Sabkha Ecosystems

Tasks for Vegetation Science 46

SERIES EDITOR

H. Lieth, *University of Osnabrück, Germany*

For other titles published in this series, go to
www.springer.com/series/6613

M. Öztürk • Benno Böer
Hans-Jörg Barth • Siegmar-Walter Breckle
Miguel Clüsener-Godt • M. Ajmal Khan
Editors

Sabkha Ecosystems

Volume III: Africa and Southern Europe

Springer

Editors
Prof. M. Öztürk
Ege University
Fen Fakultesi A Blok
Botany Department E Blok
35100 Bornova, Izmir
Turkey
munirozturk@gmail.com

Benno Böer
UNESCO Office
Doha
Qatar

Hans-Jörg Barth
Department of Physical Geography
University of Regensburg
Regensburg
Germany
hans-joerg.barth@geographie.uni-regensburg.de

Siegmar-Walter Breckle
University of Bielefeld
Department of Ecology
Wasserfuhr 24-26
D-33619 Bielefeld
Germany
sbreckle@gmx.de

Miguel Clüsener-Godt
Acting Chief, Ecological Sciences & Biodiversity Section
Division of Ecological and Earth Sciences
UNESCO
1 rue Miollis
F-75732 PARIS Cedex 15
France
m.clusener-godt@unesco.org

M. Ajmal Khan
University of Karachi
Institute of Sustainable Halophyte Utilization
75270 Karachi
Pakistan
majmalk@uok.edu.pk

ISBN 978-94-007-3449-4 ISBN 978-90-481-9673-9 (eBook)
DOI 10.1007/978-90-481-9673-9
Springer Dordrecht Heidelberg London New York

Printed on acid-free paper

Springer is part of Springer Science+Business Media (www.springer.com)

Foreword

Africa is one of UNESCO's priorities. This has a lot to do with the availability of critical resources that are absolutely essential for the daily livelihood, or shall we say, survival of families, such as drinking water, fuel and food. It has also to do with the availability and accessibility of scientific knowledge as well as education, both of which are critical factors for empowering people to enhance their living standards and escape the trap of poverty.

The Sabkha Ecosystem series deals with the globally important scientific problem of how to utilise abundant saline soils and water for fodder and food production, and is consistent with science-based ecosystem management.

Sabkha – the Arabic term for flat salt deserts, a term which has been accepted by multi-disciplinary scientists – occur globally in abundance, especially in the northern and southern dry desert belts. These areas are hundreds of thousands of square kilometres in size, and people make little use of them for agricultural production due in part to their excessive salinity. On the one hand this is good, because it allows for these geologically interesting ecosystems to remain largely untouched. They do actually have an ecosystem function and biodiversity value, even though they may appear on first sight as just barren wastelands. They also have a scientific, education, and heritage value. At the same time, there is evidence that biosaline agriculture in coastal sub-tropical and tropical dry desert sabkhat can make use of abundant seawater for the irrigation of halophyte cash crops, such as mangroves, seagrasses, algae, and other types of plants. The editors have demonstrated this clearly in Volumes I and II, as well as during the "1st International Symposium on Sabkha Ecosystems" which was co-organised by UNESCO in Tunis in 2006.

Using seawater for biosaline agriculture is certainly most challenging. Highly complex questions arise on issues such as practicability, marketing, corrosion, soil salinisation, irrigation techniques, drainage, seed ecology, root systems, element concentration analyses of soil, water, and biota, and this ranges all the way to salinity conversion tables, and soil fertility, nutrients, as well as farmers' cultural acceptance of science-based innovations, and finally to investment, and profit.

This means biosaline agriculture using high-saline, and even full-strength seawater, as an irrigation source is not an easy task. However, it offers multiple promising prospects. Herbivorous seacows, and various species of marine turtles and fish for example, use seagrasses as their staple food supply. The feeding ecology of terrestrial livestock however, does not allow them to graze on those valuable plants under the seawater surface. Is it possible then to produce them in land-based coastal systems, similar to rice-fields, harvest them, rinse the salt off, and use them as a cash fodder crop for camels, sheep, goats, and cattle?

There are many scientific questions that need to be addressed and experiments that need to be conducted, in order to provide answers to these questions. Mangrove leaves and seeds can provide good fodder for camels. Certain other halophytes from among the more than 2,000 known salt-tolerant plant species have huge economic potential. This can be of benefit for African Members States that suffer from a lack of freshwater resources, as well as for other countries.

It is with the above thoughts in mind that I encourage the stakeholders in biosaline agriculture development as well as in nature conservation and environmental management related to salt deserts to read and study the contributions from this new volume on sabkha ecosystems. It is a small, yet significant science-based contribution to find an answer to an old question: How can we use saline water, and salty soils for agricultural productivity?

Prof. Dr. Walter Erdelen
UNESCO Assistant Director-General for Natural Sciences

Kingdom of Saudi Arabia
Ministry Of Defense & Aviation
Presidency of Meteorology & Environment
P M E

Preface

We, as the people of the world, are facing real environmental challenges today which must be taken seriously. These challenges are numerous and can no longer be ignored. Some of the most pressing challenges are freshwater scarcity, groundwater depletion and salinization, climate change, loss of biodiversity and fertile soils, and food supply in competition with biofuels. This is not science-fiction – this is real and present danger to us, and to our children.

UNESCO has already produced two volumes on Sabkha Ecosystems, dealing with the sabkhas of the Arabian Peninsula and adjacent countries, as well as West and Central Asia. This is a laudable effort, trying to make scientific information available on various aspects of sabkhas including how to utilize sabkhas for biosaline productivity, and describing the educational, heritage, and ecosystem value of sabkhas.

Sabkhas belong naturally to many dry areas in the world, including the dry zones of Africa and Southern Europe, and their agricultural potential is considered quite low. However, with research and development in saline irrigation and good drainage it is possible to convert at least coastal sabkhas into agro-systems with mangrove, seagrass, and algal biomass. This can contribute to land-based fish and shrimp production, livestock fodder, carbon sequestration, and production of charcoal and biofuels on currently non-productive soils. It would also provide jobs and income in dryland agriculture using high-saline irrigation and drainage, assisted by a non-corrosive irrigation system and solar energy for seawater pumps.

There are examples and success stories with biosaline productivity. However, more needs to be done to obtain knowledge on seawater-based seagrass productivity in coastal seagrass-terraces, as well as mangrove and salt marsh crops as livestock fodder. It is therefore important to continue research and experiments into this field since it can potentially make hundreds of thousands of square kilometers of salt-deserts green using saline soils and seawater for the production of cash-crop-halophytes.

In this context, sabkha development can make a significant contribution to reduce dependency and wastage of freshwater in agriculture, which is globally the highest freshwater consumer.

I highly welcome this new volume entitled "*Sabkha Ecosystems Vol III: Africa and Southern Europe*", and encourage further studies into this subject, with a particular view towards enhancing professional environmental management, and impact assessments.

Turki Bin Nasser Bin Abdulaziz Al-Saud
President, Presidency of Meteorology and Environment (PME)
Chairman, Council of Arab Ministers Responsible for Environment (CAMRE)

P.O.Box 1358 Jeddah 21341. Saudi Arabia Tel:6536000, Fax:6511424 www.pme.gov.sa

Acknowledgements

Invaluable contributions of 40 scientists from Africa and Southern Europe covering both scientific as well as practical concepts regarding Sabkhas is highly appreciated, their sincere efforts made it possible to produce this volume. We gratefully acknowledge the financial assistance given to us by the UNESCO Office Doha-Qatar. Our special thanks go to the production editorial team of Springer-Verlag for their close collaboration, flexibility as well as professional handling during the compilation of this book.

Contents

Disclaimer

United Nations
Educational, Scientific and
Cultural Organization

The designations employed and the presentation of the material throughout this publication do not imply the expression of any opinion whatsoever on the part of UNESCO or the editorial team concerning legal status of any country, territory, city or area of its authorities, or concerning the delimination of its frontiers or boundaries.

Sabkha Regions of Tropical East Africa

Shahina A. Ghazanfar and Henk J. Beentje

Abstract Sabkhat in the region of tropical East Africa (treated here as Uganda, Kenya and Tanzania) are comparatively small in area and limited mostly to lake basins in the Eastern Rift and a few coastal areas. A relatively large inland sabkha lies in North Kenya, west of Lake Turkana, associated with the Chalbi Desert (a former lake). The chief plants of saline flats that surround the saline lakes in Kenya and Uganda include *Cyperus laevigatus*, *Sporobolus spicatus* and *Dactyloctenium* spp. The coastal sabkhat, flooded only at spring tides, are occupied by monospecific stands of stunted *Avicennia marina*. At the more open parts of the *Avicennia* fringe *Arthrocnemum indicum*, *Paspalum vaginatum*, *Sesuvium portulacastrum*, *Sporobolus virginicus*, *S. spicatus*, *S. kentrophyllus*, *Pedalium murex* and *Suaeda monoica* form the main associates. In the inland sabkhat low rainfall and high potential evaporation have resulted in an arid and saline landscape dominated by grasses *Aristida adscensionis*, *A. mutabilis*, *Drake-Brockmania somalensis*, *Sporobolus consimilis*, *S. virginicus and Psilolemma jaegeri*, and the subshrubs *Duosperma eremophilum* and *Indigofera spinosa*. *Lagenantha nogalensis* occurs on gypsophilous soils and *Dasysphaera prostrata* on saline soils at the edges of Lake Turkana and the Chalbi Desert. Stunted woody vegetation is dominated by *Acacia reficiens* and in drainage channels by *Salvadora persica*. Floristically the coastal sabkhat of tropical East Africa fall in the Zanzibar-Inhambane regional mosaic and the inland sabkhat in the Somalia-Masai regional centre of endemism. There are no endemic genera in the coastal sabkhat, but all of the nine East African mangrove species occur in Kenya, Tanzania and Mozambique. In the inland sabkhat, the Somalia-Masai regional centre of endemism includes the endemic genera *Drake-Brockmania* and *Dasysphaera*. There is no arid-zone agriculture in the inland sabkhat region and nomadic pastoralists, depending on their livestock for subsistence, are the main occupants; *Duosperma* and *Indigofera* are amongst some of the important food plants of livestock (camels). Salt deposits are harvested from the extensive flats surrounding the saline and soda lakes, and the mangrove is an important economic resource as a nursery for fish and crustaceans, as well as a source of poles, timber and firewood. Over-harvesting of wood and conversion to saltpans and aquaculture, housing and industry is a threat to many parts of the mangrove area. There are no strict nature reserves in the inland sabkhat of tropical East Africa designated for the protection of arid landscapes and its flora; however the Mount Kulal Biosphere Reserve in northern Kenya covers the salt desert and lake ecosystems; Lake Manyara and Amboseli Biosphere Reserves also partly cover the saline and alkaline ecosystems. Mangrove areas are included in Watamu Marine National Park, Kiunga National Marine Reserve and Ras Tenewi Marine National Park in Kenya, and in Mafia Island Marine Park, Jozani National Park and Sadaani Game Reserve in Tanzania. Other areas of East African mangrove are included in forest reserves, with varying degrees of protection.

1 Introduction

Sabkhat in the region of tropical East Africa (treated here as Uganda, Kenya and Tanzania) are comparatively

S.A. Ghazanfar (✉) and H.J. Beentje
Royal Botanic Gardens, Kew, Richmond, Surrey, TW9 3AB
e-mail: s.ghazanfar@kew.org

M. Öztürk et al. (eds.), *Sabkha Ecosystems*, Tasks for Vegetation Science 46,
DOI 10.1007/978-90-481-9673-9_1, © Springer Science+Business Media B.V. 2011

small in area when compared to the extensive arid and saline areas of the Sahara desert and the Karoo-Namib region in southern Africa. In tropical East Africa sabkhat are limited mostly to lake basins in the Eastern Rift (mainly in the Kenyan Rift Valley) and a few coastal areas. A relatively large inland sabkha lies in north Kenya, west of Lake Turkana, associated with the Chalbi Desert (a former lake) (Fig. 1).

Lakes in the Eastern Rift were created by tectonic and volcanic activity associated with the formation of the Rift Valley (±2 myBP). Most of these are fed by springs and surface drainage and are thus rich in solutes; as most of these are closed lakes and water loss is only through evaporation, they are saline or alkaline and fluctuate seasonally in their solute concentration and water levels. Soda lakes in the eastern Rift valley are among the most biologically productive ecosystems in the world (Melack 1996) with phytoplankton and cyanobacteria attracting flocks of flamingos and other avifauna. The monospecific blooms of *Spirulina platensis* characteristic of soda lakes forms the primary graze for flamingos; it has also been a diet item for the people of Chad for many years (Léonard and Compère 1967)

The distribution of saline soils in East Africa is mainly determined by salt deposits in lake basins and valleys where the mean annual rainfall is between 250 and 1,000 mm. Some areas may not be sufficiently saline to have halophytic vegetation, such as the sodium soils created by rocks containing soda feldspar perthite, nevertheless they support a distinct vegetation. In this chapter we will describe the vegetation, the main plant communities and associated species, their local uses and economic importance, and conservation

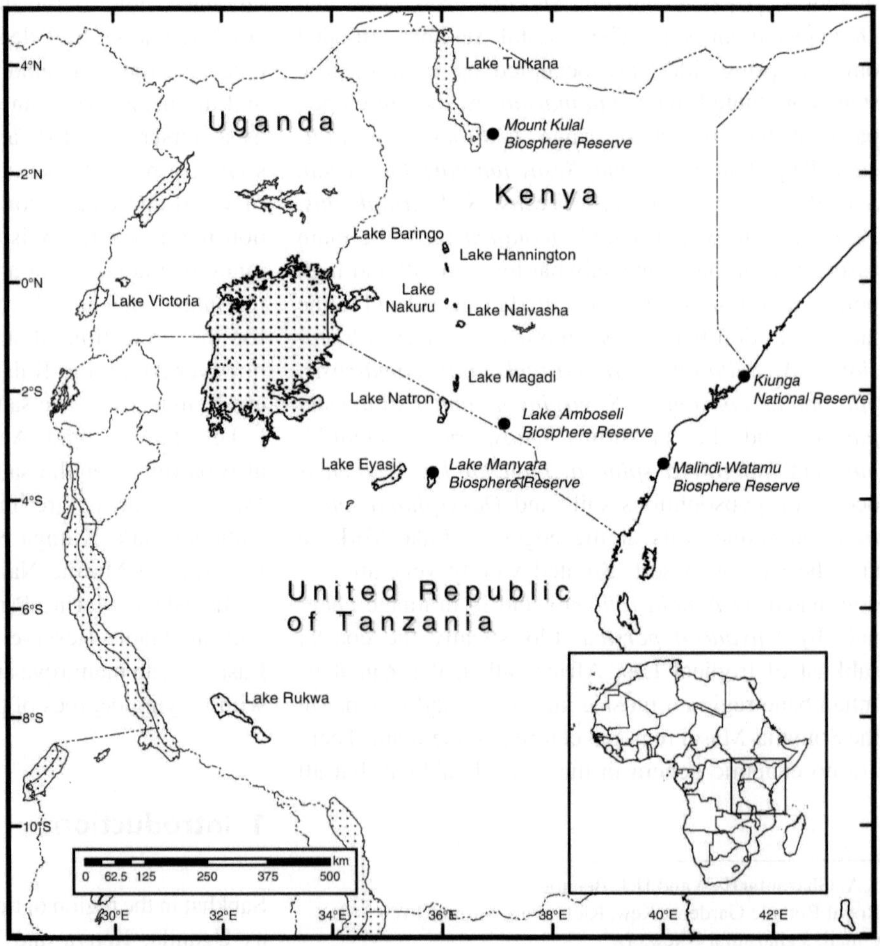

Fig. 1 Tropical East Africa, showing the location of the lakes and the Biosphere Reserves

status of coastal, true saline (soluble salts 0.2–0.3%) and alkaline (pH 8.0–9.6) ecosystems in tropical East Africa.

2 Inland Sabkhat

2.1 *Saline Sabkhat*

In tropical East Africa, the inland saline sabkhat are not extensive and occupy relatively small areas. A large area in northern Kenya in the Marsabit District, the Chalbi Desert and its immediate surrounding plains can be classified as an inland sabkhat.

The Chalbi desert is composed of saline and alkaline sediments and stabilised sand dunes associated with the old bed of Lake Chalbi. During the Pleistocene (1.8 mya to 10,000 years BP) this lake drained the volcanoes at Marsabit, but has since dried up to form the extensive Chalbi desert. The area lies in a depression between 430 and 500 m above sea level surrounding a large central plain. Precipitation is low with an average of between 100 and 250 mm a year for the region, occurring between March–May and October–December. Much of the drainage from the surrounding mountains (Mt Marsabit, Hurri Hills, Mt Kulal, Mt Nyiru, Ol Doinyo Mara and Ndotos) is lost by evaporation and never reaches the Chalbi Desert.

2.2 *Vegetation*

The vegetation of the inland sabkhat in tropical East Africa (northern Kenya) falls under the Somalia-Masai regional centre of endemism which includes NE Uganda (Karamoja), most of Kenya in the North and between the Highlands and the coastal belt, and the dry lowlands of north and central Tanzania; the region is characterised by an *Acacia-Commiphora* deciduous bushland and thicket and secondary and wooded grassland. Two genera, *Drake-Brockmania* and *Dasysphaera* that are found in the inland sabkhat, are endemic to this floristic region.

The major part of the Chalbi desert itself is barren except for at the edges where there is subsurface water flow from springs from the surrounding mountains. Due to extensive evaporation the area is hyper-arid, and after seasonal rains and flood, the soil becomes highly saline supporting a few halophytic grasses. These too grow only near runnels and water outlets. In the surrounding plains and parts of the Chalbi Desert the most extensive vegetation type is the semi-desert annual grassland. Second to this is the semi-desert dwarf shrubland which occupies the stabilized sand dunes (White 1983). The main halophytic grass which grows in these locations is *Drake-Brockmania somalensis*. This is a mat-forming annual, spreading by stolons and rooting at nodes, and with leaves covered with stiff hairs. It occupies seasonally flooded places in silty and saline soils. *Drake-Brockmania somalensis* is distributed from Tanzania through to NE Africa (Sudan, Somalia, Ethiopia; excluding Uganda); it is also found on the Farasan Island (Saudi Arabia) in the Red Sea (Phillips 1997).

By far the most common and widespread in the desert and semi-desert are the annual grasses, *Aristida adscensionis* and *A. mutabilis*. They occupy the driest areas, extending up to elevations of 1,000 m on the drier parts of the surrounding hills. They occur on poor, shallow soils and during periods of drought may be absent for as long as the drought lasts (White 1983). Associated with these are a few species of subshrubs which form the perennial woody components of the sabkhat. Dominant amongst these are *Duosperma eremophilum* (Family Acanthaceae) and *Indigofera spinosa* (Family Leguminosae); the former found on relatively moister soils, and the latter on the drier soils. *Indigofera spinosa* is a small intricate subshrub found throughout the semi-desert grassland in East Africa and is distributed through Sudan, Somalia and Ethiopia to Arabia. It is also present in the dry *Acacia-Commiphora* deciduous bushland (Gillett et al. 1971). In northern Kenya, around settlements where there is heavy utilisation of vegetation through grazing and where soils are degraded, *Indigofera spinosa* is the main species. Its cover is 8.5% (Lusigi et al. 1986), and it forms the main graze for livestock.

Amongst other shrubs, *Lagenantha nogalensis* (Family Chenopodiaceae) a succulent that is tolerant of gypsophilous soils, forms almost pure stands on white calcareous soils of the old Chalbi Lake bed; *Dasysphaera prostrata* (Family Amaranthaceae) occurs on saline or alkaline soils at the margins of Lake Turkana and at the edges of the Chalbi Desert (White 1983).

There are a few annuals which come up after rain but are grazed out as soon as they appear. These form a very low cover (0–1%) and constitute only a few species such as *Cenchrus ciliaris*, *Dactyloctenium aegyptium*, *Psilolemma jaegeri*, *Sporobolus consimilis*, *S. virginicus*, *Gisekia pharnaceoides*, *Hermannia* sp. and *Limeum praetermissum*.

Only a few trees or large shrubs are associated with arid and saline soils. These are usually stunted, with *Acacia reficiens* as the dominant species; other species include *A. seyal*, *A. tortilis*, *A. horrida*, *A. senegal* and *Commiphora* spp. In areas degraded by intense grazing pressure the unpalatable *Calotropis procera* is widespread and *Salvadora persica* occurs on saline soils associated with the Chalbi drainage system.

2.3 Alkaline Sabkhat

The alkaline sabkhat are associated with lakes of the Rift Valley. Except for Lakes Baringo and Naivasha, the others on the Eastern Rift (Lake Bogoria, Nakuru, Elementeita, Magadi, Natron, Manyara, Eyasi and Rukwa) are highly alkaline and show salt deposits around their shores. Several are centres for the commercial harvesting of soda ash (Lind and Morrison 1974).

The main species around the saline lakes in Kenya and Uganda are *Cyperus laevigatus*, *Sporobolus spicatus* and *Dactyloctenium* spp. The vegetation of the grassland on the flats close to Kiboko river is described by Bogdan (1958) which shows different dominant species inhabiting soils with different concentrations of salts: thus in slightly alkaline conditions the grass cover is mainly *Cenchrus ciliaris*, but as alkalinity increases *Chloris gayana* becomes dominant. With increasing moisture and alkalinity *Sporobolus consimilis* becomes the dominant species and where soils are shallow, *Cynodon dactylon* appears, and in highly alkaline soils that are waterlogged in the rainy season, *Sporobolus spicatus* takes over and forms dense growth. In the flat valleys in the drier parts of Tanzania, such as the Pangani River valley, where the waters are saline, the flood plains are dominated by grasses *Sporobolus consimilis*, and shrubs such as *Suaeda monoica*, *Sesbania sesban*, *Salvadora persica* and *Triplocephalum holstii*.

The vegetation of the Lake Rukwa basin in Tanzania, which is chiefly grassland has been described by Vesey-FitzGerald (1963); this is summarised here: the vegetation can be divided into three zones from the fringe of the lake to the lake bed itself. The edges of the lake are occupied by almost pure stands of *Sporobolus consimilis*. This species does not grow in water, but colonizes the alkaline soils on the beach of the lake. The shallowly flooded alkaline swamp present on the extensive flat lake bed is vegetated by the grass *Diplachne fusca*. This is a rhizomatous perennial rooting and branching from the lower nodes forming dense mats over large areas. *Diplachne fusca* is found throughout East Africa and is distributed right through the tropics and subtropics of the Old World including Australia (Clayton et al. 1974). On the lake bed, the alkaline flats are occupied by two species: when the lake is dry *Sporobolus spicatus* grows associated with *Psilolemma jaegeri*; when the flats are flooded with the highly alkaline water, *Psilolemma jaegeri* takes over and replaces *Sporobolus*.

Other saline and alkaline patches of vegetation exists around Lake Amboseli in Kenya, where the major vegetation type is the *Commiphora-Acacia* bushland, and the saline and alkaline plains are dominated by *Suaeda monoica* and *Salvadora persica*. Alkaline grassland dominated by *Sporobolus spicatus* is also present around Lake Manyara.

3 Coastal Sabkhat

The coastal plain consists of deposits of corals and sands with occasional dune formations (Lind and Morrison 1974). The climate is controlled by the monsoonal currents of the Indian Ocean, with long rains from April to June and short rains from October or November to January. Annual rainfall in the coastal strip decreases from ±1,425 mm in the South of Tanzania to ±1,075 mm in northern coastal Kenya. The mean annual temperature is 25°C and does not vary much (Dahdouh-Guebas et al. 2004).

Floristically the coastal sabkhat of tropical East Africa fall in the Zanzibar-Inhambane regional mosaic. There are no endemic genera in the coastal sabkha, but all of the nine East African mangrove species occur in Kenya and Tanzania: *Avicennia marina*, *Bruguiera gymnorhiza*, *Ceriops tagal*, *Heritiera littoralis*, *Lumnitzera racemosa*, *Pemphis acidula*, *Rhizophora mucronata*, *Sonneratia alba* and *Xylocarpus granatum* (Beentje and Bandeira 2007).

The mangrove is an important economic resource as a nursery for fish and crustaceans, as well as a source of poles, timber and firewood. Both mangrove trees and mangrove vegetation are decreasing in East Africa, particularly around urban areas and large villages. This is due to increasing populations and continuing poverty, leading to more tree cutting for fuelwood, building material and urban development; large mangrove areas are also converted into salt-pans or to agricultural land, and to a lesser extent for aquaculture purposes (oyster, shrimp and fish ponds). Coastal developments further threaten mangrove areas, and the main reasons are those of infrastructure and tourism; increased population has led to increasing pollution from industry and agriculture (e.g. crude oil, heavy metals, pesticides) (Semesi 1998).

3.1 Vegetation

Along the coast fringing coral reef is more or less continuous, and where it is lacking, mangroves occur. Much of the rocky coral shore is overlaid with sand (Lind and Morrison 1974). Mangrove forest is found in suitable sites all along the East African coast, from Egypt to South Africa, but in East Africa it reaches relatively high diversity, with the mouth of the Rufiji River in Tanzania among the richest mangroves in Africa.

The true mangrove is flooded regularly, but an area inland of the mangrove is flooded only twice a year (during the equinoctal spring tides), and the salt content of the sandy soil is so high through evaporation that often vegetation is almost absent. During the rainy season the salt is leached out, and so even halophytes have to struggle to cope with this changing habitat. The vegetation cover is usually low, in the region of 10% on the edge of the mangrove, just about the High Water Mark (HWM), to virtually absent in the area only flooded by the spring tides (Walter and Steiner 1936; Knapp 1973; White 1983; Table 1).

Diversity is low, with twelve species reported (see the list below). *Avicennia marina* and *Lumnitzera racemosa* both are mangrove trees, but extend into the sabkha zone, becoming more stunted as the salt concentration in the soil increases. *Avicennia* extends into the sabkha zone thanks to its capacity to withstand high salt concentrations; McCusker (1977) reports that the upper limit of soil osmotic potential for other East African mangrove trees is between 38.8 and 50.1 atm, but that of *Avicennia* is up to 97.8 atm. The pneumatophores of *Avicennia* are colonized by red algae (*Bostrychia* sp.). *Suaeda monoica* may form large stands on the edge of the barren sand flats, while *Arthrocnemum indicum*, *Sporobolus virginicus* and *Suaeda monoica* grow on slightly elevated soil between the HWM and the barren zone. *Salicornia pachystachya* may form carpets in summer.

Table 1 List of coastal salt marsh halophytes (Data from Walter and Steiner 1936; Knapp 1973; White 1983)

Chenopodiaceae	
Suaeda monoica J.F.Gmel	Between the stunted Avicennia zone and the vegetation-free sandflats, about the HWM
Arthrocnemum indicum (Willd.) Moq.	On slightly raised parts just above HWM
Salicornia pachystachya Ungern.-Sternb.	On slightly raised parts just above HWM
Gramineae	
Sporobolus spicatus (Vahl) Kunth	No detail
Sporobolus kentrophyllus (K.Schum.) W.D.Clayton	On slightly raised parts just above HWM
Sporobolus virginicus (L.) Kunth	No detail
Paspalum vaginatum Sw.	No detail
Dactyloctenium geminatum Hack.	No detail
Pedaliaceae	
Pedalium murex L.	No detail
Aizoaceae	
Sesuvium portulacastrum (L.)L.	About HWM, in moister sites
Verbenaceae/(Avincenniaceae)	
Avicennia marina (Forssk.)Vierh.	Next to mangrove,the more inland ones becoming more stunted
Combretaceae	
Lumnitzera racemosa Willd.	Next to mangrove in slightly moister sites, the more inland ones becoming more stunted

4 Utilisation and Economic Importance

There is no arid-zone agriculture in the inland sabkhat of tropical East Africa. Nomadic pastoralists are the main occupants in this area, and the region is used as a basic grazing resource for their livestock (camels, cattle and goats). *Duosperma* and *Indigofera* are amongst some of the important food plants, and wood from the larger shrubs is used as a source of fuelwood and materials for livestock fencing. The Range Resource Assessment and Management Strategies for South-Western Marsabit, Northern Kenya (Lusigi et al. 1986) showed that the rangelands were heavily utilised, well beyond their carrying capacity leaving the soil barren and encouraging erosion. The annuals provide little cover, but are grazed very soon after they come up.

Harvesting of salt from the saline and soda lakes is an important industry in tropical East Africa. In western Uganda, Lake Katwe, a closed crater lake fed by brine springs is an important source of salt and revenue. Soda ash, harvested from Lake Magadi in Kenya is also of much economic importance to the country.

Uses of plants of coastal regions are minimal, apart from some casual grazing or fodder collecting (reported for *Avicennia marina*) and use as vegetables (reported for *Pedalium murex*, *Sesuvium portulacastrum*); plus some firewood collection. Hence, economic importance is virtually nil.

5 Conservation and Designated Protected Areas

There are no strict designated protected areas in the inland sabkhat of tropical east Africa for the protection of the flora per se. However several Biosphere Reserves are designated as part of the UNESCO Man and the Biosphere Programme which are essentially aimed at the conservation and sustainable development of terrestrial and coastal ecosystems and are recognised within the framework of UNESCO (Clüsener-Godt 2002). Amongst these is The Mount Kulal Biosphere Reserve in northern Kenya, established in 1978 and situated on the eastern side and southern end of Lake Turkana, with Mount Kulal in its centre. The area comprises a variety of landscapes and habitats, including brackish water at the southern end of the Lake, a volcanic landscape with lava flows, an extensive lava desert and a volcanic island within the lake, hot springs, the occasionally flooded Chalbi salt desert, sand dunes and seasonal water courses. The Amboseli Biosphere Reserve located in the south of Kenya and the Lake Manyara Biosphere Reserve in Tanzania offers protection to the saline and alkaline ecosystems (Fig. 1) (UNESCO 1977; UNEP-WCMC and the IUCN World Commission on Protected Areas (WCPA)).

There is increasing anthropogenic pressure on the coastal sabkhat ecosystems in East Africa. Fuelwood exploitation, urban development, tourism development, and conversion to saltpans and aquaculture are on the increase. Pollution from an increasing population and industry also contribute to the threats; for an overview of the various factors (see Semesi 1998). In Kenya, around 70% of the people living on the coast rely on mangrove poles for construction (Beentje and Bandeira 2007) and so access to the mangroves, which necessarily goes through the sabkha, would result in trampling and compaction.

Mangrove areas, including coastal sabkha areas, are included in Watamu Marine National Park, Kiunga National Marine Reserve and Ras Tenewi Marine National Park in Kenya, and on Mafia Island Marine Park, Jozani National Park and Sadaani Game Reserve in Tanzania. The Malindi-Watumu Biosphere Reserve (IUCN Management Category II, VI, IX: National Park, Resource Reserve, Biosphere Reserve) located south of Malindi was established in 1981. This includes intertidal rocks, sand and mud flats and sublittoral areas; the Mida Creek mud flats with fringing mangrove swamps with *Rhizophora mucronata, Bruguiera gymnorrhiza* and *Ceriops spp.* are also included. It also offers protection to the nesting grounds for the roseate tern, *Sterna dougalii* and the bridled tern, *S. anaethetus*. Several species of shore birds and crabs, and sea grasses *Thallasia chemprechii, Syringodium* spp., *Cymodocea* spp., and *Halodule wrightii* on the intertidal sand and mud are also included in the Reserve (UNEP WCMC, Protected Areas Programme). Other areas of East African mangrove are included in forest reserves, with varying degrees of protection.

6 Conclusions

The sabkhat regions of tropical East Africa are few and limited in overall area, nevertheless constitute an important ecosystem that provides habitats for a number of plants and animal species, and contributes to the economy of the region. As most arid regions in Africa, the sabkhat are also over-utilized and are in need of management and protection. This is especially relevant for the coastal sabkhat, which are constantly subject to damage due to coastal development and over-use of coastal resources. The conservation and the sustainable use of coastal sabkhat should be a priority for any conservation and management plan for a country with setting up more Biosphere Reserves essential for the protection and proper management of coastal regions. Programmes on establishing living halophytic collections and genetic material for restoration, and undertaking research projects to understand the dynamics of coastal and inland sabkhat ecosystems are also necessary to preserve these important ecosystems.

Acknowledgements We thank Justin Moat for providing Fig. 1.

References

Beentje HJ, Bandeira S (2007) A field guide to the mangrove trees of Africa and Madagascar. Royal Botanic Gardens, Kew

Bogdan AV (1958) Some edaphic vegetatonal types at Kiboko, Kenya. J Ecol 46:115–116

Clayton WD, Phillips SM, Renvoize SA (1974) Gramineae. In: Milne-Redhead E, Polhill RM (eds) Flora of Tropical East Africa. Crown Agents, London, pp 281–282

Clüsener-Godt M (2002) UNESCO's work for the protection of arid zones with particular emphasis to sabkhat. In: Barth H-J, Boer B (eds) Sabkha ecosystems, vol 1, The arabian peninsula and adjacent countries. Kluwer Academic, Dordrecht, Boston, London, pp 347–352

Dahdouh-Guebas F, Van Pottelbergh I, Kairo J, Cannicci S & Koedam N (2004) Human-impacted mangroves in Gazi (Kenya), predicting future vegetation based on retrospective remote sensing, social surveys, and tree distribution. Marine Ecology Progress Series 272:77–92

Gillett JB, Polhill RM, Verdcourt B (1971) Leguminosae. In: Polhill RM (ed) Flora of Tropical East Africa. Crown Agents, London, p 253

Knapp R (1973) Die Vegetation von Afrika. Fisher, Stuttgart

Léonard J, Compère P (1967) *Spirulina platensis* (Gom.) Geitl., algue bleue de grande valeur alimentaire par sa richesse en proteines. Bulletin du Jardin Naturel de Belge 37(Supplement 1)

Lind EM, Morrison MES (1974) East African vegetation. Longman, London

Lusigi WJ, Nkurunziza ER, Awere-Gyekye K, Masheti S (1986) Range resource assessment and management strategies for south-western Marsabit, Northern Kenya. UNESCO-FRG-MAB integrated Project in Arid Lands, Nairobi, p 230

McCusker A (1977) Seedling establishment in mangrove species. Geo Ecol Trop 1:23–33

Melack JM (1996) Saline and freshwater lakes of the Kenyan rift. In: McClanahan TR, Young TP (eds) East African ecosystems and their conservation. Oxford University Press, New York, Oxford, pp 171–190

Phillips S (1997) Gramineae. In: Hedberg Inga, Edwards Sue (eds) Flora of ethiopia. The National Herbarium, Biology Department, Addis Ababa University, Ethiopia and The Department of Systematic Botany, Uppsala University, Sweden, pp 108–110

Semesi AK (1998) Mangrove management and utilization in Eastern Africa. Ambio 27:620–626

UNEP-WCMC. Protected Areas Programme, UNEP-WCMC: www.unep-wcmc.org

UNESCO (1977) Man and the environment in Marsabit District. Proceedings of a symposium held at Mt Kulal. Nairobi, Regional Office for Science and Technology for Africa

Vesey-FitzGerald DF (1963) Central African grasslands. J Ecol 51:243–273

Walter H, Steiner M (1936) Der Okologie der ostafrikanischen Mangroven. Z Bot 30:65–193

White F (1983) The vegetation of Africa. A descriptive memoir to accompany the UNESCO, AETFAT, UNSO vegetation map of Africa. UNESCO, Paris

A Review of Fauna and Flora Associated with Coastal and Inland Saline Flats from Namibia with Special Reference to the Etosha Pan

Peter L. Cunningham and Willem Jankowitz

Abstract Coastal saline flats are often frequented by a plethora of migratory birds, mainly waders, and mammals such as the elusive Brown Hyaena (*Hyaena brunnea*) and Black-backed Jackal (*Canis mesomelas*). Inland saline flats, particularly Etosha Pan, which is fringed by broad- and fine-leafed arid savanna woodlands, teem with a variety of game. Greater Flamingo (*Phoenicopterus ruber*) and Lesser Flamingo (*Phoeniconaias minor*) also occasionally breed on the Etosha Pan depending on local rainfall. Only vegetation adapted to halophytic conditions occur in association with saline systems in Namibia. Grasses such as *Sporobolus salsu* and *Sporobolus spicatus* may occasionally occur on the Etosha Pan while the succulent-like perennial herbs such as *Salicornia natalensis* and *Anthrocnemum africanum* may grow in the coastal saline conditions. Although no fauna and flora are exclusively known to be associated with saline flats in Namibia, these habitats are important for wildlife mainly due to the vegetation (i.e. source of food and shelter) surrounding these flats, fountains (i.e. source of water) often associated with these flats and breeding/feeding/resting sites for migrant waders and other birds.

1 Introduction

Saline flats are widely distributed throughout Namibia with numerous relatively small flats located between "dune streets" or in open dry woodland areas from the Kalahari Desert, in the east, to coastal saline flats to one of the largest and most well known flats such as the Etosha Pan in the central north of Namibia.

1.1 Etosha Pan

The Etosha Pan is a flat saline depression approximately 120 km long × 55 km wide and covers an area of 6,133 km² at an altitude of 1,100 m (Berry 1972) and one of the most remarkable geomorphological landforms in the world (Buch 1997). The pan is situated in the north of Namibia (Lat. 19°S, Long. 16°E) mostly within the Etosha National Park which covers a total area of 22,270 km² and is the largest protected saline wetland in Africa (Lindeque and Archibald 1991). The Etosha Pan is classified as a wetland of international importance and listed as a Ramsar site (Kolberg).

The Etosha Pan forms part of the Cuvlai drainage system that drains the Sierra Encoco Mountains – some 300 km inside Angola – and eventually ends up in the Etosha Pan under exceptional flood years. The saline wetlands of the Etosha Pan are ephemeral in nature (Berry 1972; Clarke 1998; Lindeque and Archibald 1991; Lowery 2001), holding water for several months after extensive flooding although usually rarely more than a month. The pan holds large sheets of shallow water, usually not exceeding 1 m in depth and when dry the area is a hostile stretch of salt-crusted wasteland. Flooding of the lower reaches of the pan seldom occurs while the main inlet on the eastern side of the pan – Fishers Pan – is flooded almost every year although the amount of water received and the duration of the wetland varies annually. Water entering the pan turns brackish towards the end of the rainy season through

P.L. Cunningham(✉)
Environment & Wildlife Consulting Namibia, P. O. Box 90717, Windhoek, Namibia, 9000
e-mail: pckkwrc@yahoo.co.uk

W. Jankowitz
Department of Nature Conservation, Polytechnic of Namibia, School of Natural Resources and Tourism, Private Bag 13388, Windhoek 9000, Namibia

M. Öztürk et al. (eds.), *Sabkha Ecosystems*, Tasks for Vegetation Science 46,
DOI 10.1007/978-90-481-9673-9_2, © Springer Science+Business Media B.V. 2011

evaporation and contact with previous evaporites with a suspected salinity gradient increasing from north to south (Lindeque and Archibald 1991).

The climate is semi-arid and harsh with highly unpredictable rainfall mainly between October and April varying from 300 mm in the west to 500 mm in the east (Berry 1972; Clarke 1998; Mendelsohn et al. 2000). The average maximum temperature for December is 39°C and the lowest mean temperature of 6°C occurs during July (Baker 1996). Mendelsohn et al. (2000) states that September to December have the highest maximum temperatures followed by slightly cooler months associated with greater cloud cover and rain. Evaporation is high as Mendelsohn et al. (2000) and Wellington (1938) measured annual evaporation of between 2,500 mm and 2,700 mm per year, respectively, while wind speeds of up to 30 knots have been measured on the pan.

The vegetation associated with the Etosha Pan is broadly classified as Saline Desert with a Dwarf Shrub Savannah fringe (Giess 1971) and surrounded by salt-tolerant shrubs and spiny grasses (Strohbach 1996).

The wetlands in northern Namibia are most important in terms of supporting human populations and biodiversity as approximately 60% of Namibia's total population lives in this area of which 75% live alongside the perennial and/or ephemeral wetlands (Hines and Kolberg 1996). About 44% of the people living adjacent to the wetland system live within the Cuvelai drainage system, where a mixed economy is dependant on the flood regime for water, grazing and food – mainly in the form of fish. The Etosha Pan is also important as a tourist destination as the Etosha National Park is the most visited conservation area in Namibia (Baker 1996) with more than 140,000 visitors recorded annually (Berry 1997). The starkness of the Etosha Pan during the dry season(s) does not decrease the importance of the area as wildlife habitat and tourist attraction.

1.2 Coastal Flats

Namibia is well endowed with coastal saline flats along its entire coast from the Orange River mouth (bordering South Africa) in the south to the Kunene River mouth (bordering Angola) in the north. These flats vary in size from approximately 10,000 m^2 to 60 km^2 (Grűnert 2000) with most of the flats occurring close to the sea. The Namibian coastline is extremely arid (Namib Desert), however the shallow, brackish or saline wetlands make a major contribution to the ecology of the coastal zone. Three of these sites (Walvis Bay lagoon, Sandwich Harbour and mouth of the Orange River) have been registered as wetlands of international importance under the Ramsar Convention (Kolberg n.d.). The Namib Desert is one of the oldest deserts with most of the area protected by either the Namib-Naukluft Park or the Skeleton Coast Park (Baker 1996). The Desert biome is an extreme desert with highly irregular precipitation – mean annual varying between 5 and 85 mm annually – while fog is experienced throughout the year and a crucial life-support system (Lovegrove 1993). The region is generally cold and windy with a mean annual temperature of 16.3°C (Baker 1996). The vegetation is characterized by a dominance of therophytes and lichens (Lovegrove 1993). Very little work has been conducted on the fauna and flora associated with these coastal and inland saline flats within Namibia. This paper attempts to collate the known information on this subject.

2 Methods

An extensive literature survey was conducted on the fauna and flora associated with saline flats – with special reference to the Etosha Pan as an inland saline system as well as coastal flats – from Namibia. Personal observations as a result of numerous field excursions throughout the country over a number of years added to this information, as did anecdotal information gathered from people currently living and working in close proximity to the saline flats.

3 Results and Discussion

3.1 Fauna

3.1.1 Etosha Pan

3.1.1.1 Mammals

No mammal species are specifically associated and/or unique to the Etosha Pan system although a wide variety of species make use of the pan in some or other way. Etosha National Park – which includes the largest portion of the Etosha Pan system – has 114 mammal species (Baker 1996) with most of these (especially the water dependant species) congregating at waterholes at the edge of the pan during the dry season, which usually extends from May to September. Mendelsohn et al. (2000) states that the grass and dwarf-shrub plains surrounding the Etosha Pan are favoured by grazing antelope and zebra with high densities of wildlife generally found to the east, south and west of the pan. Mammals utilize the pan in the following ways:

- Obtain water at natural seepage's associated with the edge of the pan
- Source of food
- Movement and migration routes
- Shelter and resting
- Natural licks

Ungulate species that frequent these natural waterholes (seepage) include most of the herd forming "plains game" such as Blue Wildebeest (*Connochaetes taurinus*), Burchell's Zebra (*Equus zebra burchelli*), Oryx (*Oryx gazella*) and Springbok (*Antidorcas marsupialis*) (C. Brain Pers. comm., B. Kötting Pers. comm. and W. Versveld Pers. comm.). Elephant (*Loxodonta africana*) also frequent this seepage at the edge of the pan. Predator species such as Lion (*Panthera leo*) use these waterholes as ambush sites especially during the dry season. Although many species are water independent they do utilize water when available and are subsequently attracted to the waterholes at the edge of the pan. According to Griffin and Grobler (1991) the following mammal species – all water dependant i.e. require permanent drinking water – are associated with the seasonal wetlands of Owambo and Etosha: Elephant, Black Rhino (*Diceros bicornis*), Burchell's Zebra, Blue Wildebeest, Black-faced Impala (*Aepyerus melampus petersi*), Kudu (*Tragelaphus strepsiceros*) with the Serval (*Felis serval*) not verified, but expected. Buffalo (*Syncerus caffer*) also used to occur on the eastern fringes of the pan during the 1950s although currently do not occur in the region (Berry et al. 1997).

When stands of *Sporobolus salsus*, a short-lived micro-perennial grass, occur on the pan after good rains, grazers converge on the pan to feed on these grasses (Berry et al. 1997). A small island aptly named "Haas Eilande" (Hare Island) refers to Scrub Hare (*Lepus saxatilis*) that are found on these grassy shrubland islands located in the pan (Berry et al. 1997). Few predatory animals venture far out into the pan (Berry 1972). Other than waterholes used as ambush sites, predators such as Black-backed Jackal (*Canis mesomelas*) and Spotted Hyena (*Crocuta crocuta*) venture out onto the pan to feed on Greater Flamingo (*Phoenicopterus ruber*) and Lesser Flamingo (*Phoeniconaias minor*) chicks and eggs (Brain Pers. comm., Kötting Pers. comm. and Versveld Pers. comm.). Flamingo breeding does not however occur on an annual basis, but rather only in years of good rainfall. Lion have also been known to venture out onto the pan to prey on ungulates that utilize grass growing at the edge of the pan, especially after good rains. A spring on the western tip of a peninsula named "Leeunes" (Lion Den) is due to lions using a copse of *Salvadora persica* as an ambush site (Berry et al. 1997).

Animal footpaths are often located out on the pan often as far as a few hundred meters from the edge of the pan. It would seem that these footpaths are used as a result of predator pressure close to the edge of the pan (e.g. predators lying up in denser vegetation at the edge of the pan) and/or easier walking (Brain Pers. comm., Kötting Pers. comm. and Versveld Pers. comm.). Ungulate species known to utilize these footpaths include Blue Wildebeest, Burchell's Zebra, Oryx and occasionally Giraffe (*Giraffa cemelopardalis*). Elephant often cross the pan following age old migration routes especially across some peninsula's although they also cross the pan during the wet season when the pan may be flooded (Brain Pers. comm., Kötting Pers. comm. and Versveld Pers. comm.). Giraffe are also known to cross the pan regularly in certain areas.

Many animals use the pan as a lying up place due to the precise nature of the area – i.e. makes the approach of predators obvious. Species known to utilize the pan as a resting-place include Blue Wildebeest, Burchell's Zebra, Oryx and Springbok. Oryx even go so far as to leave their young (predator avoidance) on their own – often several hundred meters from the edge of the pan – whilst foraging in the area. This even occurs during the hottest part of the day during summer (Brain Pers. comm., Kötting Pers. comm. and Versveld Pers. comm.). Predators such as lion and cheetah (*Acinonyx jubatus*) are also known to hide their young in the sparse vegetation at the pan's edge whilst the adults are away foraging. The saline clays associated with the pan is often used by ungulates as a natural lick to supplement their salt requirements (Brain Pers. comm., Kötting Pers. comm. and Versveld Pers. comm.).

3.1.1.2 Birds

According to Baker (1996) 340 bird species are in the Etosha National Park area although not all are associated with the pan. Although no bird species are specifically associated and/or unique to the Etosha Pan system, birds frequent the pan after the rains. Birds utilize the pan in the following ways:

- Breeding
- Feeding
- Migratory stopover

The Etosha Pan is the only known breeding reservoir of Greater and Lesser Flamingo in Southern Africa (Berry 1972). Up to a million or more birds have been known to congregate at the pan when sufficient water allows for breeding. Berry (1972) supplies a detailed report on the breeding of flamingo's at the Etosha Pan including a rescue operation conducted to save some of the chicks when the water dried up prior to them fledging. Interesting enough it has been determined that flamingos migrate between the coastal saline flats on the west coast of Namibia and the Etosha Pan depending on water availability in the latter location. The largest island on the pan (2.5 km × 700 m in size) is named Pelican Island due to White Pelicans (*Pelecanus onocrotalus*) occasionally breeding on this island and foraging in the vicinity after good rains (Berry et al. 1997).

Other species known to breed at the pan, often in association with flamingos, include Grey-headed Gulls (*Larus cirrocephalus*), Grey Heron (*Ardea cinerea*), Sacred Ibis (*Threskiornis aethiopicus*) and Glossy Ibis (*Plegadis falcinellus*). Spoonbill (*Platalea alba*) has also been documented as breeding on a small island on the pan albeit away from the flamingo colony. A variety of duck species and Dabchick (*Tachybaptus ruficollis*) have been documented as breeding on the pan depending on the water level (Brain Pers. comm., Kötting Pers. comm. and Versveld Pers. comm.). According to Williams (1991) it is probable that species such as the Whiskered Tern (*Chlidonas hybridus*) and Openbilled stork (*Anastomas lamelligerus*), which breed in seasonally flooded wetlands in Bushmanland (eastern Namibia), also breed in the seasonally flooded wetlands in Owamboland (includes Etosha Pan). It is expected that some of the smaller waders (e.g. plovers and sandpipers, etc.) also nest in the vicinity of the pan. Globally near-threatened species that occur at Etosha Pan include the Slaty Egret (*Egretta vinaceigula*), Lesser Kestrel (*Falco naumanni*), Wattled Crane (*Bugeranus carunculatus*) and Blue Crane (*Anthropoides paradiseus*) (Simmons et al. 1998).

A species that breeds on the pan, but not necessarily dependant on water to do so is the Ostrich (*Struthio camelus*) (Berry 1972; Brain Pers. comm., Kötting Pers. comm. and Versveld Pers. comm.). Berry (1972) even mentions an Ostrich nest as far as 10 km from the edge of the pan on raised ground surrounded by water. Vulture species are also known to breed in trees on the vegetated islands in the pan (Brain Pers. comm., Kötting Pers. comm. and Versveld Pers. comm.).

Wetland associated birds (to numerous to mention here) including the species breeding at the site migrate to the Etosha Pan after good rains to feed on aquatic invertebrates. Predatory and scavenger avian species – usually associated with the flamingo breeding attempts – such as Tawny Eagle (*Aquila rapax*), Grey-headed Gull, Pied Crow (*Corvus alba*), Black Crow (*Corvus corvus*) and Lappet-faced Vulture (*Torgos tracheliotus*) have been known to prey on flamingo nestling and eggs (Berry 1972).

A wetland such as Etosha Pan – albeit temporary of nature – would serve as a stopover for migratory bird species. This has been documented for flamingos (Berry 1972) and would be expected for any other aquatic migratory species, of which many different species visit Namibia annually.

3.1.1.3 Herpetofauna

According to Baker (1996) 110 reptile species and 16 amphibian species are in the Etosha National Park area although not all are associated with the pan. According to Griffin and Channing (1991) relatively few (only eight species) Namibian reptiles are dependent on wetlands due to the country being primarily arid and semi-arid in nature. Furthermore Griffin and Channing (1991) state that 12 species (verified) and 4 species (not verified, but expected) of reptiles and amphibians are associated with the seasonal wetlands of Owambo and the Etosha Pan area. Of these, six (verified) and three (not verified, but expected) belong to the

Bufonidae and Ranidae (Amphibians). Jurgens (1979) states that of the 127 anuran species found in Southern Africa only 10 are found within the boundaries of the Etosha National Park. This scarcity of anuran species is probably as a result of the low number of different habitats and also the chemical composition of the water (pH levels vary from 8.92 to 10.3 in Etosha with high concentrations of dissolved chemicals making the water unsuitable for anuran use) (Jurgens 1979).

A Namibian endemic, *Agama etoshae* (Etosha Agama), occurs in sandy areas in the Etosha National Park area although not specifically associated with the pan. A Botswana endemic, *Agama makarikarica* (Makgadikadi Spiny Agama), on the other hand is restricted to the Nata and Makgadikadi salt flats in northeastern Botswana (Branch 1998). Other than the occasional *Psammophis* (Sand Snake) species crossing the pan no retiles are specifically associated with the pan itself (M. Griffin Pers. comm.).

3.1.1.4 Fish

According to Baker (1996) only one fish species occurs in the Etosha National Park area. Bethune and Robberts (1991) state that at least 49 species of fish are present – albeit temporarily after flooding episodes – in the Owambo and Etosha area. Fish of these temporary pans in northern Namibia are dominated by the genera *Barbus*, *Clarias* and *Oreochromis* (Van der Waal 1991). Fish species that do enter the Etosha Pan under flood conditions are usually located in the Ekuma area (north central part of the pan) which is the major inlet from the Cuvelai drainage system further north and usually consist of *Barbus* species (Brain Pers.comm., Kötting Pers.comm. and Versveld Pers.comm.). According to S. Bethune (Pers. Comm.) only the hardiest species survive the "migration" during occasional flooding episodes into the Etosha Pan and don't survive for long either due to the temporary nature of the pan.

3.1.2 Coastal Saline Flats

3.1.2.1 Mammals

Mammals are not common along Namibia's extremely arid coastal areas more commonly known as the Namib Desert, including the northern section aptly termed the "Skeleton Coast". Rainfall is irregular with as little as 15–100 mm precipitating annually (Baker 1996). Mammals that do occur and that undoubtedly utilize the saline flats whilst foraging include predators such as the Black-backed Jackal, Brown Hyaena (*Hyaena brunnea*) and occasionally Lion – in the far northern reaches of the Namibian coast (Baker 1996; Lovegrove 1993; O'Toole 1996). Cape Fur Seals (*Arctocephalus pusillus*) have breeding colonies at Sandwich Harbour and Cape Cross (both areas well endowed with saline flats) and consequently support high densities of Black-backed Jackal and Brown Hyaena. Black-backed Jackal's are masters at foraging for beached prey originating from the sea and very rarely rely on the desert for food (Lovegrove 1993). Oryx and Springbok are two ungulate species also often seen along the inhospitable Namibian coast and obtain their water at springs emerging close to the coast at the mouth of some of the westward flowing ephemeral rivers. The authors personally observed tracks of the above-mentioned species crossing coastal flats probably in search of food or moving from one area to another.

3.1.2.2 Birds

The combination of nutrient rich ocean waters, tidal shoreline and coastal wetlands along the Namibian coast provides a habitat and feeding ground for many species of seabirds and waders, that congregate in very large numbers at certain times of the year (Lowery 2001). The coastal region south of Swakopmund, including Walvis Bay and Sandwich Harbour (both including saline estuaries and flats situated in the central western part of Namibia's coastline), offers sanctuary to hundreds of thousands of wading birds (Baker 1996). Of the 21 sites identified as important for birds in Namibia, 10 are along the Namibian coast and all someway associated with the saline flats systems (Simmons et al. 1998). The Greater and Lesser Flamingos – which occasionally migrate to and breed at the Etosha Pan – extensively use these coastal wetlands for feeding (Berry 1972). It has been estimated that the Swakopmund coastal salt flats support at least 1% of the Southern African population of the aforementioned two species and up to 250,000 pairs of Cape Cormorants (*Phalacrocorax capensis*) (Lowery 2001). The artificial saltworks (salt extracted by means of evaporation of seawater) at Swakopmund supports more than 20,000 shore-birds (Williams 1988). Globally near-threatened species that occur at the Swakopmund saltworks include the Lesser Flamingo, Damara Tern (*Sterna balaenarum*) and African Black Oystercatcher (*Haematopus moquini*) (Simmons et al. 1998).

Overall populations of wetland birds range between 37, 500 and 78,200 (peak counts up to 150,000) at the Walvis Bay saline flats (Noli-Peard and Williams 1991). Globally near-threatened species that occur at the Walvis Bay saline flats include the Lesser Flamingo, Damara Tern, African Black Oystercatcher and Bank Cormorant (*Phalacrocorax neglectus*) (Simmons et al. 1998).

Sandwich Harbour seasonally supports up to 25% of the estimated world population of the Chestnutbanded Plover (*Charadrius pallidus*) and >3% of the sucontinental populations of the Blacknecked Grebe (*Podiceps nigricollis*) (Noli-Peard and Williams 1991). Globally near-threatened species that occur at Sandwich Harbour saline flats include the Lesser Flamingo, Damara Tern, African Black Oystercatcher and Bank Cormorant (Simmons et al. 1998).

There is also a rich variety of waders in the Cape Cross Seal Reserve further north (saline flats are common throughout this area) with the endemic and rare Damara Tern breeding in the vicinity (Baker 1996) as well as other localities along the Namibian coast, often in conjunction and/or close to saline flats (Clinning 1978) although more commonly on gravel plains (Frost 1976). Williams (1991) shows that the Cape Cross saline lagoons support between 6–16% of the non-breeding populations of the endemic Southern African race of the Blacknecked Grebe (*Podiceps nigricollis gurneyi*) and between 3,000 and 7,500 birds of any 30 species at any time. Globally near-threatened species that occur at the Cape Cross saline flats include the Lesser Flamingo and Damara Tern (Simmons et al. 1998).

The Namibian feeding grounds are essential to the survival of a large variety of Palearctic migrants, which include species such as the Ruff (*Phylomachus pugnax*) and the Curlew Sandpiper (*Calidris ferruginea*) (Lowery 2001).

3.1.2.3 Herpetofauna

The herpetofauna associated with coastal saline flats is extremely limited at best with few reptiles specifically associated with any type of wetland from Namibia (Griffin and Channing 1991). The mouth of the Cunene River (river separating Angola from Namibia in the north western part of the country) is of particular interest due to it being the southernmost (disjunct) range of the Nile Softshelled Turtle (*Trionyx triunguis*) (Branch 1998; Griffin and Channing 1991). Green turtles (*Chelonia midas*) and even the Nile Crocodile (*Crocodylus niloticus*) have been reported in this same vicinity (Branch 1998; Griffin and Channing 1991) although not exclusively associated with the saline flats abounding in the area. A roadkill of a *Psammophis* (Sand Snake) species (unconfirmed) in the vicinity of Cape Cross, adjacent to coastal saline flats in the area, suggests that other reptiles do occur there although not necessarily specifically associated with these areas (Pers. obs.). Amphibians would generally not be expected in these saline areas due to them not typically being suited morphologically to inhabit saline areas (Loughland and Cunningham 2002).

3.1.2.4 Fish

Although no fish species are exclusively associated with coastal saline flats in Namibia, the larger saline estuaries and flats occasionally inundated during high tides may hold fish temporarily. Bethune and Robberts (1991) have documented 36 estuarine and marine fish species occurring in coastal wetlands along the Namibian coast.

3.2 Flora

3.2.1 Etosha Pan

The high sodium content of most of the pans results in vegetation being non-existent or very rare with few species adapted to cope with halophytic soil conditions. According to Le Roux (1980) the Etosha Pan is largely without any vegetation due to the high sodium content. The whitish clay soils have a sodium contents in excess of 30,000 ppm and the pH varies between 8.8 and 10.2, which is also very high. In good rain years some grass species such as *Sporobolus salsus* and *Sporobolus spicatus* may occur on the pan (Berry et al. 1997; Le Roux 1980; Mendelsohn et al. 2000).

Along the pan margins the dominant perennial grasses are *Odyssea paucinervis*, *Sporobolus spicatus*, *S. ioclados*, *S. tenellus* and the sedge *Cyperus marginatus* (Mendelsohn et al. 2000). Woody species are generally absent but salt-loving woody dwarf-shrubs such as *Suaeda articulata* and *Sporobolus salsus* can be found on sand hummocks, which occur as little "islands" in the pan and on the pan margins (Le Roux 1980; Mendelsohn et al. 2000). *Salvadora persica* is occasionally found on some of the larger islands (Berry et al. 1997).

Several different vegetation types, in most cases well adapted to halophytic conditions, occur at the very edge of the pan. These vegetation types are classified by Le Roux (1980) as follows.

The **Andoni Veld** vegetation type is situated to the north east of the Etosha Pan. The vegetation in this vegetation type is dominated by an almost mono-specific stand of the perennial course grass species *Sporobolus spicatus*. Other prominent grass species in this vegetation type include *Crassipedorachis sarmentoasa*, *Digitaria setrivalva*, *Eragrostis sabine* and *Odyssea paucinervis*. Shrubs such as *Dichrostachys cinerea* are currently encroaching into this vegetation type.

To the southwest and south of the Etosha Pan, sweet grasses are associated with lime soils and classified as the **Sweet grass veld**. The vegetation in this area can generally be described as treeless plains although three tree species that do however occur in this area include *Acacia mellifera*, *Acacia nebrownii* and *Acacia reficiens*. *Albizia anthelmintica* may also be present in certain areas. Thickets of *A. nebrownii* occur on the southern side near the edge of the pan.

To the west of the pan small shrubs, such as *Monechma tonsum*, *Monechma qenistifolium* and *Petalidium enqelerianum* dominate the vegetation. Closer to the edge of the pan *Salsola* species such as *Salsola aphylla* and *Suaeda articulata* are found on the halophytic soils. Some perennial grass species such as *Cenchrus ciliaris* and *Stipagrostis hochstetteriana* may occur, but more common grasses such *Aristida adensionis*, *Enneapogon cenchroides*, *Enneapogon desvauxii* and *Eragrostis nindensis* are more typical for this vegetation type.

On the western side of the Etosha Pan shrubs such as *Leucosphera bainesii* and *Salsola tuberculatec* dominate the vegetation type known as **Ondeka Duneveld** and grow right up to the edge of the pan. Some of the important grasses are *Enneapogon desvauxii*, *Eragrostis sabinea* and *Eragrostis porosa*. Herbs such as *Monechma dwaricatum*, *Herpicium gazansoides* and *Zygophyllum simplex* also occur in this vegetation type.

On the northern side of the pan a geographic feature known as Poachers Peninsula – peninsula extending southwards into the pan – is dominated by shrubs such as *Catophraetes alexanderii* and *Dichrostachys cinerea* while the grass layer is dominated by *Stipagrostis uniplumis*.

To the east of Poachers Peninsula a relative small vegetation type known as the **Ekuma grass lands** extends down to the edge of the pan from the north. The vegetation can be described as a tall grass veld with palatable grasses such as *Anthephora pubescens*, *Cenchrus ciliaris*, *Schmidtia pappophoroides* and *Stipagrostis uniplumis*. The most important shrub in this area is *Leucosphera bainesii*.

On the northeastern side of the pan the vegetation is characterized by a **Sandveld vegetation type**. The vegetation in this area is dominated by tree species such as *Acacia erioloba*, *Acacia mellifera*, *Philenoptera nelsii*, *Terminalia sericea* and *Terminalia prunioides*. Several *Grewia* species for example *Grewia flava* are typical of the shrub components that occur in this area. The grass layer is poorly developed with *Schmidtia kalaharienses* as the most prominent species.

Clarke (1998) describes the vegetation associated with saline pools and flats slightly to the north of the Etosha Pan, but still in the Cuvelai drainage basin, as sparse and poorly vegetated. Species that are characteristic of saline pools in this area include the small grass *Sporobolus coromandelianus* and large clumps of *Sporobolus ioclados*. Sandy edges support *Odyssea paucinervis* and *Sporobolus spicatus*. Trees are absent possibly due to the waterlogged and saline soils in this area.

3.2.2 Coastal Saline Flats

The coastal saline flats are devoid of vegetation (except for some algae species) except on the edges of the flats. Similar to the Etosha Pan, only vegetation adapted to halophytic conditions occurs around these flats. In general the density of vegetation is low in the coastal areas, but does increase further away from the flats.

Species such as *Anthrocnemum africanum* and *Sarcornia natalensis* are succulent-like perennial herbs (small ±30 cm in height) that succeed in growing in these very saline conditions. Small shrubs such as *Galenia papulasa*, *Lycium cinereum*, *Salsola* species, *Suaeda* species, *Psilocaulon* species, *Tamarix usineoides* and *Zygophyllum* species occur slightly further away (i.e. inland). Members of the grass family that succeeded in establishing themselves in this area are *Odyssea paucinervis*, *Sporobolus nebulosus* and the well-known reed, *Phragmites australis*, which occur in these halophytic conditions especially near an area called "Jakkalsputs" north of Swakopmund. The salt marsh and adjacent brackish waters at Sandwich Harbour support four communities – *Sarcocornia natalensis* var. *affinis*, *Sporobolus virginicus*, *Typha capensis* and the *Odyssea paucinervis* communities, whose distribution is determined by the salinity gradient between the fresh water seepage and the marine system (Robinson 1976).

In many cases the vegetation grows on sand hummocks created by the wind. On the gravel/gypsum plains the most prominent plant is a perennial shrub (±30–40 cm high) known as *Arthraerua leubnitzii*. A large number of endemic lichens are also found on these gravel plains. These plants are a symbiosis between fungi and algae and are dependent on the fog-regime along the coast for their moisture requirements.

4 Conclusion

As far as could be determined no specific fauna (mammals, birds, reptiles and fish) and flora are specifically associated with inland and coastal saline flats in Namibia. This however, should not detract from the immense importance such systems have in terms of supporting a wide variety of life especially in an arid country such as Namibia. Not only are the saline flats important from a national perspective such as the tourism potential of Etosha National Park, Cape Cross Seal Reserve, Walvis Bay lagoon, etc., but also from an international perspective. The fact that these saline flats especially the coastal flats host Palearctic waders, underscores its international importance.

The ecological importance of saline flats in Namibia is often overlooked and certainly not well understood and usually deemed worthless from a farming and tourism perspective. Secondary benefits of these saline flats such as the importance for migratory waders that in their turn attract tourists (birders) should be encouraged. The importance of these saline flats should not be underestimated and it is suggested that an awareness campaign, similar to the awareness campaigns regarding lichens and Damara Terns, should be initiated in Namibia. Further research into the saline flats system(s) from Namibia is warranted especially regarding the invertebrates, which form the basis of the food chain for migratory birds. This little understood and unique landform deserves further attention.

Acknowledgements We would like to convey our sincere appreciation to the following people for their assistance in gathering information towards this paper:

Ministery of Environment and Tourism staff from the Okaukeujo Research Centre in the Etosha National Park with special thanks to Nad Brain, Birgitte Kötting, Wilfred Versveld and Mike Griffin.

References

Baker L (1996) An overview of Namibia's game parks and recreation areas. Namibia Environ 1:32–50

Berry HH (1972) Flamingo breeding on the Etosha Pan, South West Africa, during 1971. Madoqua 1(5):5–31

Berry HH (1997) Historical review of the Etosha region and its subsequent administration as a National Park. Madoqua 10(1):3–12

Berry HH, Rochèr CJV, Paxton M, Cooper T (1997) Origin and meaning of place names in the Etosha National Park, Namibia. Madoqua 10(1):13–35

Bethune S, Robberts KS (1991) Checklist of the fishes of Namibia for each wetland region. Madoqua 17(2):193–199

Branch B (1998) Field guide to snakes and other reptiles of Southern Africa. Struik Publishers, Cape Town, South Africa, pp 399

Buch MW (1997) Etosha Pan – the third largest lake in the world? Madoqua 10(1):49–64

Clarke NV (1998) Guide to the common plants of the Cuvelai wetlands. Southern African Botanical Diversity Network, Windhoek, Namibia, pp 43

Clinning CF (1978) The biology and conservation of the Damara tern in South West Africa. Madoqua 11(1):31–39

Frost PGH (1976) Breeding adaptations of the Damara Tern *Sterna balaenarum*. Madoqua 9(3):33–39

Giess W (1971) A provisional vegetation map of South West Africa. Dinteria 4:5–112

Griffin M, Channing A (1991) Wetland-associated reptiles and amphibians of Namibia – a national review. Madoqua 17(2):221–225

Griffin M, Grobler HWJ (1991) Wetland-associated mammals of Namibia – a national review. Madoqua 17(2):233–237

Grünert N (2000) Namibia fascination of geology. Klaus Hess, Germany, pp 176

Hines C, Kolberg H (1996) Importance of wetland management in arid regions. Namibia Environ 1:75–78

Jurgens JD (1979) The anura of the Etosha National Park. Madoqua 2(3):185–208

Kolberg H Preliminary inventory of Namibia's wetlands. Directorate of Scientific Services, Ministry of Environment and Tourism, Windhoek, Namibia

Le Roux CJG (1980) Vegetation classification and related studies in the Etosha National Park. D.Sc. thesis, University of Pretoria, Pretoria, South Africa, pp 364

Lindeque M, Archibald TJ (1991) Seasonal wetlands in Owambo and the Etosha National Park. Madoqua 17(2):129–133

Loughland RA, Cunningham PL (2002) Vertebrate fauna of sabkhat from the Arabian Peninsula: a review of mammalia, reptilia & amphibia. In: Barth, Böer (eds) Sabkha ecosystems. Kluwer Academic, The Netherlands

Lovegrove B (1993) The living deserts of Southern Africa. Fernwood Press, Vlaeberg, South Africa, pp 224

Lowery R (Ed) (2001) Wetlands in Namibia. A resource book for colleges and schools in Namibia. Unpublished report, Swedish International Development Co-operation Agency (SIDA)

Mendelsohn J, El Obeid S, Roberts C (2000) A profile of north-central Namibia. Directorate of Environmental Affairs, Windhoek, pp 79

Noli-Peard KR, Williams AJ (1991) Wetlands of the Namib Coast. Madoqua 17(2):147–153

O'Toole M (1996) Namibia's marine environment. Namibia Environ 1:51–55

Robinson ER (1976) Phytosociology of the Namib Desert Park South West Africa. Unpublished M.Sc. thesis, University of Pietermaritzburg, Pietermaritzburg, South Africa

Simmons RE, Boix-Hinzen C, Barnes KN, Jarvis AM, Robertson A (1998) Important bird areas of Namibia. In: Barnes KN (ed) The important bird areas of Southern Africa. BirdLife South Africa, Johannesberg

Strohbach M (1996) Exciting world of Namibian flora. Namibia Environ 1:79–82

Van der Waal BCW (1991) Fish life of the oshona delta in Owambo, Namibia, and the translocation of Cunene species. Madoqua 17(2):201–209

Wellington JE (1938) The Kunene River and the Etosha plain. S Afr J Geogr 20:150

Williams AJ (1988) Walvis Bay and other coastal gems. Afr Wildl 42:82–85

Williams AJ (1991) Numbers and conservation importance of coastal birds at the Cape Cross Lagoons, Namibia. Madoqua 17(2):239–243

Introducing the Namib Desert Playas

Frank D. Eckardt and N. Drake

Abstract The hyper arid Central Namib gravel plains to the north of the Kuiseb River feature a series of playa clusters which are nested in shallow drainage channels. The playas form primarily as the result of groundwater water ponding at bedrock obstructions which produces surface salts such as halite and gypsum on top of thin stream sediments. They differ considerable in size, shape and composition when compared with playas elsewhere in the region such as the Kalahari and Karoo.

1 Introduction

The Namib Desert is mainly associated with dunes, inselbergs, escarpments, gravel plains and coastal sabkhas. This paper aims to add desert playas to the list of landforms of the coastal region. The pans occur in a number clusters, in particular in the central region between the Ugab and Kuiseb Rivers on the broad, 120 km wide, shallow gradient (1%) gravel plains below the escarpment (Fig. 1).

The playas or pans are nowhere near as well known, as the Etosha pan complex or their counterparts in the Kalahari but deserve attention for a number of reasons. The pans produce some of the most saline waters in southern Africa (Day 1993). Furthermore, while currently occupying less than 5% of the Namib gravel plains, given the age of the desert, the pans must have played an important role in the denudation of the inselberg landscape. The aggressive salt weathering environment in the Namib (Goudie et al. 1997; Goudie and Parker 1998) may be amplified by salt deflation from pans and the high relative humidity and fog advection for which the region is known. The deflation of gypsum dust from the pans may also have created the widespread pedogenic gypsum crusts (Eckardt et al. 2001; Eckardt and Spiro 1999). Many of the sites described here feature small salty springs. Most of the pans described here have no name. Some names given here are not official but refer to nearby place names which are used as a means to distinguish between the different systems.

The environment and geomorphology of the Namib Desert has been described previously (Meigs 1966; Logan 1969). Rainfall gradually decreases from the semi-arid environment found in the highlands (250 mm), towards the Pro Namib Desert (70 mm) and on to the hyper arid coast, which receives less than 20 mm of rainfall per annum. The Namib Desert is considered to have been hyper-arid for the last 5 million years (Ward et al. 1983).

The playas are found on the gravel plains, which exhibit an extensive but subdued and poorly developed drainage system. This dentritic drainage network provides a conduit for sparse Pro Namib rain towards the hyper-arid Atlantic coast. Playas occur at numerous locations, clusters have been identified in Landsat Thematic Mapper imagery and aerial photography. The ones visited in the field were to the northeast of Swakopmund, on the Welwitschia Flats and towards the east of the Okahandja Lineament and will be described in turn. In addition to some basic field validation and XRD analyses of salts, water was collected and analysed for its chemical composition. About ten pans have been visited throughout the area and are considered representative of what can be expected in the coastal region.

F.D. Eckardt (✉)
University of Cape Town, Department of Environmental and Geographical Science, 7701 Rondebosch, South Africa
e-mail: frank.eckardt@uct.ac.za

N. Drake
Department of Geography, Kings College, Strand, London, UK

M. Öztürk et al. (eds.), *Sabkha Ecosystems*, Tasks for Vegetation Science 46,
DOI 10.1007/978-90-481-9673-9_3, © Springer Science+Business Media B.V. 2011

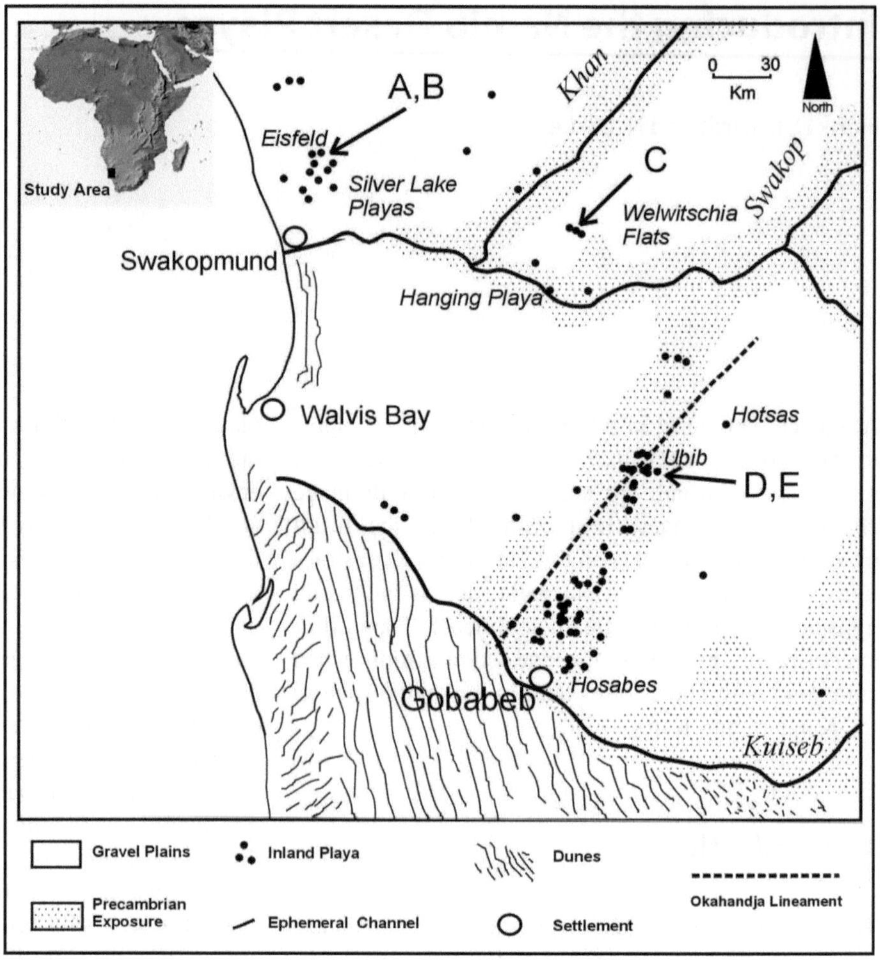

Fig. 1 Map of Study Area and major pans in the Central Namib (A, B, C, D, E Locations of Sections in Figure 3)

2 Results

2.1 The Silver Lake Playas

The playas are found 10 km to the NE of Swakopmund and form part of the Silver Lake Pans (Roper undated; Table 1, Fig. 2). Here Ehrhorn's playa has briefly been mined for halite in the 1930s, while Eisfeld, Kaliman and Kalifrau seem to have been subjected to some initial prospecting (Gevers and Van der Westhuyzen 1931; Schneider and Genis 1992; Table 1). Judging by the rusty diggers left at those pans, salt mining has been abandoned a long time ago. Salt production has entirely shifted to the nearby coastal lagoons where marine water is evaporated in a series of shallow ponds such as Walvis Bay and Swakopmund. These are among the largest salt productions sites in Africa.

Two short soil profiles have been obtained at Eisfeld which reveal halite, gypsum and moist, clay rich sediments, superimposed onto a shallow Precambrian bedrock (Fig. 3a, b). The pans form due to relatively impermeable dolerite outcrops traversing drainage channels that appear to carry groundwater, as is suggested by spring discharge at a number of points. This results in groundwater ponding behind the dense network of dolerite dyke outcrops which results in the rise of the water table and which promotes evaporation (Plate 1).

2.2 The Welwitschia Flats Playas

A second and less pronounced cluster of playas exists on the Welwitschia Flats (Table 2). Again these are the result of groundwater ponding. While these pans are

Table 1 Silver Lake Playas as identified in Landsat Thematic Mapper image

Playa	Size (ha)	Latitude	Longitude
1	21	22° 26′ 09″	14° 35′ 09″
2	69	22° 28′ 22″	14° 34′ 33″
3 Eisfeld	76	22° 31′ 12″	14° 32′ 56″
4 Ehrhorn	63	22° 32′ 24″	14° 34′ 19″
5	43	22° 33′ 28″	14° 37′ 37″
6	20	22° 29′ 17″	14° 37′ 16″
7	25	22° 28′ 01″	14° 37′ 22″
8	19	22° 27′ 09″	14° 38′ 43″
10 Kaliman	10	22° 35′ 09″	14° 33′ 00″
11 Kalifrau	16	22° 34′ 21″	14° 32′ 39″

comparable in terms of size to the Silver Lake playas, they appear to form due to Precambrian marble obstructions rather than Karoo dolerite dykes. The morphology of the playas is comparable to the Silver Lake pans at Swakopmund and features thin, moist sediment overlying Precambrian rocks (Fig. 3c).

An unusually perched hanging playa can be found at the nearby Swakop Canyon rim, upstream from the Swakop/Khan confluence (Lat 22° 42′ 34″ S, Lon 14° 59′ 09″ E). This has created a massive contemporary pan-type salt deposit 50 m high, covering a length of nearly 200 m on the north side of the steep the Canyon

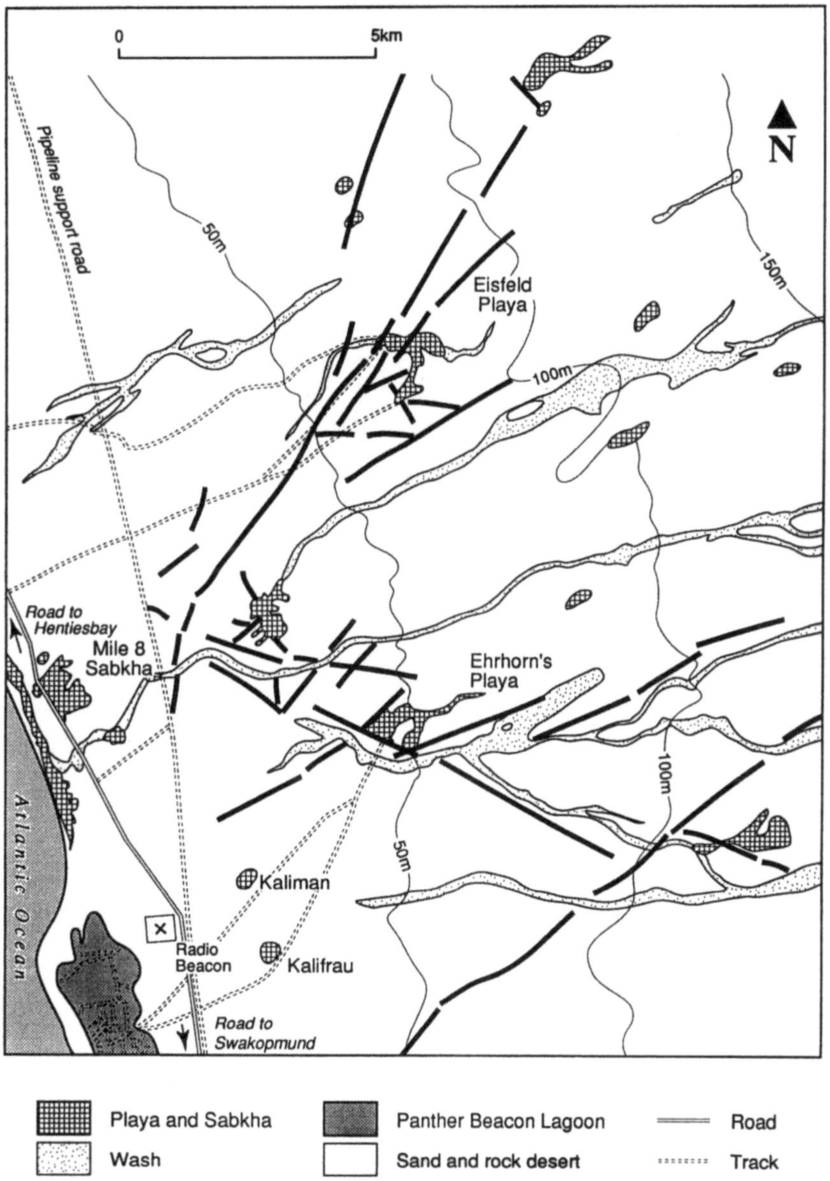

Fig. 2 Map of Silver Lake Playas and dolerite ridges near Swakopmund

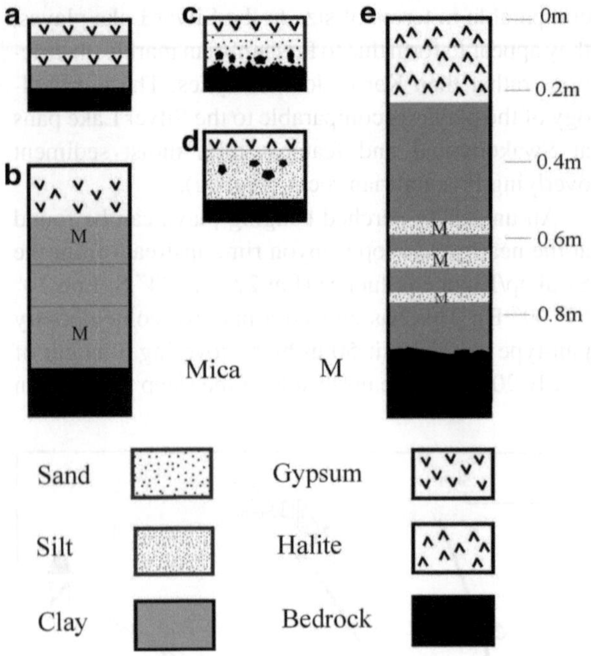

Fig. 3 Section of playas taken in the Namib. (**a**, **b**) Eisfeld, (**c**) Welwitschia Flats, (**d**, **e**) Ubib

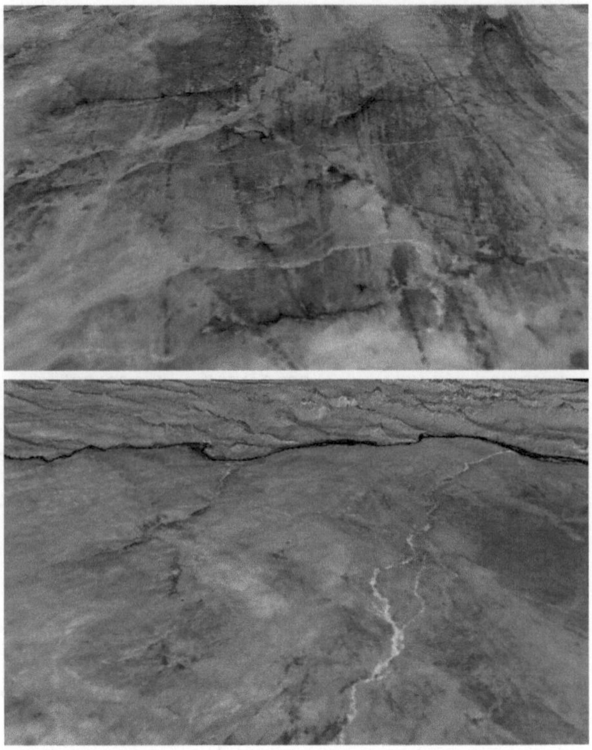

Plate 2 Three dimensional terrain Visualisation. Landsat Enhanced Thematic Mapper image (Display: *Red, Green, Blue, Bands 7,4,2*) draped of Shuttle Radar Topography Mission data depicting playas at Okahandja Lineament. Salt and gypsum appear as dark patches. (top) Playas at Ubib, note gravel road network. (bottom) Playas at Hosabes, note Kuiseb River and dunes to the south

Plate 1 Low level aerial photo of Eisfeld playa surrounded by gravel plains. (**A**) Dolerite Ridges, (**B**) Pan Surface, (**C**) Wash (note tyre tracks for scale)

Table 2 Playas on the Welwitschia Flats as identified in LandsatThematic Mapper image

	Size (ha)	Latitude	Longitude
1	31	22° 38′ 46″	15° 03′ 59″
2	32	22° 39′ 17″	15° 04′ 04″
3	40	22° 37′ 43″	15° 06′ 41″

wall. Halite stalactites and stalagmites have coated material such as ostrich feathers and other detritus on the dry river bed.

2.3 The Okahandja Lineament Playas

The third and largest accumulation of playas can be established through the careful examination of aerial photography, which highlights the presence of between 50–60 small playas (<5 ha). Pans are found within the gravel plain drainage and can be identified on aerial photography by their dark appearance, caused by the presence of silt and clays, polygonal thrust faults and phreatophyte mounds. It appears that these playas are usually the result of Damara bedrock outcrops associated with the Tinkas schists and the Salem granites in the vicinity of the Okahandja lineament (Fig. 1) and fault lines to its east (Plate 2). They form a pronounced belt which runs from Hosabes playa near Gobabeb on to Ubib Spring and terminates to the west of Hotsas water hole which approximately coincides with the eastern margin of appreciable gypsum occurrences in the Central Namib Desert.

The presence of clay, silt and some sand sized material in most of the playas suggests surface water input to take place. This is likely to occur during sporadic high magnitude rain events, which will cause surface run-off and erode material from inselbergs and the gravel plain into playa depressions. The deepest profile (Fig. 3d, e) is found in a playa, which does not have an outflow but instead two distinct inflows, with different types of source materials from different sized catchments producing alternating layers in its profile. Other playas, which are positioned in more pronounced drainage channels, promote coarser and thinner profiles.

2.4 Water Chemistry

Water can be found at most Namib playas (Table 3). Some feature permanent springs and some water holes may be maintained by frequenting gemsboks and springboks. In some cases water may only be assessed by digging a shallow pit.

3 Discussion

The Namib playas feature thin halite (NaCl) and gypsum crusts ($CaSO_4{\cdot}2H_2O$) have phreatophyte mounds, thrust polygon surfaces or other irregular microtopagraphy. The playas form, as sporadic surface water and sparse groundwater traverse the 150 km wide desert from east to west. The water is subject to evaporative concentration in particular when the porous tertiary colluvium of the plains is punctuated by impermeable bedrock such as dolerite, marble or granite. All playas presented here occupy drainage channels and depression, which have accumulated sediments that are often less than 1 m deep. The crusts are replenished by capillary rise of shallow groundwater stored in moist, fine-grained clays and silt. Permanent surface water is present at a few small pools where it reaches extreme salinities in an environment of very high potential evaporation rates and moderate flow rates.

The small surface water pools appears perennial as they were present in 1994 and 1995, and some had even been sampled in the 1980s (Day 1993). The salinity changes at Welwitschia Playa (−18%), the Hanging Playa (−11%) and Eisfeld (+3%) are noted to be minimal, suggesting that evaporation, inflow and crystallisation have attained equilibrium. Hosabes samples show considerable temporal variability (+72%), which could be controlled by a number of factors, including sampling distance from discharge point as well as mixing of stratified bottom and surface water.

The hypersaline spring at Hosabes was investigated in detail by Day and Seely (1988) revealing flow rate maxima at 0.023 m/s with discharge ranging from 0.22 to 0.75 Conductivity was observed to double over a flow length of 80 m, with concentration being controlled by channel depth and distance from source.

Other evaporates such as sylvite (KCl) and humberstonite ($K_3Na_7Mg_2(SO_4)6(NO_3)2 \cdot H_2O$) have also been identified at Hosabes (Lat 23° 30′ 15″ S, Lon 15° 03′ 11″ E). Humberstonite is a rare hydrous sulphate-nitrate of potassium, sodium and magnesium, which has for the first time been found outside the Atacama Desert (Mrose et al. 1970).

The composition of Namib Playas is comparable to the coast parallel sabkhas, which stretch from Angola into South Africa. More than 20 well-established sabkhas with a cumulative area of 300 km² are known. The coast

Table 3 Playa water samples obtained in 1994 and 1995 (all figures in ppm)

Water Samples	date	pH	TDS	SO_4	Cl	Na	Ca	Mg	K	HCO_3
Ubib playa east	23.5.94	8.1	11,213	800	4,800	2,850	330	885	270	176
Ubib playa east	2.10.95	7.8	16,104	12	7,400	3,900	387	1,153	330	178
Eisfeld playa	23.5.94	7.4	46,728	3,000	26,000	15,300	3,496	3,089	215	170
Eisfeld playa	10.10.95	7.6	48,444	19	34,600	18,200	4,545	3,377	460	286
Hosabes playa	24.5.24	8.1	111,203	10,000	46,000	32,500	190	4,859	580	200
Hosabes playa	10.10.95	8.1	192,060	25,200	96,000	67,000	55	7,659	1,940	226
Hanging playa	23.5.94	7.7	20,328	2,040	11,400	5,000	2,597	1,977	194	148
Hanging playa	10.10.95	7.6	18,216	2,120	10,400	5,500	2,447	1,771	185	104
Welwitschia playa	17.5.94	7.7	26,994	2,000	13,600	7,050	3,646	2,553	185	112
Welwitschia playa	10.10.95	7.9	22,242	2,420	13,100	6,800	3196	2,059	280	108

Table 4 Playas identified in the Central Namib, Keetmanshoop District and Owamboland

Playa	Location	Size (ha)	Evaporites
Central Namib Desert			
Reserve 0, 1, 2, 3	Henties Bay area	25	Halite (NaCl) Gypsum ($CaSO_4{\cdot}2H_2O$)
Rottraut	Henties Bay area	60	Halite (NaCl) Gypsum ($CaSO_4{\cdot}2H_2O$)
Marko	Hentiesbaai area	50	Halite (NaCl) Gypsum ($CaSO_4{\cdot}2H_2O$)
Eisfeld	17 km north east of Swakopmund	20	Halite (NaCl) Gypsum ($CaSO_4{\cdot}2H_2O$)
Kaliman	2 km east of Panther Beacon	8	Halite (NaCl) Gypsum ($CaSO_4{\cdot}2H_2O$)
Kalifrau	2 km east of Panther Beacon	11	Halite (NaCl) Gypsum ($CaSO_4{\cdot}2H_2O$)
Ehrhorn	16 km north-northeast of Swakopmund	30	Halite (NaCl) Gypsum ($CaSO_4{\cdot}2H_2O$)
Arandis	13 km north of Arandis	11	Halite (NaCl) Gypsum ($CaSO_4{\cdot}2H_2O$)
Tsondab Vlei	Namib Naukluft Park Tsondab River (size obtained from Landsat Thematic Mapper imagery)	660	Magnesium calcite (MgCa), Dolomite ($CaMg(CO_3)2$) Aragonite ($CaCO_3$), Halite (NaCl)
Soutrivier	Namib Naukluft Park, Central Namib	3–4	Halite (NaCl) Gypsum ($CaSO_4{\cdot}2H_2O$)
Keetmanshoop District			
Keetmanshoop District Pans	Keetmanshoop District		Glauber salt (Na_2SO_4)
Verwall Pan	15 km east of Koes	550	Epsom salt ($MgSO_4$)
	Keetmanshoop district		Halite (NaCl) Thenardite (Na_2SO_4) Sylvite (KCl)
Owamboland			
Odangwa Pan	Owamboland Etosha (Aminius)	2,100	Halite (NaCl) Thenardite (Na_2SO_4) Sylvite (KCl) Anhydrite ($CaSO_4$) Calcium carbonate ($CaCO_3$) Magnesium chloride ($MgCl_2$)
Otjivalundu Pans No. 1 and 2	Owamboland Etosha	40 and 400	Thenardite (Na_2SO_4) Trona ($NaH(CO_3)_2{\cdot}11H_2O$) Sulphohalite ($2Na_2SO_4{\cdot}NaCl{\cdot}NaF$) Pirssonite ($CaCO_3{\cdot}NaCO_3{\cdot}2H_2O$)

Source: Forshag 1933; Teller and Last 1990; Schneider and Genis 1992.

parallel, flat, salt encrusted surfaces, often less than a meter above mean sea level, are found near the high water mark between the Kunene and Orange Rivers. Some are well documented (Schneider and Genis 1992) and have been described already in the 1930s. Gypsum and halite exists at Swakopmund, Cape Cross (Gevers and Van der Westhuyzen 1931), Conception Bay (Kaiser and Neumaier (1932) and Torra Bay (Torien 1964).

These playas and sabkhas in the hyper arid Namib differ significantly from those found elsewhere in

Namibia (Forshag 1933; Schneider and Genis 1992). The well-known scenic Vleis or lake deposits associated with the Namib Sand Sea such as Sossus, Koichab and Tsondab Vlei are recharge points in the permeable sand sea. Shortlived highland floods terminate in the dunes where they produce temporary lakes and leave behind fine calcified sediments and shorelines (Teller and Last 1990). In the Kalahari and Etosha area, playas hold Epsom salts ($MgSO_4$), halite (NaCl), thenardite (Na_2SO_4), anhydrite ($CaSO_4$), calcium carbonate ($CaCO_3$), magnesium chloride ($MgCl_2$), and sylvite (KCl). The playas of the Etosha Pan in northern Namibia hold, thenardite (Na_2SO_4), trona ($NaH(CO_3)_2 \cdot 11H_2O$), sulphohalite ($2Na_2SO_4 \cdot NaCl{\cdot}NaF$), and pirssonite ($CaCO_3{\cdot}NaCO_3 \cdot 2H_2O$; Table 4.

4 Conclusion

There are numerous playas throughout the Central Namib which precipitate halite and gypsum on the gently sloping drainage of the gravel plains. The playas are best developed in areas where bedrock obstructions run perpendicular to the drainage pattern and groundwater flow. Ground water appears to be present all year round despite conditions of hyper aridity. In most cases playas have a distinct in and outflow channel and do not appear to produce major aeolian features such as lunette dunes. In that regard they differ considerably from playas in the Kalahari and the Karoo region. Their main role in shaping the landscape of the Namib would be enhanced denudation by salt weathering. Their contribution to the regional ecology still needs to be determined.

Acknowledgement We would like to thank Dr. Mary Seely from the Desert Ecological Research Unit of Namibia for hosting us and Mrs. Conradie from Department of Water Affairs, Windhoek, Namibia who kindly assisted in the analyses of water data. Research funding was provided by the NERC Environment Research Council and the Trapnell Fund. The Namibian Ministry of Environment and Tourism provided research permits. Additional thanks has to go to Mark Barry, Christine Hänel, Dr. Peter and Kathy Jacobsen, Stephanie Wolters and Juliane Zeidler. Final thanks also goes to Dr Kevin White and Andrew Goudie.

References

Day JA (1993) The major ion chemistry of some southern African saline systems. Hydrobiologia 267(1–3):37–59

Day JA, Seely MK (1988) Physical and chemical conditions in an hypersaline spring in the Namib Desert. Hydrobiologia 160:141–153

Eckardt FD, Spiro B (1999) The origin of sulphur in gypsum and dissolved sulphate in the Central Namib Desert, Namibia. Sed Geol 123(3–4):255–273

Eckardt FD, Drake NA, Goudie AS, White K, Viles H (2001) The role of playas in the formation of pedogenic gypsum crusts of the Central Namib Desert. Earth Surf Process Land 26:1177–1193

Forshag WF (1933) Sulfohalite and other minerals from the Otjiwalundo salt pan, SWA. Am Mineralog 18:431–434

Gevers TW, Van der Westhuyzen JP (1931) The occurrences of salt in the Swakopmund area, South West Africa. Trans Geol Soc SA 34:61–80

Goudie AS, Parker AG (1998) Experimental simulation of rapid rock block disintegration by sodium chloride in a foggy coastal desert. J Arid Environ 40:347–355

Goudie AS, Viles HA, Parker AG (1997) Monitoring of rapid salt weathering in the central Namib Desert using limestone blocks. J Arid Environ 37:581–598

Kaiser E, Neumaier F (1932) Sand- steinsalz- kristallskelette aus der Namib Südwestafrikas. Zentralblatt für Mineralogie 6A:177–188

Logan RF (1969) Geography of the Central Namib Desert. In: Arid lands in perspective. University of Arizona Press, Tucos, AZ, pp 129–143

Meigs P (1966) Geography of coastal deserts, UNESCO arid zone research. UNESCO, Paris, pp 79–114

Mrose ME, Fahey JJ, Ericksen GE (1970) Mineralogical studies on the nitrate deposits in Chile III Humberstonite. Am Mineralog 55:1518–1533

Roper H (undated) Salt in the coastal strip between Swakopmund and the Huab River S.W.A. Geological survey of South Africa, Internal Report

Schneider GIC, Genis G (1992) Salt Namibian ministry of mines and energy, geological survey Namibia, Windhoek, Open File Report MRS 21

Teller JT, Last WM (1990) Paleohydrological indicators in playas and salt lakes, with examples from Canada, Australia, and Africa. Davis, Owen K. Palaeoenvironments of arid lands; a selection of papers presented at the Twelfth congress of the International Quaternary Association. University of Arizona, Department of Geoscience, Tucson, AZ, United States. Palaeogeogr Palaeoclimatol Palaeoecol 76(3–4): 215–240

Torien DK (1964) Geologiese Verslag Oor Die Soutpanne Langs Die Weskus Tussen Swakopmund En Die Kunenemund Suidwest-Africa. Namibian Ministry of Mines and Energy, Geological Survey Namibia, Windhoek, Open File Report EG 069

Ward JD, Seely MK, Lancaster N (1983) On the antiquity of the Namib. S Afr J Sci 79:175–183

Quantitative Eolian Transport of Evaporite Salts from the Makgadikgadi Depression (Ntwetwe and Sua Pans) in Northeastern Botswana: Implications for Regional Ground-Water Quality

Warren W. Wood, Frank D. Eckardt, Thomas F. Kraemer, and Ken Eng

Abstract Eolian salts from the evaporite-covered Makgadikgadi Depression in Botswana were observed in the soil as far as 150 km downwind from the depression. Over three million metric tons of chloride, sodium, and bicarbonate are transported each year from the basin to the adjacent land. Infiltering soil water mobilizes and transports these soluble salts to the ground water, where they degrade the water quality. A relation between the size of the evaporative floor area and the length of the downwind salt "footprint" on the soil was established. This permits use of readily available topographic maps to estimate the area and length of potential degradation of ground water without extensive field sampling. Significant mass of naturally-occurring radioactive 226-radium is present in the eolian-transported salts that may have deleterious health consequences for individuals residing in the downwind area.

1 Introduction

Solute concentrations of ground water in many arid and semi-arid regions are so elevated and spatially variable that they inhibit development for many domestic, livestock, irrigation, or industrial uses. In a significant number of these regions, the concentration of ground-water solutes appears to be impacted by eolian transport of evaporite salts from local or regional saline pans (Wood and Sanford 1995). Salts from these pans are transported tens of kilometers by wind, where they fall to the ground and are dissolved in recharge water to become part of the ground-water system. The resulting affect on ground water is a plume of high-solute water that reflects a combination of source, dominant wind, and ground-water flow directions.

Eolian-transport processes in arid and semi-arid areas are well documented in the geologic record by the presence of dunes, loess, volcanic ash, and other features (Péwé 1981). The observation that dust can travel great distances has been shown by detection of dust from Asia in the atmosphere over Alaska (Rahn et al. 1981) and of dust from the Sahara over the North Atlantic (Schütz et al. 1981). Eolian transport of sea aerosols on the continents also is well established (Jung and Werby 1958; Mandel and Shiftean 1981; Hingston and Gailitis 1976), as are the effects on surface water and soils (Person and Fisher 1971; Dethier 1979; Eisenreich 1980; Wiman and Agren 1985; Wiman et al. 1990). The presence of ^{3}H, ^{14}C, and ^{36}Cl isotopes from above-ground nuclear testing in many water-table aquifers worldwide clearly demonstrates the existence of eolian transport of aerosols on a global scale. However, specific contributions of solutes to ground water by eolian processes, other than the isotopes indicated above, have received relatively little attention in the literature.

Heim and Giuliana (1988) showed that radioactive material from the Chernobyl accident has contaminated the ground water regionally; Lehman and Hansen (1988) documented uranium derived from smoke-stack

W.W. Wood (✉)
Department of Geological Sciences, Room 206 Natural Sciences building, Michigan State University, East Lansing, MI, 48824
e-mail: wwwood@msu.edu

F.D. Eckardt
University of Cape Town, Department of Environmental and Geographical Science, 7701 Rondebosch, South Africa

T.F. Kraemer and K. Eng
U. S. Geological Survey, MS 430, Reston, VA 20192, USA

M. Öztürk et al. (eds.), *Sabkha Ecosystems*, Tasks for Vegetation Science 46,
DOI 10.1007/978-90-481-9673-9_4, © Springer Science+Business Media B.V. 2011

release in the ground water in parts of Ohio; and Robertson et al. (1989) demonstrated that eolian transport of sulfur from smelters in Sudbury, Ontario resulted in high sulfate levels in aquifers hundreds of kilometers from the source. Brown and Sharp (1992) summarized much of the previous work and developed a generic numerical model to simulate mass transport from a point source at the surface, through the unsaturated zone, and into the ground water.

Eolian transport of soluble evaporite minerals from a small 4.7-km^2 pan in western Texas (Double Lake) created a "footprint" of soluble salts that could be identified in the soil as far as 35 km downwind from the pan (Wood and Sanford 1995), negatively impacting the quality of ground water. For efficient water management, it is important to develop a methodology to identify and quantify these eolian inputs to the ground water. It was hypothesized that the length of the soluble footprint derived from a pan is a function of the size of the source area. That is, the length of the solute footprint is limited by the mass of the evaporite minerals, if all other factors remain the same. If this "size/length of footprint" hypothesis can be demonstrated to be true, existing maps of pans coupled with prevailing wind directions could then be used to estimate the area of eolian salt influence on regional ground-water quality. This knowledge would be useful in ground-water resource evaluation by helping to minimize drilling and sampling necessary for water-quality characterization as current maps generally define the locations and area of these saline pans.

The experimental design used to test the "size/length of solute footprint" hypothesis was to evaluate the concentration of various soluble salts from soil samples collected at known distances from a pan and compare these concentrations with similar distance/area relations developed for the western Texas pan. The Makgadikgadi Depression in northeastern Botswana (Fig. 1) was identified as a possible site to test the hypothesis. The Makgadikgadi Depression exhibits wind and precipitation patterns similar to those in the western Texas study area. Both sites exhibit significant ground-water evaporation that contributes solutes to the pan. It was assumed that both sodium sulfate (western Texas) and sodium carbonate (Makgadikgadi Depression) evaporative salts would form similar sized eolian particles with similar density and thus have similar transport properties.

1.1 Hydrogeology

An estimated 8,000 pans cover Botswana, which is at the heart of the Kalahari basin. The Makgadikgadi Depression represents by far the largest pan system in the country, occupying more than 30,000 km^2. The Makgadikgadi consists of several dozen connected pans, but it is dominated by the Ntwetwe Pan to the west and Sua Pan to the east. The Ntwetwe Pan is fed by surface water from the Boteti River, an overflow of the Okavango Delta, but surface water from this source has failed to reach the pan since 1991 (Personal communication, Botswana Department of Water Affairs). The Sua Pan receives surface water mainly from the Nata River, which drains the wetter eastern regions of Botswana and Zimbabwe. The Nata River discharges seasonally but is subject to a considerable degree of flood variability. The Sua Pan represents the lowest point of the southern African inland basin, at 890 m above sea level. Because of its low altitude, the Makgadikgadi is the natural discharge point for regional ground-water flow. The water table is typically less than a meter below the surface, permitting evaporativc discharge from the capillary fringe.

The eastern sides of the pans feature elevated terrain. Here, the metamorphic basement of Achaean and Precambrian material forms the so-called hardveld. The south of the pan system is home to a 100-m sandstone escarpment, part of the Carboniferous-Jurassic Karoo sequence (Thomas and Shaw 1991). To the west and north, the pans open to the Kalahari sandveld. Bedrock exposure on the surface is limited due to the extensive cover of Tertiary Kalahari beds. These eolian and fluvial sands are poorly stratified and are cemented by calcrete and silcrete, particularly along drainage lines and pans (Thomas and Shaw 1991). This terrain features limited surface-flow input and is dominated by numerous shorelines, indicative of past lake levels between the 912- and 945-m contours. Western pan margins have preserved sand ridges (lunettes) produced by prevailing winds and past wave action. The hardveld is home to tree savanna, while the remaining sandveld, which dominates Botswana, consists largely of open shrub savanna. The Makgadikgadi pans themselves lack vegetation because they are covered with evaporative salts. Grasslands extend beyond the perimeters of the pans.

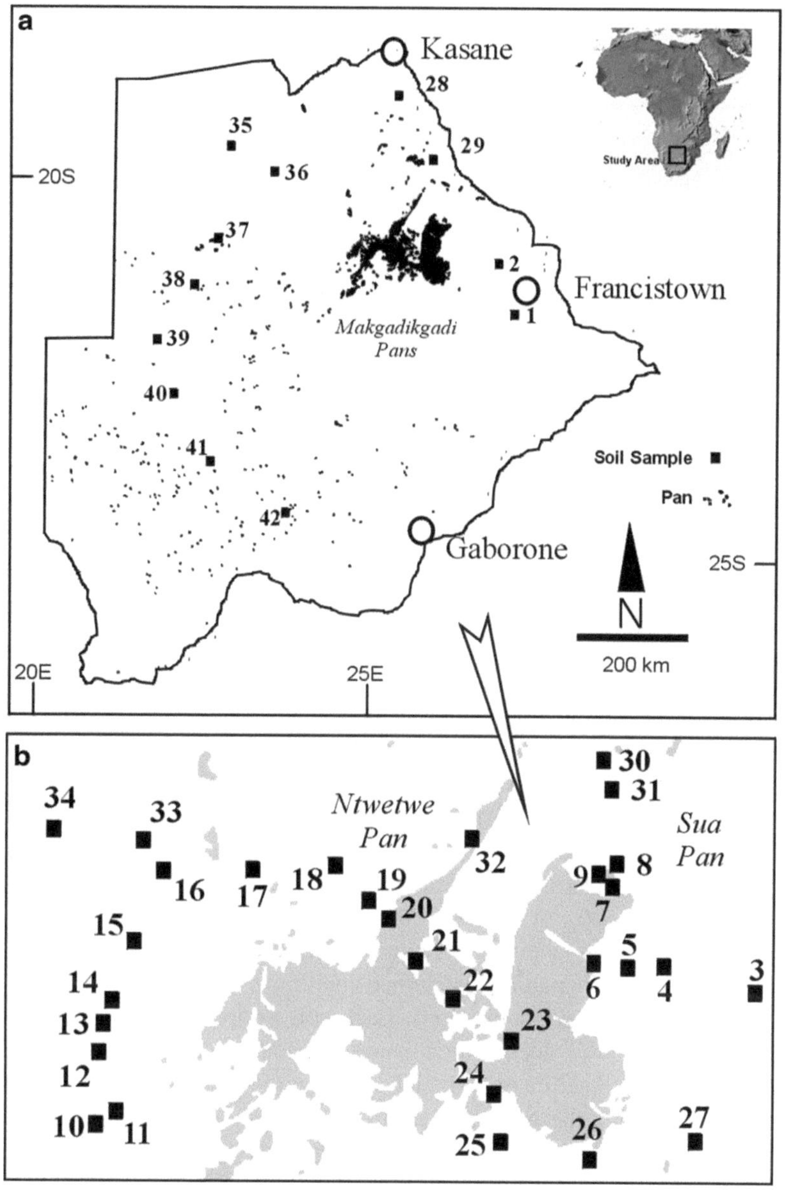

Fig. 1 Map showing location of Makgadikgadi Depression (**b**) and soil sampling in Botswana (**a**)

The climate in the study area is semi-arid and subject to considerable annual and inter-annual rainfall variability. Generally, most of the rainfall occurs between the months of November and March, while April through October are dry (Thomas and Shaw 1991). Rainfall gradient in Botswana ranges from 600 mm/year in the northeast to 250 mm/year in the southwest (Thomas and Shaw 1991), with between 400 and 450 mm/year at the Makgadikgadi Depression. Winds are dominated by the Eastern Trade Winds that have an average speed of 5.9 m/s, with maximum gusts to 33 m/s (Fig. 2) recorded just south of the pan complex. The input of eolian salts, rainfall, and wind direction in the Makgadikgadi Depression appears to have been relatively constant over the last several hundreds of years, representing a steady-state condition of soluble salts in the soil comparable to those documented for western Texas (Wood and Sanford 1995). Steady-state concentration of soluble salts in the soil is directly related to solute flux entering the ground-water table. This concentration can be calculated if the ground-water recharge rate and soil moisture content are known.

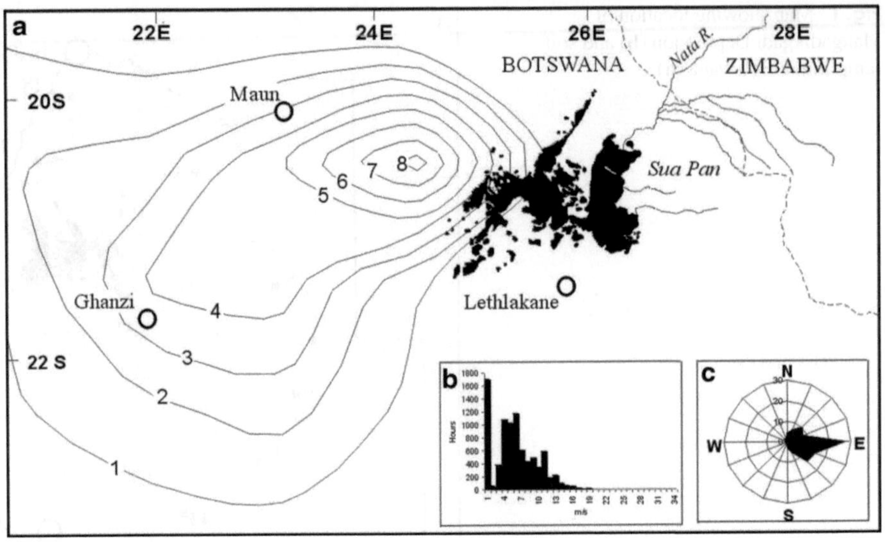

Fig. 2 Map showing wind direction associated with a map of contours of Aerosol Index (optical opaqueness) in the atmosphere. Aerosol Index data from the TOMS (Total Ozone Mapping Spectrometer)

2 Methods

Under an assumption of steady-state soluble salt and rainfall input, surface soil samples represent the concentration profile in the unsaturated zone (Wood and Sanford 1995). Thus, soil samples were collected at varying horizontal distances from the pan (Fig. 1). Approximately 500 g of soil were collected at each site by excavating shallow pits (10–20 cm). Samples were selected to represent natural, un-farmed conditions and both up- and downwind locations. Each soil sample collection site was documented using a hand-held Global Position Satellite receiver. Distances of the collection sites from the nearest basin floor (defined by the presence of evaporites on the surface) were determined from an enlargement of Fig. 1 as follows: Samples falling between the northern and southern latitude boundary of the basin floor were measured directly east or west to the floor margins (Samples 2–25, plus samples 27, 37, and 38, Fig. 1). Distances of samples north or south of the bounding latitudes (1, 39, and 40, Fig. 1) were measured to the nearest floor margin of the pan.

Samples were transported to the U.S. Geological Survey in Reston, Virginia where they were disagraded and dried at 50°C for 24 h. 50 g of the dried soil were placed into 100 ml of high-purity distilled water. This suspension was gently stirred for 24 h and after decanting, filtered through a 0.45-μm filter. The resulting solutions were analyzed for soluble ions utilizing the methods of Fishman and Friedman (1989). Major ions are reported as micrograms/gram dry weight (PPM) (Table 1). Radium was analyzed by gamma-ray spectrometry based on the emission from the soil of the 186-Kev gamma ray from 226-radium. A weighted sample of dried soil was placed in a metal ointment can and sealed with clear tape. The can was then placed on a high-purity germanium detector for counting for periods of from one to five days. The detector was previously calibrated by analyzing a NIST river sediment standard of known radium activity (National Bureau of Standards 4350). Results are reported as disintegrations per minute per gram (dpm/g) of dried soil (Table 2). Trace elements were determined by ICPMS (inductively coupled plasma mass spectrometry) using the U.S.E.P.A. method 200.8 (U.S. Environmental Protection Agency 1994). Data for the trace elements are reported in nanograms/gram (PPB) (Table 3).

3 Results

Samples 7, 20, 21, 23, and Sua Pan crust were assumed to represent input conditions, suggesting that the dominant ions available for eolian transport are sodium, bicarbonate, and chloride (Table 1). The logarithmic concentration of these ions as a function of logarithm

Table 1 Chemical analysis of major ions (micrograms/gram dried weight) derived from a distilled-water leach of soil samples

Sample ID and Map number	South	East	Distance Km	Ca μg/g	Mg μg/g	Na μg/g	K μg/g	Sr μg/g	Si μg/g	HCO_3 μg/g	Cl μg/g	SO_4 μg/g	Br μg/g	NO_3 μg/g	PO_4 μg/g
B-12/7/99-1	21.6150	27.3684	140	14	2.7	2.3	12	0.11	11	52	15	1.2	0.07	0.00	3.7
B-12/7/99-2	21.0691	27.4188	115	4.0	1.5	1.5	35	0.04	9.5	43	15	1.4	0.00	0.00	0.00
B-12/7/99-3	20.7198	26.8315	58	68	9.6	1.1	41	0.16	11.8	285	17	2.4	0.00	0.00	0.00
B-12/7/99-4	20.5198	26.4077	26	46	10	2.3	25	0.21	13	205	17	0.0	0.30	0.00	0.00
B-12/7/99-5	20.5531	26.1543	16	72	7.3	38	29	0.66	34	343	25	10	0.00	0.00	0.00
B-12/7/99-6	20.5525	26.2816	2	3.8	2.3	1.6	11	0.06	16	30	14	0	0.00	0.00	0.00
B-12/7/99-7	20.3486	26.2433	0	3300	68	84200	4200	48	87	14100	126400	10200	99	0.00	0.00
B-12/7/99-9	20.3238	26.2362	9	7.7	0.1	63	211	0.15	51	180	0.0	400	8.5	0.00	9.7
B-12/7/99-10	21.0890	24.2978	44	17	3.0	7.2	8.1	0.12	19	67	16	0	0.00	0.00	0.00
B-12/7/99-11	21.0864	24.3503	35	44	3.6	10	18	0.33	25	175	16	0	0.00	0.00	0.00
B-13/7/99-12	20.9981	24.3600	56	67	8.9	11	16	0.63	27	275	18	1	0.00	0.00	0.00
B-13/7/99-13	20.8481	24.3603	39	47	4.3	6.8	17	0.32	27	175	16	0	0.00	0.00	0.00
B-13/7/99-14	20.6303	24.4142	39	19	2.9	2.0	11	0.19	9.9	69	15	0	0.00	0.00	0.00
B-13/7/99-15	20.4509	24.5175	88	17	3.9	2.2	14	0.22	9.4	69	15	0	0.06	0.00	0.00
B-13/7/99-16	20.2131	24.5942	105	13	1.2	0.7	3.0	0.14	8.5	46	14	0	0.00	0.00	0.00
B-13/7/99-17	20.2350	24.9215	74	12	3.1	4.3	21	0.11	8.7	130	15	5	0.00	0.00	8.6
B-13/7/99-18	20.1950	25.2239	44	6.7	1.5	3.2	4.5	0.08	10	34	17	0	0.00	0.00	0.00
B-14/7/99-19	20.4201	25.3533	9	42	4.6	33	68	1.1	38	265	30	3	0.00	3.1	0.00
B-14/7/99-20	20.4556	25.3819	0	10	1.2	62100	1350	0.90	81	20000	67700	13200	110	0.00	0.00
B-14/7/99-21	20.5711	25.5108	0.3	42	0.7	680	118	0.63	220	610	550	210	3.4	88	0.00
B-14/7/99-22	20.6886	25.6067	4	48	4.9	58	23	0.52	48	315	30	1.9	0.00	0.00	0.00
B-14/7/99-23	20.8886	25.8289	0	5.0	0.1	38800	2100	0.27	375	8200	54400	5800	47	1900	0.00
B-14/7/99-24	21.0787	25.6497	9	37	5.2	23	34	0.37	31	160	54	2	0.00	0.00	0.00
B-14/7/99-25	21.3094	25.5706	23	52	7.5	8.1	28	0.35	24	215	16	1	0.00	0.00	0.00
B-14/7/99-26	21.3248	26.0449	9	40	2.3	11	16	0.11	14	140	21	0	0.00	0.00	0.00
B-14/7/99-27	21.1850	26.5831	39	28	1.9	1.7	11	0.06	18	105	14	0	0.00	0.00	0.00
B-16/7/99-29	19.1656	25.7686	105	3.4	1.5	1.1	8.4	0.01	1.6	9.3	14	0	0.00	0.00	0.00
B-16/7/99-30	19.9172	26.1456	49	2.4	0.5	0.4	3.2	0.02	1.5	6.6	13	0	0.00	0.00	0.00
B-16/7/99-31	19.9494	26.1542	38	92	4.7	11	19	0.93	32	330	15	0	0.00	0.00	0.00
B-16/7/99-32	20.1417	25.7111	26	26	2.0	20	14	0.21	22	120	24	0	0.00	0.00	0.00
B-16/7/99-33	20.1861	24.3231	114	35	3.4	0.7	6.6	0.24	10	120	15	0	0.00	0.00	0.00
B-16/7/99-34	20.1639	23.7717	158	26	3.3	2.7	10	0.20	15	72	15	0	0.00	11	0.00
B-18/7/99-36	19.8078	23.6317	150	22	5.9	0.4	39	0.22	15	99	13	0	0.00	0.48	22
B-18/7/99-37	20.6216	22.7417	220	3.6	0.6	0.6	13	0.02	6.0	16	13	0	0.00	0.55	0.00
B-18/7/99-38	21.0597	22.4011	260	13	3.8	1.7	25	0.13	7.0	30	15	5	0.00	25	0.00
B-19/7/99-39	22.0081	21.3739	370	1.8	0.6	0.6	13	0.08	9.1	14	16	0	0.01	0.00	0.00
B-19/7/99-40	22.9911	22.1764	400	1.4	0.3	0.2	4.7	0.00	3.1	4.2	14	0	0.00	0.00	0.00
Sua Pan Crust			0	3600	60	61800	2850	51	170	15300	108000	0	0.00	0.00	0.00

Table 2 Analysis of 226-radium and 238-uranium, in dpm/g of dried weight

Map No.	238-uranium	226-radium
1	4.41	4.88
2	1.4	0.48
3	–	0.36
4	1.23	0.43
5	0.41	1.48
6	1.28	0.26
9	1.09	2.13
10	0.71	0.57
11	0.96	1.29
12	1.56	1.12
13	0.47	0.35
14	0.14	0.11
15	0.14	0.24
16	0.32	0.24
17	0.55	0.36
18	0.63	0.36
19	2.98	1.67
21	1.93	3.76
22	3.25	3.31
24	0.42	1.62
25	1.58	0.63
26	–	0.82
27	2.23	0.87
28	–	0.53
29	0.34	0.49
30	0.24	0.34
31	1.17	1.71
32	0.49	0.54
33	0.37	0.39
34	0.05	0.48
36	0.27	0.24
37	0.32	–
38	0.49	0.74
39	1.22	0.76
40	0.7	0.74

of distance from the source is shown in Figs. 3–5. The activity of radioactive element 226-radium (Fig. 6; Table 2) is plotted arithmetically as a function of the logarithm of the distance. Trace elements concentrations (Table 3) failed to yield a significant correlation with distance. Samples 3, 4, 5, 6, 27, 30, and 31 were considered upwind and not plotted in Figs. 3–7. Samples 1 and 2 were used as background for the major ions. Background for 226-radium was assumed to be the lowest values observed. Extrapolation of the lines in Figs. 3–6 to the background values suggests a solute influence distance of approximately 125 km downwind. The length is consistent with data from Total Ozone Mapping Spectrometer (TOMS), 2003 (Fig. 2). The data in Fig. 2 are contoured as Aerosol Index, which is a measure of optical opaqueness. October–February, corresponding to the driest period, are the months of the largest index values and hence, highest aerosol concentration.

The hypothesis that the solute footprint would extend farther downwind in a regular fashion as a function of the larger area of the source is consistence with these observations. In this case, the larger area led to a significant increase in the length of the solute footprint over that observed in western Texas (35 km), suggesting that eolian transport of evaporative salts is proportional to the amount of salt available, all other factors remaining constant. The relation between the two locations is shown in Fig. 7 and is expressed as:

$$x = 27.94y^{0.145}, \quad (1)$$

where x is the length of the footprint in km, and y is the area of the pan in km^2. Thus, a pan with an area of 1,000 km^2 would be expected to yield a solute footprint approximately 76 km long, given the same wind conditions and mineral source. Systems dominated by halite (sodium chloride) would likely have a smaller eolian footprint than those dominated by sodium sulfate or sodium carbonate. Halite tends to form in interlocking grains that create "ice-like" surfaces compared to the fine, "fluffy," more easily mobilized texture of sodium sulfate and sodium carbonate minerals.

Average concentrations over distance downwind from the source were used to provide estimates of bulk-deposited amounts of sodium, bicarbonate, chloride, and 226-radium. These concentrations were determined by first establishing a linear regressed relationship between the sample concentrations and the distance from the source (Figs. 3–6). The form of the equation for sodium, bicarbonate, and chloride is:

$$y = a\,x^{b}, \quad (2)$$

where x is the distance from source, in km, and y is the concentration, a is the intercept, and b is the slope (Table 4). The expression for calculating the distance vs. concentration relation for 226-radium is of the form:

$$y = b\,\ln x + a, \quad (3)$$

Table 3 Chemical analysis of trace ions (nanograms/gram dried weight) derivedfrom a distilled-water leach of soil samples

Co (ng/g)	Cr (ng/g)	Cs (ng/g)	Cu (ng/g)	Dy (ng/g)	Er (ng/g)	Eu (ng/g)	Fe (ng/g)	Ga (ng/g)	Gd (ng/g)	Ge (ng/g)	Hf (ng/g)	Hg (ng/g)	Ho (ng/g)	I (ng/g)	In (ng/g)	Ir (ng/g)	La (ng/g)	Li (ng/g)	Lu (ng/g)	Mn (ng/g)
2	1	0	17	0	0	0	35	0	0	0	0	0	0	5	0	0	0	2	0	97
6	6	0	8	0	0	0	45	0	0	0	0	0	0	1	0	0	0	0	0	900
0	0	0	0	0	0	0	38960	0	0	0	0	0	0	0	0	0	0	7670	0	0
5	17	0	32	0	0	0	0	0	0	0	0	0	0	9	0	0	6	16	0	5
1	2	0	3	0	0	0	38	0	0	0	0	0	0	1	0	0	0	2	0	62
0	0	0	0	0	0	0	25450	0	0	0	0	0	0	0	0	0	0	9930	0	0
0	5	0	3	0	0	0	0	0	0	0	0	0	0	1	0	0	0	11	0	2
0	3	0	8	0	0	0	3	0	0	0	0	0	0	2	0	0	0	0	0	11
1	31	0	5	0	0	0	115	0	0	0	0	0	0	0	0	0	0	0	0	36
1	0	0	2	0	0	0	75	0	0	0	0	0	0	0	0	0	0	1	0	57
0	1	0	3	0	0	0	0	0	0	0	0	0	0	1	0	0	0	1	0	4
0	2	0	4	0	0	0	17	0	0	0	0	0	0	0	0	0	0	1	0	1
0	1	0	3	0	0	0	30	0	0	0	0	0	0	1	0	0	0	5	0	18
0	0	0	0	0	0	0	34590	0	0	0	0	0	0	0	0	0	0	12630	0	0

Re (ng/g)	Rh (ng/g)	Ru (ng/g)	Sb (ng/g)	Sc (ng/g)	Se (ng/g)	Sm (ng/g)	Sn (ng/g)	Ta (ng/g)	Tb (ng/g)	Te (ng/g)	Th (ng/g)	Ti (ng/g)	Tl (ng/g)	Tm (ng/g)	U (ng/g)	V (ng/g)	W (ng/g)	Y (ng/g)	Yb (ng/g)	Zn (ng/g)	Zr (ng/g)
0	0	0	0	0	5	0	3	0	0	0	0	0	0	0	0	23	0	1	0	5	0
0	0	0	0	0	1	0	0	0	0	0	0	0	0	0	0	3	0	1	0	14	0
0	0	0	0	0	2710	0	0	0	0	0	0	0	0	0	1680	4210	0	0	0	0	0
0	0	0	0	0	0	0	0	0	0	0	2	0	0	0	2	330	8	3	0	4	0
0	0	0	0	0	4	0	0	0	0	0	0	0	0	0	0	4	0	0	0	18	0
0	0	0	0	0	1860	0	0	0	0	0	0	0	0	0	4250	0	0	0	0	0	0
0	0	0	3	0	3	0	1	0	0	0	0	0	0	0	1	4	0	0	0	4	0
0	0	0	0	0	5	0	0	0	0	0	0	0	0	0	0	6	0	0	0	6	0
0	0	0	0	0	1	0	0	0	0	0	0	0	0	0	0	10	0	2	0	4	3
0	0	0	0	0	3	0	0	0	0	0	0	0	0	0	0	1	0	0	0	8	0
0	0	0	0	0	0	0	0	0	0	0	0	0	0	0	0	7	0	0	0	3	0
0	0	0	6	0	0	0	0	0	0	0	0	0	0	0	0	3	0	0	0	3	0
0	0	0	0	0	0	0	0	0	0	0	0	0	0	0	0	0	0	0	0	0	0
0	0	0	0	0	0	0	0	0	0	0	0	0	0	0	0	0	0	0	0	18350	0

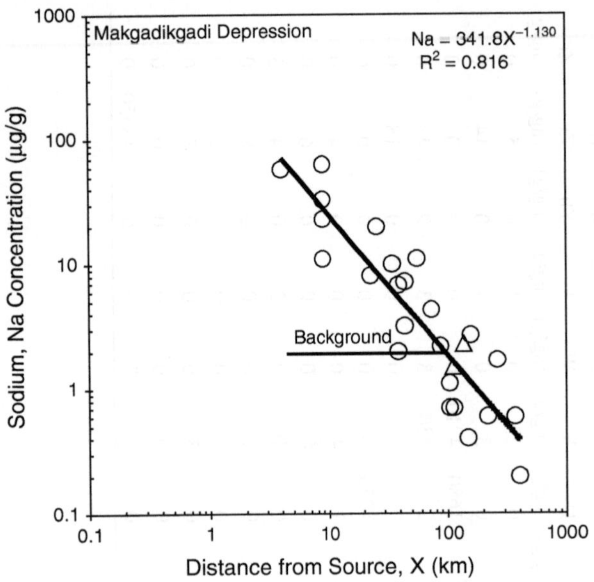

Fig. 3 Graph showing soluble sodium concentration in the soil as a function of distance from the edge of the Makgadikgadi Depression

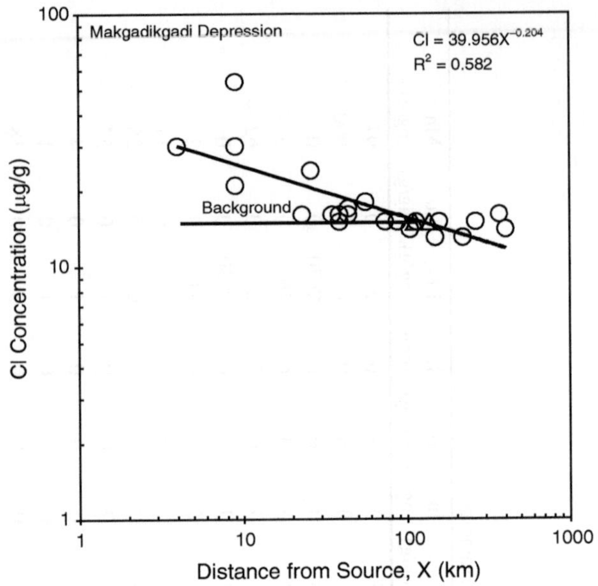

Fig. 5 Graph showing soluble chloride concentration in the soil as a function of distance from the edge of the Makgadikgadi Depression

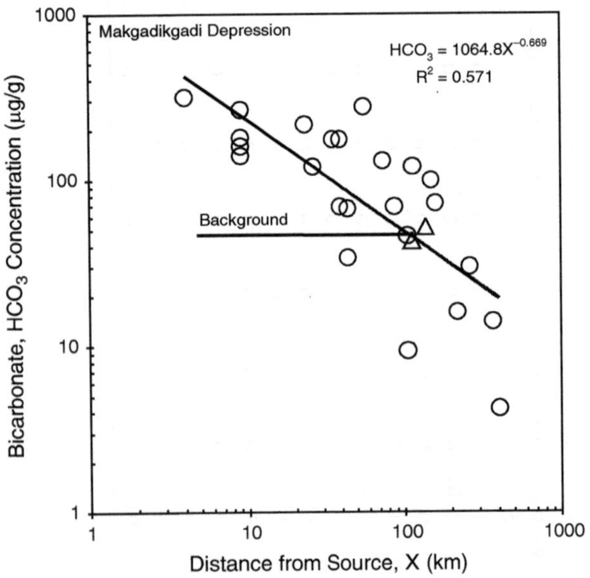

Fig. 4 Graph showing soluble bicarbonate concentration in the soil as a function of distance from the edge of the Makgadikgadi Depression

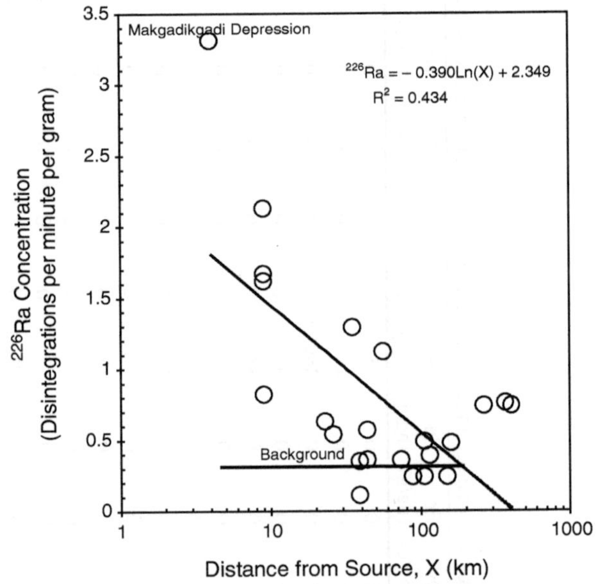

Fig. 6 Graph showing total radium activity in the soil as a function of distance from the edge of the Makgadikgadi Depression

where y is activity of the ion of interest, and ln is the natural logarithm of the distance x; other parameters are defined above (Table 4).

The linear regressed parameters were determined by ordinary least squares and are listed in Table 4. Lilliefors tests (Wilks 1995) at the 20% level strongly indicate that all these variables can be described by lognormal distributions. However, the distribution for chloride was found not to follow any type of theoretical distribution. The distribution for sodium is given in Figs. 8 and 9.

The idealized area receiving eolian input is assumed to be a rectangular area 140 by 125 km. That is, the N/S dimension of the Makgadikgadi Depression is approximately 140 km, and the downwind distance for solutes is approximately 125 km. Thus, the total area receiving soluble eolian input is approximately 17,500 km^2. Because of the complex shape of the shoreline of the Makgadikgadi Depression, the wind dispersion addition used in the study of the west Texas pan, where the shoreline is regular, was ignored. The area beneath the curves (Figs. 3–6) was determined by integration from the closest site at 4 km to approximately 125 km downwind. The distance where the sampled concentrations reached a background value was identified as the upper integration limit. Average concentrations for sodium, bicarbonate, chloride, and 226-radium were obtained by dividing the area underneath the curve by the effective distance of transport (125 km), they are listed in Table 4. The background concentrations were then subtracted from these average concentrations before any bulk estimates were calculated. Assuming soil moisture content remains at steady state with a recharge flux of 4 mm/year, similar to that of western Texas, estimates in metric tons/year are sodium 309,905; bicarbonate 2,737,266; chloride 176,615; and 226-radium 0.0154 (Table 4). The aerosol chloride flux in the small western Texas pan represented 450 metric tons/year of chloride.

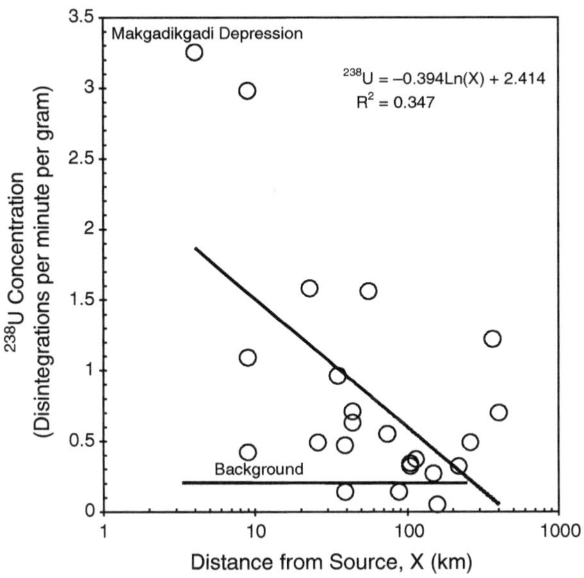

Fig. 7 Graph showing total 238-uranium activity in the soil as a function of distance from the edge of the Makgadikgadi Depression

It is suspected that the regional topography may play a role in eolian transport of particles. The wind is down the topographic gradient in the western Texas site, while the dominant wind in the Makgadikgadi Depression site is up the topographic gradient. This condition tends to minimize the length of the Makgadikgadi footprint. Because the wind direction (east to west) is opposite the regional ground-water flow direction (west to east), the amount of ground water impacted by degraded quality is lower than it would be if the gradients were the same. Further, it is clear that because of the opposing gradients, much of the solute mass is recycled back into the depression.

An additional assumption is that flux calculation assumed that the ground-water recharge is the same in both the Botswana and western Texas systems. This assumption has no observational basis. The only similarities are the amount of annual rainfall and seasonal deposition. Thus, values could be significantly different if actual values of ground-water recharge were used. Another assumption is that the estimates of radium in the calculations are based on the total amount of 226-radium in a sample above a background level (based on the lowest values seen in the data set) rather

Table 4 Mass of salts, in metric tons/annum, deposited on the surface down wind of the Makgadikgadi depression

Solute	a	b	R^2	RMSE	Area under curve	Background concentration	Average concentration over 125km	4 mm/year or 4 L/m^2 year of recharge bulk amount deposited metric tons/year
Sodium	341.8	−1.1305	0.816	0.285	790,902	1.9	4,427	310,000
Bicarbonate	1064.8	−0.6687	0.571	0.309	10825,476	47.5	39,104	2,740,000
Chloride	39.956	−0.2042	0.582	0.089	2190,206	15	2,522	177,000
226-Radium	2.3491	−0.3895	0.434	1.757	98,452	0.3	0,488	0.0154
238-Uranium	2.4142	−0.3939	0.347	1.848	104,230	0.275	0,559	52,000

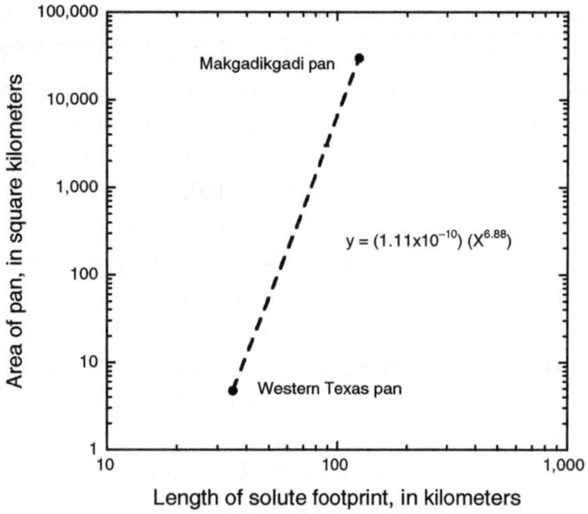

Fig. 8 Graph showing relation between length of solute footprint and evaporite covered area of the contributing pan

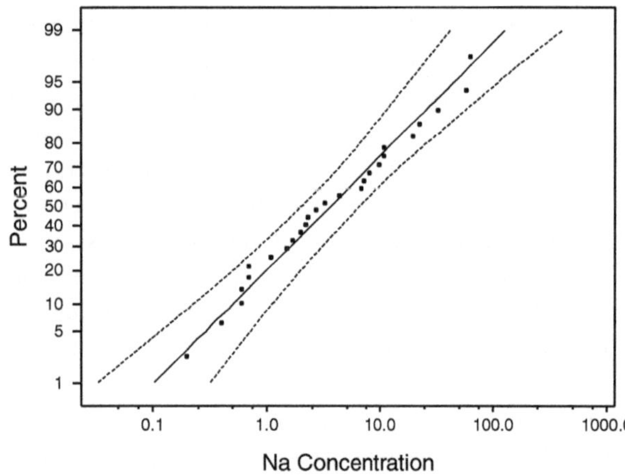

Fig. 9 Probability of non-exceedence of sodium concentration. The solid straight line represents the lognormal distribution with the same mean and standard deviation as the data; the dashed limes are the 95% confidence interval

than a soluble leach extract, as was done for the other components examined. This was done because the relative instrumental insensitivity of the technique used for quantification (gamma ray spectroscopy) would not have produced measurable responses from leach solutions used for the other analytes. Even though the radium value exhibits the same general trends as the other constituents and are consistent with the described hypothesis, it must be recognized that there is the possibility that different processes of concentration in the solute footprint of the soils may be operative for this element, which may affect the interpretation of the total quantity transported to the ground-water system. An explicit assumption in this analysis is that the flux of salts generated is the same in similar settings; thus, the total mass available for eolian transport is a function of the area of the pan. In this approach, the area of the pan is used as a surrogate for the mass of evaporite salts available, as annual mass generation is difficult to calculate.

4 Conclusions

Eolian salts from the evaporite-covered Makgadikgadi Depression in Botswana were observed in the soil between 125 and 150 km downwind from the basin. This salt input is almost certain to have an adverse affect on the underlying ground-water quality. Over three million metric tons of chloride, sodium, and bicarbonate are transported each year from the basin to adjacent land, where these soluble salts are dissolved and transported to the ground water. The relation between the size of the evaporative floor area and the length of the salt footprint on the soil was established. This relation permits use of readily available maps depicting areas of an evaporite-covered depression to estimate the area of potential degradation of ground water. This relation minimizes the amount of field sampling necessary to define the area of potential degradation. It is clear that similar eolian transport processes are occurring in the Makgadikgadi Depression and the western Texas pan and that significant eolian transport of soluble salts to the water table is likely to occur. These data should be considered in water-quality management decisions. Additional work on different basins is necessary to improve this relation and to add more control to area vs. length of the solute footprint (Fig.7). Significant mass of radioactive 226-radium is present in the eolian-transported salts and may have deleterious health consequences for individuals residing in the downwind area.

References

Brown TJ, Sharp JM (1992) A model for the effects of point-source emission of aerosols on groundwater systems. Appl Hydrogeol 1(3/92):33–46

Dethier DP (1979) Atmospheric contributions to stream-water chemistry in the North Cascade Range, Washington. Water Resour Res 15:787–794

Eisenreich SJ (1980) Atmospheric input of trace metals to Lake Michigan. Water Air Soil Pollut 13:287–301

Fishman MJ, Friedman LC (1989) Methods for determination of inorganic substances in water and fluvial sediments. In: Techniques of water-resources investigations of the United States geological survey, 3rd edn. Wiley, New York, Chapter A1, Book 5, pp 545

Heim GE, Giuliana P (1988) Chernobyl provides lessons regarding ground-water contamination. In: Moore JE, Zaporozec AA, Csallany SC, Varney TV (eds) Recent advances in ground-water hydrology. American Institute of Hydrology, Minneapolis, MN, pp 195–202

Hingston FJ, Gailitis V (1976) The geographic variation of salt precipitation over Western Australia. Aust J Soil Res 14: 319–335

Jung CE, Werby RT (1958) The concentration of chloride, sodium, potassium, calcium and sulfate in rainwater over the United States. J Meteorol 15(5):417–425

Lehman L, Hansen H (1988) Secondary concentration of air-released uranium through watershed runoff at the Feed Materials Production Center, Fernald, Ohio. In: Post R.G. (ed) Waste processing, transportation, storage and disposal. Technical Programs and Public Education 1, Low-Level Waste, Waste Management Eighty Eight, pp 781–784

Mandel S, Shiftean ZL (1981) Groundwater resources. Academic, New York, pp 269

Person FJ.Jr, Fisher DW (1971) Chemical composition of atmospheric precipitation in the northeastern United States, U.S. Geological Survey Water Supply Paper 1535-P, 1–23

Péwé TL (1981) Desert dust: an overview. In: Péwé TL (ed) Desert dust: origin, characteristics, and effect on man. Geological Society of America, Boulder, CO, pp 1–10, Special Paper 186

Robertson WD, Cherry JA, Shiff SL (1989) Atmospheric sulfur deposition 1959–1985 inferred from sulfate in ground water. Water Resour Res 25:1111–1123

Rahn KA, Borys RD, Shaw GE (1981) Asian desert over Alaska: anatomy of an Arctic haze episode. In: Péwé TL (ed) Desert dust: origin, characteristics, and effect on man. Geological Society of America, Boulder, CO, pp 37–70, Special Paper 186

Schütz L, Jaenicke R, Pietrek H (1981) Saharan dust transport over the North Atlantic Ocean. In: Péwé TL (ed) Desert dust: origin, characteristics, and effect on man Special Paper 186. Geological Society of America, Boulder, CO, pp 87–100

Thomas DSG, Shaw PA (1991) The Kalahari Environment. Cambridge University Press, Cambridge, pp 284

Total Ozone Mapping Spectrometer (TOMS) (2003) http://jwocky.gsfc.nasa.gov/aerosols/today last revised 31 Mar

U. S. Environmental Protection Agency (1994) Methods for the determination of metals in environmental samples – Supplement 1, EPA-600/R-94–111

Wiman BLB, Agren GL (1985) Aerosol depletion and deposition in forests – a model analysis. Atmos Environ 19:335–347

Wiman BLB, Unsworth MH, Lindberg SE, Bergkvist B, Jaenicke R, Hanson HC (1990) Perspectives on aerosol deposition to natural surfaces: interactions between aerosol residence times, removal processes, the biosphere and global environmental change. J Aerosol Sci 21:313–338

Wilks DS (1995) Statistical methods in the atmosphere science. Academic, San Diego, CA, pp 467

Wood WW, Sanford WE (1995) Eolian transport saline lake basins, and groundwater solutes. Water Resour Res 31: 3121–3129

Palaeo-Ecological Aspects of Farafra Oasis (Egyptian Sabkha) During the Mid-Neolithic Period (7130–6190 BP): A Multivariate Analysis

Ahmed Gamal-El-Din Fahmy and Monier M. Abd El-Ghani

Abstract Studies have been carried out on carbonized plant macro-remains recovered from the Hidden Valley, a Neolithic settlement located on ancient playa shoreline deposits near Farafra Oasis, Egypt. Site contexts have been radiocarbon-dated to 7130–6190 BP. A total of 63 soil samples were collected the total volume of these samples is 81 1 collected in two seasons (1996 and 1997), during which a total area of 61 m^2 was excavated. Soil samples were processed by dry-screening, Recovered plant macroremains were dominated by grains of wild grasses. Multivariate and correspondence analyses were employed to explore the temporal distribution of plant macro-remains and their relationship to archaeological features, including hearths, milling stones, and pot-holes. Two-Ways Indicator Species Analysis (TWINSPAN) of a data matrix of 40 samples × 36 taxa using density values resulted in the recognition of seven floristic groups. Canonical Correspondence Analysis (CCA) was used to examine the relationships between floristic groups and 18 archaeological variables. The highest densities of plant macro-remains characterised sections of the site that had been occupied by human Neolithic inhabitants for longer periods of time (horizons 1 and 2). The highest species diversity indices were that of pot holes, while sediments collected near milling stones had the lowest values. The study indicated that the playa deposition on the archaeological site provided a unique opportunity to preserve the botanical remains underneath. Reconstruction of the past vegetation around the Hidden Valley settlement increased our knowledge on palaeoecological aspects of the Farafra Oasis during the Mid-Neolithic period.

A. Gamal-El-Din Fahmy (✉)
Botany Department, University of Helwan, Cairo, Egypt
e-mail: afahmyleg@yahoo.com

M.M. Abd El-Ghani
The Herbarium, University of Cairo, Giza 12613, Egypt

1 Introduction

Early Neolithic human populations of the Eastern Sahara are known to have inhabited the shorelines of ancient playa lakes as early as 8,000 years ago (Wendorf and Schild 1980). Previous macro botanical studies in Nabta playa by Wasylikowa et al. (1995) have shed the light on their significance for palaeo-environmental studies in the Western Desert of Egypt. Remnants of these settlements are preserved today in playa deposits, such as those found at the site of the Hidden Valley located 65 km north of Farafra Oasis. Excavations conducted at this locality in 1996 and 1997 by Prof. Barbara Barich (University of Rome "La Sapienza") have produced the remains of a permanent or semi-permanent Neolithic human settlement preserved in playa shoreline deposits. Long-term continuous occupation of this site is indicated by a deep stratigraphy (1 m), which revealed a clear succession of occupation floors containing many hearths. According to Barich (1996) radiocarbon determinations for the site have been done in Silesian Technical University, Gliwice, Poland, they are: 7030 BP (Gd-1170)/6910 BP (Gd- 9629) and 7130 BP (Gd-10505)/6710 (Gd-7820). Carbon dating was followed by radiocarbon laboratory number between practice for documentation. Also, she recognizes three horizons of occupation, all of which have produced notable traces of human activity, including rock-lined hearths, small pot holes, and milling stones. In 1996, 38 m^2 were excavated using a 10 × 10 m grid, and the work concentrated on the archaeological sectors 96 E

M. Öztürk et al. (eds.), *Sabkha Ecosystems*, Tasks for Vegetation Science 46,
DOI 10.1007/978-90-481-9673-9_5, © Springer Science+Business Media B.V. 2011

and 96 F/1. In 1997, excavations were extended further east to cover sectors 97 E and 97 F trench, and a total area of 23 m^2 was uncovered. A corridor was cut in order to connect the eastern sector with the central area of the settlement, which contained circular structures used by Neolithic inhabitants as dwellings.

At present, the climate of this region is extremely arid. Precipitation records from 1969 to 1989 indicate a mean of 2 mm/year consequently the area around the Hidden Valley is completely devoid of plant cover. The present study aims at finding correlation between the distribution of the recovered plant macro-remains and the excavated archaeological features including hearths, milling stones and pot holes using multivariate analysis techniques. Another goal of the study is to reconstruct past vegetation around the site in question.

2 Materials and Methods

A total of 63 soil samples amounting to 81 l was collected from hearths, pot-holes, milling stone and living floors in 1996–1997. Hearths are spherical holes, about 40 cm in diameter and 20 cm in depth. The inhabitants used these features as primitive ovens. Pot holes are small depressions in which vessels containing food are placed for cooking by piling hot ash around them (Mitka and Wasylikowa 1995). Milling stones are rounded, concave, stony artefacts used to grind grains of wild grasses.

Quantities of individual soil samples ranged from 0.75 to 4.00 l. Ashy layers were targeted for the recovery of plant macro remains including, seeds, grains and culm fragments. All samples were dry-screened in the field using a 3 mm mesh to separate stones, potsherds, and charcoal fragments, after which they were transported in plastic bags to the laboratory. Although samples collected in 1997 were entirely examined, those excavated in 1996 were sub-sampled in 100 ml units using a 5 ml spatula. Each sample (or sub-sample) was sieved through 1.0 and 0.5 mm mesh sieves and preliminary sorting was completed using a binocular objective magnification (8–25 X). Plant macro-remains were examined in more detail using a Wild stereoscopic microscope with a range of magnification from 20 to 50 X. Identifications were made using descriptions and illustrations in publications concerned with the flora of Egypt (Täckholm 1974; Boulos 1999; Cope and Hosni 1991).

Some illustrations in these contributions are provided with drawings of spikelets, grains and seeds of living plants, which have been used by the first author to narrow the scope of identification into definite species or genera. On the other hand, illustrations, photos and descriptions of plant macro-remains from the Nabta playa by Wasylikowa (1992, 1997) and Wasylikowa and Kubiak-Martens (1995). According to Wasylikowa (1997) identification categories include the qualification "type" following plant name, which means that morphological resemblance to a taxon named but without excluding the possibility that similar fruits, seeds may be found in other taxa which were not examined by the first author. On the other hand, the abbreviation "cf" means closer identification than type. Identifications were confirmed through comparison with modern reference collections of Egyptian plants housed at the Department of Botany, University of Helwan, Cairo.

Two-Way Indicator Species Analysis (TWINSPAN; Hill 1980) a polythetic divisive classification method; was applied on a data matrix comprising 40 samples × 36 taxa using their density values. Here, density is the number of plant macro remains per liter of soil sediment retrieved from the archaeological site in question TWINSPAN program was run using the default options, with the exception of the pseudospecies cut levels which were altered to: 0, 5, 10, 20, 40, 60 and 80. The computer program CANOCO 3.12 (ter Braak 1991) is used for all ordinations, and plots were drawn using CANODRAW 3.0 (Smilauer 1993).

Preliminary analyses were made by applying Detrended Correspondence Analysis (DCA) to check the magnitude of change in species composition along the first ordination axis (i.e., gradient length in standard deviation units). All default settings were used for CCA, except the samples scores that are the weighted averages of species scores. The variables in the CCA biplots are represented by arrows pointing in the direction of maximum variation, with their length proportional to the rate of change (ter Braak 1986) Each arrow determines an axis on which the species points can be projected (ter Braak and Prentice 1988). In general, these projection points estimate the optima of species distribution for each archaeological variable. Intra-set correlations were used to assess the importance of the archaeological variables. All data variables were assessed for normality prior to the CCA analysis (Berk 1994), and appropriate transformations were performed when necessary to improve normality according to (Zar 1984) In total 18 archaeological variables were used: 4 archaeological artefacts (AF), including hearths (H), pot holes (PH), milling stones (MS) and corridors (C); and 13 depths below surface (D), including D_1 (10 cm), D_2 (15 cm), D_3 (20 cm), D_4 (25 cm), D_5 (30 cm), D_6 (35 cm), D_7 (40 cm), D_8 (45 cm), D_9 (50 cm), D_{10} (55 cm), D_{11} (60 cm), D_{12} (65 cm), D_{13} (70 cm), D_{14} (75 cm) and D_{15} (80 cm). We checked all archaeological variables for multicollinearity problems, and two depths: D_{10} and D_{15} are removed from the analysis. Monte Carlo permutation tests (99 permutations) were performed to test the significance of the first canonical axis.

The species diversity within each recognized floristic group was assessed using two indices expressing species richness and relative evenness of species abundance. Species richness was expressed as the average number of species per sample, while relative evenness was calculated using the Shannon–Wiener index ($H` = -\sum pi \log pi$) where p*i* is the relative density of the *i*-th species (Pielou 1975).

3 Results

Seven floristic groups (A–G) were recognized after the application of TWINSPAN technique. Group A includes three samples dominated by *Sorghum* grains, and represent the corridor in sectors 97 E corridor and 97 F trench as well as a hearth in sector 97 E. It is. Group B comprised seven samples dominated by *Echinochloa* and *Setaria* grains belonging to sectors 97 E, 97 E corridor, 96 E and 96 F/1, and were all retrieved from hearths. Group C includes 13 samples, nine of which are dominated by *Setaria* while *Echinochloa* prevailed in four. Specimens originated from hearths and pot holes from sectors 97 E, 97 E corridor, 96 E and 96 F/1. Group D includes samples dominated by seed fragments of *Acacia* type, recovered from sectors 97 E, 97 E corridor and 97 F. Group E includes six samples dominated by grains of *Panicum* and *Sorghum*, and collected from hearths at 97 E, 97 F corridor and 96 E. Group F includes one sample that had fruits (siliqua) of *Coronopus* which was recovered from a hearth in sector 96 F/1. Group G had four samples clustered in this group, dominated by Leguminosae type-seeds. The sediments were excavated near a milling stone and from hearths in sector 96 F/1 and sector 97 E. The highest densities of plant macro remains occur in sectors 96 E and 96 F/1, both of which were inhabited for longer periods than sectors 97 E, 97 E corridor and 97 F trench.

4 Discussion

The present study shows that plant macro-remains appear in higher concentrations in the stratigraphic levels 10, 25, 35, and 40 cm below surface. In particular, the remains of *Echinochloa*, *Setaria*, and *Brachiaria* appear to cluster at a depth of 40 cm (D_7). Stratigraphic levels 10–40 cm correspond to the first and second horizons of occupation recognized through artifactual evidence by Barich (1996). Consequently, densities of plant macro-remains appear to support the archaeological interpretation that the first and second horizons were occupied for a longer periods than others, perhaps due to the prevalence of favourable environmental conditions, such as summer rainfall associated with low temperatures.

The associations of macro-remains with archaeological features including hearths, pot holes, milling stones, and corridors. Grains of *Digitaria* and *Sorghum* are clearly associated with hearths at 10 cm below the surface (D_1). Perhaps these grains were deposited in the hearths as the result of cooking accidents. In addition, there is a high correspondence between the seeds of Cruciferae-type and *Schouwia* with recovered pot holes (AF_2). This pattern could suggest that these vessels were used to prepare foodstuffs, which included wild grasses and *Schouwia* seeds. There is another possibility that parts of these seeds could have been introduced into the deposits via being burned as ash for fuel.

The results of DCA indicate a reasonable segregation among groups along the ordination plane of axes 1 and 2. The eigenvalue for the first DCA axis 1 was relatively high ($\lambda = 0.675$), indicating that it captured the greater proportion of the variation in the taxa composition among the floristic groups. Statistical analysis of the samples under study reveal that wild grasses are highly represented in most of the studied samples. The 289 caryopses identified are attributed to 8 taxa: *Digitaria* type, *Sorghum*, *Echinochloa* cf. *colona*, *Setaria*, *Stipagrostis* type, *Brachiaria*, *Panicum* cf. *turgidum*, and *Cenchrus* type. In particular, *Sorghum* grains show a highly significant p value (0.001), indicating that this taxon must have been a common plant in the area during the various periods of habitation.

Canonical correspondence analysis (CCA) was performed on 40 samples, 34 taxa and 16 archaeological variables. The eigenvalues for the first two CCA axes (1 = 0.395 and 2 = 0.350) indicate acceptable levels of separation of sample scores along the measured archaeological variables. The successive eigenvalues of the first three axes (1 = 0.395, 2 = 0.350 and 3 = 0.330) of the CCA decrease rapidly, suggesting a well-structured data set. In general, intra-set correlations are low.

Although medicinal uses of *Schouwia* species have considerable antiquity (Fahmy 1995), based on this evidence alone, it is not possible to argue convincingly that this species was used as a medicine during the Neolithic occupation of Hidden Valley. Indeterminate grains and milling stones (AF_3) are highly correlated (results not shown). This pattern could be attributed to the fact that grains are destroyed in the grinding process, making them difficult to identify to definite taxonomic categories.

The highest species diversity indices are recorded from pot holes while sediments recovered near milling stones show the lowest. Because of their association with cooking (Mitka and Wasylikowa 1995), it is not surprising that pot holes produced highest species diversity values. Possible sources of seeds in these features are foods lost during cooking as well as plant materials in the ashes used to heat the pots. The ash layer surrounding the pot holes would have provided a gentle charring environment where grains could fall into the hot ashes surrounding the vessels. In case of hearths, grains are more apt to fall directly into fire resulting in their destruction. This could explain the high number of taxa recorded from pot holes compared to hearths. The low species diversity indices of milling stones are attributed to the fact that grinding results in the deposition of fragmentary grains, making them difficult to identify. The relatively high number of species recorded from the corridor is probably related to the large quantities of soil samples (17 l) examined from this context.

5 Palaeo-Economic Interpretations

The recovery of several wild grasses from hearths, pot holes, and milling stones strongly suggests that Neolithic inhabitants of the Hidden Valley gathered them for food consumption. The eight wild grasses identified in site deposits are: *Brachiaria*, *Cenchrus* type, *Digitaria* type, *Echinochloa colona*, *Panicum turgidum*, *Setaria*, *Sorghum*, and *Stipagrostis* type. Many of these taxa have been identified at other Saharan sites, and this, in addition to ethnohistoric evidence, further indicates the widespread use of these species by Saharan populations from ancient to recent times (Barakat and Fahmy 1999) *Sorghum* represents 41.5% of the total number of grasses identified in the Hidden Valley samples. The caryopses are well preserved by carbonization. They are dorsally flattened, obovate to oval elongate in outline, and some are still enclosed by coriaceous glumes. Morphologically, these *Sorghum* grains compare very well with those recovered from Nabta Playa (8,000 BP), Wasylikowa and Kubiak-Martens (1995), Wasylikowa et al. (1995) and Wasylikowa (1997) attribute the Nabta remains to *Sorghum bicolor* (L.) Moench. subsp. *arundinaceum* (Desv.) De Wet and Harlan, a species whose distribution extends across the African savanna (Wendorf and Schild 1980; De Wet 1978). Further investigations are underway to more precisely identify the Hidden Valley *Sorghum* grains. Neolithic inhabitants may have intensively collected its panicles for food due to its big grains in comparison to the associated wild grasses.

From an ecological point of view, the dominance of *Sorghum* during the Neolithic period is expected, as this grass is characterized by wide ecological amplitude due to its remarkable physiological features. *Sorghum* can cope with hot and dry conditions, and it is a high salt tolerance. Furthermore, it can withstand high rainfall, and even temporary water-logging. Its penetrating multi-branched roots are very useful to compete for water with other plants at different depths of the soil (National Research Council 1996).

The macro-remains of eight wild species are represented at Hidden Valley in relatively low frequencies: *Acacia*, *Coronopus niloticus*, *Juncus*, *Phragmites*, *Tamarix aphylla*, *Typha*, *Carex*, and *Cyperaceae* type. This assemblage of plants probably formed part of the natural vegetation in the vicinity of the site during the mid-Holocene. Although many of these wild plant remains have not yet been identified to the species level, some ecological inferences are possible. The relative profusion and assortment of the recovered specimens may indicate that regional plant cover was rich in species and that compared to modern times, a much wetter environment was in existence during the Neolithic. Rainfall is estimated to have been between 100 and 250 mm annually during the wet phases of the early and middle Holocene (Neumann 1989). Also it is very interesting to note that the assemablage of wild grasses identified from the Hidden Valley are recorded in studies on recent vegetation of North and Central parts of the Sudan (Schulz 1988; Wilson 1978). These areas of the Eastern Sahara enjoys precipitation ranges between 100 and 250 mm (Walter 1979).

The present data indicate that two types of vegetation existed in the vicinity of the playa, one dominated by reeds and the other by thorny vegetation. Reed vegetation probably existed along the playa margins and may have experienced some fluctuation in water levels, but for the most part remained waterlogged throughout the year. *Typha* and *Phragmites* dominated this vegetation type. The wet fringes around the playa probably provided a favorable habitat for the growth of associated species, such as *Tamarix* and *Juncus*, as well as water loving species such as *Carex* and *Coronopus niloticus*. This reconstruction is based on modern vegetation analysis around water bodies (e.g. springs) in the oases of the Western Desert (Zahran and Willis 1992). Thorny vegetation probably occurred on the plain surrounding the playa. Trees of *Acacia* grew on the upstream sections of water runnels that dissected the plateau towards the playa. Several wild grasses including, *Digitaria sanguinalis*, *Sorghum*, *Echinochloa cf. colona*, *Setaria*, *Stipagrostis type*, *Brachiaria*, *Panicum cf. turgidum, and Cenchrus* type would have dominated the herb layer.

6 Conclusions and Recommendations

The present study provides archaeobotanical evidence on the relationship between man and their Neolithic environment in the Egyptian Sahara. Also, it increases our knowledge regarding economy and ecology of settelements, where, the settlers lived close to sabkhas collecting wild grasses and herding animals.

Results obtained from Nabta and the Hidden Valley (Fahmy 2001; Barakat and Fahmy 1999) add important dimension to the sabkha ecosystem research. Understanding the subsistence strategy of these early groups of people can be achieved only through archaeological and archaeobotanical studies.

References

Barakat H, Fahmy AG (1999) Wild grasses as "Neolithic" food resources in the eastern Sahara. In: Van der Veen M (ed) The exploitation of plant resources in ancient Africa. Kluwer/Plenum, New York, pp 33–45

Barich B (1996) Early to Mid-Holocene occupation at Farafra (Western Desert, Egypt): a social approach. The Italian Archaeological Mission in Cairo An official report to the Egyptian Antiquities organ pp 45

Berk KN (1994) Data analysis with student SYSTAT Windows Edition. Course Technology, Inc. pp 595

Boulos L (1999) Flora of egypt, vol I. Al Hadara Publishing, Cairo, Egypt, pp 287

Cope T, Hosni H (1991) A key to Egyptian grasses. Royal Botanic Gardens, Kew, London, pp 75

De Wet J (1978) Systematics and evolution of Sorghum sect. Sorghum (Gramineae). Am J Bot 65(4):477–484

Fahmy AG (1995) Historical Flora of Egypt. Ph.D. Thesis, Cairo University. pp 248

Fahmy AG (2001) Palaeoethnobotanical studies of the Neolithic settlement in Hidden Valley, Farafra Oasis, Egypt. Vegetation Hist Archaeobot 10:235–246

Hill MO (1980) TWINSPAN – a fortran program for arranging multivariate data in an ordered two-way table of classification of individuals and attributes. Cornell University, Ithaca, NY, pp 90

Mitka J, Wasylikowa K (1995) Numerical analysis of charred seeds and fruits from an 8000 years old site at Nabta Playa, western desert, south Egypt. Acta Palaeobotanica 35(1):175–184

National Research Council (1996) Lost crops in Africa. National Academy Press, Washington, DC

Neumann K (1989) Holocene vegetation of the Eastern Sahara, *The African Archaeological Review* 7:97–116

Pielou EC (1975) Ecological diversity. New York: John Wiley. pp 165

Schulz E (1988) Der Südrand der Sahara. Würzburger Geographische Arbeiten 69:167–210

Smilauer P (1993) CANODRAW 3.0, user's guide. Microcomputer Power, Ithaca, NY, pp 118

Täckholm V (1974) Student's flora of Egypt. Cairo University, Giza, Egypt, pp 888

ter Braak CJF (1986) Canonical correspondence analysis: a new eigenvector technique for multivariate direct gradient analysis. J Ecol 67:1167–1179

ter Braak CJF (1991) CANOCO version 3.12. Agricultural mathematics group, Wageningen, The Netherlands, p 35

ter Braak CJF, Prentice IC (1988) A theory of gradient analysis. Adv Ecol Res 18:271–317

Walter H (1979) Vegetation und Klimazonen. Ulmer, Stuttgart, pp 342

Wasylikowa K (1992) Exploitation of wild plants by prehistoric peoples in the Sahara. Würzberger Geographische Arbeiten 84:247–262

Wasylikowa K (1997) Flora of the 8000 years old archaeological site E-75-6 at Nabta playa, Western Desert, southern Egypt. Acta Palaeobot 37(2):99–205

Wasylikowa K, Kubiak-Martens L (1995) Wild sorghum from the early neolithic site Nabta Playa, S. Egypt. In: Kroll H, Pasternak R (eds) Rescherches Archaeobotanicae, 9th Symposium IWGP, pp 345–358

Wasylikowa K, Schild R, Wendorf F, Krolik H, Kubiak-Martens L, Harlan JR (1995) Archaeobotany of the Early neolithic site E-75–6 at Nabta Playa, Western Desert, South Egypt (preliminary results). Acta Palaeobotanica 35(1):133–155

Wendorf F, Schild S (1980) Prehistory of the Eastern Sahara. Academic, New York

Wilson RT (1978) The "Gizu" winter grazing in the South Libyan Desert. J Arid Environ 1:327–344

Zar JH (1984) Biostatistical analysis, 2nd edn. Prentice-Hall, Englewood Cliffs, NJ, pp 718

Zahran M, Willis J (1992) The vegetation of Egypt. Chapman and Hall, London, pp 424

Overview of Halophytic Plants of the Sudanese Red Sea Salt Marsh

Omyma Osman Karrar

Abstract Halophytes of the Sudanese Red Sea coast are considered to be one of the important key habitats in the region providing many important ecological functions and useful services. At present, these communities are exposed to coastal disturbances and wide spread degradation due to combination of factors. To this effect, this review was organized in order to illustrate the extent and magnitude of these effects if exists .The results of the effects on the status of these communities have been enlisted followed by discussion leading to conclusions and recommendations that are hoped to contribute towards future conservation and management.

1 Introduction

Sudan is the largest country in the African continent with a total gross area of 2.505,800 sq km including deserts and water-bodies. It lies between 4–22ºNC and 22–36ºE. This location has given Sudan distinctive characteristic reflected in variety of environments and hence variety of flora and fauna. Ecologically Sudan is divided into six zones. These are the desert zone constituting 29% of the total land area, semi-desert zone covering 20%, low rain Savanna on sand and low rainfall Savanna on clay constituting 13% and 14% of the total arid area respectively; and the heavy rainfall Savanna of a total area of 13.8%.The sixth zone is comprised of what is known as Special habitats. This includes the Erkawit plateau 900 meters above the sea in the Red Sea Hills in the eastern part of the country, Imatong and Dongotoren mountains in the south. Special habitat zones are characterised by unique botanical composition (Kassas 1957).

The river Nile is the most significant feature in the country forming a prominent incision in the terrain. The Red Sea is the second dominant water body. Sudanese Red Sea extends for 750 km including embayments and inlets. It constitutes a unique ecosystem with high biological diversity and its natural resources provide substantial economic support for the region (Schoreder 1982). The climate along the coast is very dry with annual rainfall around 111 mm varying from 39 mm at Haliab (Northern Sudanese Red Sea coast) and 164 mm at Suakin harbour (southern coast). The Sudanese coast lies in the intermediate region of the Red Sea which is characterized by a definite wind regime between the North–north west at the northern half and the south east winds. This area is characterized by relatively low pressure and temperatures which shows progressive increase in southward direction. Fundamental movement of surface waters follows the winds pattern. (Sheppard et al. 1992). Salinity varies from39% to 41% and surface water temperature ranges from 26°C to30.5°C and it may be higher in shallow and enclosed areas.

2 Study Area

2.1 Red Sea State

The most important land forms of the Red Sea state are the coastal strip, The Red Sea hills and the plateau stretching west of the hills with an elevation of up to 1,000 m. The most dominant topographical feature is the rocky hills which are intersected by the seasonal

O.O. Karrar(✉)
Arab Open University, P.O. Box: 3322 Safat, 13033 Kuwait, The State of Kuwait
e-mail: Omymak@gmail.com; O.karrar@arabou.edu.kw

M. Öztürk et al. (eds.), *Sabkha Ecosystems*, Tasks for Vegetation Science 46,
DOI 10.1007/978-90-481-9673-9_6, © Springer Science+Business Media B.V. 2011

water courses draining east into the Sea and west into the Nile. The region is characterized by a number of varying habitats, these are:

The Wadis or water courses (local name denoting the presence of ridges and elevations with low lying runnels).

The mist, low temperature oasis of Erkawait.

Inland river deltas.

Salt marshes along The Red Sea.

All these features are reflected in the structure and composition of flora and fauna of the region which is classified as the semi-desert grass and shrubs.

2.1.1 Basic data

The Red Sea state, as the name implies borders the Red Sea, yet few people earn a living from the sea. The population of the state live at the centres located on the coast including Port-Sudan, Suakin, Toker and the Sinkat sub-region. The population consists of rural nomadic, rural settles and urban people who represent 48%, 26% and 26% of the total population respectively. People to the north of Port-Sudan are generally classified as being more dependent upon the sea and those to the south on nomadic pastoralism and small scale seasonal agriculture in the Toker Delta. Port-Sudan is the major sea port for Sudan. The city has a population of about 30,0000 which is approximately 30% of the estimated population for the Red Sea (CIDA 1995). Most of the industrial and economic activities are concentrated in Port-Sudan. The nearest populated centre is Port Suakin 57 km south of Port Sudan. It is the second largest sea port on the Sudanese coast. It handles mostly passenger vessels and small vessels. Most of the newly planned socio-economic projects are located within Suakin and Port-Sudan.

Coastal and marine habitats of Sudan include sandy, muddy and rocky beaches, salt marshes, sub-tidal soft bottom areas and extensive coral reefs. A typical feature of the 720 km coastline are "Mersas" which are sheltered embayments often fringed with mangroves. Significant mangrove areas are located south of Port-Sudan and Suakin with great number of species and habitat diversity (Krupp et al. 1994).

However based on recent studies, it is reported that the range of plant species have declined greatly due to the increasing pressure from the surrounding region during the recent recurrent drought and the unfriendly human activities. At present there is a great concern about the rapidly deteriorating conditions along the coast from various hazards resulting from coastal activities. (Suga 1999; Khider 2000; Bashir 2001).

3 Objectives

To provide a review of the current status of halophytic communities of the coastal region.

To identify possible threats to these resources.

To provide some suggestions for conservation programmes.

4 Literature Review

This review has prompted the need for consultation of literature on the subject matter with reference to topics of relevance. However biology and taxonomy of the Sudanese Red Sea marine and coastal habitats have been investigated by relatively few scientists. Andrews (Andrews 1948, 1950; 1952; 1956) gave a passing reference on the taxonomy of the halophytes of the Red Sea region. His outstanding pioneer work gave merit to his work as a leading accomplishment. Kassas (1957) and Kassas and Zahrah (1967) provided detailed description of these natural communities which stand out beyond any other in the region at that time. The references for the last three decades include SMSD (1950–1980); Chapman (1974); Mohamed (1984); UNDP, UNESCO (1986); Eltom (1989); Sheppard et al. (1992); Abdel Latif (1993a); Krupp et al. (1994); FNC (1994); CIDA (1995); UNEP/PERSGAA (1997); Elhag and Abdel Gadir (1998); Suga (1999); Abu Gideiri (1990); Ali et al. (2000); Khider (2000); Bashir (2001); Karrar (2003). It is worth mentioning that Bashir (2001) in Sudan Country Study on Biodiversity dedicated many parts to this topic and contains more information on the subject that has been accumulated in any subsequent publication.

5 Methodology

The assessment consisted of inventory of all related elements on the main subject matter. These include information on the main habitats and their characteristics, species composition, environmental threats and pressures to which they are currently subjected. Sources of information consulted include published literature, departmental reports and results of previous questionnaires and field surveys of selected sites of the study area by the author, in addition to discussions with the staff of the ministry of agriculture-Red Sea State and community leaders.

6 Results

The following brief summaries together with tables and diagrams describe the current status of salt marsh communities in terms of species distribution, diversity and related threats and environmental impacts. The data obtained are assessed and utilized in the overall assessment of the current status as basis for the prediction of the future.

The sea salt marsh covers approximately 7 km in width, it consists of shallow waters interrupted by wide bays, lagoons and drainage extremities of seasonal streams originating at the Red Sea Hills and mainly covered by salt tolerant grasses and shrubs.

6.1 Floristic Composition

The flouristic composition of the Sudanese salt marsh constitutes 11 major families with a varying number of representative species (Table 1). Families Chenopodiaceae and Gramineae are represented by a wide range of species while the rest are represented by fewer species; two to three species at most. The distribution of the Halophytic communities along the different zones of the Salt Marsh is shown in Table 2.

The halophytic communities dominate three Marsa or gulf habitats, namely Heidub, Ata and Arus. In these habitats halophytic communities exhibit clear and distinct zonation according to soil type and salinity gradient.

Marsa Heidub is rich Red Sea Mangrove; halophytic communities include *Arthrocnemum glaucum, Halopoplis perfoliata, Suaeda fruticosa* and *Suaeda monoiaca.*

Marsa Ata is an open bay protected by coral islets. Its location is a typical representative of salt marsh sector. Most dominant halophytes are *Artrocnemum glaucum, Halopoplis perfoliata. Acluropus lagopoides, Suaeda fruticosa and Suaeda vermiculata.*

Table 1 Floristic composition of the Red Sea Salt Marsh

Family	Species
Chenopodiaceae	*Arthrocnemum glaucum, Atriple farinose, Halopolis perfoliata, Salicornia fruticosa, Salsola baryosma, Suaeda fruticosa, S. monaica, S. vermiculata, Acluropus lagopoides*
Gramineae	*Aristida adscensionis, Conchrus setigrus, Cyperus glomeratus, C. rotundus, Dactylotenium aegyptiun, D. scindicum, Echinochola colonum, Juncus arabicus, Lasiurus hirsutus, Panicum turgidum, Sporobolus spicatus, Targus beteronianus*
Mimosaceae	*Acacia tortilis, Prospis chillensis, Limonium axillare*
Resedaceae	*Ochradenus baccatus*
Rhamnaceae	*Ziziphus spina-christi*
Salvadorceae	*Salvadora persica*
Solanaceae	*Datura stramonium, Soanum dubium*
Tamariacea	*Tamarixx aphylla*
Tiliacea	*Corochorus depressus*
Verbenaceae	*Avicennia marina*
Zygophyllaceae	*Tribulus terrestris, Zygophyllum album and Z. simplex*

After Bashir (2001)

Table 2 Distribution of Halophytes in the Sudanese Red Sea Salt Marsh

Location	Species
Marsa Heidub	*Avicennia mania, Arthrocnemum glaucum, Halopoplis perfoliata, Suaeda fruticosa, suaeda monoica*
Marsa Ata	*Avicennia mania, Arthrocnemum glaucum*
	Halopoplis perfoliata, Acluropus lagopoides, Suaeda fruticosa, Suaeda vermiculata
Marsa Arus	*Limonim axillare*
	Salicornia fruticosa, Artiplex farinose
Khour Arus	*Suaeda monoica*
Transition zone	*Jatropha aceroidesAcacia Trotilis, Tamarix aplylla, Calotropis procera.Zygophyllum album,Limonium axillare,Sporobolus spicatus,Aeluropus*

Marsa Arus represents the salt marsh in the North with varying habitat from the southern part due to the presence of the water edge which in direct contact with the coral reefs. It is characterized by hard calcareous soils and consequently different halophytic species such as *Limonium axillare, Salicornia fruticosa and Atriplex farinosa.*

Khor Arus is characterized by salinity gradient that decreases from the water edge towards the upper reaches of the Khor and consequently it supports different types of vegetation. Halophytes are dominated by *Suaeda monoica* which thins out at the upper reaches of the Khor where the semi desert plants prevail. These include *Jatropha aceroides, Calotropis procera, Acacia trotilis and Tamarix aphylla.*

Other halophytic communities include patches dominated by *Zygophylum album, Limonium axillare, Sporobolus spicatus* and *Aeluropus lagopoides.*

Mangroves: are the most conspicuous of the coastal vegetation. Thirteen mangrove sites are recorded along the Sudanese southern coastline south of Port-Sudan and Suakin Harbours mainly fringing coastal lagoons. Three of the four mangrove species known to occur in the Red Sea are found in the Sudanese coastline. These are *Avicenia marina, Rhizophora muncronata, Bruguiera gymnorhiza. Avicenia* is the most porominant while *Rhizophora* is restricted to few locations in the southern coast. *Bruguiera* is the least prevalent mangrove and mainly found in the northern part of the coast. Apart from the above mentioned localities, other important mangrove stands include Kilo-Tammania south of Port-Sudan, Klianeeb and Heidub south of Suakin.

7 Potential Threats

The coastal vegetation is subjected to various environmental threats which are expected to augment as further coastal development occurs. These are summarized in Table 3. One of the most significant environmental problems is related to coastal land use and human activities. These include filling and dredging which is usually carried out in shallow inter-tidal waters resulting in the destruction of a number of biologically important habitats including halophytes and mangroves through removal or accumulation of silt. Another threat is the pollution from various sources of the marine and

Table 3 Threats to halophytic communities; impacts and causes

Issue	Impacts	causes
Habitat destruction		
Coastal development	Dredging and filling (siltation and burial), destruction of mangroves	Urban, Industrial and port development
Waste disposal	Pollution of coastal areas around Port Sudan and Suakin harbors	Improper disposal facilities
Coastal tourism	Disturbance to vegetation	Treading, collection of shells
Salt pans	Damage of mangroves and associated flora and fauna	Construction and consequent removal of vegetation cover
Sedimentation	Mobilization of sand	Deforestation grazing of livestock-fuel wood collection
Urban development		
Coastal construction	Damage to coastal and marine biota	Infrastructure development
Discharge of untreated or insufficient treated sewage and solid wastes	Eutrophication and alteration of coastal environment	Lack of sewage treatment plants, lack of maintenance of existing ones and improper disposal facilities
Industrial activities		
Industrial pollution	Decline in water quality	Chemical pollution, organic pollution and thermal pollution
Maritime risks		
Marine vessel sewage and discharges of solid wastes	Localized marine and beach pollution	Discharges from ships
Petroleum development and transport		
Oil spills	Beach contamination, damage to coastal and marine biota	Discharges from terminals accidents, ballasts and bilge water

coastal environment. Pollution is becoming a problem of growing magnitude especially along the coasts of Port-Sudan and Suakin harbours. This includes oil and thermal pollution, industrial effluents, chemicals, pesticides and solid wastes from industrial and domestic sources (Elhag and Abdel Gadir 1998). Coastal tourism if uncontrolled is expected to result in disturbance to vegetation. Another concern is related to planned socio-economic development of the coastal region which includes a number of projects and activities, each with a different set of requirements and impacts. A major threat is attributed to one of the oldest and major activity along the Sudanese coast i.e. salt pans construction. Construction results in the removal of halophytic plants including well established mangroves.

8 Discussion

The Red Sea Coast is an area of intense activity, an area of interchange within and between physical, biological, social and economic processes. Its shallow coastal area have traditionally supported high productive ecosystems from which resources are harvested, in addition to natural protective and defensive fluctuations against coastal erosion and flooding. Red Sea marine coastal habitats are potentially subject to various interactions, not only physical interactions but also with regard to human population inhabiting the area with wide variety of activities leading to certain changes in coastal environment.

The combined action of the basic natural constraints and patterns of human exploitation may eventually stress coastal ecosystems that might in the future lead to complete degradation, risks to become endangered or even extinct (Karrar 2003). The expected change in climatic patterns and trends of natural deficiency seem to enhance environmental challenges to sustainability. The findings of the present review contribute supporting evidence to the above mentioned factors.

Halophytes – Saline zones include the salt marshes that fringe the land masses, coastal swamps and bays. Maritime salt marshes are relatively more spread. The dominant natural vegetation in saline habitats consists essentially of micro flora, phanerogamic herbs, some shrubs and trees. In the warm marshes there is a predominance of tropical true halophytes (Chapman 1974). These include succulents such as *Arthocnemum sp*, *Sueda spp* and *halopeplis sp*. The common salt tolerant types that are non-fleshy include *Limonium sp*, a number of *wiry grasses* such as *Sporobolus sp, Aleuropus sp* in addition to mangroves which are considered as special group of perennial halophytes (Suga 1999). Coastal areas are fragile in view of ecological hazards to which they have been subjected. These hazards include active erosion and periodical inundation of sea water leading to removal of top fertile soil and increase in salinity levels (UNDP, UNESCO 1986). Other factors include microclimatic conditions of the region (Eltom 1989).Vegetation densities and diversity are mainly governed by physiographic, edaphic and biotic factors. It is generally known that rainfall and relative humidity are higher in the southern part of the province (SMSD 1950–1980). All these factors act in a mosciac pattern creating favourable conditions for some where biological impacts such as human interference and overgrazing of more palatable species leading to destruction of some species being intense in the area (Mohamed 1984). Local administrators claim that the danger of herding domestic live-stock lies in the traditional pastoral nomads. Nomadic overgrazing of vegetation causes considerable damage to the natural plant covers and consequently inflicts barrenness of the landscape. Intermittent droughts which reflect upon the annual rainfall averages periodically limit vegetation cover to small pockets (Khider 2000; Bashir 2001). Cutting woods for domestic fuel, poses another local hazard (NEA 1983). Supplementing poor income in dry years is solved by the recent increasing demand for commercial fuel-wood, charcoal and bakery wood. This practice is likely to continue as an important source of income generation. The need for fuel wood is progressively increasing due to a clear trend of rural–urban migration and the network of fast roads to congested areas. The advancing recent aridity and excessively severe browsing of the more palatable species may account for the increasingly restricted presence of some species (FNC 1994).

Indigenous knowledge claims a strong invasion of the more competitive *Prosopis chilensis* trees being more drought tolerant and undesirable by animals.

In salt marshes where halophytes dominate in a zonal pattern of pure stands, camels are the sole animals which browse on twigs, leaves and young branches. These areas are highly degraded now (Abdel Latif 1993a; Khalil 1994). In general there is a decrease in the number of communities between the Red Sea

Coast and desert plain regarding halophytic and non-halophytic species when compared to Kassas (1957) records (Mohamed 1984; Suga 1999; Khider 2000). The present day vegetation and flora of The Red Sea Salt Marsh were reduced to poor diffuse tri-zonal assemblage of halophytes in the marginal saline habitat along the coast. The floristic composition consists of medium sized stands and shrubs in open formation. Pure stands were reduced and sporadic (Bashir 2001). Previously existing zones of halophytic communities of Marsa Ata do not exist anymore in the ecological sense. Absence of zone boundaries and overall morphological deformity is quite evident. Deterioration is a direct outcome of salt pans construction which involves the total removal of vegetation cover.

Mangroves–Mangroves are extremely important forms of coastal vegetation providing many important ecological functions in stabilizing sediments, protecting the coastline as well as providing feeding and nesting grounds for commercially important fisheries in addition to other economically valuable products. However harsh environmental conditions in the region where temperature and salinity are near upper limits of their existence make them very sensitive to disturbances and probably limits their ability to recover (UNEP/PERSGAA 1997). Most of the recent studies on Sudanese mangroves have confirmed that some stands are severely damaged. The magnitude of damage is variable from severe or moderate to light. The most affected mangroves are those of Kilo-Tammania, Shuokh and Halout (Abdel Latif (1993b); Suga 1999). Indications of mangrove destruction is also evident in Marsa Heidub which used to be typically rich Red Sea mangroves (Bashir 2001). This is quite evident when shoreline mangroves are compared with the inaccessible stands on the islands. The current condition is mainly attributed to the general degradation of the region rangeland. Mangroves offered an alternative substitute for grazing animals and important source for wood for fuel and housing construction. Some stands are heavily stressed or deteriorated. This situation is expected to intensify due to further pressure imposed by the newly planned socio-economic projects (Abu Gideiri 1990; Ali et al. 2000). Degradation of mangroves world wide is still attributed to active rehabilitation, expansion of ports, construction of new ones and human activities. The aims of reclamation are socio-economic growth in industry, urban and rural settlement. Such human manipulations have seriously threatened mangroves. Simultaneous with the above man-made destruction to extensive fragile and vulnerable ecosystems is a natural considerable degradation stress that have been observed in the last two decades.

9 Managerial Aspects

Halophytes have low priority index in the minds of the policy – makers, development planners and local administrative bodies. They are generally considered as resources with limited economic potential. The importance of halophytes is admittedly indirect and therefore their value tends to be underestimated and often limited to uses for fuel wood and sources of fodder. Halophyte ecosystems are unique whose habitat support diverse flora and fauna. They play a vital role in coastal stabilization and in maintaining the Sudanese Red Sea biodiversity heritage which would be lost if they perished. Other problems are related to data collection systems which lack proper planning and transparency, another factor is that management policies are not well defined and not based on reliable data base due to inadequate human and financial resources. Modern guidelines for responsible management is not yet part of the legal national framework. This requires an approach less focused on short term considerations and more concerned with long term sustainability of resources and the environment. Within this context, it is worth mentioning that UNESCO Regional Office-Doha is establishing a bio-saline pilot farm in the Red Sea State for the utilization of sea water to grow mangroves and other palatable halophytes as part of sustainable development of the region to address local communities basic needs as well as restoring coastal habitats.

10 Conclusions

The Sudanese Red Sea salt marsh shows signals and trends of some degree of degradation. Recent surveys revealed that halophytes belt along the Red Sea coast including mangroves are shrinking gradually under natural and man-made stress upon this unique ecosystem. Damage to coastal habitat by pollution from different sources of Port-Sudan and Suakin harbours is expanding.

It follows from the above that there is a great need for a well organized systematic monitoring programmes where research plans are well tied to development programmes. A closer interaction is required amongst different agencies in the region through a central body for the co-ordination of efforts to restore coastal habitats through integration of environmental and natural resources issues into planning and management of the region.Community participation is a must.

References

Abdel Latif EM (1993a) Factors threatening the marine and coastal environment of the Red Sea in Sudan. Sudan Environment Conservation Society, Khartoum

Abdel Latif EM (1993b) Environment view for development in Sudan (in Arabic). Sudan Environment conservation Society, Khartoum

Abu Gideiri YB (1990) Biodiversity in the Red Sea. National Biodiversity Strategy and Action Plan (NBSAP), Khartoum-Sudan

Ali SM, Elhag AD, Eltayeb MM, Goiballa AJ, Mohammed MM, Ibraheem MY, Elamin SM, Mohamed MO, Mohamed KA, Mohamed LH (2000) Environmental Evaluation of Eastern Part of Port Sudan Harbor. Presented to Shipping Ports Corporation, Port Sudan-Sudan

Andrews FW (1948) Vegtation of the Sudan. In: Tothill JD (ed) Agriculture in the Sudan. .Oxford University Press, London

Andrews FW (1950; 1952; 1956) The Flowering plants of the Anglo – Egyptian Sudan (Vol. II) and I. The flowering plants of Sudan (Vol. III). T. Buncle and Co, Arbroath, Scotland

Bashir M (ed) (2001) Sudan Country Study ON Biodiversity: prepared by the National Strategy and Action plan (NBSAP) Project No. Sud/97/G3/.Sudan

Chapman VJ (1974) Salt marshes and salt deserts of the world. In: Reimold RJ, Queen (eds) Ecology of halophytes. Academic, London

CIDA (1995) Preliminary Environmental Assessment, Sudan Integrated Fisheries Project: project NO: B535 Halifax, Nora Scotia, Hull, Canada

Elhag EA, Abdel Gadir SA (1998) The effect of oil pollution on the distribution of marine faura at Port Sudan and Suakin Harbours. Sudan. J Sci III

Eltom M (1989) An outline of the climate of the Red Sea Region of The Sudan. Proceedings of the Red Sea Area program (RESAP).Khartoum

FNC (1994) Annual Reports. Forests National Corporation, Khartoum

Karrar O (2003) Aspects of effects of hydro-carbons on selected target groups in the vicinity of Bashaier Oil Terminal. PhD Thesis – Department of Zoology, University of Khartoum-Khartoum

Kassas M (1957) On the ecology of the Red Sea coastal-land. J Ecol Sudan 3:696–741

Kassas M, Zahrah MA (1967) On the ecology of the Red Sea Littoral salt marsh. Egypt. Ecological monographs, Egypt

Khalil AS (1994) Ecological studies on fishes of the mangrove Ecosystem. M.Sc. Thesis (Zoology Dept.), University of Khartoum, Khartoum, Sudan

Khider S (2000) A study on the ecology of The Khour Arbaat Valley. Red Sea State, Sudan. M.Sc. Thesis (Botany Dept.), University of Khartoum. Khartoum, Sudan

Krupp F, Turkay M, Elhag AGD, Nasr D (1994) Introduction to: Comparative ecological analysis of biota and habitats in littoral and shallaow sub-littoral waters of the Sudanese Red Sea. In: Friedhelm Krupp, Michael Turkay, Abdel Gadir D. ElHag Dirar Nasr (eds) Forschungs institute Senchenberg, Germany and Faculty of Marine Science and Fisheries, Sudan

Mohamed BF (1984) Ecological observations on Mangroves of the Red Sea shores of the Sudan. Hydrobiologia, Khartoum, Sudan

National Energy Administration (1983) Base year Energy supply, demand balances and demand projection methodology. National Energy Administration, Khartoum

Schoreder JH (1982) Aspects of coastal zone management at the Sudanese Red Sea.Envi Research Report: 3, IKS, Khartoum extracts, Khartoum

Sheppard CR, Price A, Roberts C (1992) Marine ecology of the arabian region. Patterns and processes in extreme tropical environments. Academic, London

Sudan Metrological Services Department (SMSD) (1950–1980). Climatalogical normals of Khartoum, Halaib, Port Sudan, Suakin and Toker, Sudan

Suga AAB (1999) On the Ecology of Sudanese Red Sea Coastal Vegetation with Emphasis on Mangrove Ecosystem. Ph.D. Thesis, Faculty of Science, University of Khartoum, Sudan

UNEP/PERSGAA (1997) Assessment of land-based on sources and activities affecting the Marine Environment In the Red Sea and Gulf of Aden. UNEP Regional Seas Reports and Studies

A General Assessment of Marine Turtle Nesting Activity on the Libyan Saline Coast

Daw A. Haddoud and Hisham M. El Gomati

Abstract Marine turtle nesting sites in Libya have remained unknown until 1992/1993 for a long time due to the shortage of information with the exception of Kouf national protected area. A nesting activities survey were first conducted in 1993 between the Egyptian border and Zweetina. The first phase to assess nesting activity was conducted in 1995 between the Egyptian borders to Sirte. The Second phase was carried out in 1996 and covered the coast from Sirte to Misratah, from May to August 1996. All the tracks identified as *Caretta caretta*, and the green turtle *Chelonian mydas* was observed during this phase. The third phase was conducted in July of 1998 between the Tunisian border and Misratah, 15 crawl tracks of nesting were recorded and identified as *Caretta caretta* later in July 1999. Three beaches were resurveyed in eastern part of Libya. A general assessment of marine turtle nesting activity on the whole coast was carried out to allow us to implement a sound conservation strategy for the conservation of this endangered species.

1 Introduction

The current number of species of marine turtles that live in seas and oceans amounts to eight types, three of which have existence logged in the Mediterranean sea water, namely: *Caretta caretta*, *Chelonia mydes* and *Dermochelys coricea*. The existence of nests of the first two types was proved to be at the Mediterranean shores, and the studies conducted in the South Mediterranean countries proved the continuous decline for these two types due to environmental and natural causes (Kasparek 1993; Laurent 1990; Broderick et al. 2002, 2003). These studies have been conducted at the shores of Algeria, Morocco, and Egypt and the nest ratios logged in these countries do not represent any addition to the deteriorating stock of marine turtles in the Mediterranean sea, while at the northern and eastern parts of Mediterranean the activity of marine turtle nesting is concentrated on Greece, Cyprus and Turkey.

Relevant institutions and researchers in these countries gave much concern in setting certain programs to study such creatures and protect their nest locations (Demetropoulos and Hadjichristophorou 1988; Margaritoulis et al. 1992; Broderick and Godley 1996). Hazards which threaten marine turtles are outlined in the different pollution factors, and exploitation of sandy beaches in setting up tourism buildings increase and expand year by year, and which are concurrent with turtles' egg-laying on such shores.

Libya, being one of the Mediterranean States, and at the same time occupying a large space of the Mediterranean southern shore; deserves to have some studies conducted on its more significant shores as an attempt to estimate the real stock of marine turtles latent in the Mediterranean. The recent available information lacked sufficient and adequate accuracy

D.A. Haddoud (✉) and H.M. El Gomati
Marine Biology Research Center, Tripoli, Tajura, Libya

M. Öztürk et al. (eds.), *Sabkha Ecosystems*, Tasks for Vegetation Science 46,
DOI 10.1007/978-90-481-9673-9_7, © Springer Science+Business Media B.V. 2011

and details. Part of such information was logging of existence of loggerhead marine turtles on the shores of natural protectorate at Al-Kuf Valley (Herbert 1979; Armsby 1980; Schleich 1987) as well as the information already logged about the locations of marine turtles nesting within the field visits paid in the years 1992–1993 for the shores of the eastern region (Hadoud D and Assigier 1995). As a completion for this exerted effort, a complete study was made along the Libyan coast as a part of an integrated program for determining the nesting locations of marine turtles, their species and possible dangers threatening turtles and nests. The study further included the determination of the shores that is in need for special protection programs being the most susceptible to pollution or in which predatory acts by natural hostilities are excessive. The aim of this study is a general assessment of marine turtle nesting activity on the hole Libyan coast to allow us in putting a good strategy for future conservation programs for this endangered marine organism.

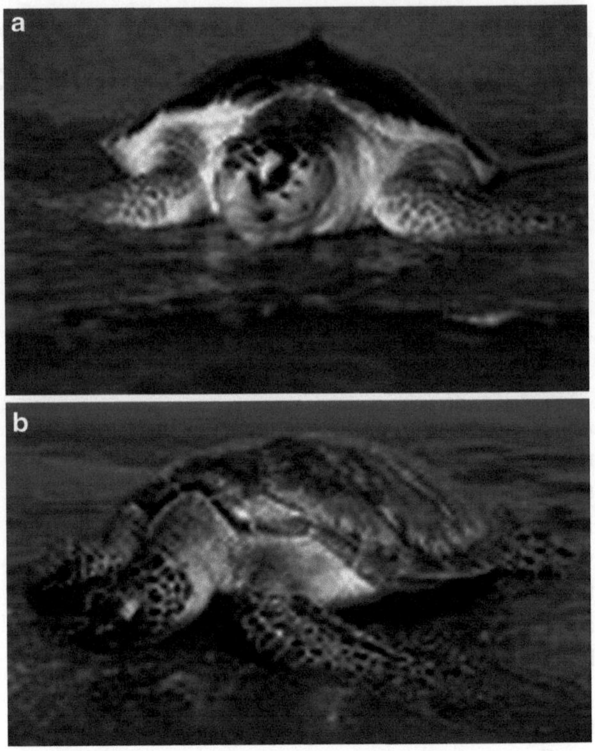

Fig. 1 (**a**) *Caretta caretta*. (**b**) *Chelonia mydas*

2 Material and Methods

This 3-year lasting study concerning of the nesting sites of marine turtle has been accomplished on three phases where the Libyan shore was subdivided into three areas (Fig. 1) such as follows:

1. Phase I: The area located between the Egyptian borders and Sirte city.
2. Phase II: The area located between Sirte and Misratah city.
3. Phase III: The area located between Misratah city and the Tunisian borders In all three phases different survey method were applied such as "walk (w), motorbike (m), vehicle (v), Quad (q)" and by such means the targeted sandy beaches were surveyed once in Phase (I) and Phase (III) and several times for some shores selected in Phase (II) in addition to re-survey of three high density nesting shores located within the boundaries of Phase (I), in Summer 1999. As with respect to classifying nesting signs applied in all phases, they were within the following five classifications: UCT, FCT, NCT, CT, N "Laurent et al. 1995".

3 Results

The total length of the Libyan coast is 1,975 km, of which a total of 1,144 km was for the sandy beaches. A large number of halophytes are distributed in the country such as the species of *Acacia, Tamarix* and *Calotropis procera.* The coastal zone vegetation strip is dominated by dunes covered by *Ammophila arenaria, Agropyron junceum, Aeluropus littoralis,* bulbous plants like *Urginea maritima* and *Pancratium maritimum*, along with scattered shrubs *Lycium europaeum*, *Rhus tripartit,* woody plants like *Limoniastrum monopetalum, Tamarix nilotica*. The seasonal mud flats include the halophytes such as *Suaeda fruticosa, Juncus maritimus, Mesembryanthemum nodiflorum* and *Cakile maritima*. *Phragmites communis, P. australis* and *Typha capensis*, together with the species of *Limonium, Salicornia, Sarcocornia, Suaeda*, and *Juncus maritimus* are scattered over the coast (Boulos 1975). There are several thousand hectares of mixed stands of *Acacia saligna*

Table 1 Loggerhead turtle nesting activity during the three phase of study

Phase of study	Coordinate of phase		Coastline (km)	Sand Beach Length (km)	Number of sites sampled	Total length of sand beach sampled (km)	Total crawl tracks (N + NC)
1995	31° 57 69′	31° 13 00′	1,195	797.7	50	141.65	342 (176)
Phase I	24° 59 28′	16° 40 00′				(17.76)	
1996	31° 13 28′	32° 12 13′	209	186.3	8	87	58 (27)
Phase II	16° 22 98′	15° 19 26′				(46.70)	
1998	32° 25 06′	33° 07 58′	407	160	23	105	15 (7)
Phase III	14° 59 62′	11° 41 15′				(65.63)	
Total			1,811	1,144	81	333.65	415(210)
						(29.17)	

Table 2 Crawl tracks survey at the east coast of Libya

	Prospected beaches				
	Coordinator		Crawl tracks		
Coastal area	Latitude	Longitude	1992/1993	1995	1999
Ayn el Ghazalah 1	32° 12 30′	23° 20 70′	24 (15/8/1992)	2 (20/6/95)	7 (23/7/99)
Ayn el Ghazalah 2	32° 12 36′	32° 20 56′	33 (23/7/1993)	11 (20/6/95)	23 (20/7/99)
Abu el Frais	32° 16 00′	23° 13 00′	18 (21/7/1993)	8 (19/6/95)	17 (23/7/99)

Table 3 Dead Turtles recorded from 1987 to1998

	1990–1992	1995 phase I	1996 phase II	1998 phase III	Total
Ayn el Ghazalah (1)	2	19	7	9	37
Ayn el Ghazalah (2)	2	2	2	–	6
Abu el Frais (3)	1	–	–	–	1

and *Atriplex canescens* found in the country. All these habitats form seclusive areas for the turtles.

Within the three phases a total of 81 shores with 333.65 km were surveyed which they represent a percentage of 29.16% of the total length of the sandy beaches along with the Libyan coast. The total crawl tracks which recorded in three phases is 415, 210 of which were tracks of turtle nests "loggerhead type", namely the percentage of 50.60% of the total sum of tracks so logged (Table 1).

A three shores (1, 2, 3) located at the eastern part of the Libyan coast were re-surveyed in summer of 1999 these were: Ayn Al-Ghazalah (1), Ayn Al-Ghazalah (2), and Abu El Frais and the number of tracks logged were 7, 23, and 17 respectively (Table 2). The number of dead turtles logged along the Libyan coast before and within the 3 phases of survey was 44 turtles (Table 3) distributed such as follows.

Thirty seven *Caretta caretta* loggerhead, six *Chelonia mydas* (green turtle), one *Dermochelys coriacea* (leather back turtle). Two turtles of *Dermochelys coriacea* type were caught alive at Tajura shore by means of small coastal fishing gear. The length of the first turtle was 137 cm (1996), while the length of the second turtle was 1.40 cm (November 2000). The tagged turtles within the past years 1995–2002 were recoded; all being the loggerhead type. Five turtles were tagged with Phase I, one turtle was tagged in Phase II and ather most of them tagged in the sea (Table 4). The percentages of the predator nests within survey duration were 44.8%, 37.0% and 45.4% in the three phases respectively.

Table 4 Tagged Turtles *Caretta caretta* recorded from 1995 to 2002

Name of beach	Coordinates	Date	Carapace length (cm)	Number of tag
Deriana	32° 17 60′	25/6/1995	79.5 cm	F 3031
	20° 13 31′			
	31° 02 23′			
Sultan	20° 09 25′	29/6/1995	74.5 cm	F 3021
		29/6/1995	74.3 cm	F 3011
		29/6/1995	81 cm	F 3051
		29/6/1995	86.3 cm	F 3001
Al Gbiba	31° 13 28′	23/6/1996	83 cm	F 3101
	16° 22 98′			F 3121
Tajura	32° 53 73′	15/12/1996	72.5 cm	F 3022
	13° 22 54′			
North Musratha	32° 31′	20/4/2002	SCCL 74	F 3104
	15° 52′		CW 52	
North Benghazi	32° 08′	24/4/2002	CCL 71	F3172
	19° 15′		CCW 62	
North Benghazi	32° 35′	3/5/2002	CCL 90	F3123
	19° 36′			
North Benghazi	32° 51′	8/6/2002	CCL 76	F3128
	20° 10′		CCW 71	
North Bengazi	32° 27′	17/6/2002	CCL 61	F3105
	19° 51′		CCW 55	
South musratha	31° 29′	20/6/2002	CCL 68	F3106
	16° 12′		CCW 56	
North Benghazi	32° 21′	28/6/2002	CCL 62	F3114
	15° 40′		CCW 52	

4 Discussion

Based on the currently emphasized information, a number of three types of marine turtles in the Libyan waters was logged, namely: *Caretta caretta, Chelonia mydas,* and *Dermochelys coriacea.* Numerous researchers agreed upon the existence of *Caretta caretta* along the Libyan coast and with large (Laurent et al. 1995), nesting number and the findings of this comprehensive study confirms this fact and further support the already logged findings prior to this study, by other researchers during their study conducted for the natural protectorates shores at Al-Kuf Valley (Herbert 1979; Armsby 1980; Schleich 1987). The existence of, for *Chelonia mydas*, its existence in Libyan waters was emphasized for the first time, but its recorded in Ayn Al-Ghazala lagoon in the far eastern shore of Libya (Hadoud D and Assigier 1995) and the same existence was re-logged again in one of Phase (II) shores, namely: Abuwirat Al-Hasson shore (Hadoud and El Gomati 1997).

The third type "*Dermochelys coriacea*" has an existence, which was logged for the first time in the Libyan waters in years 1927, 1928 in Benghazi and Tripoli Coasts (Capra 1949). Recently two turtles of this type were suspended at Tajura area during 1996 and 2000. However, the nesting of this type at the Libyan shores has not been proved yet up to this date. Fretey (1986) presumed the existence of nesting activity of this species of turtles on the Libyan shores although there is nothing to enhance the correctness of this presumption. The density of crawl tracks in the three phases was 2.16, 0.66, and 0.113/km respectively with overall average of 1.24/km for tracks and 0.62/km as nesting density. Such ratios give intense significance to the Libyan shores and render them the most important nesting locations for Caretta caretta type worldwide.

In contrast with the logged density in the shores of the neighboring African States, it will be noticeable that in Morocco and Algeria no tracks were reported during the multiple survey made for 97.9 km. of the shores (Laurent 1990), while in Tunisia the average tracks was 0.16 km (Bradai 1993) and a calculated density of 0.039 km. witnessed by Egypt during the study conducted on the western Egyptian shores (Kasparek 1993). The large increase in the activity of marine turtles nesting on the Libyan shores in comparison with the neighboring States is the length of its sandy beaches which are still conserving their natural condition as well as the few number of turtles that subject to intentional catching and that is mainly attributed to the fact that their meat does not constitute a foodstuff as it is the case in the neighboring states. The density of the tracks logged in the three phases were not convergent, as they range from zero in some shores and frequently reach 5.8/km. in Abu El Frais shore in Phase I: (Laurent et al. 1999) and 4.4/km. In the second shore in Phase II (Hadoud and El Gomati 1997). The imperfection of density in some sandy beaches could not be perfectly interpreted rather than some probabilities would be made on the grounds of the scientific observations already logged within the study.

For instance, in phase I: that could easily be attributed to the apparent pollution in some shores with tarball, while in the shores of Phase II: interpretation of non-existence of nesting density in some shores could be made despite the fact that they are sandy beaches, with low pollution rate and far from whatever population activity. However, one fact can be taken into account, namely the concurrence of turtle nesting season with the season of egg-laying of some species of sharks which approach in large quantities towards these shores the thing which cause leaving of turtles far away from these areas. As regards to Phase III: the least nesting density area is because of the high population density and centering of the main cities of Libya within the limits of this area.

The risks that threaten the marine turtles are much diversified. For instance, turtles in the sea water are suspended in nets of fisherman, like Tunara – fisheries of tuna fish – also there are some fisheries for sharks which are gill nets called "Khanagates" in which marine turtles are suspended. What increase the number of suspended turtles in these two types of fisheries is the concurrence of tuna fishing season "June" and the shark fishing season (February – June) with the season of turtles reproduction and their approaching to the shores. And most dangerous impact of sea turtle in the Libyan coast through the season of tuna by the foreign fisheries in April to July every year many turtle catch by long line and surrounding nets. Likewise the female turtles are liable to predatory when moving out for egg-laying and the most important predators is "Jackal *Canis aurens*". As for nests, the Predatory percentages were too high and the predators reported within the study were "fox *Vulpes vulpes*, sand crab, Ocypode cursor, Jackal *Canis aurens*".

Comparing the logged percentages for marine turtle nest, predatory acts at the Libyan shores, with other shores, we will find them 44.8% in phase I (Laurent et al. 1995), 37.0% in phaseII, (Hadoud and El Gomati 1997), 45.4% in phase (3) (Laurent et al. 1999), 41.2% at Turkish Shores (Brown and Macdonald 1995) and 70% at the southern shores of Cyprus (Demetropoulos and Hadjichristophorou 1988). That will definitely make it obvious that the phenomenon of nest ravening is deemed to be a natural phenomenon of which both increase and decrease will be controlled by environmental mode of each shore and the existence or on-existence of predators. Within the past years a number of seven *Caretta caretta* were tagged while moving out to lay eggs. This process has become a routine style for data collection and clarification of some vague points concerning the biology and behavior of this animal such as study of rates of growth of this creature and migration (Carr 1967) as well as the nutritional locations and the places in which turtles spend the winter season (Margaritoulis 1988) and finally the ages of the turtles (Mrosovsky 1983).

The occupation of Libya to the longest shore southwards the Mediterranean sea enabled it to acquire a special importance and in particular after the findings obtained in this three-phase study which showed a significant nesting density in the Libyan shores and which in turn requires focusing on the intensive re-surveying specially in the high density shares in order to reach realistic findings and estimations on nesting density and the types of turtles having nesting activity in the Libyan shores. Setting a permanent protection program in some important shores such as Ayn El Ghazalah, Al-Kuf, Abu Frais shares within the limits of Phase I, and shore No. (1), within the limits of phase II as an attempt to preserve and develop the stock of this marine organism which is really threatened with extinction.

References

Armsby JK (1980) Koufnational park marine survey final report: Apr–July 1980 ACSAD

Boulos L (1975) The Mediterranean element in the flora of Egypt and Libya. In La flore du bassin méditerranéen: essai de systématique synthétique No. 235CRNS, pp 119–124

Bradai MN (1993) La nidificatiol de la Tortue marine Caretta caretta dans le sudest de la Tunisie. Rapport pour le RAC/ SPA et I, Association de la Protection de la Nature et de I, Environnement a Sfax (APNES)

Broderick AC, Glen F, Godley BJ, Hays GC (2002) Estimating the number of green and loggerhead turtles nesting annually in the Mediterranean. Oryx 36:227–235

Broderick AC, Glen F, Godley BJ, Hays GC (2003) Variation in reproductive output of marine turtles. J Exp Mar Biol Ecol 288:95–109

Broderick AC, Godley BJ (1996) Population and nesting ecology of the Green Turtle, *Chelonia mydas,* and the Loggerhead Turtle, *Caretta caretta,* in northern Cyprus. Zool Middle East 13:27

Brown L, Macdonald D (1995) Predation on Green turtle *Chelonia mydas* nests by wild canids at Akyatan beach Turkey. Biol Conserv 71(1):55–60

Capra F (1949) La *Dermochelys coriacea* (L.) nel Golfo di Genova e nel Mediterraneao. Ann Mus Civ Stor Nat Genovo 63:270–282

Carr A (1967) So excellent afishe: a natural history of sea turtles. Natural History Press, New York, pp 248

Demetropoulos A; Hadjichristophorou M (1988) Turtles and turtle conservation in Cyprus (handout to visitors to the Lara Turtle Station). Department of Fisheries. pp 3

Fretey J (1986) Les reptiles de franco metropolitaino et des les Satellites. Tortues et lezards, Paris, Hatier, pp 128

Hadoud D, Assigier F (1995) Survey of sea turtle in eastern part of Libya. Premier Congres Magrepin des Sciences e la Mer. Hammamet-Tunisie, 20–22 Nov 1995

Hadoud DA, El Gomati H (1997) The coast survey of marine turtle activity along the coast of Libya. Phase 2: between Sirte and Misratah. Deuxiemes Journees Maghrebines des sciences de la Mer. ISTPM

Herbert J (1979) Kouf national park. Wildlife survey and development. The Arab center for the studies of arid zones and dry lands

Kasparek M (1993) Marine turtles in Egypt. Phase I. Survey of the Mediterranean coast between Alexandaria and EL-Salum. Report for Medasset and RAC SPA

Laurent L (1990) Les tortues marines en Algerie et au Maroc (Mediterranee) Bulletin de la Societe Herpetologique de France 55-1-23.

Laurent L, Bradai M, Haddoud D, El Gomati H (1995) Marine turtle nesting activity assessment on Libyan coasts. Phase 1 Survey of the coasts between the Egyptian border and Sirte. RAC/SPA

Laurent L, Bradai M, Haddoud D, El Gomati H, Abdelmola AH (1999) Marine turtle nesting activity assessment on Libyan coasts. Phase 3 Survey of the coasts between the Musrata to Tunisian border. RAC/SPA

Margaritoulis D (1988) Post-nesting movements of loggerhead sea turtle tagged in Greece. Rapp Commun Int Mer Medit 31(2):284

Mrosovsky N (1983) Conserving sea turtles. British erpetological society, pp 176

Schleich HH (1987) Contributions to the herpetology of kouf national Park (NE Libya) and adgacent areas. Spixiana 10(1):37–80

Current Challenges and Future Opportunities for a Sustainable Utilization of Halophytes

Ahmed Debez, Bernhard Huchzermeyer, Chedly Abdelly, and Hans-Werner Koyro

Abstract Increasing salinisation has significant and detrimental impacts on land, water and vegetation quality, wildlife environments, agronomy and ecosystem functioning. This is particularly true for arid and semiarid areas where high evapo-transpiration rates expose plants to further adaptive pressure. Unlike conventional crops, halophytes are plants that survive and are able to reproduce in environments (coasts, wetlands, and inland deserts) with higher salinity levels. These species, which represent about 1% of the world's flora, have evolved complex mechanisms at different levels (whole plant, cellular, and molecular) enabling them to successfully cope with these hostile conditions. There are about a billion ha of salt-affected land world wide, which are unsuitable for agriculture and may therefore provide unique opportunities for "halo-biotechnologies". Taking into account the increasing pressure on fresh water resources and considerable diversity of potentially useful halophytes, such an approach may help in the mid- and long-term to rehabilitate these marginal zones and create sustainable production systems. Agriculture on saline soils is an alternative agriculture under a range of salinity levels in groundwater, and/or soils. A precondition for its use is the economic value added. Yet, fundamental prerequisites have to be considered to ensure that this promising approach would be cost-effective and environmentally safe. It must yield economically viable crops at yields high enough to be accepted by the farmers. This should be concomitant with the development of agronomic techniques relevant for growing saline, water-irrigated crops in a sustainable manner. Most importantly, these practices should be sustainable, ecologically well-tolerated and not lead to further damage of natural environments. If applied successfully, such an approach may lead to domestication of wild, salt-tolerant plants to be used as food, forage, oilseed crops, as well as pharmaceutical or ornamental plants. Soil desalination represents also important tasks for the so called cash crop halophytes. The successful rehabilitation of saline marginal zones by introduction of halophytes largely depends on collecting reliable data on salt-tolerance limits during life cycle of the respective candidate species. In this contribution, we present an overview of new data gained under saline conditions during the last years with respect to halophytes of interest and discuss their likely implications at applied level.

1 Introduction

Growth of the human population by 50%, from 6.1 billion in mid-2001 to 9.3 billion by 2050 (http://www.unfpa.org/swp/2001/), requires a significant and concomitant increase of crop production to ensure food security, especially in the developing countries. Recent estimations of the Food and Agriculture Organization of the United Nations (FAO) indicate that more than 850 million people are suffering from chronic malnutrition. Additionally, this problem becomes enhanced

A. Debez (✉) and B. Huchzermeyer
Institut für Botanik, Leibniz Universität Hannover, Herrenhäuser Str. 2, 30419 Hannover, Germany
e-mail: ahmed_debez@yahoo.com

C. Abdelly and A. Debez
Laboratoire des Plantes Extrêmophiles, Centre de Biotechnologie de Borj-Cedria (CBBC), BP 901, Hammam-lif, 2050, Tunisia

H.-W. Koyro
Institut für Pflanzenökologie, Justus-Liebig Universität Giessen (JLU), Heinrich-Buff-Ring 26-32, D-35392 Giessen, Germany

M. Öztürk et al. (eds.), *Sabkha Ecosystems*, Tasks for Vegetation Science 46,
DOI 10.1007/978-90-481-9673-9_8, © Springer Science+Business Media B.V. 2011

by the current climate change. Among the most important challenges are the increase in temperatures, desertification, and salinisation of soil and water resources. Salinity can be either natural or related to human activity (inaccurate irrigation management or fertilizer use) (Galvani 2007; Breckle 2009).

Salinity is actually an ever-present limiting factor to crop yields, notably in countries where agriculture relies on irrigation, which allows farmers to be less dependent on seasonal rainfall and the weather uncertainties (Galvani 2007). Yet, inappropriate irrigation strategies increase salinity of soils and water by depositing ions like sodium, calcium, magnesium, potassium, sulphate and chloride. In arid and semi-arid regions, solutes from irrigation water can highly accumulate, reaching levels that are harmful for the vigour and productivity of crops. According to Szabolcs (1994), 7% (1 billion hectares) of all land area is salt-effected, and about 10 million ha of agriculturally productive soils are being lost annually worldwide because of irrigation-induced salinity. Crop production is restricted by salinity on 40% of the world's irrigated land. In addition, 19.5% of the currently 230 million ha irrigated land, and ca. 5% of 1,500 million ha under dry land agriculture are salt-affected to varying degrees (Munns et al. 1999). It is estimated that salinisation of irrigated lands causes annual global income loss of about US $12 billion, severely impacting aggregate national incomes in countries affected by degradation of salt-affected land and saline water resources (Ghassemi et al. 1995). Generally, the worst salinity impacts occur where farming communities are relatively poor and face economic difficulties. In Australia, the annual expansion rate of dry land salinity is 3–5%, resulting in losses of Australian $270 million. Future projections estimate that 17 million ha of Australia's agricultural land will be significantly affected by salinity by 2050 (http://audit.ea.gov.au). The situation in Australia is even worsened by the extensive secondary salinity caused by substituting superficial rooted annual species for the deep-rooted perennial native vegetation with. As a result, rising water tables are observed that bring salt stored deep in the profiles close to the soil surface (Barrett-Lennard 2002).

Despite considerably varying in their response when salt-challenged, crop species are generally intolerant to one-third of the concentration of salts found in seawater (Flowers 2004). Several attempts to improve salt tolerance of crops through conventional breeding programs have yielded very limited success, mainly because salinity is a multifactorial problem (Koyro et al. 2008). Halophytes are plants that can complete their life cycle in soils with salinity concentrations above 200 mm NaCl (Flowers and Colmer 2008). Halophytes, which represent 1% of the world's flora, thrive in a wide range of habitats, from arid regions to coastal marshes. Several species grow in waterlogged or flooded soils, even withstanding total immersion in seawater (Munns 2008). Some halophytes require fresh water for germination and early establishment but can tolerate higher salt levels during vegetative and reproductive stages, other's may germinate at high salinities but require lower salinity for maximum growth (Debez et al. 2003, 2004; Koyro and Eisa 2008). In extreme cases (obligate or euhalophytes), increased biomass production occurs only under increased salinity. Further, some plants grow well on permanently wet areas; others grow best where the soil dries out in the summer (Galvani 2007). Mechanisms that allow such an extraordinary adaptation are still largely unknown.

Given that oceans contain most of the water on earth, the concept of sustainable agriculture using the so-called "cash crop halophytes" irrigated with saline waters up to seawater salinity is gradually emerging (Lieth et al. 1999). Although economic consideration of halophytes and other salt-tolerant plants is just beginning, they are now receiving increased attention in arid regions where intensive irrigation has led to saline soils or where water shortages are forcing use of marginal resources such as brackish underground water (O'Leary 1987; Öztürk et al. 2006, 2008). According to Yensen (2006) twenty-first century will likely be the century of halophyte agriculture expansion, as diminishing fresh water resources put pressure on civilization to utilize the vast saline soils and aquifers. While much of this land occurs in the Middle East, Central Asia, Northern Africa and Australia it seems that many other countries will face salinity issues" if steps are not taken. More recently, Galvani (2007) claimed: "To meet future agriculture needs, a solution is an environmentally friendly technique which controls salinisation, uses salt removing crops, chooses halophytes crops for direct salt water irrigation, or selects species that have high salt-removing capacity and commercial value. Some plants can be integrated into rotation programs or planted as intercrops for perennial plants to control salinisation". As hypothesized by this author, the unsuitability of sand and saline water for conventional crops may actually be advantageous when considering salt-tolerant plants, owing to the mineral composition of seawater (among the 13 mineral nutrients required by plants, 11 are present in seawater in sufficient amounts) and the texture of sandy soils, which favours the infiltration of water and prevent the harmful salt build up in the zone of root development. If applied carefully, the combination of sand, seawater, sun and salt-tolerant plants constitutes hence a valuable and realistic opportunity for many developing countries.

Several halophytes have been evaluated as potential crops for direct seawater or brackish water irrigation. Their potentialities cover a wide range of applications: the improvement of soil characteristics (desalination, heavy metal extraction), biomass production, food, fuel, fodder, fiber. Although direct consumption of halophytes by humans and animals may be limited, the seeds of many of them are being considered as new sources of grains or vegetable oils (Hinman 1984). Salt-tolerant plants can also be used to produce materials with high economical value, such as essential oils, flavours, fragrances, gums, resins, oils, pharmaceuticals, and fibers (Galvani 2007; Ksouri et al. 2007). They may also be marketed for use for ornamentation for their foliage or flowers (Messedi et al. 2004; Slama et al. 2006). Fuel-wood and building materials may also be manufactured from salt-tolerant species using land and water unsuitable for conventional crops.

The development of successful saline agriculture necessitates a better understanding of the potential of plant species to withstand ambient salinity and sodicity levels in soil and water, and also of the potential uses and markets for the agricultural products. In this contribution, we address the most recent information available with respect to the potential utilizations of halophytes in the context of halo-biotechnology, and present some practical recommendations to be considered for a economically viable and environmentally compatible sustainable biosaline agriculture.

2 Soil Bioremediation by Desalination-Efficient Halophytes

In the arid to semi-arid regions, water quality is a major factor limiting crop production (Cantero-Martínez et al. 2007). Interestingly, the natural ecosystems of the Mediterranean-climate areas are characterized by a mixture of rustic perennial shrubs and trees (mainly xerophytes and halophytes) with annual crops or fodder plants (Turner 2004). Growth activity of the annuals is governed by the rainfall regime, the rainy season extending from autumn to spring, while the slowly growing perennials are able to cope with water shortage and soil/water salinity. Variations in biomass and productivity among and within natural ecosystems may be attributed mostly to differences in water and nutrient (especially N) availability and salinity (Sherman et al. 2003). Understanding the functioning of plant saline ecosystems capable of significant productivity under salt stress is of paramount importance in perspective of improvement of soil characteristics using halophytes.

One of the most fascinating natural saline ecosystems are coastal or continental sabkhas, which are typical of the semi-arid and arid areas around the world. Because of the extreme environmental (edaphic and climatic) conditions characterizing these biotopes, they are considered as non productive wastelands. Sabkhas could actually be valorised by implementing agro-ecosystems using "alternative" cash-crop halophytes (Böer and Gliddon 1998). In Tunisia for instance, the annual fodder plant *Hordeum maritimum* (Poaceae) is commonly found in the saline depressions in close association with strict halophytes, such as *Arthrocnemum indicum* and *Halocnemum strobilaceum* (Hafsi et al. 2007). In a 2-year survey, Abdelly et al. (2006) have been monitoring the impact of halophyte-fodder species association on biomass production and soil salinity level in the inland sabkha of Enfidha (100 km south-east from Tunis), a semi-arid bioclimatic zone with moderate winters and mean annual rainfall rarely above 300 mm. The vegetation of studied area was characterized by perennial halophytes tufts in association with fodder annuals, mainly *Medicago* species, highly appreciated for their high fodder value, and ensuring most of to the ecosystem primary production in the absence of water shortage. Parallel field and laboratory investigations revealed that *Medicago* growth was restricted by salt (Abdelly et al. 1995; Abdelly et al. 1996), and nutrient (mainly nitrogen and phosphorus) deficiencies (Abdelly 1997). Halophytes displayed slow growth activity associated with poor accumulation of mineral nutrients (e.g. nitrogen, inorganic Phosphate, and Potassium) and high sodium concentrations in their shoots. On the contrary, annuals were characterized by high growth activity (up to 40% of the plant annual biomass production of the ecosystem), depending on the precipitation and sustained by a high capacity for nutrient acquirement (up to 70% of the total nutrient uptake).

The annuals were almost exclusively clustered under the halophyte tufts, or at their immediate vicinity, where soil nitrogen and inorganic phosphate levels were significantly higher, and salinity significantly lower than between the halophyte tufts. Furthermore, the shoots of the annuals growing in association with halophyte species contained relatively low Na^+ concentrations. These findings indicate that the upper horizon enclosing the halophyte tufts (where sensitive annuals grow) is fertile and contains low salt levels. This was also corroborated by the study of soil samples taken from the upper profile in the tuft centers. However, further studies are needed to support this by following the water movement in different soil horizons in different seasons. Desalination of the upper horizon by the superficial roots of halophytes may be responsible for this micro-gradient of salinity (Caldwell 1974). Further, the litter formed by halophyte fallen organs and by organic debris accumulated by the wind at the vicinity of halophyte tufts, could contribute to localized soil enrichment in N and P. Hence, the upper soil profile, where these plants grew, was fertile and contained (relatively) low salt levels, as corroborated by the results of soil analysis (upper horizon in the tuft centre was always less saline than when taken at the tuft periphery). Soil aeration near the annual glycophytes would be improved by this organic matter and by the higher soil level under the tufts, leading to better drainage capacity. Improved soil aeration is favourable for nitrification and N_2 fixation, which in turn sustains the colonization of the halophytes tufts by the *Medicago* spp. Halophytes may also play an indirect role by developing deep root systems exploiting the more saline horizons, as shown by the presence of halophyte roots at 1 m depth and by the vertical increasing salinity gradient. So, the halophytes directly contributed to maintenance of a relatively low salinity and high fertility in upper horizon, enabling the growth of annuals.

These findings were confirmed using a biological test of soil fertility. Some tufts of *Salicornia arabica* were removed for sampling soil in the upper profile (0–20 cm), where roots of annual plants developed. Other samples were taken between halophytes tufts, in zones devoid of vegetation or weakly populated. Four annual *Medicago* species (*M. ciliaris*, *M. polymorpha*, *M. truncatula* and *M. minima*) were grown on these soil samples, without mineral fertilization, in a greenhouse under controlled conditions. The plants were harvested at flowering stage. In the four species, total biomass production (dry matter per pot) was higher on soils sampled under the halophyte tufts than on soils from nude zones (Fig. 1a). Although these studies revealed that perennial halophytes improve soil characteristics by lowering salt content and by increasing nitrogen and phosphorus concentrations but further studies are needed in this connection by comparing with the salt secreting halophytes.

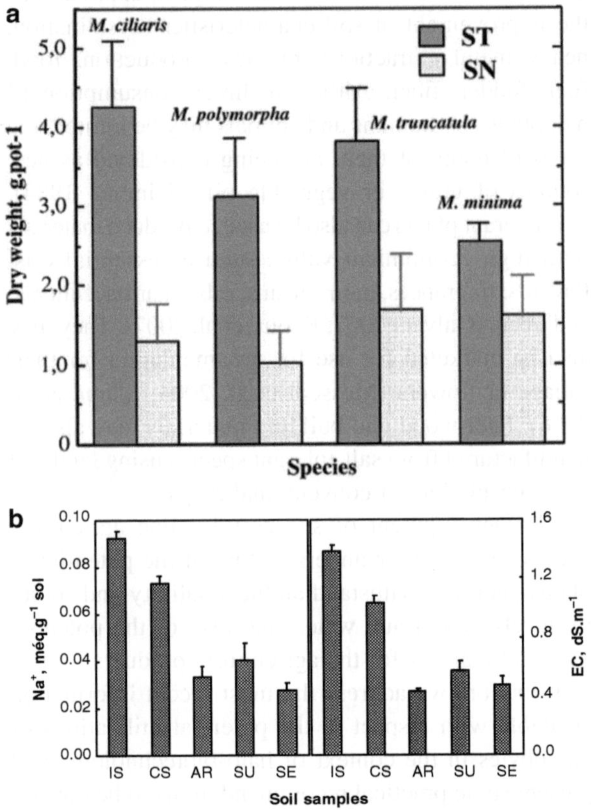

Fig. 1 Capacity of halophyte species for soil desalination. (**a**) Changes in biomass production of *Medicago ciliaris*, *M. polymorpha*, *M. truncatula*, and *M. minima* (g.pot^{-1}) depending on the soil origin ($n = 20 \pm$ SE at $p < 0.05$). ST: culture carried out on soil sampled under halophytes tufts, SN: cultures carried out on soil sampled in outside of halophytes tufts. (**b**) Soluble sodium content and electrical conductivity (EC) of 10% (w/w) aqueous extract of soil samples ($n = 6 \pm$ SE at $p < 0.05$) taken from the upper profile (15 cm deep) of pots (8 kg capacity), in which three halophytes: *A. indicum* (AR), *S. fruticosa* (SU), and *S. portulacastrum* L. (SE) were grown for 170 days. Plants were regularly irrigated with tap water. IS and CS stand for the initial and control (irrigated without plantation) soils respectively (adapted from Abdelly et al. 2006)

Involvement of halophytes in creation of micro-habitats favouring development of *Medicago* spp. and their micro-symbionts has been documented by Bekki (1995) who showed that *M. ciliaris* plants growing in combination with *Suaeda fruticosa*, had higher growth rates and better nodulation and nitrogen fixation potentialities than isolated ones. Several studies have highlighted the advantageous role of halophytes in soil desalination processes, especially for the most productive species, such as *Salsola salsa* (20 t · ha^{-1} among which, 3–4 t of salt exported from the soil), *Batis maritima* (3 t · ha^{-1}) (Le Houérou 1993), and the succulent *Sesuvium protulacastrum* (Pasternak and Nerd 1996). Although salt removal from soil is a common feature latter would be particularly interesting because of its high salt tolerance (growth stimulation up to 800 mM NaCl) despite accumulating high salt levels in the shoots (6 mmol · g^{-1} DW, representing ca. 35% of the whole plant biomass) (Messedi et al. 2001). In a 4-year study, Keiffer and Ungar (2002) observed a significant decline in brine-affected soils, following introduction of halophytes (*Atriplex prostrata*, *Spergularia marina*, and *Suaeda calceoliformis*), so that glycophytes could successfully establish. Recent findings of Kiliç et al. (2008) showed that purslane (*Portulaca oleracea*) may be cultivated as an intercrop all year round in one growing season, with a threshold value of salinity at 6.5 dS · m^{-1}. It removed considerable amounts of salt (up to 210 kg · ha^{-1} of Cl^- and 65 kg · ha^{-1} of Na^+) from the soil. However,it will better to use it for phyto-reclamation if efficiency is enough in comparison with total soil salinity, thus salt balance calculations are needed for this purpose, because it will take 500 years to get rid of the salt if there are 5,000 t of salt in soil per ha in the physiological and ecological sense.

The capacity of desalination of saline soil by halophytes was also evaluated in strictly controlled conditions, using *Arthrocnemum indicum, Suaeda fruticosa*, and *Sesuvium portulacastrum* (Rabhi et al. 2008). The plants, also originating from the edge of a sabkha, were cultivated for 6 months on saline soil and irrigated with tap water without losses by drainage. Salt export by plants was assessed by the difference between the amounts of Na^+ and Cl^- initially measured in the culture substrate, and those found at the end of the experiment in the soil. After 170 days, the soil electric conductivity (EC) in the pots used for the culture of the three halophytes decreased by more than 50% in average (0.6 dS · m^{-1} vs. 1.4 dS · m^{-1}) (Fig. 1b). Among the three studied species, *S. portulacastrum* was most productive and showed highest capacity of sodium extraction. It accumulated up to 26% of the original Na^+ content of the soil after 170 days, while both, *A. indicum* and *S. fruticosa* plants, accumulated 8% (Fig. 1b). Based on these findings, the calculated Na^+ export rates per ha were 2.50, 0.80 and 0.71, respectively, for *S. portulacastrum*, *A. indicum*, and *S. fruticosa* (Table 1). This study demonstrates once again that the halophyte traits displayed by *S. portulacastrum*, namely high growth rate and high capacity for salt accumulation, contribute efficiently to the soil desalination, even in short-term cultures. Hence, this species would be promising for rehabilitation of saline lands, especially those located in arid and semi-arid regions. Ravindran et al. (2007) evaluated the desalination capacity of six halophytes (*Suaeda maritima*, *S. portulacastrum*, *Clerodendron inerma*, *Ipomoea pes-caprae*, *Heliotropium curassavicum*, and the tree *Excoecaria agallocha*), in terms of fast growth rate associated with high salt accumulation. All tested species decreased the EC of the saline soil used, in concomitance with an increase of the EC of plant samples. Salt removal capacity calculated over in 4 months of time was 504, 473.9, 396.3, 359.5, 325.2 and 301.5 kg · ha^{-1} of NaCl in *Suaeda*, *Sesuvium*, *Excoecaria*, *Clerodendron*, *Ipomoea* and *Heliotropium*, respectively.

Table 1 Estimation of the capacity of soil desalination by three perennial halophytes grown during 170 days on a soil originating from a sabkha, and constantly irrigated with tap water

Species	*S. portulacastrum*	*S. fruticosa*	*A. indicum*
DW per pot (g)	30.00	9.25	10.50
Quantity of Na^+ exported per pot (g)	11.32	3.63	3.22
Quantity of Na^+ exported (t · ha^{-1})	2.50	0.80	0.71

Source: Adapted from Rabhi et al. (2008)

3 Heavy Metal Phytoextraction

Saline depressions, which are less populated, often constitute sites of accumulation of industrial and urban effluents contaminated by heavy metals. The plants from saline habitats can be evaluted to some extent in their phytoextraction potential as an environmental friendly alternative for the remediation of such contaminated areas (Cunningham and Berti 1993; Brooks 1998; Djebali et al. 2002). The first step in this direction will be to assess the potential interest in a plant species for phytoextraction and quantify, under fully-controlled conditions, the optimal level of toxic metal accumulation in relation to the growth rate (Lutts et al. 2004). Hyperaccumulators (plants able to concentrate high levels of heavy metals in their aerial parts without showing any symptom of injury) have to be identified. Plants are usually considered as hyperaccumulators if they contain more than 10,000 mg · kg^{-1} Zn or 100 mg · kg^{-1} Cd (Brooks 1998). Several studies demonstrated different tolerance mechanisms operating at the whole-plant level. For instance, salt excluding mechanisms are not always specific to sodium, so that other toxic elements such as copper, zinc, or cadmium may accumulate in salt glands, as in *Armeria maritima* and *Avicennia marina* (Lutts et al. 2004), or in trichomes, as found in *Tamarix aphylla* (Hagemeyer and Waisel 1988). Several halophyte species of the genus *Atriplex*, which are naturally salt- and drought-tolerant and mostly typical "includers" (Reimann and Breckle 1993), have been suggested as potential candidates for a phytoremediation approach (Salo et al. 1996; Glenn et al. 2001). A recent study reported that *Atriplex halimus* displayed high resistance to heavy-metals (Cd and Zn) in contaminated mining sites and that it has a strong potential for the rehabilitation of heavy metal-polluted lands (Lutts et al. 2004). This was strongly correlated to the over-accumulation of phytochelatins in combination with the induction of detoxification mechanisms, such as the co-precipitation of Cd and/or Zn with Ca in oxalate structures (oxalic acid is commonly produced in *Atriplex* species) within the stems, to prevent their toxic accumulation in the photosynthetic tissues (Karimi and Ungar 1986). Given that this species is able to produce 4–5 t of dry matter.ha^{-1} · $year^{-1}$ (Ben Ahmed et al. 1996) and that it may accumulate up to 0.083% Cd and 0.044% Zn on a dry weight basis in the shoots, the amounts expected to be removed could reach 4.15 kg · ha^{-1} Cd and 2.2 kg · ha^{-1} Zn (Lutts et al. 2004).

Halimione portulacoides was also suggested as a suitable species for the phytoremediation owing to the high translocation rates of Cd and Cu towards the aboveground tissues (Reboreda and Caçador 2007). Ghnaya et al. (2005) compared growth, cadmium accumulation and mineral nutrition in two halophytes from Aizoaceae: *S. portulacastrum* (perennial) and *Mesembryanthemum crystallinum* (annual) under the combination of mild salinity (100 mM NaCl) and Cd-stress (up to 300 μM $CdCl_2$). Cd-exposure had no impact on the growth of *S. portulacastrum*, while that of *M. crystallinum* was significantly decreased at the lowest Cd level (50 μM $CdCl_2$) (Fig. 2a). The better behavior of *S. portulacastrum* was partly ascribed to its

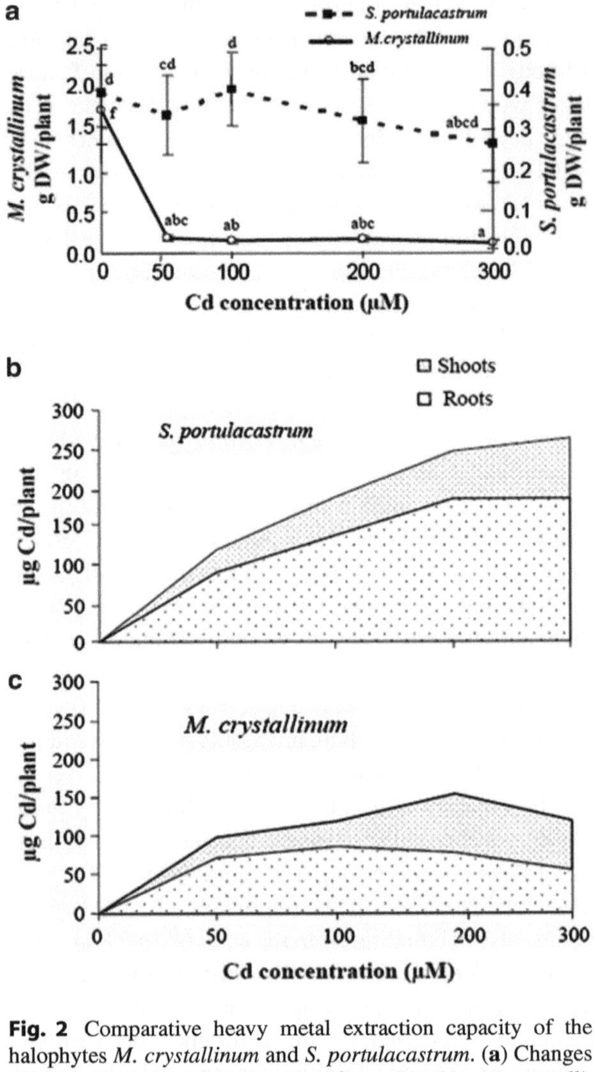

Fig. 2 Comparative heavy metal extraction capacity of the halophytes *M. crystallinum* and *S. portulacastrum*. (**a**) Changes in whole plant dry matter (g · $plant^{-1}$) produced by *M. crystallinum* and *S. portulacastrum* treated by various Cd concentrations. Means ($n = 8 \pm$ SE at $p < 0.05$) marked with same letter are not significantly different at $p < 0.05$. (**b**) Distribution of the cadmium absorbed by the whole plant (mg · $plant^{-1}$) between roots and shoots of *S. portulacastrum* and *M. crystallinum*. (adapted from Ghnaya et al. 2005)

better ability to avoid oxidative stress, when challenged with Cd (Nouairi et al. 2006). Indeed, leaves of *S. portulacastrum* showed lower decrease in their lipid content as compared to *M. crystallinum*, and cadmium treatment did not induce changes in the fatty acid composition of the total membrane lipids. Both halophytes accumulated more Cd in the roots than in the shoots. Yet, values of Cd concentrations in shoots (350–700 $\mu g \cdot g^{-1}$ DM) were characteristic of Cd hyperaccumulator plants (Fig. 2b). Hence, *S. portulacastrum* could extract up to 5950 g Cd · $ha^{-1} \cdot year^{-1}$, this performance being even improved in the calcareous soils (Ghnaya et al. 2007).

4 Halophytes for Livestock Production

The amount of edible biomass produced is an essential aspect with respect to livestock production in biosaline agriculture. Even at high salinities a range of halophytic grasses and shrubs with appreciable crude protein and digestible fibber contents are still able to produce 0.5–5 t of edible dry matter per year. Valuable species include *Leptochloa fusca* (Ahmad et al. 1990), *Brachiaria mutica* (Kumar and Abrol 1984), *Sesbania bispinosa* (Ahmad et al. 1990), *Cynodon dactylon* (Oster et al. 1999), *Kochia scoparia* (Garduno 1993), *Echinochloa crusgalli* (Aslam et al. 1987), and *P. oleracea* (Grieve and Suarez 1997). Among the most used halophytes for grazing and forage production under both saline and non-saline conditions are the shrubs from the Chenopodiaceae family, including the saltbushes, small-leafed bluebush (*Maireana brevifolia*), *Kochia* spp., *Tamarix* spp., glassworts (*Salicornia* spp.), and *Suaeda* spp. (Le Houérou 1994; Masters et al. 2001). All these species are well known to be drought and salt tolerant. According to Le Houérou (1992), 5–10 kg · DM $ha^{-1} \cdot year^{-1}$ of edible forage production from a range of saltbush species could be produced for each mm of rainfall in areas with low salinity. Under favorable conditions in terms of soil structure and rainfall (between 200 and 400 mm), yields of 2–4 t DM $ha^{-1} \cdot year^{-1}$ could be reasonably attained. Regular cutting practices associated with heavy grazing of woody shrubs such as saltbush favors the regeneration of less woody and more digestible shoots. Significant attention should also be paid to the appropriate cutting height, a critical point to avoid die-off of the existing plant material. *Atriplex lentiformis*, *Batis maritima*, *Atriplex canescens*, *Salicornia bigelovii* and *Distichlis palmeri* yielded over 10 t · $ha^{-1} \cdot year^{-1}$ following successful establishment (Glenn and O'Leary 1985), whereas *Atriplex* spp. irrigated with saline drainage water were able to yield 2.2–5.3 t DM · $ha^{-1} \cdot year^{-1}$ of forage (Watson and O'Leary 1993). Recent studies reported that highly saline groundwater (30 dS · m^{-1}) allowed to produce much larger amounts of hay from irrigated and intensively managed inland salt grass (*Distichlis spicata*) and marine couch (*Sporobolus virginicus*) (International Center for Biosaline Agriculture 2004). Abdelly et al. (2006) investigated the salt-response of two Poaceae halophytes native to Tunisia, with fodder potential: *Aeluropus littoralis* (perennial) and *Catapodium rigidum* (annual). The latter was slightly more productive than *A. littoralis* in the absence of salt. Increasing salinity decreased the growth activity of both species, but *A. littoralis* was more tolerant than *C. rigidum* (Fig. 3a). In the salinity range not exceeding 400 mM NaCl, the relative growth rate (RGR) of both species remained between 0.04–0.06 day^{-1}. Under non saline conditions, the following data were found for other plant species: *Medicago* spp. (0.08–0.09 day^{-1}) (Abdelly 1997), *Suaeda fruticosa* (0.07–0.09 day^{-1}) (Sleimi and Abdelly 2002), *Spartina alterniflora* (0.03 day^{-1}) (Sleimi and Abdelly 2002), *Spartina anglica* and *Puccinellia maritima* (0.02–0.05 day^{-1}) (Rozema and Van Diggelen 1991).

Barhoumi et al. (2007) reported that *A. littoralis* was able to survive up to 800 mM NaCl, with only a 50% reduction in leaf water content. Salt glands were observed on both leaf sides at high density in salt-treated plants. Interestingly, these morphological structures showed to be selective for Na^+ and Cl^-, and owing to their high density, enabled the elimination of about 50% of absorbed NaCl (Fig. 3b). If the results on production, yield, and reproduction are significantly high the species can be evaluted properly. In a recent survey, Laudadio et al. (2009) assessed the nutritional value of fourteen halophytes (*Aeluropus littoralis*, *Artemisia campestris*, *Atriplex halimus*, *Frankenia thymifolia*, *Imperata cylindrica*, *Limoniastrum guyonianum*, *Nitraria retusa*, *Reaumuria vermiculata*, *Salicornia Arabica*, *Salsola tetragona*, *Salsola tetrandra*, *Suaeda mollis*, *Tamarix gallica*, and *Zygophyllum album*) naturally growing in the arid regions of Tunisia. The contents of ash, crude protein, crude lipid, structural carbohydrate and nitrogen-free extract indicated nutritional

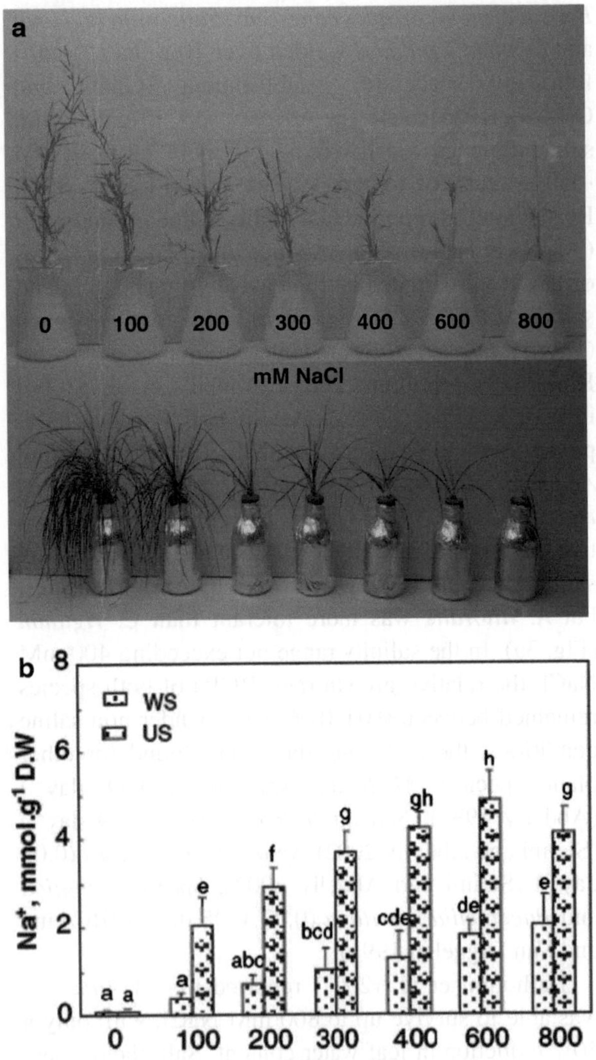

Fig. 3 (**a**) Comparative salt response of two fodder halophytes: *Aeluropus littoralis* (*above*) and *Catapodium rigidum* (below) to increasing salinity. (**b**) Importance of salt glands in salt excretion for *A. littoralis*: Na^+ contents (mmol · g^{-1} DW) mainly in washed (WS) and unwashed shoots (US) of plants exposed for 2 months to 0–800 mM NaCl NaCl. Values ($n = 7 \pm$ SE at $p < 0.05$) followed by the same letter are not significantly different at $p < 0.05$ (adapted from Barhoumi et al. 2007)

characteristics close to those found in the essence of the typical pastures of the Mediterranean region and compatible with the digestive physiology of the ruminant species.

Despite varying in both palatability and nutritive value, most halophytes contain sufficient levels of crude proteins and essential nutrients, covering the nutritional requirements of animals, particularly during the wet season (El Shaer 2006). In summer and autumn (dry season), they need to be supplemented with other feed ingredients, particularly with energy feed resources (El Shaer 1997). On the field, the mineral composition of the plants may be considerably affected by both the concentration and type of salts in the soil and/or water. For instance, sodium, potassium, chloride, calcium and magnesium may accumulate above the maximum tolerable levels for livestock (Masters et al. 2007). The high concentrations of sodium chloride in particular may decrease feed intake and under some conditions even threaten animal health. Furthermore, plants growing in saline environments often accumulate secondary compounds, which can have either beneficial (e.g., vitamin E and betaine) or toxic (oxalates, coumarins, saponines, and nitrites) impact on grazing livestock. Therefore, further investigations are essentially needed to ensure the well-tolerated application of halophytes as fodder plants.

Several treatments have been applied in order to improve the palatability and nutritive values of halophytic forage species (El Shaer and Kandil 1990). Chopping is an effective process to increase the palatability of succulent species and hence enable a more efficient utilization of whole shrubs. Appropriate conservation such as hay making and ensiling of halophytes could improve their utilization as good quality fodder and their acceptability by sheep and goats (El Shaer and Kandil 1990). This was ascribed to the effect of anaerobic fermentation during the ensiling process on some anti-nutritional factors such as tannins and other phenolic compounds. Additionally, the ash content and fibber materials are lowered. Feeding these silages to animals represents an economic alternative, feed costs being 30–50% lower in comparison with conventional diets such as berseem hay, which is of high importance in developing countries (El Shaer and Kandil 1990). The problem of preparing proper feed supply from halophytes is far from being solved. There are several reports showing mineral imbalances and shortage of essential mineral intake by animals in arid areas. It was found, for instance, that additional supply of the essential trace element Selenium may overcome nutritional constraints limiting animal productivity (see Khan et al. 2005 and citations therein).

Further combinations with other feed ingredients such as broiler litter, crushed date seeds, fodder beet and other forages could be also considered. In this way, voluntary feed intake was shown to increase by ensiling a mixture of some halophytic species, *A. halimus*, *H. strobilaceum*, *Tamarix mannifera* and *Zygophyllum album* with some agro-industrial byproducts such as ground date seeds and olive pulp (El Shaer 1997).

5 Halophytes for Oil Production

Seeds of several halophytes may contain appreciable quantities of edible oil (Weber et al. 2001). Oil extracted from the seeds (Up to 30% of seed DW) of *Salicornia bigelovii*, a highly salt tolerant annual halophyte, was found to be of good quality with unsaturation degree comparable to oils from conventional oil seeds (Glenn et al. 1991). Interestingly, *S. bigelovii* (Glenn et al. 1991) and several accessions (Mexico, Egypt, and United Arab Emirates) of the stem succulent *Salicornia europaea* appear to be highly productive, with a biomass production of 20 t · ha^{-1} among which 2 t are seeds (Goodin, et al. 1990). According to Yajun et al. (2003), *Descurainaia sophia* collected from saline soil (0.4% NaCl) contained higher amounts of linolenic acid in their seeds as compared to plants originating from non-saline soil (<0.1% salt) (53.7% and 36% linolenic acid, respectively). Weber et al. (2007) analysed the seeds of *Arthrocnemum indicum*, *Alhaji maurorum*, *Cressa cretica*, *Halopyrum mucronatum*, *Haloxylon stocksii* and *Suaeda fruticosa* to assess their potential as source of edible oil. Seed-oil content was 22–25%, and amounts of unsaturated fatty acids were high (65–74%) excepting *Alhaji maurorum*. In addition, seed lipids contained 12 unsaturated and four saturated fatty acids.

Zarrouk et al. (2003) retained *Zygophyllum album* (Zygophyllaceae), *Crithmum maritimum* (Apiaceae), and *Cakile maritima* (Brassicaceae) as potential oilseed halophytes. The dry weight of 100 seeds ranging from 133 mg in *Z. album* to 774 mg in *C. maritima*. Interestingly, the value for the latter species is nearly two times higher than that for rape (*Brassica napus*) seeds, a conventional oleaginous plant. Seeds of *C. maritima* and *C. maritimum* were rich in oil, (up to 42% and 30% of the seed FW, respectively), whereas seed-oil content was much lower in *Z. album* (ca. 6%). As for olive oil, fatty acid composition of *C. maritimum* seeds was characterised by a high level of oleic acid (81%), whereas that of *Z. album* seeds showed a high percentage of linoleic acid (64%), similar to sunflower seed-oil composition. Therefore, both species contained oil of good quality which may be used without any further modification. Erucic acid (22:1) was prominent in fatty acid spectrum (over 25% of total fatty acids) of *C. maritima* seeds, suggesting the possibility of its utilisation for industrial purposes (nylon, emollients, lubricants etc.) (Bhardwaj and Hamama 2003).

In more recent papers, the natural variability of fruit and seed oil characteristics was investigated in twelve local accessions of *C. maritima*, harvested along the Tunisian coasts, and covering different bioclimatic stages (humid to arid) (Ghars et al. 2006). Seed-oil amount was highly variable among accessions (25–39%) (Fig. 4a) and triacylglycerols (TAG) represented 80–97% of the lipid categories. Erucic acid (22:1), found only in TAG, represented 25–35% of total fatty acids. Such a variability was consistent with previous studies on Australian and Moroccan ecotypes of *C. maritima*, whose seed-oil content ranged from ca. 35–48%, respectively (Hocking 1982; Kumar and Tsunoda 1978). This intra-specific variability in the amount and the quality of the extracted oil could be ascribed to environmental factors, such as precipitation (Rana and Ahmed 1981; Pannelli et al. 1994), temperature (Champolivier and Merrien 1996), and salinity (Zarrouk and Cherif 1983). Fruit morphology and NaCl accumulation also were accession-dependent. Furthermore, the repartition pattern of Na^+ and Cl^- between the plant tissues (stems, siliques, and seeds) suggested that some protective mechanisms prevent excessive salt accumulation in reproductive organs, which is essential for the establishment and survival of *C. maritima* in saline biotopes. The successful rehabilitation of saline marginal zones by introduction of halophytes for the creation of sustainable production systems largely depends on the germination capacity at each site as stressed by Debez et al. (2004, 2006a, 2008) (Fig. 4b) and Ben Amor et al.(2006) and Ksouri et al.(2007).

At fructification stage, the number of seeds per plant was also significantly augmented at salt concentrations optimal for plant growth (50–100 mM) (Debez et al. 2004). Concentrations exceeding 200 mM NaCl strongly reduced both seed mass and seed viability, likely owing to salt-induced restriction of seed nutrition and filling. While seed-oil content was only slightly increased at the extreme salt levels (25%), qualitative changes of seed-oil were more obvious: erucic acid (C22:1) was the major fatty acid (29%) in seed-oil of plants grown at NaCl above 200 mM to the detriment

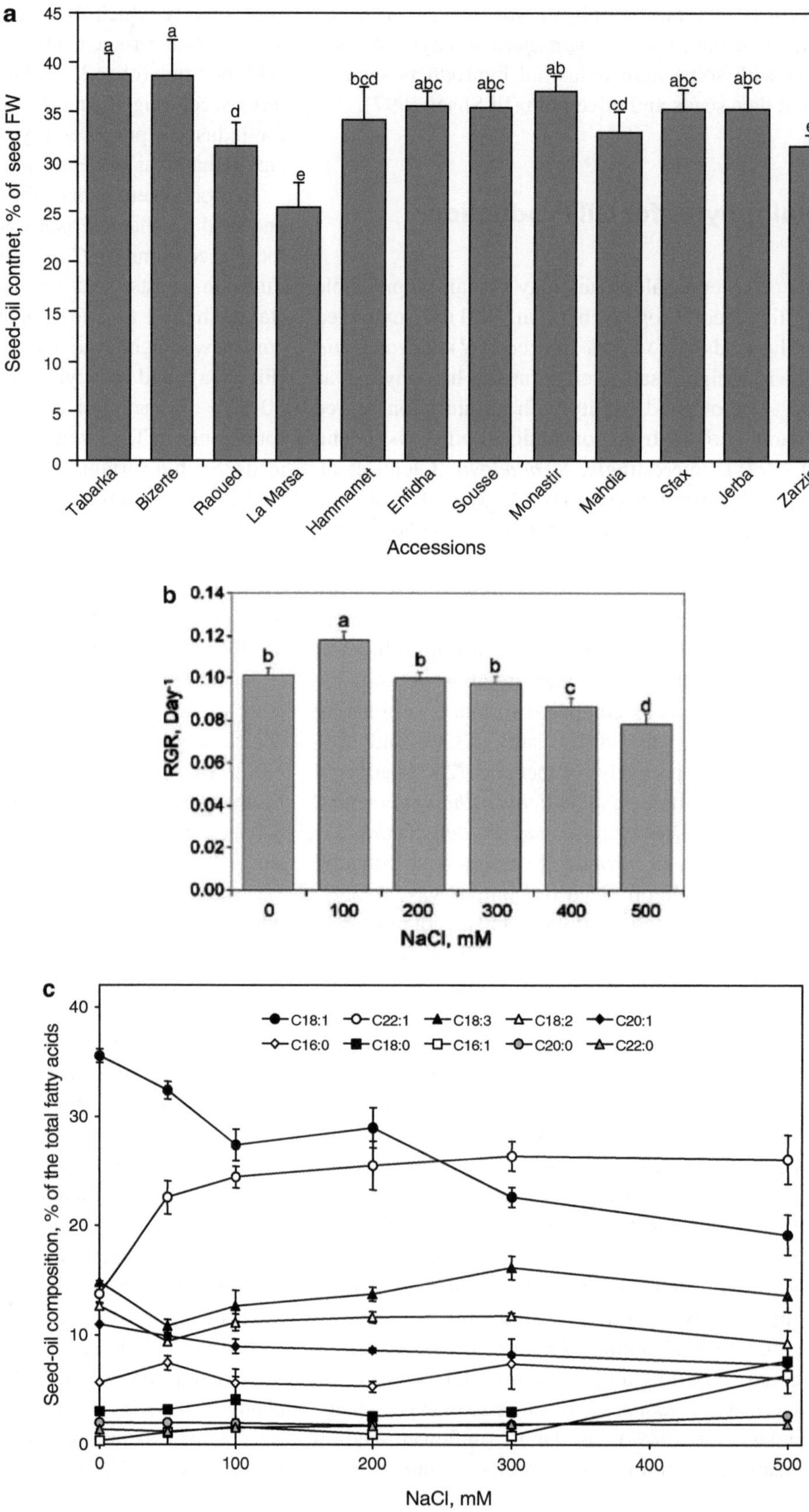

Fig. 4 Seed-oil characteristics and salt-responses of the halophyte *C. maritima* (**a**) Variability of seed-oil content in seeds of *C. maritima* Tunisian accessions. Means ($n = 3 \pm$ SE at $p < 0.05$) followed by at least one same letter are not significantly different between the accessions at $p < 0.05$ (adapted from Ghars et al. 2006). (**b**) Changes in relative growth rate (RGR) of plants exposed for 6 weeks to 0–500 mM NaCl salinities ($n = 10 \pm$ SE at $p < 0.05$) (adapted from Debez et al. 2006). (**c**) Effect of increasing NaCl on the seed-oil composition ($n = 3 \pm$ SE at $p < 0.05$)

of oleic acid (C18:1), which was prominent in the control (36%) (Debez et al. 2006b). Such a trend likely mediated elongases, which are known to catalyze the formation of long fatty acids (such as erucic acid), using oleic acid as initial substrate (Katavic et al. 2002). The search for accessions with a better match between salt concentrations optimal for seed production, seed oil content, and erucic acid concentration, is therefore of high interest. Seed-oil content seemed also to be unaffected by salinity in the oleaginous halophyte *Lesquerella fendleri* (Dierig et al. 2003).

Another promising oilseed halophyte is *Kosteletzkya virginica*, native to the American salt marsh. This species was introduced into China as a candidate species to improve tideland and develop ecologically sound saline agriculture Ruan et al. (2008). A 10 year-long field study revealed that *K. virginica* was flooding-tolerant, displaying multiple eco-benefits, such as landscape beautification, revegetation, and representing a food source for migratory birds. Besides, seed yields of unselected mixed and bred lines were 621 kg · ha^{-1} and 957 kg · ha^{-1}, respectively. Oil contents in the seeds of the unselected mixed and bred lines were 18% and 21%, respectively. Gallagher (1985) reported that seed yield of *K. virginica* in the USA was 800–1,500 kg · ha^{-1} with a seed-oil content of 20–30%. Unsaturated fatty acids (70%) in the seed oil predominated over saturated ones (30%). Oil content in the seeds of *K. virginica* was equivalent to the soybean, whose oil content was about 5%–20% (Yang et al. 2000). The content of unsaturated fatty acids was slightly less than soybean (85%), peanut (82%) and gingili (86%). All these indicated that seed of *K. virginica* has a high potential to become a oil resource at saline sites.

6 Halophytes with Medicinal Potential and for Cosmetics

The use of medicinal plants is common in many developing countries. In addition, there is a renewed interest in developed countries in using medicinal plants to treat humans, pets and livestock. In the late 1990s, the world market for herbal remedies was estimated at US $19.4 billion (Qadir et al. 2008). Environmental stresses (salinity, drought, heat/cold, luminosity and other hostile conditions) often trigger in plants a significant oxidative stress, generating the formation of ROS, leading to cellular damage, metabolic disorders, and senescence processes (Wang et al. 2004). Increasing ROS concentrations eventually affect biological molecules, such as DNA, proteins, or lipids, resulting in mutations and membrane damage, and finally lead to cell and tissue injuries (Abdil and Ali 1999). Enhanced synthesis of protective secondary metabolites under stressful conditions is believed to save the cellular structures from oxidative effects (Buchanan et al. 2000). Halophyte ability to withstand salt-triggered oxidative stress is governed by multiple biochemical mechanisms that facilitate retention and/or acquisition of water, protect chloroplast functioning, and maintain ion homeostasis. Most essential ones include the synthesis of osmotically active metabolites, specific proteins, and antioxidant compounds (Ksouri et al. 2008). This might explain (i) the utilization of some halophytes as traditional medicinal plants and (ii) the increasing interest for their potential for industrial applications (e.g. agro-food, cosmetic and medicine). Natural antioxidants occur in all plant parts, and the typical compounds that exhibit antioxidant activities include phenols, carotenoids and vitamins (Chanwitheesuk et al. 2005). Among various kinds of natural antioxidants, polyphenols constitute the main powerful compound. Accordingly they are widely applied in food industry, cosmetic, pharmaceutical and medicinal materials (Maisuthisakul et al. 2007). In plants, polyphenol synthesis and accumulation is generally stimulated in response to biotic/abiotic stresses (Naczk and Shahidi 2004), such as salinity (Navarro et al. 2006). In addition to their role as antioxidant, these compounds exhibit a wide spectrum of physiological properties, such as anti-allergic, anti-arthero-genic, anti-inflammatory, anti-microbial, anti-thrombotic, cardio-protective and vasodilatory effects (Balasundram et al. 2006).

The nonhalophytic species like medicinally important *Catharanthus roseus* can perform successfully at 10 dS · m^{-1} salinity without showing substantial decline in biomass production (Anwar et al. 1988). *Lycium barbarum* (Solanaceae) is a perennial halophytic species, which is important in traditional medicine in arid and semi-arid areas of China. It has been widely used as health-giving food for about 2,300 years (Cui and Xing 1999). Furthermore, *L. barbarum* fruits are used owing to their nourishing virtues for the kidney, liver, and for brightening eyes (Peng and Tian 2001). Wei et al. (2006) reported that moderate salinity (100 mM NaCl) was optimal for the plant biomass production (+30% as compared to the plants cultivated in non-saline conditions). This was associated with higher shoot succulence, significant accumulation of osmotically active compounds (i.e., betaine and sugar) and the absence of depressive effect on chlorophyll content, gas exchanges (CO_2 assimilation rate and stomatal conductance) and maximum quantum efficiency of photosystem II. Recent findings of Meot-Duros et al. (2008) confirmed the multiple potential of halophytes as natural food (or cosmetic) preservatives. The authors assessed both antioxidant and antimicrobial activities (against 12 bacterial and yeast strains) of three halophytes (*Eryngium maritimum*, *C. maritimum*, and *C. maritima*) collected from the Brittany (France) shoreline. The chloroformic and methanolic extracts were tested for their antimicrobial activities. Interestingly, all bacterial strains were inhibited by plant extracts with exception of *Listeria monocytogenes*. Both phenol content and the antioxidant activity were variable among the tested extracts, *E. maritimum* showing the lowest phenolic level as well as the lowest radical scavenging activity. On the contrary, *C. maritimum* exhibited the highest total phenol content and ABTS (2,2'-azinobis (3-ethylbenzothiazoline-6-sulfonic acid)) radical scavenging activity. Total antioxidant capacity was strong and ranged from 32 mg · g^{-1} DW in *E. maritimum* to 48 mg · g^{-1} DW in *C. maritima*. Antimicrobial activity of the apolar fraction was appreciable in the three investigated species, as most of the extracts showed a MIC (minimum inhibitory concentrations) of 100 μg · ml^{-1}, being even active at 1 μg · ml^{-1} in some cases. Polar fraction was effective as well. *E. maritimum* showed a strong antibacterial activity against *Pseudomonas aeruginosa* and *P. fluorescens*, while *C. maritima* was very effective against *Salmonella arizonae*. A good antimicrobial activity was found in *C. maritimum* against *Pseudomonas aeruginosa* and *Candida albicans*.

Besides its potential as oilseed, *C. maritima* could be of interest for production of secondary metabolites and antioxidant compounds. A comparative survey of two Tunisian accessions (Jerba and Tabarka, respectively, sampled from arid and humid bioclimatic stages) of this halophyte highlighted the presence of an accession-dependent capacity to induce antioxidant mechanisms in response to salt (up to 400 mM NaCl). This may result in a corresponding variability for growth sustainability (Ksouri et al. 2007). Indeed, Tabarka growth (shoot biomass, leaf expansion) was significantly restricted by increasing salinity whereas Jerba growth increased at 100 mM NaCl before declining at 400 mM NaCl. The better behaviour of Jerba was closely related to higher polyphenol content (+56% and 30% of the control values at 100 and 400 mM NaCl, respectively) and better antioxidant activity (lower IC_{50} values for both 1,1-diphenyl-2-picrylhydrazyl and superoxide scavenging), associated with lower leaf MDA (malonyldialdehyde) accumulation (*ca.* −66% of the control at 100 mM NaCl) (Table 2). The parallel stimulations of shoot biomass and polyphenol concentration in tissues in Jerba at 100 mM NaCl hence support the assumption that stress-tolerant plants (such as halophytes) are potentially interesting systems for production of secondary metabolites useful for food and medicinal applications. A similar trend was documented by Falleh et al. (2008) in *Cynara cardunculus*, (Asteraceae). Moderate salinity (25–50 mM NaCl) had no significant impact on leaf growth (biomass, length and number), but these parameters showed a severe reduction at 150 mM NaCl (−30% to −90% as compared to the control). Leaf phenolic content was significantly higher at 25–50 mM NaCl, before declining at 150 mM NaCl. The superoxide anion scavenging capacity of leaf extracts was proportional to the external NaCl concentration. *C. cardunculus* occurs mainly in Mediterranean arid regions, characterized by high

Table 2 MDA and total polyphenol and contents and antioxidant activities (IC_{50} values) in leaves of two *C. maritima* accessions irrigated for 28 days with a nutrient solution containing 0, 100, or 400 mM NaCl

NaCl (mM)	MDA (nmol · g^{-1} FW)		Total polyphenol (mg of GAE · g^{-1} DW)		DPPH· scavenging activity IC_{50} (μg · ml^{-1})		Superoxide scavenging activity IC_{50} (μg · ml^{-1})	
	Jerba	Tabarka	Jerba	Tabarka	Jerba	Tabarka	Jerba	Tabarka
0	4.30 ± 0.2e	4.64 ± 0d	43.02 ± 4c	42.84 ± 8c	0.67 ± 0.1c	0.62 ± 0.2c	3.10 ± 0.1bc	3.57 ± 0.9bc
100	1.48 ± 0.2f	7.48 ± 0.2b	66.93 ± 4a	37.73 ± 8c	0.76 ± 0.0c	1.34 ± 0.2b	1.70 ± 0.3c	5.10 ± 1.1b
400	5.61 ± 0.1c	12.03 ± 0.0a	56.04 ± 1b	31.58 ± 8c	0.89 ± 0.5c	1.95 ± 0.1a	3.90 ± 0.1bc	14.90 ± 3.0a

Source: Adapted from Ksouri et al. (2007)

Values ($n = 8 \pm$ SE at $p < 0.05$) of each parameter followed by at least one same letter are not significantly different at $p < 0.05$

temperature, elevated salinity and drought in summer (Gominho et al. 2000). It is noteworthy that *C. cardunculus* flowers are used for cheese preparation (Valentao et al. 2002), while leaves are traditionally known for their therapeutic virtues as diuretic, antidiabetic, and antimicrobial agent (Fratianni et al. 2007).

Saïdana et al. (2008) focused on the chemical composition of the essential oil and the antibacterial and antifungal activities from the flowering parts of *Suaeda fruticosa* (Chenopodiaceae) and *Limonium echioides* (Plumbaginaceae) sampled from the Sahel region in Tunisia, where semi-arid Mediterranean climate prevails. The first species is known for its hypoglycaemic action, while *Limonium* spp. is known as an antioxidant medicinal herb. 65 compounds were identified in *L. echioides* among which 48 were common with *S. fruticosa*. The main components in *L. echioides* were hexacosane (10.7%), palmitic acid (9.8%), nonacosane (8.4%), (E,E)-farnesyl acetate (7.0%) and vanillin (6.5%). Palmitic acid (15.2%) was also prominent in *S. fruticosa*, followed by methyl linoleate (10.8%), phytyl acetate (8.8%), hexacosane (7.4%) and methyl decanoate (7.0%). Oil of both species was effective against bacteria such as *Staphylococcus aureus*, *S. epidermidis*, *Micrococcus luteus*, *Escherichia coli*, and *Salmonella typhimurium*, but no antifungal activity was detected, presumably because of the low concentrations used in this study.

Recently, more exhaustive studies revealed large variability in total polyphenol contents and antioxidant activities (DPPH and superoxide radicals scavenging activities, and iron chelating and reducing powers) of several halophytes (*C. maritima*, *Limoniastrum monopetalum*, *Mesembryanthemum crystallinum*, *M. edule*, *Salsola kali*, and *Tamarix gallica*), depending on biological (species, organ and developmental stage), environmental (original habitat), and technical (extraction solvent) factors (Ksouri et al. 2008; Ksouri et al. 2009; Falleh et al. 2009). Such a variability might be of great importance in the perspective of valorising these halophytes as a source of naturally secondary metabolites.

7 Agroforestry Ecwosystems and Carbon Sequestration

Implementing salt-tolerant tree plantations while utilising saline drainage or groundwater represents a promising alternative of using abandoned lands. Of the most promising species are *Prosopis juliflora*, *Acacia nilotica*, *Casuarina equisetifolia*, and *Eucalyptus camaldulensis* (Qadir et al. 2008). Widely used *Acacia* species in agroforestry on saline soils include *A. stenophylla*, *A. nilotica*, *A. ampliceps*, *A. tortilis*, *A. maconochieana*, and *A. cyclops*. Although most *Acacia* species are not halophytes but some are highly salt-tolerant and at the same time sensitive to waterlogging (Craig et al. 1990). *Casuarina*, a fast-growing evergreen tree native to Asia and Australia, was successfully used in the coastal regions of Africa due to its ability to cope with extreme environmental conditions. Marcar et al. (2000) reported a 77% survival rate in *C. cunninghamiana* growing in soils with 11.5 dS $\cdot$ m^{-1} EC for 30 months and 55% for more than 7 years. The evergreen tree *E. camaldulensis* is a medium to tall (20–45 m) tree native to Australia with several uses, such as essential oil and paper production, landscaping, sand dune stabilization, and to lower high water tables of saline environments. In a mid-term (4-year) field study, Oster et al. (1999) found significant restriction of both growth rate and water use of *Eucalyptus* when irrigated with saline-sodic waters (20–22 dS $\cdot$ m^{-1} salinities in the root zone). Whereas a long-term (9 year-long) investigation with 31 tree species challenged with saline water at 8.5–10.0 dS $\cdot$ m^{-1} EC showed that *Tamarix articulata*, *A. nilotica*, *P. juliflora*, *E. tereticornis*, *A. tortilis* and *Cassia siamea* were performing best (Tomar et al. 2002).

The process of carbon sequestration, or flux of carbon, into soils forms part of the global carbon cycle. Movement of carbon between the soil and the above ground environment is bidirectional and consequently carbon storage in soils reflects the balance between the opposing processes of accumulation and loss. Carbon in soils comes from CO_2 in the atmosphere, which is captured by plants *via* photosynthesis, and ultimately stored to soil by direct decomposition of plant material. Halophytic plants adapted to saline dry lands offer potential for sequestering carbon in this hostile ecosystems. Investment in anti-desertification measures in the world's drylands appears to be an economical method to offset CO_2 build-up in the atmosphere while accomplishing a major international objective of re-vegetating drylands. This alternative relies on the assumption that salinity of a halophyte agro-ecosystem may slow the processes of residue decay, and thereby increases the potential for sequestering halophyte carbon. According to Glenn et al. (1993), 0.6–1.2 Gt C $\cdot$ $year^{-1}$ could be assimilated by halophytes, among which 50% of C might enter long-term storage in soil (Glenn et al.

1993). Taking into account the irrigation requirements (based on the irrigation demand which is related to the plant biomass production), the authors consider the rate of 22.5–30% as realistic. Yet, this assumption needs to be corroborated by further field surveys, as the potential for storing organic carbon in agronomic soils depends on the type of organic material in the system. Goodfriend et al. (1998) addressed the decomposition of three promising halophytic crops (*S. bigelovii*, *Suaeda* sp., and *B. maritima*) in order to assess their potential for carbon storage in a coastal desert agro-ecosystem, near Sonora (Mexico). Several parameters including decay rates, incorporation of residues into the soil, microbial decomposition activity, the abundance of bacteria, fungi were determined. Results showed that after 1 year, the rate of halophyte residue decay and the number of microbial decomposers associated with the residues are similar to those of fresh water systems. There appears to be no advantage in storing carbon as plant residue in this saline agricultural system.

Mangroves are woody tress or shrubs located at the interface between land and sea, presently occupying ca. 181,000 km^2 of tropical and subtropical coastline. Over the past 50 years, approximately one-third of the world's mangrove forests have been lost, mostly because of urban development, aquaculture, mining and overexploitation for timber, fish, crustaceans and shellfish (Alongi 2002). Mangroves grow in sheltered shores, penetrating into the estuaries of rivers, tidal creeks, backwaters, salt marshes and coastal mudflats. They are open systems since they are constantly subject to tidal flow and seasonal flooding. As halophytes, mangroves thrive well in saline waters but require fresh water to a certain extent in order to maintain an optimum salinity balance and to get nutrients. Mangrove ecosystems contribute to stabilize sediments and may augment local sediment accumulation and enrich soil with nitrogen, phosphorus and sulphur (Hussain and Badola 2008). They not only create a living buffer between neighbouring river, marine and terrestrial communities, but also play an important role in supporting the productivity of the coastal environments. Among the most common species there are *Avicennia officinalis*, *A. alba*, *Aegiceras majus*, *Aegialitis rotundifolia*, *Acanthus ilicifolius*, *Kandelia rheedii*, *Bruguiera caryophylloides*, *Sonneratia griffithii*, *Laguncularia racemosa* and *Carapa obovata*. Ecophysiological (salt-tolerant, highly efficient nutrient retention) and morphological features (respiratory aerial roots, viviparous embryos, absence of growth rings, wood with narrow densely distributed vessels rings) make mangroves structurally and functionally unique (Alongi 2002).

Because mangroves fix and store significant amounts of carbon, their loss may drastically affect global carbon budgets. Cebrian (2002) calculated that a loss of about 35% of the world's mangroves led to a net loss of $3.8 \cdot 10^{14}$ g · C stored as mangrove biomass. The ecological significance of mangroves in protecting the coastal areas was recently highlighted by Danielsen et al. (2005) who showed that mangroves and *Casuarina* plantations attenuated the tsunami-induced waves and protected shorelines against damage: indeed, villages located within the *Casuarina* plantations remained undamaged except for rows of 5–10 trees nearest to the shore, which were uprooted. In addition, based on analytical models, mangrove plantations at 30 trees per 100 m^2 in a 100-m wide belt may reduce the maximum tsunami flow pressure by more than 90%. Measurement of wave forces and modeling of fluid dynamics suggested that tree vegetation may protect coastlines from tsunami damage by reducing wave amplitude and energy.

8 Conclusions and Perspectives

Crop diversification and production systems based on halophytes are likely to be the key to both socially and economically rehabilitate, and thereby valorise, the salt-affected-regions. Such an approach will be particularly relevant for developing arid and semi-arid countries, where most farmers exploiting these marginal lands are resource-limited and communities face severe unemployment, poverty, and population migration (Qadir et al. 2008). Revolutions in agriculture rely on empirical approaches. Looking towards the future, sometimes means taking a step to the past, in order to discover guided by an open mind thousands of possibilities hidden by nature, and enlarged by cultural evolution (Galvani 2007). Historical evidence suggests that farmers always shift from more sensitive to more tolerant crops as salinity in their fields rises (Flowers 2004). Ending this infernal series could be achieved by the sustainable use of the wide range of halophytes available, whose large potential as crops is still under-explored (Koyro and Huchzermeyer 2004). The use of salt-affected land and saline water resources by crop diversification could also be an alternative to move from subsistence farming to higher income-generating ventures.

The emerging data supporting the feasibility of successfully cultivating salt-tolerant plants in saline agroecosystems offer unexpected opportunities for researchers, farm advisors, and farmers to identify the most appropriate cash crop halophytes and their combinations to optimize the input/output ratios. Besides, since salt-affected soils and ground waters cross national boundaries, co-operation and coordination at regional and global levels is of high importance to elaborate and apply effective salinity strategies. With this respect, it is crucial to initially involve politicians, institutions, select farmers, water user associations, and other potential beneficiaries, so that (when the critical time for expansion comes) all parties will be familiar with their role. (Yensen 2006). This means changing the general opinion of the affected farming communities and policy makers about the questionability of using salt-affected soils.

Preliminary assessments suggest that there are 26 salt-tolerant plant species capable of producing 13 products (or services) of value to agriculture in Australia (Barrett-Lennard 2002). As suggested by Koyro et al. (2006), once the selection of halophytic species suitable for a particular climate and for a particular utilization has been achieved, a progressive realization of the following steps may bring to establish useful cash-crop halophytes:

(a) Performing greenhouse experiments at the local substrates (and climatic conditions) to select and propagate promising sites (Fig. 5)
(b) Using Lysimeters on the field to gather reliable and long-term data about the water consumption and ion movements
(c) Developing a sustainable production system in plantations at coastal areas or at inland sites (for example for economical use)
(d) Assessing the plant yield and the financial outcome for the farmers
(e) Evaluating the economic acceptance of the product The development of successful industries for salt-land will require pertinent information regarding the capabilities of land and water, plants and animals, and markets
(f) Protecting the environment by using appropriate culture and irrigation practices and finally
(g) A major task is the need to change perception in the community of biosaline agriculture

Fig. 5 On the left side gravel/hydroponic quick check system (QCS) with automatic drip irrigation under photoperiodic conditions in a growth cabinet (plant species: *Atriplex nummularia*) and on the right side hydroponic culture of *Sesuvium portulacastrum*. Controls are visible in the foreground, the sea water salinisation treatment (100% sws ≅ 500 mol * m^{-3} NaCl) in the background

References

Abdelly C (1997) Mécanismes d'une association de Luzernes spontanées et de halophytes pérennes en bordure de sebkha. Ph.D. Thesis, Faculty of Sciences of Tunis, Tunisia

Abdelly C, Lachâal M, Grignon C, Soltani A, Hajji M (1995) Association épisodique d'halophytes stricts et de glycophytes dans un écosystème hydromorphe salé en zone semi-aride. Agronomie 15:557–568

Abdelly C, Zid E, Hajji M, Grignon C (1996) Biomass production and nutrition of *Medicago* species associated to halophytes on the edge of a sebkha in Tunisia. In: Malcolm CV, Hamdy A, Choukr-Allah R (eds) Halophytes and Biosaline Agriculture. Marcel Dekker, NewYork, pp 313–324

Abdelly C, Barhoumi Z, Ghanya T, Debez A, Ben Hamed K, Sleimi N, Ouerghi Z, Smaoui A, Huchzermeyer B, Grignon C (2006) Potential utilisation of halophytes for the rehabilitation and valorisation of salt-affected areas in Tunisia. In: Öztürk M, Waisel Y, Khan MA, Görk G (eds) Biosaline agriculture and salinity tolerance in plants. Birkhäuser Verlag, Switzerland, pp 163–172

Abdil S, Ali A (1999) Role of ROS modified human DNA in the pathogenesis and ethiology of cancer. Cancer Lett 142:1–9

Ahmad N, Qureshi RH, Qadir M (1990) Amelioration of a calcareous saline-sodic soil by gypsum and forage plants. Land Degrad Dev 2:277–284

Alongi DM (2002) Present state and future of the world's mangrove forests. Environ Conserv 29:331–349

Anwar M, Singh DV, Subrahmanyam K (1988) Safe limits of salinity for three important medicinal plants. Int J Trop Agric 6:125–128

Aslam Z, Saleem M, Qureshi RH, Sandhu GR (1987) Salt tolerance of *Echinochloa crusgalli*. Biol Plant 29:66–69

Balasundram N, Sundram K, Samman S (2006) Phenolic compounds in plants and agri-industrial by-products: antioxidant activity, occurrence, and potential uses. J Food Chem 99: 191–203

Barrett-Lennard EG (2002) Restoration of saline land through revegetation. Agric Water Manage 53:213–226

Barhoumi Z, Djebali W, Smaoui A, Chaïbi W, Abdelly C (2007) Contribution of NaCl excretion to salt resistance of *Aeluropus littoralis* (Willd) Parl. J Plant Physiol 164:842–850

Bekki A (1995) Effet de la salinité sur la fixation d'azote chez l'association *Rhizobium-Medicago*: sensibilité du partenaire bactérien à l'état libre et sous forme de micro-symbiote. Tentative d'essai au champ. In: Drevon JJ (ed) Facteurs Limitant la Fixation Symbiotique de l'Azote dans le Bassin Méditerranéen. Les Colloques n° 77. INRA, Paris, pp 139–148

Ben Ahmed H, Zid E, El Gazzah M, Grignon C (1996) Croissance et accumulation ionique chez *Atriplex halimus* L. Cah Agric 5:367–372

Ben Amor N, Jiménez A, Megdiche W, Lundqvist M, Sevilla F, Abdelly C (2006) Response of antioxidant systems to NaCl stress in the halophyte *Cakile maritima*. Physiol Plant 126:446–456

Bhardwaj HL, Hamama AA (2003) Accumulation of glucosinolate, oil, and erucic acid in developing *Brassica* seeds. Ind Crop Prod 17:47–51

Böer B, Gliddon D (1998) Mapping of coastal ecosystems and halophytes (case study of Abu Dhabi, United Arab Emirates). Mar Freshwat Res 49:297–301

Breckle SW (2009) Is sustainable agriculture with seawater irrigation realistic? In: Ashraf M, Öztürk M, Athar HR (eds) Salinity and water stress-improving crop efficiency, tasks for vegetation science-44. Springer, the Netherlands, pp 187–196

Brooks RR (1998) Plants that hyperaccumulate heavy metals: their role in phytoremediation, microbiology, archeology, mineral explo ration and phytomining. CABI, Wallingford, UK

Buchanan BB, Gruissem W, Jones R (2000) Biochemistry and molecular biology of plants. American Society of Plant Physiologists, Beltsville, MD

Caldwell MM (1974) Physiology of desert halophytes. In: Reimold RJ, Queen WH (eds) Ecology of halophytes. Academic Press, New York

Cantero-Martínez C, Angás P, Lampurlanés J (2007) Long-term yield and water use efficiency under various tillage systems in Mediterranean rainfed conditions. Ann Appl Biol 150:293–305

Cebrian J (2002) Variability and control of carbon consumption, export, and accumulation in marine communities. Limnol Oceanogr 47:11–22

Champolivier L, Merrien A (1996) Evolution de la teneur en huile et de sa composition en acides gras chez deux variétés de tournesol (oléique ou non) sous l'effet de températures différentes pendant la maturation des graines. Oléagineux, Corps Gras, Lipides 3:140–144

Chanwitheesuk A, Teerawutgulrag A, Rakariyatham N (2005) Screening of antioxidant activity and antioxidant compounds of some edible plants of Thailand. Food Chem 92:491–497

Craig GF, Bell DT, Atkins CA (1990) Response to salt and waterlogging stress of ten taxa of Acacia selected from naturally saline areas of Western Australia. Austr J Bot 38: 619–630

Cui KR, Xing GS (1999) Effect of hydrogen peroxide on somatic embryogenesis of *Lycium barbarum* L. Plant Sci 146:9–169

Cunningham DS, Berti RW (1993) Remediation of contaminated soils with green plants: an overview. In Vitro Cell Dev Biol 29:207–212

Danielsen F, Sørensen MK, Olwig MF, Selvam V, Parish F, Burgess ND, Hiraishi T, Karunagaran VM, Rasmussen MS, Hansen LB, Quarto A, Suryadiputra N (2005) The asian tsunami: a protective role for coastal vegetation. Science 310:643

Debez A, Ben Hamed K, Abdelly C (2003) Some physiological and biochemical aspects of salt tolerance in two oleaginous halophytes. In: Lieth H, Moschenko M (eds) Tasks for vegetation science. Cash crop halophytes recent studies, vol 38, Ten years after Al Ain meeting. Kluwer Academic, Dordrecht, the Netherlands, pp 31–39

Debez A, Ben Hamed K, Grigon C, Abdelly C (2004) Salinity effects on germination, growth, and seed production of the halophyte *Cakile maritima*. Plant Soil 262:179–189

Debez A, Saadaoui D, Ramani B, Ouerghi Z, Koyro H-W, Huchzermeyer B, Abdelly C (2006a) Leaf H^+-ATPase activity and photosynthetic capacity of *Cakile maritima* under increasing salinity. Environ Exp Bot 57:285–295

Debez A, Taamalli W, Saadaoui D, Ouerghi Z, Zarrouk M, Huchzermeyer B, Abdelly C (2006b) Salt effect on growth, photosynthesis, seed yield and oil composition in the potential crop halophyte *Cakile maritima*. In: Öztürk M, Waisel Y, Khan MA, Görk G (eds) Biosaline agriculture and salinity tolerance in plants. Birkhäuser Verlag, Switzerland, pp 55–63

Debez A, Koyro H-W, Grignon C, Abdelly C, Huchzermeyer B (2008) Relationship between the photosynthetic activity and the performance of *Cakile maritima* after long-term salt treatment. Physiol Plant 133:373–385

Dierig DA, Grieve CM, Shannon MC (2003) Selection for salt tolerance in *Lesquerella fendleri* (Gray) S.Wats. Ind Crop Prod 17:15–22

Djebali W, Chaïbi W, Ghorbel MH (2002) Croissance, activité peroxydasique et modifications ultrastructurales induites par le cadmium dans la racine de tomate. Can J Bot 80:942–953

El Shaer HM (1997) Practical approaches for improving utilization of feed resources under extensive production system in Sinai. Proceedings of the international symposium on systems of sheep and goat production, Bella, Italy, 25–27 Oct 1997

El Shaer MH (2006) Halophytes as cash crops for animal feeds in arid and semi-arid regions. In: Öztürk M, Waisel Y, Khan MA, Görk G (eds) Biosaline agriculture and salinity tolerance in plants. Birkhäuser Verlag, Switzerland, pp 117–128

El Shaer HM, Kandil HM (1990) Comparative study on the nutritional value of wild and cultivated Atriplex halimus by sheep and goat in Sinai. Com Sci Dev Res 29:81–90

Falleh H, Ksouri R, Megdiche W, Trabelsi N, Boulaaba M, Abdelly C (2008) Effect of salinity on growth, leaf-phenolic content and antioxidant scavenging activity in *Cynara cardunculus* L. In: Abdelly C, Öztürk M, Ashraf M, Grignon C (eds) Biosaline agriculture and high salinity tolerance. Birkhäuser Verlag, Switzerland, pp 335–343

Falleh H, Ksouri R, Oueslati S, Guyot S, Magné C, Abdelly C (2009) Interspecific variability of antioxidant activities and phenolic composition in Mesembryanthemum genus. Food Chem Toxicol 47:2308–2313

Flowers TJ, Colmer TD (2008) Salinity tolerance in halophytes. New Phytol 179:945–963

Flowers TJ (2004) Improving crop salt tolerance. J Exp Bot 55:307–319

Fratianni F, Tucci M, De Palma M, Pepe R, Nazzaro F (2007) Polyphenolic composition in different parts of some cultivars of globe artichoke (*Cynara cardunculus* L. var. *scolymus* (L.) Fiori). Food Chem 104:1282–1286

Gallagher JL (1985) Halophytic crops for cultivation at sea water salinity. Plant Soil 89:323–336

Galvani A (2007) The challenge of the food sufficiency through salt tolerant crops. Rev Environ Sci Biotechnol 6:3–16

Garduno MA (1993) Kochia: a new alternative for forage under high salinity conditions of Mexico. In: Lieth H, Al Masoom A (eds) Towards the rational use of high salinity tolerant plants, vol 1. Kluwer Academic, Dordrecht, pp 459–464

Ghars MA, Debez A, Smaoui A, Zarrouk M, Grignon C, Abdelly C (2006) Variability of fruit and seed-oil characteristics in Tunisian accessions of the halophyte *Cakile maritima* (Brassicaceae). In: Ajmal Khan M, Weber DJ (eds) Ecophysiology of high salinity tolerant plants, vol 40, Series: tasks for vegetation science. Springer, Berlin, Heidelberg, New York, pp 55–67

Ghassemi F, Jakeman AJ, Nix HA (1995) Salinisation of land and water resources: human causes, extent, management and case studies. University of New South Wales Press Ltd., Sydney

Ghnaya T, Nouairi I, Slama I, Messedi D, Grignon C, Abdelly C, Ghorbel MH (2005) Cadmium effects on growth and mineral nutrition of two halophytes: *Sesuvium portulacastrum* and *Mesembryanthemum crystallinum*. J Plant Physiol 162: 1133–1140

Ghnaya T, Slama I, Messedi D, Grignon C, Ghorbel MH, Abdelly C (2007) Effects of Cd^{2+} on K^+, Ca^{2+} and N uptake in two halophytes *Sesuvium portulacastrum* and *Mesembryanthemum crystallinum*: consequences on growth. Chemosphere 67:72–79

Glenn EP, O'Leary W (1985) Productivity and irrigation requirements of halophytes grown with seawater in the sonora desert. J Arid Environ 9:81–91

Glenn EP, O'leary JW, Watson MC, Thompson TL, Kuehl RO (1991) *Salicornia bigelovii* Torr.: an oilseed halophyte for seawater irrigation. Science 251:1065–1067

Glenn E, Squires V, Olsen M, Frye R (1993) Potential for carbon sequestration in drylands. Water Air Soil Poll 70: 341–355

Glenn EP, Waugh WJ, Moore D, McKeon C, Nelson SG (2001) Revegetation of an abandoned uranium millsite on the Colorado Plateau, Arizona. J Environ Qual 30:1154–11621

Gominho JJ, Pereira FH (2000) *Cynara cardunculus* L., a new fiber crop for pulp and paper production. Ind Crop Prod 13:1–100

Goodfriend WL, Olsen MW, Frye RJ (1998) Decomposition of seawater-irrigated halophytes: implications for potential carbon storage. Plant Soil 202:241–250

Goodin JR, Epstein E, McKell CM, O'Leary JW (1990) Saline agriculture, salt tolerant plants for developing countries. National Academy, Washington, DC

Grieve CM, Suarez DL (1997) Purslane (*Portulaca oleracea* L.): a halophytic crop for drainage water reuse systems. Plant Soil 192:277–283

Hafsi C, Lakhdar A, Rabhi M, Debez A, Abdelly C, Ouerghi Z (2007) Interactive effects of salinity and potassium availability on growth, water status, and ionic composition of *Hordeum maritimum*. J Plant Nutr Soil Sci 170:469–473

Hagemeyer J, Waisel Y (1988) Excretions of ions (Cd^{2+}, Li^+, Na^+, and Cl^-) by *Tamarix aphylla*. Physiol Plant 73:541–546

Hinman CW (1984) New crops for arid lands. Science 225:1445–1448

Hocking PJ (1982) Salt and mineral nutrient levels in fruits of two strand species, *Cakile maritima* and *Arctotheca populifolia*, with special reference to the effect of salt on the germination of *Cakile*. Ann Bot 50:335–343

Hussain SA, Badola R (2008) Valuing mangrove ecosystem services: linking nutrient retention function of mangrove forests to enhanced agroecosystem production. Wetlands Ecol Manage 16:441–450

ICBA (International Center for Biosaline Agriculture) (2004) Development of salinity-tolerant sorghum and pearl millet cultivars for enhanced productivity on saline lands. Progress Report 2003–2004. Dubai, United Arab Emirates: ICBA

Karimi SH, Ungar IA (1986) Oxalate and inorganic ion concentration in *Atriplex triangularis* Willds. organs in response to salinity, light level, and aeration. Bot Gaz 147:65–70

Katavic V, Mietkiewska E, Barton DL, Giblin EM, Reed DW, Taylor DC (2002) Restoring enzyme activity in nonfunctional low erucic acid *Brassica napus* fatty acid elongase1 by a single amino acid substitution. Eur J Biochem 269:5625–5631

Keiffer CH, Ungar IA (2002) Germination and establishment of halophytes on brine-affected soils. J Appl Ecol 39:402–415

Khan ZI, Ashraf M, Hussain A, Koyro H-W, Huchzermeyer B (2005) Seasonal variation in selenium status of different classes of grazing goats in a semi-arid region of Pakistan. Dtsch Tieraerztl Wschr 112:460–465

Kiliç CC, Kukul YS, Anaç D (2008) Performance of purslane (*Portulaca oleracea* L.) as a salt-removing crop. Agr Water Manage 95:854–85

Koyro H-W, Huchzermeyer B (2004) Ecophysiological mechanisms leading to salinity tolerance – screening of cashcrop halophytes. Recent Res Dev Plant Sci 1:187–207

Koyro H-W, Geissler N, Hussin S, Huchzermeyer B (2006) Mechanisms of cash crop halophytes to maintain yields and reclaim saline soils in arid areas. In: Khan MA, Weber DJ (eds) Ecophysiology of high salinity tolerant plants, vol 40, Series: tasks for vegetation science. Springer, Berlin, Heidelberg, New York, pp 345–366

Koyro H-W, Eisa SS (2008) Effect of salinity on composition, viability and germination of *Chenopodium quinoa* Willd. seeds. Plant Soil 302:79–90

Koyro H-W, Geißler N, Hussin S, Debez A, Huchzermeyer B (2008) Strategies of halophytes to survive in a salty environment. In: Khan NA, Singh S (eds) Abiotic stress and plant responses. IK International, pp 83–104. ISBN: 978-7319-883-0

Ksouri R, Megdiche W, Debez A, Falleh H, Grignon C, Abdelly C (2007) Salinity effects on polyphenol content and antioxidant activities in leaves of the halophyte *Cakile maritima*. Plant Physiol Biochem 45:244–249

Ksouri R, Falleh H, Megdiche W, Trabelsi N, Mhamdi B, Chaieb K, Bakrouf A, Magné C, Abdelly C (2009) Antioxidant and antimicrobial activities of the edible medicinal halophyte *Tamarix gallica* L. and related polyphenolic constituents. Food Cheml Toxicol 47:2083–2091

Ksouri R, Megdiche W, Falleh H, Trabelsi N, Boulaaba M, Smaoui A, Abdelly C (2008) Influence of biological, environmental and technical factors on phenolic content and antioxidant activities of Tunisian halophytes. C R Biologies 331:865–873

Kumar A, Abrol IP (1984) Studies on the reclaiming effect of Karnal-grass and para-grass grown in a highly sodic soil. Indian J Agric Sci 54:189–193

Kumar PR, Tsunoda S (1978) Fatty acid spectrum of Mediterranean wild Cruciferae. J Am Oil Chem Soc 5:320–323

Laudadio V, Tufarelli V, Dario M, Hammadi M, Seddik MM, Lacalandra GM, Dario C (2009) A survey of chemical and nutritional characteristics of halophytes plants used by camels in Southern Tunisia. Trop Anim Health Prod 41:209–215

Le Houérou HN (1992) The role of saltbushes (*Atriplex* spp.) in arid land rehabilitation in the Mediterranean Basin: a review. Agroforest Syst 18:107–148

Le Houérou HN (1993) Salt-tolerant plant for the arid regions of the Mediterranean isoclimatic zone. In: Lieth H, Al Masoom A (eds) Towards the rational use of high salinity tolerant plants. Kluwer Academic, Dordrecht, pp 403–422

Le Houérou HN (1994) Forage halophytes and salt-tolerant fodder crops in the Mediterranean Basin. In: Ayoub AT, Squires VR (eds) Halophytes as a resource for livestock and for rehabilitation of degraded lands. Kluwer Academic, Dordrecht, pp 123–137

Lieth H, Moschenko M, Lohmann M, Koyro H-W, Hamdy A (1999) Halophyte uses in different climates I. Ecological and ecophysiological studies. In: Lieth H (ed) Progress in biometeoroogy. Backhuys, Leiden, the Netherlands, pp 1–258

Lutts S, Lefèvre I, Delpérée C, Kivits S, Dechamps C, Robledo A, Correal E (2004) Heavy metal accumulation by the halophyte species Mediterranean saltbush. J Environ Qual 33:1271–1279

Maisuthisakul P, Suttajit M, Pongsawatmanit R (2007) Assessment of phenolic content and free radical-scavenging capacity of some Thai indigenous plants. Food Chem 100:1409–1418

Marcar NE, Hossain AKMA, Crawford DF, Nicholson AT (2000) Evaluation of tree establishment on saline seeps near Wellington and Young in New South Wales. Austr J Exp Agric 40:99–106

Masters DG, Norman HC, Dynes RA (2001) Opportunities and limitations for animal production from saline land. Asian-Aust J Anim Sci 14:199–211

Masters DG, Sharon EB, Norman HC (2007) Biosaline agriculture for forage and livestock production. Agr Ecosyst Environ 119:234–248

Meot-Duros L, Le Floch G, Magné C (2008) Radical scavenging, antioxidant and antimicrobial activities of halophytic species. J Ethnopharm 116:258–262

Messedi D, Sleimi N, Abdelly C (2001) Salt tolerance in *Sesuvium portulacastrum*. In: Horst WJ (ed) Plant nutrition-food security and sustainability of agro-ecosystems. Kluwer Academic, Dordrecht, pp 406–407

Messedi D, Laabidi N, Grignon C, Abdelly C (2004) Limits imposed by salt to the growth of the halophyte *Sesuvium portulacastrum*. J Plant Nutr Soil Sci 167:1–6

Munns R (2008) Living with salinity. New Phytol 179:903–905

Munns R, Cramer GR, Ball MC (1999) Interactions between rising CO_2, soil salinity and plant growth. In: Luo Y, Mooney HA (eds) Carbon dioxide and environmental stress. Academic, London, pp 139–167

Naczk M, Shahidi M (2004) Extraction and analysis of phenolics in food. J Chromatogr A 1054:95–111

Navarro JM, Flores P, Garrido C, Martinez V (2006) Changes in the contents of antioxidant compounds in pepper fruits at ripening stages, as affected by salinity. Food Chem 96:66–73

Nouairi I, Ghnaya T, Ben Youssef N, Zarrouk M, Ghorbel MH (2006) Changes in content and fatty acid profiles of total lipids of two halophytes: *Sesuvium portulacastrum* and *Mesembryanthemum crystallinum* under cadmium stress. J Plant Physiol 163:1198–1202

O'Leary W (1987) Halophytic food crops for arid lands. In: Strategies for classification and management of native vegetation for food production in arid areas, Report RM-150, Forest Service, USDA, CO

Oster JD, Kaffka SR, Shannon MC, Grattan SR (1999) Saline-sodic drainage water: A resource for forage production? In Proceedings of the 17th International Congress on Irrigation and Drainage, Granada, Spain, pp 67–79

Öztürk M, Guvensen A, Gork G, Gork C (2006) An overview of the coastal zone plant diversity and management strategies in the Mediterranean region of Turkey. In: Öztürk et al (eds) Biosaline agriculture and salinity tolerance in plants. Birkhauser Verlag, Basel, Switzerland, pp 89–100

Öztürk M, Guvensen A, Gucel S (2008) Ecology and economic potential of halophytes – a case study from Turkey. In: Kafi, Khan (eds) Crop and forage production using saline waters. NAM S & T Centre, Daya Publishing House, Delhi, India, pp 255–264

Pannelli G, Servili M, Selvaggini R, Baldioli M, Montedoro GF (1994) Effect of agronomic and seasonal factors on olive (*Olea europaea* L.) production and on the qualitative characteristics of the oil. Acta Hort 356:239–243

Pasternak D, Nerd A (1996) Research and utilization of halophytes in Israel. In: Choukr-Allah R, Malcolm CV, Hamdy A (eds) Halophytes and biosaline agriculture. Marcel Dekker, New York, pp 325–348

Peng X, Tian G (2001) Structural characterization of the glycan part of glycoconjugate LbGp2 from *Lycium barbarum* L. Carbohyd Res 331:95–99

Qadir M, Tubeileh A, Akhtar J, Larbi A, Minhas PS, Khan MA (2008) Productivity enhancement of salt-affected environments through crop diversification. Land Degrad Develop 19:429–453

Rabhi M, Talbi O, Atia A, Abdelly C, Smaoui A (2008) Selection of a halophyte that could be used in the bioreclamation of salt-affected soils in arid and semi-arid regions. In: Abdelly C, Öztürk M, Ashraf M, Grignon C (eds) Biosaline agriculture and high salinity tolerance. Birkhäuser Verlag, Switzerland, pp 241–246

Rana MS, Ahmed AA (1981) Characteristics and composition of Libyan olive oil. J Am Oil Chem Soc 58:630–631

Ravindran KC, Venkatesan K, Balakrishnan V, Chellappan KP, Balasubramanian T (2007) Restoration of saline land by halophytes for Indian soils. Soil Biol Biochem 39:2661–2664

Reboreda R, Caçador I (2007) Halophyte vegetation influences salt marsh retention capacity for heavy metals. Environ Poll 146:147–154

Reimann C, Breckle SW (1993) Sodium relations in Chenopodiaceae: a comparative approach. Plant Cell Environ 16:323–3283

Rozema J, Van Diggelen J (1991) A comparative study of growth and photosynthesis of four halophytes in response to salinity. Acta Ecol 12:673–681

Ruan C-J, Lia H, Guo Y-Q, Qin P, Gallagher JL, Seliskar DM, Lutts S, Mahy G (2008) *Kosteletzkya virginica*, an agroecoengineering halophytic species for alternative agricultural production in China's east coast: Ecological adaptation and benefits, seed yield, oil content, fatty acid and biodiesel properties. Ecol Eng 3:320–328

Saïdana D, Mahjoub S, Boussaada O, Chriaa J, Mahjoub MA, Cheraif I, Daami M, Mighri Z, Helal AN (2008) Antibacterial and antifungal activities of the essential oils of two Saltcedar species from Tunisia. J Am Oil Chem Soc 85:817–826

Salo LF, Artiola JF, Goodrich-Mahoney JW (1996) Plant species for revegetation of a saline flue gas desulfurization sludge pond. J Environ Qual 25:802–808

Sherman RE, Fahey TJ, Martinez P (2003) Spatial patterns of biomass and aboveground net primary productivity in a mangrove ecosystem in the Dominican Republic. Ecosystems 6:384–398

Slama I, Messedi D, Ghnaya T, Savoure A, Abdelly C (2006) Effects of water deficit on growth and proline metabolism in *Sesuvium portulacastrum*. Env Exp Bot 56:231–238

Sleimi N, Abdelly C (2002) Growth and mineral nutrition of some halophytes under seawater irrigation. In: Ahmad R, Malik KA (eds) Prospects for saline agriculture. Academic, Dordrecht, pp 403–410

Szabolcs I (1994) Soil and salinization. In: Pessarakli M (ed) Handbook of plant and crop stress. Marcel Dekker, New York, pp 3–11

Tomar OS, Minhas PS, Sharma VK, Singh YP, Gupta RK (2002) Performance of 32 tree species and soil conditions in a plantation established with saline irrigation. Forest Ecol Manage 177:333–346

Turner NC (2004) Sustainable production of crops and pastures under drought in a Mediterranean environment. Ann Appl Biol 144:139–147

Valentao P, Fernandes E, Carvalho F, Andrade PB, Seabra RM, Bastos ML (2002) Antioxidative properties of cardoon (*Cynara cardunculus* L.) infusion against superoxide radical, hydroxyl radical, and hypochlorous acid. J Agric Food Chem 50:4989–4993

Wang B, Lüttge U, Ratajczak R (2004) Specific regulation of SOD isoforms by NaCl and osmotic stress in leaves of the C3 halophyte *Suaeda salsa* L. J Plant Physiol 161:285–293

Watson MC, O'Leary JW (1993) Performance of *Atriplex* species in the San Joaquin Valley, California, under irrigation and with mechanical harvests. Agr Ecosyst Environ 43:255–266

Weber DJ, Gul B, Khan MA, Williams T, Wayman P, Warner S (2001) Comparison of vegetable oil from seeds of native halophytic shrubs. In: Mc Arthur ED, Fairbanks DJ (eds) Proceeding of shrubland ecosystem genetics and biodiversity. RMRS-P-21. USDA Forest Service, Rocky Mountain Research Station, Ogden, UT, USA, pp 287–290

Weber DJ, Ansarib R, Gulb B, Khan MA (2007) Potential of halophytes as source of edible oil. J Arid Environ 68:315–321

Wei Y, Xu X, Tao H, Wang P (2006) Growth performance and physiological response in the halophyte *Lycium barbarum* grown at salt-affected soil. Ann Appl Biol 149:263–269

Yajun B, Xiaojing L, Weiqiang L (2003) Primary analysis of four salt tolerant plants growing in Hai-He plain, China. In: Lieth H, Moshenko M (eds) Cash crop halophytes: recent studies. Kluwer Academic, Dordrecht, pp 135–138

Yang G-Y, Wang Y, Ma X-P (2000) Research on fat content and composition of fatty acid of wild soybean (*G. soja*) in China. Soybean Sci 19:258–262

Yensen NP (2006) Halophyte uses for the twenty-first century. In: Khan MA, Weber DJ (eds) Ecophysiology of high salinity tolerant plants, vol 40, Series: tasks for vegetation science. Springer, Berlin, Heidelberg, New York, pp 367–396

Zarrouk M, Cherif A (1983) Teneur en lipides des halophytes et résistance au sel. Z Pflanzenphysiol 112:373–380

Zarrouk M, El Almi H, Ben Youssef N, Sleimi N, Smaoui A, Ben Miled D, Abdelly C (2003) Lipid composition of local halophytes seeds: *Cakile maritima, Zygophyllum album* and *Crithmum maritimum*. In: Lieth H, Moschenko M (eds) Cash crop halophytes recent studies. Kluwer Academic, Dordrecht, pp 121–126, Ten years after the Al Ain Meeting

The Salt Marsh (Sabkha) in the Western Part of Libya

M.O. El-Magsodi and Daw A. Haddoud

Abstract The total coastline of Libya is about 1975 km long. Numerous salt marshes bordered by sabkhas occur along the coast. These sabkhas cover some a very large area on the coast line and most of them are completely dry during the summer season. One of the larger sabkhas is located in the western part of Libya and is called "the Abu Kammash salt marsh (Abu Kammash subkha)" and it is found along the coastal Libyan-Tunisian highway about 150 km west of Tripoli in the Abu Kammash area near the Tunisian border, at an altitude of 2.5–4.0 m below sea level. This is a typical salt marsh of coastal Mediterranean region. It is generally a flat sabkha with a total area of about 50 km^2. There are 38 wells in the sabkha, from which brine water is pumped through pipelines to Abu Kammash factory (chemical plant) which was established on the area near the seashore for the mass production of pure salt and bleaching agents. The saline ponds and pools are formed when such low land area floods in rainy winters, and 12 km long and 4 km wide small seasonal lake is formed. In summer, however, most of these lakes dry up. The high salinity and evaporation lead to precipitation of a thick layer of salt on the sabkha floor.The phytoplankton density is quite low. The sabkha has been identified as a natural biotope for brine shrimp (*Artemia*). Next to the *Artemia* comes the birds which are also found in moderate number. Majority of the birds are migratory, as such the sabkha may serve as good refuge for such birds provided certain measures are taken. There is a very poor vegetation cover composed mainly of halophytes and desert plant taxa around the sabkha due to coarse sand texture and prevailing semiarid conditions.

1 Introduction

Libya has a total area of 1,759,540 km^2 and a coast line over 1,975 km long. Numerous salt marshes (sabkhas) occur along the coast. These sabkhas cover an area of hundreds of square kilometer all along the coast and most dry up during the summer season. The huge areas of isolated marsh lands are unique habitats of great significance. Therefore an evaluation of their biodiversity will be highly fruitful. The biodiversity inhabiting such habitst plays important role in this type of ecosystem and human population living around these areas depend on them for food. They are also a source of industrial raw materials such as salt, chemicals and *Artemia*. One of the main marshes among these is Abu Kammash salt marsh "Abu Kammash sabkha" extending from the western coast from the Tunisian border to Musratha. It shows natural topographical structure all along the Libyan-Tunisian coastal highway lying about 150 km west of Tripoli in the Abu Kammash area close to the Tunisian border. Major part of the sabkha is located in the middle of coast from Musratha to Benghazi. The coves and marine lagoons are found in the eastern part of this coast from Benghazi upto the Egyptian border. These are Ain Ziyana, Bumba and Ain Gazala lagoon. The Farwa lagoon is found on the western part near Libyan-Tunisian boarder (Kerambrun 1986; Haddoud 1999; Gashout and Haddoud 2001). Several experimental trials have been done to use some of these lagoons and sabkhas in aquaculture activities

M.O. El-Magsodi (✉) and D.A. Haddoud
Marine Biology Research Center, Tajura, Tripoli, Libya

M. Öztürk et al. (eds.), *Sabkha Ecosystems*, Tasks for Vegetation Science 46,
DOI 10.1007/978-90-481-9673-9_9, © Springer Science+Business Media B.V. 2011

especially in Farwa, Ain Gazala, but no progress has been acheived due to several reasons (Gashout and Haddoud 2001).

2 Material and Methods

2.1 Study Area

The Abu Kammash subkha (salt marsh) is located c. 33° 48' 00' N, 1 1° 36' 00' E, approximately 150 km west of Tripoli, along Tripoli–Tunisia coastal highway close to the Tunisian border (Fig. 1). The sabkha has an elongated basin extending from west to east with an axial length of c. 12 km, a maximum width of 4 km, and a total area of c. 50 km^2. The depth of the lagoon varies from 10 to more than 80 cm. The actual basin (Abu Kammash sabkha) is separated from the sea by a narrow sand bar of c. 3 km width. It is a typical salt marsh of coastal Mediterranean region, generally a flat with an altitude varying between 2.5 and 4.0 m below sea level. The floods in rainy winters form saline ponds and pools on such low land areas and a small seasonal lake is formed. In summer however, most of these lakes dry up.

The Abu Kammash area is characterized mainly by semiarid climate, coarse sand and poor vegetation cover made up mostly of halophytes and desert plants. The climate of the area is mostly maritime which does not differ from that on the nearest mainland. The temperature of the Abu Kammash area reaches a maximum of 43°C in summer and a minimum of 13°C during winter. The relative humidity ranges from 64–88% (x = 73.6%).

In 2001, a pilote workplan was proposed by MBRC to study Libyan sabkhas and wetlands covering; biological, chemical and socio-economic features and use of each site. The information was gathered by using a questionaire prepared for this purpose. The scientific information was collected from different sites by using portable instruments. The results from previous studies were also taken into consideration during the evaluation.

Artemia cysts (eggs) were collected in summer 2001, when the pond was completely dry, together with debris and dry muddy-sand which formed a dark gray thin layer at the edges of the dried pond. All samples were processed at the *Artemia* Reference Center (Belgium). Cysts were cleaned by means of the bi-flotation technique as described by Sorgeloos et al. (1978). The lake samples produced only limited amount of full cysts because of considerable amounts of salt and impurities. The diameters of hydrated non-decapsulated, hydrated decapsulated cysts, length of instar-I nauplii and the thickness of chorion were measured according to the technique of Vanhaecke and Sorgeloos (1980). The incubation of cysts for collection instar-I nauplii and the decapsulation of cysts were performed according to Sorgeloos et al. (1986). The cyst diameters and the length of nauplii were measured using a light microscope equiped with a micrometer eyepiece previously calibrated with a stage graticule. A random sample of 100 cysts and nauplii was evaluated.

Artemia sibling species were defined for Abu Kammash brine shrimp in cross-breeding tests performed following procedures outlined by with *Artemia* from Sanfrancisco Bay (California-USA) and Lamaca Lake (Cyprus) were used as reference material for *A, franciscana* (Bowen et al. 1978) and *A. salına* (*tunisiana*) respectively.

In November 2001 the same sabkha was rain flooded forming a pond of about 12 km length and a maximum 4 km width, and four study trips were organized to collect samples and scientific information from the area. During these study trips the *Artemia* biomass was collected by using a standard plankton net of 200 μm mesh size, with mouth diameter of 0.25 m for further analysis. Also, samples of water were collected and brought to the laboratory for chemical analysis. Moreover, glass jars were used to collect water samples for salinity, oxygen and pH determination. Jars were also used to skim water and top soil layer for analysis of microscopic flora. Temperature and humidity were recorded in all study trips, by using the following equipments: Maximum and minimum thermometer, wet and dry bulb hygrometer and whirling hygrometer.

The sediment samples for sediment analysis were collected from pond periphery and near its center. These samples were analyzed for the determination of soil texture. The technique adapted and used in the analysis was that described by Folk (1974).

Samples of the plant cover in the area arround the sabkha were collected and identified following Tackholm (1956) and Keith (1965).

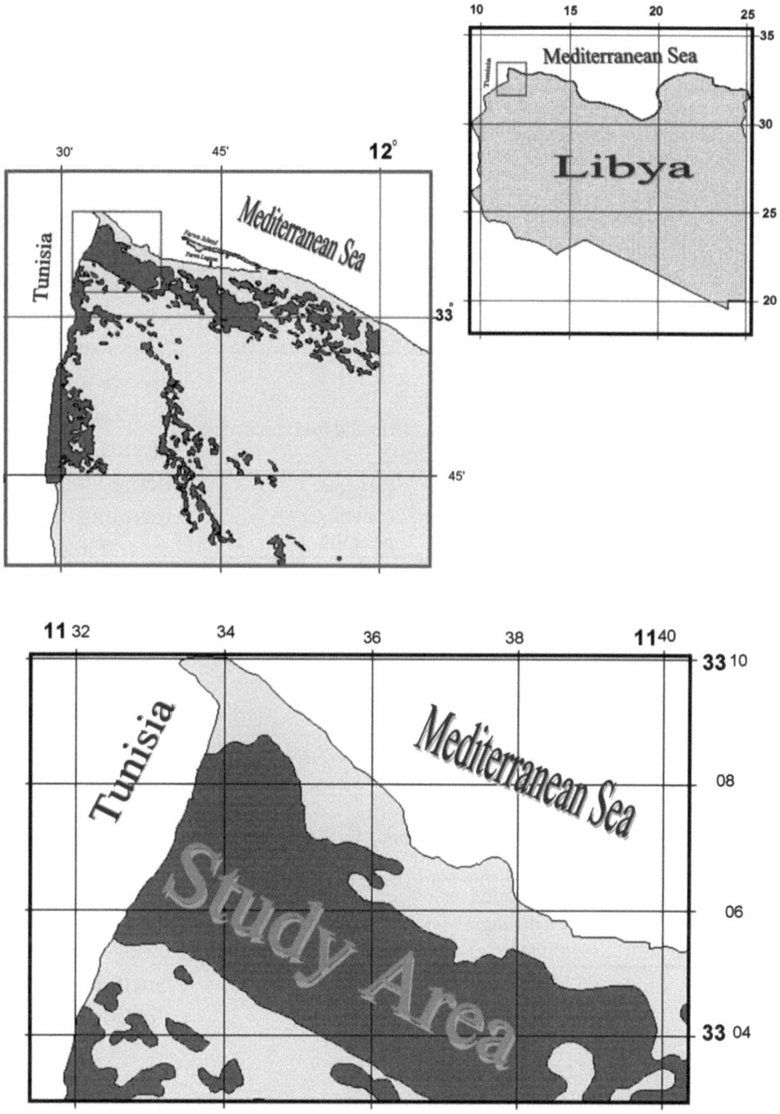

Fig. 1 A map showing location of the study area (Abu-Kammash subka)

Birds were mainly observed and identified by field glasses (Binoculars). Also we followed Heinzel et al. (1972), Bundy (1976) and Hollom et al. (1988) for identification of bird fauna on the area.

3 Results

Phytoplanktons: Planktonic organisms found in the water and top soil surfaces of the pond are listed in Table 1. There are seven genera and ten species of diatoms, one genus of dinoflagellates and one genus and one species of blue green algae. The analyses of the sabkha substrate indicated that it has sandy texture; composed of coarse, medium and very fine sand. The silt and clay were around 35% in most samples. Fine sediment covered the largest area of the sabkha especially near the edge, while the area around it shows mainly coarse sand texture.

Table 1 List of phytoplanktons in pond water and soil samples

A. Diatoms: *Amphora coffeaeformis* (C. Agardh) Kutzing , *A. coffeaeformis* var. *perpmilla* Grunow., *A. lineolata* Ehrenberg, *A. cstata* W. Smith, *Nitzschia vivax* W, Smith, *Navicula minuscitlata Grunow*, *Fragilaria* sp., *Mefosira* sp., *Synedra* sp., and *Surirella* sp.

B. Blue green algae: *Phormidium fragile* (Menegh) Gomont

C. Dinoflagellates: *Gymnodinium* sp.

Table 2 Summary of biometrical measurements

Sample	Biometrical measurements	Sample size	Mean (μm)	Standard deviation
Libyan cysts (ARC code number 1437)	Non decapsulated cysts	100	256.8	13.7
	Decapsulated cysts	100	227.8	11.2
	Instar-I nauplin	100	478.0	21.0
	Chorion thicknes	–	14.5	–

Table 3 Temperature, salinity, oxygen content and (pH) recorded during the four study trips in 2001 and 2002

Field trip dates	Temprature (°C)		Salinity (o/oo)	Oxygen content (parts per million)	pH
	Air	Water			
August 2001	43	ND	ND	ND	ND
November 2001	19	18	312.4	2.2	6.83
January 2002	13	12	302.0	3,6	7.72
April	28	25	317.4	1.63	7.31

ND No data recorded due to the drying of the sabkha

Table 4 Summary of the average values of the individual constituent of salts

Chemicals composition	Concentration in g/1 during (summer)	Concentration in g/1 during (winter)
NaCl	237.0	287.6
$MgCl_2$	31.0	5.6
$MgSO_4$	23.5	9.4
KCL	11.0	4.7
Br	0.4	0.1
$CaSO_4$	1.7	2.8

Artemia: The results obtained for the measurement of hydrated non-decapsulated and decapsulated cyst diameters and instar-I naupliar length are presented in Table 2. Although the cross-breeding tests between Abu Kammash *Artemia* and Larnaca (Cyprus) brine shrimp yielded viable fl offsprings, the Abu Kammash and San Francisco Bay matings were sterile. Abu Kammash *Artemia* can be classified within the *A. salina* (*Tunisiana*) sibling species complex. The *Artemia* biomass was found to be the predominant member of the plankton. It was present in such prolific abundance as to impart red colour to the surface of water of the sabkhas.

Physical and chemical parameters: The results of physical and chemical parameters of the Abu Kammash sabkha waters revealed that during the four study trips temperature, salinity, oxygen content and pH varied between; 17–33°C, 302–317‰, 1.63–3.6 part per million and 6.83–7.72 (pH) respectively (Table 3). The results of brine analysis during summer and winter seasons are presented in Table 4.

Birds: A large number of birds were observed to visit the area. Out of these 65% were migratory and 35% were resident species. Most of the birds observed are passerines. The details are given in Table 5.

Plant diversity: The vegetation comprises of xerophytes and halophytes with few date palms, the rest being coarse sand. Most of the plants in the area (arround the sabkha) are halophytes and desert species. These are not favoured by animals. The dominant species are *Salicornia fruticosa* and *Halimione porlulacoides*. There is no great variation between the flora of the mainland (area around the sabkha) and that of the nearest island (Farwa island) (Annajar and Hussain 1984).

4 Discussion

Salt evaporating ponds of varying sizes can be found along the coastal deserts in "many parts of the world. Such areas are traditionally used by man for salt production. The production of food in these salterns could be an important by-product. The natural productivity of these ponds per unit surface area was reported to approach that of the open sea, even though the ponds are only 30–50 cm deep.

The results of present work indicate that ponds formed during rainy seasons in sabkha and marshes along the north-western coast of Libya contain a number of diatoms and other microscopic algae that can thrive and grow to support food availability to newly hatched larvae (nauplii) and adults of *Artemia*. The soil texture of such ponds may provide more aeration for these microscopic organisms. Also, the nature of the substrata seem to be a suitable habitat for both soil diatoms and *Artemia* cysts (Fig. 2a–c).

The brine shrimp (*Artemia salina*) was found to be the predominant member of the plankton. It was present in such prolific abundance as to impart red colour

Table 5 List of birds observed on the sabkhas

Species	Common name	Habitat	Status
Sterna hirundo	Common tern	A	B
Sterna anaethetus	Bridded tern	A	B
Tringa totanus	Red shank	A	B
Pholacrocorax carbo	Cormorant	A	M
Ardea cinerea	Grey heron	A	M
Egretta garzetta	Little egert	A	M
Fulica atra	Coot	A	M
Larus argentatus	Herring gull	A	M
Alceda atthis	King fisher	A	M
Luscinia megarhynchos	Nightingale	T	M
Erithacus rebecula	Robin	T	M
Apus melba	Alpine swift	T	M
Motacilla alba	White wagtail	T	M
Ficedula hypoleuca	Pied fly catcher	T	M
Passer hispaniolensis	Spanish sparrow	T	R
Fringilla coelebs	Chaffinch	T	R
Upupa epops	Hoopoe	T	R
Columba livia	Rock dove	T	R
Sfreptopelia turtur	Turtle dove	T	R
Lanius excubitor	Great grey shrike	T	R

A aquatic, *T* terrestrial, *Mr* migratory, *R* resident, *B* breeding. Heinzel et al. (1972), Bundy (1976), Hollom et al. (1988) were followed for identifications

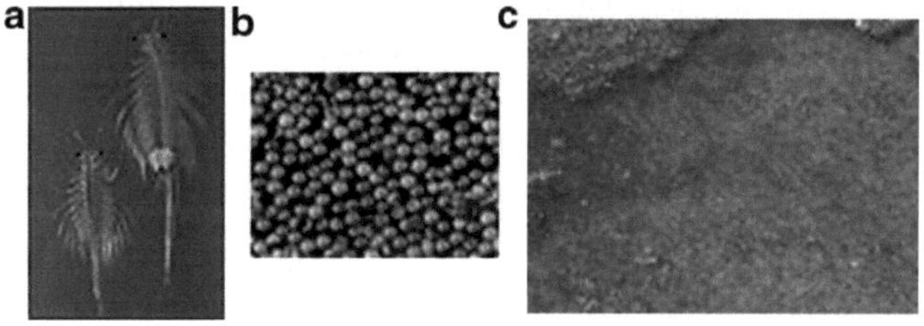

Fig. 2 *Artemia salina* general view (**a**), cysts (**b**), and imparting red colour(**c**)

to the surface of water of the sabkha. The *A. salina* is an arthropod which belongs to class Crustacea, subclass Branchiopoda, and order Anostraca, commonly known as fairy shrimp, which are the most primitive crustacean of our recent fauna. They are elongated, somewhat cylindrical, delicate animals without the carapace. *Artemia* is an extreme example of euryhalinity, since it can inhabit salt solution containing as little as 0.26% NaCl and concentrated brine having 30% salt (Croghan 1959), and in high salinity waters of over 200 parts per thousand (Allee et al. 1949). In the present study we found it in the sabkha with a salinity of more than 300 parts per thousand. This salinity is higher than that of the Great Salt Lake and that of the Dead Sea which have a salinity of 203 and 192 parts per thousand respectively. The present preliminary quantitative study of the Abu Kammash sabkha also revealed that its fauna has low diversity of species as well as number of planktons. This may be attributed to the scarcity of food, high salinity, soil texture and lack of different habitats, or all might be considered as limiting factors for the presence, abundance, diversity and stability of the planktons on the sabkha.

Next to the *Artemia* come the birds which are found in moderate numbers. Majorities of these are migratory depicting that the Abu kammach area (including the sabkha) may serve as a good refuge for such birds provided certain measures are taken. Moreover, the effect of distance as a barrier is less effective for birds than mammals. This is in accordance with our results where birds are the largest and the most abundant group of vertebrates recorded on the area. However, only few species of mammals were observed here; Gerbil, House mouse, Red fox, Jackal, Wild rabibit and three species of reptiles: Skink, Chamaeleon, Montpellier-snake and 24 species of insects (Annajar and Hussain 1984). Amphibians are not represented in the area due to the lack of moisture and high salinity that could be considered as limiting factors for their existence. Darlington (1957) staded that distribution of amphibians on these areas is limited because of high salinity. High salinity,coarse sand texture and semiarid condition lead to poor vegetation cover mainly halophytes and desert plant species.

5 Conclusions

From this quantitative and qualitative study we conclude that the Abu Kammash area (including the Abu-Kammash sabkha) has low general faunal diversity and number of mammals too is low. This may be due to the scarcity of food, high salinity, soil texture and lack of different habitats, even all might be considered as limiting factors for the presence, abundance, diversity and stability of the animals in the area. Moreover semiarid conditions are prevealing on the area. However, the sabkha could serve as a source of brine shrimp *Artemia* for research and development purposes.

References

Allee WC, Emerson AE, Park O, Park T, Schmidt KP (1949) Principles of animal ecology. W. B. Saunders, Philadlphia

Annajar BB, Hussain MG (1984) Preliminary survey of the Farwa Island ecosystem. BSc. thesis Faculty of Scinece, Department of Zoology, Al-Fatah University, Tripoli, Libya, pp 23

Bowen ST, Durkin JP, Sterling G, Clark LS (1978) *Artemia* hemoglobins: genetic variation in parthenogenetic and zygogenetic populations. Biol Bull 155:273–287

Bundy G (1976) The birds of Libya. British Ornithologist Union London, pp 102

Croghan PC (1959) Osmotic and ionic regulations in *Artemia*. J. Exp. Biol., 35

Darlington PJ (1957) Zoogeography: the geographical distribution of animals. Wiley, New York, pp 675

Folk RL (1974) Petrology of sedimentary rocks. Hemphill publishing Co, Austin

Gashout SF, Haddoud AD (2001) Libyan Lagoon Management. Proceedings of the Fifth Internationl Conference On the Mediterranean Coastal Environment, Medcoast Ozkan (ed), Tunis, pp 1033–1042

Haddoud AD (1999) Study of bio-ecological factor at Ain-El-Gazala Lagoon M.Sc. thesis

Halophyte-Fodder Species Association May Improve Nutrient Availability and Biomass Production of the Sabkha Ecosystem

Chedly Abdelly, Ahmed Debez, Abderrazak Smaoui, and Claude Grignon

Abstract Sabkhas are often regarded as marginal non productive areas. Yet, a wide range of halophytes together with less salt-tolerant fodder species thrive in these ecosystems. In the present study, the potential of halophytes in improving biomass production and nutrient uptake of an un-grazed parcel edging an inland sabkha (100 km south-east of Tunis) was addressed. The vegetation of studied area was characterised by perennial halophytes tufts in association with fodder annuals, mainly *Medicago* species. Halophytes grew slowly and accumulated low contents of mineral nutrients (N, Pi, K^+) and high Na^+ concentrations in their shoots. Depending on the precipitation, the annual fodder species showed high growth activity (up to 40% of the plant annual biomass production of the ecosystem), which was sustained by a high capacity of nutrient uptake (up to 70% of the total nutrient uptake). The annuals were almost exclusively clustered under the halophyte tufts, or at their immediate vicinity, where soil N and Pi levels were significantly higher, and salinity significantly lower than between the tufts. The perennial halophytes provided optimal conditions for the growth of these annual Leguminosae, which valorised the nutrients accumulated around the halophytes, owing to their high growth rate, nutrient-absorption rate and use-efficiency. These data are of high ecological significance and suggest that the salt tolerant-fodder species association may constitute a low-cost and valuable approach for the restoration and reclamation of the marginal saline ecosystems.

C. Abdelly and A. Smaoui
Laboratoire des Plantes Extrêmophiles, Centre de Biotechnologie de Borj-Cedria, BP 901, Hammam-lif, 2050, Tunisia

A. Debez (✉)
Laboratoire des Plantes Extrêmophiles, Centre de Biotechnologie de Borj-Cedria, BP 901, Hammam-lif, 2050, Tunisia
and
Institut für Botanik, Leibniz Universität Hannover, Herrenhäuser Str. 2, 30419 Hannover, Germany
e-mail: ahmed_debez@yahoo.com

C. Grignon
B&PMP, CNRS-INRA-SupAgro-Université Montpellier 2, France

1 Introduction

Frequently associated with drought, salinity leads to considerable loss of fertile soils and threatens the world nutritional balance (Malcolm et al. 2003; Reynolds et al. 2005). About 1.5 billion ha of soils suffer excessive salt level throughout the world (Choukr-Allah 1996). Annually 10 million ha of arable lands are abandoned because of salt accumulation following the utilisation of saline irrigation water for agriculture. In Tunisia, 1.5 million ha (10% of the whole territory and 18% of its arable land) are affected by salt. Besides impacting plant water status, increasing salinity enhances plant nutrient requirements and may restrict the uptake of essential nutrients such as K^+, Ca^{2+}, and Mg^{2+} (Hu and Schmidhalter 2005), resulting in salt-induced nutrient stress. Such resource limitations control ultimately plant community properties (e.g. structure and species distribution), as well as ecosystem functions (i.e. productivity and nutrient cycling) (James et al. 2005).

Unlike conventional crops, halophyte plants are able to survive and to reproduce in environments (coasts, wetlands, inland deserts) where salt concentration reaches or even exceeds seawater salinity level. These species, which represent about 1% of the world's flora, have evolved complex mechanisms at different levels (whole plant, cellular, and molecular) enabling them to successfully cope with these hostile conditions.

M. Öztürk et al. (eds.), *Sabkha Ecosystems*, Tasks for Vegetation Science 46,
DOI 10.1007/978-90-481-9673-9_10,

Restoration of saline areas using salt-tolerant vegetation is progressively emerging as a sustainable and low-cost approach (Aschenbach 2006; Ravindran et al. 2007; Sargeant et al. 2008; Khan et al. 2009). Concomitantly, better understanding of plant ecosystems and their dynamics that are capable of significant productivity under salt stress is of paramount importance. For instance, the analysis of plant association in their natural environments may provide key information about their intrinsic potentialities and the likely ways of their utilisation in order to restore marginal salt-affected areas.

Coastal or continental sabkhas are saline ecosystems frequently or occasionally subjected to flooding, which are typical of the semi-arid and arid areas around the world. Because of the extreme environmental (edaphic and climatic) conditions prevailing there, these biotopes are considered as non productive wastelands (Böer and Gliddon 1998). However, vegetation in sabkhas is diverse and heterogeneous, consisting in numerous perennial halophytic plants in association with annual species, more or less sensitive to salt and mineral deficiency stresses. In Tunisia for instance, the annual fodder plant *Hordeum maritima* (Poaceae) is common in the saline depressions in close association with strict halophytes, such as *Arthrocnemum indicum* and *Halocnemum strobilaceum* (Hafsi et al. 2007). Sabkhas could be valorised by implementing agro-ecosystems using "alternative" cash-crop halophytes (Böer and Gliddon 1998; Keiffer and Ungar 2002). Indeed, among the 2600 recognised halophytic species, some present economic (human food, fodder, materials of high economic values) or ecological (soil desalinisation, dune fixation, phytoremediation, landscaping and ornament) potentials (Abdelly et al. 2006; Ghars et al. 2006; Weber et al. 2007).

Despite sabkhas are common in Tunisia these unique biotopes remain poorly investigated with respect to their spatial and temporal dynamics, vegetation structure, nutrient relations, and their agro-economical potentialities. In the framework of the sustainable saline ecosystem valorisation, the present study investigates the role of halophytes on soil physico-chemical characteristics and nutrient relations, enabling enhanced primary production of fodder species. The chosen area borders a sabkha located in a semi-arid region of Tunisia (mean annual rainfall ranging from 250 to 700 mm). The flora of this site is composed of strict halophytes (*Arthrocnemum indicum*, *Halocnemum strobilaceum*, *Salsola cruciata*, *Suaeda fruticosa*) and of glycophytes (mainly *Medicago* species).

2 Material and Methods

2.1 Experimental Site and Sampling Periods

An experimental parcel covering about 1,000 m^2 was delimited and protected from livestock grazing, at the border of the sabkha of Enfidha (100 km south-east from Tunis), situated in the semi-arid bioclimatic zone (moderate winters and mean annual rainfall rarely above 300 mm). Biomass production of the plant cover (perennial halophytes and annual glycophytes) was assessed three times annually during two successive years. The first year was characterised by irregular and scarce rainfall (235 mm), and the second one was relatively rainy (423 mm). The first sampling took place shortly after of the establishment of the annuals, the second at their flowering stage (generally corresponding to the period of halophyte maximum growth activity), and the third one following ripening of the annuals. At the final sampling, biomass may have been underestimated since most of the desiccated organs were lost.

2.2 Plant Sampling Method

Perennial plants. A non-destructive, indirect sampling method was used to evaluate the contribution of each species to total biomass production. In the protected study area (Fig. 1) observations were made in a representative (with respect to the plant cover) homogeneous 5 m × 5 m plot. Within this experimental area, the halophytes were thriving in isolated tufts of different sizes. At each sampling, diameter and height of the perennial tufts inside the observation plot were measured. For each species, size classes were defined, and their populations were counted. Then, tufts with sizes representative for each class, outside the reference square, were sampled (green twigs and dry twigs separately), and fresh weight (FW), dry weight (DW) after drying for 3 days at 80°C, and mineral content determined. For each species *j*, with size classes *i* = 1, 2, …, k, each containing $n_{i,j}$ plants of mass $m_{i,j}$, dry weight was calculated as:

$$DW_j = \sum_{i=1}^{i=k} n_{i,j}\, m_{i,j}$$

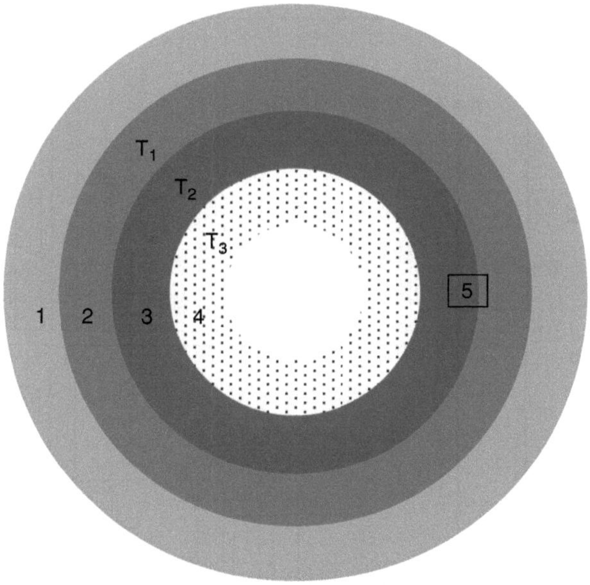

Fig. 1 Schematic representation of the studied parcel. T_1, T_2 and T_3 indicate the harvesting sites of tufts 1, 2, and 3 respectively. Numbers 1–4 represent the different vegetation zones from the periphery to the centre of the sabkha of Enfidha. 1: Most external zone, containing mainly fodder shrubs (*Acacia cyanophylla* and *Medicago arborea*), Poaceae and other annual glycophytes. 2: Transitional zone, between abundant and poor flora areas, characterised by *Atriplex halimus* and *Suaeda fruticosa* tufts, and other species typical of more saline (*Hordeum maritimum* and *Triglochin bulbosa*) or less saline (*Trifolium tomentosum* and *Limoniastrum guyonianum*) hydromorphic soils. 3: The third zone, colonised mainly by perennial halophytes, indicating temporarily inundated saline soils. The sssociation *Salsola tetrandra-Suaeda fruticosa* was frequent, in addition to *Arthrocnemum indicum* and *Phragmites cummunis*. 4: The fourth zone, characterised by the prominence of two species indicating high hydromorphy and salinity : *Halocnemum strobilaceum* and *Arthrocnemum indicum*. 5: Location of the studied parcel, delimited at the border of zones 2 and 3

Annual plants. Six 0.5 m × 1 m plots were selected inside the protected parcel, representative for the distribution of the annual species within the experimental area. Sampled shoots from each small area were partitioned into Leguminosae and Poaceae, and their respective FW and DW determined.

2.3 Soil Sampling Method

Three *Arthrocnemum indicum* tufts growing in association with annuals were selected. Tuft 1 was close to the sabkha periphery, whereas tufts 2 and 3 were close to its centre (Fig. 1). Soil samples (n = 4) were taken at the centre of each tuft, and along two orthogonal axes at 0.5, 1 and 2 m from the centre. At each position, five 20-cm wide soil horizons (from 0 to 1 m depth) were sampled.

2.4 Mineral Elements

Plants. Plant samples were finely grounded. Na^+, K^+ and inorganic phosphate (Pi) were extracted in 0.5% HCl and assayed by flame emission photometry and colorimetry respectively (Fleury and Leclerc 1943). For Pi, the extracts were first treated with organic carbon. Reduced nitrogen was determined by the Kjeldahl Ũmethod.

Soil. Assimilable phosphorous was extracted and estimated using the fine soil fraction according to the method described by Bonneau and Souchier (1979). Electrical conductivity (EC) was measured and soluble cations (Na^+ and K^+) were determined by flame emission photometry (Corning, UK) in aqueous extracts (soil/water: 1/10 w/w).

2.5 Statistical Analysis

A one way analysis of variance (ANOVA) at the $P < 0.05$ significance level was performed using the SPSS programme.

3 Results

3.1 Vegetation Inventory

Four vegetation zones were distinguished from the periphery to the centre of the sabkha (Fig. 1). The outer zone, which was partially cultivated, contained barley and fodder shrubs (*Acacia cyanophylla* and *Medicago arborea*), in addition to a mixture of naturally-growing vegetation, mainly comprising Poaceae (*Lolium regidum*, *Lagurus ovatus*, *Aeluropus littoralis*, *Cynodon dactylon*), Leguminosae (*Medicago ciliaris*, *Medicago truncatulata*, *Medicago polymorpha*, *Medicago hispida*, *Medicago minima*, *Astragalus hamosus*, *Trifolium tomentosum*, *Scorpirus muricatus*), and Compositeae (*Calendula arvensis*, *Anacyclus clavatus*, *Pteranthus dichotomus*, *Pallenis spinosa*, *Scolymus hispanicus*). The majority of these species is known to be salt-sensitive. However, *Aeluropus littoralis* is an indicator of moderately hydromorphic zones with variable EC (from 10 to 60 dS m^{-1}).

The second zone represents a transitional zone species-rich to species-poor areas. It was characterised by tufts of Chenopodiaceae, mostly *Atriplex halimus* and *Suaeda fruticosa*, tolerating a wide range in salinity (EC between 10 and 70 dS m^{-1}). Species typical of more saline (*Hordeum maritima* and *Triglochin bulbosa*) or less saline (*Trifolium tomentosum* and *Limoniastrum guyonianum*) hydromorphic soils, were also growing in this area. Indicator species of low EC soils (less than 25 dS m^{-1}: *Medicago ciliaris* and *Trifolium scabrum*) were also found.

The third zone (EC ranging between 10 and 70 dS m^{-1}) was mainly characterised by perennial halophytes, especially the association *Salsola tetrandra-Suaeda fruticosa*, also characteristic of low to moderate hydromorphic soils. Other species like *Arthrocnemum indicum* indicated temporarily flooded saline soils (EC up to 80 dS m^{-1}) or pronounced hydromorphy (*Phragmites cummunis*) or high salinity (*Halocnemum strobilaceum*). Four *Medicago* species, as well as *Trigonella maritima* and *Trifolium tomentosum* were also found.

Vegetation in the fourth zone distinguished by prominence of two species indicating strong hydromorphy and salinity: *Halocnemum strobilaceum* and *Arthrocnemum indicum*. The tufts of *H. strobilaceum* formed a suitable micro-habitat for *Spergularia salina* and *Trifolium tomentosum*.

Thus, the variations in species composition among the four zones indicated increasing salinity and hydromorphy gradients from the outer boundary towards the centre of the sabkha. Interestingly, the annual and perennial species were spatially associated, annuals thriving at the vicinity of the halophyte tufts, while almost no vegetation was found between these tufts.

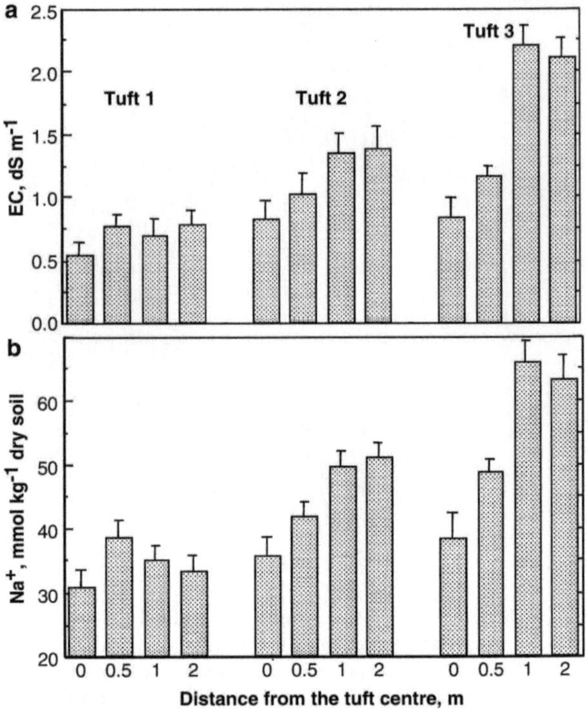

Fig. 2 Variation in soil (0–20 cm depth horizon) salinity with distance from the perennial halophytes. (**a**) EC of aqueous extracts (1 g de soil for 10 ml H_2O). (**b**) soluble sodium. Tuft 1 situated at the sabkha periphery, tufts 2 and 3 near its centre. Means of 4 replicates ± S.E. at $*P < 0.05$

3.2 Soil Salinity

Soil salinity (as EC, dS m^{-1}) in the centre of the tufts showed non-significant (at $P < 0.05$) spatial variation (0.55–0.88 dS m^{-1}) within the sabkha (Fig. 2a). These values correspond to a salinity ranging between 55 and 88 meq kg^{-1} dry soils, i.e. a relatively low salt level. EC generally increased with distance from the tufts, exceeding 1 dS m^{-1}, and even 2 dS m^{-1} (200 meq kg^{-1} dry soil), at 2 m from the tuft centre, i.e. characteristic for a highly saline soil.

EC was significantly higher (at $P < 0.05$) for tufts 2 and 3 (located inside the sabkha) than in tuft 1, closer to its periphery. This is consistent with the increasing salinity gradient from the periphery to the sabkha centre derived from vegetation analyzes. A horizontal salinity micro-gradient was thus observed from the centre towards the exterior of the halophyte tufts. Na^+ concentration varied in parallel with the EC (Fig. 2b), along both large scale gradients and micro-gradients, indicating its large contribution to the total soil salinity.

3.3 Soil Fertility

Potassium concentration in the soil (2–4 mmol kg^{-1} dry soil) did not vary systematically across the sabkha,

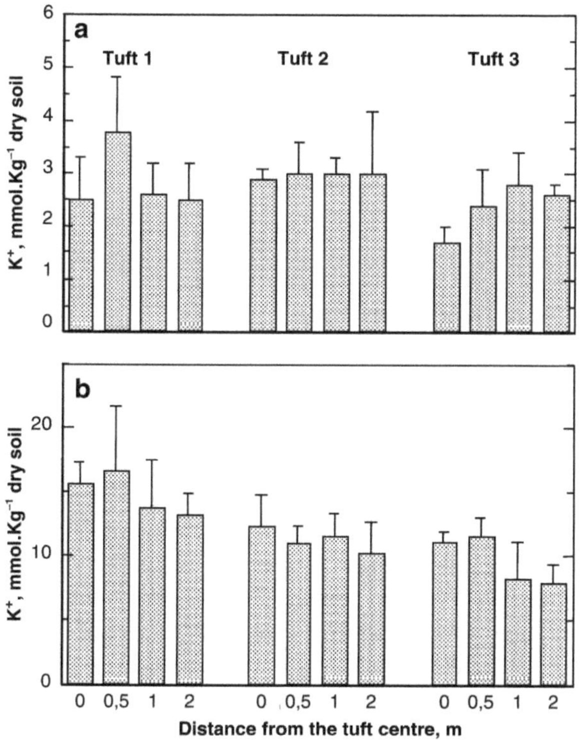

Fig. 3 Variation in soil (0–20 cm depth horizon) potassium content with distance from *Arthrocnemum indicum* tufts (**a**) soluble K^+. (**b**) exchangeable K^+. Tuft 1 situated at the sabkha periphery, tufts 2 and 3 near its centre. Means of 4 replicates ± S.E. at $^*P < 0.05$

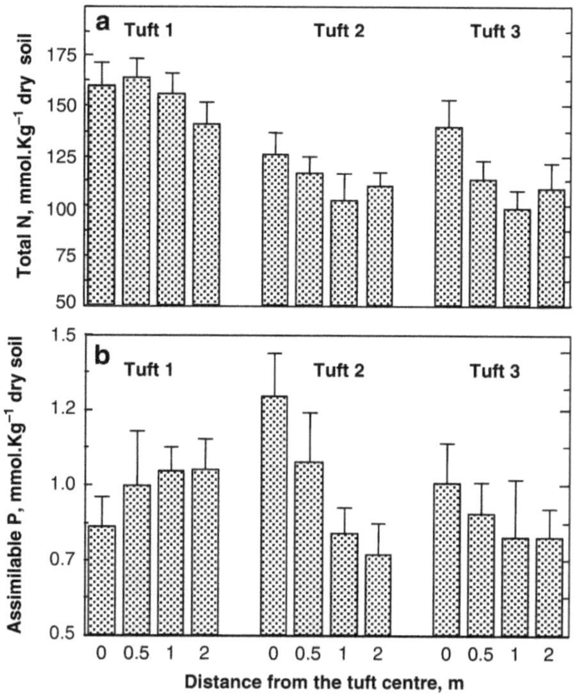

Fig. 4 Variations in soil N and P with distance from *Arthrocnemum indicum* tufts. (**a**) total N. (**b**) assimilable P. The samples were extracted from 0–20 cm deep horizon. Tuft 1 was situated at the sabkha periphery, tufts 2 and 3 were near its centre. Means of 4 replicates ± S.E. at $^*P < 0.05$

nor with distance from the tufts (Fig. 3a). Exchangeable potassium concentration decreased from the periphery towards the sabkha centre (Fig. 3a) and with distance from the halophytes, though non significant. This gradient could be ascribed to substitution of K^+ by Na^+ in clay, which was abundant outside the tufts. Total soil nitrogen ranged from 1 to 2.5 mg N g^{-1} soil, equivalent to 71–178 mmol N kg^{-1} dry soil (Fig. 4a). Generally, significantly (at $P < 0.05$) larger amounts were found in tuft 1, i.e. higher nitrogen availability at the sabkha outer boundary. Moreover, total nitrogen decreased with distance from the tuft centre. These gradients, more or less pronounced according to the tuft size, reveal the importance of the litter for the soil nitrogen balance. Phosphorus concentration was variable among tufts (Fig. 4b). The upper soil horizon at the centre of tufts 2 and 3 contained more assimilable phosphorus than that farther from the tuft, suggesting that its availability was increased by soil enrichment with organic matter.

3.4 Ecosystem Productivity

Biomass production. Two plant groups contributed to ecosystem biomass production: the perennial halophytes (*Arthrocnemum indicum*, *Salsola cruciata* and *Suaeda fruticosa*) and the annuals, mainly the Leguminosae (*Medicago* sp.), the Poaceae (*Hordeum* sp.) and the Compositeae. Maximal biomass production of the whole ecosystem was about 50% higher in the rainy year than in the dry year, especially in samplings 5 and 6 (Fig. 5a).

The perennial halophytes consisted of dry as well as green twigs throughout the year. The green organ contribution to the total biomass production was predominant only during the spring, reaching 50–65% in *Arthrocnemum indicum*, *Salsola cruciata* and *Suaeda fruticosa*. In the remainder of the year, desiccated twigs part could make up 75% of the biomass (data not shown). While the primary production of *Arthrocnemum indicum* and *Salsola cruciata* was strongly improved by the rainfall (+23% and +45%, respectively), no impact was observed for *Suaeda fruticosa* (Fig. 5b). The rainfall effect was more pronounced for the annuals, especially the Leguminosae, which produced six times as much in the wet year. The proportion of these species to total ecosystem (perennials and annuals) biomass production increased from 15% in the dry year to 40% in the rainy year, of which the Leguminosae (mainly *Medicago ciliaris*, *Medicago polymorpha*, *Medicago truncatula*, and *Medicago minima*) contributed up to 80%.

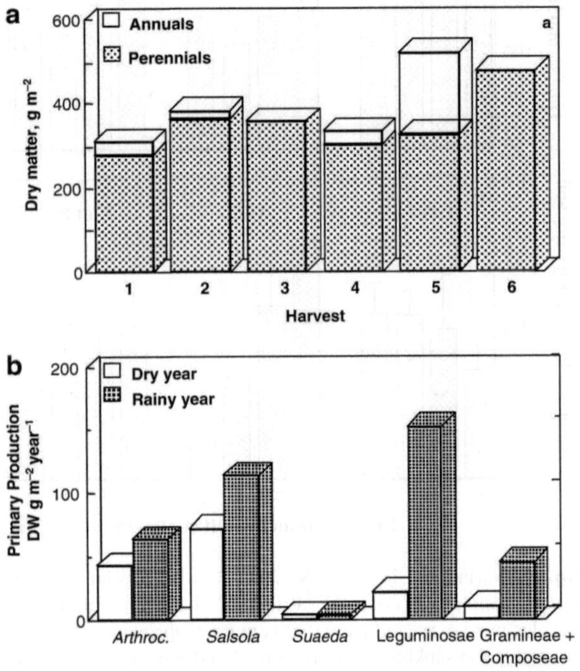

Fig. 5 Biomass production in the sabkha. (**a**) shoot dry weight (DW) per surface area unit as a function of the sampling dates. Annuals: essentially Leguminosae; perennials: halophytes. Samplings 1, 2 and 3: during the dry year, respectively in April, June and September. Samplings 4, 5 and 6: during the rainy year, respectively in January, March and June. (**b**) Primary production (DW) per surface area unit of the different plant types. For the halophytes, the annual production is estimated by the difference between the biomasses at the beginning (January) and the end (June) of the growing period. For the annuals, it corresponds to the spring harvests. Means of 4 replicates ± S.E. at $*P < 0.05$

3.5 Nutrient Uptake

K^+ concentrations ranged between 0.3 and 0.8 mmol · g^{-1} DW in perennial halophyte shoots, being generally higher in the dry year (Fig. 6). Potassium concentration in the green twigs of halophytes was three times that in the desiccated twigs, suggesting a remobilisation or leaching from the former. In annuals, K^+ concentrations ranged from 0.4 to 1 mmol g^{-1} DW, and were higher in the rainy year. Halophyte green twigs contained more nitrogen than the desiccated ones (Fig. 6). The highest N concentrations in *Suaeda fruticosa* (up to 3 mmol g^{-1} DW) were always lower than those in the annuals, especially in the Leguminosae. For the latter, the fluctuations in N concentration were large (a factor two over the two years), depending on rainfall. For phosphorus, Pi concentrations in shoots of all species were relatively low (30–200 μmol g^{-1} DW). In halophytes, and in contrast to K^+ and N, Pi was uniformly distributed between the green and the desiccated twigs (data not shown). Pi concentrations of the annuals, particularly the Leguminosae, were four times higher in the rainy year than in the dry year. In spring, when the annuals attained maximal development, the Leguminosae (mainly including the four *Medicago* species), representing *ca.* 32% of the

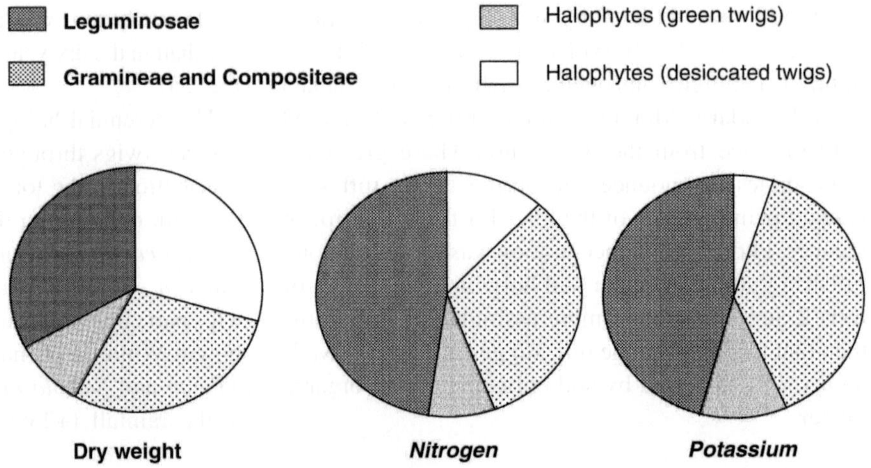

Fig. 6 Contribution of the plant groups to the biomass production and to N and K^+ uptake in the sabkha. Data from the spring harvest during the rainy year, corresponding to the maximal development of the annuals. Each sector area is proportional to concentrations in shoots, per soil surface area unit

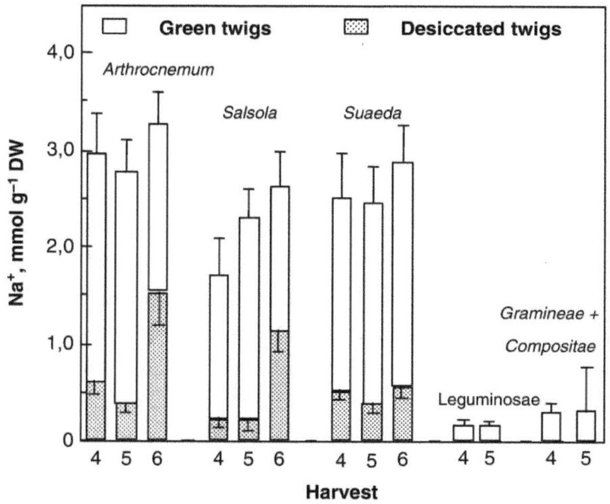

Fig. 7 Sodium contents of the shoots of different species growing in the sabkha. Harvests 4, 5 and 6 were made in January, March and June of the rainy year, respectively. Means of 6 replicates ± S.E. at $*P < 0.05$

total biomass, contained 45% of the total nitrogen and 48% of the potassium (Fig. 6). Thus, the *Medicago* species were more efficient than the other species in the uptake of nutrient resources.

3.6 Na^+ Accumulation

Most of the sodium transported to the plant cover was found in the halophytes. Na^+ strongly accumulated in the halophyte green twigs (up to 3 mmol g^{-1} DW) (Fig. 7), where it likely contributed to their osmotic adjustment. The tissue Na^+ concentration ranged from 0.5 to 1.5 M, while Na^+ concentrations in the desiccated organs represented only 20% of those accumulated in the green ones. Leguminosae and Poaceae tissues contained much lower (at $P < 0.05$) Na^+ levels (0.2–0.3 mmol g^{-1} DW), suggesting that (i) these plants may be provided with mechanisms for protection against sodium invasion, and that (ii) they exploited the low-salt levels superficial zones, while the halophytes owing to deeper roots, exploit more saline soil horizons.

4 Discussion

As reported by Turner (2004), the natural ecosystems of the Mediterranean-climate areas are usually characterised by a mixture of perennial shrubs and trees (mainly xerophytes and halophytes) and annual crops or fodder plants. Growth activity of the annual is governed by the rainfall regime, the rainy season extending from autumn to spring, while the perennials grow slowly, but are able to cope with water shortage and soil/water salinity. Variations in biomass and productivity among and within natural ecosystems may be attributed mostly to differences in water and nutrient (especially N) availability and salinity (Sherman et al. 2003). In the arid to semi-arid regions bordering the Mediterranean Sea, water quality is a major factor limiting crop production (Cantero-Martínez et al. 2007).

In the present study, both soil analyses and vegetation distribution indicated increasing salinity from the periphery to centre of the sabkha. Vegetation cover was more abundant in the peripheral zones, containing the less salt-tolerant perennial halophytes and the annuals. Plant diversity, lower in the depression centre, was dominated by two highly salt-tolerant halophytes: *Arthrocnemum indicum* and *Halocnemum strobilaceum*. Similar heterogeneous distribution of salinity has been reported in other studies (Loveland and Ungar 1983; Bekki 1995), and has been ascribed to salt transport by rainwater towards the basin centre. Our data showed an inverse fertility gradient (notably of nitrogen), decreasing from the periphery to the sabkha centre. The low nitrogen availability in the depression centre could be due to the depressive effect of salinity on soil organic matter accumulation (subsequent to the low plant biomass production) and on microbial activity (Rosenberg et al. 1986; Mc Clung and Frankenberger 1987).

We found that the spontaneous *Medicago* spp., characterised by high nutritive value, largely contributed to the ecosystem primary production in the absence of water shortage. These annuals mainly thrived within or very close to the halophytes tufts, while the soil was almost bare between the tufts. Parallel studies performed in our laboratory pointed out that these spontaneous *Medicago* are particularly sensitive to salinity (Abdelly et al. 1995), and to N (Abdelly 1997) and P (Abdelly et al. 1991) deficiency. Furthermore, shoots of the annuals growing in association with the *Medicago* species contained relatively low Na^+ concentrations (Fig. 7). Hence, the upper soil horizon, where these plants grew, was fertile and contained (relatively) low salt levels, as corroborated by the results of soil analysis (upper soil horizon in the tuft centre was always less saline than when taken at the tuft periphery). Desalinisation of the upper soil horizon by the superficial roots of halophytes could be responsible for this micro-gradient of salinity. Moreover, litter from the halophyte fallen organs and organic debris accumulated by the wind at the feet of tufts, may have contributed to localised soil enrichment in N and P (Fig. 4a, b). Soil aeration near these plants would be improved by this organic matter and by the higher soil level under the tufts, leading to better drainage capacity. Improved soil aeration is favourable for nitrification and N_2 fixation, favouring the colonisation of the halophytes tufts by the *Medicago* spp. Therefore, halophytes directly contributed to maintenance of a relatively low salinity and high fertility in the upper soil horizon, enabling the growth of annuals.

The halophytes may also play an indirect role by developing deep root systems exploiting the more saline soil horizons, as shown by the presence of halophyte roots at 1 m depth and by the vertical increasing salinity gradient (Table 1). The halophytes could thus limit rise of the salty water-table, allowing rainfall water infiltration into the upper soil layers. Several studies have demonstrated the role of halophytes in the soil desalinisation. Zhao (1991) showed that *Salsola salsa* biomass production could reach 20 t ha^{-1}, including 3 to 4 tonnes of salt exported from the soil. Less productive *Batis maritima* displays a similar desalinisation capacity (3 t ha^{-1}) (Le Houérou 1993), as does the succulent *Sesuvium protulacastrum* (Pasternak and Nerd 1996). The latter would be particularly interesting because of its high salt tolerance (growth stimulation up to 800 mM NaCl) despite accumulating high salt levels in the shoots (6 mmol g^{-1} DW, representing ca. 35% of the whole plant biomass) (Messedi et al. 2001). In a 4-year study, Keiffer and Ungar (2002) observed a significant decline in brine-affected soils, following introduction of halophytes (*Atriplex prostrata*, *Spergularia marina*, and *Suaeda calceoliformis*), so that glycophytes could successfully establish. On the other hand, involvement of halophytes in the creation of micro-habitats favouring the development of *Medicago* spp. and their microbial symbionts (namely, root-nodulating and nitrogen-fixing rhizobia), has been reported by Bekki (1995) who showed that *Medicago ciliaris* plants growing in combination with *Suaeda fruticosa*, had a higher growth rates and better nodulation and nitrogen fixation potentialities than the isolated ones.

Recently, Ravindran et al. (2007) assessed the desalinisation capacity of six halophytes (*Suaeda maritima*, *S. portulacastrum*, *Clerodendron inerma*, *Ipomoea pes-caprae*, *Heliotropium curassavicum*, and the tree *Excoecaria agallocha*), in terms of fast growth rate associated with high salt accumulation. All tested species decreased the EC of the saline soil used, in concomitance with an increase of the EC of plant samples. Salt removing capacity calculated over in 4 months of time was 504, 473.9, 396.3, 359.5, 325.2 and 301.5 kg ha^{-1} of NaCl in *S. maritima*, *S. portulacstrum*, *E. agallocha*, *C. inerma*, *I. pescaprae*, and *H. curassavicum* respectively. Similar to our findings (Fig. 7), the desalinisation potential of the species tested was mostly associated with their salt accumulation capacity within their above ground tissues, without adversely impacting the plant growth activity. Salt compartmentalisation in the leaf vacuoles and osmotic adjustment are key mechanisms of plant salt tolerance. Both of these processes enabling the use of Na^+ as a cheap osmoticum rather than its harmful accumulation in the cytosol, often take place in succulent halophytes lacking salt-excreting structures at their leaf surface (Tester and Davenport 2003; Debez et al. 2006).

5 Conclusions

This study revealed substantial soil and vegetation heterogeneity and diversity in the sabkha. In this area, the plant biomass production is largely determined by that of the forage annuals, especially the Leguminosae, growing in localised habitats surrounding bushy halophytes tufts. Thus, ecosystem micro-heterogeneity, combined with large-scale radial heterogeneity, is an important factor of its productivity, in addition to plant diversity. The salt-tolerant perennials improved the soil properties, mainly by decreasing the soil salinity owing to their high salt accumulation capacity. In addition, they provided "islands of fertility", enabling the growth

Table 1 Soil ionic characteristics in the central zone of the studied area in the Enfidha sabkha

	Horizon depth (cm)				
Parameter	0–20	20–40	40–60	60–80	80–100
EC (dS m^{-1})	0.13 ± 0.01	0.13 ± 0.02	1.21 ± 0.15	2.33 ± 0.2	2.62 ± 0.32

EC electric conductivity of the aqueous extract (1 g soil per 10 ml water)

of fodder annuals with high agronomic value. The latter species used the mineral resources accumulated in the upper soil horizons, near the halophytes, owing to a sustained fats growth, associated to a high efficiency of nutrient availability. These data constitute a strong argument in favour of the practical utilisation of halophytes towards the restoration of marginal and/or cultivated salt-affected lands, by reducing the soil salinity level and improving its physical and chemical properties. Long-term data are required to better assess the positive impact of halophytes on the sabkha ecosystem.

References

Abdelly C (1997) Mécanismes d'une association de luzernes spontanées et d'halophytes pérennes en bordure de sabkha. Es-Doctoral Thesis. Science Faculty of Tunis, Tunis, pp 290

Abdelly C, Soltani A, Hajji M, Grignon C (1991) Effet de la déficience en phosphore sur la croissance et la nutrition minérale de trois espèces de *Medicago* provenant de zones salées. Bulletin de la Société des Sciences Naturelles de Tunisie 91(92):1–7

Abdelly C, Lachaâl M, Grignon C, Soltani A, Hajji M (1995) Association épisodique de halophytes strictes et de glycophytes dans un écosystème naturel. Agronomie 15:557–568

Abdelly C, Barhoumi Z, Ghanya T, Debez A, Ben Hamed K, Sleimi N, Ouerghi Z, Smaoui A, Huchzermeyer B, Grignon C (2006) Potential utilisation of halophytes for the rehabilitation and valorisation of salt-affected areas in Tunisia. In: Öztürk M, Waisel Y, Khan MA, Gork G (eds) Biosaline agriculture and salinity tolerance in plants. Birkhäuser Verlag, Switzerland, pp 163–172

Aschenbach TA (2006) Variation in growth rates under saline conditions of *Pascopyrum smithii* (Western wheatgrass) and *Distichlis spicata* (Inland saltgrass) from different source populations in Kansas and Nebraska: implications for the restoration of salt-affected plant communities. Restor Ecol 14:21–27

Bekki A (1995) Effet de la salinité sur la fixation d'azote chez l' association *Rhizobium-Medicago*: sensibilité du partenaire bactérien à l'état libre et sous forme de micro-symbiote. Tentative d'essai au champ. In: Drevon JJ (ed) Facteurs limitant la fixation symbiotique de l'azote dans le Bassin Méditerranéen, vol 77, Les Colloques. INRA, Paris, pp 139–148

Bonneau M, Souchier B (1979) Méthodes pratiques d'analyse du sol. Pédologie. 2, Constituants et Propriétés du Sol. Masson, France

Böer B, Gliddon D (1998) Mapping of coastal ecosystems and halophytes (case study of Abu Dhabi, United Arab Emirates). Mar Freshwater Res 49:297–301

Cantero-Martínez C, Angás P, Lampurlanés J (2007) Long-term yield and water use efficiency under various tillage systems in Mediterranean rainfed conditions. Ann Appl Biol 150:293–305

Choukr-Allah R (1996) The potential of halophytes in the development and rehabilitation of arid and semi-arid zones. In: Choukr-Allah R, Malcolm CV, Hamdy A (eds) Halophytes and biosaline agriculture. Marcel Dekker, New York, pp 3–13

Debez A, Saadaoui D, Ramani B, Ouerghi Z, Koyro HW, Huchzermeyer B, Abdelly C (2006) Leaf H^+-ATPase activity and photosynthetic capacity of *Cakile maritima* under increasing salinity. Environ Exp Bot 57:285–295

Fleury P, Leclerc M (1943) La méthode nitro-vanado-molybdique de Misson pour le dosage colorimétrique du phosphore. Son intérêt en biochimie. Bulletin de la Société de Chimie Biologique 25:201–205

Ghars MA, Debez A, Smaoui A, Zarrouk M, Grignon C, Abdelly C (2006) Variability of fruit and seed-oil characteristics in Tunisian accessions of the halophyte *Cakile maritima* (Brassicaceae). In: Khan MA, Weber DJ (eds) Ecophysiology of high salinity tolerant plants, vol 40, Series: tasks for vegetation science. Springer, Berlin, Heidelberg, New York, pp 55–67

Hafsi C, Lakhdhar A, Rabhi M, Debez A, Abdelly C, Ouerghi Z (2007) Interactive effects of salinity and potassium availability on growth, water status, and ionic composition of *Hordeum maritimum*. J Plant Nutr Soil Sci 170:469–473

Hu Y, Schmidhalter U (2005) Drought and salinity: a comparison of their effects on mineral nutrition of plants. J Plant Nutr Soil Sci 168:541–549

James JJ, Tiller RL, Richards JH (2005) Multiple resources limit plant growth and function in a saline-alkaline desert community. J Ecol 93:113–126

Keiffer CH, Ungar IA (2002) Germination and establishment of halophytes on brine-affected soils. J Appl Ecol 39:402–415

Khan MA, Ansari M, Ali H, Gul B, Nielsen BL (2009) *Panicum turgidum*, a potentially sustainable cattle feed alternative to maize for saline areas. Agric Ecosyst Environ 129:542–546

Le Houérou HN (1993) Salt-tolerant plant for the arid regions of the Mediterranean isoclimatic zone. In: Lieth H, Al Masoom A (eds) Towards the rational use of high salinity tolerant plants. Kluwer Academic, Dordrecht, pp 403–422

Loveland DG, Ungar IA (1983) The effect of nitrogen fertilisation on the production of halophytes in an inland salt marsh. Am Midland Nat 109:346–354

Malcolm CV, Lindley VA, O'Leary JWO, Runciman HV, Barrett-Lennard EG (2003) Halophyte and glycophyte salt tolerance at germination and the establishment shrubs in saline environments. Plant Soil 253:171–185

Mc Clung G, WTJr F (1987) Nitrogen mineralisation rates in saline vs. salt amended soils. Plant Soil 104:13–21

Messedi D, Sleimi N, Abdelly C (2001) Salt tolerance in *Sesuvium portulacastrum*. In: Horst WJ (ed) Plant nutrition-food security and sustainability of agro-ecosystems. Kluwer Academic, Dordrecht, pp 406–407

Pasternak D, Nerd A (1996) Research and utilization of halophytes in Israel. In: Choukr-Allah R, Malcolm CV, Hamdy A (eds) Halophytes and biosaline agriculture. Marcel Dekker, New York, pp 325–348

Ravindran KC, Venkatesan K, Balakrishnan V, Chellappan KP, Balasubramanian T (2007) Restoration of saline land by halophytes for Indian soils. Soil Biol Biochem 39:2661–2664

Reynolds MP, Mujeeb-Kazi A, Sawkins M (2005) Prospects for utilising plant-adaptive mechanisms to improve wheat and

other crops in drought- and salinity-prone environments. Ann Appl Biol 146:239–259
Rosenberg RJ, Christensen NW, Jackson TL (1986) Chloride, soil solution osmotic potential, and soil pH effects on nitrification. Soil Sci Soc Am J 50:941–945
Sargeant MR, Tang C, Sale PWG (2008) The ability of *Distichlis spicata* to grow sustainably within a saline discharge zone while improving the soil chemical and physical properties. Aust J Soil Res 46:37–44
Sherman RE, Fahey TJ, Martinez P (2003) Spatial patterns of biomass and aboveground net primary productivity in a mangrove ecosystem in the Dominican Republic. Ecosystems 6:384–398
Tester M, Davenport R (2003) Na^+ tolerance and Na^+ transport in higher plants. Ann Bot 91:503–527
Turner NC (2004) Sustainable production of crops and pastures under drought in a Mediterranean environment. Ann Appl Biol 144:139–147
Weber DJ, Ansari R, Gul B, Khan MA (2007) Potential of halophytes as source of edible oil. J Arid Environ 68:315–321
Zhao K (1991) Desalinisation of saline soils by *Sueada fruticosa*. Plant Soil 135:303–305

Potential Role of Salt Marshes in the Sabkhas of Egypt

Hassan M. El Shaer

Abstract Salinization is one of the major problems for food production in Egypt. It has significantly impaired agricultural productivity. The total area of salt affected soils is estimated to be 1.8 million ha. These are mainly located along the Red sea coast, in the Mediterranean part, some areas in middle, western and eastern parts of the Nile delta, El Fayoum, Wadi El Natroun and the Oasis in the western desert area. Out of these saline areas salt marshes are important habitats for grazing animals, waterfowl and fish. They are integral components of the coastal and inland ecosystems of the country with a great potential as a source for many raw materials like food for human, animal feeds, fiber materials, habitat for fish, insects,etc.). However these habitats are under threat due to the uncontrolled human and orther biotic interferences. An attempt will be made here to give some evaluations on these salt marshes and their potential role in the Sabkhas of Egypt.

1 Introduction

Salt marshes cover considerable areas in the Middle East region with altitudes ranging from – 300 m below sea level to almost 200 m above sea level. In Egypt these occur along the Red Sea coastal belt in South Sinai, East of the eastern desert, and in the vicinity of lakes in the Mediterranean part. There are large depressions and Siwa, Moghra, Wadi EL-Natroun and Bahariya, Kharga, Dakhla and Dungul oases in the Western Desert, which are accepted as the most productive ecosystems in Egypt. These are mainly inhabited by halophytes, therefore revegetation of these habitats with halophytes could be highly beneficial for all living beings in the area and lessen the risk of degradation (El Shaer and Zahran 2002; El Shaer 2009). This paper will focus on the characteristics of salt marshes and their potential role in the sabkhas of Egypt.

2 Salt Marshes in Egypt

There are three major types of salt marshes in Egypt namely; Red Sea, Mediterranean, and Oases (El-Khouly and Zahran 2002). The area from Suez to Mersa Halaib (near Sudan) is covered by the Red Sea salt marshes. The precipitation in the form of rains varies here between 4 and 25 mm per year. The soils are arid with insufficient leaching and salts form crusts on the surface (Younes et al. 1983), which thus effects the growth of plants, their zonation and distribution along the littoral part. The plant cover here is mainly composed of mangroves, reed swamps and salt marshes (Zahran 1993), dominated by *Avicennia marina, Rhizophora mucronata, Phragmites australlis, Typha domingensis, Arthrocnemum macrostachyum, Halocnemum strobilaceum, Halopeplis perfoliata, Zygophyllum album, Limonium axillare, Sporobulus spicatus, Halopyrum mucronaum, Nitraria retusa, Suaeda monoica, Atriplex farinosa, Juncus rigidus, Imperata cylindrica, Anabasis setifera and Tamarix nilotica.*

H.M. El Shaer (✉)
Desert Research Center, 1 Mathaf El Mataria Street, P.O. Box 11753, Mataria, Cairo, Egypt
e-mail: drc_elshaer@hotmail.com

M. Öztürk et al. (eds.), *Sabkha Ecosystems*, Tasks for Vegetation Science 46,
DOI 10.1007/978-90-481-9673-9_11,

The Mediterranean salt marshes are found in the vicinity of Bardawil, Manzalah, Mariout and Buroulos lakes occurring along the coast of Mediterranean Sea. These are generally oligotrophic (El Shaer 2009). Some like Bardawil are hypersaline. Biotic interferences, salinity and some other environmental parameters are the most important factors affecting the distribution of plants here with large zonation (Zahran and Willis 1992). More than 25 taxa of halophytes have been recorded from these habitats. The dominating ones are *Salsola tetrandra, Nitraria retusa, Zygophyllum album, Halocnemum strobilaceum, Arthrocnemum macrostachym, Suaeda aegyptiaca,* and *Sarcocornia fruiticosa* (Khedr 1999; Khedr and Zahran 2002). The areas with 2–5% of salinity in general lack any plant cover but wet sabkhas are coverede by some vegetation (Shaheen 1998). These lake systems are important habitat for a lrge number of migratory birds. These salt marshes are found in a region with extreme drought and are at the same time under a heavy presure due to grazing and trampling by Bedouin herds as well as severe sand shifting.

The Oases salt marshes characterized by large depressions occur in the Western Desert including northern oases like Siwa, Moghra, Wadi EL-Natrun and Bahariya and southern oases like Kharga, Dakhla and Dungul. The soils are calcareous with high $CaCO_3$ content and lower organic carbon (El-Khouly 2001; El-Khouly and Zahran 2002). These are dominated by different types of plant covers like iced swamp, salt marshes, sand dunes and desert plains. A total of 32 taxa have been recorded from here and 46.9% of these belong to the families Poaceae and Chenopodiaceae. The most common genera was *Tamarix* spp., *Juncus* spp., *Cyperus laevigatus, Alhagi graecorum, Sporobolus spicatus, Zygophyllum album, Suaeda monica, Salsola imnbricate, Nitraria retusa, Sarcocornia fruticosa, Desmostachya bipinnata, Cressa cretica,* and *Aeluropus lagopoides*. These can be grouped under nine growth forms such as; rhizomatous (31.3%), succulents (25%), non-succulents (12.5%), stoloniferous (9.4%), non succulent herbs (6.3%), thorny shrub, leafless trees and succulent herb (3.1%). Many of these are characterised by salt lands and some have modified their organs for water storage (El-Khouly and Khedr 2000).

3 Salt Marshes as Potential Habitats

The wetlands like river Nile, estuaries (Damietta and Rosetta), coastal and inland swamps are dominated by reeds and rushes like *Phragrmites*, *Typha* and *Juncus* species. All these taxa tolerate salinity to different degrees (Serag et al. 2002). They provide nesting sites, substratum, and feeding materials for a wide range of organisms and are important part in the food chain of freahwater ecosystems as well as a source of nutrient rich freshwater habitats in river Nile and other lakes, drainage and irrigation canals. Average growth of these plants helps in the oxygenation of water and thus permits survival of fishes and invertebrates (Zahran and Willis 2002; El Shaer 2009). *Phragmites* is widely distributed in Nile, Oases, Mediterranean and Red Sea coastal zones, Desert area and Sinai (Boulos 1995). It is valued as animal fodder and fuel, and can be used to renovate wastewaters (Öztürk et al. 2005). The two species of *Typha* (*T. Domingensis, T. Elephantina*) found in Egypt are important as fish habitats if an overgrowth does not take place.They too play an important role in the nutrient cycling and renovation of wastert-ers (Serag et al. 1999; Öztürk et al. 2005). In addition to these two dominating taxa there are nine species of *Juncus* distributed in the salt marshes (Mohammed 1980; El Shaer 2009), majority being salt tolerant flourishing under hot and dry climate. These cab be used to produce fiber for a good quality paper; for making mats, baskets and sandals; leaves and culms have been used as pens for writing on papyrus (Zahran 1993; Snogerup 1993; Öztürk et al. 2008 ; El Shaer 2009). They can be evaluted at the industrial scale because their seeds are rich in oils, proteins, amino-acids, and carbohydrates (Serag et al. 2002; El Shaer 2009). The obligate halophytes *Inula crithmoides, Suaeda vera, Arthrocnemum macrostachyum, Halocnemum strobilaceum* and *Atriplex portulacoides* accompany these rush species in different habitats. These halophytes can provide 'indirect' and economical benefits like reclamation of degraded lands, CO_2 sequestration, and can be used as forage/fodder, oil-seed crops, fertilizers, in soap industry, building materials, wood for furniture, timber, paper production, herbal tea, sea floor fixation, in lanscape architectural projects, fire wood and medicinal plants (Menzel and Leith 1998; Squires and Ayoub 1994; El Shaer 2009).

4 Prospects of Halophytes as Animal Feeds

The halophytes serve as important areas of primary production for coastal food chains and important habitat for the production of grazing animals, waterfowls and fish. The palatable halophytic taxa like the species of *Suaeda*, *Atriplex*, *Salsola* and *Nitraria retusa* are very common in the salt marshes and disappear fast due to overgrazing (El Shaer 1981). They are valuable fodders during drought season. Their nutritive value differs from species to species and several factors are invloved in this, the highest forage values are found during the wet season of the year. The nutritive value decreases with growth and maturation.The digestibility is generally higher in grazing season than in drought season (El Shaer 2009). Although many halophyte taxa are high in protein content they should not be offered as sole diet to animals for long time.The performance of animals on the halophytic ranges depends on factors like animal species, season of the year, forage abundance, and nutritional values of forage species. In addition to some physical and chemical defenses of the plants, main factors involved in poor intake are high Na, Ca and silica contents and higher levels secondary metabolites (El Shaer and Attia- Ismail 2002). In general feeding small ruminants with good quality halophytes can prove economical for the farmers.

The salt marshes in Egypt are facing a severe threat from the biotic pressures. Some of the elements in the biodiversity of these habitats has vanished and some are endangered. There is a great need for the protection of these valuable habitats through suitable conservation measures and sustainable use of its biodiversity in particular the halophytes (Szabolcs 1994; Öztürk et al. 2006). For this purpose several approaches can be used like gene bank preservation system, in-situ conservation of endangered species, and cultivation of palatable and other multi-purposes halophytic taxa on saline marshes by using saline and brackish waters.

References

Boulos L (1995) Flora of Egypt checklist. Al Hadara, Gairo, pp 203

El-Khouly AA (2001) Plant diversity in the dry land habitats of Siwa Oasis in the Western desert of Egypt. J Env Sci, Mansoura University 22:125–143

El- Khouly AA, Zahran MA (2002) On the ecology of the halophytic vegetation of the oases in Egypt. Proceedings of the International Symposium on the optimum resources utilization in salt-affected ecosystems in arid and semi-arid regions, pp 277–286, 8–10 Apr 2002, Cairo, Egypt

El-Khouly AA, Khedr AA (2000) Species diversity and phenology in the wetland vegetation of Sewa Oasis, in the Western desert of Egypt. Desert Inst Bull Egypt 50:239–258

El Shaer HM (1981) A comparative nutrition study on sheep and goats grazing Southern Sinai desert range with supplements. Ph.D. Thesis, Fac. Agric. Ain Shams Univ. Egypt

El Shaer HM (2009) Potential role of Sabkhas in Egypt: an overview. In: Ashraf et al (eds) Salinity and water stress – improving crop efficiency, Tasks for vegetation Science – 44. Springer Science and Business Media BV, the Netherlands, pp 221–228

El Shaer HM, Attia Ismail SA (2002) Halophytes as animal feeds: potentiality, constraints, and prospects. Proceedings of international symposium on the optimum resources utilization in salt-affected ecosystems in arid and semi-arid regions, pp 411–418, 8–10 Apr 2002, Cairo, Egypt

El Shaer HM, Zahran MA (2002) Utilization of halophytes in Egypt: an overview. In: Proceedings of the international conference on Halophyte utilization and regional sustainable development of agriculture. Huanghua, Shijiazhnag, China, pp 14–20, Sept 2001

Khedr AA (1999) Floristic composition and phytogeography in a Mediterranean deltaic lake (Lake Burollos), Egypt. Ecologia Meditternea 25:1–11

Khedr AA, Zahran MA (2002) The salt marsh visitation of lake Bardawil, North Sinai, an overview. Proceedings international symposium on the optimum resources utilization in salt-affected ecosystems in arid and semi-arid regions, pp 339–345, 8–10 Apr 2002, Cairo, Egypt

Menzel U, Leith H (1998) Tabulation of halophytes reported as utilized in different publications and handbook. In: Hamdy A, Leith Fl, Todorovic M, Moschenko M (eds) Halophytes uses in different climates, Biometerology Ii. Bbackuys, Leiden, pp 127–133

Mohammed NE (1980) Studies in the genus *Juncus* in Egypt. M.Sc. Thesis, Fac. Sci., Cairo Univ. Egypt

Öztürk M, Alyanak I, Sakcali S, Guvensen A (2005) Multipurpose plant systems for renovation of waste waters. Arab J Sci Eng 30(2C):17–28

Öztürk M, Guvensen A, Gork G, Gork C (2006) An overview of the coastal zone plant diversity and management strategies in the Mediterranean region of Turkey. In: Öztürk et al (eds) Biosaline agriculture and salinity tolerance in plants. Birkhauser Verlag, Basel Switzerland, pp 89–100

Öztürk M, Guvensen A, Gücel S (2008) Ecology and economic potential of halophytes – a case study from Turkey. In: Kafi, Khan (eds) Crop and forage production using saline waters. NAM S & T Centre, Daya, House-Delhi, India, pp 255–264

Serag MS, Khedr AA, Zahran MA, Willis AJ (1999) Ecology of some aquatic plants in polluted water courses, Nile Delta, Egypt. J Union Arab Biol 9(B):85–97

Serag MS, Zahran MA, Khedr AA (2002) Ecology and economic potentialities of the dominant salt-tolerant reeds and rushes in the nile delta. Proceedings of International Symposium on the optimum resources utilization in salt-affected ecosystems in arid and semi-arid regions, pp 245–256, 8–10 Apr 2002, Cairo, Egypt

Shaheen SE (1998) Geoenvironmental studies on Bardawil lagoon and its surroundings, North Sinai, Egypt. PhD Thesis, Faculty of Science, Mansoura University, Mansonra, Egypt

Snogerup S (1993) A revision of *Juncus* subgen. *Juncus* (*Juncaceae*). Willdenowia 23:23–73

Squires VR, Ayoub AT (1994) Halophytes as resource for livestock and for rehabilitation of degraded lands. In: Proceedings of international workshop on halophytes for reclamation of saline wasteland and as a resource for livestock: problems and prospects. Kluwer Academic, London, pp 315

Szabolcs I (1994) Salt affected soils as ecosystem for halophytes. In: Squires VR, Ayoub AT (eds), Halophytes as a resource for livestock and for rehabilitation of degraded lands, Kluwer Academic. pp 19

Zahran MA (1993) *Juncus* and *Kochi* fiber- and fodder-producing halophytes under salinity and aridity stress. In: Pessarakli M (ed) Plant and crop stress. New York, Basal, Hong Kong, pp 505–528

Zahran MA, Willis AJ (2002) The plant life of the River Nile in Egypt. Marc, Reyadh, Suadi Arabia

Halophyte Plant Diversity, Coastal Habitat Types and Their Conservation Status in Cyprus

M. Öztürk, S. Gucel, A. Guvensen, C. Kadis, and C. Kounnamas

Abstract This chapter focuses on the identification, description and determination of the conservation status of the halophytic plant diversity and the coastal habitat types of Cyprus. The chapter presents the results of a study that was undertaken during 2006–2008 in the coastal zone of Cyprus, which revealed that a total of 457 plant taxa with distinct features are distributed along a 770 km long coastline; 129 being typical halophytes. The taxa dominating the major intact habitats along the coast are *Limonium* spp., *Mesembryanthemum* spp., *Euphorbia paralias, Pancratium maritimum, Eryngium maritimum, Medicago marina, Taraxacum aphrogenes, Otanthus maritimus* and *Cakile maritima.* The saline habitats along the coast and alongside the internationally important salt lakes of Larnaca and Akrotiri (both included in the Ramsar list of Wetlands of International Importance) abound in halophytic taxa like *Salicornia* spp., *Arthrocnemum microstachyum, Suaeda vera, Juncus* spp., *Halimione portulacoides* and *Inula crithmoides*. These salt lakes serve as migration stations for a large number of migratory birds during their route from Europe to Africa and back. Their conservation value is related not only to their function as important bird habitats but also to their significant floristic diversity. The coastal waters of the island give shelter to 197 fish species and various species of crabs, sponges and echinoderms. The Green and the Loggerhead turtles breed regularly on the island's sandy beaches, some of which are strictly protected. The coastal zone has been seriously degraded over the last 25 years. The habitats in this zone face serious degradation problems due to economic, recreational and intensive tourist development. An attempt has been made here to present the information on the life forms and other features of the major plant taxa occuring within the coastal zone. The conservation status of the plant taxa has been identified according to the Red Data Book of the Flora of Cyprus and the criteria set by the IUCN: 6 taxa critically endangered, 12 endangered and 14 vulnerable. Moreover, in the coastal zone of Cyprus 16 habitats types that are included in the Habitats Directive (92/43/EEC) Annex I have been identified out of which 3 are classified as priority ones and need immediate conservation measures. Information on coastal endemics has also been included so that it can be utilised for future effective conservation action.

M. Öztürk and A. Guvensen
Ege University, Fen Fakultesi, A Blok Botany Department, E Blok 35100, Bornova, Izmir Turkey
e-mail: munirozturk@gmail.com; aguvensen@hotmail.com

S. Gucel (✉)
Near East University, Institute of Environmental Sciences, Nicosia, Cyprus
e-mail: sgucel@hotmail.com

C. Kadis and C. Kounnamas
Frederick University, Nature Conservation Unit, Nicosia, Cyprus
e-mail: pre.kc@fit.ac.cy; ckounnamas@hotmail.com

1 Introduction

Cyprus is an island lying at a strategic location in the East Mediterranean with an area of 9,250 km^2 and 770 km long coastline, having Turkey in the north, Syria and Lebanon in the west and Egypt in the south. The indented coastline of the island -the third largest in the Mediterranean - is full of cliffy or rocky parts interchanging with small or large sandy or gravelly beaches. It is 240 km long from end to end and 100 km

M. Öztürk et al. (eds.), *Sabkha Ecosystems*, Tasks for Vegetation Science 46,
DOI 10.1007/978-90-481-9673-9_12,

wide at its widest point. The name of the island is said to have originated from the Greek word for the Mediterranean Cypress tree or from the Greek name of the henna plant. However, some historians report that the name stems either from the Eteocypriot or Sumerian word for copper (zubar) or for bronze (kubar), because large deposits of copper ore are found in the country. Cyprus has given its name to the classical latin word for copper via overseas trade through the phrase *aes Cyprium*, i.e. metal of Cyprus, later shortened to *Cuprum – Copper*.

The extinction of dwarf hippos and elephants depicts the arrival of first humans (Simmons 2001) on the island. The earliest site of human activity is said to lie in the southern coastal part Aetokremnos. The site shows remains of hunter-gatherers dating back to 10000 BC (Mithen 2005). The remains of the oldest known settlements on the island date from 7000 to 3900 BC around the well preserved neolithic village Khirokitia (Wade 2007; Walton 2004). These remains bear witness to the existence of an ancient civilization which developed along the North and South coasts. Most Chalcolithic settlements (3900–2500 BC) are found in Western Cyprus. One comes across here the oldest water wells in the world dating back to 9,000–10,500 years (BBC 2009).

The data published on the biodiversity of Cyprus reveals that 7, 26 and 357 species of land mammals, amphibians/reptiles and birds, respectively, distribute on the island. The coastal waters abound of various species of crabs, sponges and echinoderma and give shelter to 197 fish species. The largest wild animal that still lives on the island is the Cyprus moufflon (*Ovis aries ophion*), a rare type of wild sheep found only in Cyprus. The sandy beaches on the coastal zone are typical breeding habitats for Green turtle (*Chelonia mydas*) and Loggerhead turtle (*Caretta caretta*). The island is used by millions of birds during their migration from Europe to Africa and back, the main reason being the presence of Larnaca and Akrotiri Salt Lakes. Cyprus is one of the few places on earth where the geological processes and microclimate have played an important role in the development of its habitat diversity and thus the rich biodiversity (BBC 2009).

With its approximately 2000 taxa of flowering plants, it is an extremely interesting island, which make it a botanist's paradise. The island's isolation has allowed an evolution of a strong endemic flora. This richness is further strenghtened by an incorporation of botanical elements from the neighbouring land masses, where 8% of the indigenous plants (145 taxa) are endemics. The aim of this review is to enlighten the environmental threats faced by the coastal zone plant diversity in Cyprus and present information for a geographical/taxonomical/ecological database of European coastal zone plants, thus contributing towards effective conservation action during the future planning programmes. The list of threatening factors has also been compiled.

2 Geology, Hydrology, Soils and Climate

Geomorphologically, the island is characterised by the ophiolitic complex of large Troodos mountain range displaying a variable relief with a predominance of slopes and rounded peaks as well as prevalence of vertical cliffs, steep slopes, gorges and deep gullies. It covers most of the southern and western parts of the island and accounts for roughly half of its area. The highest peak is Mount Olympus (1952 m above sea level), located in the center of this range. The small Kyrenia mountain range extends along the northern coastline. Both are dominating the central Mesaoria plain. There are six districts in the country (IPI 2007); Nicosia (Lefkosia-Lefkosa), Famagusta (Ammochostos-Gazimagusa), Kyrenia (Keryneia-Girne), Larnaca (Larnaka-Larnaka/Iskele), Limassol (Lemesos-/Leymosun) and Paphos (Pafos/Bafos-Baf/Gazibaf) (Fig. 1). The country is traversed by rivers and small and large streams, the largest being Pediaios, Gialias and Serrachis which rise from Troodos. The significant natural salt lakes of Larnaca and Akrotiri accummulate large volumes of water. There are also smaller lagoons like Paralimni and Galateia-Mehmetcik. The largest water reservoirs are Kouris and Asprokremmos, the former being at the homonymous river with a capacity of 115 million m^3 and the later on the Xeros river with a capacity of 51 million m^3. The indented coastline is full of small or extensive cliffy or rocky parts interchanging with small or large sandy or gravely beaches (Pantelas et al. 2003).

The area of arable land of the island is approximately 15% and the irrigated land is about 400 km^2. For the purposes of this study soil samples collected randomly and analysed at a soil laboratory in 2008.

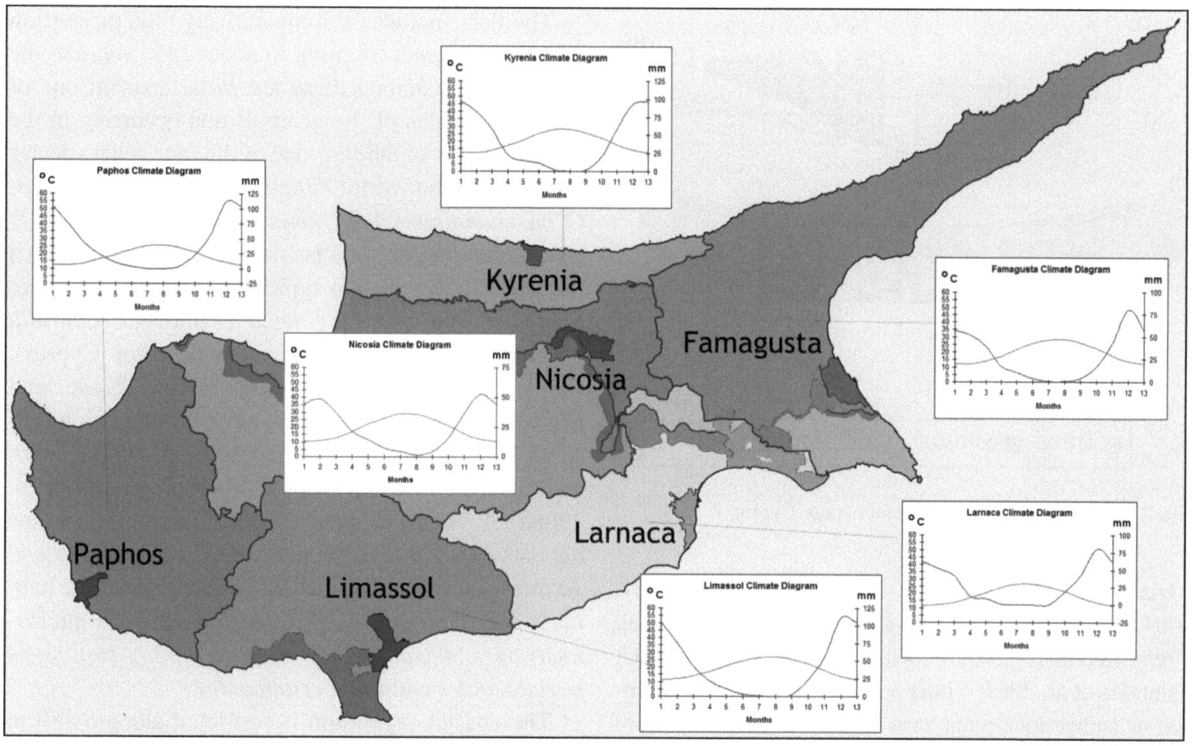

Fig. 1 Map of Cyprus with districts and their climate diagrams

The results obtained showed that the water potential of soils around coastal zone varies between 26.42–61.9%. The pH was found to be between 7.1 and 8.9. The texture is mainly sandy, soluble salts vary between 0.01 and 2.56%, organic matter from trace level to 1.94%, $CaCO_3$ between 22.2 and 52.5%, cation exchange capacity between 1.63–50 me/100 gr and Sodium between 50 and 6,300 ppm.

The climate is typical mediterranean with average daytime temperature lying around 32°C from June to September, 16°C from December to February and 25°C in the other 5 months. Rainfall is common in winter and precipitation in the form of snow occurs from January to March on the mountain peaks (Fig. 1).

3 Plant Diversity

Cyprus was known as the "Green Island" because most of it was covered with forests. Over time, this forest vegetation has been destroyed, mainly due to human activities, such as expansion of cultivations and human settlements, misuse and over exploitation, intensive grazing, fires and recently, climatic change. The varied microclimate and geology of the island has resulted in a great variety of habitats, which together with its isolation and location near the big neighbouring land masses has contributed a lot to its rich biodiversity. These factors have allowed the evolution of strong endemic flora which makes it an extremely interesting place for botanists (Gucel and Yildiz 2008). Nearly 8% (145 taxa) of the indigenous plants are endemics. The earliest systematic studies on the island were carried out by Sibthorp (1758–1796), Hawkins (1758–1841) and Bauer (1760–1826), and were followed by many others (cf. Yildiz and Gucel 2006). However, detailed studies were undertaken by Meikle (1977, 1985) who compiled the collected data in two volumes. He was followed by a large number of local investigators (Alziar 1985; Alziar and Guittonneau 2004; Alziar and Christodoulou 2006a, b; Christodoulou 2003; Christodoulou 2006a, b; Christodoulou and Hadjikyriakou 2006; Christodoulou et al. 2006; Chrtek and Slavik 1993; Costa et al. 1984a, b; Della and Iatrou 1995; Della 1994, 1996; Georgiades 1994; Georgiades and Hadjikyriakou 1993; Gucel and Yildiz 2008; Hadjichambis and Della 2007; Hadjichambis et al. 2003a, b, c; 2004; Hadjikyriakou et al. 1996; Hadjikyriakou et al. 2003; Hadjikyriakou and

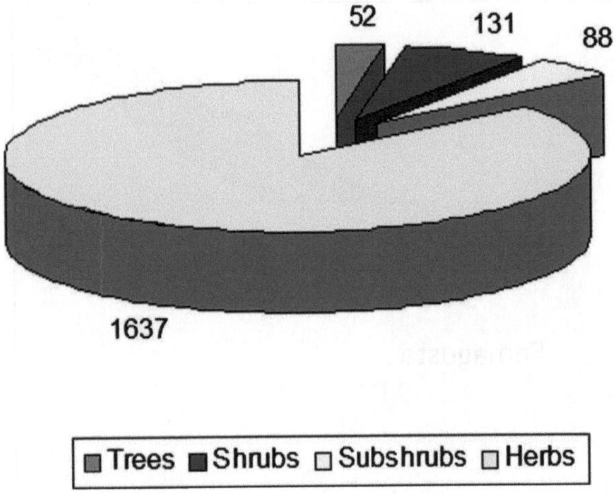

Fig. 2 Number of indigenous plant taxa in Cyprus

Alziar 2006; Hadjikyriakou 2009; Hand 2000, 2001, 2004, 2006; Hand and Hadjikyriakou 2009; Iacovou et al. 1995; Kephalas 2005, 2006a, b; Pantelas et al. 1997; Tsintides et al. 2007; Yildiz and Gucel 2006). The number of indigenous plant taxa recorded until now from Cyprus exceeds 1900 (Fig. 2).

The plant cover in general is composed of Forest, Maquis, Garrigue and Phrygana, accounting for 42.3% of the total land area and other high forests found on the Troodos and Pentadaktylos mountain ranges and along the coast of the Akamas peninsula in the west, the Akrotiri peninsula in the south, Cape Gkreko in the southeast, the Karpasia peninsula in the east, as well as along the northern coasts make up the 43.8%. These include species very common in the Mediterranean basin like *Pinus brutia* and *P. nigra* ssp. *pallasiana* accompanied by species like *Juniperus phoenicea, J. foetidissima, Olea europaea, Ceratonia siliqua, Arbutus andrachne, Myrtus communis, Rhus coriaria, Pistacia lentiscus, P. terebinthus, Berberis cretica, Sorbus aria* ssp. *cretica, Rosa chionistrae, Cotoneaster racemiflorus* var. *nummularia, Smilax aspera* and *Cistus* species. The most important are the natural forests of *Cedrus brevifolia* and *Quercus infectoria* ssp. *veneris*. The common broadleaved species are *Platanus orientalis, Alnus orientalis, Salix alba, Nerium oleander, Arundo donax, Laurus nobilis* and *Rubus sanctus*. The forests of the *Cupressus sempervirens* occur on mountainous areas, chiefly on limestone and marls and rarely on igneous formations. The most common shrubs are *Ziziphus lotus, Crataegus azarolus, Sarcopoterium spinosum* and *Thymus capitatus*.

The flora includes a comparatively high proportion of endemic plants counting to about 145. Some of the most eminent among these are *Bosea cypria*; one of the three species of the genus Bosea occurring in the world; *Cedrus brevifolia;* one of the four cedar species occurring in the world; *Tulipa cypria, Crocus cyprius, C. hartmannianus* and *Quercus alnifolia.*

Another important constituent of the indigenous flora of Cyprus are the typical plants of the Eastern Mediterranean like *Pinguicula crystallina, Cyprinia gracilis* (a monotypic genus named after Cyprus), *Colchicum troodi, Glaudosciadium cordifolium,* and *Euphorbia thompsonii*. The halophytic vegetation is of special interest and high conservation value (Guvensen and Öztürk 2003: Guvensen et al. 2006; Öztürk et al. 2006a, b). This type of vegetation flourishes on saline habitats along the coast and around the Larnaca and Akrotiri salt lakes. Characteristic species of these habitats are *Salicornia* spp., *Arthrocnemum microstachyum, Suaeda vera, Juncus* spp., *Halimione portulacoides* and *Inula crithmoides*.

The coastal vegetation is restricted along a 150 m wide narrow belt, where the dominant plants are *Limonium* spp., *Mesembryanthemum* spp., *Euphorbia paralias, Pancratium maritimum, Eryngium maritimum, Medicago marina, Taraxacum aphrogenes, Otanthus maritimus, Cakile maritima, Crithmum maritimum* and *Matthiola tricuspidata* (Gehu et al. 2000; Uslu 1998, 2001). The wetland vegetation that colonizes areas of the salt lakes, river mouths and other marshy places consists of *Arthrocnemum macrostachyum, Salicornia fruticosa, S. europaea, Halocnemum strobilaceum, Phragmites australis* and other species. The dominating halophytic taxa, their ecological features and a list of the families with the highest number of taxa are presented in Table 1 and Fig. 3.

The families with the highest number of taxa are given in Fig. 3. Out of a total of 129 taxa, *Poaceae* has the highest number. The families with lower taxa are; Apiaceae (5), Caryophyllaceae (5), Juncaceae (5), Brassicaceae (4), Gentianaceae (4), Liliaceae (4), Aizoiaceae (3), Convolvulaceae (2), Euphorbiaceae (3), Papaveraceae (2), Tamaricaceae (3), Frankeniaceae (2), Plantaginaceae (2), Agavaceae (1), Amaryllidaceae (1), Apocynaceae (1), Geraniaceae (1), Illecebraceae (1), Juncaginaceae (1), Linaceae (1), Neuradaceae (1), Orobanchaceae (1), Polygonaceae (1), Zannichelliaceae (1), Zygophyllaceae (1).

Table 1 List of halophytes and their ecological features

Families	Taxa	Life Form	Ecotype	Altitude (m above sea-level)	Flowering Time
Aizoaceae	*Aizoon hispanicum* L.	T	PH	0–70	3–5
	Mesembryanthemum crystallinum L.	C	X	SL	3–5
	Mesembryanthemum nodiflorum L.	C	X	SL	3–5
Agavaceae	*Agave sisalana* Perrine ex Engelm.	C	X	SL	8
Amaryllidaceae	*Pancratium maritimum* L.	C	PH	SL	8–10
Apiaceae	*Crithmum maritimum* L.	C	X	SL	4–7
	Cachyrs scabra (Fenzl) Meikle	CH	PH	SL-30	4–6
	Eryngium maritimum L.	C	PH	SL	6–8
	Ferulago syriaca Boiss.	CH	X	150–300	5–6
	Pseudorlaya pumila (L.) Graunde	T	X	SL	3–6
Apocynaceae	*Trachomitum venetum* (L.) Woodson	C	PH	SL	6–8
Asteraceae	*Achillea cretica* L.	CH	X	15–150	4–7
	Achillea maritima (L.) Ehrend. & Y.P. Guo ssp. *maritima*	CH	PH	SL	6–8
	Aetheorhiza bulbosa (L.) Cass.	C	X	0–400	4–6
	Ambrosia maritima L.	CH	X	SL	7–12
	Chlamydophora tridentata (Delile) Less.	T	HG	0–300	2–5
	Helichrysum conglobatum (Viv.) Steud.	CH	X	0–700	3–5
	Inula crithmoides L.	CH	HG	SL	6–8
	Launea resedifolia (L.) O.Kuntze	CH	X	SL	3–5
	Reichardia picroides (L.)Roth	H	X	SL-30	2–11
	Taraxacum aphrogenes Meikle	H	X	SL	10–12
Brassicaceae	*Cakile maritima* Scop.	T	PH	SL	2–7
	Enarthrocarpus arcuatus Labill.	T	X	SL-30	3–5
	Malcolmia nana (DC.) Boiss. var. *glabra* Meikle	T	PH	SL	3
	Matthiola tricuspidata (L.) R. Br.	T	PH	SL	2–5
Caryophyllaceae	*Silene colorata* Poiret var. *decumbens* (Bjv.) Rohrb.	T	X	0–800	3–6
	Silene discolor Sibth. et Sm.	T	PH	SL	3–5
	Silene kotschyi Boiss. var. *maritima* Boiss.	T	PH	SL	6–7
	Spergularia bocconii (Scheele) Aschers. et Graebn.	T	X	0–760	2–6
	Spergularia marina (L.) Griseb.	T	XH	SL	3–6
Chenopodiaceae	*Arthrocnemum macrostachyum* (Moric.) C. Koch	CH	HA	SL	5–7
	Arthrocnemum perenne (Miller) Moss	CH	HA	SL	4–7
	Atriplex halimus L.	CH	XH	0–150	7–10
	Atriplex patula L.	T	X	SL	5–7
	Atriplex prostrata Boucher ex DC.	T	X	SL	7–10
	Bassia hirsuta (L.) Ascherson	T	X	SL	7–9

(continued)

Table 1 (continued)

Families	Taxa	Life Form	Ecotype	Altitude (m above sea-level)	Flowering Time
	Beta vulgaris L. ssp. *maritima* (L.) Arcangeli	T	X	0–150	2–5
	Halimione portulacoides (L.) Aellen.	CH	HA	SL	6–10
	Halocnemum strobilaceum (Pall.) Bieb.	CH	HA	SL	8–10
	Halopeplis amplexicaulis (Vahl.) Ung.-Sternb. ex Ces.	T	HA	SL	7–9
	Salicornia europaea L.	T	HA	SL	5–10
	Salicornia fruticosa L.	CH	HA	SL	9–11
	Salsola inermis Forssk.	T	XH	0–150	9–11
	Salsola kali L.	T	PH	SL	5–9
	Salsola soda L.	T	PH	SL	5–9
	Suaeda aegyptiaca (Hasselq.) Zohary	CH	HA	SL	8–10
	Suaeda maritima (L.) Dumort.	T	HA	SL	8–10
	Suaeda vera Forssk.	CH	HA	0–30	10–5
Convolvulaceae	*Cressa cretica* L.	H	XH	SL	6–10
	Ipomoea imperati (Vahl) Griseb.	H	P	SL	7–11
Cyperaceae	*Bolboschoenus maritimus* (L.) Palla	C	HG	0–1500	4–9
	Carex divisa Huds.	C	HG	0–1700	3–5
	Carex extensa Good.	H	HG	SL	6–8
	Carex flacca Schreber ssp. *serrulata* (Bir.) Greuter	C	HG	0–1675	3–9
	Cyperus capitatus Vand.	C	HG	SL	2–5
	Eleocharis palustris (L.) Roem. & Schult.	C	HG	0–1500	3–6
	Isolepis cernua (Vahl) Roem. & Schult.	T	HG	SL-250	4–5
	Scirpoides holoschoenus L.	C	HG	0–1900	4–9
	Scirpus lacustris L. ssp. *tabernaemontani* (C.C.Gmel.) Syme	C	H	SL-1300	3–6
	Schoenoplectus litoralis (Schrad.) Palla	C	HG	SL	4–7
	Schoenus nigricans L.	H	HG	0–500	3–7
Euphorbiaceae	*Euphorbia paralias* L.	CH	PH	SL	2–11
	Euphorbia peplis L.	T	PH	SL	5–7
	Euphorbia terracina L.	CH	PH	0–300	2–6
Fabaceae	*Acacia saligna* (Labill.) Wendl. fil.	P	HG	0–100	4–7
	Alhagi maurorum Medik var. *Turcorum*	CH	PH	SL	6–8
	Argyrolobium uniflorum Jaub. & Spach	CH	X	SL	3–4
	Coronilla repanda (Poir.) Guss. ssp. *repanda*	T	PH	SL	2–4
	Lotus halophilus Boiss et Spruner	T	PH	SL	3–4
	Medicago littoralis Rohde ex Loisel. var. *littoralis*	T	X	0–200	3–5
	Medicago marina L.	H	PH	SL	2–4
	Ononis diffusa Ten.	T	PH	SL	3–6

Frankeniaceae	*Frankenia hirsuta* L. var. *hispida* (DC.) Boiss.	CH	PH	SL	4–10
	Frankenia pulverulenta L.	T	PH	0–200	4–10
Gentianaceae	*Blackstonia perfoliata* (L.) Hudson	T	HG	0–700	6–7
	Centaurium tenuiflorum (Hoffmanns. & Link) Fritsch	T	HG	0–800	6–8
	Centaurium maritimum (L.) Fritsch	T	X	SL-400	4–5
	Centaurium erythraea Rafn. ssp. *rhodense* (Boiss. et Reut.) Melderis	T	HG	0–1065	5–7
Geraniaceae	*Erodium crassifolium* L'Her.	CH	HG	SL	3–5
Illecebraceae	*Paronychia argentea* Lam.	H	PH	0–1500	12–7
Juncaceae	*Juncus acutus* L.	C	HG	0–1150	3–5
	Juncus heldreichianus Marsson ex Parl.	C	HG	0–1800	4–8
	Juncus littoralis C.A. Mey.	C	HG	0–150	4–6
	Juncus maritimus Lam.	C	HG	SL	7–8
	Juncus rigidus Desf.	C	HG	0–150	6–8
Juncaginaceae	*Triglochin bulbosa* L.	C	HG	SL	3–5
Liliaceae	*Allium curtum* Boiss. & Gaillard	C	PH	0–30	4–5
	Allium trifoliatum	C	XH	0–300	3–5
	Asparagus stipularis Forssk.	C	X	0–700	3–6
	Asphodelus tenuifolius Cav.	T	X	SL	3–4
Linaceae	*Linum maritimum* L.	H	HG	SL	5–10
Neuradaceae	*Neurada procumbens* L.	T	P	SL	3–6
Orobanchaceae	*Cistanche helypaea* (L.) P. Coutinho	H	HA	SL	3–5
Papaveraceae	*Glaucium flavum* Crantz	H	X	SL	5–7
	Hypecoum procumbens L.	T	X	0–300	3–6
Plantaginaceae	*Plantago coronopus* L. ssp. *commutata* (Guss.) Pilger	T	PH	0–800	2–7
	Plantago maritima L. ssp. *crassifolia*	T	PH	0–150	3–10
Plumbaginaceae	**Limonium albidum* (Guss.) Pignatti ssp. *cyprium* Meikle	H	X	SL	6–11
	Limonium echioides (L.) Mill. ssp. *echioides*	T	HA	0–230	4–6
	Limonium echioides (L.) Mill. ssp. *exaristatum*	T	HA	0–200	4–7
	Limonium meyeri (Boiss.) O.Kuntze	T	HA	SL	7–10
	Limonium mucronulatum (H. Lindb. fil.) Greuter&Burdet	H	HA	SL	6–9
	Limonium sinuatum (L.) Mill.	H	HA	SL	3–7
	Limonium virgatum (Willd.) Fourr.	H	HA	0–160	5–8

(continued)

Table 1 (continued)

Families	Taxa	Life Form	Ecotype	Altitude (m above sea-level)	Flowering Time
Poaceae	*Ammophila arenaria* (L.) Link	C	PH	SL	5–7
	Arundo donax L.	C	HG	0–600	6–10
	Aeluropus lagopoides (L.) Trin.	C	PH	0–160	4–7
	Aeluropus littoralis (Gouan) Parl.	C	PH	SL	4–7
	Brachypodium distachyon (L.) P. Beauv.	T	X	0–400	3–6
	Bromus madritensis L.	T	X	0–1000	3–7
	Catapodium marinum (L.) *C.E.* Hubb.	T	PH	0–100	4
	Elymus farctus (Viv.) Runemark ssp. *farctus* var. *farctus*	C	PH	SL	6–8
	Hordeum marinum Hudson	T	PH	0–100	5–6
	Imperata cylindirica L.	C	PH	0–300	4–6
	Lagurus ovatus L.	T	PH	0–50	4–6
	Parapholis incurva (L.) C.E. Hubb.	T	HG	SL	4–5
	Parapholis marginata Runemark	T	HG	0–230	4–5
	Phragmites australis (Cav.) Trin. ex Steudel	H	HG	0–1500	8–10
	Polypogon maritimus Willd.	T	PH	0–150	5–6
	Sporobolus virginicus (L.) Kunth.	C	PH	SL	6–9
	Trachynia distachya (L.) Link.	T	PH	0–1500	3–5
	Triplachne nitens (Guss.) Link	T	P	SL	4
	Vulpia brevis Boiss & Kotschy	T	PH	SL	3–4
Polygonaceae	*Polygonum maritimum* L.	H	PH	SL	6–11
Tamaricaceae	*Tamarix smyrnensis* Bunge	P	HA	0–1000	4–8
	Tamarix tetragyna Ehrenb.	P	HA	0–300	2–4
	Tamarix tetrandra Pall. ex M. Bieb.	P	HA	0–1300	5
Zannichelliaceae	*Zannichellia palustris* L.	T	HG	0–1500	3–6
Zygophyllaceae	*Zygophyllum album* L.	CH	XH	SL	3–5

Life Forms: *C* = Cryptophytes, *CH* = Chamaephytes, *H* = Hemicryptophytes, *P* = Phanerophytes, *T* = Therophytes
Ecological Types: *HA* = Halophyte, *HG* = Hygrohalophytes, *PH* = Psammohalophytes, *X* = Xerophyte, *XH* = Xerohalophytes
* Endemic
SL = Sea-level
Flowering Time: Month 1 is January and month 12 is December

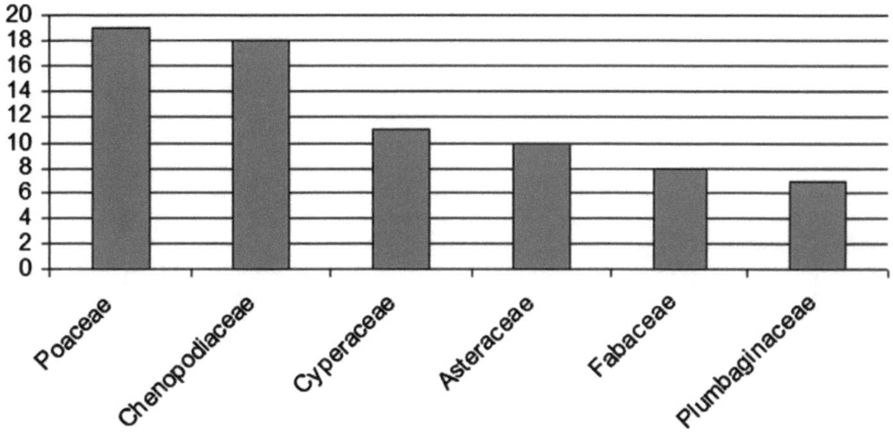

Fig. 3 Families with the highest number of taxa

4 Endemic, Rare and Threatened Plants of Cyprus

Biodiversity in Cyprus is threatened by a number of factors, the principal ones being the expansion of built-up areas, tourism development, heavy tourist use of sandy coasts, trampling of beaches with machinery installation of tourism infrastructure, sand removal, habitat degradation, recreation activities, stochastic events, demographic changes, fires, deforestations, agricultural activities, agricultural expansion, over-grazing, road maintenance, off-road driving, invasion by exotic species like *Acacia saligna*, changes in water balance, drought, habitat degradation and loss, changes in land use, drying out of watercourses and climatic change (Fig. 4a).

A Red Data List for the Cyprus flora compiled by Tsintides et al. (2007) indicates that about 300 plant species are threatened and need immediate protection measures. In this list *Achillea cretica* (Fig. 4b) was recorded as vulnerable (Table 2), but no mention was made for *Pancratium maritimum* (Fig. 4c) and *Trachomitum venetum* (Fig. 4d) which, in our opinion, should be included in the list as endangered species.

Approximately 145 taxa are endemics and some of these are very rare. Twenty-three endemic and other native plants are listed in Appendix I of the Convention for the Conservation of the European Wild Life and the Natural Habitats (Council of Europe 2000) and they are charactewrised as strictly protected. 18 species (16 endemics and 2 other native plants) are of community importance and are listed in Appendices II & IV of the European "Habitats Directive" (92/43/EEC), which provides for the conservation of species through the conservation of their habitats (Council of Europe 1992). It is noted that Appendix II includes the threatened animal and plant species of European Union interest, whose conservation requires the designation of special areas of conservation, while Appendix IV includes animals and plants of European Union interest in need of strict protection. Also, 51 species are included in *the Red List of the Threatened Plants* prepared by IUCN (Walter and Gillett 1998). Moreover, six endemic plants of Cyprus are included in the IUCN & SSC - Species Survival Commission edition " *The Top 50 Plants of the Mediterranean Islands*" which presents fifty of the most threatened plant taxa of the Mediterranean (Montmollin and Strahm 2005). The above information indicades that a significant number of endemic and other plants on the island are part of world's natural heritage.

The situation on the coastal habitats in the country is not different. During the studies carried out between 2006–2008 along 770 km long coastal zone, a total of 457 taxa were found in this area. Out of these plant taxa, 129 halophytic taxa were surveyed at length and their life forms, ecological types, altitudinal distribution and phenology were recorded. The conservation status of 32 of these species is described according to the Red Data Book of the Flora of Cyprus (Tsintides et al. 2007), which evaluates the conservation status of the Cyprus flora based on the criteria set by the International Union for Conservation of Nature (IUCN 2003); 6 were found to be critically endangerd, 12 endangered and 14 vulnerable (Table 2, Fig. 5).

Regarding the habitats diversity observed within the coastal zone, a total of 16 habitat types were

Fig. 4 Land degradation along the coast-Bafra (**a**), *Achillea cretica*-Vulnerable species (**b**), *Pancratium maritimum*-Endangered species (**c**), *Trachomitum venetum*-Endangered species (**d**)

Table 2 Conservation status of the Halophytes of Cyprus

Latin name	Threatened category: IUCN Criteria (IUCN 2003)
Achillea cretica	VU: D2
A. maritima subsp. *maritima*	VU: A4C
Aizoon hispanicum	EN: D1
Ambrosia maritima	CR: B1 ab (i–v) + 2ab(i–v); C2a(i); D1
Ammophila arenaria	EN: B1ab(iii,v) + 2ab(iii,v)
Argyrolobium uniflorum	CR: B1ab(iii,v) + 2ab(iii,v)
Cachyrs scabra	EN: B1ab(iii,v) + 2ab(iii,v)
Centaurium maritimum	EN: B1ab(iii) + 2ab(iii)
Cistanche phelypaea	CR:D1
Coronilla repanda subsp. *repanda*	VU: D2
Enarthrocarpus arcuatus	VU:D2
Erodium crassifolium	VU:D2
Euphorbia paralias	EN:A4ac
Ferulago syriaca	VU:D2
Ipomoea imperati	EN:C2a(i)
Isolepis cernua	EN:D1
Juncus littoralis	VU:D2
Juncus maritimus	VU: D1+2
Limonium mucronulatum	CR: B1ab(iii,v) + 2ab(iii,v); C2a(ii)
Linum maritimum	VU:D2
Malcolmia nana var. *glabra*	CR: B1ab(iii,iv) + 2ab(iii,iv); C2a(ii)
Mesembryanthemum crystallinum	VU:D2
Neurada procumbens	EN:B1ab(iii,v) + 2ab(iii,iv)
Ononis diffusa	EN:B1ab(iii,v) + 2ab(iii,v)
Reichardia picroides	VU:D1+2
Salsola soda	VU:D1+2
Scirpus lacustris ssp. *tabernaemontani*	EN:D1
Silene kotschyi var. *maritima*	EN:B1ab(iii,v) + 2ab(iii,v)
Suaeda aegyptiaca	EN:B1ab(iii,v) + 2ab(iii,v)
Taraxacum aphrogenes	VU:B1ab(iii,v) + 2ab(iii,v)
Triplachne nitens	VU:C2a(i)
Vulpia brevis	CR:B1ab(i–v) + 2ab(i–v)

Fig. 5 Percentage of endangered coastal plants among endangered plants of Cyprus

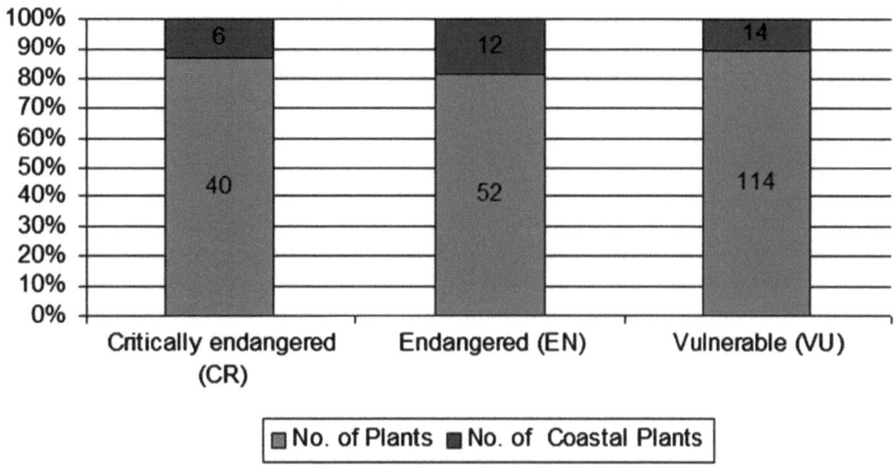

Table 3 Description of habitats

No	Code Name	Importance
1	1110 Sandbanks which are slightly covered by sea water all the time	HD I
2	1140 Mudflats and sandflats not covered by seawater at low tide	HD I
3	1150 Coastal lagoons	HD I*
4	1210 Annual vegetation of drift lines	HD I
5	1240 Vegetated sea cliffs of the Mediterranean coasts with endemic *Limonium* spp.	HD I
6	1310 Salicornia and other annuals colonising mud and sand	HD I
7	1410 Mediterranean salt meadows (Juncetalia maritimi)	HD I
8	1420 Mediterranean and thermo-Atlantic halophius scrubs (Arthrocnemetalia fructicosae)	HD I
9	1430 Halo-nitrophilous scrubs (Pegano-Salsoletea)	HD I
10	1510 Mediterranean salt steppes (*Limonietalia*)	HD I*
11	2110 Embryonic shifting dunes	HD I
12	2190 Humid dune slacks	HD I
13	2230 Malcolmietalia dune grasslands	HD I
14	2240 Brachypodietalia dune grasslands with annuals	HD I
15	2250 Coastal dunes with Juniperus spp.	HD I*
16	2260 Cisto-Lavendulatalia dune sclerophyllous scrubs	HD I

classified as habitat types of the Annex I of the Habitats Directive, according to the Interpretation Manual of European Union Habitats - EUR 25. Out of these habitats, three were found to be priority ones (Table 3). It is noted that Annex I includes the natural habitat types of European Union interest, whose conservation requires the designation of special areas of conservation. As priority habitat types are defined the natural habitat types in danger of disappearance, which are present on the EU territory and for the conservation of which the EU has particular responsibility in view of the proportion of their natural range which falls within its territory.

Code_Name: Included in Annex I of Directive 92/43/EEC, HD I – included in Annex I, HD I* – priority habitat of Annex I.

In the ecological survey of coastal shores, dunes and cliffs in Europe, under the title "Dry Coastal Ecosystems of the World" a lot of information has been presented on the occurrence of endemic littoral plant species (van der Maarel 1996). An evaluation of this data as well as other detailed surveys together with present studies were used for an evaluation of the coastal zone plant diversity in Cyprus.

This evaluation includes plant taxa from salt marshes, tidal flats and saline waters i.e. lagoons. Although species like *Ammophila arenaria* and *Hippophae rhamnoides* ssp. *maritima* and *Puccinellia maritima* are important for the dunes and salt marshes, many coastal species have limited role in the composition and structure of littoral plant communities. Moreover, many species dominating coastal zone habitats do not belong to these habitats at all such as *Phragmites australis*, *Salix* spp., *Juniperus* and *Quercus* species.

5 Conclusion

We do not always realize how relatively small the areas occupied by coastal ecosystems are and how vulnerable the coastal flora is. We hope that this survey will prove helpful in the preservation of coastal zone plant diversity and habitats. The coastal ecosystems harbour numerous littoral plant taxa of which the majority can be considered endemics in a strict sense. For most of these species their future existence seems to very uncertain, since initiatives to protect rare and threatened coastal species are insufficient. The data presented here will serve as a reference source to build up a detailed database for littoral plant diversity in Cyprus.

References

Alziar G (1985) Contribution a l'histoire naturelle de l'ile de Chypre. La flore. Biocosme Mésogéen 2:1–20

Alziar G, Guittonneau GG (2004) Compte rendu des journées d'études de la Société Botanique de France à Chypre (5-12 avril 2001 et 2-9 mai 2002). J Bot Soc Bot France 25:5–25

Alziar G, Christodoulou C (2006a) Geum urbanum. In: Hand R (ed) Supplementary notes to the flora of Cyprus V. Willdenowia 36(2):770

Alziar G, Christodoulou C (2006b) Phillyrea latifolia. In: Hand R (ed) Supplementary notes to the flora of Cyprus V. Willdenowia 36(2):785

BBC (2009) Stone Age wells found in Cyprus. Retrieved 2009-07-31. http://news.bbc.co.uk/1/hi/world/europe/8118318.stm

Christodoulou C (2006a) Pinguicula crystallina. In: Hand R (ed) Supplementary notes to the flora of Cyprus V. Willdenowia 36(2):787

Christodoulou C (2006b) Salsola soda. In: Hand R (ed) Supplementary notes to the flora of Cyprus V. Willdenowia 36(2):792

Christodoulou C, Hadjikyriakou G (2006) Juncus littoralis. In: Hand R (ed) Supplementary notes to the flora of Cyprus V. Willdenowia 36(2):798

Christodoulou C, Hadjikyriakou G, Papachristophorou T (2006) Najas marina subsp. armata. In: Hand R (ed) Supplementary notes to the flora of Cyprus V. Willdenowia 36(2):800

Christodoulou CS (2003) The impact of Acacia saligna invasion on the autochthonous communities of the Akrotiri salt marshes. BSc Thesis, University of Central Lancashire, Cumbria Campus Newton Rigg, Penrith:0020100 pp

Chrtek J, Slavík B (1993) Contribution to the flora of Cyprus 2. Fl Medit 3:239–259

Costa M, Géhu JM, Peris JB, Biondi E, Arnold N (1984a) Donées sur la végétation maritime des cotes meridionales de l' île de Chypre (plages, dunes, lacs sales et falaises). Documents Phytosociologiques 8:343–364

Costa M, Géhu JM, Peris JB, Biondi E, Arnold N (1984b) Sobre la vegetation thermomediterranea littoral de la Isla del Chipre. Documents Phytosociologiques 8:365–376

Council of Europe (1992) Council Directive 92/43/EEC of 21 May 1992 on the conservation of natural habitats and of wild fauna and flora. Official Journal L 206, 007–050

Council of Europe (2000) Convention on the Conservation of European Wildlife and Natural Habitats (ETS No. 104), Appendices to the Convention. Status in force since 3 March 2000

Della A (1994) On the cultivated flora of Cyprus. Nicosia. doi:10.2307/4110645

Della A, Iatrou G (1995) New plant records from Cyprus. Kew Bull 50:387–396

Della A (1996) Cyprus flora in mythology, history and beauty, pp 53–57. In: Proceedings of the consultation meeting on neglected plant genetic resources with a landscape and cultural importance for the Mediterranean region, Naples, Italy, 7–9 November 1996

Gehu JM, Costa M, Uslu T (2000) Phytosociological research of coastal vegetation of North Cyprus with concern for preservation. 2. Int Cyprus Stud Congr 4:365–373

Georgiades C (1994) I epigenis chlorida tis Kyprou, taxinomiki, chloridiki, fytogeografiki, oikofysiologiki meleti [The adventive flora of Cyprus, taxonomic, floristic, phytogeographic, ecophysiological study]. Ph.D Thesis, Athens University

Georgiades T, Hadjikyriakou G (1993) Centaurea akamantis (Compositae), a new species from Cyprus. Willdenowia 23:157–162

Gucel S, Yildiz K (2008) Morphological investigations and transplantation attempts on some endemic species of Northern Cyprus. Pakistan J Bot 40(4):1399–1410

Guvensen A, Gork G, Öztürk M (2006) An overview of halophytes in Turkey. In: Khan A, Boer B, Kust GS, Barth HJ (eds) Sabkha ecosystems, Vol II. Springer, Netherlands, pp 9–30

Guvensen A, Öztürk M (2003) Halophytic plant diversity of South Aegean coastal zone in Turkey. Pakistan J Bot 35:853–864

Hadjichambis ACh, Della A (2007) Ecology of threatened coastal ecosystems of Cyprus. Agricultural research institute – research promotion foundation, Nicosia, Cyprus, pp 412

Hadjichambis ACh, Baker RM, Della A (2003a) Contribution to the knowledge of the flora and vegetation of coastal sand dunes of Cyprus, pp 348–349. In: Proceedings of hellenic society for biological sciences 25th annual conference, Mytilene, Greece, May 2003

Hadjichambis ACh, Dimopoulos P, Georghiou K (2003b) Conservation Biology of threatened coastal sand dune habitats of Cyprus, pp 350–351. In: Proceedings of hellenic society for biological sciences 25th annual conference, Mytilene, Greece, May 2003

Hadjichambis ACh, Paraskeva-Hadjichambi D, Della A, Dimopoulos P, Georghiou K (2003c) Vulnerable coastal habitats of Cyprus, pp 352–353. In: Proceedings of hellenic society for biological sciences 25th annual conference, Mytilene, Greece, May 2003

Hadjichambis ACh, Della A, Paraskeva-Hadjichambi D, Georghiou K, Dimopoulos P (2004) Flora of the sand dune ecosystems of Cyprus. In: Arianoutsou M, Papanastasis VP (eds) Proceedings of the 10th MEDECOS conference, Rhodes, Greece, 25 April–1 May 2004. Millpress, Rotterdam

Hadjikyriakou G, Orphanos G, Georghiou K, Kadis CC (1996) Cyprian flora ten years after the edition of "Flora of Cyprus. Proceedings of the 6th scientific meeting of the hellenic botanical society, Paralimni, Cyprus, pp 213–218

Hadjikyriakou G, Orphanos J, Alziar G (2003) Ranunculus repens. In: Hand R (ed) Supplementary notes to the flora of Cyprus III. Willdenowia 33:306

Hadjikyriakou G, Alziar G (2006) "2005": peucedanum kyriakea (Apiaceae), a new species from Cyprus. Biocosme Mésogéen 22:177–183

Hadjikyriakou G (2009) Symvoli sti meleti tis chloridas tis Kyprou 13. Dasoponos 39:7–10

Hand R (ed) (2000) Contributions to the flora of Cyprus I. Willdenowia 30:53–65

Hand R (ed) (2001, 2004, 2006) Contributions to the flora of Cyprus. II, IV, V. Willdenowia 31:383–409, 34:427–456. doi:10.3372/wi.34.34210, 36:761-809, doi:10.3372/wi.36.36211

Hand R, Hadjikyriakou G (2009) Cynara makrisii (Asteraceae, Cardueae), a new artichoke species in Cyprus. Willdenowia 39:77–81. doi:10.3372/wi.39.39108

Iacovou NG, Loizidou XI, Hulsbergen CH, Hoozemans FMJ (1995) Coastal zone management for Cyprus. Delft Hydraulics Publication no. 489. Waterloopkundig Laboratorium Delft Hydraulics, Emmeloord, The Netherlands, 12 pp

IPI (2007) Deeply concerned over criminal defamation charges brought against daily newspaper in Northern Cyprus. http://www.freemedia.at/cms/ipi/statements_detail.html?ctxid=CH0055&docid=CMS1168350896599

IUCN (2003) Guidelines for application of IUCN Red List Criteria at regional levels: Version 3.0. Gland. IUCN Species Survival Commission, Switzerland and Cambridge, UK

Kephalas K (2005) *Trachomitum venetum* (L.) Woodson, a rare plant of Cyprus. Dasoponos (Forester) 24:11

Kephalas K (2006a) The species *Eryngium campestre L. in Cyprus*. Dasoponos (Forester), 29: 5.

Kephalas K (2006b) Sympliromatika Stoicheia gia ti Chlorida stin Karpasia. Dasoponos 27:13–16

Meikle RD (1977, 1985). Flora of Cyprus, Vol. 1, 2. The Bentham,-Moxon Trust Royal Botanic Gardens, Kew

Mithen S (2005) After the Ice: a global human history, 20000 BC–5000 BC. Harvard University Press, Boston

Montmollin Bde, Strahm W (2005) The top 50 mediterranean island plants: wild plants at the brink of extinction, and what is needed to save them. IUCN/SSC Mediterranean Islands Plant Specialist Group. IUCN, Gland, Switzerland and Cambridge, UK, x + 110 pp

Öztürk M, Guvensen A, Gork C, Gork G (2006a) An overview of coastal zone plant diversity and management strategies in the mediterranean region of Turkey. In: Öztürk M et al (eds) Biosaline agriculture & salinity tolerance in plants, pp. 89–100. Birkhauser Verlag (Springer Science), Basel

Öztürk M, Waisel Y, Khan MA, Gork G (2006b) Biosaline agriculture and salinity tolerance in plants. Birkhauser Verlag, Basel, Switzerland, 205 p

Pantelas V, Hadjikyriakou G, Christodoulou S. Ch, Papachristophorou T, Makris Ch, Orphanos J, Kadis CC (1997) Additions to the flora of Cyprus. Abstracts of the first Balkan Botanical Congress

Pantelas V, Papachristophorou T, Christodoulou P (2003) Cyprus flora in color. The Endemics, 2nd edn. Kailas Printers and Lithographers Ltd, Athens

Simmons AH (2001) In: Swiny S (ed) The earliest prehistory of Cyprus from colonization to exploitation. Cyprus American archaeological research institute monograph series, Vol. 2. American Schools of Oriental Research Publications, Boston

Tsintides T, Christodoulou CS, Delipetrou P, Georghiou K (2007) The red data book of the flora of Cyprus. Cyprus Forestry Association, Nicosia, Cyprus

Uslu T (1998) The coastal areas of biological richness of international value in the Northern Cyprus. Turkish Foundation Publication, Ankara, pp 278–279

Uslu T (2001) Research of ethnobotanical and medical plants of Northern Cyprus forest and coastal flora. Prime Ministers Office Project Turkey, Ankara, pp 258

van der Maarel E, van der Maarel-Versluys M (1996) Littoral vascular plant species and habitat diversity of Cyprus and their conservation status. J Coastal Conserv 2:73–92

Wade N (2007) Study traces cat's ancestry to Middle East. The New York Times

Walter KS, Gillett HJ (1998) 1997 IUCN red list of threatened plants. Compiled by the World Conservation Monitoring Centre. IUCN – The World Conservation Union, Gland, Switzerland and Cambridge, UK, lxiv + 862pp

Walton M (2004) Ancient burial looks like human and pet cat. CNN. http://edition.cnn.com/2004/TECH/science/04/08/cats.cyprus/index.html

Yildiz K, Gucel S (2006) Chromosome numbers of 16 endemic plant taxa from northern Cyprus. Turk J Bot 30:181–192

Salt-Affected Soils and Their Native Vegetation in Hungary

Tibor Tóth

Abstract Approximately 13% of Hungary is considered to be salt-affected and with this large extent it is unique in Europe. There are large areas of naturally saline and sodic soils, but also secondary salinization is known to occur. Due to the geological and hydrological conditions, the country demonstrates the most characteristic features of natural continental (not marine) salinization, sodification and alkalinization. Since the most important direct source of soil salinization is the shallow groundwater level below the lowland surface, there is a chance of irrigation-related salinization in two dominant situations: when the abundant use of river waters causes waterlogging and rise of saline groundwater (salinization from below); and when typically saline tubewell-waters are used for irrigation (salinization from above). The spatial assessment of salt-affected areas began with the systematic mapping of salt-affected areas. There are a series of ten maps describing different aspects (salt-affected soil types, vegetation types, salt-efflorescences) of the salinity-status nation-wide from 1897 onward, with the latest survey finished last year. Besides the national scale of 1:500,000, soil salinity is also mapped at the scale of 1:100,000 on the "AGROTOPO" map sheets and 1:25,000 in the "Kreybig"-practical soil information (spatial vector data for maps and database for profiles and borings) systems. In spite of the two systems being digitally available, the most detail information collected at the scale of 1:10,000 is available only for 2/3 of the country and is not digitised. Very early maps at field scale, later at regional scale showed numerical salinity/sodicity values. At present field scale numerical maps are analysed in order to optimise salinity mapping in space and time. Parallel to soil studies, the assessment of the vegetation of saline and sodic lands is a traditional topic of Hungarian botanists. The vegetation of these areas is used for millennia by grazing and provides medicine and raw material for several purposes.

T. Tóth (✉)
Research Institute for Soil Science and Agricultural Chemistry of the Hungarian Academy of Sciences (RISSAC), Budapest, Hungary
e-mail: tibor@rissac.hu

1 Introduction

Traditionally the study of salt-affected soils (SAS) is one of the most popular topics among Hungarian soil scientists. The origin, properties and reclamation of these soils (in Hungarian "szik") were investigated thoroughly during the last two centuries. A complete list of the 22 monographs on salt-affected soils is reported by Tóth and Szendrei (2006). The mapping of these soils started in 1897, mapping at the scale of 1:25,000 was finished by the 1950s, and their last assessment, now of the areas covered with native halotolerant vegetation, was carried out in the years 2003–2006 (Bölöni et al. 2003), see Fig. 2. This summary is based on Tóth (2008).

2 Environmental Conditions in Hungary

About one third of the soils on the Great Hungarian Plain (N 46–48.5° and E 19–22.5°) are affected by salinity/sodicity, mainly by sodification, one third of the territory is covered by potential SAS, and one third does not have such soils. Potential SAS are defined as

M. Öztürk et al. (eds.), *Sabkha Ecosystems*, Tasks for Vegetation Science 46,
DOI 10.1007/978-90-481-9673-9_13, © Springer Science+Business Media B.V. 2011

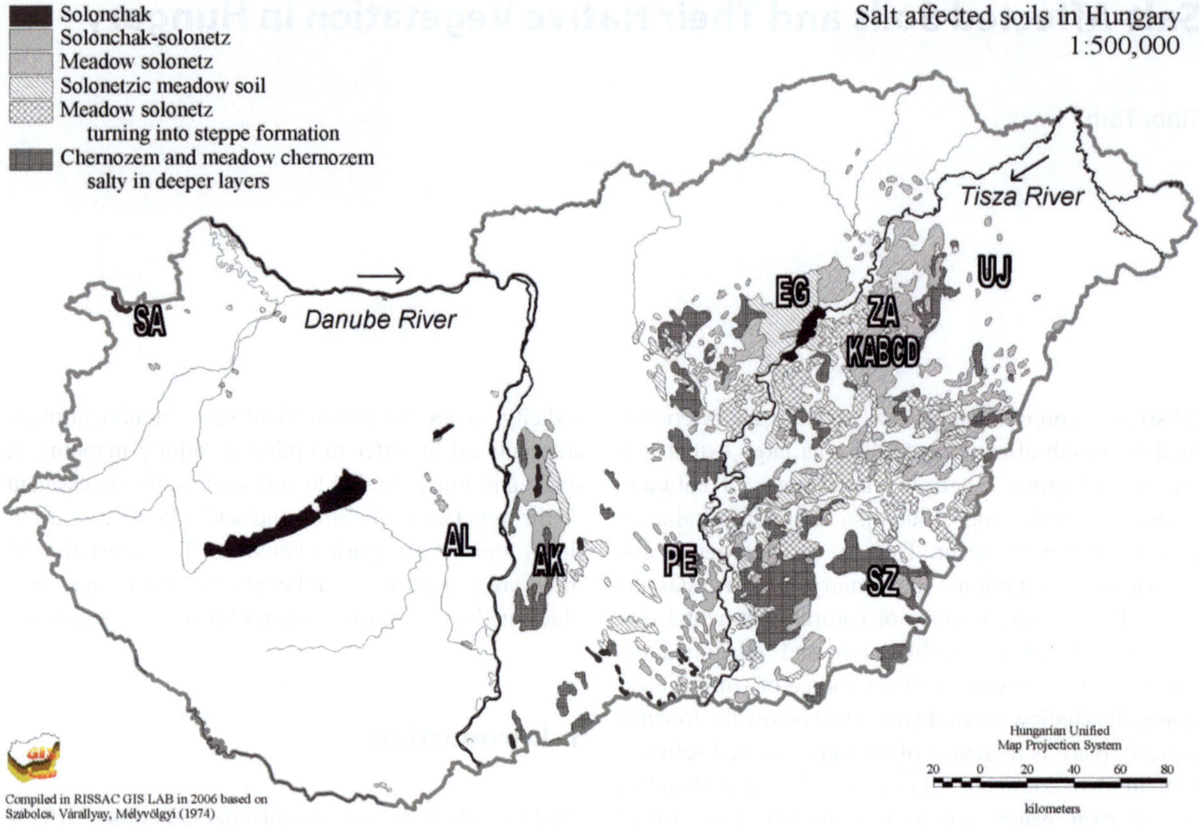

Fig. 1 Map of Hungarian salt-affected soils as published by Szabolcs (1974) and the location of the profiles described in Tables 2–4

soils, which are not salt-affected at present, but which could become considerably saline or sodic as a consequence of irrigation (Szabolcs 1974). The territorial segregation of some types of SAS is evident (Fig. 1). Soil types Solonchak and Solonchak-Solonetz are concentrated mainly in the Danube-Tisza Interfluve, types "Meadow Solonetz" and "Deep Mollic Solonetz"[1] are more typical in the Tisza Plain.

3 Meteorological Conditions

The Great Hungarian Plain is the hottest and driest region of the Carpathian Basin, which is otherwise characterized by temperate climate. In the central region, where SAS are most common, data describing annual averages and dynamism is summarized in Table 1. The area of Hungarian SAS, is located at an elevation of 80–90 m above sea level, under temperate continental climate, with 10°C mean annual temperature of −2°C in January and +21°C in July, 527 mm average annual precipitation (June is the most rainy month with 71 mm, March has the least precipitation with 28 mm), and 900 mm mean annual pan evaporation.

4 Hydrological Conditions

The Great Hungarian Plain is a basin filled with sediments deposited by rivers and wind. Therefore, the position of surface waters had an important impact on soil formation. These rivers, as typical lowland rivers, affected a vast territory by the periodic floods, creating huge marshlands. According to their origin, sediments deposited from the rivers differ much and the base materials of soil formation reflect these differences.

[1]In this report the term "Meadow Solonetz Turning into Stepp Formation" of Szabolcs, 1966 and Szabolcs, 1989 p 245 has been replaced by the term "Deep Mollic Solonetz" for reasons of clarity.

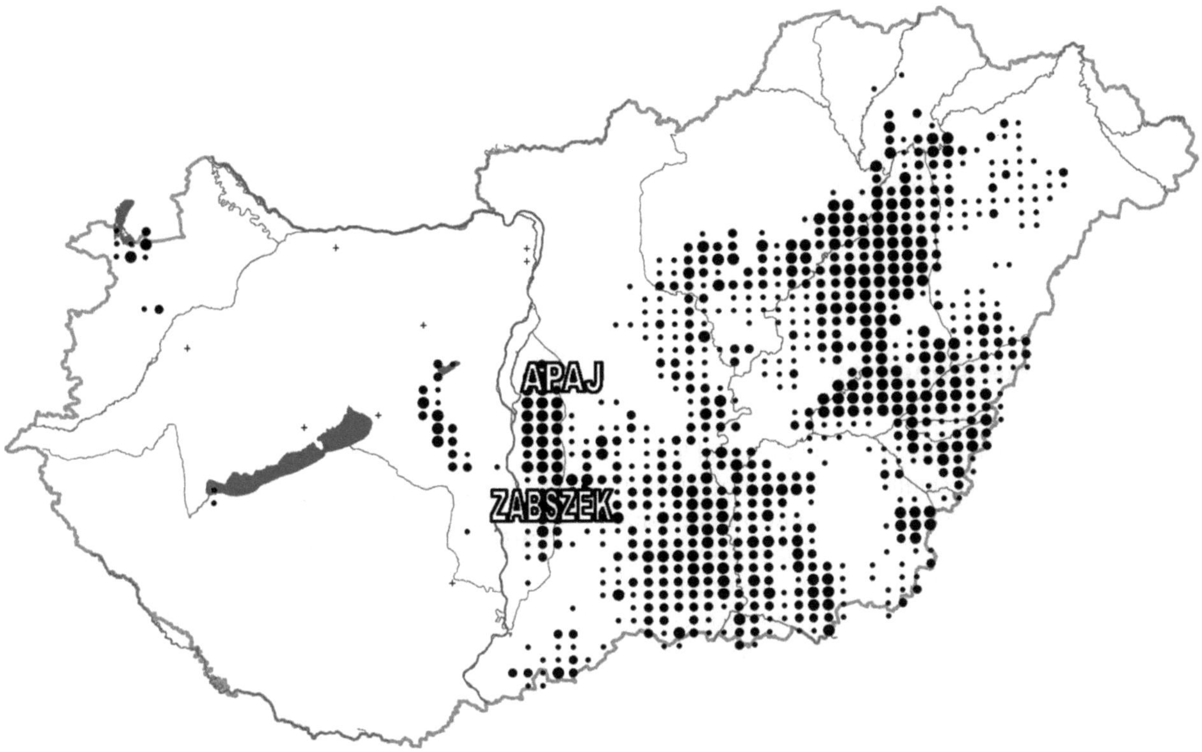

Fig. 2 The distribution of F2, "Salt meadow" habitat, copied from Molnár et al. (2008) and the location of study sites (Tables 5–6)

Table 1 Average meteorological parameters in the middle of the Great Hungarian Plain

Parameter (monthly)	1	2	3	4	5	6	7	8	9	10	11	12	Year
Precipitation (mm)	30	30	28	41	51	71	53	50	34	33	46	46	527
Potential Evaporation	12	19	40	78	112	136	156	144	106	58	25	14	900
Drought Index	0.40	0.63	1.43	1.90	2.20	1.92	2.94	2.88	3.12	1.64	0.54	0.30	1.71
Actual evap (mm)	11	15	27	63	102	91	76	58	35	21	16	12	527
Air temp (°C)	1.8	0.5	5.2	10.9	16.0	19.7	21.3	20.5	16.4	10.7	5.3	0.6	10.0

In the formation of salt affected soils a decisive role is played by saline groundwater, so the different types of SASs in the Hungarian soil classification system are closely related to distinct groundwater table depths. There are regional and local differences in the composition and concentration of groundwater that resulted in the wide variety of salt-affected soils.

5 Soil Conditions

According to the general Hungarian classification of soils, there are soils of the Atlantic region (Brown Forest soils) in the hilly marginal regions of the plain and of the steppe region (Chernozems) in the inner plateaus of the plain. Important "azonal" soils are the salt-affected soils and Meadow soils. These, together with the "intrazonal" Alluvial soils, form a catena. As the parent material between the Danube and $\text{Tisza}_{\text{Hungarian}}$ ($=\text{Theiß}_{\text{German}}$ or $\text{Tisa}_{\text{Slovakian\&Ukrainian}}$) rivers is rich in calcium-carbonate, the Solonchak and Solonchak-Solonetz soils developed on the alluvial sandy soils are classified as "Calcareous Sodic soils", whereas the more or less leached Solonetz-like soils that were developing on the more acidic sediments of the Tisza River (loamy and clayey parent material) are frequently referred to as "Non Calcareous Sodic Soils". The latter is characterized by higher clay-content and unfavourable hydrophysical properties, high ESP (Exchangeable Sodium Percentage) and high pH in the columnar B horizon and, as a rule, low salt content.

The unfavourable properties that limit the fertility of these soils are the consequence of the high clay content, high ESP, high pH and the resulting special moisture regime. The climatic conditions, e. g. the unequal distribution of the precipitation, the high aridity index and the high fluctuating saline groundwater call for a complex approach for improvement for agricultural purposes.

6 Groundwater Conditions

The Great Hungarian Plain consists of a variable layered and textured deep aquifer where the groundwater table varies between 0.5 and 4.0 m below surface, with an average fluctuation of 0.5–2.0 m. The shallow water table often causes waterlogging on the lower parts of the fields. Surface waterlogging appears also on the low-lying, low permeability plots at the end of winter, after snowmelt and/or during high-precipitation periods. The high salt content of the groundwater and its high $Na^+/(Ca^{++} + Mg^{++})$ ratio often result in salinization and alkalinization of the soils.

7 The Formation of the SAS of Hungary

At the beginning of the Miocene geological Era (23–5.3 million years before present) between the ancient "Carpathian" and "Dinaric" Mountains a vast gulf of the "Tethys" ancient sea flowed in to create the "Parathethys". This sea gulf later became detached from the Tethys and – known as the "Sarmathian" or "Pannonic" Sea – by the end of the Pliocene Era (5.3–1.8 million years before present) has been filled up with several hundred meter thick alluvial sediment. During the Pleistocene Era (1.8 million to 11,550 years before present) this process continued and loess formation took also place on the previously deposited alluvial sand. In some areas the sand was blown into dunes.

On the parent materials formed during the Pleistocene Era the influence of shallow fluctuating, saline-sodic groundwaters, as well as the permanent or temporary waterlogging created the conditions of SAS formation. The sodium ions, being considered as the most important factors, either dissolved from the Tertiary Era (65–1.8 million years before present) deposits into the groundwater (supported by data of Mádlné Szőnyi et al. 2005) or concentrated during consecutive drying and wetting of infiltrated water (as argued by Bakacsi and Kuti 1998). Szöőr et al. 1991 have shown that salinization has been present in the Great Hungarian Plain at least 30,000 years before. Among the anions in the groundwater and soil solutions there were plenty of bicarbonates, carbonate and other ions with alkaline hydrolysis and these caused almost irreversible sodium exchange processes.

8 Classification of Hungarian Salt-Affected Soils

In the late US classification of SAS the term "white alkali soil" stood for Solonchak soils and "black alkali" for Solonetz soils. The modern Hungarian soil classification is based on these categories as well. The categories like saline, sodic and saline-sodic as suggested by Richards 1954 and de Sigmond 1938, are also still in use. In agronomic practice the limit for a soil to be called SAS is 0.1% soluble salt content, as suggested by de Sigmond 1938 and Richards 1954.

The current classification system of Hungarian SAS meets two requirements: it fits the general principles of genetic soil classification, first developed in Russia (described in Gerasimov 1960) and later further developed in Europe (Kubiena 1953) and USA (Marbut 1927; Soil Survey Staff 1951) up to the middle of the twentieth century, and it keeps the traditional categories of Hungarian SAS.

The Hungarian SAS, belonging to the "Main soil type" of "Halomorphic soils" of the national soil classification system (Szabolcs 1966) are divided into five soil types: Solonchak soils, Solonchak-Solonetz soils, Meadow Solonetz soils, Deep Mollic Solonetz, Secondary Salt-Affected soils. The following list shows the current "official" classification of the main types of "salt-affected soils" of Hungary (Szabolcs 1966; Guidelines 1989). The acreage of the soil types was calculated from the Agrotopographical Map Database (in short AGROTOPO) database, (described by Várallyay et al. 1985). The map of salt-affected soils as shown by Fig. 1 was compiled by Szabolcs (1974).

9 Solonchak Soils (Total Area 47 km^2)

These soils are *per definitio* the saline soils, which are mainly located in low-lying areas, typically shorelines of saline/sodic lakes, in the region between the Danube and Tisza Rivers, but also occur in patches east of the Tisza River. These soils are characterized with 60–80 cm deep groundwater table and an average total soluble salt content of 0.3–0.5% at the surface. Dominant salts are sodium-carbonate, bicarbonate, sulphate and chloride. There is calcium-carbonate in the whole profile. It is difficult to distinguish horizons in the profile of this soil. These soils are not cropped, but sustain native halophyte vegetation which is grazed.

10 Solonchak-Solonetz Soils (Total Area 659 km^2)

These saline-sodic soils are also located mostly between the Danube and Tisza Rivers, but above deeper groundwater level, ca at 1 m. In the profile a weakly developed columnar/prismatic natric (= solonetzic) B horizon can be distinguished. There is calcium-carbonate in the whole profile. These soils sustain native halophyte vegetation which is grazed.

11 Meadow Solonetz Soils (Total Area 2,749 km^2)

The typical "solonetz" soils of Hungary are the typical sodic soils on the Great Hungarian Plain, mostly east of the Tisza River, but also west of the Danube River. These soils are characterized by large exchangeable sodium percent and not high salt content. This latter can be low enough in the "A" horizon to permit cultivation on these soils. Otherwise these soils sustain native halophyte vegetation which is grazed. The fertility of the soil is proportional to the thickness of slightly saline "A" horizon. In the characteristic columnar/prismatic natric B_1 horizon, where the maximum of the sodium adsorption can be encountered, the value of exchangeable sodium percent (ESP) is at least 20–25%. The maximum of salt accumulation can be found in the B_2 horizon, where the soil structure is prismatic or "nutty" (= large subangular blocky). Calcium-carbonate is generally absent from the A and B_1 horizons. The depth to the groundwater table is between 150 and 350 cm.

12 Deep Mollic Solonetz (Total Area 2,122 km^2)

When the groundwater table is lower (3–4 m below the soil surface) the leaching reduces the soluble salt and calcium-carbonate content of the upper horizons of these sodic soils. "Turning into steppe formation", the term originally used for this soil type by Szabolcs (1966) denotes soil forming processes similar to those of the steppe (Chernozem) soils. These soils are typically ploughed.

13 Secondary Salt-Affected Soil (Not Distinguished on the Agrotopo Database as Polygons)

This soil type comprises all soils, which were originally not salt-affected, but due to human influence became salt-affected. Besides the mentioned SAS there are other SAS types which belong not to the main type of "Halomorphic soils" but to other main soil types, such as "Solonetzic Meadow soils" with total area of 2,419 km^2, and "Chernozem soils with saline/sodic subsoil" with total area of 4,185 km^2. These soils are typically ploughed.

In Hungary the total area of salt-affected soils, based on the AGROTOPO database (printed on the map sheets by the Kartográfiai Vállalat in 1983), is 12,181 km^2. With this acreage the overall area of SAS covers 13% of the national territory. The map provided by the Hungarian Central Statistical Office, 2006 is almost the same as published by Stefanovits (1963), based on his map of the soils of Hungary at the scale of 1:200,000. Although there are differences in the acreages of the distinct SAS types between the two sources, but the total acreages, as 11,087 (Stefanovits 1963) or 12,181 km^2 (AGROTOPO) are close.

14 The Utilization of SAS in Hungary

Though the improvement (reclamation or amelioration) of these soils is scientifically well founded, it is a rather costly operation. This is a reason why large tracts of these soils are kept as grazeland or hayfield, land for afforestation, paddy field or fishpond. Most of the Hungarian National Parks have salt-affected grasslands, hayfields, marshes, reedlands, lakes and these provide habitat for protected animals (mainly birds), plants and attract lots of tourists. Many of the protected animals barely find a place for feeding and breeding on other soil types, since most of those are cropped or otherwise intensively utilized. In total some 88% of the surface of the country has no natural vegetation cover (cropland, tree plantations with exotic species, orchards, vineyards, settlements, roads, etc.).

Among the crops that may be grown economically on these soils the most important is winter wheat. It covers above the half of the area of the SAS. Other important crops are winter barley, sunflower, Sudangrass, vetch, rice and sometimes maize, sugarbeet and pea.

15 Information Available on the Spatial Extent of SAS

There is an outstanding record of collecting soil information in Hungary. The historical past is summarized in several publications (Ballenegger and Finály 1963). Just like in other countries in the early period of soil mapping, before the First World War there were two tendencies: special mapping of selected, usually small areas and preparation of very small-scale maps, based on scarce observations and continental-scale conceptual models. In Hungary the first Hungarian soil (that time called agrogeologic) map was compiled in 1861 (Szabó 1861) for the area of two counties at the scale of 1: 576,000. Few years later the soil map of Tokaj-Hegyalja intended to improve the production of the famous Tokaj wine in the region (Szabó 1865–6). A major achievement was the first complete soil map of Hungary prepared by Timkó in 1914. During the pre-war and after-war periods of 1935–1951 the "Kreybig" practical soil mapping was completed and displayed on maps at 1:25,000. From the 1960s the 1:10,000 scale mapping of the agricultural lands was performed. From 1989 no systematic large-scale soil mapping is carried out.

16 Agrotopo Map Database

It was the 1:100,000 scale AGROTOPO soil spatial database, which became available first digitally. This database, which was developed in 1990s, integrates the dominantly small-scale soil related data into digital format and is organized into spatial soil information systems (AGROTOPO: Várallyay 1989; HunSOTER: Várallyay et al. 1994; MERA: Pásztor et al. 1998). Information in the AGROTOPO is provided for nine properties, such as soil types, soil parent material, soil texture, clay-mineral composition, soil water regime category, soil reaction and carbonate status, soil organic matter stock, depth of solum and soil bonitation value. There are altogether 3,312 polygons for the total area of 93,000 km^2 of the country. As background to the soil polygons there is a general topographic sheet with landuse categories, elevation contour lines, settlements, waterways, roads, etc.

16.1 The 1:25,000 Soil Information System (Kreybig Digital Soil Information System)

The national soil mapping project initiated and led by L. Kreybig was unique, being a national survey based on both field and laboratory soil analyses and at the same time serving practical purposes (Kreybig 1937). Due to inactivity during the Second World War it was carried out between 1935 and 1951 in several stages. In the fifties, when the mapping was successfully completed, Hungary was among the first countries in the world to have such detailed soil information for the whole country. These maps still represent a valuable treasure of soil information. The soil and land use conditions were shown jointly on the maps. Altogether three characteristics were attributed to soil mapping units and displayed on each mapsheet. First feature distinguished was land use, both ploughland and grassland was not distinguished. Second was the chemical reaction shown by colours and third feature was the physical soil properties of the soil root zone. Some further soil properties were determined and measured in soil profiles. A very remarkable feature of the map series is that it distinguishes three different categories of SAS by colour codes:

Reddish purple colour: SAS suitable for cropping.

Light purple colour: SAS potentially suitable for cropping, can be reclaimed with $CaCO_3$.

Dark purple colour: SAS not suitable for cropping, which cannot be reclaimed with $CaCO_3$.

The GIS adaptation of soil information originating from the soil maps displayed on 1:25,000 scale is still under construction (Pásztor et al. 2006). There is much more utilizable information originating from this survey, than it was processed earlier and published on the map series and in reports, and what is provided by simply archiving them digitally. The surplus information should be exploited by the new technologies provided by GIS and DSM (digital soil mapping) and provide the basis of improved management of the soils.

16.2 Genetic Soil Maps of 1:10,000 Scale

In the early 1960s a mapping technology was elaborated by the Hungarian soil scientists, soil surveyors and soil-mapping specialists for the large-scale soil survey to satisfy the practical needs of soil information of large farming units (state and co-operative farms), which characterized the Hungarian agriculture between 1950 and 1990. Such maps were prepared for about one-third of the area of Hungary, representing two thirds of the cropland (ca 35,000 km^2). The mapping reports consist of four main parts: (i) genetic soil map, indicating soil taxonomy units, and the parent material; (ii) thematic soil maps on the most important physical and chemical soil properties; (iii) thematic maps, indicating recommendations for rational land use, cropping pattern, amelioration, tillage practice and fertilization; (iv) explanatory booklets including a short review on the physiographical conditions; description of soils, recommendations for their rational utilization; field description of soil profiles; results of field observations or measurements and data of laboratory analyses (Szabolcs 1966). These maps were widely and successfully used in Hungary and became an easily applicable scientific basis of intensive, large-scale agricultural production, in spite of the fact that generally these maps were not published in printed form and are available only as manuscripts at the given farming units or at the Plant and Soil Conservation Stations. The large-scale soil-mapping programme was restarted in 1986 within the framework of the National Land Evaluation Programme (Guidelines 1989). The aim of this Programme was to valuate the agricultural land based on soil survey at the scale of 1:10,000, but was also left uncompleted. These huge archives provide appropriate raw material for recent digitally based applications. Spatial soil information systems based on these data could be efficiently used in numerous fields.

Szabolcs (1966) described the methodology to be used in the detailed mapping of soils. For example in the case of SASs this method at the scale of 1:10,000 can be illustrated best with the set of individual map sheets which might make up a complete soil mapping document.

17 Exemplary Data of Hungarian Salt-Affected Soils

Twelve soil profiles were selected to demonstrate the characteristics of Hungarian salt-affected soil types from Jozefaciuk et al. (2006). Basic properties of the studied soils are presented in Tables 2–4. The studied profiles were described in different regions of the Hungarian Plain, a floodplain with varying thickness of alluvial sediments. Occurrence of salt-affected soils is closely related to groundwater depth and salinity, being the most important factors of salinization, and also to surface waters – the frequency and time of waterlogging. During the last 200 years great changes in the hydrological situation in the Plain took place and the distribution of salt-affected soils reflects these changes closely (Tóth et al. 2001). At most of the sampling sites the groundwater level has been sufficiently close to the surface and enough saline to cause salt accumulation.

Very slightly salt affected Karcag (K) profiles represent the best stands for crops: Chernozems (KA (irrigated), KB and KD) and a Vertisol (lower elevated site KC, irrigated). The native vegetation on higher areas is thick grasses and the soil is fertile. In the lowest patches fine particles have been settled and the soil is cracking, but also shows frequent waterlogging. As it is typical for such soils, sodicity is observed only in deeper layers of the subsoil (where the capillary water-rise affects the chemical features) and the topsoil is salt-free (Table 2).

Among the other soils taken from native sites, some occurred at relatively higher elevation in the toposequence (Egerlovo EG, Szabadkigyos SZ and Alap AL)

Table 2 Characteristics of the sites of the profiles

Site	Sarród	Egerlövő	Szabadkígyós	Zám	KarcagM1	KarcagM1K	KarcagM2	KarcagM2K
Code in Table 4	SA	EG	SZ	ZA	KA	KB	KC	KD
Description	Grassland	Grassland	Grassland	Grassland	Cropland	Cropland	Cropland	Cropland
Geographic coordinates northern	47°39.253′	47°43.600′	46°36.232′	47°31.669′	47°18.628′	47°18.650′	47°18.606′	47°18.655′
Eastern	16°48.764′	20°35.706′	21°5.815′	21°2.381′	20°48.737′	20°49.969′	20°48.632′	20°48.633′
Elevation m	118	95	91	94	88	87	86	87
Soil type	Stagnic	Vertic	Mollic	Haplic	Haplic	Haplic	Haplic	Haplic
World reference base	Solonchak	Solonetz	Solonetz	Solonetz	Chernozem	Chernozem	Vertisol	Chernozem
Hungarian classification	Solonchak	Medium Meadow Solonetz	Crusty Meadow Solonetz	Crusty Meadow Solonetz	Meadow Chernozem	Lowland Chernozem	Meadow Soil	Lowland Chernozem
Native plants	Salicornea europea Suaeda salsa Plantago maritima	Suaeda salsa Limonium gmelini Artemisia santonicum	Camphorosma annua Puccinellia limosa Artemisia santonicum	Salicornea europea Suaeda salsa Plantago maritima	na	na	na	na
Groundwater depth (cm)	190	180	204	250	200	200	220	220
EC	6.0	1.5	5.0	7.5	7.4	5.7	2.8	4.1
pH	7.7	7.8	8.0	8.1	8.5	8.4	7.7	8.2
SAR	19	4	68	65	49	43	11	24
Ca (me/l)	5	5	0.62	0.77	1.08	1.04	2.89	0.6
Mg	18	3	1	4	4.07	3.74	3.91	2.28
Na	66	9	61	100	79.4	67.2	20.95	28.8
K	0.25	0.24	0.07	1	0.04	0.03	0.04	0.03
SO_4	68	9	9	1	3.55	5.2	2.1	1.85
Cl	16	1	27	73	35.7	14.0	4.41	3.07
CO_3	4	2	7	2	4.66	2.54	0	1.2
HCO_3	13	7	20	26	19.44	27.31	8.44	17.70

Site	Újfehértó	Péteritó	Alap	Akasztó
Code in Table 4	UJ	PE	AL	AK
Description	Grassland	Grassland	Grassland	Grassland
Geographic coordinates northern	47°47.795′	46°34.635′	46°48.927′	46°40.578′
Eastern	21°41.250′	19°53.984′	18°39.503′	19°9.121′
Elevation (m)	114	93	92	94
Soil type world reference base	Stagnic Solonchak	Gleyic Solonchak	Calcic Solonchak	Salic Solonchak
Hungarian classification	Solonchak	Solonchak	Solonchak	Solonchak-solonetz
Native plants	Kochia scoparia	Puccinellia limosa	Camphorosma annua	Lepidium crassifolium
	Puccinellia limosa	Lepidium crassifolium Aster tripolium	Limonium gmelini	Puccinellia limosa
Groundwater depth (cm)	150	80	200	95
EC	4.7	11.0	11.0	7.5
pH	8.15	8.3	8.2	9.6
SAR	45	90	80	145
Ca (me/l)	1	0.1	1.4	0.05
Mg	2.5	7	6	1
Na	60	169	154	105
K	0.2	2.3	0.3	0.5
SO_4	2.3	0.9	121	7.2
Cl	5.2	31	37	16
CO_3	7.5	35	5.3	45
HCO_3	60	56	17.3	79

na – not applicable

Table 3 Colour, texture and structure of the horizons distinguished in the studied soil profiles

Site and date of sampling	Code of horizon	Depths (cm)	Wet colour	Texture	Structure
Sarród 2000 Aug 29	1	0–10	10YR5/2	SL	s-SUB
Sarród 2000 Aug 29	2	10–18	10YR3/2	C	s-SUB
Sarród 2000 Aug 29	3	18–52	2.5Y7/3	CL	s-SUB
Sarród 2000 Aug 29	4	52–54	10YR4/3	LS	nps
Sarród 2000 Aug 29	5	54–75	2.5Y7/2	CL	nps
Újfehértó 1999 Aug 16	1	0–4	5Y5/3	SL	PLA
Újfehértó 1999 Aug 16	2	4–14	5Y5/3	CL	ABL
Újfehértó 1999 Aug 16	3	14–40	2.5Y5/4	L	SAB
Újfehértó 1999 Aug 16	4	40–85	2.5Y5/6	L	SAB
Péteritó 2000 Jul 28	1	0–3	2.5Y4/2	SL	PLA
Péteritó 2000 Jul 28	2	3–13	2.5Y5/2	LS	s-SAB
Péteritó 2000 Jul 28	3	13–39	2.5Y4/2	LS	s-SAB
Péteritó 2000 Jul 28	4	39–58	2.5Y5/2	LS	s-SAB
Péteritó 2000 Jul 28	5	58–77	2.5Y6/1	LS	s-SAB
Alap 2OOO Aug 24	1	0–2 (efflorescence)	7.5YR2/2	SC	nps
Alap 2OOO Aug 24	2	2–28	7.5Y2/2	SC	COL
Alap 2OOO Aug 24	3	28–57	7.5Y3/2	SC	SAB
Alap 2OOO Aug 24	4	57–71	2.5Y7/4	SG	SAB
Alap 2OOO Aug 24	5	71–82	2.5Y7/6	SL	SAB
Akasztó 1998 Jul 30	A	0–3	5Y5/2	SL	s-SAB
Akasztó 1998 Jul 30	B	3–19	5Y6/2	CL	COL
Akasztó 1998 Jul 30	BC	19–38	5Y6/3	L	SAB
Akasztó 1998 Jul 30	C1	38–58	5Y6/3	L	SAB
Akasztó 1998 Jul 30	C2	58–90	5Y6/3	L	SAB
Akasztó 1998 Jul 30	C3	90–92	5Y5/3	LS	SAB
Egerlövö 2000 Aug 9	A/e	0–8	2.5Y3/1	L	PLA
Egerlövö 2000 Aug 9	B1	8–21	2.5Y2/1	CL	PRI
Egerlövö 2000 Aug 9	B2	21–46	2.5Y3/1	CL	l-SAB
Egerlövö 2000 Aug 9	B3	46–78	2.5Y2/1	C	ABL
Egerlövö 2000 Aug 9	BC	78–118	2.5Y3/1	C	ABL
Egerlövö 2000 Aug 9	C	118–130	2.5Y4/1	LC	SAB
Szabadkígyós 2000 Sep 8	A	0–2	10YR4/2.5	L	PLA
Szabadkígyós 2000 Sep 8	B1	2–17	10YR2/1	CL	COL
Szabadkígyós 2000 Sep 8	B2	17–36	10YR2/2	CL	SAB
Szabadkígyós 2000 Sep 8	BC	36–67	2.5Y4/3	CL	PRI
Szabadkígyós 2000 Sep 8	C	67–90	2.5Y4.5/5	CL	PRI
Zám puszta 2001 Aug 23	A/e	0–7	2.5Y3/2	L	PRI
Zám puszta 2001 Aug 23	B1	7–22	2.5Y3/2	C	s-SAB
Zám puszta 2001 Aug 23	B2	22–38	2.5Y3/2	C	ABL
Zám puszta 2001 Aug 23	BC	38–68	2.5Y4/2	C	ABL
Zám puszta 2001 Aug 23	C1	68–75	2.5Y5/3	C	SAB
Zám puszta 2001 Aug 23	C2	100–120	2.5Y5/3	C	SAB
Karcag_M1 1998 Oct 22	Ap	0–18	10YR3/1	CL	ABL
Karcag_M1 1998 Oct 22	A	18–32	10YR3/1	CL	ABL
Karcag_M1 1998 Oct 22	B	32–51	2.5Y3/1	CL	s-SAB
Karcag_M1 1998 Oct 22	BC	51–92	10YR3/2	CL	s-SAB
Karcag_M1 1998 Oct 22	C	92–100	10YR4/3	CL	s-SAB
Karcag_M1K 1998 Oct 22	Ap	0–21	10YR2/2	CL	s-SAB
Karcag_M1K 1998 Oct 22	A	21–46	10YR3/1	CL	s-SAB
Karcag_M1K 1998 Oct 22	B	46–68	10YR3/2	CL	SAB
Karcag_M1K 1998 Oct 22	BC	68–111	2.5YR4/2	CL	ABL
Karcag_M1K 1998 Oct 22	C	111–118	2.5YR4/3	CL	ABL

(continued)

Table 3 (continued)

Site and date of sampling	Code of horizon	Depths (cm)	Wet colour	Texture	Structure
Karcag_M2 1998 Oct 22	Ap	0–20	10YR2/1	CL	ABL
Karcag_M2 1998 Oct 22	A	20–36	10YR2/1	CL	PRI
Karcag_M2 1998 Oct 22	B	36–75	10YR2/1	CL	ABL
Karcag_M2 1998 Oct 22	BC	75–95	10YR2/1	CL	PRI
Karcag_M2 1998 Oct 22	C	95–105	10YR3/2	CL	ABL
Karcag_M2K 1998 Oct 22	Ap	0–28	10YR2/2	CL	ABL
Karcag_M2K 1998 Oct 22	A	28–45	10YR2/2	CL	ABL
Karcag_M2K 1998 Oct 22	B	45–68	10YR2/2	CL	ABL
Karcag_M2K 1998 Oct 22	BC	68–105	10YR3/3	CL	ABL
Karcag_M2K 1998 Oct 22	C	111–115	2.5YR4/3	L	ABL

SL sandy loam, *L* loam, *CL* clay loam, *C* clay, *PLA* platy, *SAB* subangular blocky, *ABL* angular blocky, *COL* columnar, *PRI* prismatic, *l* large, *s* small, *nps* no pedological structure has developed

Table 4 Selected properties of the studied profiles

Sample	pH	OM	Carb.	Clay	CEC	SP	EC	Na	SAR
SA5	8.98	0.89	22.1	15	6.1	33	27	345	45.1
SA14	8.93	0.93	33.7	38.1	12.9	62	16.2	167	33.5
SA35	8.94	0.56	57.9	45	10.6	55	15.3	162	30.8
SA53	9.21	0.48	26.6	17.8	7.9	33	11.2	110	26.5
SA65	9.13	0.42	53.7	36.6	8.6	57	11.6	111	26.2
EG4	8.08	4.31	1.3	29.4	25.4	66	25.2	440	69.7
EG14	8.66	2.33	9	50.6	32.3	93	21	342	64.2
EG34	9.11	1.55	14.8	50	32	128	14	192	73.6
EG62	9.14	1.3	13.3	46.9	31.4	143	8	97	71
EG98	9.24	0.99	12.6	42.2	28.3	140	2.7	28	40.8
EG124	9.39	0.71	28.2	38.1	22.3	98	2.5	11.2	21.8
SZ1	9.26	2.2	4.24	30	23.6	53	21	380	223
SZ9	9.56	2.37	12.7	44.5	26.3	67	25.4	456	357
SZ26	9.8	1.74	18.2	41.1	23.7	69	19.5	345	361
SZ53	9.92	1.18	20.4	38.5	20.4	73	11.8	187	258
SZ78	9.97	0.59	22.9	37.1	16	86	6.9	100	171
ZA3	9.21	2.05	0.3	28.5	18.2	52	36.2	489	409
ZA14	9.68	1.56	0.3	34.7	21.8	66	22.8	289	282
ZA30	9.75	1.41	0.8	44.9	25.6	93	16.3	200	249
ZA53	9.73	1	5.6	48.2	25.7	116	10.5	117	137
ZA71	9.72	0.84	21.4	44.3	19.5	98	8.3	87	110
ZA110	9.48	0.49	20.6	33.6	16.6	88	7.2	75	77.4
UJ2	10.2	0.79	6.72	10.3	6.3	37	20	405	684
UJ9	10.4	0.4	11.13	21.8	9	44	8	138	223
UJ27	10.3	0.38	36.8	28.8	11	57	4.7	82	160
UJ62	10.3	0.21	26.25	22.8	11	50	3.5	62	111
PE2	9.95	1.29	15.2	7.1	5.2	37	71	1080	620
PE8	10.2	0.69	20.2	10.7	3.7	33	7.9	100	197
PE26	10.4	0.3	13.5	9.8	3.9	33	6.3	80	169
PE48	10.4	0.22	20.2	15.8	3.8	47	6.2	75	180
PE68	10.3	0.22	31.6	20.4	3.9	44	6.2	74	136
AL1	9.87	1.18	2.5	14.2	8.3	33	78	1750	2060
AL15	9.84	1.17	4.2	22.7	15.2	48	25	386	387
AL43	10	0.68	16.9	21.7	9.6	46	16.2	220	339
AL64	10.2	0.06	13.9	4.2	3.8	33	10	124	560
AL76	9.87	0.16	23.2	17.2	10.5	73	9.5	114	260

(continued)

Table 4 (continued)

Sample	pH	OM	Carb.	Clay	CEC	SP	EC	Na	SAR
AK2	9.88	1.19	29.3	9.2	5.4	35	70	1000	2100
AK11	10.2	0.63	33.9	20.1	7.3	39	29	398	3370
AK28	10.1	0.2	38.2	17.8	7.3	43	27.5	374	7780
AK48	10.1	0.23	33.9	13	7.5	40	24.3	329	10,000
AK74	10.5	0.23	26.3	3	5.8	35	12.7	120	10,000
AK92	10.5	0.13	18.7	3.6	4.9	30	12.8	134	484
KA9	7.33	5.32	0	41.3	26.7	47	1.1	3.8	2.05
KA25	6.91	4.08	0	40.8	33.3	46	0.7	2.3	1.69
KA42	7.67	3.15	1.4	41.4	34.5				
KA72	8.51	2.59	7.2	41.7	33.2	52	0.5	3.6	5.19
KA96	9.52	0.94	8.8	39.3	15.7	51	1.2	12	19
KB11	7.4	6.48	0	40.2	24.3	48	0.8	2.7	1.62
KB33	7.84	2.99	1.2	39.5	24.4	52	0.6	2.5	1.94
KB57	8.38	2.54	5.8	40.3	27.3	53	0.5	2.9	3.08
KB89	9.25	1.83	9.5	40.3	17	47	1	9.3	13.4
KB114	9.72	0.07	9.9	35.9	14	56	2.1	22	30.7
KC10	7.47	5.61	0	39	24.1	48	0.6	3.5	3.46
KC28	7.67	3.87	0	40.1	21.6	45	0.7	5.6	6.55
KC56	8.76	2.91	0	49	31.8	46	1.2	11	15.9
KC85	8.95	1.99	0	47.8	29.8	49	2.6	27	31.1
KC100	8.97	1.22	0	46.6	19.6	55	2.6	27	33.2
KD14	7.53	5.42	0	40.1	27.7	48	1.1	9.1	10
KD26	7.39	3.37	0	40.3	31.5	53	0.6	3	2.92
KD56	8.05	2.71	4.9	38.5	22.3	55	0.5	2.2	2.05
KD87	8.28	2.02	8.8	39.8	18.8	52	0.5	2.9	3.34
KD113	9.03	1.54	11.4	37.8	20.3	46	0.7	6	8.81

Sample is identified by horizon depth mid-point (cm). pH H_2O, *OM* organic matter %, *Car.* carbonates %, *Clay* clay %, *SP* saturation percentage, *EC* electric conductivity of the soil saturation extract (mS cm^1), *Na* sodium concentration in the soil saturation extract (mM dm^{-3}), *SAR* sodium adsorption ratio

where the salt efflorescenses can be found in small patches only. Solonetz is the typical soil type in this situation, but Solonchak limited to small patches is found also. The latter soil type is the most typical one in waterlogged areas and beside saline lakes. The above soils have *Artemisia santonicum* as the dominant plant species. Intermediate elevation is occupied at (Akaszto AK), which was earlier a temporary lake, already being dried out and covered by sparse vegetation stands of *Camphorosma annua*. From lower elevations we sampled Ujfeherto (UJ) and Zam (ZA) profiles, temporary waterlogged sites, where *Kochia scoparia* and *Salicornia europaea* are characteristic plant species, respectively. The lowest elevation sites (Sarrod SA and Peterito PE) are close to typical saline lakes, where *Suaeda salsa* and *Aster tripolium* plus *Puccinellia limosa* (more hydrophytic vegetation) dominate, respectively, although these soils are not usually covered by lake water. The drier, upland sites of highly saline soils are grazed by sheep and the lowland ones, especially those beside lakes, are grazed by wild geese.

Chernozem and Vertisol profiles at Karcag are more brown (10YR) than salt-affected soils. Profiles close to lakes are paler in colors. The variable layering of the subsoil of Sarrod profile, caused by depositions, is evidenced in color and texture as well (Table 4).

The texture of the studied soils ranges from loamy sand to clay. The eluvial A/E horizons of Solonetz soils have coarser texture and a lower clay content than the illuvial horizons. In Solonetz soils prismatic and columnar structure reflects high sodicity. The amount of clay in soils of higher salinity is usually the highest in illuvial B horizons while in less saline or non-saline soils this is rather uniform throughout the whole profile.

The presence of calcium carbonate is common all over the Hungarian Plain because of the frequent occurrence of loessial and carbonatic deposits of River Danube. All studied profiles contain carbonates except the Vertisol profile KC. As it is typical for chernozems, calcium carbonate occurs in deeper layers in KA, KB and KD. Leached (low carbonate) topsoil layers are characteristic for Solonetz soils, whereas Solonchak contain rather high amounts of carbonates right at the surface (Table 4).

Electrical conductivity of the saturated soil extracts ranges between 2 and 78 dS/m for salt-affected soils and this is significantly lower for slightly salt-affected soils. Solonchak soils show continuously decreasing EC values down from the topsoil. Solonetz soils show small electric conductivities at the surface and a maximum in B2 horizons. In non saline soil types (Chernozem and Vertisol) largest EC values are found in the deepest horizon, affected mostly by saline groundwater. The electric conductivity of the saturated soil extracts decreases in general with the profile depth for all soils which is also noted for the sodium adsorption ratio (Table 4).

Except one horizon all samples are alkaline and mostly reach pH values higher than 9. More leached horizons of slightly saline and Solonetz profiles can have neutral pH at the surface. Soil pH increases down the profiles, similarly as the carbonate content (Table 4).

As a minimal value, soil organic matter of 0.8% was determined. Solonchak soils typically do not have SOM larger than 1.5%. SOM content around 3% can be found in the uppermost horizon of Solonetz soils. Chernozem and Vertisol profiles show SOM above 5%. The amount of soil organic matter accumulated in the profiles studied is closely related to their pH. Presence of such dependence was very surprising for us, because amount of soil organic matter in soils depends on many environmental factors, conditions of soil genesis and physical soil properties, and not on a single parameter. However all studied soils are located within rather small area and, as far as the pH is a main parameter responsible for organic matter leaching, one may suspect that at higher pH its accumulation has been prevented by the extreme conditions for plants and microbes. Where there is better plant growth there is larger soil organic matter content, that dissolves $CaCO_3$, plants use the Ca and it results in a decrease in pH.

18 Vegetation of Saline and Sodic Areas

The list of plant species of saline and sodic habitats is best represented by the floras of the National Parks. Out of the ten National Parks, there are five parks possessing saline and sodic areas. The first flora was written about the first Hungarian National Park, the Hortobágy National Park (Szujkó-Lacza et al. 1982) and others followed.

Soó (1980) listed the 25 saline and sodic plant associations, belonging to the division "*Puccinellio-Salicornea*" in two association class, 4 association series, 7 association groups. The most frequent of these are the following: *Salicornietum prostratae, Suaedetum pannociae, Salsoletum sodae, Crypsidetum aculeatae, Puccinellietum limosae, Pholiuro-Plantaginetum tenuiflorae, Hordeetum hystricis, Camphorosmetum annuae, Lepidio crassifolii-Puccinellietum limosae, Agrostio-Alopecuretum pratensis, Eleochariti-Alopecuretum geniculati, Agrostio-Beckmannietum, Achilleo-Festucetum pseudovinae, Artemisio-Festucetum pseudovinae.*

There are other associations, which are related to saline and sodic soils, such as *Bolboschoenetum maritimae continentalis, Glycerietum maximae*, typical for salt marshes.

There is a sodic woody association, the *Galatello-Quercetum roboris* and on its clearings the *Peucedano-Galatelletum.*

Recently a full list of associations was given by Molnár and Borhidi (2003).

Molnár et al. (2008) summarizes the actual distribution of saline vegetation. Here we copy their map (Fig. 2) only on one habitat, the salt meadows (Code F2 according to Bölöni et al. 2003), which are the most frequent grassy habitats in the country. Most characteristic associations belonging to this category are *Agrostio-Alopecuretum pratensis, Agrostio-Caricetum distantis, Eleochari-Alopecuretum geniculati, Agrostio-Glycerietum poiformis, Agrostio stoloniferae-Beckmannietum eruciformis, Rorippo kerneri-Ranunculetum lateriflori, Loto-Potentilletum anserinae.*

19 Example of the Soil and Vegetation Conditions in Two Characteristic Kiskunság Sites

The soil properties and vegetation at two nearby sites is described here based on Tóth et al. (2003). The study region is the Danube valley where there was a drop in groundwater level and soil desalinization was observed together with a shift in vegetation (Table 5). Data on more than five years monitoring are presented here.

The "Artemisia saline puszta" vegetation at Apaj is a result of gradual drying and groundwater sinking. Right after the drainage of the area probably Bolbo schoenetum maritimi, after which Lepidio crassifolii-Puccinellietum limosae plant associations was characteristic, of these two characteristic species was found in the quadrats. The official nomenclature of the present plant association is Artemisio–santonici–Festucetum pseudovinae Soó (1933) 1947 corr. Borhidi 1996 (Molnár and Borhidi 2003).

This plant community is named as ***Artemisia* saline puszta (15.A113)** by Devillers and Devillers-Terschuren (1996) and we use this name in the text. Its characteristics are described under the code of F1a by Molnár et al. (2003) and can be summarized as:

- Typical soil is Solonetz, the soil is affected by shallow groundwater and surface waterlogging as well.
- There are no shrubs or trees among these grasses, neither tall grasses.
- Most characteristic species are *Artemisia santonicum* subsp. *monogyna* subsp. *patens*, *Festuca pseudovina*, *Limonium gmelini*, *Podospermum canum*, *Trifolium retusum*, *Trifolium angulatum*, *T. parviflorum*, *Ranunculus pedatus*, *Bupleurum tenuissimum*, *Gypsophila muralis*, *Lotus tenuis* (*L. glaber*), *Cerastium dubium.*
- They occur in an elevation zone between Pannonic *Puccinellia limosa* hollow (lower neighbor) and slightly saline **Grassy saline puszta** or **nonsaline Pannonic loess steppic grasslands** (upper neighbor) (names according to Devillers and Devillers-Terschuren 1996).
- Regenerative potential is good, but the details are not known.

At present the site is grazed by sheep.

The "Pannonic Puccinellia limosa hollow" vegetation at Zabszék lake is the result of continous drying of the lake and the shifting of lake margin towards the bottom of the lake. The previous stage might have been lake-margin vegetation and *Bolboschoenetum maritimi* plant association. The official nomenclature of the plant association is Lepi diocrassi folii–*Puccinellietumlimosae* Soó 1947–puccinellietosum.

This plant community is named as **Pannonic Puccinellia limosa hollows (15.A131)** by Devillers and Devillers-Terschuren (1996) and EUNIS(2002) and as used in this text. Its characteristics are described under the code of F4 by Bagi and Molnár (2003) and can be summarized as:

- Typical soil is Solonchak. A condition of its occurrence is shallow saline groundwater and repeated waterlogging, typical on the bank of saline lakes. The spring and late summer aspects might be very different.
- Physiognomy is determined by waterlogging. If there is no waterlogging the *Puccinellia limosa* is small, on wet places it grows high, forms tussocky patches.
- Most characteristic species are *Puccinellia limosa*, *P. festuciformis* subsp. *intermedia*, sometimes important species are *Lepidium crassifolium* (*L. cartilagineum*), *Aster tripolium* subsp. *pannonicus, Artemisia santonicum* subsp. *santonicum* and subsp. *patens, Plantago maritima.*
- They occur in the neighbourhood of saline lakes.
- Regenerative potential is very good.

At present the only major grazing animals are wild geese, mostly *Anser anser*.

The two sites represent the most typical salt-affected habitats of Kiskunság region, Hungary.

Although the clay percent is similar at the two sites (Table 6), the "Pannonic Puccinellia limosa hollow" site has less sand and more silt particles, which is the result of the effect of the sedimentation onto the lake bottom. The same process might have resulted in the larger $CaCO_3$ content of the soil in the same community. These differences are reflected in the Cation Exchange Capacity values, water retention values and soil hydraulic conductivity values as well. The surface soil of "Pannonic Puccinellia limosa hollow" site binds the water with stronger force and permits its movement less (Tóth and Kuti 2002).

Due to the complete plant cover at the "Artemisia saline puszta" site, the soil organic matter content is larger. The surface salinity is shown by the electrical conductivity (EC) of the saturation extract (Richards 1954). This standard indicator of soil salinity was analyzed for the profile characterization from the soil genetic horizons one time and showed three times as large EC at the "Pannonic Puccinellia limosa hollow"

Table 5 Some characteristics of the main species of the two exemplary associations based on Horváth et al. 1995 and the sources indicated

Name	Festuca pseudovina Hack. ex Wiesb.	Artemisia santonicum L. (ssp.)	Bromus hordaceus L. ssp hordaceus	Plantago maritima L.	Podospermum canum C. A. Mey.	Puccinellia limosa (Schur.) Homberg.	Aster tripolium ssp. pannonicus (Jacq.) Soó
Synonym	–	Artemisia maritima L.	–	–	Scorzonera cana (Mey.) Griseb.	–	–
Family	Poaceae	Compositae	Poaceae	Plantaginaceae	Compositae	Poaceae	Compositae
Mean (min–max) cover%	50.5 (5–76)	11.7 (0–29)	1 (0–9)	0.95 (0–4)	0.50 (0–2.5)	36 (0–80)	0.22 (0–4.75)
Site	Artemisia saline puszta	Artemisia saline puszta	Artemisia saline puszta	Artemisia saline puszta	Artemisia saline puszta	Pannonic Puccinellia limosa hollow	Pannonic Puccinellia limosa hollow
Raunkiaer life-form	Hemicryptophyton	Chamaephyton-Hemicryptophyton	Therophyton	Hemicryptophyton	Hemicryptophyton-Hemitherophyton	Hemicryptophyton	Hemicryptophyton
Vertical distribution	Plain–mountain	Plain	Plain-subalpine	Plain–mountain	Plain–mountain	Plain	Plain–hill
Flora element	Continental	Eurasian	Cosmopolitan	Circumpolar	Pontic-Submediterranean	Pannonian endemism	Eurasian
Vegetative propagation (Csontos 2001)	Bunching grass	Bunching grass	None	None	None	Bunching grass	Bunching grass
Dispersion of seeds (Thompson et al. 1997, Csontos et al. 2002)	Endozoochory	Anemo-, endozoo-, epizoochory	Endozoochory	Anemo-, endozoochory	Anemochor	Endozoochort	Anemo-, zoochory
Flowering (months)	6, 7	7, 8, 9, 10	5, 6, 7,	6, 7, 8, 9	5, 6, 7, 8, 9, 10,	6, 7	7,8,9
Seed ripening (months)	7	8, 9, 10,11	6, 7, 8	8, 9, 10	6, 7, 8, 9, 10	7, 8	8.9
Naturalness (Simon, 1988, 1992)	Disturbance tolerant native species	Native accessorial species	Disturbance tolerant native species	Native accessorial species	Native accessorial species	Native accessorial species	Native accessorial species
Strategy (Borhidi, 1995)	Competitor	Competitor	Disturbance tolerant	Generalist	Generalist	Competitor	Specialist
Heat supply (Borhidi, 1995)	Submediterranean woodland and grassland belt	In accordance with thermophilous forest or woodland belt	In accordance with submontane broad leafed forest belt	In accordance with mesophilous broad-leaved forest belt	Submediterranean woodland and grassland belt	In accordance with thermophilous forest or woodland belt	In accordance with thermophilous forest or woodland belt

(continued)

Table 5 (continued)

Name	Festuca pseudovina Hack. ex Wiesb.	Artemisia santonicum L. (ssp.)	Bromus hordaceus L. ssp hordaceus	Plantago maritima L.	Podospermum canum C. A. Mey.	Puccinellia limosa (Schur.) Homberg.	Aster tripolium ssp. pannonicus (Jacq.) Soó
Soil moisture (Borhidi, 1995)	Xero-tolerants, but occasionally occurring on fresh soils	Xero-tolerants, but occasionally occurring on fresh soils	Plants of semi-humid habitats under intermediate conditions	Plants of fresh soils	Plants of semi-dry habitats	Plants of moist soils not drying out and well aerated	Xero-indicators on habitats with long dry period
Soil reaction (Borhidi, 1995)	Basiphilous plants	Basifrequent plants, mostly on basic soils	Mostly on neutral soils, but also on acidic and basic ones, generally widely tolerant more or less indifferent plants	Basiphilous plants	Basiphilous plants	Basifrequent plants, mostly on basic soils	Basiphilous plants
Nitrogen supply (Borhidi, 1995)	Plants of moderately oligotrophic habitats	Plants of sub-mesotrophic habitats	Plants of mesotrophic habitats	Plants of moderately oligotrophic habitats	Plants of habitats very poor in nitrogen	Plants of mesotrophic habitats	Plants of moderately oligotrophic habitats
Light intensity (Borhidi 1995)	Full light plants of open habitats	Full light plants of open habitats	Halflight plants mostly livingin full light, but shadow tolerant	Light plants	Light plants	Full light plants of open habitats	Full light plants of open habitats
Continentality (Borhidi, 1995)	Subcontinental, area in eastern CE	Continental species, reaching only eastern part of CE	Oceanic–suboceanic species area in whole Central Europe (CE)	Continental–subcontinental species, main area in Eastern Europe	Continental–subcontinental species, main area in Eastern Europe	Continental–subcontinental species, main area in Eastern Europe	Eucontinental species, main aera in Siberia and Eastern Europe
Salinity tolerance (Borhidi, 1995)	Beta-mesohaline plants living in soils of little chloride concentration 0.3–0.5%	Alfa-mesohaline plants living in soils of intermediate chloride concentration 0.7–0.9%	Species not occurring in saline or alkali soils	Polyhaline plants living in soils with high chloride concentration 1.2–1.6%	Alfa-mesohaline plants living in soils of intermediate chloride concentration 0.7–0.9%	Euhaline plants living in soils of very high chloride concentration 1.6–2.3 %	Polyhaline plants living in soils with high chloride concentration 1.2–1.6%

Table 6 Abiotic characteristics of the studied sites

Plant community	Artemisia saline puszta	Pannonic Puccinellia limosa hollow
Site	Apaj	Zabszék
Coordinates latitude	N 47° 05′14.0″	N 46° 50′ 34.8″
Coordinates longitude	E 19° 05′ 54.7″	E 19° 10′ 37.1″
Soil type	Solonetz	Solonchak
Soil clay % in 0–40 cm depth	30	34
Soil sand % in 0–40 cm depth	13	8
Soil physical properties in the uppermost layer of 3–9 cm		
Soil bulk density (g/cm^3)	1.7	1.76
Soil water retention at 0.1 kPa (saturation) (gravimetric %)	22	23
Soil water retention at 20 kPa (field capacity) (gravimetric %)	18	20
Soil water retention at 1500 kPa (wilting) (gravimetric %)	4.5	13
Soil saturated hydraulic conductivity (cm/day)	0.07	0.028
Chemical properties of 0–10 cm layer		
Soil organic matter in 0–10 cm layer %	2.2	0.9
Soil $CaCO_3$ content %	12	32
Exchangeable Na %	23	39
Cation exchange capacity-me/100 g soil	23	31
pH	8.3	8.5
Electrical conductivity of saturation extract (mS/cm)	1.8	5.3
Equivalent Na in cation composition of saturation extract (%)	76	98
Equivalent Ca in cation composition of saturation extract (%)	13	1
Equivalent Mg in cation composition of saturation extract (%)	1	1
Equivalent Cl in cation composition of saturation extract (%)	23	34
Equivalent SO_4 in cation composition of saturation extract (%)	18	14
Equivalent HCO_3 in cation composition of saturation extract (%)	53	44
Equivalent CO_3 in cation composition of saturation extract (%)	6	8
Average dynamic groundwater and lakewater properties		
Average (minimum, maximum) depth of groundwater level (m)	1.62 (0.6–2.3)	0.95 (0.7–1.5)
Average (minimum, maximum) EC of groundwater (mS/cm)	4.0 (0.92–5.5)	3.19 (1.45–3.82)
Average (minimum, maximum) pH of groundwater	8.7 (8.1–9.2)	8.2 (7.5–8.9)
Average (minimum, maximum) elevation of lake water level (m)	–	−0.26(−0.55+0.27)
Average (minimum, maximum) EC of lake water (mS/cm)	–	8.3 (2.2–31.9)
Average (minimum, maximum) pH of lake water	–	9.3 (8.1–9.9)
Average dynamic properties of 0–40 cm layer		
Average (minimum, maximum) $EC_{2.5}$ of soil in 0–40 cm layer-mS/cm	0.85 (0.45–1.17)	1.0 (0.75–1.78)
Average (minimum, maximum) pH of soil in 0–40 cm layer	9.6 (9.0–10.2)	10.0 (9.7–10.4)
Average (minimum, maximum) pNa of soil in 0–40 cm layer	2.36 (1.99–2.89)	2.2 (1.69–2.9)
Average (minimum, maximum) soil moisture of soil in 0–40 cm layer (%)	17.6 (5.2–38.3)	20.9 (14.6–29.3)

site than at the "Artemisia saline puszta" site in the surface layer. The reason for the difference is the contrastive soil types, which is Solonchak at the more saline "Pannonic Puccinellia limosa hollow" site. This soil type has the maximum of salt accumulation right at the surface. On the other hand, the Solonetz of the "Artemisia saline puszta" site has less saline, and alkaline, somewhat leached surface horizon, and reaches its maximum of salt concentration at greater depth. Consequently plants with low salt tolerance can live on Solonetz soils. On the other hand the differences in the temporal, depth average values (0–40 cm depth, 67 months) were not great.

The chemical composition of saturation extract of the surface layer (Table 6) shows that there is more Ca ion and less Cl at "Artemisia saline puszta" site, indicating the tendency of evaporative concentration of solutes. With increasing concentration of solutes Ca has a tendency of precipitating and Cl the tendency of increasing concentration, since this anion is the last to precipitate (Tanji 1990).

The groundwater was shallower at the lakeside "Pannonic Puccinellia limosa hollow" site and both average levels corresponded well to the characteristic groundwater depths in Solonchak (<1 m) and Solonetz (1.5–3 m) soils. The salinity of groundwater is only slightly higher at "Artemisia saline puszta" site probably due to less effect of atmospheric water. During the 5.5 years study period extreme low values were observed in one winter and were related to shallow groundwater. The average pH of soil at surface and in 0–40 cm layer is slightly higher at lakeside "Pannonic Puccinellia limosa hollow" site. There is an opposite tendency in the pH values of the groundwater, since it fluctuates in the more saline layers of "Artemisia saline puszta" site (Solonetz) and less saline layers of "Pannonic Puccinellia limosa hollow" site (Solonchak). The average value of lake elevation shows that it was 0.26 m deeper than the surface of the "Pannonic Puccinellia limosa hollow" site. During the driest period, when the salinity of lake water approached the concentration of sea, the level of lake was 0.55 m deeper. On the other hand the maximum level was 0.27 m above the surface, when the data collection was suspended for 16 months in the year 1999 due to extreme large precipitation. Lake water pH showed a clearly positive linear relationship to the EC. The dilution of the lake water by precipitation results in the lower dominance of carbonates (Dvihally 1960; Tóth et al. 2003).

Average dynamic properties of 0–40 soil layers showed the tendencies expected because of soil type (salinity = $EC_{2.5}$, pH) and presence of lake (moisture). The largest average moisture percentages measured at the two sites were within the limit of "available moisture range" defined between the field capacity and wilting point. On the other hand in the surface 0–10 cm layer soil moisture decreased below wilting capacity from time to time.

The plant composition of the sites, being dominated by Poaceae, Compositae and Plantaginaceae as shown in Table 5 is very similar to other temperate inland saline habitats and reflect the abiotic conditions. Compared to "Artemisia saline puszta" site the plants in "Pannonic Puccinellia limosa hollow"site prefer wet soil, light places and tolerate high salinity.

At "Artemisia saline puszta" site out of the five dominant plants four are typical for Pannonic saline grasslands. *Bromus hordaceus* is a common weed and its spreading is a relatively new feature in these vegetation types (Bagi 1989). Most of the ecological characteristics of *B hordaceus* are very different from the other four dominant species and show that the site is in change. At the surface where the plant roots live the conditions has become permissible for this plant.

At "Pannonic Puccinellia limosa hollow" site *Puccinellia limosa* is an endemic species, and the accompanying *Aster tripolium* is represented by an endemic subspecies. As Table 6 indicates, the surfaces of the quadrats were not covered completely with green vegetation during most of the study period. The bare surface changed between 0% when large stalks and leaves covered the soil surface and 100% during and immediately after waterlogging.

20 Utilization of Halophytes in Hungary

The first and most important utilization of the native vegetation is the grazing by domestic animals. Earlier grazing was typical for each of the saline and sodic vegetation types ranging from the tall grass occurring in the highest lying areas down to the lowest lying marshes. At present grazing is not typical in the wetlands. Loess steppes and taller grass sodic grasslands were and are grazed by horses and cattle. Shortgrass sodic grasslands were and are grazed by sheep. In the salt meadows typically there is no grazing, but hay cutting. Water buffalo and pigs were grazing the wetlands earlier. Also the Hungarian Grey Cattle are able to graze wet habitats, including reeds.

Besides grazing and hay cutting, reed (*Phragmites australis*) cutting is also typical in the wetlands belonging to saline and sodic vegetation. Also cattails (*Typha* sp) are collected here.

Bare sodic habitats have *Matricaria chamomilla* (chamomile) stands, which are collected as medicinal plant. Also *Achillea* sp, *Artemisia* sp, *Symphytum officinale, Althaea officinalis* are collected for medicinal purposes. *Limonium gmelini* and *Aster sedifolius* are collected from time to time as decorative plants.

Among mushrooms it is the *Agaricus bernardii* and *Marasmius oredes* which are collected most.

References

Bagi I (1989) The pedological reasons and ecological implications of the occurrence of *Gypsophila muralis* L. in the salt-affected areas of Kiskunság. Botanikai Közlemények 76:51–63, in Hungarian

Bagi I, Molnár Zs (2003) Habitat F4. In: Bölöni J, KunA, Molnár Zs (eds) Comparative survey and evaluation of the vegetation of Hungary. Working Group for data quality. Manual of habitats 2.0. Vácrátót (in Hungarian)

Bakacsi, Zs, Kuti L (1998) Agrogeological investigation on a salt affected landscape in the Danube Valley, Hungary. Agrokémia és Talajtan 47:29–38

Ballenegger R, Finály I (1963) The history of soil science until 1944. Budapest, Akadémiai Kiadó (in Hungarian)

Bölöni J, Kun A, Molnár Zs (2003) Habitat guide. Institute of Ecology and Botany of the HAS, Vácrátót, Hungary, p 186, (in Hungarian)

Borhidi A (1995) Social behaviour types, the naturalness and relative ecological indicator values of the higher plants in the Hungarian Flora. Acta Botanica Hungarica 39:97–181

Csontos P (2001) Research methods of the study of natural seedbanks. Scientia Kiadó, Budapest, in Hungarian

Csontos P, Tamás J, Tobisch T (2002) Presentation of the database of the way of seed distribution of Hungarian flora. Exemplary evaluations and analysis of social behaviour. In: Salamon-Albert É (ed) A 70 éves Borhidi Attila köszöntése. PTE kiadványa, Pécs, pp 557–569 (in Hungarian)

de Sigmond A (1938) The principles of soil science. Thomas Murby, London

Devillers P, Devillers-Terschuren J (1996) A classification of Palaearctic habitats. Nature and environment, No. 78. Council of Europe. See http://www.botanika.hu/project/habhun/

Dvihally Zs (1960) Seasonal changes of the chemical composition of saline lakes. Hidrológiai Közlemények 40:313–316 (in Hungarian)

EUNIS Habitat Classification (2002) Version 2.3. 2002. http://mrw.wallonie.be/dgrne/sibw/EUNIS/home.html

Gerasimov IP (1960) The soils of Central Europe and related geographical problems. Izd. AK. Nauk SSSR, Moscow (in Russian)

Guidelines (1989) Guidelines to the field-scale mapping of soils. Agroinform, Budapest

Horváth F, Dobolyi K, Morschhauser T, Lõkös L, Karas L, Szerdahelyi T (1995) "Flora" database V1.2. Taxonlist and attribute file. Vácrátót (in Hungarian)

Hungarian Central Statistical Office (2006) nvironmental statistical yearbook of Hungary 2005. Hungarian Central Statistical Office, Budapest, 2006 November

Jozefaciuk G, Tóth T, Szendrei G (2006) Surface and micropore properties of saline soil profiles. Geoderma 135:1–15

Kreybig L (1937) The methods of soil survey and mapping used at the geological insitute. MÁFI, Budapest, in Hungarian

Kubiena WL (1953) Bestimmungsbuch and Systematik der Böden Europas. Ferdinand Enke Verlag, Stuttgart

Mádlné Szőnyi J, Simon Sz, Tóth J, Pogácsás Gy (2005) Connection between surface and groundwaters in the case of Kelemen-lake and Kolon-lake. Általános Földtani Szemle 30:93–110, in Hungarian

Marbut CF (1927) A scheme for soil classification. Proceedings and papers of the I. Intern. Congr. of Soil Sci. Vol IV. Washington

Molnár Zs, Borhidi A (2003) Hungarian alkali vegetation: origins, landscape history, syntaxonomy, conservation. Phytocoenologia 33:377–408

Molnár Zs, Bagi I, Tímár G (2003) Habitat F1.a. In: Bölöni J, Kun A, Molnár Zs (eds) Comparative survey and evaluation of the vegetation of Hungary. Working Group for data quality. Manual of habitats 2.0, Vácrátót (in Hungarian)

Molnár Zs, Biró M, Bölöni J, Horváth F (2008) Distribution of the (semi-)natural habitats in Hungary I.: marshes and grasslands. Acta Botanica Hungarica 50(Suppl):59–105

Pásztor L, Szabó J, Bakacsi Zs, László P, Dombos M (2006) Large-scale soil maps improved by digital soil mapping and GIS-based soil status assessment. Agrokémia és Talajtan 55:79–88

Pásztor L, Szabó J, Németh T (1998) GIS-based stochastic approach for mapping soil vulnerability. Agrokémia és Talajtan 47(1–4):87–96

Richards LA (ed) (1954) Diagnosis and improvement of saline and alkali soils. US Salinity Laboratory Staff. Soil and Water Conservation Research Branch. Agricultural Research Service, Washington DC, p 160

Simon T (1988) Categorization of the higher plants of the Hungarian Flora according to naturalness. Abstracta Botanica 12:1–23, in Hungarian

Simon T (1992) Keys to the Hungarian Flora. Tankönyvkiadó, Budapest, in Hungarian

Soil Survey Staff (1951) Soil survey manual. In: USDA-SCS Agriculture Handbook 18. U. S. Gov. Print. Office, Washington, DC

Soó R (1980) Conspectus associationum regionis Pannonicae. In Soó: A magyar flóra és vegetáció rendszertaninövényföldrajzi kézikönyve. Synopsis systematico-geobotanica florae vegetationisque Hungariae. Akadémiai Kiadó Budapest, VI: 525–538 (in Hungarian)

Stefanovits P (1963) The soils of Hungary, 2nd edn. Akadémiai Kiadó, Budapest, in Hungarian

Szabó J (1861) Counties of Békés and Csanád. Description of geological conditions and soil types accompanied with a coloured geological map. Magyar Gazdasági Egyesület, Budapest (in Hungarian)

Szabó J, Molnár J (1865–6) Description and classification of the soils of Tokaj-Hegyalja. Matematikai és Természettudományi Közlemények IV (in Hungarian)

Szabolcs I (ed) (1966) Handbook of genetic soil mapping of fields. OMMI, Budapest, in Hungarian

Szabolcs I (1974) Salt-affected soils in Europe. Martinus Nijhoff. Hague, the Netherlands

Szabolcs I (1989) Salt-affected soils. CRC Press, Boca Raton, FL

Szöőr Gy, Sümegi P, Balázs É (1991) Sedimentological and geochemical facies analysis of Upper Pleistocene fossil soil zones discovered in the Hajdúság region, NE Hungary. pp. 47–59. In: Pécsi M, Schweitzer F (eds). Quaternary environment in Hungary. Studies in geography in Hungary 26. Akadémiai Kiadó, Budapest, p 103

Szujkó-Lacza J, Fekete G, Kováts D, Szabó L, Siroki Z (1982) The vascular plants of the Hortobágy National Park. In: Szujkó-Lacza J (ed) Natural history of the Hungarian National Parks, vol 3, The flora of the Hortobágy National Park. Akadémiai Kiadó, Budapest

Tanji KK (ed) (1990) Agricultural salinity assessment and management. American Society of Civili Engineers, New York

Thompson K, Bakker JP, Bekker RM (1997) The soil seed banks of North West Europe: methodology, density and longevity. Cambridge University Press, Cambridge

Tóth T (2008) Salt-Affected Soils in Hungary, pp 75–81. In: Needs and priorities for research and education in biotechnology applied to emerging environmental challenges in see countries. Workshop Proceedings. NOVI SAD, SERBIA 2008. UNESCO Office in Venice-Bresce, Italy and Institute of Lowland Forestry and Environment, Novi Sad, Serbia

Tóth T, Kuti L (2002) Factors of the changes in soil salinity at Apaj, Kiskunság region, pp. 106–116. In: Kátai J, Jávor A (eds) Soil and environment. Debreceni Egyetem Agrártudományi Centrum, Debrecen (in Hungarian)

Tóth T, Kuti L, Fügedi U (2003) Monthly observations at Zab-szék saline lake. Temporal changes of lake water and groundwater, soils and vegetation. Természetvédelmi Közlemények 10:191–206, in Hungarian

Tóth T, Szendrei G (2006) Types and distribution of salt affected soils in Hungary, and the characterisation of the processes of salt accumulation. Topographia Mineralogica Hungariae IX:7–20, in Hungarian

Tóth T, Kuti L, Kabos S, Pásztor L (2001) Use of digitalized hydrogeological maps for evaluation of salt-affected soils of large areas. Arid Land Res Manage 15:329–346

Várallyay Gy (1989) Soil mapping in Hungary. Agrokémia és Talajtan 38:696–714

Várallyay Gy, Szabó J, Pásztor L, Michéli E (1994) SOTER (Soil and Terrain Digital Database) 1:500,000 and its application in Hungary. Agrokémia és Talajtan 43:87–108

Várallyay Gy, Szücs L, Zilahy P, Rajkai K, Murányi A (1985) Soil factors determining the agro-ecological potential of Hungary. Agrokémia és Talajtan 34(Suppl):90–94

Halophytes in Republic of Macedonia

S. Mitrev, Lj. Mihajlov, F. Trajkova, B. Kovacevikj, and V. Zlatkovski

Abstract Republic of Macedonia is situated in Balkan Peninsula with a total area of 25,942 km^2 and 1.26 million hectares of agricultural land, where 560,000 ha are arable land (44.2%) mainly located in the valleys and 700,000 ha are pastures (55.6%), located in hilly and mountainous regions of the country. There are approximately 11,000 ha saline soils, mainly found in four valleys (Ovche Pole, Pelagonija, Skopska and Strumica). These soils are of natural origin, but anthropogenic factors play important direct and indirect role in their genesis. The halophyte vegetation on these soils has been studied at length. This vegetation is facing great biotic interferences as a result of this some species or associations are threatened with extinction. This paper will therefore try to enlighten the current situation of this vegetation in Macedonia.

1 Introduction

Republic of Macedonia is situated in Balkan Peninsula with a total area of 25,713 km^2 (Fig. 1). Around the half of the total area (1.26 million hectares) belongs to the agricultural land with 560,000 ha arable land (44.2%) and 700,000 ha pastures (55.6%). From the total arable land, 461,000 ha are (82%) cultivated area and horticultural crops, 58,000 ha meadows (10%), 26,000 ha vineyards (5%) and 15,000 ha orchards (3%). Presently, the cultivated area in the country has been recorded as 0.625 ha per person, which is higher than the standard in EU-25 (National Strategy for Agriculture and Rural Development for Republic of Macedonia 2007–2013 (2007)). According to the statistics, 1/3 of the total cultivated area is abundaned every year because of the migration of rural people into the towns and industrial developments. The agricultural land has declined from 633,000 ha in 1999 to 560,000 ha in 2004.

The country enjoys two major types of climate due to its specific natural and geographical characteristics; Mediterranean and continental. There are two seasons cold, wet winters and dry, hot summers. However, the high mountainous areas are characterized by a mountainous climate with short, cool summers and considerably cold and moderately wet winters, with precipitation being mainly in the form of snow. In spite of the fact that Macedonia lays relatively close to the Aegean and Adriatic Seas, the influence of the Mediterranean climate does not reach very deeply into the country, except within a few valleys. This is a result of the high mountains which rise up in the west and south of the country. The average annual temperature is 11.3°C. The towns of Valandovo and Gevgelija show the highest average temperatures laying around 14.5°C and 14.3°C, respectively. In the mountainous areas, the mean annual temperatures are 4.7°C in Popova Shapka, 6.8°C in Lazaropole and 8.2°C in Krushevo.

The average precipitation in the country is 683.7 mm/year. Humidity is higher in October and December, less in March and May. Rain splash in spring and autumn is very often exuberant, and contribute much to the soil erosion and local overflow. During the vegetative period, very often dry spills occur, signifying that water is limited factor for intensive agricultural production. Late spring frosts, and early autumn rimes happen very often. The areas of highest precipitation occur in

S. Mitrev (✉), Lj. Mihajlov, F. Trajkova, B. Kovacevikj, and V. Zlatkovski
Goce Delchev University – Shtip, Krste Misirkov b.b., P.O. 201, 2000 Shtip, Republic of Macedonia
e-mail: sasa.mitrev@ugd.edu.mk

M. Öztürk et al. (eds.), *Sabkha Ecosystems*, Tasks for Vegetation Science 46,
DOI 10.1007/978-90-481-9673-9_14, © Springer Science+Business Media B.V. 2011

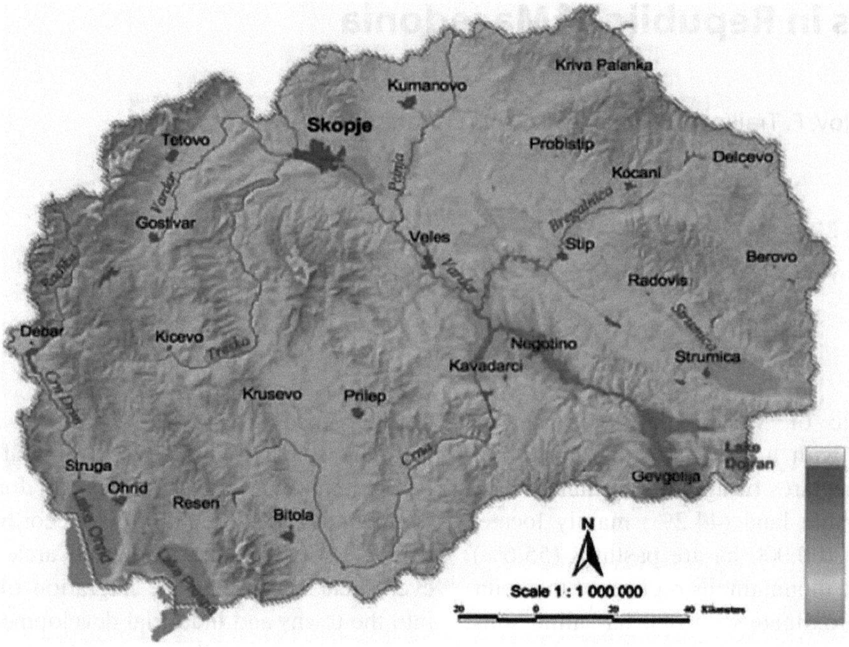

Fig. 1 Map of Republic of Macedonia

Mavrovi Anovi and Resen, with 1,197 mm and 757.9 mm, respectively, and the least in Ovche Pole Plain with only 490.3 mm.

2 Genesis and Classification of Saline Soils

The landform in the Republic of Macedonia is characterized by complex geotectonic features, which produce developed relief, complex geology and diversity of soil types: automorphous, hydromorphic, halomorphic, subaquatic) (The Country Study for Biodiversity of the Republic of Macedonia 2003). There are over 30 soil types, due to the diversity of the relief, climate, vegetation, geological formation and the anthropogenic factors. We come across eight climate–vegetative–soil areas, very heterogenic in climate, vegetation and soil profile. The arable area is located mostly in sub-Mediterranean and warm continental area between 50 and 900 m above the sea level (Filipovski 1985).

The area of saline soils is 0.42% of the total country area. There are approximately 11,000 ha of saline soils, mainly placed in four valleys: Ovche Pole, Pelagonija, Skopje and Kochani, where halomorphic soils extend on large areas, and in Valandovo and Strumica, where the halomorphic soils can be found in the form of small patches of several square meters located on the arable lands (Filipovski 1985). According to Second National Ecological Action Plan (2006) irrigation, especially in arid areas is leading to increased soil salinization, but because of the absence of monitoring or research, the intensity and current situation of salinization cannot be defined precisely.

According to Filipovski (1985), halomorphic soils are formed under certain climatic, relief, hydro-geographic and vegetation conditions. They mostly occur in arid and semiarid areas, halomorphic soils in the region of Ovche Pole, belongs mostly to the type of solonchak and solonetz. The concentration of salts in the underground waters of the solochak soils in this valley is very high 25–100 g/L. In other valleys the concentration of salts in underground waters does not exceed 2–4 g/L. The typical solonchak soils are formed under the influence of shallower underground waters, which contain more salts, but the neutral salts are predominant (chlorides and sulfates).

The bedrock plays very important role in the genesis of halomorphic soils. These soils are formed only on the top of sediments (alluvium, lakesides, marine, eolic), at accumulation grounds, where soils have been precipitated together with sediments in the past. The anthropogenic factors are very important for genesis and evolution of these soils. The irrigation is contributing towards the expansion of such soils. The erosion contributes by filling the rivers and through an alleviation of underground waters. Intensive grazing and destruction of natural vegetation, especially grasses, encourage the process of evaporation of saline underground waters and accumulation of soils on the surface horizon.

3 Plant Diversity of Saline Soils

The halophyte plant communities show a restricted distribution in the country. These develop on small area in Ovche Pole and in the steppe-like area between Veles, Shtip and Negotino. The halophyte vegetation in the steppe-like areas is categorized as center of high floristic diversity. This vegetation is found on small areas. Although detailed studies have been undertaken, discussions on definition on certain halophytic associations is still continuing (Tzonev et al. 2008).

Very few botanists have worked on the saline soil plant diversity in Macedonia. The earliest research on the taxonomy and phytosociology of halophytes has been conducted by Micevski (1965). He has given a detailed description of vegetation and floral classes, orders and communities in Ovche Pole. Although the area is very small, Micevski (1965) has defined 11 floral associations with a mosaic distribution of the halophyte communities conditioned by the relief and heterogenic chemical composition of the soil (Fig. 2).

The following associations have been recorded by Micevski (1965) on the saline soils in Ovche Pole.

1. Class *Salicornietea* on saline soils (solonchaks and soloneotzs) Br. Bl. et Tx. 1952. The halophytic associations presented here are:
   - Ass. *Suaedetum maritimae balcanicum* is very poor in plant taxa characterised by *Suaeda maritime* Dum. and individual specimens of *Punccinellia convolute* (Horn.) Hay. The association is accompanied by *Hordeum maritimum* Huds. and *Matricaria chamomilla* L. This association develops only on solonchak soils.
   - Ass. *Camphorosmetrum annuae balcanicum* is represented by the characteristic species of *Camphorosma ovata* W.K. and characteristic species for the alliance *Punccinellion convolutae* is *Punccinellia convoluta* (Horn.) Hay. The association shows poor species composition and because of high salt concentrations, only two species are growing on solonchak soil, related to the eroded solonetz rich in sodium.
   - Ass. *Puccinellietum convolutae* is represented by the characteristic species of *Puccinellia convoluta* (Horn.) Hay. and its companion species are *Hordeum marinum* Huds., *Lepidium ruderale* L., *Matricaria chamomilla* L. and where water is dwelled *Phragmites communis* Trin.
   - Ass. *Hordeo-Trifolietum parviflori* is represented by the characteristic species of *Trifolium parviflorum* Ehrh. and *Podospermum canum* C.A.Mey. This association is rich in companion taxa such as *Hordeum marinum* Huds., *Cynodon dactylon* Pers., *Matricaria chamomilla* L., *Trifolium nigrescens* Viv., *Poa bulbosa* L., *Trifolium echinatum* M.B., *Lepidium ruderale* L., *Trifolium resupinatum* L., *Crepis setosa* Hall., etc. Although this association is under an impact of heavy grazing it is richer in species because of the favourable chemical conditions of the soils.
   - Ass. *Camphorosmetum monspeliaceae* is represented by the characteristic species of *Camphorosma monspeliaca* L. and *Plantago coronopus* L. var *commutata* (Guss.) Bég. The association develops on solonetz soils in small patches with dimensions 1–3 m. *Camphorosma monspeliaca* L. is protecting the solonetz from erosion.
   - Ass. *Pholiureto-Plantaginetum balcaninicum* is represented by the characteristic species *Lepturus pannonicus* Kunth. and *Plantago tenuiflora* W.K. This association is rarely found and developes on very small areas, connected to microdepressions with stagnant waters.
   - Ass. *Crypsidetum aculeatae balcanicum* is represented by the characteristic species *Crypsis aculeata* Ait. and *Spergularia marginata* DC. It is typical for wet lands rich in salts. In the area of Ovche Pole it is developing on small depressions and around the water channels where water stays during spring and summer. The association flourishes during the last part of summer, mostly in August.

Fig. 2 Typical halophytic vegetation next to agricultural fields in Ovche Pole

Matevski et al. (2008) have conducted a study in the central part of the country, in the triangle between Veles, Shtip and Negotino. This is a special region used to be called steppe or semi-desert by some researchers who worked there. The bedrock in this area is composed of Paleogenic and more rarely Neogenic sediments. All Paleogenic sediments are salt-rich, thus halophytic vegetation develops at the site. Matevski et al. (2008) has given floristic valorization of endemics and rare species, halophytes, and steppe species. The survey on halophytes shows endemic taxa like *Artemisa martima* L., *Krascheninnikova ceratoides* (L.) Gueldenst., *Camphorosma monspeliaca* L., *Camphorosma annua* Pall. distributed in the steppe region. Their ecology differs significantly from the halophytes species growing on solonchaks in Ovche Pole. The halophytes in the steppe region grow on undeveloped soils, where the erosive processes are even more intense due to the completely degraded forest ecosystems (Matevski et al. 2008).

The halophytic vegetation communities in Macedonia are considered as communities with restricted distribution and they are seriously endangered and threatened with extinction. The most threatened is Ass. *Camphorosmetum monspeliacae*, and there is a great probability that other associations (e.g., Ass. *Crypsidetum aculeatae balcanicum* and Ass. *Pholiureto-Plantaginetum balcanicum*), which develop in small, shallow depressions, will also disappear if steps are not taken to preserve these.

The cultivation on the saline soils of the Ovche Pole Plain has endangered some halophytic species and communities. This especially refers to the species *Camphorosma monspeliaca* L., but also to other halophytes such as: *Crypsis aculeatus* (L.) Aiton, *Puccinellia convolute* (Horn.) Hay, *Suaeda maritima* L. etc. The species *Allium obtusiflorum* DC is now considered to be extinct (Fig. 3).

Heavy agricultural activity is posing a severe threat to the biodiversity of Macedonia, especially due to the current unfavorable conditions and negative developmental trends like draining of wetlands, destruction of valuable habitats like halophytic vegetation, pollution of soils, destruction of bio-corridors, etc. (Capacity self-assessment within the thematic area of biodiversity in Republic of Macedonia 2004). Filipovski (1985) has estimated the agricultural production value of saline soils in the country. The production characteristics of solonchak and solonetz soils depends on the total salt content, Na-ions content, physical characteristics, etc. The worst solonchak soils are without vegetation (so called bald places) or just with halophytic vegetation. Some of the solonchak/solonetz soils are used as poor pastures which are green only in spring when the soil is wet and salts are diluted. Also, they are

Fig. 3 Saline soils in Ovche Pole under severe anthropogenic influence

used in the production of field crops such as barley, sunflower, alfalfa, but without amelioration only chamomile can be grown, but the productivity and crop yields are very low.

4 Conclusion

The distribution of halophytic vegetation is restricted to saline soils and the total area under saline soils in the Republic of Macedonia is 11,000 ha. Thus, the halophytic vegetation has limited distribution, mainly in Ovche Pole and between Veles, Shtip and Negotino. The species composition of the halophytic vegetation is represented by plants typical for continental and Mediterranean climate types, specific soil composition, hydrological and relief characteristics. Although the taxonomy of halophytes has been studied in details, there is a need for future studies on the ecological and higher rank syntaxa. This vegetation type is under significant anthropogenic pressures and some of the halophytic taxa and associations are seriously endangered and threatened with extinction.

References

Braun-Blanquet J (1952) Les groupements végétaux de la France Méditerranéenne C.N.R.S., Paris, p 297

Capacity self-assessment within the thematic area of biodiversity in Republic of Macedonia (2004) UNDP, Skopje, Jan 2004, p 1–80

Country Study for Biodiversity of the Republic of Macedonia (First National Report) (2003) Ministry of Environment and Physical Planning, Skopje, p 1–217

Filipovski Gj (1985) Pedologija, IIIth edn. Univerzitet Kiril i Metodij, Skopje, p 1–600

Matevski V, Čarni A, Kostadinovski M, Košir P, Šilic U, Zelnik I (2008) Flora and Vegetation of the Macedonian Sttepe. ZRC SAZU, Biološki inštitut Jovana Hadžija, Ljubljana, p 1–96

Micevski K (1965) Halofitska vegetacija Ovčeg polja. Musei Macedonici Scentarum Naturalium Skopje 10:67–90

National Strategy for Agriculture and Rural Development for Republic of Macedonia 2007–2013 (2007) Ministry of Agriculture, Forestry and Water Management, Skopje, June 2007, p 1–156

Second National Ecological Action Plan (National Action Plan for Environment) (2006) Ministry of Environment and Physical Planning, Skopje, p 1–230

Tzonev R, Lysenko T, Gussev C, Zhelev P (2008) The halophytic vegetation in South-East Bulgaria and along the Black Sea Coast. Hacquetia 7(2):95–121. doi:10.2478/v10028-008-0006-3

Low-Cost Mapping of Sabkha-Ecosystems Using Satellite Imagery: An Example from SE Qatar

Jörg Beineke and Andreas Kagermeier

Abstract The article tries to show up an easy way to acquire satellite images (SPOT monochromatic/grayscale) for free, using the internet portal "Geoengine", maintained by the National Imagery and Mapping Agency (NIMA) of the United States. In a second step, these images (in this case sabkha-dune ecosystems in south-eastern Qatar) are loaded into a GIS (Geographical Information System)-program (ArcView® 3.2) and analysed for reflectance values, correlated with the values for known sabkha areas, that were taken during earlier field trips in Qatar.

The combination of satellite imagery and ground control points within areas of known geomorphological and geoecological characteristics produces an approximate map of the spatial extent and distribution of sabkha ecosystems in south-eastern Qatar.

1 Introduction

Systematic mapping of ecosystems using satellite imagery – especially for education purposes – is often hindered by the availability of adequate metadata and sufficient financial support to acquire the necessary scenes. Even though this is still true for most of the actual and high resolution data, there are few possibilities to acquire satellite images through the internet free of charge, except online-fees.

The online distribution of satellite images was limited by the data transfer rate for a long time. Up to the end of the 1990s, only enterprises, universities or other governmental institutions could afford fast internet access. In the meantime, the demand for high-speed internet connection has risen, so that a complete coverage system of fast internet access is available in most industrial countries.

Following the spread of fast and reliable internet access, some companies – like Microsoft®'s Terraserver – developed systems that allowed the user to view and print small parts of satellite images for private purposes. However, the user could not save whole images to disk or print large scenes, nor were the images ready to use in Geographical Information Systems or other professional digital imaging tools.

2 The Geoengine: Data Acquisition

One of the rare opportunities to acquire images at a larger scale is provided by the Geospatial Engine (http://geoengine.nima.mil) (Fig. 1). Managed by the National Imagery and Mapping Agency (NIMA), this internet portal offers satellite images, digital elevation data and vectorized basemaps for many parts of the world.

The data coverage includes the western central part of the United States, Europe/Eurasia excluding Portugal, Ireland and Scandinavia, major parts of the Arabian Peninsula including the Sinai and some scattered areas (e.g. Central America, Libya, Korea and South Eastern Asia). South America, major parts of Asia and Africa, as well as Australia and Antarctica are not covered (Fig. 2).

The article was submitted in 2003.

J. Beineke (✉)
German Academy of Sciences Leopoldina,
Emil-Abderhalden-Strasse 37, 06108 Halle (Saale),
Germany
e-mail: joerg.beineke@leopoldina.org

A. Kagermeier
Freizeit- und Tourismusgeographie,
Fachbereich VI Geographie/Geowissenschaften,
Campus II - Behringstrasse,
University of Trier,
54286 Trier, Germany

M. Öztürk et al. (eds.), *Sabkha Ecosystems*, Tasks for Vegetation Science 46,
DOI 10.1007/978-90-481-9673-9_15, © Springer Science+Business Media B.V. 2011

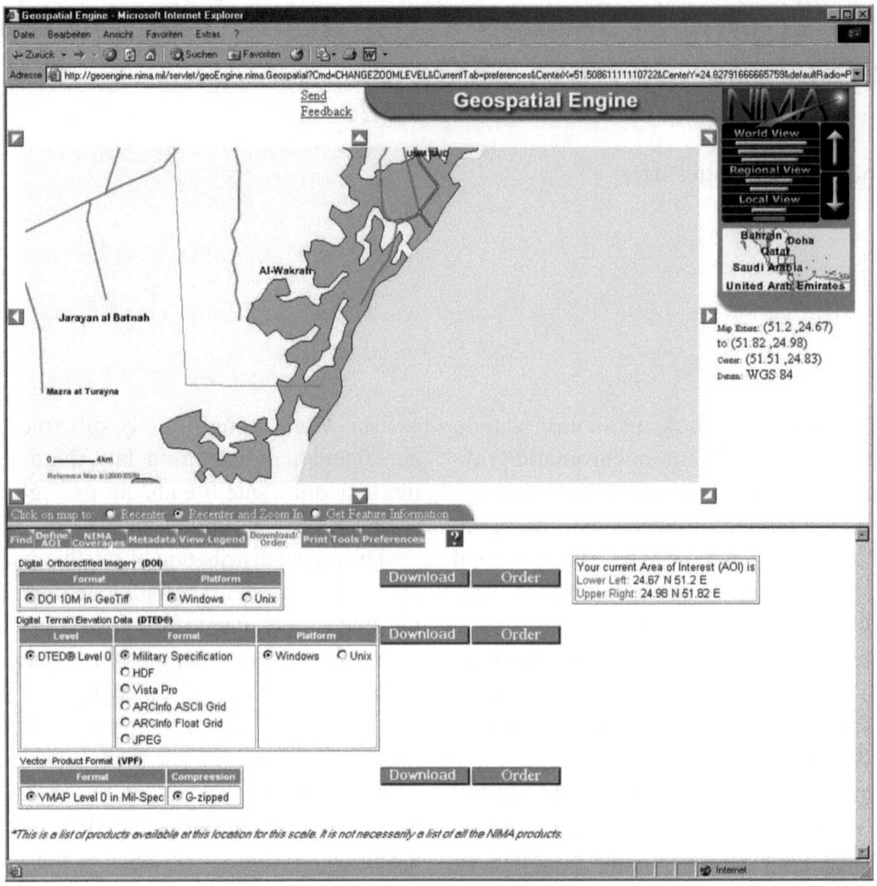

Fig. 1 The Geoengine website

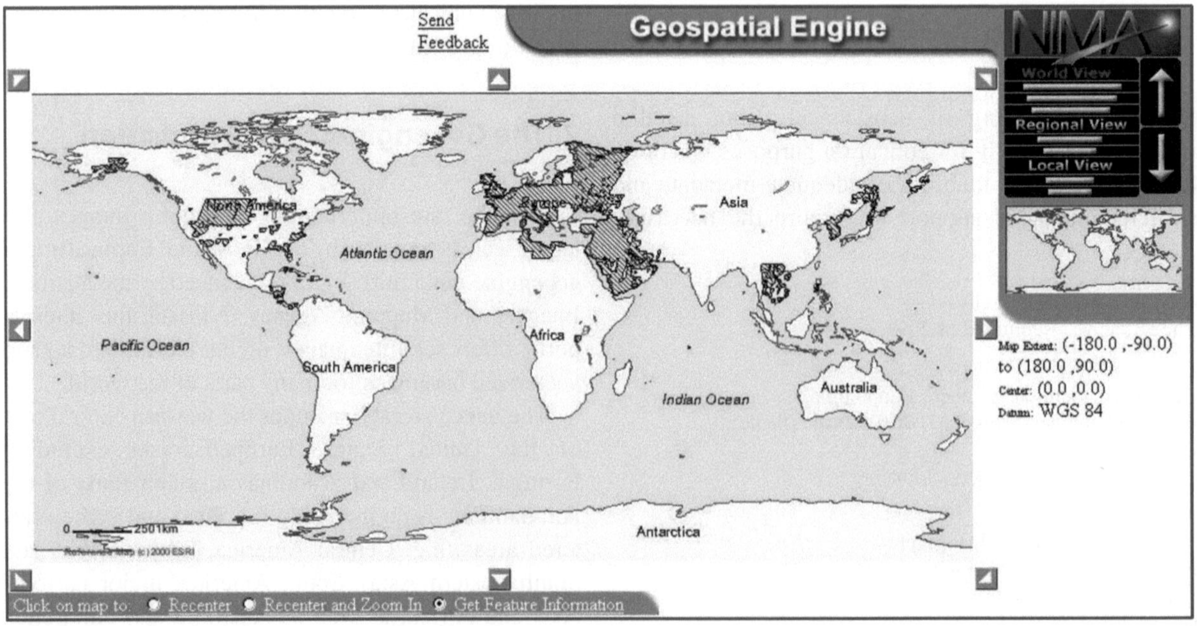

Fig. 2 Spatial coverage of DOI 10M satellite images

The offered satellite images were originally obtained from the SPOT Image Corporation under an unrestricted license. They are delivered as 10 m digital resolution orthorectified imagery (DOI-10M) and were acquired between 1992 and 1994. Depending upon operating system, the images are delivered in the georeferenced GeoTiff-format for Windows or UNIX and can be used in professional GIS-Software as well as in conventional low cost image processing systems without making full use of all advantages of the given format.

Data access is very simple. The website is divided into two frames: the upper frame contains a browsable map of the world, based on ESRI®'s mapserver technology. The window allows users to choose their area of interest via browsing and zooming on a "world" to "local view" level. The lower frame is subdivided into several categories regarding file format, data source, date and extend of data acquisition, download proceedings and data find, which allows users to search images via spatial coordinates.

When the reference map in the upper frame is centred on the appropriate area of interest, the data can be downloaded in the required format from the "download" button in the lower frame. If the user is not sure, whether the requested spatial extent is covered by the DOI-10M or needs a preview of the area, the "view" button gives a quick (low quality) overview of the availability and quality of the existing imagery. The initial download request is followed by the determination of the estimated download size, which has to be below 50 MB to reduce the server load of the Geoengine-Website. If the file-size is bigger than 50 MB, the user is asked to reduce the spatial extent and an optional part of the area is presented. To speed up the download time, the satellite imagery is automatically compressed to a ZIP-archive afterwards, which reduces the file-size to about 50% of the original image. This process takes several minutes (depending on server load), but the building progress is documented on the website every few seconds. When the build of the compressed file is finished, the server indicates the final size of the ZIP-archive and draws attention to the licence conditions that are implied in the use of these images. The download time of the compressed archive depends on the size of the original image, the type of the internet connection and the server load of the NIMA Geoengine. Given a high speed internet access, download time should be in the range of minutes with maximum transfer rates of about 400–500 KB per second. Users with slower access can use download managers to stabilize the data connection, because the Geoengine-Server supports the "resume"-function of most of these programs. After download is complete, the ZIP-file has to be decompressed using an unzip utility. The resulting GeoTiff-file can now be viewed or analysed in the corresponding software programs.

A similar procedure is necessary for the download of digital elevation data (DTED®), that is available in various specifications like ARCInfo ASCII or Float Grid, JPG or HDF etc. The Digital Terrain Elevation Data (DTED®) Level 0 was developed by the NIMA to provide the scientific community as well as the interested public with basic quantitative terrain data on a digital basis. In contrast to other high resolution elevation data that is intended for military purposes, the DTED® Level 0 data was deliberately "thinned" from the original NIMA DTED® Level 1 to a nominal resolution of 1 km per pixel (= 30 Arc Second Terrain Data). This specification is suitable for the modeling of whole regions or countries, but only of limited use on a local level. The diverse specifications of the data permits the use of the data in most GIS-systems and makes it accessible to the general public.

Apart from the described satellite imagery and digital terrain data, the Geoengine-website offers many other features that can be used in a geographical context, but are not listed here for clarity reasons. One example for additional data is the availability of vectorized basemaps (VMAP Level 0 in military specification) in the Vector Product Format that can be used in most GIS programs. The map preview in the upper frame is based on the VMAP Level 0 and gives a first impression of the quality of the downloadable files available on a CD-ROM sized basis.

3 Geographical Aspects of the Study Area

The study area is located in the southeastern part of Qatar, south of Umm Said Industrial City (Fig. 3). It is part of an approximately 60 km long and 30 km wide belt of active dunes, that are moving from NNW to SSE (Embabi and Ashour 1993). The up to 40 m high dunes are orientated perpendicular to the dominant "Shamal" winds from NNW and increase their density towards the SSE (Al-Sheeb 1998). In upwind direction the sand accumulations mainly consist of single barch-

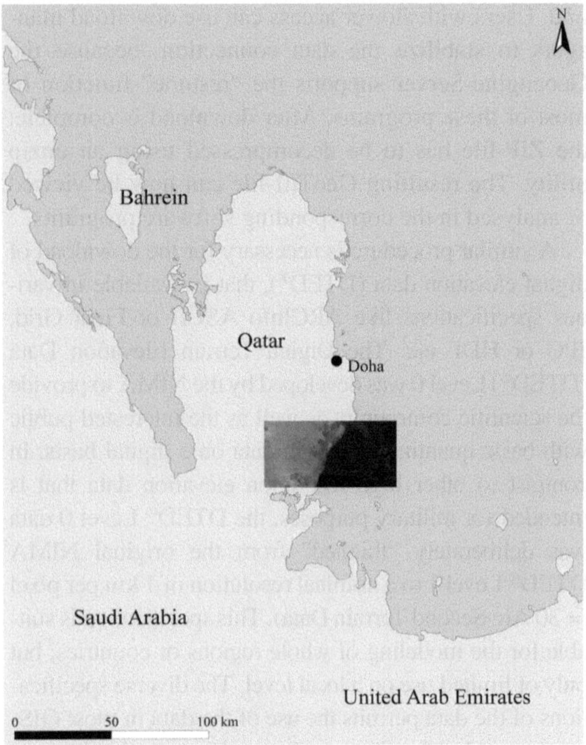

Fig. 3 Location of the study area

ans, whereas they coalesce into complex/compound barchans and transverse dunes further downwind due to a higher sediment supply.

Most dunes – especially in the inland area – are underlain by Eocene limestones and dolomites of the Upper Damman Formation, that crops out in major parts of Qatar peninsula. Near the southeastern coastline, these Tertiary rock formations are overlain by an up to 30 m thick sequence of sabkha deposits, mainly composed of siliciclastic sands and minor parts of dolomite or other carbonate fragments (Shinn 1973). This approximately 500 km² large area of sabkha is partly covered by sandsheets and active sand dunes that are migrating over the damp surface towards the sea. In contrast to most coastal dunefields, where dunes are driven inland by onshore winds, the Umm Said Dunes are prograding offshore, thereby creating small pools between dunearms that are frequently flooded during high tides.

The origin of the dunes in southeastern Qatar is still uncertain. Most researchers believe that the siliciclastic sands of the Umm Said dunefield are of external origin and were not generated on Qatar Peninsula (Shinn 1973; Ashour 1987). The mineralogy of the dune sands indicates to an origin on the Arabian mainland, that nowadays is separated from Qatar peninsula by the Gulf of Salwa. It is speculated that during late glacial lowstands of the Arabian Gulf, combined with the enhanced "Shamal"-Winds of this time period, dunes were piled up in the Eastern Province of Saudi Arabia and moved onwards in SSE direction towards the Gulf of Salwa (Shinn 1973). Sandbars (former transverse dunes today covered by coastal waters) which are detectable on satellite images of this area (Al-Hinai et al. 1987), point to the fact that the dunes moved over the then dried out Gulf and reached the northwestern coast of Qatar. As sea-level rose in the early Holocene, the dunes were cut off from the sediment source areas in the NW and were eroded by undersatured winds, thereby moving further to the SSE. Due to an almost flat territory, the sands passed the Eocene limestone formations of central Qatar without major disturbances and reached the southeastern coast, where huge amounts of sand were spilled into the low-lying coastal areas and the sea.

Interspersed with small portions of local carbonate grains, these siliciclastic sands were responsible for the development of the sabkha sequence, whose thickness is controlled by the sea-level as an absolute erosion basis. Given the limited to exhausted sand supply and steady winds from the NNW, the sand dunes of the Umm Said area will be gone in a matter of 1,000 years (Shinn 1973), thereby producing larger sabkha areas, that will be overformed by coastal processes after the depositions of quartz sands ends.

4 Methods

This study uses Esri's ArcView 3.2 GIS program to analyze the satellite images. Although most image processing programs are able to execute the basic manipulation processes of this study like grouping of color values, GIS-systems simplify and enlarge the possible modifications to a certain degree. To make full use of ArcView's image processing capabilities, it was necessary to load the Spatial Analyst 2.0 extension in addition to the core program.

After downloading the selected sample imagery of southeastern Qatar (Fig. 4) from the Geoengine website, the GeoTIFF satellite image was extracted from the ZIP-archive and loaded into ArcView using the

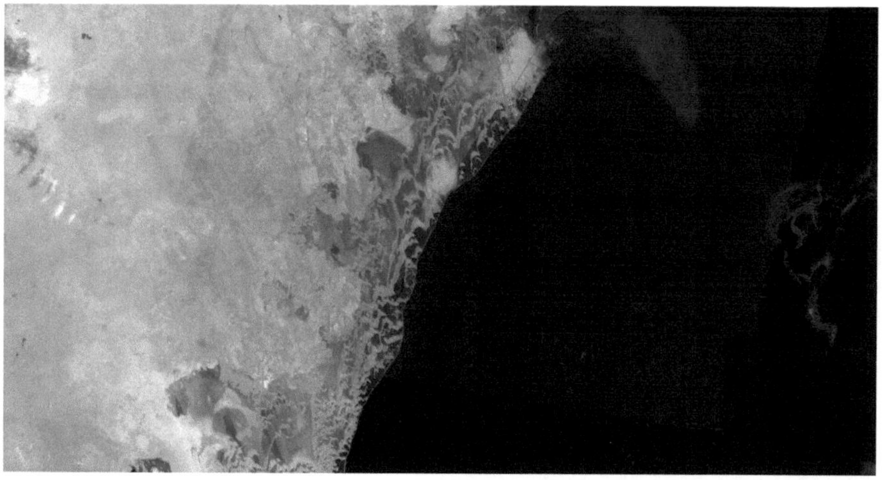

Fig. 4 Unchanged original satellite image from the Geoengine website. See Fig. 3 for scale (© CNES/SPOT Image 1992–1994)

TIFF 6.0 Image Support module. Due to its data format that contains spatial references, the image is directly placed in its correct geographical position. This feature makes it possible to avoid problems with the Geoengine's limitation of 50 MB per image because related files can be placed in the right position without further efforts, thereby facilitating the analysis of larger areas.

Afterwards, the GeoTiFF was converted into an ArcView Grid using the Spatial Analyst extension. Even though basic information like gray values can be directly derived from the original image, this transformation is necessary to classify single color values into distinct groups and to mark areas with special characteristics. By default, the image conversion produces nine distinct classes (equal distance) of color values and one field for missing data using a random color ramp. Although no intentional modifications of the image were made at this point, the combination of single values into groups results in the basic structuring of dominant landscape elements. This effect can be recognized extremely well in the northeast of the grid, where underwater features like longshore-drift of sediments are detectable, that are barely visible in the original satellite image (Fig. 5). Something similar applies to the sabkha areas that are marked by dark (original) values due to their high water content, whereas the sea is characterized by its unmistakable black value.

To refine this random classification, ground control points were set in the study area during an earlier field trip. These points were taken in areas that were visually identified in the field as sabkha areas using a handheld GPS (Garmin etrex, 5 m average resolution). They were converted into shapes (point theme) and loaded into the ArcView project. Comparison of the ground control points with the color values of the grid theme showed that all known sabkha areas have values in the range of 30–90 (based on the original color scheme of the gray scale image). If applied to the random grouping, setting values below 30 as sea and above 90 as other terrestrial environments, this classification produces an approximation of the sabkha areas in southeastern Qatar. To avoid the inclusion of marine elements like shallow waters in the "sabkha ecosystems group" (compare with Fig. 5), only pixels west of the coastline were included in the reclassification, whereas all pixels east of the coastline were classified as sea (Fig. 6).

Due to the simple processing measurements, this classification of the sabkha ecosystems in southeastern Qatar is far from perfect. As can be seen in the upper right corner of the random classification, reflectance values of sabkha areas with a high moisture content are very similar to those of very shallow waters. Sabkhas in a more continental setting are often covered by thin veneers of blown sand that falsify the readings as well. Taken all these things together, this method is not appropriate (and not intended) for high resolution interpretation of satellite images with the concentration on single elements, but for a quick low-cost overview of landscape elements with similar reflectance characteristics.

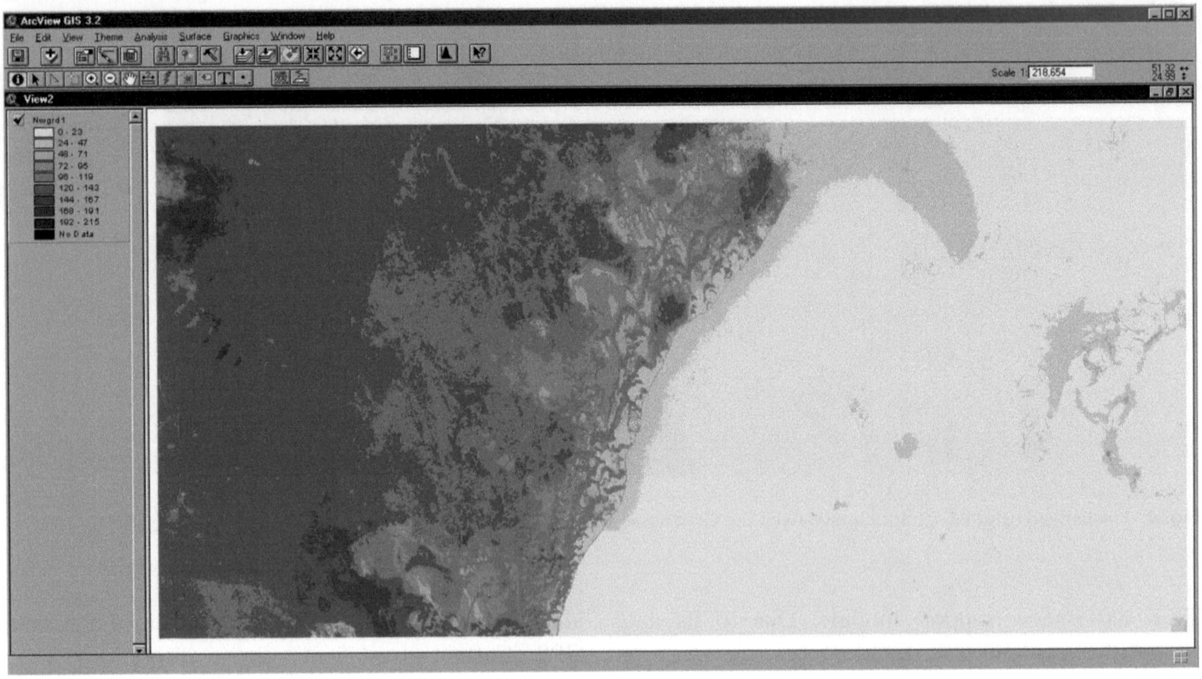

Fig. 5 Random classification after conversion into ArcView Grid. Please note the visibility of shallow water in the NE of the image that was barely recognizable in the original image

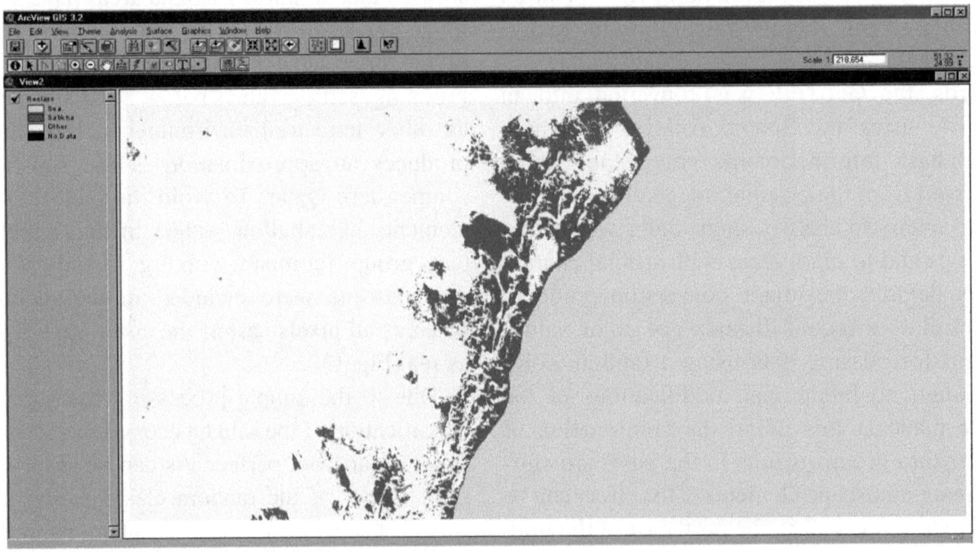

Fig. 6 Approximate map of sabkha ecosystems in southeastern Qatar after reclassification

5 Conclusion

The presented study shows that it is possible to acquire high quality satellite images for free, using the Geoengine internet portal. The images are not appropriate for the mapping and study of short-term processes like infrastructure planning, because of its pre-determined date, but are suitable for landscape classification purposes. This classification can be done by the use of GIS programs like ArcView without major efforts. The combination of image processing and comparison with reflectance values of known geoecological specification produces an approximate map of the existing landscape units.

References

Al-Hinai KG, McMahon Moore J, Bush PR (1987) LANDSAT image enhancement study of possible submerged sand-dunes in the Arabian Gulf. Int J Remote Sens 8:251–258

Al-Sheeb AI (1998) Sand movement in the state of Qatar – The problem and solution. In: Omar SAS, Misak R, Al-Ajmi D (eds) Sustainable development in arid zones, vol 1, Assessment and monitoring of desert ecosystems. Balkema, Rotterdam, pp 223–239

Ashour MM (1987) Surficial deposits of Qatar Peninsula. In: Frostick L, Reid I (eds) Desert sediments. Ancient and modern (= Geological Society Special Publication No. 35). The Geological Society, London, pp 361–367

Embabi NS, Ashour MM (1993) Barchan dunes in Qatar. J Arid Environ 25:49–69

Shinn EA (1973) Sedimentary accretion along the Leeward, SE Coast of Qatar Peninsula, Persian Gulf. In: Purser BH (ed) The Persian Gulf, Holocene carbonate sedimentation in a shallow epicontinental sea. Springer, New York, pp 199–209

Index

M. Öztürk et al. (eds.), *Sabkha Ecosystems*, Tasks for Vegetation Science 46,
DOI 10.1007/978-90-481-9673-9, © Springer Science+Business Media B.V. 2011

MIX
Papier aus verantwortungsvollen Quellen
Paper from responsible sources
FSC® C105338

If you have any concerns about our products,
you can contact us on
ProductSafety@springernature.com

In case Publisher is established outside the EU,
the EU authorized representative is:
Springer Nature Customer Service Center GmbH
Europaplatz 3, 69115 Heidelberg, Germany

Printed by Libri Plureos GmbH
in Hamburg, Germany

Computational Methods for Rational Drug Design

Combinatorial Methods for Rational Drug Design

Computational Methods for Rational Drug Design

Edited by

Mithun Rudrapal
Department of Pharmaceutical Sciences,
School of Biotechnology and Pharmaceutical Sciences,
Vignan's Foundation for Science, Technology & Research,
Guntur, Andhra Pradesh,
India

WILEY

Published by John Wiley & Sons, Inc., Hoboken, New Jersey.
Published simultaneously in Canada.

Library of Congress Cataloging-in-Publication Data
Names: Rudrapal, Mithun, editor.
Title: Computational methods for rational drug design / edited by Mithun Rudrapal.
Description: Hoboken, New Jersey. : Wiley, [2025] | Includes index.
Identifiers: LCCN 2024045117 (print) | LCCN 2024045118 (ebook) | ISBN 9781394249169 (hardback) | ISBN 9781394249183 (adobe pdf) | ISBN 9781394249176 (epub)
Subjects: MESH: Drug Design | Models, Molecular
Classification: LCC RM301.25 (print) | LCC RM301.25 (ebook) | NLM QV 745 | DDC 615.1/9–dc23/eng/20241009
LC record available at https://lccn.loc.gov/2024045117
LC ebook record available at https://lccn.loc.gov/2024045118

Cover Design: Wiley
Cover Image: © Grafissimo/Getty Images

Set in 9.5/12.5pt STIXTwoText by Straive, Pondicherry, India

Contents

List of Contributors

Raghu Ram Achar
Division of Biochemistry
School of Life Sciences
JSS Academy of Higher Education & Research
Mysuru, Karnataka
India

Department of Biotechnology
JSS Science and Technology University
Mysuru, Karnataka
India

Richie Rashmin Bhandare
Department of Pharmaceutical Sciences
College of Pharmacy and Health Sciences
Ajman University, Ajman
United Arab Emirates

Centre of Medical and Bio-allied Health Sciences Research
Ajman University, Ajman
United Arab Emirates

Abanish Biswas
Department of Pharmaceutical Sciences and Technology
Birla Institute of Technology
Ranchi, Jharkhand
India

Diptanil Biswas
Department of Statistical Genomics
National Institute of Biomedical Genomics
Kalyani, West Bengal
India

Ulviye Acar Çevik
Department of Pharmaceutical Chemistry
Faculty of Pharmacy
Anadolu University, Eskişehir
Turkey

Debarupa Dutta Chakraborty
Royal School of Pharmacy
The Assam Royal Global University
Guwahati, Assam
India

Prithviraj Chakraborty
Royal School of Pharmacy
The Assam Royal Global University
Guwahati, Assam
India

Shrimanti Chakraborty
Department of Pharmaceutical Sciences and Technology
Birla Institute of Technology
Ranchi, Jharkhand
India

Soumi Chakraborty
Department of Pharmaceutical Sciences and Technology
Birla Institute of Technology
Ranchi, Jharkhand
India

N. Chandana
Division of Biochemistry, School of Life Sciences
JSS Academy of Higher Education & Research
Mysuru, Karnataka
India

Department of Biotechnology
JSS Science and Technology University
Mysuru, Karnataka
India

Tabsum Chhetri
Department of Bioinformatics
University of North Bengal
Darjeeling, West Bengal
India

Sohan S. Chitlange
Department of Pharmaceutical Chemistry
Dr. D. Y. Patil Institute of Pharmaceutical Sciences and Research
Pune, Maharashtra
India

Jay Mukesh Chudasama
Department of Pharmacy
Sumandeep Vidyapeeth Deemed to be University
Vadodara, Gujarat
India

Bhrigu Kumar Das
Pharmacology and Toxicology Laboratory
School of Pharmaceutical Sciences
Girijananda Chowdhury University, Guwahati
India

Riddhi Dave
Gujarat Arts and Science College
Ahmedabad, Gujarat
India

Shine Devarajan
School of Biotechnology and Bioinformatics
D Y Patil Deemed to be University
Navi Mumbai, Maharashtra
India

Sneha Dokhale
Department of Biotechnology
B. K. Birla College of Arts, Science & Commerce
Kalyan, Maharashtra
India

Samiksha Garse
School of Biotechnology and Bioinformatics
D Y Patil Deemed to be University
Navi Mumbai, Maharashtra
India

John J. Georrge
Department of Bioinformatics
University of North Bengal
Darjeeling, West Bengal
India

Abhirup Ghosh
School of Biosciences and Technology
Vellore Institute of Technology
Vellore, Tamil Nadu
India

Rahul Ghosh
Department of Pharmaceutical Sciences and Technology
Birla Institute of Technology
Ranchi, Jharkhand
India

Ashis Kumar Goswami
Department of Pharmaceutical Sciences
Faculty of Science and Engineering
Dibrugarh University
Assam
India

Shikha Goswami
Department of Pharmacology
Delhi Pharmaceutical Sciences and Research University
New Delhi
India

Anmol Gupta
Department of Biosciences
Integral University, Lucknow
Uttar Pradesh, India

Ayşen Işik
Department of Biochemistry
Faculty of Science, Selçuk University, Konya
Turkey

Vanesa James
Department of Regulatory Affairs and Quality Assurance
LJ Institute of Pharmacy
Ahmedabad, Gujarat
India

Risy Namratha Jamullamudi
Department of Pharmacy
Koneru Lakshmaiah Education Foundation
Vaddeswaram, AP
India

Priyanka Kamaria
Department of Pharmaceutical Chemistry
KLE College of Pharmacy
Bangalore
India

Koyel Kar
Department of Pharmaceutical Chemistry
BCDA College of Pharmacy and Technology
Kolkata, West Bengal
India

Abdüllatif Karakaya
Department of Pharmaceutical Chemistry
Faculty of Pharmacy, Zonguldak Bulent Ecevit University, Zonguldak
Turkey

Priya Kashav
Department of Pharmacy
Sumandeep Vidyapeeth Deemed to be University
Vadodara, Gujarat
India

Ankita Kashyap
Institute of Pharmacy
Assam Medical College and Hospital
Dibrugarh
India

Monalisa Kesh
Centre for Digital Health
Indian Institute of Technology Bombay
Mumbai
India

Kevser Kübra Kırboğa
Faculty of Engineering, Bioengineering Department
Bilecik Seyh Edebali University
Bilecik
Türkiye

Shaunak Kolhapure
School of Biotechnology and Bioinformatics
D Y Patil Deemed to be University
Navi Mumbai, Maharashtra
India

Sathish Kumar Konidala
Department of Pharmaceutical Sciences
School of Biotechnology and Pharmaceutical Sciences
Vignan's Foundation for Science, Technology and Research
Guntur, AP
India

Nigam Jyoti Maiti
Department of Pharmaceutical Sciences and Technology
Birla Institute of Technology
Ranchi, Jharkhand
India

Manshi Mishra
Department of Pharmaceutical Sciences and Technology
Birla Institute of Technology
Ranchi, Jharkhand
India

Saurav Kumar Mishra
Department of Bioinformatics
University of North Bengal
Darjeeling, West Bengal
India

Maitreyee Mukherjee
Department of Pharmaceutical Technology
NSHM Knowledge Campus – Group of Institutions
Kolkata 700053,West Bengal
India

Podila Naresh
Department of Pharmaceutical Sciences
School of Biotechnology and Pharmaceutical Sciences
Vignan's Foundation for Science, Technology and Research
Guntur, AP
India

Chandan Nayak
School of Pharmaceutical Education and Research
Berhampur University
Berhampur, Odisha
India

André M. Oliveira
Department of Environment Studies
Federal Centre of Technological Education of Minas Gerais
Contagem, Minas Gerais
Brazil

Ipsa Padhy
Department of Pharmaceutical Chemistry
School of Pharmaceutical Sciences
Siksha 'O' Anusandhan (Deemed to be University)
Bhubaneswar, Odisha
India

Ipsita Panigrahi
Division of Biochemistry
School of Life Sciences
JSS Academy of Higher Education & Research
Mysuru, Karnataka
India

Department of Biotechnology
JSS Science and Technology University
Mysuru, Karnataka
India

Ghanshyam Parmar
Department of Pharmacy
Sumandeep Vidyapeeth (Deemed to be University)
Vadodara, Gujarat
India

Vaishali Patel
Department of Pharmaceutics
Laxminarayandev College of Pharmacy
Bharuch, Gujarat
India

Mukesh Kumar Patwa
Department of Microbiology
King George Medical University
Lucknow, Uttar Pradesh
India

Sathiaseelan Perumal
Department of Chemistry
Bishop Heber College
Tiruchirappalli, Tamil Nadu
India

Shantanu Raj
Department of Pharmaceutical Sciences and Technology
Birla Institute of Technology
Ranchi, Jharkhand
India

Gourav Rakshit
Department of Pharmaceutical Sciences & Technology
Birla Institute of Technology
Ranchi, Jharkhand
India

Sanket S. Rathod
Department of Pharmaceutical Chemistry
Bharati Vidyapeeth College of Pharmacy
Kolhapur, Maharashtra
India

Sharanya Roy
Department of Pharmaceutical Sciences and Technology
Birla Institute of Technology
Ranchi, Jharkhand
India

Sneha Roy
Department of Bioinformatics
University of North Bengal
Darjeeling, West Bengal
India

Mithun Rudrapal
Department of Pharmaceutical Sciences
School of Biotechnology and Pharmaceutical Sciences
Vignan's Foundation for Science, Technology & Research
Guntur, Andhra Pradesh
India

Biprajit Sarkar
Department of Pharmaceutical Sciences and Technology
Birla Institute of Technology
Ranchi, Jharkhand
India

Anupam Sarma
Advanced Drug Delivery Laboratory
School of Pharmaceutical Sciences
Girijananda Chowdhury University
Guwahati
India

Ashish Shah
Department of Pharmacy
Sumandeep Vidyapeeth Deemed to be University
Vadodara, Gujarat
India

Afzal Basha Shaik
Department of Pharmaceutical Sciences
School of Biotechnology and Pharmaceutical Sciences
Vignan's Foundation for Science, Technology and Research
Guntur, AP
India

Tripti Sharma
Department of Pharmaceutical Chemistry
School of Pharmaceutical Sciences
Siksha 'O' Anusandhan (Deemed to be University)
Bhubaneswar, Odisha, India

School of Pharmaceutical Sciences and Research
Chhatrapati Shivaji Maharaj University
Navi Mumbai, Maharashtra
India

Sonali S. Shinde
Department of Pharmaceutical Chemistry
Dr. D. Y. Patil Institute of Pharmaceutical Sciences and Research
Pune, Maharashtra
India

Irum Siddiqui
IIRC-1, Department of Bioengineering
Integral University
Lucknow, Uttar Pradesh
India

Aditi Singh
Division of Biochemistry
School of Life Sciences
JSS Academy of Higher Education & Research
Mysuru, Karnataka
India

Department of Biotechnology
JSS Science and Technology University
Mysuru, Karnataka
India

Kratika Singh
Department of Microbiology
King George Medical University
Lucknow, Uttar Pradesh
India

Nisha Kumari Singh
Department of Pharmaceutical Sciences and Technology
Birla Institute of Technology
Ranchi, Jharkhand
India

Urmila Singh
Department of Microbiology
King George Medical University
Lucknow, Uttar Pradesh
India

Ashapurna Sinha
Department of Biosciences
Integral University
Lucknow, Uttar Pradesh
India

Kamma Harsha Sri
Department of Pharmaceutical Sciences
Vignan Pharmacy College
Guntur, AP
India

Shivananju Nanjunda Swamy
Department of Biotechnology
JSS Science and Technology University
Mysuru, Karnataka
India

Vaishnavi Thakur
School of Biotechnology and Bioinformatics
D Y Patil Deemed to be University
Navi Mumbai, Maharashtra
India

Rajiv Kumar Tonk
Department of Pharmaceutical Chemistry
Delhi Pharmaceutical Sciences and Research University
New Delhi
India

Sridhar Vemulapalli
Department of Pharmaceutical Sciences
University of Nebraska Medical Center
Omaha, NE
United States

Vivek Yadav
Department of Pharmaceutical Chemistry
Delhi Pharmaceutical Sciences and Research University
New Delhi
India

Neha Zachariah
Swami Shraddhanand College
University of Delhi, New Delhi
India

Preface

The growing incidence of diseases has created a dire need for proper medications, which are readily available, potent and, cost-effective. To meet up this demand, the discovery of novel drug molecules that would be clinically effective and safe is need of the hour. Traditionally, drug discovery requires many years of development and a huge expenditure. However, using modern computational approaches a drug molecule can be identified by investing less time at reduced cost of discovery. It facilitates scientists to find potent and safe therapeutic molecules in more rational way as compared to conventional approaches. In computational drug discovery, various tools, methods/techniques, and software are used in designing better drug molecules with predictive modeling and biophysical studies ranging from physicochemical/ADMET assessments, understanding the molecular mechanisms and toxicity screening. The computational approaches and methods in drug designing include structure-based methods, ligand-based methods, virtual screening, predictive analytics, informatics tools, artificial intelligence (AI) approaches, machine learning methods, and multi-database methods.

This book mainly delves into recent advances in drug designing tools and techniques and their applications pertaining to the discovery of novel therapeutic molecules. It includes armamentarium of latest approaches and advanced methodologies available for drug design and their practical applications in diverse cutting-edge therapeutic areas including cancer, multidrug-resistant bacterial infections, neurodegenerative disorders, inflammatory diseases, and viral infections. It comprises twenty-three chapters in unique topics contributed from experts all around the globe. Some of the key features of the book are as follows:

- Computer-assisted methods and tools for structure- and ligand-based drug design, virtual screening and lead discovery, artificial intelligence and machine learning approaches for drug design and ADMET and physicochemical assessments
- *In silico* and pharmacophore modeling, fragment-based design, *de novo* drug design and scaffold hopping, network-based methods and drug discovery
- Rational design of natural products, peptides, enzyme inhibitors, drugs for neurodegenerative disorders, anti-inflammatory therapeutics, antibacterials for multidrug-resistant infections, and antiviral and anticancer therapeutics

The content of the book has been crafted in such a way that it would serve as useful resource materials for postgraduate and doctoral courses in multidisciplinary academic and/or research disciplines such as Pharmaceutical Sciences, Medicinal Chemistry, Pharmacology, Biomedical Sciences, Biochemistry, Microbiology, Biotechnology, Drug Discovery, Computational Drug Design, and Allied Sciences. This book will be particularly useful to drug developers, pharmaceutical scientists (R&D), discovery scientists, biomedical scientists, healthcare professionals, biochemists, medicinal chemists, pharmacologists, research students, professors, and other researchers working in the field of drug design and discovery. In addition, scientists involved in diverse areas of medicinal chemistry, molecular modeling, drug design and discovery, computational drug discovery (CADD), *in silico* drug discovery. and related areas are also expected to be the wider audience, users, or readers of this book, regardless whether they are engaged in basic (chemistry, computational chemistry, biophysical techniques, molecular modeling, and drug design) or applied (such as pharmaceutical, medicinal chemistry, and drug discovery) research.

I extend my heartfelt gratitude to the contributors and publication team of Wiley for their invaluable contributions and tireless efforts, supports, and dedication, which have enabled the successful compilation of the book volume.

October, 2024

Mithun Rudrapal
Guntur, India

1

Molecular Modeling and Drug Design

Monalisa Kesh[1], Abhirup Ghosh[2], and Diptanil Biswas[3]

[1] *Centre for Digital Health, Indian Institute of Technology Bombay, Mumbai, India*
[2] *School of Biosciences and Technology, Vellore Institute of Technology, Vellore, Tamil Nadu, India*
[3] *Department of Statistical Genomics, National Institute of Biomedical Genomics, Kalyani, West Bengal, India*

1.1 Introduction

1.1.1 What Is Molecular Modeling?

Molecular modeling is at the cutting edge of scientific innovation and discovery, providing a paradigm-shifting method for comprehending the microscopic world of atoms and molecules. Researchers in various disciplines, including chemistry, biology, materials science, and pharmacology, are given newfound strength by this computational technique that enables them to precisely model, examine, and forecast the behavior of molecules and molecular systems. At its core, molecular modeling uses computers' computational capacity to reveal the nanoscale world's well-kept secrets, providing knowledge that is essential for expanding our understanding of the natural world and fostering technological advancements [1]. The attraction of molecular modeling is its capacity to connect theory and experiment. It offers a comprehensive understanding to the researchers to examine the dynamics of large molecular assemblies, the creation of chemical bonds, and the delicate description of atoms. Researchers can now examine issues that are frequently difficult, expensive, or even impractical to solve using only conventional experimental methods but much easier by this computational playground [2].

The significance of molecular modeling is most notable in drug development, where it has completely changed how pharmaceutical molecules are created and optimized. Researchers can quickly screen and prioritize prospective treatments by modeling the interactions between drug candidates and their target proteins, considerably speeding up the drug development process. This has contributed to personalized medicine, in which treatments are matched to specific genetic profiles as well as the development of novel therapeutics. The design of novel materials with specific features is aided in materials research by molecular modeling. Computational modeling directs the development of materials for various applications, from electronics to aerospace, whether it is optimizing the structure of innovative polymers, investigating the behavior of sophisticated composites, or comprehending the properties of nanomaterials [3, 4].

Beyond these areas, molecular modeling provides a flexible tool for understanding chemical processes, researching the principles of protein folding, and evaluating environmental effects. It is essential to education because it aids in the visualization of intricate molecular relationships and

Computational Methods for Rational Drug Design, First Edition. Edited by Mithun Rudrapal.

structures, which deepens comprehension of the underlying ideas that control the natural world. By enabling researchers to solve the riddles of molecules and molecular systems, molecular modeling acts as a light of scientific discovery in this era of computational inquiry, advancing us to new horizons of knowledge and innovation [5].

1.1.2 Software Used for Molecular Modeling

In the discipline of molecular modeling, software is crucial because it allows researchers to carry out intricate simulations, see molecular structures, and quickly process data. There are numerous software programs accessible, each suited to particular modeling methodologies and study goals [6]. These famous applications are frequently used for molecular modeling.

1.1.2.1 Schrodinger

For molecular modeling and drug development, Schrodinger provides a complete range of software tools. A user-friendly interface is offered by Maestro, one of its flagship products, for various modeling activities, including molecular dynamics (MD) simulations, virtual screening, and structure-based drug creation [7].

1.1.2.2 GROMACS

GROMACS is a potent simulation tool for MD that is typically employed to examine the behavior of biomolecules like proteins and lipids. It is a well-liked option in both academia and business because of its speed and scalability [8].

1.1.2.3 Amber

Amber is a second extensively used program for modeling MD, with a significant emphasis on biomolecular systems. It is appropriate for various research applications since it has tools for modeling proteins, nucleic acids, and tiny compounds [9].

1.1.2.4 CHARMM

Known for its prowess in simulating intricate biomolecular systems and researching protein–ligand interactions, CHARMM (Chemistry at HARvard Molecular Mechanics) is a leading name in the field. Drug discovery and structural biology both make substantial use of it [10].

1.1.2.5 AutoDock

A well-liked molecular docking program, AutoDock, forecasts how tiny compounds will interact with protein receptors. It is useful for identifying prospective medication candidates during virtual screening [11].

1.1.2.6 VMD

The versatile molecular visualization program VMD (visual molecular dynamics) is used to examine and display molecular structures, trajectories, and data from numerous simulation programs. It is especially helpful for producing gorgeous molecular graphics [12].

1.1.2.7 PyMOL

PyMOL is a popular molecular visualization program that provides an easy-to-use interface for developing professional-grade 3D molecular images and animations. It is helpful for both academic and research endeavors [13].

1.1.2.8 Open Babel

A useful tool for data preparation and software package compatibility, Open Babel, is an open-source chemical toolkit that enables users to convert between multiple chemical file formats [14].

1.1.2.9 Avogadro

Avogadro is a user-friendly, open-source tool for building and analyzing molecular structures. It is a molecular editor and visualization tool [15].

1.1.2.10 Discovery Studio

Developed by BIOVIA (formerly Accelrys), Discovery Studio is a complete set of molecular modeling and simulation tools used in materials science and drug development [16].

To conduct cutting-edge molecular modeling studies, advance our understanding of molecular systems, and hasten discoveries in areas like drug development, materials science, and structural biology, these software tools are indispensable aids for researchers in various scientific disciplines. The unique study objectives, available computing power, and user skills all play a role in the software selection process.

1.1.3 Molecular Mechanics

Due to the invaluable insights of molecular mechanics into the interactions between medications and their biological targets, typically proteins, molecular mechanics plays a significant role in the development of new drugs. This computational method helps researchers identify and optimize prospective medication candidates by assisting in the study of the structural and energetic elements of molecular interactions. Here is how molecular mechanics assists in the search for new medicines [17].

1.1.3.1 Prediction of Binding Affinity

Molecular mechanics can calculate the binding affinity between a potential medication and its intended protein target. Researchers can forecast how tightly the medicine will bind to the target by estimating the potential energy of the drug–receptor combination. When choosing or creating molecules with higher affinity, which increases medicinal efficacy, this information is essential.

1.1.3.2 Conformational Analysis

Molecular mechanics that aids in the investigation of various conformations of a drug molecule may take up in the binding site of its target. This is important because proteins are dynamic structures, and knowing how a medicine interacts with different conformations might help formulators create more potent drugs.

1.1.3.3 Virtual Screening

To quickly assess the binding of hundreds of chemicals to a target protein, high-throughput virtual screening uses molecular mechanics. To find possible drug candidates, virtual screening makes use of molecular mechanics simulations, which optimize molecular interactions for effective drug discovery. The prioritization of compounds with favorable binding energies speeds up the identification of prospective therapeutic candidates [18].

1.1.3.4 Lead Discovery

Molecular mechanics helps refine lead compounds during drug development. It aids in the decision-making process for scientists when deciding whether chemical alterations will improve a drug's binding affinity, selectivity, and overall pharmacological qualities [19].

1.1.3.5 Mechanism of Action

Computer simulations of molecular mechanics shed light on the ways in which medicines affect their intended protein targets. By modeling molecular interactions, molecular mechanics sheds light on drug processes and the finer points of a drug's mode of action. Researchers can use this information to create medications with more on-target effects and precise mechanisms of action. Absorption, distribution, metabolism, and excretion (ADME) and possible toxicity of drug candidates can also be predicted using molecular mechanics. Evaluation of a drug's safety and bioavailability is aided by knowledge of how it interacts with cellular and physiological elements [20].

In drug development, molecular mechanics is a crucial tool for logically designing and perfecting therapeutic medicines. Reducing the number of prospective candidates saves time and money for researchers while increasing the likelihood that successful medications with focused effects and few adverse effects will be developed.

1.2 Types of Molecular Models

Molecular models are critical in clarifying complicated structures at the molecular level, assisting both scientists and students in picturing three-dimensional (3D) atom configurations. Three important models stand out among the numerous types: the ball-and-spoke model, the space-filling model, and the crystal lattice model (Figure 1.1).

The ball-and-spoke model represents atoms as spheres and bonds as sticks, simplifying chemical structures. This basic method explains connectedness and geometry. In contrast, the space-filling model depicts atoms as spheres that occupy available space, displaying relative sizes and molecular packing. Crystal lattice models play an important role in crystallography because they depict the recurring atomic configurations in crystals while stressing symmetry and periodicity. These models, each with its own set of strengths, contribute to a full understanding of molecular structures by providing different viewpoints on the microscopic world [21, 22].

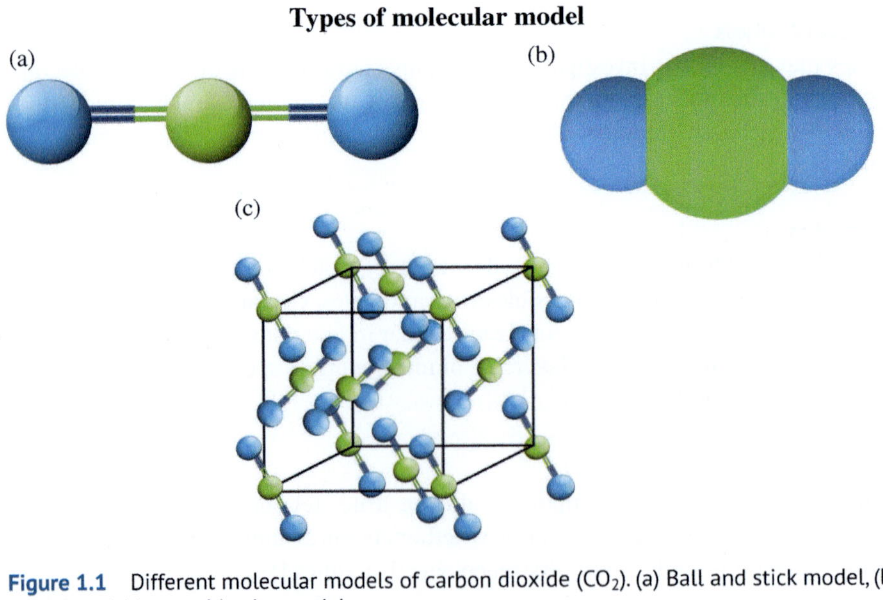

Figure 1.1 Different molecular models of carbon dioxide (CO_2). (a) Ball and stick model, (b) space-filling model, and (c) crystal lattice model.

1.2.1 Ball-and-Spoke Model

In molecular modeling, ball-and-spoke models are crucial visual representations that simplify the complex 3D structures of molecules for analysis and comprehension. This paradigm, which depicts atoms as spheres and bonds as sticks, has found widespread applicability in various scientific disciplines. The ball-and-spoke model, often known as the stick model, is a molecular depiction in which atoms are represented as spheres and the bonds that connect them as sticks. This straightforward approach visualizes molecular geometry in a practical and simple manner, assisting researchers in comprehending the spatial arrangement of atoms within a molecule [23].

Applications of this model:

- **Structural analysis**: The ball-and-spoke model is extensively used in structural elucidation because it allows researchers to easily determine bond angles, bond lengths, and overall molecular shapes.
- **Teaching tool**: The model is an effective teaching tool in educational settings, reducing complex chemical structures for pupils and facilitating a better understanding of key chemistry principles.
- **Communication**: The model is a universal language in scientific communication, commonly used to transmit molecular structures to varied audiences in research publications, presentations, and textbooks.

Pros of this model:

- **Simplicity**: The model's simplicity makes it approachable to both academics and students, promoting a clear understanding of molecular geometry.
- **Intuition**: The visual depiction aids in the rapid understanding of molecular structures, allowing for instant insights into atom connectivity and arrangement.
- **Communication effectiveness**: The ball-and-spoke paradigm promotes clarity in research publications and presentations, making it an effective means of communication in the scientific community.

Cons of this model:

- **Oversimplification**: In some circumstances, the model may oversimplify molecular activity, ignoring nuances that more advanced models may capture.
- **Electron density representation is limited**: The model does not directly express information regarding electron density or orbital features, which can be critical for a thorough understanding of molecular properties [23, 24].

1.2.1.1 Future Directions

As molecular modeling techniques advance, researchers are exploring ways to integrate the strengths of the ball-and-spoke model with more advanced computational methods, aiming for a more holistic representation of molecular structures.

In conclusion, the ball-and-spoke model remains a cornerstone in molecular modeling, offering a balance between simplicity and effectiveness. While it has its limitations, its widespread applications and continued use underscore its significance in the scientific community's pursuit of understanding molecular complexities.

1.2.2 Space-filling Models

Space-filling models, often known as CPK models (after the initials of the creators Corey, Pauling, and Koltun), describe molecules by portraying atoms as spheres with radii proportional to the van der Waals radii of the respective elements. These models try to provide a realistic representation of molecular structures by demonstrating how atoms pack together efficiently in 3D space [25, 26].

Applications of this model:

- **Steric considerations**: Space-filling models are very useful for comprehending steric hindrance and molecule packing. They explain how atoms inhabit space and impact the orientations of one another.
- **Drug design**: Space-filling models in pharmaceutical research help in drug design by visualizing how compounds can interact within active areas of biological macromolecules, allowing researchers to optimize binding interactions.
- **Molecular size visualization**: The models excel in conveying the relative sizes of atoms within a molecule, allowing for a better understanding of molecular dimensions and overall structure.

Pros of this model:

- **Steric considerations**: Space-filling models can help you understand steric hindrance and molecular packing. They describe how atoms inhabit space and influence one another's orientations.
- **Drug design**: In pharmaceutical research, space-filling models aid in drug design by visualizing how chemicals interact within active areas of biological macromolecules, allowing researchers to optimize binding interactions.
- **Visualization of molecular size**: The models excel in communicating the relative sizes of atoms within a molecule, allowing for a better comprehension of molecular dimensions and overall structure.

Cons of this model:

- **Complexity for large molecules**: Space-filling models for large and complicated molecules may appear cluttered, making it difficult to detect individual characteristics.
- **Inadequate bond information**: Space-filling models, unlike ball-and-spoke models, do not directly depict bond information, instead focus on atomic configurations [27].

1.2.2.1 Future Directions

Advances in processing power and visualization techniques provide opportunities to improve space-filling models. The incorporation of interactive and dynamic aspects may alleviate some of the difficulties involved with seeing complicated chemical structures.

Finally, space-filling models provide a detailed and realistic description of molecular structures, revealing important information about molecular size, steric effects, and overall packing. Despite significant limitations, their applications in various scientific disciplines highlight their importance as indispensable tools for both molecular modelers and researchers.

1.2.3 Crystal Lattice Models

In molecular modeling, crystal lattice models describe the repeated arrangement of atoms within a crystal structure. These representations highlight crystals' 3D periodicity, emphasizing the regular arrangement of atoms and molecules inside the lattice. These models help researchers grasp the relationship between microscopic arrangements and macroscopic features by offering a macroscopic picture of the crystalline structure [28].

Applications of this model:

- **Material science**: Crystal lattice models are important in materials science because they guide researchers in the discovery and design of novel materials. Predicting material qualities requires an understanding of the crystal structure.
- **Drug development**: Crystallography aids in the elucidation of the crystal structures of medicinal molecules in pharmaceutical research, providing critical information for optimizing medication formulation and increasing bioavailability.
- **Electronic properties**: In order to understand the electronic properties of semiconductors and other electronic materials, crystal lattice models are required [29, 30].

Pros of this model:

- **Periodic insight**: Crystal lattice models depict the periodic arrangement of atoms, allowing researchers to identify patterns and symmetries within the crystal lattice.
- **Macroscopic understanding**: These models bridge the gap between microscopic and macroscopic scales, providing a comprehensive view of the crystal structure and assisting with experimental data interpretation.
- **Predictive capability**: Understanding the crystal lattice is critical for forecasting material properties and driving the creation of novel materials with specific functions [31].

Cons of this model:

- **Oversimplification**: While crystal lattice models are excellent at depicting periodicity, they may oversimplify certain aspects of molecular activity, ignoring dynamic and nonperiodic processes.
- **Noncrystalline material complexity**: Because crystal lattice models are inherently intended for crystalline materials, their application for examining noncrystalline or amorphous structures is limited [31].

1.2.3.1 Future Directions

Crystal lattice models are becoming more useful as computational tools and experimental procedures advance. Integrating real-time simulations and dynamic elements could improve the accuracy and application of these models, allowing them to represent a wider range of materials.

Finally, crystal lattice models continue to be vital in molecular modeling, providing a powerful tool for understanding the periodicity and structure of crystalline materials. Their applications in materials science, drug development, and electronics highlight their value as valuable tools for researchers attempting to unravel the mysteries of molecular configurations in the solid state.

1.3 Computational Methods in Drug Discovery

1.3.1 What Is Drug Discovery?

By combining computational, mathematical, experimental, translational, and clinical models, possible novel medicinal entities can be identified through the process of drug development. It is the technique of locating and analyzing compounds that have the ability to govern illness in a secure way, with the aim of developing medications that can prolong patients' lives. Drug development is still an arduous, expensive, time-consuming, and ineffective process with a high dropout rate of novel therapeutic discovery, despite advancements in biotechnology and physiological system understanding. Knowing a biological target while employing creativity to find novel drugs can be referred to as drug designing. Creating molecules that complement the molecular target they interact and bind to in terms of charge and shape is the fundamental aspect of drug design. In the big data era, drug design typically, but not always, depends on computer modeling tools and bioinformatics methodologies [32].

Apart from small molecules, biopharmaceuticals, particularly therapeutic antibodies, have emerged as a significant class of medications [33]. Significant progress has also been made in computational methods for enhancing the stability, selectivity, and affinity of these protein-based treatments. Preclinical research on animal and cell models as well as human clinical trials are all part of the process of developing and discovering new drugs. Afterward, the drug must receive regulatory approval before being put on the market. In contemporary drug discovery, screening hits are identified, medicinal chemistry is applied, and those hits are optimized to improve the hit's affinity, specificity, efficacy, metabolic stability, and oral bioavailability. Before conducting clinical trials, a molecule that satisfies all of these criteria will be found, and the drug development process will start [34].

An illness or clinical condition for which there are not any appropriate pharmaceuticals on the market leads to the start of a drug discovery program, and this unmet clinical need serves as the project's primary source of motivation. The preliminary study, which is frequently conducted in academic settings, produces information to support a hypothesis that a protein's or pathway's activation or inhibition will have a beneficial impact in the context of a medical condition. The result of this action is the identification of a target that, in order to support a drug development endeavor, may need additional validation before moving forward into the lead discovery phase. It is a drawn-out, resource-intensive procedure that calls for close collaboration among various disciplines. The pharmaceutical industry is extremely interested in optimizing the drug discovery process since identifying and selecting appropriate drug candidates promptly can have a significant impact on the price and profitability of new medications [35, 36].

1.3.2 Computational Platforms for Drug Discovery

In the field of drug development, computational platforms and algorithms are invaluable resources that have revolutionized the conventional and tedious approaches. These tools utilize artificial intelligence (AI), chemical simulations, and advanced algorithms to accelerate multiple phases of the drug development process. Computational models aid in the identification of pharmaceuticals with the best pharmacokinetic profiles by forecasting important pharmacological aspects including absorption, distribution, metabolism, excretion, and toxicity. By combining distinct datasets, these technologies not only expedite decision-making processes but also promote collaboration. This leads to the simplification of drug development pipelines and a notable increase in the efficiency of generating innovative and effective medicines. Some most important computational tools and platforms for drug development are enumerated below. These tools aid in the analysis of enormous volumes of biological data, the prediction of drug–target interactions, and the optimization of drug candidates.

1.3.2.1 NCBI

The National Center for Biotechnology Information or NCBI, a division of the National Library of Medicine (NLM) of the National Institutes of Health (NIH), which was founded in 1988, is a comprehensive molecular biology information repository. It offers access to many biological databases, including PubMed, BLAST, GenBank, NCBI Gene, Genome Resources, and NCBI Bookshelf [37].

1.3.2.2 Chemical Databases

Comprehensive chemical data and compound databases are available through platforms such as PubChem and ChemSpider. Both databases provide a wealth of chemical data, but PubChem is especially useful for biological data, such as bioassays, because it is part of the larger NCBI resources. In contrast, ChemSpider places a strong emphasis on using a community-driven methodology to create an extensive library of chemical structures. ChemSpider fosters a collaborative atmosphere by encouraging user contributions and the curation of chemical data. Community data entry is also permitted through PubChem. Chemical information can be accessed and contributed to by academics, researchers, and professionals working in the field of chemistry pharmacology and allied sectors using PubChem and ChemSpider. When used in tandem, they can offer a more thorough comprehension of chemical molecules and their biological functions [38, 39].

1.3.2.3 PDB

Protein Data Bank or PDB acts as the hub for 3D structural data of biological macromolecules and is a cornerstone of structural biology. The PDB is a massive collection of structures that have been determined through experimentation. It captures the complex architecture found in nucleic acids, proteins, and complex molecular organizations. These structures are explored by techniques such as crystallographic X-rays, NMR spectroscopy, and cryo-EM, providing crucial insights into the spatial arrangement of atoms within those biomolecules. Researchers can study, examine, and understand the structural nuances of biomolecules with the help of this publicly available resource. The PDB continues to be an essential tool in expanding our knowledge of the connections between structure and function, making a substantial contribution to areas like drug development, molecular biology, and bioinformatics. This is made possible by its cooperative efforts and interaction with other databases [40].

1.3.2.4 AutoDock, AutoDock Vina, DOCK, PatchDock, HADDOCK, SwissDock, Glide, Gold, FlexX, UCSF Chimera, and DockThor

These programs help uncover possible therapeutic candidates by predicting the binding modes of small compounds with the target proteins [41].

1.3.2.5 UniProt

The Universal Protein Resource, or UniProt, is a comprehensive and open-access resource that offers details on the functional annotations of proteins. UniProt is a centralized platform that unifies data from many protein databases. It contains references to scholarly publications as well as data on post-translational modifications, structural details, functional annotations, and protein sequences. It is a priceless tool for scientists studying molecular biology, computational biology, and related subjects. It provides a standardized, carefully curated repository that makes it easier to find and examine data about proteins. In order to guarantee that the database continues to be an extensive and trustworthy resource for the international scientific community, the UniProt Consortium is dedicated to continuously updating and growing it [42].

1.3.2.6 QSAR

Quantitative structure–activity relationship or QSAR uses a compound's chemical structure to predict its biological activity. The foundation of QSAR techniques is the statistical correlation between target drug interactions and other molecular descriptors. The QSAR approach is based on the observation that compounds with similar structures typically exhibit comparable biological activities. These models provide a mathematical explanation of how the structural characteristics of a ligand affect the activity responses of a target that binds it. The association between various features of tiny ligand binders and biological activity acquired through experimentation is used to calculate QSAR. It is possible to forecast the activity of novel drug molecule analogs using QSAR relationships [43].

1.3.2.7 GROMACS, AMBER, NAMD, PLUMED, LAMMPS, CHARMM, GROMOS, OpenMM, Orac, XMD, YASARA, Ms2, MacroModel, and Avizo

These software packages replicate the motions of atoms and molecules over a period of time, providing an understanding of the dynamic behavior of biological systems [44].

1.3.2.8 Desmond

Specifically created for mimicking biomolecular systems. This cutting-edge MD modeling program Desmond, created by DE Shaw Research, has been developed for examining the shifting dynamics of biological systems in great detail down to the atomic level. Desmond distinguishes out for its effectiveness and flexibility, making it especially appropriate for simulating large and complicated biomolecular systems over longer scales. Desmond facilitates the investigation of the conformation dynamics, interactions, and the thermodynamic laws of macromolecules including nucleic acids, proteins, and lipids by means of advanced algorithms and parallel processing capabilities. Advanced capabilities in the software, such as parallel processing and flexible integration time steps, enable faster and more accurate simulations. Desmond has contributed significantly to the advancement of our knowledge of biological processes, aiding in the search for new drugs, and offering insightful information on the structural dynamics underlying various physiological and pathological occurrences [45].

1.3.2.9 OpenBabel

An open-source chemical package named OpenBabel has been developed for molecular structure manipulation and chemical file format conversion. OpenBabel is a flexible and effective tool in the field of cheminformatics, created by an interconnected group of computational chemists. Its extensive chemical file format compatibility enables users to transform molecular representations between various platforms and programs with ease. Activities including ligand docking, structure-based virtual screening, and chemical database analysis are made easier by OpenBabel. Its extensive feature set includes descriptor computations, chemical fingerprinting, and structural perception. It is an important technology in computational chemistry that improves accessibility and interoperability among various applications and tools utilized in the field of chemical informatics [46].

1.3.2.10 DeepChem and Cheminformatics for Python (RDKit)

A machine learning package named DeepChem has been developed especially for cheminformatics and drug discovery applications. It provides an extensive collection of tools and methods for the creation and use of models for deep learning in biology and chemistry. DeepChem makes it easier for researchers to use machine learning for drug development projects by facilitating processes like compound screening, molecular featurization, and property prediction. Conversely, a popular open-source cheminformatics toolbox for Python is called RDKit (Cheminformatics for Python). RDKit provides a comprehensive variety of cheminformatics features, including molecular structure processing, biochemical informatics, and descriptor computation. It is a useful tool for researchers in computational biology and medicinal chemistry since it is widely used in the investigation and alteration of chemical data. DeepChem and RDKit work together to provide a powerful blend of traditional cheminformatics tools with deep learning capabilities, enabling scientists to investigate, evaluate, and optimize chemical structures for drug development applications [47–49].

1.3.2.11 SBML

Systems biologists can represent and exchange computational models in a uniform format by using the Systems Biology Markup Language (SBML). Researchers can describe mathematical models representing processes in biology, such as pathways of metabolism, signaling pathways, and gene regulatory networks, using SBML, an open, XML-based language. The power of SBML resides in its capacity to enable model integration and interchange between different platforms and computational tools. In the ever-evolving field of systems biology, it ensures interoperability and collaboration by offering a uniform framework for model representation. SBML is now a vital tool for modelers, allowing them to share, replicate, and expand upon the work of one another, ultimately leading to a more thorough knowledge of complex biological systems. It has gained widespread acceptance with encouragement from the systems biology community [50].

1.3.2.12 Virtual Screening

A widely used computational method in drug development is virtual screening, which helps find promising drug candidates rapidly from vast compound libraries. Using an in silico study of chemical databases, this method predicts and ranks the compounds that have the best chance of binding to a target, usually a protein linked to a specific disease. The two types of virtual screening techniques are structure based and ligand based. The goal of ligand-based virtual screening is to find compounds with comparable features by using the understanding of existing ligands and their effects on biology. Conversely, structure-based virtual screening determines the affinity of ligand binding by evaluating how well a target's structure complements the chemical structures of putative ligands [51].

1.3.3 Applications of Computer-Based Methods in Steps of Drug Discovery

Throughout the drug development process, computer-based techniques are essential for increasing productivity, cutting expenses, and finding new therapeutic targets [52]. In order to identify and validate pharmacological targets, various disciplines including bioinformatics databases, systems biology, pathway analysis, network modeling, and computational biology work together [53]. By describing gene and protein functions, bioinformatics databases like NCBI and UniProt offer a plethora of genomic and proteomic data that makes it easier to explore potential targets. These databases function as biological information archives, providing researchers access to a wide range of molecular information such as protein structures, genetic sequences, and functional annotations. By enabling scientists to examine the function of certain genes and proteins in a range of illnesses, these databases prove to be extremely beneficial in the identification of possible therapeutic targets [54]. Identification of important components and possible therapeutic targets is facilitated by systems biology approaches, which provide an understanding of the intricate interconnections within biological systems. Systems biology techniques use experimental data integration to generate a comprehensive picture by simulating biological system dynamics using computational models. The identification of important biological processes and possible targets for medication intervention is made easier by this comprehensive understanding [55]. Pathway analysis helps researchers identify critical nodes for therapeutic intervention by examining the interlinked biochemical pathways linked to a disease using tools such as PathVisio and Cytoscape [56]. This is furthered by network analysis, which pinpoints interactions between molecules and signaling cascades to discover possible targets inside the framework of cellular networks [57, 58]. Targeting certain proteins is evaluated as a viable option using computational biology methods. Drug targets are validated in a substantial way by computational methods including molecular docking, MD modeling, and structural bioinformatics. Molecular docking facilitates the identification and refinement of putative therapeutic candidates by forecasting the orientation and binding affinity of small compounds to target proteins. Protein structures are analyzed using structural bioinformatics approaches in order to determine binding sites and comprehend their activities. Furthermore, network pharmacology methodologies employ computational techniques to investigate the interplay between pharmaceuticals and biological networks, offering perspectives on the wider consequences of focusing on particular proteins [59–61]. By combining these methods, a thorough framework for locating and confirming potential therapeutic targets is created, which improves the efficacy and precision of the drug development steps.

1.4 Potential Use and Application of AI in Drug Designing

In terms of drug discovery and design, AI has emerged as a game-changer, providing previously unheard-of chances to speed up the creation of innovative therapies. Here are a few prospective AI applications and uses in medication development.

1.4.1 Target Identification and Validation

AI is able to find prospective therapeutic targets linked to diseases by analyzing enormous biological information, such as genomes, proteomics, and clinical records. AI is essential to drug discovery because it can quickly identify and evaluate therapeutic targets from a wide range of biological data, which speeds up the process of selecting viable candidates for additional research and development. Machine learning algorithms forecast the possibility that a given protein or metabolic pathway would be an effective therapeutic target. AI can validate targets by identifying their contribution to the development of diseases through network analysis [62].

1.4.2 Drug Screening and Lead Optimization

By modeling interactions between millions of chemicals and target proteins, AI speeds up high-throughput virtual screening. With data-driven screening, AI revolutionizes drug discovery by quickly finding promising molecules. In drug development, it helps in lead optimization by forecasting and improving molecular structures for increased safety and efficacy. It forecasts pharmacological and binding affinities, assisting researchers in finding potential lead compounds. To improve therapeutic efficacy and safety, molecular structures are optimized using machine learning models [62] (Figure 1.2).

1.4.3 *De Novo* Drug Design

AI creates brand-new compounds with desired pharmacological characteristics. By creating new compounds, tailoring structures, and forecasting chemical interactions, AI speeds up the *de novo* drug discovery process. It offers specialized treatments for particular targets and disorders,

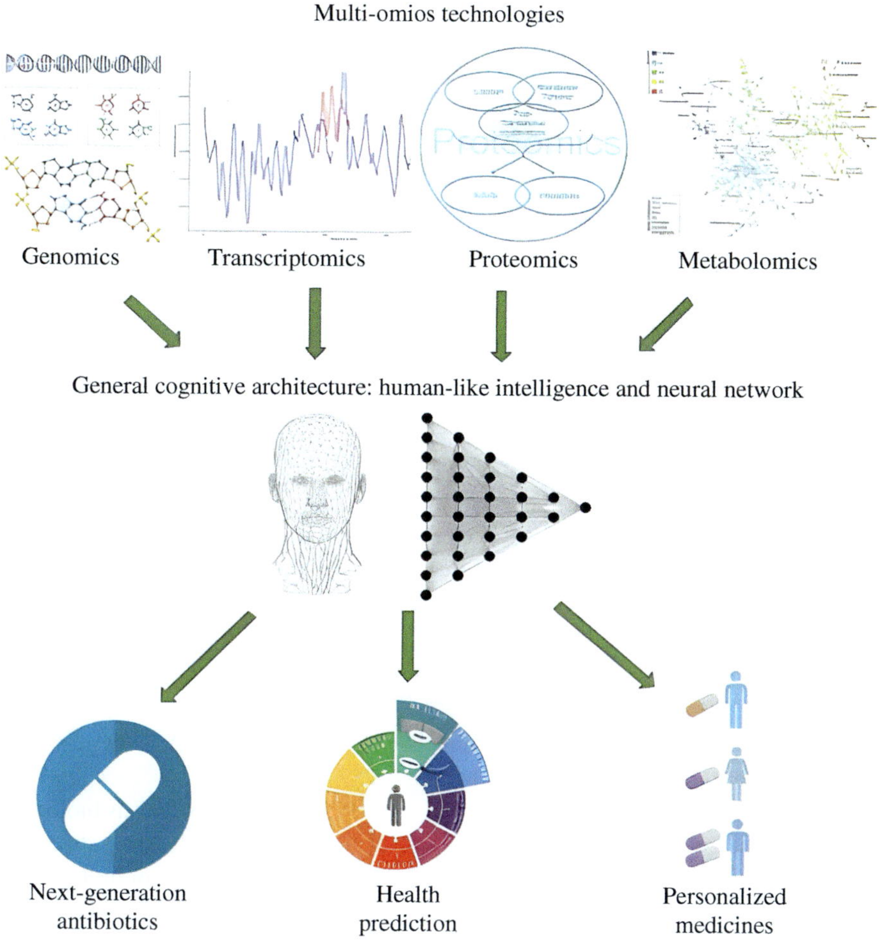

Figure 1.2 AI in drug development.

speeding up the drug development process. To improve hit-to-lead efficiency, generative adversarial networks and reinforcement learning produce drug-like molecules.

1.4.4 Predictive Toxicology and ADMET

Early in the drug development process, AI evaluates potential drug toxicity and forecasts adverse effects. AI predicts absorption, distribution, metabolism, excretion, and toxicity (ADMET) characteristics and toxicity, which improves drug discovery. It streamlines the creation of safer and more potent medications by analyzing enormous databases, spotting possible hazards, and optimizing molecules. Drug ADMET qualities are predicted by models, which enhance compound choice [63].

1.4.5 Clinical Trial Optimization

AI examines patient data to find trial candidates who are a good fit and forecast patient reactions. AI enhances clinical trials for drug discovery by evaluating patient data, forecasting results, and selecting eligible subjects. It streamlines operations, lowers expenses, and quickens the entire drug development process. Adaptive trial designs optimize trial efficiency and success rates by modifying protocols in response to real-time data [64].

1.4.6 Drug Repurposing

AI finds current medications that can be used for alternative therapeutic purposes. AI expedites the repurposing of medications by uncovering possible new uses for existing drugs through analysis of large datasets. It speeds up research and provides affordable treatments for a range of illnesses by reallocating already-existing molecules. It compares medications with comparable modes of action or protein targets to find new applications [65].

1.4.7 Concept of Personalized Medicine

In drug discovery, AI evaluates enormous amounts of biological and chemical data to provide quick insights into possible drug candidates, optimize structures, and forecast interactions. It simplifies the entire process of developing new drugs. It analyses massive omics data, aiding in the identification of biomarkers and the comprehension of disease mechanisms. For better compound design, it locates pertinent chemical features and patterns in chemical libraries. Thus, AI can enable the discovery of patient subpopulations that respond well to particular medications. AI in medication development analyses patient data to enable individualized medicine. By customizing care based on clinical, molecular, and genetic data, it improves accuracy and effectiveness for certain patient populations. It tailors treatments to individual patient profiles, taking into account genetic, genomic, and clinical data [66].

1.4.8 Drug Combination Optimization

AI analyses intricate relationships and forecasts synergistic effects to optimize medication combinations in drug discovery. It reduces adverse effects and speeds up the process of finding combinations that work well together. AI analyses synergistic effects by taking into account the intricate interactions of different pharmaceuticals, and it optimizes drug combinations to maximize therapeutic effectiveness and minimize side effects [66].

To summarize, target identification, chemical screening, clinical trial optimization, and personalized medicine are just a few of the areas where AI is being used in drug design. By drastically decreasing the time and expense associated with drug discovery, raising the success rate of clinical trials, and ultimately providing patients with safer and more effective therapies, these technologies have the potential to revolutionize the pharmaceutical business. AI's influence on medication development and healthcare is expected to increase as it develops, creating new opportunities for invention and research.

1.5 Limitations of Current Methods

The subject of molecular modeling and drug design has shown tremendous potential for AI, which is revolutionizing how researchers approach drug development. To fully utilize the potential of present AI techniques, however, a number of constraints must be overcome [67, 68].

1.5.1 Data Restrictions

The availability and caliber of data are two of the biggest problems. When it comes to molecular modeling, such data can be rare or lacking. AI models need huge, diverse datasets for training. A fundamental bottleneck in the subject is the creation of high-quality, well-annotated datasets. Additionally, experimental data frequently lags behind the rate of AI advancement, making it challenging to maintain AI models current with the most recent data.

1.5.2 Interpretability

Due to their intricacy, deep learning techniques in particular are sometimes referred to as "black boxes" in many AI-driven models. For regulatory and scientific validation in drug design, it is essential to understand how these models make their predictions. The development of interpretable AI techniques is still difficult.

1.5.3 Generalization

AI models frequently find it difficult to extrapolate beyond the data that was used to train them. This is a challenge for molecular modeling because chemical and biological processes might vary greatly. An AI model's utility may be constrained if it overfits its training data and performs badly on new data.

1.5.4 Resources and Computation

AI-driven molecular modeling can have high computational requirements. For smaller research groups or organizations with limited resources, the requirement for significant computer power to simulate and predict molecular interactions at a high level of precision can be a barrier.

1.5.5 Ethical Considerations

AI has the potential to unintentionally reinforce biases found in the data it was trained on, which raises ethical questions about medication design and molecular modeling. Predictions that are biased may result in the creation of medicines that are less efficient or secure for particular demographic groups.

1.5.6 Validation and Experimentation

Although AI may forecast possible therapeutic candidates, actual biological system validation and testing remain crucial. Validating AI-generated ideas can be expensive and time-consuming, and not all forecasts will result in useful medicines.

1.5.7 Regulatory Obstacles

Regulating authorities, like the FDA, have not yet established precise standards for the application of AI to the production of pharmaceuticals. The incorporation of AI in pharmaceutical research may be hampered by the absence of regulatory clarity and standardization.

AI has the potential to greatly speed up molecular modeling and drug development. However, its full potential is now constrained by restrictions on data accessibility, interpretability, generalization, resource requirements, ethical issues, validation, and regulatory difficulties. To ensure the ethical and successful application of AI in the creation of novel medications and treatments, researchers in the field must address these concerns.

1.6 Case Studies

1.6.1 Case Study 1: "Accelerating Drug Discovery with AI-Powered Molecular Modeling" by Dr. Jane Mitchell [69]

Background: To tackle germs that are resistant to antibiotics, Dr. Jane Mitchell, a senior researcher at BioTech Innovations, led a team in the search for novel antibiotics. A more effective strategy was required because previous drug discovery techniques were inefficient and expensive.

AI solution: To put state-of-the-art molecular modeling methods into practice, Dr. Mitchell's team worked with AI startup DrugAI Solutions. A deep learning model by DrugAI Solutions had been created and trained using large chemical datasets and well-known antibiotic structures. The drug development process could be considerably sped up by using this model to estimate the binding affinities of prospective drugs to bacterial targets.

Results: The AI-enhanced methodology significantly sped up the identification of potential antibiotic candidates. Within a short period of time, the researchers had discovered a number of highly effective chemicals. Further research and development on these substances resulted in the invention of a potential new antibiotic.

1.6.2 Case Study 2: "AI-Driven Drug Design for Rare Genetic Disorders" by Prof. David Reynolds [70]

Background: A rare genetic condition that only affects a tiny patient population was the focus of Prof. David Reynolds' research at the GenoMed Research Institute. The intricacy and genetic variety of this illness were difficult for conventional medication development techniques to address.

AI solution: Prof. Reynolds' group worked with GenoAI Therapeutics' AI specialists. They used deep learning and AI models that were trained on genomes, molecular structures, and well-known disease pathways. Personalized medication candidates were recommended by these models based on the distinct genetic profiles of each patient.

Results: The AI-driven method offered individuals with the rare illness personalized treatment options. The specially formulated medications specifically addressed the genetic defects, improving symptom control and enhancing the quality of life for those who were impacted. The research of Prof. Reynolds demonstrated the promise of AI in creating specialized treatments for rare diseases.

1.6.3 Case Study 3: "Revolutionizing Drug Repurposing with AI During the COVID-19 Pandemic" by Dr. Maria Fernandez [71]

Background: During the epidemic, Dr. Maria Fernandez, a virologist at the Global Health Institute, was presented with the pressing task of locating potential COVID-19 therapeutics. To effectively address the situation, the traditional medicine repurposing procedure proved too sluggish.

AI solution: Dr. Fernandez's team worked with the AI and data analytics firm DataRx Solutions. In order to assess the possibility of existing medications to block the virus, DataRx Solutions used AI algorithms that combined molecular docking simulations and data from thousands of different drugs. The AI system quickly located medications having COVID-19 antiviral characteristics.

Results: The AI-driven drug repurposing strategy made it possible to identify viable COVID-19 therapies quickly. A number of already-approved medications were repurposed for emergency use, aiding in the pandemic response. Dr. Fernandez's research demonstrated how AI can speed up the drug development process in times of public health crisis.

These three case studies highlight the crucial role of AI in molecular modeling and medication design, from expediting drug development to personalizing therapies for uncommon diseases and responding to global health crises. They were written by well-known researchers and subject matter experts. The field of pharmaceutical research and drug development is still being shaped by AI.

1.7 Molecular Docking

Molecular docking is a computer approach that predicts the interaction of tiny compounds with target proteins. It entails simulating molecule binding in order to find possible medication candidates.

Molecular docking is extremely useful in medication design. It aids in estimating binding affinity, elucidating binding locations, and optimizing therapeutic candidates by investigating the interactions between ligands and target proteins. Understanding enzyme–substrate interactions to designing treatments for complex disorders are some of the applications [72, 73].

Among the notable breakthroughs in drug design achieved using molecular docking is the discovery of anti-HIV medications such as Ritonavir [74] and Lopinavir [75]. In the case of COVID-19, molecular docking was used to identify possible inhibitors of the viral primary protease. Such examples demonstrate molecular docking's critical role in expediting drug discovery, proving its ability to speed the identification of new medicines with substantial clinical significance.

1.7.1 What Is Molecular Docking?

Molecular docking is a cornerstone of computational biology because it elucidates the delicate dance between tiny molecules and target proteins, providing crucial insights for drug discovery and design. Predicting the best shape and binding affinity of a ligand within the binding region of a target protein provides a virtual platform for exploring prospective therapeutic options.

1.7.1.1 Procedure

The creation of 3D structures of the ligand and target protein is the first step in molecular docking. Ligand structures can be determined experimentally or predicted computationally. Various docking techniques are then used to systematically examine the ligand's conformational space within the binding site. Autodock, Autodock Vina, Glide, and GOLD are popular software solutions that each use its own search algorithms and score methods [41].

1.7.1.2 Biophysical Laws

The energy changes during molecule docking are governed by biophysical rules, namely the laws of thermodynamics. These principles take into account enthalpy, entropy, and Gibbs free energy, offering a theoretical foundation for understanding molecule binding thermodynamics [76].

1.7.1.3 Rigid and Flexible Docking

Rigid docking assumes that during the binding process, both the ligand and the protein maintain fixed conformations. While rigid docking is computationally efficient, it may overlook induced-fit effects, which occur when binding causes structural changes. Flexible docking recognizes the dynamic nature of molecular interactions by enabling ligand, protein, or both flexibility. This method represents the binding process more realistically, but it requires more computational resources [77].

1.7.1.4 Types of Docking

There are different types of docking, including protein–ligand docking, which is commonly used in drug development, investigates the interactions between a tiny molecule (ligand) and a target protein, assisting in the identification of possible therapeutic candidates, and protein–protein docking, which investigates the interactions between two proteins, revealing information on protein–protein interactions that are important in various cellular processes [78].

1.7.1.5 Challenges and Future Perspectives

Despite its effectiveness, molecular docking poses obstacles such as accurately forecasting binding affinities and taking protein flexibility into consideration. To improve accuracy and efficiency, ongoing research focuses on improving scoring systems, utilizing quantum physics, and leveraging machine learning approaches. Finally, molecular docking functions as a virtual laboratory, allowing researchers to explore and anticipate intricate molecular interactions. This computational technique is becoming increasingly important in expediting drug development and enhancing our understanding of molecular recognition [79].

1.7.2 Applications of Molecular Docking in Drug Designing

Molecular docking is an important computational approach with numerous applications in drug design, providing vital insights into the interactions of small compounds and target proteins. One of its key applications is virtual screening, which involves screening enormous libraries of compounds to uncover prospective therapeutic candidates. Molecular docking accelerates the identification of compounds with therapeutic promise by anticipating the binding affinity and preferred conformations of ligands within a target protein's active site [80]. Furthermore, molecular docking is important in lead optimization, aiding medicinal chemists in altering existing compounds to improve binding affinity and bioavailability. The method aids in the knowledge of structure–activity interactions, which is critical for refining medication candidates and enhancing their efficacy [81]. Furthermore, molecular docking is critical for investigating protein–ligand interactions in various illnesses, including cancer, infectious diseases, and neurological disorders. Researchers can build more targeted and effective medications by gaining insights into the molecular pathways behind diseases. To summarize, molecular docking applications in drug design range from initial virtual screening to lead optimization, making it a crucial tool in the search for novel and effective therapeutic agents [59].

1.7.3 Success of Molecular Docking Cases in Drug Designing

Several significant case studies demonstrate the efficacy of molecular docking in drug design, demonstrating its efficacy in finding and optimizing therapeutic candidates. One interesting example is the creation of HIV protease inhibitors, in which molecular docking was critical. The HIV medications Ritonavir and Lopinavir were discovered and improved using molecular docking experiments, revealing their capacity to bind well to the viral protease active site [82, 83]. In terms of developing diseases, the COVID-19 pandemic triggered intensive molecular docking attempts to uncover possible SARS-CoV-2 virus inhibitors. Computational approaches were used to evaluate existing medication libraries and create new compounds that target the viral primary protease. These efforts resulted in the identification of viable candidates, demonstrating molecular docking's speedy and crucial role in responding to global health crises [84]. Furthermore, molecular docking has made major contributions to the field of cancer treatments. Understanding the binding interactions between the inhibitors and the kinase domain was facilitated by molecular docking in the creation of tyrosine kinase inhibitors, such as imatinib for the treatment of chronic myeloid leukemia. This understanding aided in the rational design of targeted medications, resulting in increased selectivity and efficacy [85]. The success of molecular docking extends to anti-inflammatory medications as well. Molecular docking experiments were utilized to optimize celecoxib, a selective COX-2 inhibitor used in the treatment of arthritis. Understanding the particular interactions between celecoxib and the target protein enabled the development of a medicine with fewer adverse effects than typical nonsteroidal anti-inflammatory treatments [86].

These case examples show the critical importance of molecular docking in drug creation. Molecular docking accelerates drug development by offering insights into the binding interactions of candidate compounds with target proteins, resulting in the identification and optimization of therapeutically useful molecules. As computational approaches evolve, the success stories of molecular docking in drug design are likely to grow, helping in the development of creative and effective treatments for a wide range of ailments [87].

1.8 Conclusion and Future Works

As technology keeps transforming the domains of drug development and molecular modeling, both of these fields have enormous promise toward the future. AI and computational methods are driving molecular modeling, which has the potential to completely transform the way drugs are discovered. Researchers are able to forecast molecular interactions, examine large datasets, and find novel therapeutic ideas faster, thanks to the incorporation of machine learning methods and big data analytics. The transition to in silico techniques minimizes the need for lengthy experimental trials by enabling more effective and economical chemical screening. The precision and rapidity of molecular simulations should also be improved by developments in quantum computing, allowing researchers to study intricate biological systems in previously unheard-of detail. Molecular modeling is projected to have an important role in the development of personalized medicine, which will be driven by biological data and precision medicine in order to customize drug discovery to meet the specific needs of each patient. All things considered, the field of molecular modeling and the development of drugs hold immense promise for a paradigm change in the future. Multidisciplinary teams working with state-of-the-art technologies are going to generate therapeutic solutions, which are safer, more efficient, and more specifically targeted. MD simulations and structural bioinformatics collectively are additionally offering novel approaches to investigate the dynamics of biological macromolecules. The design of more precise and potent medications is made achievable by this greater understanding of protein folding, interactions, and changes in conformation. High-resolution structures can also be obtained by applying sophisticated techniques like cryo-electron microscopy, which has served to provide significant insights for drug design and development. Preclinical testing might experience a revolution with the introduction of 3D bioprinting and organ-on-a-chip technologies, which would allow researchers to more closely replicate human physiology and forecast drug reactions in physiologically relevant scenarios. Partnerships among academia, business, and government organizations are becoming more and more essential to creating an atmosphere that encourages innovation and streamlines the drug development process. Furthermore, the emphasis on data sharing and open science is generating enormous databases that researchers may use to collaborate more quickly. Concerns over confidentiality of information, intellectual property rights, and the responsible application of AI in drug research will become increasingly important as the field develops. A drive toward exploring cutting-edge therapeutic modalities including personalized medicine and RNA-based medications is becoming more and more prominent in the setting of drug discovery. Molecular modeling methods are crucial for the development and enhancement of these novel therapies. Additionally, the understanding of medication efficacy and safety aspects in various groups may be improved by incorporating patient-generated data and real-world evidence into drug development processes. In conclusion, an integrated strategy that incorporates technology developments, interdisciplinary cooperation, and a dedication to ethical practices is what will characterize the next phase of molecular simulation and drug development. This method will address complicated healthcare concerns. This all-encompassing viewpoint sets the area at the center of medical innovation and has the potential to completely transform the healthcare system by providing more individualized, accurate, and readily accessible options for treatment [88–91].

References

1 Francoeur, E. (1997). The forgotten tool: the design and use of molecular models. *Social Studies of Science* 27 (1): 7–40.

2 Doi, M. and Takimoto, J. (2003). Molecular modelling of entanglement. *Philosophical Transactions of the Royal Society of London. Series A: Mathematical, Physical and Engineering Sciences* 361 (1805): 641–652.

3 Kumar, S., Plotnikov, N.V., Rouse, J.C., and Singh, S.K. (2018). Biopharmaceutical informatics: supporting biologic drug development via molecular modelling and informatics. *Journal of Pharmacy and Pharmacology* 70 (5): 595–608.

4 Nadendla, R.R. (2004). Molecular modeling: a powerful tool for drug design and molecular docking. *Resonance* 9 (5): 51–60.

5 Giannantoni, C. (2010). Protein folding, molecular docking, drug design – the role of the derivative "drift" in complex systems dynamics. *First International Conference on Bioinformatics* 2: 193–199.

6 Barril, X. and Soliva, R. (2006). Molecular modelling. *Molecular BioSystems* 2 (12): 660–681.

7 Subhani, S., Jayaraman, A., and Jamil, K. (2015). Homology modelling and molecular docking of MDR1 with chemotherapeutic agents in non-small cell lung cancer. *Biomedicine & Pharmacotherapy* 71: 37–45.

8 Khalili, S., Rasaee, M.J., Mousavi, S.L. et al. (2017). *In silico* prediction and in vitro verification of a novel multi-epitope antigen for HBV detection. *Molecular Genetics, Microbiology and Virology* 32: 230–240.

9 Salomon-Ferrer, R., Case, D.A., and Walker, R.C. (2013). An overview of the Amber biomolecular simulation package. *Wiley Interdisciplinary Reviews: Computational Molecular Science* 3 (2): 198–210.

10 Mallajosyula, S.S., Jo, S., Im, W., and MacKerell, A.D. (2015). Molecular dynamics simulations of glycoproteins using CHARMM. *Glycoinformatics* 407–429.

11 Butt, S.S., Badshah, Y., Shabbir, M., and Rafiq, M. (2020). Molecular docking using chimera and autodock vina software for nonbioinformaticians. *JMIR Bioinformatics and Biotechnology* 1 (1): e14232.

12 Spivak, M., Stone, J.E., Ribeiro, J. et al. (2023). VMD as a platform for interactive small molecule preparation and visualization in quantum and classical simulations. *Journal of Chemical Information and Modeling* 63 (15): 4664–4678.

13 Lill, M.A. and Danielson, M.L. (2011). Computer-aided drug design platform using PyMOL. *Journal of Computer-Aided Molecular Design* 25: 13–19.

14 Field, M.J. (2008). The pDynamo program for molecular simulations using hybrid quantum chemical and molecular mechanical potentials. *Journal of Chemical Theory and Computation* 4 (7): 1151–1161.

15 Hanwell, M.D., Curtis, D.E., Lonie, D.C. et al. (2012). Avogadro: an advanced semantic chemical editor, visualization, and analysis platform. *Journal of Cheminformatics* 4 (1): 1–17.

16 Ranjbar, A., Torabi, S., Nematollahi, D. et al. (2021). Molecular modelling of the therapeutic agents for COVID-19 treatment. *Infectious Disease Research* 2 (1): 1.

17 Pissurlenkar, R.R., Shaikh, M.S., Iyer, R.P., and Coutinho, E.C. (2009). Molecular mechanics force fields and their applications in drug design. *Anti-Infective Agents in Medicinal Chemistry* 8 (2): 128–150.

18 De Vivo, M., Masetti, M., Bottegoni, G., and Cavalli, A. (2016). Role of molecular dynamics and related methods in drug discovery. *Journal of Medicinal Chemistry* 59 (9): 4035–4061.

19 Kesh, M. and Goel, S. (2023). Target-based screening for lead discovery. *CADD and Informatics in Drug Discovery* 103–119.

20 Tripathi, M.K., Sharma, A., and Singh, S.K. (2022). Fundamentals of molecular modeling in drug design. *Computer Aided Drug Design (CADD): From Ligand-based Methods to Structure-based Approaches*, 125–155.

21 Goodman, R.E. (1970). Molecular models in biology. *The American Biology Teacher* 32 (1): 34–38.

22 Kozlíková, B., Krone, M., Falk, M. et al. (2016). Visualization of biomolecular structures: state of the art revisited. *Computer Graphics Forum* 36 (8): 178–204.

23 Turner, M. (1971). Ball and stick models for organic chemistry. *Journal of Chemical Education* 48 (6): 407.

24 Chiu, M., Chou, C.C., Chen, Y.H. et al. (2019). Model-based learning about structures and properties of chemical elements and compounds via the use of augmented realities. *Chemistry Teacher International* 1 (1): 20180002.

25 Joseph, V.R. (2016). Space-filling designs for computer experiments: a review. *Quality Engineering* 28: 28–35.

26 Corey, R.B. and Pauling, L. (1953). Molecular models of amino acids, peptides, and proteins. *Review of Scientific Instruments* 24 (8): 621–627.

27 Koltun, W.L. (1965). Precision space-filling atomic models. *Biopolymers* 3 (6): 665–679.

28 Scattergood, A. (1937). The making of crystal lattice and unit cell models. *Journal of Chemical Education* 14 (3): 140.

29 Wing, R.M., Drew, H., Takano, T. et al. (1980). Crystal structure analysis of a complete turn of B-DNA. *Nature* 287 (5784): 755–758.

30 Teif, V.B. and Rippe, K. (2010). Statistical–mechanical lattice models for protein–DNA binding in chromatin. *Journal of Physics: Condensed Matter* 22 (41): 414105.

31 Acharya, K.R. and Lloyd, M.D. (2005). The advantages and limitations of protein crystal structures. *Trends in Pharmacological Sciences* 26 (1): 10–14.

32 Zhong, W.Z. and Zhou, S.F. (2014). Molecular science for drug development and biomedicine. *International Journal of Molecular Sciences* 15 (11): 20072–20078.

33 Luu, K.T., Kraynov, E., Kuang, B. et al. (2013). Modeling, simulation, and translation framework for the preclinical development of monoclonal antibodies. *The AAPS Journal* 15 (2): 551–558.

34 Xiao, X., Min, J.L., Lin, W.Z. et al. (2015). iDrug-target: predicting the interactions between drug compounds and target proteins in cellular networking via benchmark dataset optimization approach. *Journal of Biomolecular Structure & Dynamics* 33 (10): 2221–2233.

35 Bateman, T.J. (2022). Drug discovery. In: *Atkinson's Principles of Clinical Pharmacology* (ed. S.-M. Huang, J.J.L. Lertora, P. Vicini, and A.J. Atkinson), 563–572. Elsevier.

36 Hughes, J.P., Rees, S., Kalindjian, S.B., and Philpott, K.L. (2011). Principles of early drug discovery. *British Journal of Pharmacology* 162 (6): 1239–1249.

37 NCBI Resource Coordinators (2013). Database resources of the National Center for biotechnology information. *Nucleic Acids Research* 41: D8–D20.

38 Brown, N., Cambruzzi, J., Cox, P.J. et al. (2018). Big data in drug discovery. In: *Progress in Medicinal Chemistry*, vol. 57 (ed. D.R. Witty and B. Cox), 277–356. Elsevier.

39 Pence, H.E. and Williams, A. (2010). ChemSpider: an online chemical information resource. *Journal of Chemical Education* 87 (11): 1123–1124.

40 Burley, S.K., Berman, H.M., Bhikadiya, C. et al. (2019). RCSB protein data Bank: biological macromolecular structures enabling research and education in fundamental biology, biomedicine, biotechnology and energy. *Nucleic Acids Research* 47 (D1): D464–D474.

41 Pagadala, N.S., Syed, K., and Tuszynski, J. (2017). Software for molecular docking: a review. *Biophysical Reviews* 9 (2): 91–102.

42 The UniProt Consortium (2017). UniProt: the universal protein knowledgebase. *Nucleic Acids Research* 45 (D1): D158–D169.

43 Verma, J., Khedkar, V.M., and Coutinho, E.C. (2010). 3D-QSAR in drug design – a review. *Current Topics in Medicinal Chemistry* 10 (1): 95–115.

44 Hollingsworth, S.A. and Dror, R.O. (2018). Molecular dynamics simulation for all. *Neuron* 99 (6): 1129–1143.

45 Bowers, K.J., Chow, E., Xu, H. et al. (2006). Scalable algorithms for molecular dynamics simulations on commodity clusters. *Proceedings of the 2006 ACM/IEEE Conference on Supercomputing*, 84.

46 O'Boyle, N.M., Banck, M., James, C.A. et al. (2011). Open babel: an open chemical toolbox. *Journal of Cheminformatics* 3: 33.

47 Vora, L.K., Gholap, A.D., Jetha, K. et al. (2023). Artificial intelligence in pharmaceutical technology and drug delivery design. *Pharmaceutics* 15 (7): 1916.

48 Wu, Z., Ramsundar, B., Feinberg, E.N. et al. (2017). MoleculeNet: a benchmark for molecular machine learning. *Chemical Science* 9 (2): 513–530.

49 Bento, A.P., Hersey, A., Félix, E. et al. (2020). An open source chemical structure curation pipeline using RDKit. *Journal of Cheminformatics* 12 (1): 51.

50 Hucka, M., Bergmann, F.T., Dräger, A. et al. (2018). The systems biology markup language (SBML): language specification for level 3 version 2 core. *Journal of Integrative Bioinformatics* 15 (1): 20170081.

51 Dhingra, A.K., Chopra, B., Bhardwaj, S., and Grewal, A.S. (2020). Chapter 8: Virtual screening. *Computational Approaches in Drug Discovery, Development and Systems Pharmacology* Elsevier.

52 Kapetanovic, I.M. (2008). Computer-aided drug discovery and development (CADDD): *in silico*-chemico-biological approach. *Chemico-Biological Interactions* 171 (2): 165–176.

53 Yue, R. and Dutta, A. (2022). Computational systems biology in disease modeling and control, review and perspectives. *NPJ Systems Biology and Applications* 8 (1): 37.

54 Hutchins, J.R. (2014). What's that gene (or protein)? Online resources for exploring functions of genes, transcripts, and proteins. *Molecular Biology of the Cell* 25 (8): 1187–1201.

55 Chen, B.S. and Wu, C.C. (2013). Systems biology as an integrated platform for bioinformatics, systems synthetic biology, and systems metabolic engineering. *Cells* 2 (4): 635–688.

56 Greenblum, S.I., Efroni, S., Schaefer, C.F., and Buetow, K.H. (2011). The PathOlogist: an automated tool for pathway-centric analysis. *BMC Bioinformatics* 12: 133.

57 Dasika, M.S., Burgard, A., and Maranas, C.D. (2006). A computational framework for the topological analysis and targeted disruption of signal transduction networks. *Biophysical Journal* 91 (1): 382–398.

58 Navlakha, S., Gitter, A., and Bar-Joseph, Z. (2012). A network-based approach for predicting missing pathway interactions. *PLoS Computational Biology* 8 (8): e1002640.

59 Agu, P.C., Afiukwa, C.A., Orji, O.U. et al. (2023). Molecular docking as a tool for the discovery of molecular targets of nutraceuticals in disease management. *Scientific Reports* 13 (1): 13398.

60 Zhang, X., Wu, F., Yang, N. et al. (2022). *In silico* methods for identification of potential therapeutic targets. *Interdisciplinary Sciences: Computational Life Sciences* 14: 285–310.

61 Salo-Ahen, O.M.H., Alanko, I., Bhadane, R. et al. (2021). Molecular dynamics simulations in drug discovery and pharmaceutical development. *Processes* 9 (1): 71.

62 Duch, W., Swaminathan, K., and Meller, J. (2007). Artificial intelligence approaches for rational drug design and discovery. *Current Pharmaceutical Design* 13 (14): 1497–1508.

63 Kumar, A., Kini, S.G., and Rathi, E. (2021). A recent appraisal of artificial intelligence and *in silico* ADMET prediction in the early stages of drug discovery. *Mini Reviews in Medicinal Chemistry* 21 (18): 2788–2800.

64 Lim, J.J., Goh, J., Rashid, M.B.M.A., and Chow, E.K.H. (2020). Maximizing efficiency of artificial intelligence-driven drug combination optimization through minimal resolution experimental design. *Advanced Therapeutics* 3 (4): 1900122.

65 Mohanty, S., Rashid, M.H.A., Mridul, M. et al. (2020). Application of artificial intelligence in COVID-19 drug repurposing. *Diabetes & Metabolic Syndrome: Clinical Research & Reviews* 14 (5): 1027–1031.

66 Awwalu, J., Garba, A.G., Ghazvini, A., and Atuah, R. (2015). Artificial intelligence in personalized medicine application of AI algorithms in solving personalized medicine problems. *International Journal of Computer Theory and Engineering* 7 (6): 439.

67 Sacan, A., Ekins, S., and Kortagere, S. (2012). Applications and limitations of *in silico* models in drug discovery. *Bioinformatics and Drug Discovery* 87–124.

68 Qing, X., Yin Lee, X., De Raeymaeker, J. et al. (2014). Pharmacophore modeling: advances, limitations, and current utility in drug discovery. *Journal of Receptor, Ligand and Channel Research* 81–92.

69 Lu, H. and Tonge, P.J. (2008). Inhibitors of FabI, an enzyme drug target in the bacterial fatty acid biosynthesis pathway. *Accounts of Chemical Research* 41 (1): 11–20.

70 Wang, D.R., Ai, P., Sun, Y. et al. (2022). Gene signatures associated with temporal rhythm as diagnostic markers of major depressive disorder and their role in immune infiltration. *International Journal of Molecular Sciences* 23 (19): 11558.

71 Fernandez, D., Marabotti, A., Facchiano, A., and Tagliaferri, R. (2021). A review on drug repurposing applicable to COVID-19. *Briefings in Bioinformatics* 22 (2): 726–741.

72 Morris, G.M. and Lim-Wilby, M. (2008). Molecular docking. *Methods in Molecular Biology* 443: 365–382.

73 Vidal-Limón, A., Aguilar-Toalá, J.E., and Liceaga, A.M. (2022). Integration of molecular docking analysis and molecular dynamics simulations for studying food proteins and bioactive peptides. *Journal of Agricultural and Food Chemistry* 70 (4): 934–943.

74 Dawood, A.A. (2023). The efficacy of Paxlovid against COVID-19 is the result of the tight molecular docking between Mpro and antiviral drugs (nirmatrelvir and ritonavir). *Advances in Medical Sciences* 68 (1): 1–9.

75 Rafi, M.O., Bhattacharje, G., Al-Khafaji, K. et al. (2020). Combination of QSAR, molecular docking, molecular dynamic simulation and MM-PBSA: analogues of lopinavir and favipiravir as potential drug candidates against COVID-19. *Journal of Biomolecular Structure & Dynamics* 40 (8): 3711–3730.

76 Torres, P.H.M., Sodero, A.C.R., Jofily, P., and Silva-Jr, F.P. (2019). Key topics in molecular docking for drug design. *International Journal of Molecular Sciences* 20 (18): 4574.

77 Alogheli, H., Olanders, G., Schaal, W. et al. (2017). Docking of macrocycles: comparing rigid and flexible docking in glide. *Journal of Chemical Information and Modeling* 57 (2): 190–202.

78 Hernández-Santoyo, A., Tenorio-Barajas, A.Y., Altuzar, V. et al. (2013). *Protein–Protein and Protein–Ligand Docking*. InTech eBooks.

79 Saikia, S. and Bordoloi, M. (2019). Molecular docking: challenges, advances and its use in drug discovery perspective. *Current Drug Targets* 20 (5): 501–521.

80 Bartuzi, D., Kaczor, A.A., Targowska-Duda, K.M., and Matosiuk, D. (2017). Recent advances and applications of molecular docking to G protein-coupled receptors. *Molecules* 22 (2): 340.

81 Asiamah, I., Obiri, S., Tamekloe, W. et al. (2023). Applications of molecular docking in natural products-based drug discovery. *Scientific African* 20: e01593.

82 Ghosh, A.K., Osswald, H.L., and Prato, G. (2016). Recent progress in the development of HIV-1 protease inhibitors for the treatment of HIV/AIDS. *Journal of Medicinal Chemistry* 59 (11): 5172–5208.

83 Razzaghi-Asl, N., Sepehri, S., Ebadi, A. et al. (2015). Effect of biomolecular conformation on docking simulation: a case study on a potent HIV-1 protease inhibitor. *Iranian Journal of Pharmaceutical Research* 14 (3): 785–802.

84 Subhaswaraj, P. and Siddhardha, B. (2022). Molecular docking and molecular dynamic simulation approaches for drug development and repurposing of drugs for severe acute respiratory syndrome-Coronavirus-2. *Computational Approaches for Novel Therapeutic and Diagnostic Designing to Mitigate SARS-CoV-2 Infection*, 207–246.

85 Gagic, Z., Ruzic, D., Djokovic, N. et al. (2020). *In silico* methods for design of kinase inhibitors as anticancer drugs. *Frontiers in Chemistry* 7: 873.

86 Shi, M., Wang, L., Li, P. et al. (2021). Dasatinib-SIK2 binding elucidated by homology modeling, molecular docking, and dynamics simulations. *ACS Omega* 6 (16): 11025–11038.

87 Pinzi, L. and Rastelli, G. (2019). Molecular docking: shifting paradigms in drug discovery. *International Journal of Molecular Sciences* 20 (18): 4331.

88 Borhani, D.W. and Shaw, D.E. (2012). The future of molecular dynamics simulations in drug discovery. *Journal of Computer-Aided Molecular Design* 26 (1): 15–26.

89 Doytchinova, I. (2022). Drug design-past, present, future. *Molecules* 27 (5): 1496.

90 Bassani, D. and Moro, S. (2023). Past, present, and future perspectives on computer-aided drug design methodologies. *Molecules* 28 (9): 3906.

91 AlRawashdeh, S. and Barakat, K.H. (2024). Applications of molecular dynamics simulations in drug discovery. *Methods in Molecular Biology* 127–141.

2

Bioactive Small Molecules and Drug Discovery

Ashish Shah[1], Vaishali Patel[2], Sathiaseelan Perumal[3], Riddhi Dave[4], Neha Zachariah[5], Ghanshyam Parmar[1], and Jay Mukesh Chudasama[1]

[1] *Department of Pharmacy, Sumandeep Vidyapeeth (Deemed to be University), Vadodara, Gujarat, India*
[2] *Department of Pharmaceutics, Laxminarayandev College of Pharmacy, Bharuch, Gujarat, India*
[3] *Department of Chemistry, Bishop Heber College, Tiruchirappalli, Tamil Nadu, India*
[4] *Gujarat Arts and Science College, Ahmedabad, Gujarat, India*
[5] *Swami Shraddhanand College, University of Delhi, New Delhi, India*

2.1 Introduction

2.1.1 Introduction to Drug Design and Discovery

The process of finding possible new therapeutic entities through the application of computational, experimental, translational, and clinical models is known as drug discovery. Despite improvements in biotechnology and our understanding of biological systems, the process of finding new therapeutics is still exceedingly time-consuming, expensive, challenging, and inefficient [1]. The entire process of creating a new medication, from conception to market release, can take 12–15 years and exceed one billion dollars. Based on the knowledge of a biological target, the inventive process of finding new medications is known as drug design [2]. Preclinical research on animal and cell models as well as human clinical trials is all part of the process of developing and discovering new drugs [3]. Afterward, the drug must receive regulatory approval before being put on the market. Identification of screening hits, medicinal chemistry, and optimization of those hits to boost their affinity, selectivity (to lower the risk of adverse effects), efficacy/potency, metabolic stability (to lengthen the half-life), and oral bioavailability are all important components of modern drug research. Following the discovery of a chemical meeting each of these criteria, the process of developing a medicine will start before any clinical trials are conducted [4].

Drug design, in its most basic form, is the design of molecules that complement the molecular target that they interact and bind to in terms of charge and shape. In the big data era, bioinformatics methods and computer modeling techniques are often, but not always, used in drug design. Apart from small molecules, biopharmaceuticals, particularly therapeutic antibodies, are becoming a significant class of medications. Significant progress has also been made in computational methods for enhancing the stability, selectivity, and affinity of these protein-based treatments [5].

Computational Methods for Rational Drug Design, First Edition. Edited by Mithun Rudrapal.

2.1.2 Brief History of Small-Molecule Drug Discovery

Numerous small-molecule medications have enhanced patient outcomes and advanced medicine throughout history. While some of these pioneering medications remain on the market today, others have vanished from use but played a significant role in opening the door for more effective therapies within their particular use. For almost a century, small molecules have been crucial in producing medical advancements for human ailments. The quick development of biopharmaceutical research and technology creates opportunities for fresh, imaginative methods of creating small-molecule medications [6].

For almost a century, small-molecule medications have served as the foundation of the pharmaceutical business. Small-molecule medications, which are defined as any organic substance with a low molecular weight, offer several unique therapeutic benefits. The majority are oral medications that can cross cell membranes and reach intracellular targets [7]. Because of these qualities, small-molecule medicines are becoming more popular in the pharmaceutical industry and are excellent candidates for therapeutic development. Because of their inadequate efficacy or severe side effects, the medications now employed in clinical practice are insufficient, even with the developments in molecular genetics and the creation of new, efficient drug discovery techniques. For this reason, creating novel, effective medications is crucial to the prevention and treatment of disease [8].

2.1.3 Importance of Bioactive Small Molecules in Drug Discovery

Their inherent qualities, such as their capacity to pass through biological barriers and affect various distinct biological targets, have played a significant role in their effectiveness as medications. Majority of small-molecule medications have oral bioavailability, which allows for conventional oral administration as a tablet [9]. When compared to biologics, this convenience factor is a significant benefit, even though the pharmacological impact size may be comparable. For example, recombinant biologic erythropoietin (EPO) can be treated for renal anemia with small-molecule inhibitors (HIF-PH), which can be taken as a pill instead of an injection or infusion. The modular structure and chemical synthesis accessibility of tiny molecules are further essential features [10]. Because of this, their chemical structures may be quickly varied, and their attributes can be improved in a systematic way. The unaltered tiny molecule is the primary component of the majority of pharmaceutical preparations. The unaltered small molecule is typically the active ingredient in drug formulations. Prodrugs, or appropriate reversible chemical changes, can be created to compensate for possible drawbacks such as low water solubility. Targeted activation or transport of a pharmaceutical component toward a target tissue can also be accomplished using a comparable method. Lastly, most medication formulations and administration routes are compatible with small molecules, and they often have a high shelf life [11]. For a more comprehensive overview of small-molecule properties, see Table 1.

2.2 Importance of Computational Methods in Bioactive Small-Molecules Discovery

Computational techniques play a crucial role in the search for bioactive small molecules, making them a powerful and efficient strategy for drug development. Molecular docking, synthetic screening, molecular dynamics (MD) models, and quantitative structure–activity relationship (QSAR) models are all components of these methodologies. The ability of computational methods to predict potential bioactive compounds and their interactions with target proteins, thereby decreasing the number of candidates that need to be experimentally validated, is a significant advantage in the process of drug discovery [12]. The intricate molecular mechanisms underlying drug–target interactions can be comprehended more effectively through the utilization of these approaches, thereby facilitating the rational development of novel compounds possessing enhanced pharmacological attributes [13]. Computer programs also assist in the search for novel chemical entities with favorable therapeutic characteristics by enabling the identification of structurally diverse molecules. Integrating computational methodologies provides an efficient and cost-effective method for identifying potential therapeutic candidates, as traditional drug development can be resource-intensive and time-consuming. Computational methodologies are essential in the detection of bioactive small molecules since they enhance the velocity, efficacy, and cost-efficiency of pharmaceutical research and development. Consequently, this enables us to tackle medical demands that have not been fulfilled and enhance the results for patients [14].

2.2.1 Structure-Based Methods

Computational techniques for the discovery of small bioactive compounds rely heavily on structure-based methods. To aid in the development of pharmaceuticals, these techniques make use of the three-dimensional structures of biological macromolecules, such as proteins or nucleic acids. One of the most well-known methods in this area is molecular docking, which uses computational methodologies to model the binding geometry of small molecules (ligands) with target proteins in order to anticipate how these two would interact [15]. To aid in the selection of potential medication candidates, this method makes it easier to identify potential binding sites and evaluate the binding strength of ligands [16, 17]. In order to find compounds that can bind to a certain target protein, structure-based virtual screening methodically searches through large compound libraries [18, 19]. This approach helps reduce the time and resources needed for hit identification by prioritizing compounds for experimental testing. By showing how biomolecular complexes behave dynamically over time, MD simulations improve structure-based methods and provide a better understanding of ligand binding and stability [14]. Also, structure-based methods are useful for developing fragment-based drugs or designing novel small compounds through de novo ligand design. In *de novo* ligand design, new compounds with desired properties are generated computationally, whereas in fragment-based approaches, smaller molecular fragments are found and optimized to form a ligand with high binding affinity. Finding bioactive tiny compounds relies heavily on structure-based approaches. By utilizing these tools, we may make informed predictions about ligand–target interactions, virtually screen compound libraries, analyze molecules' dynamic behaviors, and rationally create new therapeutic candidates [20]. The use of these computational techniques considerably improves the efficacy and efficiency of medication development (Table 2.1).

2.2.2 Ligand-Based Methods

In order to infer information about possible medication candidates, ligand-based approaches to bioactive small-molecule discovery examine the features and qualities of already-known bioactive molecules. An alternative to using target protein structures, these strategies make use of ligand-derived data [21]. The process of discovering new drugs often makes use of a few essential ligand-based methods enlisted in Table 2.2.

Table 2.1 Structure-based methods for drug discovery.

Structure-based methods	Description
Docking process	Uses simulated binding geometry to predict how a tiny chemical (ligand) will interact with a target protein. A useful tool for determining binding affinity and locating possible binding locations
Structure-based virtual screening	Systematically screens large compound libraries to identify molecules with the potential to bind to a specific target protein. Prioritizes compounds for experimental testing, saving time and resources
Molecular dynamics simulations	Provides insights into the dynamic behavior of biomolecular complexes over time. Enhances understanding of ligand binding and stability, contributing to a more comprehensive analysis
De novo ligand design	Generates new molecules computationally with desired properties. Facilitates the design of novel small molecules for specific targets
Fragment-based drug discovery	Identifies and optimizes smaller molecular fragments that can be combined to form a high-affinity ligand. A strategic approach to drug design

Table 2.2 Essential ligand-based methods for drug discovery.

Ligand-based methods	Description
QSAR	Statistical approaches are employed to establish a relationship between the chemical structure and biological activity of a set of ligands. By analyzing the structural properties of novel compounds, QSAR models can predict their activity
Pharmacophore modeling	Discovers shared spatial or structural characteristics needed for ligand–target interactions. New compounds with comparable pharmacological characteristics can be designed using this three-dimensional arrangement of features as a template
Molecular fingerprints	Creates a binary fingerprint using molecular characteristics like substructures or chemical patterns. To help with ligand similarity searches and virtual screening, fingerprints that are structurally similar are useful
Machine learning approaches	Uses various machine learning algorithms to draw conclusions about ligand bioactivity from data, including neural networks and random forests. These techniques improve predicting abilities by handling complicated interactions among massive datasets

QSAR models are used to create mathematical relationships between the chemical structures of ligands and their observed biological activities. QSAR models evaluate datasets of known bioactive chemicals and predict the activity of new molecules to aid in the design of more potent drugs or those with fewer adverse effects. Pharmacophore models are extremely useful for determining the essential structural and spatial components involved in ligand binding to a target [22]. These models guide medicinal chemists in constructing compounds that imitate the required interactions, thereby facilitating the design of new molecules by emphasizing crucial characteristics. Molecular fingerprints enable the comparison of ligand structures by converting molecular information into a binary format. Comparable fingerprints facilitate the virtual screening of potential drug candidates and the search for similar ligands by highlighting structural similarities. Neural networks and random forests are two machine learning methods that analyze large amounts of ligand data to identify patterns that can provide insights about the behavior of chemicals in the body. Ligand-based approaches offer valuable tools for the discovery of bioactive small molecules by utilizing data from known ligands to forecast the activities and characteristics of novel compounds [23, 24]. These methodologies are highly effective in forecasting the pharmacological characteristics of novel drugs and in identifying intricate patterns and correlations in intricate datasets. When there is an absence or inadequate amount of structural data available for the target protein, numerous strategies seem to be effective [25].

2.2.3 Network-Based Methods

Network-based approaches utilize network theory and analysis to comprehend and forecast the interactions among biological entities, such as proteins, genes, or tiny molecules. These methods are especially effective for investigating the intricacy of biological systems and revealing correlations that contribute to the identification of drugs [26]. Various essential network-based methodologies are employed in the investigation of bioactive small molecules (Table 2.3).

Protein–protein interactions (PPIs) are illustrated in PPI networks, providing a comprehensive view of biological activity at a systems level. These networks are useful in drug development as they help identify core proteins or hubs that could serve as therapeutic targets or play significant roles in disease pathways. Drug–target interaction networks refer to networks that illustrate the connections between drugs and the proteins they specifically operate upon. Through the exploration of drug–target interaction networks, researchers can get insights into the polypharmacology of medications, enabling them to comprehend the mechanisms of therapeutic action by detecting potential off-target effects. Chemical similarity networks were created to facilitate the identification of compounds that share similar structures. These networks were built using the chemical similarity of small molecules. Chemical similarity networks facilitate the discovery of novel drug concepts and the identification of alternative applications for existing ones. A network that displays the interconnectedness among different biological processes is referred to as a

Table 2.3 Network-based methods for bioactive small molecules.

Network-based methods	Description
Protein–protein interaction (PPI) networks	Represent interactions between proteins in a network format. Analyzing PPI networks helps identify key proteins or hubs that play crucial roles in cellular processes, aiding in the identification of potential drug targets
Drug–target interaction networks	Model the interactions between drugs and their target proteins. Analyzing these networks provides insights into the polypharmacology of drugs, helping understand their multiple interactions and potential side effects
Chemical similarity networks	Constructed based on the similarity of chemical structures between small molecules. These networks assist in identifying structurally related compounds, guiding the exploration of chemical space for drug discovery
Biological pathway networks	Depict the relationships between biological pathways, illustrating how different pathways interact. Analyzing these networks aids in understanding the broader context of a target's involvement in cellular processes

biological pathway network. Analyzing these networks allows researchers to get a more comprehensive foundation for drug discovery projects by better understanding the potential involvement of a single target in several pathways. Hence, considering the many interconnections among biological systems, network-based techniques offer a comprehensive methodology for identifying bioactive small compounds. These methods provide insights into chemical similarities, biological pathways, PPIs, drug–target links, and drug–target linkages, enabling a better understanding of potential drug candidates and their effects on cellular processes [27–30].

2.3 Natural Products in Bioactive Small-Molecule Discovery

Since ancient times, natural products (NPs) have been the main source of treatment. Traditional medicines like Ayurveda, Chinese medicine, Siddha, and Unani all use NPs as a source of medicine. The key component of these NPs is bioactive small molecules. Bioactive small molecules are soluble molecules that can interact and moderate the activity of a cell. Many of these bioactive molecules are derived from primary and secondary metabolites of plants. They are also obtained from marine sources. These bioactive molecules are mainly used as anticancer, antiviral, and antimalarial agents [31].

2.3.1 Plant Primary and Secondary Molecules as Bioactive Molecules

Chemicals that are fundamental for activities such as respiration, photosynthesis, and translocation are known as primary metabolites. Although they are not directly engaged, plants may need secondary metabolites for growth and development [32].

Primary metabolites include amino acids, nucleosides, and carbohydrates. Branched chain amino acids like valine, leucine, and isoleucine have been found to enhance muscle health. The consumption of glycine improved the quality of sleep compared to chemical substances like benzodiazepines. Supplementation of cysteine and theanine improved the overall immunity by antigen-specific antibody production after encountering an antigen. Plants also contain a wide range of vitamins, which are essential for metabolism. Several vitamins originating from plants, including pro-vitamins A, B complex, C, E, and K1, have antioxidant properties. Supplements containing these vitamins are available [33].

Secondary metabolites include phenolics, terpenes, steroids, alkaloids, and flavonoids. Phenolics have antioxidant and anti-inflammatory properties that can moderate inflammatory signaling pathways and can serve as bioactive small molecules for anti-inflammatory drugs used in the treatment of various diseases like cancer. Phenolics also serves as antimicrobial and anticancer compounds.

Flavonoids have antioxidative, anti-inflammatory, antitumoral, and cardioprotective activities. These are mainly found in plants like cabbage, cauliflower, berries onion, apples, and pears [34]. Terpenes contain many bioactive small molecules like eugenol (from *Syzygium aromaticum*), Germacrone (from *Eryngium maritimum*), and Santalol (from *Santalum album*), which acts as antiviral drugs against influenza infections. They also serve as anti-inflammatory compounds by inducing the release of pro-inflammatory cytokines and leukotrienes [35]. Steroids have a great affinity toward nuclear receptors. This led to the development of drugs derived from steroids for receptor-mediated diseases. Several anticancer molecules and antihormonal drugs have been synthesized from steroids. Alkaloids are a major part of Chinese herbal medicine. Alkaloids are majorly found in higher plants, especially in dicots. Berberine obtained from *Coptidis rhizoma* acts as an antibacterial and anticancer drug. Matrine from *Sophora flavescens* acts as an anti-inflammatory, anti-arrhythmic, and antifibrotic

Table 2.4 Pharmacological effects of bioactive compounds.

Bioactive compound	Source	Effect	Reference
Preussiafuran Cissetin Asterric acid	*Preussia* sp.	Antimalarial	[37]
Piliformic acid	*Xylaria* sp.	Antimalarial	[37]
Fumoquinone Pseurotin	*Aspergillus* sp.	Antimalarial	[37]
Bostrycin Nigrosporin B Fusarubin	*Fusarium* sp.	Antimalarial	[37]
Gallic acid	*Vitis vinifera* and *Camellia sinensis*	Antioxidant, antimicrobial, anti-inflammatory, and antitumoral	[38]
Luteolin	*Brassica oleracea* and *Apium graveolens*	Antioxidant, anti-inflammatory, and antitumoral	[38]
Anthocyanins	*Vitis vinifera* and *Daucus carota*	Antimicrobial, anti-inflammatory, antioxidant, and anti-proliferative	[38]
Curcumin	*Curcuma longa*	Anti-inflammatory	[39]
Paclitaxel	*Taxus brevifolia*	Antimitotic agent – various cancers	[39]
Artemisinin	*Artemisia annua*	Malignant cerebral malaria	
Tomatine	*Solanum lycopersicum*	Anticancer, immune effects, and antifungal	[39]
Oblongolide	*Phomopsis* sp.	Antiviral	[39]
Illudin S	*Omphalotus illudens*	Antiviral	[39]
Fucoidans	*Fucus vesiculosus and Lonicera japonica*	Antioxidative	[40]
Fucoidans	*Fucus vesiculosus*	Antithrombin, anti-inflammatory	[41]
Nisin A	*Lactococcus lactis*	Cancers of the head and neck, breast, liver, and blood cells (acute T cell leukemia)	[41]
Pediocin	*Pediococcus acidilactici*	Colon adenocarcinoma	[42]
N-butanol fraction	*Acacia arabica*	Anti-HIV activity	[43]
Methanolic extract	*Ocimum basilicum*	Inhibition of viral replication	[43]
Aqueous extra	*Olea europaea*	Inhibition of cell-to-cell transmission of HIV	[43]

drug. Piperine from *Piper nigrum* has antioxidant, anti-asthmatic, anticarcinogenic, anti-inflammatory, anti-ulcer, and anti-amoebic properties. Rhyncophylline obtained from *Uncaria rhynchophylla* can be used as a substitute for amphetamine to treat psychological disorders [36] (Table 2.4).

2.3.2 Anticancer Agents as Bioactive Molecules

Developing less toxic anticancer medications is crucial because cancer is one of the top causes of mortality in the modern era. Bioactive compounds with anticancer properties regulate metabolic and signaling pathways, thereby limiting and preventing the growth and division of aberrant cancer cells. Untailored natural materials or their synthetic versions constituted 47.1% of the 155 anticancer medications that were clinically authorized between 1961 and 2006. Majority of the bioactive compounds with anticancer properties originate from microbes and plants. The presence of bioactive molecules like anthraquinones, polysaccharides, and chromones in *Aloe vera* makes it an anticancer and antiproliferative drug [44]. Luteolin, a flavonoid acts as an anticancer agent by inhibiting the histone acetyltransferase (HAT) pathway of cancer cells. HAT is an enzyme that causes acetylation of chromatin to isomers of histone. In most cases of cancer, elevated levels of HAT are observed. Vinblastine, Vincristine (from *Catharanthus roseus*), and Colchicine (from *Colchium autumnale*) are alkaloids that inhibit cancer cell migration and metastasis by binding to microtubulins. Microtubules are cytoskeletal structures that are involved in various processes like cell migration, maintaining the cell shape, and cell division [45]. Podophyllotoxin is a lignan (secondary metabolite) derived from *Podophyllum peltatum* L. It prevents the growth of tumor cells by preventing polymerization of tubulin thereby arresting mitosis. Etoposide, teniposide, and etopophos are semisynthetic derivatives of Podophyllotoxin [46].

Certain antibiotics like Actinomycin D, Bleomycin, Doxorubicin, and Mitomycin C are used to treat cancers like sarcomas, leukemia, and carcinomas. They are derived from *Actinomycetes* such as *Streptomyces antibioticus*, *Streptomyces verticillus*, *Streptomyces peucetius*, and *Streptomyces caespitosus*, respectively. Actinomycin D treats tumors by inducing p53-independent apoptosis. Bleomycin cleaves the DNA and causes DNA damage. Doxorubicin blocks replication and transcription process and causes oxidative damage to the cell components. Mitomycin C crosslinks the DNA strands [42, 47].

2.3.3 Antiviral Agents as Bioactive Molecules

Antiviral drugs are important for treating clinically significant viruses like HIV (human immunodeficiency virus), HSV (herpes simplex virus) influenza, and hepatitis C. Most of the existing viral drugs like zidovudine, acyclovir, and ribavirin have serious adverse effects on the body. Therefore, development of antiviral drugs with the least side effects is necessary for which natural bioactive molecules are a great choice. Methanol extracts from various plants like *Alchornea laxiflora*, *Panax notoginseng*, and *Senecio scandens* have inhibitory effects on HIV integrase. Ethanol extracts from *Justicia gendarussa* and *Bauhinia variegata* and hexane extracts from *Aerva lanata*, *Acorus calamus*, and *Rhinacanthus nasutus* had inhibitory effects on HIV reverse transcriptase [48]. Anti-HSV activity was shown by ethanol, methanol, and chloroform extracted from plants like *Aglaia odorata*, *Pistacia vera*, and *Quercus brantii*. They halted viral growth by inhibiting viral replication, viral DNA synthesis, and virus entry. Dichloromethane and methanol obtained from plants like *Mussaenda elmeri* and *Baccaurea angulata* in 1:1 ratio inhibited hemagglutination due to influenza infection [20]. An anti-influenza drug, oseltamivir synthesized from shikimic acid and quinine acid was found to decrease the mortality rate in COVID-19-infected patients [49].

Several bioactive compounds from fungi are also known to have antiviral properties. Ganodeeric acid in *Ganoderma lucidum* and adenosine in *Cordyceps militaris* show inhibitory effects on HIV protease. Beta glucan protein of *Agaricus subrufescens* limits the replication of HSV [50].

2.3.4 Antimalarial Agents as Bioactive Molecules

Antimalarial bioactive molecules help combat the malarial infection caused by various *Plasmodium* species. Malaria caused by *Plasmodium falciparum* is considered the most fatal one due to its nature of causing cerebral infections. The most successful antimalarial drugs, quinine and artemisinin, are secondary metabolites of plants [51]. Bisbenzylisoquinoline from *Stephania erecta* can restrict the growth of chloroquinone-resistant strains of *Plasmodium falciparum*. Quassinoids from *Brucea javanica*, *Simaba guianensis*, and *Quassia indica* are effective against *Plasmodium falciparum*. Certain flavonoid compounds like lanaroflavona, volkensiflavone, aetonin B, and cyclohetrophyllin are effective against malaria. Xanthones like allanxanthone, torophyllin A, and symphonin have *in vitro* class antiplasmodial activity against sensitive and resistant strains [52]. Asterric acid and preussiafuran found in *Preussia* sp. are fungal bioactive compounds, which are known to be effective against malaria. Some of the other antimalarial compounds found in fungi are cytochalasin, acremonisol, fumoquinone, and pseurotin [53, 54].

2.3.5 Marine Bioactive Products

In terms of the discovery of bioactive chemicals, marine resources have been the least utilized. The pharmaceutical, cosmetic, and food industries could all benefit from the substances it contains. These biomolecules found in the ocean have a stellar reputation for fighting cancer, inflammation, and free radicals. The phycobilin pigments present in cyanobacteria and red algae have hepatoprotective anti-inflammatory properties. Marine algae have great antioxidative properties due to the presence of compounds like phlorotannin, beta carotene, and astaxanthin [55]. Red and brown algae and sea cucumber (*Ludoigothurea grisea*) are rich in sulfated polysaccharides that have antithrombotic and anticoagulant activities. The anticoagulant activity is due to the presence of fucoidins. These algae also inhibit replication of enveloped viruses. *Nostocmuscorum* and *Oscillatoria* have antitumor activity against hepatocellular cancer cell lines. HeLa cancer cells were inhibited by compounds present in *Calothrix*. Marine cyanobacterial depsipeptide gallinamide A was found to inhibit cathepsin L, which is used by coronaviruses to cause infection inside the cells [56]. Plitidepsin obtained from *Aplidium albicans* reduces the proliferation of cancer and reduces lung inflammation and viral load in COVID-19 infections. It has also been proved that plitidepsin has greater therapeutic potential in treating COVID-19 in comparison to remdesivir [57]. Marine bioactive molecules can be used as drug targets and as dietary supplements with proper research to rule out the possible chances of toxicity.

2.4 Role of Density Functional Theory (DFT) Studies in Bioactive Small-Molecule Discovery

2.4.1 Importance of DFT in Small-Molecule Drug Discovery

If medicinal chemists want to improve their drug discovery and design procedures, they need better computational methods. If we want to learn important things about the electrical characteristics of small-molecule pharmaceuticals and biological systems in a short amount of time, we need to use the correct quantum mechanical (QM) methods [13]. In QM calculations, conventional Hartree–Fock (HF) methods may not be able to accurately describe certain properties owing to a lack of electron correlation, leading to the development of post-HF methods such as the Moller–Plesset perturbation theory (MPn) and configuration interaction (CI), which consider electron correlation [58]. However, these methods are also computationally expensive. As a result, DFT has gained popularity as it offers a high level of accuracy with reduced computational time, making it a cost-effective option compared to other methods. In the drug design and discovery process, DFT is used to study isolated drug molecules to provide information related to chemical reactivity and global reactivity parameters. DFT geometry optimization is the key method for small molecules, as it directly determines the electronic probability density and the geometry of molecule is optimized to provide low energy conformation which is useful for the ligand of greater possible conformations [59].

Extensive DFT studies of diverse molecular properties have demonstrated the accuracy of DFT. Moreover, a comprehensive study comparing DFT calculations with experimental data has confirmed its suitability. The B3LYP functional has been widely utilized in this field of study. Drug design often focuses on the energetic properties of drug molecules, including their ionization energies, relative energies, and electron affinities. While DFT can effectively study these properties, it does not provide reliable information on atomization energies [60]. The process of designing new drugs requires an understanding of the different types of interactions that occur between drug molecules and their target sites. DFT calculations accurately predict the strength of hydrogen bonds and reveal the strength of ionic and covalent interactions between drug-like compounds and their targets. However, dispersion interactions that affect hydrogen bonds to hydrophobic interactions are not predictable by DFT calculations [59]. DFT calculations cannot provide a reliable description of systems with low-energy forces, such as those between 1 and 5 kcal/mol [61]. Therefore, when analyzing the interactions between a drug and its target, it is imperative not to depend exclusively on DFT. The transition states of small pharmacological compounds can be accurately determined using DFT calculations. Calculations employing B3LYP functionals offer insights into the drug–target interaction process.

DFT is additionally used with molecular docking and MD simulations to predict drug–receptor interactions more precisely. In molecular docking, the drug candidate is considered flexible, whereas the macromolecule (active site) is considered rigid. So, in order to mimic an actual biological system in which the protein is flexible, the ligand–macromolecule complex is simulated through MD simulations, which are based on classical Newtonian physics, to address the unrealistic rigidity of a system [61].

The docking and MD simulations reveal drug–target binding mechanisms and potential noncovalent interactions between small-molecule drugs and their targets. However, these methods are expensive. To overcome these disadvantages, integrated QM/molecular mechanics (MM) hybrid modeling has been utilized. DFT part in QM/MM method studies the enzyme active site where different possible bonding interactions take place while MM method takes care of the rest of macromolecules. Because of this, force field-based MM methods can effectively simulate the complex macromolecular portion, which consists of hundreds of atoms, while QM methods calculate electronic properties like charge transfer at protein target [62]. Some selected DFT functionals used in molecular modeling are listed in Table 2.5.

2.5 Application of DFT to Bioactive Small Molecules

2.5.1 HOMO–LUMO Calculation

Based on the predictions made from the highest occupied molecular orbital (HOMO) and the lowest unoccupied molecular orbital (LUMO), DFT calculates the Frontier molecular orbital (FMO) to find out how many electrons a molecule may take or donate. A large HOMO–LUMO gap suggests that the pharmacological molecule is relatively unresponsive or inactive. The HOMO–LUMO energy gap (EHOMO–ELUMO) is inversely proportional to the compounds' stabilizing interactions in protein–ligand interactions. A number of reactivity metrics are computed for drug

Table 2.5 Summary of selected DFT functionals used in molecular modeling studies of small-molecule drug design and discovery.

Functional	Type	χ	Basis set	Correlation functional	Description
wB97X-D	GGA RSH	100 22	6-31+G(d,p) def2-TZVP	Becke97	The most suitable DFT functional for accurately representing hydrogen bonding interactions
B3LYP	Meta GGA Hybrid GGA	20	def2-TZVP 6-311G* 6-311G(d,p) 6-31G* 6-31G	Lee-Yang-Parr (LYP)	Computationally less expensive and gives accurate results for a vast number of drug molecules. It is suitable for small organic molecules and it poorly performs with transition meta complexes and very large organic molecules
CAM-B3LYP	RSH	85	def2-TZVP 6-311G* 6-311G(d,p) 6-31G* 6-31G	Lee-Yang-Parr (LYP)	When compared to B3LYP, CAM-B3LYP does better for transferring charge excitations in a dipeptide model; it also beats B3LYP when it comes to nucleophilic substitutions and subsets of barrier height
LC-BLYP	RSH	100	def2-TZVP 6-311G* 6-311G(d,p) 6-31G* 6-31G	Lee-Yang-Parr (LYP)	LC-BLYP is resistant to spurious oscillations in molecular properties
B3PW91	Hybrid GGA	20	6-311++G**	Perdew-Wang91 PBE	It is suitable for the systems with uniform density
PBE0-D3	Semi-empirical GGA	25	6-31G*	PBE	It is used to study flexible drug–target complex

χ, the percentage of HF exchange functional.
Abbreviations: GGA, generalized gradient approximation; RSH, range separation hybrid.

molecules, including chemical hardness (ε), chemical potential (μ), chemical softness (σ), ionization potential (I), electron affinity (A), electrophilicity index ($\ddot{\upsilon}$), electronegativity (χ), and dipole moment (Debye). Extensive calculations are performed in accordance with the theory in order to assess the stability and reactivity of drugs [63].

2.5.1.1 Molecular Electrostatic Potential (MEP) Map

A valuable electronic property in molecular is the MEPs, which has been studied alongside other theories such as FMO analysis, chemical reactivity, and local descriptors. MEP surface analysis reveals the molecular charge distribution and related properties of small-molecule drugs [64]. The MEP map is a three-dimensional picture that can be used to understand the electronic nature of reactive sites, such as electron-deficient (most positive) and electron-rich (most negative) regions.

The MEP domains can be shown using various colors; for instance, blue can be used to represent the most positive regions, red to represent the most negative, and yellow or green to represent the intermediate regions. This picture helps understand possible covalent connections upon binding to macromolecules and other interactions such as hydrogen bonding or weak London forces [65].

2.5.1.2 The Two Main Methods Used in Population Statistics Are the Mulliken and Natural Population Analyses

According to MEP, the type of charge on each atom in the drug molecule can be revealed by using natural population analysis (NPA) and Mulliken population analysis (MPA). A number of possible interactions between the drug molecule and the enzyme active site can be anticipated if we suppose that the molecule contains a nucleophilic and electrophilic core.

2.5.1.3 Natural Bond Orbital (NBO) Analysis

DFT uses a chemical bond analysis technique called NBO analysis (DFT). It investigates all possible combinations of "empty" (acceptor) non-Lewis NBOs and "full" (donor) Lewis-type NBOs. Second-order perturbation theory is then applied to estimate the energy of these interactions. Hydrogen bonds between molecules, intramolecular hydrogen bonds, and hyperconjugative interactions can all be identified by NBO analysis. The donor–acceptor stabilization energy (E2) is linked to the *i*/*j* delocalization between the acceptor NBO (*j*) and donor NBO (*i*). The stabilizing energy (E2) is greater, allowing for a more thorough conjugation over the entire molecular system. Furthermore, the medicinal molecule's electronic spectra's π–π^* and n–π^* transitions are validated using the results of NBO [66].

Researchers can also learn new things about the catalytic mechanism of active site enzymes thanks to the QM/MM technique, which is important for mechanism-based, logical drug design: For the past 20 years, QM clusters have been in use for enzyme active site reaction mechanism modeling. This approach, which uses simple QM as opposed to QM/MM, provides high-level chemical precision, especially at the catalytic regions of enzymes, allowing for a thorough examination of protein–ligand interactions and reaction processes. The enormous computational cost of this approach is a drawback [67].

In summary, electron density, electronic charge, and other reactivity parameters can be studied for any isolated drug molecule using DFT as a standalone method; DFT is also utilized to study the type of interactions at the target site of enzyme by MM method. Finally, QM/MM method is combinedly used to evaluate the mechanisms of inhibition reactions by a small-molecule drug [68].

2.5.1.4 Implementations and Tools

Thoms and Fermi started the work on the DFT in the 1920s but the broad usage of DFT was established by Hohenberg, Kohn, and Sham in the 1960s. Later, the field booms after the development of popular B3LYP functionals in the early 1990s that the application of DFT in molecular modeling came into reality. Some of the most widely used quantum chemical programs are listed in Table 2.6, including Gaussian, GAMESS, Jaguar, and ORCA [69, 70].

2.6 Factors Affecting the Choice of Bioactive Molecules in Drug Discovery

Drug discovery is a complex, multifaceted process that involves identifying, designing, and developing compounds that exhibit therapeutic properties. The selection of bioactive molecules plays a pivotal role in this process, and several factors influence this choice. Bioactive molecules are

Table 2.6 Widely used quantum chemical programs.

Quantum chemical programs	License	URL
Gaussian	Commercial	https://gaussian.com/
Jaguar	Commercial	https://newsite.schrodinger.com/platform/products/jaguar/
GAMESS	Academic free	https://www.ameslab.gov/gamess-open-source-quantum-chemistry-software
ORCA	Academic free	https://orcaforum.kofo.mpg.de/app.php/portal
Quantum Espresso	Open source	https://www.quantum-espresso.org/Doc/user_guide/node8.html
MOLPRO	Commercial	https://www.molpro.net/
Psi4	Academic free	https://psicode.org/

Table 2.7 Common factors that typically affect the choice of bioactive molecules in drug discovery.

Factor	Description
Target specificity	The degree of selectivity of a bioactive molecule toward the intended drug target
Safety profile	Evaluation of the molecule's safety, encompassing toxicity and potential adverse effects
Pharmacokinetics	Study of how the molecule is absorbed, distributed, metabolized, and excreted in the body
Structural characteristics	Analysis of the molecular structure and its relevance to interactions with the target
Bioavailability	Measurement of the amount of the administered dose that reaches the intended target site
Potency	Assessment of the strength or efficacy of the molecule in inducing the desired biological effect

compounds capable of interacting with biological systems, often exhibiting specific biological activity that can be harnessed for therapeutic purposes. These molecules serve as the foundation for potential drugs, affecting target specificity, pharmacokinetics, and safety profiles. The choice of bioactive molecules profoundly influences the success of drug discovery due to their direct impact on efficacy, safety, and feasibility in pharmaceutical development. This factor plays a crucial role in determining a molecule's suitability for therapeutic development. Table 2.7 gives the summary of common factors that typically affect the choice of bioactive molecules in drug discovery [1].

Here are some key factors (Table 2.8) and the corresponding reference articles that discuss these aspects.

2.6.1 Target Identification and Validation

The starting point in drug discovery is the identification and validation of a specific molecular target associated with a disease. This target could be a protein, enzyme, receptor, or nucleic acid. The choice of bioactive molecules revolves around their ability to interact with and modulate these targets

Table 2.8 A concise overview of the factors influencing the selection of bioactive molecules in drug discovery.

Factor	Description	References
Target identification and validation	Target identification methods, encompassing affinity-based, genetic, computational, and chemical proteomics approaches, delineating their pros, cons, and applications, and categorizing methodologies by execution steps to aid strategic selection in drug discovery	[71]
	Comparison of chemogenomic compound evaluations, affinity-based chemoproteomics, activity-based protein profiling, label-free techniques, and computational approaches, emphasizing their components, advantages, limitations, and applications to assist researchers in effective target identification for drug discovery	
Target specificity	Discusses the importance of molecules interacting selectively with intended biological targets while minimizing off-target interactions	[72]
Bioavailability and pharmacokinetics	Barriers (e.g., enzymatic degradation and low permeability) affecting oral peptide and protein delivery, along with strategies (e.g., formulation techniques and encapsulation methods) categorized by efficacy in overcoming specific challenges, providing valuable insights for researchers developing optimized oral delivery systems for protein therapeutics	[72]
	Explores the evolution and relevance of the Rule of Five in drug discovery, how the Rule of Five impacts protein targets and drug binding sites, emphasizing distinctive features and providing examples of exceptions predominantly found in natural products (NPs) and synthetic compounds, ultimately contributing to a comprehensive understanding of structural influences on bioactive molecule selection in drug discovery for strategic decision-making	
Chemical structure and drug-likeness	A structured overview of essential physiochemical properties influencing drug development; categorizing methods for screening, diagnosis, and analysis; and offering a comprehensive guide for scientists and students in medicinal chemistry to assess and optimize drug-like properties throughout drug discovery and development	[73]
Safety and toxicity	Models incorporate gene disruption, structural information (MOLD2), toxicity data (TOX21), and adverse event reporting (FAERS) to predict drug-induced liver injury (DILI). Models display various performances in this regard. The article showcases the use of ensemble voting methods to improve predictive capacities by highlighting accuracies and Matthews correlation coefficients (MCCs) for different clinical DILI subtypes. Highlighting the intricacy of predictive modeling for DILI, this all-inclusive resource provides insights into model performance for safety evaluation during drug development	[74]
Toxicity and side effects	Discusses strategies to filter out problematic compounds early in the drug discovery process to minimize toxicity and adverse effects	[74]
Cost-effectiveness, synthetic feasibility, and scalability	Explores economic considerations in pharmaceutical production and emphasizes cost-effective synthesis methods. Focuses on the importance of synthetic accessibility on a large scale for clinical production and scalability for practical manufacturing processes	[75]
Structural diversity and novelty	Diversity-oriented synthesis (DOS) aims to generate diverse pure compounds for drug lead discovery, blending strategies from nature and synthetic chemistry. It tackles limitations of relying solely on natural sources, employing molecular descriptor analysis and a freely available tool to assess structural diversity in synthesis	[76]
Patentability and intellectual property	Considerations in protecting pharmaceutical innovation	[77]

effectively. The process of identifying and validating potential drug targets is pivotal in drug discovery, particularly concerning the selection of bioactive molecules [71]. Several aspects affecting target identification and validation are critical to the selection of bioactive compounds. Important decisions regarding which bioactive compounds to pursue in further stages of drug development are made during target identification and validation. Several strategies have been suggested for identifying targets, including computational methods, genetic methods, phenotype-based methods, and affinity-based methods. Thermal shift assays, drug affinity responsive target stability (DARTS), and photoaffinity labeling (PAL) are all affinity-based methods that directly identify targets for small molecules. PAL is one example; it employs a photoreactive moiety on a chemical probe that, when triggered by light of a certain wavelength, forms a covalent bond with its target [78]. DARTS, on the other hand, identifies interactions without prior modification of the drug, relying on protease digestion and subsequent analysis of stabilized proteins. Thermal shift assays, like cellular thermal shift assay (CETSA), investigate protein stabilization upon ligand binding. These methods offer advantages in sensitivity and specificity but may face challenges in terms of scalability and resource requirements. Additionally, phenotype-based assays, exemplified by MorphoBase, utilize high-content image analysis to classify drug effects based on cell morphology. Genetic approaches, particularly CRISPR-Cas9, revolutionize drug target discovery by enabling precise genome editing, as demonstrated in the identification of novel therapeutic targets in diseases like acute myeloid leukemia. Computational methods, involving analyses of genomics, transcriptomics, proteomics, and metabolomics data, contribute to a comprehensive understanding of drug interactions. While these diverse approaches offer valuable insights into target identification, each comes with its own set of advantages and disadvantages, emphasizing the importance of a multifaceted approach in drug discovery [79].

2.6.2 Target Specificity

Molecules must interact selectively with the intended biological target (e.g., protein, enzyme, and receptor) to exert the desired therapeutic effect while minimizing off-target effects. Target specificity stands as a pivotal factor in the selection of bioactive molecules during the intricate process of drug discovery. This ensures efficacy and safety. It is believed that stress ligand efficiency is a key metric for lead selection, emphasizing the importance of molecule effectiveness in binding to intended targets for drug specificity. Specificity in targeting biomolecules or receptors significantly impacts a molecule's efficacy and safety profile. Balancing potency and selectivity poses a fundamental challenge in drug development, highlighting the need for researchers to carefully consider target specificity in the intricate process of selecting bioactive molecules [80].

2.6.3 Bioavailability and Pharmacokinetics

Bioavailability and pharmacokinetic properties determine the molecule's efficacy and safety in drug discovery. The strategies to improve bioavailability, particularly for biologics, proteins, and peptidesare essential considerations in drug discovery. Bioavailability and pharmacokinetics are critical factors influencing the choice of bioactive molecules in drug discovery, particularly when considering the oral delivery of complex proteins. The introduction of proteins as therapeutic agents brings forth advantages in terms of high bioactivity and specificity, but challenges arise in ensuring their effective delivery through the gastrointestinal (GI) tract. The GI environment presents hurdles such as enzymatic degradation, low permeability, and weak absorption, impacting the bioavailability of orally administered proteins and affecting clinical outcomes. Overcoming these barriers is crucial for improving patient compliance and reducing production costs. Current oral delivery strategies, including formulation approaches and the utilization of delivery systems, aim to enhance the stability and absorption of proteins (Peng et al. 2023). However, translating these strategies from *in vitro* and *in vivo* animal studies to human applications requires careful consideration of safety and limitations. The shortage of clinical data poses a challenge, emphasizing the need for innovative design strategies, novel model systems, and a deeper understanding of bioavailability and pharmacokinetics to advance oral protein drug delivery in the field of drug discovery [81].

Selecting bioactive compounds with good absorption, distribution, metabolism, and excretion rates is essential. The ADME pathway allows molecules to reach their target location at effective concentrations; molecules should have appropriate characteristics for this pathway. The chemical space is enormous, yet biologically active molecules tend to congregate in tiny areas. The number of protein folds and pockets also limits the options for drug design. The biophysics of cavity creation in protein ligand-binding sites is in good agreement with the physicochemical characteristics of pharmaceuticals, as stated by the Rule of Five. Experimental solubility is greatly affected by the parameter lipophilicity, which has nothing to do with binding site biology. Lipinski [82] discusses the evolution of the "Rule of Five" as a guideline for drug discovery, emphasizing its alignment with structural constraints in protein targets and ligands. Ro5, derived from physicochemical profiles of phase II drugs, guides ligand chemistry and design philosophy, influencing drug-likeness concerning bioavailability and pharmacokinetics. Lipinski in 2016 explores two extremes in ligand construction: synthetically derived ligands and NP metabolites shaped by evolutionary selection. Cyclosporine A, a chemical chameleon and an outlier of the NP Ro5 population, is a good example of how NPs can serve outliers owing to their peculiar chemistry and the effects of natural selection. The chameleon-like nature of NPs, caused by intramolecular hydrogen bonding, is demonstrated by exceptions to Ro5, particularly among NPs like cyclosporine A. Navitoclax and other non-NP Ro5 outliers were found by a team of very competent and coordinated individuals, with special emphasis on the complex interaction between medicinal chemists and in vivo bioassay specialists. Discovering Navitoclax – a non-NP Ro5 outlier – required a great deal of skill, resources, and unconventional thinking. In spite of confirmed accomplishments, the cost-effectiveness of finding oral medications that do not originate from NPs outside the Ro5 space is still up in the air. This highlights the persistent difficulties in effectively locating such compounds during the drug development process [83].

2.6.4 Chemical Structure and Drug-likeness

Chemical structure and drug-likeness play a crucial role in determining the success of drug development. Drug-likeness describes the chemical and physical properties of a molecule that make it suitable for use as a drug. Factors like molecular weight, lipophilicity, solubility, and structural complexity influence a molecule's drug-like properties. It underscores the critical role of selection of drug candidates in drug discovery, emphasizing the importance of compounds with favorable absorption, distribution, metabolism, elimination (ADME), and toxicity profiles [84]. It navigates physiochemical properties, offering in-depth exploration of screening, diagnosis, and modification strategies, and serves as a valuable guide for scientists and students involved in understanding, discovering, and developing clinically optimal candidates in drug development. In drug discovery, only a fraction of newly invented compounds with therapeutic target binding capabilities possess sufficient ADME properties and acceptable toxicology profiles to progress through human Phase I clinical trials. Kerns and Di's book provides scientists and students with the background and tools to comprehend, discover, and develop optimal clinical candidates [85]. The resource explores physiochemical properties, including solubility and permeability, and covers a wide variety of current methods for screening, diagnosis, and in-depth analysis of drug properties. Through case studies, structure–property relationship descriptions, and modification strategies, it equips scientists with tools and methods for all aspects of drug research, discovery, design, development, and optimization [5].

2.6.5 Safety and Toxicity

Understanding a molecule's safety profile is crucial to avoid adverse effects. Assessing potential toxicities and side effects early in the drug discovery process helps in selecting molecules with better safety profiles. Safety and toxicity assessments are critical in drug development to mitigate potential side effects. The study conducted by Adeluwa et al. [74] sheds light on the challenges and complexities in predicting DILI, a pivotal concern in regulatory drug clearance. One major issue in regulatory clearance is the prediction of DILI, which is the subject of the study by Adeluwa et al. [74]. In order to create prediction models for various DILI classes, the study used a wide variety of datasets, including gene expression data, structural information, toxicity profiles, and reports of adverse events [86]. The findings indicate varying accuracies and MCCs among models based on expression signatures, FAERS, MOLD2, and TOX21 data for predicting clinical DILI subtypes. Integrating these data types through ensemble voting methods enhanced predictive capabilities, achieving balanced accuracies up to 0.60 for clinically relevant DILI subtypes. The challenges in employing traditional machine learning for optimal classification with the current dataset emphasize the complexity of safety prediction in drug development [87]. The limitations of traditional machine learning in handling imbalanced datasets prompt the use of resampling techniques. The findings underscore the ongoing need for improvement in both model construction and the development of robust predictive data to comprehensively capture the complex landscape of DILI. As drug discovery continues to evolve, the study advocates for a broader focus on predicting biomarkers specific to DILI using advanced omics data, such as single-cell and metabolomics signatures, to enhance the safety assessment of bioactive molecules. Understanding a molecule's safety profile remains paramount in mitigating adverse effects and advancing the overall success of drug development efforts [88].

2.6.6 Toxicity and Side Effects

It is of utmost importance to ensure that a molecule is safe, with minimum toxicity and side effects in medication candidates. Research by Baell and Holloway [89] explains how to identify potentially harmful substances at an early stage of drug development and eliminate them or at least reduce their impact. In biochemical high-throughput screens, it draws attention to substructural properties that are critical for finding frequent hitters (promiscuous chemicals), which helps in identifying and excluding pan-assay interference compound (PAINS) from screening libraries and bioassays [90]. These compounds, despite evading detection by common filters for reactive compounds, exhibit activity across various assays, raising concerns about their prevalence as starting points in drug discovery. The accompanying table likely details specific problematic substructural features, their prevalence in diverse assay systems, and potential toxicity and side effects. This comprehensive guide aids researchers in stringent filtering and exclusion criteria, enhancing the reliability of screening libraries and bioassays in drug discovery [89].

2.6.7 Cost-Effectiveness, Synthetic Feasibility, and Scalability

Cost-effectiveness, synthetic feasibility, and scalability are critical considerations in the selection of bioactive molecules for drug discovery. The cost of synthesizing and developing bioactive compounds significantly influences their practical application in drug development. The synthesis processes must be economically feasible to enhance the viability of drug production. Moreover, the scalability of synthesis processes is vital to ensure that bioactive molecules can be produced on a large scale for clinical applications [91]. Factors such as the complexity of molecules, the availability of raw materials, and the chosen synthetic routes play a pivotal role in determining both the cost and scalability of production. These considerations underscore the importance of balancing cost-effectiveness and scalability when exploring bioactive molecules for drug discovery. Achieving economically viable synthesis processes is essential for advancing drug development and making these molecules accessible for large-scale production in clinical settings [92].

2.6.8 Structural Diversity and Novelty

Structural diversity enhances the discovery of novel bioactive compounds in drug development. Diversity-oriented synthesis aims to efficiently create a wide array of pure compounds suitable for lead generation. Structural diversity and novelty are pivotal factors influencing the choice of bioactive molecules in drug discovery, as highlighted by the focus on diversity-oriented synthesis (DOS) discussed by Bender et al. [76]. While nature inherently produces a vast array of structurally diverse small molecules, relying solely on NP sources for drug discovery has its limitations. The goal of DOS is to chemically synthesize small compounds with diverse and complicated structures as efficiently as possible so that phenotypic, high-throughput screening methods can provide leads [93]. It compares and contrasts the methods used by synthetic chemists with those used by nature, highlighting the significance of learning from the former. Examining molecular characteristics for chemical classification and the subjective nature of comparing DOS approaches are also covered in detail [94]. To address this, the article introduces freely available software for evaluating structural diversity (www.cheminformatics.org/diversity) in combinatorial synthesis, emphasizing the need for a systematic and objective approach to enhance the selection of diverse and novel bioactive molecules in drug discovery programs. It underscores the importance of molecular descriptor analysis and provides a tool for assessing structural diversity in combinatorial synthesis (www.cheminformatics.org/diversity).

2.6.9 Patentability and Intellectual Property

The considerations of patentability and intellectual property (IP) play a crucial role in determining the selection of bioactive molecules in drug discovery. The potential for obtaining patent protection becomes a significant factor, especially when dealing with unique chemical structures or novel applications. This aspect is pivotal in safeguarding the substantial investments made in research and development [95]. The works of Yu (2016) underscore the importance of IP in drug discovery, emphasizing the need for a comprehensive understanding of the interface between competition policy and IP rights, particularly in collaborative pharmaceutical research [96]. Yu (2016) delves into the intricate relationship between these factors and their impact on the drug discovery process, assessing the compliance of collaboration agreements with Article 101. Evaluation of patentability and IP rights in such collaborations is fundamental for ensuring adherence to legal frameworks and optimizing the protection of innovative bioactive molecules in drug development [96].

In summary, the intricate process of selecting bioactive molecules in drug discovery involves careful consideration of diverse factors, encompassing target identification, pharmacokinetics, chemical structure, safety, cost, and IP considerations. Achieving a delicate balance among these elements is crucial for identifying and advancing promising candidates that hold the potential for effective pharmaceutical development. The interplay of these factors underscores the complexity and nuanced decision-making required in the pursuit of successful drug discovery. Researchers must balance these factors to identify promising candidates for further development into effective pharmaceuticals [77, 97].

2.7 Conclusion

In conclusion, a vibrant and exciting area of pharmaceutical research is the investigation of bioactive small molecules in drug discovery. The enormous variety of tiny molecules present in nature, along with developments in computational techniques and synthetic chemistry, have completely changed how we approach medication creation. Researchers have learned a great deal about disease causes, biological processes, and possible treatment targets by studying bioactive small molecules. NPs are still a great source of inspiration for drug discovery because of their wide range of chemical structures and biological activity. Furthermore, the development of high-throughput screening methods has sped up the process of developing new drugs by facilitating the quick discovery and optimization of lead compounds. The development of multidisciplinary methods such as computer modeling, medicinal chemistry, and structural biology has accelerated the search for bioactive small molecules with improved pharmacokinetic, potent, and selective characteristics. Additionally, the development of target-based screening techniques and fragment-based drug design has made it easier to identify innovative treatment candidates with previously unheard-of accuracy.

Drug development has a bright future ahead of it as we work to understand biological systems and realize the potential of tiny compounds. Through the utilization of bioactive small-molecules' extensive chemical variety, we can tackle unfulfilled medical requirements, counteract medication resistance, and enhance patient results. In the quest for new treatments, bioactive small molecules will surely continue to be essential instruments when working together and utilizing creative approaches.

References

1 Zhou, S.F. and Zhong, W.Z. (2017). Drug design and discovery: principles and applications. *Molecules (Basel, Switzerland)* 22 (2): 279.

2 Bonam, S.R., Sekar, M., Guntuku, G.S. et al. (2021). Role of pharmaceutical sciences in future drug discovery. *Future Drug Discovery* 3 (3): FDD64.

3 Romano, J.D. and Tatonetti, N.P. (2019). Informatics and computational methods in natural product drug discovery: a review and perspectives. *Frontiers in Genetics* 10: 368.

4 Talevi, A. and Bellera, C.L. (2020). Challenges and opportunities with drug repurposing: finding strategies to find alternative uses of therapeutics. *Expert Opinion on Drug Discovery* 15 (4): 397–401.

5 Daina, A., Michielin, O., and Zoete, V. (2017). SwissADME: a free web tool to evaluate pharmacokinetics, drug-likeness and medicinal chemistry friendliness of small molecules. *Scientific Reports* 7 (1): 42717.

6 Gu, M., Sun, S., You, Q., and Wang, L. (2023). Forward or backward: lessons learned from small molecule drugs approved by FDA from 2012 to 2022. *Molecules (Basel, Switzerland)* 28 (24): 7941.

7 Li, H., Yang, Y., Hong, W. et al. (2020). Applications of genome editing technology in the targeted therapy of human diseases: mechanisms, advances and prospects. *Signal Transduction and Targeted Therapy* 5 (1): 1.

8 Seyhan, A.A. and Carini, C. (2019). Are innovation and new technologies in precision medicine paving a new era in patients centric care? *Journal of Translational Medicine* 17 (1): 114.

9 Stielow, M., Witczyńska, A., Kubryń, N. et al. (2023). The bioavailability of drugs – the current state of knowledge. *Molecules (Basel, Switzerland)* 28 (24): 8038.

10 Alqahtani, M.S., Kazi, M., Alsenaidy, M.A., and Ahmad, M.Z. (2021). Advances in oral drug delivery. *Frontiers in Pharmacology* 12: 618411.

11 Mitchell, M.J., Billingsley, M.M., Haley, R.M. et al. (2021). Engineering precision nanoparticles for drug delivery. *Nature Reviews Drug Discovery* 20 (2): 101–124.

12 Carpio, L.E., Sanz, Y., Gozalbes, R., and Barigye, S.J. (2021). Computational strategies for the discovery of biological functions of health foods, nutraceuticals and cosmeceuticals: a review. *Molecular Diversity* 25 (3): 1425–1438.

13 Sliwoski, G., Kothiwale, S., Meiler, J., and Lowe, E.W. Jr. (2014). Computational methods in drug discovery. *Pharmacological Reviews* 66 (1): 334–395.

14 Adelusi, T.I., Oyedele, A.-Q.K., Boyenle, I.D. et al. (2022). Molecular modeling in drug discovery. *Informatics in Medicine Unlocked* 29: 100880.

15 Shah, A.P., Parmar, B.M., Ghodawala, M.A., and Seth, A. (2018). *In silico* drug discovery of novel small lead compounds targeting Nipah virus attachment glycoprotein. *Journal of Integrated Health Sciences* 6 (2): 60.

16 Sadybekov, A.V. and Katritch, V. (2023). Computational approaches streamlining drug discovery. *Nature* 616 (7958): 673–685.

17 Shah, A., Sanghvi, K., Sureja, D., and Seth, A. (2018). *In silico* drug design and molecular docking studies of some natural products as tyrosine kinase inhibitors. *International Journal of Pharmaceutical Research* 10 (2).

18 Shah, A., Choudhary, A., Jain, M. et al. (2024). Discovery of novel anticancer flavonoids as potential HDAC2 inhibitors: virtual screening approach based on molecular docking, DFT and molecular dynamics simulations studies. *3 Biotech* 14 (3): 83.

19 Shah, A. and Seth, K.A. (2021). *In silico* identification of novel flavonoids targeting epidermal growth factor receptor. *Current Drug Discovery Technologies* 18 (1): 75–82.

20 Manigrasso, J., Marcia, M., and De Vivo, M. (2021). Computer-aided design of RNA-targeted small molecules: a growing need in drug discovery. *Chem* 7 (11): 2965–2988.

21 Raschka, S. and Kaufman, B. (2020). Machine learning and AI-based approaches for bioactive ligand discovery and GPCR-ligand recognition. *Methods (San Diego, Calif)* 180: 89–110.

22 Agu, P.C., Afiukwa, C.A., Orji, O.U. et al. (2023). Molecular docking as a tool for the discovery of molecular targets of nutraceuticals in diseases management. *Scientific Reports* 13 (1): 13398.

23 Bissantz, C., Kuhn, B., and Stahl, M. (2010). A medicinal chemist's guide to molecular interactions. *Journal of Medicinal Chemistry* 53 (14): 5061–5084.

24 Vázquez, J., López, M., Gibert, E. et al. (2020). Merging ligand-based and structure-based methods in drug discovery: an overview of combined virtual screening approaches. *Molecules (Basel, Switzerland)* 25 (20): 4723.

25 Cereto-Massagué, A., Ojeda, M.J., Valls, C. et al. (2015). Molecular fingerprint similarity search in virtual screening. *Methods (San Diego, Calif)* 71: 58–63.

26 Fiscon, G., Conte, F., Farina, L., and Paci, P. (2018). Network-based approaches to explore complex biological systems towards network medicine. *Genes* 9 (9): 437.

27 Ginsberg, S.D., Sharma, S., Norton, L., and Chiosis, G. (2023). Targeting stressor-induced dysfunctions in protein–protein interaction networks via epichaperomes. *Trends in Pharmacological Sciences* 44 (1): 20–33.

28 Basar, M.A., Hosen, M.F., Kumar Paul, B. et al. (2023). Identification of drug and protein–protein interaction network among stress and depression: a bioinformatics approach. *Informatics in Medicine Unlocked* 37: 101174.

29 Kuzmanov, U. and Emili, A. (2013). Protein–protein interaction networks: probing disease mechanisms using model systems. *Genome Medicine* 5 (4): 37.

30 Kenanova, D.N., Visser, E.J., Virta, J.M. et al. (2023). A systematic approach to the discovery of protein–protein interaction stabilizers. *ACS Central Science* 9 (5): 937–946.

31 Yuan, H., Ma, Q., Ye, L., and Piao, G. (2016). The traditional medicine and modern medicine from natural products. *Molecules (Basel, Switzerland)* 21 (5): 559.

32 Dias, D.A., Urban, S., and Roessner, U. (2012). A historical overview of natural products in drug discovery. *Metabolites* 2 (2): 303–336.

33 Zhang, S., Zeng, X., Ren, M. et al. (2017). Novel metabolic and physiological functions of branched chain amino acids: a review. *Journal of Animal Science and Biotechnology* 8: 10.

34 Wink, M. (2015). Modes of action of herbal medicines and plant secondary metabolites. *Medicines (Basel, Switzerland)* 2 (3): 251–286.

35 Shahidi, F. and Ambigaipalan, P. (2015). Phenolics and polyphenolics in foods, beverages and spices: antioxidant activity and health effects – a review. *Journal of Functional Foods* 18: 820–897.

36 Mahadik, N., Bhattacharya, D., Padmanabhan, A. et al. (2022). Targeting steroid hormone receptors for anti-cancer therapy – a review on small molecules and nanotherapeutic approaches. *Wiley Interdisciplinary Reviews Nanomedicine and Nanobiotechnology* 14 (2): e1755.

37 Ibrahim, S.R.M., Mohamed, G.A., Al Haidari, R.A. et al. (2018). Potential anti-malarial agents from endophytic fungi: a review. *Mini Reviews in Medicinal Chemistry* 18 (13): 1110–1132.

38 Albuquerque, B.R., Heleno, S.A., Oliveira, M.B.P.P. et al. (2021). Phenolic compounds: current industrial applications, limitations and future challenges. *Food & Function* 12 (1): 14–29.

39 Costa, G., Francisco, V., Lopes, M.C. et al. (2012). Intracellular signaling pathways modulated by phenolic compounds: application for new anti-inflammatory drugs discovery. *Current Medicinal Chemistry* 19 (18): 2876–2900.

40 Al-Salihi, S.A.A. and Alberti, F. (2021). Naturally occurring terpenes: a promising class of organic molecules to address influenza pandemics. *Natural Products and Bioprospecting* 11 (4): 405–419.

41 Suleria, H.A.R., Gobe, G., Masci, P., and Osborne, S.A. (2016). Marine bioactive compounds and health promoting perspectives; innovation pathways for drug discovery. *Trends in Food Science & Technology* 50: 44–55.

42 Karpiński, T.M. and Adamczak, A. (2018). Anticancer activity of bacterial proteins and peptides. *Pharmaceutics* 10 (2): 54.

43 Akram, M., Tahir, I.M., Shah, S.M.A. et al. (2018). Antiviral potential of medicinal plants against HIV, HSV, influenza, hepatitis, and coxsackievirus: a systematic review. *Phytotherapy Research* 32 (5): 811–822.

44 Choudhari, A.S., Mandave, P.C., Deshpande, M. et al. (2019). Phytochemicals in cancer treatment: from preclinical studies to clinical practice. *Frontiers in Pharmacology* 10: 1614.

45 Asma, S.T., Acaroz, U., Imre, K. et al. (2022). Natural products/bioactive compounds as a source of anticancer drugs. *Cancers* 14 (24): 6203.

46 Dhyani, P., Quispe, C., Sharma, E. et al. (2022). Anticancer potential of alkaloids: a key emphasis to colchicine, vinblastine, vincristine, vindesine, vinorelbine and vincamine. *Cancer Cell International* 22 (1): 206.

47 Gao, Y., Shang, Q., Li, W. et al. (2020). Antibiotics for cancer treatment: a double-edged sword. *Journal of Cancer* 11 (17): 5135–5149.

48 Razonable, R.R. (2011). Antiviral drugs for viruses other than human immunodeficiency virus. *Mayo Clinic Proceedings* 86 (10): 1009–1026.

49 De Clercq, E. (2004). Antiviral drugs in current clinical use. *Journal of Clinical Virology* 30 (2): 115–133.

50 Suwannarach, N., Kumla, J., Sujarit, K. et al. (2020). Natural bioactive compounds from fungi as potential candidates for protease inhibitors and immunomodulators to apply for coronaviruses. *Molecules (Basel, Switzerland)* 25 (8): 1800.

51 Shibeshi, M.A., Kifle, Z.D., and Atnafie, S.A. (2020). Antimalarial drug resistance and novel targets for antimalarial drug discovery. *Infection and Drug Resistance* 13: 4047–4060.

52 Cui, L., Mharakurwa, S., Ndiaye, D. et al. (2015). Antimalarial drug resistance: literature review and activities and findings of the ICEMR network. *The American Journal of Tropical Medicine and Hygiene* 93 (3 Suppl): 57–68.

53 Azmi, W.A., Rizki, A.F.M., Djuardi, Y. et al. (2023). Molecular insights into artemisinin resistance in *Plasmodium falciparum*: an updated review. *Infection, Genetics and Evolution* 112: 105460.

54 Shah, A.P., Parmar, G.R., Sailor, G.U., and Seth, A.K. (2019). Antimalarial phytochemicals identification from *Euphorbia hirta* against Plasmepsin protease: an *in silico* approach. *Folia Medica* 61 (4): 584–593.

55 Parmar, G., Chudasama, M.J., Shah, V. et al. (2023). A mini-review on marine bioactive compounds as a source of antiviral drug lead. *Current Traditional Medicine* 9 (5): 116–119.

56 Shah, A.P., Doshi, K., Fatepurwala, H. et al. (2023). Naturally occurring diterpenes as anti-hypertensive agents: a review. *Multidisciplinary Approaches in Pharmaceutical Sciences* 201–213.

57 Amdal, C.D., Falk, R.S., Singer, S. et al. (2023). A multicenter international prospective study of the validity and reliability of a COVID-19-specific health-related quality of life questionnaire. *Quality of Life Research* 32 (2): 447–459.

58 Lin, X., Li, X., and Lin, X. (2020). A review on applications of computational methods in drug screening and design. *Molecules (Basel, Switzerland)* 25 (6): 1375.

59 Shah, A., Parmar, G., Shah, U., and Perumal, S. (2023). Virtual screening, molecular docking studies and DFT calculations of novel anticancer flavonoids as potential VEGFR-2 inhibitors. *Chemistry Africa* 6 (4): 1847–1861.

60 De Souza, L.A., Nogueira, C.A.S., Lopes, J.F. et al. (2013). DFT study of cisplatin@carbon nanohorns complexes. *Journal of Inorganic Biochemistry* 129: 71–83.

61 Abkari, A., Chaabane, I., and Guidara, K. (2016). DFT (B3LYP/LanL2DZ and B3LYP/6311G+(d,p)) comparative vibrational spectroscopic analysis of organic–inorganic compound bis(4-acetylanilinium) tetrachlorocuprate(II). *Physica E: Low-Dimensional Systems and Nanostructures* 81: 136–144.

62 El-Demerdash, S.H., Halim, S.A., El-Nahas, A.M., and El-Meligy, A.B. (2023). A density functional theory study of the molecular structure, reactivity, and spectroscopic properties of 2-(2-mercaptophenyl)-1-azaazulene tautomers and rotamers. *Scientific Reports* 13 (1): 15626.

63 Mumit, M.A., Pal, T.K., Alam, M.A. et al. (2020). DFT studies on vibrational and electronic spectra, HOMO–LUMO, MEP, HOMA, NBO and molecular docking analysis of benzyl-3-N-(2,4,5-trimethoxyphenylmethylene)hydrazinecarbodithioate. *Journal of Molecular Structure* 1220: 128715.

64 Vidhya, V., Austine, A., and Arivazhagan, M. (2019). Quantum chemical determination of molecular geometries and spectral investigation of 4-ethoxy-2, 3-difluoro benzamide. *Heliyon* 5 (11): e02365.

65 Pandey, J., Prajapati, P., Srivastava, A. et al. (2018). Spectroscopic and molecular structure (monomeric and dimeric model) investigation of Febuxostat: a combined experimental and theoretical study. *Spectrochimica Acta Part A: Molecular and Biomolecular Spectroscopy* 203: 1–12.

66 Renj, D.P.L., Geetha, R.R.J., Seles, R.F. et al. (2024). Density functional theory calculations, vibrational spectral analysis and topological analysis of 1-acethyl-2(4-isopropoxy-3-methoxyphenyl) cyclopropane with docking studies. *Chemical Physics Impact* 100524.

67 Frenking, G. and Froehlich, N. (2000). The nature of the bonding in transition–metal compounds. *Chemical Reviews* 100 (2): 717–774.

68 Ayisha Begam, K., Kanagathara, N., Marchewka, M.K., and Lo, A.Y. (2022). DFT, Hirshfeld and molecular docking studies of a hybrid compound – 2,4-diamino-6-methyl-1,3,5-triazin-1-ium hydrogen oxalate as a promising anti-breast cancer agent. *Heliyon* 8 (8): e10355.

69 Hirschi, J.S., Bashirova, D., and Zuehlsdorff, T.J. (2023). Opening the density functional theory black box: a collection of pedagogic Jupyter notebooks. *Journal of Chemical Education* 100 (11): 4496–4503.

70 Haunschild, R., Barth, A., and French, B. (2019). A comprehensive analysis of the history of DFT based on the bibliometric method RPYS. *Journal of Cheminformatics* 11 (1): 72.

71 Schenone, M., Dančík, V., Wagner, B.K., and Clemons, P.A. (2013). Target identification and mechanism of action in chemical biology and drug discovery. *Nature Chemical Biology* 9 (4): 232–240.

72 Zhao, Z., Ukidve, A., Kim, J., and Mitragotri, S. (2020). Targeting strategies for tissue-specific drug delivery. *Cell* 181 (1): 151–167.

73 Batool, M., Ahmad, B., and Choi, S. (2019). A structure-based drug discovery paradigm. *International Journal of Molecular Sciences* 20 (11): 2783.

74 Adeluwa, T., McGregor, B.A., Guo, K., and Hur, J. (2021). Predicting drug-induced liver injury using machine learning on a diverse set of predictors. *Frontiers in Pharmacology* 12: 648805.

75 Vora, L.K., Gholap, A.D., Jetha, K. et al. (2023). Artificial intelligence in pharmaceutical technology and drug delivery design. *Pharmaceutics* 15 (7): 1916.

76 Bender, A., Fergus, S., Galloway, W.R. et al. (2006). Diversity oriented synthesis: a challenge for synthetic chemists. *Ernst Schering Research Foundation Workshop* 58: 47–60.

77 Saha, C.N. and Bhattacharya, S. (2011). Intellectual property rights: an overview and implications in pharmaceutical industry. *Journal of Advanced Pharmaceutical Technology & Research* 2 (2): 88–93.

78 Anderson, A.C. (2003). The process of structure-based drug design. *Chemistry & Biology* 10 (9): 787–797.

79 Salo-Ahen, O.M.H., Alanko, I., Bhadane, R. et al. (2020). Molecular dynamics simulations in drug discovery and pharmaceutical development. *Processes* 9 (1): 71.

80 Agamah, F.E., Mazandu, G.K., Hassan, R. et al. (2020). Computational/*in silico* methods in drug target and lead prediction. *Briefings in Bioinformatics* 21 (5): 1663–1675.

81 Bashir, S., Fitaihi, R., and Abdelhakim, H.E. (2023). Advances in formulation and manufacturing strategies for the delivery of therapeutic proteins and peptides in orally disintegrating dosage forms. *European Journal of Pharmaceutical Sciences* 182: 106374.

82 Barman, P., Joshi, S., Sharma, S. et al. (2023). Strategic approaches to improvise peptide drugs as next generation therapeutics. *International Journal of Peptide Research and Therapeutics* 29 (4): 61.

83 Zhang, Z. and Tang, W. (2018). Drug metabolism in drug discovery and development. *Acta Pharmaceutica Sinica B* 8 (5): 721–732.

84 Schneider, G. (2013). Prediction of drug-like properties. In: *Madame Curie Bioscience Database [Internet]*. Landes Bioscience.

85 Bickerton, G.R., Paolini, G.V., Besnard, J. et al. (2012). Quantifying the chemical beauty of drugs. *Nature Chemistry* 4 (2): 90–98.

86 Hughes, J.P., Rees, S., Kalindjian, S.B., and Philpott, K.L. (2011). Principles of early drug discovery. *British Journal of Pharmacology* 162 (6): 1239–1249.

87 Taniguchi, C.M., Armstrong, S.R., Green, L.C. et al. (2008). Drug toxicity. In: *Principles of Pharmacology: The Pathophysiologic Basis of Drug Therapy*, 63–69. Lippincott Williams & Wilkins.

88 Pognan, F., Beilmann, M., Boonen, H.C.M. et al. (2023). The evolving role of investigative toxicology in the pharmaceutical industry. *Nature Reviews Drug Discovery* 22 (4): 317–335.

89 Baell, J.B. and Holloway, G.A. (2010). New substructure filters for removal of pan assay interference compounds (PAINS) from screening libraries and for their exclusion in bioassays. *Journal of Medicinal Chemistry* 53 (7): 2719–2740.

90 Baell, J.B. and Nissink, J.W.M. (2018). Seven year itch: pan-assay interference compounds (PAINS) in 2017 – utility and limitations. *ACS Chemical Biology* 13 (1): 36–44.

91 Yan, X., Liu, X., Zhao, C., and Chen, G.-Q. (2023). Applications of synthetic biology in medical and pharmaceutical fields. *Signal Transduction and Targeted Therapy* 8 (1): 199.

92 Patra, J.K., Das, G., Fraceto, L.F. et al. (2018). Nano based drug delivery systems: recent developments and future prospects. *Journal of Nanobiotechnology* 16 (1): 71.

93 Galloway, W.R., Isidro-Llobet, A., and Spring, D.R. (2010). Diversity-oriented synthesis as a tool for the discovery of novel biologically active small molecules. *Nature Communications* 1: 80.

94 Pavlinov, I., Gerlach, E.M., and Aldrich, L.N. (2019). Next generation diversity-oriented synthesis: a paradigm shift from chemical diversity to biological diversity. *Organic & Biomolecular Chemistry* 17 (7): 1608–1623.

95 Atanasov, A.G., Zotchev, S.B., Dirsch, V.M. et al. (2021). Natural products in drug discovery: advances and opportunities. *Nature Reviews Drug Discovery* 20 (3): 200–216.

96 Giugni, D. and Giugni, V. (2010). Intellectual property: a powerful tool to develop biotech research. *Microbial Biotechnology* 3 (5): 493–506.

97 Paul, D., Sanap, G., Shenoy, S. et al. (2021). Artificial intelligence in drug discovery and development. *Drug Discovery Today* 26 (1): 80–93.

3

Novel Drug Targets for Small Molecule-based Drug Discovery

Raghu Ram Achar[1,2], Ipsita Panigrahi[1,2], Aditi Singh[1,2], N. Chandana[1,2], and Shivananju Nanjunda Swamy[2]

[1] *Division of Biochemistry, School of Life Sciences, JSS Academy of Higher Education & Research, Mysuru, Karnataka, India*
[2] *Department of Biotechnology, JSS Science and Technology University, Mysuru, Karnataka, India*

3.1 Introduction

A drug target can be defined as a molecule, usually a protein that is associated with a specific disease process in the body and can be addressed by drugs to produce therapeutic effects. The drug target must exhibit the following properties: participate in the main pathway of the biological system; have functional features and characteristics; and have efficacy (its ability to bind to small molecules means there is a site for binding). Traditionally, analytical methods have been used to find effective drug targets, leading to the concept of "druggability." Drug targets are generally defined as proteins with protein structures that facilitate interaction with drugs. Many proteins can be drugged based on their structure, but their combination does not provide medical assistance. Over the past two decades, many efforts have been made to analyze and classify drug targets. The most common proteins include proteases, kinases, G protein-coupled receptors (GPCRs), and nuclear hormone receptors. Druggability is not a necessary characteristic that defines a "good" drug. In fact, Scientists often choose initial drug candidates based on their involvement in biological processes important for disease. Groups such as enzymes, substrates, metabolites, proteins, receptors, ion channels, transporters, DNA, RNA, and ribosomes are targets of monoclonal antibodies. Although proteins have historically formed the basis of the most effective drugs, there are now many emerging drug targets such as nucleic acids, control DNA content, and noncoding RNAs (ncRNAs). Its importance in medicine and precision medicine is rapidly increasing. In fact, drugs that target nucleic acids already exist, especially in the fields of antibacterial and anticancer therapy. RNA is now recognized as an important factor in many regulatory processes, just like proteins. In fact, RNA plays an important role in transcriptional regulation, translation regulation, catalysis, protein function, protein transport, peptide bonds, and RNA splicing. Compared to DNA, RNA may provide better therapeutic results because RNA exhibits higher diversity patterns and fewer repairs. Like proteins, RNA has three-dimensional folds that form complex structures that enable specific binding of drug molecules. RNA targeting has been used successfully in the field of antibiotics. Additionally, with the emergence of new RNA classes and properties in animal control systems, their applications are rapidly expanding. Among these new drug targets, ncRNA is attracting increasing attention. ncRNA refers to a broad class of

Computational Methods for Rational Drug Design, First Edition. Edited by Mithun Rudrapal.

endogenous RNA molecules that cannot code for proteins but have specific biological functions. Although ncRNA cannot code for proteins, they can influence the expression of other genes through various mechanisms. In some cases, the methods of operation are well known and the strategies for managing these activities are correct. The ability of ncRNAs to regulate gene expression makes them a target for drug development. But uncertainty about how (or whether) ncRNAs work makes the drug development process even more difficult. Targeting RNA has the opportunity to modulate many cellular processes, including targeting of "weak" proteins. Examples include proteins with many important functions that are difficult to block using a single molecule or proteins that resemble other proteins, making them difficult to adequately select. Drug targets are important substances that affect biological pathways specific to a disease or an organism. The difference between drug targets and other biomolecules involved in the same pathway is only their location and action. The target of the putative drug should be disease related, meaning it should have minimal effects on other diseases. However, human diseases are often complex and involve many interacting pathways, which can hinder the identification of molecular targets. However, new drug targets are biomolecules whose roles are not fully understood and for which established drugs do not exist. These targets deserve more attention because they could lead to new treatments. The focus of preclinical drug discovery is often on the delivery of biomolecules (e.g., DNA, RNA, proteins, and peptides), including known and novel targets. Candidate drug targets should be characterized by a balance of efficacy and safety ("unsafe drugs"). Efficacy should focus on how well the biomolecule works for the drug target. The "druggability" target should be taken into account in the evaluation [1]. Identifying drug–target interactions (DTIs) is an important step in the development of new drugs and understanding their side effects. Two experimental methods are commonly used to identify DTIs: affinity chromatography and protein microarrays. Due to the increasing number of synthetic antibodies produced against many proteins and biological processes, identification of DTIs using biological assays is time consuming and expensive, and very few true DTIs have been discovered through such a process. Therefore, in recent years, researchers have tried to use methods to analyze DTI. There are many ways that drugs interact with receptors; different relationships lead to different biochemical reactions. For example, agonists generally have higher activity and can cause biological effects, while antagonists have lower activity in the body and then do not bind to receptors and cause biological effects, but they can block the binding of agonists The type of relationship between the drug and the protein is important in determining the drugs [2]. The term "drug" used here refers only to medical devices and includes all small molecules and biologics currently (or previously approved) approved by the US FDA (prior to June 2015). Health and antimalarial drugs are approved elsewhere in the world. The word "prescription" does not include prescription drugs, nutritional supplements, sunscreens, or vaccines. James Black's famous quote from 2000: "The best way to find a new drug is to start with an old drug." If possible, a deeper understanding of the effects of vaccines will continue to inform drug discovery, clinical trials, and efforts to address drug resistance. Therefore, maintaining accurate and up-to-date information about proposed drugs and their effective targets (e.g., prevention targets at which the drug exerts its therapeutic effect) is important to guide future drug development and innovation. Possibly, the first attempt to compile the list dates back to 1996, when Drews and Ryser estimated the number of human targets for approval of small drugs. From this paper and subsequent reviews, the concept of "essential" protein families with a good history and value in drug discovery began to emerge. Many repositories now provide information on drug use, and each has a different focus. The first is Treatment Target Database Drug Bank, which is the most advanced drug database that uses data to match drugs to the proteins they are reported to bind to; Super Target, on the other hand, is a text mining-based collection of direct and indirect targets of drugs. Recently, RaskAndersen and colleagues have provided a new perspective on the current drug and the human target thought to be responsible for its results. In addition, Munos outlines the main trends in drug class and target innovation over the past decade and examines the overlap and specificity of the types of drug plans pursued. However, although many useful drugs are available online, it is still a challenge to be consistent and visionary about the target of approved drugs (including small molecules and biologics) and their associated molecular performance-intended purpose through medicine (human and pathogenic). In addition, although the concept of purpose is a natural concept for researchers working in the field, there are significant problems in practice in ongoing studies on the concept of purpose for seeds and products [3]. Target identification is an important step in new drug discovery and development as it allows researchers to understand the mechanism of drug action. Therefore, much of the progress made in drug discovery and development over the past few hundred years can be attributed to advances in target identification. Target identification is also important to optimize drug selection and reduce side effects. There are many types of biomolecules that can serve as therapeutic targets, including enzymes, cell receptors, ion channels, DNA, and transcription factors (TFs). Determining the biological target of a drug can be tedious as there are vast reserves of potential protein and nonprotein targets; therefore, recent advances in bioinformatics overtake the traditional target discovery methods. In addition to a large research database containing different physical and chemical properties of various ligands and targets, recent advances in the gene/vaccine field provide various methods of drug identification. This article will review various new drug targets for small-molecule drug discovery [4].

3.2 Drug Target Identification

Drug target is a biological protein or nonprotein component that interacts with its potential drug candidate to bring about physiological changes usually targeting a disease.

An ideal drug target candidate would withhold the following properties [5]:

- Functionality of target on binding of drug.
- DTI should be able to bring about pathophysiological modifications in favor of the disease.
- The target should be druggable and favor simpler assayability methods for enabling its high-throughput screening.
- Favorable prediction of target binding sites of drugs for efficient interaction.
- Target structure and function should be drug driven.

Target-based drug discovery is the basic strategy for drug development. Recent development in computational biology has made several open web databases and bioinformatics tools available for linking molecular drug targets to diseases. Analysis of "omics" datatypes like genomics, transcriptomics, proteomics, and metabolomics has increased the efficacy of drug target discovery more than the traditional methodologies [1]. Cellular or genetic chemicals in the system usually interact with drug molecules, therefore acting as drug targets [6]. Phenotype screening is another target identification method in which the drug target is unknown; yet, the validated drug candidate is discovered on the basis of 3D modeling method. Most first-in-class drugs discovered before 1980 were based on phenotypic screening like aspirin that was synthesized in 1897 but its mechanism of action and target was identified in 1976 [7].

Various tools and technologies are used for target identification and validation. Protein and gene expression profiling to identify novel targets and validate the existing ones can be performed using microarray techniques, whereas validation can be performed using gene knockout model and antisense RNA technology [8].

Drug target identification imposes another challenge to druggability of the target. An ideal druggable target binds to ligands that obey Ro5; however, the absence of Ro5-associated ligand does not make a target undruggable. Therefore, structure-based druggability assessment is performed by identification of potential binding sites, function discrimination based on the physiological property of binding pocket, and by assessing the difficulty of druggability [9] (Figure 3.1).

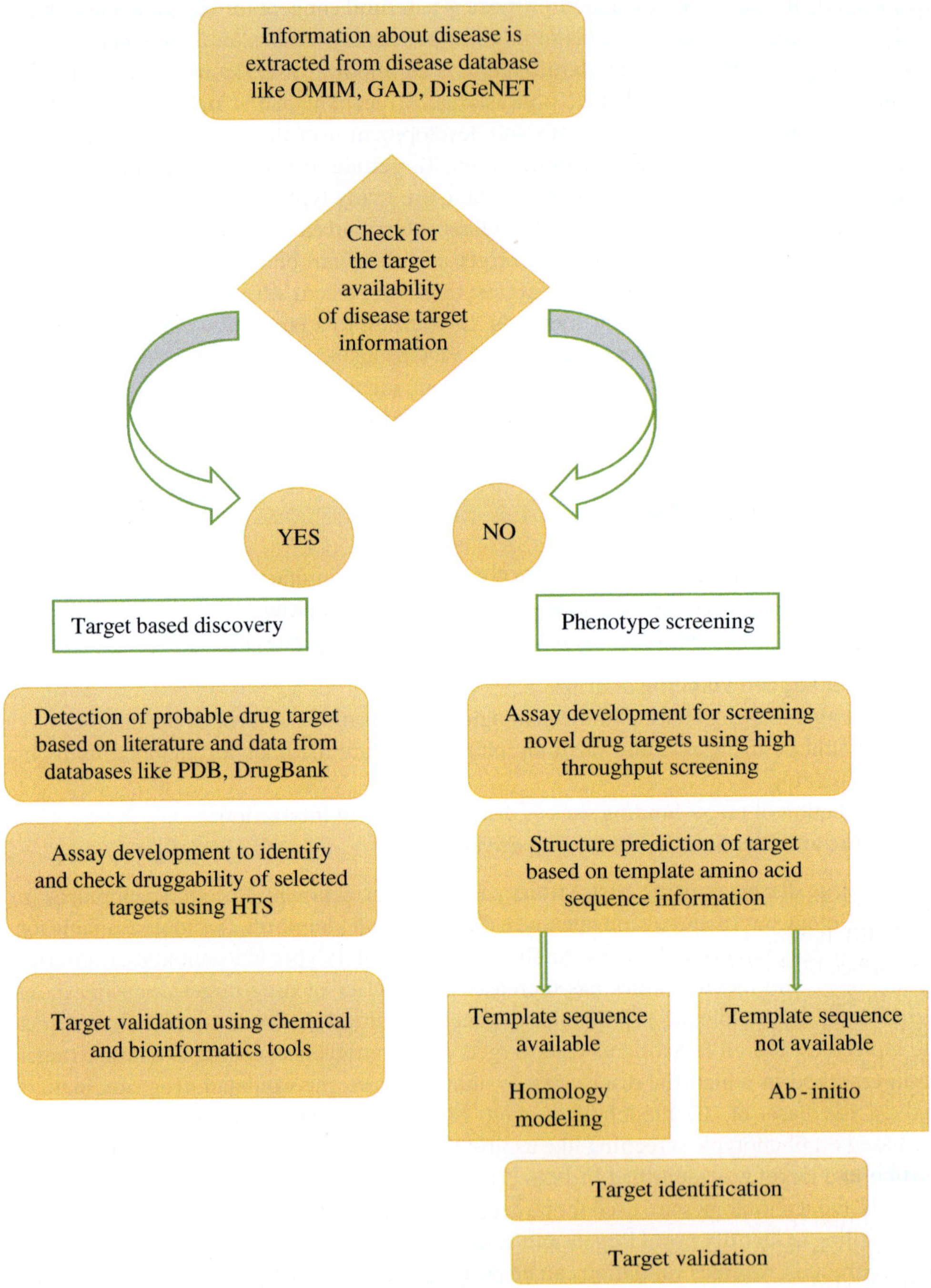

Figure 3.1 Canonical steps in drug target identification.

3.3 Classification of Novel Drug Targets

Drug target identification is a crucial stage in drug discovery as it enables the researchers to trace the mechanism of action of drug molecules and also ensures an efficient understanding of the cryptic DTIs [4]. Selection of the right drug target is essential as majority of drug failure in the R&D of the therapeutic industry is attributed to lack of drug efficacy. Proteins were initially considered major druggable targets that could be modulated by drugs such as small molecular weight chemical compounds or biologics such as an antibody or recombinant protein [5]. Almost all approved drugs are bound to protein targets and about 80% of the targets were either enzymes or receptors. The protocol for lead identification also follows Lipinski's rule of 5 that is valid only for protein drug targets. However, the number of protein drug targets is limited as only 600 disease-modifying protein targets exist out of 25,000 genes that is encoded by the human genome [10]. Therefore, the focus on target selection has now shifted to other biomolecules that serve as novel drug targets. Macromolecules including enzymes and cellular receptors along with novel targets like cytokines, DNA, RNA, chaperones, antigens, peptides, TFs, and transporters are being exploited for drug discovery.

3.3.1 Transcription Factors

TFs constitute the most essential classes of proteins that bind to specific sites on DNA and play a fundamental role in regulation of genome. TFs have been found to be associated with diseases and thus are capable of being a therapeutic target. Although it is an interesting target, only a few TFs are targeted by small molecules due to the absence of small-molecule binding pocket and intrinsic disordered arrangement of TFs [11]. Novel TFs are emerging as a drug target for cancers, cardiovascular diseases (CVDs), and COVID-19. NF-κB is a TF found to play an important role in pathogenesis of atherosclerosis. Vinpocetine and metformin are certain novel small molecules that are under preclinical trials that target NF-κB [12]. Autophagy is a lysosome-dependent catabolic process that takes place during starvation and helps in breaking down the product to form energy. It also plays an important role in lipid metabolism by shuttling the lipid droplets to lysosomes to form free fatty acids and glycerol [13]. During tumor growth, the process of autophagy is notably upregulated making it a therapeutic target for cancer. Transcription factor EB (TFEB) is the key regulator for lysosomal biogenesis and autophagy. TFEB is found to be highly expressed in multiple types of cancer, thus making it a therapeutic target for cancer. US FDA-approved drug eltrombopag is used for the treatment of thrombocytopenia and is a potent TFEB inhibitor. Eltrombopag binds to TFEB and creates a physical barrier between TFEB and the DNA, thus impeding the regulatory function of TFEB. It binds to the basic helix–loop–helix–leucine zipper domain of TFEB, targeting the bottom surface of the helix–loop–helix region. This interaction interferes with DNA recognition, leading to disruption of TFEB–DNA interaction. The way eltrombopag interacts with TFEB does not cause any severe neurological effects, indicating its potential as an autophagy inhibitor. Eltrombopag has been found to block autophagy and increase sensitivity to temozolomide in glioblastoma cells [14, 15].

3.3.2 Cytokines

Cytokines constitute 50% of ligands targeted by FDA-approved drugs and 40% of the novel ligands for which agents are in development. Cytokines have been approved as drug targets because of their specificity, size, and target modulation. In particular, they are soluble ligands that can be targeted by fusion proteins, mAbs, and aptamers that can block protein ligand interaction eliciting therapeutic response [16]. Several novel cytokine targets such as interleukin-6 (IL-6) were identified in inflammatory cardiomyopathy that were responsible for myocardial infiltration, tissue degradation, and T-cell activation. Novel cytokines that were previously not associated with cardiomyopathies included COLEC 12 – a scavenger receptor that plays a role in host defense, LAIR1 – a protein expressed on human leukocytes, LILRB4 – a leukocyte immunoglobulin-like receptor, and FSTL3 was recently identified as a relevant prognostic marker for major adverse cardiovascular events in stable coronary artery disease [17]. IL-17 is an important drug target for autoimmune diseases like psoriasis. Two molecules targeting IL-17 namely secukinumab and ixekizumab have been approved by FDA in 2015 and 2016, respectively. IL-23 regulates the T-helper cells that produce IL-17; therefore, IL-23 is also a target for a wide range of mAb drugs like guselkumab, tildrakizumab, and risankizumab approved in 2017, 2018, and 2019, respectively, for the treatment of psoriatic disease and other immune-inflammatory disease [16]. Cytokine storm syndrome (CSS) is an implication of COVID-19 prognosis that leads to dysregulation of synthesis of cytokines. IL-6 plays an important role in the pathogenesis of CSS. The significant role of IL-6 in pathogenesis of COVID-19 was confirmed in a range of studies, which showed that the plasma concentration of IL-6 was increased in patients with severe COVID-19. IL-6 inhibitors like tocilizumab and sarilumab suppress the overactive cytokine in COVID-19-affected individuals but are used as an off-label drug [18].

3.3.3 Chaperones

Molecular chaperones play an important role in the proteostasis network. Proteostasis network maintains the health of the proteome by controlling protein folding, misfolding, intracellular protein transport, protein aggregation, and trafficking. Dysregulation of this network leads to various neurodegenerative diseases and can also be a probable cause of cancer. Two of the most significant chaperone families, heat shock protein 70 kDa (Hsp70) and 90 kDa (Hsp90), rely on adenosine triphosphate (ATP) hydrolysis for energy and cooperation with their co-chaperones to create the dynamic complex [19]. Hsp-90 is an evolutionary conserved molecular chaperone that plays a major role in cell signaling; therefore, it attracts cancer cells to prevent oncoprotein mutation and degradation and is thus regarded as a potential anticancer target. Inhibition of Hsp-90 triggers the release of heat shock inhibition factors, Hsp70 and Hsp27 chaperones that can degrade normal proteins along with oncoproteins, thus reducing Hsp-90 inhibitor efficacy. Therefore, a protein kinase cochaperone CDC37 that promotes oncogenesis on interacting with Hsp-90 is chosen as a potential therapeutic target as the anticancer drugs acting on this protein–protein interaction (PPI) system destroy the oncoproteins specifically retaining the conformational integrity of normal proteins. The inhibition of these PPIs is facilitated by novel small-molecule drugs such as DDO-5942. NEN (FDA-approved antihelminthic drug) was proved to be an efficient drug against hepatocellular carcinoma by inhibiting CDC37 interactions and DCZ3112 that were found to disrupt PPI and degrade the Hsp-90 client proteins [20].

Small molecular chaperones are found to be potential stabilizers of the transthyretin–Aβ interaction. Transthyretin is a homotetrameric protein that prevents aggregation of amyloid beta peptides, thus inhibiting the prognosis of Alzheimer's disease. Being a tetramer, its stability is a key factor in interacting with Aβ peptide as the degradation of transthyretin to a monomeric form leads to amyloid aggregation. Therefore, small-molecule chaperones such as sulindac, olsalazine, flufenamic acid (marketed drugs), and luteolin (investigational drug) are used as novel therapeutic drugs to stabilize transthyretin–Aβ interaction [21].

3.3.4 Viral Targets

The emergence of various viral infections such as HIV, hepatitis, HPV infection, and the recent emergence of severe acute respiratory syndrome (causative agent being coronavirus) has led to the investigation of novel drug targets to combat the same. Antiviral drugs mostly comprise small molecules that target the viral cellular machinery. HIV that is responsible for causing AIDS is a deadly viral infection that was countered with Isentress approved by FDA in 2007 as the first integrase inhibitor that blocks integration of viral DNA with the host genome. The first peptide (36 amino acids)-based HIV inhibitor to be approved was enfuvirtide (Fuzeon) in 2003. Enfuvirtide imitates the helix in heptad region 2 of the viral glycoprotein 41 (GP41) to prevent HIV from fusing with the host cell membrane. Ibalizumab (Trogarzo), a monoclonal antibody that attaches to T-cell surface CD4 receptors to stop HIV from entering and reproducing, was approved in 2018. Dolutegravir and lamivudine (Dovato), as a two-drug combination, were authorized by the FDA in April 2019 to treat HIV-1 infection. Dolutegravir, an integrase strand transfer inhibitor (INSTI) INSTI, and lamivudine, an nucleoside reverse transcriptase inhibitor (NRTI), work together to inhibit HIV-1 replication and protease activity [22].

The severe acute respiratory syndrome coronavirus-2 (SARS-COV-2) is the cause of the COVID-19 pandemic, which is continuing to spread quickly around the world. More than 16.2 million confirmed cases from 213 countries were recorded as of July 26, 2020. Deaths have exceeded 648,866 to date [23]. SARS-CoV-2 belongs to the family Coronaviridae, which also includes the virulent strains SARS-CoV and Middle East respiratory syndrome CoV (MERS-CoV) that infect both humans and animals [24]. Remdesivir triphosphate (RDV-TP) is a direct antagonist of ATP and blocks the function of RNA dependant RNA polymerase (RdRP) (facilitates viral replication) when it takes its place upon elongation of the positive sense RNA of the virus by RdRP. Therefore, RdRP is considered a potential drug target to counter the virus. Favipiravir is also another inhibitor of RdRp that suppresses replication of SARS-COV-2 RNA [25]. Ivermectin is another FDA-approved drug that shuts down the ability of transporter importin to transport viral proteins into the nucleus of the host [26].

3.3.5 G Protein-coupled Receptors

GPCRs are a class of seven-transmembrane proteins that have garnered significant attention in academia and the pharmaceutical industry for many years due to their physiological and therapeutical significance as well as their accessibility for small-molecule drug discovery. GPCRs represent the largest family of druggable proteins in the human genome and unsurprisingly are targeted by more than 30% of marketed drugs worldwide [27]. Recent developments in GPCR structure illumination reveal that ligands may bind to distant allosteric sites in addition to the conventional orthosteric site, opening up a wide range of novel possibilities for drug discovery [28]. The contribution of GPCRs to the druggable genome has been approximated in earlier research. We have discovered 134 GPCRs that are targets for presently licensed medications, namely FDA-approved medications, by gathering data from open sources. About two-thirds of currently druggable GPCRs control cAMP, and many drug targets linked to GPCRs affect cAMP signaling (e.g., phosphodiesterases (PDEs) or proteins associated with ligand synthesis/transport/regulation whose receptors control cAMP synthesis). As a result, the cAMP pathway is probably the signaling route that currently approved treatments target the most frequently [29]. Phosphodiesterase-4 (PDE4) is a key enzyme that breaks down cyclic 3′,5′-adenosine monophosphate (cAMP, or cyclic AMP), an extremely effective inhibitor of PDE4 is roflumilast that increases intracellular cAMP. Roflumilast was approved by FDA in 2018 to manage chronic obstructive pulmonary disease as per PubChem data. Tiotropium, a long-acting muscarinic receptor antagonist, was included in 2015 in the Global Initiative for Asthma (GINA) guidelines. It is the only LAMA that is presently licensed for the treatment of asthma, whereas aclidinium, glycopyrronium, tiotropium, and umeclidinium are approved for the treatment of COPD [30]. Calcitonin gene-related peptide (CGRP) receptors are majorly exploited to manage migraines. Some novel small molecules targeting these receptors are Ubrogepant and Rimegepant that was approved by FDA in 2019 and 2020, respectively [31].

Class A GPCR CC chemokine receptor 5 interacts with the viral glycoprotein gp120 of HIV to form a co-receptor complex (PPI) that is responsible for the infectivity of the virus. Maraviroc is an anti-HIV drug that can disrupt CCR5-gp120 interaction by attaching to CCR5 allosterically; therefore, it acts as a chemokine agonist and inhibitor of GPCR [32].

3.3.6 Transporters

Transporters are proteins that traverse membranes helping move ions and biological substances across the phospholipid bilayer. Solute carrier (SLC) transporters and ATP binding cassette (ABC) transporters are two primary kinds of transporters. SLC transporters are involved in passive transport, whereas ABC transporters are involved in active transport. Cystic fibrosis transmembrane transporter coded by CFTR gene belongs to ABCC7 class of molecules [33]. Tezacaftor is a novel CFTR modulator belonging to the class of correctors that was approved by FDA in 2018. Correctors target the defectively folded protein due to the most common CFTR mutation F508del. Tezacaftor improves processing and trafficking of F508del-CFTR and increases chloride transport from 2.8% to 8.1% in bronchial epithelium [34]. Neonatal FcRn is another transcytosis receptor that is coded by FCGRT gene that mediates transport of IgG from mother to infant as proposed by Brambell in the 1960s. In autoimmune disorders like myasthenia gravis, pathogenic IgG autoantibodies are produced that bind to immune antigenic checkpoints and organ targets leading to organ damage. FcRn was recognized as a novel yet crucial transporter that is responsible for increasing the half-life of IgG autoantibodies in plasma by salvaging it from lysosomal degradation [35]. Therefore, FcRn targeting small-molecule drugs like Efgartigimod alfa, which is a monoclonal antibody acts as a potential therapeutic approved by FDA [36].

3.3.7 Enzymes

Protein phosphorylation plays an important role in regulation of metabolic processes. Protein kinases are a group of enzymes that carry out phosphorylation and dysregulation of these enzymes plays an important role in development of autoimmune, inflammatory, cardiovascular, and neurological disorders. Protein kinases are the second most prominent drug targets after GPCRs [37, 38]. The search for new treatments for various diseases like CVDs and cancer has led to the discovery of novel protein kinase targets. Recent studies have shown Janus kinase, which is activated by various cytokines, to be a prominent target in treatment of Systemic Lupus Erythematosus (SLE) [39]. Tofacitinib which targets JAK 1/3 is a US FDA-approved drug for RA and ulcerative colitis, which showed good results during phase I and I trials. Brepocitinib selectively inhibits JAK1/Tyk2 and shows a promising treatment in SLE. It is in phase II trials targeting nonrenal SLE patients [40]. DRAK1 is a negative regulator of TRAF6 protein, which is involved in inflammatory pathways that induce tumorigenesis. Thus, targeting DRAK1 can be used as a therapeutic target to antagonize TRAF6 protein overexpression [41]. Cell death is an essential process that helps in maintaining development and homeostasis of an organism. Apoptosis is the process of programmed cell death, which is important for various cellular processes. Cell necrosis, unlike apoptosis, is the process of unregulated cell death that happens due to external stress either physical or chemical. Necroptosis is a signal transduction pathway that regulates the process of programmed necrosis via the RIPK1 class of kinase protein and its downstream mediators: RIPK3 kinase and pseudokinase MLKL [42, 43]. RIPK1 is the central regulator of necroptosis, which is carried out via its kinase and scaffolding activities [44]. Thus, RIPK1 has become an important drug target for various neurodegenerative, autoimmune, and inflammatory diseases. GFH312 is a small molecule that inhibits RIPK1 and shows an antinecroptosis effect in neurological and inflammatory disease models [45]. GFH312 is now under phase II trials for inflammatory conditions after proving to be safe in the phase I trials.

ATG4 is a cysteine protease required for autophagosome formation. It has four homologs that are ATG4A, ATG4B, ATG4C, and ATF4D. ATG4B is overexpressed in tumor cells of colorectal cancer, glioblastoma, gastric cancer, and other cancers. Advanced gastric cancer is usually fatal due to lack of potential therapeutics. Azalomycin F4a is found to tightly bind and inhibit ATG4B, thus blocking the metastatic progression of advanced gastric cancer and sensitizing tumors to chemotherapy [46]. S130 is a novel molecule that is found to bind tightly to ATG4B and inhibits only ATG4B and not to other proteases [47].

3.3.8 RNA Targets

RNA plays an important role in the transfer of information in biological systems; thus, it can be an important therapeutic target for small molecules. Most of the small molecules available are target proteins; among these proteins, some disease-related proteins are termed as undruggable. This problem could be solved by directly targeting mRNAs, which can modulate the proteins before their biogenesis [48].

70% of the human genome is pervasively transcribed into ncRNAs; therefore, it acts as an emerging future target for drug discovery. The FDA approved the first direct RNA-splicing modifier for the treatment of spinal muscular atrophy (SMA) in 2020 using the smallmolecule risdiplam. Exon 7 inclusion and increased expression of the motor neuron protein are facilitated by risdiplam's direct binding to the survival of the mRNA-spliceosome complex of motor neuron 2 (SMN2) [49].

3.4 Small Molecules as Drugs

Small molecules are referred to as compounds that have a molecular weight of <900 Da. 90% of the drug molecules available in the pharmaceutical industry are small molecules [50]. These molecules are capable of affecting functions of various proteins by forming complexes with their targets, thus affecting the PPIs [51]. For more than a century, small molecules have been providing medical breakthroughs for human diseases. The features that make small molecules a good drug molecule are oral bioavailability, modular structure, and their ease of accessibility via chemical synthesis. It typically does not show immunogenicity and can bind to targets with high selectivity. The metabolic stability of small molecules ranges from low to high. Other than these features, it also shows high shelf life stability and is compatible with different drug formulations and administration, thus proving to be a good drug molecule [52]. Table 3.1 consists of some of the small-molecule drugs that interact with novel therapeutic target.

Table 3.1 Small molecule drugs targeting novel drug targets.

Small molecule drug	Nature of drug	Drug target	PDB ID	Target class	Indication	Approval status
Eltrombopag	Small-molecule nonpeptide TPO-R agonist	TF-EB	7Y62	Transcription factor	Glioblastoma	Preclinical trials
Guselkumab	mAb	IL23	3DUH	Cytokine	Psoriatic disease and immune-inflammatory disorder	FDA-approved
NEN (Niclosamide ethanolamine)	Small-molecule inhibitor	CDC37-DCZ112 PPI	2W0G	Chaperone	Hepatocellular carcinoma	FDA-approved – antiheminthic drug
Flufenamic acid, Luteolin, Olsalazine	Small molecular chaperones	Transthyretin-amyloid-β aggregation	6C3T	Protein–protein interaction	Alzheimer's disease	Preclinical trials
Remdesivir triphosphate (RDV-TP), Favipiravir	Small-molecule adenosine analog Small-molecule substrate analog	RNA-dependent RNA polymerase	6M71	Viral enzyme	COVID-19	FDA-approved
Roflumilast	Small-molecule long-acting inhibitor	PDE-4 (phosphodiesterase-4)	4NW7	Enzyme	Chronic obstructive pulmonary disease (COPD)	FDA-approved
Tiotropium	Small molecule antagonist (LAMA)	Long-acting M3 muscarinic acetylcholine receptor	4DAJ	GPCR	Asthma	FDA-approved
Ubrogepant, Rimegepant	Small-molecule inhibitor	Calcitonin gene-related peptide receptor (CGRPR)	5N7S	GPCR	Migraine	FDA-approved
Maraviroc	Small-molecule inhibitor	CCR5 (allosteric inhibitor)	4MBS	GPCR	HIV infection	FDA-approved
Tezacaftor	Small-molecule CFTR corrector	Misfolded CFTR due to F508 del mutation	1XMI	ABC transporter	Cystic fibrosis	FDA-approved

Efgartigimod alfa	mAb	Neonatal FcRn receptor	1I1A	Transcytosis receptor	Myasthenia gravis	FDA-approved
Tofacitinib	Small-molecule inhibitor	JAK 1/3	[a]6N7A [a]6NY4	Enzyme	SLE	Phase II trials
Brepocitinib	Selective inhibitor	JAK1/Tyk2	[a]6N7A [a]6X8G	Enzyme	SLE	Phase II trials
GFH312	Small-molecule inhibitor	RIPK1	[a]6NW2	Enzyme	Inflammation	Phase II trials
Azalomycin F4a	Selective inhibitor	ATG4B	2D1I	Enzyme	Advanced gastric cancer	Preclinical study
Risdiplam	Small-molecule mRNA splicing modifier	mRNA spliceosome complex of motor neuron 2 (SMN2)	5OR0	Long noncoding RNA	Spinal muscular atrophy	FDA-approved

[a] The structures present are complex with another molecule.

3.5 Conclusion

The discovery and identification of novel targets and drugs is the current requirement of the pharmaceutical industry as well as in biomedical research [53]. Novel therapeutic drug target identification is a tedious, intensive, and complex approach that demands integration of various biochemical, genetic, and bioinformatics tools and information. Identification of effective novel drug target serves as a potential progress in the drug discovery process. Recent advancements in the target identification approach have enabled researchers to focus on nonprotein molecules as novel targets apart from the age-old extensively exploited protein targets. Small-molecule drugs have been used as potential targeting molecules for druggable targets over the years. Small molecules hold advantages in the drug market due to their bioavailability, target specificity, simpler manufacturing and characterization, and nontoxic and nonimmunogenic properties. However, the exploitation of small molecules has made many targets undruggable to small molecules as they are merely biological mimics. As stated by Zhong et al. [54], crizotinib, a small-molecule drug inhibitor of ALK in various cancer types, showed drug resistance in patients with L1196M and G1269A mutation within 12 months post-administration in randomized phase 3 trials. Many cancer and bacterial cells have developed small-molecule drug resistance; therefore, small molecule fails to be a cosmopolitan drug (Table 3.2).

Table 3.2 Structure of small molecules.

Small molecule	Structure	Pubchem CID
Eltrombopag	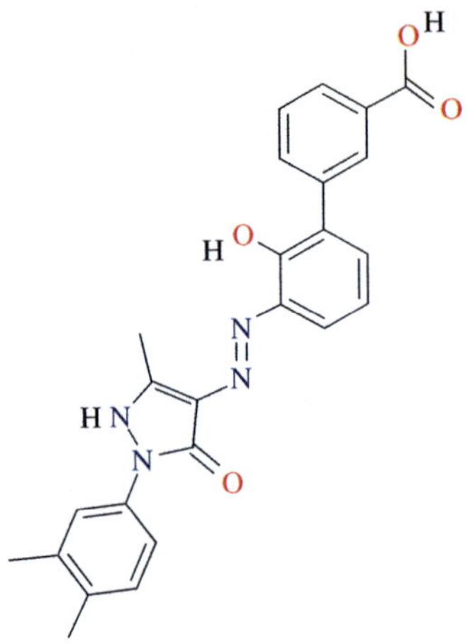	135449332
NEN	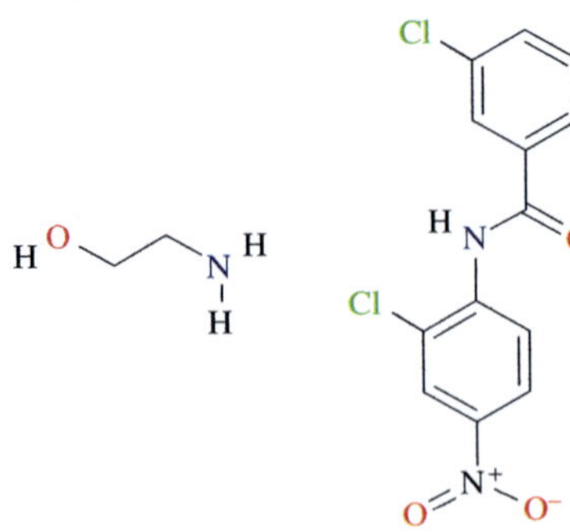	14992

Table 3.2 (Continued)

Small molecule	Structure	Pubchem CID
Flufenamic acid	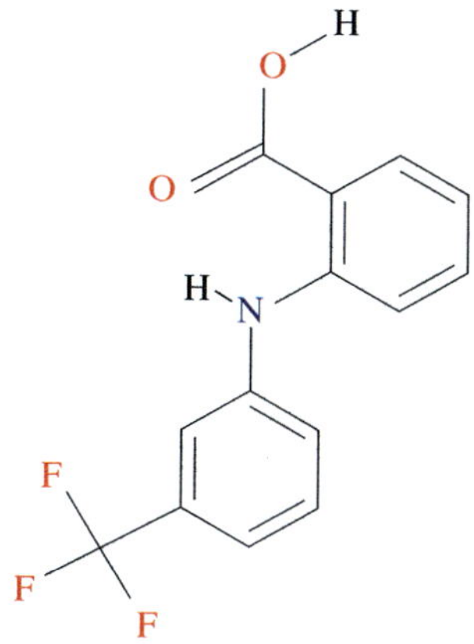	3371
Luteolin	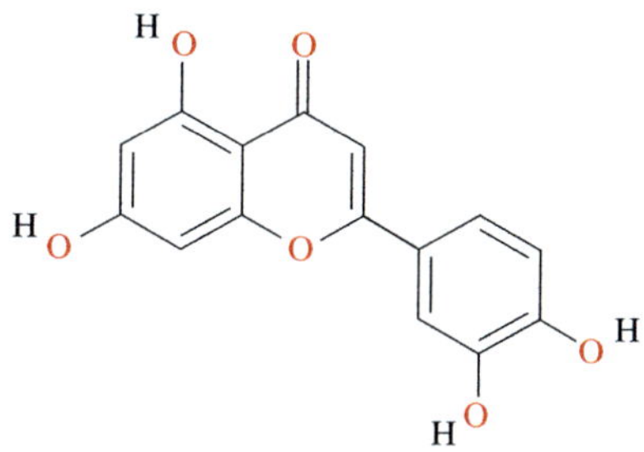	5280445
Olsalazine	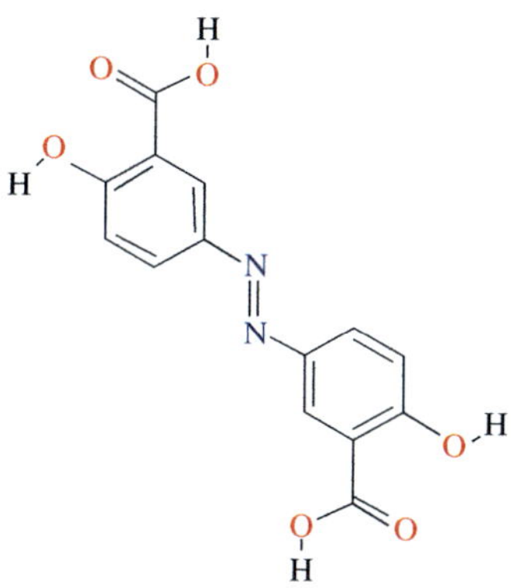	22419
Remsedivir triphosphate	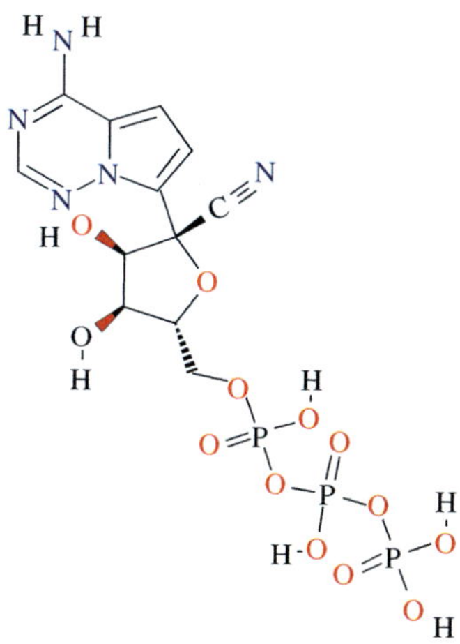	56832906

(*Continued*)

Table 3.2 (Continued)

Small molecule	Structure	Pubchem CID
Favipiravir	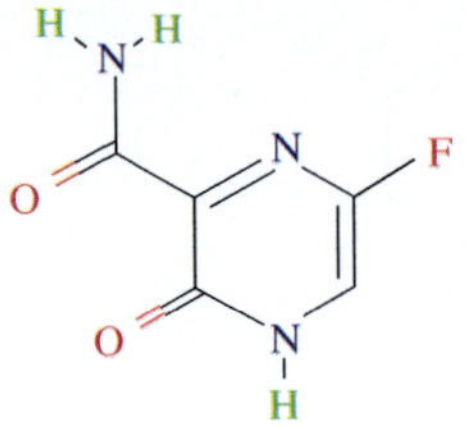	492405
Roflumilast	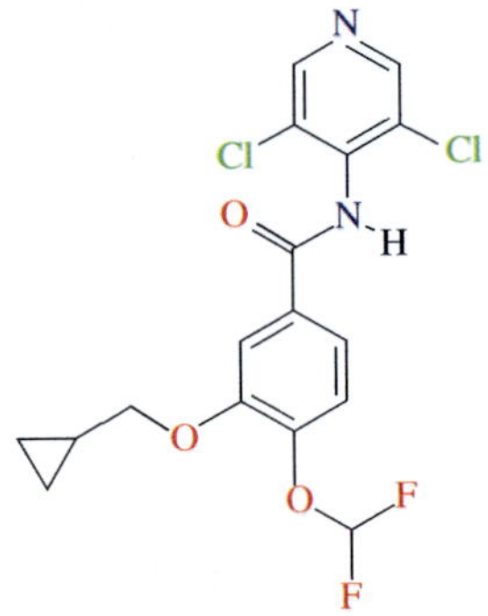	449193
Tiotropium	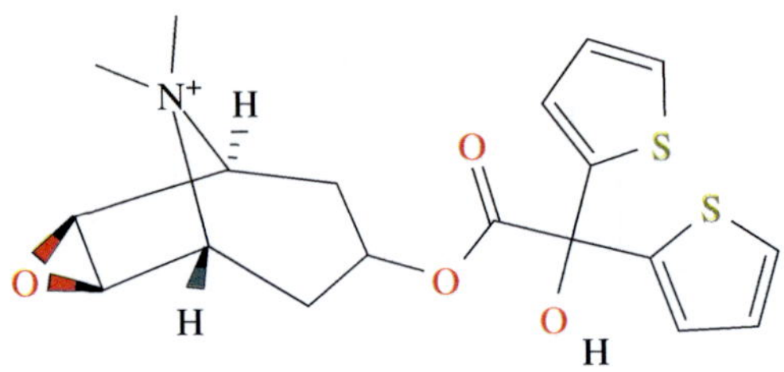	5487427
Ubrogepant	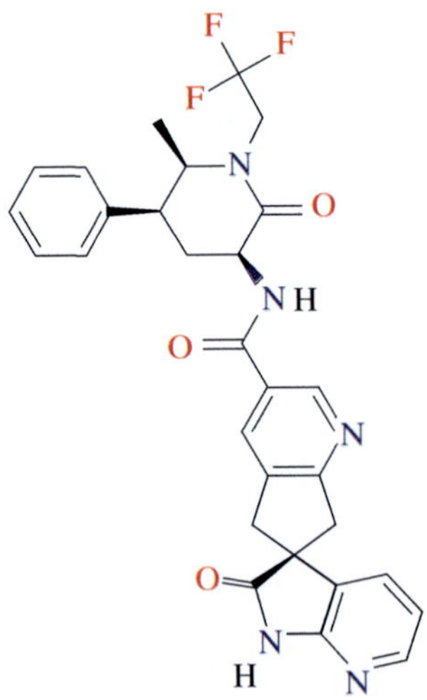	68748835

Table 3.2 (Continued)

Small molecule	Structure	Pubchem CID
Rimegepant	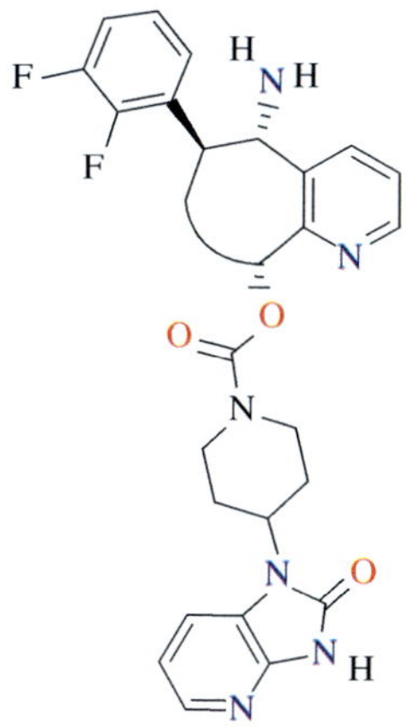	51049968
Maraviroc	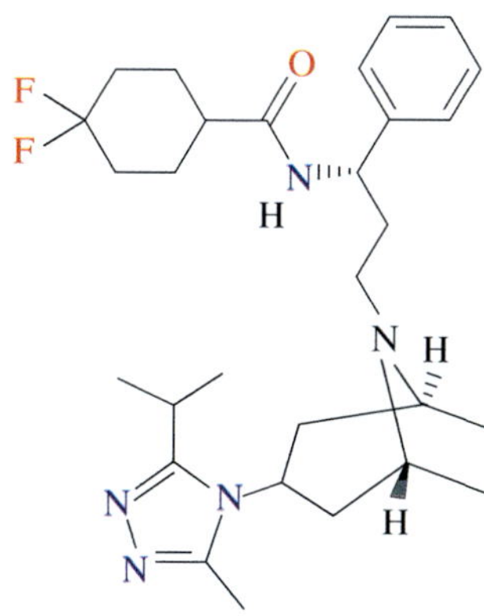	3002977
Tezacofactor	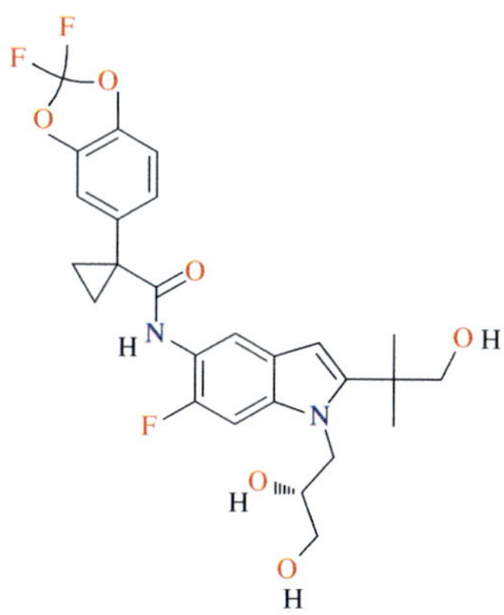	46199646
Tofacitinib	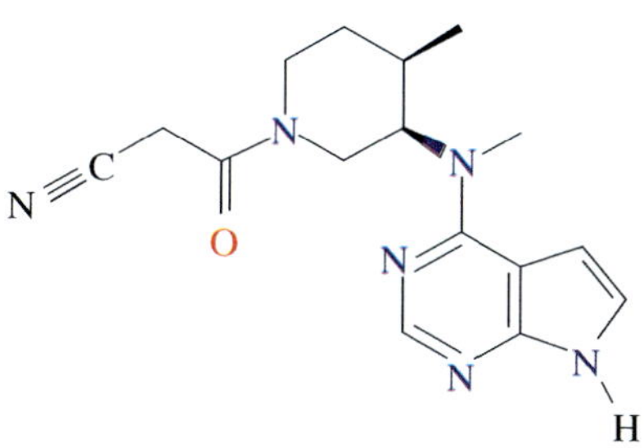	9926791

(Continued)

Table 3.2 (Continued)

Small molecule	Structure	Pubchem CID
Brepocitinib	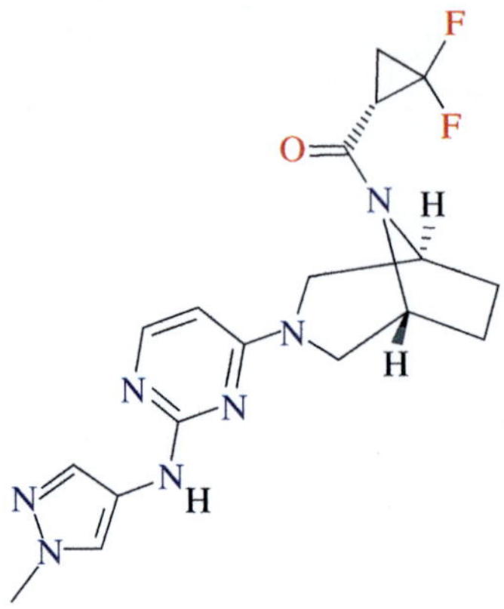	118878093
Azalomycin F4a	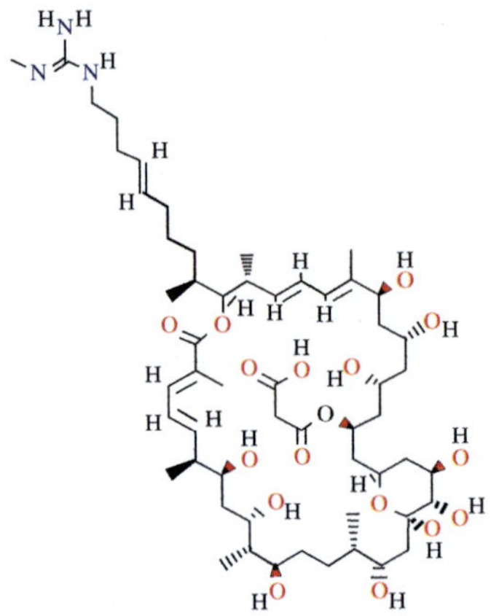	76963406
Risdiplam	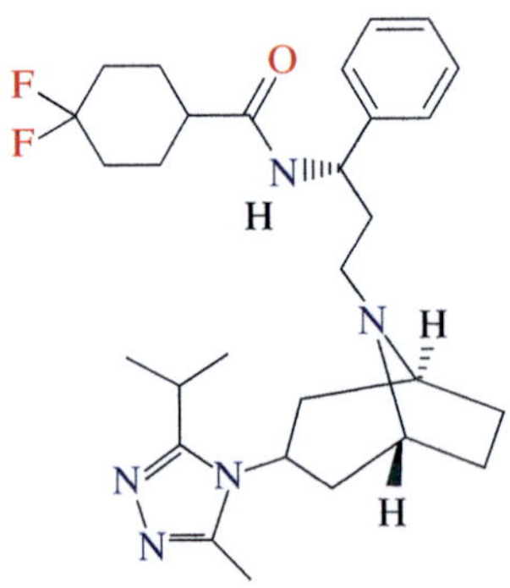	118513932

Although small molecules are extensively used in the pharmaceutical industry, they will soon be dominated by biologics. The only advantage of small molecules over biologics is it cost-effectiveness and ease of manufacture. Biologics as therapeutic candidates fulfill the unmet clinical needs and are considered an important drug of the future especially targeting the undruggable molecules [55]. In summary, small molecules are effective drug candidates for modulating drug targets. However, there is scope for further identification of druggable novel targets and there is also a need for modifying small molecules to interact effectively with the target and bring about desired physiological changes.

References

1 Paananen, J., Fortino, V. (2020). An omics perspective on drug target discovery platforms. *Briefings in Bioinformatics* 21(6): 1937–53.

2 Zhao, T., Hu, Y., Valsdottir, L.R. et al. (2021). Identifying drug–target interactions based on graph convolutional network and deep neural network. *Briefings in Bioinformatics* 22(2): 2141–50.

3 Santos, R., Ursu, O., Gaulton, A. et al. (2017). A comprehensive map of molecular drug targets. *Nature Reviews. Drug Discovery* 16(1): 19–34.

4 Tabana, Y., Babu, D., Fahlman, R. et al. (2023). Target identification of small molecules: an overview of the current applications in drug discovery. *BMC Biotechnology* 23(1): 44.

5 Gashaw, I., Ellinghaus, P., Sommer, A., and Asadullah, K. (2011). What makes a good drug target? *Drug Discovery Today* 16(23–24): 1037–43.

6 Phoebe Chen, Y.P. and Chen, F. (2008). Identifying targets for drug discovery using bioinformatics. *Expert Opinion on Therapeutic Targets* 12(4): 383–9.

7 Van Den Broeck, W.M.M. (2015). Drug targets, target identification, validation, and screening. In: *The Practice of Medicinal Chemistry [Internet].* Elsevier. [cited 2024 Feb 12]. p. 45–70. Available from: https://linkinghub.elsevier.com/retrieve/pii/B9780124172050000031.

8 Wang, S., Sim, T.B., Kim, Y.S., and Chang, Y.T. (2004). Tools for target identification and validation. *Current Opinion in Chemical Biology* 8(4): 371–7.

9 Fauman, E.B., Rai, B.K., and Huang, E.S. (2011). Structure-based druggability assessment – identifying suitable targets for small molecule therapeutics. *Current Opinion in Chemical Biology* 15(4): 463–8.

10 Schmidt, M.F. (2014). Drug target miRNAs: chances and challenges. *Trends in Biotechnology* 32(11): 578–85.

11 Henley, M.J. and Koehler, A.N. (2021). Advances in targeting 'undruggable' transcription factors with small molecules. *Nature Reviews. Drug Discovery* 20(9): 669–88.

12 Gan, J., Guo, L., Zhang, X. et al. (2023). Anti-inflammatory therapy of atherosclerosis: focusing on IKKβ. *Journal of Inflammation* 20(1): 8.

13 Settembre, C., De Cegli, R., Mansueto, G. et al. (2013). TFEB controls cellular lipid metabolism through a starvation-induced autoregulatory loop. *Nature Cell Biology* 15(6): 647–58.

14 Lin, Y., Shi, Q., Yang, G. et al. (2023). A small-molecule drug inhibits autophagy gene expression through the central regulator TFEB. *Proceedings of the National Academy of Sciences of the United States of America*. 120(7): e2213670120.

15 Zhang, L., Li, Z., Zhang, L. et al. (2023). Dissecting the multifaced function of transcription factor EB (TFEB) in human diseases: from molecular mechanism to pharmacological modulation. *Biochemical Pharmacology* 215: 115698.

16 Attwood, M.M., Jonsson, J., Rask-Andersen, M., and Schiöth, H.B. (2020). Soluble ligands as drug targets. *Nature Reviews. Drug Discovery* 19(10): 695–710.

17 Suwalski, P., Golpour, A., Weiner, J. et al. (2023). Cytokines as potential novel therapeutic targets in severe inflammatory cardiomyopathy [Internet]. *Cardiovascular Medicine.* [cited 2024 Jan 27]. Available from: doi:10.1101/2023.07.27.23293253

18 Nasonov, E. and Samsonov, M. (2020). The role of Interleukin 6 inhibitors in therapy of severe COVID-19. *Biomedicine & Pharmacotherapy* 131: 110698.

19 Wang, L., Bergkvist, L., Kumar, R. et al. (2021). Targeting chaperone/co-chaperone interactions with small molecules: a novel approach to tackle neurodegenerative diseases. *Cells* 10(10): 2596.

20 Wang, L., Zhang, Q., and You, Q. (2022). Targeting the HSP90–CDC37–kinase chaperone cycle: a promising therapeutic strategy for cancer. *Medicinal Research Reviews* 42(1): 156–82.

21 Cotrina, E.Y., Santos, L.M., Rivas, J. et al. (2021). Targeting transthyretin in Alzheimer's disease: drug discovery of small-molecule chaperones as disease-modifying drug candidates for Alzheimer's disease. *European Journal of Medicinal Chemistry* 226: 113847.

22 Tompa, D.R., Immanuel, A., Srikanth, S., and Kadhirvel, S. (2021). Trends and strategies to combat viral infections: a review on FDA approved antiviral drugs. *International Journal of Biological Macromolecules* 172: 524–41.

23 Wondmkun, Y.T. and Mohammed, O.A. (2020). A review on novel drug targets and future directions for COVID-19 treatment. *Biologics: Targets and Therapy* 14: 77–82.

24 Dai, L. and Gao, G.F. (2021). Viral targets for vaccines against COVID-19. *Nature Reviews. Immunology* 21(2): 73–82.

25 Lei, S., Chen, X., Wu, J. et al. (2022). Small molecules in the treatment of COVID-19. *Signal Transduction and Targeted Therapy* 7(1): 387.

26 Panoutsopoulos, A.A. (2020). Known drugs and small molecules in the battle for COVID-19 treatment. *Genes & Diseases* 7(4): 528–34.

27 Zhou, J. and Wild, C. (2019). GPCR drug discovery: emerging targets, novel approaches and future trends. *Current Topics in Medicinal Chemistry* 19(16): 1363–4.

28 Chan, H.C.S., Li, Y., Dahoun, T. et al. (2019). New binding sites, new opportunities for GPCR drug discovery. *Trends in Biochemical Sciences* 44(4): 312–30.

29 Sriram, K. and Insel, P.A. (2018). G protein-coupled receptors as targets for approved drugs: how many targets and how many drugs? *Molecular Pharmacology* 93(4): 251–8.

30 Thomson, N.C. (2017). New and developing non-adrenoreceptor small molecule drugs for the treatment of asthma. *Expert Opinion on Pharmacotherapy*; 18(3): 283–93.

31 Bhutani, P., Joshi, G., Raja, N. et al. (2021). U.S. FDA approved drugs from 2015–June 2020: a perspective. *Journal of Medicinal Chemistry* 64(5): 2339–81.

32 Lu, S. and Zhang, J. (2019). Small molecule allosteric modulators of G-protein-coupled receptors: drug–target interactions. *Journal of Medicinal Chemistry* 62(1): 24–45.

33 Carmichael, N. and Day, P.J.R. (2022). Cell surface transporters and novel drug developments. *Frontiers in Pharmacology* 13: 852938.

34 Sala, M.A. and Jain, M. (2018). Tezacaftor for the treatment of cystic fibrosis. *Expert Review of Respiratory Medicine* 12(9): 725–32.

35 Zhu, L.N., Hou, H.M., Wang, S. et al. (2022). FcRn inhibitors: a novel option for the treatment of myasthenia gravis. *Neural Regeneration Research* 18(8): 1637–1644.

36 Avram, S., Halip, L., Curpan, R., and Oprea, T.I. (2022). Novel drug targets in 2021. *Nature Reviews. Drug Discovery* 21(5): 328–328.

37 Bazan-Socha, S. and Jakieła, B. (2023). Editorial to the special issue "Pathogenesis and Novel Therapeutics in Asthma" *Biomedicine* 11(2): 268.

38 Roskoski, R. (2024). Properties of FDA-approved small molecule protein kinase inhibitors: a 2024 update. *Pharmacological Research* 200: 107059.

39 Nikolopoulos, D. and Parodis, I. (2023). Janus kinase inhibitors in systemic lupus erythematosus: implications for tyrosine kinase 2 inhibition. *Frontiers in Medicine* 10: 1217147.

40 Mok, C.C. (2023). Targeted small molecules for systemic lupus erythematosus: drugs in the pipeline. *Drugs* 83(6): 479–96.

41 Park, Y., Pang, K., Park, J. et al. (2020). Destablilization of TRAF6 by DRAK1 suppresses tumor growth and metastasis in cervical cancer cells. *Cancer Research* 80(12): 2537–49.

42 Chen, L., Zhang, X., Ou, Y. et al. (2022). Advances in RIPK1 kinase inhibitors. *Frontiers in Pharmacology [Internet]* 13 [cited 2024 Feb 11]. Available from: doi: 10.3389/fphar.2022.976435.

43 Degterev, A., Ofengeim, D., and Yuan, J. (2019). Targeting RIPK1 for the treatment of human diseases. *Proceedings of the National Academy of Sciences* 116(20): 9714–22.

44 Berger, S.B., Kasparcova, V., Hoffman, S. et al. (2014). Cutting edge: RIP1 kinase activity is dispensable for normal development but is a key regulator of inflammation in SHARPIN-deficient mice. *Journal of Immunology (Baltimore, Md.: 1950)* 192(12): 5476–80.

45 Lickliter, J., Wang, S., Zhang, W. et al. (2023). A phase I randomized, double-blinded, placebo-controlled study assessing the safety and pharmacokinetics of RIPK1 inhibitor GFH312 in healthy subjects. *Clinical and Translational Science* 16(9): 1691–703.

46 Zhong, L., Yang, B., Zhang, Z. et al. (2022). Targeting autophagy peptidase ATG4B with a novel natural product inhibitor Azalomycin F4a for advanced gastric cancer. *Cell Death & Disease* 13(2): 161.

47 Fu, Y., Hong, L., Xu, J. et al. (2019). Discovery of a small molecule targeting autophagy via ATG4B inhibition and cell death of colorectal cancer cells in vitro and in vivo. *Autophagy* 15(2): 295–311.

48 Warner, K.D., Hajdin, C.E., and Weeks, K.M. (2018). Principles for targeting RNA with drug-like small molecules. *Nature Reviews. Drug Discovery* 17(8): 547–58.

49 Nguyen, L.D., Chau, R.K., and Krichevsky, A.M. (2021). Small molecule drugs targeting non-coding RNAs as treatments for Alzheimer's disease and related dementias. *Genes* 12(12): 2005.

50 Govardhanagiri, S., Bethi, S., and Nagaraju, G.P. (2019). Small molecules and pancreatic cancer trials and troubles. In: *Breaking Tolerance to Pancreatic Cancer Unresponsiveness to Chemotherapy* [Internet]. Elsevier, 117–31. [cited 2024 Feb 11]. Available from: https://linkinghub.elsevier.com/retrieve/pii/B9780128176610000081

51 Li, Q. and Kang, C. (2020). Mechanisms of action for small molecules revealed by structural biology in drug discovery. *International Journal of Molecular Sciences* 21(15): 5262.

52 Beck, H., Härter, M., Haß, B. et al. (2022). Small molecules and their impact in drug discovery: a perspective on the occasion of the 125th anniversary of the Bayer Chemical Research Laboratory. *Drug Discovery Today* 27(6): 1560–74.

53 Rask-Andersen, M., Almén, M.S., and Schiöth, H.B. (2011). Trends in the exploitation of novel drug targets. *Nature Reviews. Drug Discovery* 10(8): 579–90.

54 Zhong, L., Li, Y., Xiong, L. et al. (2021). Small molecules in targeted cancer therapy: advances, challenges, and future perspectives. *Signal Transduction and Targeted Therapy* 6(1):201.

55 Ngo, H.X. and Garneau-Tsodikova, S. (2018). What are the drugs of the future? *MedChemComm* 9(5): 757–8.

4

Computer-assisted Methods and Tools for Structure- and Ligand-based Drug Design

Saurav Kumar Mishra, Sneha Roy, Tabsum Chhetri, and John J. Georrge

Department of Bioinformatics, University of North Bengal, Darjeeling, West Bengal, India

4.1 Introduction

In the present situation, computer-aided drug design is evolving worldwide, and researchers are also keenly interested in it. Before the advancement in the CADD, the conventional drug design was in the authentic methods that were utilized; however, using this concept, target identification, inhibitor identification, and many more were time-consuming and less accurate. Several studies were performed in this field to develop and design the crucial steps useful in the drug discovery process [1]. Recently, to overcome the challenges of drug discovery, the CAAD came as a new hope that scaled the drug design process with high accuracy [2]. It comprises target retrieval, structure prediction, active site identification, ligand library preparation, and virtual screening for the novel inhibitor identification toward the target structure. These compounds were optimized after the crucial steps of validation. Interestingly, the virtual screening process reduces the costs of the novel discovery. Based on the computational screening, the research easily finds the promising one and utilizes it for further experimental evaluation, such as *in vivo* and *in vitro*. The computer-assisted drug designs lie on several important aspects and among these tools and databases are two main pillars that are handling and revolutionizing this CADD method in modern drug discovery [3–5]. The accurate methodology lies in different concepts, such as the selection of targets and whether their structure is available, as it is essential for inhibitor identification. The computer-assisted drug design mainly lies in the concept of structure-based drug discovery (SBDD) and ligand-based drug discovery (LBDD). Several methodologies were employed utilizing the concept of SBDD and LBDD in the present situation [6, 7]. In this chapter, the list of tools and databases that are available for the SBDD and LBDD were briefly described. Moreover, the SBDD- and LDBB-based case studies were also mentioned, which can help the readers understand the concept of computer-assisted study related to the drug discovery process.

4.2 Structure-Based Drug Discovery Concept

As the number of emerging and remerging cases of pathogens increased daily [8], the requirement for inhibitors through computer assistance also increased, along with identifying novel compounds that had promising interaction with the pathogen target. SBDD is the most impactful approach for

Computational Methods for Rational Drug Design, First Edition. Edited by Mithun Rudrapal.

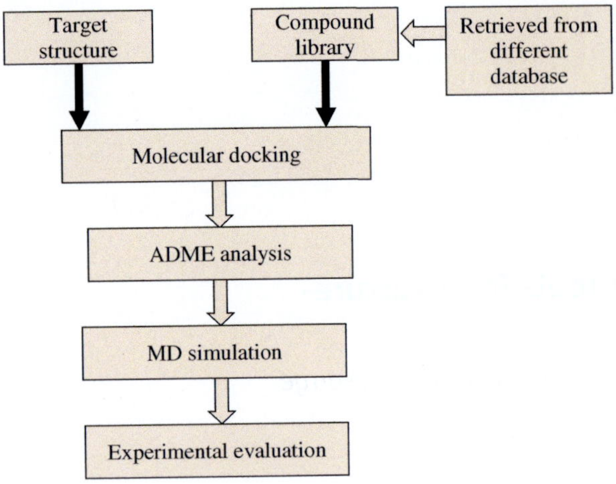

Figure 4.1 Illustration of basic concepts that are involved in the SBDD.

the identification of therapeutics for pathogens. In this aspect, computational capacity is one of the pillars that can help achieve accuracy in less time [6]. The initial steps are target identification and ligand library retrieval from a different set of databases to the molecular docking examination and the ADME (absorption, distribution, metabolism and excretion) properties of the identified compound to reduce the number of compounds for the molecular dynamic simulation [9]. However, based on crucial validation steps, the final one was experimentally validated before administration. The basic concept behind the SBDD is illustrated in Figure 4.1. In the case of the SBDD, the target structure is most essential, and if no information is available, the structure is modeled followed by homology modeling, *ab initio* modeling, threading methods, and many more [3, 6]. Only after the accurate structure of the target is available or retrieved, the concept of SBDD can be further processed.

4.2.1 Structure Generation of the Target

The starting phase of SBDD procedure consists of these steps. The 3D structure is created in this stage; nevertheless, it may or may not be approachable in the PDB. Without discovery or solution, the structure can be predicted through a sequence search of a similar nature. Three distinct modeling methods, homology modeling, fold recognition, and *ab initio* modeling, are introduced. Homology modeling, which implies 3D models and templates with a high degree of sequence similarity, is the most prevalent technique in the interim. Many tools and resources are available [6]. These tools are including MODELER, SWISS-MODEL, PHYRE2, INTFOLD, RAPTORX, HHPRED, ALPHA FOLD, SNPWEB, MODBASE, MOULDER, and MODLOOP. All these tools are for homology modeling with different algorithms and specific uses. But if a similar sequence or structure is unavailable in the comparative modeling database, fold recognition and *ab initio* or *de novo* modeling are used. This method is based on the energy functions, and it is given a number of possible confirmations, which depend on the thermodynamic stability, low energy states, and native-like models. Some of the web servers, including ROBETTA, QUARK, and I-TASSER, are available for initio-based prediction. The list of tools that are associated with it is shown in Table 4.1.

4.2.1.1 The Detailed Description of Each Tool

Modeller: Primarily on the basis of its alignment (templates), the program MODELLER can be utilized to accomplish this by predicting the 3D structure of a particular protein sequence. In

Table 4.1 List of associated tools for the target structure generation.

Sl. no.	Name	URLs	References
1	Modeller	https://salilab.org/modeller/	[10]
2	Swiss model	https://swissmodel.expasy.org/	[11]
3	Phyre2	http://www.sbg.bio.ic.ac.uk/~phyre2/html/page.cgi?id=index	[12]
4	IntFOLD	https://www.reading.ac.uk/bioinf/IntFOLD/	[13]
5	RaptorX	http://raptorx6.uchicago.edu/	[14]
6	HHpred	http://protevo.eb.tuebingen.mpg.de/hhpred https://toolkit.tuebingen.mpg.de/tools/hhpred	[15]
7	Robetta	https://robetta.bakerlab.org/	[16]
8	AlphaFold	https://alphafold.ebi.ac.uk/	[17]
9	SNPWEB	http://salilab.org/SNPWeb	[18]
10	QUARK	https://zhanggroup.org/	[19]
11	I-TASSER	https://zhanggroup.org/I-TASSER/	[20]
12	MODBASE	http://salilab.org/modbase	[21]
13	MOULDER	http://salilab.org/modweb	[21]
14	MODLOOP	http://salilab.org/modloop	[21]
15	MODWEB	http://salilab.org/modweb	[21]

addition to facilitating the recalculation of models in the event that the alignment is modified, MODELLER also offers tools for integrating pre-existing information of the target, including cross-linking constraints and secondary structures, and enables the calculation of models based on multiple templates. *Ab initio* modeling of insertions is an additional capability provided by MODELLER, which is often essential when annotating functions. Important to comparative modeling, particularly in between 30% and 50% sequence identity, the program MODELLER is also practical for loop modeling. Whereas the core regions remain relatively conserved and precisely aligned, the loops among the homologs vary in this range. MODELLER offers the requisite instruments to accomplish loop modeling, which can be considered a miniature protein-folding problem. In addition, the efficiency and practicality of protein structure models can be assessed using MODELLER. It provides insights into the accuracy and applicability of the models by enabling the docking of ligands into comparative models and the identification of putative binding regions [10]. Swiss Model: Starting with primary sequences of partners, the SWISS-MODEL server constructs three-dimensional models of protein complexes. Through homology modeling, the server subsequently deduces the structure. Extrapolating experimental data from protein structures that are evolutionarily related and function as templates for the target sequences constitutes this procedure. Providing input data, selecting a template, constructing the model, and estimating the grade of the model are all components of the default modeling workflow. ProMod3 was additionally developed to facilitate flexible and rapid prototyping for forthcoming modeling developments in SWISS-MODEL. In addition, a novel approach based on an original description and evolutionary distance has been implemented in ProMod3. This enhancement enables more accurate modeling of the stoichiometry and the overall structure of protein complexes when simulating the quaternary structures of homo- and hetero-oligomers of proteins. The accuracy and dependability of the 3D models produced by the SWISS-MODEL server have been enhanced due to the progress made with the ProMod3 modeling engine [11]. Phyre2: About its methodologies and constraints, the Phyre2 protein structure prediction application is similar to other servers designed for protein structure prediction. With an intuitive interface that enables access to state-of-the-art bioinformatics techniques, Phyre2 is primarily distinguished by its approach to methods. Phyre2 constructs three-dimensional models that protein sequence and forecast ligand binding sites using sophisticated remote homology detection techniques. Model quality assessment, alignment confidence, conflicts and rotamers analysis, pocket detection, and mutational analysis are a few of the supplementary tools it provides for the comprehensive management and analysis of protein structure modeling projects. When considering constraints, Phyre2 is comparable to other servers in its category. The detection of homology between a sequence supplied by the user and a sequence whose structure is known constitutes one limitation. The persistent challenge of the protein-folding problem is reflected in the fact that modeling will be either unattainable or highly unreliable in the absence of homology detection. Phyre2 is a highly regarded protein structure prediction tool that distinguishes itself through its intuitive interface and extensive collection of protein structure analysis and prediction tools. Although alternative robust structure prediction servers are also accessible, Phyre2 is the most widely utilized and dependable for various modeling tasks. The prediction center's website does not currently offer access to CASP11 data regarding the average quality of models. Non-bioinformatician usability is the principal distinction between these servers and Phyre2 rather than accuracy. Predicting the structural consequences of point mutations constitutes the second constraint, which reapplies to all commonly employed methodologies. Prediction capabilities are present in Phyre2 [12]. ntFOLD: The web resource known as the IntFOLD server facilitates the prediction of protein structure and function. The server's recent utilization in CASP experiments, user-friendly interfaces, and benchmarked performance are emphasized. The server provides predictions regarding natively unstructured regions, protein tertiary structures, structural domain boundaries, and protein–ligand interactions. Graphical outputs of predicted models are now available; they have been updated to enhance performance. Protein structural domain boundaries, natively unstructured or disordered regions in proteins, protein–ligand interactions, and estimates of model accuracy (EMA) are all included in the unified resource provided by the IntFOLD server, enabling the automated prediction of protein tertiary structures. Access to an integrated suite comprising six component methods is facilitated through the server. The server's recent utilization in CASP experiments, user-friendly interfaces, and benchmarked performance are emphasized. The server provides predictions regarding natively unstructured regions, protein tertiary structures, structural domain boundaries, and protein–ligand interactions. Graphical outputs of predicted models are now available; they have been updated to enhance performance. It has consistently exhibited outstanding performance. According to independent official evaluation metrics, the IntFOLD server has positioned itself among the top-performing publicly accessible servers in protein structure prediction [13]. RaptorX: A bioinformatics application known as the RaptorX server, which predicts the structure of proteins based exclusively on their sequence or sequence profile. In the case of proteins lacking near homologs in the Protein Data Bank (PDB) or possessing sparse sequence profiles, it exhibits superior performance compared to alternative servers. The server simultaneously predicts the solvent accessibility, disordered regions, and proteins' secondary structure, utilizing a deep learning model known as DeepCNF. The quality of the training data for secondary structure and disorder prediction is a determinant of the accuracy of the predictions. In addition to representing the intricate relationship between structure and sequence through a deep hierarchical structure, this model also depicts the interdependence among neighboring property labels. Processing time on the RaptorX server is contingent on sequence length and prediction mode. It returns a unique job ID and URL for monitoring the progress of submitted jobs and provides a web interface for job submissions. The server can be utilized for genome-scale structure property prediction and is designed to respond promptly to users who only require structure property prediction. Additional research could enhance prediction accuracy and decrease execution time during sequence profile construction. Furthermore, the accuracy and utility of the server in protein analysis could be significantly improved by augmenting the quality of training data utilized for secondary structure and disorder prediction [14]. HHpred: A bioinformatics application known as HHpred detects protein homology and predicts its structure. The tool's rapidity, user-friendliness, and sensitivity in detecting remote homology are emphasized. HHpred is a rapid option for structure prediction, as it can search the PDB in approximately one minute using a 300-amino acid sequence. It facilitates the exploration of an extensive assortment of structure and protein family databases that are consistently updated. Such databases comprise the PDB, SCOP, Pfam, SMART, COG, and CDD. HHpred is suited for predicting domain boundaries because it has been designed to function effectively with query sequences. HHpred employs profile hidden Markov models (HMMs) to detect protein homology and compares the query HMM with every HMM in the chosen database. Precalculated secondary structure information, either predicted by PSIPRED or assigned by DSSP from the 3D structure, is contained in the database HMMs. The HHsearch software is utilized to conduct the database search in order to compare HMM–HMM. This software enhances sensitivity through the implementation of position-specific gap penalties. In addition, HHpred can compute secondary structure similarity scores, thereby significantly improving the sensitivity toward homologous proteins. Nevertheless, structurally analogous proteins that are not homologous may yield scores that are only marginally significant. HHsearch, which utilizes profile HMMs and facilitates HMM–HMM comparison, enables HHpred to conduct a thorough and sensitive search for homologous proteins, thereby augmenting its capability to identify protein homology. Investigating the homologous relationship among distinct protein families is facilitated by the useful bioinformatics utility Hhpred [15]. Robetta: The Robetta protein structure prediction server predicts and analyses structures using multiple computational approaches. Through RosettaNMR, the Robetta server derives protein structures from experimental NMR limitations. If available, we generate fragment libraries that match chemical shifts, NOE constraint data, and residual dipolar couplings. After using these fragment libraries, the RosettaNMR *de novo* fragment insertion approach generates decoys using constraints data in its scoring function. In Robetta, computational interface alanine scanning detects energetically relevant side chains in protein–protein complexes to anticipate mutation effects on protein–protein interactions. Lab-developed protein design and protein–protein docking methods will be added to improve the Robetta server. To expand the Robetta server, the server will deliver high-quality structural information to enhance research, infer function, and aid medication creation [16]. AlphaFold: AlphaFold accurately predicts protein structures. It describes the AlphaFold network's training, construction, performance, and bioinformatics applications. AlphaFold improves structure prediction accuracy by training and inferring with raw MSAs and attention methods. AlphaFold uses a new architecture to include MSAs and paired features to learn from sequence data evolutionary constraints and patterns in the training process. Bioinformatics inter-residue contact prediction requires linked mutations and evolutionary sequence variation, which the system captures. AlphaFold uses the predicted confidence score to select the best model per target to improve structure prediction and runs the network trunk numerous times with different random MSA cluster center choices. Structured bioinformatics and biological applications benefit from the AlphaFold system's atomic-accuracy protein structure prediction. This innovation could transform protein structure and function research and speed up drug discovery and protein engineering. Accurate protein structure predictions can illuminate protein function pathways and shed light on molecular biology. Identifying pharmacological targets and designing more effective therapies can result. Large-scale protein structural studies could benefit structural bioinformatics if protein structures can be accurately predicted. This helps us comprehend biological systems by identifying novel protein folds, characterizing protein–protein interactions, and studying protein dynamics. Accurate protein structure prediction can also help develop novel proteins with specific functionalities, such as enzymes with improved catalytic activity or proteins with increased stability, which have biotechnology and industrial uses. In structural bioinformatics and biological applications, the AlphaFold system's atomic-accuracy protein structure predictions could affect drug discovery, protein engineering, and our understanding of biological systems [17]. SNPWeb: The SNPWEB web server simulates the effects of a single amino acid residue substitution. It inputs the wild-type protein's structure and substitution. The server predicts the mutant's function and justifies its influence based on wild-type and mutant structures. Without locating the wild-type structure in PDB or MODBASE, modeling with MODWEB is attempted. The server computes characteristics for wild-type and mutant proteins based on their sequences and structures, such as accessible surface area, rigidity, variations in residue volume, charge, hydrophobicity, evolutionary conservation, and replacement likelihood. In exceptional circumstances, supplementary characteristics, such as the structural significance and location of substituting established functional sites, may be incorporated. A decision tree categorizes the mutation as either harmful or neutral. According to the protocol, a mutation is detrimental in two ways: it may significantly alter functional sites' composition when exposed to the solvent or impede the changes in the native fold when submerged in the core [18]. QUARK: QUARK predicts protein structure using energy concepts and algorithms. QUARK protein structure prediction uses energy concepts and algorithms in many ways. Energy terms and conformational change motions are easier to calculate and apply. Predicting structural properties with neural networks and assembling small fragments with REMC simulations aid the development of force fields and search engines. A composite knowledge-based force field leads REMC simulations, improving the force field and search engine with unique energy terms and Monte Carlo movements. QUARK uses multiple feature predictions, fragment synthesis, REMC simulation using the semi-reduced protein model, decoy structure clustering, and full-atomic refinement. QUARK improves *ab initio* protein structure prediction accuracy and efficiency by tackling force field design and conformational search with energy concepts and algorithms. QUARK outperforms Rosetta in protein structure prediction in numerous instances. In a study of 145 test sequences, QUARK models outperformed Rosetta models in RMSD (root mean square deviation), TM (template modeling) score, and HB score for 96 targets. Small- to medium-sized globular protein structures are also well-predicted by QUARK. QUARK beat Rosetta in both sets of proteins, with a higher TM and HB score, indicating better folding of tiny proteins. Due to its more properly generated potentials for low-resolution simulations, QUARK has been shown to generate structural models for protein targets more accurately than Rosetta. Force field design and conformational search issues complicate *ab initio* protein structure prediction. Protein structure prediction requires accurate force fields and efficient conformational space exploration [19]. I-TASSER: The sequence-to-structure-to-function paradigm used by I-TASSER predicts protein structure and function. I-TASSER generates 3D atomic models from amino acid sequences utilizing multiple threading alignments and iterative structure assembly simulations. For forecasts, the I-TASSER server utilizes the C-score, a confidence score. The C-score depends on threading template alignments and structure assembly simulation convergence settings. Based on its substantial C-score correlation, the first I-TASSER model estimates quality in large-scale benchmark testing. The

C-score and quality of lower-ranked models are less associated, making it impossible to determine their quality. Despite numerous benchmark tests, automated structure and function prediction quality estimates can be unpredictable and error-prone. Thus, user-collected experimental data is necessary to validate predictions. The I-TASSER service offers Threading, *Ab Initio* Modeling, C-score, Sequence Submission, External Restraints, and result output for protein structure prediction [20]. MODBASE: The relational database MODBASE contains annotated comparative protein structural models. Applying MODPIPE to all SWISS-PROT protein sequences yielded datasets for all protein sequences matched to at least one known protein structure. MODBASE models domains in 415-937 of 733-239 distinct SWISS-PROT protein sequences. MySQL permits direct SQL queries [21]. MOULDER: MODWEB's optional MOULDER protocol optimizes alignment and implied model via genetic algorithms. The protocol optimizes model evaluation scores by aligning, realigning, constructing, and assessing models. New alignments, spatial restraint-based comparative models, and composite criterion assessments are made in iteration. The iterative method blurs comparative modeling and threading. MOULDER in a Linux cluster takes a day on 100 CPUs to compute a 150-residue target sequence [21]. MODLOOP: MODLOOP is a web server for precise protein structure loop modeling using coordinate files and residue positions. Users can optimize several loops simultaneously, especially coupled loops. Starting with random beginning conformations and optimizing nonhydrogen atom locations, the service generates 300 loop predictions. The lowest objective function score determines the final loop prediction. The server submits calculations to a Linux cluster to reply faster and can only calculate 300 loop conformations. Loops not longer than 20 residues are allowed [21]. MODWEB: MODWEB, an automatic comparison structure of proteins for modeling purposes, accepts FASTA sequences and produces models using the best PDB template structures. This method helps structural genomics examine how a newly established structure affects sequence space modeling. MODPIPE, a software pipeline uses template structures and sequence–structure alignments to construct protein sequence comparison models. By matching the PSI-BLAST sequence profile to each PDB template sequence and scanning the target sequence against IMPALA's template profile database, sequence–structure matches are produced. We evaluate sequence–structure relationships after building and evaluating the model [21].

4.2.2 Active Binding Site Within the Target

Moreover, based on the final target-generated structure before the docking analysis, it is essential that the ligand molecule will strongly bind with the target residue there, for the active residue within the target is most promising for the docking purpose. The binding site of a protein is a region where a molecule binds to produce the desirable product. Understanding the structure of ligands with proteins can help in structure-based drug design. A lack of structural information on binding pockets can lead to possible issues. Binding pockets anticipated using *in silico* approaches. While these methods are useful for predicting binding sites, their accuracy is affected by features, including template similarity and pocket size [6]. There are various tools with different algorithms and specifications, including CASTp, Prankweb, BSpred, DoGSitesScorer, Consurf, GRaSP, Metapocket, COACH, Pockdrug, SiteMap, FPocket, PocketDepth, and Caver. The list of available tools is illustrated in Table 4.2.

4.2.2.1 The Detailed Description of Each Tool

CastP: The computed atlas of surface topography of proteins lets you find, draw, and measure the geometric and topology characteristics of protein forms online. Using the alpha shape approach, the server finds physical characteristics, counts area and volume,

Table 4.2 List of associated tools for the active site prediction.

Sl. no.	Name	URLs	References
1	CASTp	http://sts.bioe.uic.edu/castp/index.html?1ycs	[22]
2	Prankweb	https://prankweb.cz/	[23]
3	DoGSitesScorer	https://proteins.plus/	[24]
4	ConSurf	https://consurfdb.tau.ac.il/	[25]
5	GRaSP	https://grasp.ufv.br	[26]
6	PockDrug	https://pockdrug.rpbs.univ-paris-diderot.fr/cgi-bin/index.py?page=home	[27]
7	Fpocket	https://mobyle2.rpbs.univ-paris-diderot.fr/cgi-bin/portal.py#forms::fpocket	[28]

determines impose, and retrieves secondary structure data from UniProt and SIFTS. The method gives a full and accurate numerical summary of the topographic aspects of proteins and has been demonstrated to be useful for various research [22]. Prankweb: PrankWeb is a web-based application that helps you use P2Rank, regarded as the most current way to predict ligand sites' binding. PrankWeb employs a template-free machine learning approach that predicts the capacity of near compounds to attach to peptides according to places on their surface that agents can access. PrankWeb helps viewers estimate and observe which ligands and proteins will bind and match their predictions to real binding locations and areas that stay the same [23]. DoGSitesScorer: Identifying probable interaction sites and subpockets of a particular protein of interest is possible with DoGSitesScorer. After that, it looks at these pockets' geometric and physicochemical characteristics while employing a support vector machine (SVM) to determine how druggable they are [24]. ConSurf: Per the PDB, ConSurf-DB is a location where the evolutionary conservation investigation into peptides via established structures can be viewed. A grouping of conserved results is currently calculated based on MSAs taken from the HSSP database. It uses Bayesian analysis in order to figure out conservation scores, which are then broken down into levels of conservation using 1–9 codes of color [25]. GRaSP: The GRaSP method is a new, scalable way to predict the residues that will bind to ligands. It works better than other methods, can be used on a large scale, and can predict the binding site for a protein complex in 10–20 seconds on average. It is a way to learn with supervision that shows a residue and its structure neighbors as a graph, storing them as a feature vector. The atoms and interactions of each residue are used to determine the relative solvent accessibility, physicochemical traits, and interactions [26]. PockDrug: PockDrug is a strong model for studying druggability in pockets that can handle unknown pocket boundaries. The PockDrug-Server consistently gives druggability results when using various pocket measurement methods. Because it is robust against pocket boundary and estimation uncertainties, it can be used effectively with apo pockets that are hard to predict. It used various estimation methods to separate druggable and less druggable pockets and did better than recent druggability models for apo pockets [27]. Fpocket: This is an open-source package for finding pockets. It uses Voronoi tessellation and alpha spheres, also available to the public. Fpocket is a solid starting point for quicker, more reliable, and freely available protein pocket identification, pocket descriptor of the extraction process, and druggability prediction [28].

4.2.2.2 Molecular Docking Analysis

The approach to designing drugs based on shape discusses the docking of molecules. The molecular docking process finds how the ligand and target molecule will mix. Determining the preferred position of the low free binding energy shows how well the ligand will bind to the protein and form a stable complex. Several noncovalent relationships occur in this reaction, including hydrogen bonds, ionic bonds, hydrophobic interactions, and interactions mediated by van der Waals. Investigations into molecular docking can be done among proteins, between proteins and ligands, and between proteins and nucleotides [6]. A ligand is identified with high binding affinity when the structure and binding site are determined during the drug design process. The molecular docking process can predict the ligand's perfect orientation with the protein's binding site in this ligand identification method. The docking analysis is anticipated to be of three types: rigid, flexible, and semiflexible docking. Semiflexible docking is mainly used for multiple protein structures, and rigid docking can only show the target's and ligand's geometry. Additional flexible docking is used for the refinement [1]. Docking can be manual or automated, and it is found to fit the ligand best. There are two ways of docking protocol, i.e., one ligand with different orientations and the binding affinity score of the ligand. Various tools and web servers are available for analysis. These tools (mentioned in Table 4.3) are different in their algorithm, scoring, and strategy.

Table 4.3 List of available tools and databases for the docking analysis.

Sl. no.	Name	URLs	References
1	Cluspro	https://cluspro.bu.edu/login.php	[29]
2	PatchDock	https://bioinfo3d.cs.tau.ac.il/PatchDock/	[30]
3	SymmDock	https://bioinfo3d.cs.tau.ac.il/SymmDock/php.php	[30]
4	GRAMM-X	https://gramm.compbio.ku.edu/request	[31]
5	RosettaDock	https://rosie.graylab.jhu.edu/	[32]
6	FoXSDock	https://modbase.compbio.ucsf.edu/foxsdock/	[33]
7	HDOCK	http://hdock.phys.hust.edu.cn/	[34]
8	HADDOCK	https://wenmr.science.uu.nl/	[35]
9	ZDOCK	https://zdock.umassmed.edu/	[36]
10	FRODOCK	https://chaconlab.org/modeling/frodock/frodock-donwload	[37]
11	MEGADOCK	https://www.bi.cs.titech.ac.jp/megadock/	[38]
12	CombDock	http://bioinfo3d.cs.tau.ac.il/CombDock/download/	[39]
13	FibreDock	http://bioinfo3d.cs.tau.ac.il/FiberDock/php.php	[40]
14	F2Dock	https://www.cs.utexas.edu/~bajaj/cvc/software/f2dock.shtml	[41]
15	GOLD	https://www.ccdc.cam.ac.uk/solutions/software/gold/	[42]
16	pyDock	https://life.bsc.es/pid/pydock/	[43]
17	MOE	https://www.chemcomp.com/Products.htm	[44]
18	AutoDock Vina	https://autodock.scripps.edu/	[45]
19	SurflexDock	https://www.biopharmics.com/	[46]
20	GEMDOCK	http://gemdock.life.nctu.edu.tw/dock/	[47]

4.2.2.3 The Detailed Description of Each Tool

ClusPro: ClusPro is an online docking tool that works autonomously and employs the clustering technique to find the bound complexes of proteins with the best breakdown and electrostatic free energy properties. You may modify your search criteria using several advanced methods with ClusPro. With the help of desolvation and electrostatic energies (calculated using a Coulombic potential), it quickly filters the output of the Fourier correlation method. This method lets many structures very close to being native using the filter while eliminating many erroneous results [29]. PatchDock and SymmDock: PatchDock and SymmDock represent two online docking applications that use shape matching and geometry-based docking algorithms to determine the structure of amino acid compounds. These two tools are used to figure out the molecular makeup of PPI complexes and a homo multimer with cyclic symmetry. Both use the Shape Similarity Principles and the symmetrical limitations method for docking [30]. GRAMM-X: GRAMM-X is a docking web-based user interface, which employs the Fast Fourier transformation (FFT) method. FFT incorporates based on knowledge evaluation, smooth potential, and improvement. The global search FFT stage uses a smooth Lennard-Jones possible on an extremely fine grid. This is followed by refined optimization in continuous dimensions and rescoring with several possible phrases based on knowledge [31]. RosettaDock: RosettaDock is an online docking assistance that sorts bound protein molecules by rigid-body alignment and side chain geometry. It uses a Monte Carlo-based algorithm for local protein–protein interaction [32]. FoXSDock: FoXSDock is an internet-based coupling technique employing rigid-body docking and the SAXS (small angle X-ray scattering) models to generate better complexes of proteins docked with a small amount of vitality. It has rigid global modeling and flexible docking methodology for rigid body docking to show the geometry [33]. HDOCK: HDOCK is a web server that uses an alternate docking method to make good docked complexes. It does both modeling and docking with provided PDB frameworks [34]. HADDOCK: HADDOCK is an online tool using an information-driven coupling method to guess how protein complexes are assembled. Biomolecular interaction with high inconsistency is what it does. The flexible docking method is used to show the binding strength between the object being studied and its receptor [35]. ZDOCK: ZDOCK is an easy-to-use, rigid docking-based protein docking program that guesses how complexes of proteins and uniform complexes are put together. The FFT method is used to predict PP complexes and symmetric multimers [36]. FRODOCK: FRODOCK serves as a docking framework, which assists in the two proteins binding collectively through its complimentary based on knowledge possibilities. This platform does rapid circular docking by using added knowledge-based capability [37]. MEGADOCK: MEGADOCK is a speedy protein–protein docking application built on FFT. Various types of supercomputers are used to speed up the docking procedure. The system uses various types of supercomputers to accomplish outstanding efficiency [38]. CombDock: Randomized techniques are employed by CombDock to do protein–protein docking. The function numerous peptides make in how they communicate alongside each other is put jointly and anticipated. Sequential development was done by docking more than one item at a time [39]. FibreDock: FiberDock is the initial web server for dock biomolecules that takes into consideration the adaptability of both sides of chains and the core to enhance the docked molecules' strength. In addition to performing rigid-body refining and rescoring, this application performs adaptable induced-fit backbone improvement [40]. F2Dock: F2Dock is a quick Fourier transform-based tool for docking biomolecules to one another. It boosts up the procedure by using numerous threads and Lennard-Jones' perspective to rescore the docked complexes of proteins depending on their desolvation activity [41]. GOLD: To predict how easily malleable compounds will bind to protein molecules, GOLD is a docking equipment. Protein–ligand docking in GOLD is carried out using an algorithm based on genetics technique, and the corresponding ligand and the protein can be either entirely or partially flexible [42]. pyDock: The pyDockWEB website host forecasts the rigid-body docking of intricate protein structures, employing an updated version of the pyDock scoring method. The pyDockWEB website serves the pyDock rigid-body docking and assessment technique while making it easy for scholars to use through a web-based user interface [43]. MOE: The MOE drug discovery application platform combines visualization, simulation, modeling, and method design. The MOE command, scripting, and application programming syntax is SVL. MOE uses SVL, a flexible, high-performance programming language [44]. AutoDock Vina: AutoDock Vina is a quick and popular open-source docking technology. The computerized docking apparatus uses a simple scoring process and quick gradient-optimization conformation exploration. These universal computational docking tools accept receptor and ligand coordinates data and suggest optimum docked compliance [45]. SurflexDock: Surflex is a completely automatic and adaptable molecular docking method that uses the Hammerhead docking system's scoring mechanism and a surface-based molecular homology search tool to create molecular segment poses quickly. Surflex docking efficiency was equal to the best [46]. GEMDOCK: GEMDOCK, a protein–ligand docking software, uses the differential evolution algorithm to compute elegantly physiologically. A docking program like GEMDOCK uses a search algorithm and a scoring system to identify target's binding site. Differential evolutionary algorithm piecewise potential energy function is used in the GEMDOCK scoring function [47].

4.2.3 Molecular Dynamic Simulations

Molecular dynamics (MD) simulations have been adopted as a substitute for computationally forecasting protein binding pockets and ligands. Protein structural characteristics and stability of protein–ligand complexes contribute to the drug design. It can facilitate the development of more viable drugs by identifying additional druggable binding sites and the virtual screening of chemical compounds. Typically, simulations are performed to validate promising complexes [6]. Various simulation applications are utilized to simulate the dynamics of protein and ligand topologies generated with AMBER CharmGUI force fields, including GROMOS, GROMACS, YASARA, LAMMPS, GENESIS, NAMD, Gaussian software, Discovery Studio, Chimera, TINKER, and OpenMM as shown in Table 4.4 [59]. MD simulations require a more exact anatomical force field, which may result in a greater computing overhead.

Table 4.4 List of available tools and databases for the MD simulation analysis.

Sl. no.	Name	URLs	References
1	AMBER	https://ambermd.org/	[48]
2	CHARMM	https://www.charmm.org/	[49]
3	GROMACS	https://www.gromacs.org/	[50]
4	YASARA	http://www.yasara.org/mdanalysis.htm	[51]
5	LAMMPS	https://www.lammps.org/#gsc.tab=0	[52]
6	GENESIS	https://www.r-ccs.riken.jp/labs/cbrt/	[53]
7	NAMD	https://www.ks.uiuc.edu/Research/namd/	[54]
8	Gaussian software	N/A	[55]
9	Discovery Studio	https://www.3ds.com/products/biovia/discovery-studio/simulations	[56]
10	TINKER	https://dasher.wustl.edu/tinker/	[57]
11	OpenMM	https://openmm.org/	[58]

4.2.3.1 The Detailed Description of Each Tool

Amber: Amber is a collection of codes that work together for molecular simulations. It consists of separate programs for system preparation, simulation, and trajectory analysis, allowing for flexibility, modularity, and compatibility with different coding practices. While this code separation has advantages, such as easy upgrades and portability, it also has disadvantages, such as a lack of consistent user interface and limited interaction between different program components. The main MD program in the Amber suite is called Sander. It is a parallel program written in Fortran 90 that uses the MPI (Message Passing Interface) programming interface for communication among processors. The program divides force-field tasks among processors and performs MD simulations by computing potential energy and gradients, communicating force vectors, and updating positions. An optimized version called pmemd has also been developed to improve performance by efficiently sharing only the necessary coordinate information for energy calculations. The suite also includes the nmode code for analyzing nonperiodic simulations and computing thermodynamic quantities, primarily simulations [48]. CHARMM: CHARMM-GUI is a web application that helps users assemble complicated simulation systems without software or modeling expertise. CHARMM input scripts can be downloaded and reused because it generates and executes them using the CHARMM binary. CHARMM-GUI uses the model-view-controller (MVC) architecture pattern to divide the program into three interrelated parts: a model for data and logic, a view for rendering information, and a controller for user input and model or view control. This modular approach lets new modules be developed quickly without extensive module knowledge. CHARMM-GUI also uses Python scripts and third-party applications for difficult instances that CHARMM cannot handle. Users can inspect their system at each step using JSmol to visualize three-dimensional molecule structures. The viewing modes include protein-only complexes containing band illustrations, protein/membrane complexes with lipids and proteins, orientation view of the bilayer hydrophobic core, packing picture view for lipid-like pseudo atoms, and electrostatic potential view. Users can see and analyze structures in CHARMM-GUI without installing additional applications, improving workflow [49]. GROMACS: GROMACS is a popular free and open-source chemistry program for biomolecule dynamical simulations. GROMACS is a popular MD simulation tool that provides spatial and temporal resolution that is impossible in experiments. It runs on supercomputers, embedded systems, and laptops and is optimized for performance and efficiency. GROMACS is a nearly two-million-line software project. GROMACS 5 uses C++ and improves code modularity, memory handling, and error handling. Balanced hardware works well with GROMACS 5; however, unbalanced hardware may slow performance. GROMACS 5 and the Copernicus ensemble framework support ensemble-level parallelism. GROMACS evaluates short-ranged nonbonded interactions using domain-level parallel decomposition and a new algorithm. It supports OpenMP-based multithreading and uses SIMD and GPU acceleration [50]. YASARA: YASARA represents atomic visual modeling, and YASARA uses PVL, a revolutionary programming environment that beats conventional software. PVL uses GPUs' 1993 molecular simulation application for Windows, Linux, MacOS, and Android to view even the biggest molecules and do full active real-time calculations using extremely accurate force fields on regular PCs. VR-enabled chemical modeling program YASARA Model blends efficiency and enjoyment. It interacts with VR headsets via OpenVR and broadcasts front-camera footage for viewing a mouse and keyboard. The alpha gradient on the billboard material combines the image when looking or moving aside [51]. LAMMPS: MD's ability to model huge systems over extended timeframes has made it popular. Advanced computer hardware has made MD simulations faster and more efficient, enabling complex phenomenon analysis. Complex potentials like many-body and machine learning have also enhanced material property predictions. A popular open-source MD algorithm, LAMMPS allows parallel computations and includes many material models and customization possibilities. A parallel MD code for materials and biomolecular modeling, LAMMPS, was developed in the mid-1990s. It offers 200 pair styles, hundreds of fix and compute styles, and hybrid models from various models. Model simulation may determine the best GPU exploit approach. Spatial decomposition can divide the simulation box into subdomains, with each processor handling a subset of atoms and their interactions [52]. GENESIS: Genesis is a novel software for general-purpose supercomputers that efficiently simulates huge biomolecular systems with all-atom MD. GENESIS includes ATDYN and SPDYN MD simulators plus analysis and setup tools. This version of GENESIS supports CHARMM force-fields and FFTE for 3D FFT (Fast Fourier Transform). Basic structural parameters and advanced analysis functions like PCA (Principal Component Analysis) are also available [53]. NAMD: MD simulations on all-atom models have improved, but sampling uncommon occurrences remains an issue. Charm++ manages NAMD's hybrid spatial/force decomposition, which is highly scalable. Charm++ supports many NAMD instances, and its versatile Tcl interface makes scripting easy. Another set of MPI communicators allows independent NAMD instances to communicate. Charm++ can be broken into numerous local subcommunicators [54]. Gaussian software: Gaussian accelerated molecular dynamics (GaMD) computes biomolecule-free energy and unconstrained increased sampling. It speeds up and accurately reconstructs biomolecule-free energy landscapes. This enhanced sampling method uses Gaussian boost potentials to calculate biomolecule-free energy landscapes accurately. GaMD smooths the potential energy surface using monotonicity, minor potential difference, and threshold energy range. The dual-boost GaMD accelerates more than the other two simulations. Simulation settings include harmonic force constants and threshold energy values [55]. Discovery Studio: Discovery Studio is an easy-to-use, single-graphical interface for drug design and protein modeling research. Additionally, Discovery Studio Standalone provides a complete molecular modeling environment for independent modelers. DS CHARMm can reduce receptor atoms and analyze vast numbers of ligands in Discovery Studio. It optimizes docked poses and calculates molecular system entropic energy [56]. TINKER: Tinker offers the latest molecular mechanics and dynamic programs and tools for modeling molecules and biopolymers that attempt to make Tinker user-friendly without a GUI [57]. OpenMM: OpenMM, a multilayer computer package, mimics molecules on robust computing systems. It is expandable to accommodate new hardware architectures and add functionality. The molecular simulation code OpenMM solves previous systems' reusability, extension, and availability issues. Intended for easy integration into programs, it allows adding new functionality without altering the OpenMM library. With its open-source license, OpenMM can be used in any application. Covalent and noncovalent interactions, implicit solvent models, integrators, thermostats, and barostats are OpenMM's main features. OpenMM is expandable so that users can add functionality. OpenMM plugins add new functionalities. Overall, OpenMM's extensibility simplifies simulation technique prototyping and testing [58].

4.3 Ligand-Based Drug Discovery Concept

When the target's three-dimensional structure is not known, LBDD is a handy technique during the drug discovery process. When a target structure is not comparable, the known ligand molecule is employed to identify the target by leveraging its physiochemical, structural, and pharmacological characteristics. The correlation between the target compound's biological activity and the compounds' chemical information is established. LBDD approaches predict novel drug compounds with comparable biological effects using prior knowledge of active drugs' structural, physical, and chemical characteristics, as shown in Figure 4.2. The high structural and physicochemical

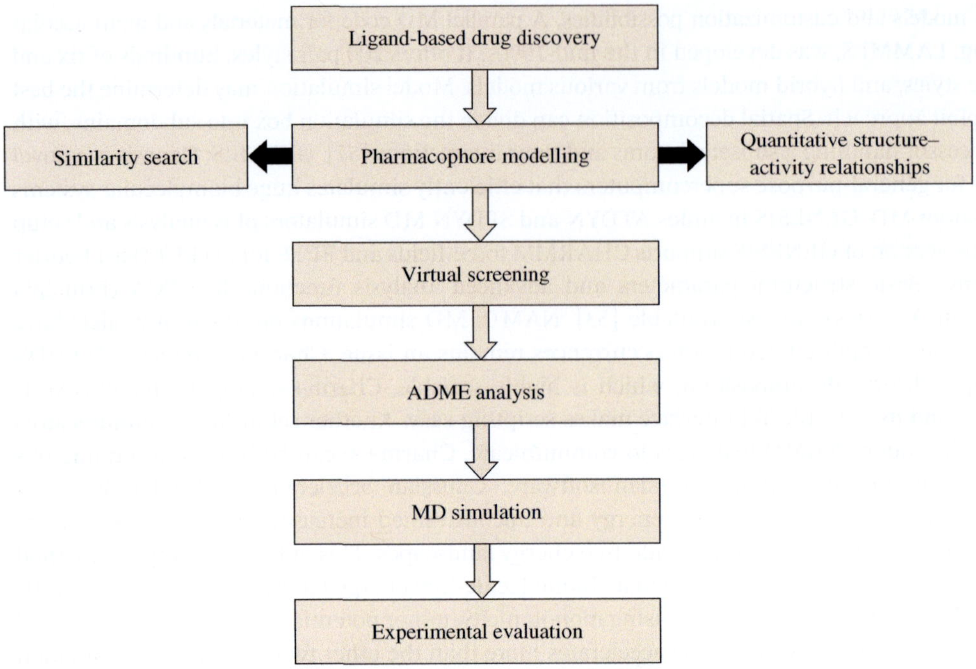

Figure 4.2 Illustration of basic concepts that are involved in the LBDD.

similarity between chemical compounds signifies a greater biological similarity upon which these predictions are founded. Several techniques are used in the drug discovery process, one of which is similarity search. This approach identifies compounds that share a ligand molecule through their structural and functional similarities. Pharmacophore and quantitative structure–activity relationship (QSAR) methodologies are implemented after comparable datasets have been identified. These are the two most prevalent methodologies employed in the LBDD [1, 6, 45]. The technique known as QSARs is utilized to determine the relationship between a compound's chemical structure and its function. In order to determine the biological activity of compounds, the QSAR model is utilized to optimize them. In contrast, pharmacophore is also a valuable method for describing the essential characteristics that enable a compound to exert its biological activity. It enhances knowledge regarding the interactions between ligands and proteins.

By utilizing the structural data of the ligand molecule, it is possible to construct it. However, within AI and machine learning, several databases and tools (DrugBank, Chembl, Zinc, ChemBank, and TCM) are accessible to facilitate LBDD-related procedures. Furthermore, because the LBDD is predominately founded upon a collection of ligand libraries, certain databases are accessible. These databases encompass a vast quantity of compound data, spewing specific information regarding a natural product, FDA-approved drug, phytochemical, small molecular, and numerous others. The worldwide researcher is utilizing several impactful tools and databases, which are detailed in Table 4.5.

4.3.1.1 The Detailed Description of Each Tool

LigandScout: LigandScout is a comprehensive, integrated platform that employs 3D chemistry feature pharmacophore models for exact virtual screening. The program supports streamlined

Table 4.5 List of tools and databases available in the case of LBDD.

Sl. no.	Name	URLs	References
1	LigandScout	http://www.inteligand.com/ligandscout/	[60]
2	MOE	http://www.chemcomp.com/MOE-Pharmacophore_Discovery.htm	[61]
3	MolSign	http://www.vlifesciences.com/products/Functional_products/Molsign.php	[62]
4	PharmaGist	http://bioinfo3d.cs.tau.ac.il/pharma/	[63]
5	Pharmer	http://smoothdock.ccbb.pitt.edu/pharmer/	[64]
6	PharmMapper	http://59.78.96.61/pharmmapper	[65]
7	Phase	http://www.schrodinger.com/Phase/	[66]
8	QSAR-Co	https://sites.google.com/view/qsar-co	[67]
9	QSAR ToolBox	https://qsartoolbox.org/	[68]
10	McQSAR	http://users.abo.fi/mivainio/mcqsar/index.php	[69]

procedures that begin with ligand- and structure-based pharmacophore modeling. It integrates inventive outstanding performance alignment methods that guarantee outstanding prediction quality while allowing screen at a level never before possible [60]. MOE: MOE offers three approaches for generating conformers. These algorithms, which are analyzed in this work, are summarized, and their main default settings, which were employed unless otherwise specified, are also described [61]. MolSign: MolSign is a comprehensive package for pharmacophore modeling. Finding pharmacophoric properties allows for a comprehensive mapping of the molecule, which may be utilized to search databases using pharmacophores [62]. PharmaGist: PharmaGist detects pharmacophores. The applied technique is ligand-based. Specifically, it does not require the target receptor's structure. Instead, the input consists of drug-like compounds with known receptor-binding properties. The output includes potential pharmacophores computed using several flexible alignments of the input ligands. The method directly and systematically handles the input ligands' flexibility during the alignment phase [63]. Pharmer: Pharmer is a revolutionary computational pharmacophore search method that grows with query breadth and complexity rather than database capacity. Geometric hashing and the generalized Hough transform inspire Pharmer, although they differ. Pharmer can search millions of structures for a given pharmacophore in minutes, extending pharmacophore screening [64]. PharmMapper: The PharmMapper online tool is a web server that uses reversed pharmacophore matching to identify possible drug targets by comparing the query chemical to an inside pharmacophore database [65]. Phase: PHASE was developed to improve pharmacophore-based lead discovery and optimization methods, which increasingly depend on models and 3D databases [66]. QSAR-Co: QSAR is a computational technique in the field of drug design. QSAR-Co is an open-source standalone software for model development and predicts the module. This provides a robust multitarget model using linear discriminant analysis [67]. QSAR ToolBox: QSAR Toolbox is a software application to find the QSAR models. This incorporates data from various sources and computational tools to identify targeted chemical molecules [68]. McQSAR: It utilized the descriptor for the multirepresentation of the molecules based on their variability and structure changes [69].

4.4 Structure- and Ligand-Based Assisted Studies

In this current situation, drug design through structure-based and ligand-based revolutions is the process for identifying novel targets and inhibitors employing multiple screening steps, cross-validation, and many more. Most of the researchers utilized the SBDD, which states the importance of the available target structure and how it can be helpful for novel inhibitor identification. Like the ligand of the chemical compound, natural products and approved-investigated libraries are available, which can help get the novel ligands through the ligand-based assisted study. Within this concept, the high-throughput virtual screening process also lies, which has been utilized to get the top inhibitor that can be further precisely investigated. Apart from that, various pivotal steps occurred during the successful structure- and ligand-based concept, successfully performed worldwide by comprehensive researchers, and their concept improvement was ongoing daily. So in this section, 10 selective studies of the structure-based and ligand-based were discussed, which can be helpful for readers to get a better understanding, i.e., how this concept was lying and how it leads to successful study design for the novel discovery. The structure-based assisted successful studies and the other computational aspects are shown in Table 4.6; similarly, the ligand-based assisted study is illustrated in Table 4.7. Moreover, the detailed description of the mentored study in Tables 4.6 and 4.7 was discussed in the section.

Table 4.6 List of successful structure-based assisted studies.

Sl. no.	Target	Remarks	References
1	SARS-CoV-2 Mpro	Identification of protease inhibitors employing structure-based methods	[70]
2	COX-2	Identification of novel inhibitors through structure-based methods along with advanced computation	[71]
3	SARS-CoV-2 Mpro	Identification of druggable compounds through structure-based methods	[72]
4	TMPK	Identification of natural product-based inhibitors through a structure-based approach	[73]
5	CTLA4, CDK8, EGFR, mTOR, p53R2, and PR	Identification of phytochemical-based inhibitors through a structure-based approach	[74]
6	α-glycosidase	Identification of α-glucosidase inhibitors employing structure-based methods	[75]
7	Protein-polymerase basic protein	Identification of natural inhibitors compound-based employing structure-based methods	[76]
8	Secreted aspartic protease 2 (SAP2)	Identification of novel inhibitors through structure-based methods	[77]
9	SARS-CoV-2 Mpro	Identification of novel nonpeptide inhibitors through structure-based methods	[78]
10	CYP1B1	Identification of novel inhibitors through a structure-based approach along with experimental validation	[79]

Table 4.7 List of successful ligand-based assisted studies.

Sl. no.	Target	Remarks	References
1	DNA Topoisomerase I (Top1)	Identification of potential inhibitors through ligand-based approach	[80]
2	Histone deacetylase 3 (HDAC3)	Identification of potential inhibitors through ligand based along with pharmacophore	[81]
3	Ld-PriS	Identification of novel drug compounds through ligand-based approach	[82]
4	CatD-statin	Assessment of chemical features of the selected inhibitor through a ligand-based approach	[83]
5	TcTIM	Identification of benzimidazoles potential inhibitors through ligand-based approach	[84]

4.4.1 The Detailed Description of Each Tool

In this situation, several successful therapeutics have been developed to reduce the transmission and infection rate of COVID-19. A study by Kolybalov et al. targeted the Mpro of the SRAS-CoV-2, the most important therapeutic target, as it plays an essential role in viral mechanisms. In this study, the author manually investigated four ligands, i.e., L1–L4, prepared by altering the binding groups, and the docking were executed through the GOLD database. The structure was retrieved from PDB with ID:7NG6 and prepared using authentic approaches. The binding site was identified and set, followed by a 10 Å spherical cavity. The docked complex was further investigated through the MD simulation followed by a 100 ns period. Based on the energy calculation and several results validation, the author suggested that the LK2, L3, and L4 can be utilized for further optimization. The study suggested that these identified compounds can be a potential antiviral candidate against SRAS-CoV-2 [70]. Another study by Krishnendu et al. targeted cyclooxygenase (COX) to identify its potential inhibitors employing the pharmacophore structure-based approaches and many more validation steps. COX is an enzyme essential for the body's reaction to inflammation and for making prostaglandins, lipid substances that participate in many bodily functions. Moreover, it plays an essential role in severe gastrointestinal irritation and cardiac toxicities and targeting this can be helpful for a better tomorrow. In this study, a total of 190 inhibitors were reported and studied in the year 2022, a part of 18 extensive research articles followed by various chemical structures drawn by the ChemDraw Ultra 12.0. Further, the pharmacophore model was generated with the help of the PHASE module available in Schrodinger, considering different characteristics. A set of 130 compounds were incorporated using the PHASE module. Using automated random selection, these compounds were utilized to generate a 3D QSAR model. The structure of COX-2 was computed from PDB having ID: 5KIR, and the models' docking analysis was performed, followed by the virtual screening. Additionally, the ligands were also examined for the drug-like properties analysis. The overall study suggests that the identified molecules have better promising results than standard rofecoxib drugs, and apart from that, the designed methodology can also be utilized for further study, and the reliable approach shows a promising accuracy in terms of novelty [71]. Different ligand libraries were developed to revolutionize the drug discovery process, with numerous compound collections that can be helpful for novel inhibitor identification. To keep the exploring concept, Ghufran et al. performed another study, which utilized a reliable approach to find a druggable target of Mpro protease. In this study, the Mpro protease was retrieved from the PDB database with the ID 6LU7 for further investigation. The downloaded targets were further prepared and minimized before further processing. Similarly, a set of ligands was retrieved from different databases followed by ZINC, ChemBridge, and in-house designed structural analogs comprising 1600 compounds. These compounds were minimized and optimized using the molecular operating environment (MOE) framework. Afterward, the authors performed structure-based virtual screening followed by molecular docking analysis, and the promising compound was further evaluated through the MD simulation process. The binding free energy was also calculated to better understand the ligand and target interaction throughout the simulation computation. The overall study shows that the three potential active inhibitors that were identified from each set of libraries have a promising bonding with the selected target. The authors also suggested that these compounds can be considered the most active antiviral compounds [72]. Another study was executed by Khan et al. to identify natural product-based promising inhibitors of the Monkeypox virus. Monkeypox is one of the leading re-emerging diseases that will be noticed in 2022 in different countries. In this study, authors targeted the thymidylate kinase as the structure was not available; the target was modeled considering the vaccinia virus having PDB ID: 2V54. Further, the structure quality was validated using the subsequent servers, Procheck and ProSA-web. Further, the natural products library was screened, followed by traditional Chinese medicines, NPASS, SANCDB, and Coconut. Further, all retrieved ligand data were screened based on Ro5 to identify promising ones. Afterward, the compounds were screened on the active site of the target structure utilizing the AutoDock vina. The stability of the ligand–target complex must be crucial in the drug design process, so the top three compounds from each dataset screened based on the target structure were further examined through the simulation process. The simulation was performed over 300 ns, and the authors calculated the binding energy. Based on the overall study, the top three compounds from the TCM (TCM26463, TCM2079, TCM29893), NPASS (NPC474409, NPC278434, NPC158847), SANCDB (SANC00240, SANC00984, SANC00986), and coconut (CNP0404204, CNP0262936 and CNP0289137) databases were found as most promising based on the crucial steps of validation. Further, the authors suggested that these compounds need experimental validation followed by *in vitro* and *in vivo* before final administration [73]. A study by Jha et al. on identifying phytochemicals for breast cancer employs molecular docking. In the present situation, it is one of the leading infections worldwide, and their cases are increasing daily; moreover, there are different cancer therapeutics available; however, their adverse effect is found to be less accurate. Therefore, in this study, the authors retrieved the target protein associated with breast cancer from the PDB database followed by its ID: 4OAR, 2J6M, 4DRH, 3MF1, IDQT, and 6T41. These targets were preprocessed with the help of Discovery Studio 2020, followed by standard protocol. Moreover, the active site within the target was identified using the CASTp server, which identified more pivotal residue that can be considered a target active site followed by 3–5 in residue. Afterward, the ligand was retrieved from Dr. Duke's phytochemical and ethnobotanical databases, and their required structure was retrieved from the PubChem databases. The identified ligand must have all the drug-like properties; therefore, the author performed the initial screening steps to reduce the ligand library and identify drug-like molecules. Moreover, the docking was performed using AutoDock Vina. The identified compounds through the docking analysis were utilized for the bioavailability and toxicity analysis. The overall study shows that among the top selected compounds with each target, only ursolic acid, enterolactone, parthenolide, and berberine have promising activity based on the selection criteria. However, the authors suggested that MD simulation and experimental validation can further investigate these compounds [74]. Another study performed by Liu et al. targeted α-glycosidase for the identification of novel inhibitors. In the current situation, diabetes is one of the leading metabolic disorders, which leads to severe medical complications in human health. In this study, the authors employed different crucial computational aspects followed by the experimental approach. The retrieved target was further prepared through the Protein Preparation Wizard, followed by preprocessing, energy minimization, and many more. The active site within this target was designed based on the residues within 15 Å centered in the structure. The docking analysis was performed, followed by the standard protocol using the Specs database to identify the inhibitors. A total of 52 compounds were purchased in this study, and the inhibitor assay analysis was performed through the standard protocol. Afterward, the authors performed the quenching, kinetic assay, and cell viability assay. The overall study shows that the four compounds were found to be the most promising ones based on molecular docking screening and experimental validation. The authors suggest that these compounds are the most promising hit [75]. Of the four types of influenza virus, the two types, i.e., types A and B, affect the human and are commonly known as the flu. Several natural compounds that showed inhibitory activity toward the influenza virus target were reported. A study performed by Jin et al. targeted the RNA polymerase to identify novel inhibitors toward it. The structure was extracted from the PDB database using PDB ID: 2ZNL and prepared using Schrödinger's software through the protein preparation wizard considering standard protocol followed by preprocess energy minimization and elimination of steric hindrance. Furthermore, the grid centers were generated through the position of the PB1 subunit. Afterward, a total of 70,000 natural compounds were collected from the different databases and utilized for the virtual scanning process, followed by the HTVS, SP, and XP, considering the top 10% compound as a selection criterion. The screened compounds were examined through the Swiss ADME calculation for a detailed investigation of drug-like properties. The stability of the identity compound with the target was validated through the simulation process. Based on the binding energy score obtained, the overall finding suggests that compound 1 is the most stable bonding among the most promising compounds. The authors suggest that no inhibitors are available in the market based on the target utilized in this study, so this investigation will help understand the molecular aspect between the target and identified novel inhibitors [76]. Another study performed by Li et al. targeted secreted aspartic protease 2 (SAP2) for the identification of pivotal inhibitors followed by experimental validation. In this study, a total of 713,000 compounds were utilized for the virtual screening process and ranked by the GOLD. Within this screening, a total of 7851 molecules were identified through the steps, and a further 582 molecules were utilized for the 100 structural diverse candidates and further purchased. Moreover, out of these compounds, 50 were utilized for the *in vivo* experimental validation. The inhibitory assay shows that among the 50 compounds, a total of 8 compounds showed a high inhibitory rate, and within that four compounds have a large IC50 value. Within the screened compound, one compound was reported as a Shp-2 inhibitor, and further, the authors performed the structural activity relationship of identified pyrazolone, along with its chemistry. The identified compounds were biologically evaluated, and their expression was analyzed on their corresponding cell. The authors also analyzed the antifungal activity to evaluate the inhibitor's activity. The overall study suggests that the one derivative named 24a showed promising inhibitor activity and was confirmed through the experimental evaluation. The authors reported that the mentioned compound has antifungal activity, and their optimization is ongoing [77]. SARS-CoV-2 is one of the leading causes of COVID-19 infection, and to overcome this, several approaches were employed, and different therapeutic approaches were developed. Similarly, another study was performed by Yang et al., which targeted Mpro for the identification of nonpeptide inhibitors. In the viral mechanism of SARS-CoV-2, Mpro is the essential one that plays a major role in pathogenesis. The authors retrieved the target structure with PDB ID: 6LU7 and prepared it using the protein preparation wizard followed by the standard protocol. Further, the receptor grid generation program identified the receptor sites within the target using the Glide program. Afterward, 8960 commercially available compounds were utilized in this study and prepared using the LigPrep module available at Schrodinger, followed by the standard protocol. Further, the ensemble docking approach was utilized in this study through the high-throughput virtual screening; moreover, the selection criteria were set at 10%. Finally, 49 compounds screened through the virtual screening were purchased from TargetMol for the experimental evaluation. Afterward, the biological assay experiment, MD simulation, and binding calculation were performed. The overall study shows that the two compounds (Z1244904919 and Z1759961356) were the most promising throughout the experimental validation. The study suggests that these protocols and identified antiviral drugs could be an initial point to combat COVID-19 [78]. Wang et al. targeted cytochrome P450 (CYP) to identify its inhibitors through structure-based virtual screening and experimental validation in another study. The target structure was retrieved from the PDB database with ID: 6iq5, having an inhibitor with an IC50 score of 4.4 nm. The author employed a docking validation protocol in which 201 reported inhibitors and 5025 inactive compounds were extracted through the DUD-E databases. Moreover, considering the collected external dataset and FDA databases, the LigPrep module was employed for the ligand library preparation. Further, for accurate identification, reverse docking analyses were performed to validate the docking protocol using different methods, CDOCKER, Libdock, and LigandFit in Discovery Studio, Autodock, and Vina. The Glide program was utilized for the virtual screening analysis on the active site of the CYP1B1 through the standard precision (SP), followed by the MM/GBSA method. Further, the identified inhibitors were examined through EROD (ethoxyresorufin-O-deethylase) assay. Moreover, the authors also perform the MD simulation over 100 ns, along with the density functional investigation. The cell culture and statistical analysis were also performed during the investigation. Furthermore, the overall study suggests that carvedilol could be a potential repurposed inhibitor with IC50 of 1.11 μM [79].

4.4.2 The Detailed Description of Each Tool

In the case of anticancer and antibacterial drug development, targeting topoisomerase could be a potential lead as it involves resolving the complication associated with the supercoiling of DNA and its further mechanisms. In these aspects, a study performed by Pal et al. targeted topoisomerase for the identification of its potential inhibitors employing the pharmacophore model along with the virtual screening. In this study, 62 molecules were retrieved from the literature review for the dataset generation and split into the test and training set, followed by their biological activity. Further, the structure of the selected ligands was drawn through the ChemDraw Ultra software and optimized through the CHARMM force fields. Utilizing this set of data, the 3D QSAR pharmacophore model was generated; however, the designed model was validated through the Cost analysis, Test set analysis, and Fischer's randomization test. Apart from the set of ligand libraries, they were designed for the virtual screening process followed by the drug-like properties analysis. For the docking purpose, the topoisomerase structure was retrieved from the PDB database using its ID: 1T81. Finally, the MD simulation study evaluated the most promising identified compounds. The findings show that ZINC68997780, ZINC15018994, and ZINC38550809, based on excessive steps of validation followed by docking, simulation, and ADME analysis, have the most promising profile and can be further utilized for the potential inhibitors [80]. Similarly, another study was performed by Kumbhar et al. using the concept of pharmacophore modeling toward the HDAC3 to identify potential inhibitors. In this study, 84 reported inhibitors were utilized for the dataset generation, followed by training and test set data. Moreover, their structure was designed using ChemDraw. Moreover, the model was generated using this dataset, and their validation was performed, followed by Fisher's randomization. Afterward, the virtual screening was performed using the diverse chemical datasets (NCI, Asinex, Chembridge, and Maybridge chemical databases). For the virtual screening, the Fast and Flexible methodology was employed along with the ADME analysis. The compound identified through the virtual screening was utilized for the docking analysis with the HDAC3 (PDB code: 4A69) target through the GOLD software. Moreover, the most pressing ones were utilized for the simulation study over 50 ns and the binding energy calculation. Additionally, the PCA was also performed for a better understanding. The overall finding shows that the 22 compounds show a promising interaction on the selected target. However, the top three hits were further analyzed for the simulation analysis. The study shows that the two compounds have stable interaction through the energy calculation. The authors suggest that the reported screened compounds can be potential inhibitors and an experimental evaluation is required [81]. A study by Bhowmik et al. utilized the concept of ligand-based virtual screening to identify potential drugs toward the catalytic subunit of *Leishmania donovani*. In this study, the authors performed the BLAST to get the gene sequence of the Ld-PriS, and the homology modeling was performed, along with their validation, to get an understanding of the quality of the designed model. Moreover, the structure was minimized through the YASARA. Furthermore, the active site within the target was completed through the COFACTOR and COACH software. Finally, the virtual screening and socking analysis were performed using the PyRx and iGEMDOCK software. Further, the ligands were evaluated for drug-like properties analysis and toxicity analysis. Afterward, the most promising hits were utilized for the simulation analysis over 50 ns. The overall finding shows that the ZINC000009219046, ZINC000025998119, and ZINC000004677901 show promising molecular activity with the target, and their simulation confirmed that the compound has stable binding. The authors suggest that these compounds can be considered the best inhibitors [82]. Another study by Sakkiah et al. targeted Cathepsin D, a significant component of lysosomes and an important component of catabolism and degenerative conditions. The author performed the ligand-based virtual screening in this study to identify promising inhibitors. Initially, the authors employed the pharmacophore modeling using the DiscoveryStudio v2.5 software. The compound utilized for the modeling was drawn using the ChemSketch V12 software. Fit the model generation, a total of 76 reported inhibitors were utilized for the test and training set data generation using these compounds. The pharmacophore model was validated. The virtual screening was also performed, utilizing 300,000 molecules for the process, which was evaluated through the ADME properties analysis. Afterward, the molecular docking analysis was performed, and for that, the target structure was received from the PDB database with the PDB ID 1LYB. The overall finding suggests that the 49 hit compounds were found promising based on the utilized selection criteria. The authors suggest that Hypothesis 1 can also be utilized for another study as it shows a new molecule for the selected target as an inhibitor [83]. Another one by Jimenez et al. employed the virtual screening methods along with the docking analysis for inhibitors of triosephosphate isomerase assessment. In this study, a total of 10 compounds were utilized as a control, and their structure was drawn through the ChemDraw software. The target structure was prepared through the UCSF Chimera 1.14.1. Further, the molecular docking analysis was performed through PyRx software, followed by the blind docking analysis. The benzimidazole scaffold was utilized for the virtual screening steps, available at the ZINC15 database, and analyzed through the

OpenBabel program. Further, the MD analysis was performed over 100 ns, followed by the binding energy calculation. The overall finding shows that the BP5 compound has a promising profile and a lower docking score on the human TIM compared to the utilized control. The identified compound shows reliable activity toward drug-like properties [84].

4.5 Advancement and Challenges in SBDD and LBDD

The structure- and ligand-based drug design revolutionized the concept of drug design. It shows very promising results in terms of accuracy and high processing activity. Through virtual screening within a bulk library, users can easily identify the most promising one. Structure-based analysis can easily help understand the target structure and its activity, which can be useful before processing the work. However, apart from the basic concept of the drug design, the fragment-based method shows the promising concept behind the drug design in terms of identifying and developing the drug in a simple form. To top up the concept, the researcher also employed machine learning methods through which they trained a model that can easily help to identify promising inhibitors. Moreover, the 3D QSAR concept was also evolving, which is helpful for the arrangement of the reported ligand for the novel discovery. The structure and ligand base can also be helpful in metabolic prediction, which means the candidate can be helpful in inhibiting metabolic disease and reducing its effects. Moreover, with time by time, many advances were occurring daily to get a more insightful impact. Similarly, as the advancement occurred, the structure- and ligand-based study had certain drawbacks and challenges. In the case of structure based, the binding site within the protein is difficult as it easily adopts multiple confirmations. The target selection is also important during the drug resistance mechanism in the structure against the identified inhibitors. As the nature of the target is hydrophobic and complicated, the targeting of the membrane protein is still challenging; moreover, as the work is based on the computer, it requires a high-performance computing system for better study. Similarly, in the ligand-based drug design case, the ligand library's limitations are challenging as it generates the concept of the limited collected compounds. Moreover, as the QSAR model is based on the data retrieved during the study, it will lead to challenges in identifying false positive leads. Moreover, several studies employed several concepts for model generation, so the impact of the identified compound also varies. In the case of the ligand-based study, the author explained the ADME calculation for the drug-like properties analysis; the accurate prediction of the properties and remains is challenging. Apart from this, severe advantages and challenges occur daily, and to overcome these, worldwide researchers apply various extensive concepts to deal with and handle these challenges easily.

4.6 Conclusion

The computer-assisted study has the most impactful result in the present situation in the case of drug design. Several studies were performed along with the structure- and ligand-based drug designs. To enhance the study, the drug target and inhibitor identification server tools and database were also designed and developed, which can be helpful for numerous studies. Using these developed databases, researchers are utilizing the concept of structure- and ligand-based drug design by enhancing the novel discovery. Moreover, it is essential that before the drug discovery process, the author explores the available databases and tools that are associated with the study. This study explores the in-detail concept behind the drug design, followed by structure- and ligand-based drug design. The mentioned tools and database can be helpful for the researcher to understand inhibitor identification, virtual screening, statistical evaluation, and many more. Moreover, several databases have certain limitations, and their available dataset is low, which requires an excessive updation to improve the tools and databases for better accuracy.

References

1 Sulimov, A.V., Ilin, I.S., Tashchilova, A.S. et al. (2021). Docking and other computing tools in drug design against SARS-CoV-2. *SAR and QSAR in Environmental Research* 35 (2): 91–136.
2 Karaman, B. and Sippl, W. (2019). Computational drug repurposing: current trends. *Current Medicinal Chemistry* 26 (28): 5389–5409.
3 Isert, C., Atz, K., and Schneider, G. (2023). Structure-based drug design with geometric deep learning. *Current Opinion in Structural Biology* 79: 102548.
4 Vinjoda, P., Mishra, S.K., Sharma, K., and Georrge, J.J. (2024). *In silico* identification of novel drug target and its natural product inhibitors for herpes simplex virus. *Nanotechnology and In Silico Tools* 2024: 377–383.
5 Vaghasia, V.V., Sharma, K., Mishra, S.K., and Georrge, J.J. (2024). *In silico* identification of natural product inhibitor for multidrug resistance proteins from selected gram-positive bacteria. *Nanotechnology and In Silico Tools* 2024: 309–317.
6 Shaker, B., Ahmad, S., Lee, J. et al. (2021). *In silico* methods and tools for drug discovery. *Computers in Biology and Medicine* 137: 104851.
7 Vakhariya Sakina, S., Mishra, S.K., Sharma, K., and Georrge, J.J. (2023). Designing of a novel curcumin analogue to inhibit mitogen-activated protein kinase: a cheminformatics approach. *Journal of Phytonanotechnology and Pharmaceutical Sciences* 3 (1): 37–47.
8 Mishra, S.K., Priya, P., Rai, G.P. et al. (2023). Coevolution based immunoinformatics approach considering variability of epitopes to combat different strains: a case study using spike protein of SARS-CoV-2. *Computers in Biology and Medicine* 163: 107233.
9 Vakhariya, S., Kumar Mishra, S., and Georrge, J.J. (2023). *Identification of Novel Curcumin Analogues to Inhibit Mitogen-Activated Protein Kinase 1 Using In Silico Combinatorial Library Design and Molecular Docking Approach*. Available at SSRN 4649193.
10 Webb, B. and Sali, A. (2016). Comparative protein structure modeling using MODELLER. *Current Protocols in Bioinformatics* 54: 561–567.
11 Waterhouse, A., Bertoni, M., Bienert, S. et al. (2018). SWISS-MODEL: homology modelling of protein structures and complexes. *Nucleic Acids Research* 46 (W1): W296–W303.
12 Kelley, L.A., Mezulis, S., Yates, C.M. et al. (2015). The Phyre2 web portal for protein modeling, prediction and analysis. *Nature Protocols* 10 (6): 845–858.
13 McGuffin, L.J., Adiyaman, R., Maghrabi, A.H.A. et al. (2019). IntFOLD: an integrated web resource for high performance protein structure and function prediction. *Nucleic Acids Research* 47 (W1): W408–W413.
14 Wang, S., Li, W., Liu, S., and Xu, J. (2016). RaptorX-Property: a web server for protein structure property prediction. *Nucleic Acids Research* 44 (W1): W430–W435.
15 Soding, J., Biegert, A., and Lupas, A.N. (2005). The HHpred interactive server for protein homology detection and structure prediction. *Nucleic Acids Research* 33 (Web Server Issue): W244–W248.

16 Kim, D.E., Chivian, D., and Baker, D. (2004). Protein structure prediction and analysis using the Robetta server. *Nucleic Acids Research* 32 (Web Server Issue): W526–W531.

17 Jumper, J., Evans, R., Pritzel, A. et al. (2021). Highly accurate protein structure prediction with AlphaFold. *Nature* 596 (7873): 583–589.

18 Karchin, R., Diekhans, M., Kelly, L. et al. (2005). LS-SNP: large-scale annotation of coding non-synonymous SNPs based on multiple information sources. *Bioinformatics* 21 (12): 2814–2820.

19 Xu, D. and Zhang, Y. (2012). *Ab initio* protein structure assembly using continuous structure fragments and optimized knowledge-based force field. *Proteins* 80 (7): 1715–1735.

20 Roy, A., Kucukural, A., and Zhang, Y. (2010). I-TASSER: a unified platform for automated protein structure and function prediction. *Nature Protocols* 5 (4): 725–738.

21 Pieper, U., Webb, B.M., Dong, G.Q. et al. (2014). ModBase, a database of annotated comparative protein structure models and associated resources. *Nucleic Acids Research* 42 (Database Issue): D336–D346.

22 Tian, W., Chen, C., Lei, X. et al. (2018). CASTp 3.0: computed atlas of surface topography of proteins. *Nucleic Acids Research* 46 (W1): W363–W367.

23 Jendele, L., Krivak, R., Skoda, P. et al. (2019). PrankWeb: a web server for ligand binding site prediction and visualization. *Nucleic Acids Research* 47 (W1): W345–W349.

24 Volkamer, A., Kuhn, D., Rippmann, F., and Rarey, M. (2012). DoGSiteScorer: a web server for automatic binding site prediction, analysis and druggability assessment. *Bioinformatics* 28 (15): 2074–2075.

25 Goldenberg, O., Erez, E., Nimrod, G., and Ben-Tal, N. (2009). The ConSurf-DB: pre-calculated evolutionary conservation profiles of protein structures. *Nucleic Acids Research* 37 (Database Issue): D323–D327.

26 Santana, C.A., Silveira, S.A., Moraes, J.P.A. et al. (2020). GRaSP: a graph-based residue neighborhood strategy to predict binding sites. *Bioinformatics* (Suppl_2): 36, i726–i34.

27 Borrel, A., Regad, L., Xhaard, H. et al. (2015). PockDrug: a model for predicting pocket druggability that overcomes pocket estimation uncertainties. *Journal of Chemical Information and Modeling* 55 (4): 882–895.

28 Le Guilloux, V., Schmidtke, P., and Tuffery, P. (2009). Fpocket: an open source platform for ligand pocket detection. *BMC Bioinformatics* 10: 168.

29 Kozakov, D., Hall, D.R., Xia, B. et al. (2017). The ClusPro web server for protein-protein docking. *Nature Protocols* 12 (2): 255–278.

30 Schneidman-Duhovny, D., Inbar, Y., Nussinov, R., and Wolfson, H.J. (2005). PatchDock and SymmDock: servers for rigid and symmetric docking. *Nucleic Acids Research* 33 (Web Server Issue): W363–W367.

31 Tovchigrechko, A. and Vakser, I.A. (2006). GRAMM-X public web server for protein–protein docking. *Nucleic Acids Research* 34 (Web Server Issue): W310–W314.

32 Chaudhury, S., Berrondo, M., Weitzner, B.D. et al. (2011). Benchmarking and analysis of protein docking performance in Rosetta v3.2. *PLoS One* 6 (8): e22477.

33 Schneidman-Duhovny, D., Hammel, M., Tainer, J.A., and Sali, A. (2016). FoXS, FoXSDock and MultiFoXS: single-state and multi-state structural modeling of proteins and their complexes based on SAXS profiles. *Nucleic Acids Research* 44 (W1): W424–W429.

34 Yan, Y., Tao, H., He, J., and Huang, S.Y. (2020). The HDOCK server for integrated protein–protein docking. *Nature Protocols* 15 (5): 1829–1852.

35 Dominguez, C., Boelens, R., and Bonvin, A.M. (2003). HADDOCK: a protein–protein docking approach based on biochemical or biophysical information. *Journal of the American Chemical Society* 125 (7): 1731–1737.

36 Pierce, B.G., Wiehe, K., Hwang, H. et al. (2014). ZDOCK server: interactive docking prediction of protein–protein complexes and symmetric multimers. *Bioinformatics* 30 (12): 1771–1773.

37 Ramirez-Aportela, E., Lopez-Blanco, J.R., and Chacon, P. (2016). FRODOCK 2.0: fast protein–protein docking server. *Bioinformatics* 32 (15): 2386–2388.

38 Ohue, M., Shimoda, T., Suzuki, S. et al. (2014). MEGADOCK 4.0: an ultra-high-performance protein-protein docking software for heterogeneous supercomputers. *Bioinformatics* 30 (22): 3281–3283.

39 Inbar, Y., Benyamini, H., Nussinov, R., and Wolfson, H.J. (2003). Protein structure prediction via combinatorial assembly of sub-structural units. *Bioinformatics* 19 (Suppl 1): i158–i168.

40 Mashiach, E., Nussinov, R., and Wolfson, H.J. (2010). FiberDock: a web server for flexible induced-fit backbone refinement in molecular docking. *Nucleic Acids Research* 38 (Web Server Issue): W457–W461.

41 Bajaj, C., Chowdhury, R., and Siddavanahalli, V. (2011). F2Dock: fast Fourier protein–protein docking. *IEEE/ACM Transactions on Computational Biology and Bioinformatics* 8 (1): 45–58.

42 Verdonk, M.L., Cole, J.C., Hartshorn, M.J. et al. (2003). Improved protein–ligand docking using GOLD. *Proteins* 52 (4): 609–623.

43 Jimenez-Garcia, B., Pons, C., and Fernandez-Recio, J. (2013). pyDockWEB: a web server for rigid-body protein–protein docking using electrostatics and desolvation scoring. *Bioinformatics* 29 (13): 1698–1699.

44 Vilar, S., Cozza, G., and Moro, S. (2008). Medicinal chemistry and the molecular operating environment (MOE): application of QSAR and molecular docking to drug discovery. *Current Topics in Medicinal Chemistry* 8 (18): 1555–1572.

45 Eberhardt, J., Santos-Martins, D., Tillack, A.F., and Forli, S. (2021). AutoDock Vina 1.2.0: new docking methods, expanded force field, and python bindings. *Journal of Chemical Information and Modeling* 61 (8): 3891–3898.

46 de Almeida Filho, J.L. and Fernandez, J.H. (Eds). (2022). HTP SurflexDock: a web tool for structure-based virtual screening analysis based on the ensemble docking protocol. *Anais do XVI Brazilian e-Science Workshop*, SBC.

47 Bitencourt-Ferreira, G. and de Azevedo, W.F. Jr. (2019). Docking with GemDock. *Methods in Molecular Biology* 2053: 169–188.

48 Case, D.A., Cheatham, T.E. 3rd, Darden, T. et al. (2005). The Amber biomolecular simulation programs. *Journal of Computational Chemistry* 26 (16): 1668–1688.

49 Brooks, B.R., Brooks, C.L. 3rd, Mackerell, A.D. Jr. et al. (2009). CHARMM: the biomolecular simulation program. *Journal of Computational Chemistry* 30 (10): 1545–1614.

50 Abraham, M.J., Murtola, T., Schulz, R. et al. (2015). GROMACS: high performance molecular simulations through multi-level parallelism from laptops to supercomputers. *SoftwareX* 1: 19–25.

51 Ozvoldik, K., Stockner, T., and Krieger, E. (2023). YASARA model-interactive molecular modeling from two dimensions to virtual realities. *Journal of Chemical Information and Modeling* 63 (20): 6177–6182.

52 Thompson, A.P., Aktulga, H.M., Berger, R. et al. (2022). LAMMPS – a flexible simulation tool for particle-based materials modeling at the atomic, meso, and continuum scales. *Computer Physics Communications* 271: 108171.

53 Jung, J., Mori, T., Kobayashi, C. et al. (2015). GENESIS: a hybrid-parallel and multi-scale molecular dynamics simulator with enhanced sampling algorithms for biomolecular and cellular simulations. *Wiley Interdisciplinary Reviews: Computational Molecular Science* 5 (4): 310–323.

54 Jiang, W., Phillips, J.C., Huang, L. et al. (2014). Generalized scalable multiple copy algorithms for molecular dynamics simulations in NAMD. *Computer Physics Communications* 185 (3): 908–916.

55 Wang, J., Arantes, P.R., Bhattarai, A. et al. (2021). Gaussian accelerated molecular dynamics (GaMD): principles and applications. *Wiley Interdisciplinary Reviews: Computational Molecular Science* 11 (5).

56 Pawar, S.S. and Rohane, S.H. (2021). Review on discovery studio: an important tool for molecular docking. *Asian Journal of Research in Chemistry* 14 (1): 86–88.

57 Foscato, M., Deeth, R.J., and Jensen, V.R. (2015). Integration of ligand field molecular mechanics in tinker. *Journal of Chemical Information and Modeling* 55 (6): 1282–1290.

58 Eastman, P., Friedrichs, M.S., Chodera, J.D. et al. (2013). OpenMM 4: a reusable, extensible, hardware independent library for high performance molecular simulation. *Journal of Chemical Theory and Computation* 9 (1): 461–469.

59 Wang, Y., Lamim Ribeiro, J.M., and Tiwary, P. (2020). Machine learning approaches for analyzing and enhancing molecular dynamics simulations. *Current Opinion in Structural Biology* 61: 139–145.

60 Wolber, G. and Langer, T. (2005). LigandScout: 3-D pharmacophores derived from protein-bound ligands and their use as virtual screening filters. *Journal of Chemical Information and Modeling* 45 (1): 160–169.

61 Chen, I.J. and Foloppe, N. (2008). Conformational sampling of druglike molecules with MOE and catalyst: implications for pharmacophore modeling and virtual screening. *Journal of Chemical Information and Modeling* 48 (9): 1773–1791.

62 VLife M. 3.5 (2008). *Molecular Design Suite*, 2010. Pune: Vlife Sciences Technologies Pvt Ltd.

63 Schneidman-Duhovny, D., Dror, O., Inbar, Y. et al. (2008). PharmaGist: a webserver for ligand-based pharmacophore detection. *Nucleic Acids Research* 36 (Web Server Issue): W223–W228.

64 Koes, D.R. and Camacho, C.J. (2011). Pharmer: efficient and exact pharmacophore search. *Journal of Chemical Information and Modeling* 51 (6): 1307–1314.

65 Wang, X., Shen, Y., Wang, S. et al. (2017). PharmMapper 2017 update: a web server for potential drug target identification with a comprehensive target pharmacophore database. *Nucleic Acids Research* 45 (W1): W356–W360.

66 Dixon, S.L., Smondyrev, A.M., and Rao, S.N. (2006). PHASE: a novel approach to pharmacophore modeling and 3D database searching. *Chemical Biology & Drug Design* 67 (5): 370–372.

67 Ambure, P., Halder, A.K., Gonzalez Diaz, H., and Cordeiro, M. (2019). QSAR-Co: an open source software for developing robust multitasking or multitarget classification-based QSAR models. *Journal of Chemical Information and Modeling* 59 (6): 2538–2544.

68 Dimitrov, S.D., Diderich, R., Sobanski, T. et al. (2016). QSAR Toolbox – workflow and major functionalities. *SAR and QSAR in Environmental Research* 27 (3): 203–219.

69 Vainio, M.J. and Johnson, M.S. (2005). McQSAR: a multiconformational quantitative structure-activity relationship engine driven by genetic algorithms. *Journal of Chemical Information and Modeling* 45 (6): 1953–1961.

70 Kolybalov, D.S., Kadtsyn, E.D., and Arkhipov, S.G. (2024). Computer aided structure-based drug design of novel SARS-CoV-2 main protease inhibitors: molecular docking and molecular dynamics study. *Computation* 12 (1): 18.

71 Krishnendu, P.R., Benny, S., Kumar, S. et al. (2024). A structure-based approach to explore novel COX-2 inhibitors using pharmacophore modelling, 3D-QSAR analysis, virtual screening and dynamics simulation study. *Journal of Molecular Structure* 1295: 136634.

72 Ghufran, M., Ullah, M., Khan, H.A. et al. (2023). In-silico lead druggable compounds identification against SARS COVID-19 main protease target from in-house, chembridge and zinc databases by structure-based virtual screening, molecular docking and molecular dynamics simulations. *Bioengineering (Basel)* 10 (1).

73 Khan, A., Adil, S., Qudsia, H.A. et al. (2023). Structure-based design of promising natural products to inhibit thymidylate kinase from Monkeypox virus and validation using free energy calculations. *Computers in Biology and Medicine* 158: 106797.

74 Jha, V., Devkar, S., Gharat, K. et al. (2022). Screening of phytochemicals as potential inhibitors of breast cancer using structure based multitargeted molecular docking analysis. *Phytomedicine Plus* 2 (2): 100227.

75 Liu, S.K., Hao, H., Bian, Y. et al. (2021). Discovery of new alpha-glucosidase inhibitors: structure-based virtual screening and biological evaluation. *Frontiers in Chemistry* 9: 639279.

76 Jin, Z., Wang, Y., Yu, X.F. et al. (2020). Structure-based virtual screening of influenza virus RNA polymerase inhibitors from natural compounds: molecular dynamics simulation and MM-GBSA calculation. *Computational Biology and Chemistry* 85: 107241.

77 Li, C., Liu, Y., Wu, S. et al. (2020). Targeting fungal virulence factor by small molecules: structure-based discovery of novel secreted aspartic protease 2 (SAP2) inhibitors. *European Journal of Medicinal Chemistry* 201: 112515.

78 Yang, J., Lin, X., Xing, N. et al. (2021). Structure-based discovery of novel nonpeptide inhibitors targeting SARS-CoV-2 M(pro). *Journal of Chemical Information and Modeling* 61 (8): 3917–3926.

79 Wang, Y., He, X., Li, C. et al. (2020). Carvedilol serves as a novel CYP1B1 inhibitor, a systematic drug repurposing approach through structure-based virtual screening and experimental verification. *European Journal of Medicinal Chemistry* 193: 112235.

80 Pal, S., Kumar, V., Kundu, B. et al. (2019). Ligand-based pharmacophore modeling, virtual screening and molecular docking studies for discovery of potential topoisomerase I inhibitors. *Computational and Structural Biotechnology Journal* 17: 291–310.

81 Kumbhar, N., Nimal, S., Barale, S. et al. (2022). Identification of novel leads as potent inhibitors of HDAC3 using ligand-based pharmacophore modeling and MD simulation. *Scientific Reports* 12 (1): 1712.

82 Bhowmik, D., Jagadeesan, R., Rai, P. et al. (2021). Evaluation of potential drugs against leishmaniasis targeting catalytic subunit of *Leishmania donovani* nuclear DNA primase using ligand based virtual screening, docking and molecular dynamics approaches. *Journal of Biomolecular Structure & Dynamics* 39 (5): 1838–1852.

83 Sakkiah, S., Thangapandian, S., and Lee, K.W. (2012). Ligand-based virtual screening and molecular docking studies to identify the critical chemical features of potent cathepsin D inhibitors. *Chemical Biology & Drug Design* 80 (1): 64–79.

84 Vazquez-Jimenez, L.K., Juarez-Saldivar, A., Gomez-Escobedo, R. et al. (2022). Ligand-based virtual screening and molecular docking of benzimidazoles as potential inhibitors of triosephosphate isomerase identified new trypanocidal agents. *International Journal of Molecular Sciences* 23(17).

5

Virtual Screening and Lead Discovery

Nisha Kumari Singh, Nigam Jyoti Maiti, Manshi Mishra, Shantanu Raj, Gourav Rakshit, Rahul Ghosh, and Sharanya Roy

Department of Pharmaceutical Sciences and Technology, Birla Institute of Technology, Ranchi, Jharkhand, India

5.1 Introduction to Virtual Screening and Lead Discovery

5.1.1 Overview of Drug Discovery Process

The process of drug discovery is intricate and financially demanding, encompassing the identification and advancement of novel therapeutic agents. This multifaceted endeavor unfolds through various phases, starting with target identification, followed by lead discovery, lead optimization, preclinical evaluations, and ultimately, clinical trials. Each stage plays a crucial role in the development and validation of potential pharmaceutical candidates, culminating in the delivery of safe and effective treatments to address diverse medical needs [1].

Efforts to discover new chemical entities and novel structural scaffolds for therapeutic applications are central to pharmaceutical chemistry. Traditionally, this involves the laborious synthesis of various potential structures or the screening of natural products, often leading to chance discoveries rather than rational approaches. However, recent advancements in computational techniques and protein crystallography have spurred the adoption of *in silico* approaches [2], which have become integral to drug design and discovery in both industrial and academic settings. These approaches, encompassing genomics, proteomics, bioinformatics, and chemoinformatics, facilitate the identification and development of potential ligands tailored for specific protein targets, thereby forming the cornerstone of drug discovery processes. Once potential ligands, or hits, are identified, the next step involves refining them into active compounds with nonpromiscuous binding behavior, known as leads. This process, known as lead identification and optimization, is crucial in drug discovery programs and typically involves medicinal chemists modifying substituents to enhance activity against a given receptor or disease. Ultimately, these efforts aim to yield compounds with improved potency and efficacy, often requiring further study of pharmacokinetics and toxicological profiles through *in vivo* screening procedures [3–5]. However, challenges persist, particularly in identifying the target protein and active site, designing inhibitors for membrane proteins, and accounting for various factors influencing drug interaction. Despite the advances in computational methods, such as high-throughput screening and virtual screening (VS) approaches, there remains a need to develop databases and refine concepts to categorize chemical entities as drug-like molecules. High-throughput screening methods, which involve testing large libraries of compounds for

Computational Methods for Rational Drug Design, First Edition. Edited by Mithun Rudrapal.

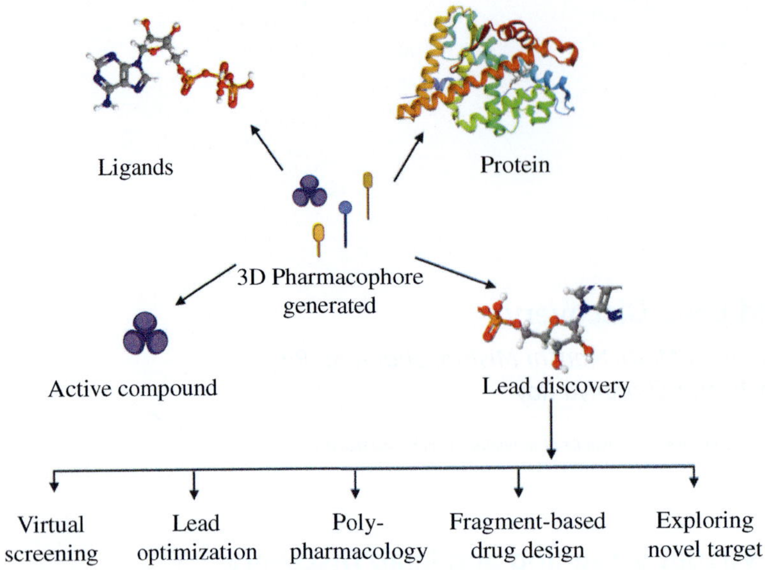

Figure 5.1 Depicting the drug discovery process.

biological activity, have revolutionized drug discovery but also pose challenges in data analysis and interpretation.

In recent years, researchers have shifted toward "smart" approaches for identifying compounds with desired biological activity, leveraging innovative computational techniques to process large datasets effectively. While the initial excitement surrounding combinatorial screening methods has waned, there is a growing focus on adopting more strategic approaches to identify compounds with therapeutic potential (Figure 5.1).

5.1.2 Role of Virtual Screening

VS serves as a pivotal computational methodology within drug discovery, aimed at pinpointing promising drug candidates amidst extensive chemical repositories. This technique significantly streamlines the compound selection process for experimental evaluation, thereby conserving both time and resources. VS methodologies are typically categorized into two main approaches: structure-based virtual screening (SBVS) and ligand-based virtual screening (LBVS) [6, 7], each offering distinct strategies for identifying compounds with potential therapeutic activity. The experimental efforts required for the biological screening of billions of compounds remain substantial, leading to increased interest in computer-aided drug design approaches. VS has emerged as a dynamic and lucrative technology for identifying novel drug-like compounds, or "hits," in the pharmaceutical industry [8]. Utilizing bioactive conformers obtained from structural methods or molecular modeling, VS employs ligand-based and structure-based approaches to rank novel ligands by 3D similarity searching or pharmacophore pattern matching. These computational methods enable the prediction of putative binding affinities between small molecules and biological receptors of interest in pharmaceutical research [9–11]. Ligand-based VS, also known as neighborhood behavior search, involves searching databases of chemical structures to find compounds similar to known actives or possessing common pharmacophores or substructures. Meanwhile, structure-based VS automates the docking of numerous chemical compounds against protein-binding or active sites, leveraging the increasing number of available protein 3D structures. These VS methodologies are categorized based on their principles, with a focus on introducing the underlying theories and principles behind these procedures. Additionally, this review presents recent case studies illustrating the application and effectiveness of various VS methods.

5.1.3 Importance of Lead Discovery

Lead discovery is a critical phase in drug development, involving the identification of potential drug candidates. For instance, VS techniques, such as molecular docking simulations, efficiently sift through large chemical databases to highlight promising compounds. Examples include identifying a small molecule that binds to a specific protein target implicated in a disease or discovering a natural product with therapeutic properties through screening plant extracts [12]. Early identification of these lead compounds, like a novel kinase inhibitor for cancer treatment or an antiviral peptide derived from marine organisms, accelerates drug development and reduces costs by expediting subsequent optimization and testing processes.

5.2 Molecular Targets and Biomolecular Structures

The study of molecular targets and biomolecular structures is fundamental to various fields, including molecular biology, biochemistry, pharmacology, and structural biology. It provides insights into how biological processes occur at the molecular level and informs the development of therapies and interventions for a wide range of diseases and conditions [7, 13]. Researchers continue to explore these concepts to deepen our understanding of life processes and to advance medical science. Molecular targets are specific biological molecules, like proteins or DNA sequences, crucial for various cellular functions and often associated with diseases. Biomolecular structures refer to the three-dimensional arrangements of these molecules, determining their functions. Understanding these structures is vital for scientific fields like molecular biology, pharmacology, and structural biology [14], as it enables the development of treatments and enhances our knowledge of life processes. Techniques like X-ray crystallography and NMR spectroscopy help in determining these structures at different levels, from the linear sequence to the overall three-dimensional folding of biomolecules.

5.3 Virtual Screening Approaches

VS approaches are essential tools in drug discovery, allowing researchers to identify potential drug candidates from large chemical libraries using computational methods. There are three primary VS approaches.

5.3.1 Structure-based Virtual Screening

SBVS is a computational drug discovery approach that utilizes the three-dimensional structures of target proteins to screen and identify potential ligands from large compound libraries. SBVS relies on molecular docking algorithms to predict the binding affinity and orientation of small molecules within the active site of a target protein. By analyzing the interactions between the ligand and the target protein, SBVS helps in identifying lead compounds with high binding affinities and favorable pharmacological properties, thus accelerating the drug discovery process [15–18].

SBVS relies on the knowledge of a target protein's three-dimensional structure, often obtained through techniques like X-ray crystallography or homology modeling. Key points include:

- Utilization of the target protein's 3D structure.
- Prediction of potential drug compounds' interaction with the protein's active site.
- Screening of compound libraries to identify molecules that fit well with the protein's structure and have therapeutic potential.
- Valuable when the protein's structure is known or can be accurately predicted.

5.3.2 Ligand-based Virtual Screening

LBVS is a computational drug discovery approach that focuses on the chemical properties and structural features of known ligands (or active compounds) rather than the three-dimensional structure of the target protein. LBVS methods analyze the similarities between the molecular descriptors or fingerprints of reference ligands and those of compounds in a chemical database. The goal is to identify new molecules with similar structural and chemical features to known active compounds, thus potentially possessing similar biological activities [19, 20].

LBVS does not require the target protein's structure but instead focuses on known ligands (small molecules) that bind to the target. Key aspects are:

- Utilization of data from known ligands or active compounds.
- Analysis of structural and chemical properties to identify common features or patterns.
- Search for molecules in chemical libraries with similar properties, potentially having similar biological activity.
- Particularly useful when the protein structure is unavailable or hard to obtain.

5.3.3 Hybrid Approaches

Hybrid approaches in SBVS and LBVS refer to integrating methodologies from both SBVS and LBVS to improve the accuracy and efficiency of VS in drug discovery [21, 22]. These hybrid approaches leverage the strengths of both SBVS, which utilizes the three-dimensional structure of the target protein, and LBVS, which focuses on the chemical properties of ligands, to overcome limitations associated with each method when used independently [23]. One common hybrid approach involves using structural information from SBVS to refine the results obtained from LBVS or vice versa. For example, after identifying potential hits through LBVS, these compounds can be docked into the active site of the target protein using SBVS techniques to assess their binding modes and interactions. Another hybrid strategy involves combining molecular descriptors derived from ligand-based methods with structural information obtained from target proteins to develop more accurate predictive models for VS [24].

Hybrid approaches combine elements of SBVS and LBVS to enhance VS accuracy. These approaches involve:

- Integration of information from both protein structure (SBVS) and ligand data (LBVS).
- Improvement of the screening process by considering multiple factors, including protein–ligand interactions and ligand similarity.
- Increased likelihood of identifying potential drug candidates with desired properties [25] (Figure 5.2).

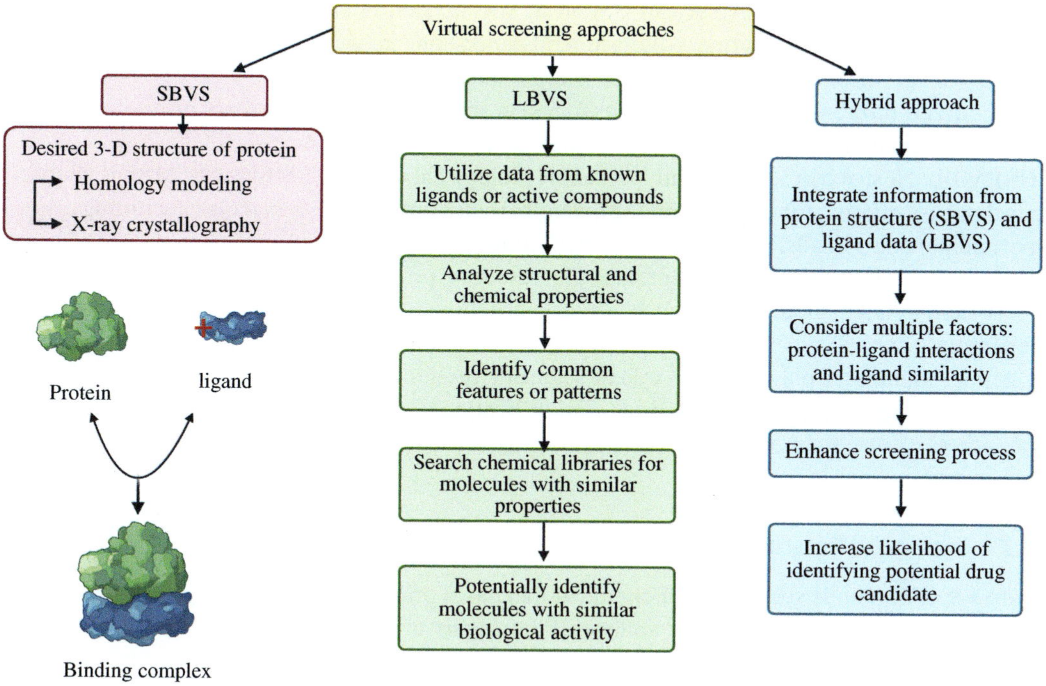

Figure 5.2 Workflow for the virtual screening approaches in the drug discovery process.

5.4 Databases and Compound Collections

5.4.1 Overview of Chemical Databases

"Chemical Databases" encompass repositories of structured chemical information, including molecular structures, properties, reactions, and associated data, utilized across various scientific disciplines for research, education, and industrial applications. These databases serve as vital resources for chemists, biochemists, pharmacologists, and other researchers, enabling efficient storage, retrieval, and analysis of chemical data. An overview of chemical databases includes various types and examples [8]:

Public databases: These freely accessible databases provide extensive collections of chemical information to the scientific community and the general public. Examples include:

PubChem: Maintained by the National Center for Biotechnology Information (NCBI), PubChem is a comprehensive chemical database containing information on millions of chemical compounds, including their structures, properties, bioactivities, and references to the literature.

ChemSpider: Curated by the Royal Society of Chemistry, ChemSpider is a chemical structure database offering access to over 67 million chemical compounds, along with their properties, spectral data, and literature references.

Commercial databases: These databases, typically requiring a subscription or license, offer curated chemical data and specialized tools for research and industry. Examples include:

Reaxys: Provided by Elsevier, Reaxys is a web-based chemistry database offering access to a vast collection of organic, inorganic, and organometallic compounds, along with their properties, reactions, and synthesis information.

Scifinder: Offered by the American Chemical Society, Scifinder is a comprehensive research tool providing access to chemical literature, patents, and substance information, facilitating searches for chemical compounds, reactions, and related data [26–30].

Specialized databases: These databases focus on specific areas of chemistry or research interests, providing tailored data and tools. Examples include:

ChEMBL: Hosted by the European Molecular Biology Laboratory (EMBL), ChEMBL is a bioactivity database offering information on the biological activities of small molecules, including drugs, experimental compounds, and natural products, along with their targets and assay data.

DrugBank: DrugBank is a comprehensive resource providing information on drugs, drug targets, and related data, including chemical structures, pharmacological actions, and drug interactions.

Chemical databases play a crucial role in scientific research, education, and industrial applications, facilitating the exploration, analysis, and utilization of chemical compounds and their properties across diverse fields of study.

5.4.2 Compound Filtering and Preparation

Before VS, compounds sourced from chemical databases often undergo filtering and preparation steps to enhance their suitability for analysis. Filtering involves the exclusion of compounds that do not meet specific criteria, such as drug-likeness or safety considerations, ensuring that only relevant candidates are retained for further investigation. Subsequently, compound preparation tasks are carried out, including the generation of three-dimensional structures and energy minimization to optimize molecular geometries. These preparatory measures ensure that the compounds are in an optimal state for subsequent VS simulations, thereby facilitating more accurate predictions and efficient identification of potential drug candidates [31, 32].

5.4.3 Diversity and Size of Compound Collections

The diversity of compound collections plays a crucial role in conducting comprehensive VS processes. Such collections, encompassing a broad spectrum of chemical structures, significantly enhance the likelihood of identifying novel drug candidates. These collections can vary greatly in size, ranging from thousands to millions of compounds, depending on the database or resource utilized. While larger collections offer more extensive screening possibilities, they also demand substantial computational resources to execute effectively [33].

5.5 Molecular Docking

5.5.1 Principles of Molecular Docking

Molecular docking stands as a vital computational tool in drug discovery, aiming to predict the interaction between a small molecule (known as a ligand) and a biological target, commonly a protein (receptor). The primary objective is to ascertain the most energetically favorable binding pose of the ligand within the binding site of the receptor. Leveraging principles of molecular physics and chemistry, molecular docking simulations simulate these interactions, offering insights into the potential binding modes and affinities between the ligand and receptor [34, 35].

5.5.2 Docking Algorithms and Scoring Functions

Docking algorithms serve as computational methodologies designed to explore the conformational space of ligands within the binding site of receptors. These algorithms meticulously search for ligand poses that optimize binding interactions, taking into account crucial factors such as hydrogen bonding, electrostatics, and van der Waals forces. To assess and rank the quality of docking poses, scoring functions are employed, which evaluate predicted binding energy or affinity. Lower energy scores typically indicate more favorable binding interactions, guiding researchers toward the most promising ligand–receptor complexes for further investigation [34, 36–41].

5.5.3 Validation of Docking Results

Validation is integral to evaluating the accuracy and reliability of docking outcomes in drug discovery. Experimental validation involves laboratory testing of predicted ligand–receptor interactions, often utilizing techniques like X-ray crystallography or binding assays to confirm the predicted binding modes. Additionally, cross-validation procedures involve docking studies on datasets with known binding affinities, enabling assessment of the predictive capabilities of the docking method. Visualization and analysis of docking poses and interactions further contribute by offering insights into the specific ligand–receptor binding modes, facilitating the refinement and optimization of drug candidate designs [6, 7, 36, 42–48] (Figure 5.3).

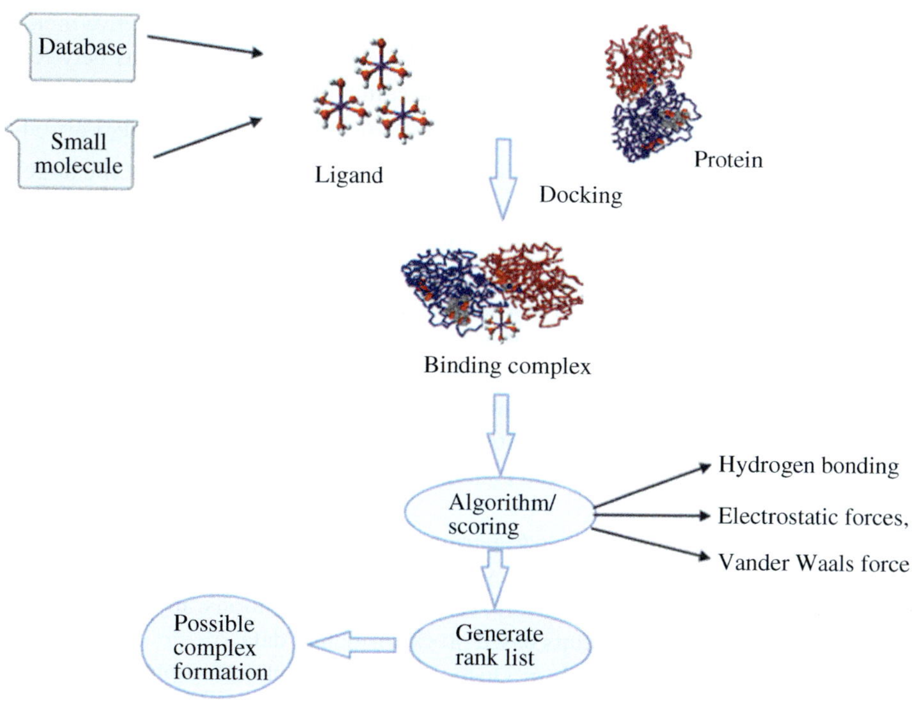

Figure 5.3 Workflow for the docking in drug discovery process.

5.6 Pharmacophore Modeling

5.6.1 Concept of Pharmacophores

The concept of pharmacophores refers to a spatial arrangement of atoms or functional groups within a molecule that is responsible for its biological activity by interacting with specific structural features of a target receptor or enzyme. Pharmacophores represent the essential elements of a molecule that contribute to its pharmacological effects, such as binding affinity and selectivity. Understanding pharmacophores is crucial in drug design and discovery [41], as it helps in identifying and optimizing compounds with desired biological activities [46].

A pharmacophore model typically consists of specific features, including hydrogen bond donors and acceptors, hydrophobic regions, aromatic rings, and positively or negatively charged groups, which mimic the complementary binding sites on the target protein or receptor. These features are often represented as points in three-dimensional space, defining the spatial arrangement and orientation required for optimal binding to the target.

Pharmacophore-based drug design involves the identification of pharmacophoric features in known active compounds or ligands and the subsequent use of these features to search chemical databases for potential new drug candidates with similar pharmacological properties. Pharmacophore modeling techniques, such as ligand-based or structure-based approaches, play a vital role in rational drug design by guiding the selection and optimization of lead compounds with improved potency and selectivity [49].

5.6.2 Generating Pharmacophore Models

Generating pharmacophore models is a crucial step in computer-aided drug design, where the aim is to identify and characterize the key structural and chemical features of biologically active molecules. Pharmacophore models provide valuable insights into the molecular interactions between ligands and their target receptors, guiding the design of novel compounds with optimized pharmacological properties. The process of generating pharmacophore models involves several steps [50–52].

1) **Data collection:** Biologically active compounds with known activities against a specific target are collected from experimental assays or databases.
2) **Structural alignment:** The three-dimensional structures of the collected ligands are aligned or superimposed to identify common structural motifs or pharmacophoric features.
3) **Feature mapping:** Pharmacophoric features such as hydrogen bond donors, acceptors, aromatic rings, and hydrophobic regions are identified within the aligned ligand structures. These features represent the essential chemical interactions required for binding to the target receptor.
4) **Pharmacophore generation:** Based on the identified pharmacophoric features, a pharmacophore model is generated. This model represents a spatial arrangement of these features in three-dimensional space, defining the essential chemical properties necessary for ligand binding and biological activity [53].
5) **Validation and optimization:** The generated pharmacophore model is validated using various techniques, such as cross-validation or decoy testing, to assess its predictive power and robustness. Optimization of the model may involve refining feature positions, adjusting feature types, or incorporating additional constraints based on experimental data.
6) **Virtual screening and lead identification:** Once validated, the pharmacophore model is used for VS of compound databases to identify potential lead compounds that match the pharmacophoric features of the model. Hits obtained from VS can then be further evaluated experimentally for their biological activity and drug-like properties.

By elucidating the structural requirements for ligand–receptor interactions, pharmacophore models play a crucial role in rational drug design, facilitating the discovery of novel therapeutic agents with improved potency and selectivity.

5.6.3 Applications in Lead Discovery

Pharmacophore modeling is a pivotal tool in lead discovery and drug design, facilitating the identification and optimization of lead compounds with desired pharmacological properties by elucidating the essential structural and chemical features necessary for ligand–receptor interactions. This computational technique plays a crucial role in rational drug design, guiding the selection and optimization of candidate compounds with improved potency, selectivity, and pharmacokinetic profiles. Pharmacophore modeling involves the systematic identification and characterization of key chemical features within a set of biologically active compounds known as ligands or inhibitors. These features represent the essential interactions between the ligands and their target receptors or proteins, such as hydrogen bond donors and acceptors, hydrophobic regions, aromatic rings, and positively or negatively charged groups. By generating pharmacophore models, researchers can gain valuable insights into the structural requirements for ligand binding and biological activity. These models serve as three-dimensional representations of the spatial arrangement and orientation of pharmacophoric features, guiding the design and optimization of novel compounds with similar pharmacological properties.

Pharmacophore modeling is particularly useful in lead discovery, where it aids in the identification of structurally diverse compounds that share common pharmacophoric features with known active compounds. These models are also valuable in the VS of compound databases, enabling the rapid identification of potential lead compounds that match the pharmacophoric features of the model.

Overall, pharmacophore modeling serves as a powerful tool in lead discovery and drug design, offering insights into ligand–receptor interactions and guiding the rational design of novel therapeutic agents with optimized pharmacological properties.

Pharmacophore modeling plays a crucial role in lead discovery and drug design [18]:

- **Virtual screening:** Pharmacophore models are used to screen compound databases to identify molecules with matching pharmacophoric features, potentially serving as lead compounds.
- **Lead optimization:** During lead optimization, pharmacophores guide structural modifications to enhance a compound's affinity and selectivity for the target.
- **Poly-pharmacology:** Pharmacophores can aid in designing compounds that interact with multiple targets, enabling the development of multitarget drugs.
- **Fragment-based drug design:** Pharmacophore analysis helps in fragment-based drug design by identifying critical fragments to build upon.
- **Exploring novel targets:** Pharmacophores can be used to investigate the potential of existing drugs against new targets by matching pharmacophoric features.

5.7 Quantitative Structure–Activity Relationship (QSAR)

5.7.1 Basics of QSAR

QSAR is a computational modeling technique used in drug discovery, environmental chemistry, and toxicology to establish quantitative relationships between the chemical structure of compounds and their biological activities or properties. QSAR models predict the biological activity or property of a compound based on its structural features, enabling the rational design and optimization of new compounds with desired activities.

The basics of QSAR involve several key concepts and steps [54–56]:

Data collection: QSAR analysis begins with the collection of a dataset comprising chemical structures of compounds (ligands) and their corresponding biological activities or properties, typically obtained from experimental assays or literature.

Descriptor calculation: Molecular descriptors, which quantitatively represent the structural and physicochemical properties of compounds, are calculated for each compound in the dataset. Descriptors can include molecular size, shape, electronegativity, hydrophobicity, and various other molecular properties.

Data preprocessing: The dataset may undergo preprocessing steps such as removing duplicates, handling missing values, and scaling descriptors to ensure uniformity and improve model performance.

Model development: QSAR models are built using statistical or machine learning (ML) techniques to correlate the calculated molecular descriptors with the observed biological activities or properties. Common modeling methods include multiple linear regression, partial least squares regression, support vector machines (SVMs), random forests, and neural networks [57].

Model validation: QSAR models are validated using techniques such as cross-validation, external validation with an independent test set, or bootstrapping to assess their predictive performance and generalization ability.

Model interpretation: QSAR models provide insights into the relationship between chemical structure and activity or property by identifying the most important descriptors contributing to the model's predictions. Interpretation of QSAR models helps in understanding the underlying mechanisms governing the observed biological activities or properties.

Model application: Once validated, QSAR models can be used to predict the activities or properties of new compounds not present in the training dataset. These predictions guide the selection and optimization of lead compounds with desired activities or properties in drug discovery and other applications.

QSAR models are valuable tools in computational chemistry and drug design, enabling the efficient screening and optimization of large compound libraries and reducing the need for costly and time-consuming experimental assays.

5.7.2 Model Development and Validation

In QSAR model development, the initial step involves selecting relevant molecular descriptors, chosen based on chemical intuition and prior knowledge of their expected influence on biological activity. Following descriptor selection, various mathematical techniques, including linear regression, multiple linear regression, or ML algorithms, are employed to construct QSAR models that relate these descriptors to biological activity. Model validation is crucial to ensure prediction reliability, achieved through partitioning the dataset into training and testing sets [58]. The model is trained on the training set and then evaluated on the testing set to assess its predictive accuracy. Cross-validation techniques, such as *k*-fold cross-validation, further enhance model robustness by partitioning the dataset into multiple subsets for comprehensive performance evaluation. QSAR models also provide insights into structure–activity relationships by identifying which molecular features contribute most significantly to observed activity, thereby guiding further compound design endeavors.

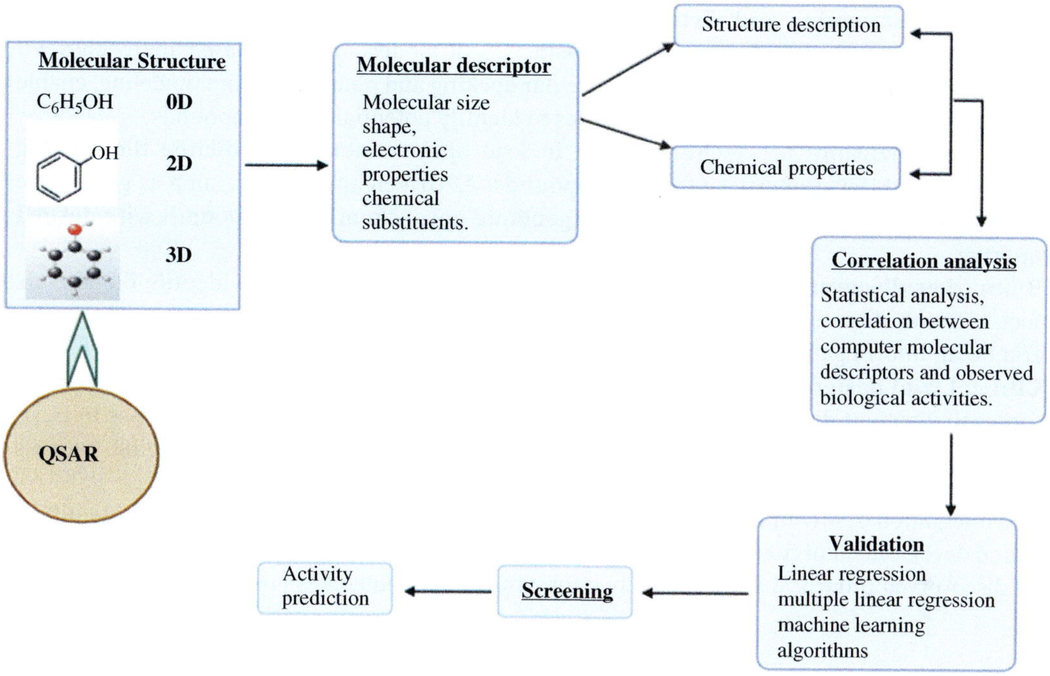

Figure 5.4 Workflow for the QSAR in the drug discovery process.

5.7.3 QSAR in Virtual Screening

QSAR models are versatile tools utilized across various stages of drug discovery and development. Integrated seamlessly into VS workflows, they aid in prioritizing compounds from vast chemical libraries based on predicted activity against specific targets. During lead optimization, QSAR guides chemical modifications to enhance lead compound activity and selectivity. Furthermore, QSAR models predict absorption, distribution, metabolism, excretion, and toxicity (ADMET) properties, facilitating the selection of drug candidates with favorable pharmacokinetic profiles. Additionally, these models assess compounds' potential to interact with multiple targets, facilitating the design of multitarget drugs with improved therapeutic efficacy and reduced side effects [57] (Figure 5.4).

5.8 Machine Learning and AI in Virtual Screening

5.8.1 Introduction to Machine Learning and AI

Introduction to ML and artificial intelligence (AI) in drug discovery involves the application of computational techniques to analyze biological and chemical data, accelerate the drug development process, and discover novel therapeutics. ML and AI methods play a crucial role in various stages of drug discovery, from target identification and validation to lead optimization and clinical trial design. In the context of drug discovery, ML and AI techniques are used to:

Target identification and validation: ML algorithms analyze biological data, such as genomics, proteomics, and transcriptomics data, to identify potential drug targets involved in diseases. AI methods help prioritize and validate these targets based on their relevance and druggability [57].

Compound screening and virtual screening: ML models predict the biological activity of chemical compounds against target proteins using QSAR models or ligand-based approaches. VS methods, powered by AI algorithms like molecular docking and pharmacophore modeling, enable the rapid screening of large compound libraries to identify potential lead compounds.

Lead optimization: ML techniques aid in lead optimization by predicting the binding affinity and ADMET properties of lead compounds. AI-driven approaches, such as generative models and *de novo* design algorithms, help generate novel chemical compounds with desired properties.

Biomarker discovery: ML and AI methods analyze multiomics data to identify biomarkers associated with disease progression, treatment response, or adverse drug reactions. These biomarkers can inform patient stratification and personalized medicine approaches.

Clinical trial design and patient recruitment: ML algorithms optimize clinical trial design by analyzing patient data to identify optimal trial parameters, predict patient responses to treatment, and stratify patient populations. AI-powered recruitment platforms match eligible patients with clinical trials based on their clinical and molecular profiles.

The integration of ML and AI in drug discovery holds the promise of accelerating the identification and development of novel therapeutics, reducing the time and cost associated with traditional drug discovery approaches, and ultimately improving patient outcomes [16].

5.8.2 Feature Selection and Model Training

Feature selection: In the context of VS, feature selection involves choosing relevant molecular descriptors or characteristics that best represent chemical compounds. These descriptors can include structural, physicochemical, or biological properties. Feature selection is a crucial step as it influences the model's performance.

In ML and AI model development for VS, after selecting relevant features, models are trained using datasets containing information on compound structures and their corresponding biological activities. Commonly employed algorithms include random forest, a versatile ensemble learning method capable of handling diverse data types and providing feature importance rankings; SVMs, effective for classification tasks such as distinguishing active and inactive compounds; and deep learning, particularly utilizing neural network architectures like convolutional neural networks (CNNs) and recurrent neural networks, which have demonstrated effectiveness in analyzing molecular structures and properties. Model performance is then evaluated using various metrics such as accuracy, sensitivity, specificity, and the area under the receiver operating characteristic curve. To ensure robustness, cross-validation techniques are often employed [59].

5.8.3 Applications in Virtual Screening

- **Enhanced virtual screening:** ML and AI techniques can significantly improve the efficiency and accuracy of VS. They can prioritize compounds for experimental testing, reducing the time and resources required.
- **Multitarget screening:** ML models can be designed to predict compound activities against multiple targets simultaneously, facilitating the discovery of multitarget drugs.
- **ADMET prediction:** ML models are employed to predict ADMET properties of compounds, aiding in the selection of drug candidates with favorable pharmacokinetic profiles.

- **Drug repurposing:** ML and AI can identify existing drugs that may have potential for new therapeutic uses by analyzing their chemical structures and known activities.
- **Toxicity prediction:** ML models can predict the toxicity of compounds, helping to eliminate potentially harmful drug candidates early in the drug discovery process [60].

5.9 Hit-to-Lead Optimization

5.9.1 Prioritizing Hits from Virtual Screening

In the drug discovery process, after conducting VS to identify potential drug candidates from a pool of compounds, the next critical step is hit-to-lead optimization. This phase involves prioritizing and refining the initial "hits" obtained from VS to develop promising lead compounds. Here is an explanation with an example:

Example: Imagine a pharmaceutical company is searching for a new drug to treat a specific cancer type. After VS, the company has identified a set of chemical compounds that have shown potential for inhibiting a key protein involved in cancer growth [61].

5.9.2 SAR Analysis and Iterative Design

5.9.2.1 SAR Analysis (Structure–Activity Relationship)

SAR analysis is a fundamental approach in drug discovery, focusing on elucidating the connection between the chemical structure of compounds and their biological activity. Researchers meticulously investigate how minor alterations to the chemical structure influence crucial pharmacological properties such as potency and selectivity. By discerning these relationships, SAR analysis aids in pinpointing structural features pivotal for eliciting the desired biological effects, thereby informing rational drug design and optimization strategies [62].

Example: In the cancer drug discovery scenario, SAR analysis involves systematically modifying the chemical structure of the initially identified compounds. Researchers change functional groups, alter substituents, or adjust the compound's scaffold while assessing the impact on its ability to inhibit the cancer-associated protein. Through SAR analysis, they may find that specific chemical groups enhance the compound's potency, leading to the identification of more promising lead candidates [7].

5.9.2.2 Iterative Design

Hit-to-lead optimization typically follows an iterative process guided by SAR insights, where compounds undergo successive modifications. These modifications are based on SAR findings aimed at enhancing the compounds' biological activity. The modified compounds are subsequently subjected to biological assays and testing to evaluate their improved potency and selectivity. This cycle of design, synthesis, testing, and analysis iterates until lead compounds exhibiting the desired efficacy, safety, and pharmacokinetic properties are identified. This iterative approach ensures the refinement and optimization of lead candidates for further development as potential drug candidates.

Example: In our cancer drug discovery case, researchers synthesize and test multiple versions of the initially identified compounds with structural changes guided by SAR insights. They discover that a specific modification significantly enhances the compound's ability to inhibit the cancer protein while maintaining low toxicity. This compound becomes a lead candidate and moves forward in the drug development process [63].

5.9.3 ADME/Tox Considerations

5.9.3.1 ADME (Absorption, Distribution, Metabolism, Excretion)

ADME stands for absorption, distribution, metabolism, and excretion, and it represents a set of pharmacokinetic properties that collectively determine the fate of a drug within the body. Understanding ADME is crucial in drug discovery and development as it influences the efficacy, safety, and pharmacokinetic profile of a drug candidate.

Absorption: Absorption refers to the process by which a drug enters the bloodstream after administration. It involves the movement of the drug across biological barriers such as the gastrointestinal tract (oral administration), skin (topical administration), or mucous membranes (intranasal or sublingual administration). Factors affecting absorption include the drug's physicochemical properties (e.g., solubility and lipophilicity), formulation, and the presence of transporters or efflux pumps.

Distribution: Distribution involves the transportation of a drug from the bloodstream to various tissues and organs in the body. Factors influencing drug distribution include blood flow, tissue permeability, protein binding, and the drug's lipophilicity. Distribution can impact the concentration of a drug at its target site and its efficacy.

Metabolism: Metabolism, also known as biotransformation, refers to the enzymatic conversion of a drug into metabolites, typically in the liver. Metabolism can lead to the activation or inactivation of a drug and plays a crucial role in determining its duration of action and potential for toxicity. The major enzyme systems involved in drug metabolism are the cytochrome P450 enzymes (CYPs) and phase II conjugation enzymes.

Excretion: Excretion involves the removal of a drug and its metabolites from the body, primarily through the kidneys (urine) and liver (bile). Other routes of excretion include feces, sweat, saliva, and exhalation. Renal excretion, facilitated by glomerular filtration and tubular secretion, is particularly important for hydrophilic and water-soluble drugs, while hepatic excretion is more relevant for lipophilic drugs.

In drug discovery and development, assessing ADME properties early in the process helps identify potential issues that may impact a drug candidate's pharmacokinetic profile, efficacy, or safety. Optimization of ADME properties is often pursued to enhance the bioavailability, distribution, metabolism, and elimination of a drug, thereby improving its overall therapeutic profile.

Understanding ADME is essential for predicting and optimizing the pharmacokinetic behavior of drug candidates, ultimately contributing to the selection of promising candidates for further development and clinical evaluation [64, 65].

5.9.3.2 Toxicity Considerations

Toxicity considerations are paramount in drug discovery, as they directly impact the safety and success of potential drug candidates. Understanding and mitigating toxicity risks is crucial to advancing compounds through preclinical and clinical development stages.

Predictive toxicology: Employing predictive toxicology approaches early in drug discovery helps identify potential toxicities before compounds enter clinical trials. Computational models, such as QSAR models and ML algorithms, analyze chemical structures to predict toxicological endpoints, such as mutagenicity, carcinogenicity, hepatotoxicity, cardiotoxicity, and nephrotoxicity. These predictive models enable the prioritization of safer compounds for further development.

***In vitro* toxicity screening:** *In vitro* assays assess the toxicological effects of drug candidates on cellular and molecular targets, providing valuable data on mechanisms of toxicity.

High-throughput screening platforms evaluate compound cytotoxicity, genotoxicity, and organ-specific toxicity using cell-based assays, organ-on-a-chip models, and biochemical assays. These assays help identify compounds with potential safety liabilities early in the drug discovery process.

***In vivo* toxicology studies:** Preclinical toxicology studies conducted in animal models evaluate the safety profile of drug candidates before initiating human clinical trials. Acute, subacute, and chronic toxicity studies assess systemic toxicity, organ toxicity, and potential adverse effects following repeated dosing. Pharmacokinetic studies determine compound exposure levels and distribution in tissues, guiding dose selection and formulation strategies.

Safety pharmacology: Safety pharmacology studies assess the potential adverse effects of drug candidates on vital physiological systems, including the cardiovascular, central nervous, and respiratory systems. These studies evaluate compound effects on cardiac electrophysiology, respiratory function, and central nervous system activity to identify safety concerns that may impact patient safety.

Regulatory requirements: Regulatory agencies, such as the Food and Drug Administration (FDA) and the European Medicines Agency (EMA), require comprehensive toxicity data as part of the Investigational New Drug (IND) application for clinical trials. Toxicity studies conducted in compliance with regulatory guidelines ensure the safety of human subjects participating in clinical trials and support drug approval processes.

By integrating toxicity considerations throughout the drug discovery process, researchers can identify and mitigate potential safety risks early, leading to the development of safer and more effective drug candidates for patient use [64, 66, 67] (Figure 5.5).

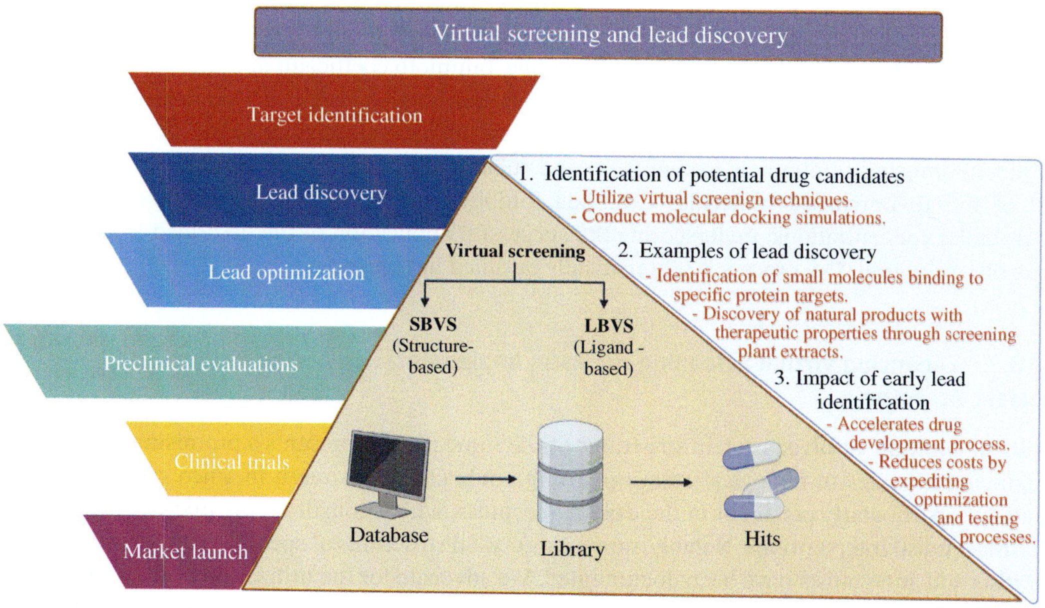

Figure 5.5 Refined workflow for employing virtual screening approaches in the drug discovery process.

5.10 Case Studies and Examples

5.10.1 Exploration Protocol for Mutant-targeted PI3K Inhibitors

The PIK3CA gene stands out as one of the frequently mutated oncogenes in various human cancers. It is responsible for encoding the p110α catalytic subunit of phosphatidylinositol 3-kinase, isoform α (PI3Kα). This particular protein plays a pivotal role in orchestrating signaling cascades that govern critical cellular processes such as proliferation, survival, and growth. Notably, a common mutation observed in PI3Kα involves a histidine-to-arginine substitution at position 1047 (H1047R) [68] within exon 20. Initially, the mutant H1047R PI3Kα [69] crystal structure (PDB ID: 3HIZ) was built using a combination of homology and loop modeling techniques. Subsequently, the atomistic model of the full-length H1047R mutant was generated utilizing Modeller 9v8. This resultant model was then solvated in water and subjected to molecular dynamics (MD) simulations lasting approximately 70 ns, employing the NAMD package. Following the simulation, the final snapshot was extracted and utilized for binding site identification calculations through the SiteMap module of Schrödinger v2.4. Notably, a non-ATP binding site proximal to the H1047R mutation emerged as one of the top-ranking sites and was thus selected for the present study. Subsequent to binding site identification, VS was executed using the docking program Glide 5.7 [49]. The drug-like subset of compounds from the HitFinder collection within the Maybridge database (accessible via www.maybridge.com) served as the basis for the VS exercise [69]. Employing the glide standard precision (SP) mode, all structures were docked and scored. Subsequently, the top 10,000 ranked structures from the SP filter underwent redocking and rescoring utilizing the glide extra precision (XP) mode. Complexes corresponding to the top-ranked 1000 compounds resulting from the XP processing were then subjected to further post-processing via the ChemBioServer. To eliminate poses distant from the energy minimum, the van der Waals filter of the server was utilized to exclude compounds with steric clashes. The remaining compounds underwent physicochemical property filtering according to the "Jorgensen rule of 3" [70], alongside toxicity filtering based on a database available in ChemBioServer, which cataloged known toxic moieties. Subsequently, hierarchical clustering was executed for the remaining compounds utilizing the Tanimoto coefficient and the ward clustering linkage, aiming to maximize chemical diversity while minimizing visual inspection efforts. Finally, the resulting cluster representatives underwent visual inspection, leading to the selection of the 10 most promising compounds for purchase and subsequent *in vitro* assay testing. Remarkably, 4 out of the 10 purchased molecules exhibited inhibition of PI3Kα protein activity *in vitro* at micromolar concentrations, underscoring the successful application of the described workflow in drug discovery [69]. The process is schematically outlined in Figure 5.6.

5.10.2 Enhancing Virtual Screening Hit Rate: Implementation on the RXRα Nuclear Receptor

Utilizing multiple resolved protein structures of the same protein presents a promising avenue for enhancing hit rates in biological assays of compounds chosen through in silico methods. This strategy is particularly pertinent in the case of receptors, which constitute the primary targets for pharmaceutical interventions. Notably, numerous crystal structures of specific receptors with both agonists and antagonists have been documented. We advocate for the utilization of these available structures in conjunction with the below-developed protocol to elevate the efficiency of identifying compounds capable of modulating the target receptor effectively. In this context, we utilize the nuclear receptor retinoid X receptor alpha (RXRα) [71, 72] as a model for a selective SBVS screening

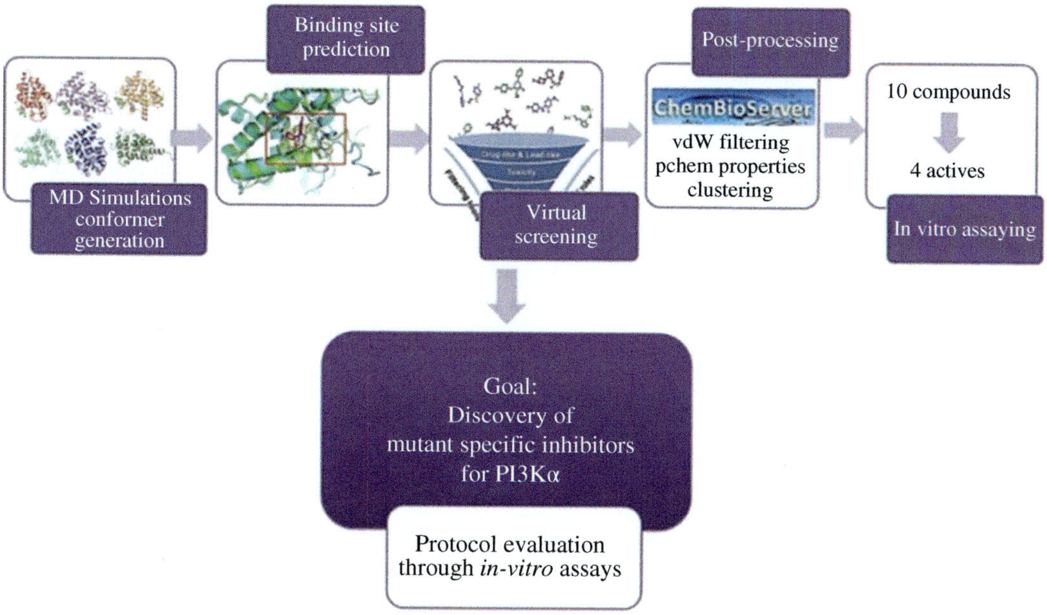

Figure 5.6 Workflow for the discovery of mutant-specific PI3K inhibitors based on an SBVS protocol involving conformer generation, binding site prediction, and compound post-processing.

protocol, employing ensemble docking to identify new selective agonists. RXR acts as a heterodimeric partner for approximately one-third of the 48 human nuclear receptor superfamily members, implying potential alterations in its ligand binding pocket upon interaction with different heterodimerization partners. This heterodimerization significantly enhances DNA binding and transcriptional activation, with reports indicating that increased expression of RXRα augments the transcriptional response to ligand binding. Selective RXR agonists have demonstrated the ability to enhance the antiproliferative and apoptotic responses of breast cancer cell lines to PPAR ligands. Recent discoveries have unveiled RXR-selective ligands that not only inhibit the proliferation of breast cancer cells resistant to all-trans RA *in vitro* but also induce regression of the disease in animal models. Hence, RXR emerges as a promising therapeutic target for cancer treatment. Considering the presence of multiple RXR receptor conformations due to binding with different heterodimerization partners, our objective is to enhance the specificity of identified binders for a given heterodimer partner by carefully selecting the RXR receptor structure for SBVS. Our method entails selecting RXR conformations for SBVS based on four criteria:

- Conducting pairwise comparison of receptor conformations using RMSD calculations.
- Analyzing and clustering RXR structures based on binding-site shape and volume using SiteMap (Figure 5.7).
- Docking a small database of known actives for a specific heterodimer partner to the resulting shape-diverse subset of binding sites from steps (a) and (b) using Glide 5.8 SP and XP [73].
- Retrieving representative protein conformations for the structure of interest from MD simulations using GROMACS [73].

Performing VS on three different subsets of RXR receptor conformations, arising from binding to different heterodimerization partners selected as mentioned above, may increase the success

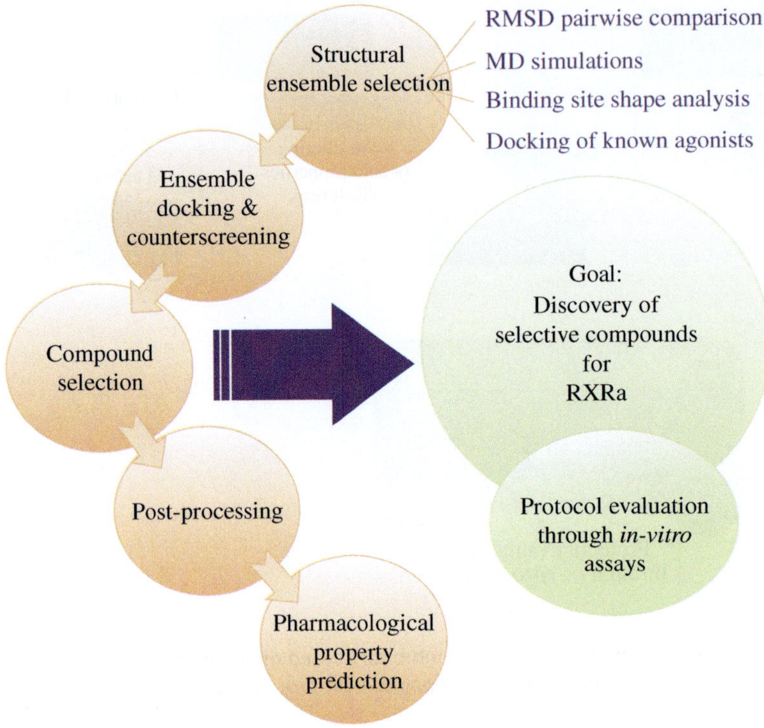

Figure 5.7 Workflow for the discovery of selective RXR ligands based on an SBVS protocol of ED and counter-screening.

rate of the process. Preprocessing for docking compound databases in this SBVS exercise involves assigning protonation and tautomeric states as described previously. Compounds for assaying are chosen based on the following criteria: Molecules that exhibit high scores when docked in the RXR protein ensemble binding to the heterodimer partner of interest, while simultaneously displaying low scores for RXR structures binding to heterodimer partners of no interest, are selected to achieve selectivity. Subsequently, a post-processing step is applied to the top-scoring compounds using ChemBioServer [74] and FAF-Drugs2 filtering tools [75] pharmacological property prediction using the QikProp software. The workflow of this protocol is depicted in Figure 5.7.

5.11 Challenges and Future Directions

5.11.1 Limitations of Virtual Screening

VS is a computational technique used in drug discovery to identify potential drug candidates by screening large chemical libraries, typically consisting of thousands to millions of compounds, against a target of interest, such as a protein receptor implicated in a disease pathway. While VS has become an integral part of the drug discovery process, it also comes with several limitations.

Scoring functions and accuracy: VS methods often rely on scoring functions to estimate the binding affinity between a ligand and a target protein. However, these scoring functions may not always accurately predict the binding affinity due to the complex and dynamic nature of protein–ligand interactions. A study by Gohlke et al. [16] highlights the challenges in accurately predicting binding affinities, particularly for flexible binding sites and conformational changes upon ligand binding.

Ligand flexibility: Conformational flexibility of ligands and proteins adds another layer of complexity to VS. Traditional rigid docking approaches may fail to account for the conformational changes required for binding, leading to false negatives or inaccurate predictions. Flexible docking methods aim to address this issue by considering the flexibility of both ligands and receptors.

Library size and diversity: While VS allows for the rapid screening of large chemical libraries, the quality and diversity of the compounds in these libraries greatly influence the success of the screening process. Limited structural diversity within the chemical library can result in a biased representation of chemical space and may overlook potential drug candidates. An analysis highlights the importance of chemical diversity in compound libraries for successful VS campaigns.

Computational resources and speed: VS methods often require significant computational resources, especially when using computationally intensive techniques such as molecular dynamics simulations or ensemble docking. Limited computational resources and time constraints may restrict the thoroughness of the screening process or necessitate the use of simplified models, potentially compromising accuracy. A study by Kitchen et al. [7] discusses the impact of computational resources on VS strategies.

Experimental validation and false positives: VS often generates a list of potential drug candidates that require experimental validation to confirm their biological activity and pharmacological properties. False positives, where compounds exhibit activity in silico but fail in experimental assays, can occur due to limitations in the scoring functions, incomplete representation of protein flexibility, or neglect of relevant physicochemical properties. A study discusses the importance of experimental validation in VS studies.

In conclusion, while VS is a valuable tool in drug discovery, researchers must be aware of its limitations and continuously strive to improve methods and algorithms to overcome these challenges. Addressing issues related to scoring accuracy, ligand flexibility, library diversity, computational resources, and experimental validation will enhance the effectiveness of VS in identifying promising drug candidates.

5.11.2 Emerging Technologies and Trends

Emerging technologies like AI and ML are increasingly crucial in VS. Deep learning models, such as CNNs, are gaining prominence for analyzing complex molecular structures. For example, Alpha-Fold, a deep learning-based tool, has demonstrated remarkable accuracy in predicting protein structures. Integrating these technologies can significantly enhance the precision and efficiency of VS.

5.11.3 Integration with High-throughput Experimentation

A promising trend is the integration of VS with high-throughput experimentation. In drug discovery, VS can identify potential drug candidates quickly. These candidates are then subjected to high-throughput experimental tests to validate their efficacy. For example, during the COVID-19 pandemic, VS identified existing drugs as potential treatments and high-throughput experiments were used to rapidly confirm their effectiveness. This integration accelerates the drug discovery process, especially in urgent situations like pandemics, where time is critical.

5.12 Ethical and Regulatory Considerations

5.12.1 Intellectual Property and Patents

Importance: Intellectual property (IP) and patents are critical in drug discovery. Companies invest heavily in research, and IP protection incentivizes innovation by granting exclusive rights to develop and market new drugs for a specified period.

Example: Consider a pharmaceutical company that discovers a novel compound through VS. To protect their investment, they file a patent for the compound. This patent gives them the exclusive right to develop and commercialize the drug for a certain number of years. IP protection encourages pharmaceutical companies to invest in expensive drug development processes.

5.12.2 Ethical Use of Computational Tools

Responsible research: Ethical considerations are vital in using computational tools. Researchers must ensure that their methods and data sources are ethical and accurate. Misuse of data or algorithms can lead to erroneous results or unintended consequences.

Example: When using VS to identify potential drug candidates, it is essential to ensure that the data sources are reliable and that the research is conducted with transparency and integrity. Misrepresenting results or manipulating data unethically can have serious consequences.

5.12.3 Regulatory Approval Process

Safety and efficacy: Regulatory approval is a crucial step before a drug can be marketed and prescribed to patients. Regulatory agencies like the FDA in the United States evaluate a drug's safety and efficacy based on clinical trial data.

Example: A pharmaceutical company that has successfully identified a lead compound through VS must conduct extensive preclinical and clinical trials to demonstrate its safety and effectiveness. The regulatory approval process ensures that drugs reaching the market meet rigorous standards and do not pose undue risks to patients.

5.13 Conclusion

VS and lead discovery have indeed revolutionized the drug discovery process by offering efficient and cost-effective methods for identifying potential drug candidates. The integration of computational techniques with experimental approaches has significantly accelerated the early stages of drug development, allowing researchers to sift through vast chemical libraries and prioritize compounds with higher chances of success. Moreover, advancements in technology, such as increased computational power, improved algorithms, and enhanced molecular modeling tools, have further bolstered the capabilities of VS methods.

5.13.1 Future Prospects in Virtual Screening and Lead Discovery

The integration of advanced ML and AI is anticipated to significantly enhance the accuracy and predictive capabilities of VS models, consequently leading to more successful lead discoveries. Furthermore, the integration of Big Data and Omics data, including genomics, proteomics, and metabolomics, into VS approaches promises to provide a deeper understanding of drug–target interactions. VS's potential role in personalized medicine, tailoring drug treatments to individual patients based on their genetic profiles, is expected to increase drug efficacy and minimize side effects. Additionally, VS will continue to be instrumental in drug repurposing efforts, identifying existing drugs for new therapeutic uses and potentially expediting the availability of treatments for various diseases. The advancement of high-throughput computational screening methods will enable faster and more efficient screening of larger compound libraries, thus broadening the scope of lead discovery endeavors.

5.13.2 Summary of Key Points

In summary, VS and lead discovery involve using computational techniques to identify potential drug candidates, optimizing them, and bringing them to market. Key points include:

- **Virtual screening:** The process of using computer simulations to identify compounds with potential therapeutic activity.
- **Lead discovery:** The process of refining initial hits from VS into promising lead compounds.
- **SAR analysis:** Studying the relationship between the chemical structure of compounds and their biological activity.
- **ADME/Tox considerations:** Evaluating how compounds are absorbed, distributed, metabolized, excreted, and assessing toxicity.
- **Ethical and regulatory considerations:** Ensuring ethical research practices and adhering to regulatory standards in drug development.
- **Future prospects:** Advancements in AI, big data integration, and personalized medicine are expected to shape the future of VS and lead discovery.

VS and lead discovery continue to drive innovation in drug development, offering hope for the discovery of novel therapies for a wide range of diseases.

References

1 Bajorath, J. (2002). Integration of virtual and high-throughput screening. *Nature Reviews. Drug Discovery* 1 (11): 882–894.

2 Ewing, T.J.A. and Kuntz, I.D. (1997). Critical evaluation of search algorithms for automated molecular docking and database screening. *Journal of Computational Chemistry* 18 (9): 1175–1189.

3 Schneider, G. and Fechner, U. (2005). Computer-based de novo design of drug-like molecules. *Nature Reviews. Drug Discovery* 4 (8): 649–663.

4 Cronk, D. and Shearer, J. (2012). High-throughput screening. In: *Drug Discovery and Development-E-Book: Technology in Transition* (ed. R.G. Hill and H.P. Rang), 95. Churchill Livingstone.

5 Lipinski, C.A., Lombardo, F., Dominy, B.W., and Feeney, P.J. (1997). Experimental and computational approaches to estimate solubility and permeability in drug discovery and development settings. *Advanced Drug Delivery Reviews* 23 (1–3): 3–25.

6 Shoichet, B.K. (2004). Virtual screening of chemical libraries. *Nature* 432 (7019): 862–865.

7 Kitchen, D.B., Decornez, H., Furr, J.R., and Bajorath, J. (2004). Docking and scoring in virtual screening for drug discovery: methods and applications. *Nature Reviews. Drug Discovery* 3 (11): 935–949.

8 Wang, R., Fang, X., Lu, Y. et al. (2005). The PDBbind database: methodologies and updates. *Journal of Medicinal Chemistry* 48 (12): 4111–4119.

9 Ferreira, L.G., Dos Santos, R.N., Oliva, G., and Andricopulo, A.D. (2015). Molecular docking and structure-based drug design strategies. *Molecules* 20 (7): 13384–13421.

10 Alberts, B., Johnson, A., Lewis, J. et al. (2003). *Molecular Biology of the Cell*, 4e. New York: Garland Science.

11 Berg, J.M., Tymoczko, J.L., and Stryer, L. (2002). Protein structure and function. *Biochemistry* 262: 159–173.

12 Branden, C.I. and Tooze, J. (2012). *Introduction to Protein Structure*. Garland Science.

13 Schneider, G. (2010). Virtual screening: an endless staircase? *Nature Reviews. Drug Discovery* 9 (4): 273–276.

14 Shoichet, B.K. and Kobilka, B.K. (2012). Structure-based drug screening for G-protein-coupled receptors. *Trends in Pharmacological Sciences* 33 (5): 268–272.

15 Teague, S.J. (2003). Implications of protein flexibility for drug discovery. *Nature Reviews. Drug Discovery* 2 (7): 527–541.

16 Gohlke, H. and Klebe, G. (2002). Approaches to the description and prediction of the binding affinity of small-molecule ligands to macromolecular receptors. *Angewandte Chemie, International Edition* 41 (15): 2644–2676.

17 Irwin, J.J. and Shoichet, B.K. (2005). ZINC – a free database of commercially available compounds for virtual screening. *Journal of Chemical Information and Modeling* 45 (1): 177–182.

18 Medina-Franco, J.L., Giulianotti, M.A., Welmaker, G.S., and Houghten, R.A. (2013). Shifting from the single to the multitarget paradigm in drug discovery. *Drug Discovery Today* 18 (9–10): 495–501.

19 Bohacek, R.S. and McMartin, C. (1994). Multiple highly diverse structures complementary to enzyme binding sites: results of extensive application of a de novo design method incorporating combinatorial growth. *Journal of the American Chemical Society* 116 (13): 5560–5571.

20 Kuntz, I.D., Blaney, J.M., Oatley, S.J. et al. (1982). A geometric approach to macromolecule–ligand interactions. *Journal of Molecular Biology* 161 (2): 269–288.

21 Ripphausen, P., Nisius, B., Peltason, L., and Bajorath, J. (2010). Quo vadis, virtual screening? A comprehensive survey of prospective applications. *Journal of Medicinal Chemistry* 53 (24): 8461–8467.

22 Friesner, R.A., Banks, J.L., Murphy, R.B. et al. (2004). Glide: a new approach for rapid, accurate docking and scoring. 1. Method and assessment of docking accuracy. *Journal of Medicinal Chemistry* 47 (7): 1739–1749.

23 Warren, G.L., Do, T.D., Kelley, B.P. et al. (2012). Essential considerations for using protein–ligand structures in drug discovery. *Drug Discovery Today* 17 (23–24): 1270–1281.

24 Cheng, T., Li, Q., Zhou, Z. et al. (2012). Structure-based virtual screening for drug discovery: a problem-centric review. *The AAPS Journal* 14: 133–141.

25 Wang, Y., Xiao, J., Suzek, T.O. et al. (2009). PubChem: a public information system for analyzing bioactivities of small molecules. *Nucleic Acids Research* 37 (suppl_2): W623–W633.

26 Pence, H.E. and Williams, A. (2010). *ChemSpider: An Online Chemical Information Resource*. ACS Publications.

27 Gaulton, A., Bellis, L.J., Bento, A.P. et al. (2012). ChEMBL: a large-scale bioactivity database for drug discovery. *Nucleic Acids Research* 40 (D1): D1100–D1107.

28 Law, V., Knox, C., Djoumbou, Y. et al. (2014). DrugBank 4.0: shedding new light on drug metabolism. *Nucleic Acids Research* 42 (D1): D1091–D1097.

29 Muller, K., Faeh, C., and Diederich, F. (2007). Fluorine in pharmaceuticals: looking beyond intuition. *Science (1979)* 317 (5846): 1881–1886.

30 Kim, S., Chen, J., Cheng, T. et al. (2021). PubChem in 2021: new data content and improved web interfaces. *Nucleic Acids Research* 49 (D1): D1388–D1395.

31 Hawkins, P.C.D., Skillman, A.G., and Nicholls, A. (2007). Comparison of shape-matching and docking as virtual screening tools. *Journal of Medicinal Chemistry* 50 (1): 74–82.

32 Schneider, N., Lowe, D.M., Sayle, R.A., and Landrum, G.A. (2015). Development of a novel fingerprint for chemical reactions and its application to large-scale reaction classification and similarity. *Journal of Chemical Information and Modeling* 55 (1): 39–53.

33 Kim, S., Thiessen, P.A., Bolton, E.E. et al. (2016). PubChem substance and compound databases. *Nucleic Acids Research* 44 (D1): D1202–D1213.

34 Morris, G.M., Huey, R., Lindstrom, W. et al. (2009). AutoDock4 and AutoDockTools4: automated docking with selective receptor flexibility. *Journal of Computational Chemistry* 30 (16): 2785–2791.

35 Trott, O. and Olson, A.J. (2010). AutoDock Vina: improving the speed and accuracy of docking with a new scoring function, efficient optimization, and multithreading. *Journal of Computational Chemistry* 31 (2): 455–461.

36 Bissantz, C., Folkers, G., and Rognan, D. (2000). Protein-based virtual screening of chemical databases. 1. Evaluation of different docking/scoring combinations. *Journal of Medicinal Chemistry* 43 (25): 4759–4767.

37 Guedes, I.A., de Magalhães, C.S., and Dardenne, L.E. (2014). Receptor–ligand molecular docking. *Biophysical Reviews* 6: 75–87.

38 Sousa, S.F., Fernandes, P.A., and Ramos, M.J. (2006). Protein–ligand docking: current status and future challenges. *Proteins: Structure, Function, and Bioinformatics* 65 (1): 15–26.

39 Jones, G., Willett, P., Glen, R.C. et al. (1997). Development and validation of a genetic algorithm for flexible docking. *Journal of Molecular Biology* 267 (3): 727–748.

40 Perola, E. and Charifson, P.S. (2004). Conformational analysis of drug-like molecules bound to proteins: an extensive study of ligand reorganization upon binding. *Journal of Medicinal Chemistry* 47 (10): 2499–2510.

41 Eldridge, M.D., Murray, C.W., Auton, T.R. et al. (1997). Empirical scoring functions: I. The development of a fast empirical scoring function to estimate the binding affinity of ligands in receptor complexes. *Journal of Computer-Aided Molecular Design* 11: 425–445.

42 Halgren, T.A. (1996). Merck molecular force field. I. Basis, form, scope, parameterization, and performance of MMFF94. *Journal of Computational Chemistry* 17 (5-6): 490–519.

43 Klebe, G. (2000). Recent developments in structure-based drug design. *Journal of Molecular Medicine* 78: 269–281.

44 Rarey, M., Kramer, B., Lengauer, T., and Klebe, G. (1996). A fast flexible docking method using an incremental construction algorithm. *Journal of Molecular Biology* 261 (3): 470–489.

45 Sotriffer, C. and Klebe, G. (2002). Identification and mapping of small-molecule binding sites in proteins: computational tools for structure-based drug design. *Il Farmaco* 57 (3): 243–251.

46 Meng, E.C., Shoichet, B.K., and Kuntz, I.D. (1992). Automated docking with grid-based energy evaluation. *Journal of Computational Chemistry* 13 (4): 505–524.

47 Muegge, I. and Martin, Y.C. (1999). A general and fast scoring function for protein–ligand interactions: a simplified potential approach. *Journal of Medicinal Chemistry* 42 (5): 791–804.

48 Gehlhaar, D.K., Verkhivker, G.M., Rejto, P.A. et al. (1995). Molecular recognition of the inhibitor AG-1343 by HIV-1 protease: conformationally flexible docking by evolutionary programming. *Chemistry & Biology* 2 (5): 317–324.

49 Dixon, S.L., Smondyrev, A.M., Knoll, E.H. et al. (2006). PHASE: a new engine for pharmacophore perception, 3D QSAR model development, and 3D database screening: 1. Methodology and preliminary results. *Journal of Computer-Aided Molecular Design* 20: 647–671.

50 Koes, D.R. and Camacho, C.J. (2011). Pharmer: efficient and exact pharmacophore search. *Journal of Chemical Information and Modeling* 51 (6): 1307–1314.

51 Wolber, G. and Langer, T. (2005). LigandScout: 3-D pharmacophores derived from protein-bound ligands and their use as virtual screening filters. *Journal of Chemical Information and Modeling* 45 (1): 160–169.

52 Sliwoski, G., Kothiwale, S., Meiler, J., and Lowe, E.W. (2014). Computational methods in drug discovery. *Pharmacological Reviews* 66 (1): 334–395.

53 Elokely, K.M. and Doerksen, R.J. (2013). Docking challenge: protein sampling and molecular docking performance. *Journal of Chemical Information and Modeling* 53 (8): 1934–1945.

54 Tropsha, A. (2010). Best practices for QSAR model development, validation, and exploitation. *Molecular Informatics* 29 (6–7): 476–488.

55 Golbraikh, A. and Tropsha, A. (2002). Beware of q2! *Journal of Molecular Graphics & Modelling* 20 (4): 269–276.

56 Cramer, R.D., Patterson, D.E., and Bunce, J.D. (1988). Comparative molecular field analysis (CoMFA). 1. Effect of shape on binding of steroids to carrier proteins. *Journal of the American Chemical Society* 110 (18): 5959–5967.

57 Cherkasov, A., Muratov, E.N., Fourches, D. et al. (2014). QSAR modeling: where have you been? Where are you going to? *Journal of Medicinal Chemistry* 57 (12): 4977–5010.

58 Hansch, C., Leo, A., and Hoekman, D. (1995). *Exploring QSAR: Hydrophobic, Electronic, and Steric Constants*, vol. 2. Washington, DC: American Chemical Society.

59 LeCun, Y., Bengio, Y., and Hinton, G. (2015). Deep learning. *Nature* 521 (7553): 436–444.

60 Rupp, M., Tkatchenko, A., Müller, K.R., and Von Lilienfeld, O.A. (2012). Fast and accurate modeling of molecular atomization energies with machine learning. *Physical Review Letters* 108 (5): 058301.

61 Congreve, M., Carr, R., Murray, C., and Jhoti, H. (2003). A 'rule of three' for fragment-based lead discovery? *Drug Discovery Today* 8 (19): 876–877.

62 Lavecchia, A. and Di Giovanni, C. (2013). Virtual screening strategies in drug discovery: a critical review. *Current Medicinal Chemistry* 20 (23): 2839–2860.

63 Di, L. and Kerns, E.H. (2003). Profiling drug-like properties in discovery research. *Current Opinion in Chemical Biology* 7 (3): 402–408.

64 Hughes, J.D., Blagg, J., Price, D.A. et al. (2008). Physiochemical drug properties associated with in vivo toxicological outcomes. *Bioorganic & Medicinal Chemistry Letters* 18 (17): 4872–4875.

65 Ekins, S., Mestres, J., and Testa, B. (2007). In silico pharmacology for drug discovery: methods for virtual ligand screening and profiling. *British Journal of Pharmacology* 152 (1): 9–20.

66 Kola, I. and Landis, J. (2004). Can the pharmaceutical industry reduce attrition rates? *Nature Reviews. Drug Discovery* 3 (8): 711–716.

67 Cook, D., Brown, D., Alexander, R. et al. (2014). Lessons learned from the fate of AstraZeneca's drug pipeline: a five-dimensional framework. *Nature Reviews. Drug Discovery* 13 (6): 419–431.

68 Mandelker, D., Gabelli, S.B., Schmidt-Kittler, O. et al. (2009). A frequent kinase domain mutation that changes the interaction between PI3Kα and the membrane. *National Academy of Sciences of the United States of America* 106 (40): 16996–17001.

69 Gkeka, P., Eleftheratos, S., Kolocouris, A., and Cournia, Z. (2013). Free energy calculations reveal the origin of binding preference for aminoadamantane blockers of influenza A/M2TM pore. *Journal of Chemical Theory and Computation* 9 (2): 1272–1281.

70 Spitzer, R. and Jain, A.N. (2012). Surflex-Dock: docking benchmarks and real-world application. *Journal of Computer-Aided Molecular Design* 26: 687–699.

71 Crowe, D.L. and Chandraratna, R.A.S. (2004). A retinoid X receptor (RXR)-selective retinoid reveals that RXR-α is potentially a therapeutic target in breast cancer cell lines, and that it potentiates antiproliferative and apoptotic responses to peroxisome proliferator-activated receptor ligands. *Breast Cancer Research* 6 (5): 1–10.

72 Tanaka, T. and De Luca, L.M. (2009). Therapeutic potential of "rexinoids" in cancer prevention and treatment. *Cancer Research* 69 (12): 4945–4947.

73 Van Der Spoel, D., Lindahl, E., Hess, B. et al. (2005). GROMACS: fast, flexible, and free. *Journal of Computational Chemistry* 26 (16): 1701–1718.

74 Athanasiadis, E., Cournia, Z., and Spyrou, G. (2012). ChemBioServer: a web-based pipeline for filtering, clustering and visualization of chemical compounds used in drug discovery. *Bioinformatics* 28 (22): 3002–3003.

75 Lagorce, D., Sperandio, O., Galons, H. et al. (2008). FAF-Drugs2: free ADME/tox filtering tool to assist drug discovery and chemical biology projects. *BMC Bioinformatics* 9 (1): 1–9.

6

ADMET and Physicochemical Assessments in Drug Design

Ulviye Acar Çevik[1], Ayşen Işik[2], and Abdüllatif Karakaya[3]

[1] Department of Pharmaceutical Chemistry, Faculty of Pharmacy, Anadolu University, Eskişehir, Turkey
[2] Department of Biochemistry, Faculty of Science, Selçuk University, Konya, Turkey
[3] Department of Pharmaceutical Chemistry, Faculty of Pharmacy, Zonguldak Bulent Ecevit University, Zonguldak, Turkey

6.1 ADMET

6.1.1 Absorption

Obtaining a candidate that exerts a therapeutic effect at the site of action and is eliminated within a specified time period is the primary objective of drug discovery [1]. The majority of drugs are formulated for oral administration due to their convenience for patients to self-administer. Oral administration of a drug necessitates traversing numerous cellular membranes in order to access the site of action. The mechanism that controls how drugs cross biological membranes is essentially the same. For instance, drug permeability across the blood–brain barrier (BBB), the gastrointestinal tract barrier, and the human colon carcinoma cell line Caco-2 monolayers all utilize this principle [2]. Two prevalent categories of mechanisms underlie the transmembrane transport of drugs. The absorption of small-molecule drugs (with a molecular weight of around 150–1500 Da) is most commonly achieved through passive transcellular diffusion, which refers to the process of diffusion through the lipid bilayers. The process is propelled by concentration gradients, starting in compartments with high concentrations and moving down to compartments with low concentrations. Drugs with a high affinity for lipids tend to accumulate in the membrane, resulting in a greater difference in concentration between the membrane and the cytosolic fluid. The process of diffusion from the membrane to the cytosolic solution happens quickly because of the large gradient that exists between the two. When the steady state is established, the concentration of drugs that are not attached to the membrane should be the same on both sides of the membrane [3, 4]. The physiology of the organ, including surface area, transit time, and pH gradient, as well as the molecular characteristics of a drug, such as passive permeability, pK_a, and other physicochemical parameters, all have an impact on how well it absorbs. In the intestine, compounds with a molecular weight of less than 200 can be taken in through the paracellular pathway, which is made up of the spaces between cells [5]. No energy is required for the trans-membrane transportation of nonpolar and uncharged drugs, which is why passive transport is the most common route for these types of drugs. Predicting medication absorption and permeability primarily focuses on this channel since passive diffusion is more significant for both the intestinal epithelium and the BBB [6, 7]. "Active

Computational Methods for Rational Drug Design, First Edition. Edited by Mithun Rudrapal.

transport" is the second mechanism by which drugs are absorbed. Active transport is a key process in drug efflux and the absorption of hydrophilic substances into cells. It is energy dependent (ATP dependent) and can operate against a concentration gradient. For highly polar or charged drugs, carrier proteins are essential for facilitating active transport. At its core, active transport is defined by saturation, movement against an electrochemical gradient, and very high selectivity [8].

Permeability and solubility can be used to describe drug absorption, and the biopharmaceutics classification system (BCS) has identified them as key parameters [9, 10]. Drug classification is based on three dimensionless parameters: the absorption number, the dissolution number, and the dose number. It also uses the basic drug properties of solubility and permeability. Highly soluble drugs are those with a dose-to-solubility ratio of less than 250 mL in media with a pH range of 1.0–6.8 at 37°C. On the other hand, drug candidates that demonstrate an extent of absorption of at least 85% in humans are classified as "highly permeable" [11]. Four distinct BCS classes have been defined: BCS class 1 substances have high solubility and permeability, and gastric emptying or drug dissolution limits drug absorption. BCS class 2 drugs have low solubility and high permeability, limiting absorption. Permeability limits drug absorption in BCS class 3 compounds with high solubility and low permeability. Low solubility and permeability compounds in BCS class 4 make oral administration of drugs difficult. The BCS is useful for guiding the development of formulations once a drug has been categorized. As an illustration, if the solubility of a component is troublesome, it may be necessary to make further efforts in the development of formulations consisting of, for instance, enabling formulations (Figure 6.1).

Based on research into the World Drug Index, Lipinski came up with the concept of inadequate absorption and penetration. This concept is known as the "rule of five." This criterion specifies that if a molecule surpasses two or more of the following restrictions, the medication will have poor oral absorption or permeability. Molecular weight must be more than 500, the computed octanol–water partition coefficient (log P) must be more than 5, the number of hydrogen bond (HB) donors (OH and NH groups) must be more than 5, and the number of HB acceptors (N and O atoms) must be more than 10 [12].

Last but not least, the characteristics of the drug, elements related to the formulation, and physiological conditions all have an impact on the very complicated process of drug absorption following oral administration. Before a drug reaches other body organs, oral bioavailability or drug exposure in the systemic circulation depends on its capacity to be absorbed across the membranes of the gastrointestinal tract and to be metabolized via the intestines and the liver. Immensely important, then, is the development of drugs capable of penetrating the intestinal epithelium throughout the course of drug discovery. As a result, given that the topic of drug absorption will be covered in more depth, it is essential to underline the fact that solubility and membrane permeability are genuinely critical factors in the process of absorption.

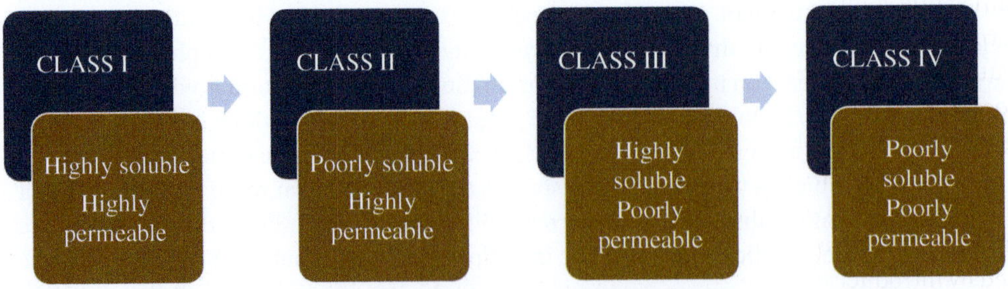

Figure 6.1 Biopharmaceutical classification system (BCS) of drugs.

6.1.1.1 Solubility and Dissolution

The term "solubility" describes a compound's capacity to dissolve in a certain solvent, such as water, a buffer with a particular pH, a fasted-state intestinal fluid simulation, or a fed-state intestinal fluid simulation [13]. The assessment of a drug's oral absorption has been shown to depend significantly on solubility despite the fact that this parameter is not always regarded as an ADMET feature. Since solubility determines bioavailability and drugs must be in solution to pass through the GI tract's membrane, it is clear that this property is an important determinant of absorption rates [14]. Aqueous solubility, which controls medication absorption, transfer, and excretion from the body, is an essential characteristic that is essentially engaged in all phases of drug development [15–17]. The effectiveness of drugs is mostly determined by their ability to dissolve in water; hence, drugs that have a low dissolution rate or a low solubility will be destroyed before they enter the blood circulation, and as a result, they will not provide the necessary pharmacological effects [18, 19]. It is possible to evaluate a drug's solubility in water by determining its capacity to partition from lipid environments to aqueous settings. This ability is dependent on the ionization of the drug that is being experimented with. The majority of drugs consist of weakly basic or acidic compounds that are incapable of complete ionization in aqueous solutions and thus ionize only partially. The solvent's pH is typically considered to determine the above partition behavior, and pK_a is a common parameter used to characterize a compound's dissolution process. This is because drug ionization is very sensitive to the pH of the solvent. As a rule, ionized medicines have significantly higher water solubility than their nonionized equivalents [14]. Recently, there has been an increase in the number of poorly soluble drugs, and concerns such as poor absorbability, food effects, and the absence of pharmacokinetic linearity have also emerged [20]. Numerous basic and acidic drugs frequently exhibit solubility profiles that depend on pH levels. Consequently, their solubilities in the stomach and small intestine are distinct. When the pH of the stomach is between 1 and 2, for instance, moderately basic substances may have extremely good solubility. However, when the pH of the intestine is reached, the solubility may be significantly lower than it is in the stomach [21]. Furthermore, the consumption of food can influence solubility through its effects on stomach acidity and the elevation of bile acid levels. Given the wide range of physicochemical and physiological parameters that influence drug solubility, several strategies can be employed during the early stages of drug design to enhance the oral drug's solubility. One of these processes entails ionization into various salts: basic drugs are ionized into salts using hydrochloride, citrate, tartrate, or other acid complexes, whereas acidic drugs are ionized into sodium, calcium, and potassium salts. Better solubility usually follows salt production. Sodium salts are the most prevalent form of basic drug salts, with 43% of the top 200 best-selling medications being in this form, and HCl salts are the most common form of acidic drug salts, with 11% of the top 200 best-selling drugs in this form [22].

The fact that 75% of candidates for drug development are categorized as BCS classes 2 or 4 and are not sufficiently soluble [23] provides further evidence that drug solubility is a limiting factor in drug absorption, as stated previously. In order to minimize the risk of progressing drug candidates throughout the development stage, it is critical to evaluate their solubility and any potential problems as soon as feasible. Limited solubility may impair absorption and, thus, drug likelihood.

During the early stages of the discovery process, kinetic and thermodynamic solubility are two methods that can be utilized to quantify solubility [14]. **The kinetic solubility** approach is the preferred method since it uses a reasonably high-throughput technology and is measured when a tiny volume of DMSO is added to the aqueous buffer [24]. **Thermodynamic solubility** is determined by introducing an aqueous buffer directly to the solid crystalline material and allowing sufficient time for equilibrium to be established between the dissolved and solid components.

In developing new formulations, this is essential since it represents the compound's actual solubility [25]. Kinetic solubility measures can be used to check for insolubility that happens in standard *in vitro* potency and ADME tests. On the other hand, measurements of kinetic solubility are not very relevant to the problems that arise *in vivo*. It is possible to achieve thermodynamic solubility by adding the aqueous buffer directly to the solid crystalline material and then waiting for a considerable amount of time for the dissolved and solid materials to reach a state of equilibrium. The measurement process is time consuming and requires more material, even though thermodynamic solubility is more important.

Poorly soluble drugs may have their absorption limited not only by their solubility but also by their dissolution rate [26, 27]. A molecule's solubility can be evaluated by measuring its dissolution rate. A chemical with a rapid rate of dissolution will enter the solution quickly, leading to an absorption phase that happens swiftly. This increases the likelihood that the molecule will be absorbed within the GI transit time, and its solubility will remain constant. The dissolving rate of the substance depends on its physical and salt form as well as its particle size. Decreasing the size of the particles enhances the amount of solid material that comes into contact with the solvent, hence accelerating the rate at which it dissolves [3]. One other approach to modifying the solubility characteristics is to employ the principle of physical form [14]. In drug discovery, amorphous solids are used most often. These are solids whose molecules are not organized in a specific way, unlike crystals, which have molecules that are very well ordered. In comparison to their amorphous counterparts, the compound's crystalline forms are less soluble. Various crystalline structures, such as polymorphs, salts, and solvates, can display varying levels of solubility. Water solubility can be enhanced via the cocrystallization of drugs with solvents like ethanol or water, a process known as hydrates [22].

To provide a brief summary, low oral bioavailability and a poor oral absorption rate are often the outcomes of a low dissolution rate and poor water solubility. There is a correlation between the lipophilicity of the substance and its low solubility in water. Accordingly, knowing a drug's lipophilicity is essential for figuring out its permeability and oral absorption.

6.1.1.2 Lipophilicity

It has been reported that lipophilicity has a significant influence on a variety of pharmacokinetic properties, including the absorption, distribution, and permeability of a pharmaceutical, as well as the routes of drug clearance [28–30]. As a result, lipophilicity plays an important role in the evaluation of the therapeutic suitability of a pharmaceutical. Between the point of application and the location of the target protein, an active agent must pass through a number of lipid membrane barriers and aqueous compartments in order to reach its destination. A sufficient amount of lipophilicity is required in order to accomplish the desired distribution. This makes perfect sense given that lipid membranes divide the many watery phases that make up biological systems. It is consequently necessary for lipophilicity to be a factor in the transportation and distribution of tiny molecules in such systems. These requirements mostly stem from the fact that biological targets are lipid-based. For instance, neurotransmitter route, anatomical, and intracellular targets require lipophilic agonists [31, 32].

When it comes to absorption, lipophilic chemicals and molecules with a molecular weight of more than 500–600 Da are often well absorbed; yet, they are subject to biliary clearance at a relatively rapid rate. This typically occurs immediately following absorption from the intestines during the initial hepatic passage [33]. Only compounds that have moderate lipophilicity have a fair probability of "migrating" into the aqueous phase as well as the lipid phase in order to arrive at enough concentrations in the tissue that can be targeted. The reason is that highly lipophilic compounds are both more rapidly digested and potentially harmful from a toxicological standpoint. On the other side, drugs that have a very low lipophilicity are likewise undesirable since they will not have the impact that is required. When ions or membrane-bound receptor-interacting substances are enriched in the surrounding membrane, they can frequently reach their targets more readily. In order to achieve this, the compounds must possess lipophilic properties or possess a substantial lipophilic group that can be used to secure them within the membrane. At the same time, appropriate drug formulations need to demonstrate a high degree of water solubility and an acceptable degree of lipophilicity to evaluate the most effective oral absorption in addition to the necessary deposition and activity. Lipophilicity and solubility have an inverse relationship, meaning that increased lipophilicity results in decreased water solubility. This is characterized by the partition coefficient that exists between the aqueous phase and the lipid phase. A measurement is taken of the distribution of octanol and water in the model that is the most basic.

The partition coefficient (P):

$$P_{\mathrm{o/w}} = C_{n\text{-octanol}} \,/\, C_{\mathrm{water}}$$

The partition coefficients are calculated by taking the logarithm (log P) of the ratio of equilibrium concentrations of the aqueous water to the organic solvent [34]. Log P is included in Lipinski's "rule of 5" because, at high log P values (>5), solubility is frequently poor, impeding absorption, and/or molecules partition into membranes, rendering them impervious to enterocytes [21]. This is the reason why log P is included in the rule. To provide a more comprehensive understanding of the connection between the log P value (lipophilicity) and the solubility/permeability [14]:

A drug's log P value:

i) Between 0 and 3 is considered optimal for passive diffusion.
ii) <1 indicates that a compound's solubility will be high due to its hydrophilic nature, but its permeability will be low.
iii) A value greater than 3 suggests that a molecule has a high affinity for lipids, may have limited solubility, and is likely to undergo metabolism and/or be excreted through the bile.

The distribution coefficient, log D, represents the proportion of the compound's total concentrations (including both ionized and unionized forms) in two solvents (namely, the aqueous phase at a given pH and n-octanol). It is important to note that the log P will change depending on the conditions under which it is measured as well as the partitioning solvent that is selected. On the other hand, log D represents the log distribution coefficient at a specific pH degree. The equilibrium between ionized and nonionized species in the two solvents is considered in the analysis [35].

$$\text{Distribution coefficient, } D = \frac{[\text{Unionized}]_{(\mathrm{o})}}{[\text{Unionized}]_{(\mathrm{aq})}} + [\text{Ionized}]_{(\mathrm{aq})}, \quad \text{Log}\, D = \log_{10}(\text{Distribution coefficient})$$

When taking into consideration the connection between lipophilicity and ADME, log D is utilized more frequently than log P for ionizable compounds [13]. It is common practice to use log D at a pH of 7.4 as a means of providing an indicator of the lipophilicity of a drug at the pH of blood plasma.

Calculating log P is a straightforward and rather reliable process. However, determining log D pH is more complex as it necessitates knowledge of the pK_a. Drug solubility, lipophilicity, permeability, and absorption are among the physicochemical and ADME characteristics that are significantly impacted by the ionization state of compounds, as indicated by the pK_a value [36, 37]. The ionization constant (pK_a) is traditionally represented by the Henderson–Hasselbalch equation, which quantifies the proportion of ionized and unionized chemicals at a specific pH.

$$\mathrm{pH} = pK_a + \log \frac{[\text{Ionized compound}]}{[\text{Unionized compound}]} \Rightarrow \text{for weak acid}$$

$$\mathrm{pH} = pK_a + \log \frac{[\text{Unionized compound}]}{[\text{Ionized compound}]} \Rightarrow \text{for weak base}$$

6.1.1.3 Permeability

Permeability is a significant physicochemical characteristic of compounds that has a direct impact on the pace at which drugs are absorbed. One of the primary factors that determine the ADME/PK of a drug and its metabolites, as well as their exposures to various tissues, is its permeability. Permeability is the factor that influences the molecule's capacity to pass through the biological membrane. The drug's interdependent physicochemical qualities, including its molecular size, polarity, and lipophilicity, are a significant factor in determining its effectiveness of permeability. As previously stated, lipophilicity is commonly assessed using the partition coefficient. In general, a molecule with a high partition coefficient has a greater ability to move between an aqueous fluid and a biological membrane, which results in higher permeability. Conversely, a high partition coefficient value leads to the drug being trapped in the membrane, preventing it from further distributing across the membrane and blood. Concerning molecular size, molecules with high molecular weight exhibit low permeability. The permeability of a molecule depends on the equilibrium of the intermolecular forces between the drug and both water and the membrane. There are a number of research that have discussed the hydrogen bonding parameter as a means of describing drug permeability. The majority of the findings have demonstrated that HB acceptor descriptors are connected with less value when it comes to predicting the permeability of the human intestinal epithelium [38, 39]. Furthermore, there exists an inverse link between the polarity of a molecule and its permeability. Polar molecules that have several HB acceptors and donors interact better with water [40].

Multiple investigations have demonstrated that drug absorption varies across different regions. The pH gradient in the digestive system, ranging from acidic (pH 2–3) to basic (pH 8–9), has been identified as a significant determining factor. Most of the process of absorbing nutrients from the intestines takes place in the small intestine, specifically in the duodenum, jejunum, and ileum. The pH in this part of the intestine typically ranges from 4 to 7 [41, 42]. In order to provide an explanation for this occurrence, the pH-partition hypothesis [43] proposes that the lipophilic membrane can only allow uncharged molecules to pass through it. The pH-partition hypothesis states that only free drugs, which are unionized and unbound, are able to passively pass across the cell membrane. It is widely acknowledged that the unbound unionized molecule exhibits a greater permeability, typically between three and four log units, in comparison to its ionized counterpart [44]. Therefore, in order to pass through the majority of membranes, the drug must possess a relatively low polarity. At a pH level found in living organisms, ionized molecules have a tendency to engage with lipid membranes that carry a negative charge. This interaction often leads to a low level of permeability, with the molecules arranged in the following order: neutrals, bases, zwitterions, and finally, acids [45].

Passive permeability impacts all aspects of drug candidate ADME properties, including as gastrointestinal absorption, first-pass extraction, tissue distribution, cell penetration, clearance mediated by enzymes and transporters, renal reabsorption, enterohepatic circulation, and drug–drug interactions. The advantage of passive permeability is that it is not reliant on the concentration of the substrate; as a result, it is nonsaturable, which results in linear absorption. Compounds with high passive permeability can easily pass through cell membranes. This property has several advantages, including effective absorption when taken orally, the ability to be used in modified-release dosage forms, increased exposure to the brain, reduced clearance by the kidneys, and a lower risk of drug–drug interactions mediated by transporters [39]. For most tissues, cell types, and animal species, scaling factors allow for the translation of passive permeability. Scaling factors are required to account for variations in lipid compositions [46]. Some examples of *in vitro* cell-based or artificial membrane assays that can be used for permeability measurements are Caco-2, LLC-PK1, MDCK, and PAMPA. Caco-2 is a cell line derived from human epithelial colon adenocarcinoma that exhibits characteristics of both enterocytes and colonocytes [47]. Furthermore, the findings produced in these cells show a strong association with the permeability and bioavailability of the human jejunum *in vivo* [48, 49]. MDCK cells, derived from the distal renal tissue of canines, emerged thereafter as a more expeditious and economically advantageous substitute for Caco-2 cells [50]. PAMPA, or parallel artificial membrane permeability assay, is a widely used and cost-effective screening tool that does not require the use of cells. It is commonly used alongside cell-based assays [51]. Although PAMPA is unable to simulate active transport, this limitation is counterbalanced by the fact that over 90% of medicines are absorbed through passive diffusion [52, 53]. The permeability of PAMPA at both pH 7.4 and 5.5 shows a strong correlation with the permeability of Caco-2 in small datasets [54]. Because of its adaptability to high-throughput as well as its flexibility with experimental circumstances (various lipid compositions/range of pH conditions), PAMPA is an effective screening approach that can be utilized in the early stages of drug discovery [55]. In summary, membrane permeability is essential for the absorption of medications via the biomembrane of the gastrointestinal system and is also vital for their distribution to all other bodily regions. Hence, it is imperative to establish principles that facilitate the attainment of favorable passive permeability, as well as other essential qualities, in order to effectively facilitate drug design.

6.1.2 Distribution

Another important process that ends with the target organ being exposed to the drug is drug distribution, which occurs after the drug is absorbed from the gastrointestinal luminal fluid into the systemic circulation. This involves the drug moving from the blood into different compartments within tissues and organs, such as the interstitial space and the intracellular space [56]. Due to the fact that it has the potential to influence the quantity of drug that is ultimately delivered to the active sites, the process of drug distribution is of great significance [57]. Fat, muscle, and brain tissue are among the many areas that a drug can reach after it has been absorbed. The distribution of drugs is typically very quick, and the majority of small lipophilic compounds are able to quickly move across lipid membranes through a process that is passive. Passive diffusion does not allow large or highly polar molecules to flow across lipid membranes. Instead, these compounds need particular transporters to enter the tissue [58]. A drug's concentration in tissue may be significantly greater than its plasma concentration if it enters a tissue through an active transporter mechanism. The drug molecules can be transported to several organs for action, including the skin, tumors, brain, and other tissues that contain receptors. Additionally, some drugs can be disseminated to the liver and kidneys, which are responsible for disposal. A number of factors, including the drug's interaction with blood components like plasma proteins and the rate of blood flow, as well as its molecular size and lipophilicity, impact the distribution process, which mostly takes place in the systemic circulation, or blood. Several *in vitro* assays, including as lipophilicity, solubility, plasma stability, and plasma protein binding, have been established by researchers in order to evaluate the distribution process [59]. The enormous volumes of distribution that occur as a consequence of lipophilic positively charged compounds partitioning into biological membranes are a consequence of their positive interactions with the phospholipid bilayer. Neutral compounds do not exhibit any electrostatic interaction with the surface of membranes. As a result, the rise in lipophilicity will be the primary factor that determines their capacity to partition into membranes. Even though negatively ionized substances bind strongly to serum albumin – the most abundant plasma protein – they have a poor affinity for membranes and, as a result, a low volume of distribution [60]. When it comes to neutral and basic compounds, lipophilicity is once again the most important factor in albumin binding [61]. There is a correlation between a drug's distribution and its binding strength to tissue proteins, with higher log$D$ values indicating stronger binding [62]. Excessive lipophilicity not only enhances the dispersion volume but also the oral absorption rate. Lipophilic compounds, on the other hand, are more susceptible to metabolism, which results in a higher clearance [63]. This, in turn, reduces the drug's half-life and decreases its bioavailability.

Another crucial metric for determining the distribution process is the assessment of aqueous solubility at physiological pH. Insoluble compounds can have a negative impact on the ADME analysis, as a portion of the molecule will form a precipitate, potentially rendering the action sites unavailable [59].

Plasma protein binding is a critical component that significantly impacts drug distribution and, by extension, the pharmacokinetic and pharmacodynamic behavior of drugs in general. There are 7% proteins in plasma, with human serum albumin being the most significant protein for drug binding, followed by α1-acid glycoprotein and lipoproteins [64]. The amount of drugs that are freely available in the plasma and can be distributed to the active sites is determined by analyzing the plasma protein binding data. As a result of the majority of the drug fraction binding to the plasma proteins during the process of drug absorption into the bloodstream, the amount of drug that is accessible to reach the target is decreased. This, in turn, determines the efficacy of the drug, as well as its metabolism and other pharmacokinetic parameters. Conversely, drugs that have a lower affinity for plasma proteins have a greater ability to be distributed to organs and tissues. On the other hand, drugs that strongly bind to plasma proteins are limited to the blood vessels and have a relatively small volume of distribution [59]. The temporary localization, reduced or eliminated pharmacological effects or hazardous actions, and difficulty of the attached molecule to penetrate biological membranes or diffuse into tissues make plasma protein binding of foreign compounds, and drugs in particular, an essential process. Among the plasma proteins that have the ability to attach to foreign compounds, albumin exhibits a broad capacity to bind several molecules. Nevertheless, it is important to note that other plasma proteins, such as globulins, also play a role in this process. Numerous factors, including variations in species, shifts in pH, and hormonal impacts, have the potential to modify the binding of foreign molecules. The interaction between drugs or chemicals and plasma proteins is occasionally crucial, as a chemical that forms a strong bond with a plasma protein will exhibit an extended duration of effect. Suramin, a potent therapeutic drug for trypanosomiasis, exhibits a significant affinity for plasma proteins. The drug provides protection for a period of at least three months with just a single dose [65]. Permeability barriers allow concentration gradients to be formed and sustained, which in turn supports biological specialization and complexity. Because of this, diffusion across membranes via extracellular gaps is more likely to occur in most tissues than in those with continuous membrane barriers due to the structure of cells. It is at the level of the capillary endothelium that the effective tight junction takes place in permeability barriers, such as those found in the brain, the placenta, and among the thymus. Blood–organ barriers are the name given to these barriers. In comparison to other organs, the brain has unique characteristics that affect a drug's penetration. A crucial characteristic of the brain is its total isolation from the bloodstream by the BBB [66]. It is necessary for the drug to get through the BBB in order for it to enter the brain through simple diffusion. Hence, the capacity of drugs to permeate the BBB is crucial in the process of drug development. For drugs that are active in the central nervous system (CNS), a high level of penetration is required, whereas a low level of penetration may be desirable in order to reduce the adverse effects on the CNS that are associated with pharmaceuticals that have a peripheral site of action [67]. Since passive diffusion allows some drugs to pass through the BBB, physicochemical features, including charge, lipophilicity, and molecular weight, are crucial for substances discovered in the CNS [68]. The distribution of a drug can be determined either by analyzing its systemic exposure or by directly measuring its presence in tissues. Typically, the majority of data is acquired by the former approach. It is possible to determine the extent of drug distribution throughout the body by using a theoretical volume known as the apparent volume of distribution at the steady state (V_{dss}) of a molecule. Despite the fact that V_{dss} does not have any physiological significance, it is a major pharmacokinetic parameter that is directly related to the elimination half-life ($T_{1/2}$) and protein binding. Furthermore, it is not dependent on the total clearance (CL) of a drug. A liter per kilogram of bodyweight (L/kg) is a common unit of measurement for V_{dss}. $V_{dss} > 1$ L/kg (>100% of body weight) suggests that the drug is concentrated in specific tissue compartments. As a result, V_{dss} is a valuable parameter that may be used when determining the relative amount of drug that is present in the tissues or outside of the central compartment [69].

$$T_{1/2} = \frac{0.693 \times V_{dss}}{CL}$$

$T_{1/2}$ = Half life
V_{dss} = Volume of distributions
CL = Clearance

6.1.3 Metabolism

Drugs are compounds that are not native to the body and are converted by a wide range of enzymes into other products that are referred to as metabolites. Metabolites typically exhibit more polarity and higher solubility in water, and are more easily eliminated through urine. Metabolism, also known as biotransformation, refers to the process by which a drug undergoes a chemical transformation in order to become a metabolite. After being absorbed in the intestines, the drugs next travel through the hepatic portal system to the liver, which is where the majority of the drug's metabolism takes place. Mostly, the liver is responsible for the metabolism of a wide variety of xenobiotics, which includes the metabolism of drugs and hazardous chemicals [70]. There is a correlation between the physicochemical properties of medicines and their metabolic clearance, for example, a high $\log D$ 7.4 for lipophilicity and a high metabolic clearance [71]. Many metabolizing enzymes make xenobiotics more soluble in water, which helps the body get rid of them.

A significant family of enzymes that are responsible for the metabolism of drugs is known as the microsomal cytochrome P450 (CYP450) family [72]. The enzyme CYP3A4, which is the most prevalent P450 enzyme in the liver, is accountable for the metabolism of around 40–50% of the drugs that are utilized in clinical practice. Because they reveal useful information regarding the possibility of drug exposure, the fate of metabolites in preclinical species to humans, and clinical interactions caused by inhibition and induction, CYP inhibition assays have become an important factor in the exploration and development of new medications [73]. The following reactions can be catalyzed by CYPs: oxidation of carbon and oxygen, carboxylation of nitrogen and oxygen, aromatic hydroxylation, oxidation of nitrogen and sulfur, deamination, and dehalogenation. Humans have around 100 isoforms of CYPs. The primary site of action is predominantly in the liver. The human body contains 12 different CYPs, all of which are recognized to play a significant role in the metabolism of xenobiotics. CYP2C, CYP2D, and CYP3A subfamilies are the most active with regard to the metabolism of drugs. CYP3A4 is the enzyme that is most frequently expressed and is involved in the metabolism of nearly 50% of all drugs for treatment [74]. Following CYP3A4 in terms of the number of medicines that are metabolized by P450 enzymes, CYP2D6 is responsible for the metabolism of the second biggest number of drugs. The brain and liver are two places where this enzyme is expressed. CYP2D6 has a tendency to oxidize substrates that have a protonated basic nitrogen at a pH that is typical for the body and a flat aromatic ring [75]. Monoamine oxidases (MAOs) are enzymes that contain flavin and are found in the mitochondria. They play a crucial role in catalyzing the process of deamination for both biogenic and xenobiotic amines. Most tissues have two isoforms of MAO, called MAO-A and MAO-B [76, 77]. Flavin-containing monooxygenases, also known as FMOs, are found in a wide variety of tissues, including those found in the liver, kidney, lungs, skin, brain, and other tissues. However, the liver, kidney, and lungs in particular have the highest concentrations of these enzymes [78]. There are now five distinct isoforms of FMOs that have been identified and described. FMO1 and FMO3 have a significant involvement in the adult human liver. The primary substrates for FMO1 are imipramine and orphenadrine. In addition to inhibiting FMO-dependent pathways, methimazole also inhibits CYP2B6, CYP2C9, and CYP3A4 [79]. Many nucleophilic molecules with nitrogen, sulfur, phosphorous, or selenium atoms are oxidized by FMOs using NADPH [80]. While nonbasic molecules are more likely to be oxidized by CYPs, the degree of N-substitution also plays a significant effect. Basic amines are often FMO substrates, but there is some overlap with CYPs (e.g., CYP2D6). In contrast to primary amines, which are typically better CYP substrates, tertiary amines are typically FMO substrates. Secondary amines, on the other hand, are less clear-cut [45]. For N-oxidation, if a molecule has many nitrogen atoms, the more basic nitrogen center is typically the chosen location.

The primary purpose of metabolic transformation is to enhance the drug's hydrophilicity. Metabolism is commonly categorized into two distinct stages: Initially, in a phase known as Phase I, the molecule undergoes functionalization, resulting in the incorporation or release of polar groups such as alcohols, phenols, and amines by processes such as dealkylation, hydroxylation, heteroatom oxidation, and oxidative dehalogenation. In addition, both esters and amides undergo hydrolysis. Phase I enzymes modify drugs' functional groups. Because of this, phase I metabolism is frequently referred to as the functionalization phase of the drug metabolism process. If the parent molecule has sufficient solubility and/or transporter selectivity, both unmodified parent compound and Phase I metabolites are eliminated through bile and urine. Phase II reactions mostly encompass conjugation reactions, including glucuronide, sulfate, and glutathione conjugations. Phase II reactions typically yield very hydrophilic molecules. For instance, the presence of the strongly polar glucuronide moiety typically results in the swift elimination of a drug from the body. Nevertheless, not all phase II reactions produce a metabolite that is more polar. The biotransformation reactions are facilitated by enzymes that are found in several organs and tissues, with the liver being the most crucial one overall [81–84].

To provide a brief summary, it has been recently estimated that drug metabolism is one of the primary characteristics that has been demonstrated to be taken into serious consideration during the process of discovering, developing, and designing potential drug candidates. To better understand a compound's pharmacokinetic properties, optimize PK parameters, and choose the most drug-like compounds for development, *in vitro* biotransformation experiments are an important tool [85, 86].

6.1.4 Excretion

The process by which toxic compounds and/or their metabolites are transported from the body to the environment outside the body in an irreversible manner is referred to as excretion. Therefore, excretion is one of the key systems that protect the body from the poisonous effects of toxicants. The process of excretion allows for the removal of these substances from the body. This process decreases the concentration of the drug at its specific location, where it exerts its effects. A gradual elimination mechanism facilitates accumulating the required drug concentration to sustain the desired therapeutic effects. The two main forms of drug metabolites that are excreted from the body are their unaltered forms and their structurally distinct forms. The energy-dependent transport pathways that regulate drug clearance by the kidneys and liver are discussed in this section. The kidney plays a crucial role in eliminating drugs and their metabolites from the body by excreting them through urine. Renal clearance is the kidneys' process of getting rid of drugs through urine. It is composed of multiple components, including active tubular secretion, passive and active glomerular filtration, and reabsorption [87]. In the process of glomerular filtration, an ultrafiltrate of the body's plasma is produced. This ultrafiltrate contains foreign substances and their metabolites in amounts that are roughly equivalent to those that are found in the blood. After filtration at the glomeruli, the molecule can either be ejected or passively reabsorbed into the bloodstream by crossing the tubular cells of the nephron. The chemicals that are reabsorbable are not very common in urine. However, lipid-insoluble ionized molecules, like conjugates of foreign chemicals, are less able to be reabsorbable and are therefore more easily flushed out of the body by the kidneys than their precursors [65]. Transport mechanisms across membranes are involved in both passive and active tubular secretion and reabsorption. Drug filtering is a passive process that occurs through diffusion. Drug excretion by tubular secretion is facilitated by many active transporters that uptake substances from the interstitium of the kidney and release them into the tubular lumen [88]. The majority of renal transporters comprise OAT1, OAT3, OCT2, MATE, P-gp, and BCRP. These transporters play an active role in facilitating the elimination of xenobiotics from the bloodstream into the urine [59]. The rate of filtration and secretion can be added to determine the renal excretion of medicines, and the rate of reabsorption can be subtracted. The amount of medication expelled over a given period can be divided by the time interval and the average plasma concentration to calculate renal clearance [69].

$$\mathbf{CL_R}\left(\text{mL/min}\right) = \frac{\text{excreted amount time interval}}{\text{plasma concentration}}$$

The liver has developed a number of different systems that allow it to take in nutrients and endogenous substances that are essential to its function and to the maintenance of homeostasis in general. An excretory pathway exists in the liver for the removal of nonionized compounds that contain lipophilic and polar groups. Furthermore, apart from the aforementioned transport systems for organic chemicals, the liver likely possesses an additional transport mechanism specifically designed for the elimination of metals. Efflux mechanisms are present on the canalicular and sinusoidal surfaces of hepatocytes, and they have developed to detoxify and eliminate byproducts or metabolites. Multiple transporters are accountable for the absorption of medicines into the hepatocytes through the sinusoidal (basolateral) membrane [89, 90]. The OATP (SLC21 family) transporters facilitate the transportation of endogenous compounds, including thyroid hormones and bile acids. These transporters have demonstrated the ability to transport drugs such as fexofenadine, rifampicin, pravastatin, and rifamycin. Additional uptake transporters comprise the organic anion transporters (OATs) and organic cation transporters (OCTs) belonging to the SLC22 family, as well as the sodium taurocholate co-transporting polypeptide (NTCP) encoded by the SLC10A1 gene, and the peptide transporter (PGT) encoded by the SLC21A2 gene [91].

$$\mathbf{CL_B}\left(\mathrm{mL/min}\right) = \frac{\text{bile concentration} \times \text{bile flow}}{\text{plasma concentration}}$$

Liver cells have the ability to discharge chemicals into bile, which then enters the small intestine. A process known as the enterohepatic cycle can occur when substances or their metabolites have characteristics that facilitate intestinal reabsorption. This cycle continues with biliary secretion and intestinal reabsorption until the compounds are excreted from the body by the kidneys. Biliary clearance can be determined by monitoring the rate of bile flow and the concentration of drugs in both plasma and bile. When a drug's concentration in bile is significantly higher than its concentration in plasma, resulting in a clearance value greater than bile flow (0.5–0.8 mL/min), biliary clearance becomes significant [69].

6.1.5 Toxicity

During the period of development that occurs in the womb or from postnatal to adolescent, drugs and other xenobiotics have the ability to disrupt homeostasis as well as proper growth, differentiation, and developmental processes among the developing organism. The normal development of the organism that is developing is disrupted as a result of this disorder, which could result in birth defects, teratogenesis, structural abnormalities, and ultimately the death of the creature. Developmental toxicity and acute oral toxicity are assessed by numerous international regulatory agencies. An estimated 100 million rodents are subjected to toxicity studies every year, with the majority of these animals belonging to the cosmetics and pharmaceutical industries [92]. Because of this, drug toxicity is a significant challenge in the process of drug development. The following criteria must be satisfied in order to conduct an *in vitro* assessment of the toxicity of a drug:

i) Human-derived cells should be utilized in the process of determining the toxicity of drugs to humans.
ii) It is recommended that cells generated from the organ in issue, such as hepatocytes, be utilized in order to assess the level of toxicity that is exhibited by particular organs, such as the liver.
iii) Various cell types can be found in an organ. When dealing with particular kinds of organ toxicity, it is important to use the ideal target cell.
iv) The testing panel may also include enzyme-based and receptor-binding assays as additional criteria.
v) Toxicogenomics is a highly effective method for identifying drug toxicity in people due to its comprehensive assessment of several causes of drug toxicity. As a result of the fact that toxicogenomics, which is the evaluation of the effects of drugs on the expression of genes that are toxicologically relevant, enables the simultaneous evaluation of multiple toxicity mechanisms, it is most likely to be useful for toxicity evaluation, particularly for the elucidation of mechanism and, consequently, significance [93–95].

6.2 Physicochemical Assessments

6.2.1 Partition Coefficient

Lipophilicity is regarded as a primary physicochemical descriptor in several contexts, including drug absorption, plasma protein binding, hydrophobic drug–receptor interactions, and, to some extent, the pharmacokinetic behavior and toxicological characteristics of drug molecules, as well as formulation attributes like solubility [96].

The partition coefficient, which measures the fat-soluble tendency of the drug, is one of the most important factors affecting its activity. The partition coefficient (P), defined as the ratio of the concentrations of a drug in equilibrium in two immiscible phases (polar aqueous phase and nonpolar organic phase), is formulated as follows:

$$P = \frac{[A]_{\text{octanol}}}{[A]_{\text{water}}}$$

The above equation is the general partition coefficient equation based on octanol and water. The relative lipophilicity of a medicine can be determined by measuring the octanol–water partition coefficient. Octanol was partly chosen for the octanol–water system due to its flexibility and polar head and nonpolar tail resemblance to biological membrane components. Therefore, it is believed that a medication's propensity to exit the aqueous phase and partition into octanol is a measure of how well the drug will diffuse through and partition into biological barriers like the intestinal membrane [97].

Lipophilicity is the result of all intermolecular interactions involved in the partitioning of a solute between two phases, a property that combines contributions from polarity as well as hydrophobicity, which is the tendency of nonpolar groups or molecules to associate in an aqueous environment [98].

Currently, log $P_{0/w}$ is arguably the most successful and instructive physicochemical property utilized in medicinal chemistry; it is a critical component of many quantitative structure–activity relationships (QSAR) created for pharmacological, environmental, biochemical, or drug design applications. It has been widely used to explain drug–receptor and drug–biological membrane interactions [99].

A compound's physicochemical characteristics, such as its lipophilicity, are closely correlated with its chemical structure and can be quantitatively linked to its pharmacological activity. The change in the log P value as a result of the replacement change is displayed in Figure 6.2. Therefore, both the drug's physicochemical and biological qualities are determined by its structure [100].

If we look at the research done by Meyer and Overton, they established a linear relationship, within certain bounds, between the lipophilic nature of a group of related medications and their biological activity.

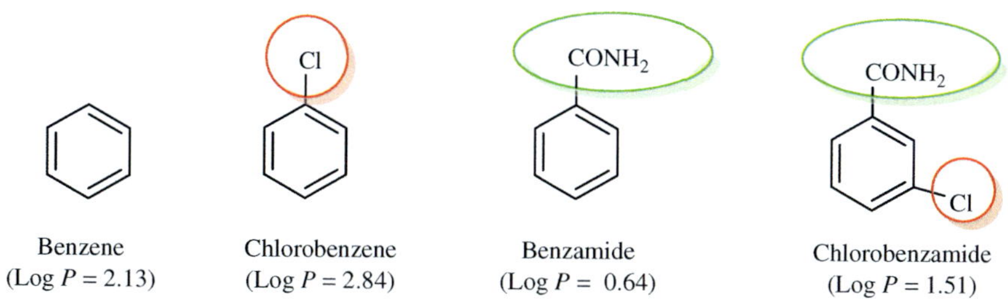

Figure 6.2 Impact of substitution on a molecule's log P value.

$$\log(1/C) = 0.94 \log P + 0.87$$

This formula expresses the link between the octanol–water partition coefficient and the molar concentration (*C*) that causes isonarcosis as a linear free energy relationship [101].

The groups a substance contains determine its lipophilic; compounds with high partition coefficients are more lipophilic. The partition coefficient value rises as groups that offer the molecule lipophilic characteristics are added. Hansch established the hydrophobic bond constant, which is the numerical representation of the substituent's contribution to the partition coefficient, to establish quantitative correlations between activity and the partition coefficient. In the equilibrium below, Hammett σ values are used to monitor electronic effects, Taft E values are employed to monitor steric effects, and 1-octanol–water $\log P$ uses σ to indicate the difference in $\log P$ between the substituted and parent molecules.

$$\log(1/C) = \rho\sigma + a\pi + b\pi^2 + cE_s$$

where π is determined by the variation in $\log P$ between the substituted compound and the parent, and σ is the Hammett constant for the substituent's electronic impact [102].

$$\pi = \log P_x - \log P_H$$

6.2.2 Log *D*: Ionizable Compound Lipophilicity

When characterizing the lipophilicity of possible medications, it is important to include the chemical's ionic state because a molecule's ionization reduces lipophilicity relative to the neutral state [103]. A discernible difference exists between the distribution ratio *D*, determined for the total analytical concentration, and the partition coefficient *P*, which represents the partition of neutral species between octanol and water.

$$P = \frac{[HA]_o}{[HA]_w} \qquad D = \frac{[HA]_o + [A^-]_o}{[HA]_w + [A^-]_w}$$

the concentrations of neutral and anionic species are represented by [HA] and $[A^-]$, where the subscripts *o* and *w* stand for the concentrations in the aqueous and octanol phases, respectively [104].

P and *D* are the same since there are only neutral species in the case of nonionizable compounds. In the pH range where ionic species exist, *D* should be used instead of *P* for ionizable chemicals. For example, it was shown that the "Rule of 5" can be significantly affected when employing $\log D$ rather than $\log P$. But in many systems, ionic species go into the nonaqueous phase, necessitating a more sophisticated strategy [105].

6.2.2.1 Methods for Calculating Lipophilicity

For medicinal chemists, determining lipophilicity has become a standard procedure [106]. Two types of experimental procedures are used to measure lipophilicity: direct methods and indirect approaches. When utilizing direct experimental techniques, the concentration ratio in the equilibrium of a substance divided into the nonaqueous and aqueous phases is utilized to calculate the partition coefficient directly. Through correlations, such as those between a compound's partition coefficient and retention factor in a reverse-phase chromatographic system, the partition coefficient is calculated via indirect experimental approaches. Liquid chromatography (LC) is a significant tool for directly and indirectly determining lipophilicity [107].

6.2.2.2 Direct Experimental Determination of Lipophilicity

In silico determination of lipophilicity: Numerous silico techniques have been developed to predict lipophilicity from a molecule's molecular structure since the seminal work of Hansch et al., who discovered that the partition coefficients had an additive constitutive nature [108]. *In silico* calculations are made based on a set of experimental data and evaluated with predicted results. Sometimes, the calculated $\log P$ value may need to be corrected. Additionally, *in silico* methods do not draw definitive conclusions as there can be differences of up to two different $\log P$ values depending on the method used [109].

Shake-flask method: Quantification of the chemical concentration in the aqueous and non-aqueous phases is necessary for the direct assessment of lipophilicity. The basic process applied in determining the partition coefficient is based on shaking a certain amount of chemical compound with measured amounts of water-saturated octanol and octanol-saturated water or buffer adjusted to pH 7.4. After sufficient time for the equilibrium between the two phases to be established, the amount of substance in one or both phases is measured by an appropriate analytical method. The ratio of the substance concentration in the octanol phase to the substance concentration in the aqueous phase gives the partition coefficient. Although various analysis methods can be used to determine the amount of substance in the phases, the most preferred is the spectrophotometric method. Since P values are generally quite large, logarithmic P values are used and the partition coefficient equation is written as follows:

$$\log P = \log C_{\text{octanol}}/C_{\text{water}} = \log C_{\text{octanol}} - \log C_{\text{water}}$$

C_{octanol} and C_{water} = drug concentration in the octanol and water phase at equilibrium [110, 111].

One disadvantage of this approach is that it takes a long time to attain equilibrium (up to two or three days) while stirring [112].

6.2.2.3 Indirect Experimental Determination of Lipophilicity

The need to create techniques that can estimate lipophilicity without requiring measurement is growing as a result of the drawbacks of conducting direct trials. In addition to being more accurate and versatile than direct procedures, these techniques provide higher throughput and environmental benefits [113].

Reversed-phase thin layer chromatography: In reversed-phase systems, the partitioning between the stationary and mobile phases primarily controls retention. Therefore, the RM value may be used to calculate the lipophilicity index as determined by RP-TLC.

$$\text{RM} = \log\left[\left(\frac{1}{R_{\text{f}}}\right) - 1\right]$$

where R_{f}, the retention factor, is calculated by dividing the sample's distance traveled by the mobile phase's distance traveled [114].

Table 6.1 Programs based on fragmental methods.

Program	Method	Platform	Commercial
∑f-SYBYL	Rekker, revised version	Unix	Tripos
PCMODELS	Hansch/Leo	Unix	Daylight
CLGOP	Hansch/Leo	PC, Mac, Vax	BioByte
∑f	Rekker	—	—
PROLOGP_cdr	Rekker	PC	Compudrug

Table 6.2 Programs based on atomic contributions.

Program	Method	Platform	Commercial
GLGOP	Atomic fragments	—	—
ATOMIC5	Atomic values	PC	Compudrug
Tsar 2.2	Atomic values	Unix	Oxford molecular
SMILOGP	Atomic contributions	PC	—
MOLCAD	Atomic values	PC	—

Reversed-phase high-performance liquid chromatography: In order to measure lipophilicity experimentally, reversed-phase high-performance liquid chromatography (RP-HPLC) is the most used indirect approach. The OECD has officially suggested a standard protocol for determining log *P* [115].

***Programs for Calculation of Log* P** The initial methods to calculate lipophilicity were fragmental approaches, which involved splicing molecules into suitable segments and using correction criteria in conjunction with molecular connectivity. Atom-based techniques, on the other hand, do not require correction factors and specify an enormous number of atom-types; lipophilicity is simply measured by adding up the values of each atom-type. A satisfactory treatment of the effect of stereochemistry on lipophilicity is not provided by fragmentary and atom-based methods. Therefore, the most recent generation of calculation techniques aims to supply the user hydrophobic fields for use in 3D QSAR and to reflect conformational features. The calculating techniques and programs are briefly outlined in the following based on the subclassification above [116] (Tables 6.1 and 6.2).

6.2.3 Acid–Base Properties and Ionization

The acid–base ionization constant is an essential physicochemical parameter of a compound. Because it has an impact on biological activity, absorption, distribution, metabolism, and excretion (ADME), it is particularly important in medicinal chemistry [117–119].

According to Brønsted and Lowry, substances that donate protons are acids, and substances that accept protons are bases [120]. pK_a and pK_b, as ionization coefficients, indicate that an acid donates a proton and a base accepts a proton [121] (Figure 6.3).

For acids:

$$AH + H_2O \longleftrightarrow A^- + H_3O^+$$

$$pK_a = -\log\left([A^-][H_3O^+]/[AH]\right)$$

For bases:

$$BH^+ + H_2O \longleftrightarrow B + H_3O^+$$

$$pK_a = -\log\left([B][H_3O^+]/[BH^+]\right)$$

Knowing that $pH = -\log\left[H_3O^+\right]$

For acids:

$$pH - pK_a = \log\left[A^-\right] - \log[AH]$$

For bases:

$$pH - pK_a = \log[B] - \log\left[BH^+\right]$$

Figure 6.3 Acid and base equations.

Two parameters determine a drug's degree of ionization in solution: pH and pK_a. It is understood whether a compound is a strong or weak acid or base by its pK_a value [122, 123].

$pK_a < 2$: Strong acid
pK_a 4–6: Weak acid
pK_a 8–10: Very weak acid
$pK_a > 12$: strong base

Generally, the functional groups determine the pK_a of a drug molecule. A molecule may have more than one pK_a if it has more than one functional group. A chemical with several functional groups, like ciprofloxacin, can have both basic and acidic characteristics [122] (Figure 6.4).

At physiological pH, 95% of pharmaceutical substances ionize to some degree. The majority of medications have weak bases or weak acids. Permeability and solubility, two characteristics frequently utilized in pharmaceutical research to forecast the pharmacokinetic profile of a molecule, are controlled by the degree of ionization [122, 123].

The certain equation calculates the concentration of ionized and nonionized forms of a drug with known pK_a at a Henderson–Hasselbach pH [110, 124].

For weak acid: $pH - pK_a = \log [i]/[ni]$
For weak base: $pH - pK_a = \log [ni]/[i]$

Weak acids will be more unionized at pH values below their pK_a and more ionized at pH values above their pK_a. For weak bases, the opposite is true [100] (Figure 6.5).

The solubility, medication absorption, and distribution are significantly impacted by the pK_a value due to the assumption that only neutral species can pass through the lipophilic

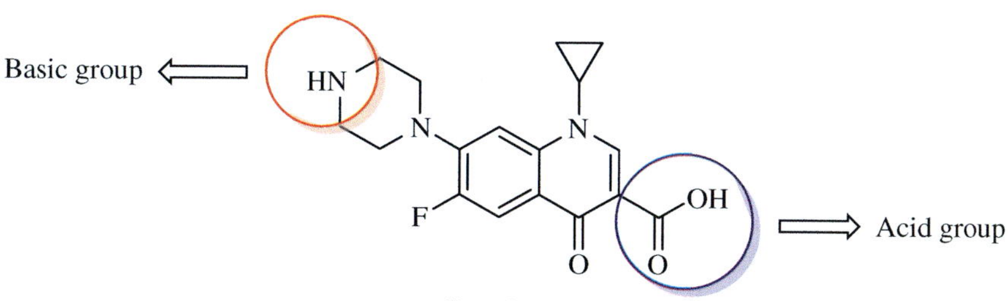

Ciprofloxacin

Figure 6.4 Acidic and basic groups of ciprofloxacin.

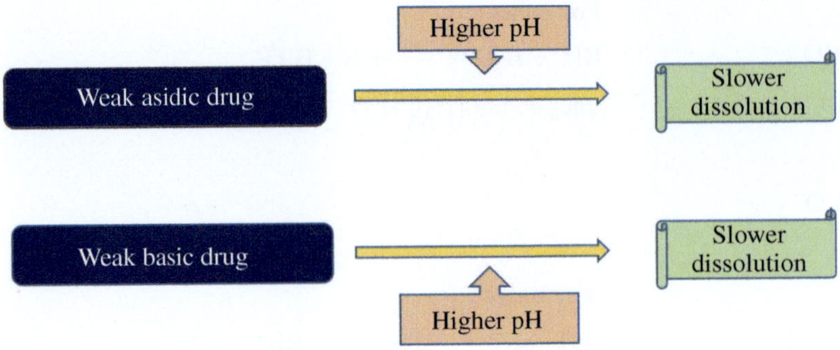

Figure 6.5 The impact of pH on weak bases and acids.

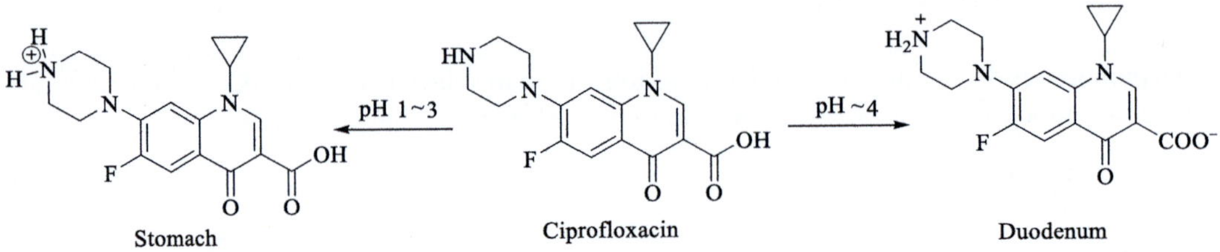

Figure 6.6 Chemical change of ciprofloxacin depending on pH.

membrane [125]. With some exceptions, drugs pass through nonpolar membranes such as cell membranes and the BBB in a nonionized manner. They generally show their effects in an ionized form through electrostatic interactions such as ionic and ion–dipole bonds, as seen in drug–receptor interactions. Therefore, the environment's pH determines the substance's ionized and nonionized state, its passage through cell membranes, and the compound's activity. Accordingly, acidic drugs are generally more active at low pH. Because they will be nonionized at higher rates at these pHs, their passage through membranes, that is, their absorption, will be easier, and the amounts reaching the site of action will be higher [110, 126]. This also explains why organic acids and bases make up so many medications. A neutral chemical, an acid, or a base can be well absorbed at a certain point in the gastrointestinal tract due to the significantly different pH values in the stomach and intestines. Absorption may become troublesome if the pK_a values are excessively high compared to the physiological pH values, such as amidines or guanidines. This also holds for molecules with several acidic or basic groups inside their structure, such as amino acids and zwitterionic chemicals [127].

Figure 6.6 illustrates how the zwitterionic molecule ciprofloxacin changes in the gastrointestinal tract. The strongest base in stomach fluid is the secondary amine, which can take up a proton. The carboxylic acid is the most acidic functional group in the duodenum; it produces a carboxylate anion by giving off a proton. A zwitterion is a molecule that has both positive and negative charges. In the duodenum, ciprofloxacin is a zwitterion [122].

6.2.4 Solubility

A solid material's solubility is defined as the concentration at which, at a certain temperature and pressure, the solution phase and a particular solid phase are in equilibrium [124]. A growing number of recently developed pharmaceuticals have low aqueous solubility, inadequate and variable bioavailability, and gastrointestinal mucosal irritation. These characteristics are anticipated to present difficulties when these medications are developed into oral medicinal formulations. For medications taken orally, solubility is the most important rate-limiting component because it enables the drug to reach the appropriate concentration in the systemic circulation for a pharmacological reaction. Low solubility of the pharmaceutically active constituent suggests a greater likelihood of medication development and innovation failure. In addition, pharmacodynamics, pharmacokinetics, drug distribution, absorption, and protein binding are all adversely affected by poor solubility, which has a substantial impact on drug delivery. Furthermore, poorly soluble medicines reduce the rate of *in vitro* dissolution and have a major negative effect on drug delivery [128, 129].

Solubility is affected by the properties of the solvent and many molecular parameters [121] (Figure 6.7). These:

pK_a: This is important for the solubility of drugs. It may affect a drug's solubility in water. The charge state of the drug molecules can be changed by adjusting the pH of the solution. If the pH of the solution is such that a particular molecule bears no net electric charge, the solute frequently has limited solubility and precipitates out of the solution [130]. Variations in gastrointestinal pH brought on by food intake can dramatically increase or reduce drug solubility in the case of weakly acidic or basic medicines, which reside in the aqueous GI environment in ionized and unionized form. The pH of the stomach in healthy individuals usually ranges from 1 to 3 when they are fasting, but it can momentarily increase to 4–6 when they eat. Food consumption may accelerate the dissolving of drugs in the stomach because it increases the amount of ionization and, thus, the solubility of a weakly acidic medicine at higher pH levels. Conversely, at higher stomach pH, a weakly basic medication's degree of ionization will decrease, which will lead to less drug dissolution and/or possible precipitation of already-dissolved drug molecules [131, 132].

Lipophilicity: The term hydrophilic or lipophobic refers to solubility in polar environments, and the term lipophilic or hydrophobic refers to solubility in nonpolar environments. Theoretically, a substance can dissolve in both polar and nonpolar environments, but their degree of dissolution is different. As a general rule, similar substances dissolve in similar solvents. In other words, polar molecules dissolve more easily in polar solvents, and nonpolar molecules dissolve more easily in

Figure 6.7 Factors affecting solubility.

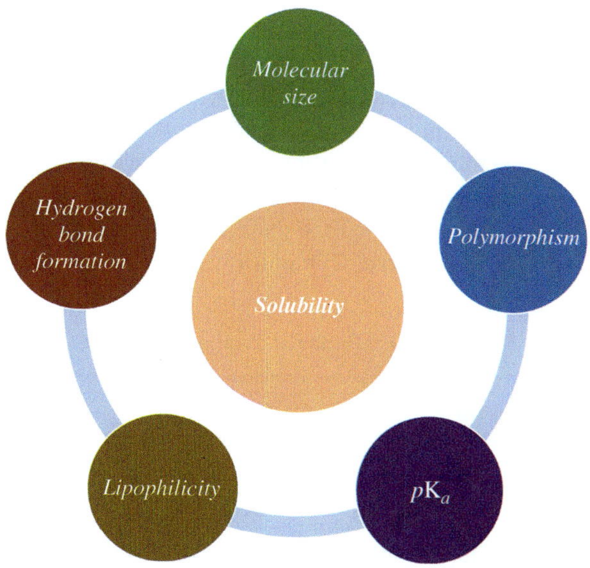

nonpolar solvents. Alkyl chains give the molecule oil solubility, and as the alkyl chain lengthens, oil solubility increases. The introduction of heteroatoms such as oxygen and nitrogen or functional groups containing them into the molecule increases the polarity of the molecule, that is, its solubility in water [110].

Hydrogen bond formation: While polar compounds that can make HBs with water molecules dissolve in water, Van der Waals and hydrophobic bonds cause nonpolar molecules to dissolve in organic solvents [110].

Molecular size: The drug's particle size determines how soluble it is. Because they have a lower surface area, large particles interact with the solvent less. One of the approaches to increase the drug's surface area is to minimize its particle size, which improves its dissolving property [130, 133].

Polymorphism: A polymorph is a substance that has a different crystal structure but the same chemical content. As a result, polymorphs exhibit various physicochemical properties because of their distinct network architectures and molecular conformations. Many medications can crystallize into different polymorphic forms to boost solubility through a frequent phenomenon called polymorphism [130].

One way to find solubility is to use the shake-flask method. This procedure involves adding a component to a solvent in a known volume and shaking the mixture until equilibrium is established. Next, using a variety of analytical methods, including high-performance LC, the concentration of the dissolved component is ascertained [134].

6.2.5 Polymorphism

When studying calcium carbonate in 1788, Klaporoth and associates first investigated polymorphism. "The ability of a given elementor compound to crystallize more than one distinct crystal species" is how McCroneas described polymorphism in 1965 [135]. The ability of a drug to exist in multiple crystal forms, or polymorphism, is widely used in the pharmaceutical sector. When a substance is in solution, all of its polymorphs behave chemically in the same way, but when it is solid, the behaviors can differ greatly. There may be variations in characteristics like surface free energy, shape, solubility, and melting point [136–139].

Since they include solvent molecules that are integrated into the crystal lattice in addition to the specific medicinal molecule, hydrates, and solvates are frequently referred to as "pesudopolymorphs." Drug compounds can crystallize into a variety of polymorphs and pseudopolymorphs quite frequently [124].

One of the known examples of differences in solubility of polymorphs is Ritonavir, the antiretroviral drug used in the treatment of HIV. It was discovered by Abbott Laboratories in 1992, and the FDA approved it in 1996 for use in both liquid and semisolid capsule formulations. However, in 1998, the product failed the dissolution test. The novel polymorph Form II, which is less soluble (Table 6.3) and thermodynamically more stable than Form I, was identified as the precipitated

Table 6.3 Solubility of Ritonavir in ethanol/water (mg/mL).

Ethanol/water	100/1	75/25
Form I	90 mg/mL	170 mg/mL
Form II	19 mg/mL	30 mg/mL

Source: Brog et al. [140]/with permission of Royal Society of Chemistry.

form during this test. The FDA approved the two forms of the medication, and in 1999, it was eventually made available for sale [140].

Mebendazole is an anthelmintic drug with a benzimidazole structure. Mebendazole has three different forms, and the difference in solubility of these three different forms causes changes in the activity of the drug. In physiological media, the solubility order is Form B > Form C > Form A. Form B is toxic due to its high solubility and consequent bioavailability, while form A nonanthelmintic activity due to its lesser solubility. Form C is the favored form in pharmaceuticals because of its high solubility, which guarantees good absorption without the risk of toxicity [141].

Different methods are used for the characterization of polymorphs. These are:

- Hot stage microscopy.
- **Thermal analysis:** Differential thermal analysis (DTA), differential scanning calorimetry (DSC), and thermogravimetric analysis (TGA).
- Spectroscopy.
- X-ray [140, 142].

6.2.6 Molecular Weight

The atomic weights (or atomic masses) of every atom in a molecule add up to the molecular mass (MW) of a substance, which is also referred to as the molar mass or molecular mass and is usually given in Daltons (Da). The molecular weight (MW) of a substance influences its membrane permeability, which in turn influences its intestinal absorption and tissue distribution, especially when it comes to permeability via the BBB. The atomic weights of each atom in the chemical formula are summed up to determine the MW. Mass spectrometers and other analytical tools are useful for obtaining accurate MWs [134].

There is often a general trend (even sometimes correlation) toward increased lipophilicity as MW increases [143]. Comprehensive research has demonstrated that permeability decreases as MW increases, leading Lipinski to suggest a 500 Da cutoff for certain medications. Some molecules, MW > 500 Da, are absorbed, though. In certain instances, MW is purposefully raised (and may surpass 500 Da) through the creation of a prodrug to enhance permeability. For instance, the dianionic medication that is the active ingredient is poorly absorbed, while the ester prodrug olmesartan medoxomil is effectively absorbed [125] (Figure 6.8).

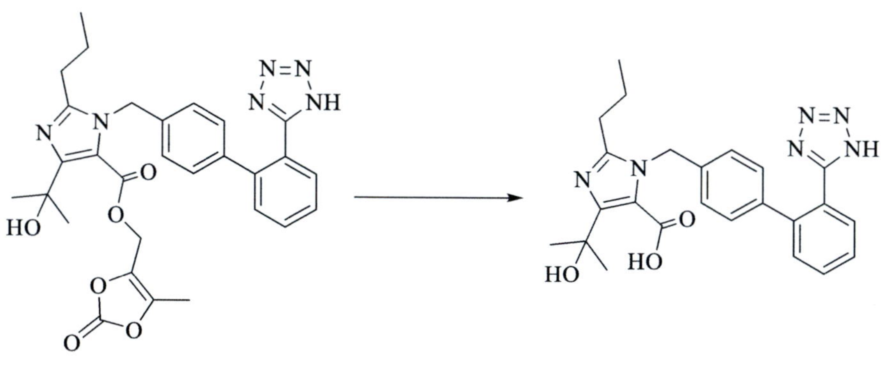

Olmesartan medoxomil

Figure 6.8 Olmesartan's conversion from prodrug to active medication.

6.2.7 Number of Hydrogen Bond Donors (HDB) and Acceptors (HDA)

The drug's distribution, metabolism, excretion, and absorption (ADME) are all impacted by HBs [144]. A drug's potency in forming HBs with water is reflected in how easily it may be transported from water to a nonpolar environment. This process has a significant impact on both passive permeability and *in vivo* distribution. Drugs that are in their neutral form must have HBs with water that are energetically more favorable than HBs between their solid-state molecules in order for them to have good aqueous solubility. The fact that most medicines' intracellular unbound concentrations cannot be determined *in vivo* is a substantial difficulty in the design of new therapies. HBDs can result in extremely poor solubility, permeability, and bioavailability, especially for substances in the upper end of the molecular weight property range. On the other hand, compounds that solely consist of HBAs and do not have any formal donor groups can nevertheless maintain good permeability, which in turn allows for BBB penetration and bioavailability [145].

References

1 Fedi, A., Vitale, C., Ponschin, G. et al. (2021). *In vitro* models replicating the human intestinal epithelium for absorption and metabolism studies: a systematic review. *Journal of Controlled Release* 335: 247–268.
2 Hou, T. and Wang, J. (2008). Structure–ADME relationship: still a long way to go? *Expert Opinion on Drug Metabolism & Toxicology* 4 (6): 759–770.
3 Zhong, H.A. (2017). ADMET properties: overview and current topics. In: *Drug Design: Principles and Applications* (ed. A. Grover), 113–133. Springer.
4 Pocock, K., Delon, L., Bala, V. et al. (2017). Intestine-on-a-chip microfluidic model for efficient *in vitro* screening of oral chemotherapeutic uptake. *ACS Biomaterials Science & Engineering* 3 (6): 951–959.
5 Di, L., Artursson, P., Avdeef, A. et al. (2020). The critical role of passive permeability in designing successful drugs. *ChemMedChem* 15 (20): 1862–1874.
6 Fagerholm, U., Nılsson, D., Knutson, L., and Lennernäs, H. (1999). Jejunal permeability in humans *in vivo* and rats *in situ*: investigation of molecular size selectivity and solvent drag. *Acta Physiologica Scandinavica* 165 (3): 315–324.
7 Di, L. and Kerns, E.H. (ed.) (2015). *Blood–Brain Barrier in Drug Discovery: Optimizing Brain Exposure of CNS Drugs and Minimizing Brain Side Effects for Peripheral Drugs*. John Wiley & Sons.
8 Fowler, S., Chen, W.L.K., Duignan, D.B. et al. (2020). Microphysiological systems for ADME-related applications: current status and recommendations for system development and characterization. *Lab on a Chip* 20 (3): 446–467.
9 Lamberti, G., Cascone, S., Marra, F. et al. (2016). Gastrointestinal behavior and ADME phenomena: II. *In silico* simulation. *Journal of Drug Delivery Science and Technology* 35: 165–171.
10 Amidon, G.L., Lennernäs, H., Shah, V.P., and Crison, J.R. (1995). A theoretical basis for a biopharmaceutic drug classification: the correlation of *in vitro* drug product dissolution and *in vivo* bioavailability. *Pharmaceutical Research* 12: 413–420.
11 Charoo, N.A., Mohamed, E.M., Eltigani, M. et al. (2024). Drug and formulation development processes. In: *From Current to Future Trends in Pharmaceutical Technology*, 257–292. Academic Press.
12 Benet, L.Z., Hosey, C.M., Ursu, O., and Oprea, T.I. (2016). BDDCS, the Rule of 5 and drugability. *Advanced Drug Delivery Reviews* 101: 89–98.

13 Komura, H., Watanabe, R., and Mizuguchi, K. (2023). The trends and future prospective of *in silico* models from the viewpoint of ADME evaluation in drug discovery. *Pharmaceutics* 15 (11): 2619.

14 Yanni, S.B. (2015). *Translational ADMET for Drug Therapy: Principles, Methods, and Pharmaceutical Applications*. John Wiley & Sons.

15 Balakin, K.V., Savchuk, N.P., and Tetko, I.V. (2006). *In silico* approaches to prediction of aqueous and DMSO solubility of drug-like compounds: trends, problems and solutions. *Current Medicinal Chemistry* 13 (2): 223–241.

16 Prajapati, R.N., Tekade, R.K., Gupta, U. et al. (2009). Dendrimer-mediated solubilization, formulation development and *in vitro in vivo* assessment of piroxicam. *Molecular Pharmaceutics* 6: 940950.

17 Kayat, J., Gajbhiye, V., Tekade, R.K., and Jain, N.K. (2011). Pulmonary toxicity of carbon nanotubes: a systematic report. *Nanomedicine* 7: 4049.

18 Jain, N.K. and Tekade, R.K. (2013). *Dendrimers for Enhanced Drug Solubilization. Drug Delivery Strategies for Poorly Water-Soluble Drugs*, 373409. John Wiley & Sons Ltd.

19 Ghanghoria, R., Tekade, R.K., Mishra, A.K. et al. (2016). Luteinizing hormone-releasing hormone peptide tethered nanoparticulate system for enhanced antitumoral efficacy of paclitaxel. *Nanomedicine (London)* 11: 797816.

20 Kuentz, M. and Imanidis, G. (2013). *In silico* prediction of the solubility advantage for amorphous drugs – are there property-based rules for drug discovery and early pharmaceutical development. *European Journal of Pharmaceutical Sciences* 48 (3): 554562.

21 Khojasteh, S.C., Wong, H., and Hop, C.E. (2011). *Drug Metabolism and Pharmacokinetics Quick Guide*. Springer Science & Business Media.

22 Zhong, H.A., Mashinson, V., Woolman, T.A., and Zha, M. (2013). Understanding the molecular properties and metabolism of top prescribed drugs. *Current Topics in Medicinal Chemistry* 13 (11): 1290–1307.

23 Di, L., Kerns, E.H., and Carter, G.T. (2009). Drug-like property concepts in pharmaceutical design. *Current Pharmaceutical Design* 15 (19): 2184–2194.

24 Könczöl, Á. and Dargó, G. (2018). Brief overview of solubility methods: recent trends in equilibrium solubility measurement and predictive models. *Drug Discovery Today: Technologies* 27: 3–10.

25 Sou, T. and Bergström, C.A. (2018). Automated assays for thermodynamic (equilibrium) solubility determination. *Drug Discovery Today: Technologies* 27: 11–19.

26 Kostewicz, E.S., Abrahamsson, B., Brewster, M. et al. (2014). *In vitro* models for the prediction of *in vivo* performance of oral dosage forms. *European Journal of Pharmaceutical Sciences* 57: 342–366.

27 Wang, Q., Fotaki, N., and Mao, Y. (2009). Biorelevant dissolution: methodology and application in drug development. *Dissolution Technologies* 16 (3): 6–12.

28 Smith, D.A., Beaumont, K., Maurer, T.S., and Di, L. (2018). Clearance in drug design: miniperspective. *Journal of Medicinal Chemistry* 62 (5): 2245–2255.

29 Tekade, R.K., Dutta, T., Gajbhiye, V., and Jain, N.K. (2009). Exploring dendrimer towards dual drug delivery: pH responsive simultaneous drug-release kinetics. *Journal of Microencapsulation* 26 (4): 287–296.

30 Dwivedi, P., Tekade, R.K., and Jain, N.K. (2013). Nanoparticulate carrier mediated intranasal delivery of insulin for the restoration of memory signaling in Alzheimer' disease. *Current Nanoscience* 9: 4655.

31 Bergström, C.A. (2005). Computational models to predict aqueous drug solubility, permeability and intestinal absorption. *Expert Opinion on Drug Metabolism & Toxicology* 1 (4): 613627.

32 Bergström, C., Charman, W., and Porter, C. (2016). Computational prediction of formulation strategies for beyondrule-of-5 compounds. *Advanced Drug Delivery Reviews* 101: 6.

33 Gerhard, K. (2013). *Drug Design: Methodology, Concepts and Mode-of-Action*. Springer 850 p.

34 Liu, X., Testa, B., and Fahr, A. (2011). Lipophilicity and its relationship with passive drug permeation. *Pharmaceutical Research* 28: 962–977.

35 Vogel, H.G. (ed.) (2006). *Drug Discovery and Evaluation: Safety and Pharmacokinetic Assays; with 125 Tables*. Springer Science & Business Media.

36 Jelfs, S., Ertl, P., and Selzer, P. (2007). Estimation of pK_a for druglike compounds using semiempirical and information-based descriptors. *Journal of Chemical Information and Modeling* 47 (2): 450–459.

37 Milletti, F., Storchi, L., Sforna, G., and Cruciani, G. (2007). New and original pK_a prediction method using grid molecular interaction fields. *Journal of Chemical Information and Modeling* 47 (6): 2172–2181.

38 Refsgaard, H., Jensen, B., Brockhoff, P. et al. (2005). *In silico* prediction of membrane permeability from calculated molecular parameters. *Journal of Medicinal Chemistry* 48 (3): 805811.

39 König, J., Müller, F., and Fromm, M.F. (2013). Transporters and drug–drug interactions: important determinants of drug disposition and effects. *Pharmacological Reviews* 65 (3): 944–966.

40 Veber, D.F., Johnson, S.R., Cheng, H.Y. et al. (2002). Molecular properties that influence the oral bioavailability of drug candidates. *Journal of Medicinal Chemistry* 45 (12): 2615–2623.

41 Avdeef, A. (2012). *Absorption and Drug Development: Solubility, Permeability, and Charge State*. John Wiley & Sons.

42 Vertzoni, M., Augustijns, P., Grimm, M. et al. (2019). Impact of regional differences along the gastrointestinal tract of healthy adults on oral drug absorption: an UNGAP review. *European Journal of Pharmaceutical Sciences* 134: 153–175.

43 Shore, P.A., Brodie, B.B., and Hogben, C.A.M. (1957). The gastric secretion of drugs: a pH partition hypothesis. *The Journal of Pharmacology and Experimental Therapeutics* 119 (3): 361–369.

44 Trapp, S., Rosania, G.R., Horobin, R.W., and Kornhuber, J. (2008). Quantitative modeling of selective lysosomal targeting for drug design. *European Biophysics Journal* 37: 1317–1328.

45 Charifson, P.S. and Walters, W.P. (2014). Acidic and basic drugs in medicinal chemistry: a perspective. *Journal of Medicinal Chemistry* 57 (23): 9701–9717.

46 Treyer, A., Mateus, A., Wisniewski, J.R. et al. (2018). Intracellular drug bioavailability: effect of neutral lipids and phospholipids. *Molecular Pharmaceutics* 15 (6): 2224–2233.

47 Grasset, E., Pinto, M., Dussaulx, E. et al. (1984). Epithelial properties of human colonic carcinoma cell line Caco-2: electrical parameters. *The American Journal of Physiology* 247 (3): C260–C267.

48 Artursson, P. and Karlsson, J. (1991). Correlation between oral drug absorption in humans and apparent drug permeability coefficients in human intestinal epithelial (Caco-2) cells. *Biochemical and Biophysical Research Communications* 175 (3): 880–885.

49 Lennernäs, H., Palm, K., Fagerholm, U., and Artursson, P. (1996). Comparison between active and passive drug transport in human intestinal epithelial (Caco-2) cells *in vitro* and human jejunum *in vivo*. *International Journal of Pharmaceutics* 127 (1): 103–107.

50 Putnam, W.S., Pan, L., Tsutsui, K. et al. (2002). Comparison of bidirectional cephalexin transport across MDCK and Caco-2 cell monolayers: interactions with peptide transporters. *Pharmaceutical Research* 19 (1): 27–33.

51 Kerns, E., Di, L., Petusky, S. et al. (2004). Combined application of parallel artificial membrane permeability assay and Caco-2 permeability assay in drug discovery. *Journal of Pharmaceutical Sciences* 93 (6): 1440–1453.

52 Oja, M. and Maran, U. (2018). pH-permeability profiles for drug substances: experimental detection, comparison with human intestinal absorption and modelling. *European Journal of Pharmaceutical Sciences* 2018 (123): 429–440.

53 Oja, M. and Maran, U. (2015). The permeability of an artificial membrane for wide range of pH in human gastrointestinal tract: experimental measurements and quantitative structure–activity relationship. *Molecular Informatics* 34 (6–7): 493–506.

54 Verma, R.P., Hansch, C., and Selassie, C.D. (2007). Comparative QSAR studies on PAMPA/ modified PAMPA for high throughput profiling of drug absorption potential with respect to Caco-2 cells and human intestinal absorption. *Journal of Computer-Aided Molecular Design* 21 (1–3): 3–22.

55 Williams, J., Siramshetty, V., Nguyễn, Đ.T. et al. (2022). Using *in vitro* ADME data for lead compound selection: an emphasis on PAMPA pH 5 permeability and oral bioavailability. *Bioorganic & Medicinal Chemistry* 15 (56): 116588.

56 Chen, M., Lan, H., Jin, K., and Chen, Y. (2023). Responsive nanosystems for targeted therapy of ulcerative colitis: current practices and future perspectives. *Drug Delivery* 30 (1): 2219427.

57 Meijer DKF, & Groothuis GMM. Hepatic transport of drugs and proteins W. Graham, E. Cooksley, Geoffrey McCaughan, Lawrie Powell, Richard Smallwood. In *Oxford Textbook of Clinical Hepatology* Oxford University Press, vol. 1 (N. McItyre, J.-P. Benhamou, J. Bircher, M. Rizzetto, and J. Rodes, eds.). 1991.

58 Chung, T.D., Terry, D.B., Smith, L.H. et al. (2015). *In vitro* and *in vivo* assessment of ADME and PK properties during lead selection and lead optimization – guidelines, benchmarks and rules of thumb. 2015 Sep 9. In: *Assay Guidance Manual [Internet]* (ed. S. Markossian, A. Grossman, M. Arkin, et al.). Bethesda, MD: Eli Lilly & Company and the National Center for Advancing Translational Sciences.

59 Chunduri, V. and Maddi, S. (2023). Role of *in vitro* two-dimensional (2D) and three-dimensional (3D) cell culture systems for ADME-Tox screening in drug discovery and development: a comprehensive review. *ADMET and DMPK* 11 (1): 1–32.

60 Smith, D.A., Beaumont, K., Maurer, T.S., and Di, L. (2015). Volume of distribution in drug design: miniperspective. *Journal of Medicinal Chemistry* 58 (15): 5691–5698.

61 Endo, S. and Goss, K.U. (2011). Serum albumin binding of structurally diverse neutral organic compounds: data and models. *Chemical Research in Toxicology* 24 (12): 2293–2301.

62 Stevens, J.C., Shipley, L.A., Cashman, J.R. et al. (1993). Comparison of human and rhesus monkey *in vitro* phase I and phase II hepatic drug metabolism activities. *Drug Metabolism and Disposition* 21 (5): 753–760.

63 Waterbeemed, H.V.D., Smith, D.A., and Jones, B.C. (2011). Lipophilicity in PK designs: methyl, ethyl, futile. *The Journal of Computer-Aided Molecular Design* 15: 273–286.

64 Caldwell, G.W. and Yan, Z. (2014). *Optimization in Drug Discovery*. New York: Springer.

65 Gupta, P.K. (2016). *Fundamentals of Toxicology: Essential Concepts and Applications*. Academic Press.

66 Lin, J.H. and Lu, A.Y. (1997). Role of pharmacokinetics and metabolism in drug discovery and development. *Pharmacological Reviews* 49 (4): 403–449.

67 Norinder, U. and Haeberlein, M. (2002). Computational approaches to the prediction of the blood–brain distribution. *Advanced Drug Delivery Reviews* 54 (3): 291–313.

68 Raney, K.D., Coles, B., Guengerich, F.P., and Harris, T.M. (1992). The endo-8,9-epoxide of aflatoxin B1: a new metabolite. *Chemical Research in Toxicology* 5 (3): 333–335.

69 Lin, J., Sahakian, D.C., De Morais, S.M. et al. (2003). The role of absorption, distribution, metabolism, excretion and toxicity in drug discovery. *Current Topics in Medicinal Chemistry* 3 (10): 1125–1154.

70 De Kanter, R., Olinga, P., De Jager, M.H. et al. (1999). Organ slices as an *in vitro* test system for drug metabolism in human liver, lung and kidney. *In Vitro Toxicology* 13 (4–5): 737–744.

71 Van De Waterbeemd, H. and Gifford, E. (2003). ADMET *in silico* modelling: towards prediction paradise? *Nature Reviews. Drug Discovery* 2 (3): 192–204.

72 Hu, B., Zhou, X., Mohutsky, M.A., and Desai, P.V. (2020). Structure–property relationships and machine learning models for addressing CYP3A4-mediated victim drug–drug interaction risk in drug discovery. *Molecular Pharmaceutics* 17 (9): 3600–3608.

73 Valicherla, G.R., Mishra, A., Lenkalapelly, S. et al. (2019). Investigation of the inhibition of eight major human cytochrome P450 isozymes by a probe substrate cocktail *in vitro* with emphasis on CYP2E1. *Xenobiotica* 49 (12): 1396–1402.

74 Zhao, M., Ma, J., Li, M. et al. (2021). Cytochrome P450 enzymes and drug metabolism in humans. *International Journal of Molecular Sciences* 22 (23): 12808.

75 de Groot, M.J., Wakenhut, F., Whitlock, G., and Hyland, R. (2009). Understanding CYP2D6 interactions. *Drug Discovery Today* 14 (19–20): 964–972.

76 Lang, D. and Kalgutkar, A.S. (2003). Non-P450 mediated oxidative metabolism of xenobiotics. In: *Drug Metabolizing Enzymes: Cytochrome P450 and Other Enzymes in Drug Discovery and Development* (ed. J. Lee, R.S. Obach, and M.B. Fisher), 483–540. Taylor & Francis.

77 Wen, B. and Nelson, S.D. (2011). Common biotransformation reactions. In: *Mass Spectrometry in Drug Metabolism and Disposition: Basic Principles and Applications* (ed. M.S. Lee and M. Zhu), 13–41. Wiley.

78 Başaran, R. and Benay, C. (2017). Flavin containing monooxygenases and metabolism of xenobiotics. *Turkish Journal of Pharmaceutical Sciences* 14 (1): 90.

79 Wynalda, M.A., Hutzler, J.M., Koets, M.D. et al. (2003). *In vitro* metabolism of clindamycin in human liver and intestinal microsomes. *Drug Metabolism and Disposition* 31 (7): 878–887.

80 Krueger, S.K. and Williams, D.E. (2005). Mammalian flavin-containing monooxygenases: structure/function, genetic polymorphisms and role in drug metabolism. *Pharmacology & Therapeutics* 106 (3): 357–387.

81 Richardson, S.J., Bai, A., Kulkarni, A.A., and Moghaddam, M.F. (2016). Efficiency in drug discovery: liver S9 fraction assay as a screen for metabolic stability. *Drug Metabolism Letters* 10 (2): 83–90.

82 Kumar, G.N. and Surapaneni, S. (2001). Role of drug metabolism in drug discovery and development. *Medicinal Research Reviews* 21 (5): 397–411.

83 Lin, J., Sahakian, D.C., De Morais, S.M. et al. (2003). The role of absorption, distribution, metabolism, excretion and toxicity in drug discovery. *Current Topics in Medicinal Chemistry* 3 (10): 1125–1154.

84 Gaohua, L., Miao, X., and Dou, L. (2021). Crosstalk of physiological pH and chemical pK_a under the umbrella of physiologically based pharmacokinetic modeling of drug absorption, distribution, metabolism, excretion, and toxicity. *Expert Opinion on Drug Metabolism & Toxicology* 17 (9): 1103–1124.

85 Eddershaw, P.J., Beresford, A.P., and Bayliss, M.K. (2000). ADME/PK as part of a rational approach to drug discovery. *Drug Discovery Today* 5 (9): 409–414.

86 Masimirembwa, C.M., Bredberg, U., and Andersson, T.B. (2003). Metabolic stability for drug discovery and development: pharmacokinetic and biochemical challenges. *Clinical Pharmacokinetics* 42: 515–528.

87 Paine, S.W., Barton, P., Bird, J. et al. (2010). A rapid computational filter for predicting the rate of human renal clearance. *Journal of Molecular Graphics & Modelling* 29 (4): 529–537.

88 Dresser, M.J., Leabman, M.K., and Giacomini, K.M. (2001). Transporters involved in the elimination of drugs in the kidney: organic anion transporters and organic cation transporters. *Journal of Pharmaceutical Sciences* 90 (4): 397–421.

89 Hagenbuch, B. (1997). Molecular properties of hepatic uptake systems for bile acids and organic anions. *The Journal of Membrane Biology* 160 (1): 1–8.

90 Jetter, A. and Kullak-Ublick, G.A. (2020). Drugs and hepatic transporters: a review. *Pharmacological Research* 154: 104234.

91 Ayrton, A. and Morgan, P. (2001). Role of transport proteins in drug absorption, distribution and excretion. *Xenobiotica* 31 (8–9): 469–497.

92 Drwal, M.N., Banerjee, P., Dunkel, M. et al. (2014). ProTox: a web server for the *in silico* prediction of rodent oral toxicity. *Nucleic Acids Research* 42 (Web Server issue): W53–W58.

93 Li, A.P. (2002). A review of the common properties of drugs with idiosyncratic hepatotoxicity and the "multiple determinant hypothesis" for the manifestation of idiosyncratic drug toxicity. *Chemico-Biological Interactions* 142 (1–2): 7–23.

94 Lloyd, S., Hayden, M.J., Sakai, Y. et al. (2002). Differential *in vitro* hepatotoxicity of troglitazone and rosiglitazone among cryopreserved human hepatocytes from 37 donors. *Chemico-Biological Interactions* 142 (1–2): 57–71.

95 Li, A.P. (2004). *In vitro* approaches to evaluate ADMET drug properties. *Current Topics in Medicinal Chemistry* 4 (7): 701–706.

96 Mannhold, R. and Dross, K. (1996). Calculation procedures for molecular lipophilicity: a comparative study. *Quantitative Structure–Activity Relationships* 15 (5): 403–409.

97 Leo, A., Hansch, C., and Elkins, D. (1971). Partition coefficients and their uses. *Chemical Reviews* 71 (6): 525–616.

98 Soares, J.X., Santos, Á., Fernandes, C., and Pinto, M.M. (2022). Liquid chromatography on the different methods for the determination of lipophilicity: an essential analytical tool in medicinal chemistry. *Chem* 10 (8): 340.

99 Buchwald, P. and Bodor, N. (1998). Octanol-water partition: searching for predictive models. *Current Medicinal Chemistry* 5 (5): 353–380.

100 Pal, Y., Deb, P.K., Bandopadhyay, S. et al. (2018). Role of physicochemical parameters on drug absorption and their implications in pharmaceutical product development. In: *Dosage Form Design Considerations* (ed. R.K. Tekade), 85–116. Academic Press.

101 Hansch, C. (1969). Quantitative approach to biochemical structure–activity relationships. *Accounts of Chemical Research* 2 (8): 232–239.

102 Martin, Y.C. (2018). How medicinal chemists learned about log *P*. *Journal of Computer-Aided Molecular Design* 32: 809–819.

103 Bhal, S.K., Kassam, K., Peirson, I.G., and Pearl, G.M. (2007). The rule of five revisited: applying log D in place of log *P* in drug-likeness filters. *Molecular Pharmaceutics* 4 (4): 556–560.

104 Nowosielski, B.E. and Fein, J.B. (1998). Experimental study of octanol–water partition coefficients for 2,4,6-trichlorophenol and pentachlorophenol: derivation of an empirical model of chlorophenol partitioning behaviour. *Applied Geochemistry* 13 (7): 893–904.

105 Westall, J.C., Leuenberger, C., and Schwarzenbach, R.P. (1985). Influence of pH and ionic strength on the aqueous–nonaqueous distribution of chlorinated phenols. *Environmental Science & Technology* 19 (2): 193–198.

106 Hitzel, L., Watt, A.P., and Locker, K.L. (2000). An increased throughput method for the determination of partition coefficients. *Pharmaceutical Research* 17: 1389–1395.

107 Liang, C. and Lian, H.Z. (2015). Recent advances in lipophilicity measurement by reversed-phase high-performance liquid chromatography. *TrAC Trends in Analytical Chemistry* 68: 28–36.

108 Bergström, C.A., Norinder, U., Luthman, K., and Artursson, P. (2002). Experimental and computational screening models for prediction of aqueous drug solubility. *Pharmaceutical Research* 19: 182–188.

109 Hanson, K.B., Hoff, D.J., Lahren, T.J. et al. (2019). Estimating *n*-octanol-water partition coefficients for neutral highly hydrophobic chemicals using measured *n*-butanol-water partition coefficients. *Chemosphere* 218: 616–623.

110 Akgün, H., Aytemir, M., Balkan, A. et al. (2019). *Farmasötik ve Medisinal Kimya*. Hacettepe University Press.

111 Sharapova, A., Ol'khovich, M., Blokhina, S., and Perlovich, G.L. (2022). Experimental examination of solubility and lipophilicity as pharmaceutically relevant points of novel bioactive hybrid compounds. *Molecules* 27 (19): 6504.

112 Brook, D.N., Dobbs, A.J., and Williams, N. (1986). Octanol–water partition coefficients (P): measurement, estimation, and interrelation, particularly for chemicals with P > E + 5. *Ecotoxicology and Environmental Safety* 11: 257–260.

113 Godard, T. and Grushka, E. (2011). The use of phospholipid modified column for the determination of lipophilic properties in high performance liquid chromatography. *Journal of Chromatography A* 1218 (9): 1211–1218.

114 Komsta, Ł., Skibiński, R., Berecka, A. et al. (2010). Revisiting thin-layer chromatography as a lipophilicity determination tool – a comparative study on several techniques with a model solute set. *Journal of Pharmaceutical and Biomedical Analysis* 53 (4): 911–918.

115 Abbas, N.S., Mohamed, Y.A.S., Derayea, S.M. et al. (2021). Simple TLC–spectrodensitometric method for studying lipophilicity and quantitative analysis of hypoglycemic drugs in their binary mixture. *Biomedical Chromatography* 35 (11): e5154.

116 van De Waterbeemd, H. and Mannhold, R. (1996). Programs and methods for calculation of log *P*-values. *Quantitative Structure–Activity Relationships* 15 (5): 410–412.

117 Manallack, D.T. (2007). The pK_a distribution of drugs: application to drug discovery. *Perspectives in Medicinal Chemistry* 1: 1177391X0700100003.

118 Alqahtani, S. (2017). *In silico* ADME-Tox modeling: progress and prospects. *Expert Opinion on Drug Metabolism & Toxicology* 13 (11): 1147–1158.

119 Barbour, N.P. and Lipper, R.A. (2008). Introduction to biopharmaceutics and its role in drug development. In: *Biopharmaceutics Applications in Drug Development* (ed. R. Krishna and L. Yu), 1–25. Boston, MA: Springer US.

120 Beale, J.M. Jr. and Block, J.H. (2010). *Wilson and Gisvold's Textbook of Organic Medicinal and Pharmaceutical Chemistry*. New Delhi: The Point 1010 p.

121 Kerns, E.H. and Di, L. (2004). Physicochemical profiling: overview of the screens. *Drug Discovery Today: Technologies* 1 (4): 343–348.

122 Zhong, H.A. (2017). ADMET properties: overview and current topics. *Drug Design: Principles and Applications* 113–133.

123 Cruciani, G., Milletti, F., Storchi, L. et al. (2009). *In silico* pK_a prediction and ADME profiling. *Chemistry & Biodiversity* 6 (11): 1812–1821.

124 He, X. (2009). Integration of physical, chemical, mechanical, and biopharmaceutical properties in solid oral dosage form development. In: *Developing Solid Oral Dosage Forms* (ed. Y. Qiu, Y. Chen, G.G.Z. Zhang, et al.), 407–441. Academic Press.

125 Khojasteh, S.C., Wong, H., and Hop, C.E. (2011). *Drug Metabolism and Pharmacokinetics Quick Guide*. Springer Science & Business Media.

126 Malan, S.F., Chetty, D.J., and Du Plessis, J. (2002). Physicochemical properties of drugs and membrane permeability. *South African Journal of Science* 98 (7): 385–391.

127 Tekade, R.K. (2018). *Basic Fundamentals of Drug Delivery*. Academic Press.

128 Kapoor, D.U., Singh, S., Sharma, P., and Prajapati, B.G. (2023). Amorphization of low soluble drug with amino acids to improve its therapeutic efficacy: a state-of-art-review. *AAPS PharmSciTech* 24 (8): 253.

129 Williams, R.O. III, Watts, A.B., and Miller, D.A. (ed.) (2012). *Formulating Poorly Water Soluble Drugs*, vol. 3. New York: Springer.

130 Bhalani, D.V., Nutan, B., Kumar, A., and Singh Chandel, A.K. (2022). Bioavailability enhancement techniques for poorly aqueous soluble drugs and therapeutics. *Biomedicine* 10 (9): 2055.

131 Dressman, J.B., Vertzoni, M., Goumas, K., and Reppas, C. (2007). Estimating drug solubility in the gastrointestinal tract. *Advanced Drug Delivery Reviews* 59 (7): 591–602.

132 Koziolek, M., Schneider, F., Grimm, M. et al. (2015). Intragastric pH and pressure profiles after intake of the high-caloric, high-fat meal as used for food effect studies. *Journal of Controlled Release* 220 (Part A): 71–78.

133 Jambhekar, S.S. (2002). Physicochemical and biopharmaceutical properties of drug substances and pharmacokinetics. *Foye's Principles of Medicinal Chemistry* 61–105.

134 Komura, H., Watanabe, R., and Mizuguchi, K. (2023). The trends and future prospective of *in silico* models from the viewpoint of ADME evaluation in drug discovery. *Pharmaceutics* 15 (11): 2619.

135 Zhou, Y., Lv, C., Liu, X. et al. (2023). An overview on polymorph preparation methods of active pharmaceutical ingredients. *Crystal Growth & Design* 24 (1): 584–600.

136 Lieberman, H. and Vemuri, N.M. (2015). Chemical and physicochemical approaches to solve formulation problems. In: *The Practice of Medicinal Chemistry*, 767–791. Academic Press.

137 Zhou, Y., Wang, J., Xiao, Y. et al. (2018). The effects of polymorphism on physicochemical properties and pharmacodynamics of solid drugs. *Current Pharmaceutical Design* 24 (21): 2375–2382.

138 Yao, C., Zhang, S., Wang, L., and Tao, X. (2022). Recent advances in polymorph discovery methods of organic crystals. *Crystal Growth & Design* 23 (1): 637–654.

139 Pindelska, E., Sokal, A., and Kolodziejski, W. (2017). Pharmaceutical cocrystals, salts and polymorphs: Advanced characterization techniques. *Advanced Drug Delivery Reviews* 117: 111–146.

140 Brog, J.P., Chanez, C.L., Crochet, A., and Fromm, K.M. (2013). Polymorphism, what it is and how to identify it: a systematic review. *RSC Advances* 3 (38): 16905–16931.

141 Saha, B.K., Nath, N.K., and Thakuria, R. (2023). Polymorphs with remarkably distinct physical and/or chemical properties. *Chemical Record* 23 (1): e202200173.

142 Chistyakov, D. and Sergeev, G. (2020). The polymorphism of drugs: new approaches to the synthesis of nanostructured polymorphs. *Pharmaceutics* 12 (1): 34.

143 Tinworth, C.P. and Young, R.J. (2020). Facts, patterns, and principles in drug discovery: appraising the rule of 5 with measured physicochemical data. *Journal of Medicinal Chemistry* 63 (18): 10091–10108.

144 Rankovic, Z. (2015). CNS drug design: balancing physicochemical properties for optimal brain exposure. *Journal of Medicinal Chemistry* 58 (6): 2584–2608.

145 Kenny, P.W. (2022). Hydrogen-bond donors in drug design. *Journal of Medicinal Chemistry* 65 (21): 14261–14275.

7

In Silico Modeling and Drug Design

Sonali S. Shinde[1], *Sanket S. Rathod*[2], *and Sohan S. Chitlange*[1]

[1] *Department of Pharmaceutical Chemistry, Dr. D. Y. Patil Institute of Pharmaceutical Sciences and Research, Pune, Maharashtra, India*
[2] *Department of Pharmaceutical Chemistry, Bharati Vidyapeeth College of Pharmacy, Kolhapur, Maharashtra, India*

7.1 Introduction

Target recognition and lead compound discovery are currently leading developments in drug discovery, with an emphasis on the causes of diseases and their comprehension. A personalized medicine system dependent on the conversion of molecular phases (and developments, from DNA to RNA to protein) has become essential in the development of drugs in the era of personal medicine and more updated economic healthcare outcomes [1, 2]. The molecular identification of illness is an essential requirement for developing a system like this, and environmental factors and the gut microbiota must also be taken into account. Safety regulations are becoming more stringent at the same time [3, 4]. We believe that to deal with the previously discussed interaction in a high-throughput form, data technology and chemoinformatic methods should be used in concert with human intelligence and artificial interplay. While people are limited in their ability to identify patterns by data-intensive and intellectually challenging environments and procedures computer algorithms may not be able to do so [5]. This coordination will support group data techniques and help in direct judgment calls, and proper and quick effective data output. Large and varied data sets necessitate stringent screening as well as in-depth analysis and interpretation. Scientists working in biomedicine must simultaneously collaborate while making selections in an efficient and effective manner. Large-scale volumes of intricate, multifaceted data must be meaningfully gathered, mined, and analyzed to achieve this in such an environment; computer-aided drug discovery will be more efficient when methodologies for selecting targets and verification are dependable. Additionally, new computational frameworks grounded in systems biology and networks incorporate omics databases and improve combinational drug development strategies. In such an environment, more effective computer-aided drug discovery will be made possible by reliable target choice and verification combined with drug discovery methodologies. Furthermore, new computational models based on networks and systems biology integrate omics databases and enhance combinational drug development approaches [6].

Computational Methods for Rational Drug Design, First Edition. Edited by Mithun Rudrapal.

7.2 Target Identification

A biological entity, typically a protein, that can modify the symptoms of a disease is referred to as a target for drugs [7]. Therefore, the first and most crucial phase in the drug discovery system is the recognition of prime drug targets. Traditional methods of discovering drug targets involve conducting experiments to find genes that are expressed differently in healthy and infected cells or tissues, as well as proteins that have strong connections to proteins associated with the disease.

7.2.1 Experimental Approaches

Target recognition using traditional experimental techniques necessitates biochemical and molecular investigations of disease pathophysiology. Although these investigations depend on the information on different disorders, there may be long-duration methods of locating possible treatment targets. Genome-scale testing methods like target deconvolution, haploinsufficiency profiling (HIP), and solid isotope identifying by amino acids in cell culture (SILAC) have been introduced recently to improve target recognition. HIP is an assay where genome-wide investigation is done to determine possible drug targets, finds gene products linked to the sensitivity of disease cell lines, and exposes cells to chemical compounds [8]. One benefit of the HIP assay is that it can assess thousands of genes simultaneously and does not require any prior understanding of the pathophysiology of the disease. With traditional drug discovery methods, it is difficult to recognize viable therapeutic targets because of the complex pathophysiologies of several illnesses. In such an example, a different way might be employed: drugs that modify the symptoms of a disease could be found through screening, which would reveal target proteins matching criteria [9]. A variety of techniques, including affinity chromatography, biochemical suppression, and protein microarrays, are used for target deconvolution [10]. Figure 7.1 gives an idea of strategies for target identification. The effective reverse screening method known as SILAC

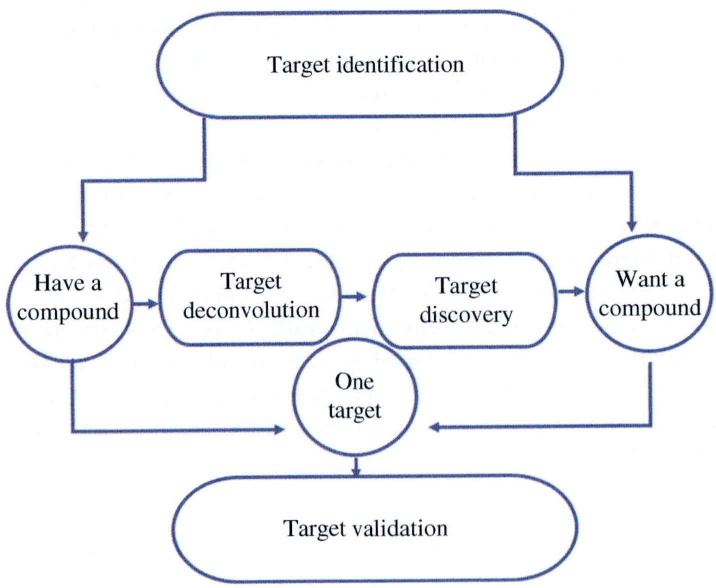

Figure 7.1 An overview of target identification strategies.

makes it possible to identify target proteins that are bound to medications and small-molecule probes in an objective, thorough, and reliable manner [11]. More precise determination of drug–protein interactions is now possible thanks to the recent integration of this method with affinity chromatography and quantitative mass spectrometry-based proteomics. Although SILAC has many advantages, its wide-ranging and practical use is limited by a number of drawbacks such as: (i) labeling of an isotope is expensive; (ii) complex equipment, such as (HRMS) high-resolution mass spectrometers, is needed; and (iii) it takes time to generate chemically immobilized drugs and ensure their biological activity [12].

7.2.2 Computational Target Identification

Experimental techniques are costly and usually applied at lower throughput scales because of their complex nature. To overcome these hurdles, computational techniques have been created to determine the best pharmacological targets [13]. The targeted proteins can be calculated computationally using actual experiment results [14, 15], anticipated from protein networks, or extracted from the literature survey [16]. Several web servers provide the names of possible drug targets that have been calculated using various databases, like Harmonizome and the Open Target Platform [17]. Instead, the concept that similar-structured ligands could attached to the same protein with the same binding energies and exhibit the same biological properties can be applied to find putative protein targets through the use of a reverse docking technique [18, 19]. One popular method for identifying drug targets is association-based recognition. As an illustration, the Open Targets Platform [20] incorporates information from a variety of sources, such as text-mined information from the scientific literature, animal model outcomes of experiments, and omics data. After that, the platform ranks genes based on how closely they are linked to illness [21]. Several machine learning and statistical methods such as the Similarity Ensemble method [22], PharmMapper, Tarfisdock, and TargetHunter have been established to determine the biological target of a drug in a pharmaceutical formulation [23–25]. A common method of protein target discovery is ligand-based when there is nonavailability of drug pathophysiology information. Lavecchia investigated several machine-learning methods created to process ligand finding by utilizing fingerprints and molecular descriptors that correspond to the physical–chemical characteristics of a drug [26]. Often applied in the making of descriptors, fingerprints, and predictive models offer a representation of the physical and chemical characteristics of a substance [27]. A subtractive approach might help achieve planned objectives more accurately. For example, the best therapeutic targets for the treatment of *Helicobacter pylori* infection can identify removing extracellular enzymes from respective humans and or gut flora, nonessential proteins, redundant enzyme, and materials from the *H. pylori* proteome [28] (Table 7.1).

7.2.3 Target Validation

Verifying if altering a biological function of a target influences the disease phenotype comes next after a target has been determined [37]. Modulating biological processes and assessing estimated targets can be done in a number of ways. Small interfering RNAs (siRNAs) [38] are the most popular of these techniques because they may replicate the effects of drugs by repressing translation, which causes the target protein to be temporarily suppressed [39, 40]. With siRNAs, target inhibition can be studied without the need for inhibitors or prior protein structure information [41]. However, the level of suppression by siRNAs can impact cellular physiologies variously and may

Table 7.1 Protein target databases.

Database	Description	URL	Reference
Drug bank	A collection of 13,857 drug entities including 2661 approved drug molecules and 1425 approved biologics (peptides, proteins, and vaccines)	https://go.dru http://gbank.com	[29]
ChEMBL	A collection of 13,382 drug targets and 1.9 million drug molecules	https://www.ebi.ac.uk/chembl	[30]
ChemBank	Information on hundreds of biomedical assays and millions of small drug compounds	https://data.broa http://dinstitute.org/chembank/assay	[31]
Therapeutic Target Database	Experimental validation data on 37,316 drug compounds and 3419 drug targets	http://db.idrblab.net/ttd	[32]
Traditional Chinese Medicines	Information on 37,170 unique compounds from 352 different herbs, minerals, and animal products	http://tcm.cmu.edu.tw/	[33]
MATADOR	Information on manually annotated drugs and targets from drug bank and super targets	http://matador.embl.de/	[34]
ChemSpider	Structural and text information on over 67 million chemical compounds	http://www.ch http://emspider.com	[35]
Chem2BioRDF	Information on chemical compounds, biological targets, and phenotypic data	http://chem2bio http://2rdf.org	[35]
Promiscuous	Information on 991,805 small molecules, 9430 drug targets and 2,727,520 drug–target interactions	http://bioinfo http://rmatics.charite.depromiscuous2/index.php	[36]

therefore result in incompatible effects for conditions with complicated pathophysiology such as neurological disorders [42]. In these situations, animal models with mutated or deleted target genes may provide more useful information for target approval.

7.3 Computer-Aided Drug Design

Drug discovery aims to find small substances so those can change the roles of an identified target protein; thus consequently, there is a change in the disease phenotype. Additionally, it is imperative to find small molecules with good pharmacokinetic parameters and minimum cell toxicity. A lengthy, expensive, and tedious series of complex steps, including pharmacokinetics, preclinical toxicity assessments, applicant confirmation, and drug candidate identification, are involved in the search for new pharmaceuticals. Traditional pharmaceutical research and development (R&D) is expensive and has a long duration. A drug specifically requires 10–14 years to come into the market, and each successful drug discovery is thought to have cost between $900 million and $1.9 billion USD [43]. Finding chemical compounds that are pharmacologically operational is the first challenge in the drug discovery process. In general, experimental HTS has a hit rate of 0.01–0.14% [44]. Approximately 40–70% of drug failures afterward are caused by deficiencies in ADME-Tox, which

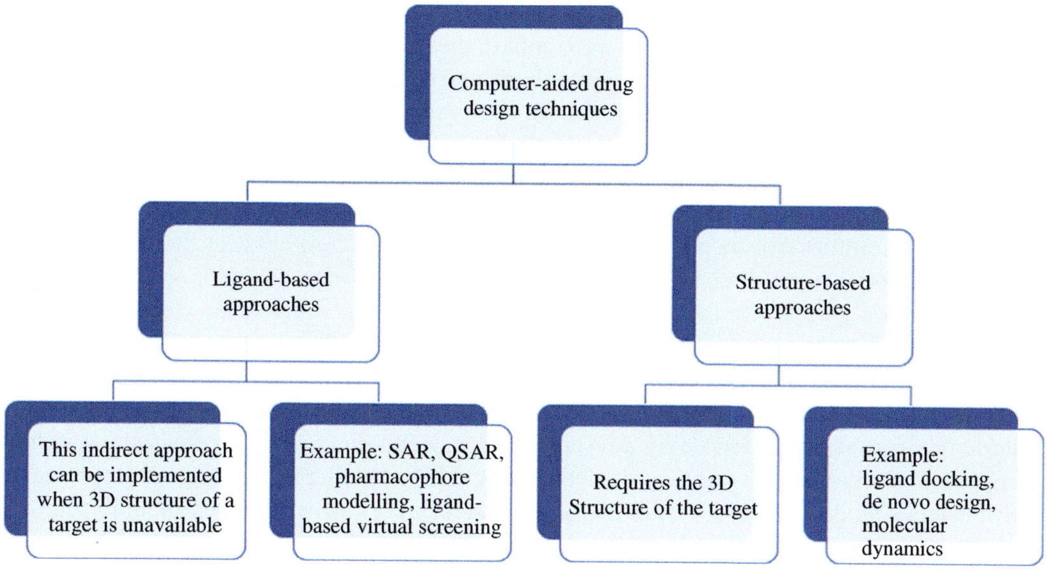

Figure 7.2 Classification of computer-aided drug design (CADD) techniques and mainly applied techniques. SAR, structure–activity relationship; QSAR, quantitative structure–activity relationship. *Source:* Koszła et al. [50]/MDPI/CC BY 4.0.

is another major obstacle [45, 46]. The use of *in silico* drug development technology has proven to be very beneficial to any pharmaceutical company for a long time [47]. The two key advantages of computational drug designing are it is cost effectiveness and time-saving capabilities. Moreover, it can be used in every stage, from drug testing to preclinical–clinical phases [48], greatly reducing the likelihood that drug procedures will not succeed. Thanks to recent developments in machine learning algorithms, gathered databases, and readily procured synthetic drug substance lists, *in silico* methods can practically test a wide range of chemical molecules and then immediately assess their ADMET aspects and identify potential pharmaceuticals with good potency and less cell toxicity [49]. Figure 7.2 shows the main two ways of computer-aided drug design techniques: (i) ligand-based approach and (ii) structure-based approach.

7.3.1 Ligand-based CADD

Ligand-based CADD techniques search for novel, more effective compounds by utilizing information about small molecules which interact with the target in concern. This data covers physicochemical characteristics, chemical structure, binding affinities, etc. These approaches are thought to be more effective than structure-based approaches in some circumstances. Selecting novel compounds based on their chemical resemblance to well-established active ones is one ligand-based strategy. Numerous fingerprint techniques are available for this purpose, enabling the effective visualization of a molecular structure in comparison to other molecules [51]. These techniques, which are based on chemical-based information about compounds, provide an extremely qualitative method for finding novel, best-effective ligands [29]. A suitable method for the determination of T-type blockers of calcium channels has been linked to neuropathic pain and seizures. This work was presented by Ijjaali and peers [30]. In this work, a ligand-based online testing procedure was carried out on a two million compound database by using Chem Axon's PF and CCG's

GpiDAPH3 fingerprints tools to evaluate 38 molecules for their good functionality on recombinant human CaV3.2. Half of the 38 molecules were potent, meaning that these hits were blocking the T-type current mediated by CaV3.2. The quantitative structure–activity relationship (QSAR) is another crucial method, where a QSAR model can clarify the connection between a group of molecules' structures and their intended response. Gathering a number of both inactive and active compounds against a targeted protein and creating descriptors that characterize their fundamental and physicochemical parameters make up the standard QSAR workflow. A forecasting tool for novel molecular entities can then be produced by using the model to connect these descriptors with the laboratory activity [52]. QSAR algorithms, which are always changing, can be used to implement a variety of 2D and 3D descriptors, including volume, molecular weight, distance between two atoms, types of atoms, rotatable bonds, electronegativity, atom distribution, solving characteristics, and aromaticity. There are multiple levels of definition for these descriptors, each with increasing complexity. This strategy works well for finding hit compounds that act as mGlu5 allosteric modulators, as evidenced by a recent study. A well-known pharmacological target for schizophrenia, Parkinson's disease, and anxiety is mGlu5 [53]. Mueller and colleagues generated a QSAR model using data from an earlier mGlu5 HTS screen [54], which allowed them to find 28 active pharmaceutical compounds that alter the signaling pathways of proteins. While HTS had a 0.2% hit rate, the QSAR model had an overall success rate of 3.6% [55, 56]. Proteochemometrics (PCM) and polypharmacology modeling is another area receiving attention in the context of computational drug discovery [57]. Using a single predictive model, PCM modeling integrates target and ligand data to forecast an outcome diverse of interest [58, 59]. Personalized medicine is made achievable by integrating information from targeted protein and ligand information into a unique machine-developing model, as opposed to relying solely on ligand- or protein-only approaches to determine the optimal drug therapy for a given genotype [60]. Applying PCM techniques, Frimurer et al. were able to find roughly 60 compounds for the prostaglandin D-2 receptor 2 (CRTH2) after testing of a list of 1.4 million structures [61, 62].

7.3.2 Structure-Based CADD

The use of 3D protein and DNA data in drug design started almost 30 years ago. The Protein Databank (PDB) is the biggest source of biomolecule structure data, primarily from X-ray crystallography and NMR methods. 1998 saw the addition of 2058 structures in all to the protein data bank. Till now, almost there have been 105,465 structures total as of 2014, an increase in depositions of roughly 7.5% per year. For the past few years, the pharmaceutical industry and academia have been using this abundance of structural data to inform their drug design strategies. Dynamic macromolecules by nature are proteins. An interaction of protein with a small molecule cannot be fully understood from a structured snapshot, nor can its binding site be determined. Molecular dynamics (MD) simulations represent one of the most significant facets of protein behavior research [34, 63]. In MD simulations, dihedral angles are modeled applying a sinusoidal function, while chemical interactions and atomic angles are modeled applying straightforward online springs. Since the hydrophobic and electrostatic interactions are computed employing Lennard-Jones potential and Coulomb's law, nonbonded forces like van der Waals interactions. These simulations have a big influence on the drug discovery industry when combined with experimental data. Historically, these computations have been carried out on CPU clusters using software that can parallelize complex system simulation processes [64]. The computations needed for these simulations have been established recently so that computer graphics and video game applications can handle them. MD simulations have eventually been accelerated by the graphics processing units

(GPUs), which were initially created to speed up video games, typically by magnitude range [33, 65]. Many MD functions are applied for the free energy calculation to compare the estimated values with the determined by experiment binding capacities of small drugs to a targeted protein. Molecular mechanisms Poisson–Boltzmann surface area (MM/PBSA), and linear interaction energy (LIE), free energy perturbation (FEP) techniques are some of these software programs [36, 66]. They can then be used for *in silico* affinity binding prediction after that. Among the applications of these techniques are the prediction of binding free energies, the forecasting of the binding mode of efavirenz to HIV-1 RT by MM-PBSA, the application of LIE models to predict the binding mode of β-secretase (BACE) inhibitors, and calculations of relative binding free energy of biotin and the interactions of analogs with streptavidin using FEP strategies [67]. Sometimes the structure of targeted protein for a drug design project remains a mystery. Here are some predicted tools for developing comparative models. Comparative modeling is utilized to determine the structure of a protein from a structural template because protein molecules with the same sequences typically have the same structures. The main applied *in silico* technique to accomplish this is the homology modeling, which creates a protein model by identifying a structurally related template protein with the same series, aligning their sequences, applying aligned region coordinates, predicting the construction of the model, and fine-tuning the target's missing atom coordinates. MODELER is one of the most popular programs for homology modeling [68] and SWISS-MODEL [69]. A chemo-attractant model of the receptor OXE-R, which is founded on the CXCR4 crystal structure [35] as a template, is one instance of how homology modeling is used in drug design. Consequently, a small-molecular modulator called Gue1654 was found via virtual screening techniques to inhibit a particular GPCR signaling pathway [70]. An additional instance is the prediction of the binding energy mechanism of antihypertensive medications to the type 1 angiotensin-II receptor (AT1). The receptor with a homology model and an investigation of the pharmaceutical binding mode utilizing pharmacophore and MD modeling were later confirmed upon the discovery of the crystal structure. When molecules are bound together to create a stable complex, docking [71] techniques are employed to predict which orientation a molecule will prefer to occupy on a protein. The versatility of the compounds or the targeted protein via the docking process is taken into account in different ways depending on the method [72, 73]. The most widely used approach holds that the protein docking site is rigid and that the ligand is flexible, typically after applying force fields from MD. Several software programs, including SURFLEX, Autodock, Gold [74], DOCK, AutoDock Vina, GLIDE [75], and others, are available for docking. Scoring functions are used to calculate the docking score, which is an assessment of the energetic affinity of the complex. These scoring functions may be empirical, knowledge-based, consensus-based, or based on molecular mechanics. For instance, SURFLEX employs an empirical function to assess the binding energetics, whereas DOCK applies the AMBER force field. Consensus scoring is a widely researched technique that predicts the binding energies of a compound for a specific target by combining multiple scoring algorithms. In one case study [76], Tuchinardi et al. determined the overall docking and scoring of multiple algorithms along 83 ligand–receptor X-ray structures. One more familiar technique for determining how likely it is for a molecule to attach to a targeted protein binding location is the pharmacophore modeling method. A pharmacophore is the set of electronic and steric properties needed to confirm the best supramolecular interaction with a typical structure of a biological target [77]. The complex of ligand and target protein can be mapped using pharmacophore features, which show the geometrical and chemical properties. These parameters cover basic and acidic groups, hydrogen bond donors and acceptors, and aliphatic and aromatic hydrophobic moieties, partial charge [78]. These features involve hydrogen bond donors and acceptors, acidic and basic groups, partial charge, and aromatic and aliphatic hydrophobic molecules [79]. This depiction can be utilized in online testing assignments to find effective binding sites on the basis of the interaction. Pharmacophore modeling has been incorporated into many software applications. For example, The Pocket v.2 and Ligandscout map ligand–target interactions using algorithms found in complex protein–ligand data targets.

7.4 ADMET Assessment

The next step after finding a drug candidate is to evaluate its pharmacokinetic characteristics, like ADMET. Thanks to developments in machine learning algorithms and gathered datasets, computational methods can also be used to predict ADMET [80]. Preclinical studies are thought to withdraw 40–60% of drug candidates due to ADMET concerns. To reach their biological targets and produce their pharmacological effects, drug compounds must pass through several physiological barriers, including the blood–brain barrier, the gastrointestinal barrier, and microcirculatory barriers. They might need to be metabolically changed to be activated, or they might change into a hazardous substance with unfavorable effects [81]. For ADMET evaluations, conventional experimental techniques are still time-consuming and expensive. Lipinski's called as "rule of five," which is a molecular weight <500 Da, lipophilicity <5, number of rotatable bonds <10, hydrogen bond donors <5, and hydrogen bond acceptors <10, is a straightforward guideline for determining whether a chemical compound is drug-like [82]. More sophisticated prediction techniques are being used more frequently these days to forecast drug likeness in the form of ADME-Tox characteristics, as opposed to relying solely on this basic guideline. Several models based on machine learning have been created to forecast the pharmacokinetic characteristics of chemical substances. ALOGPS, PreADMET, DrugMint, and SwissADME are a few of these models. For instance, Schyman et al. used the variable nearest neighbor (vNN) approach to build 15 models for the prediction of ADMET properties [83]. Additionally, Abdul et al. used a decision tree algorithm to create a model for predicting chemical toxicity. The method was applied for toxicity screening after they selected an ideal number of characteristics from a list of parameters. Furthermore, Yu et al. developed a model for predicting androgen receptor toxicity using a coevolutionary neural network algorithm [84]. Remarkably, they trained the model using 2D chemical structure photos of molecules rather than the widely used molecular descriptors. Their model effectively categorized substances that are inactive and agonists of androgen receptors. Ultimately, Lee et al. utilized a neural network algorithm to create a trustworthy cardiotoxicity prediction model for the human ether-a-go-go-related gene (hERG).

7.5 Conclusion

The efficiency and accuracy of *in silico* drug targets and therapeutic drug identification have increased over the last few decades. The accumulation of freely accessible biological data and the quick development of computational techniques have recently led to an acceleration in *in silico* drug discovery. The biological functions of targets are explained by chemical biology, and to help identify promising drug candidates, CADD techniques use structural data about either ligands or structure based or with referred to bioactivity (ligand based). Thanks to its capacity to expedite the process of finding new drugs by utilizing pre-existing information on ligand–receptor interactions, structural optimization, and synthesis, CADD techniques have become an indispensable component of the drug discovery process. The key pharmacological characteristics for a successful drug development strategy are adsorption, distribution, metabolism, excretion, and toxicity. A growing number of models based on machine learning have also been created using biological data. As of this writing, a number of drugs discovered through CADD techniques are available to consumers and have reached the market. Further developments are needed, though, especially in the areas of molecular docking scoring functions, identifying receptors in the absence of structural data, adding solvent effects and molecular adaptability to force fields used in MD simulations, and increasing computational efficiency. By taking care of these limitations of CADD approaches, all the possibilities of CADD may be realized.

References

1 Barabási, A.L., Gulbahce, N., and Loscalzo, J. (2011). Network medicine: a network-based approach to human disease. *Nature Reviews Genetics [Internet]* 12 (1): 56–68. https://doi.org/10.1038/nrg2918.

2 Chen, B. and Butte, A.J. (2013). Network medicine in disease analysis and therapeutics. *Clinical Pharmacology and Therapeutics* 94 (6): 627–629.

3 Culligan, E.P., Sleator, R.D., Marchesi, J.R., and Hill, C. (2014). Metagenomics and novel gene discovery: promise and potential for novel therapeutics. *Virulence* 5 (3): 37–41.

4 Wang, L. and Xie, X.Q. (2014). Computational target fishing: what should chemogenomics researchers expect for the future of *in silico* drug design and discovery? *Future Medicinal Chemistry* 6 (3): 247–249.

5 Agrawal, D., Bernstein, P., Bertino, E. et al. (2011). *Challenges and Opportunities with Big Data 2011-1*. 1–16. Available from: https://docs.lib.purdue.edu/cgi/viewcontent.cgi?article=1000&context=cctech.

6 Tsiliki, G., Karacapilidis, N., Christodoulou, S., and Tzagarakis, M. (2014). Collaborative mining and interpretation of large-scale data for biomedical research insights. *PLoS One* 9 (9): e108600.

7 Katsila, T., Spyroulias, G.A., Patrinos, G.P., and Matsoukas, M.T. (2016). Computational approaches in target identification and drug discovery. *Computational and Structural Biotechnology Journal [Internet]* 14: 177–184. https://doi.org/10.1016/j.csbj.2016.04.004.

8 Giaever, G., Shoemaker, D.D., Jones, T.W. et al. (1999). Genomic profiling of drug sensitivities via induced haploinsufficiency. *Nature Genetics* 21 (3): 278–283.

9 Kubota, K., Funabashi, M., and Ogura, Y. (2019). Target deconvolution from phenotype-based drug discovery by using chemical proteomics approaches. *Biochimica et Biophysica Acta (BBA) – Proteins and Proteomics [Internet]* 1867 (1): 22–27. https://doi.org/10.1016/j.bbapap.2018.08.002.

10 Ong, S.E., Schenone, M., Margolin, A.A. et al. (2009). Identifying the proteins to which small-molecule probes and drugs bind in cells. *Proceedings of the National Academy of Sciences of the United States of America* 106 (12): 4617–4622.

11 Ong, S.E., Blagoev, B., Kratchmarova, I. et al. (2002). Stable isotope labeling by amino acids in cell culture, SILAC, as a simple and accurate approach to expression proteomics. *Molecular & Cellular Proteomics [Internet]* 1 (5): 376–386. https://doi.org/10.1074/mcp.M200025-MCP200.

12 Chan, J.N.Y., Nislow, C., and Emili, A. (2010). Recent advances and method development for drug target identification. *Trends in Pharmacological Sciences* 31 (2): 82–88.

13 Jenkins, J.L., Bender, A., and Davies, J.W. (2006). *In silico* target fishing: Predicting biological targets from chemical structure. *Drug Discovery Today: Technologies* 3 (4): 413–421.

14 Ezzat, A., Wu, M., Li, X.L., and Kwoh, C.K. (2018). Computational prediction of drug–target interactions using chemogenomic approaches: an empirical survey. *Briefings in Bioinformatics* 20 (4): 1337–1357.

15 Cichonska, A., Ravikumar, B., Parri, E. et al. (2017). Computational-experimental approach to drug–target interaction mapping: a case study on kinase inhibitors. *PLoS Computational Biology* 13 (8): 1–28.

16 Zheng, S., Dharssi, S., Wu, M. et al. (2019). Text mining for drug discovery. *Methods in Molecular Biology* 1939: 231–252.

17 Agamah, F.E., Mazandu, G.K., Hassan, R. et al. (2020). Computational/*in silico* methods in drug target and lead prediction. *Briefings in Bioinformatics* 21 (5): 1663–1675.

18 Rouillard, A.D., Gundersen, G.W., Fernandez, N.F. et al. (2016). The harmonizome: a collection of processed datasets gathered to serve and mine knowledge about genes and proteins. *Database: The Journal of Biological Databases and Curation* 2016: 1–16.

19 Ochoa, D., Hercules, A., Carmona, M. et al. (2021). Open targets platform: supporting systematic drug-target identification and prioritisation. *Nucleic Acids Research* 49 (D1): D1302–D1310.

20 Byrne, R. and Schneider, G. (2019). *In silico* target prediction for small molecules. *Methods in Molecular Biology* 1888: 273–309.

21 Chen, Y.Z. and Zhi, D.G. (2001). Ligand–protein inverse docking and its potential use in the computer search of protein targets of a small molecule. *Proteins: Structure, Function, and Genetics* 43 (2): 217–226.

22 Paul, N., Kellenberger, E., Bret, G. et al. (2004). Recovering the true targets of specific ligands by virtual screening of the protein data bank. *Proteins: Structure, Function, and Genetics* 54 (4): 671–680.

23 Li, H., Gao, Z., Kang, L. et al. (2006). TarFisDock: a web server for identifying drug targets with docking approach. *Nucleic Acids Research [Internet]* 34 (Suppl_2): W219–W224. https://doi.org/10.1093/nar/gkl114.

24 Carvalho-Silva, D., Pierleoni, A., Pignatelli, M. et al. (2019). Open targets platform: new developments and updates two years on. *Nucleic Acids Research* 47 (D1): D1056–D1065.

25 Wang, L., Ma, C., Wipf, P. et al. (2013). Targethunter: an *in silico* target identification tool for predicting therapeutic potential of small organic molecules based on chemogenomic database. *The AAPS Journal* 15 (2): 395–406.

26 Liu, X., Ouyang, S., Yu, B. et al. (2010). PharmMapper server: a web server for potential drug target identification using pharmacophore mapping approach. *Nucleic Acids Research* 38 (Suppl. 2): 5–7.

27 Keiser, M.J., Setola, V., Irwin, J.J. et al. (2009). Predicting new molecular targets for known drugs. *Nature [Internet]* 462 (7270): 175–181. https://doi.org/10.1038/nature08506.

28 Geppert, H., Vogt, M., and Bajorath, J. (2010). Current trends in ligand-based virtual screening: molecular representations, data mining methods, new application areas, and performance evaluation. *Journal of Chemical Information and Modeling* 50 (2): 205–216.

29 Wishart, D.S., Knox, C., Guo, A.C. et al. (2008). DrugBank: a knowledgebase for drugs, drug actions and drug targets. *Nucleic Acids Research* 36 (Suppl. 1): 901–906.

30 Kim, S., Chen, J., Cheng, T. et al. (2019). PubChem 2019 update: improved access to chemical data. *Nucleic Acids Research* 47 (D1): D1102–D1109.

31 Seiler, K.P., George, G.A., Happ, M.P. et al. (2008). ChemBank: a small-molecule screening and cheminformatics resource database. *Nucleic Acids Research* 36 (Suppl. 1): 351–359.

32 Wang, Y., Zhang, S., Li, F. et al. (2020). Therapeutic target database 2020: enriched resource for facilitating research and early development of targeted therapeutics. *Nucleic Acids Research* 48 (D1): D1031–D1041.

33 Chen, C.Y.C. (2011). TCM Database@Taiwan: the world's largest traditional Chinese medicine database for drug screening *in silico*. *PLoS One* 6 (1): 1–5.

34 Günther, S., Kuhn, M., Dunkel, M. et al. (2008). SuperTarget and Matador: resources for exploring drug-target relationships. *Nucleic Acids Research* 36 (Suppl. 1): 919–922.

35 Hettne, K.M., Williams, A.J., Van Mulligen, E.M. et al. (2010). Automatic vs. manual curation of a multi-source chemical dictionary: the impact on text mining. *Journal of Cheminformatics* 2 (1): 10–11.

36 Von Eichborn, J., Murgueitio, M.S., Dunkel, M. et al. (2011). PROMISCUOUS: a database for network-based drug-repositioning. *Nucleic Acids Research* 39 (Suppl. 1): 1060–1066.

37 Lavecchia, A. (2015). Machine-learning approaches in drug discovery: methods and applications. *Drug Discovery Today [Internet]* 20 (3): 318–331. https://doi.org/10.1016/j.drudis.2014.10.012.

38 Danishuddin, K.A.U. (2016). Descriptors and their selection methods in QSAR analysis: paradigm for drug design. *Drug Discovery Today [Internet]* 21 (8): 1291–1302. https://doi.org/10.1016/j.drudis.2016.06.013.

39 Nettles, J.H., Jenkins, J.L., Bender, A. et al. (2006). Bridging chemical and biological space: "Target fishing" using 2D and 3D molecular descriptors. *Journal of Medicinal Chemistry* 49 (23): 6802–6810.

40 Hert, J., Willett, P., Wilton, D.J. et al. (2004). Comparison of fingerprint-based methods for virtual screening using multiple bioactive reference structures. *Journal of Chemical Information and Computer Sciences* 44 (3): 1177–1185.

41 Raymond, J.W. and Willett, P. (2002). Effectiveness of graph-based and fingerprint-based similarity measures for virtual screening of 2D chemical structure databases. *Journal of Computer-Aided Molecular Design* 16 (1): 59–71.

42 Gao, K., Nguyen, D.D., Sresht, V. et al. (2020). Are 2D fingerprints still valuable for drug discovery? *Physical Chemistry Chemical Physics* 22 (16): 8373–8390.

43 Ibrahim, K.A., Helmy, O.M., Kashef, M.T. et al. (2020). Identification of potential drug targets in *Helicobacter pylori* using *in silico* subtractive proteomics approaches and their possible inhibition through drug repurposing. *Pathogens* 9 (9): 1–21.

44 Gfeller, D., Grosdidier, A., Wirth, M. et al. (2014). SwissTargetPrediction: a web server for target prediction of bioactive small molecules. *Nucleic Acids Research* 42 (W1): 32–38.

45 Dunkel, M., Günther, S., Ahmed, J. et al. (2008). SuperPred: drug classification and target prediction. *Nucleic Acids Research* 36 (Web Server Issue): 55–59.

46 Awale, M. and Reymond, J.L. (2017). The polypharmacology browser: a web-based multi-fingerprint target prediction tool using ChEMBL bioactivity data. *Journal of Cheminformatics* 9 (1): 1–10.

47 Liu, X., Vogt, I., Haque, T., and Campillos, M. (2013). HitPick: a web server for hit identification and target prediction of chemical screenings. *Bioinformatics* 29 (15): 1910–1912.

48 Peón, A., Li, H., Ghislat, G. et al. (2019). MolTarPred: a web tool for comprehensive target prediction with reliability estimation. *Chemical Biology & Drug Design* 94 (1): 1390–1401.

49 Alberga, D., Trisciuzzi, D., Montaruli, M. et al. (2019). A new approach for drug target and bioactivity prediction: the multifingerprint similarity search algorithm (MuSSeL). *Journal of Chemical Information and Modeling* 59 (1): 586–596.

50 Koszła, O., Stępnicki, P., Zięba, A. et al. (2021). Current approaches and tools used in drug development against Parkinson's disease. *Biomolecules* 11 (6): 897.

51 Piñero, J., Ramírez-Anguita, J.M., Saüch-Pitarch, J. et al. (2020). The DisGeNET knowledge platform for disease genomics: 2019 update. *Nucleic Acids Research* 48 (D1): D845–D855.

52 Wang, J., Wolf, R.M., Caldwell, J.W. et al. (2004). 20035_Ftp. *Journal of Computational Chemistry* 56531 (9): 1157–1174.

53 Van Westen, G.J.P., Wegner, J.K., Ijzerman, A.P. et al. (2011). Proteochemometric modeling as a tool to design selective compounds and for extrapolating to novel targets. *MedChemComm* 2 (1): 16–30.

54 Singh, U.C., Brown, F.K., Bash, P.A., and Kollman, P.A. (1987). An approach to the application of free energy perturbation methods using molecular dynamics: applications to the transformations of methanol .fwdarw. ethane, oxonium .fwdarw. ammonium, glycine .fwdarw. alanine, and alanine .fwdarw. phenylalanine in aqueous solution and to $H_3O+(H_2O)_3$ .fwdarw. $NH_4+(H_2O)_3$ in the gas phase. *Journal of the American Chemical Society [Internet]* 109 (6): 1607–1614. https://doi.org/10.1021/ja00240a001.

55 Miyamoto, S. and Kollman, P.A. (1993). Absolute and relative binding free energy calculations of the interaction of biotin and its analogs with streptavidin using molecular dynamics/free energy perturbation approaches. *Proteins: Structure, Function, and Bioinformatics* 16 (3): 226–245.

56 Cortés-Ciriano, I., Ain, Q.U., Subramanian, V. et al. (2015). Polypharmacology modelling using proteochemometrics (PCM): recent methodological developments, applications to target families, and future prospects. *MedChemComm* 6 (1): 24–50.

57 Zhang, S. (2011). Computer-aided drug discovery and development. *Methods in Molecular Biology* 716: 23–38.

58 Lemkul, J., Genheden, S., Ryde, U. et al. (2015). Assessing the performance of the MM_PBSA and MM_GBSA methods. 1. The accuracy.pdf. *Journal of Chemical Information and Modeling [Internet]* 10 (7): 449–461. Available from: http://www.gromacs.org/@api/deki/files/198/=gmx-tutorial.pdf.

59 Liu, W., Schmidt, B., Voss, G., and Müller-Wittig, W. (2008). Accelerating molecular dynamics simulations using graphics processing units with CUDA. *Computer Physics Communications* 179 (9): 634–641.

60 Yuriev, E., Agostino, M., and Ramsland, P.A. (2011). Challenges and advances in computational docking: 2009 in review. *Journal of Molecular Recognition* 24 (2): 149–164.

61 Matsoukas, M.T., Cordomí, A., Ríos, S. et al. (2013). Ligand binding determinants for angiotensin II type 1 receptor from computer simulations. *Journal of Chemical Information and Modeling* 53 (11): 2874–2883.

62 Wu, B., Chien, E.Y.T., Mol, C.D. et al. (2010). Structures of the CXCR4 chemokine GPCR with small-molecule and cyclic peptide antagonists. *Science (80–)* 330 (6007): 1066–1071.

63 Schwede, T., Kopp, J., Guex, N., and Peitsch, M.C. (2003). SWISS-MODEL: an automated protein homology-modeling server. *Nucleic Acids Research* 31 (13): 3381–3385.

64 Myers, S. and Baker, A. (2001). Drug discovery – an operating model for a new era. Despite the advent of new science and technologies, drug developers will need to make radical changes in their operations if they are to remain competitive and innovative. *Nature Biotechnology* 19 (8): 727–730.

65 Hecker, N., Ahmed, J., Von Eichborn, J. et al. (2012). SuperTarget goes quantitative: update on drug–target interactions. *Nucleic Acids Research* 40 (D1): 1113–1117.

66 Davis, A.P., Grondin, C.J., Johnson, R.J. et al. (2021). Comparative toxicogenomics database (CTD): update 2021. *Nucleic Acids Research* 49 (D1): D1138–D1143.

67 Chen, B., Dong, X., Jiao, D. et al. (2010). Chem2Bio2RDF: a semantic framework for linking and data mining chemogenomic and systems chemical biology data. *BMC Bioinformatics* 11.

68 Wishart, D., Arndt, D., Pon, A. et al. (2015). T3DB: the toxic exposome database. *Nucleic Acids Research* 43 (D1): D928–D934.

69 Lindsay, M.A. (2003). Target discovery. *Nature Reviews. Drug Discovery* 2 (10): 831–838.

70 Noori, H.R. and Spanagel, R. (2013). *In silico* pharmacology: drug design and discovery's gate to the future. *Silico Pharmacology* 1 (1): 1–2.

71 Na, D., Rouf, M., O'Kane, C.J. et al. (2013). NeuroGeM, a knowledgebase of genetic modifiers in neurodegenerative diseases. *BMC Medical Genomics* 6 (1): 1–14.

72 Drews, J. (2000). Drug discovery: a historical perspective. *Science (80–)* 287 (5460): 1960–1964.

73 Venkatesh, S. and Lipper, R.A. (2000). Role of the development scientist in compound lead selection and optimization. *Journal of Pharmaceutical Sciences* 89 (2): 145–154.

74 Kennedy, T. (1997). Managing the drug discovery/development interface. *Drug Discovery Today [Internet]* 2 (10): 436–444. https://doi.org/10.1016/S1359-6446(97)01099-4.

75 Zhu, T., Cao, S., Su, P.C. et al. (2013). Hit identification and optimization in virtual screening: practical recommendations based on a critical literature analysis. *Journal of Medicinal Chemistry* 56 (17): 6560–6572.

76 Tuccinardi, T., Poli, G., Romboli, V. et al. (2014). Extensive consensus docking evaluation for ligand pose prediction and virtual screening studies. *Journal of Chemical Information and Modeling* 54 (10): 2980–2986.

77 Moustakas, D.T., Lang, P.T., Pegg, S. et al. (2006). Development and validation of a modular, extensible docking program: DOCK 5. *Journal of Computer-Aided Molecular Design* 20 (10–11): 601–619.

78 Jain, A.N. (2003). Surflex: fully automatic flexible molecular docking using a molecular similarity-based search engine. *Journal of Medicinal Chemistry* 46 (4): 499–511.

79 Daina, A., Michielin, O., and Zoete, V. (2017). SwissADME: a free web tool to evaluate pharmacokinetics, drug-likeness and medicinal chemistry friendliness of small molecules. *Scientific Reports* 7 (March): 1–13.

80 Friesner, R.A., Murphy, R.B., Repasky, M.P. et al. (2006). Extra precision glide: docking and scoring incorporating a model of hydrophobic enclosure for protein–ligand complexes. *Journal of Medicinal Chemistry* 49 (21): 6177–6196.

81 Houston, D.R. and Walkinshaw, M.D. (2013). Consensus docking: improving the reliability of docking in a virtual screening context. *Journal of Chemical Information and Modeling* 53 (2): 384–390.

82 Wolber, G. and Langer, T. (2005). LigandScout: 3-D pharmacophores derived from protein-bound ligands and their use as virtual screening filters. *Journal of Chemical Information and Modeling* 45 (1): 160–169.

83 Allouche, A. (2012). Software news and updates gabedit – a graphical user interface for computational chemistry softwares. *Journal of Computational Chemistry* 32: 174–182.

84 Mueller, R., Rodriguez, A.L., Dawson, E.S. et al. (2010). Identification of metabotropic glutamate receptor subtype 5 potentiators using virtual high-throughput screening. *ACS Chemical Neuroscience* 1 (4): 288–305.

8

Pharmacophore Modeling in Drug Design

Rahul Ghosh, Sharanya Roy, Gourav Rakshit, Nisha Kumari Singh, and Nigam Jyoti Maiti

Department of Pharmaceutical Sciences & Technology, Birla Institute of Technology, Ranchi, Jharkhand, India

8.1 Introduction

The landscape of drug discovery and development unfolds as a multifaceted and resource-intensive journey spanning a duration exceeding a decade [1]. The intricacies of drug design and exploration demand a holistic embrace of interdisciplinary strategies. Within this context, computer-aided drug design (CADD) methodologies come to the fore, predominantly orchestrating the early to intermediate phases of the drug discovery continuum. The expansion of computational prowess, data reservoirs, software innovations, and algorithmic sophistication [2–4] has synergistically propelled the significant amplification of CADD's role within the realms of drug discovery.

The CADD's imprint resonates across diverse domains encompassing the pursuit of target identification, the validation of prospective targets, the pursuit of promising hits, the judicious curation of lead compounds, and the meticulous refinement of their therapeutic potential [5]. This discourse converges its focus on the meticulous scrutiny of pharmacophore modeling situated within the pantheon of CADD methodologies, thereby accentuating its prominence and implications.

8.1.1 The Role of Pharmacophore Modeling in Drug Design

A pharmacophore delineates the inherent molecular attributes of a small molecule that govern its effective engagement within the cellular environment for biological or pharmacological functions. These attributes encompass facets such as geometry, spatial orientation, physiological relevance, and various other characteristics that collectively contribute to its functional role [6]. Serving as a prime exemplar, a pharmacophore offers a tangible link between the three-dimensional (3D) structure and the activity of a molecule. Investigating the pharmacokinetic, pharmacodynamic, and ADMET (absorption, distribution, metabolism, excretion, and toxicity) properties of a compound constitute an indispensable stride toward establishing its candidacy as a potential drug-like entity. The assessment of ADMET profiles holds pivotal sway, wielding significance for both synthetic and endogenous molecules alike. Essential criteria encompassing favorable, adverse, and inhibitory responses of ligands – small molecules – constitute pivotal landmarks for achieving targeted ligand–receptor interactions [7].

Computational Methods for Rational Drug Design, First Edition. Edited by Mithun Rudrapal.

In drug development, there is a pivotal phase dedicated to the thorough investigation of interactions between ligands and proteins [8]. The application of pharmacophore modeling enhances the precision of this approach. Pharmacophores were initially proposed in 1970 as a method to gain a better understanding of the interactions between small molecules (ligands) and receptors, particularly in cases where detailed structural information was unavailable. Pharmacophore modeling plays a prominent role in CADD, seamlessly integrating with similarity analysis and quantitative structure–activity relationship (QSAR) studies [9].

Fundamentally, the objective of pharmacophore characterization is to shed light on the spatial arrangement of functional groups or fragments within an active compound that primarily contributes to its biological efficacy. The identification of pharmacophore properties derived from ligand–receptor interactions at binding sites enables the screening of extensive chemical libraries, encompassing numerous potential novel entities. The construction of a robust pharmacophore model hinges on the accurate depiction of compounds with biological activity [10–12].

A distinguishing hallmark of the pharmacophore concept lies in the amalgamation of atom-centric models with their pharmacophoric properties. This amalgamation intentionally incorporates critical atoms or groups essential for biological interactions. It encompasses features such as positively or negatively charged moieties, aromatic motifs, hydrophobic clusters, hydrogen bond acceptors (HBAs), and hydrogen bond donors (HBDs). To facilitate comprehension, pharmacophore aspects are often represented as points, such as the centroid of five- or six-membered rings. Inter-feature distances, crucial for the structural definition of pharmacophore models, delineate these feature sites. The amalgamation of different features and the distances between them elucidate the chemical properties and spatial configurations of a pharmacophore.

Pharmacophore modeling stands as a pivotal stride within the intricate tapestry of drug design, serving as a potent tool for sifting through potential inhibitors by leveraging the intricate tapestry of their pharmacophore attributes [13]. The journey of drug design traverses a multifaceted landscape that commands substantial financial investments, often reaching into the millions, in pursuit of the elusive "magic bullet" capable of targeting and remedying specific diseases. Pioneers within the biopharmaceutical sphere are driven to harness the formidable potential embedded within the pharmacophore modeling paradigm, aiming to discern bioactive entities through a substrate of foundational biological insights. This endeavor not only augments the precision of the process but also compresses the temporal horizons of the drug design narrative [8, 14–18].

At the core of drug design, the pharmacophore model is delineated into two primary facets: structure-based pharmacophore modeling and ligand-based pharmacophore modeling, which is depicted in Figure 8.1. The former hinges upon preexisting cognizance of ligand properties or documented inhibitor–receptor complexes. In contrast, the latter thrives on the diversity of pharmacophore attributes, encompassing hydrogen bonding, hydrophobic interactions, aromatic contacts, metal-mediated associations, and charged interactions, all marshaled in the pursuit of fashioning novel therapeutic agents [19].

A wealth of online bioinformatics tools stands ready to empower researchers in their pharmacophore modeling ventures, making these sophisticated methods widely accessible. Present-day, a generalized pharmacophore-based computational approach unfolds through several sequential steps. This journey commences with the quest for the 3D structure of a biological target implicated in a specific ailment. Pharmacophore modeling enters the scene, complemented by virtual screenings (VSs) of compound databases. This dual strategy endeavors to unearth molecules displaying propitious attributes for future therapeutic endeavors. Hits are then subjected to thorough physicochemical evaluations, as these attributes emerge as pivotal determinants in establishing their potential as potent inhibitors [20].

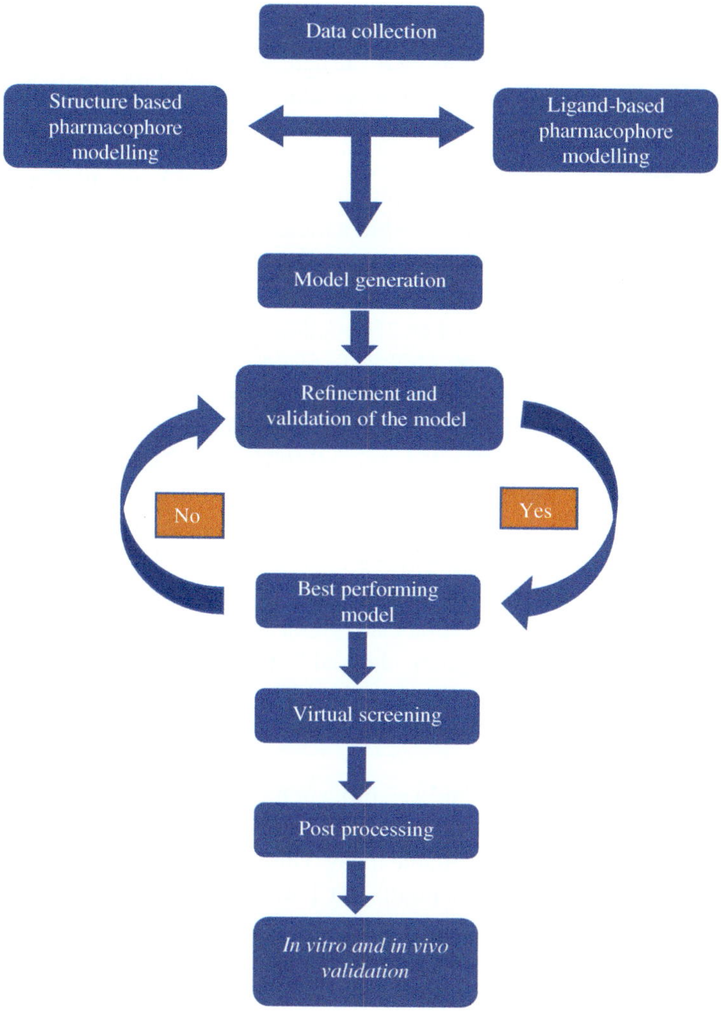

Figure 8.1 Flowchart illustrating the procedure of computational drug design.

Expanding the scope, the biological activities of the chosen molecules undergo scrutiny using predictive servers that illuminate enzyme-catalyzed metabolic pathways, facilitating the identification of high-scoring hits. The following assessments use the Lipinski Rule of Five to examine bioavailability and ADME-Tox characteristics. As a result, compounds with promising bioavailability, bioactivity, and ADMET features should be given priority for *in vitro* testing. The pharmacophore concept thus catalyzes hastening the drug design trajectory [2].

Aligning medications with individual genetic profiles is one area where the pharmacophore method has expanded into modern personalized medicine. The pharmacophore idea has several practical uses, including but not limited to target discernment, off-target prediction, VS, and ADME-Tox modeling. The use of the pharmacophore paradigm in molecular docking simulations improves the accuracy of binding posture prediction even more. A vital subset within the sphere of CADD, pharmacophore modeling encapsulates innovation and potential at the crossroads of modern scientific advancement [21].

8.1.2 Historical Perspective and Evolution of Pharmacophore Concepts

The origin of the pharmacophore concept is attributed to Paul Ehrlich, who introduced a novel approach to developing dyes by examining chromophores, the molecular components responsible for coloration. In 1890, Ehrlich provided the initial definition of a pharmacophore as "a molecular framework that bears (phoros) the essential attributes contributing to the biological activity (pharmacon) of a drug." In contemporary scientific terminology, a pharmacophore is defined, as established by Peter Günd, as "a collection of structural features within a molecule that is recognized at a receptor site and is accountable for the molecule's biological activity" [22]. Peter Günd has further explored the evolutionary history of the pharmacophore concept in his review [22].

The pharmacophore concept reached its full potential when 3D database search software became available in the 1990s. The pioneering computer program MOLPAT [6], designed to identify pharmacophore patterns, was created by Günd, Wipke, and Langridge at Princeton University in 1974. The demand for 3D structure search software grew alongside the development of rapid 3D structure generation programs like CONCORD [23], CORINA [24, 25], AIMB [26], and WIZARD [27]. Pharmaceutical companies contributed to the development of 3D search software, with examples such as ALADDIN [28] (originally by Abbott Laboratories, later commercialized by Daylight Chemical Information Systems, Inc.) and 3D-Search [29] (developed by Lederle Laboratories), while academic and government institutions introduced CAST-3D [30] (Chemical Abstract Services), DOCK [31] (University of California at San Francisco), and CAVEAT (University of California at Berkeley). Subsequently, the Marshall group devised a pharmacophore method rooted in ligand structures, known as the "active analog" approach [22]. They applied this approach to a group of ACE inhibitors [32] and validated the pharmacophore model against available experimental data, demonstrating a strong correlation [33].

The advent of commercial 3D searching systems marked a significant milestone in the field with the release of MACCS-3D by Güner et al. [22]. Over the subsequent four years, critical advancements were made, paving the way for the technology available today. This period saw the development of key 3D searching technologies, including ChemDBS-3D [34] (Chemical Design Inc., USA), UNITY [35] (Tripos Inc., USA), and Catalyst [36] (Accelrys Inc., USA). The demand for pharmacophore development software surged as these 3D searching technologies became widely accessible. While many of these 3D searching software tools included built-in query generation capabilities, specialized pharmacophore generation software also emerged. Notable examples included DISCO [37] by Martin et al. (Tripos Inc., USA), HipHop [36] by Barnum et al. (Accelrys Inc., USA), and GASP [38] by Jones et al. (Tripos Inc., USA). Simultaneously, predictive models rooted in QSAR, such as CoMFA (Tripos Inc., USA) [39] by Cramer et al., Apex-3D (Accelrys Inc., USA) by Golander and Vorpagel, and HypoGen [40] by Teig et al. (Accelrys Inc., USA), also came into existence. A comprehensive exploration of the usage and validation of pharmacophore development software can be found in the pharmacophore book [8, 28, 41–43]. In Table 8.1, we have given commonly employed server names with descriptions.

8.2 Essential Concepts in Pharmacophore Hypothesis Generation

Ideally, IC50 values serve as a reliable metric derived from *in vitro* experimentation, encompassing target-based and cell-free methodologies, with certain factors mitigated, such as cell efflux, cellular uptake, and metabolic influences. Cell-based assays, on the other hand, rely on discerning target interactions with ligands, which may induce modifications in protein expression, protein

Table 8.1 Programs and servers used in pharmacophore modeling.

Server name	Description
CATALYST-HipHop [36]	CATALYST has been incorporated into the BIOVIA Discovery Studio and comprises essential algorithms for pharmacophore generation, specifically HipHop and HypoGen. HipHop is employed for aligning active ligands concerning a particular target, enabling the identification of 3D conformations featuring common pharmacophoric elements through the superimposition of diverse molecular structures
CATALYST-HypoGen [40]	By assimilating data derived from biological analyses, pharmacophore modeling establishes hypotheses that facilitate the quantitative estimation of molecular activity. This integration enables a more streamlined approach to pharmacophore modeling by establishing meaningful connections between structural features and activity data
GASP [44]	GASP is a component of the SYBYL package that employs a genetic algorithm to identify pharmacophores. Unlike conventional pharmacophore determination methods, GASP conducts conformational searches in real time and seamlessly integrates this process into its program workflow. Before superimposing them onto each input chemical, the analysis involves the examination of conformational changes utilizing a singular low-energy structure and random spinning
LigandScout [45]	While LigandScout offers the capability to conduct structure-based and ligand-based pharmacophore modeling, it stands out as one of the pioneering software applications primarily designed for specialized structure-based pharmacophore modeling. It is particularly favored for applications where the 3D structure of the target protein in complex with its ligands is available
GALAHAD [46]	This software utilizes an adapted genetic algorithm that addresses specific limitations found in the GASP program, resulting in enhanced performance. It accelerates computational processing by employing preconstructed molecular structures as an initial reference point
MOE [47]	MOE possesses the capability to conduct both ligand-based and structure-based pharmacophore modeling. The process of constructing these models involves pairwise alignment of active ligands. For optimal results, it is advisable to reduce the size of the training dataset by clustering molecules with similar characteristics
PHASE [48]	This tool is included in the Schrodinger package. It serves as a practical method employed in the field of drug discovery, whether the receptor structure is available. It generates a hypothesis based on one or more ligands, protein–ligand complexes, and apoproteins. It features a specialized algorithm specifically tailored for optimizing lead compounds and conducting virtual screening
PharmaGist [49]	This is an openly accessible web server utilized for the generation of ligand-based pharmacophores. This online tool identifies pharmacophores through numerous flexible alignments of the input molecules
Pharmer [50]	In contrast to conventional molecular library screening methods, this pharmacophore technique conducts searches based on the breadth and intricacy of the query. The source code is accessible under an open-source license, and the technique is well-regarded for its exceptional speed
PharmMapper [51]	This publicly accessible web service is employed to identify potential targets for input ligands. Utilizing semirigid pharmacophore mapping, the methodology involves the calculation of pharmacophores

degradation, protein trafficking, and perturbations in cell membrane stability. When dealing with inactive compounds, it is imperative to elucidate the underlying reasons for their lack of activity, whether attributed to the absence of target interactions, metabolic processes, or efflux pump activity. It is noteworthy that *in vitro* data, often acquired through fluorometry-based or spectrophotometry-based techniques, may be prone to inaccuracies, partially due to the presence of chromophoric groups. Therefore, meticulous scrutiny is essential when interpreting such data. In the contemporary scientific landscape, an immense volume of data, spanning both *in silico* and *in vitro* realms, has been exponentially accumulating and populating various databases. For instance, activity data can be readily accessed from publicly available sources, including but not limited to open PHACT [52] and PubChem [53]. However, when extracting activity data from these sources, caution must be exercised, as a single compound may exhibit activity against multiple targets and be the result of various bioassays. It is essential to acknowledge that minor discrepancies exist in virtually every database; for example, a typical release of chEMBL may contain approximately 5% inaccuracies in chemical structures, 3% erroneous target information, and 1% errors in data deposition. Hence, users should exercise due diligence when selecting and utilizing data from these repositories.

8.2.1.1 Partitioning Initial Data into Distinctive Datasets

In accordance with contemporary scientific practices, the raw initial dataset is meticulously partitioned into three distinct subsets, namely, the training set, the test set, and the decoy set. This systematic division serves as a pivotal step in the process of constructing a robust pharmacophore model, as elucidated by some research (Figure 8.2).

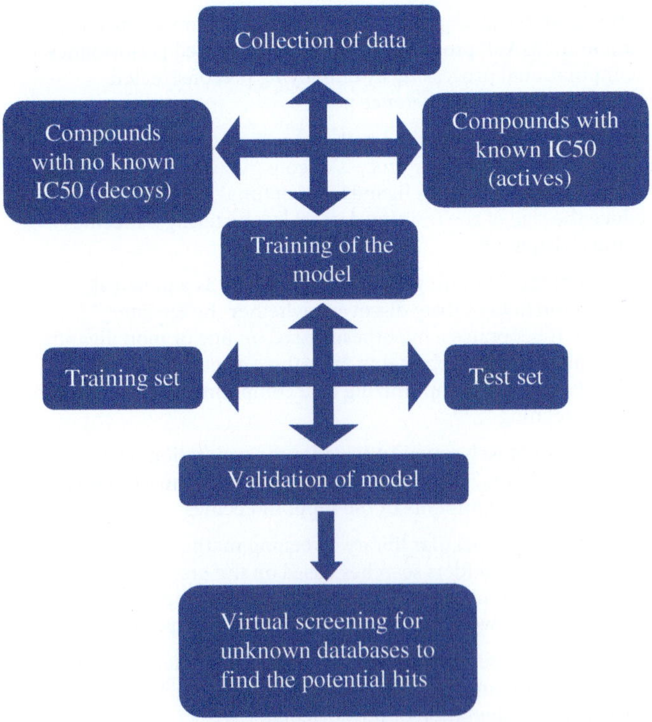

Figure 8.2 An overarching procedure for partitioning the data into many datasets.

Training Set In the context of ligand-based pharmacophore modeling, a designated ensemble of compounds exhibiting known activity serves as the foundational training dataset. Typically, this dataset comprises a minimum of two active molecules. Within a specific training dataset, compounds are categorized according to their activity levels, which may include classifications such as "very potent," "moderately potent," and "inactive," particularly in the context of semiquantitative pharmacophore models. These pharmacophore models are constructed based on shared chemical attributes found among highly active molecules, characterized by their diverse 3D structures, as exemplified by studies conducted by Leach et al. [54], as well as Seidel et al. [55].

Test Set A test set in pharmacophore modeling is a distinct subset of compounds, separate from the training set, and it serves a pivotal role in assessing the performance of a constructed pharmacophore model. Its primary objective is to evaluate how effectively the model can predict the bioactivity or properties of compounds that it has not encountered during the model-building process. The test set typically comprises compounds with known bioactivity, which can be categorized into two main groups: active compounds and inactive compounds. Active compounds are those with confirmed biological activity, while inactive compounds serve as controls. Both are essential for evaluating the model's ability to discriminate between active and nonactive molecules [56].

Decoys Set Decoys in pharmacophore modeling are negative control molecules used to differentiate between active and inactive compounds, assisting in model validation and optimization. They aid in assessing the model's ability to distinguish between active compounds and chemically similar yet inactive decoys, often measured through sensitivity and specificity metrics. In addition, decoys play a vital role in refining pharmacophore models to accurately capture essential features for biological activity. Once validated, these models are applied in VS to identify potential drug candidates, with decoys helping prioritize compounds likely to be active against the target. Proper selection of diverse decoys representative of the chemical space is crucial for meaningful discrimination tests, making decoys an integral component of pharmacophore modeling in drug discovery [53].

8.3 Diverse Approaches to Pharmacophore Modeling

There are two distinct approaches for generating pharmacophores: (i) the ligand-based method [2, 57] and (ii) the structure-based method [58]. The ligand-based approach is indirect, particularly applicable when a 3D structure of the target protein is unavailable. This method relies on known active molecules or ligands (inhibitors) for the target and searches for new inhibitors based on their functional attributes, referred to as features, which are utilized to construct a model [59]. Initially, the model is trained using shared pharmacophoric features from multiple known compounds, followed by model assessment and validation. On the other hand, a structure-based pharmacophore model is developed when the crystallographic structure of the target receptor is accessible [58]. The 3D structure of the protein complexed with a potential inhibitor is highly desirable as it provides detailed information about residues and the structural characteristics of the active site pocket. Consequently, the interaction between the ligand and the receptor serves as the foundation for constructing the structure-based model. In cases where the crystal structure is unavailable, a homology model of the target protein can be employed [47, 50, 59, 60].

Pharmacophore models comprise diverse chemical features such as HBDs and HBAs within ligand-receptor complexes. The central position and the tolerance sphere play pivotal roles in assessing the effectiveness of a pharmacophore model. In essence, a pharmacophore model encapsulates three fundamental aspects: feature "type," representing the specific chemical characteristic; "location," signifying the precise spatial coordinates of the feature; and "weight," which denotes the degree of significance attributed to each feature. The weight factor assumes significance during the computation of a geometric fit value for compounds [13].

8.3.1 Ligand-Based Pharmacophore Modeling

In scenarios where structural information for a macromolecular target is unavailable, the utilization of ligand-based pharmacophore modeling emerges as a crucial computational tool in the field of drug development. The standard procedure for this approach is to choose a set of known ligands that interact critically with a specific macromolecular target and then extract their shared chemical characteristics from their 3D structures. There are essentially two stages to the pharmacophore synthesis process that starts with a collection of ligands, or training set compounds:

Step One: Generating a Conformational Ensemble: To capture the natural conformational flexibility of the training set ligands, it is necessary to construct a complete conformational space for each of them.

Step Two: The process of identifying pharmacophore features and aligning ligands involves the critical step of aligning diverse ligands within the training set and discerning their fundamental shared chemical properties. Following the generation of conformational ensembles, pharmacophore models are constructed based on these identified properties. The fundamental approaches and main obstacles in the field of ligand-based pharmacophore modeling are accurately dealing with the conformational flexibility of ligands and performing accurate molecular alignment.

The domain of ligand-based pharmacophore modeling has undergone considerable advancements, although it continues to encounter notable challenges, particularly in simulating ligand flexibility. Addressing this issue involves two distinct approaches. The first method, referred to as the pre-enumeration approach, entails the prior calculation and storage of numerous conformations for each molecule in a database [61]. The second method, known as the "on-the-fly" approach, involves concurrent conformational analysis during the pharmacophore modeling process [61]. Despite the possible need for large amounts of storage space, the pre-enumeration method has computational benefits due to its ability to lower the cost of molecular alignment. On the other hand, while optimizing on the fly gets rid of bulk storage, it could increase CPU usage. According to the data, the pre-enumeration method is more effective than the on-the-fly calculating method [62].

In ligand-based pharmacophore modeling, molecular alignment is the second major obstacle. Both point-based and property-based techniques are basic types of alignment methods [62]. Aligning points – which could represent atoms, fragments, or chemical properties – through least-squares fitting is a common practice in point-based algorithms. When working with different ligands, it might be difficult to set specified anchor points, which is the main drawback of these methods. Alignments are made easier by property-based methods, which use molecular field descriptors, which are usually represented by sets of Gaussian functions. The goal function in alignment optimization is the intermolecular overlap of Gaussian functions, which are similarity measurements. Previous reviews of traditional molecular alignment algorithms have covered a lot of ground [62], and new alignment methods are being introduced all the time as a result of continuous discoveries. New developments include grid-based interaction energies [63], fuzzy pattern recognition [64], atomic property fields [65], and stochastic proximity embedding [66].

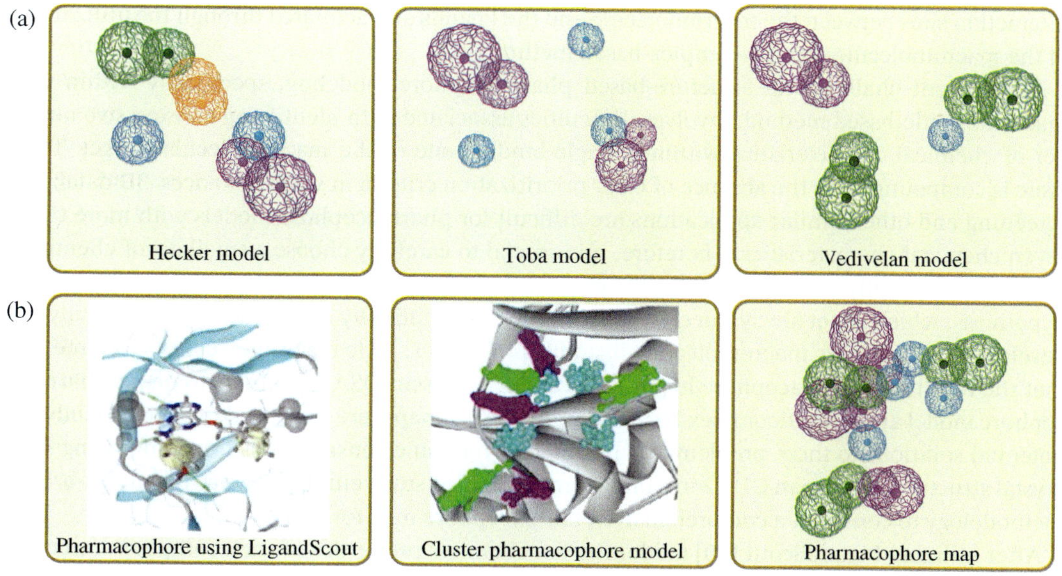

Figure 8.3 Pharmacophore models delineating inhibitors targeting cyclin-dependent kinase 2 (CDK2). (a) Utilizing Catalyst software developed by Hecker, Toba, and Vadivelan, pharmacophore models specific to CDK2 inhibitors were formulated. (b) The fundamental methodology entails generating a comprehensive pharmacophore map of CDK2 inhibitors employing a multicomplex-based strategy. Chemical attributes are depicted using specific color codes: green signifies HBAs, magenta denotes hydrogen-bond donors, light blue represents hydrophobic features, and orange indicates aromatic rings. *Source:* Yang [70]/with permission of Elsevier.

Choosing the right compounds for the training set is another major obstacle; this may appear like a simple but confusing process to both new and seasoned users. Several factors influence the ultimate pharmacophore model, encompassing the nature of ligand molecules, the size of the dataset, and the chemical diversity inherent in the compounds under consideration [61]. For ligands that interact with the same macromolecular target, using the same method and software with different training sets can sometimes produce completely different pharmacophore models. As an example, three separate groups of researchers used the Catalyst software with different training sets to independently construct pharmacophore models for cyclin-dependent kinase 2 (CDK2) inhibitors. These groups were Hecker et al. [67], Toba et al. [68], and Vadivelan et al. [69]. The models showed significant differences in the kind of features and the amount of space they had to work with (Figure 8.3a), an issue that will be covered more in the review's later parts.

8.3.2 Structure-Based Pharmacophore Modeling

This approach to pharmacophore modeling is known as structure-based pharmacophore modeling, and it makes direct use of the 3D structure of the target macromolecule or the complex it forms with a ligand. An in-depth examination of the spatial correlations and complementary chemical properties inside the target's active site is the first step in the methodology for doing structure-based pharmacophore modeling. Then, a pharmacophore model is assembled using these features. Two subsets of structure-based pharmacophore modeling methods exist those that focus on macromolecules with ligand complexes and those that focus on macromolecules alone. Locating the ligand-binding site within the macromolecular target and identifying pivotal interaction sites between the macromolecule and the ligands are facilitated through the utilization of the macromolecule–ligand–complex-based method [71].

A persistent challenge in structure-based pharmacophore modeling, specifically within the macromolecule-based method, involves difficulties associated with identifying an excessive number of chemical characteristics within a single-binding site of the macromolecular target. This issue is compounded by the absence of clear prioritization criteria in some instances. 3D database screening and other similar applications are difficult for pharmacophore models with more than seven chemical characteristics. Therefore, it is crucial to carefully choose a small set of chemical properties – usually between three and seven features – to build a workable pharmacophore hypothesis, which is not always an easy process. A further difficulty is that the model may only be developed from a single macromolecule–ligand complex or a single macromolecule, which means that the resultant pharmacophore hypothesis cannot represent QSARs. A most frequent pharmacophore model and a multicomplex-based comprehensive map were suggested in a recent study as potential solutions to these problems [71]. Drawing from an extensive repository comprising 124 crystal structures of human CDK2 inhibitor complexes, this study employed a multicomplex-based methodology to construct a comprehensive pharmacophore map for CDK2.

After utilizing LigandScout [70] to identify the chemical properties of each compound, a complete pharmacophore map was generated by clustering all these features. Figure 8.3b shows that the known map included almost all the important chemical characteristics necessary for CDK2–inhibitor interactions. Except for an aromatic ring feature (depicted in orange) present in the Hecker model, all pharmacophore features identified in the published ligand-based models, including the Hecker model, Toba model, and Vadivelan model (refer to Figure 8.3a), are encompassed within the comprehensive map. This is an important finding. Accordingly, it appears that these ligand-based models stand in for parts of the whole map. It appears that ligand scaffolds relate to the aromatic ring characteristic in the Hecker model, which is the sole exception. A structure-based pharmacophore modeling technique is unable to detect a chemical property present in tiny molecular scaffolds that functions as a pseudo-pharmacophore feature as it does not represent an interaction between ligands and macromolecules. A simplified model is necessary for real-world applications since the comprehensive pharmacophore map is too limited and not good for VS. One workable approach is to focus on the most common characteristics, those that are known to be critical for the efficacy of CDK2 inhibitors. A most-frequent-feature pharmacophore model was thus constructed by meticulously selecting and merging the seven most prominent characteristics, which were present in 124 complexes with a probability greater than 25%. In addition to showing that this most-frequent-feature model can successfully distinguish between known CDK2 inhibitors and molecules in an inactive focused dataset, validation studies have also shown that it can accurately predict the activities of a diverse range of CDK2 inhibitors in an active external dataset [71].

8.4 Application of Pharmacophore Modeling

A 3D pharmacophore can be constructed either by utilizing a set of active ligands or by employing complexes derived from apo-state proteins (targets) and ligands. Subsequently, VS is applied to assess a chemical library (ligands) against these 3D pharmacophores. Compounds that meet the criteria outlined in the query pharmacophore are subjected to further validation through experimental techniques, as outlined by Schaller et al. [72].

8.4.1 Applications of Pharmacophore-Based Virtual Screening

A VS process involves the systematic application of a pharmacophore model to screen extensive chemical libraries. This results in the creation of a virtual hit list containing molecules that align with the model's specifications. Subsequent experimental testing often confirms the activity of these compounds. Consequently, VS serves as a valuable tool for efficiently screening large compound collections, thereby augmenting the pool of active molecules in chemical databases available for further experiments. There is a consensus within the scientific community that VS is a valuable strategy to complement traditional high-throughput screening (HTS) initiatives [73, 74]. This preference is attributed to its substantially higher true positive hit rates compared to random testing procedures [75–77]. Although various prospective trials have been documented [78, 79], reported hit rates from prospective pharmacophore-based VS typically range from 5% to 40%. In contrast, a random selection of test compounds to identify active molecules often yields hit rates lower than 1%, exemplified by rates such as 0.075% for peroxisome proliferator-activated receptor (PPAR) [80], 0.55% for glycogen synthase kinase-3, and 0.021% for protein tyrosine phosphatase-1B [77].

8.4.1.1 Drug Discovery

Pharmacophore-based VS proves particularly advantageous in the initial categorization of compound classes and the refinement of compound properties, offering utility across various stages of the drug development pipeline, as expounded upon in subsequent sections.

Lead Identification The "cherry-picking" technique, which prioritizes lead identification, represents a prevalent application of pharmacophore-based VS. These programs are commonly employed to minimize the number of compounds subjected to biological screens, thereby optimizing the selection of molecules for experimental testing. The overarching objective is the discovery of novel lead compounds tailored to target specific diseases, with the potential for further development into therapeutic drugs addressing the identified illnesses. This application has been substantiated in numerous recent research studies [81–86]. For instance, Ha et al. utilized a ligand-based pharmacophore modeling strategy to uncover new ligands targeting the CXCR2 chemokine receptor [86]. Lepailleur et al. [87] identified novel compounds with enhanced binding affinity to the 5HT4 receptor through pharmacophore-based VS, aiming to discover new histamine H3 receptor antagonists. The first example of a compound exhibiting a dual mode of action was deemed advantageous for the treatment of Alzheimer's disease [87].

Structure–Activity Relationship A pharmacophore model encapsulates essential functionalities for a substance to exhibit bioactivity, delineating the potential binding mode of active compounds to their specific targets. The primary objective of a pharmacophore model is to discriminate between active and inactive molecules, ideally within the same chemical series. This characteristic makes pharmacophore modeling particularly valuable for elucidating complex structure–activity relationships (SARs). The presence or absence of specific chemical groups, denoted as pharmacophore characteristics, within individual molecules may account for observed variations in the biological activities of a given compound set in experimental settings. The process of constructing a pharmacophore model provides an opportune moment to commence the elucidation of SARs, shedding light on the underlying mechanisms driving biological activity. For a deeper understanding of how chemicals interact with the P-glycoprotein drug binding site, Ferreira et al. employed pharmacophore models [88]. Notably, active compounds isolated in-house from

Euphorbia species did not incorporate the nitrogen atom, a component postulated to play a crucial role in modulators according to previous studies. The researchers utilized a dataset comprising modulators from literature, compounds from their collection, and inactive molecules to assess the numerous refined pharmacophore models they developed. Their final model indicated that P-glycoprotein modulators must exhibit hydrophobic interactions and an HBA characteristic. Furthermore, even with the addition of another HBA/HBD characteristic, the mapping of active chemicals remained consistent [88].

Moreover, pharmacophore models play a pivotal role in recapitulating established SARs, thereby contributing to the discovery of novel bioactive compounds. In 2002, Flohr et al. conducted experimental investigations utilizing the natural peptide urotensin II and synthetic analogs to identify crucial interactions necessary for binding to the urotensin II receptor [89]. Leveraging existing SAR information, pharmacophore models were constructed and employed to screen a chemical library comprising small, drug-like molecules. Subsequent experimental testing of the virtual hits led to the identification of six new scaffold classes, notably featuring nonpeptidic molecules in prominent roles [89].

Scaffold Hopping Pharmacophore features delineate general chemical functions rather than specific functional groupings. Furthermore, the identification of virtual hits occupying 3D locations critical for biological activity is integral to pharmacophore models, primarily relying on localized functional similarities among active chemicals. It is noteworthy that the exact 2D architectures of the mapping molecules are not constrained in any particular manner. Pharmacophore models exhibit flexibility to accommodate structurally diverse hits, even as their composition is influenced by the 2D structures of the molecules utilized in their generation and refinement. This characteristic of pharmacophore modeling extends its utility to both natural products and synthetic chemicals, enabling exploration across a broad spectrum of chemical space. Crucially, this flexibility facilitates the implementation of the "scaffold hopping" strategy, allowing for the discovery of new scaffolds not previously associated with the target. A comprehensive examination of pharmacophore modeling within the framework of scaffold hopping has been extensively covered in a prior article [90]. Recent research has leveraged pharmacophore modeling to identify new ligands for the TRPV1 channel, representing the transient receptor potential vanilloid type 1 [91]. While initial hits may exhibit weak target contacts, they serve as intriguing starting points for optimizing the chemical process. Such studies often prioritize the discovery of novel chemical scaffolds over the identification of highly active molecules in theoretical validation datasets, resulting in lower overall similarity scores [91].

The significance of scaffold hopping in the pharmaceutical industry is underscored by its capacity to explore compounds without encountering intellectual property concerns. This attribute is particularly relevant in a broader context, as scaffold hopping enables the discovery of compounds with limited available data. This scenario frequently arises when examining chemicals found in consumer goods and environmental contaminants, both of which often possess characteristics distinct from traditional drugs.

8.4.2 Applications in Drug Target Fishing

Computer-Aided Pharmacological Design (CADD) emerges as a potent approach for unraveling the mechanisms of action of pharmacological compounds, particularly in cases where these mechanisms are not well-established. Chemometrics-based similarity search algorithms are employed to identify related compounds with known action mechanisms [92]. While pharmacophore queries prove beneficial for discovering specific chemicals, pharmacophore modeling encompasses various other applications. A notable use case involves employing the compound itself as a query to identify an appropriate pharmacophore model. These models can either be sourced from existing databases or manually crafted [93]. Moreover, this strategy facilitates the identification of a biological target for a specific chemical with an unknown biological function. A notable example is the discovery of numerous potential therapeutic targets through the analysis of various plant metabolites [79].

8.4.3 Applications in Ligand Profiling

Within ligand profiling, pharmacophore modeling serves to predict potential targets, anticipate side effects, and propose novel therapeutic targets. Both structure-based and ligand-based pharmacophore modeling are employed in ligand profiling, although the former is more prevalent [94]. Instead of molecular docking limitations, pharmacophore modeling offers an alternative for ligand profiling [95]. A study utilizing ligand profiling based on pharmacophore models achieved accurate target assignments. The investigation involved screening a dataset with 2108 pharmacophore models using 16 metabolites from *Ruta graveolens*. Experimental trials were conducted to validate computational results, with a specific focus on acetylcholinesterase (AChE), the coat protein of the human rhinovirus (HRV), and cannabinoid receptor type 2 (CB2). The binding patterns elucidated through pharmacophore modeling consistently aligned with experimental data [79].

8.4.4 Applications in Docking

Various strategies exist for the integration of molecular modeling techniques based on docking and pharmacophores, offering the potential to mitigate certain limitations inherent in each method and enhance overall outcomes. In the realm of docking, pharmacophore models can be employed as guiding constraints based on pharmacophoric properties. Within the docking process, these models serve to restrict the number of compounds subjected to docking, and post-docking, they function as filters for ligand selection and pose ranking [57]. Prior to initiating molecular docking, these pharmacophore models are applied as filters to assess whether molecules meet the fundamental chemical and structural requirements specified by the query [96]. To pre-filter the ligand database, pharmacophores are employed as search queries preceding the docking assessment. Certain molecular docking programs permit the inclusion of specific types of ligand–protein interactions in docking poses, facilitating pharmacophore-constrained docking [97]. Another approach to enhance the quality of top-ranking docking results involves the utilization of pharmacophore models after docking to evaluate the optimal binding mode of a chemical [98].

Illustratively, a search for potential inhibitors of Chk-1 (checkpoint kinase-1) within a chemical library employed a hybrid approach combining pharmacophore and docking methodologies. Initially, small molecules possessing HBA and HBD properties, capable of interacting with the adenine region of Chk-1, were screened using a pharmacophore model. Subsequently, docking procedures were executed to identify poses, considering these specific interactions [99]. For each ligand, multiple poses were recorded and subsequently rescored. Ultimately, the methodology identified chemical entities with potential binding affinities [100]. In a distinct study, new dual adenosine A1/A2A receptor antagonists were discovered through the application of docking and pharmacophore modeling [101].

8.4.5 Applications in ADMET

When drugs exhibit deviations from their intended effects within the body, it is often attributed to issues related to ADMET [102]. Consequently, there is a consensus that the early assessment of ADMET characteristics plays a crucial role in the drug development process. In the initial stages of drug development, pharmacophore modeling approaches are employed to predict ADMET features to reduce the incidence of unsuccessful attempts in the search for new medications [22].

Through the comparison of the chemical characteristics of tested compounds with medicines possessing well-documented ADMET profiles, pharmacophore models can be utilized to identify potential interactions with drug-metabolizing enzymes [103]. For instance, interactions between known medicines and cytochrome P450 (CYP) enzymes have been leveraged to develop a pharmacophore model estimating the binding of drug-like compounds to specific CYP enzymes, with a focus on potential degradation by these enzymes [104]. Similarly, an ADMET pharmacophore approach has been employed to model 5′-diphospho-glucuronosyl transferase enzymes due to their involvement in drug excretion processes [105].

8.4.6 Modulation of the Immune System

Small compounds capable of influencing or modulating immune responses are termed modulators. Toll-like receptors (TLRs) represent one such example, playing a pivotal role in eliciting innate immune responses in reaction to recognized infection-related molecular patterns. According to Murgueitio et al. [106], an auspicious strategy for addressing various ailments, including cancer, autoimmune diseases, neurological disorders, and allergies, involves the development of modulator molecules targeting TLRs. In a study, structure-based VS was employed in conjunction with experimental validation to identify a novel TLR8 modulator. Molecular docking was utilized to analyze the diverse binding mechanisms of the screened molecule. Subsequently, molecules were categorized based on optimal binding positions, and the preferred position was employed to construct a 3D model of the pharmacophore. This pharmacophore model was then applied to VS processes.

8.5 Emerging Trends in Pharmacophore Model Development

Pharmacophore modeling, encompassing both structure-based and ligand-based techniques, stands as a computational tool widely employed in various computer-assisted biological applications. Numerous CADD initiatives heavily rely on its utilization. Furthermore, VS and investigations into protein function extensively depend on the integration of 3D pharmacophore models. Recent advancements have showcased the application of 3D pharmacophore models in combination with molecular dynamics (MDs) simulations, providing dynamic insights into macromolecular–ligand interactions within a simulated cellular physiological environment. The integration of newly developed 3D pharmacophores and machine learning methods has facilitated the screening of diverse types of novel compounds [107].

8.5.1 Involvement of Machine Learning

The field of drug discovery has experienced substantial advancements owing to the availability of extensive datasets and the enhancement of computational capabilities. Over recent years, there has been a growing integration of pharmacophore modeling with machine learning, a branch of artificial intelligence (AI). Machine learning methods, particularly deep learning, have demonstrated increased efficiency, especially in the era of large-scale data, leading to the emergence of numerous pharmacophore-based machine learning algorithms. Machine learning has also played a pivotal role in refining scoring functions [72].

Prominent machine learning methods trained on pharmacophores include DeepSite, PharmIF (Pharmacophore-based Interaction Fingerprint), and Hot Spots guided Receptor-based Pharmacophores (HSPharm). HSPharm employs random forest decision trees trained with pharmacophoric descriptors to reduce the number of pharmacophore attributes [108]. PharmIF, on the other hand, utilizes a Support Vector Machine (SVM) trained with pharmacophoric fingerprints to rank docking poses of small compounds [107]. DeepSite employs a Convolutional Neural Network (CNN) trained with pharmacophoric descriptions to detect cavities and determine binding affinities [37].

8.5.2 Prediction of Pharmacokinetic Properties

Accurately forecasting the ADMET features of a compound is imperative in evaluating its drugability and forms a critical prerequisite for the rational design of effective therapeutic compounds. Pharmacophore modeling serves as one approach to predict ADMET characteristics, allowing the identification of drugs with similar profiles as other studied compounds. This involves precise predictions of ADMET properties and binding affinities using a set of pharmacophore models [4].

In a recent study, a novel strategy amalgamating pharmacophore modeling and QSAR approaches were employed, focusing on the hexadecahydro-1*H*-cyclopenta[*a*]phenanthrene framework (HHCPF). This innovative method facilitated a comprehensive understanding of the intricate interplay between the characteristics of target binding sites and specific pharmacophore features within a group of medicinal compounds. The investigation incorporated drug compounds based on HHCPF and their target proteins through docked complexes. Subsequently, various target structures obtained from renowned databases such as UniProt, DrugBank, and the Protein Data Bank (PDB) were meticulously compared with the diverse ADMET characteristics of these medications [4].

8.5.3 Structural Biology and Protein Functionality Studies

Pharmacophore modeling serves as a prominent approach for gaining mechanistic insights into protein functionality. It is particularly valuable for dissecting the ligand-dependent functions of receptors. Notably, G-protein-coupled receptors (GPCRs), a class of vital drug targets implicated in various diseases, exhibit intricate pharmacophore attributes and display flexibility in their ligand-dependent activities. An effective study delved into the receptor-specific binding patterns within closely related GPCR members of the same protein family. This analysis involved scrutinizing binding patterns and identifying key residues located in the loop regions. In addition, some investigations have explored the concept of ligand bias, where ligands demonstrate preferences for specific signaling pathways mediated by their target GPCRs. Certain ligands have displayed conformational constraints, particularly in the extracellular regions [23].

8.5.4 Integration with MDs Simulations

The active sites residing on the target molecule exhibit substantial flexibility. Consequently, relying solely on a structure-based pharmacophore model featuring a solitary conformation of the drug–target interaction may fail to capture the full spectrum of involved features in this interaction. To address this challenge posed by active site flexibility, dynamic pharmacophore models prove invaluable. MDSs simulations capitalize on the inherent dynamism of protein–ligand complexes within the cellular milieu. These MD simulation approaches can construct dynamic pharmacophore models that align with the adaptable nature of target active sites or binding pockets. An advanced strategy in 3D pharmacophore modeling involves the integration of multiple conformations into a pharmacophore model using MD simulations. Given the dynamic nature of both the target and ligands, the resulting target–ligand complex and their interaction patterns mirror this dynamism. MD simulations offer an effective means of mapping the dynamics inherent to these complexes [109].

Advancements in the field have led to the creation of various software tools that combine MD simulations and the generation of 3D pharmacophore models. One notable approach is the Hydration Site-Restricted Pharmacophore (HSRP), which operates on the principle of constraining pharmacophore features to hydration sites identified on protein molecules. These hydration sites are characterized by unfavorable thermodynamic properties, as determined through MD simulations. HSRP effectively serves as a filter, enabling the selection of ligands that exhibit greater relevance and favorable entropic characteristics when binding to the target protein [109].

The Dynophore methodology, with a focus on spatiotemporal feature analysis through statistical approaches, integrates 3D pharmacophore models with MDs simulations. Utilizing conformers from MD simulations, the Dynophore program assesses the chemical interaction characteristics of protein–ligand complexes. It evaluates various interaction properties, including charges, hydrogen bonds, and lipophilic contacts, using the MD trajectory data. The statistical ranking of these properties is determined based on their interaction patterns within the protein complex. Subsequently, a 3D pharmacophore model is generated using the best-fitting statistical model.

For optimizing 3D pharmacophores derived from MD simulations, two prevalent approaches are MYSHAPE and the Common Hits Approach. In the Common Hits Approach, 3D pharmacophore models obtained from MD simulations are categorized based on their interaction patterns, and VS utilizes these models to construct a virtual ligand library. In contrast, MYSHAPE employs MD simulations of protein–ligand complexes to optimize shape-dependent pharmacophore characteristics. The GRAIL method elucidates the molecular interaction fields of pharmacophore models generated in MD simulations. This method can calculate atomic densities of water, ligand, and protein molecules constituting the interaction complex. The PyRod algorithm, incorporating heuristic scoring methods for water molecules near the protein's binding site, generates a 3D model of the pharmacophore. It also examines site-specific hydration-related pharmacophore models, which are subsequently employed in VS [110].

Proteins are simulated using cosolvent-based MD in the presence of ethanol and water molecules employing the AutoDock Bias with Solvent Sites method. This approach analyzes the hydroxyl and methyl groups of ethanol to identify their binding locations or hotspots using MD trajectory data. In docking simulations, AutoDock incorporates the computed ethanol-free energies [72], extensively studied and contrasted with earlier approaches. A novel method utilizing cosolvent simulations to create 3D pharmacophore models is Pharmmaker. VS of ligand molecules makes use of these pharmacophore models, and the druggability of several probe compounds is evaluated using DruGUI in a case study following cosolvent simulations. Pharmmaker assesses binding sites within protein–ligand interactions to identify the most druggable binding site, constructing a 3D pharmacophore model for VS using snapshots of these interactions [111].

8.6 Case Studies

8.6.1 Case 1

The X-linked inhibitor of apoptosis protein (XIAP) belongs to the inhibitor of apoptosis protein (IAP) family and is responsible for counteracting the activity of caspases-3, caspases-7, and caspases-9. The upregulation of the protein led to a reduction in the occurrence of apoptosis within the cellular environment, thereby impeding the progression of cancer. Various classes of XIAP antagonists are commonly employed to rectify the impaired apoptotic pathway, hence facilitating the eradication of cancer within organisms. The currently identified chemically synthesized substances that function as XIAP inhibitors have been found to have adverse effects, hence posing challenges in the context of chemotherapy treatment. The project was undertaken to identify novel natural chemicals capable of inducing apoptosis by activating caspases while exhibiting low toxicity. Therefore, to find natural compounds, a structure-based pharmacophore model was constructed for the protein's active site cavity. This was followed by VS, molecular docking, and MDs simulation. In the initial stage, a total of seven hit compounds were obtained. Subsequently, a molecular docking approach was employed to assess the compounds, resulting in the selection of four compounds for subsequent examination. The stability of the chosen drug candidate to the target protein was validated using the MDs simulation technique, which successfully demonstrated the stability of all three compounds. According to the results, three recently acquired compounds, specifically Caucasicoside A (ZINC77257307), Polygalaxanthone III (ZINC247950187), and MCULE-9896837409 (ZINC107434573), have been identified as potential lead compounds for combating XIAP-related cancer treatment. The compounds that were chosen for analysis exhibited a greater degree of binding affinity, with values ranging from −6.9 to −8.0 kcal/mol, toward the XIAP protein of interest. The *in silico* toxicity test revealed a reduced level of toxicity, while the ADME study indicated that the substance can be rapidly absorbed by the tissue site due to its high-fat solubility. The study commenced by constructing a structure-based pharmacophore model, which was subsequently employed for VS, molecular docking, ADMET analysis, and MD simulation. During the final step of the MD simulation, four compounds were examined. However, it was seen that the compound ZINC1070004335 exhibited unfavorable stability with the protein XIAP. As a result, this compound was rejected. The A-to-Z VS technique may yield three prominent natural compounds that have the potential to function as lead molecules in the battle against cancer [112]. All the details of this case study are mainly depicted in Figures 8.4–8.6.

8.6.2 Case 2

The present study employed an innovative computational methodology that combined ligand-based pharmacophore filtering and molecular docking approaches to discover prospective compounds with the ability to simultaneously inhibit the tyrosine kinase activities of both EGFR and VEGFR2. The results of the study revealed that six compounds demonstrated a strong alignment with the characteristics of the designated pharmacophore models. These compounds also showed higher docking scores when compared to erlotinib and axitinib, which were used as reference medications. Moreover, a comprehensive examination of the binding mechanisms revealed that these compounds exhibited analogous interactions with the reference medications. Nevertheless, by conducting a thorough analysis of the stability of their binding modes during MDs simulations, it was revealed that two distinct compounds, namely, ZINC16525481 and ZINC38484632, exhibited consistent hydrogen bonding interactions (with an occupancy surpassing 50%) with essential residues of both targets. Moreover, these compounds demonstrated favorable binding free energies, suggesting their capacity to strongly bind to the binding pockets of both EGFR and VEGFR2,

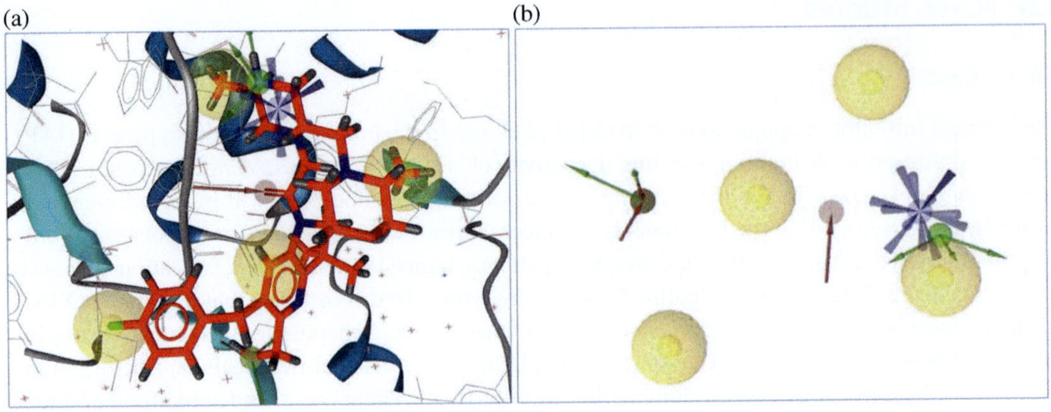

Figure 8.4 (a) The pharmacophore model of XIAP protein bound to the 46781908 ligands, derived from the crystallographic structure of XIAP protein (PDB ID: 5OQW), is founded on its 3D conformation. (b) Following intricate molecular interactions, multiple pharmacophore attributes are delineated by four yellow spherical representations. The interaction between the protein–ligand complex is typified by the presence of a hydrophobic interaction, depicted by a blue star shape indicating a positively ionizable group with a tolerance of 2 Å. Moreover, three red arrows and spherical representations signify HBAs with a tolerance of 1.5 Å. In addition, five HBDs are identified within the interaction, depicted by green spherical or arrow shapes. Notably, the schematic illustration does not encompass the 15 exclusion volumes generated during pharmacophore modeling.

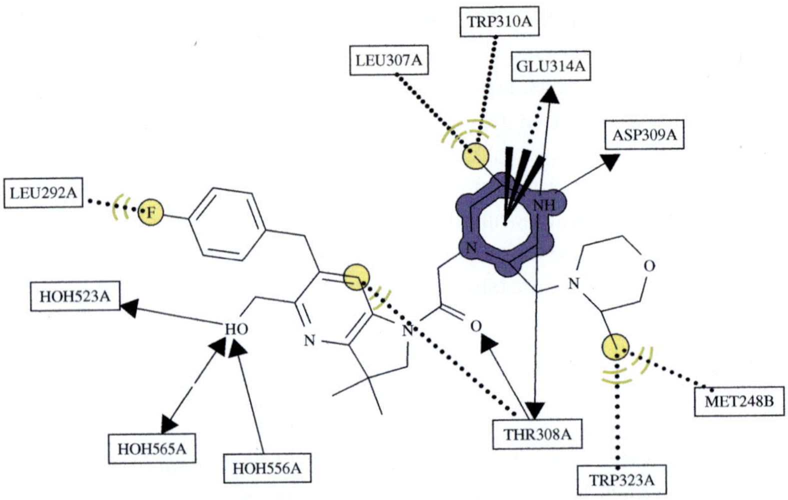

Figure 8.5 The pharmacophore modeling yielded a 2D representation highlighting the hydrophobic interactions, depicted in yellow, involving specific amino acid residues within the XIAP protein. The prevalent HBD properties engaged in ligand–protein interactions are delineated in green. The coloration reflects the interactions between HBAs and the oxygen and nitrogen atoms of the benzene ring, along with its diverse side chains. Notably, the depiction does not encapsulate the morphology and positioning of the binding pocket, which are upheld by hydrogen atoms and delimited regions.

hence having a potent inhibitory impact. Consequently, the study proposed that ZINC16525481 and ZINC38484632 exhibit promise as prospective contenders for the simultaneous suppression of EGFR and VEGFR2. This finding underscores the need for additional research and exploration in future investigations [113]. All above explanations are depicted in Figures 8.7 and 8.8.

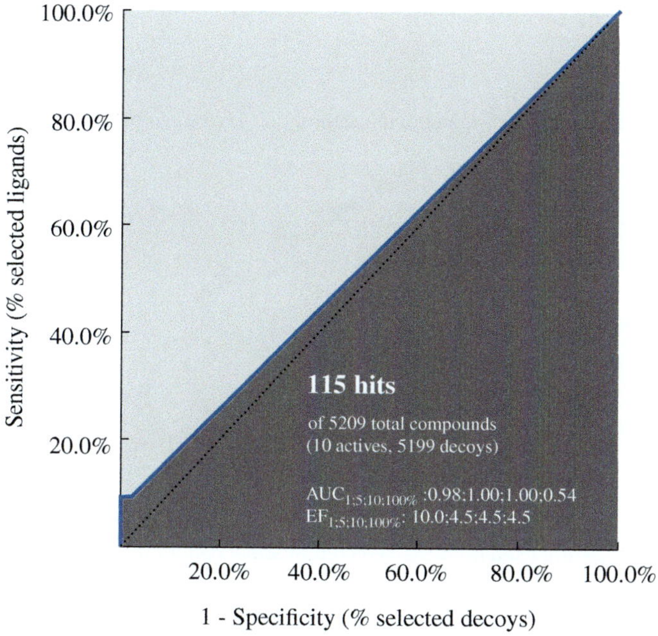

Figure 8.6 A receiver operating characteristic (ROC) curve was generated to evaluate the discriminative capability of the active molecule in distinguishing decoy compounds, employing the structure-based pharmacophore model. The validation of the pharmacophore model was conducted using a dataset comprising 10 XIAP active compounds and 5199 decoy compounds.

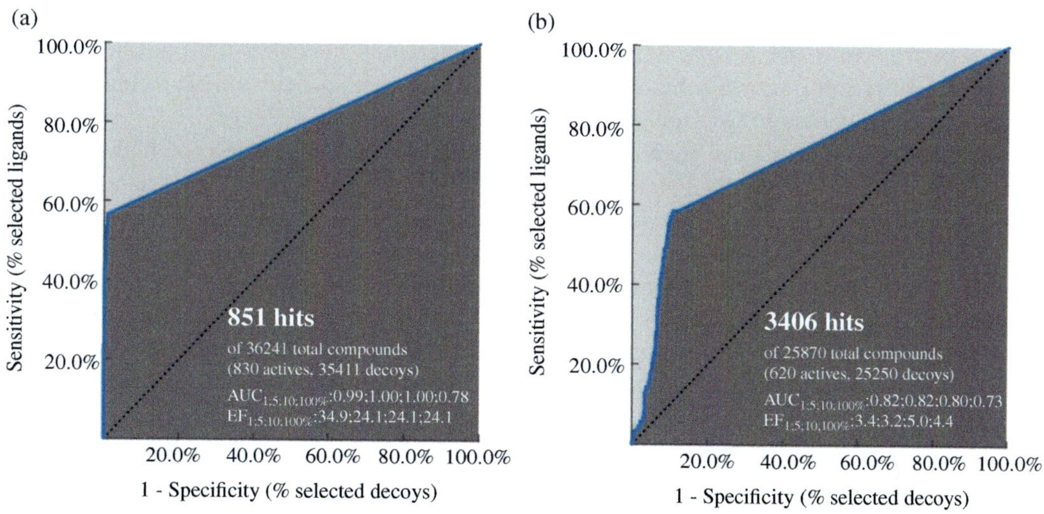

Figure 8.7 A receiver operating characteristic (ROC) curve was generated to evaluate the discriminative capability of the active molecule in distinguishing decoy compounds, employing the structure-based pharmacophore model. The validation of the pharmacophore model was conducted using a dataset comprising 10 XIAP active compounds and (a) 35411 and (b) 25250 decoy compounds.

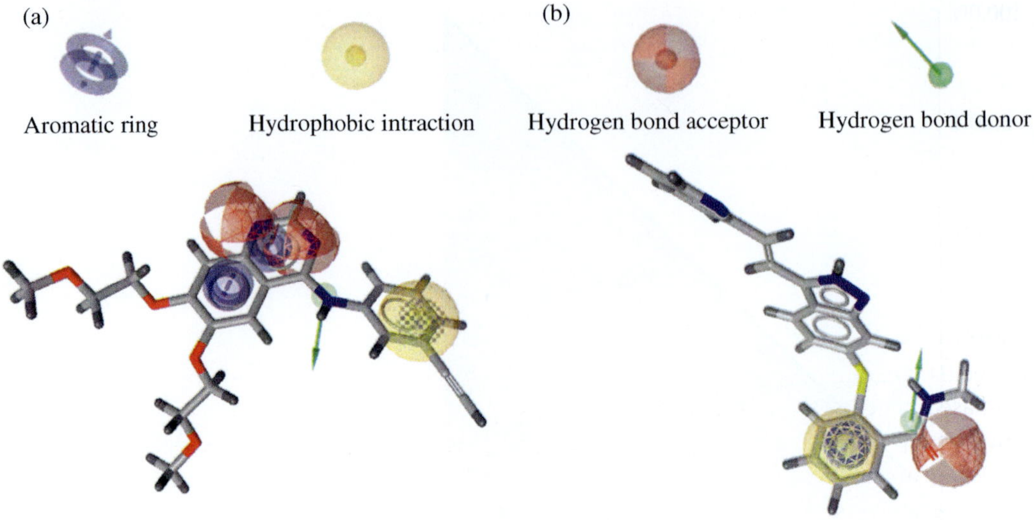

Figure 8.8 Erlotinib (a) and axitinib (b) are superimposed onto the selected pharmacophore model of EGFR. These compounds serve as inhibitors of VEGFR2.

8.7 Challenges in Pharmacophore Modeling

Pharmacophore modeling encounters several challenges that require resolution to enhance modeling quality [114].

One application area of pharmacophore modeling is VS using pharmacophores. However, effective scoring functions are deficient for this purpose [115]. Typically, the degree of alignment between a ligand and the pharmacophore query is expressed through RMSD, which assesses the similarity between the query patterns and the compound's atoms. Unfortunately, this measurement doesn't consider the likeness to known inhibitors, making it incapable of estimating the overall similarity with the receptor. Consequently, compounds matching the pharmacophore query might differ from known inhibitors and contain functional groups unable to bind to the receptor-binding site, rendering them inactive despite perfect matches [116].

Another formidable challenge in pharmacophore-based VS is the presence of higher "false positive" rates, where virtual hit ligands may lack biological activity [70]. This limitation can be attributed to insufficient hypothesis quality, the pharmacophore model's accuracy, and deviations from actual biological conditions. To overcome this drawback, it's essential to incorporate expertise, comprehensive validation, including relevant target information, and integration with other computational methods [12].

Modeling ligand flexibility presents another significant challenge. To address this, structural analysis based on predetermined structure databases or during the pharmacophore modeling process may be employed. Notably, the method reliant on predetermined structural databases has demonstrated superior performance [62]. However, limitations exist in VS by pharmacophore using this database-dependent approach. These databases typically contain only a few low-energy structures per molecule, potentially missing the structure of an active ligand. This is especially relevant for structures with rotatable bonds in small molecular functional groups like hydroxyl, where distinguishing between various rotations based on RMSD value differences during structure generation can be challenging [117]. In general, pharmacophore search tools can account for bond rotations during the matching process to identify the correct directional conformations of small, flexible, polar functional groups. Nevertheless, creating a pharmacophore query remains a challenge without a clearly defined approach [118].

Similarly, in structure-based pharmacophore modeling, addressing protein flexibility and ligand conformational flexibility poses major challenges. These challenges can be mitigated by generating the pharmacophore model from a docked complex created through flexible docking or by generating and aligning models from protein-ligand MDs simulations simultaneously. Combining the structure-based approach with flexible docking and MDs simulations may alleviate these limitations [119]. In addition, the generation of pharmacophore models in the structure-based approach is not straightforward, particularly when various combinations of features are possible. Each pharmacophore model may lead to a different set of compounds [120].

Molecular alignment is a complex aspect of pharmacophore modeling and can be categorized as point-based or feature-based approaches based on their fundamental nature. Point-based algorithms overlap double atoms, fragments, or chemical pattern points using least squares matching. However, a major drawback of this approach is the requirement for predetermined connection points. Feature-based algorithms, on the other hand, employ molecular domain determinants, often represented by Gaussian function sets, to create alignments. Ongoing developments aim to introduce new alignment methods [42].

Another practical challenge lies in the selection of the appropriate training set molecules. While nontechnical, this issue can perplex users, and the choice of ligand molecule type, dataset size, and chemical diversity has been shown to significantly influence the final pharmacophore model generated [70].

8.8 Conclusion

The term "pharmacophore" refers to a 3D configuration of chemical attributes that are essential for the biological functionality of a given molecule. Different software applications are deployed in the construction of pharmacophore models, which are then used to identify new compounds that meet the required pharmacophoric criteria, showing potential for biological activity.

The application of pharmacophore modeling is widely beneficial throughout multiple stages of the drug discovery process. VSs have gained significant popularity due to their ability to aid in the discovery of chemicals that can produce the appropriate biological reactions. Pharmacophore models are utilized as efficient tools to discriminate compounds that meet pharmacophoric criteria before, during, and after docking simulations. Furthermore, these entities assume crucial functions in several activities, such as drug target prediction, ligand screening, and the anticipation of ADMET features.

The inclusion of modern computational tools has aided the overcoming of inherent problems in pharmacophore modeling. Significantly, the combination of pharmacophore modeling and MDs simulations has promise in addressing challenges arising from ligand flexibility. Furthermore, the issue of inadequate scoring functions employed in VS through pharmacophore methods presents a significant problem, which could potentially be mitigated by the application of machine learning techniques. Therefore, current developments in pharmacophore modeling offer the potential to generate models with enhanced characteristics.

In conclusion, pharmacophore modeling plays a crucial role in the field of drug discovery. The improvement of computing capabilities, increased access to data, integration with other computational approaches, and the use of advanced algorithms have all contributed to the enhancement of the quality of pharmacophore models that are developed. As the development and refinement of these models progress, their capacity to enhance drug discovery is positioned for additional growth. Therefore, it is crucial to continue the efforts in improving pharmacophore models to further drug development activities.

Acknowledgments

The authors are thankful to the Department of Pharmaceutical Sciences and Technology, Birla Institute of Technology, Mesra, Ranchi, Jharkhand, India, for facilitating the required infrastructure, and computing facilities to carry out the chemical stimulations.

References

1 Deore, A.B., Dhumane, J.R., Wagh, R., and Sonawane, R. (2019). The stages of drug discovery and development process. *Asian Journal of Pharmaceutical Research and Development* 7 (6): 62–67.
2 Sliwoski, G., Kothiwale, S., Meiler, J., and Lowe, E.W. (2014). Computational methods in drug discovery. *Pharmacological Reviews* 66 (1): 334–395.
3 Le, A.V., Phan, T.C.Q., and Nguyen, T.H. (2015). In silico drug design: prospective for drug lead discovery. *International Journal of Engineering and Science Invention* 4 (10): 2319–6726.
4 Macalino, S.J.Y., Gosu, V., Hong, S., and Choi, S. (2015). Role of computer-aided drug design in modern drug discovery. *Archives of Pharmacal Research* 38: 1686–1701.
5 Prachayasittikul, V., Worachartcheewan, A., Shoombuatong, W. et al. (2015). Computer-aided drug design of bioactive natural products. *Current Topics in Medicinal Chemistry* 15 (18): 1780–1800.
6 Langer, T. and Hoffmann, R.D. (2006). Pharmacophore modelling: applications in drug discovery. *Expert Opinion on Drug Discovery* 1 (3): 261–267.
7 Mason, J.S., Good, A.C., and Martin, E.J. (2001). 3-D pharmacophores in drug discovery. *Current Pharmaceutical Design* 7 (7): 567–597.
8 Guner, O.F. (2002). History and evolution of the pharmacophore concept in computer-aided drug design. *Current Topics in Medicinal Chemistry* 2 (12): 1321–1332.
9 Kaalia, R., Kumar, A., Srinivasan, A., and Ghosh, I. (2015). An ab initio method for designing multi-target specific pharmacophores using complementary interaction field of aspartic proteases. *Molecular Informatics* 34 (6–7): 380–393.
10 McGregor, M.J. and Muskal, S.M. (1999). Pharmacophore fingerprinting. 1. Application to QSAR and focused library design. *Journal of Chemical Information and Computer Sciences* 39 (3): 569–574.
11 Hughes, J.P., Rees, S., Kalindjian, S.B., and Philpott, K.L. (2011). Principles of early drug discovery. *British Journal of Pharmacology* 162 (6): 1239–1249.
12 Bradshaw, R.A. and Stahl, P.D. (2015). *Encyclopedia of Cell Biology*. Academic Press.
13 Khedkar, S.A., Malde, A.K., Coutinho, E.C., and Srivastava, S. (2007). Pharmacophore modeling in drug discovery and development: an overview. *Medicinal Chemistry (Los Angeles)* 3 (2): 187–197.
14 Ansari, F., Ghasemi, J.B., and Niazi, A. (2020). Three dimensional quantitative structure activity relationship and pharmacophore modeling of tacrine derivatives as acetylcholinesterase inhibitors in Alzheimer's treatment. *Medicinal Chemistry (Los Angeles)* 16 (2): 155–168.
15 Muegge, I. (2002). Pharmacophore features of potential drugs. *Chemistry – A European Journal* 8 (9): 1976–1981.

16 Sakkiah, S., Thangapandian, S., John, S. et al. (2010). 3D QSAR pharmacophore based virtual screening and molecular docking for identification of potential HSP90 inhibitors. *European Journal of Medicinal Chemistry* 45 (6): 2132–2140.

17 Schwab, C.H. (2010). Conformations and 3D pharmacophore searching. *Drug Discovery Today: Technologies* 7 (4): e245–e253.

18 Acharya, C., Coop, A., Polli, J.E., and MacKerell, A.D. (2011). Recent advances in ligand-based drug design: relevance and utility of the conformationally sampled pharmacophore approach. *Current Computer-Aided Drug Design* 7 (1): 10–22.

19 Lu, X., Yang, H., Chen, Y. et al. (2018). The development of pharmacophore Modeling: generation and recent applications in drug discovery. *Current Pharmaceutical Design* 24 (29): 3424–3439.

20 McGregor, M.J. and Muskal, S.M. (2000). Pharmacophore fingerprinting. 2. Application to primary library design. *Journal of Chemical Information and Computer Sciences* 40 (1): 117–125.

21 Fourches, D., Muratov, E., and Tropsha, A. (2010). Trust, but verify: on the importance of chemical structure curation in cheminformatics and QSAR modeling research. *Journal of Chemical Information and Modeling* 50 (7): 1189.

22 Güner, O.F. (2000). *Pharmacophore Perception, Development, and Use in Drug Design*, vol. 2. Internat'l University Line.

23 Tyagi, R., Singh, A., Chaudhary, K.K., and Yadav, M.K. (2022). Pharmacophore modeling and its applications. In: *Bioinformatics*, 269–289. Elsevier.

24 Sadowski, J. (1997). A hybrid approach for addressing ring flexibility in 3D database Searching. *Journal of Computer-Aided Molecular Design* 11: 53–60.

25 Gasteiger, J. (2012). *Software Development in Chemistry 4: Proceedings of the 4th Workshop "Computers in Chemistry" Hochfilzen*, Tyrol, November 22–24, 1989, Springer Science & Business Media.

26 Wipke, W.T. and Hahn, M.A. (1988). AIMB: analogy and intelligence in model building. System description and performance characteristics. *Tetrahedron Computer Methodology* 1 (2): 141–167.

27 Dolata, D.P., Leach, A.R., and Prout, K. (1987). WIZARD: AI in conformational analysis. *Journal of Computer-Aided Molecular Design* 1: 73–85.

28 van Drie, J.H. (2003). Pharmacophore discovery-lessons learned. *Current Pharmaceutical Design* 9 (20): 1649–1664.

29 Sheridan, R.P., Nilakantan, R., Rusinko, A. III et al. (1989). 3DSEARCH: a system for three-dimensional substructure searching. *Journal of Chemical Information and Computer Sciences* 29 (4): 255–260.

30 Fisanick, W., Cross, K.P., and Rusinko, A. III (1992). Similarity searching on CAS registry substances. 1. Global molecular property and generic atom triangle geometric Searching. *Journal of Chemical Information and Computer Sciences* 32 (6): 664–674.

31 Kuntz, I.D. (1992). Structure-based strategies for drug design and discovery. *Science (1979)* 257 (5073): 1078–1082.

32 Mayer, D., Naylor, C.B., Motoc, I., and Marshall, G.R. (1987). A unique geometry of the active site of angiotensin-converting enzyme consistent with structure-activity studies. *Journal of Computer-Aided Molecular Design* 1: 3–16.

33 Kuster, D.J. and Marshall, G.R. (2005). Validated ligand mapping of ACE active site. *Journal of Computer-Aided Molecular Design* 19: 609–615.

34 Murrall, N.W. and Davies, E.K. (1990). Conformational freedom in 3-D databases. 1. Techniques. *Journal of Chemical Information and Computer Sciences* 30 (3): 312–316.

35 Hurst, T. (1994). Flexible 3D searching: the directed tweak technique. *Journal of Chemical Information and Computer Sciences* 34 (1): 190–196.

36 Barnum, D., Greene, J., Smellie, A., and Sprague, P. (1996). Identification of common functional configurations among molecules. *Journal of Chemical Information and Computer Sciences* 36 (3): 563–571.

37 Jiménez, J., Doerr, S., Martínez-Rosell, G. et al. (2017). DeepSite: protein-binding site predictor using 3D-convolutional neural networks. *Bioinformatics* 33 (19): 3036–3042.

38 Jones, G., Willett, P., and Glen, R.C. (1995). A genetic algorithm for flexible molecular overlay and pharmacophore elucidation. *Journal of Computer-Aided Molecular Design* 9: 532–549.

39 Cramer, R.D.I., Patterson, D.E., and Bunce, J.D. (1988). Comparative molecular field analysis (CoMFA). 1. Effect of shape on binding of steroids to carrier proteins. *Journal of the American Chemical Society* 110 (18): 5959–5967.

40 Li, H., Sutter, J., and Hoffmann, R. (2000). HypoGen: an automated system for generating 3D predictive pharmacophore models. *Pharmacophore Perception, Development, and Use in Drug Design* 2: 171.

41 Kurogi, Y. and Guner, O.F. (2001). Pharmacophore modeling and three-dimensional database searching for drug design using catalyst. *Current Medicinal Chemistry* 8 (9): 1035–1055.

42 Dror, O., Shulman-Peleg, A., Nussinov, R., and Wolfson, H.J. (2004). Predicting molecular interactions in silico: I. A guide to pharmacophore identification and its applications to drug design. *Current Medicinal Chemistry* 11 (1): 71–90.

43 Glennon, R.A., Dukat, M., and Liao, L. (2004). Musings on A4β2 nicotinic acetylcholine (NACh) receptor pharmacophore models. *Current Topics in Medicinal Chemistry* 4 (6): 631–644.

44 Patel, Y., Gillet, V.J., Bravi, G., and Leach, A.R. (2002). A comparison of the pharmacophore identification programs: catalyst, DISCO and GASP. *Journal of Computer-Aided Molecular Design* 16: 653–681.

45 Wolber, G. and Langer, T. (2005). LigandScout: 3-D pharmacophores derived from protein-bound ligands and their use as virtual screening filters. *Journal of Chemical Information and Modeling* 45 (1): 160–169.

46 Richmond, N.J., Abrams, C.A., Wolohan, P.R.N. et al. (2006). GALAHAD: 1. Pharmacophore identification by hypermolecular alignment of ligands in 3D. *Journal of Computer-Aided Molecular Design* 20: 567–587.

47 Vilar, S., Cozza, G., and Moro, S. (2008). Medicinal chemistry and the molecular operating environment (MOE): application of QSAR and molecular docking to drug discovery. *Current Topics in Medicinal Chemistry* 8 (18): 1555–1572.

48 Dixon, S.L., Smondyrev, A.M., and Rao, S.N. (2006). PHASE: a novel approach to pharmacophore modeling and 3D database searching. *Chemical Biology & Drug Design* 67 (5): 370–372.

49 Schneidman-Duhovny, D., Dror, O., Inbar, Y. et al. (2008). PharmaGist: a webserver for ligand-based pharmacophore detection. *Nucleic Acids Research* 36 (suppl_2): W223–W228.

50 Koes, D.R. and Camacho, C.J. (2011). Pharmer: efficient and exact pharmacophore search. *Journal of Chemical Information and Modeling* 51 (6): 1307–1314.

51 Liu, X., Ouyang, S., Yu, B. et al. (2010). PharmMapper server: a web server for potential drug target identification using pharmacophore mapping approach. *Nucleic Acids Research* 38 (suppl_2): W609–W614.

52 Gaulton, A., Bellis, L.J., Bento, A.P. et al. (2012). ChEMBL: a large-scale bioactivity database for drug discovery. *Nucleic Acids Research* 40 (D1): D1100–D1107.

53 Maho, N. (2015). The PubChemQC project: a large chemical database from the first principle calculations. *AIP Conference Proceedings*, AIP Publishing, Vol. 1702.

54 Leach, A.R., Gillet, V.J., Lewis, R.A., and Taylor, R. (2010). Three-dimensional pharmacophore methods in drug discovery. *Journal of Medicinal Chemistry* 53 (2): 539–558.

55 Seidel, T., Ibis, G., Bendix, F., and Wolber, G. (2010). Strategies for 3D pharmacophore-based virtual screening. *Drug Discovery Today: Technologies* 7 (4): e221–e228.

56 Huang, N., Shoichet, B.K., and Irwin, J.J. (2006). Benchmarking sets for molecular docking. *Journal of Medicinal Chemistry* 49 (23): 6789–6801.

57 Caporuscio, F. and Tafi, A. (2011). Pharmacophore modelling: a forty year old approach and its modern synergies. *Current Medicinal Chemistry* 18 (17): 2543–2553.

58 Griffith, R., Luu, T.T.T., Garner, J., and Keller, P.A. (2005). Combining structure-based drug design and pharmacophores. *Journal of Molecular Graphics & Modelling* 23 (5): 439–446.

59 Patel, S., Modi, P., and Chhabria, M. (2018). Rational approach to identify newer caspase-1 inhibitors using pharmacophore based virtual screening, docking and molecular dynamic simulation studies. *Journal of Molecular Graphics & Modelling* 81: 106–115.

60 Lin, X., Li, X., and Lin, X. (2020). A review on applications of computational methods in drug screening and design. *Molecules* 25 (6): 1375.

61 Poptodorov, K., Luu, T., and Hoffmann, R.D. (2006). Pharmacophore model generation software tools. *Pharmacophores and Pharmacophore Searches* 32: 15–47.

62 Wolber, G., Seidel, T., Bendix, F., and Langer, T. (2008). Molecule-pharmacophore superpositioning and pattern matching in computational drug design. *Drug Discovery Today* 13 (1–2): 23–29.

63 Baroni, M., Cruciani, G., Sciabola, S. et al. (2007). A common reference framework for Analyzing/comparing proteins and ligands. Fingerprints for ligands and proteins (FLAP): theory and application. *Journal of Chemical Information and Modeling* 47 (2): 279–294.

64 Nettles, J.H., Jenkins, J.L., Williams, C. et al. (2007). Flexible 3D pharmacophores as descriptors of dynamic biological space. *Journal of Molecular Graphics & Modelling* 26 (3): 622–633.

65 Totrov, M. (2008). Atomic property fields: generalized 3D pharmacophoric potential for automated ligand superposition, pharmacophore elucidation and 3D QSAR. *Chemical Biology & Drug Design* 71 (1): 15–27.

66 Bandyopadhyay, D. and Agrafiotis, D.K. (2008). A self-organizing algorithm for molecular alignment and pharmacophore development. *Journal of Computational Chemistry* 29 (6): 965–982.

67 Hecker, E.A., Duraiswami, C., Andrea, T.A., and Diller, D.J. (2002). Use of catalyst pharmacophore models for screening of large combinatorial libraries. *Journal of Chemical Information and Computer Sciences* 42 (5): 1204–1211.

68 Toba, S., Srinivasan, J., Maynard, A.J., and Sutter, J. (2006). Using pharmacophore models to gain insight into structural binding and virtual screening: an application study with CDK2 and human DHFR. *Journal of Chemical Information and Modeling* 46 (2): 728–735.

69 Vadivelan, S., Sinha, B.N., Irudayam, S.J., and Jagarlapudi, S.A.R.P. (2007). Virtual screening studies to design potent CDK2-cyclin A inhibitors. *Journal of Chemical Information and Modeling* 47 (4): 1526–1535.

70 Yang, S.-Y. (2010). Pharmacophore Modeling and applications in drug discovery: challenges and recent advances. *Drug Discovery Today* 15 (11–12): 444–450.

71 Zou, J., Xie, H.-Z., Yang, S.-Y. et al. (2008). Towards more accurate pharmacophore modeling: multicomplex-based comprehensive pharmacophore map and most-frequent-feature pharmacophore model of CDK2. *Journal of Molecular Graphics & Modelling* 27 (4): 430–438.

72 Schaller, D., Šribar, D., Noonan, T. et al. (2020). Next generation 3D pharmacophore modeling. *Wiley Interdisciplinary Reviews: Computational Molecular Science* 10 (4): e1468.

73 Bajorath, J. (2002). Integration of virtual and high-throughput screening. *Nature Reviews. Drug Discovery* 1 (11): 882–894.

74 Tanrikulu, Y., Krüger, B., and Proschak, E. (2013). The holistic integration of virtual screening in drug discovery. *Drug Discovery Today* 18 (7–8): 358–364.

75 Schuster, D., Spetea, M., Music, M. et al. (2010). Morphinans and Isoquinolines: acetylcholinesterase inhibition, pharmacophore modeling, and interaction with opioid receptors. *Bioorganic & Medicinal Chemistry* 18 (14): 5071–5080.

76 Polgár, T., Baki, A., Szendrei, G.I., and Keserűu, G.M. (2005). Comparative virtual and experimental high-throughput screening for glycogen synthase kinase-3β inhibitors. *Journal of Medicinal Chemistry* 48 (25): 7946–7959.

77 Doman, T.N., McGovern, S.L., Witherbee, B.J. et al. (2002). Molecular docking and high-throughput screening for novel inhibitors of protein tyrosine phosphatase-1B. *Journal of Medicinal Chemistry* 45 (11): 2213–2221.

78 Vuorinen, A. and Schuster, D. (2015). Methods for generating and applying pharmacophore models as virtual screening filters and for bioactivity profiling. *Methods* 71: 113–134.

79 Rollinger, J.M., Schuster, D., Danzl, B. et al. (2009). *In silico* target fishing for rationalized ligand discovery exemplified on constituents of Ruta Graveolens. *Planta Medica* 75 (03): 195–204.

80 Wu, B., Gao, J., and Wang, M. (2005). Development of a complex scintillation proximity assay for high throughput screening of PPARγ modulators. *Acta Pharmacologica Sinica* 26 (3): 339–344.

81 Murgueitio, M.S., Henneke, P., Glossmann, H. et al. (2014). Prospective virtual screening in a sparse data scenario: design of small-molecule TLR2 antagonists. *ChemMedChem* 9 (4): 813–822.

82 Krautscheid, Y., Senning, C.J.Å., Sartori, S.B. et al. (2014). Pharmacophore modeling, virtual screening, and in vitro testing reveal haloperidol, eprazinone, and fenbutrazate as neurokinin receptors ligands. *Journal of Chemical Information and Modeling* 54 (6): 1747–1757.

83 Joung, J.Y., Lee, H.Y., Park, J. et al. (2014). Identification of novel Rab27a/melanophilin blockers by pharmacophore-based virtual screening. *Applied Biochemistry and Biotechnology* 172: 1882–1897.

84 Lu, P., Wang, Y., Kai Ouyang, P. et al. (2015). 3d-QSAR based pharmacophore modeling and virtual screening for identification of novel G protein-coupled receptor40 agonists. *Current Computer-Aided Drug Design* 11 (1): 51–56.

85 Singh, N., Tiwari, S., Srivastava, K.K., and Siddiqi, M.I. (2015). Identification of novel inhibitors of mycobacterium tuberculosis PknG using pharmacophore based virtual screening, docking, molecular dynamics simulation, and their biological evaluation. *Journal of Chemical Information and Modeling* 55 (6): 1120–1129.

86 Temml, V., Voss, C.V., Dirsch, V.M., and Schuster, D. (2014). Discovery of new liver X receptor agonists by pharmacophore modeling and shape-based virtual screening. *Journal of Chemical Information and Modeling* 54 (2): 367–371.

87 Lepailleur, A., Freret, T., Lemaître, S. et al. (2014). Dual histamine H3R/serotonin 5-HT4R ligands with antiamnesic properties: pharmacophore-based virtual screening and polypharmacology. *Journal of Chemical Information and Modeling* 54 (6): 1773–1784.

88 Ferreira, R.J., dos Santos, D.J.V.A., Ferreira, M.-J.U., and Guedes, R.C. (2011). Toward a better pharmacophore description of P-glycoprotein modulators, based on macrocyclic diterpenes from euphorbia species. *Journal of Chemical Information and Modeling* 51 (6): 1315–1324.

89 Flohr, S., Kurz, M., Kostenis, E. et al. (2002). Identification of nonpeptidic urotensin II receptor antagonists by virtual screening based on a pharmacophore model derived from structure–activity relationships and nuclear magnetic resonance studies on urotensin II. *Journal of Medicinal Chemistry* 45 (9): 1799–1805.

90 Hessler, G. and Baringhaus, K.-H. (2010). The scaffold hopping potential of pharmacophores. *Drug Discovery Today: Technologies* 7 (4): e263–e269.

91 Goldmann, D., Pakfeifer, P., Hering, S., and Ecker, G.F. (2015). Novel scaffolds for modulation of TRPV1 identified with pharmacophore modeling and virtual screening. *Future Medicinal Chemistry* 7 (3): 243–256.

92 Medina-Franco, J.L. (2013). Advances in computational approaches for drug discovery based on natural products. *Revista Latinoamericana de Quimica* 41 (2): 95–110.

93 Koutsoukas, A., Simms, B., Kirchmair, J. et al. (2011). From in silico target prediction to multi-target drug design: current databases, methods and applications. *Journal of Proteomics* 74 (12): 2554–2574.

94 Rognan, D. (2010). Structure-based approaches to target fishing and ligand profiling. *Molecular Informatics* 29 (3): 176–187.

95 Rella, M., Rushworth, C.A., Guy, J.L. et al. (2006). Structure-based pharmacophore design and virtual screening for novel angiotensin converting enzyme 2 inhibitors. *Journal of Chemical Information and Modeling* 46 (2): 708–716.

96 Paliwal, S., Mittal, A., Sharma, M. et al. (2015). Pharmacophore and molecular docking based identification of novel structurally diverse PDE-5 inhibitors. *Medicinal Chemistry Research* 24: 576–587.

97 Peach, M.L. and Nicklaus, M.C. (2009). Combining docking with pharmacophore filtering for improved virtual screening. *Journal of Cheminformatics* 1 (1): 1–15.

98 Hindle, S.A., Rarey, M., Buning, C., and Lengauer, T. (2002). Flexible docking under pharmacophore type constraints. *Journal of Computer-Aided Molecular Design* 16: 129–149.

99 Mobley, D.L., Liu, S., Lim, N.M. et al. (2014). Blind prediction of HIV integrase binding from the SAMPL4 challenge. *Journal of Computer-Aided Molecular Design* 28: 327–345.

100 Lyne, P.D., Kenny, P.W., Cosgrove, D.A. et al. (2004). Identification of compounds with nanomolar binding affinity for checkpoint kinase-1 using knowledge-based virtual screening. *Journal of Medicinal Chemistry* 47 (8): 1962–1968.

101 Wang, M., Hou, S., Wei, Y. et al. (2021). Discovery of novel dual adenosine A1/A2A receptor antagonists using deep learning, pharmacophore Modeling and molecular docking. *PLoS Computational Biology* 17 (3): e1008821.

102 Alavijeh, M.S. and Palmer, A.M. (2004). The pivotal role of drug metabolism and pharmacokinetics in the discovery and development of new medicines. *Current Opinion in Investigational Drugs* 5: 755–763.

103 Mohan, C.G. (2019). *Structural Bioinformatics: Applications in Preclinical Drug Discovery Process*, vol. 27. Springer.

104 de Groot, M.J. and Ekins, S. (2002). Pharmacophore modeling of cytochromes P450. *Advanced Drug Delivery Reviews* 54 (3): 367–383.

105 Miners, J.O., Smith, P.A., McKinnon, R.A., and Sorich, M.J. (2004). Multiple pharmacophores for the investigation of human UDP-glucuronosyltransferase isoform substrate selectivity. *Molecular Pharmacology* 65 (2): 301–308.

106 Murgueitio, M.S., Rakers, C., Frank, A., and Wolber, G. (2017). Balancing inflammation: computational design of small-molecule Toll-like receptor modulators. *Trends in Pharmacological Sciences* 38 (2): 155–168.

107 Sato, T., Honma, T., and Yokoyama, S. (2010). Combining machine learning and pharmacophore-based interaction fingerprint for in silico screening. *Journal of Chemical Information and Modeling* 50 (1): 170–185.

108 Barillari, C., Marcou, G., and Rognan, D. (2008). Hot-spots-guided receptor-based pharmacophores (HS-pharm): a knowledge-based approach to identify ligand-anchoring atoms in protein cavities and prioritize structure-based pharmacophores. *Journal of Chemical Information and Modeling* 48 (7): 1396–1410.

109 Schuetz, D.A., Seidel, T., Garon, A. et al. (2018). GRAIL: grids of pharmacophore interaction fields. *Journal of Chemical Theory and Computation* 14 (9): 4958–4970.

110 Perricone, U., Wieder, M., Seidel, T. et al. (2017). A molecular dynamics-shared pharmacophore approach to boost early-enrichment virtual screening: a case study on peroxisome proliferator-activated receptor α. *ChemMedChem* 12 (16): 1399–1407.

111 Lee, J.Y., Krieger, J.M., Li, H., and Bahar, I. (2020). Pharmmaker: pharmacophore modeling and hit identification based on druggability simulations. *Protein Science* 29 (1): 76–86.

112 Opo, F.A.D.M., Rahman, M.M., Ahammad, F. et al. (2021). Structure based pharmacophore Modeling, virtual screening, molecular docking and ADMET approaches for identification of natural anti-cancer agents targeting XIAP protein. *Scientific Reports* 11 (1): 4049.

113 Sangande, F., Julianti, E., and Tjahjono, D.H. (2020). Ligand-based pharmacophore modeling, molecular docking, and molecular dynamic studies of dual tyrosine kinase inhibitor of EGFR and VEGFR2. *International Journal of Molecular Sciences* 21 (20): 7779.

114 Hassan Baig, M., Ahmad, K., Roy, S. et al. (2016). Computer aided drug design: success and limitations. *Current Pharmaceutical Design* 22 (5): 572–581.

115 Hamza, A., Wei, N.-N., and Zhan, C.-G. (2012). Ligand-based virtual screening approach using a new scoring function. *Journal of Chemical Information and Modeling* 52 (4): 963–974.

116 Scior, T., Bender, A., Tresadern, G. et al. (2012). Recognizing pitfalls in virtual screening: a critical review. *Journal of Chemical Information and Modeling* 52 (4): 867–881.

117 Gimeno, A., Ojeda-Montes, M.J., Tomás-Hernández, S. et al. (2019). The light and dark sides of virtual screening: what is there to know? *International Journal of Molecular Sciences* 20 (6): 1375.

118 Kirchmair, J., Wolber, G., Laggner, C., and Langer, T. (2006). Comparative performance assessment of the conformational model generators omega and catalyst: a large-scale survey on the retrieval of protein-bound ligand conformations. *Journal of Chemical Information and Modeling* 46 (4): 1848–1861.

119 Drwal, M.N. and Griffith, R. (2013). Combination of ligand- and structure-based methods in virtual screening. *Drug Discovery Today: Technologies* 10 (3): e395–e401.

120 Vancraenenbroeck, R., De Raeymaecker, J., Lobbestael, E. et al. (2014). In silico, in vitro and cellular analysis with a kinome-wide inhibitor panel correlates cellular LRRK2 dephosphorylation to inhibitor activity on LRRK2. *Frontiers in Molecular Neuroscience* 7: 51.

9

Scaffold Hopping and *De Novo* Drug Design

Shrimanti Chakraborty, Soumi Chakraborty, Biprajit Sarkar, Rahul Ghosh, Sharanya Roy, Nisha Kumari Singh, and Gourav Rakshit

Department of Pharmaceutical Sciences and Technology, Birla Institute of Technology, Ranchi, Jharkhand, India

9.1 Introduction

Drug discovery is a process where a novel lead compound is present, which binds to a specific macromolecular receptor and exhibits pharmacological activity. The aim is to find novel lead compounds with desirable pharmacological properties. To find this compound *de novo* design is applied in various cases. A supportive approach is provided by computer-assisted *de novo* design is the screening of virtual molecules. For this, the process of compound synthesis and testing must be emulated by smart algorithms [1], so that software can suggest chemically feasible substances which exhibit drug-like properties.

Computer-assisted *de novo* drug design has an immense advantage in the field of compound diversity which is theoretically unlimited and innovative also [2]. By *de novo* drug design, we can resolve two main problems: first is how to construct chemically and synthetically feasible drug-like molecules (the assembly problem) and second one is how to assess the quality of the designed compounds (the scoring problem) [1]. For the solution of the first problem, fragment-based methods are a good approach. Fragment-based drug discovery is an important method for finding potential drugs and improving their effectiveness. With the vast possibilities in the world of potential drugs, computers help us explore small building blocks of chemicals and find good starting points for designing effective drugs [3].

In the early stage *de novo* drug design was basically based on receptor-based scoring schemes. There are two main ways to deal with the scoring problem. One way is to make better scoring functions for designing based on the receptor (called "*in situ*" design). The other way is to not focus on the receptor's structure and instead rely on ligand-based scoring. In the first approach, you need a model of the receptor's binding cavity. In the second approach, you need at least one known reference ligand to use as a guide for designing compounds [1]. Due to significant improvement in the area of *de novo* design methods, it will defiantly help in drug discovery more precisely in the near future.

Computational Methods for Rational Drug Design, First Edition. Edited by Mithun Rudrapal.

9.2 Scaffold Hopping

Scaffold hopping (Figure 9.1) refers to a procedure focused on discovering molecular structures with different molecular backbones while maintaining iso functionality [4]. Also known as lead hopping, leapfrogging, chemotype switching, and scaffold searching [5]. Scaffold-hopping approaches typically initiate with recognized active compounds and conclude by altering the central core structure of the molecule to generate a novel chemotype [6]. At times, adjusting side chains is enough to mitigate the undesirable properties linked to the parent molecule, while in other cases, it becomes necessary to modify the core structure or scaffold of the parent molecule [6].

There exist various motives for pursuing a diverse array of molecular frameworks: diverse chemotypes provide options for chemical accessibility and opportunities for lead optimization. Having multiple lead structures reduces the risk of drug development setbacks in the presence of undesirable ADMET (absorption, distribution, metabolism, excretion, and toxicity) properties. Scaffold-hopping can additionally be employed to transition from natural substrates to more drug-like agents or enhance receptor-subtype selectivity [4]. This method necessitates having a template, which is a chemical structure showcasing the desired biological activity. It operates on the assumption that other compounds, structurally different from the template but retaining certain essential features, can exhibit the same biological activity [7]. Attaining this involves modifying factors such as mode of action, selectivity, potency, and other relevant aspects of activity, administration, availability at the target site, or other metabolic properties. This process encompasses not only bioisosteric replacement but also substantial changes and substitutions to the foundational structure, aiming to introduce both novelty and the desired response [5]. From a chemist's viewpoint, two scaffolds may differ if they are constructed through distinct synthetic routes. Numerous cases

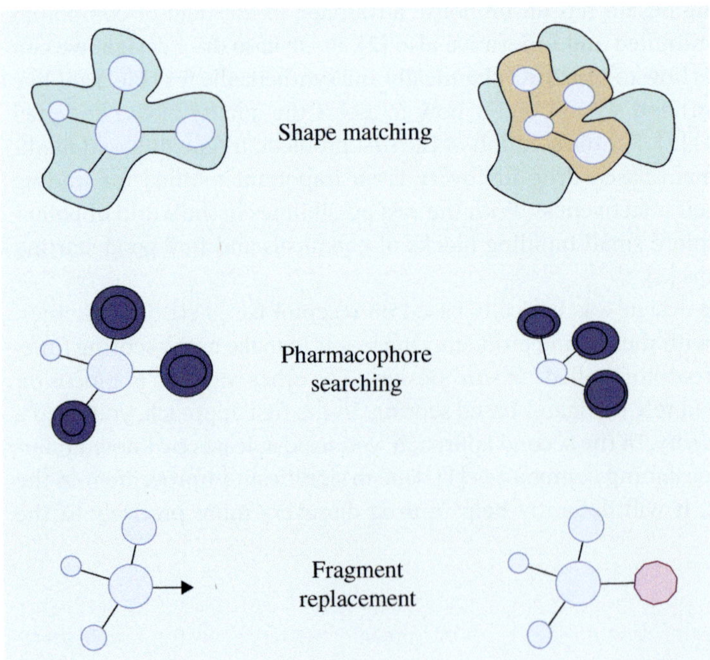

Figure 9.1 Illustration of scaffold hopping.

illustrate that minute alterations can lead to dramatic changes in molecular properties. Therefore, a pharmacologist may differentiate between an agonist and an antagonist based on the compound's function, even if the compounds vary by just one minor substituent [7].

A prevalent strategy in scaffold hopping involves seeking replacements for a fragment within an active compound rather than entire compounds. An early example of a 3D database searching program, CAVEAT, was specifically designed to address this challenge [7]. Several other programs, employing variations of the methodology used by CAVEAT or targeting the common objective of substituting a specified scaffold in a 3D structure, include Brood, sparkV10, Core Hopping, Scaffold Replacement, SHOP, ParaFrag, and ReCore. These programs take as input a 3D structure of an active compound or a scaffold portion, with designated bonds marked for use as attachment vectors linking the scaffold to substituents (R-groups). All these programs conduct searches within predefined fragment databases and provide matching 3D structures that can replace the original scaffold. The proposed new scaffolds are then organized based on criteria such as size, predicted physicochemical properties, pharmacophoric features, common structural framework, or other attributes that assist users in reviewing the results and evaluating the potential of the suggested structural replacements [8].

9.2.1 Classification of Scaffold Hopping

Boehm et al. established a classification criterion for distinguishing two scaffolds: they deemed them different if synthesized through distinct synthetic routes, irrespective of the magnitude of the alteration. This assertion has found support in numerous instances where closely related chemical structures, exhibiting sufficient disparity to warrant separate patent claims or US Food and Drug Administration (US FDA) approvals for new drug applications, coexist. A notable illustration involves the phosphodiesterase enzyme type 5 (PDE5) inhibitors sildenafil and vardenafil, where a mere exchange of a carbon atom and a nitrogen atom in the 5–6 fused ring constitutes the primary structural difference, yet it suffices for distinct patent coverage. Similarly, the cyclooxygenase 2 (COX-2) inhibitors Rofecoxib (Vioxx™) and valdecoxib (Bextra™) exhibit minimal dissimilarity, limited to the five-member hetero rings connecting the two phenyl rings, but were marketed separately by Merck and Pharmacia/Pfizer. The rationale behind the concept of scaffold hopping hinges on evaluating the extent of change relative to the original parent molecule [6].

9.2.1.1 1° Hop: Heterocycle Replacement

Minor alterations, such as substituting or interchanging carbon and heteroatoms within a backbone ring, fall under the classification of a 1° hop. Heterocycles, serving as the cores of drug molecules, typically present multiple vectors projecting in various directions. Introducing novel scaffolds can be achieved by replacing carbon (C), nitrogen (N), oxygen (O), and sulfur (S) atoms in a heterocycle while preserving the outward vectors. Enhanced binding affinity is often attainable when the heterocycle directly engages in interactions with the target protein [6].

Illustratively, nonsteroidal anti-inflammatory drugs (NSAIDs) operate by inhibiting the enzyme cyclooxygenase (COX), which is responsible for catalyzing the biosynthesis of prostaglandins (PGs) from arachidonic acid (AA). In humans, two enzymes, COX-1 and COX-2, catalyze the initial step in PG biosynthesis. Despite sharing the same catalytic activity, COX-1 and COX-2 differ in sequence (60% identity), tissue distribution, and physiological function. COX-1 plays a role in gastroprotection and vascular homeostasis, while COX-2 is primarily associated with inflammatory processes. Selectively inhibiting the COX-2 isozyme can mitigate the ulcerogenic effects linked to traditional NSAIDs like aspirin and ibuprofen. Despite a 60% sequence homology between COX-1 and COX-2, their protein backbones, particularly the ligand-binding sites, closely resemble each other. Nevertheless, subtle structural distinctions at the ligand-binding sites prove sufficient to generate COX-2 selective inhibitors [6].

9.2.1.2 2° Hop: Ring Opening and Closure: Pseudo Ring Structures

Manipulating a molecule's flexibility can be achieved through ring opening and closure, exerting control over the total number of free rotatable bonds. Intramolecular hydrogen bonds (HBs) often guide where to initiate ring closure, as exemplified in the following instances. The GlaxoSmithKline group, intrigued by the potential intramolecular HB between the o-alkoxy group and biaryl NH, synthesized a series of indole compounds as PG EP1 receptor antagonists. The strategic design of ring closure effectively constrained the molecule into a bioactive conformation. One of the resulting indole compounds, featuring iso-butyl as R, exhibited noteworthy activities in binding and functional EP1 antagonist assays at low nM and sub-nM levels.

While ring closure positively impacts binding free energy, it can potentially have adverse effects on solubility and other ADME properties. To mitigate these effects and enhance drug-likeness, medicinal chemists may opt for ring opening. An illustrative example is the pyridopyrimidinone moiety found in protein kinase inhibitors. PD166285, a broad-spectrum tyrosine kinase inhibitor with a 6-aryl substituted pyridopyrimidinone structure, served as a template for designing novel tyrosine kinase inhibitors. Furet et al. performed ring opening by relocating the nitrogen atom in position 1 of the pyrimidine ring to position 5, forming a pseudo-six-member ring with the adjacent urea through intramolecular hydrogen bonding. *Ab initio* calculations and data mining confirmed the favorable pseudo cyclic conformation, supported by assay results showing submicromolar inhibition against several tyrosine kinases, including c-Src, EGFR, and c-Abl, for the pyrimidinyl urea compound [6].

9.2.1.3 3° Hop: Pseudopeptides and Peptidomimetics

The strategic design of small molecules to emulate the structural attributes of peptides, with active peptide conformations serving as templates, has yielded promising outcomes for challenging targets. This approach has been expansively employed in targets associated with protein–protein interactions, aiming to create small molecules that mimic the interacting components of proteins. In peptide-based drug discovery, the primary objective is to diminish peptide characteristics to bolster resistance against proteolysis, all the while preserving essential chemical features for molecular recognition. A common method employed for the transition from peptides to small molecules is scaffold hopping. Frequently observed secondary structures like α-helices, β-sheets, and β/γ-turns at peptide-protein and protein–protein interfaces have prompted the design of synthetic structures to replicate these secondary structures.

Tools such as Recore and CAVEAT prove invaluable in devising suitable scaffolds to substitute specific segments of peptides. Additionally, pharmacophore modeling packages offered by Chemical Computing Group, Accelrys, and Schrodinger find widespread application in the design of peptidomimetics [6].

9.2.1.4 4° Hop: Topology/Shape-Based Scaffold Hopping

Instances of successful topology/shape-based scaffold hopping are infrequently documented in the literature. One potential explanation is that numerous attempts have been made, but the majority ended in failure and remained unpublished. Another scenario arises when the new chemotype significantly diverges from its template; scientists may categorize the process as virtual screening (VS) rather than scaffold hopping.

The Cambridge Structural Database (CSD) extends beyond being a mere repository of small-molecule crystal structures; it boasts a robust search engine and tools for query construction and structure mapping. The CSD package serves as an effective topology hopping tool, empowering users to specify diverse topological requirements, encompassing dihedral angle, point-to-plane distance, and plane-to-plane angle. Additionally, ROCS stands out as a potent topology-based scaffold-hopping tool. The grid-based method SHOP has also been employed to identify novel scaffolds featuring significantly distinct chemotypes from their source queries [6].

9.2.2 Advantages of Scaffold Hopping

- **Diversity in chemical space:** Scaffold hopping allows for the exploration of diverse chemical structures, potentially leading to the discovery of novel compounds with unique biological activities.
- **Improved selectivity:** Modifying the scaffold can enhance the selectivity of a compound for a specific biological target, reducing off-target effects and improving the overall safety profile.
- **Overcoming limitations:** If a scaffold has limitations such as poor pharmacokinetics or toxicity issues, scaffold hopping can be used to address these challenges and improve the overall drug-like properties.
- **Structure–activity relationship (SAR) exploration:** Scaffold hopping facilitates the exploration of SAR, providing insights into how specific structural changes influence the biological activity of a compound.
- **Intellectual property opportunities:** Developing new scaffolds may offer opportunities for obtaining new patents, providing a competitive advantage and exclusivity in the market.
- **Lead optimization:** When a lead compound shows promising activity but has undesirable properties, scaffold hopping can be employed to optimize the compound for better drug-like characteristics.
- **Avoidance of patent issues:** Scaffold hopping can help navigate patent landscapes, allowing researchers to design compounds with unique structures that do not infringe on existing intellectual property.
- **Leveraging previous knowledge:** If there is existing knowledge about the biological activity of a certain scaffold, scaffold hopping can leverage this information to design new compounds with similar or improved properties.

9.2.3 Disadvantages of Scaffold Hopping

- **Loss of activity:** Modifying the scaffold may lead to a loss of biological activity, and there is a risk that the newly designed compounds may not exhibit the desired therapeutic effects.
- **Unforeseen toxicity:** Introducing structural changes through scaffold hopping may result in unexpected toxicity issues, and the safety profile of the new compounds must be thoroughly evaluated.
- **Synthetic challenges:** Some scaffold modifications may pose synthetic challenges, making the production of new compounds difficult or impractical.
- **Limited predictability:** The success of scaffold hopping is not always predictable, and the relationship between structural changes and biological activity can be complex and difficult to anticipate.
- **Time and resource intensive:** Designing, synthesizing, and evaluating new compounds through scaffold hopping can be time consuming and resource-intensive.
- **Biological relevance:** The newly designed scaffolds may not always be biologically relevant or may not interact favorably with the target, leading to a lack of efficacy.

9.2.4 Reasons for Scaffold Hopping

In recent days, the bridge between scaffolds, compounds, and biological activities has been looked into everywhere. The development of new scaffolds and their guiding search calculations are very useful in representing new potent compounds. For various reasons, scaffold hopping is applied in medicinal chemistry as for example, one might be interested in circumventing an intellectual property position by identifying novel chemical entities having a desired activity, replacing a chemically complex natural product with a synthetically accessible molecule, or improving pharmacological properties of known actives [9]. Some substances can be hard to absorb because of certain qualities they have. For example, if a molecule does not dissolve easily or if it mostly exists in a charged form like strong acids or bases, it might not be absorbed well because it cannot easily pass through certain types of membranes [10]. It is mostly used in the context of virtual or computational screening. The replacement of scaffold from a chemical perspective is by ingenious design or systematic chemical modifications of core structures representing compound series [9].

In systemic scaffold hopping applications, computational approaches are important [9]. By this method we can generalize the structures of known reference molecules and separate from them to identify compounds having different scaffolds. This method explores the idea of molecular similarity and dissimilarity in various ways. On the other hand, docking can also be utilized for searching of novel active compounds. There is another reason for the necessity of scaffold hopping is the metabolic transformation of the molecule. As for example, amide bonds are often hydrolyzed rapidly by proteases, but peptides often occur as natural ligands of many targets. In order to derive a lead structure from such a peptide, the scaffold often needs to be replaced by a less peptidic one [10].

For the prediction of the affinity between targets and small molecules is another interest for choosing scaffold hopping. In simple terms, when we do not know what a substance does, we can predict its effects by looking at other substances that are similar to it chemically and already have known effects [10]. Target profile prediction and prediction of new ligands for individual targets at the early discovery stage.

9.2.5 Properties and Key Methods of Scaffold Hopping

- **Pharmacophore searching:** The concept of pharmacophore searching has been preferred for a long time. A pharmacophore is a three-dimensional (3D) arrangement of molecular features or chemical groups that are essential for a compound to interact with a biological target and exhibit a specific biological activity. If the pharmacophores are transferred from the reference to the test molecule, then the scaffolds will also be replaced [9]. Mostly, a pharmacophore query molecule has a limited number of interaction properties like H-bond donors and acceptors, and lipophilic or aromatic groups, which have a particular position and orientation in space [10]. There are some specific examples: Catalyst and Unity.
- **Fragment replacement:** In the context of drug design, the fragment replacement technique is equivalent to bioisosteric replacement, which defines a pair of compounds that differ only by a single-site substructure change [8]. CAVEAT is a groundbreaking software that is mostly used in fragment replacement. This program looks through specific databases for pieces and finds 3D shapes that can be swapped with the original structure. The new suggestions are sorted based on their size, expected properties, key features, shared structure, or other factors. This helps users see and understand the potential of the suggested changes [8]. This method can be performed on 2D or 3D structures and has a high success rate [7].

- **Similarity searching:** When the information regarding 3D structure is not available, this similarity searching is applied. Earlier, this is performed by using one or multiple query structures and searching against a database of existing (or virtual) molecules [11]. These methods have a downside. To search for a molecule in the database, it must already be clearly listed. This is usually fine when looking through small compound collections or corporate libraries, which have around 10^5–10^8 molecules. However, with large combinatorial libraries, like the virtual FE library mentioned earlier, there can be many more molecules, making the search more challenging [11].

9.3 *De Novo* Drug Design

The Latin word "*De novo*" means "from the beginning," "afresh," or "anew." The term "*De novo*" refers to the process of designing therapeutic molecules from scratch using computer-aided drug design (CADD) technology. Better chemical space exploration and the discovery of novel chemical structures for medication development are made possible by it; these structures are not available in any known chemical database, whether they are discovered in synthetic chemical libraries or natural compounds. As a result, the chemicals or medications that are found will be special [12, 13]. To create an ideal solution, *de novo* drug design necessitates the simultaneous and independent optimization of multiple goals. This renders the evolutionary algorithm – a soft computing methodology – a problem-independent search technique that may be used for a variety of optimization problems, ranging from robot behavior to medication discovery and design [14].

De novo drug design involves building pharmacological compounds from scratch using either atoms or molecular fragments [13, 15]. There is a vast amount of search space and many novel chemical structures could be found. Nevertheless, the atom-based approach produces some compounds that are not synthesizable in addition to taking longer to create these unique structures (Figure 9.2).

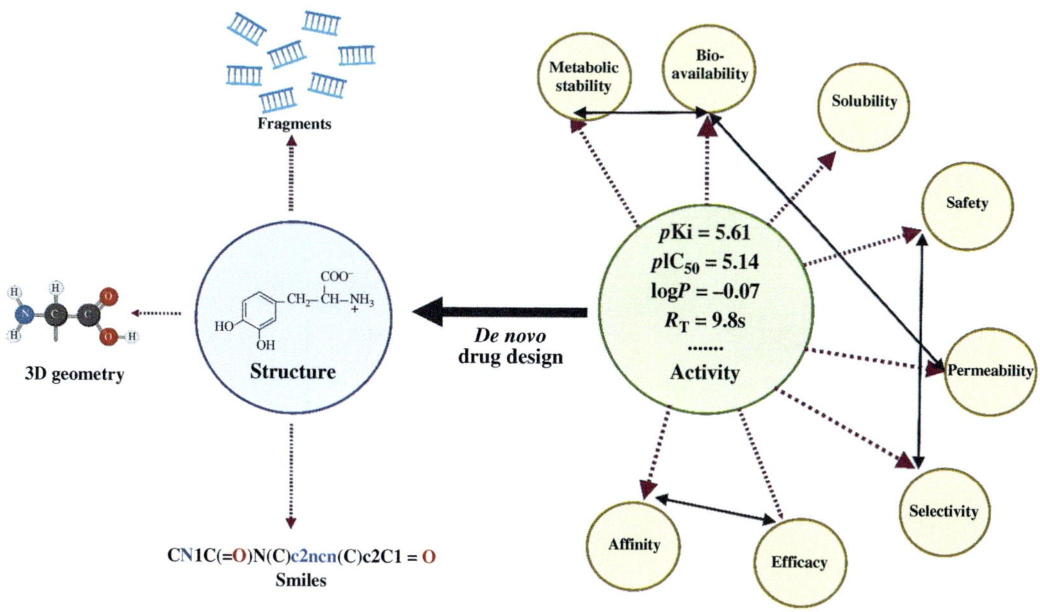

Figure 9.2 Basic principle of *de novo* drug design.

Assembling fragments is the favored drug synthesis strategy in many *de novo* drug design tools, known as the fragment-based method. Better ADME (absorption, distribution, metabolism, and excretion) qualities and significant diversity are found in the solutions that are found, and the molecular search space is relatively smaller. The goal of *de novo* drug design is to create novel molecules with the desired biological activity while matching the biological target's binding pattern. There are two distinct *de novo* design approaches based on the binding pattern: ligand-based and structure-based approaches [13, 15].

With structure-based *de novo* drug design, novel ligands are created without previous knowledge of other ligands by utilizing data from the 3D structure of a protein target. Constructing the molecule directly in the target protein's binding site and measuring the molecule's interaction energy with the protein are popular structure-based design techniques. Conversely, in ligand-based *de novo* design, the structure of the protein target is unknown and the new molecule is built based on its resemblance to an existing ligand molecule.

9.3.1 Classification of *De Novo* Drug Design

De novo drug design is a computational approach used in drug discovery to design novel molecules with desired pharmacological properties. The classification of *de novo* drug design can be broadly categorized into two main approaches. The classification has also been presented pictorially in Figure 9.3.

9.3.1.1 Structure-based Drug Design

1) **Rational Drug Design**
   i) **Construction:** Designing molecules based on the 3D structure of the target.
   ii) **Example:** Designing HIV protease inhibitors by understanding the structure of the viral protease and creating molecules that fit its active site.
2) **Fragment-based Drug Design**
   i) **Construction:** Identifying small, bioactive fragments and building them into larger molecules.
   ii) **Example:** Developing antibiotics by assembling fragments that individually interact with essential bacterial enzymes.

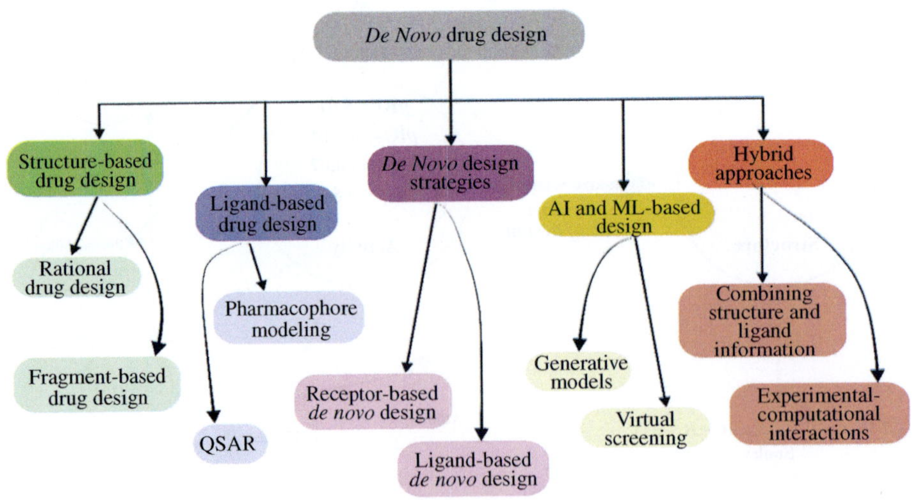

Figure 9.3 Pictorial representation of overall classification of *de novo* drug design.

9.3.1.2 Ligand-based Drug Design

1) **Quantitative Structure–Activity Relationship (QSAR)**
   i) **Construction:** Establishing statistical relationships between chemical structure and biological activity.
   ii) **Example:** Predicting the activity of new antihypertensive compounds based on the quantitative relationships derived from known structure–activity data.
2) **Pharmacophore Modeling**
   i) **Construction:** Identifying key spatial arrangements of functional groups essential for biological activity.
   ii) **Example:** Designing kinase inhibitors by aligning molecules with a pharmacophore representing the critical features for binding to the kinase active site.

9.3.1.3 *De Novo* Design Strategies

1) **Receptor-based *De Novo* Design**
   i) **Construction:** Generating entirely new molecules that interact with a specific biological target.
   ii) **Example:** Designing novel anticancer drugs by considering the structure of a unique cancer-specific protein.
2) **Ligand-based *De Novo* Design**
   i) **Construction:** Designing molecules based on the properties of known active ligands.
   ii) **Example:** Designing analogs of an existing painkiller with modifications to improve efficacy and reduce side effects.

9.3.1.4 Artificial Intelligence (AI) and Machine Learning-based Design

1) **Generative Models**
   i) **Construction:** Using AI to generate novel molecular structures.
   ii) **Example:** Employing a deep learning model to generate potential drug candidates with desired properties.
2) **Virtual Screening**
   i) **Binding Pattern:** Predicting the likelihood of compounds binding to a target.
   ii) **Example:** Screening a large chemical database using machine learning algorithms to identify potential inhibitors for a specific enzyme.

9.3.1.5 Hybrid Approaches

1) **Combining Structure and Ligand Information**
   i) **Construction and Binding Pattern:** Integrating structural and ligand-based considerations.
   ii) **Example:** Designing new antibiotics by combining features of known antibiotic structures with insights from the target's 3D structure.
2) **Experimental–Computational Interactions**
   i) **Construction and Binding Pattern:** Iteratively refining designs based on computational predictions and experimental results.
   ii) **Example:** Adjusting the chemical structure of a designed compound after assessing its activity in experimental assays.

9.3.2 Basic Principle of *De Novo* Drug Design

De novo drug design is an approach that generates new chemical entities solely from data about a biological target, such as a receptor or its known active binders, which are ligands with strong binding or inhibitory activity against the receptor. The description of the receptor active site or ligand pharmacophore modeling, molecule creation (sampling), and molecule evaluation are the

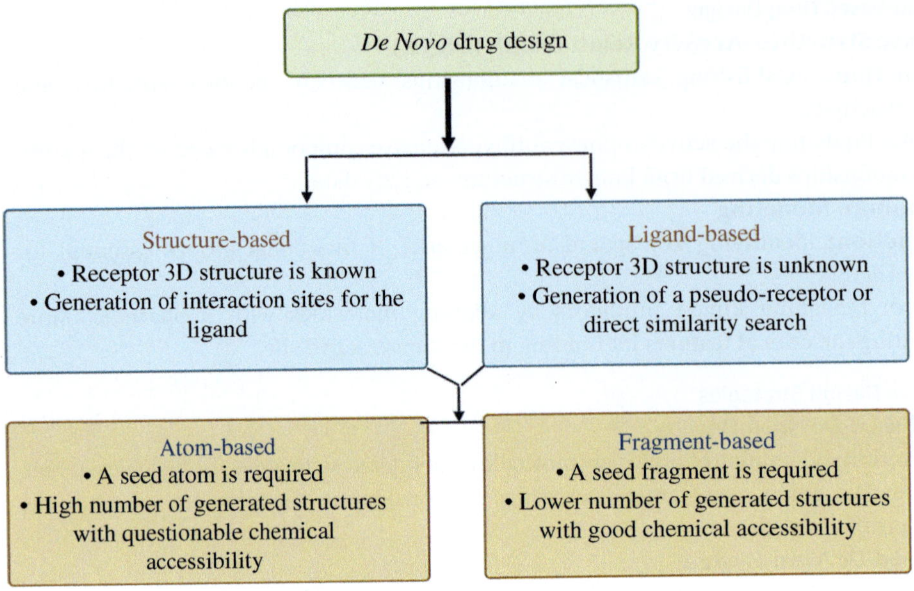

Figure 9.4 *De novo* drug-design approach schematic illustration.

main steps in the *de novo* drug design process. Two main *de novo* drug design strategies are accessible, such as design based on ligands and structure. X-ray crystallography, nuclear magnetic resonance, or electron microscopy are typically used to obtain the 3D structures of receptors (Figure 9.4). When the receptor's structure is unknown, homology modeling can be used to find a good structure for *de novo* medication design. On the other hand, sequence similarity and template structure quality are what determine how good a homology model is. When one or more active binders are known, but no structural data for the biological target are available, the ligand-based method is typically employed.

9.3.3 Application of *De Novo* Drug Design

De novo drug design has numerous applications in the field of drug discovery and development. Some of the key applications include:

1) **Target-based drug discovery**
   De novo drug design is often used in conjunction with information about a specific molecular target (such as a protein implicated in disease) to design molecules that interact selectively and with high affinity.
2) **Orphan disease**
   For rare or orphan diseases where existing drug treatments may not be available, *de novo* drug design can be a valuable approach to discovering new compounds targeting specific molecular pathways related to the disease.
3) **Drug resistance**
   De novo drug design can be applied to address the challenges of drug resistance. By designing novel compounds that interact with a target differently than existing drugs, it may be possible to overcome resistance mechanisms.

4) **Polypharmacology**
   Designing compounds that interact with multiple targets can be important in addressing complex diseases with multiple contributing factors. *De novo* drug design allows for the creation of multitarget drugs.
5) **Fragment-based drug design**
   De novo approaches, particularly fragment-based design, are valuable for identifying small-molecular fragments that can be assembled to form larger, more complex compounds. This approach is especially useful for targets with large binding sites.

9.3.4 Historical Overview of Scaffold Hoping and *De Novo* Drug Design

The pharmaceutical industry began in the 1880s–1930s when chemical corporations set up research facilities to create novel medications and separate active ingredients from natural ingredients, and their biological activity was examined [16]. Schneider et al. presented the idea of scaffold hopping in 1999 as a method to find isofunctional molecular structures with noticeably distinct molecular backbones. In the brief definition, the new compounds' different core structures and similar biological activity compared to the parent compounds were highlighted as two important aspects of scaffold hopping [17].

The concise definition highlights two essential elements of scaffold hopping: the new compounds' comparable biological activity to those of their parent compounds and their distinct core structures. The similarity property concept states that substances with comparable chemical structures typically have similar physicochemical qualities and biological activities [18]. These two requirements seem to contradict this idea. The possibility of structurally distinct molecules binding to the same target is not ruled out by this concept. The concept of the similarity property is widely recognized because, in order for the ligands to fit in the same pocket, they should have some structural similarities, such as a similar shape and electropotential surface, even though they may belong to different chemotypes. The Lennard–Jones potential's repulsive penalty is highly sensitive to interatomic distances, so the activity landscape is not linear. For example, adding or removing a small methyl group can cause significant changes in biological activity [19], and similarity metrics are insufficient to capture this nonlinearity. Furthermore, the flexibility of small molecules and proteins further complicates the relationship between similarity and activity. Scaffold hopping is not an exception to the similarity property principle, which is still the cornerstone of contemporary drug discovery techniques like SARs.

The Latin word "*De novo*" means "from the beginning," "afresh," or "anew." The term "*De novo*" refers to the process of designing therapeutic molecules from scratch using CADD technology. Better chemical space exploration and the discovery of novel chemical structures for medication development are made possible by it; these structures are not available in any known chemical database, whether they are discovered in synthetic chemical libraries or natural compounds. As a result, the chemicals or medications that are found will be special [12, 13]. In 1991, the first *de novo* drug design tool, called LEGEND, was published. As of right moment, there are more than 20 of these tools available. Structure-based drug design approaches are used by the tools LigMerge [19], LEGEND [20], LUDI [21, 22], SPROUT [23], BREED [24], and PRO LIGAND [25] to generate novel compounds. LEGEND is one of these instruments that construct the molecule or medicine atom by atom, while the other technologies employ fragment-by-fragment construction. To generate novel compounds, the ligand-based drug design technique is used by the tools NEWLEAD [26] and PhDD [27]. Apart from these tools, the simulated annealing technique has also been applied to tools like AMOSA [28], CONCERTS [29], SkelGen [30], and MOLig [31]. LigBuilder [32], LEA [33], ADAPT [34], PEP [35], SYNOPSIS [36], LEA3D [37], GANDI [38], TOPAS [39], Flux [1, 40],

FLUX [41], MEGA [42], and EvoMD [43] are a few examples of instruments. The process of finding new drugs is heavily dependent on the continuous supply of fresh lead compounds that bind to macromolecular receptor molecules that are relevant to the disease and display the appropriate pharmacological profile. There are several approaches to identifying these leads, the majority of which rely on high-throughput screening (HTS) campaigns that involve the experimental testing of physically accessible molecules [44, 45]. By screening virtual molecules, computer-assisted *de novo* design offers a supplementary method that significantly increases the number of compounds that are easily accessible for HTS (usually 1–3 million compounds). In computer-aided *de novo* design, intelligent algorithms must simulate the steps involved in compound manufacturing and testing in order for the program to suggest chemically plausible, drug-like compounds. It has been demonstrated that fragment-based virtual synthesis techniques are very useful for this task [46, 47]. *De novo* ligand design was introduced in 1991 as a purely structure-based method to suggest ligands for synthesis and was later augmented by ligand-based approaches. Pharmacophores and shape restrictions are identified by both structure-based and ligand-based approaches, which then match them with complimentary features included in small-molecule topologies. *De novo* ligand design has received a lot of interest lately when combined with biophysical fragment screening and X-ray structural clarification. From a specialized application of *de novo* design, scaffold hopping has developed into a rapidly growing software toolkit with a wide range of applications in the pharmaceutical sector. While both *De Novo* design and scaffold hopping-based approaches are crucial in drug design, they exhibit distinct differences as illustrated in Figure 9.5.

9.3.5 Methodological Approaches in *De Novo* Drug Design

On the basis of the information regarding the biological target or the known active target site, ligand pharmacophore modeling, and construction of the sample, *de novo* drug design creates

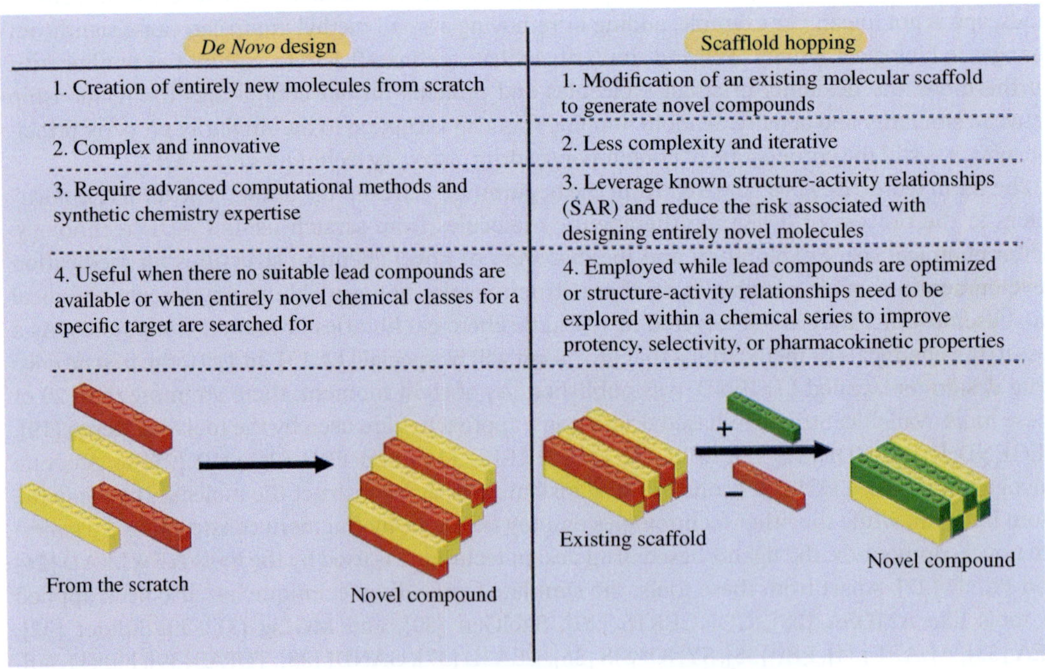

Figure 9.5 The basic difference between *de novo* drug design and scaffold hopping.

novel chemical compounds. Ligands are found to have good binding or inhibitory activity against the target site [48]. In the structure-based design, the structure of the target molecule is unknown; therefore, homology modeling is introduced. When the structure of the target molecule is available, but here, one or more active binders are also known, then the ligand-based approach is followed.

9.3.5.1 Structure-based *De Novo* Drug Design

The active site's physical–chemical properties and molecular shape are analyzed to determine the shape resistance and the noncovalent interactions to a specific binding of a ligand [13]. Distinction of the receptor's active site is described by receptor-based *de novo* drug design. During the interaction of the ligand with an active site containing a receptor, there may be noncovalent interactions like hydrogen bonds, electrostatic, and hydrophobic. Based on the interaction on the active site, it will increase the structure selectivity [48] (Figure 9.6).

Several methods are used to describe the interaction site for the receptor's active site. A few examples are HSITE [49], LUDI and PRO_LIGAND [49, 50], and HIPPO [50], respectively, consider hydrogen-bond donors and acceptors, hydrophobic interaction sites and interactivity with covalent and metal bonds.

9.3.5.2 Ligand-based *De Novo* Drug Design

The quality of the pharmacophore model holds a considerable role in ligand-based *de novo* drug design. For the evaluation of the pharmacophore model, a quantitative SAR model is used parallelly. In the absence of 3D structure of a biological target, the known active binders give an

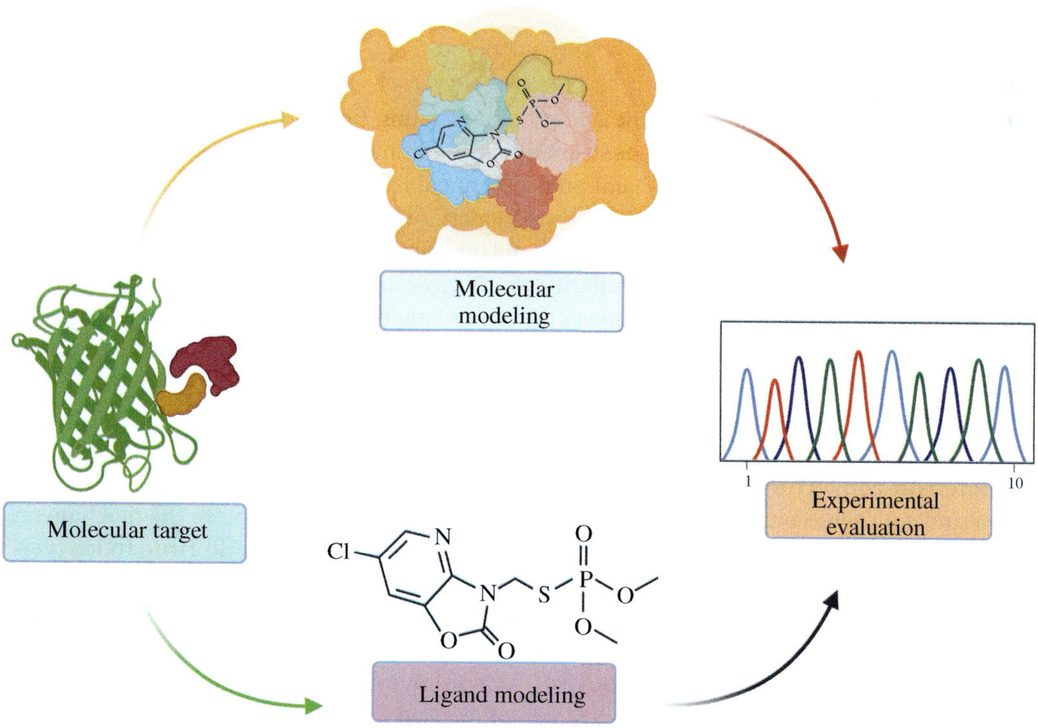

Figure 9.6 Structure-based *de novo* drug design.

alternative strategy for *de novo* design [49]. By SAR studies or screening, these data are accessible from the literature [50]. ChEMBL is a database where we can find active binders that contain bioactive molecules with drug-like properties [51]. When finding a crystal structure is challenging, this method is applied to finding biological targets. For one or more active binders, a ligand pharmacophore model is applied and used to generate novel structures [52]. Specifically, the ligand pharmacophore model can be employed for generating a pseudo-receptor or for directly executing similarity design [53]. There are a few tools that are used for ligand-based drug design are – DOGS [54], TOPAS [54], and SYNOPSIS [54].

9.3.5.3 Generation of Drug-Like Molecular Fragments

For the computerized molecular design algorithms are used. These algorithms are based on the concepts of reproduction, evaluation, mutation, crossover, and selection. This concept or algorithms are applied to the already existing molecules to create novel molecules. During this operation, mutation and crossover play a pivotal role in generating new molecules [55]. On the basis of the Kawai et al. study the atom-based method applied for the mutation purpose. This method usually resulted in a lot of unusual structures that contained invalid hetero–hetero atom bonds [55]. This strategy aims to explore possible options that show structural similarities to the reference molecule but introduce differences in both their scaffolds and peripheral chains. The technique moves across a large chemical space by using a known active molecule of interest as a guide. The fragment library should ideally be put together using a set of well-known compounds that are connected to the reference molecule's target. Every fragment in the collection can be used for mutation, depending on fragmentation. Three categories – side chains, linkers, and rings – are used to group these fragments. By employing the fragment library and crossover, new entities are created through fragment-based mutation. Computational experiments conducted with our proprietary fragment library, derived from GPCR SARfari, have validated the feasibility of our approach to drug discovery [55].

9.3.5.4 Similarity Searching

When considering any compounds to be added to an existing library by purchasing, it is very important as well as hard to find whether structurally similar molecules have the same biological activity shown or not. Within chemical and pharmaceutical research, similarity searching stands as a traditional and extensively employed method to identify compounds with specific properties from databases. Various techniques have been devised for small-molecule similarity searching [56]. Regardless of the method employed, a similarity search typically begins with one or more known active compounds serving as reference molecules. It fundamentally comprises three key components: a selected molecular representation, the search algorithm, and a metric to measure the similarity among these represented molecules [56].

9.3.5.5 Selection of Target Reference Structure

- **Angiotensin-converting enzyme inhibitors**

Krueger and colleagues and their team conducted a database search to explore the mechanism of action associated with the keyword "ACE-Inhibitor" in the Derwent World Drug Index (version July 2006). The software used for this search was ISIS/base version 2.5 developed by MDL Information Systems. Lisinopril (1), a nonprodrug ACE inhibitor, was chosen as the template structure for evolutionary *de novo* design, along with the smaller Captopril (2), which, to the best of knowledge, is the only other nonprodrug ACE inhibitor available on the market [57]. In a second *de novo* approach, the team utilized a set of focused reference structures with a similar

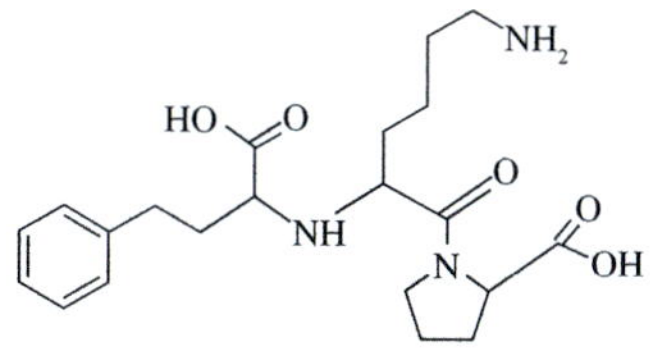

1. Lisinopril

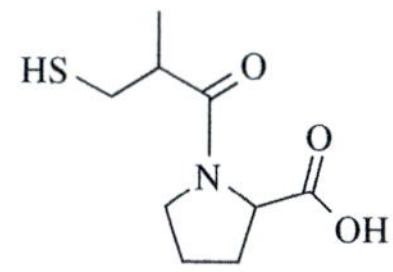

2. Captopril

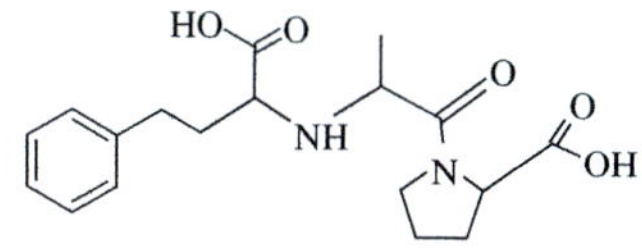

3. Enalapril

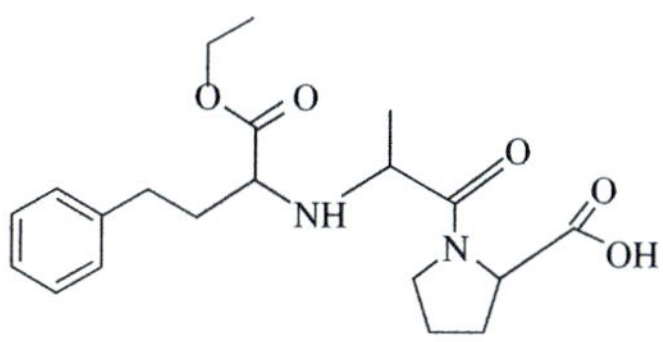

4. Enalaprilmaleate

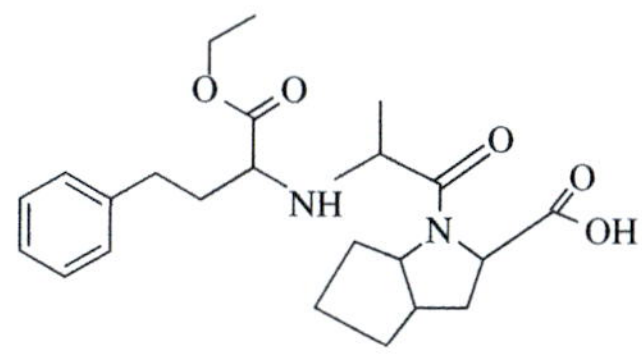

5. Ramipril

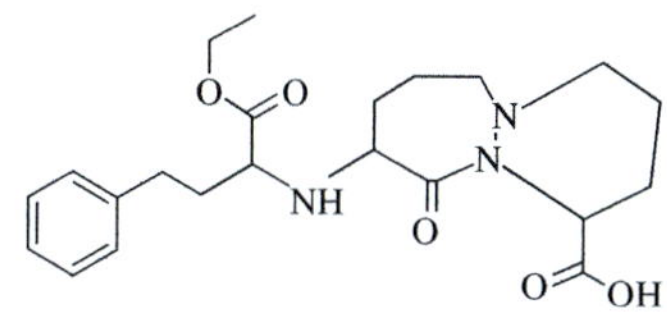

6. Cilazapril

Figure 9.7 Reference structures for ACE inhibitors.

chemotype. A similarity search in the reference actives for Lisinopril was conducted using the FCFP-4 descriptor and Tanimoto index. The resulting set included the four most similar substances (3–6) along with Lisinopril (1), forming a set of five "focused" reference structures (Figure 9.7).

- **Angiotensin-II receptor antagonists**

To assemble a collection of active compounds as antagonists for the angiotensin-II receptor (AT2), the team conducted a database search in the WDI using the mechanism of action keyword "Angiotensin-2-inhibitor." This was followed by a filtering step employing a custom drug-like filter. The team opted for the "first-in-class" drug Losartan (7) as the reference structure for optimization, employing a single-reference approach. Additionally, for parallel-focused optimization with five similar reference structures, four substances (8–11) from the compiled actives were identified to be most similar to Losartan using the FCFP-4 descriptor and Tanimoto index [57] (Figure 9.8).

- **Aldose reductase inhibitors**

A collection of active compounds for a second target class was found by running a WDI database search with the mechanism of action keyword "Aldose-reductase inhibitor" (ARI). After that, this collection was filtered using our drug-likeness standards. A single template structure, Epalrestat (12), was chosen as the template for *de novo* design. Four related compounds (13–16) were found using a similarity search (FCFP-4, Tanimoto) for Epalrestat within the active chemicals list. These compounds were then employed as templates for focused *de novo* design [57] (Figure 9.9).

9.3.5.6 Similarity Analysis of *De Novo*-generated Compounds

A target class-specific "best-split" similarity value can be computed and used as a similarity radius or threshold around each known active from the previously built collection to build a preliminary screening. A *de novo* compound may be classified as "potentially active" according to the "similar property principle" or the "neighborhood behavior principle" if its similarity radius is less than the threshold determined for any of the known actives. Once identified as possible actives, compounds can be subjected to additional testing. *De novo* compounds with a high probability of inactivity may be rejected if they do not show at least the threshold degree of resemblance to any recognized active [40].

7. Losartan

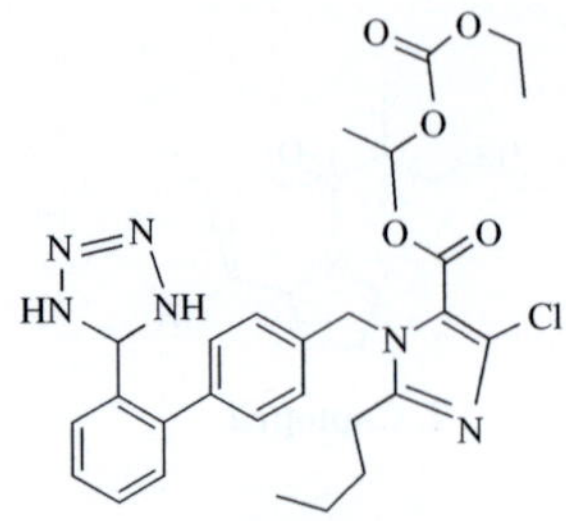

8. Elisartan

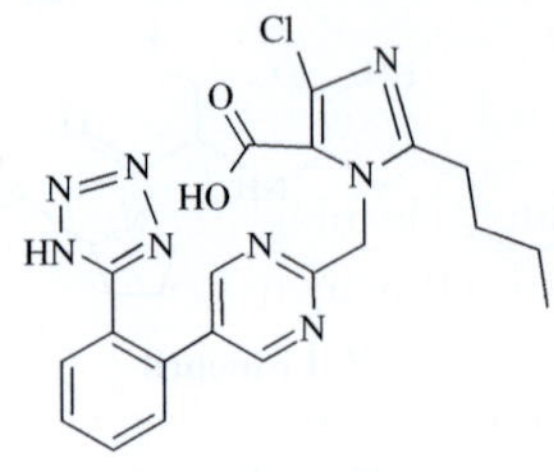

9. CP161418

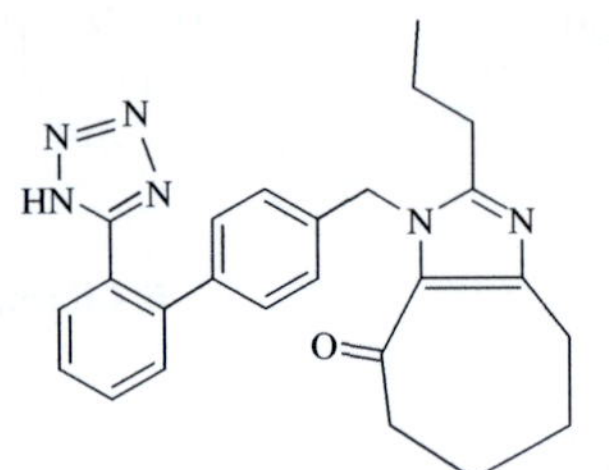

10. Pratosartan

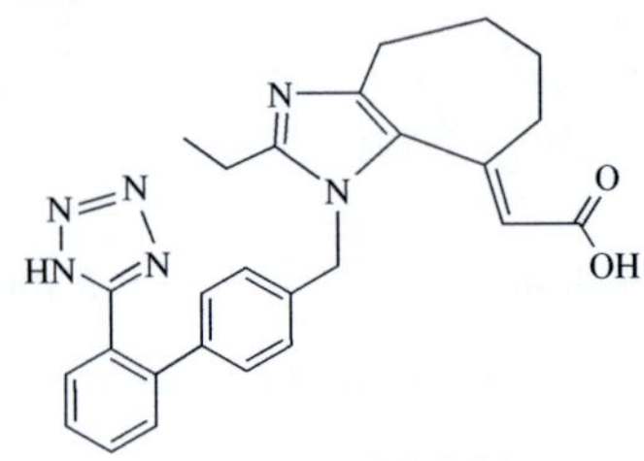

11. KT-3-866

Figure 9.8 Reference structures for angiotensin-II receptor antagonists.

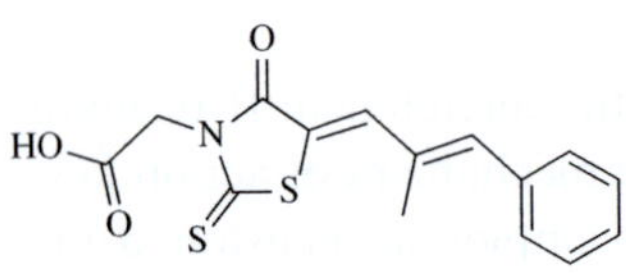

12. Epalrestat

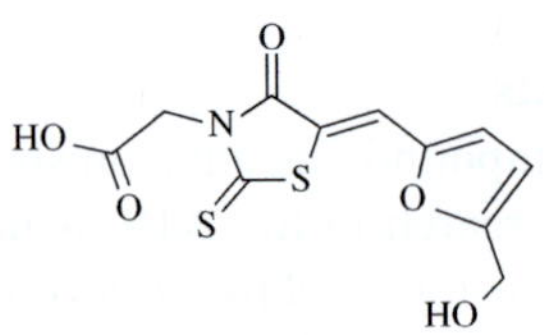

13. ONO-1

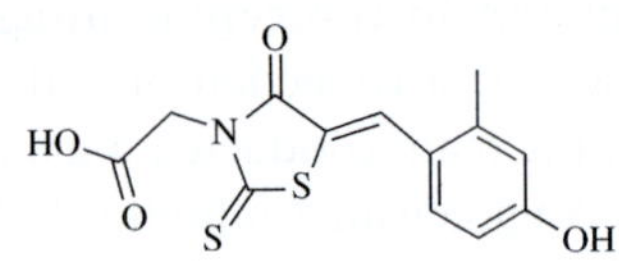

14. ONO-3

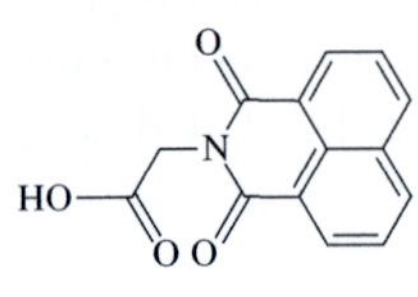

15. Alrestatin

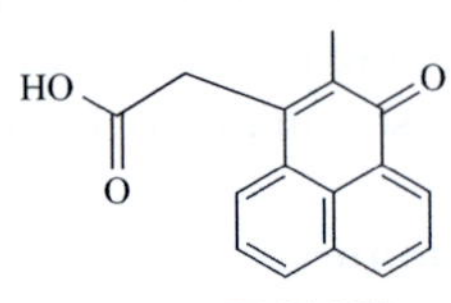

16. AY-31358

Figure 9.9 Reference structures for Aldose reductase inhibitors.

9.3.5.7 Evaluation of Scaffold Diversity

The research team, along with Krueger et al., computed Murcko scaffolds for every *de novo* compound and every member of the active set in order to evaluate the diversity of the recently formed *de novo* structures. A customized MOE script (Molecular Operating Environment, version 08.2006, Chemical Computing Group) was used to do this. Only the cyclic substructures of a molecule and their shortest linkers are captured by Murcko scaffolds; these indicate the shortest bond path between two rings. Two types of Murcko scaffolds were created: atom-based scaffolds that kept information about hetero atoms in rings and linkers, and graph-based scaffolds that removed all information about hetero items in rings and linkers. For each scaffold, both MACCS keys and FCFP-4 fingerprints were computed and subsequently clustered based on maximum dissimilarity, employing the "complete linkage" method [57].

9.4 Results and Discussion

9.4.1 Generation of Drug-Like Molecular Fragments

After "cleaning" the Word Drug Index, Krueger et al. and his team identified 80,499 structures. By subjecting them to various filtering criteria (including REOS, Hann, and drug-like properties), the selection was narrowed down to 24,301 compounds. These compounds served as the foundation for our pool of drug-like molecules. Utilizing their implementation of RECAP within the Retroflux software, they dissected these compounds, resulting in 10,497 unique fragments with at least one cleaved (unvalenced) bond. These fragments were then employed as the building blocks for generating new molecules from scratch. Additionally, we gathered reference sets comprising 73 ACE inhibitors, 80 aldose reductase inhibitors, and 59 AT2 antagonists, all possessing known activity against their respective targets [57].

9.4.2 *De Novo* Design with a Single Reference Structure

- **ACE inhibitors:** Using Lisinopril (1) as a template structure, 259,829 distinct *de novo* constructions were produced in the *de novo* design studies. Although the most promising *de novo*-created construct showed a Manhattan distance of 8.8 compared to Lisinopril (Manhattan distance computed using CATS2D Fingerprints), Lisinopril itself was not reconstructed during the evolutionary optimization process. Subsequently, a similarity search was conducted across all compounds within the ACE actives set, resulting in the identification of 12 known actives (e.g., 17, 18) within the *de novo* constructs. Some of these compounds exhibited significant dissimilarity to the reference Lisinopril, as indicated by their Tanimoto scores [57] (Figure 9.10).
- **AT_2 antagonists:** In the *de novo* design process with Losartan (7) as the reference structure, a total of 972,495 unique virtual constructs were generated. Once again, the evolutionary algorithm failed to replicate the template structure. The top-performing *de novo* constructs approached Losartan with a Manhattan distance of up to 3.4 (according to CATS2D). Of these constructs, 5792 virtual entities achieved a similarity score of $\geq$0.48, meeting the target class-specific similarity threshold (calculated using FCFP-4 Tanimoto similarity) compared to any of the compounds within the known actives set. Consequently, these constructs were categorized as potential actives [57].
- **Aldose reductase inhibitors:** In the *de novo* design process with Epalrestat (12) as the sole template, a total of 250,753 virtual constructs were generated. The highest-ranking *de novo*

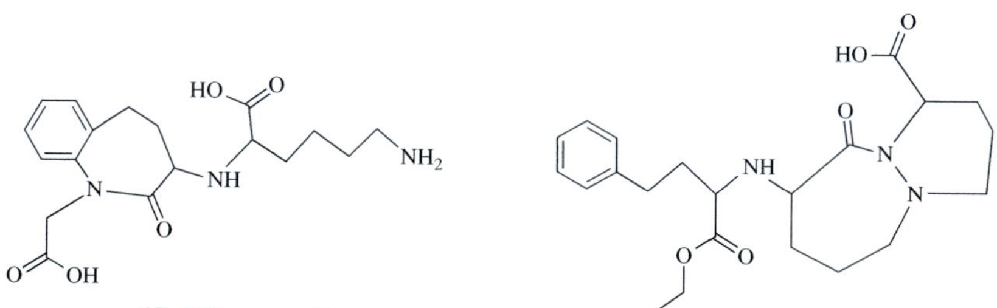

17. Libenzapril **18. Cilazapril**

Figure 9.10 Reference structures for ACE inhibitors.

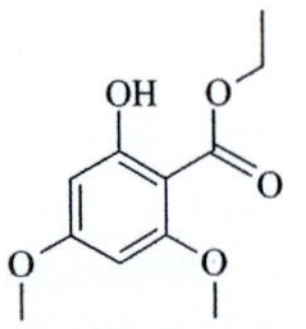

19. Xantoxylin-Carboxylate Ethylester

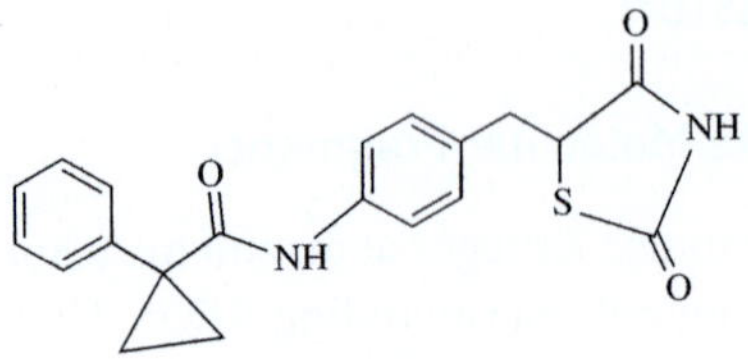

20. DN-108

Figure 9.11 Reference structures for aldose reductase inhibitors.

construct closely resembled Epalrestat, achieving a Manhattan distance of 3.7 (measured using CATS2D descriptors). During this process, two known actives, labeled as (19, 20) in the WDI, were successfully reconstructed. Furthermore, among the generated constructs, 1030 virtual entities exhibited a similarity score of ≥0.59, surpassing the target class-specific similarity threshold (determined using Tanimoto similarity with FCFP-4 fingerprints) when compared to any compound within the actives set. As a result, these constructs were identified as potentially active [57] (Figure 9.11).

9.4.3 *De Novo* Design with a Focused Set of Five Similar Templates

- **ACE inhibitor:** The Flux algorithm identified 50 more structures that performed well. These structures were close to the two molecules mentioned (44 and 45) [57], suggesting that the algorithm focused on a particular area of activity. Even though none of the suggested template designs were discovered from scratch. While looking closely at the newly created structures, it is noticed that a specific pattern often appeared in the highest-rated ones. This pattern is the same as the one found in Enalapril, which is the most common part shared by the five template compounds (1) and (3–6) [57] (Figure 9.12).
- **Angiotensin-II receptor antagonists:** 650,616 different virtual things were constructed. The best *de novo* structure had a Manhattan distance of $D = 6.5$ (CATS). It was found that 3980 constructs with a score of ≥0.58 were kind of similar to an already known active compound [57]. However, the given templates and any known active compounds are not reconstructed.

9.4.4 *De Novo* Design with a Diverse Set of Five Templates

- **ACE inhibitors:** 333,299 unique virtual designs were generated using 5 very diverse sets of starting templates (1, 2, 7–9) [57]. Each design was different, and the best one scored 7.8 and it fit well (using certain measures). This score is a bit better than starting from scratch with just one template (which scored 8.8), but not as good as using a set of similar templates (which scored 5.2) [57]. Ten known compounds were recreated. A number of 4999 *de novo* structures with a score of ≥0.52 were somewhat similar to existing active compounds, but not as many as when it started from just one template (Figure 9.13).
- **Aldose reductase inhibitors:** 376,455 unique *de novo* were constructed in this approach. Out of these, 1915 designs were found with a similarity score of ≥0.59 [57].

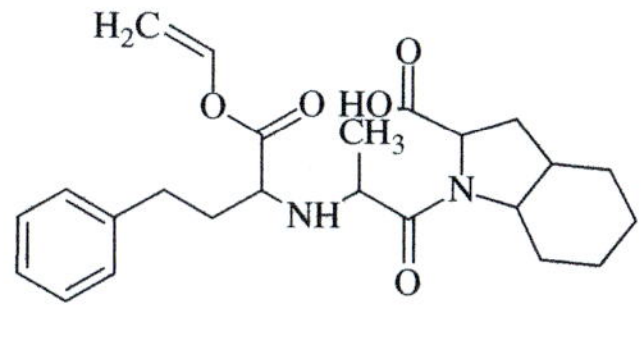

44. de novo compound 1

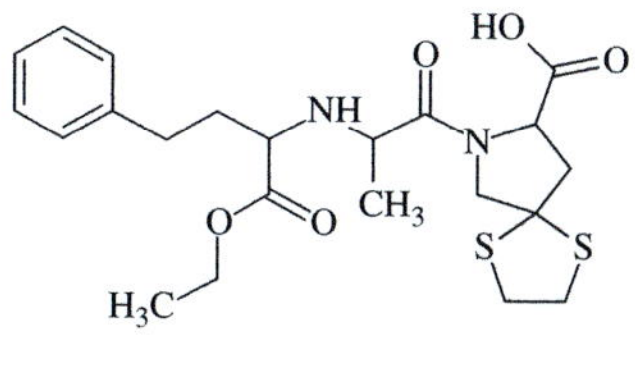

45. de novo compound 1

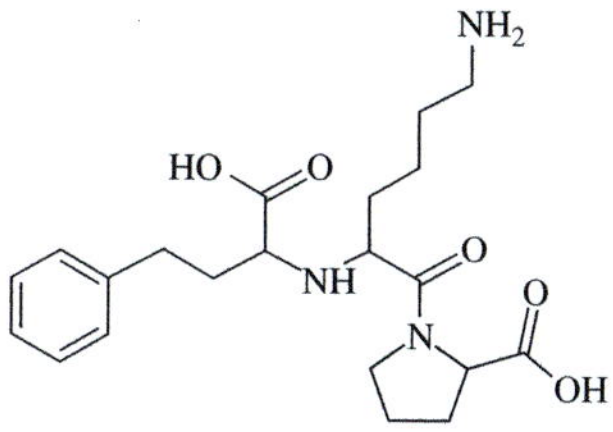

1. Lisinopril

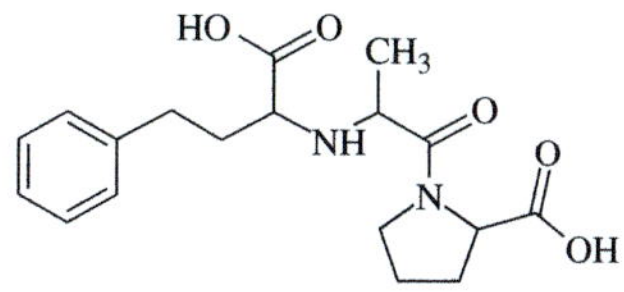

3. Enalapril

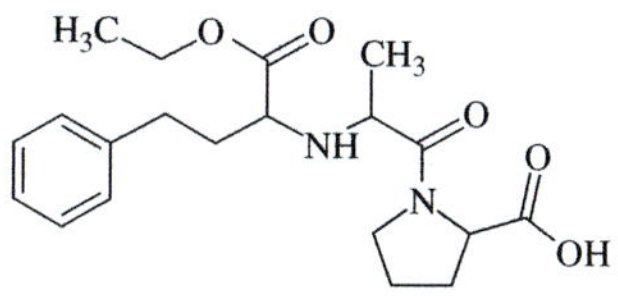

4. Enalaprilmaleate

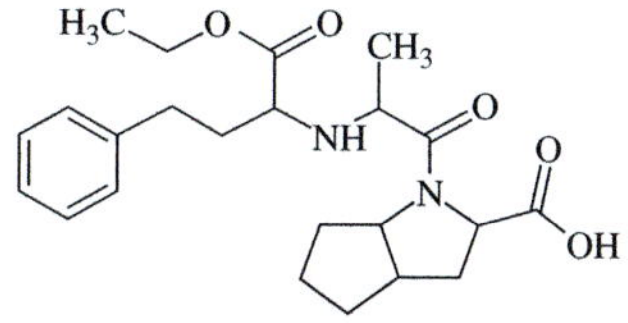

5. Ramipril

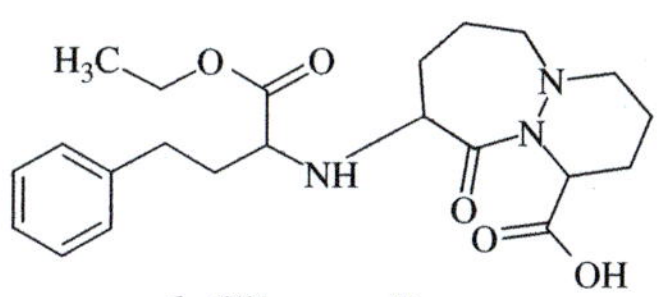

6. Cilazapril

Figure 9.12 Reference structures for ACE inhibitors.

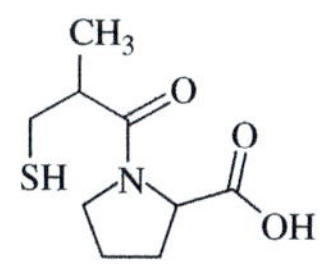

2. Captopril

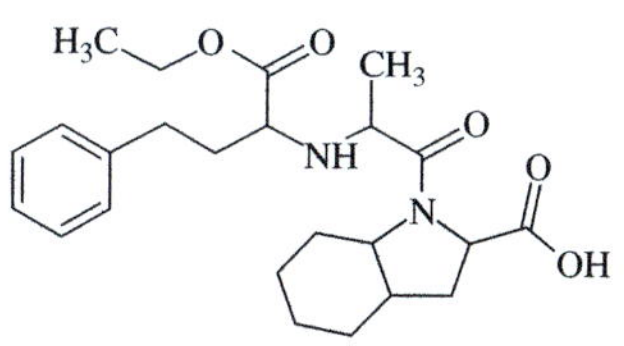

7. Perindopril

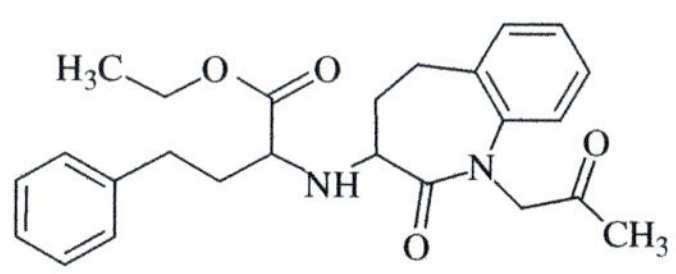

8. Benazepril

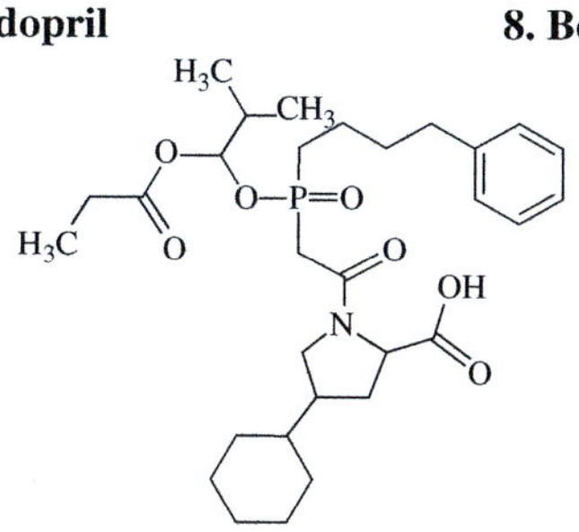

9. Fosinopril

Figure 9.13 Reference structures for ACE inhibitors.

9.5 Software Tools for SH (Scaffold Hopping) and *De Novo* Design Selection

a) **SPROUT:** The software SPROUT is one of the earliest methods for computer-aided *de novo* design of possible ligands. SPROUT's concept is to split the design process into two phases. The initial phase creates molecular skeletons that fulfill steric restrictions, also referred to as fundamental constraints. Pieces of a template are assembled step-by-step to create skeletons. Since each template fragment is made up of dummy atoms without element types and solely assigned hybridization states, it represents a set of molecular fragments. Element types are allocated in the second stage to satisfy hydrophobic and electrostatic needs (secondary constraints). The principal and secondary limitations include produced from a ligand binding site's 3D structure. As a result, SPROUT has been expanded to include other design goals, such as virtual chemical synthesis's ability to facilitate synthetic synthesis (SynSPROUT) [23, 58].
b) **FLUX/Topas:** One software program for ligand-based *de novo* design is TOPAS. The construction of candidate molecules results in a high degree of resemblance to one or more reference ligands. Measures of similarity include (i) a topological pharmacophore descriptor (CATS descriptor) and (ii) a structural 2D descriptor. Using a set of bioactive substances and the virtual retrosynthetic RECAP rules, molecular fragments utilized by TOPAS are produced. Additionally, the fragment-based assembly of novel ligand candidates follows the same set of guidelines [40].
c) **BREED:** BREED software takes a very different technique than stochastic optimization while assembling pieces. Rather, BREED recombines fragments of known ligands in their receptor-bound state overtop of them to new possible ligands – a process that essentially echoes what medicinal chemists often do. The process of aligning ligands involves covering the backbone atoms of protein structures.
d) **Skelgen:** The Skelgen software is based on an approach for computer-based *de novo* design that was proposed by Todorov and Dean in the late 1990s. Like SPROUT (see above), the creation of novel compounds is divided into two steps: (i) assigning atom types and (ii) creating skeletons. The process of assembling template fragments randomly results in the creation of molecular structures. These pieces can be used to create user-defined sets because they are made up of generic hydrogen and carbon atoms.

9.6 Case Study

9.6.1 *De Novo* Drug Design

Darolutamide (Nubeqa™) stands as a novel second-generation, nonsteroidal, selective androgen receptor (AR) inhibitor tailored for treating nonmetastatic castration-resistant prostate cancer (nmCRPC) [59]. The discovery of Darolutamide began with the identification of a potent nonsteroidal AR antagonist. This compound was initially identified by researchers at Bayer AG using a structure-guided drug design approach. The team utilized X-ray crystallography to investigate the structure of the AR ligand-binding domain in complex with various ligands. This analysis led to the identification of a unique binding pocket on the AR that could be targeted for inhibition.

The researchers then conducted a SAR study to optimize the potency and selectivity of the AR antagonist. Through the SAR study, they were able to develop a compound with a promising pharmacological profile. However, this compound had suboptimal pharmacokinetic properties, leading the researchers to explore additional modifications.

At this stage, Bayer AG entered into a collaboration with Orion Corporation, a Finnish pharmaceutical company with expertise in the development of compounds with improved pharmacokinetic properties. The collaboration aimed to optimize the pharmacokinetic profile of the AR antagonist while retaining its pharmacological properties. The model of the antagonistic AR conformer was developed with SwissModel19, drawing on the antagonistic conformation of the progesterone receptor (PDB entry 2ovh,20). A 3D low-energy conformation of Darolutamide was formed using the Discovery Studio Suite 2017. Coot version 0.8.8 was employed for receptor-ligand modeling, and PyMOL (The PyMOL Molecular Graphics System, Version 2.0 Schrödinger, LLC, Cambridge, MA) was used to prepare figures [60]. The collaborative effort of medicinal chemists, pharmacologists, and drug metabolism experts led to the development of Darolutamide, which exhibited potent and selective AR antagonism and had an improved pharmacokinetic profile compared to the initial compound.

9.6.2 Scaffold Hopping

Ledipasvir (Harvoni) is an antiviral medication used in combination with sofosbuvir for the treatment of hepatitis C virus (HCV) infection. HCV, an RNA virus belonging to the Flaviviridae family, penetrates the cell and its viral genome is translated into a polyprotein consisting of 3000 amino acids. This polyprotein is then cleaved by both host and viral proteases into several proteins. These proteins are categorized into two main types: structural proteins (such as the core, E1, and E2 proteins), which are a part of the virions, and nonstructural proteins that play a role in the replication cycle of the virus (including proteins NS2, NS3, NS4A, NS4B, NS5A, and NS5B) [61].

One of the key targets for direct-acting antivirals (DAAs) development in HCV was the NS5A protein. The NS5A protein has multiple functions in the HCV life cycle, including RNA replication, modulation of host cell signaling pathways, and assembly of new viral particles [61]. A lead compound, known as ACH-2928, was identified as a potent inhibitor of the NS5A protein. ACH-2928 exhibited favorable antiviral activity *in vitro*, but its pharmacokinetic properties were suboptimal for oral administration. To overcome this limitation, medicinal chemists at AbbVie (formerly Abbott Laboratories) focused on optimizing the structure of ACH-2928 to improve its pharmacokinetic profile, including bioavailability, metabolic stability, and safety [62]. The lead optimization process resulted in the discovery of ledipasvir, a potent and selective NS5A inhibitor with improved pharmacokinetic properties. Ledipasvir demonstrated robust antiviral activity in preclinical studies and was subsequently evaluated in clinical trials.

9.7 Conclusion

The integration of *de novo* drug design techniques into the molecular designer's toolbox represents a significant advancement in early-phase hit and lead discovery. The inherent multiobjective nature of optimizing drug molecules from atomic or fragment sets necessitates sophisticated computational methodologies. Many *de novo* drug design tools employ evolutionary techniques to simultaneously optimize multiple objectives, facilitating the identification of promising drug candidates. Furthermore, the incorporation of retrosynthetic rules and fragment-based construction ensures that the molecules generated by *de novo* drug design tools are chemically synthesizable, enhancing their translational potential. These novel drug molecules, absent from existing databases, offer unique opportunities for drug development and innovation.

In parallel, scaffold hopping has emerged as a widely accepted method in the arsenal of medicinal chemists and molecular modelers. The last decade has witnessed numerous successful scaffold hops facilitated by ligand-based virtual screening methods. The introduction of diverse ligand-based virtual screening techniques with scaffold hopping capabilities has further expanded the toolbox available to researchers, enabling the exploration of novel chemical space and the identification of promising lead compounds.

In essence, the convergence of *de novo* drug design and scaffold hopping represents a paradigm shift in drug discovery, offering unprecedented opportunities for the rapid and cost-effective identification of innovative therapeutics. Moving forward, continued advancements in computational methodologies and virtual screening techniques will further propel drug development efforts, ultimately improving patient outcomes and advancing healthcare worldwide.

References

1 Hartenfeller, M. and Schneider, G. (2011). *De novo* drug design. *Chemoinformatics and Computational Chemical Biology* 299–323.

2 Hartenfeller, M. and Schneider, G. (2011). Enabling future drug discovery by *de novo* design. *Wiley Interdisciplinary Reviews: Computational Molecular Science* 1 (5): 742–759.

3 Rodrigues, T., Reker, D., Welin, M. et al. (2015). *De novo* fragment design for drug discovery and chemical biology. *Angewandte Chemie International Edition* 54 (50): 15079–15083.

4 Schneider, G., Schneider, P., and Renner, S. (2006). Scaffold-hopping: how far can you jump? *QSAR and Combinatorial Science* 25 (12): 1162–1171.

5 Brown, N. and Jacoby, E. (2006). On scaffolds and hopping in medicinal chemistry. *Mini Reviews in Medicinal Chemistry* 6 (11): 1217–1229.

6 Sun, H., Tawa, G., and Wallqvist, A. (2012). Classification of scaffold-hopping approaches. *Drug Discovery Today* 17 (7–8): 310–324.

7 Nakano, H., Miyao, T., Swarit, J., and Funatsu, K. (2021). Sparse topological pharmacophore graphs for interpretable scaffold hopping. *Journal of Chemical Information and Modeling* 61 (7): 3348–3360.

8 Vainio, M.J., Kogej, T., Raubacher, F., and Sadowski, J. (2013). *Scaffold Hopping by Fragment Replacement*. ACS Publications.

9 Hu, Y., Stumpfe, D., and Bajorath, J. (2017). Recent advances in scaffold hopping: miniperspective. *Journal of Medicinal Chemistry* 60 (4): 1238–1246.

10 Schuffenhauer, A. (2012). Computational methods for scaffold hopping. *Wiley Interdisciplinary Reviews: Computational Molecular Science* 2 (6): 842–867.

11 Boehm, M., Wu, T.Y., Claussen, H., and Lemmen, C. (2008). Similarity searching and scaffold hopping in synthetically accessible combinatorial chemistry spaces. *Journal of Medicinal Chemistry* 51 (8): 2468–2480.

12 Loving, K., Alberts, I., and Sherman, W. (2010). Computational approaches for fragment-based and *de novo* design. *Current Topics in Medicinal Chemistry* 10 (1): 14–32.

13 Schneider, G. and Fechner, U. (2005). Computer-based *de novo* design of drug-like molecules. *Nature Reviews. Drug Discovery* 4 (8): 649–663.

14 Oduguwa, A., Tiwari, A., Roy, R., and Bessant, C. (2006). An overview of soft computing techniques used in the drug discovery process. In: *Applied Soft Computing Technologies: The Challenge of Complexity* (ed. A. Abraham, B. de Baets, M. Köppen, and B. Nickolay), 465–480. Springer.

15 Nicolaou, A. and C, Kannas C, Loizidou E. (2012). Multi-objective optimization methods in *de novo* drug design. *Mini Reviews in Medicinal Chemistry* 12 (10): 979–987.

16 Quirke, V.M. (2017). Tamoxifen from failed contraceptive pill to best-selling breast cancer medicine: a case-study in pharmaceutical innovation. *Frontiers in Pharmacology* 8: 620.

17 Schneider, G., Neidhart, W., Giller, T., and Schmid, G. (1999). "Scaffold-hopping" by topological pharmacophore search: a contribution to virtual screening. *Angewandte Chemie International Edition* 38 (19): 2894–2896.

18 Martin, Y.C., Kofron, J.L., and Traphagen, L.M. (2002). Do structurally similar molecules have similar biological activity? *Journal of Medicinal Chemistry* 45 (19): 4350–4358.

19 Lindert, S., Durrant, J.D., and McCammon, J.A. (2012). LigMerge: a fast algorithm to generate models of novel potential ligands from sets of known binders. *Chemical Biology & Drug Design* 80 (3): 358–365.

20 Nishibata, Y. and Itai, A. (1991). Automatic creation of drug candidate structures based on receptor structure. Starting point for artificial lead generation. *Tetrahedron* 47 (43): 8985–8990.

21 Böhm, H.J. (1992). LUDI: rule-based automatic design of new substituents for enzyme inhibitor leads. *Journal of Computer-Aided Molecular Design* 6: 593–606.

22 Böhm, H.J. (1992). The computer program LUDI: a new method for the *de novo* design of enzyme inhibitors. *Journal of Computer-Aided Molecular Design* 6: 61–78.

23 Gillet, V.J., Newell, W., Mata, P. et al. (1994). SPROUT: recent developments in the *de novo* design of molecules. *Journal of Chemical Information and Computer Sciences* 34 (1): 207–217.

24 Pierce, A.C., Rao, G., and Bemis, G.W. (2004). BREED: generating novel inhibitors through hybridization of known ligands. Application to CDK2, p38, and HIV protease. *Journal of Medicinal Chemistry* 47 (11): 2768–2775.

25 Clark, D.E., Frenkel, D., Levy, S.A. et al. (1995). PRO_LIGAND: an approach to *de novo* molecular design. 1. Application to the design of organic molecules. *Journal of Computer-Aided Molecular Design* 9: 13–32.

26 Tschinke, V. and Cohen, N.C. (1993). The NEWLEAD program: a new method for the design of candidate structures from pharmacophoric hypotheses. *Journal of Medicinal Chemistry* 36 (24): 3863–3870.

27 Huang, Q., Li, L.L., and Yang, S.Y. (2010). PhDD: a new pharmacophore-based *de novo* design method of drug-like molecules combined with assessment of synthetic accessibility. *Journal of Molecular Graphics & Modelling* 28 (8): 775–787.

28 Bandyopadhyay, S., Saha, S., Maulik, U., and Deb, K. (2008). A simulated annealing-based multiobjective optimization algorithm: AMOSA. *IEEE Transactions on Evolutionary Computation* 12 (3): 269–283.

29 Pearlman, D.A. and Murcko, M.A. (1996). CONCERTS: dynamic connection of fragments as an approach to *de novo* ligand design. *Journal of Medicinal Chemistry* 39 (8): 1651–1663.

30 Dean, P.M., Firth-Clark, S., Harris, W. et al. (2006). SkelGen: a general tool for structure-based *de novo* ligand design. *Expert Opinion on Drug Discovery* 1 (2): 179–189.

31 Sengupta, S. and Bandyopadhyay, S. (2012). *De novo* design of potential Reca inhibitors using multiobjective optimization. *IEEE/ACM Transactions on Computational Biology and Bioinformatics* 9 (4): 1139–1154.

32 Wang, R., Gao, Y., and Lai, L. (2000). LigBuilder: a multi-purpose program for structure-based drug design. *Molecular Modeling Annual* 6: 498–516.

33 Douguet, D., Thoreau, E., and Grassy, G. (2000). A genetic algorithm for the automated generation of small organic molecules: drug design using an evolutionary algorithm. *Journal of Computer-Aided Molecular Design* 14: 449–466.

34 Pegg, S.C.H., Haresco, J.J., and Kuntz, I.D. (2001). A genetic algorithm for structure-based *de novo* design. *Journal of Computer-Aided Molecular Design* 15: 911–933.

35 Budin, N., Majeux, N., Tenette-Souaille, C., and Caflisch, A. (2001). Structure-based ligand design by a build-up approach and genetic algorithm search in conformational space. *Journal of Computational Chemistry* 22 (16): 1956–1970.

36 Vinkers, H.M., de Jonge, M.R., Daeyaert, F.F.D. et al. (2003). Synopsis: synthesize and optimize system *in silico*. *Journal of Medicinal Chemistry* 46 (13): 2765–2773.

37 Douguet, D., Munier-Lehmann, H., Labesse, G., and Pochet, S. (2005). LEA3D: a computer-aided ligand design for structure-based drug design. *Journal of Medicinal Chemistry* 48 (7): 2457–2468.

38 Dey, F. and Caflisch, A. (2008). Fragment-based *de novo* ligand design by multiobjective evolutionary optimization. *Journal of Chemical Information and Modeling* 48 (3): 679–690.

39 Schneider, G., Lee, M.L., Stahl, M., and Schneider, P. (2000). *De novo* design of molecular architectures by evolutionary assembly of drug-derived building blocks. *Journal of Computer-Aided Molecular Design* 14: 487–494.

40 Fechner, U. and Schneider, G. (2006). Flux (1): a virtual synthesis scheme for fragment-based *de novo* design. *Journal of Chemical Information and Modeling* 46 (2): 699–707.

41 Schüller, A., Suhartono, M., Fechner, U. et al. (2008). The concept of template-based *de novo* design from drug-derived molecular fragments and its application to TAR RNA. *Journal of Computer-Aided Molecular Design* 22: 59–68.

42 Nicolaou, C.A., Apostolakis, J., and Pattichis, C.S. (2009). *De novo* drug design using multiobjective evolutionary graphs. *Journal of Chemical Information and Modeling* 49 (2): 295–307.

43 Wong, S.S.Y., Luo, W., and Chan, K.C.C. (2010). EvoMD: an algorithm for evolutionary molecular design. *IEEE/ACM Transactions on Computational Biology and Bioinformatics* 8 (4): 987–1003.

44 Bleicher, K.H., Böhm, H.J., Müller, K., and Alanine, A.I. (2003). Hit and lead generation: beyond high-throughput screening. *Nature Reviews. Drug Discovery* 2 (5): 369–378.

45 Schneider, G. and Baringhaus, K. (2013). *De novo* design: from models to molecules. *De Novo Molecular Design* 1–55.

46 Jhoti, H. (2008). Fragment-based drug discovery using rational design. In: *Sparking Signals: Kinases as Molecular Signal Transducers and Pharmacological Drug Targets in Inflammation*, 169–185. Springer.

47 Mauser, H. and Guba, W. (2008). Recent developments in *de novo* design and scaffold hopping. *Current Opinion in Drug Discovery & Development* 11 (3): 365–374.

48 Mouchlis, V.D., Afantitis, A., Serra, A. et al. (2021). Advances in *de novo* drug design: from conventional to machine learning methods. *International Journal of Molecular Sciences* 22 (4): 1676.

49 Fischer, T., Gazzola, S., and Riedl, R. (2019). Approaching target selectivity by *de novo* drug design. *Expert Opinion on Drug Discovery* 14 (8): 791–803.

50 Wang, Y., Zhao, H., Brewer, J.T. et al. (2018). *De novo* design, synthesis, and biological evaluation of 3, 4-disubstituted pyrrolidine sulfonamides as potent and selective glycine transporter 1 competitive inhibitors. *Journal of Medicinal Chemistry* 61 (17): 7486–7502.

51 Gaulton, A., Hersey, A., Nowotka, M. et al. (2017). The ChEMBL database in 2017. *Nucleic Acids Research* 45 (D1): D945–D954.

52 Wise, A., Gearing, K., and Rees, S. (2002). Target validation of G-protein coupled receptors. *Drug Discovery Today* 7 (4): 235–246.

53 Waszkowycz, B., Clark, D.E., Frenkel, D. et al. (1994). PRO_LIGAND: an approach to *de novo* molecular design. 2. Design of novel molecules from molecular field analysis (MFA) models and pharmacophores. *Journal of Medicinal Chemistry* 37 (23): 3994–4002.

54 Hartenfeller, M., Zettl, H., Walter, M. et al. (2012). DOGS: reaction-driven *de novo* design of bioactive compounds. *PLoS Computational Biology* 8 (2): e1002380.

55 Kawai, K., Nagata, N., and Takahashi, Y. (2014). *De novo* design of drug-like molecules by a fragment-based molecular evolutionary approach. *Journal of Chemical Information and Modeling* 54 (1): 49–56.

56 Stumpfe, D. and Bajorath, J. (2011). Similarity searching. *Wiley Interdisciplinary Reviews: Computational Molecular Science* 1 (2): 260–282.

57 Krueger, B.A., Dietrich, A., Baringhaus, K.H., and Schneider, G. (2009). Scaffold-hopping potential of fragment-based *de novo* design: the chances and limits of variation. *Combinatorial Chemistry & High Throughput Screening* 12 (4): 383–396.

58 Gillet, V., Johnson, A.P., Mata, P. et al. (1993). SPROUT: a program for structure generation. *Journal of Computer-Aided Molecular Design* 7: 127–153.

59 Scott, L.J. (2020). Darolutamide: a review in non-metastatic castration-resistant prostate cancer. *Targeted Oncology* 15: 791–799.

60 Sugawara, T., Baumgart, S.J., Nevedomskaya, E. et al. (2019). Darolutamide is a potent androgen receptor antagonist with strong efficacy in prostate cancer models. *International Journal of Cancer* 145 (5): 1382–1394.

61 Gentile, I., Buonomo, A.R., Borgia, F. et al. (2014). Ledipasvir: a novel synthetic antiviral for the treatment of HCV infection. *Expert Opinion on Investigational Drugs* 23 (4): 561–571.

62 Link, J.O., Taylor, J.G., Xu, L. et al. (2014). Discovery of ledipasvir (GS-5885): a potent, once-daily oral NS5A inhibitor for the treatment of hepatitis C virus infection. *Journal of Medicinal Chemistry* 57 (5): 2033–2046.

10

Fragment-based Drug Design and Drug Discovery

André M. Oliveira[1] and Mithun Rudrapal[2]

[1] *Department of Environment Studies, Federal Centre of Technological Education of Minas Gerais, Contagem, Minas Gerais, Brazil*
[2] *Department of Pharmaceutical Sciences, School of Biotechnology and Pharmaceutical Sciences, Vignan's Foundation for Science, Technology & Research, Guntur, Andhra Pradesh, India*

10.1 Introduction

Computational methods have been used in the drug discovery process for decades. In the early days, computers were used to calculate the properties of molecules and to screen large libraries of compounds for potential drug candidates. As computers have become more powerful, computational methods have become more sophisticated and have been used to address a wider range of drug discovery problems [1]. One of the most important uses of computational methods in drug discovery is to identify potential drug targets. Drug targets are proteins or other molecules that are involved in the disease process. By understanding the structure and function of drug targets, scientists can design drugs that will interact with these targets and inhibit their activity.

Molecular modeling can be used to identify potential drug targets in a number of ways. One common approach is to use molecular docking simulations, which are used to predict the binding interactions between a drug molecule and a drug target [2]. If the drug molecule can bind to the drug target in a way that inhibits its activity, then it is considered to be a potential drug candidate.

Another approach to identifying potential drug targets is to use protein structure prediction. Protein structure prediction is the process of using computational methods to predict the three-dimensional structure of a protein from its amino acid sequence. Once the structure of a protein is known, it can be used to design drugs that will interact with the protein and inhibit its activity [3].

In addition to identifying potential drug targets, computational chemistry can also be used to design new drugs. Drug design is the process of developing new molecules that have the desired pharmacological properties. One common approach is to use computer-aided drug design (CADD). CADD is a process that uses computational methods to identify new molecules that have the desired properties, such as binding affinity, selectivity, and solubility [4].

Another approach to drug design is to use artificial intelligence (AI), which deals with the creation of intelligent agents, computer programs that can learn and adapt to their environment [5]. AI can be used to design new drugs by automating the drug design process and by identifying new drug targets. Some examples of successful drugs that were developed using computational methods include Gleevec, Herceptin, and Sovaldi. Gleevec is a drug that is used to treat chronic myeloid

Computational Methods for Rational Drug Design, First Edition. Edited by Mithun Rudrapal.

leukemia [6]. Herceptin is a drug that is used to treat breast cancer [7]. Sovaldi is a drug that is used to treat hepatitis C [8].

In the context of computational chemistry, the set of techniques and strategies known as fragment-based drug design (FBDD), widely used in the modern development of new drugs, is based on the construction of new ligands for molecular targets from the combination of structural fragments chosen according to some criteria. This choice can be directed based on the structure of the ligand, the molecular target, or both. FBDD has benefited from the development of databases of structures of molecules of different sizes and statistical techniques that make it possible to explore their structural universe.

Historically, FBDD was proposed in the 1980s by Jencks (1981) and Nakamura and Abeles (1985). The methods serve as an alternative or support to high-throughput screening (HTS), which is a more expensive and exhaustive process.

The reduction of time (and consequently of costs) is a great achievement in the new pharmaceutical industry, which deals with increasing demand due to population growth and the emergence of new diseases or the resurgence of old teeth that were already believed to be eliminated.

The use of these tools rivals more traditional resources such as Combinatorial Chemistry, an object of interest to many chemists and biologists [9]. A large part of the objections to FBDD is its possible circular nature, in which a new hit obtained generates new candidates that do not necessarily become consolidated as viable drugs.

In this scenario, the development of FBDD follows the evolution of AI, machine learning, genomics, transcriptomics, proteomics, metabolomics, microbiome, and pharmacogenomics tools [10].

The process involves libraries of fragments and links obtained by X-ray crystallography, and the construction process is mediated by calculations of affinity constants (Figure 10.1). Among the most commonly used structure databases Table 10.1 presents some examples.

10.2 The Process of Finding Fragments

Finding the fragments that will serve as the basis for proposing new ligands to a molecular target is not a simple task. It is necessary to have a wide variety of fragments and connections between them (functionalized or not) and a suitable metric to evaluate the best constructions, such as interaction energies or inhibition constants obtained by theoretical means. We can summarize the main steps that guide the FBDD process: obtaining the fragment libraries, protein hot-spots identification, and computational calculation for the fragments fitting into the sites.

The first step is to obtain the fragment library. There must be a compromise between structural diversity and library complexity, and for this purpose, certain criteria are used, known as the "rule of three" (RO3) [11] (Jacquemard and Kellenberger, 2019; Brown, 2016; Kirsch et al., 2019). The rule of three is similar in design and purpose to Lipinski's rule of five, which has application in the design of drugs with good oral bioavailability (Table 10.2).

Another important factor is synthetic feasibility, which can be measured by indicators such as the number and complexity of functional groups, and the presence of fused rings and chiral centers [11].

Finding the fragments is only part of the challenge of conducting a FBDD study. It is also necessary to find suitable molecular targets and to find appropriate binding sites on those targets. The search for molecular targets is usually done by similarity to natural ligands. An example of the application of this resource is OpenTargets [12], which brings together molecular targets that can be searched according to the associated disease or genetic profile.

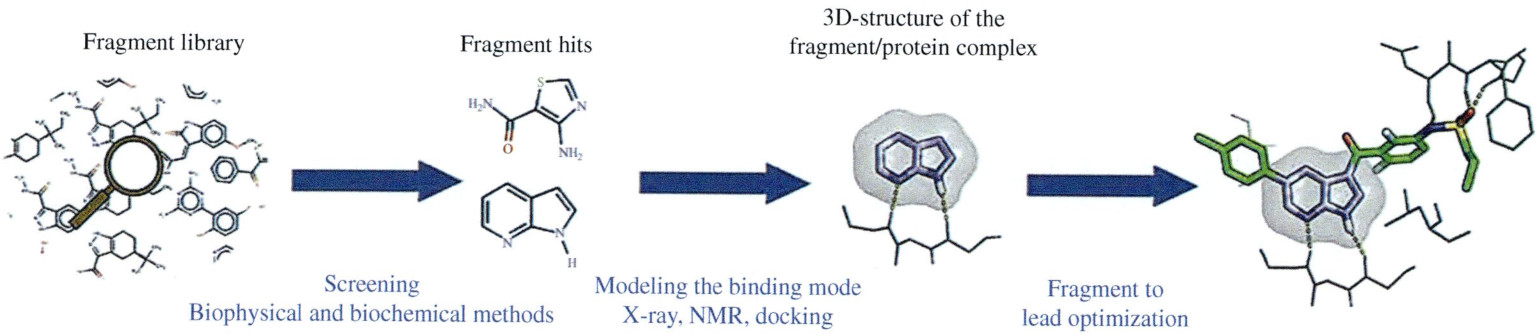

Figure 10.1 The FBDD process, exemplified on vemurafenib. The fragment 7-azaindole is a nonselective kinase inhibitor, and vemurafenib is a potent inhibitor of oncogenic B-RAF kinase activity. *Source:* Jacquemard and Kellenberger (2019)/with permission of Taylor & Francis.

Table 10.1 Some common structure databases.

Protein Databank, PDB	Protein database, containing more than 130,000 structures. http://www.rcsb.org
The Nucleic Acid Knowledgebase, NAKB	DNA and RNA database. www.nakb.org
Structural Classification of Proteins, SCOP 2	A database that lists proteins according to their structural and evolutionary similarities. https://scop.mrc-lmb.cam.ac.uk
ModBase	A database similar to SCOUP, which allows comparison between primary protein structures. http://modbase.compbio.ucsf.edu
ChemDB	Database of small molecules, containing different physicochemical data. http://cdb.ics.uci.edu
Cambridge Structural Database, CSD	One of the leading repositories of small-molecule crystallographic structures. http://www.ccdc.cam.ac.uk/products/csd
Uppsala Electron Density Server	Very complete database with crystallographic information on PDB structures. http://eds.bmc.uu.se/eds
Crystallography Open Database	Database of crystallographic structures of numerous organic, inorganic, organometallic, mineral, and biopolymer substances. http://www.crystallography.net

Table 10.2 Required and desired features of drug orally bioavailable.

Required features	Molecular weight	≤300 Da
	Number of hydrogen-bond donors	≤3
	Number of hydrogen-bond acceptors	≤3
Desired features	Number of rotatable bonds	≤3
	Polar surface area (PSA)	≤60

Another important step is the identification of possible interaction sites in the protein. Several computational methods have been developed for this task, using the topology of the macromolecule to identify pockets, such as CASTp (Tian et al. 2018) or by comparing the protein with others of the same ancestor by homology, such as SiteComp (Lin, Yoo and Sanchez, 2012). Further sites and resources are mentioned in our previous work [13].

CASTp uses the mathematical methodologies of Delaunay triangulation, alpha shape, and discrete flow [14–16], finding *pockets* (empty hollows in the protein structure that give access to solvent molecules, using a spherical probe of radius 1.4 Å), and *cavities* (empty spaces inside the structure with no access to the solvent). The CASTp method employs a triangulation algorithm using the centers of each atom situated on the walls of a concavity and their corresponding Van der Waals radii.

The following step is to make the connections among fragments and sites, which can be done through several strategies: fragment evolution, fragment linking, fragment self-assembly, and fragment optimization [17].

In *fragment evolution*, an initial fragment is optimized by adding functionality to bind to adjacent regions of the active site, as illustrated by Figure 10.2a: fragment 1 binds to the receptor at one site, and a new fragment grows into a second pocket. *Fragment linking* is done when two (or more) fragments, which bind to proximal parts of the active site, are joined together to give a larger,

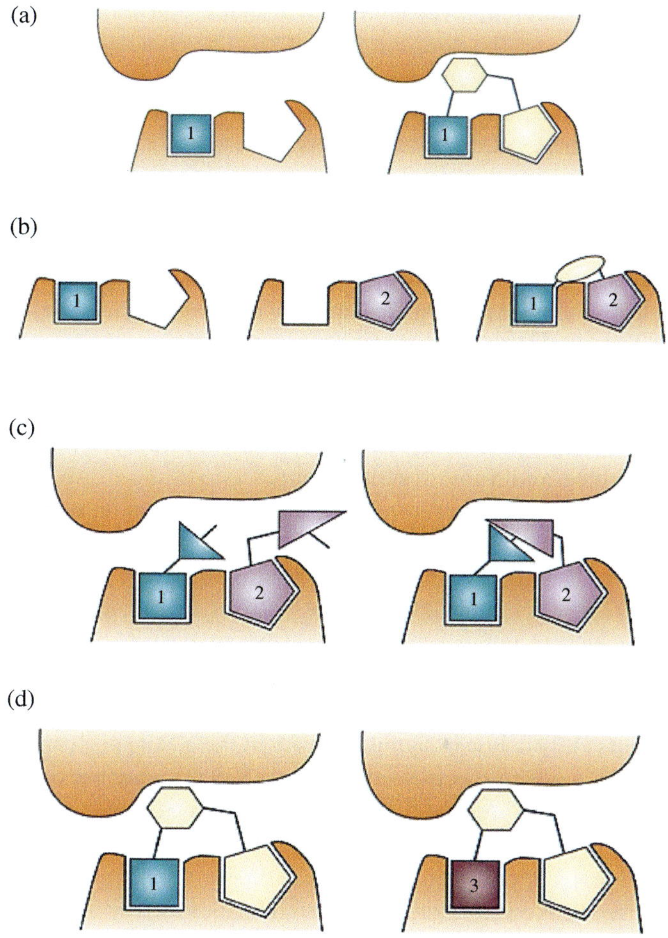

Figure 10.2 (a) Fragment evolution. (b) Fragment linking. (c) Fragment self-assembly. (d) Fragment optimization. *Source:* Ress et al. [17]/ with permission of Springer Nature.

higher-affinity-binding molecule. In Figure 10.2b, fragment 1 binds to the receptor at one site, fragment 2 to the receptor at an adjacent site, and they as linked each other. *Fragment self-assembly* occurs when complementary functional groups are built up around the protein target in order to reach the most complimentary one (Figure 10.2c). With *fragment optimization*, fragment approaches are used to optimize drug-like properties of a lead other than just binding affinity (Figure 10.2d).

It is important to warrant a wide spread of fragments in the study, a condition that can be related to molecular diversity, a measure of the variability in a set of molecules, based on statistical and chemometric techniques. One common way to measure molecular diversity is to use a similarity measure. A similarity measure is a function that takes two molecules as input and returns a value that indicates how similar they are. There are many different similarity measures available, but some of the most common include the Tanimoto coefficient and the Dice and Jaccard coefficient [18]. Once a similarity measure has been chosen, it can be used to calculate the diversity of a set of molecules. The diversity of a set of molecules is typically defined as the average similarity between all pairs of molecules in the set.

Another common way to measure molecular diversity is to use a diversity index, a function that takes a set of molecules as input and returns a value that indicates the diversity of the set. There are many different diversity indices available, but some of the most common include the Shannon index [19] and the Simpson index [20]. The choice of similarity measure or diversity index will depend on the specific application. For example, if the goal is to identify the most diverse molecules in a set, then a similarity measure or diversity index that is sensitive to small differences between molecules would be a good choice.

In addition to statistical and chemometric techniques, there are also a number of other methods that can be used to measure molecular diversity. These methods include machine learning, AI, and bioinformatics. Machine learning is a branch of AI that allows computers to learn from data. Machine learning algorithms can be used to develop models that predict the properties of molecules, such as their biological activity or their toxicity. These models can then be used to identify molecules that are likely to be diverse. Bioinformatics algorithms can also be used to identify molecules that are likely to be diverse by comparing them to other molecules in a database.

The prediction of fragments must be supported by experimental techniques that are capable of monitoring ligand–receptor interactions, such as nuclear magnetic resonance, surface plasmon resonance, and X-ray crystallography.

The most common is the use of nuclear magnetic resonance, through which changes in the chemical environments of the protein's active site are monitored with the insertion of the ligand. The most used NMR techniques for this purpose are relaxation edited 1D NMR, water–ligand observed via gradient spectroscopy (WaterLOGSY), and saturation transfer difference (STD) spectroscopy [21]. In relaxation-edited 1D NMR, ligand binding to a macromolecule is accompanied by shortened relaxation properties of the complex relative to the free ligand, and longitudinal relaxation or transverse relaxation filter is applied to the pulse sequence. Fragments that bind to the target exhibit a loss in signal intensity in the NMR spectrum, relative to fragments that do not [22].

NMR gradient spectroscopy (WaterLOGSY) makes use of the interactions between the target and the ligand that are mediated by water molecules, which cause a change in the profile of the relaxation times related to the interactions between ligand and water in the bound state or in the free state [23]. The cross-relaxation rate of dipole–dipole interaction between water and ligand is positive for free ligand and negative for bound ligands nearby water molecules associated with the protein.

The saturation-transfer difference NMR (STD-NMR) technique is based on the nuclear Overhauser effect (NOE) and is used for measuring through-space distances. The protein target is irradiated by a radiofrequency field specific for the protein nuclei and removes their magnetic polarization. The fast exchange between the free and protein-bound form causes a saturation transfer through the protein to the bound ligand and that saturation is carried on to the free ligand, where it is detected (Angulo, Enríquez-Navas and Nieto, 2010). A magnetization saturation of the target takes place, which by its turn is transferred to the bound ligand, and not to the free one. A "difference spectrum" is then obtained by subtracting the spectrum obtained with the magnetization-saturated target from the spectrum obtained with the nonirradiated target.

X-ray crystallography is another method to validate fragment generation. Obtaining crystallographic structures of macromolecules represented an important progress in the development of new drugs. X-ray structure databases have become popular, providing information at an increasing level of accuracy, such as the Brookhaven Protein Database, PDB [24]. Crystals can be obtained by using the cocrystallization method (the ligand is added to the mixture before crystal formation starts) or the soaking technique (the ligand is added directly to a mixture with pre-existing crystals).

Congreve et al. [25] explain how dynamic combinatorial chemistry (DCC) allows specific members of a combinatorial library to be selected and amplified with the use of a template. The reaction connecting the building blocks is reversible and an interchange between the different members of the dynamic combinatorial library takes place [26]. The authors report an approach in which ligands are observed directly by X-ray crystallography from their electron-density maps from crystals exposed to a dynamic combinatorial library mixture.

Thermal shift assay is a technique that supervises the changes undergone by a protein when it interacts with another molecule by measuring its denaturation temperature. Protein stability depends on factors like pH, ionic strength of the medium, presence of cofactors, and mutations. Correlated experimental parameters that are related to structural stability and that are sensitive to a temperature gradient, such as differential scanning fluorimetry, can be used to measure the ligand–macromolecule interaction [27]. The fluorescence of a protein solution depends on a temperature gradient, and the addition of fluorescence dye exhibits a low fluorescence signal in a polar environment and a high signal in a nonpolar environment [28]. The interaction of the protein with the ligand causes exposure of polar regions previously solvated by water molecules that are dispersed with the entry of the ligand, at the same time that it unbalances the distribution of nonpolar zones.

Isothermal titration calorimetry is a technique that allows the measuring of thermodynamic parameters such as enthalpy changes and Gibbs free energy, which by its turn allows for the calculation of the binding affinity [29], as well as stoichiometry in one single experiment [30]. The main advantage of this method is the discrimination it allows between enthalpic and entropic contributions to the binding.

Finally, mass spectrometry (MS) is a technique widely used in the structural determination of organic compounds. MS provides advantages over their counterpart techniques for allowing weak binding detection due to its high sensitivity. Besides, few sample amount is required, no modifications or labeling of the protein target is needed, and a direct visualization of all species in solution alongside the binding process is feasible [31, 32]. The use of milder ionization techniques, such as nondenaturing electro-spray-ionization (ESI), is always more interesting because it preserves the structural integrity of the protein: ligand complex. An example of this application is described by Liu and Quinn [33]: 643 natural products (NPs) fragment-sized library with low molecular weight were screened against 62 potential protein targets for malaria, which led to 96 low molecular NPs capable of binding and 79 fragments that could inhibit the growth of malaria parasites *in vitro*. Another worth-citing case is the discovery of a benzimidazole moiety with high affinity to a 29-mer RNA model, identified from a 180,000-fragment library using MS-based screening methods [34].

10.3 FBDD Strategies

FBDD studies can follow several strategies and approaches, which will depend on the available information and the level of refinement of the results that are desired. We can mention the following strategies: Build-up Core FBDD, HTS Complimentary, and Chemical Biology Exploration of Biological Targets. We will briefly discuss the scope of each one.

The strategy known as Build-up Core FBDD, which has applications in drug design, consists of the construction of new ligand candidates (hits) to determine targets from pure fragments. Several tools that use this path can be mentioned, the most well-known being CAVEAT [35], which employs a vector algorithm. This strategy uses molecular docking to build new candidates around the interaction site. To avoid a prohibitive number of solutions for the calculation, the pharmacophore structure is used as a guide. The choice of the docking scoring function is also important, which might include: binding energy, free energy, or interaction energies. Modern force fields group their scoring functions in force field, empirical, and knowledge-based (Guedes et al. 2014).

Another common approach is to combine FBDD with HTS. Just as large libraries of whole compounds are screened in HTS against a target, using scoring parameters such as ligand efficiency (LE, calculated from the interaction free energy), libraries of fragments can also be screened [36].

Searching for viable interaction sites in libraries of targets is also a widely used strategy, such as in the multiple solvent crystal structures method, which uses solvent molecules as bridges between the site and the ligands (Allen et al. 1996), or when guiding properties such as druggability are used to select the fragments that best fit the sites.

10.4 Case Studies

NPs are a rich source of biologically active compounds. They have been used for centuries in traditional medicine to treat a variety of diseases. In recent years, there has been a growing interest in the development of NP-inspired compounds as new drugs [12]. NPs are often complex molecules with a variety of functional groups. This complexity can make it difficult to synthesize them in the laboratory. However, there are a number of methods that can be used to synthesize NPs or their analogs. These methods include chemical synthesis, enzymatic synthesis, and combinatorial chemistry.

Once an NP-inspired compound has been synthesized, it must be tested for its biological activity. This can be done in a variety of ways, including *in vitro* assays, animal studies, and clinical trials. *In vitro* assays can be used to test the compound's activity against a variety of targets, such as enzymes, receptors, and DNA. Animal studies can be used to assess the compound's toxicity and its potential for therapeutic efficacy. Clinical trials are conducted in humans to determine the compound's safety and efficacy in treating a specific disease.

A number of NP-inspired compounds have been shown to have biological activity. Some of these compounds have been approved for use as drugs, while others are in clinical development [37]. Examples of NP-inspired drugs include paclitaxel, a chemotherapy drug used to treat cancer; artemisinin, an antimalarial drug; and rapamycin, an immunosuppressant drug used to prevent organ rejection.

The development of NP-inspired compounds is a promising area of research. These compounds have the potential to provide new treatments for a variety of diseases [38]. However, the development of new drugs is a long and expensive process. It can take many years and millions of dollars to bring a new drug to market. Despite the challenges, the development of NP-inspired compounds is an important area of research. These compounds have the potential to provide new treatments for a variety of diseases and improve the lives of millions of people.

An interesting application of FBDD is in the design of compounds inspired by NPs [39]. Due to their diversity and complexity, their biological action is accompanied by side effects associated with the ability to bind to different biological receptors. The growth in the number of NP databases signals the importance of this information source for the development of new treatments for diseases (Table 10.3).

An application of the use of FBDD/HTS in drug design is the study described by Vu et al. [32], which collected 650 compounds from various fragment libraries (Figure 10.3), following as a criterion the permitted ranges of physicochemical properties (compare Table 10.2):

- Molecular weight $\leq$250 Da
- Octanol–water partition coefficient <4
- Hydrogen bond donors $\leq$4
- Hydrogen bond acceptors $\leq$5
- Rotatable bonds $\leq$6
- Percent polar surface area <45.

Table 10.3 Some NP databases.

ANPDB	African Natural Products Database https://african-compounds.org/anpdb
AfroMalariaDB	African Antimalarial Natural Products Library https://african-compounds.org/about/afromalariadb
BIOFACQUIM	A Mexican Compound Database of Natural Products https://www.difacquim.com/d-tools
COCONUT	Natural Products Online https://coconut.naturalproducts.net
NANPDB	North American Natural Product Database http://african-compounds.org/nanpdb
NPATLAS	The Natural Products Atlas https://www.npatlas.org
NPASS	Natural Product Activity and Species Source Database https://bidd.group/NPASS
NuBBE	Nuclei of Bioassays, Biosynthesis and Ecophysiology of Natural Products https://nubbe.iq.unesp.br/portal/nubbe-search.html
SANCDB	South African Natural Compounds Database https://sancdb.rubi.ru.ac.za
StreptomeDB	Database of compounds isolated from *Streptomyces* spp. http://132.230.102.198:8000/streptomedb/
TCM-ID	Traditional Chinese Medicine Information Database https://bidd.group/TCMID
TIPdb	Database of Taiwan indigenous plants https://cwtung.kmu.edu.tw/tipdb
YaTCM	Yet another Traditional Chinese Medicine database https://github.com/jianping-grp/yatcm

These fragments were tested against 62 molecular targets associated with malaria. About 100 NP-like fragments were identified, able to interact with 32 of the molecular targets. About 80% of these compounds proved effective against *Plasmodium falciparum*.

One can also mention the search for drugs to fight COVID-19, alongside immunological treatments. FDA-approved drugs obtained from theoretical calculations, such as remdesivir, saquinavir, and darunavir, and some flavone and coumarin derivatives as potential inhibitors of human SARS-CoV2 main protease [40]. Shaffer [41] mentions some drugs being tested against SARS-Cov2, like chloroquine, lopinavir, nafamostat, hydroxychloroquine, ritonavir, camostat, corticosteroids, and sarilumab. Inhibitors for the main protease in SARS-CoV2 using FBDD have been described [42], from a fragment database derived from Auto Core Fragment *in silico* Screening (ACFIS) 2.0 web server [43] followed by docking against the enzyme using PyRx [44] and AutoDock [45], and ADMET [46] and molecular dynamics studies [47]. ADMET properties followed total polar surface area [48], water solubility, lipophilicity ($\log P$), skin permeation [49], and synthetic criteria.

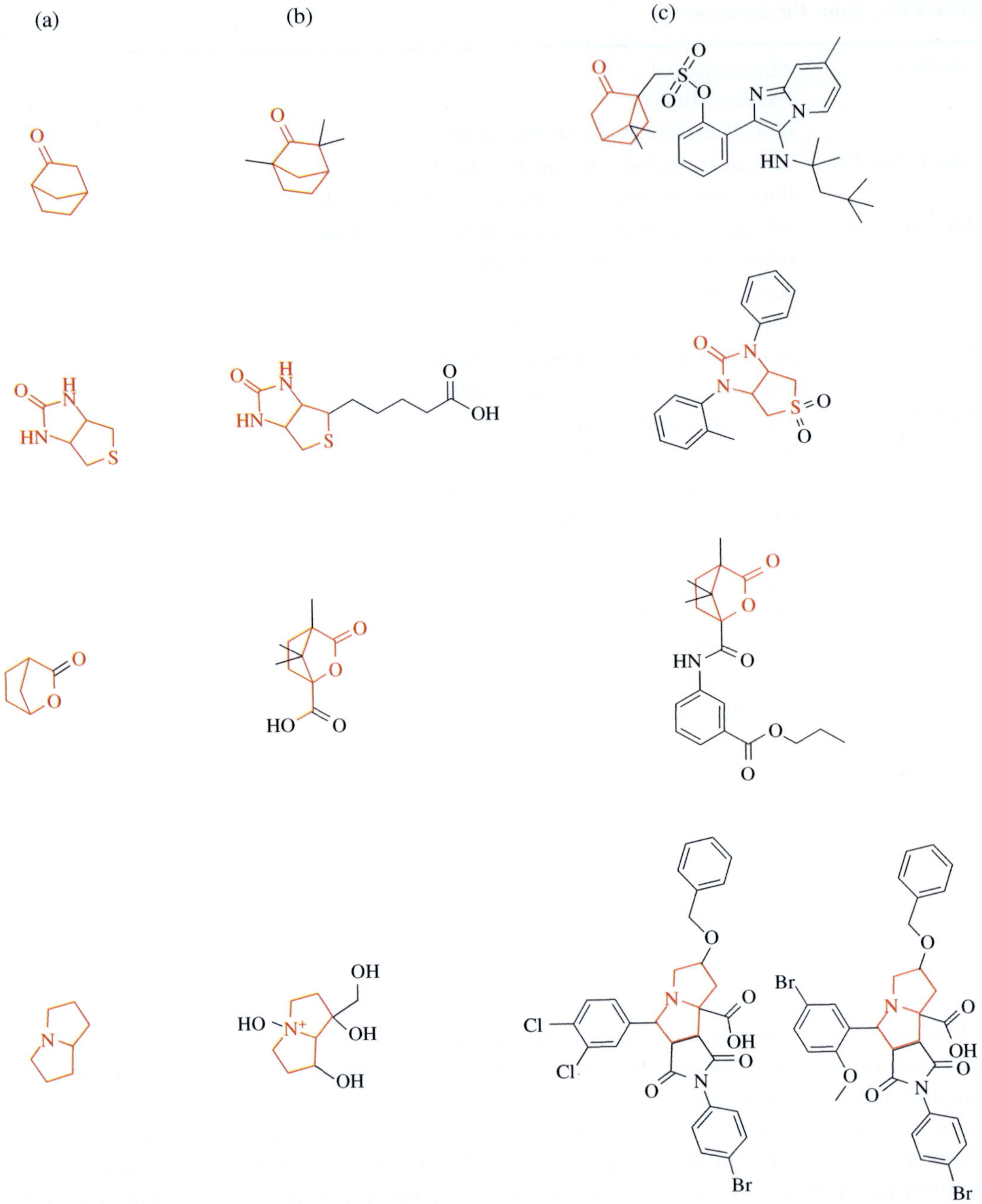

Figure 10.3 Examples of fragment hit scaffolds present among antimalarial HTS hits. (a) Fragment scaffold. (b) Original fragment. (c) Antimalarial HTS hit molecule. *Source:* Vu et al. [32]/American Chemical Society/CC BY 4.0.

10.5 Conclusion and Future Perspectives

In our previous work on FBDD [13], we listed several computational strategies employed in HTS and that serve well the purposes of FBDD studies, such as dynamic combinatorial chemistry (DCC) and *de novo* drug design approach. Dynamic combinatorial chemistry (DCC) serves as a compelling method for crafting targeted ligand libraries for macromolecular targets. Its efficacy lies in the reversible reactions of building blocks, leading to a stable thermodynamic equilibrium. An extension, termed target-directed DCC (tdDCC), identifies potent ligands for relevant pharmacological targets.

A natural progression from FBDD involves its integration with three-dimensional quantitative structure–activity relationships (3D QSAR). Unlike classic QSAR, which statistically correlates physicochemical properties with biological activity, 3D QSAR evaluates electronic and steric energy values, generating three-dimensional maps indicating favorable interaction sites. In the 1990s and 2000s, the *de novo* drug design approach using the LeapFrog (LF) program gained popularity, available in the Sybyl package. LF facilitates molecular evolution or electronic screening with three alternative modes: OPTIMIZE enhances existing leads, DREAM proposes new molecules for binding, and GUIDE supports interactive design. Binding energy calculation involves steric and electrostatic enthalpies, cavity desolvation energy, and ligand desolvation energy. Addressing the rationality of druggability and molecular shape, a concept explored by Bemis and Murcko in the 1990s, presents a challenge open to investigation. Although their work focused on 2D fragment topological features, limited graph portraits appear promising for potential drug development. This concept proves valuable in constructing fragment libraries with topological constraints.

The success of FBDD hinges on advancing computational capabilities and disease-specific fragment library availability. The COVID-19 pandemic has spurred new demands, especially in drug repurposing and toxicity studies. Despite challenges, this field offers a cost-effective avenue for synthesizing novel chemical entities, contrasting with traditional pharmaceutical combinatorial screening.

The development of novel drugs is a complex and intricate process that plays a crucial role in modern healthcare. As researchers strive to address various diseases, they encounter numerous challenges in drug planning and design. *De novo* drug design (DNDD) and FBDD are prominent methods in this endeavor. DNDD involves the construction of entirely new molecular entities. The complexity of designing molecules with desired pharmacological properties while ensuring safety and efficacy poses a significant challenge [50]. Computational tools play a crucial role in *de novo* design, but the accurate prediction of a molecule's biological activity and safety remains challenging. The reliance on computational models introduces uncertainties that need to be carefully addressed [51]. Synthesizing novel compounds is a costly and time-consuming process. Ensuring the chemical feasibility of designed molecules and their synthesis on a practical scale poses a substantial challenge [52]. Verifying the predicted biological activity of DNDD designed molecules experimentally is a critical step. Challenges arise in obtaining accurate and reliable data that can validate the effectiveness of the designed compounds [51]. The success of FBDD, by its turn, heavily relies on the quality and diversity of the fragment libraries used. Ensuring that the fragments cover a broad chemical space and possess drug-like properties is a persistent challenge [53]. Integrating fragments into a lead compound with high binding affinity is a delicate task. Achieving optimal interactions while maintaining drug-like properties can be challenging and requires careful optimization [54]. Identifying hits from fragment screening is just the initial step. Expanding these hits into viable lead compounds and linking fragments to form a cohesive, effective drug candidate presents a considerable challenge [53]. FBDD often deals with structurally complex molecules. Elucidating the three-dimensional structures of these compounds and their interactions with biological targets can be challenging using traditional methods [54].

To overcome the challenges associated with each method individually, a hybrid approach that integrates *de novo* and FBDD has gained traction [55]. By combining the strengths of designing from scratch and utilizing fragment-based information, researchers aim to enhance the efficiency and success rate of drug discovery processes. The contemporary landscape of drug planning is marked by challenges in both *de novo* and FBDD. Molecular complexity, computational limitations, chemical feasibility, and biological validation are hurdles in the *de novo* approach, while fragment library quality, binding affinity optimization, hit expansion, and structural complexity present challenges in FBDD. Researchers are continually working toward innovative solutions, and the integration of these two methods offers a promising avenue for overcoming some of these hurdles. As technology advances and our understanding of drug design deepens, the pharmaceutical industry moves closer to addressing these challenges and delivering novel therapeutics to improve global health.

References

1 Surabhi, S. and Singh, B. (2018). Computer aided drug design: an overview. *Journal of Drug Delivery and Therapeutics* 8 (5): https://doi.org/10.22270/jddt.v8i5.1894.

2 Torres, P.H.M., Sodero, A.C.R., Jofily, P., and Silva-Jr, F.P. (2019). Key topics in molecular docking for drug design. *International Journal of Molecular Sciences* 20 (18): https://doi.org/10.3390/ijms20184574.

3 Śledź, P. and Caflisch, A. (2018). Protein structure-based drug design: from docking to molecular dynamics. *Current Opinion in Structural Biology* 48: https://doi.org/10.1016/j.sbi.2017.10.010.

4 Song, C.M., Lim, S.J., and Tong, J.C. (2009). Recent advances in computer-aided drug design. *Briefings in Bioinformatics* 10 (5): https://doi.org/10.1093/bib/bbp023.

5 Wang, L., Ding, J., Pan, L. et al. (2019). Artificial intelligence facilitates drug design in the big data era. In: *Chemometrics and Intelligent Laboratory Systems*, vol. 194. https://doi.org/10.1016/j.chemolab.2019.103850.

6 Capdeville, R., Buchdunger, E., Zimmermann, J., and Matter, A. (2002). Glivec (ST1571, imatinib), a rationally developed, targeted anticancer drug. *Nature Reviews Drug Discovery* 1 (7): https://doi.org/10.1038/nrd839.

7 Shepard, H.M., Jin, P., Slamon, D.J. et al. (2008). Herceptin. *Handbook of Experimental Pharmacology* 181: https://doi.org/10.1007/978-3-540-73259-4_9.

8 Polamreddy, P., Vishwakarma, V., Arumugam, P. et al. (2020). Discovery of hit molecules targeting allosteric site of hepatitis C virus NS5B polymerase. *Journal of Biomolecular Structure and Dynamics* 38 (5): https://doi.org/10.1080/07391102.2019.1608864.

9 Drews, J. (2000). Drug discovery: a historical perspective. *Science* 287: 1960–1964. https://doi.org/10.1126/science.287.5460.1960.

10 Katsila, T., Spyroulias, G.A., Patrinos, G.P., and Matsoukas, M.T. (2016). Computational approaches in target identification and drug discovery. *Computational and Structural Biotechnology Journal* 14: 177–184. https://doi.org/10.1016/J.CSBJ.2016.04.004.

11 Schuffenhauer, A., Ruedisser, S., Marzinzik, A.L. et al. (2005). Library design for fragment based screening. *Current Topics in Medicinal Chemistry* 5 (8): 751–762. https://doi.org/10.2174/1568026054637700.

12 Vamathevan, J., Clark, D., Czodrowski, P. et al. (2019). Applications of machine learning in drug discovery and development. *Nature Reviews Drug Discovery* 18 (6): 463–477. https://doi.org/10.1038/s41573-019-0024-5.

13 Oliveira, A.M. and Rudrapal, M. (2023). Fragment-based drug design in lead discovery. In: *CADD and Informatics in Drug Discovery* (ed. J.K. Org and M. Rudrapal), 174–204. Springer Singapore.

14 Edelsbrunner, H. (1995). The union of balls and its dual shape. *Discrete & Computational Geometry* 13 (3): 415–440. https://doi.org/10.1007/BF02574053.

15 Edelsbrunner, H. and Shah, N.R. (1996). Incremental topological flipping works for regular triangulations. *Algorithmica* 15 (3): 223–241. https://doi.org/10.1007/BF01975867.

16 Facello, M.A. (1995). Implementation of a randomized algorithm for Delaunay and regular triangulations in three dimensions. *Computer Aided Geometric Design* 12 (4): 349–370. https://doi.org/10.1016/0167-8396(94)00018-N.

17 Ress, D.C., Congreve, M., Murray, C.W., and Carr, R. (2004). Fragment-based lead discovery. *Nature Reviews Drug Discovery* 3 (8): 660–672. https://doi.org/10.1038/nrd1467.

18 Ataş, İ. (2023). Performance evaluation of Jaccard-dice coefficient on building segmentation from high resolution satellite images. *Balkan Journal of Electrical and Computer Engineering* 11 (1): https://doi.org/10.17694/bajece.1212563.

19 Konopiński, M.K. (2020). Shannon diversity index: a call to replace the original Shannon's formula with unbiased estimator in the population genetics studies. *PeerJ* 2020 (6): https://doi.org/10.7717/peerj.9391.

20 He, F. and Hu, X.S. (2005). Hubbell's fundamental biodiversity parameter and the Simpson diversity index. *Ecology Letters* 8 (4): https://doi.org/10.1111/j.1461-0248.2005.00729.x.

21 Joseph-McCarthy, D., Campbell, A.J., Kern, G., and Moustakas, D. (2014). Fragment-based lead discovery and design. *Journal of Chemical Information and Modeling* 54 (3): 693–704. https://doi.org/10.1021/CI400731W/ASSET/IMAGES/LARGE/CI-2013-00731W_0005.JPEG.

22 Hajduk, P.J., Olejniczak, E.T., and Fesik, S.W. (1997). One-dimensional relaxation- and diffusion-edited NMR methods for screening compounds that bind to macromolecules. *Journal of the American Chemical Society* 119 (50): https://doi.org/10.1021/ja9715962.

23 Dalvit, C., Fogliatto, G.P., Stewart, A. et al. (2001). WaterLOGSY as a method for primary NMR screening: practical aspects and range of applicability. *Journal of Biomolecular NMR* 21 (4): https://doi.org/10.1023/A:1013302231549.

24 Berman, H.M., Westbrook, J., Feng, Z. et al. (2000). The protein data bank. *Nucleic Acids Research* 28 (1): 235–242. https://doi.org/10.1093/NAR/28.1.235.

25 Congreve, M.S., Davis, D.J., Devine, L. et al. (2003). Detection of ligands from a dynamic combinatorial library by X-ray crystallography. *Angewandte Chemie – International Edition* 42 (37): https://doi.org/10.1002/anie.200351951.

26 Seneci, P. (2000). *Solid-Phase Synthesis and Combinatorial Technologies*. Wiley-Interscience.

27 Senisterra, G., Chau, I., and Vedadi, M. (2012). Thermal denaturation assays in chemical biology. *Assay and Drug Development Technologies* 10 (2): 128–136. https://doi.org/10.1089/ADT.2011.0390/ASSET/IMAGES/MEDIUM/FIGURE5.GIF.

28 Lo, M.C., Aulabaugh, A., Jin, G. et al. (2004). Evaluation of fluorescence-based thermal shift assays for hit identification in drug discovery. *Analytical Biochemistry* 332 (1): 153–159. https://doi.org/10.1016/J.AB.2004.04.031.

29 Leavitt, S. and Freire, E. (2001). Direct measurement of protein binding energetics by isothermal titration calorimetry. *Current Opinion in Structural Biology* 11 (5): 560–566. https://doi.org/10.1016/S0959-440X(00)00248-7.

30 Damian, L. (2013). Isothermal titration calorimetry for studying protein–ligand interactions. *Methods in Molecular Biology* 1008: 103–118. https://doi.org/10.1007/978-1-62703-398-5_4/COVER.

31 Chan, D.S.H., Whitehouse, A.J., Coyne, A.G., and Abell, C. (2017). Mass spectrometry for fragment screening. *Essays in Biochemistry* 61 (5): 465–473. https://doi.org/10.1042/EBC20170071.

32 Vu, H., Pedro, L., Mak, T. et al. (2018). Fragment-based screening of a natural product library against 62 potential malaria drug targets employing native mass spectrometry. *ACS Infectious Diseases* 4 (4): 431–444. https://doi.org/10.1021/ACSINFECDIS.7B00197/ASSET/IMAGES/LARGE/ID-2017-00197K_0009.JPEG.

33 Liu, M. and Quinn, R.J. (2019). Fragment-based screening with natural products for novel anti-parasitic disease drug discovery. *Expert Opinion on Drug Discovery* 14 (12): 1283–1295. https://doi.org/10.1080/17460441.2019.1653849/SUPPL_FILE/IEDC_A_1653849_SM2190.XLSX.

34 Seth, P.P., Miyaji, A., Jefferson, E.A. et al. (2005). SAR by MS: discovery of a new class of RNA-binding small molecules for the hepatitis C virus: internal ribosome entry site IIA subdomain. *Journal of Medicinal Chemistry* 48 (23): 7099–7102. https://doi.org/10.1021/JM050815O/SUPPL_FILE/JM050815OSI20050919_010845.PDF.

35 Lauri, G. and Bartlett, P.A. (1994). CAVEAT: a program to facilitate the design of organic molecules. *Journal of Computer-Aided Molecular Design* 8 (1): 51–66. https://doi.org/10.1007/BF00124349.

36 Grädler, U., Schwarz, D., Blaesse, M. et al. (2019). Discovery of novel cyclophilin D inhibitors starting from three dimensional fragments with millimolar potencies. *Bioorganic & Medicinal Chemistry Letters* 29 (23): 126717. https://doi.org/10.1016/J.BMCL.2019.126717.

37 Atanasov, A.G., Zotchev, S.B., Dirsch, V.M. et al. (2021). Natural products in drug discovery: advances and opportunities. *Nature Reviews Drug Discovery* 20 (3): 200–216. https://doi.org/10.1038/s41573-020-00114-z.

38 Rijo, P. (2023). Drug development inspired by natural products II. *Molecules* 28 (21): 7243. https://doi.org/10.3390/MOLECULES28217243.

39 Najjar, A., Olğaç, A., Ntie-Kang, F., and Sippl, W. (2019). Fragment-based drug design of nature-inspired compounds. *Physical Sciences Reviews* 4 (9): https://doi.org/10.1515/PSR-2018-0110/MACHINEREADABLECITATION/RIS.

40 Khan, S.A., Zia, K., Ashraf, S. et al. (2021). Identification of chymotrypsin-like protease inhibitors of SARS-CoV-2 via integrated computational approach. *Journal of Biomolecular Structure & Dynamics* 39 (7): 2607–2616. https://doi.org/10.1080/07391102.2020.1751298.

41 Shaffer, L. (2020). 15 drugs being tested to treat COVID-19 and how they would work. *Nature Medicine* https://doi.org/10.1038/D41591-020-00019-9.

42 Andola, P., Pagag, J., Laxman, D., and Guruprasad, L. (2022). Fragment-based inhibitor design for SARS-CoV2 main protease. *Structural Chemistry* 33 (5): 1467–1487. https://doi.org/10.1007/S11224-022-01995-Z.

43 Hao, G.F., Jiang, W., Ye, Y.N. et al. (2016). ACFIS: a web server for fragment-based drug discovery. *Nucleic Acids Research* 44 (W1): W550–W556. https://doi.org/10.1093/NAR/GKW393.

44 Dallakyan, S. and Olson, A.J. (2015). Small-molecule library screening by docking with PyRx. *Methods in Molecular Biology (Clifton, N.J.)* 1263: 243–250. https://doi.org/10.1007/978-1-4939-2269-7_19.

45 Morris, G.M., Ruth, H., Lindstrom, W. et al. (2009). AutoDock4 and AutoDockTools4: automated docking with selective receptor flexibility. *Journal of Computational Chemistry* 30 (16): 2785. https://doi.org/10.1002/JCC.21256.

46 Daina, A., Michielin, O., and Zoete, V. (2017). SwissADME: a free web tool to evaluate pharmacokinetics, drug-likeness and medicinal chemistry friendliness of small molecules. *Scientific Reports* 7 (1): 1–13. https://doi.org/10.1038/srep42717.

47 Hess, B., Kutzner, C., Van Der Spoel, D., and Lindahl, E. (2008). GROMACS 4: algorithms for highly efficient, load-balanced, and scalable molecular simulation. *Journal of Chemical Theory and Computation* 4 (3): 435–447. https://doi.org/10.1021/CT700301Q/ASSET/IMAGES/LARGE/CT700301QF00006.JPEG.

48 Ertl, P., Rohde, B., and Selzer, P. (2000). Fast calculation of molecular polar surface area as a sum of fragment-based contributions and its application to the prediction of drug transport properties. *Journal of Medicinal Chemistry* 43 (20): 3714–3717. https://doi.org/10.1021/JM000942E/SUPPL_FILE/JM000942E_S.PDF.

49 Potts, R.O. and Guy, R.H. (1992). Predicting skin permeability. *Pharmaceutical Research* 9 (5): 663–669. https://doi.org/10.1023/A:1015810312465.

50 Shultz, M.D. (2013). Setting expectations in molecular optimizations: strengths and limitations of commonly used composite parameters. *Bioorganic & Medicinal Chemistry Letters* 23 (21): 5980–5991. https://doi.org/10.1016/J.BMCL.2013.08.029.

51 Sliwoski, G., Kothiwale, S., Meiler, J., and Lowe, E.W. (2014). Computational methods in drug discovery. *Pharmacological Reviews* 66 (1): 334. https://doi.org/10.1124/PR.112.007336.

52 Waring, M.J., Arrowsmith, J., Leach, A.R. et al. (2015). An analysis of the attrition of drug candidates from four major pharmaceutical companies. *Nature Reviews Drug Discovery* 14 (7): 475–486. https://doi.org/10.1038/nrd4609.

53 Murray, C.W. and Rees, D.C. (2009). The rise of fragment-based drug discovery. *Nature Chemistry* 1 (3): 187–192. https://doi.org/10.1038/NCHEM.217.

54 Erlanson, D.A., Fesik, S.W., Hubbard, R.E. et al. (2016). Twenty years on: the impact of fragments on drug discovery. *Nature Reviews. Drug Discovery* 15 (9): 605–619. https://doi.org/10.1038/NRD.2016.109.

55 Vijayan, R.S.K., Kihlberg, J., Cross, J.B., and Poongavanam, V. (2022). Enhancing preclinical drug discovery with artificial intelligence. *Drug Discovery Today* 27 (4): 967–984. https://doi.org/10.1016/J.DRUDIS.2021.11.023.

11

AI/ML Approaches in Drug Design

Kevser Kübra Kırboğa

Faculty of Engineering, Bioengineering Department, Bilecik Seyh Edebali University, Bilecik, Türkiye

11.1 Introduction

Historically, drug design and development have been conducted through experimental methods and chemical synthesis techniques [1, 2]. While these methods provide understanding at the molecular level, they are often time-consuming, costly, and limited, especially in the selection of starting molecules [3]. In recent years, the rise of computational methods has played a significant role in accelerating drug design processes [4–6]. These computational methods include many different technologies, such as molecular dynamics simulations, structure-based design (SBDD), and ligand-based design (LBDD) [7–9]. The adaptation of artificial intelligence (AI) and machine learning (ML) technologies in this field has enabled these methods to find a wider field of application and especially to analyze large datasets quickly and effectively [10, 11]. The contribution of AI and ML to drug design enables in-depth knowledge of particularly complex biological systems and their vast chemical space [12]. This offers an excellent opportunity for faster and more accurate identification of potential drug candidates [13] (Figure 11.1).

11.2 Traditional Drug Design Methods

Traditional drug design is mainly based on experimental laboratory methods, *in vitro* and *in vivo* testing, and chemical synthesis processes [4, 14]. These methods have formed the basis of the pharmaceutical industry over the years and have led to many essential drug discoveries [15]. Experimental methods such as high-throughput screening (HTS) make it possible to test thousands of chemical compounds [16] quickly. However, these methods are generally expensive and only suitable for small molecular changes [17]. The effectiveness and safety of drug candidates are evaluated using cell cultures or animal models [18]. These tests are critical for monitoring the biological effects of potential drugs but can be time-consuming and raise ethical issues [19]. Drug candidates are often produced through chemical synthesis [20, 21]. However, this process is often complex and multistage, leading to constraints in terms of cost and time. Although traditional methods enable many drug candidates to pass the initial stages, new and more effective methods need to be investigated due to the limitations of these methods [22].

Computational Methods for Rational Drug Design, First Edition. Edited by Mithun Rudrapal.

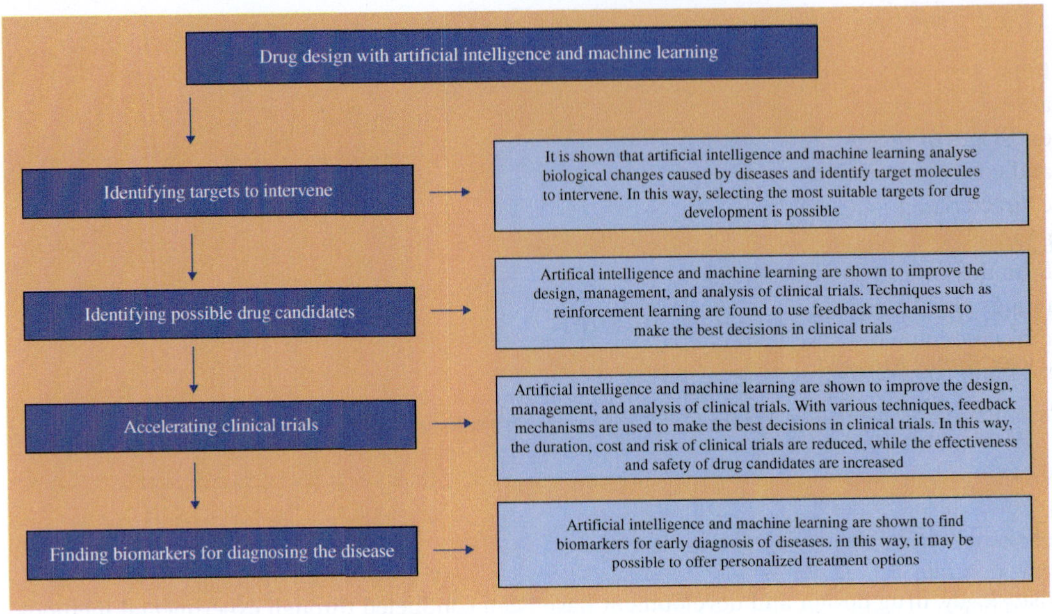

Figure 11.1 Drug design process with artificial intelligence and machine learning.

11.2.1 The Rise of Computational Methods

The limitations of traditional drug design methods have paved the way for the rise of computational methods [23, 24]. These methods include molecular modeling, simulations, and algorithm-based calculations [25].

Molecular modeling enables the study of chemical and biological systems at the atomistic level [26]. This technique is critical for understanding the interactions of potential drug molecules with target proteins [27]. Simulation techniques such as molecular dynamics simulations are used to understand the dynamics of biological systems [28]. These simulations may allow observing molecular interactions change over time [29, 30]. Simulation techniques can explain molecular systems' behavior, functions, mechanisms, and techniques to study conformational changes of drug targets, drug–binder interactions, drug–target residence times, drug resistance mechanisms, or drug side effects [31, 32].

Algorithm-based calculations analyze and model large datasets [33]. These algorithms can quickly analyze complex biological and chemical datasets, thus accelerating the drug design process [34]. Algorithm-based calculations can be applied at different levels and for different purposes in drug design. Algorithm-based calculations are used to understand complex interactions in biological systems for the identification of drug targets, perform intelligent searches in chemical space to establish molecular structures, calculate physicochemical parameters to predict molecular properties, use structural information to model molecular interactions, create chemical fingerprints for analysis of drug similarity and diversity, and can perform processes such as predicting ADMET profiles to evaluate drug safety and toxicity [35, 36]. This rise of computational methods has opened many doors for faster and more efficient drug discovery. However, these methods also have limitations and challenges, highlighting the need for future research [37].

11.2.2 The Importance of AI/ML in Modern Drug Design

In modern drug design and development processes, the contributions of AI and ML technologies are becoming increasingly evident [38, 39]. These technologies play an active role in many different stages of the drug discovery process [40]. Drug discovery and development is a long, complex, and costly process. This process has many stages, such as identifying potential drug candidates, characterizing, optimizing, and preparing for clinical tests. Each stage requires the generation, analysis, and interpretation of big data. AI/ML techniques help extract meaningful information from this data, create new hypotheses, and expand existing scientific knowledge [41]. AI/ML techniques play an essential role in identifying and selecting drug targets. Drug targets are biological molecules on which drugs act. These molecules can often be receptors, enzymes, nucleic acids, or ion channels [42, 43]. AI/ML techniques are used in identifying disease-associated genes, proteins, or metabolites, modeling target–effector relationships and target validation [44].

AI/ML techniques are also effective in designing and improving drug candidates. Drug candidates are chemical compounds that exert a biological effect by binding to drug targets. These compounds' physicochemical and pharmacokinetic properties affect factors such as bioavailability, efficacy, safety, and toxicity [45]. AI/ML techniques are used in the development of QSAR models that describe the relationships between the structural features and biological activities of drug candidates, in estimating target binding affinities of drug candidates, in optimizing the production process of drug candidates, and in determining the toxicity profiles of drug candidates [10, 22].

AI/ML techniques contribute to accelerating and reducing the cost of the drug discovery and development process. Thanks to AI/ML techniques, the number and time of experiments used in the drug discovery and development process are reduced, fewer animal experiments are performed, and the clinical success rate is increased. AI/ML techniques are of great importance in modern drug design. AI/ML techniques make drug discovery and development more efficient, safe, and economical. The accuracy and reliability of the information obtained through AI/ML techniques are constantly improved and audited.

11.3 AI/ML Landscape in Drug Design

11.3.1 AI/ML Algorithms and Methods

11.3.1.1 Machine Learning Models

ML models are a prominent category among AI algorithms and methods. It generally can learn from large and complex datasets and thus has different application areas in drug design and development processes [46, 47] (Figure 11.2). Supervised learning models are trained to predict a specific outcome. It is used to indicate the pharmacokinetic properties of a compound [48]. Supervised learning models predict outputs in new data by learning input–output relationships in training data. The success of supervised learning models depends on the quality, quantity, and diversity of training data. These models are one of the ML techniques frequently used in drug design [49, 50]. Unsupervised learning models are aimed at discovering hidden structures in datasets. It is used to understand disease mechanisms by analyzing gene expression data [51, 52]. Unsupervised learning models are used when the inputs in the training data are not labeled or the outputs are unknown. Unsupervised learning models perform similarities, differences, clustering, or dimensionality reductions in datasets. Unsupervised learning models allow the uncovering of new knowledge in drug design [53, 54].

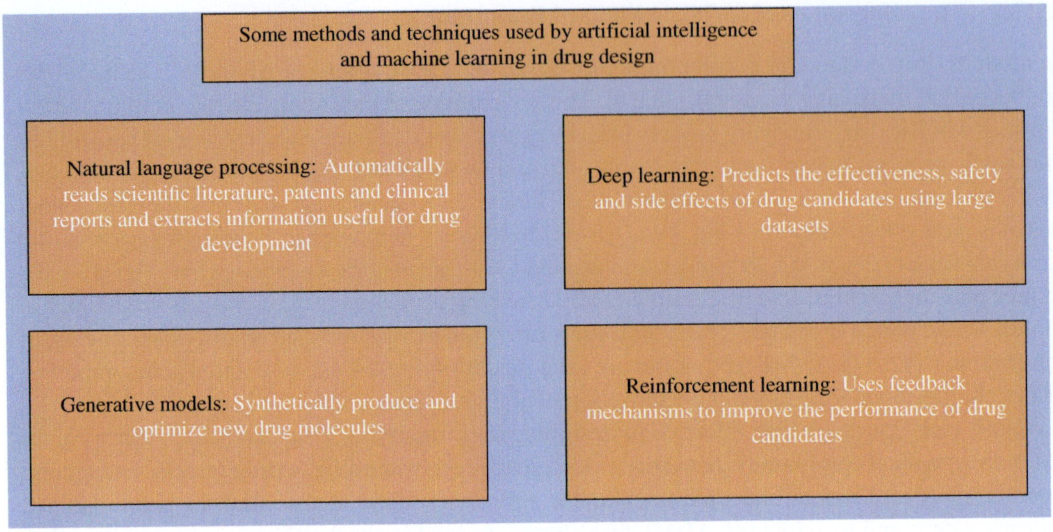

Figure 11.2 Methods and techniques used by artificial intelligence and machine learning in drug design.

Reinforcement learning algorithms learn to achieve a specific goal by solving decision-making problems. It is used in drug design, especially for compound optimization [55]. Reinforcement learning algorithms are a model in which an agent performs a specific action in an environment and receives a reward or punishment resulting from that action. The agent's goal is to find the best policy of action to achieve maximum reward in the long run. Reinforcement learning algorithms help drug design produce more effective and safe drug candidates [55, 56]. Deep learning is a subset of artificial neural networks and often uses multilayer network structures. It is used in drug design to solve complex problems such as modeling protein folding and protein–protein interactions [57, 58]. Deep learning algorithms can learn complex relationships from high-dimensional and heterogeneous data. The advantages of deep learning algorithms include high accuracy, flexibility, and generalizability. The disadvantages of deep learning algorithms include high computational cost, data requirements, and lack of interpretability [59, 60]. Random forest, SVM (support vector machine), and other algorithms are common ML algorithms often used for classification and regression problems. Such algorithms are often used to predict the biological activities of drug compounds [61]. A random forest algorithm is an ensemble method formed by combining multiple decision trees. The random forest algorithm offers high accuracy, low variance, and high interpretability [62]. The SVM algorithm is a method that tries to find a boundary that separates data points into two or more classes. The SVM algorithm provides high accuracy, low overfitting, and high generalizability [63, 64]. ML models offer increased flexibility and application potential in drug design and development processes. However, these models also need to pay attention to issues such as ethical and regulatory issues, model interpretability, and data security [65].

11.3.1.2 Neural Networks

Neural networks are among AI and ML applications' most widely used methods. In particular, deep learning, a type of neural network model, has great potential in solving complex problems. Artificial neural networks represent simple mathematical modeling of biological neural networks and contain large numbers of nodes or "neurons" [66].

Protein folding: How the three-dimensional structure of proteins is formed plays a critical role in the functioning of biological systems. However, understanding and simulating protein folding mechanisms is a complex and challenging scientific problem. Zhao et al. (2023) have shown that deep learning models are essential in understanding and simulating protein folding mechanisms [67]. Deep learning models can process large amounts of data to extract structural features of proteins from their amino acid sequences and optimize their energy functions. In this way, it allows the discovery of new and functional proteins. Deep learning models in drug discovery are important for identifying potential drug candidates for targeting and inhibiting disease-causing proteins. In addition, deep learning models produce new information on protein–protein interactions, protein–tissue relationships, and protein diseases. Therefore, deep learning models are powerful tools that revolutionize the field of protein folding and are useful in drug discovery.

Compound optimization: Neural networks are used to model complex relationships between compounds' biological activities and pharmacokinetic properties [68]. In this way, neural networks can assist in designing new compounds with desired properties or in optimizing existing compounds. Neural networks can predict the binding affinities, activity spectra, selectivity, metabolic stability, bioavailability, and toxicity of compounds. Neural networks can also show how compounds are distributed in chemical space and which regions are worth exploring [69].

Target identification: Neural networks identify potential drug targets and predict their effectiveness [70, 71]. In this way, neural networks can contribute to target validation, an essential step in drug development. Neural networks can model intracellular or extracellular signaling pathways, analyze protein–protein interactions, identify disease-associated genes or proteins, or classify drug targets [72]. Such applications of neural networks show that they are versatile and effective tools in drug design and development processes. However, these models' complexity and interpretability issues must be considered, especially in terms of ethical and regulatory issues. It may cause issues regarding confidentiality, security, and ownership of data used in drug design. In addition, neural networks can cause problems with the accuracy, reliability, and accountability of models used in drug design. Therefore, it is necessary to develop appropriate standards and guidelines for using neural networks in drug design.

Natural Language Processing (NLP) NLP is a field where the disciplines of AI and linguistics come together. NLP also has an essential place in drug design and development processes. It can be used to automatically extract information from text sources such as scientific literature, clinical reports, and patent data [73].

Information extraction: NLP algorithms can automatically extract information about drug targets or interactions from scientific literature and patent data [74, 75].

Clinical data analysis: NLP can extract data from clinical reports or patient histories and use this data to provide valuable information about drug side effects or effectiveness [76, 77].

Sentiment analysis: Social media data or patient comments can be analyzed to understand the effects of medications on patient experiences [78, 79].

These contributions to NLP drug design and development processes show that NLP is also very promising in this field. However, NLP also has some challenges, such as ethical and privacy issues, which must be addressed carefully.

11.3.2 Applications in Drug Design

11.3.2.1 Peptide Synthesis

AI and ML technologies in peptide synthesis have led to significant advances in the last five years. Nuritas has developed a platform to discover new and therapeutic peptides by integrating AI and deep learning with "omics" analysis. By integrating the AI prediction platform with in-house laboratories, it rapidly learns and preclinically validates peptide therapeutics. The Nuritas platform combines an extensive library of peptides from natural sources with publications and knowledge on disease biology, then generates peptide predictions for the selected target or disease domain. Predictions are tested in-house *in vitro*, while active peptides are tested *in vivo* by outside partners or contract research organizations. The results are fed back to the predictor at each verification stage, enabling complete optimization of the predictive feedback loop [80]. In addition, Massachusetts Institute of Technology (MIT) researchers have developed an approach combining experimental chemistry and AI that discovers nontoxic, highly active peptides to enhance drug delivery [81]. Data-driven computational methods are accelerating the discovery and development of biopeptides, and ML can rapidly and effectively predict the benefits of therapeutic peptides [82]. Developing high-throughput technologies and AI have expanded ML methods for discovering new lead peptides and incorporated them into rational drug design [83]. AI and ML have been implemented in various drug discovery processes, including peptide synthesis, streamlining drug discovery and development [84]. AI and ML technologies are accelerating the discovery and development of biological and therapeutic peptides, making these processes more efficient and cost-effective and offering new methodologies and applications in areas such as drug delivery and design. High-throughput analyses and predictions are critical to understanding and optimizing peptides' biological and therapeutic potential, providing powerful tools to accelerate drug development processes, facilitate the discovery and validation of peptides, and stimulate new and innovative research and applications in this field.

11.3.2.2 Molecular Design

Molecular design is a scientific process for discovering and designing new molecules with desired properties. Molecular design has essential applications in various fields, such as drug design and discovery, plant protection, chemical biology, materials science, and nanotechnology. Molecular design is traditionally performed using methods such as structure- and linker-based drug design, augmented drug design (ADD), multipurpose *de novo* drug design (MPO), structure–activity relationship (SAR), and big data analysis. However, these methods have some limitations. These methods are often costly, time-consuming, and complex. In addition, these methods may not fully reflect molecular systems' dynamicity, heterogeneity, and multidimensionality. AI is a technology with great potential in the field of molecular design. AI can accelerate and facilitate the discovery and design of molecules with desired properties using methods such as data analysis, ML, deep learning, machine reasoning (MR), and causal inference (CI). AI overcomes the limitations of traditional molecular design methods and enables more efficient, more reliable, and cheaper molecular design. AI can be applied to molecular design at different levels and for other purposes [85, 86].

11.3.2.3 Virtual Screening (VS)

Virtual screening (VS) enables the prioritization of compounds to identify and test new "hit" molecules, significantly reducing experimental attrition rates [87]. A fully automated AI/ML VS cascade has been implemented at a drug discovery center in Africa. An AI- and ML-based tool called ZairaChem has been developed for quantitative structure–activity/–property relationship (QSAR/QSPR) modeling. This tool requires low computational resources and can operate on various datasets. ZairaChem has been used for malaria and tuberculosis drug discovery, with 15 models forming a VS cascade. These models include a variety of decision-making tests, from whole-cell phenotypic screening to cytotoxicity, water solubility, and permeability. ZairaChem can inform the progress of frontrunner compounds in Holistic Drug Discovery and Development Center using computational profiling before synthesizing and testing compounds [87]. A technique that is iteratively trained with deep neural networks (DNNs) has been developed to enable billion-sized molecular libraries for structure-based screening. This technique introduces DNNs with small datasets and thus can screen ultra-large molecular libraries [88]. The ML models screened 3601 compounds from a specific internal library, selecting

30 for experimental bioassay validation. From this set, 10 compounds showed significant inhibitory effects on a particular enzyme [89]. ML offers a powerful way to VS and generally involves assembling a filtered training set of known active and inactive compounds. Using this data, ML algorithms develop the ability to prioritize active databases of molecules against specific protein targets [90, 91].

11.3.2.4 Quantitative Structure–Activity Relationship Models

The use of AI and ML in developing QSAR models represents a significant evolution in drug discovery and molecular design. AI and ML are used to develop and optimize QSAR models by combining wet experiments (providing experimental data and reliable validation) with molecular dynamics simulation (providing mechanistic interpretation at atomic/molecular levels) [92]. In recent years, various AI and ML algorithms have been used to analyze the current state of the drug development process and model molecular data and biological problems, especially through research on QSAR [93]. In modern drug discovery, combining chemoinformatics and QSAR modeling has emerged as a powerful alliance that enables researchers to exploit the broad potential of ML techniques for predictive molecular design and analysis [35]. In general, two of the most successful ML algorithms QSAR applications are random forests and DNNs. The work by Svetnik and colleagues, published in the *Journal of Chemical Information and Computer Sciences* in 2003, is one of the first examples of the use of random forest in QSAR and has subsequently been frequently used as a gold standard [94]. The rise of deep learning and widely accessible chemical databases has improved QSAR performance. This paper proposes a new deep learning-based method to implement QSAR prediction by combining end-to-end encoder–decoder model and convolutional neural network (CNN) architecture [95].

11.3.2.5 Drug Repurposing

AI and ML technologies are transforming the drug repositioning process and making significant contributions. Drug repurposing allows existing drugs to be reused for new medical indications, and this process becomes more cost-effective and faster, thanks to AI and ML. AI and ML approaches have been developed to identify drug repositioning candidates based on big data sources systematically. This speeds up the drug development process with computational methods and reduces risks [96]. Structure-based drug repositioning is a popular *in silico* repositioning approach. This review discusses traditional and modern AI-based computational methods and tools applied at various stages for SBDD pipelines. In addition, the role of generative models in producing molecules with skeletons is highlighted [97]. Researchers focus on supervised ML and AI methods that use publicly available databases and information sources. Although most of the exemplary applications are in anticancer drug therapies, the methods and resources reviewed are also broadly applicable to other indications, including COVID-19 treatment. Particular emphasis has been placed on comprehensive target activity profiles that will enable a systematic repositioning process by expanding the target profile of drugs [96]. Computational drug repositioning methods adapt AI algorithms to discover new applications of approved or investigational drugs [98].

These various applications and methods demonstrate how AI and ML are used in drug repositioning processes and how these technologies offer significant potential in drug discovery and development. In particular, structure-based drug repositioning and supervised ML and AI methods can make drug repositioning processes more effective, rapid, and cost-effective, enabling faster development of effective treatments for new medical indications.

The use of AI and ML technologies in drug repositioning studies in the context of the COVID-19 pandemic has a vital role in developing treatment strategies. AI- and ML-based drug repositioning strategies enable the integration of various modules to identify practical and usable drugs and drug combinations for COVID-19 treatments. These modules include virus–host interactions and protein–protein interactions [99]. Another study reviews the current status of applications of AI and ML, especially in structural biology, drug repositioning, and development, stating that it will motivate researchers to use the potential of AI in the fight against COVID-19 [100].

AI-assisted drug repositioning is seen as promising in developing new treatments for COVID-19 infection, and accumulated biological data profiles are facilitating AI-based drug repositioning efforts for developing COVID-19 therapies [101]. The drug repositioning or repositioning technique involves using existing drugs for emerging and challenging diseases, including COVID-19, and this approach has become a promising approach due to the opportunity to reduce development timelines and overall costs [102]. Some ML and AI approaches have been developed to systematically identify drug repositioning leaders based on big data sources, further accelerating and de-risking the drug development process by computational tools [96]. These studies and analyses provide valuable insights into how AI and ML technologies can support and accelerate drug repositioning efforts in the fight against COVID-19.

11.3.3 Challenges and Failures

Situations where AI and ML fail to meet expectations in drug discovery are generally related to the limitations and misapplication of these technologies. AI and ML algorithms work based on accurate and regularly prepared data. If the data is not prepared or modeled correctly, these algorithms can produce misleading results, leading to severe problems in the drug discovery process [103]. AI and ML have caused some expensive and controversial failures in the pharmaceutical industry. Some applications of AI and ML in the industry have been major disappointments, such as Watson AI's automated disease diagnosis and the clinical trial failure of Exscientia's DSP-1181 [104]. One of the main reasons why AI and ML algorithms fail in drug development programs is the lack of data and sparse data. A sufficient amount of data is needed to bring any drug to market, and the lack of this data severely limits the effectiveness of AI and ML [10]. Providing data diversity is critical for an effective model. AI platforms must be trained on various datasets representing the organization's target populations. Otherwise, these platforms may adopt fundamental biases, misread side-by-side features, and provide misleading results [103]. These limitations and misapplications may cause AI and ML to fail to meet expectations in drug discovery. To fully realize the potential of AI and ML in this field, several important factors, such as proper data management, right algorithm selection, and appropriate model validation, must be considered (Table 11.1).

11.4 Ethics, Reliability, and Regulatory Issues

The use of AI and ML in drug discovery raises various issues and controversies around ethical, reliability, and regulatory issues. Ethical and trustworthy issues generally include topics such as transparency of algorithms, data sharing, cybersecurity, and potential discrimination. Regulatory issues include the auditing, certification, and permitting processes required to ensure that AI and ML applications comply with standards, laws, and rules.

Data privacy: Patients' and participants' personal data privacy is a central issue in data-based research. Data privacy protects individuals' personal information from unauthorized access, use, or disclosure. Breach of data privacy can jeopardize individuals' privacy, security, and reputation. Therefore, institutions that collect and use data must take the necessary precautions to protect the data and ensure that the data is used following ethical rules [105].

Table 11.1 Benefits and challenges of artificial intelligence and machine learning in drug design.

Benefits	Challenges
It accelerates the drug development process, reduces its cost, and increases the success rate	Data quality, security, and privacy issues may occur
It offers personalized treatment options and provides early diagnosis of diseases	Ethical, legal, and regulatory standards may be unclear
It discovers new disease mechanisms and opens new treatment areas	May require human expertise and supervision

Benefits: The drug development process is usually very long, expensive and risky. AI and ML optimize this process, delivering better medicines to more patients. They contribute to every stage of the drug development process. By analyzing biological changes caused by diseases, it identifies target molecules to intervene. It uses large datasets to predict drug candidates' effectiveness, safety, and side effects. It synthetically produces and optimizes new drug molecules through techniques such as generative models. It improves the design, management, and analysis of clinical trials. Personalized treatment options ensure that patients receive the most appropriate medication according to their genetic, biological, and environmental characteristics. Early diagnosis of diseases increases the chance of success of treatment. AI and ML play an essential role in providing personalized treatment options. They find biomarkers for diagnosing diseases. By analyzing patients' data, it determines the best treatment protocol and monitors the effects and side effects of the treatment. Discovering new disease mechanisms helps find new treatment targets and methods. New treatment areas offer solutions to previously untreatable or difficult-to-treat diseases. They play an essential role in discovering new disease mechanisms. For instance, it detects genetic or epigenetic changes caused by diseases. It reveals the relationships between diseases and suggests new drug targets or combinations.
Challenges: Data quality, security, and privacy are critical in drug development. AI and ML require large amounts of data. This data must be accurate, up-to-date, secure, and confidential. There may be problems with data quality, security, and privacy. For example, data may be incomplete, inaccurate, or inconsistent. It may be subject to cyberattacks or misuse. Disputes may arise regarding ownership or sharing of data. Ethical, legal, and regulatory standards are the rules that must be followed in drug development. AI and ML may violate these rules or leave them unclear. There may be problems with ethical, legal, and regulatory standards. For example, who owns the patent rights to drugs produced by AI? How can we prove the safety of drugs produced by AI? Who has the ethical responsibility for drugs produced by AI? Human expertise and control are indispensable in drug development. AI and ML cannot replace humans or question their decisions. There may be problems with human expertise and control. For example, how to control the side effects of AI-generated drugs? How to explain the working mechanism of drugs produced by AI? How do we test the reliability of drugs produced by AI?

Informed consent: When collecting and using data, individuals are required to understand and consent to what type of information is being collected and how this information will be used [106]. Informed consent is essential to protect individuals' rights regarding the use of their data. Lack or inadequacy of informed consent may lead to violations or misdirection of individuals' consent. Therefore, institutions that collect and use data must inform individuals about their data's purpose, scope, duration, and consequences and obtain their consent.

Algorithm reliability: AI and ML algorithms' accuracy, reliability, and security are critical in drug discovery and development [107]. Algorithms must not produce false positive or false negative results and must identify safe and effective treatment options. To ensure algorithm reliability, the quality, quantity, and diversity of the algorithms' training data must be controlled; the performance of algorithms should be tested; errors in algorithms must be identified and corrected; and the risks of algorithms must be evaluated and minimized [108].

Algorithm transparency: It is essential for ethical and legal requirements to ensure that algorithms work and that decision-making processes are open and transparent [109]. Algorithm transparency increases the accountability of algorithms, allows algorithms' decisions to be audited, helps algorithms justify their decisions, and enables the algorithms' decisions to be challenged. To ensure algorithm transparency, sufficient information should be shared about the design, training, testing, implementation, and results of algorithms. Participation and feedback of the parties affected by the decisions of the algorithms should be ensured. The algorithms' decisions must be shown to comply with ethical and legal standards [110, 111].

Interpretability: AI and ML algorithms must be able to present their decisions and recommendations in a format that humans can understand and evaluate. Interpretability increases algorithms' reliability, transparency, and accountability; it explains the logic, significance, and consequences of algorithms' decisions. It helps evaluate the accuracy, consistency, and fairness of algorithms' decisions. Presenting the data, features, parameters, weights, coefficients, and other factors behind the algorithms' decisions in an understandable way to ensure interpretability; explaining the algorithms in a language appropriate to the parties their decisions affect. The decisions of the algorithms need to be supported by statistical criteria such as confidence interval, sensitivity, accuracy, and specificity [5].

Current regulations: Regulatory authorities such as the FDA (Food and Drug Administration) have developed regulations and established specific standards and approval processes regarding how AI and ML can be used in drug discovery and development [112].

Future prospects: A matter of great interest is how the regulatory landscape will change as AI and ML rapidly evolve, and these technologies are increasingly used in drug discovery and development. Clear standards, guidelines, and approval processes are expected in the future. These factors lead to meaningful discussions and considerations about how and when to apply AI and ML in drug discovery and development in an ethical and regulatory manner. By collaborating and discussing these issues, regulatory authorities, industry stakeholders, and the academic community are trying to determine how to apply AI and ML in this field safely, ethically, and regulatory-compliantly.

11.5 Future Directions

The rapid development of AI and ML technologies poses various directions and expectations on how they can be applied in drug discovery and development. It is thought that quantum computing, one of the emerging technologies, can be much faster and more effective than traditional molecular modeling and simulation calculations. Quantum computing can develop more effective algorithms for designing and optimizing drug molecules. In addition, blockchain technology is thought to transform drug discovery and development processes by providing data security, traceability, and transparency. In genomics and proteomics integration, AI and ML can be integrated with fields such as genomics and proteomics, providing a better understanding of disease mechanisms and more effective identification of target molecules. This integration may make it possible to develop more effective and personalized treatment strategies. AI and ML solutions must consider sustainability factors such as energy efficiency, modular design, and scalability. In addition, these solutions must be scalable to meet the requirements of large datasets and computational resources.

These guidelines and emerging technologies offer essential insights into how AI and ML can be made more effective, sustainable, and scalable in drug discovery and development. The applicability and potential of AI and ML in these fields require broad discussion and evaluation of how and when to apply these technologies ethically and regulatory-compliantly. In addition, how these technologies can be made more effective and reliable in drug discovery and development and integrated with various interdisciplinary approaches and emerging technologies will require more research and collaboration.

11.6 Conclusion

This book chapter overviews the importance, applications, methods, and challenges of AI and ML methods in drug design. AI and ML improve and accelerate drug candidates' discovery, design, optimization, and evaluation processes by providing data-based and computer-aided methods in drug design. AI and ML overcome the challenges faced by traditional methods used in drug design and enable the development of more effective, safer, and cheaper drugs. These methods can perform processes such as identifying drug targets, constructing molecular structures, predicting molecular properties, modeling molecular interactions, analyzing drug similarity and diversity, and evaluating drug safety and toxicity. AI and ML can also perform operations such as data cleaning, data integration, data visualization, data mining, and data analysis to improve the quality and reliability of data used in drug design. However, some limitations exist when using AI and ML methods in drug design. AI and ML methods often require large amounts of data, but the data used in drug design may be limited, noisy, or incomplete. AI and ML methods can also produce complex or black-box models, but interpretability and confirmability of models used in drug design are essential. Finally, AI and ML methods can cause ethical, legal, and social problems. AI and ML methods may violate data privacy, security, and ownership of copyrights or patents.

Therefore, for more effective use of AI and ML methods in drug design, it is recommended that future research should focus on the following topics:

- Leveraging new data sources or integrating existing data sources to improve the quality, coverage, and accessibility of data used in drug design.
- Developing new algorithms or frameworks or improving existing ones to increase the interpretability, confirmability, and reliability of AI and ML models used in drug design.
- Creating new policies or rules or implementing existing ones to address ethical, legal, and social issues of AI and ML methods used in drug design.

References

1 Doytchinova, I. (2022). Drug design – past, present, future. *Molecules* 27 (5): 1496.

2 Mahapatra, M.K. and Karuppasamy, M. (2022). Fundamental considerations in drug design. In: *Computer Aided Drug Design (CADD): From Ligand-Based Methods to Structure-Based Approaches* (ed. M. Rudrapal and C. Egbuna), 17–55. Elsevier.

3 Ranjan, P., Mohapatra, B., and Das, P. (2020). A rational drug designing: what bioinformatics approach tells about the wisdom of practicing traditional medicines for screening the potential of Ayurvedic and natural compounds for their inhibitory effect against COVID-19 Spike, Indian strain Spike, Papain-like protease and Main Protease protein. *Research Square* https://doi.org/10.21203/rs.3.rs-30366/v1.

4 Sliwoski, G., Kothiwale, S., Meiler, J., and Lowe, E.W. (2014). *Computational methods in drug discovery. Pharmacological Reviews* 66 (1): 334–395.

5 Kırboğa, K.K., Abbasi, S., and Küçüksille, E.U. (2023). Explainability and white box in drug discovery. *Chemical Biology & Drug Design* 102 (1): 217–233.

6 Kirboga, K.K., Kucuksille, E.U., and Kose, U. (2021). The impact of artificial intelligence on the medical area: detailed review. *Journal of Multidisciplinary Developments* 6 (1): "New Era" 6(1), 2021.

7 Durrant, J.D. and McCammon, J.A. (2011). Molecular dynamics simulations and drug discovery. *BMC Biology* 9 (1): 71.

8 Acharya, C., Coop, A.E., Polli, J.D., and MacKerell, A. (2011). Recent advances in ligand-based drug design: relevance and utility of the conformationally sampled pharmacophore approach. *Current Computer-Aided Drug Design* 7 (1): 10–22.

9 Aplin, C., Milano, S.K., Zielinski, K.A. et al. (2022). Evolving experimental techniques for structure-based drug design. *The Journal of Physical Chemistry B* 126 (35): 6599–6607.

10 Dara, S., Dhamercherla, S., Jadav, S.S. et al. (2022). *Machine learning in drug discovery: a review. Artificial Intelligence Review* 55 (3): 1947–1999.

11 Qureshi, R., Irfan, M., Gondal, T.M. et al. (2023). AI in drug discovery and its clinical relevance. *Heliyon* 9 (7): e17575.

12 Ji, Z., Yan, K., Li, W. et al. (2017). Mathematical and computational modeling in complex biological systems. *BioMed Research International* 2017: 5958321.

13 Bhattamisra, S.K., Banerjee, P., Gupta, P. et al. (2023). Artificial Intelligence in pharmaceutical and healthcare research. *Big Data and Cognitive Computing* 7: https://doi.org/10.3390/bdcc7010010.

14 Biala, G., Kedzierska, E., Kruk-Slomka, M. et al. (2023). Research in the field of drug design and development. *Pharmaceuticals (Basel)* 16 (9): 1283.

15 Lombardino, J.G. and Lowe, J.A. (2004). The role of the medicinal chemist in drug discovery – then and now. *Nature Reviews Drug Discovery* 3 (10): 853–862.

16 Szymański, P., Markowicz, M., and Mikiciuk-Olasik, E. (2012). Adaptation of high-throughput screening in drug discovery-toxicological screening tests. *International Journal of Molecular Sciences* 13 (1): 427–452.

17 Judson, R., Kavlock, R., Martin, M. et al. (2013). Perspectives on validation of high-throughput assays supporting 21st century toxicity testing. *ALTEX* 30 (1): 51–56.

18 Zhou, Z., Zhu, J., Jiang, M. et al. (2021). The combination of cell cultured technology and in silico model to inform the drug development. *Pharmaceutics* 13 (5): 704.

19 Muthuswamy, V. (2013). *Ethical issues in clinical research. Perspectives in Clinical Research* 4 (1): 9–13.

20 Katiyar, C., Arun Gupta, A.G., Satyajyoti Kanjilal, S.K., and Shefali Katiyar, S.K. (2012). Drug discovery from plant sources: an integrated approach. *Ayu* 33 (1): 10–19.

21 Schreiber, S.L. (2011). Organic synthesis toward small-molecule probes and drugs. *Proceedings of the National Academy of Sciences* 108 (17): 6699–6702.

22 Sun, D., Gao, W., Hu, H., and Zhou, S. (2022). Why 90% of clinical drug development fails and how to improve it? *Acta Pharmaceutica Sinica B* 12 (7): 3049–3062.

23 Prasad, S., Srivastava, A., Singh, N., et al. (2022). Present and future challenges in therapeutic designing using computational approaches. In *Computational Approaches for Novel Therapeutic and Diagnostic Designing to Mitigate SARS-CoV-2 Infection* (pp. 489–505). Academic Press.

24 Leelananda, S.P. and Lindert, S. (2016). Computational methods in drug discovery. *Beilstein Journal of Organic Chemistry* 12: 2694–2718.

25 Taufer, M., Estrada, T., and Johnston, T. (2020). A survey of algorithms for transforming molecular dynamics data into metadata for *in situ* analytics based on machine learning methods. *Philosophical Transactions of the Royal Society A: Mathematical, Physical and Engineering Sciences* 378 (2166): 20190063.

26 Adelusi, T.I., Oyedele, A.Q.K., Boyenle, I.D. et al. (2022). Molecular modeling in drug discovery. *Informatics in Medicine Unlocked* 29: 100880.

27 Aminpour, M., Montemagno, C., and Tuszynski, J.A. (2019). An overview of molecular modeling for drug discovery with specific illustrative examples of applications. *Molecules* 24 (9).

28 Hollingsworth, S.A. and Dror, R.O. (2018). Molecular dynamics simulation for all. *Neuron* 99 (6): 1129–1143.

29 Hospital, A., Goñi, J.R., Orozco, M., and Gelpí, J.L. (2015). Molecular dynamics simulations: advances and applications. *Advances and Applications in Bioinformatics and Chemistry* 8: 37–47.

30 Priya, P., Kesheri, M., Sinha, R.P., and Kanchan, S. (2017). Molecular dynamics simulations for biological systems. In: *Pharmaceutical Sciences: Breakthroughs in Research and Practice*, 1044–1071. IGI Global.

31 Jagger, B.R., Kochanek, S.E., Haldar, S. et al. (2020). Multiscale simulation approaches to modeling drug–protein binding. *Current Opinion in Structural Biology* 61: 213–221.

32 Amaro, R.E. and Mulholland, A.J. (2018). Multiscale methods in drug design bridge chemical and biological complexity in the search for cures. *Nature Reviews Chemistry* 2 (4).

33 Mi, T., Rajasekaran, S., and Aseltine, R. (2012). Efficient algorithms for fast integration on large data sets from multiple sources. *BMC Medical Informatics and Decision Making* 12 (1): 59.

34 Han, R., Yoon, H., Kim, G. et al. (2023). Revolutionizing medicinal chemistry: the application of artificial intelligence (AI) in early drug discovery. *Pharmaceuticals* 16 (9): 1259.

35 Niazi, S.K. and Mariam, Z. (2023). Recent advances in machine-learning-based chemoinformatics: a comprehensive review. *International Journal of Molecular Sciences* 24 (14): 11488.

36 Oliveira, T.A.D., Silva, M.P.D., Maia, E.H.B. et al. (2023). Virtual screening algorithms in drug discovery: a review focused on machine and deep learning methods. *Drugs and Drug Candidates* 2 (2): 311–334.

37 Blanco-González, A., Cabezon, A., Seco-Gonzalez, A. et al. (2023). The role of AI in drug discovery: challenges, opportunities, and strategies. *Pharmaceuticals* 16: https://doi.org/10.3390/ph16060891.

38 Paul, D., Sanap, G., Shenoy, S. et al. (2021). Artificial intelligence in drug discovery and development. *Drug Discovery Today* 26 (1): 80–93.

39 Patel, V. and Shah, M. (2022). Artificial intelligence and machine learning in drug discovery and development. *Intelligent Medicine* 2 (3): 134–140.

40 Sarkar, C., Das, B., Rawat, V.S. et al. (2023). Artificial intelligence and machine learning technology driven modern drug discovery and development. *International Journal of Molecular Sciences* 24 (3): 2026.

41 Krenn, M., Pollice, R., Guo, S.Y. et al. (2022). On scientific understanding with artificial intelligence. *Nature Reviews Physics* 4 (12): 761–769.

42 Rao, M., McDuffie, E., and Sachs, C. (2023). Artificial intelligence/machine learning-driven small molecule repurposing via off-target prediction and transcriptomics. *Toxics* 11 (10): 875.

43 Vatansever, S., Schlessinger, A., Wacker, D. et al. (2021). Artificial intelligence and machine learning-aided drug discovery in central nervous system diseases: state-of-the-arts and future directions. *Medicinal Research Reviews* 41 (3): 1427–1473.

44 Vadapalli, S., Abdelhalim, H., Zeeshan, S., and Ahmed, Z. (2022). Artificial intelligence and machine learning approaches using gene expression and variant data for personalized medicine. *Briefings in Bioinformatics* 23 (5).

45 Selvaraj, C., Chandra, I., and Singh, S.K. (2022). Artificial intelligence and machine learning approaches for drug design: challenges and opportunities for the pharmaceutical industries. *Molecular Diversity* 26 (3): 1893–1913.

46 Sarker, I.H. (2021). Machine learning: algorithms, real-world applications and research directions. *SN Computer Science* 2 (3): 160.

47 Lu, M., Yin, J., Zhu, Q. et al. (2023). Artificial Intelligence in pharmaceutical sciences. *Engineering* 27: 37–69.

48 Gill, J., Moullet, M., Martinsson, A. et al. (2022). Comparing the applications of machine learning, PBPK, and population pharmacokinetic models in pharmacokinetic drug–drug interaction prediction. *CPT: Pharmacometrics & Systems Pharmacology* 11 (12): 1560–1568.

49 Deo, R.C. (2015). Machine learning in medicine. *Circulation* 132 (20): 1920–1930.

50 Taye, M.M. (2023). Understanding of machine learning with deep learning: architectures, workflow, applications and future directions. *Computer* 12: https://doi.org/10.3390/computers12050091.

51 Chen, X.W. and Gao, J.X. (2016). Big data bioinformatics. *Methods* 111: 1–2.

52 Jafari, M., Wang, Y., Amiryousefi, A., and Tang, J. (2020). Unsupervised learning and multipartite network models: a promising approach for understanding traditional medicine. *Frontiers in Pharmacology* 11: 1319.

53 Lim, S., Tucker, C.S., and Kumara, S. (2017). An unsupervised machine learning model for discovering latent infectious diseases using social media data. *Journal of Biomedical Informatics* 66: 82–94.

54 Choudhuri, S., Yendluri, M., Poddar, S. et al. (2023). Recent advancements in computational drug design algorithms through machine learning and optimization. *Kinases and Phosphatases* 1 (2): 117–140.

55 Tan, R.K., Liu, Y., and Xie, L. (2022). Reinforcement learning for systems pharmacology-oriented and personalized drug design. *Expert Opinion on Drug Discovery* 17 (8): 849–863.

56 Tan, R., Liu, Y., and Xie, L. (2022). Reinforcement learning for personalized drug discovery and design for complex diseases: a systems pharmacology perspective. *arXiv preprint*. arXiv:2201.08894.

57 Lee, M. (2023). Recent advances in deep learning for protein–protein interaction analysis: a comprehensive review. *Molecules* 28 (13): 5169.

58 Defresne, M., Barbe, S., and Schiex, T. (2021). Protein design with deep learning. *International Journal of Molecular Sciences* 22 (21): 11741.

59 Alzubaidi, L., Zhang, J., Humaidi, A.J. et al. (2021). Review of deep learning: concepts, CNN architectures, challenges, applications, future directions. *Journal of Big Data* 8 (1): 53.

60 Ahmed, S.F., Alam, M.S.B., Hassan, M. et al. (2023). Deep learning modelling techniques: current progress, applications, advantages, and challenges. *Artificial Intelligence Review* 56 (11): 13521–13617.

61 Svetnik, V., Liaw, A., Tong, C. et al. (2003). Random forest: a classification and regression tool for compound classification and QSAR modeling. *Journal of Chemical Information and Computer Sciences* 43: 1947–1958.

62 Aria, M., Cuccurullo, C., and Gnasso, A. (2021). A comparison among interpretative proposals for random forests. *Machine Learning with Applications* 6: 100094.

63 Heikamp, K. and Bajorath, J. (2014). Support vector machines for drug discovery. *Expert Opinion on Drug Discovery* 9 (1): 93–104.

64 Rodríguez-Pérez, R. and Bajorath, J. (2022). Evolution of support vector machine and regression modeling in chemoinformatics and drug discovery. *Journal of Computer-Aided Molecular Design* 36 (5): 355–362.

65 González-Alday, R., García-Cuesta, E., Kulikowski, C.A., and Maojo, V. (2023). A scoping review on the progress, applicability, and future of explainable artificial intelligence in medicine. *Applied Sciences* 13 (19): 10778.

66 Serdar, A., Okut, H., Zackula, R. et al. (2023). Deep neural networks and applications in medical research. In: *Deep Learning – Recent Findings and Researches* (ed. M. Domínguez-Morales, J. Civit-Masot, L. Muñoz-Saavedra, and R. Damaševičius). IntechOpen: Rijeka (Chapter 1).

67 Zhao, K., Xia, Y., Zhang, F. et al. (2023). Protein structure and folding pathway prediction based on remote homologs recognition using PAthreader. *Communications Biology* 6 (1): 243.

68 Ivanciuc, O. (2009). Drug design with artificial neural networks. In: *Encyclopedia of Complexity and Systems Science* (ed. R.A. Meyers), 2139–2159. New York, NY: Springer New York.

69 Nag, S., Baidya, A.T., Mandal, A. et al. (2022). Deep learning tools for advancing drug discovery and development. *3Biotech* 12 (5): 110.

70 Kufel, J., Bargieł-Łączek, K., Kocot, S. et al. (2023). What is machine learning, artificial neural networks and deep learning? Examples of practical applications in medicine. *Diagnostics* 13 (15): 2582.

71 Wu, X., Li, Z., Chen, G. et al. (2023). Hybrid neural network approaches to predict drug–target binding affinity for drug repurposing: screening for potential leads for Alzheimer's disease. *Frontiers in Molecular Biosciences* 10: 1227371.

72 Vora, L.K., Gholap, A.D., Jetha, K. et al. (2023). Artificial intelligence in pharmaceutical technology and drug delivery design. *Pharmaceutics* 15 (7): 1916.

73 Bhatnagar, R., Sardar, S., Beheshti, M., and Podichetty, J.T. (2022). How can natural language processing help model informed drug development?: A review. *JAMIA Open* 5 (2): ooac043.

74 Vamathevan, J., Clark, D., Czodrowski, P. et al. (2019). Applications of machine learning in drug discovery and development. *Nature Reviews. Drug Discovery* 18 (6): 463–477.

75 Liu, Z., Roberts, R.A., Lal-Nag, M. et al. (2021). AI-based language models powering drug discovery and development. *Drug Discovery Today* 26 (11): 2593–2607.

76 Chen, L., Gu, Y., Ji, X. et al. (2020). Extracting medications and associated adverse drug events using a natural language processing system combining knowledge base and deep learning. *Journal of the American Medical Informatics Association* 27 (1): 56–64.

77 Hyun, S., Johnson, S.B., and Bakken, S. (2009). Exploring the ability of natural language processing to extract data from nursing narratives. *Computers, Informatics, Nursing* 27 (4): 215–223. (quiz 224–5).

78 Jiménez-Luna, J., Grisoni, F., Weskamp, N., and Schneider, G. (2021). Artificial intelligence in drug discovery: recent advances and future perspectives. *Expert Opinion on Drug Discovery* 16 (9): 949–959.

79 Greaves, F., Ramirez-Cano, D., Millett, C. et al. (2013). Use of sentiment analysis for capturing patient experience from free-text comments posted online. *Journal of Medical Internet Research* 15 (11): e239.

80 Chantelle Kiernan, P.R. (2019). *Rapid and Unrivalled AI Enabled Peptide Discovery*. Available from: https://www.nature.com/articles/d43747-020-00948-x.

81 MIT Schwarzman College of Computing (2021). Machine learning discovers new sequences to boost drug delivery. *MIT News*.

82 Basith, S., Manavalan, B., Hwan Shin, T., and Lee, G. (2020). Machine intelligence in peptide therapeutics: a next-generation tool for rapid disease screening. *Medicinal Research Reviews* 40 (4): 1276–1314.

83 Ye, J., Li, A., Zheng, H. et al. (2023). Machine learning advances in predicting peptide/protein–protein interactions based on sequence information for lead peptides discovery. *Advanced Biology* 7 (6): 2200232.

84 Gupta, R., Srivastava, D., Sahu, M. et al. (2021). Artificial intelligence to deep learning: machine intelligence approach for drug discovery. *Molecular Diversity* 25 (3): 1315–1360.

85 Brown, N., Ertl, P., Lewis, R. et al. (2020). Artificial intelligence in chemistry and drug design. *Journal of Computer-Aided Molecular Design* 34 (7): 709–715.

86 Grebner, C., Matter, H., and Hessler, G. (2022). Artificial intelligence in compound design. In: *Artificial Intelligence in Drug Design*Springer US (ed. A. Heifetz), 349–382. New York, NY.

87 Turon, G., Hlozek, J., Woodland, J.G. et al. (2023). First fully-automated AI/ML virtual screening cascade implemented at a drug discovery centre in Africa. *Nature Communications* 14 (1): 5736.

88 Gentile, F., Yaacoub, J.C., Gleave, J. et al. (2022). Artificial intelligence-enabled virtual screening of ultra-large chemical libraries with deep docking. *Nature Protocols* 17 (3): 672–697.

89 Arul Murugan, N., Priya, G.R., Sastry, G.N., and Markidis, S. (2022). Artificial intelligence in virtual screening: models versus experiments. *Drug Discovery Today* 27 (7): 1913–1923.

90 Carpenter, K.A. and Huang, X. (2018). Machine learning-based virtual screening and its applications to Alzheimer's drug discovery: a review. *Current Pharmaceutical Design* 24 (28): 3347–3358.

91 Melville, J.L., Burke, E.K., and Hirst, J.D. (2009). Machine learning in virtual screening. *Combinatorial Chemistry & High Throughput Screening* 12 (4): 332–343.

92 Mao, J., Akhtar, J., Zhang, X. et al. (2021). Comprehensive strategies of machine-learning-based quantitative structure–activity relationship models. *iScience* 24 (9): 103052.

93 Selvaraj, C., Elakkiya, E., Prabhu, P. et al. (2023). Chapter 9 – Advances in QSAR through artificial intelligence and machine learning methods. In: *QSAR in Safety Evaluation and Risk Assessment* (ed. H. Hong), 101–116. Academic Press.

94 Soares, T.A., Nunes-Alves, A., Mazzolari, A. et al. (2022). The (re)-evolution of quantitative structure–activity relationship (QSAR) studies propelled by the surge of machine learning methods. *Journal of Chemical Information and Modeling* 62 (22): 5317–5320.

95 Hu, S., Chen, P., Gu, P., and Wang, B. (2020). A deep learning-based chemical system for QSAR prediction. *IEEE Journal of Biomedical and Health Informatics* 24 (10): 3020–3028.

96 Tanoli, Z., Vähä-Koskela, M., and Aittokallio, T. (2021). Artificial intelligence, machine learning, and drug repurposing in cancer. *Expert Opinion on Drug Discovery* 16 (9): 977–989.

97 Choudhury, C., Arul Murugan, N., and Priyakumar, U.D. (2022). Structure-based drug repurposing: traditional and advanced AI/ML-aided methods. *Drug Discovery Today* 27 (7): 1847–1861.

98 Zong, N., Wen, A., Moon, S. et al. (2022). Computational drug repurposing based on electronic health records: a scoping review. *npj Digital Medicine* 5 (1): 77.

99 Lv, H., Shi, L., Berkenpas, J.W. et al. (2021). Application of artificial intelligence and machine learning for COVID-19 drug discovery and vaccine design. *Briefings in Bioinformatics* 22 (6): bbab320.

100 Prasad, K. and Kumar, V. (2021). Artificial intelligence-driven drug repurposing and structural biology for SARS-CoV-2. *Current Research in Pharmacology and Drug Discovery* 2: 100042.

101 Liu, Z., Chen, X., Carter, W., et al. (2024). AI-powered drug repurposing for developing COVID-19 treatments. *Reference Module in Biomedical Sciences*. https://doi.org/10.1016/b978-0-12-824010-6.00005-8.

102 Zhou, Y., Wang, F., Tang, J. et al. (2020). Artificial intelligence in COVID-19 drug repurposing. *The Lancet Digital Health* 2 (12): e667–e676.

103 Toivola, T. (2023). *The Evolution of Drug Discovery: Opportunities and Obstacles with ML and AI Development*. Available from: https://www.drugdiscoverytrends.com/evolution-drug-discovery-ml-ai/#:~:text=Here%20is%20a%20shortlist%20of%20the%20most%20common,to%20verify%20data%20accuracy%20or%20model%20it%20properly.#:~:text=Here%20is%20a%20shortlist%20of,accuracy%20or%20model%20it%20properly.

104 O'Leary, L. (2022). *How IBM's Watson Went From the Future of Health Care to Sold Off for Parts*. [cited 2023]; Available from: https://slate.com/technology/2022/01/ibm-watson-health-failure-artificial-intelligence.html.

105 OECD (2019). Risks and challenges of data access and sharing. In: *Enhancing Access to and Sharing of Data: Reconciling Risks and Benefits for Data Re-use across Societies*. Paris: OECD Publishing https://doi.org/10.1787/15c62f9c-en.

106 Andreotta, A.J., Kirkham, N., and Rizzi, M. (2022). AI, big data, and the future of consent. *AI & Society* 37 (4): 1715–1728.

107 Javaid, M., Haleem, A., Singh, R.P. et al. (2022). Significance of machine learning in healthcare: features, pillars and applications. *International Journal of Intelligent Networks* 3: 58–73.

108 Kelly, C.J., Karthikesalingam, A., Suleyman, M. et al. (2019). Key challenges for delivering clinical impact with artificial intelligence. *BMC Medicine* 17 (1): 195.

109 Felzmann, H., Fosch-Villaronga, E., Lutz, C., and Tamò-Larrieux, A. (2020). Towards transparency by design for artificial intelligence. *Science and Engineering Ethics* 26 (6): 3333–3361.

110 Martin, K. (2019). Ethical implications and accountability of algorithms. *Journal of Business Ethics* 160 (4): 835–850.

111 Tsamados, A., Aggarwal, N., Cowls, J. et al. (2022). The ethics of algorithms: key problems and solutions. *AI & Society* 37 (1): 215–230.

112 Cavazzoni, P. (2023). *FDA Releases Two Discussion Papers to Spur Conversation about Artificial Intelligence and Machine Learning in Drug Development and Manufacturing.* [cited 2023]; Available from: https://www.fda.gov/news-events/fda-voices/fda-releases-two-discussion-papers-spur-conversation-about-artificial-intelligence-and-machine#:~:text=The%20regulatory%20uses%20are%20real%3A,of%20therapeutic%20areas%2C%20and.

12

Network-based Methods in Drug Discovery

Ghanshyam Parmar[1], Ashish Shah[1], Jay Mukesh Chudasama[1], Priya Kashav[1], and Vanesa James[2]

[1] *Department of Pharmacy, Sumandeep Vidyapeeth (Deemed to be University), Vadodara, Gujarat, India*
[2] *Department of Regulatory Affairs and Quality Assurance, LJ Institute of Pharmacy, Ahmedabad, Gujarat, India*

12.1 Introduction

The current process of drug development, which focuses on a single target, one disease, and one drug, has serious flaws in terms of safety, efficacy, and sustainability. It has only now been realized how valuable network biology and polypharmacology (multi-targeted therapy) methods are for integrating omics data and creating medications that target numerous targets. Network pharmacology is a revolutionary paradigm that emerged from the merging of these two approaches; it studies the effects of drugs on the interactome and the disease at some levels [1–3]. The complex formulations used in Ayurveda, the traditional Indian medicine system, include a wide variety of ingredients and bioactive compounds. However, the mechanisms and scientific thinking underlying these formulations have hardly received any research attention. To better understand the possible effects, indications, and mechanisms of medications, evidence-based Ayurveda can make use of natural product approaches. In today's pharmaceutical landscape, natural products play a significant role, particularly in disease treatment [4, 5]. A wealth of valuable resources have been bestowed upon humanity by natural products throughout the millennia. To assess the pharmacological efficacy of herbal treatments, drug discovery has developed high-throughput approaches based on network pharmacology. This chapter provides a comprehensive overview of the technique, significance, and application of network pharmacology to cure a wide spectrum of complicated disorders using Indian traditional medicines [6, 7].

12.1.1 Background of Drug Discovery Future Challenges

Drug discovery is a complex process. The field encompasses the disciplines of chemistry and pharmacology and has yielded life-saving therapeutic interventions for diseases. The primary objective is to identify chemicals that fulfil unaddressed medicinal needs, particularly in critical circumstances. Notwithstanding advancements, the process of discovering new drugs remains challenging [7, 8]. The process of introducing a new pharmaceutical compound can be time-consuming and expensive. The procedure involves utilizing extensive chemical databases,

Computational Methods for Rational Drug Design, First Edition. Edited by Mithun Rudrapal.

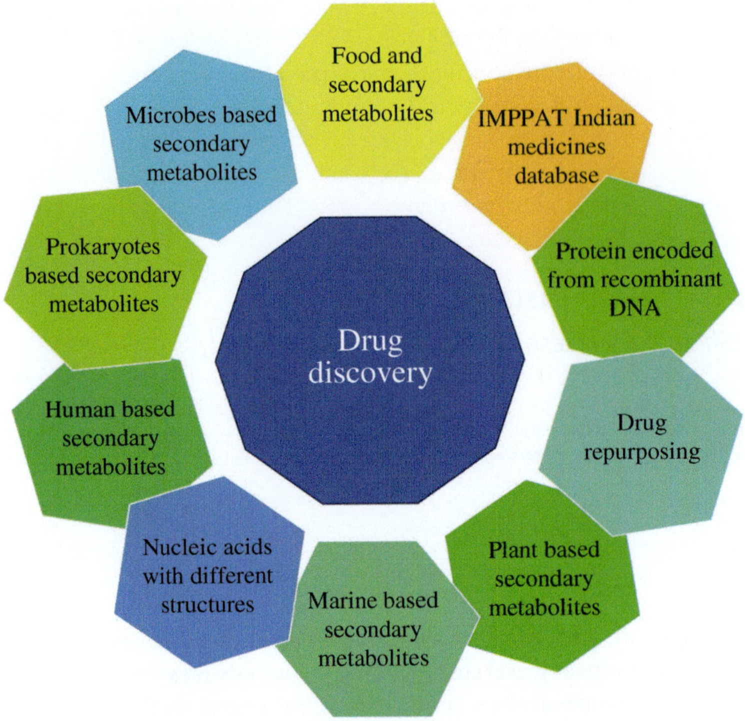

Figure 12.1 The representation of different drug applications discovery in various fields of fundamental science.

optimizing the basic constituents, and conducting comprehensive testing [9]. The creation of medication has always relied on the use of chemical synthesis and experimental biology. Nevertheless, the utilization of structural-based design and other digital tools is accelerating the process of drug discovery which is represented in Figure 12.1. There are many diseases that do not have effective therapies, and the pharmaceutical sector needs to increase output. The intricate nature of human biology poses a significant challenge in the development of drugs that have specific effects and minimal side effects [9, 10]. Notwithstanding these challenges, the discipline continues to develop. Anticipate that the integration of advanced technology and collaborative efforts will enhance the efficiency of drug research and yield a greater number of life-saving medications [10, 11].

In light of the rising costs and pricing restrictions associated with medication research, pharmaceutical corporations and other interested parties are seeking ways to increase efficiency. Three specific areas of focus include the application of big data, implementation of adaptive trial designs, and establishment of public–private partnerships. Public–private partnerships can assume various forms, each presenting distinct benefits and drawbacks, encompassing financial and legal consequences. Adequate preparation is essential for adaptive trial designs, which may involve adaptive licensing [12]. Big data can be utilized for medication development by leveraging existing information, while privacy concerns must be resolved. Overall, these novel alternatives possess the capacity to enhance the efficiency of future pharmaceutical development [13].

Drug research success in this era of rapidly evolving science and technology needs a well-thought-out plan. Scientists are discouraged, and progress is slowed by bureaucratic techniques, budget cuts, and isolated research attempts. Funding models should incorporate public and academic feedback to guarantee appropriate research methodologies. Collaboration and targeted research are impeded by the academic community's dispersion, especially among faculty members [14]. Due to the multidisciplinary nature of drug development, larger networks are necessary for collaboration beyond the confines of translational institutions, where industry, scientists, and physicians can work together. Every system in a business needs to incorporate modern technology. Recognizing and appreciating the unique skills of each individual involved in pharmaceutical creation is essential. In spite of the difficulties, open innovation and public–private partnerships are essential for making the most of everyone's abilities. The public and patients can play an active role in advancing progress through effective education and interdisciplinary training [15]. To successfully develop novel drugs, all parties involved must maintain an optimistic outlook and foster an environment of trust. There may be scholarly and business shifts brought about by the COVID-19 pandemic. By the year 2040, advances in disease biology coupled with modern technology have the potential to enhance disease prevention, management, and potential cures. To better people's health, scientists and doctors work tirelessly. A comprehensive strategy that prioritizes collaboration, foresight, and a nurturing environment can propel drug research forward [16, 17].

The efficacy of most treatments for complex disorders is negligible, and there is a decline in the process of discovering new drugs. Obstacles in fundamental and preclinical research include inadequate quality, challenges in replicating discoveries, and limited relevance to real-world applications. The focus on particular organs in the field of medicine and the belief that each disease has a singular target and therapy also impede innovation [18, 19]. The fields of systems and network medicine, in conjunction with network pharmacology, are revolutionizing our understanding, identification, management, and eradication of diseases. Distinct endotypes, identified by causative, multi-target signaling modules, explain the presence of comorbidities and supplant the need for extensive descriptions of specific symptoms. The combination of synergistic multicompound network pharmacology and medication repurposing offers a highly accurate and efficient approach to therapeutic intervention. This approach eradicates the drug discovery protocols and accelerates the translation of clinical applications [18, 20].

12.1.2 Single Target Approach Limitations

Population-based trials of commercially available medications do not demonstrate patient benefits. Indeed, the top 10 most profitable US medications do not enhance the health of the majority of patients, resulting in a high number needed to treat (NNT). The NNT is lower in high-risk individuals, indicating treatment effectiveness. However, the issue continues. As a result, it is critical to transition from continuous symptom management to a more accurate and, ideally, curative treatment that works for the majority of patients. According to Eroom's law, our ability to transfer biological knowledge into medical development has deteriorated since the 1950s [21]. To overcome this impediment, it is necessary to employ entirely novel medical approaches and identify at least two critical factors that are impeding innovation. Reproducing preclinical and foundational research is difficult due to low study quality, including insufficient statistical power and a bias in scientific journals toward favorable findings. The second issue is that we lack conceptual understanding of the majority of disease definitions. Except for viral and uncommon disorders, chronic diseases are defined by their phenotypes or organ symptoms. Nowadays, medicine focuses on organs. Our preclinical animal illness models frequently reproduce similar symptoms, but there is no conclusive evidence that the mechanism is the same in people. We focus on managing symptoms rather than treating the ailment since we do not understand the underlying causes of illness.

Primary hypertension, which has no known cause, accounts for 95% of all high blood pressure cases [22]. Until the problem goes away, blood vessel vasodilators such as calcium channel blockers or thiazide-type diuretics lower blood pressure by interacting with different proteins [23]. Because the molecular mechanism of hypertension is unknown, we prioritize preventing myocardial infarctions and strokes above treating them. Despite proper antihypertensive medication, most high-risk individuals will experience severe adverse effects. Complex disorders necessitate long-term care because present medications are neither curative nor exact. Rare diseases with precise and monogenetic ethologies are an exception to these constraints [24].

Network pharmacology finds protein–protein interactions that affect illness outcomes. Multiomics methods and computer technologies are being used to precisely document the collective metabolic response in people, enabling the study of more complicated diseases. Here, we examine biomedical applications of network pharmacology depicted in Figure 12.1 [25].

Since the discovery of monotherapy for challenging illnesses, resistance, and side effects have prompted the strategy to evolve beyond mono-triple therapy. When monotherapy is ineffective, multitarget pharmacological treatment provides hope (Figure 12.2). Polypharmacology is becoming increasingly popular among researchers and pharmaceutical corporations as a means of developing multi-target medications. Scientific research suggests that multi-target pharmacological treatment may be effective in eliminating medication resistance. First, computationally search protein networks to identify proteins that interact [26]. Proteins that assist parasites, bacteria, and fungi in surviving communicable diseases are highlighted below. Parasites such as *P. falciparum* create a large number of these proteins because they interact in hostile environments or infiltrate the host. Targeting these proteins necessitates a complete understanding of their structures and functions as they interact directly or indirectly [27, 28]. Using this

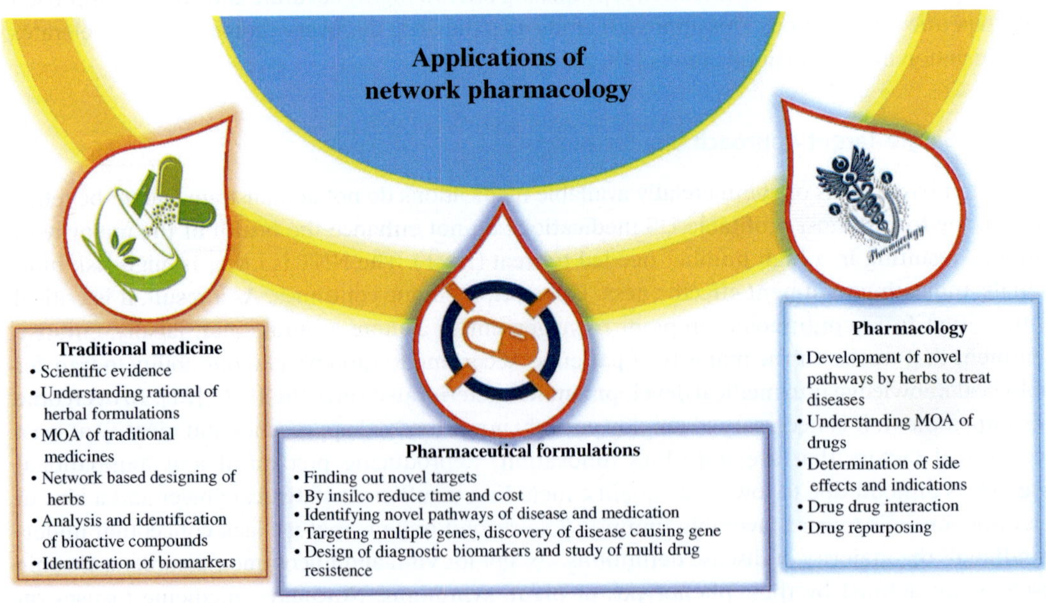

Figure 12.2 Various applications of network pharmacology, in contrast to traditional medicines and pharmaceutical formulations, and for establishing the pharmacological mechanism of drugs or phytoconstituents.

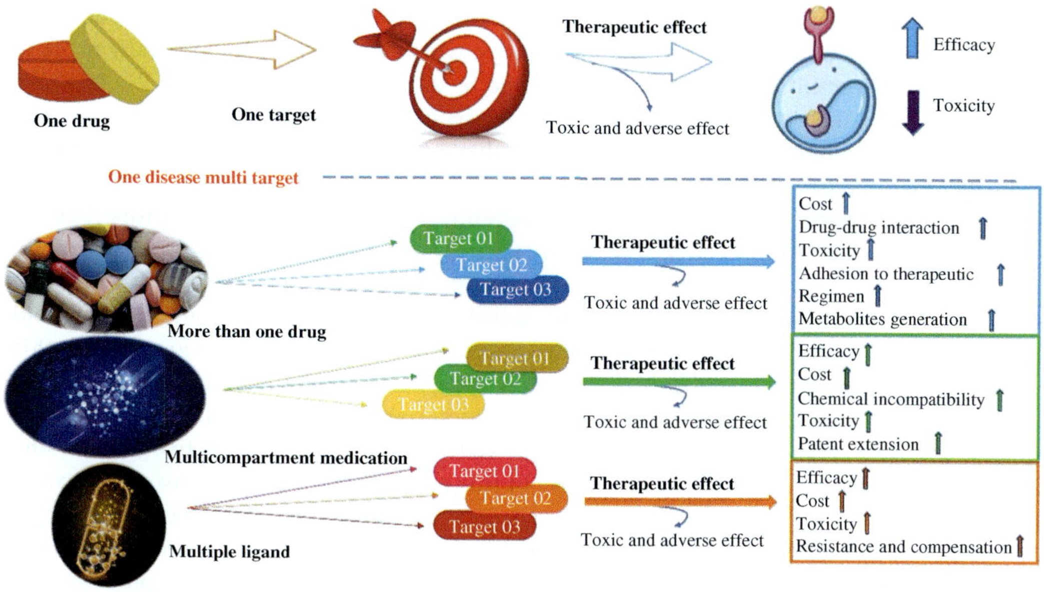

Figure 12.3 Compression between one target and multitarget approach to treat complex diseases.

information, molecules with several targets can be developed, which is represented in Figure 12.3. Since multi-targeting medications are our only option for treating complex disorders, we will most likely continue to look for them [29]. Scientists have been interested in developing nanoparticles (NPs) that target sick cells while protecting healthy tissue to reduce the side effects of chemotherapy since the development of cancer nanomedicine. Researchers are now designing biomimetic NP platforms that resist immune recognition and accumulate in tumors. Initially, they focused on bioconjugation of targeted molecules to NPs [30, 31]. Here, we examine the targeted techniques' pros and cons. First, we examine bioconjugation developments that encapsulate NPs with biomolecules for cell-specific interaction. Biomolecules can be antibodies, aptamers, peptides, or tiny compounds. Even while bioconjugated NPs have better pharmacokinetics and biodistribution than unmodified ones, the mononuclear phagocytic system (MPS) rejects them as "foreign." Several biomimetic techniques have been developed to reduce MPS clearance and immune recognition of NPs. Some biomimetic NPs imitate the structure of naturally occurring ones, while others are completely camouflaged by natural features. Overall, bioconjugated and biomimetic NPs offer great potential for future healthcare and can reduce off-target effects through site-specific delivery. We discuss each strategy, its future design, and its potential therapeutic impact on cancer treatment [32].

12.1.3 Emergence of Network Biology and Polypharmacology

There has been a continual evolution in human diseases, particularly infectious ones. Having said that, they are taking their time to establish therapeutic techniques. It is not possible to treat all diseases with the same medication because some have several causes and extremely complicated pathologies. Medication based on Traditional Indian Medicine (TIM) is an innovative approach to healthcare in India. Some of the botanicals included in TIM formulations have multi-targeting characteristics, the focus of drug research has shifted from the well known "one target, one drug" paradigm to a new "multi-target, multi-drug" strategy, which attempts to systematically regulate several targets, as a result of our growing understanding of complex disorders [33, 34]. As a new strategy to combat the declining efficiency of pharmaceutical research, polypharmacology has emerged. Still, it is not easy to find methods to evaluate multicomponent treatments and rank synergistic combinations of drugs [35].

Novel analytical approaches to study Nature's chemical and pharmacological properties are needed to find sustainable and environmentally friendly drugs to battle pandemics. Our study introduces polypharmacology-labeled molecular networking (PLMN), a new analytical technology approach. Polypharmacological high-resolution inhibition profiling and positive and negative ionization tandem mass-spectrometry-based molecular networking are used in this method. Identifying complicated extract bioactive components quickly and efficiently is the goal [36, 37]. To find antihyperglycemic and antibacterial components, the raw extract of *Eremophila rugosa* was examined using PLMN. Nodes in the molecular network show the activity of each component in the seven assays in this proof-of-concept study using polypharmacology scores, pie charts, and microfractionation variation scores. There were 27 diterpenoids that were not canonical and were generated from diphosphate. Ferruletane esters show antihyperglycemic and antibacterial properties [37]. In the fight against the epidemic and other clinically significant methicillin-resistant *Staphylococcus aureus* strains, oxacillin and a few of these esters work in tandem. A number of esters attach to the active site of protein tyrosine phosphatase 1B in a saddle-shaped fashion. The identification of drugs utilizing natural components with several pharmacological effects may be altered if PLMN's coverage of testing was to be expanded or contracted [38, 39].

12.2 Network Pharmacology: Practical Guide

12.2.1 Common Network Pharmacology Databases

Advanced technology for assessing massive datasets has enabled more compelling and efficient diagnostic and treatment solutions. Understanding how proteins disrupt the complex regulatory system is critical. The field of network biology, which postulates that complex biological systems are subject to general rules, has had a profound impact on our understanding of disease. Data mining and other computer approaches were recommended for regulatory networks in the 21st century. These strategies examined genotype-illness phenotypic relationships. Advances in network biology show that targeting individual proteins does not treat complicated diseases [40]. Drug developers realized that polypharmacology was useless and needed to be abolished to generate a viable treatment with several targets. Through network pharmacology, highly specialized drugs that target a single receptor have given way to treatments that target multiple receptors. Later, system biology and polypharmacology merged in various health areas [41].

In the last decade, indigenous people have used therapeutic plants without scientific research. Traditional treatments used many medicinal plant species. Since therapeutic plants are inexpensive and natural, they affect people's health. However, unsustainable collection and use of these plants have caused the demise of many valued plant species. Traditional remedies can treat complex ailments due to their holistic approach and extensive testing with numerous components. Herbal formulas distinguish traditional medicine. Understanding the complex combination of herbal formulae and their mechanisms can help modify conventional medicines in the age of data analysis. Modern network pharmacology allows systematic study of herbal formulations' complicated molecular makeup and interactions with complex diseases. Traditional medicine herbs have better molecular compatibility than single drugs, which may improve network responsiveness. Network-based techniques are widely used in innovative drug development research [42, 43]. The key molecule responsible for synergy and efficacy in understanding novel medicines is derived from natural sources. It has been demonstrated that these procedures are compatible with a number of herbal formulations used in traditional medicine. One cutting-edge method for identifying medicinal herbs' active ingredients and molecular targets is network pharmacology [44]. This comprehensive method provides a baseline for evaluating bioactive chemicals in medicinal plants and proposes a fresh therapeutic idea for studying their mechanisms of action in sickness therapy. Network pharmacology provides innovative approaches to discovering active compounds, biomarkers, and the scientific basis of conventional medicine by leveraging the intricate biological processes of the human body [45].

12.2.1.1 Network Pharmacology-Related Databases and Data Analysis Tools

One modern approach to drug discovery is network pharmacology, which is supported by biologically relevant databases that store massive volumes of data about biomolecule correlation (Table 12.2). Free and easy access to all of these resources is a major selling point when it comes to collecting data for studies in network pharmacology. Table 12.1 lists the major network pharmacology databases and the ones that are associated with them [46].

12.2.1.2 Exploring IMPPAT Network Pharmacology Databases

Drug development based on natural products and traditional knowledge can be accelerated by the compilation, curation, digitization, and exploration of the phytochemical space of medicinal plants used in India. In the IMPPAT, a revamped and enlarged database that includes 4010 medicinal plants from India, 17,967 phytochemicals, and 1095 medicinal uses, all of which have been handpicked by experts. Notably, IMPPAT offers a nonredundant, *in silico* stereo aware library containing 17,967 phytochemicals from medicinal plants in India, and it constructs associations at the level of plant parts. This library complies with standards. To facilitate simpler chemical space exploration, the phytochemical library has been annotated with numerous relevant features. We have also narrowed the search to 1335 phytochemicals that show promise as medications, the vast majority of which are completely unique from any already available pharmaceuticals [47]. Our team has used cheminformatics to analyze the phytochemical space of medicinal plants from India, characterizing their molecular complexity and molecular scaffold-based structural diversity. We have also compared this space to other chemical libraries. Accessible at https://cb.imsc.res.in/imppat/, IMPPAT 2.0 is a comprehensive phytochemical atlas of medicinal plants used in India that has been painstakingly compiled by hand [47].

Traditional medicine and ethnopharmacology are famous in India. Complex traditional Indian medicine compositions have many components. Their medicinal efficacy comes from empirical knowledge, not mechanistic comprehension of the mixture's active ingredients. Before recently, old Indian literature [47, 48] and old Ayurvedic literature [49] were the primary authors who documented traditional Indian medicine, particularly the use of plants and their remedies. Because it was not digitized, this material could not be used for the development of new medicinal treatments. This knowledge can be utilized to create a database that covers all aspects of Indian medicinal plants, including their phytochemistry, ethnopharmacology, and computational drug discovery [50].

Table 12.1 Data retrieval and analysis of network pharmacology data is facilitated by a variety of databases, each of which is described along with its usage.

Sr. no.	Database name	Brief description	Use of database
1	BioCarta	Pathway database	Pathway analysis and visualization
2	BioGRID	Biological general repository for interaction datasets	Protein–protein interaction dataset
3	C2Maps	Network pharmacology tool	Connectivity map for drug–target interactions
4	CB (Chemical book)	Chemical compound information	Chemical information and structure
5	ChEMBL	Bioactivity database for drug discovery	Drug discovery and pharmacology
6	ChemProt	Chemical biology resource	Protein–chemical interactions
7	ChemSpider	Chemical structure database	Chemical structure information
8	CHMIS-C	Traditional Chinese medicine database	Traditional Chinese medicine information
9	COGs	Clusters of orthologous groups	Evolutionary relationships of genes
10	CPDB	Cancer gene–drug associations database	Cancer-related gene and drug interactions
11	Cytoscape	Bioinformatics software	Network visualization and analysis
12	DAVID	Database for annotation, visualization, and integrated discovery	Functional annotation of genes
13	DIP	Database of interacting proteins	Protein–protein interactions
14	DrugBank	Drug database	Drug information and interactions
15	GeneCards	Human gene database	Information on human genes
16	Guess	Network pharmacology tool	Predicting drug–target interactions
17	HAPPI	Human annotated and predicted protein interaction database	Protein interaction information
18	HIT (Herbal ingredients' targets identification)	Herbal ingredients and their targets	Identification of targets for herbal ingredients
19	HPRD	Human protein reference database	Human protein information
20	InterPro	Protein families and domains database	Protein domain and family information
21	KEGG	Kyoto Encyclopedia of genes and genomes	Pathway and functional information
22	LookChem	Chemical marketplace	Chemical information and marketplace
23	MMsINC	Database of microRNA-mediated molecular interactions	microRNA-mediated interactions

Table 12.1 (Continued)

Sr. no.	Database name	Brief description	Use of database
24	NetMiner	Network analysis tool	Network analysis and visualization
25	NetPath	Pathway database	Signaling pathways in cancer
26	NetworkX	Python library for network analysis	Network analysis and modeling
27	OPHID	Online predicted human interaction database	Predicted human protein interactions
28	Pajek	Network analysis tool	Network analysis and visualization
29	PDB	Protein Data Bank	3D structure of biological macromolecules
30	PDTD	Potential drug target database	Drug target prediction
31	PharmGBK	Pharmacogenomics knowledge base	Pharmacogenomic information
32	PubChem	Chemical database	Chemical information and bioassay data
33	PubMed	Biomedical literature database	Biomedical literature search and retrieval
34	Reactome	Pathway database	Biological pathways and reactions
35	SignaLink	Signaling pathway database	Integration of pathway and network information
36	STITCH	Chemical–protein interaction database	Chemical–protein interactions
37	STRING	Protein–protein interaction database	Protein–protein interactions
38	SwissTargetPrediction	Target prediction for small molecules	Prediction of drug targets for small molecules
39	TCMGeneDIT	Traditional Chinese medicine gene database	Information on genes related to TCM
40	TCMSP	Traditional Chinese medicine systems pharmacology database	(Traditional Chinese Medicine) and drug information
41	IMPPAT (Indian Medicinal Plants, Phytochemistry and Therapeutics)	Indian medicinal plants database	Information of Indian medicinal plants
42	TTD	Therapeutic target database	Information on therapeutic targets
43	Ucinet	Social network analysis software	Social network analysis and visualization
44	UniProtKB	Universal protein resource	Protein sequence and functional information
45	TargetScan	miRNA target prediction	Prediction of miRNA targets in genes

Table 12.2 The table includes feature comparison tables from several databases, including IMPPAT, Polur et al., Phytochemica, KNAPSACK, and TCMID.

Feature	IMPPAT	Polur et al. (2011)	Phytochemica (2013)	KNAPSACK (2010)	TCMID (2014)
Number of Indian medicinal plants	1742	295	5	N/A	6000+
Number of phytochemicals	9596	1829	963	500,000+	N/A
Plant–phytochemical associations	Present	Present	Present	Present	Present
Plant-therapeutic use associations	Present	Present	Present	N/A	Present
Plant–medicinal formulation associations	Present	N/A	N/A	N/A	Present
Phytochemical–human target protein associations	Present	N/A	N/A	N/A	N/A
Plant part-phytochemical associations	N/A	N/A	N/A	N/A	N/A
Web interface	Present	N/A	Present	Present	Present
2D phytochemical structures	Present	N/A	Present	Present	N/A
3D phytochemical structures	Present	N/A	No	Present	N/A
Downloadable structure file formats	PDB and PDBQT, MOL2	N/A	N/A	Present	N/A
Chemical classification	Present	N/A	N/A	Present	Present
Physicochemical properties	Present	N/A	N/A	Present	No
ADMET properties	Present	N/A	N/A	N/A	N/A
Druggability properties	Present	N/A	N/A	N/A	N/A
Cytoscape network visualization	Present	N/A	N/A	N/A	N/A
Filter by physicochemical properties	Present	N/A	N/A	N/A	N/A
Filter by druggability properties	Present	N/A	N/A	N/A	N/A
Chemical similarity search	Present	N/A	N/A	Present	N/A

Well-organized databases are crucial for *in silico* drug discovery, especially for plants, their components, and chemical combinations. Databases on edible and herbaceous plant items exist; however, traditional Indian medicine databases have not been extensively developed. These databases should collect phytochemical data for drug discovery. The IMPPAT highlights Polur et al.'s [51] compendium of Ayurvedic Indian medicinal plants and their phytochemical ingredients. However, their efforts were limited compared to Chinese medicinal herb databases [52]. The authors provide the rigorously produced IMPPAT database comprising 1742 Indian medicinal plants, 9596 phytochemicals, and 1124 therapeutic applications to solve this deficiency. IMPPAT offers a nonrepetitive virtual chemical library with 2D and 3D structures to be more comprehensive than previous databases [53, 54]. Calculated physicochemical characteristics and phytochemical ADMET properties are present in the database. The researchers assessed phytochemical drug-like characteristics using cheminformatics [55]. They used Lipinski's rule of five [56], Oral PhysChem score [56, 57], Pfizer's 3/75 [58], GlaxoSmithKline's 4/400 [59], Veber rule [60], and Egan rule [61]. Based on these ratings, 960 phytochemicals may be medication candidates (Table 12.2). IMPPAT also predicts correlations between phytochemicals and STITCH human target proteins. The sentence stresses the relevance of IMPPAT, the various descriptions of the database are represented in Figure 12.4, and also known to

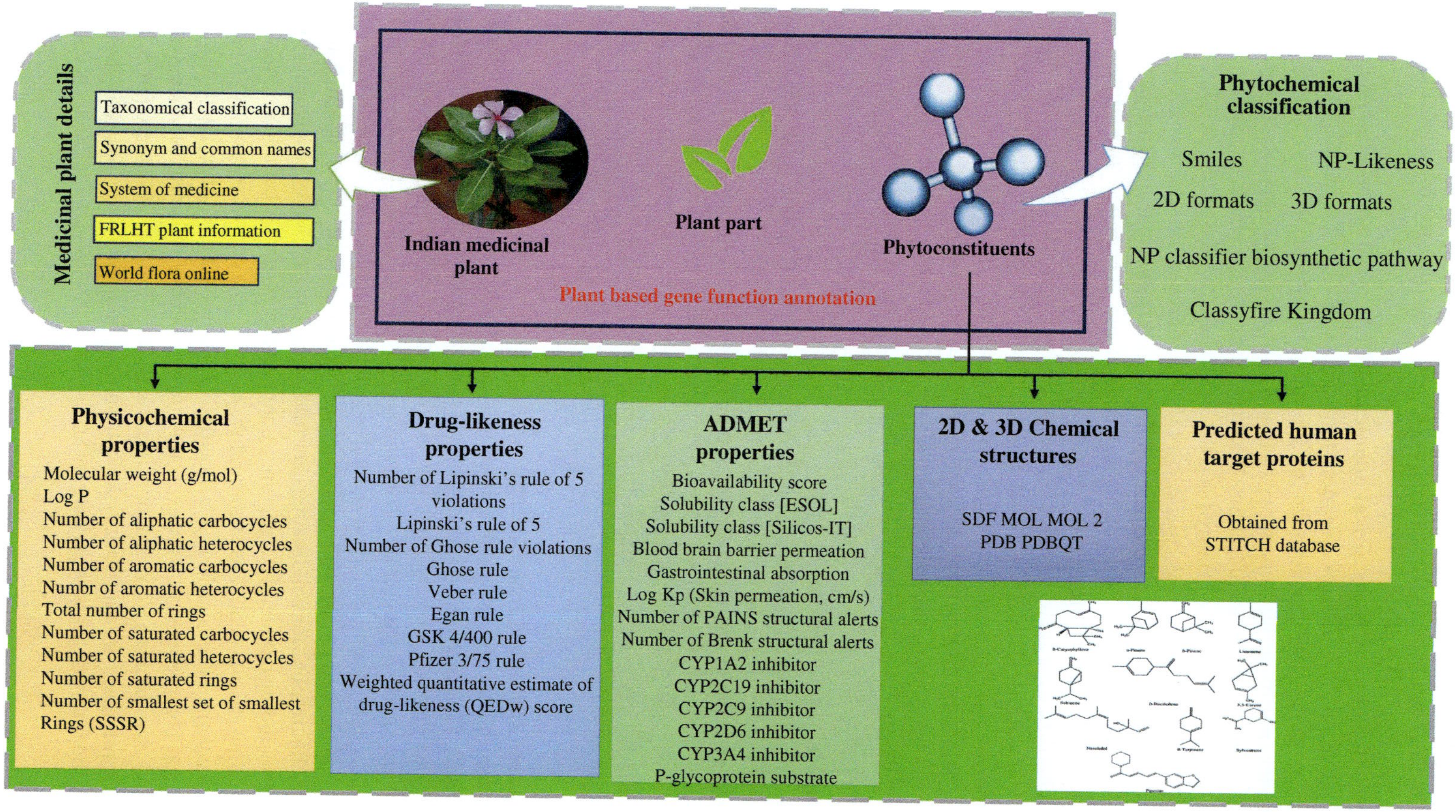

Figure 12.4 IMPPAT database features and various features of the database is highlighted in the image form.

be the largest database of Indian medicinal plant chemical compounds. It is a comprehensive platform for using cheminformatic methodologies to speed up traditional Indian medicine medication discovery [62, 63].

Therapeutic Use of Indian Botanical Species: IMPPAT also collects ethnopharmacological data on Indian medicinal herbs. We meticulously collected Indian medicinal plant therapeutic uses from Indian traditional medicine literature to achieve this goal. Polur et al. [18] created a Nutrichem9,10-derived inventory of therapeutic cures uses for 295 Ayurvedic Indian medicinal plants in addition to literature. We have chosen and evaluated Indian medicinal plant therapeutic data to ensure quality [64]. We purposely avoided automated text mining to learn more about plant-therapeutic qualities. IMPPAT database includes carefully gathered therapeutic applications of Indian medicinal plants from traditional Indian medicine books. These texts contain a plethora of medical knowledge from experience. In addition, we used identifiers from the Disease Ontology68, OMIM, UMLS, and MeSH databases to manually classify and standardize therapeutic applications of Indian medicinal plants. To the best of IMPPAT's understanding, this is the initial effort to bridge the gap between traditional Indian medicine and the scientific terminology used in contemporary medicine. The usage of OMIM, UMLS, and MeSH identifiers is crucial in databases that provide information regarding gene–disease and disease–symptom associations. It will be simple to integrate these sources' data into IMPPAT going forward [65, 66].

12.2.1.3 Target Genes of Phytoconstituents

Targets for human phytochemical proteins indicated. Using the STITCH database (http://stitch1.embl.de/), we were able to identify the human target proteins of IMPPAT phytochemicals. The STITCH database has more anticipated chemical–protein interactions than any other source. From STITCH, only interactions between human proteins and phytochemical targets with a total score of 700 or above were revealed (Figure 12.5). The STITCH database, like the TCM-Mesh16 database for traditional Chinese medicine, was utilized to anticipate IMPPAT phytochemical–target human protein interactions [62].

Recent Updates and Addons and Future Use of IMPPAT Database IMPPAT 2.0 is a rich repository of Indian medicinal plants and their valuable properties, akin to a treasure trove brimming with medicinal secrets. This extensive database reveals a quantity of plants that is four times greater than its previous version, thereby doubling the known collection of therapeutic compounds contained within them. Furthermore, it explores further by establishing correlations between distinct botanical components and their distinct chemical compositions, so uncovering the exact origins of natural treatments. IMPPAT 2.0 is not merely a compilation of names and lists; rather, it is an advanced instrument for conducting scientific investigations. Every chemical is extensively characterized, including detailed 3D structures, potential targets within the human body, and even forecasts regarding their potential as therapeutics. It serves as a connection between old wisdom and modern medicine by establishing a correlation between traditional healing terminology and Western medical comprehension. However, this treasure chest still has undisclosed mysteries. Collecting and standardizing information regarding the historic usage of these plants in formulations continues to be a difficult task. Data sharing regulations pose an additional obstacle. Nevertheless, the future appears promising [67].

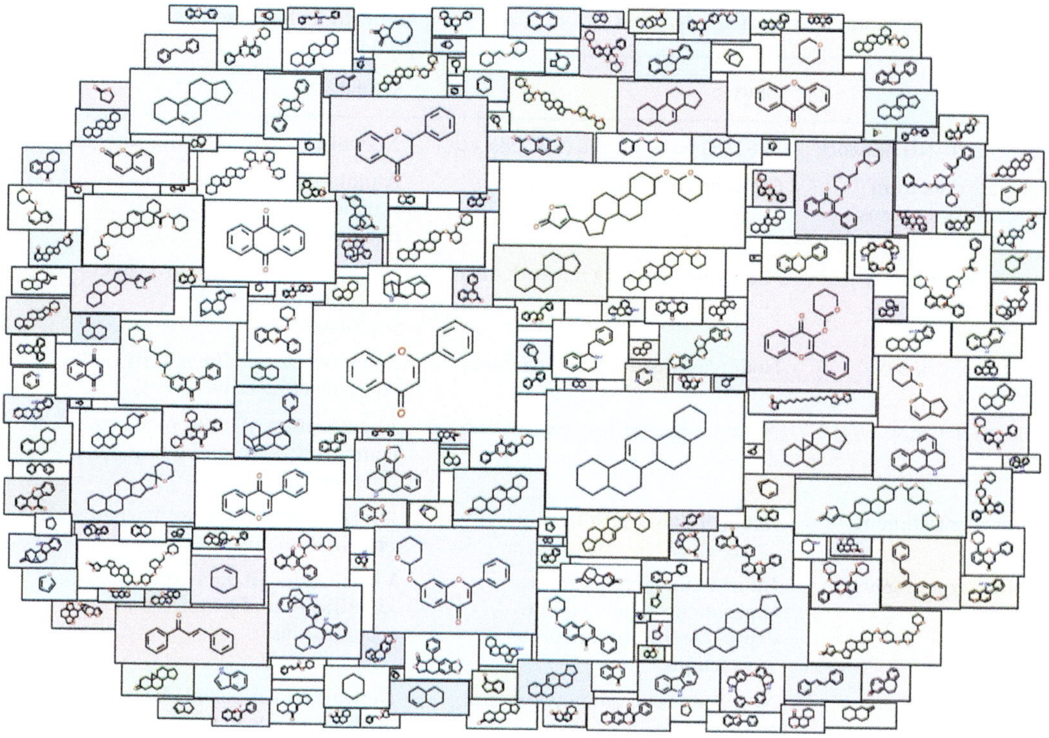

Figure 12.5 This image demonstrates the visualization of molecular clouds. When it comes to the phytochemicals of IMPPAT 2.0, the scaffold's frequency is directly proportional to the structure's size. *Source:* Adapted from Lobell et al. [57].

12.2.2 Network Analysis and Visualization

A Cytoscape StringApp for Visualizing and Analysing Proteinomics Data of Network pharmacology. Also used to visualize and analyze the compound and their related genes network, features are enlisted in Table 12.3.

The stringApp is built using the Cytoscape 3.6 App API and is implemented in Java. This app's primary purpose is to link Cytoscape to the databases and APIs of the STRING web service; second, it provides visualizations, including side panels and enrichment features, that are comparable to those on the STRING web server. Together, these two sites make it possible to use Cytoscape and its many applications to analyze the network and its associated data by integrating the STRING website into Cytoscape. As seen in the following use example, the clusterMaker2 app4 is helpful for clustering STRING networks. To retrieve data from databases and the internet, the stringApp bridge feature makes use of RESTful11 web service APIs. The application resolves protein and protein/compound query phrases to internal database identifiers using the STRING and STITCH API. To get the protein list for a given ailment, the DISEASES database API is called twice: first to convert the name of the sickness to an identifier, and again to acquire the name of the illness itself. For all three kinds of searches, StringApp lets users manually resolve confusing names. Handling PubMed queries was discussed in the preceding section. No matter what kind of query you run,

Table 12.3 This table represents the features of Cytoscape StringApp with its input and output.

Feature	Input	Output
Import STRING network	STRING network file (.txt, .xls, etc.)	Network visualization in cytoscape
Integrate data from associated databases	Gene/protein IDs or names	Annotations, pathways, functional data from associated databases
Visualize network properties	None (uses existing network data)	Network statistics, node degree distribution, clustering coefficients, etc.
Filter network by properties	Node/edge attributes, thresholds	Subnetwork containing only nodes/edges meeting criteria
Analyze network topology	None (uses existing network data)	Identification of hubs, modules, communities, potential functional relationships
Perform enrichment analysis	Gene/protein list or network cluster	Overrepresented GO terms, pathways, protein domains, etc.
Overlay experimental data	Information derived from experiments (such as gene expression or protein–protein interactions)	Visualization of data on the network, identification of patterns and relationships
Export network and analysis results	None (uses current network and results)	Network in various file formats, analysis results in tables or charts
STRING: protein query	Protein names or IDs	Network of interactions for the queried proteins
STRING: disease query	Disease name or ID	Network of proteins associated with the disease
STRING: PubMed query	PubMed ID	Network of proteins mentioned in the PubMed article
STITCH: protein/compound query	Protein or compound name or ID	Network of interactions between proteins and compounds
Expand network	Selected nodes in the current network	Expanded network including neighbors and interactions of selected nodes
Query for additional nodes	Node names or IDs	Adds specified nodes to the current network
Change confidence	New confidence score threshold	Updates network edges based on the new confidence level
Retrieve functional enrichment	Gene/protein list or network cluster	Overrepresented functional terms or pathways (requires additional plugins)
Filter enrichment	Enrichment results, selection criteria	Filtered list of enriched terms based on specified criteria
Draw charts	Network data or analysis results	Visual charts representing network properties or analysis outcomes

StringApp will be able to access the data you need from the dedicated PostgreSQL database over the web service API. To facilitate network growth, confidence cutoff decrease, or node addition, the second API retrieves data from nodes or edges [68].

12.2.3 Applications of Network Pharmacology in Drug Discovery

The current drug development paradigm involves creating highly selective ligands that interact with therapeutic targets. However, many effective drugs modulate many proteins instead of one. Recent systems biology discoveries have revealed amazing resilience in living organisms and extensive connectivity among biological components. Highly specialized compounds may be less successful in clinical settings than drugs that target several biological targets, which is undesirable [69]. The growing understanding of polypharmacology has major implications for resolving drug development's main failures, efficacy, and toxicity. Network biology and polypharmacology may expand drug development targets [70]. However, the need for creative methods to verify combinations of targets and improve structure–activity links while keeping drug-like properties hinders polypharmacology's logical progress. These advances are laying the groundwork for network pharmacology, a drug discovery paradigm change. Understanding the "drug–target–gene–disease" network is crucial to developing the network pharmacology system. We can now discover or predict how individual or mixed Chinese medicines treat diseases through their network associations. After evaluating many literature sources, the database was searched for chemical ingredients in traditional Chinese medicines and compound medications. Database searches revealed these chemical components' targets. After finding the drug's composition and target, the disease's gene target was found by reviewing the literature retrieval database again [71]. We created the "drug–target–gene–disease" relationship network through integration and analysis. Afterward, a network pharmacology system tailored to individual diseases was developed. The pharmaceutical industry has invested much in R&D, yet there has been a dramatic decline in new treatment alternatives, leading some to wonder if it is useful to focus drug discovery efforts on a single target. In these cases, the novel approach to drug discovery that is network pharmacology is useful. In contrast to conventional methods, they place an emphasis on the capacity of medications to target several disease-related proteins or networks. The development of drug–target disease network models is facilitated by bioinformatics and high-throughput screening. Drug effects on biological networks can be better understood with the use of these techniques, which assess drug–target model interactions. New insights from network biology have shed light on the mysteries of route interactions. As a result, network pharmacology is becoming increasingly important in the fight against the most lethal diseases and conditions [6]. By analyzing protein–protein interactions, network pharmacology is able to determine how a disease manifests. Researchers are now studying an increasing variety of complex disorders using multiomics and computational methods to accurately record human metabolic responses. In this chapter, we take a look at network pharmacology in the biomedical field [11, 72].

12.3 Ayurveda and Traditional Indian Medicine

12.3.1 Overview of Ayurveda and Its Complex Formulations

An increasing number of individuals are choosing Ayurveda over allopathic treatments due to its cost-effectiveness and reduced incidence of adverse effects. The significance of organic resource-based goods as supplies for pharmacotherapeutic and complementary and alternative therapies is increasing in contemporary times [73]. The extensive Ayurvedic pharmacopoeia, which has been used for centuries in different parts of India, contains minerals and metals compounds known for their therapeutic properties [74]. According to Ayurveda, any material could have medicinal properties. Mineral and metal compositions, such Lauha Kapibadkva, Rasayana, and Bhasma, are mentioned in Ayurveda. Avleha, Asavaristra, Grafa Churena, and Taila are botanical remedies. Is the use of herbs, the exclusive focus of Ayurveda, or may other therapeutic modalities that adhere to Ayurvedic principles also make use of herbs? The moment has come to deliver information that is both clear and precise. In order to advance Ayurveda, each task or project calls for a unique study methodology [74]. The four primary fields of study are literacy, fundamental science, medicines, and clinical research. The utilization of natural drug development products has been impeded by the relatively limited quantity of active constituents and the challenges associated with the extraction and isolation procedures in natural medicine. Hence, it is preferable to discover more efficient methods of extracting and isolating bioactive substances present in the natural environment. The active constituents of natural medicines are chemical compounds that possess medicinal qualities [75].

Arista and Asava: Herbs that are powdered or decocted are soaked in sugar or jaggery for a set period of time to make these sweets. Herbal active ingredients can be extracted through the slow fermentation process that produces alcohol. To make necessary liquids, the ancients suggested fermenting juices or decoctions. In the course of manufacturing, ethyl alcohol is generated. Sadhana Kalpana dose kinds (Asava, Arishta, and Kanji) are utilized in daily Ayurvedic treatment [76, 77]. The three-part mixture is then homogenized by turbidity. After the bhasma has been properly prepared, it should be rubbed between the index finger and thumb. The pulverized bhasma particles float on water and settle into finger ridges. Borax, jaggery, ghee, honey, and fruits of the Abrus precatorious Linn tree should never be allowed to return to their original mineral or metal condition once heated [78]. The main methods for preparing bhasma are shodhana, which means purification, and maarana. The terms "vira," "vati," and "gutika" describe the pill forms used in Ayurveda. Minerals, animal products, and herbal medicines go into its composition. The circular dose is known as gutika. This looks like the medicines that people take today. While Ghanavati is best chewed, Vati is best taken by tablet. The preparation methods of Agnisadhya and Anagnisadhya vati are part of Ayurveda. Sugar, jaggery, or guggulu are heated to create Agnisadhya vati. Anagnisadhya vati is rendered inert when heated [79, 80]. Avaleha, Churna products sold by Avaleha are semisolid. Sugar, jaggery, or sugar candy is added to the boiling decoction to make it sweeter. Once the mixture has cooled and been well mixed, honey is added. The ingredients and times for medication are as follows: honey, gingelly oil, salts, alkalis, milk, milk derivatives, and bhasma. To make ghrita, you just mix the decoction with ghee. Heating ghrita to 105–118 1C removes water [81]. The exudates of the Commiphora mukul plant are the source of guggulu, also known as Taila Oil. When the exudates are wrapped in cloth and cooked with Triphala or other herbs, they eventually make it into the solution. Boil until hard, and then let it dry. After drying, add ghee for a waxy paste (Table 12.4) [82].

12.3.2 Diversity of Ingredients and Bioactive Compounds in Ayurvedic Medicines

Phytobioactive substances have been utilized as traditional medicine for centuries to treat various disorders worldwide. Multiple studies have extensively documented their medicinal potential since they are widely regarded as valuable natural sources for developing novel medications that exhibit enhanced effectiveness and compatibility with living organisms. Recent publications are progressively incorporating more attributes to their already extensive range of medicinal value [83, 84]. In contemporary lifestyle, there has been a persistent rise in oxidative stress, leading to various illnesses including diabetes mellitus (DM), cancer, and cardiovascular issues. Improper utilization

Table 12.4 This table lists the many ayurvedic formulations, along with the process of their production, the medicinal advantages of these formulations, and the versions that are commercially available.

Name of process	Process of formulation	Medicinal benefits	Example of marketed formulations
Swarasa	Extraction of fresh juice from plants	Retains active principles, easily absorbed, used for various ailments based on plant properties	Amla juice for immunity, Giloy juice for fevers
Kwatha	Decoction of herbs in water	Concentrates water-soluble active constituents, good for digestive and respiratory issues	Triphala decoction for epresent, Neem decoction for skin conditions
Churna	Powdered herbs or minerals	Easy to administer, long shelf life, used for internal and external applications	Trikatu powder for digestion, Ashwagandha powder for energy
Lepa	Paste of herbs for external application	Localized action, reduces inflammation, soothes pain	Turmeric paste for wounds, Sandalwood paste for relaxation
Taila	Medicated oil prepared by infusing herbs in carrier oil	Penetrates deep tissues, nourishes skin, used for massage and pain relief	Dhanwanthari Taila for joint pain, Mahanarayan Taila for muscle weakness
Ghrita	Medicated ghee prepared by boiling herbs in ghee	Nourishes body tissues, enhances digestion, supports specific conditions	Ashwagandha Ghrita for memory, Chyawanprash for immunity
Asava/ Arishta	Fermented decoction of herbs with honey or jaggery	Enhances bioavailability of herbs, promotes gut health, supports various systems	Ashwagandha Arishta for stress management, Dashmoolarishta for digestion
Bhasma	Calcination of minerals or metals using herbal fire	Increases potency, removes impurities, used for specific conditions like anemia, infertility	Abhrak Bhasma for respiratory issues, Swarna Bhasma for mental clarity
Vatika	Pills made from powdered herbs and binding agents	Easy to administer, controlled dosage, long shelf life	Triphala Vati for constipation, Ashwagandha Vati for stress
Rasayan	Rejuvenating formulations for long life and health	Antioxidant, adaptogenic, promotes tissue repair, slows aging	Chyawanprash, Chyavanashvati

of antibiotics leads to the development of antibiotic resistance, which presents significant health concerns, particularly within hospital settings [85]. Artificial medications are inadequate to address these issues. It is imperative to investigate natural phytobioactive substances for the treatment of contemporary illnesses. A number of medical issues may be amenable to polyphenols, alkaloids, and terpenoids, including cancer, inflammation, ulcers, diabetes, platelet aggregation, microbial resistance, tumors, and oxidative stress (Table 12.5) [86, 87].

12.3.3 Indian Traditional Medicines Recently Breakthroughs in Using Different Approaches to Treat a Variety of Diseases

Indian traditional remedies have recently achieved significant advancements in disease treatment through several methodologies. By combining ancient wisdom with current science, these therapies utilize substances such as phenolics and terpenoids. Formulations rich in phenolic compounds exhibit antioxidant, anti-inflammatory, and cardiovascular advantages. Terpenoids, ranging from

Table 12.5 The several classes of phytoconstituents were listed in a table format.

Photoactive compounds	Subclass/Type	Role in treating diseases
Phenolics		
Simpler phenols	Phenolic acids and ketones	Antioxidant, anti-inflammatory, cardiovascular health
	Phenyl propanoids; coumarins, benzofuran, chromones, chromenes	Anti-inflammatory, antimicrobial, anticancer
	Phenolic quinines: benzoquinone, naphthaquinones, anthraquinones	Antimicrobial, antifungal, wound healing
Polymeric phenolics	Lignins	Antioxidant, anti-inflammatory
	Melanins	Skin protection, UV radiation absorption
	Tannins (condensed salts or gallo/epitannins)	Astringent, antidiarrheal, cardiovascular health
	Xanthones	Antioxidant, anti-inflammatory
	Stilbenoid	Antioxidant, anti-inflammatory, cardiovascular health
Flavonoids	C15 heterocyclic nucleus of flavones	Antioxidant, anti-inflammatory, immune system support
Terpenoids	Monoterpenoids	
	Iridoids	Anti-inflammatory, antimicrobial, digestive health
	Sesquiterpenoids	Anti-inflammatory, respiratory health
	Lactones	Sedative, antianxiety, digestive health
	Diterpenoid	Anti-inflammatory, anticancer
	Triterpenoid saponins	Immune system support, anti-inflammatory
	Steroid saponins	Cardiovascular health, anti-inflammatory
	Cardenolides and bufadienolides	Cardiotonic, antiarrhythmic
	Phytosterols, curcum bitacums	Anti-inflammatory, cardiovascular health
	Nortriterpenoids	Antimicrobial, anticancer
	Carotenoids	Antioxidant, eye health, immune system support
Glycosides		
Steroidal glycosides	Terpenoid aglycone	Cardiotonic, antiarrhythmic
Anthracene glycosides	Phenolic aglycone reduced anthraquinones	Laxative, anti-inflammatory
Saponin glycosides	Terpenoid aglycone	Immune system support, anti-inflammatory

Table 12.5 (Continued)

Photoactive compounds	Subclass/Type	Role in treating diseases
Cyanogenic glycosides	Hydrocyanic acid aglycone	Cytotoxic, anticancer
Flavonoid glycosides	Phenolic aglycone 2 phenylbenzopyrenes	Antioxidant, anti-inflammatory, cardiovascular health
Resinous glycosides	Resin aglycone	Anti-inflammatory, wound healing
Volatile oils		
	Terpenoids in association with alcohols, phenols, aldehydes, esters	Respiratory health, antimicrobial, aromatherapy
Gums	Amorphous polysaccharides	Gastrointestinal health, wound healing
Resins	Complex organic compounds	Anti-inflammatory, wound healing

monoterpenoids to carotenoids, provide anti-inflammatory, antibacterial, and immunological support properties [88–90]. Glycosides, volatile oils, gums, and resins are essential components that have significant impacts in several fields, ranging from promoting cardiovascular well-being to facilitating the process of wound healing. The incorporation of ancestral wisdom into modern medicine demonstrates the capacity for comprehensive healing and tackles a wide range of illnesses, representing notable progress in the discipline. Nature is widely acknowledged as a reservoir of innovative chemical compounds [91, 92]. It is critical to improve traditional medicine procedures, particularly the preservation of historical knowledge and the optimization of herbal medicine's use in human healthcare, to guarantee the efficacy, safety, and quality of traditional medicine. The following must be prioritized: postmarketing surveillance, consumer awareness, clinical risk assessment, validation methods, standardization, authentication, quality control, and regulatory elements of herbal medical products to guarantee their safe and effective usage [42, 93].

12.4 Network Pharmacology in Herbal Remedies

12.4.1 Application of Network Pharmacology in Herbal Drug Discovery

12.4.1.1 Cancer

Cancer is a major cause of illness and death worldwide. Multiple targets and carcinogenic signaling cascades must be regulated to effectively treat cancer. Similar to cancer, single-targeted drug discovery fails for severe systemic disorders with complex biological systems. Network pharmacology approaches, in contrast to more conventional drug discovery strategies, aim to inhibit disease progression by interacting with several protein or network components. When conducting clinical trials, network pharmacology is useful for addressing issues such as patient heterogeneity and disease [94, 95]. Cancers differ depending on whether therapeutic targets are present or not. Therapy choices are limited for heterogeneous populations with various malignancies. Therefore, in order to alter drug development and enhance our understanding of mechanisms of action, approaches based on network pharmacology are required. Cancer treatment therapy and medication development can both be accelerated by network pharmacology [96, 97]. There are genes that have been uncovered using network analysis that could be therapeutic targets for the reduction and treatment of cancer. Receptor 2 for human epidermal growth factor is present in HER2-positive breast cancers. Cancer cell proliferation is enhanced by protein. Amplified HER2 genes are found in 20% of breast cancer cases. To investigate the impact of Yanghe decoction on HER2-positive breast cancer, Zeng et al. employed network pharmacology [98]. Anticancer components like quercetin, luteolin, and naringenin may be present in Yanghe decoction. It was proven that Yanghe decoction works in tandem with other molecules to treat HER2-positive breast cancer. The active components and critical pathways of the Shenqi-Yi-zhu decoction were investigated by Zhen et al. for the purpose of treating stomach cancer. Network pharmacology and *in vitro* tests demonstrate the effective and advantageous molecular mechanism of Shen-qi-Yi-zhu decoction [99]. The study found that Shen-qi-Yi-zhu decoction inhibits the PI3K/AKT/mTOR pathway, which may prevent cancer. The PI3K/AKT/mTOR pathway is present in most cancers and affects cancer biology. Thus, herb/herbal formulations were crucial to antitumor research [100]. The most common cancer killer is colorectal cancer, which is quiet and lethal. Liu et al. used network pharmacology to determine how *Hedyotis diffusa* fights colorectal cancer. *Hedyotis diffusa* may cure colorectal cancer by targeting tumor growth signaling pathways, according to network analysis. This explains *Hedyotis diffusa*'s anti-colorectal cancer properties [101]. *Hedyotis diffusa* was used by Song et al. to uncover prostate cancer's multi-target pharmacological mechanism. To find the medicinal properties of herbs and herbal combinations, network pharmacology is a great tool. Bing et al. investigated the efficacy of Fuzheng Kangai's lung cancer treatment using bioinformatics and network pharmacology. The molecular mechanisms of the Kushen injection's effects on lung cancer treatment were studied by Meng et al. using molecular docking and network pharmacology. Finally, results from network pharmacology provide fresh insight into the efficacy of herb/herbal combinations in cancer treatment [102].

12.4.1.2 Cardiovascular Diseases (CVDs)

Cardiovascular and cerebral vascular diseases are significant worldwide health issues. Botanical drugs derived from traditional medicinal plants provide multiple advantages for the treatment of cardiovascular and cerebrovascular diseases (CCVDs). Nevertheless, the molecular mechanisms underlying the therapeutic effects of medicinal plants on CCVDs remain elusive. An innovative approach in systems pharmacology, network pharmacology helps us understand the pharmacological mechanisms of medicinal plants for CCVDs treatment by integrating pharmacokinetic screening, target identification, and network analysis [103]. This approach introduces an innovative methodology for investigating the mechanism by which herb/herbal formulations act against CCVDs. Yang et al. conducted a study on network pharmacology, focusing on the active compounds found in Ginkgo biloba leaf. They explored the potential targets and pathways that may be used to treat CCVD. This study provides a foundation for future research in this field. Studies suggest that Ginkgo biloba leaves have the potential to safeguard against CCVDs by modulating several biological pathways and targeting specific processes. Ginkgo biloba leaves are effective in treating CCVDs, according to this study, which also introduces a new way to find therapeutic possibilities derived from plants. To combat stroke, Ren et al. used herbal treatments [104]. Their investigation unveiled bioactive compounds that efficiently addressed stroke by specifically targeting genes associated with the condition. Tao et al. employed network pharmacology to predict the active components and possible targets of the Radix Curcumae recipe for the treatment of CCVDs. The study provides a thorough explanation of how the Radix Curcumae recipe works to treat CCVD and finds areas where herbal medicine could be improved in the future [105]. For the purpose of treating CCVD, Wang et al. investigated the active components of *Salvia miltiorrhiza* Burge and *Carthamus tinctorius* L. *Salvia miltiorrhiza* Burge and *Carthamus tinctorius* L. may increase blood flow to the brain by widening blood vessels, which in turn decreases neurotoxic damage and protects brain tissue from free radical damage, according to the study [106]. Academics can better understand the Yangxinshi tablet's complex mechanism with the use of network-based analysis. Finding the chemical and pharmacological components of other herbal compositions can also benefit from this method [107].

12.4.1.3 Diabetes Mellitus (DM)

Global DM is spreading quickly. Diabetics are more susceptible to acute diseases. Network pharmacology using natural compounds may treat DM and address the difficulties stated above. Wang et al. used network pharmacology to find the most powerful Astragaloside IV active components for type 2 diabetes [108]. Docking and network analysis can reduce screening costs and provide a complete understanding of drug development and discovery processes, according to recent research. Gu et al. used network pharmacology and molecular docking to explain Tangminling pills' T2DM therapy mechanism. The compound–compound network and compound target network showed that over 100 of the 667 compounds in the formula can target 37 T2DM proteins. Tangminling tablets include essential components, some of which have been published [109]. Procyanidin C1, Rheidin A, Rheidin C, Sennoside C, and Dihydrobaicalin have been studied as diabetes treatments. *Sorghum bicolor* is a prospective source of antidiabetic compounds due to its bioactive components. Oh et al. used network pharmacology to find diabetes treating *Sorghum bicolor* active chemicals. *Sorghum bicolor* activates PPAR signaling pathways to lower T2DM severity, according to the study [110, 111]. Four chemicals in *Sorghum bicolor* – alpha-sitosterol, 25-*oxo*-27-norcholesterol, campesterol, and propyleneglycol monoleate – are associated with the plant's antidiabetic effects. Zhou et al. investigate the mechanism of action of Xiao Ke Yin Shui's T2DM treatment using network pharmacology. Xiao Ke Yin Shui's formula significantly impacts tumor necrosis factor, phosphatidylinositol 3-kinase, and protein kinase B, according to the study. Reduced insulin resistance and antidiabetic properties are two of the Xiao Ke Yin Shui chemical's synergistic therapeutic benefits [107].

12.4.2 Screening Pharmacological Efficacy of Herbal Remedies

In developing nations, people often turn to herbal remedies, which are mixtures of plants that contain a number of different pharmacologically active chemicals. Because of advantageous "synergistic" interactions, complex combinations are more effective than individual drugs at modulating the activity of target networks linked with disease characteristics. The methods that produce their synergistic or antagonistic combinations and interactions are discovered and explained in this chapter through the application of systematic, systems pharmacology, network analysis, and machine learning methodologies. Here we take a look at a variety of studies that have investigated synergistic interactions, all using various approaches that stem from the network pharmacology paradigm [89, 90, 112]. The chapter provides a thorough analysis of evaluation mechanisms that expedite innovation in botanical research. Evaluating the pharmacological effectiveness of herbal treatments is an essential procedure in contemporary medicine. Given the increasing interest in natural remedies, it is imperative to conduct thorough scientific evaluations. Scientists utilize many techniques, including *in vitro* and *in vivo* investigations, to evaluate the bioactivity of herbal substances. *In vitro* investigations investigate molecular interactions, enzyme activity, and cellular responses, offering valuable insights into potential treatment pathways [112]. *In vivo* studies utilize animal models to enable researchers to directly evaluate the comprehensive physiological impacts and potential harm. In addition, clinical trials evaluate the effectiveness of herbal treatments in humans, establishing their safety, dose, and efficacy. Advanced methodologies such as chromatography and mass spectrometry facilitate the identification and quantification of active substances [113]. Pharmacological screening encompasses multiple factors, such as anti-inflammatory, antioxidant, antimicrobial, and immunomodulatory properties. By combining ancient knowledge with modern pharmacology, herbal therapy is able to be grounded in evidence-based practices. The rigorous screening process is essential for ensuring the safety and effectiveness of herbal treatments, enabling their incorporation into mainstream healthcare, and providing a scientific basis for their utilization in the treatment of different medical diseases [114].

12.4.3 Utilizing Network Pharmacology to Understand Complex Diseases

Network pharmacology is an advanced interdisciplinary field that has become a strong method for understanding the complicated mechanisms behind complex disorders. This novel approach integrates concepts from network science, pharmacology, bioinformatics, and systems biology to provide a comprehensive comprehension of the relationships among medications, targets, and diseases. Traditionally, drug development mostly concentrated on specific molecular targets within the traditional paradigm. Nevertheless, numerous diseases, particularly those that are intricate, encompass a multitude of interrelated pathways. Network pharmacology acknowledges the complex nature of biological systems and highlights the importance of investigating the comprehensive interactions between medications and the various networks within the body. An essential component of network pharmacology is in its capacity to elucidate the polypharmacology of medicines. Instead of acting on a single target, numerous therapies exert their effects on multiple targets [115]. Network pharmacology utilizes the integration of extensive omics data, including genomes, proteomics, and metabolomics, to identify prospective drug targets within intricate networks. The process entails the construction of biological networks that depict the associations among genes, proteins, and metabolites that are involved in illnesses. These networks might focus either on diseases or on drugs. Condition networks facilitate the integration of molecular data linked to a particular condition, aiding in the identification of crucial pathways and prospective targets for therapeutic intervention. Conversely, drug networks depict the connections between medications and the specific proteins they target [114]. Examining these networks yields valuable understanding of the biological underpinnings of diseases and the effects of drugs. By identifying network hubs, which are nodes with high connectivity, and essential pathways, researchers can prioritize prospective treatment targets. Moreover, network pharmacology assists in forecasting the adverse effects and possible off-target impacts of therapies, hence facilitating the advancement of more secure medications. The incorporation of traditional medicine into network pharmacology is remarkable. Conventional medical systems frequently include herbal treatments that consist of multiple components (Figure 12.6). Network pharmacology enables researchers to comprehend the combined effects of different chemicals inside these formulations. Through the creation of herb–drug networks, researchers can uncover the intricate connections between herbal components and pathways related to diseases. Furthermore, network pharmacology enables the investigation of the idea of "network target" [116, 117]. Instead of exclusively concentrating on particular targets, this strategy takes into account the entire network linked to a disease. Network-targeted medicines, which modulate several targets simultaneously, have the potential to be more efficacious and demonstrate reduced side effects in comparison to interventions that target a single entity. Thus, network pharmacology signifies a fundamental change in the approach to drug

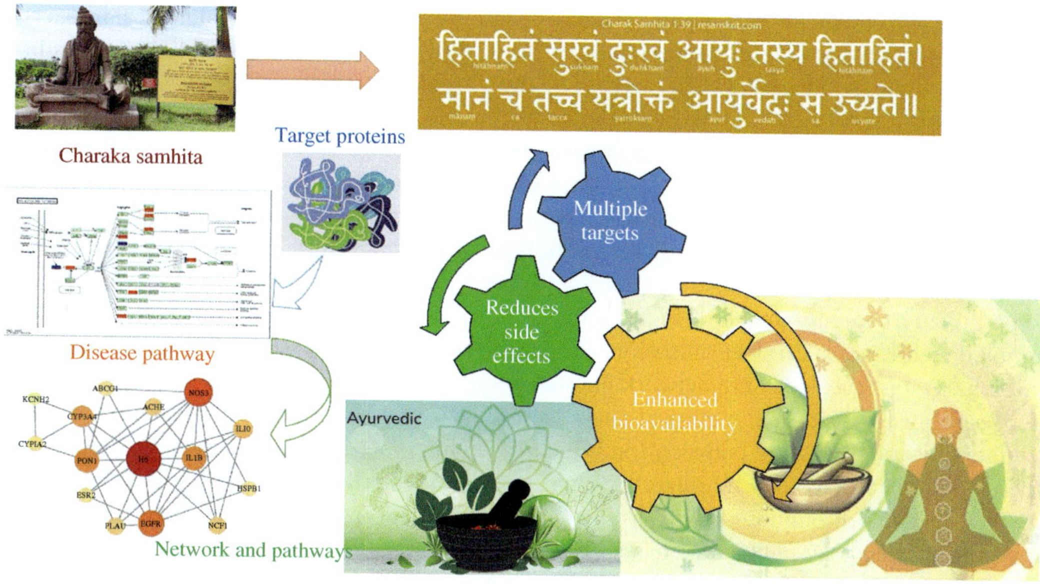

Figure 12.6 Developing the current *in silico* methods by integrating it with ancient texts of Indian traditional medicines to treat various deadly diseases.

discovery and comprehension of diseases. The holistic approach takes into account the complex interconnections within biological systems, which allows for the creation of therapies that are both more effective and safer. Network pharmacology, via the integration of various data sources and the acceptance of the intricacies of diseases, provides a complete framework for understanding complex diseases and expedites the identification of new medicines [106, 116].

12.5 Conclusion and Future Prospects

To summarize, the incorporation of Indian traditional medicines into the process of drug development via network pharmacology represents a notable advancement toward comprehensive and efficient healthcare remedies. Ancient pharmaceutical systems, firmly grounded in traditional wisdom, provide a rich collection of varied substances with therapeutic potential. Network pharmacology offers a powerful framework for comprehending the intricate relationships inside biological systems, enabling a thorough knowledge of the diverse characteristics of traditional medicines. Network pharmacology has shed light on the multitarget effects and synergistic interactions of plant-derived chemicals, including polyphenols, terpenoids, and glycosides, through a thorough investigation. This technique not only confirms the practical information contained in traditional medications but also reveals new ways to treat complicated conditions. The integration of old and new knowledge systems facilitates a fundamental change in the approach to drug discovery, placing emphasis on a comprehensive comprehension of diseases and therapies. By developing extensive networks that capture the interactions between medications, targets, and diseases, network pharmacology aids researchers in finding critical pathways, possible drug targets, and forecasting side effects. This understanding facilitates the creation of safer and more effective drugs.

The future potential of Indian traditional remedies in drug development through network pharmacology is highly exciting and diverse. Realizing these opportunities necessitates cooperative endeavors among indigenous healers, scientists, and pharmaceutical investigators. Here are some crucial paths for investigation: Omics data integration: The progress in genomes, proteomics, and metabolomics will improve the accuracy of network pharmacology. By combining omics data with traditional pharmaceutical expertise, a more intricate comprehension of the molecular mechanisms involved can be achieved. Standardization and quality control are essential for guaranteeing consistency and dependability in drug research. It involves establishing uniformity in traditional formulations and implementing strict mechanisms to monitor and evaluate the quality of the drugs. Utilizing contemporary analytical methods can enhance the uniformity and security of conventional medications. Clinical validation is crucial for confirming the effectiveness and safety of traditional drugs revealed using network pharmacology. Although *in silico* and *in vitro* researches offer useful insights, thorough clinical trials are necessary for this validation. This shift will be facilitated by collaborations between traditional medicine practitioners and modern healthcare experts. Ethnopharmacological research, which concentrates on the traditional knowledge present in diverse cultures throughout India, has the potential to reveal exceptional medicinal substances. Network pharmacology can be used to analyze the molecular pathways that are responsible for the effects of these traditional treatments.

Personalized Medicine: The customized approach of conventional medicine is well suited to the emerging concept of personalized medicine. Customizing therapies according to an individual's genetic composition and lifestyle variables, under the guidance of network pharmacology, has significant promise for improving treatment results. The interaction between Indian traditional medicines and network pharmacology gives potentially exciting prospects in drug discovery. This multidisciplinary approach not only acknowledges and values historic healing traditions but also advances them to the forefront of modern medicine, facilitating a seamless blending of the traditional and contemporary practices for the betterment of world healthcare.

References

1 Ellingson, S.R., Smith, J.C., and Baudry, J. (2014). Polypharmacology and supercomputer-based docking: opportunities and challenges. *Molecular Simulation* 40 (10–11): 848–854.

2 Hopkins, A.L. (2007). Network pharmacology. *Nature Biotechnology* 25 (10): 1110–1111.

3 Hopkins, A.L. (2008). Network pharmacology: the next paradigm in drug discovery. *Nature Chemical Biology* 4 (11): 682–690.

4 Patwardhan, B., Warude, D., Pushpangadan, P., and Bhatt, N. (2005). Ayurveda and traditional Chinese medicine: a comparative overview. *Evidence-based Complementary and Alternative Medicine* 2: 465–473.

5 Mukheriee, P.K. and Wahile, A. (2006). Integrated approaches towards drug development from Ayurveda and other Indian system of medicines. *Journal of Ethnopharmacology* 103 (1): 25–35.

6 Atanasov, A.G., Zotchev, S.B., Dirsch, V.M. et al. (2021). Natural products in drug discovery: advances and opportunities. *Nature Reviews Drug Discovery* 20 (3): 200–216.

7 Choo, M.Z.Y. and Chai, C.L.L. (2023). Chapter Two—The polypharmacology of natural products in drug discovery and development. In: *Annual Reports in Medicinal Chemistry* (ed. K.-H. Altmann), 55–100. 61: Academic Press.

8 Mohs, R.C. and Greig, N.H. (2017). Drug discovery and development: role of basic biological research. *Alzheimer's & Dementia (New York, NY)* 3 (4): 651–657.

9 Kiriiri, G.K., Njogu, P.M., and Mwangi, A.N. (2020). Exploring different approaches to improve the success of drug discovery and development projects: a review. *Future Journal of Pharmaceutical Sciences* 6 (1): 27.

10 Vora, L.K., Gholap, A.D., Jetha, K. et al. (2023). Artificial intelligence in pharmaceutical technology and drug delivery design. *Pharmaceutics* 15 (7): 1916.

11 Seyhan, A.A. (2019). Lost in translation: the valley of death across preclinical and clinical divide – identification of problems and overcoming obstacles. *Translational Medicine Communications* 4 (1): 18.

12 de Vrueh, R.L.A. and Crommelin, D.J.A. (2017). Reflections on the future of pharmaceutical public-private partnerships: from input to impact. *Pharmaceutical Research* 34 (10): 1985–1999.

13 Glicksberg, B.S., Li, L., Chen, R. et al. (2019). Leveraging big data to transform drug discovery. *Methods in Molecular Biology (Clifton, NJ)* 1939: 91–118.

14 Takebe, T., Imai, R., and Ono, S. (2018). The current status of drug discovery and development as originated in United States Academia: the influence of industrial and academic collaboration on drug discovery and development. *Clinical and Translational Science* 11 (6): 597–606.

15 Cheng, F., Ma, Y., Uzzi, B., and Loscalzo, J. (2020). Importance of scientific collaboration in contemporary drug discovery and development: a detailed network analysis. *BMC Biology* 18 (1): 138.

16 Anderson, M. (2021). How the COVID-19 pandemic is changing clinical trial conduct and driving innovation in bioanalysis. *Bioanalysis* 13 (15): 1195–1203.

17 Li, G., Hilgenfeld, R., Whitley, R., and De Clercq, E. (2023). Therapeutic strategies for COVID-19: progress and lessons learned. *Nature Reviews Drug Discovery* 22 (6): 449–475.

18 Nogales, C., Mamdouh, Z.M., List, M. et al. (2022). Network pharmacology: curing causal mechanisms instead of treating symptoms. *Trends in Pharmacological Sciences* 43 (2): 136–150.

19 Baryakova, T.H., Pogostin, B.H., Langer, R., and McHugh, K.J. (2023). Overcoming barriers to patient adherence: the case for developing innovative drug delivery systems. *Nature Reviews Drug Discovery* 22 (5): 387–409.

20 Wang, X., Hu, Y., Zhou, X., and Li, S. (2022). Network pharmacology and traditional medicine: setting the new standards by combining *in silico* and experimental work. *Frontiers in Pharmacology* 13: 1002537.

21 Nguyen, C., Naunton, M., Thomas, J. et al. (2021). Availability and use of number needed to treat (NNT) based decision aids for pharmaceutical interventions. *Exploratory Research in Clinical and Social Pharmacy* 2: 100039.

22 Ingber, D.E. (2022). Human organs-on-chips for disease modelling, drug development and personalized medicine. *Nature Reviews Genetics* 23 (8): 467–491.

23 Klein, J. (2022). Improving the reproducibility of findings by updating research methodology. *Quality & Quantity* 56 (3): 1597–1609.

24 Gao, Q., Xu, L., and Cai, J. (2021). New drug targets for hypertension: a literature review. *Biochimica et Biophysica Acta (BBA) - Molecular Basis of Disease* 1867 (3): 166037.

25 Noor, F., Rehman, A., Ashfaq, U.A. et al. (2022). Integrating network pharmacology and molecular docking approaches to decipher the multi-target pharmacological mechanism of *Abrus precatorius* L. acting on diabetes. *Pharmaceuticals* 15 (4): 414.

26 Zaman, M.A., Awais, N., Satnarine, T. et al. (2023). Comparing triple combination drug therapy and traditional monotherapy for better survival in patients with high-risk hypertension: a systematic review. *Cureus* 15 (7): e41398.

27 Kondej, M., Stępnicki, P., and Kaczor, A.A. (2018). Multi-target approach for drug discovery against schizophrenia. *International Journal of Molecular Sciences* 19 (10): 3105.

28 Talevi, A. (2015). Multi-target pharmacology: possibilities and limitations of the "skeleton key approach" from a medicinal chemist perspective. *Frontiers in Pharmacology* 6: 205.

29 Löscher, W. (2021). Single-target versus multi-target drugs versus combinations of drugs with multiple targets: preclinical and clinical evidence for the treatment or prevention of epilepsy. *Frontiers in Pharmacology* 12: 730257.

30 Parmar, G. (2022). Nanotechnology in health care. *Medknow* 10 (2): 51–52.

31 Parmar, G., Chudasama, J.M., Aundhia, C. et al. (2024). A comprehensive review on eco-friendly synthesized gold nanoparticles and its advantages. *Nanotechnology and In Silico Tools* 169–182.

32 Valcourt, D.M., Harris, J., Riley, R.S. et al. (2018). Advances in targeted nanotherapeutics: from bioconjugation to biomimicry. *Nano Research* 11: 4999–5016.

33 Church, D.L. (2004). Major factors affecting the emergence and re-emergence of infectious diseases. *Clinics in Laboratory Medicine* 24 (3): 559–586. v.

34 Zhou, Z., Chen, B., Chen, S. et al. (2020). Applications of network pharmacology in traditional Chinese medicine research. *Evidence-based Complementary and Alternative Medicine* 2020: 1646905.

35 Vitali, F., Mulas, F., Marini, P., and Bellazzi, R. (2013). Network-based target ranking for polypharmacological therapies. *Journal of Biomedical Informatics* 46 (5): 876–881.

36 Naik, B., Mattaparthi, V.S.K., Gupta, N. et al. (2022). Chemical system biology approach to identify multi-targeting FDA inhibitors for treating COVID-19 and associated health complications. *Journal of Biomolecular Structure & Dynamics* 40 (19): 9543–9567.

37 Zhao, Y., Gericke, O., Li, T. et al. (2023). Polypharmacology-labeled molecular networking: an analytical technology workflow for accelerated identification of multiple bioactive constituents in complex extracts. *Analytical Chemistry* 95 (9): 4381–4389.

38 Lee, J.H. (2003). Methicillin (Oxacillin)-resistant *Staphylococcus aureus* strains isolated from major food animals and their potential transmission to humans. *Applied and Environmental Microbiology* 69 (11): 6489–6494.

39 Barnes, E.C., Kavanagh, A.M., Ramu, S. et al. (2013). Antibacterial serrulatane diterpenes from the Australian native plant Eremophila microtheca. *Phytochemistry* 93: 162–169.

40 Cho, D.Y., Kim, Y.A., and Przytycka, T.M. (2012). Chapter 5: Network biology approach to complex diseases. *PLoS Computational Biology* 8 (12): e1002820.

41 Turnbull, L., Hütt, M.-T., Ioannides, A.A. et al. (2018). Connectivity and complex systems: learning from a multi-disciplinary perspective. *Applied Network Science* 3 (1): 11.

42 Fokunang, C.N., Ndikum, V., Tabi, O.Y. et al. (2011). Traditional medicine: past, present and future research and development prospects and integration in the National Health System of Cameroon. *African Journal of Traditional, Complementary, and Alternative Medicines: AJTCAM* 8 (3): 284–295.

43 Petrovska, B.B. (2012). Historical review of medicinal plants' usage. *Pharmacognosy Reviews* 6 (11): 1–5.

44 Eshete, M.A. and Molla, E.L. (2021). Cultural significance of medicinal plants in healing human ailments among Guji semi-pastoralist people, Suro Barguda District, Ethiopia. *Journal of Ethnobiology and Ethnomedicine* 17 (1): 61.

45 Harakeh, S., Niyazi, H.A., Niyazi, H.A. et al. (2024). Integrated network pharmacology approach to evaluate bioactive phytochemicals of *Acalypha indica* and their mechanistic actions to suppress target genes of tuberculosis. *ACS Omega* 9 (2): 2204–2219.

46 Noor, F., Tahir ul Qamar, M., Ashfaq, U.A. et al. (2022). Network pharmacology approach for medicinal plants: review and assessment. *Pharmaceuticals* 15 (5): 572.

47 Mohanraj, K., Karthikeyan, B.S., Vivek-Ananth, R.P. et al. (2018). IMPPAT: a curated database of Indian medicinal plants, phytochemistry and therapeutics. *Scientific Reports* 8 (1): 4329.

48 Rai, M.K. (1995). A review on some antidiabetic plants of India. *Ancient Science of Life* 14 (3): 168.
49 Dash, V.B. and Kashyap, V.L. (1999). *Ayurveda Materia Medica*, 711. Concept Publishing Company.
50 Lagunin, A.A., Goel, R.K., Gawande, D.Y. et al. (2014). Chemo- and bioinformatics resources for *in silico* drug discovery from medicinal plants beyond their traditional use: a critical review. *Natural Product Reports* 31 (11): 1585–1611.
51 Polur, H., Joshi, T., Workman, C.T. et al. (2011). Back to the roots: prediction of biologically active natural products from ayurveda traditional medicine. *Molecular Informatics* 30 (2–3): 181–187.
52 Chen, C.Y.-C. (2011). TCM Database@ Taiwan: the world's largest traditional Chinese medicine database for drug screening *in silico*. *PLoS One* 6 (1): e15939.
53 Vainio, M.J. and Johnson, M.S. (2007). Generating conformer ensembles using a multiobjective genetic algorithm. *Journal of Chemical Information and Modeling* 47 (6): 2462–2474.
54 O'Boyle, N.M., Banck, M., James, C.A. et al. (2011). Open Babel: an open chemical toolbox. *Journal of Cheminformatics* 3 (1): 1–14.
55 Cheng, F., Li, W., Zhou, Y. et al. (2012). *admetSAR: A Comprehensive Source and Free Tool for Assessment of Chemical ADMET Properties*. ACS Publications.
56 Lipinski, C.A., Lombardo, F., Dominy, B.W., and Feeney, P.J. (2012). Experimental and computational approaches to estimate solubility and permeability in drug discovery and development settings. *Advanced Drug Delivery Reviews* 64: 4–17.
57 Lobell, M., Hendrix, M., Hinzen, B. et al. (2006). *In silico* ADMET traffic lights as a tool for the prioritization of HTS hits. *ChemMedChem: Chemistry Enabling Drug Discovery* 1 (11): 1229–1236.
58 Hughes, J.D., Blagg, J., Price, D.A. et al. (2008). Physiochemical drug properties associated with *in vivo* toxicological outcomes. *Bioorganic & Medicinal Chemistry Letters* 18 (17): 4872–4875.
59 Gleeson, M.P. (2008). Generation of a set of simple, interpretable ADMET rules of thumb. *Journal of Medicinal Chemistry* 51 (4): 817–834.
60 Veber, D.F., Johnson, S.R., Cheng, H.-Y. et al. (2002). Molecular properties that influence the oral bioavailability of drug candidates. *Journal of Medicinal Chemistry* 45 (12): 2615–2623.
61 Egan, W.J., Merz, K.M., and Baldwin, J.J. (2000). Prediction of drug absorption using multivariate statistics. *Journal of Medicinal Chemistry* 43 (21): 3867–3877.
62 Szklarczyk, D., Santos, A., Von Mering, C. et al. (2016). STITCH 5: augmenting protein–chemical interaction networks with tissue and affinity data. *Nucleic Acids Research* 44 (D1): D380–D384.
63 Olsson, T. and Oprea, T.I. (2001). Cheminformatics: a tool for decision-makers in drug discovery. *Current Opinion in Drug Discovery & Development* 4 (3): 308–313.
64 Law, V., Knox, C., Djoumbou, Y. et al. (2014). DrugBank 4.0: shedding new light on drug metabolism. *Nucleic Acids Research* 42 (D1): D1091–D1097.
65 Hamosh, A., Scott, A.F., Amberger, J.S. et al. (2005). Online Mendelian Inheritance in Man (OMIM), a knowledgebase of human genes and genetic disorders. *Nucleic Acids Research* 33 (suppl_1): D514–D517.
66 Bodenreider, O. (2004). The unified medical language system (UMLS): integrating biomedical terminology. *Nucleic Acids Research* 32 (suppl_1): D267–D270.
67 Vivek-Ananth, R.P., Mohanraj, K., Sahoo, A.K., and Samal, A. (2023). IMPPAT 2.0: an enhanced and expanded phytochemical atlas of Indian medicinal plants. *ACS Omega* 8 (9): 8827–8845.
68 Doncheva, N.T., Morris, J.H., Gorodkin, J., and Jensen, L.J. (2019). Cytoscape StringApp: network analysis and visualization of proteomics data. *Journal of Proteome Research* 18 (2): 623–632.
69 Ebrahimi, S.B. and Samanta, D. (2023). Engineering protein-based therapeutics through structural and chemical design. *Nature Communications* 14 (1): 2411.

70 Tang, J. and Aittokallio, T. (2014). Network pharmacology strategies toward multi-target anticancer therapies: from computational models to experimental design principles. *Current Pharmaceutical Design* 20 (1): 23–36.

71 Zheng, S., Xue, T., Wang, B. et al. (2022). Application of network pharmacology in the study of mechanism of Chinese medicine in the treatment of ulcerative colitis: a review. *Frontiers in Bioinformatics* 2: 928116.

72 Askr, H., Elgeldawi, E., Aboul Ella, H. et al. (2023). Deep learning in drug discovery: an integrative review and future challenges. *Artificial Intelligence Review* 56 (7): 5975–6037.

73 Ekor, M. (2014). The growing use of herbal medicines: issues relating to adverse reactions and challenges in monitoring safety. *Frontiers in Pharmacology* 4: 177.

74 Pandey, M.M., Rastogi, S., and Rawat, A.K. (2013). Indian traditional ayurvedic system of medicine and nutritional supplementation. *Evidence-based Complementary and Alternative Medicine: Ecam* 2013: 376327.

75 Sharma, R., Bedarkar, P., Timalsina, D. et al. (2022). [Retracted] *Bhavana*, an ayurvedic pharmaceutical method and a versatile drug delivery platform to prepare potentiated micro-nano-sized drugs: core concept and its current relevance. *Bioinorganic Chemistry and Applications* 2022: 1685393.

76 Das, C., Ghosh, G., and Das, D. (2017). Ayurvedic liquid dosage form Asava and Arista: an overview. *Indian Journal Of Pharmaceutical Education And Research* 51: 169–176.

77 Savrikar, S.S. and Ravishankar, B. (2011). Introduction to 'Rasashaastra' the Iatrochemistry of Ayurveda. *African Journal of Traditional, Complementary, and Alternative Medicines: AJTCAM* 8 (5 Suppl): 66–82.

78 Gokarn, R.A., Gokarn, S.R., and Hiremath, S.G. (2014). Process standardization and characterization of Rajata Sindura. *AYU* 35 (1): 63–70.

79 Mishra, A., Mishra, A.K., Ghosh, A.K., and Jha, S. (2013). Standardization of a traditional polyherbo-mineral formulation – Brahmi vati. *African Journal of Traditional, Complementary, and Alternative Medicines: AJTCAM* 10 (3): 390–396.

80 Maurya, S.K., Seth, A., Laloo, D. et al. (2015). Śodhana: an ayurvedic process for detoxification and modification of therapeutic activities of poisonous medicinal plants. *Ancient Science of Life* 34 (4): 188–197.

81 Kaur, J., Baghel, D., Singh, S., and Mittal, A. (2019). Avaleha kalpana (medicated semisolid preparation): an synoptic overview. *International Journal of Research and Analytical Reviews* 6: 591–599.

82 Banothe, G.D., Mahanta, V., Gupta, S.K., and Dudhamal, T.S. (2018). A clinical evaluation of Kanchanara Guggulu and Bala Taila Matra Basti in the management of Mutraghata with special reference to benign prostatic hyperplasia. *AYU* 39 (2): 65–71.

83 Yuan, H., Ma, Q., Ye, L., and Piao, G. (2016). The traditional medicine and modern medicine from natural products. *Molecules (Basel, Switzerland)* 21 (5): 559.

84 Parmar, G.R., Shah, A.P., Sailor, G.U., and Seth, A.K. (2021). *In silico* discovery of novel phytoconstituents of *Amyris pinnata* as mitotic spindle kinase inhibitor. *Current Drug Research Reviews* 12 (2): 175–182.

85 Dias, D.A., Urban, S., and Roessner, U. (2012). A historical overview of natural products in drug discovery. *Metabolites* 2 (2): 303–336.

86 Riaz, M., Khalid, R., Afzal, M. et al. (2023). Phytobioactive compounds as therapeutic agents for human diseases: a review. *Food Science & Nutrition* 11 (6): 2500–2529.

87 Parmar, G. and Baile, S. (2020). A review: pharmacognosy, phytochemistry, ethnopharmacology and biological activity of inula racemose. *International Journal of Pharmaceutical Research* 12 (Spl 1): 412–417.

88 Rizvi, S.A.A., Einstein, G.P., Tulp, O.L. et al. (2022). Introduction to traditional medicine and their role in prevention and treatment of emerging and re-emerging diseases. *Biomolecules* 12 (10): 1442.

89 Parmar, G., Pundarikakshudu, K., and Balaraman, R. (2019). Anti-anaphylactic and antiasthmatic activity of *Euphorbia thymifolia* L. on experimental animals. *Journal of Traditional and Complementary Medicine* 9 (1): 60–65.

90 Parmar, G., Pundarikakshudu, K., Balaraman, R., and Sailor, G. (2018). Amelioration of anaphylaxis, mast cell degranulation and bronchospasm by *Euphorbia hirta* L. extracts in experimental animals. *Beni-Suef University Journal of Basic and Applied Sciences* 7 (1): 127–134.

91 Kumar, S., Verma, M., Hajam, Y.A., and Kumar, R. (2024). Honey infused with herbs: a boon to cure pathological diseases. *Heliyon* 10 (1): e23302.

92 Maheshwari, R.A., Parmar, G.R., Hinsu, D. et al. (2020). Novel therapeutic intervention of coenzyme Q10 and its combination with pioglitazone on the mRNA expression level of adipocytokines in diabetic rats. *Life Sciences* 258: 118155.

93 Parmar, G. and Pundarikakshudu, K. (2020). Determination and validation assay for total phenolic content in two Euphorbiaceous plants by Folin Ciocalteu method. *International Journal of Pharmaceutical Research* 12 (Suppl 1): 608–612.

94 Hashem, S., Ali, T.A., Akhtar, S. et al. (2022). Targeting cancer signaling pathways by natural products: exploring promising anti-cancer agents. *Biomedicine & Pharmacotherapy* 150: 113054.

95 Shah, A., Patel, V., Jain, M., and Parmar, G. (2023). Network pharmacology and systems biology in drug discovery. In: *CADD and Informatics in Drug Discovery*, 231–252. Singapore: Springer Nature Singapore.

96 Gonzalez, H., Hagerling, C., and Werb, Z. (2018). Roles of the immune system in cancer: from tumor initiation to metastatic progression. *Genes & Development* 32 (19–20): 1267–1284.

97 Shah, A., Parmar, G., and Seth, A.K. (2021). *In silico* discovery of novel flavonoids as poly ADP ribose polymerase (PARP) inhibitors. *Current Computer-Aided Drug Design* 17 (3): 344–350.

98 Zhong, L., Li, Y., Xiong, L. et al. (2021). Small molecules in targeted cancer therapy: advances, challenges, and future perspectives. *Signal Transduction and Targeted Therapy* 6 (1): 201.

99 Iqbal, N. and Iqbal, N. (2014). Human epidermal growth factor receptor 2 (HER2) in cancers: overexpression and therapeutic implications. *Molecular Biology International* 2014: 852748.

100 Yin, L., Duan, J.-J., Bian, X.-W., and Yu, S.-c. (2020). Triple-negative breast cancer molecular subtyping and treatment progress. *Breast Cancer Research* 22 (1): 61.

101 Harbeck, N., Penault-Llorca, F., Cortes, J. et al. (2019). Breast cancer. *Nature Reviews Disease Primers* 5 (1): 66.

102 Nitta, H., Kelly, B.D., Allred, C. et al. (2016). The assessment of HER2 status in breast cancer: the past, the present, and the future. *Pathology International* 66 (6): 313–324.

103 Kapoor, D., Vijayvergiya, R., and Dhawan, V. (2014). *Terminalia arjuna* in coronary artery disease: ethnopharmacology, pre-clinical, clinical & safety evaluation. *Journal of Ethnopharmacology* 155 (2): 1029–1045.

104 Bachheti, R.K., Worku, L.A., Gonfa, Y.H. et al. (2022). Prevention and treatment of cardiovascular diseases with plant phytochemicals: a review. *Evidence-based Complementary and Alternative Medicine: Ecam* 2022: 5741198.

105 Tao, W., Xu, X., Wang, X. et al. (2013). Network pharmacology-based prediction of the active ingredients and potential targets of Chinese herbal Radix Curcumae formula for application to cardiovascular disease. *Journal of Ethnopharmacology* 145 (1): 1–10.

106 Lu, M., Yin, J., Zhu, Q. et al. (2023). Artificial intelligence in pharmaceutical sciences. *Engineering* 27: 37–69.

107 Li, W., Yuan, G., Pan, Y. et al. (2017). Network pharmacology studies on the bioactive compounds and action mechanisms of natural products for the treatment of diabetes mellitus: a review. *Frontiers in Pharmacology* 8: 74.

108 Rogal, J., Zbinden, A., Schenke-Layland, K., and Loskill, P. (2019). Stem-cell based organ-on-a-chip models for diabetes research. *Advanced Drug Delivery Reviews* 140: 101–128.

109 Li, H., Yang, Y., Hong, W. et al. (2020). Applications of genome editing technology in the targeted therapy of human diseases: mechanisms, advances and prospects. *Signal Transduction and Targeted Therapy* 5 (1): 1.

110 Halayal, R.Y., Bagewadi, Z.K., Maliger, R.B. et al. (2023). Network pharmacology based anti-diabetic attributes of bioactive compounds from *Ocimum gratissimum* L. through computational approach. *Saudi Journal of Biological Sciences* 30 (9): 103766.

111 Parmar, G., Pundarikakshudu, K., and Balaraman, R. (2017). Antidiabetic and antihyperlipidemic activity of *Euphorbia thymifolia* L. extracts on streptozotocin-nicotinamide induced type 2 diabetic rats. *Journal of Applied Pharmaceutical Science* 7 (8): 078–084.

112 Caesar, L.K. and Cech, N.B. (2019). Synergy and antagonism in natural product extracts: when 1 + 1 does not equal 2. *Natural Product Reports* 36 (6): 869–888.

113 Thomford, N.E., Senthebane, D.A., Rowe, A. et al. (2018). Natural products for drug discovery in the 21st century: innovations for novel drug discovery. *International Journal of Molecular Sciences* 19 (6): 1578.

114 Wainwright, C.L., Teixeira, M.M., Adelson, D.L. et al. (2022). Future directions for the discovery of natural product-derived immunomodulating drugs: an IUPHAR positional review. *Pharmacological Research* 177: 106076.

115 Li, L., Yang, L., Yang, L. et al. (2023). Network pharmacology: a bright guiding light on the way to explore the personalized precise medication of traditional Chinese medicine. *Chinese Medicine* 18 (1): 146.

116 You, Y., Lai, X., Pan, Y. et al. (2022). Artificial intelligence in cancer target identification and drug discovery. *Signal Transduction and Targeted Therapy* 7 (1): 156.

117 Parmar, G., Chudasama, J.M., Shah, A., and Patel, A. (2023). *In silico* pharmacology and drug repurposing approaches. In: *CADD and Informatics in Drug Discovery*, 253–281. Singapore: Springer Nature Singapore.

13

Rational Design of Natural Products for Drug Discovery

Ankita Kashyap[1], *Anupam Sarma*[2], *Bhrigu Kumar Das*[3], *and Ashis Kumar Goswami*[4]

[1] *Institute of Pharmacy, Assam Medical College and Hospital, Dibrugarh, India*
[2] *Advanced Drug Delivery Laboratory, School of Pharmaceutical Sciences, Girijananda Chowdhury University, Guwahati, India*
[3] *Pharmacology and Toxicology Laboratory, School of Pharmaceutical Sciences, Girijananda Chowdhury University, Guwahati, India*
[4] *Department of Pharmaceutical Sciences, Faculty of Science and Engineering, Dibrugarh University, Assam, India*

13.1 Introduction

In the relentless pursuit of novel therapeutic solutions, pharmaceutical research continually explores the vast array of natural compounds. These compounds, found abundantly in nature, offer a treasure trove of chemical diversity and biological activity, making them promising candidates for drug development [1]. However, harnessing the full potential of natural products for medicinal purposes presents numerous challenges, including intricate chemical structures, limited bioavailability, and potential side effects. To overcome these hurdles and expedite the discovery of effective drugs derived from natural sources, scientists have turned to rational drug design – an approach rooted in molecular biology, biochemistry, and computational chemistry [2]. This chapter delves into the intricacies of rational drug design, exploring its methodologies, applications, and implications for advancing drug discovery.

Structure-guided computer-aided drug design is a cornerstone of rational drug design, particularly for targets with known three-dimensional structures. By elucidating the binding interactions between target proteins and potential drug candidates, this approach enables researchers to design small molecules with enhanced specificity and efficacy [3]. Utilizing techniques such as X-ray crystallography and nuclear magnetic resonance (NMR) spectroscopy, scientists can visualize the three-dimensional structure of target proteins and identify key binding sites [4]. Virtual screening (VS) of large chemical libraries further expedites the identification of promising drug candidates, laying the foundation for subsequent experimental validation and optimization.

In the era of precision medicine, state-of-the-art methodologies play a pivotal role in driving drug discovery forward. Resources such as the Protein Data Bank (PDB), the National Cancer Institute (NCI) database, and computational tools provide researchers with invaluable resources for VS and structural analysis. High-throughput screening (HTS) platforms and molecular modeling software further streamline the drug discovery process, enabling rapid iteration and optimization of lead compounds [5].

While modern drug discovery relies heavily on cutting-edge technologies, traditional medicinal practices offer valuable insights into the therapeutic potential of natural compounds. Ancient

Computational Methods for Rational Drug Design, First Edition. Edited by Mithun Rudrapal.

Asian herbal medicine, such as *Ayurveda*, emphasizes the synergistic effects of herbal combinations, paving the way for novel approaches to drug discovery. By integrating traditional wisdom with modern scientific techniques, researchers can unlock new therapeutic avenues and address unmet medical needs more effectively [6].

Global gene expression profiling provides unprecedented insights into disease pathogenesis and therapeutic responses. Platforms such as Serial Analysis of Gene Expression (SAGE) and microarray analysis enable researchers to characterize the transcriptional profiles of diseased tissues and identify dysregulated genes associated with disease progression. Bioinformatics tools facilitate the analysis and interpretation of vast genomic datasets, guiding the selection, and optimization of drug candidates with the highest therapeutic potential [7].

In conclusion, rational drug design represents a paradigm shift in drug discovery, integrating cutting-edge technologies with traditional wisdom to unlock nature's pharmacological arsenal. By harnessing the power of computational modeling, structural biology, and genomic analysis, researchers can expedite the discovery and development of safe and effective therapeutics. As we continue to unravel the complexities of disease biology and molecular interactions, rational drug design will remain at the forefront of drug discovery, shaping the future of medicine and improving patient outcomes.

13.2 Natural Products for the Development of New Drugs

Natural products constitute a diverse array of chemical compounds biosynthesized by living organisms, encompassing plants, animals, and microorganisms. Endowed with a rich history of application as curative agents dating back millennia, these compounds hold profound significance due to their inherent bioactivity and promising therapeutic attributes. Alkaloids, flavonoids, terpenes, and steroids are among the myriad subclasses of natural products, each contributing distinct molecular structures and pharmacological potential. Their pivotal role in contemporary drug discovery underscores their value as sources of novel therapeutic agents. Natural products, with their complex and intricate chemical profiles, continue to captivate scientific interest as they unveil a treasure trove of potential therapeutic preparations, perpetuating their legacy as indispensable contributors to the advancement of medicine [8]. Figure 13.1 shows the components or fundamental steps involved in developing new drugs from natural products.

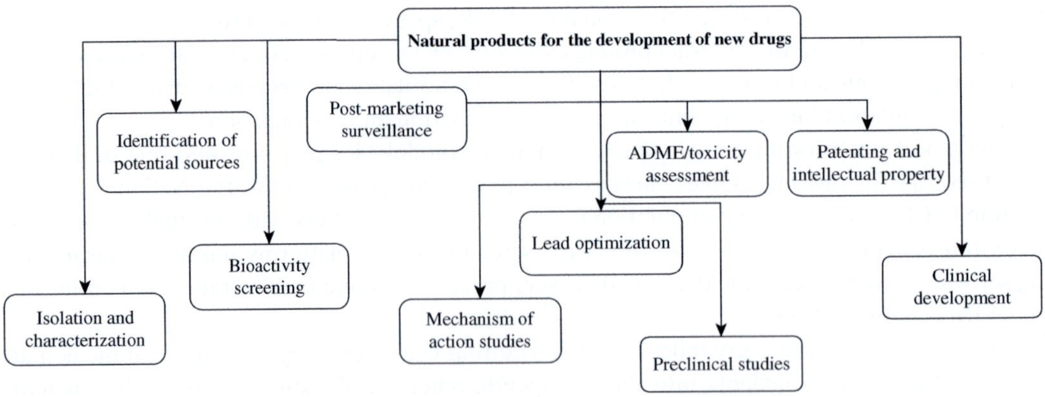

Figure 13.1 Components of the process of development of new drugs from natural products.

They are essential to the drug development process because of their unique chemical originality and deep biological activity. These substances have played a significant role in the pharmaceutical industry throughout history, accounting for around 30% of all drug sales worldwide. Natural products are unmatched resources because they have a vast reservoir of chemical variety, attractive drug-like characteristics, and the ability to interact with various biological target proteins. Their historical role as foundational lead compounds in drug discovery is augmented by an escalating recognition of their intricate structural and stereochemical attributes. In contemporary drug development initiatives, the enduring relevance of natural products persists, positioning them as invaluable reservoirs for the innovation and creation of novel therapeutic drugs [9, 10].

Sales of recombinant peptides and protein molecules are on the rise, but low-molecular-mass chemicals continue to be the gold standard for treating human diseases. This is mainly because of their greater compliance and bioavailability. Natural products have historically been the source of many novel therapeutic methods, demonstrating their creative potential in medication discovery and development. Natural products have had a significant impact on drug development, but presently, they are not given as much attention because of the significant work involved in active principle isolation and structure elucidation. To effectively compete with combinatorial chemistry, new approaches in natural product chemistry are needed to meet the requirement for large test samples in the process of HTS. Pharmaceutical companies currently spend about $350 million US on the development of new drugs. It is predicted that the course of medication development and discovery in the future will depend on advances in molecular science, the biology of cells, and genome engineering to improve therapeutic target clarification and lead discovery. The human genome is almost complete, which means that a much larger pool of possible pharmacological targets will become available for consideration during the drug discovery phase. The ability of HTS to test an increasing number of samples highlights the significance of creating large chemical collections with improved structural diversity for more successful drug development programs [11].

The landscape of natural product drug discovery confronts formidable challenges arising from the extensive chemical diversity inherent in these compounds, coupled with technical barriers. A comprehensive survey involving 52 experts underscored a perceptible dissonance between the acknowledged potential of natural products and the efficacy of prevailing drug discovery strategies. Paramount among the challenges are the constrained supply of compounds, hurdles in target identification, and the complexities associated with scale-up for commercialization. In response to these impediments, a strategic amalgamation of rational approaches, cutting-edge technologies, and computational methodologies is being orchestrated [12]. This integration encompasses structure- and ligand-based design techniques, ADMET (absorption, distribution, metabolism, excretion, and toxicity) profiling, and the utilization of cheminformatics for the systematic organization and interpretation of data. Despite the intricacies posed by these challenges, the intrinsic value of natural products in drug discovery endures as ongoing efforts strive to harness their potential through innovative methodologies and technologies [13].

Contemporary research in medicinal plant-based drug discovery adopts a multifaceted approach, integrating fractionation guided by bioassay, combinatorial chemistry, botany, phytochemistry, and biochemistry methodologies. In the quest for safe and effective drugs, natural sources continue to be explored as valuable reservoirs of pharmacological leads against severe diseases like diabetes, cardiovascular disorders, and cancer. The escalating global demand for reliable therapeutics has prompted a resurgence in exploring natural resources for potential treatments of chronic and life-threatening illnesses. However, the sector faces significant obstacles, particularly in the marketing and scaling up of active compounds, meaning that only a small percentage of lead molecules may advance to the stage of medication development. For the effective investigation and development of medicinal compounds derived from natural resources, it is essential to take a methodical and scientific approach to overcome these challenges [14].

The foundational role of conventional techniques persists as paramount in the identification of potential therapeutic agents. Historically, natural products have stood as invaluable reservoirs of drug leads, with bio-guided fractionation being a conventional approach wherein crude extracts undergo sequential separation and testing until one single active ingredient is found. However, this approach is characterized by a laborious and slow procedure. In response to these limitations, contemporary strategies, such as metabolomics and chemometrics, have emerged to enhance lead-finding efficiency by establishing correlations between chemical and biological data. Despite the inherent challenges, natural products maintain their pivotal status as a crucial source for the discovery of novel drugs, evolving alongside innovative methodologies aimed at optimizing the drug discovery process [15].

13.3 Criteria for Selecting Natural Products for Drug Design

The selection of natural products for drug design involves carefully considering several criteria to identify promising candidates with therapeutic potential. By carefully evaluating natural products based on these criteria, researchers can identify promising drug design and development candidates, ultimately discovering novel therapeutics with improved efficacy and safety profiles. These include:

1) **Bioactivity:** Natural products should exhibit specific biological activities relevant to the targeted disease or condition, such as antimicrobial, anti-inflammatory, or anticancer properties. Screening assays and bioassays are commonly used to assess the bioactivity of natural products [16].
2) **Pharmacological properties:** Natural products should possess favorable pharmacokinetic and pharmacodynamic properties, including ADMET. Compounds with suitable ADMET profiles are more likely to be developed into successful drugs [17, 18].
3) **Chemical diversity:** Natural products offer various chemical structures, representing diverse scaffolds and functional groups. Selecting compounds with structural diversity increases the likelihood of identifying novel drug leads and optimizing their pharmacological properties [19, 20].
4) **Synthetic accessibility:** The feasibility of synthesizing or obtaining natural products in sufficient quantities for further research and development is essential. Compounds that are readily accessible or amenable to chemical synthesis are preferred for drug design efforts [21, 22].
5) **Availability and sustainability:** Natural products should be obtainable from sustainable and environmentally friendly sources to ensure long-term availability. Ethical considerations regarding biodiversity conservation and traditional knowledge protection should also be considered [23].
6) **Target specificity:** Natural products should demonstrate selectivity toward the intended molecular targets associated with the diseases of interest. Understanding the mechanism of action and target engagement is crucial for assessing the therapeutic potential of natural products [24].

13.4 Importance of Biodiversity in Sourcing Natural Products

The diversity and variability of life on Earth is known as biodiversity, and it is commonly measured by differences found in species, genetic makeup, and ecosystems [25]. Natural products (NPs) are the end-products of gene expression, also known as "secondary metabolites." They are a reliable and vital source of effective medication leads from Earth's biodiverse flora and fauna [26]. Since ancient times, the development of pharmaceutical products has been aided by using natural products for medicinal purposes. Drug research and discovery depend on biodiversity conservation since its loss eliminates previously unknown genetic and natural chemical resources. The most well-known pharmacological drugs from natural sources are caffeine, acetylsalicylic acid (Aspirin), and plants' antimalarial agents Quinine and Artemisinin [25]. Similarly, the majority of known antibiotics (such as Streptomycin and Penicillin) and antifungal drugs (such as Griseofulvin) have been discovered through soil fungi [27].

Biodiversity, encompassing the vast array of life forms and ecosystems on Earth, is the cornerstone of natural product sourcing for drug discovery. It represents the collective genetic, species, and ecological diversity in different environments, ranging from tropical rainforests to marine ecosystems. Within this biological diversity lie invaluable reservoirs of secondary metabolites, known as natural products, which organisms have evolutionarily honed for various ecological purposes. These natural products exhibit a remarkable array of chemical structures and bioactivities, making them promising candidates for drug development [28]. The importance of biodiversity in sourcing natural products for drug discovery cannot be overstated. It provides researchers with an extensive library of chemical compounds with diverse pharmacological properties, offering potential solutions to various human health conditions. The loss of biodiversity poses a significant threat to drug discovery efforts, as it not only diminishes the availability of natural compounds but also erases the genetic and chemical diversity that could hold the key to novel therapeutics. As such, conservation efforts to preserve biodiversity are crucial to safeguarding these invaluable resources for future drug development endeavors [29].

Historically, natural products have played a pivotal role in developing pharmacological agents, serving as the foundation for many essential medications. Examples abound, from discovering antibiotics like penicillin to developing anticancer drugs such as vincristine and vinblastine derived from the Madagascar periwinkle. Moreover, recent advancements in drug discovery have continued to leverage biodiversity, with discoveries ranging from novel antidiabetic agents sourced from plants to marine-derived compounds with potential antiparasitic and anticancer properties [30]. *Piptoporus betulinus*, another type of fungus that grows on birches, was cooked to create charcoal, which has antibacterial and disinfecting properties, and different therapeutic cures have even been supplied by marine organisms. A drink made from the red algae *Chondrus crispus* and *Mastocarpus stellatus* was used to treat colds, sore throats, and chest illnesses, including tuberculosis. It was also reported that boiling the algae in milk or water made them effective against burns and kidney problems [31]. Also, compounds like Zamamidine D isolated from *Amphimedon* sp., Protulactone A isolated from *Aspergillus* sp., Gombasterols from *Clathria gombawuiensis*, Lamellodysidines from *Lamellodysidea herbacea*, Sulawesins from *Psammocinia* sp. have different medicinal properties [32]. In essence, preserving biodiversity is paramount for sustaining the reservoirs of natural products that hold immense promise for future drug discovery efforts. By conserving Earth's biological diversity, we can ensure continued access to the vast array of chemical compounds that nature offers, thereby advancing the frontiers of pharmaceutical research and improving global health outcomes [30].

13.5 Structural Elucidation of Natural Products

Over the past century, significant advancements have been made in elucidating the structures of natural products, leading to a profound understanding of their chemical composition and biological activity. This progress has been driven by innovative analytical techniques, methodologies, and advancements in computational chemistry and spectroscopic instrumentation [18]. Historically, the elucidation of natural product structures relied heavily on classical methods such as chemical degradation, elemental analysis, and optical rotation measurements. However, the advent of NMR spectroscopy in the mid-20th century revolutionized the field by providing detailed information about the connectivity of atoms within molecules. NMR spectroscopy enabled researchers to elucidate complex structures precisely and determine stereochemistry [33]. In parallel, mass spectrometry (MS) emerged as a powerful tool for analyzing natural products, offering insights into molecular weight, fragmentation patterns, and elemental composition. The development of high-resolution MS instruments further enhanced the accuracy and sensitivity of structural elucidation studies, enabling the characterization of compounds present in minute quantities [34]. The synergy between NMR spectroscopy and MS has facilitated the elucidation of challenging natural product structures, including complex alkaloids, terpenoids, and polyketides. Moreover, X-ray crystallography has played a pivotal role in confirming the three-dimensional arrangement of atoms in crystalline compounds, providing invaluable insights into molecular architecture and stereochemistry [35]. In recent years, advancements in computational chemistry and bioinformatics have complemented experimental approaches, enabling the prediction and modeling of natural product structures based on spectroscopic data and molecular modeling techniques. These computational tools have expedited the structural elucidation process and facilitated the discovery of novel bioactive compounds with therapeutic potential. Overall, the past century has witnessed remarkable progress in elucidating natural product structures, driven by interdisciplinary collaborations and technological innovations. Continued advancements in analytical techniques and computational methods are expected to further research in natural product chemistry, unlocking new avenues for drug discovery and development [36].

13.6 *In Silico* Computational Tools for Rational Drug Discovery from Natural Sources

13.6.1 Molecular Docking and Virtual Screening Techniques for Predicting Ligand–Receptor Interactions

Molecular docking stands as a computational cornerstone in drug discovery, employed to forecast the interaction dynamics between a small molecule (ligand) and a target protein (receptor). Within the realm of natural product drug discovery, this method assumes a pivotal role in scrutinizing potential bioactive compounds derived from diverse sources, including plants, fungi, and marine species. The primary objective is the discernment of natural products capable of interacting with specific target proteins implicated in various diseases, ranging from cancer to neurodegenerative disorders. By facilitating the screening and identification of promising lead compounds, molecular docking becomes an instrumental tool, propelling the subsequent development of these compounds into therapeutic agents. This computational approach thus serves as a strategic bridge connecting the realms of computational science and pharmaceutical innovation in the pursuit of novel and effective treatments [37].

Molecular docking emerges as an invaluable asset, facilitating the discernment of potential bioactive molecules sourced from nature and thus catalyzing the quest for innovative therapeutic agents. This method has demonstrated efficacy in pinpointing macromolecular targets for natural products, thereby contributing to the development of drugs with heightened safety and efficacy profiles [38]. Molecular docking's utility extends to the expedited identification of bioactive phytochemicals, streamlining the preliminary validation process. Its versatility is evident in medication design based on structure, exemplified by its application in the search for potential inhibitors against COVID-19 [39, 40]. In essence, molecular docking assumes a pivotal role in unlocking the therapeutic potential inherent in natural products, navigating the complex landscape of drug discovery with precision and efficiency [41].

Many natural compounds have large, pliable macrocyclic structures that interact with their targets using several sites of contact. Natural products are attractive scaffolds because of their unique combination of limited elasticity and large surface area, especially for difficult therapeutic targets and protein–protein interaction inhibition. Their preference for binding at target surfaces as opposed to deep pockets highlights why medicinal chemistry campaigns find them appealing. Their role as final therapeutic agents is no longer limited by the increasing ability to change these chemicals through semisynthetic modifications or biosynthetic processes, which turns them into suitable starting points for medicinal chemistry attempts. The logical improvement of molecular docking in modern medicinal chemistry, guided by knowledge of the complex binding mechanisms and adaptable scaffolds present in macrocyclic natural products and their analogs, greatly increases the effectiveness of drug development initiatives. This sophisticated method emphasizes how crucial molecular docking is to maximize the potential of natural macrocyclic compounds for contemporary medicinal science [42].

An essential bioinformatics method for examining and understanding ligand–receptor interactions is molecular docking, which provides information on the molecular targets of various ligands. Through its revolutionary role in chemical separation and medication development from herbal sources, this method leverages the long history of using herbs as treatments for a wide range of illnesses. Phytochemicals and drug makers are increasingly concentrating on developing innovative compounds derived mostly from herbs, hence lowering required quantities in the pursuit of safer and more effective medications. As an example of this paradigm, a plant that is commonly found in India and is well-known for its use in the treatment of urinary disorders, *Rotula aquatica* Lour was thoroughly investigated. After its root was extracted, active chemicals were produced, and molecular docking analysis was carried out between the 3-O-acetyl-11-keto-β-boswellic acid, an isolated bioactive molecule, and Tamm–Horsfall protein (THP). This investigation demonstrates the tactical application of molecular docking techniques in clarifying the medical application of natural chemicals derived from herbs, as it found possible interactions that may imply the prevention of calcium oxalate crystallization [43].

A computational method used in drug development called VS is particularly useful for finding putative bioactive chemicals. Structure-based VS utilizes two primary methodologies: (i) molecular docking, which involves docking databases of thousands of compounds onto the active site of a target of interest to pinpoint top-ranked compounds and (ii) inverse docking, wherein a compound of interest is docked against a variety of targets to unveil potential new targets, as implied by its name. Figure 13.2 shows the molecular docking mechanism and inverse docking mechanism.

The biological significance and structural diversity of natural products highlight the importance of VS in the field of natural product medication discovery. Though promising, the use of VS in this area is constrained by the availability of extensive 3D databases of natural products and issues with interoperability with HTS procedures. To mitigate these obstacles, VS must be strategically integrated with traditional pharmacognosy methodologies. This provides a means of increasing the effectiveness of natural product-based medication discovery. The integration of VS in the field of natural product medication discovery is still being researched and developed, and this could improve the accuracy and effectiveness of recognizing bioactive substances obtained from various natural sources [44].

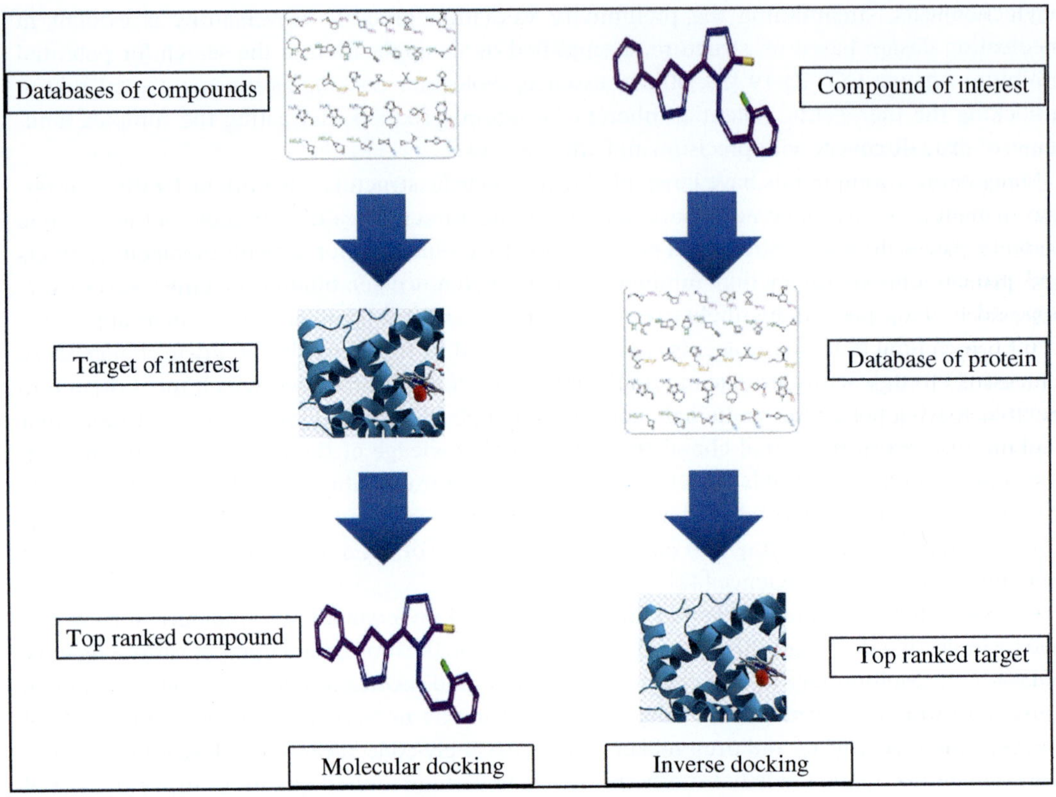

Figure 13.2 Two approaches used by virtual screening and molecular docking.

VS emerges as a pivotal computational technique employed to discern potential bioactive compounds. This technique speeds up the identification of lead compounds for later drug development by methodically analyzing large chemical databases to predict the interactions between compounds and certain biological targets. Molecular docking, ligand-based screening, and pharmacophore modeling represent prevalent techniques in VS, proving successful in recognizing macromolecular targets for natural products. The interest in natural product-oriented drug development initiatives has increased as a result of this success [45]. Moreover, cheminformatics techniques and molecular modeling techniques work together to develop drugs based on natural products at a faster rate. These tools not only organize data and interpret results but also facilitate the judicious filtering of expansive chemical databases before experimental screening, contributing to the efficiency and success of the drug discovery processes [45, 46].

VS techniques, notably molecular docking and shape-matching methodologies serve as predictive tools for illuminating ligand–receptor interactions. These computational methods play a pivotal role in expeditiously identifying potential bioactive molecules within vast compound libraries, significantly curtailing both time and financial expenditures. Particularly, cheminformatics advancements have made it easier to compare databases of natural products with modern small-molecule libraries, which has led to the successful identification of macromolecular targets for natural products. The integration of *in silico* approaches has sparked a resurgence of interest in drug discovery, offering a nuanced and efficient pathway that contributes substantive insights to the ongoing evolution of the field [47].

The ongoing exploration in the pursuit of new bioactive substances with industrial applications remains a focal point of scientific research, driven by the inherent pharmacophoric structures, pharmacokinetic attributes, and distinct chemical space exhibited by these compounds. The systematic investigation of natural sources for the extraction of valuable molecules, essential for the development of commercially and industrially relevant products, poses a substantial challenge in the field of bioprospecting. Notably, the advent of advanced VS strategies has revolutionized the discovery process by employing *in silico* analyses of expansive compound libraries. This tactical approach not only makes it easier to conduct a thorough analysis of the chemical-based space, pharmaceutical dynamics, and pharmacokinetic characteristics of these molecules but it also greatly reduces the time, infrastructure, and financial costs associated with the complex process of locating and describing novel chemical entities [48].

The principal uses for online chemical screening involve the careful selection from chemical libraries of a limited but receptor-relevant fraction with an emphasis on maximum chemical diversity. To achieve a balance between accuracy and computing efficiency throughout the hit enrichment process, our earlier research demonstrated the effectiveness of integrating ligand-centric and receptor-centric VS approaches. In this work, we present a "progressive distributed docking" methodology designed to improve the VS procedure by iteratively combining shape-matching and docking phases. First, 3D templates were used for shape comparisons within the chemical library. These templates were derived from known ligands with poor docking scores. Then, molecules with low receptor docking scores and good template shape matches were repeatedly chosen for additional rounds of shape searching and docking. This VS procedure is iterative and was verified through the enrichment of compounds from a carefully selected subset of chemical libraries that were pertinent to phosphoinositide 3-kinase and peroxisome proliferator-activated receptors. The outcomes showed how this progressive distributed docking technique improves lead-hopping procedures by increasing the chemical variety of the chosen virtual hit cohort [49].

13.6.2 QSAR (Quantitative Structure–Activity Relationship) Models for Optimizing Biological Properties

The foundation of computational science is the QSAR, which establishes relationships between the chemical structures and biological activities of various molecules. QSAR models are essential for predicting the biological activity of organic compounds against particular targets, which helps identify possible candidates for drugs. These models are useful for the effective VS of natural compounds for various pharmacological profiles, especially in the field of antiparasitic activity. Notably, nevertheless, the use of QSAR models to forecast the activity against several parasite species using a single model is still a major and continuous problem, highlighting the necessity for further developments in this computational approach [50].

Drug development activities greatly benefit from the use of QSAR models, which are crucial instruments for forecasting the biological effects of compounds. These models are useful for a wide range of pharmacological actions and toxicities, which helps find effective medications more quickly. QSAR models play a key role in the setting of natural product drug development by making it easier to anticipate the biological effects of phytochemicals, or molecules originating from plants [51]. With QSAR models' ability to simultaneously predict pharmacological activity, toxicities, and safety profiles based on the integration of varied chemical and biological data, their application becomes especially significant in the design of improved pharmaceuticals derived from phytochemicals. Consequently, by using their predictive power to identify the biological effects of phytochemicals, QSAR models become indispensable components of the seamless integration of natural product drug development to help simplify the creation of powerful yet secure medications [52].

13.6.3 High-Throughput Screening Methods for Efficient Compound Selection

The seamless integration of solid-phase extraction, high-throughput parallel preparative high-performance liquid chromatography, and automated flash chromatography highlights the crucial importance of HTS approaches. All of these approaches result in libraries that are composed of one to five compounds per well, roughly. To support this, high-throughput parallel liquid chromatography-MS-evaporative light scattering detection systems are used to do thorough library analysis before biological screening [53]. Target-based workflows in natural product drug discovery projects are significantly more efficient thanks to this novel method, which quickly identifies natural product ligands that target human drug receptors without the need for fractionation [54].

13.6.4 Molecular Dynamics Simulations for Predicting Solubility and Stability

One essential component that is considered essential for all ADME research is the evaluation of water solubility. Acknowledged as a crucial factor, water solubility defines how a medicine is absorbed orally and how bioavailable it is. Since a significant fraction of pharmaceuticals now on the market (about 40%) and a notable majority of compounds in the development stage (around 75%) have low solubility in water, it is imperative to identify this feature as soon as possible. This proactive approach illustrates the critical role that aqueous solubility plays in determining the success trajectory of drug candidates. It not only informs strategic decisions but also has the potential to minimize failures in the pharmaceutical development process [55].

A multitude of experimental techniques, such as modifications of the shake-flask method and the CheqSol methodology, have been utilized to ascertain the solubility of various substances in water. However, the experimental determination turns out to be difficult, expensive, and time-consuming, especially when dealing with the large chemical libraries that are utilized in HTS. As a result, early in the drug discovery and development process, *in silico* prediction of water solubility by quantitative structure–property relationship (QSPR) has gained significance. Although several QSPR models have been created over the years, their effectiveness on various solubility datasets has exposed some of the shortcomings of certain prediction techniques [55]. Notably, the inconsistent data sources are the primary cause of these restrictions, with an average root-mean-square error (RMSE) of 0.6–0.7 $\log S$ units. According to recent research, the insufficiency in therefore, more accurate approaches, such as advanced machine learning (ML) algorithms like random forests (RF), support vector machines (SVM), *k*-nearest neighbors (k-NN), convolutional, and recurrent networks, are recommended. Solubility prediction accuracy is not only related to the quality of experimental data. These ML techniques have been shown to perform on par with or better than conventional techniques in terms of aqueous solubility prediction [56].

Using sophisticated prediction models, especially in the field of ML, is a valuable tool for predicting a compound's solubility, which is a critical factor in the early stages of drug discovery and development. Predicting the complex characteristics of a compound's fate in a biological system is made even more accurate by adding molecular descriptors and physicochemical parameters to solubility predictions. Its usefulness is expanded by this comprehensive method to include critical parameter prediction, including ADMET [57]. Drug development initiatives can make well-informed decisions at an early stage and identify promising candidates more quickly by incorporating these predictive tools. This also makes it easier to comprehend a compound's potential within the complex field of pharmacological and toxicological research [58].

13.6.5 ADMET Attributes Predicted *In Silico*

Computational models are used *in silico* ADME-Tox prediction to evaluate a drug candidate's ADMET characteristics. Using these models is essential for strategically ranking possible compounds, which lowers development costs for new drugs and lowers attrition rates. However, the intricate nature of physiological processes poses challenges in the generation of reliable and accurate prediction models [59]. The KnowItAll system offers a comprehensive approach, providing both real-number and categorical classification predictions, thereby addressing concerns about the reliability of predictions. While *in silico* models prove invaluable for early-stage screening, it is imperative to acknowledge that they do not serve as outright replacements for *in vivo* or *in vitro* methods, emphasizing the necessity of a multifaceted approach in comprehensively assessing a drug candidate's viability [60].

13.6.6 Computational Tools for Predicting Pharmacokinetics and Pharmacodynamics (PK/PD)

A predominant challenge in drug development lies in the substantial rate of failures attributed to issues with (ADME/Tox) properties of candidate compounds, accounting for over half of the setbacks. The strategic integration of *in silico* tools to predict ADME/Tox and physicochemical properties emerges as a promising avenue to mitigate attrition rates in pharmaceutical research and development. This technology, exemplified by the KnowItAll computational environment developed by Bio-Rad Laboratories, Inc., has the potential to streamline the pharmaceutical R&D pipeline by effectively prioritizing candidate compounds. A noteworthy aspect of KnowItAll is its commitment to addressing concerns about the reliability of property predictions. Within this platform, various ADME/Tox predictors are encoded, providing the capability to verify these predictions using internal data and models, if any. Moreover, the system allows for the construction of a "consensus" model, surpassing individual predictive models in efficacy. This sophisticated environment adeptly handles both real-number and categorical classification predictions, contributing to a comprehensive and reliable approach to navigating the intricacies of drug development [61].

NPs and their derivatives are important sources for the development of new drugs; however, current *in silico* target prediction techniques frequently cannot distinguish between NPs and artificial compounds. Because NPs differ from their synthetic counterparts inherently, it is necessary to develop target prediction models that are specific to NPs. To close this gap, a study was conducted where four unique datasets were created, namely, the NP dataset, NPs and their first-class derivatives dataset, NPs and all derivatives dataset, and the ChEMBL26 compounds dataset – by curating activity data from open databases and covering a range of NP scenarios [62]. To optimize performance, systematically investigated conditions like activity thresholds and input features using eight ML techniques for NP target prediction. The most appropriate dataset for developing reliable NP-specific models turned out to be the one that included all of the derivatives of NPs. Consensus models, specifically the feedforward neural network (FNN) and SVM consensus model, and voting models were also applied, and this improved prediction performance considerably. Extensive assessments on external validation sets confirmed that these models performed better than conventional models that were trained on all of the ChEMBL26 compounds. The effectiveness of NP-specific target prediction models in boosting drug discovery efforts is highlighted by this all-encompassing strategy [62].

Predicting cytotoxicity is essential in drug development to mitigate late-stage toxicity issues. *In silico* methods, particularly deep learning, are crucial for early design processes, reducing time, costs, and reliance on animal testing. Leveraging a comprehensive dataset, a neural network model achieves a balanced accuracy exceeding 70%, demonstrating efficacy comparable to RF. While neural networks lack interpretability, the deep Taylor decomposition method is explored to identify toxicophores responsible for cytotoxic effects [63].

The escalating release of thousands of anthropogenic chemicals into the environment necessitates robust methods for rapidly screening and predicting their potential toxicity, thereby mitigating adverse impacts on human and environmental health. Computational approaches, particularly molecular docking, are evaluated here for screening the toxicity of diverse xenobiotic substances, which include chemicals produced by the chemical industry, pollution, medications, and insecticides. The methodology involves predicting binding energy between pollutants and carefully selected receptors, assuming toxicity correlates with interference in biochemical pathways. One of the method's main advantages is its speed at producing interaction maps, which provide molecular-level information on possible perturbation pathways and help choose chemicals for additional testing. When contrasted as scoring functions, Autodock Vina and the ML scoring function RF-Score-VS exhibit encouraging results but with limitations related to scoring function accuracy [64].

The VirtualToxLab is an *in silico* tool that evaluates the carcinogenicity, cardiotoxicity, and endocrine and metabolic disruption potentials of medications, chemicals, and natural items. Using an automated approach, the system provides real-time 3D/4D mechanistic interpretation in accordance with the Setubal principles for computational toxicology by simulating and quantifying the binding of small compounds to a set of 16 proteins implicated in inducing deleterious effects. Interestingly, the "*ab initio*" protocol operates in client–server mode, is globally applicable, and is free for use by academic and nonprofit institutions. It also does not require training data. Molecular dynamics simulations are made possible by the technology's thermodynamic estimate of binding affinity, which enables the investigation of the kinetic stability of ligand–protein complexes [65]. With an emphasis on *Trypanosoma brucei*, the neglected tropical disease that causes human African trypanosomiasis, the VirtualToxLab plays a key role in investigating novel medication candidates derived from natural products and utilizing their chemical diversity to address issues with current chemotherapeutic drugs [66].

Natural products have garnered increasing attention in drug discovery for their potential therapeutic advantages. The integration of ADME/Tox studies is imperative to evaluate the safety and efficacy of these compounds, given their structural diversity and distinctive biological activities. While direct evidence on the amalgamation of ADME/Tox remains limited, ensuring the safety and effectiveness of these compounds is paramount before advancing them into pharmaceutical drugs. This strategic incorporation of ADME/Tox studies contributes to the comprehensive understanding and optimization of natural products for therapeutic development [67].

For a very long time, natural products made from plants, fungi, and bacteria have been a valuable source of bioactive substances. Notably, small molecule natural products or their chemically modified analogs of FDA-approved medications; the remaining are synthetic drugs that target the same biomolecules as natural products. Despite improvements in bioinformatics analysis and HTS streamlining certain aspects of the drug discovery pipeline, the identification of specific functions, such as antibiotic activity, remains a bottleneck, requiring the laborious processes of production, purification, and assaying. The traditional sequence of assaying, production, and purification introduces the challenge of rediscovering known molecules, hindering the efficiency of natural product utilization. Although MS- and NMR-based methods offer some enhancements, they do not eliminate the fundamental requirement of having molecules for testing. Thus, the journey from the activity in a bacterial culture extract to the discovery of a novel active molecule remains a time-consuming impediment in harnessing natural products for antibiotic and therapeutic agent discovery [68].

A crucial phase in the drug development process is to effectively identify prospective biosynthetic gene clusters (BGCs) among the expanding array of bacterial genomes. However, the existing genome mining tools mostly concentrate on comparing BGCs to known natural products. This means that the present technique depends on gene-to-structure links. The problem of finding the molecules with useful functionalities has not been sufficiently addressed by prioritizing based on structural novelty, despite the fact that these techniques have identified over 147,000 BGC sequences. Many intriguing compounds from an architectural perspective have unknown functions. The potential to forecast a natural product's activity based on its BGC creates opportunities to focus search efforts on those most likely to produce products with the desired activities, thus changing the nature of natural product research [68].

13.6.7 *In Silico* Prediction of Biosynthetic Pathways and Identification of Potential Bioactive Compounds

Chemical reaction predictions without explicit rules can now be made, thanks to recent developments in deep learning techniques. Sequential models use string representations of chemicals, such as SMILES. Especially in retrosynthesis prediction tasks, these rule-free models have outperformed their rule-based counterparts in terms of performance and potential for generalization. Retrosynthetic pathway planning through methods has been made possible by single-step retrosynthesis prediction; nevertheless, the effectiveness is limited by the requirement for expensive online resource estimates. The Retro method is a recent development that combines an AND–OR tree-based searching strategy with deep learning instruction to demonstrate improved planning efficiency and solution quality. On the other hand, using multistep planning algorithms for natural product retrosynthetic planning presents particular difficulties since pathways for biosynthesis have more stages, more branching ratios, and less data available [69].

In the pursuit of understanding biosynthetic pathways NPs, where comprehensive information is often elusive, the development of a user-friendly toolkit, BioNavi-NP, proves invaluable. This toolbox uses both general organic and biosynthetic reactions to develop a single-step bioretrosynthesis prediction model using an end-to-end transformer neural network. Utilizing a planning algorithm effectively explores likely biosynthetic pathways for both NPs and chemicals that resemble them. Comprehensive tests show that BioNavi-NP is 1.7 times more accurate than current rule-based methods in identifying pathways of biosynthesis chemicals. Remarkably, the toolset effectively pinpoints biologically likely routes for intricate nanoparticles derived from current research. With the help of openly available curated datasets, trained models, and BioNavi-NP, biosynthetic pathway reconstruction for NPs can be made easier [69].

NPs stand as pivotal entities in drug discovery, contributing to over 50% of FDA-approved drugs and demonstrating distinct selectivity toward cellular targets. The unique properties of biologically active natural products make them promising candidates for influencing disease-related pathways and reshaping biological networks from a diseased to a healthy state. While large-scale network analyses, including drug–target networks, protein–protein interaction networks, metabolic networks, and disease pathways, have unveiled the action mechanisms of bioactive compounds, existing studies often focus on a limited number of molecules. With vast chemical diversity, NPs offer unparalleled potential for discovering diverse bioactive molecules. Despite sporadic analyses on aspects like chemical diversity, property distribution, molecular scaffold, and chemical space, comprehensive statistics comparing NPs with other compound types have been scarce due to the challenges of obtaining extensive datasets encompassing structures and annotations [70].

A myriad of antibiotics and therapeutic drugs, including penicillin, erythromycin, tetracycline, tacrolimus, cyclosporine (an immunosuppressant), artemisinin (an antimalarial), and acarbose (a diabetes medication), belong to the class of natural products. Originating from microorganisms or plants, these compounds are often categorized as "secondary metabolites" or "specialized metabolites" due to their biosynthetic pathways being distinct from essential growth and reproductive processes. BGCs are collections of genes essential for the synthesis of these substances in microbes and fungi. These BGCs contain all of the genetic repertoire needed for precursors in the process of biosynthesis, scaffold assembly, modification (tailoring), resistance, export, and regulation. This group makes it easier to identify entire pathways by showing how a single gene contributes to biosynthesis. On the other hand, in plants, the genes involved in specific pathways' production are dispersed throughout the genome, requiring other experimental information, like co-expression analysis, to identify [71].

Beginning in the early 2000s, the use of predictions in natural product biosynthesis was made possible by the introduction of computational tools, such as the search engine. Public tools that followed were made possible by Ecopia's 2003 introduction of its proprietary search engine and database. Most notably, SEARCHPKS automates the process of identifying polyketide synthases' (PKSs) catalytic domains. The advent of antiSMASH, an open-source genome mining platform that combined and enhanced the features of earlier tools, in 2011 marked a crucial turning point. In addition to providing an intuitive web interface, antiSMASH made larger-scale genome mining studies feasible for researchers lacking in-depth knowledge of computational biology. Considering antiSMASH has steadily grown, offering a wide range of tools and databases for comparative genomics and automated genome mining across multiple classes of secondary metabolites. The antiSMASH pipeline can be used for both fungal and bacterial genome analysis; a special branch called "fungiSMASH" is dedicated to fungus analysis. Moreover, plantiSMASH, a variation of antiSMASH, adds support for co-expression analysis and plant-specific features such as modified Hidden Markov Model profiles and cluster identification logic.

Computational approaches have become increasingly indispensable, presenting cost-effective and efficient means to identify and optimize potential drug candidates. These approaches encompass diverse methodologies such as molecular simulations, ML, and predictive modeling, collectively serving to streamline the intricate drug discovery process and curtail associated costs. Notwithstanding the inherent challenges, computational methods hold immense promise by effectively narrowing down the spectrum of compounds under consideration. These tools prove instrumental in supporting decision-makers, facilitating multiparameter optimization, and thereby guiding the judicious selection and design of compounds with optimal prospects for success [72].

13.7 Formulation Challenges with Natural Products

Phytochemicals, bioactive compounds derived from plants, have emerged as promising candidates in drug discovery due to their diverse pharmacological activities [73]. However, their translation from plant sources to effective drugs faces several pharmacokinetic limitations, impeding their progress in drug development [74].

One prominent challenge is the poor bioavailability of phytochemicals, a crucial factor influencing their efficacy. Bioavailability refers to the proportion of the administered dose that reaches the systemic circulation in an active form. Many phytochemicals exhibit low oral bioavailability owing to factors such as poor solubility, extensive metabolism, and limited absorption [75]. The hydrophobic nature of certain phytochemicals hampers their dissolution in the aqueous environment of the gastrointestinal tract, leading to reduced absorption and systemic availability [76]. In addition, first-pass metabolism in the liver further diminishes the bioavailability of phytochemicals, resulting in suboptimal therapeutic outcomes [75].

The distribution of phytochemicals within the body is also a significant hurdle in drug discovery [77]. The lipophilicity or hydrophilicity of these compounds influences their ability to traverse biological membranes and reach target tissues. Lipophilic phytochemicals may accumulate in adipose tissue, reducing their bioavailability to other organs. Conversely, hydrophilic compounds may face challenges in crossing cell membranes, limiting their distribution to specific cellular compartments [78]. Moreover, the selective permeability of physiological barriers, such as the blood–brain barrier, presents an additional obstacle for phytochemicals targeting the central nervous system [79].

Metabolism is a critical aspect of pharmacokinetics that profoundly affects the fate of phytochemicals in the body. Cytochrome P_{450} enzymes in the liver play a pivotal role in the biotransformation of many phytochemicals [80]. The rapid metabolism of these compounds can lead to a short half-life, necessitating frequent dosing to maintain therapeutic levels. Metabolites generated during this process may exhibit altered pharmacological activities, potentially influencing the overall therapeutic effect. Understanding the metabolic pathways of phytochemicals is crucial for predicting their *in vivo* behavior and optimizing drug development strategies [81].

Elimination, primarily through renal excretion, contributes to the overall pharmacokinetics of phytochemicals. The water solubility of compounds influences their renal clearance, and those with low water solubility may undergo enterohepatic recycling, further complicating their elimination kinetics. The rate of elimination affects the duration of action and the dosing frequency required for maintaining therapeutic levels [82].

To address these pharmacokinetic limitations, researchers in drug discovery have explored various strategies to enhance the pharmaceutical properties of phytochemicals. Nanoparticle-based drug delivery systems, including liposomes, micelles, and polymeric nanoparticles, have shown promise in improving solubility, protecting against degradation, and facilitating controlled release. These nanocarriers can enhance the bioavailability of phytochemicals by overcoming challenges related to their poor water solubility and promoting their transport across biological barriers [83].

Prodrug design represents another approach to mitigate pharmacokinetic challenges. Prodrugs are biologically inactive precursors that undergo enzymatic or chemical transformation *in vivo* to release the active drug. This strategy aims to improve the solubility, stability, and absorption of phytochemicals, ultimately optimizing their pharmacokinetic profile. By modifying the chemical structure of phytochemicals, prodrugs can enhance their pharmacological properties and address issues such as rapid metabolism [84].

While phytochemicals hold immense promise in drug discovery, their pharmacokinetic limitations pose substantial hurdles. Overcoming these challenges requires a multidisciplinary approach, integrating innovative drug delivery systems, prodrug strategies, and formulation approaches. The ongoing research in this field aims to unlock the full therapeutic potential of phytochemicals, paving the way for their successful integration into mainstream pharmaceuticals and contributing to improved healthcare outcomes.

13.8 Quality by Design (QbD) Approaches

In the early 20th century, Sir Ronald Fisher revolutionized research methodology by advocating for the integration of statistical analysis at the planning stages of experiments, a departure from the traditional practice of applying statistics retrospectively. This proactive approach, aligning with Deming's profound knowledge principles encompassing system thinking, variation understanding, theory of knowledge, and psychology, not only enhances the quality of the end product but also establishes a foundation for robust research [85]. Interestingly, the pharmaceutical industry lagged behind other sectors in embracing these progressive paradigms. Historically, the industry's focus centered on blockbuster drugs, and the formulation development process primarily relied on One Factor At a Time (OFAT) studies [86]. This approach contrasted with the contemporary methodologies of QbD and modern engineering-based manufacturing. Shifting toward these advanced strategies represents a crucial evolution for the pharmaceutical sector, aligning it more closely with the comprehensive and forward-thinking frameworks [87].

The integration of computational tools within QbD strategies has revolutionized the landscape of drug discovery, offering a sophisticated and systematic approach to optimize the development process. Computational tools, encompassing molecular modeling, bioinformatics, and artificial intelligence, play a pivotal role in the early stages of drug discovery by facilitating a deeper understanding of complex biological systems [88]. One key aspect of this integration is the predictive power of computational modeling in elucidating the structure–activity relationships (SARs) of potential drug candidates. Molecular docking and dynamics simulations allow researchers to explore the binding interactions between drugs and their target proteins, predicting the pharmacological effects and optimizing drug design for enhanced efficacy. In addition, QSAR analyses enable the systematic evaluation of the impact of various molecular features on drug activity, guiding the selection of lead compounds [89].

Computational tools also contribute significantly to the identification of potential drug targets through systems biology and network pharmacology approaches. By integrating diverse data sources, such as genomics, proteomics, and chemical databases, these tools help unravel complex biological pathways and identify key nodes for therapeutic intervention. This holistic understanding of the biological landscape aids in the selection of targets that are not only biologically relevant but also druggable [90].

Furthermore, the utilization of artificial intelligence (AI) and ML algorithms has emerged as a game-changer in drug discovery. These algorithms can analyze vast datasets, identify hidden patterns, and predict potential drug candidates with specific therapeutic properties. In the context of QbD, AI and ML models contribute to the systematic exploration of formulation parameters, predict drug–drug interactions, and assess the impact of different variables on product quality [91]. This accelerates the identification of optimal drug formulations, streamlining the drug development process.

The incorporation of computational tools within QbD strategies also extends to pharmacokinetic and pharmacodynamic (PK/PD) modeling. These tools enable the prediction of drug ADME properties, guiding the selection of compounds with favorable pharmacokinetic profiles. By integrating PK/PD modeling into QbD, researchers can establish design spaces that ensure a balance between therapeutic efficacy and safety [92].

Moreover, the use of computational tools in toxicity prediction and risk assessment contributes to the proactive identification of potential safety concerns during the early stages of drug development [93]. This aligns with the risk-based approach advocated by QbD, allowing for the identification and mitigation of risks associated with drug candidates, ultimately leading to the development of safer and more effective pharmaceuticals.

13.8.1 Use of Computational Models for Formulation Optimization

The pharmaceutical sector, characterized by its process-centric and quality-oriented approach, would conventionally be anticipated to promptly assimilate the aforementioned paradigms following their introduction. Contrary to this expectation, however, QbD was formally recommended by regulatory authorities such as the FDA and EMA at the onset of the new millennium. This endorsement reflects a discerning acknowledgment that quality cannot be ascribed to products through testing alone; rather, it necessitates a proactive integration during the design phase. In essence, the regulatory authorities underscored the imperative for the pharmaceutical industry to embrace a paradigm shift wherein quality is not an outcome of testing protocols but is intricately woven into the very fabric of the design process itself. This directive emphasizes a strategic and preemptive approach to quality assurance, emphasizing the necessity to embed quality principles in the early stages of product development and design [94]. This is a direct yet late acknowledgment of the significance of quality theory in pharmaceutical development, as briefly presented. The design of experiments is the main component of the statistical toolbox to deploy QbD in both research and industrial settings [95]. It is imperative to acknowledge that a multitude of mathematical modeling techniques exists to address pharmaceutical development, particularly within the QbD and process analytical technology (PAT) framework. One illustrative example is the application of multivariate data analysis (MVDA) techniques, which primarily concentrate on historical data. MVDA can serve as an initial point of departure for the design of experiments (DoEs). Consequently, the integration of MVDA in conjunction with DoE proves instrumental in the analysis of both nondesigned and designed factors. This strategic combination of techniques allows for a comprehensive approach to understanding and optimizing pharmaceutical processes, aligning with the principles of QbD and PAT. The utilization of these mathematical modeling tools contributes to a more informed and systematic exploration of the complex interplay between various factors, ultimately enhancing the efficiency and efficacy of pharmaceutical development processes [85]. Nevertheless, in light of the proactive ethos inherent in QbD, DoE emerges as the primary systematic approach necessitating the incorporation of statistical thinking at the inception of pharmaceutical development – a principle that resonates with the central tenets of Fisher's legacy. The advent of the ICH Q8 guideline has notably fostered a conducive environment for the application of DoE. Consequently, a discernible surge in relevant scientific research and industrial implementation has been observed. This trend has been further bolstered by the introduction of user-friendly software tools, which streamline the construction and analysis of experimental designs, thereby facilitating the seamless integration of DoE in pharmaceutical development processes. The combination of QbD principles, the ICH Q8 guideline, and accessible software resources collectively underscores the increasing significance and widespread adoption of DoE as a pivotal methodological framework in the pharmaceutical industry [85].

QbD is a systematic development approach that commences with predefined objectives, placing a strong emphasis on gaining a comprehensive understanding of both the product and the associated processes. This involves a meticulous focus on process control rooted in sound scientific principles and quality risk management. The core philosophy of QbD is centered around a proactive and methodical strategy, ensuring that quality is not just a desirable outcome but an integral part of the entire development process. By integrating robust scientific principles and risk management strategies, QbD facilitates a streamlined and efficient path from concept to product, fostering a culture of continuous improvement and excellence in the pursuit of high-quality outcomes [94]. Experimental design, also known as DoE, is a meticulously structured and organized methodology employed to discern the intricate relationships between factors influencing a given process and the subsequent output of that process. In essence, it serves as a systematic approach to acquiring profound process knowledge by establishing precise mathematical relationships between the various inputs and outputs of the process under consideration. This methodological framework provides a comprehensive means of understanding and optimizing processes through a strategic and well-defined exploration of the interdependencies among different factors. The utilization of experimental design, or DoE, thus contributes to a more informed and scientific decision-making process in various fields, offering a rigorous and systematic avenue for uncovering the underlying mechanisms governing complex processes and facilitating advancements in process optimization and quality enhancement [85].

A process is essentially a value-adding activity that intricately transforms a set of inputs into outputs. This transformation is intricately influenced by various factors, which can be categorized as either controlled or uncontrolled. The latter is aptly termed "noise" as these factors occur randomly and, in the grand scheme, exert a minimal impact compared to their controlled counterparts. In essence, a process involves managing and optimizing these factors to enhance efficiency and ensure that the controlled elements play a predominant role in shaping the desired outcomes [85]. However, within the realm of experimental design, Montgomery discerns process variables by categorizing them into potential design factors and nuisance factors. These are then further classified as controllable, uncontrollable, or noise. The latter pertains to uncontrolled and inevitable variations, represented by experimental error. This distinction underscores the unavoidable nature of certain factors, emphasizing the need to differentiate them within the experimental framework for a more nuanced understanding of their impact on the overall outcomes [86]. The influence of controlled factors on the quality characteristics of the end product follows a consistent pattern known as the Pareto principle. This principle, often referred to as the 20:80 rule, highlights that a relatively small number of factors contribute significantly to the overall effect. In essence, it posits that 20% of the causes (factors) account for 80% of the results (responses). Within the framework of ICH Q8, these highly impactful factors are identified as critical process parameters (CPPs). CPPs are defined as process parameters whose variability directly affects a critical quality attribute of the product, underscoring their pivotal role in determining the final product's quality [94]. Consequently, it is imperative to monitor and control CPPs to guarantee the consistent production of the desired quality. This quality is defined by the aggregation of the product's characteristics, which must consistently align with predefined ranges commonly referred to as specifications. Ensuring that these specifications are met through vigilant monitoring and control of CPPs becomes instrumental in maintaining the overall quality integrity of the process and its final outcomes.

In this context, ICH Q8 outlines critical quality attributes (CQAs) as essential physical, chemical, biological, or microbiological properties that must fall within appropriate limits, ranges, or distributions to uphold the desired product quality. DoE serves as an approach wherein the controlled input factors of the process undergo systematic and purposeful variations to discern their effects on the corresponding responses. This systematic exploration allows for a comprehensive understanding of the interplay between various factors and their impact on critical quality aspects, aiding in the refinement and optimization of the overall production process [86].

DoE surpasses OFAT in several aspects. Notably, DoE excels in maximizing process knowledge while minimizing resource usage and delivering accurate information efficiently. Its strength lies in identifying factor interactions and characterizing the relative significance of each factor, enabling predictions of process behavior within the design space [96]. In addition, DoE establishes a robust cause-and-effect relationship between CPPs and quality CQAs, facilitating the optimization of CQAs through appropriate CPP settings and defining a proven acceptable range (PAR)

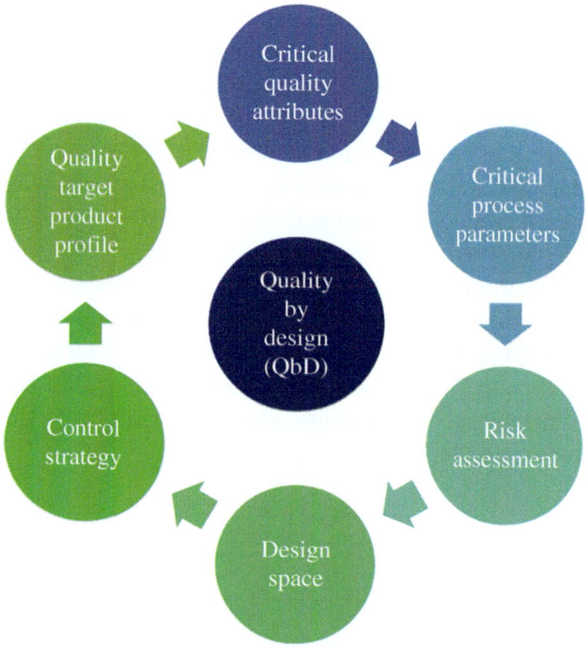

Figure 13.3 QbD in rational product development.

(Figure 13.3). This method allows for multiple response optimization, which is crucial in the pharmaceutical industry, where products have numerous CQAs requiring simultaneous enhancement. Moreover, DoE contributes to the robustness of products or processes, making them less sensitive to external influences, and ensures protection from outliers through effective identification within structured experimental matrices. In essence, DoE emerges as a comprehensive and powerful approach that not only enhances efficiency and precision but also fortifies the product or process against various challenges, reinforcing its reliability in pharmaceutical research and development [85].

In contrast, OFAT approaches focus on pinpointing local suboptimal regions by altering one factor at a time. However, this outdated and time-consuming technique lacks the capability to concurrently study multiple factor changes or identify causal relationships, including interactions. These limitations render OFAT unsuitable for QbD applications, where a more holistic and efficient approach, such as DoE, is essential for gaining a comprehensive understanding of the multifaceted relationships between various factors influencing the desired outcomes [85].

13.9 Conclusion

In conclusion, the rational design of natural products for drug discovery has seen remarkable progress with the integration of computational studies. The utilization of computational approaches offers several advantages, including the ability to expedite the identification of potential drug candidates, rationalize molecular interactions, and predict pharmacokinetic properties. By leveraging computational tools, researchers can efficiently sift through vast chemical space, thereby minimizing the time and resources required for traditional experimental methods. Moreover, computational studies facilitate the exploration of complex molecular mechanisms underlying disease pathogenesis, enabling the design of more potent and selective therapeutics. The synergy between computational and experimental techniques enhances the efficiency of drug discovery pipelines, leading to the development of novel drugs with improved efficacy and safety profiles. The pharmaceutical industry initially lagged behind in adopting progressive methodologies, notably focusing on blockbuster drugs and relying on OFAT studies for formulation development. This delayed embrace of advanced approaches like QbD and modern engineering-based manufacturing methodologies hindered the sector's overall evolution compared to other industries. However, within natural product development, the implementation of QbD has become crucial, with DoE emerging as a prominent mathematical modeling approach. DoE plays a pivotal role in QbD by establishing models that correlate CPPs and material attributes (CMAs) with CQAs, defining the design space. This comprehensive understanding of the product and process facilitates the assurance of quality in the final product, aligning with the quality target product profile. The review underscores the significance of QbD principles, tracing their evolution and emphasizing the central role of DoE as the primary tool for rational pharmaceutical development, particularly in the context of natural product development.

References

1 Mahapatra, M.K. and Karuppasamy, M. (2022). Fundamental considerations in drug design. In: *Computer Aided Drug Design (CADD): From Ligand-Based Methods to Structure-Based Approaches*, 17–55. Elsevier.

2 Simoben, C.V., Babiaka, S.B., Moumbock, A.F.A. et al. (2023). Challenges in natural product-based drug discovery assisted with *in silico*-based methods. *RSC Advances [Internet]* 13 (45): 31578.

3 Kitchen, D.B. (2017). Computer-aided drug discovery research at a global contract research organization. *Journal of Computer-Aided Molecular Design [Internet]* 31 (3): 309–318. https://doi.org/10.1007/s10822-016-9991-3.

4 Morelli, X. and Rigby, A.C. (2007). Acceleration of the drug discovery process: a combinatorial approach using NMR spectroscopy and virtual screening. *Current Computer-Aided Drug Design* 3 (1): 33–49.

5 Ejalonibu, M.A., Ogundare, S.A., Elrashedy, A.A. et al. (2021). Drug discovery for *Mycobacterium tuberculosis* using structure-based computer-aided drug design approach. *International Journal of Molecular Sciences [Internet]* 22 (24).

6 Arora, D.D. and Kumar, D.M. (2002). Rational use of ayurvedic literaature for drug development. *Ancient Science of Life [Internet]* 21 (3): 182.

7 Tuteja, R. and Tuteja, N. (2004). Serial analysis of gene expression: applications in human studies. *Journal of Biotechnology and Biomedicine [Internet]* 2004 (2): 113.

8 Bai, Y., Zhang, L., and Lei, X. (2021). Human endogenous natural products. *Progress in the Chemistry of Organic Natural Products [Internet]* 114: 313–337.

9 Tan, G., Gyllenhaal, C., and Soejarto, D. (2006). Biodiversity as a source of anticancer drugs. *Current Drug Targets [Internet]* 7 (3): 265–277.

10 Harvey, A.L., Clark, R.L., Mackay, S.P., and Johnston, B.F. (2010). Current strategies for drug discovery through natural products. *Expert Opinion on Drug Discovery [Internet]* 5 (6): 559–568.

11 Grabley, S. and Thiericke, R. (1999). Bioactive agents from natural sources: trends in discovery and application. *Advances in Biochemical Engineering/Biotechnology [Internet]* 64: 101–154.

12 Amirkia, V. and Heinrich, M. (2015). Natural products and drug discovery: a survey of stakeholders in industry and academia. *Frontiers in Pharmacology [Internet]* 6 (Oct).

13 Crane, E.A. and Gademann, K. (2016). Capturing biological activity in natural product fragments by chemical synthesis. *Angewandte Chemie International Edition [Internet]* 55 (12): 3882–3902.
14 Sharma, S.B. and Gupta, R. (2015). Drug development from natural resource: a systematic approach. *Mini-Reviews in Medicinal Chemistry [Internet]* 15 (1): 52–57.
15 Kim, H.K., Choi, Y.H., and Verpoorte, R. (2023). Natural products drug discovery: on silica or *in-silico*? *Handbook of Experimental Pharmacology [Internet]* 277: 117–141.
16 Barba-Ostria, C., Carrera-Pacheco, S.E., Gonzalez-Pastor, R. et al. (2022). Evaluation of biological activity of natural compounds: current trends and methods. *Molecules [Internet]* 27 (14).
17 Glassman, P.M. and Muzykantov, V.R. (2019). Pharmacokinetic and pharmacodynamic properties of drug delivery systems. *Journal of Pharmacology and Experimental Therapeutics [Internet]* 370 (3): 570.
18 Thomford, N.E., Senthebane, D.A., Rowe, A. et al. (2018). Natural products for drug discovery in the 21st century: innovations for novel drug discovery. *International Journal of Molecular Sciences [Internet]* 19 (6).
19 Atanasov, A.G., Zotchev, S.B., Dirsch, V.M. et al. (2021). Natural products in drug discovery: advances and opportunities. *Nature Reviews Drug Discovery [Internet]* 20 (3): 200–216.
20 Hong, J. (2011). Role of natural product diversity in chemical biology. *Current Opinion in Chemical Biology [Internet]* 15 (3): 350–354.
21 Ertl, P. and Schuffenhauer, A. (2009). Estimation of synthetic accessibility score of drug-like molecules based on molecular complexity and fragment contributions. *Journal of Cheminformatics [Internet]* 1 (1).
22 David, F., Davis, A.M., Gossing, M. et al. (2021). A perspective on synthetic biology in drug discovery and development-current impact and future opportunities. *SLAS Discovery: Advancing Life Sciences R&D [Internet]* 26 (5): 581–603.
23 Niesenbaum, R.A. (2019). The integration of conservation, biodiversity, and sustainability. *Sustainable [Internet]* 11 (17): 4676.
24 Dzobo, K. (2022). The role of natural products as sources of therapeutic agents for innovative drug discovery. *Comprehensive Pharmacology [Internet]* 2: 408.
25 Linhares, Y., Kaganski, A., Agyare, C. et al. (2023). Biodiversity: the overlooked source of human health. *Trends in Molecular Medicine [Internet]* 29 (3): 173–187.
26 Dias, D.A., Urban, S., and Roessner, U. (2012). A historical overview of natural products in drug discovery. *Metabolites [Internet]* 2 (2): 303.
27 Von Nussbaum, F., Brands, M., Hinzen, B. et al. (2006). Antibacterial natural products in medicinal chemistry – exodus or revival? *Angewandte Chemie International Edition [Internet]* 45 (31): 5072–5129.
28 Neergheen-Bhujun, V., Awan, A.T., Baran, Y. et al. (2017). Biodiversity, drug discovery, and the future of global health: introducing the biodiversity to biomedicine consortium, a call to action. *Journal of Global Health [Internet]* 7 (2).
29 Alves, R.R.N. and IML, R. (2007). Biodiversity, traditional medicine and public health: where do they meet? *Journal of Ethnobiology and Ethnomedicine [Internet]* 3.
30 Cragg, G.M. and Newman, D.J. (2013). Natural products: a continuing source of novel drug leads. *Biochimica et Biophysica Acta [Internet]* 1830 (6): 3670–3695.
31 Bernardini, S., Tiezzi, A., Laghezza Masci, V., and Ovidi, E. (2018). Natural products for human health: an historical overview of the drug discovery approaches. *Natural Product Research [Internet]* 32 (16): 1926–1950.
32 Lautié, E., Russo, O., Ducrot, P., and Boutin, J.A. (2020). Unraveling plant natural chemical diversity for drug discovery purposes. *Frontiers in Pharmacology [Internet]* 11.

33 Urban, S. and Dias, D.A. (2013). NMR spectroscopy: structure elucidation of cycloelatanene A: a natural product case study. *Methods in Molecular Biology [Internet]* 1055: 99–116.

34 Bouslimani, A., Sanchez, L.M., Garg, N., and Dorrestein, P.C. (2014). Mass spectrometry of natural products: current, emerging and future technologies. *Natural Product Reports [Internet]* 31 (6): 718–729.

35 Přichystal, J., Schug, K.A., Lemr, K. et al. (2016). Structural analysis of natural products. *Analytical Chemistry [Internet]* 88 (21): 10338–10346.

36 Sliwoski, G., Kothiwale, S., Meiler, J., and Lowe, E.W. (2014). Computational methods in drug discovery. *Pharmacological Reviews [Internet]* 66 (1): 334.

37 Morris, G.M. and Lim-Wilby, M. (2008). Molecular docking. *Methods in Molecular Biology [Internet]* 443: 365–382.

38 Wang, Y., Hu, J.S., Lin, H.Q. et al. (2016). Herbalog: a tool for target-based identification of herbal drug efficacy through molecular docking. *Phytomedicine [Internet]* 23 (12): 1469–1474.

39 Medina-Franco, J.L. and Saldívar-González, F.I. (2020). Cheminformatics to characterize pharmacologically active natural products. *Biomolecules [Internet]* 10 (11): 1–14.

40 Gogoi, N., Chowdhury, P., Goswami, A.K. et al. (2022). Integrated computational approach towards repurposing of antimalarial drug against SARS-CoV-2 main protease. *Structural Chemistry [Internet]* 0123456789. https://doi.org/10.1007/s11224-022-01916-0.

41 Puthanveedu, V. and Muraleedharan, K. (2022). Phytochemicals as potential inhibitors for COVID-19 revealed by molecular docking, molecular dynamic simulation and DFT studies. *Structural Chemistry [Internet]* 33 (5): 1423–1443.

42 Allen, S.E., Dokholyan, N.V., and Bowers, A.A. (2016). Dynamic docking of conformationally constrained macrocycles: methods and applications. *ACS Chemical Biology [Internet]* 11 (1): 10–24.

43 Vijayakumari, B., Sasikala, V., Radha, S.R., and Rameshwar, H.Y. (2016). *In silico* analysis of aqueous root extract of *Rotula aquatica* Lour for docking analysis of the compound 3-O-acetyl-11-keto-β-boswellic acid contents. *Springerplus [Internet]* 5 (1).

44 Rollinger, J.M., Stuppner, H., and Langer, T. (2008). Virtual screening for the discovery of bioactive natural products. *Progress in Drug Research [Internet]* 65: 212–249.

45 Kirchweger, B. and Rollinger, J.M. (2019). A strength-weaknesses-opportunities-threats (SWOT) analysis of cheminformatics in natural product research. *Progress in the Chemistry of Organic Natural Products [Internet]* 110: 239–271.

46 Olğaç, A., Orhan, I.E., and Banoglu, E. (2017). The potential role of *in silico* approaches to identify novel bioactive molecules from natural resources. *Future Medicinal Chemistry [Internet]* 9 (14): 1663–1684.

47 Chen, Y., De Bruyn Kops, C., and Kirchmair, J. (2017). Data resources for the computer-guided discovery of bioactive natural products. *Journal of Chemical Information and Modeling [Internet]* 57 (9): 2099–2111. https://doi.org/10.1021/acs.jcim.7b00341.

48 Santana, K., do Nascimento, L.D., Lima e Lima, A. et al. (2021). Applications of virtual screening in bioprospecting: facts, shifts, and perspectives to explore the chemo-structural diversity of natural products. *Frontiers in Chemistry [Internet]* 9.

49 Choi, J., He, N., Kim, N., and Yoon, S. (2012). Enrichment of virtual hits by progressive shape-matching and docking. *The Journal of Molecular Graphics and Modelling [Internet]* 32: 82–88.

50 Speck-Planche, A. and Cordeiro, M.N.D.S. (2015). Multitasking models for quantitative structure-biological effect relationships: current status and future perspectives to speed up drug discovery. *Expert Opinion on Drug Discovery [Internet]* 10 (3): 245–256.

51 Kar, S. and Roy, K. (2012). QSAR of phytochemicals for the design of better drugs. *Expert Opinion on Drug Discovery [Internet]* 7 (10): 877–902.

52 Kausar, S. and Falcao, A.O. (2018). An automated framework for QSAR model building. *Journal of Cheminformatics [Internet]* 10 (1).

53 Nguyen, G.T.H., Bennett, J.L., Liu, S. et al. (2021). Multiplexed screening of thousands of natural products for protein–ligand binding in native mass spectrometry. *Journal of the American Chemical Society [Internet]* 143 (50): 21379–21387.

54 Eldridge, G.R., Vervoort, H.C., Lee, C.M. et al. (2002). High-throughput method for the production and analysis of large natural product libraries for drug discovery. *Analytical Chemistry [Internet]* 74 (16): 3963–3971.

55 Falcón-Cano, G., Cabrera-Pérez, M.Á., and Molina, C. (2020). ADME prediction with KNIME: *in silico* aqueous solubility consensus model based on supervised recursive random forest approaches. *ADMET DMPK [Internet]* 8 (3): 251.

56 Deng, C., Liang, L., Xing, G. et al. (2023). Multi-channel GCN ensembled machine learning model for molecular aqueous solubility prediction on a clean dataset. *Molecular Diversity [Internet]* 27 (3): 1023–1035.

57 Ntie-Kang, F. (2013). An *in silico* evaluation of the ADMET profile of the StreptomeDB database. *Springerplus [Internet]* 2 (1): 1–11.

58 Chen, X.Q., Cho, S.J., Li, Y., and Venkatesh, S. (2002). Prediction of aqueous solubility of organic compounds using a quantitative structure–property relationship. *Journal of Pharmaceutical Sciences [Internet]* 91 (8): 1838–1852.

59 Alqahtani, S. (2017). *In silico* ADME-Tox modeling: progress and prospects. *Expert Opinion on Drug Metabolism & Toxicology [Internet]* 13 (11): 1147–1158.

60 Paul Gleeson, M., Hersey, A., and Hannongbua, S. (2011). *In silico* ADME models: a general assessment of their utility in drug discovery applications. *Current Topics in Medicinal Chemistry [Internet]* 11 (4): 358–381.

61 Abshear, T., Banik, G.M., D'Souza, M.L. et al. (2006). A model validation and consensus building environment. *SAR and QSAR in Environmental Research [Internet]* 17 (3): 311–321.

62 Liang, L., Liu, Y., Kang, B. et al. (2022). Large-scale comparison of machine learning algorithms for target prediction of natural products. *Briefings in Bioinformatics [Internet]* 23 (5).

63 Webel, H.E., Kimber, T.B., Radetzki, S. et al. (2020). Revealing cytotoxic substructures in molecules using deep learning. *Journal of Computer-Aided Molecular Design [Internet]* 34 (7): 731–746.

64 Kamerlin, N., Delcey, M.G., Manzetti, S., and Van Der Spoel, D. (2020). Toward a computational ecotoxicity assay. *Journal of Chemical Information and Modeling [Internet]* 60 (8): 3792–3803.

65 Hu, Z., Wahl, J., Hamburger, M., and Vedani, A. (2016). Molecular mechanisms of endocrine and metabolic disruption: an *in silico* study on antitrypanosomal natural products and some derivatives. *Toxicology Letters [Internet]* 252: 29–41.

66 Behera, R., Thomas, S.M., and Mensa-Wilmot, K. (2014). New chemical scaffolds for human african trypanosomiasis lead discovery from a screen of tyrosine kinase inhibitor drugs. *Antimicrobial Agents and Chemotherapy [Internet]* 58 (4): 2202.

67 Zhang, H.W., Lv, C., Zhang, L.J. et al. (2021). Application of omics- and multi-omics-based techniques for natural product target discovery. *Biomedicine & Pharmacotherapy [Internet]* 141.

68 Walker, A.S. and Clardy, J. (2021). A machine learning bioinformatics method to predict biological activity from biosynthetic gene clusters. *Journal of Chemical Information and Modeling [Internet]* 61 (6): 2560–2571.

69 Zheng, S., Zeng, T., Li, C. et al. (2022). Deep learning driven biosynthetic pathways navigation for natural products with BioNavi-NP. *Nature Communications [Internet]* 13 (1).

70 Gu, J., Gui, Y., Chen, L. et al. (2013). Use of natural products as chemical library for drug discovery and network pharmacology. *PLoS One [Internet]* 8 (4).

71 Blin, K., Kim, H.U., Medema, M.H., and Weber, T. (2019). Recent development of antiSMASH and other computational approaches to mine secondary metabolite biosynthetic gene clusters. *Briefings in Bioinformatics [Internet]* 20 (4): 1103–1113.

72 Decherchi, S. and Cavalli, A. (2020). Thermodynamics and kinetics of drug–target binding by molecular simulation. *Chemical Reviews [Internet]* 120 (23): 12788–12833.

73 Atanasov, A.G., Waltenberger, B., Pferschy-Wenzig, E.M. et al. (2015). Discovery and resupply of pharmacologically active plant-derived natural products: a review. *Biotechnology Advances* 33 (8): 1582–1614.

74 Yadav, H., Mahalvar, A., Pradhan, M. et al. (2023). Exploring the potential of phytochemicals and nanomaterial: a boon to antimicrobial treatment. *Medicine in Drug Discovery* 17: 100151.

75 Aqil, F., Munagala, R., Jeyabalan, J., and Vadhanam, M.V. (2013). Bioavailability of phytochemicals and its enhancement by drug delivery systems. *Cancer Letters* 334 (1): 133–141.

76 Singh, M., Devi, S., Rana, V.S. et al. (2019). Delivery of phytochemicals by liposome cargos: recent progress, challenges and opportunities. *Journal of Microencapsulation* 36 (3): 215–235.

77 Nasim, N., Sandeep, I.S., and Mohanty, S. (2022). Plant-derived natural products for drug discovery: current approaches and prospects. *The Nucleus* 65 (3): 399–411.

78 Testa, B., Crivori, P., Reist, M., and Carrupt, P.A. (2000). The influence of lipophilicity on the pharmacokinetic behavior of drugs: concepts and examples. *Perspectives in Drug Discovery and Design* 19 (1): 179–211.

79 Upadhyay, R.K. (2014). Drug delivery systems, CNS protection, and the blood brain barrier. *BioMed Research International* 2014.

80 Rathaur, P. and SR KJ. (2020). Metabolism and pharmacokinetics of phytochemicals in the human body. *Current Drug Metabolism* 20 (14): 1085–1102.

81 Zhang, Z. and Tang, W. (2018). Drug metabolism in drug discovery and development. *Acta Pharmaceutica Sinica B* 8 (5): 721–732.

82 Mehta, P., Shah, R., Lohidasan, S., and Mahadik, K.R. (2015). Pharmacokinetic profile of phytoconstituent(s) isolated from medicinal plants – a comprehensive review. *Journal of Traditional and Complementary Medicine* 5 (4): 207–227.

83 Chavda, V.P., Nalla, L.V., Balar, P. et al. (2023). Advanced phytochemical-based nanocarrier systems for the treatment of breast cancer. *Cancers* 15 (4): 1023.

84 Markovic, M., Ben-Shabat, S., and Dahan, A. (2020). Prodrugs for improved drug delivery: lessons learned from recently developed and marketed products. *Pharm* 12 (11): 1031.

85 Politis, S.N., Colombo, P., Colombo, G., and Rekkas, D.M. (2017). Design of experiments (DoE) in pharmaceutical development. *Drug Development and Industrial Pharmacy* 43 (6): 889–901.

86 Montgomery, D.C. *Design and Analysis of Experiments*, 8ee. Wiley.

87 Yu, L.X., Amidon, G., Khan, M.A. et al. (2014). Understanding pharmaceutical quality by design. *The AAPS Journal* 16 (4): 771–783.

88 Aminpour, M., Montemagno, C., and Tuszynski, J.A. (2019). An overview of molecular modeling for drug discovery with specific illustrative examples of applications. *Molecules* 24 (9): 1693.

89 Adelusi, T.I., Oyedele, A.Q.K., Boyenle, I.D. et al. (2022). Molecular modeling in drug discovery. *Informatics in Medicine Unlocked* 29: 100880.

90 Agamah, F.E., Mazandu, G.K., Hassan, R. et al. (2020). Computational/*in silico* methods in drug target and lead prediction. *Briefings in Bioinformatics* 21 (5): 1663–1675.

91 Han, R., Yoon, H., Kim, G. et al. (2023). Revolutionizing medicinal chemistry: the application of artificial intelligence (AI) in early drug discovery. *Pharmaceuticals* 16 (9): 1259.

92 Loisios-Konstantinidis, I., Paraiso, R.L.M., Fotaki, N. et al. (2019). Application of the relationship between pharmacokinetics and pharmacodynamics in drug development and therapeutic equivalence: a PEARRL review. *The Journal of Pharmacy and Pharmacology* 71 (4): 699–723.

93 Yang, S. and Kar, S. (2023). Application of artificial intelligence and machine learning in early detection of adverse drug reactions (ADRs) and drug-induced toxicity. *Artificial Intelligence Chemistry* 1 (2): 100011.

94 Pharmaceutical Development (2009). ICH of technical requirements for registration of pharmaceuticals for human use, pharmaceutical development Q8(R2). *International Conference on Harmonisation* https://database.ich.org/sites/default/files/Q8_R2_Guideline.pdf.

95 Box, G.E.P. and Wilson, K.B. (1951). On the experimental attainment of optimum conditions. *Journal of the Royal Statistical Society Series B-Methodological [Internet]* 13: 1–38.

96 Martins Fukuda, I., Francini, C., Pinto, F. et al. (2018). Design of experiments (DoE) applied to pharmaceutical and analytical quality by design (QbD). *Journal of Pharmaceutical Sciences* 54.

14

Design of Enzyme Inhibitors in Drug Discovery

Koyel Kar

Department of Pharmaceutical Chemistry, BCDA College of Pharmacy & Technology, Kolkata, West Bengal, India

14.1 Introduction

Enzymes are appealing targets for therapeutic intervention, given that they are crucial in numerous processes of metabolism. Designing enzyme inhibitors has become a crucial approach in the development of pharmaceuticals, giving potential cures for a variety of ailments. By suppressing specified enzymes, one may alter their activity, disrupt biochemical processes, and subsequently resume normal physiological processes. However, generating reliable enzyme inhibitors is a challenging, varied process that requires an array of strategies and confronts an assortment of hurdles [1]. This chapter attempts to serve as a foundation for the area of enzyme inhibitors, focusing on the methods adopted and challenges encountered in their development [2]. While highlighting various inhibitor types and their modes of action, the core concepts of enzyme inhibition were also highlighted. Numerous drug design approaches, including computer-aided strategies, ligand-based approaches, and structure-based discovery of drugs were also explored [3].

As many enzymes have similar biological and structural attributes, acquiring selectivity and sensitivity in enzyme inhibitor design is one of the primary difficulties [4]. To generate inhibitors that specifically target the desired enzyme while minimizing off-target effects, it is crucial to comprehend the site of activity, docking interactions, and substrate preference of the enzyme [5]. In addition, difficulties with pharmaceutical kinetics, biological availability, and possible toxic effects tend to be resolved throughout the fabrication of enzyme inhibitors [6]. The implementation of prodrugs, prodrug approaches, and formulating strategies are some of the methods being studied to boost the drug-like attributes and escalate the effectiveness of enzyme inhibitors [7]. This chapter will also provide insight into the constantly evolving field of enzyme inhibitor design, taking benefit of modern advances like fragment-based discovery of drugs, allosteric suppression, and the use of modern tools like machine learning and artificial intelligence [8].

Generating new pharmaceuticals that leverage a comprehension of the critical function of enzymes in ailments and the use of approaches to regulate their activities for medicinal purposes is a tedious and costly inventive process known as drug discovery and development. This chemotherapeutic approach requires the use of the biochemical variances between recipient and pathogen cells. These inhibitors modulate the recuperation process by limiting particular metabolic activity. As a result, pharmacology and the pharmaceutical industry are now adopting a

Computational Methods for Rational Drug Design, First Edition. Edited by Mithun Rudrapal.

fundamental approach to grasping the role of enzyme inhibitors in drug discovery. An active area of biochemical and pharmacological research is centered around the detection of intriguing robust enzyme inhibitors and the ensuing modification of these compounds. Enzyme inhibitors make up an important percentage of the orally attainable medicinal products already in use, making them ideal targets for pharmaceutical discovery [9].

The effectiveness of the inhibitor and its suitability for the target enzyme dictate how effective an enzyme inhibitor is as a drug. These traits, in turn, are impacted by the extent and form of interactions that happen involving the inhibitor and the enzyme. To design new medications, it is beneficial to screen low molecular weight molecules against both emerging and traditional protein targets. The three-dimensional arrangements of these enzymes, whether in free form or bound to inhibitors that were formerly known to exist, digging into vast chemical databases, or the incorporation of an array of computer-assisted methodologies are usually what make such a drug design process for a single enzyme target more efficient [10].

14.2 Importance of Enzyme Inhibition as a Strategy for Modulating Enzyme Activity

By emphasizing certain enzymes implicated in the development of diseases, enzyme inhibitors are essential in the medication emergence process. A successful approach for governing enzymatic activity and obstructing disease-related processes is enzyme inhibition. The enzyme's active site is in confrontation with inhibitors and the substrate. Inhibitors disrupt an enzyme's function or structure by adhering to a different location on the enzyme. Inhibitors adhere to the enzyme–substrate complex and suppress the generation of compounds [11]. It leverages the intended enzyme's three-dimensional framework to generate inhibitors that can occupy the active site. It uses data about the known ligands or substrates of the enzyme to build inhibitors with comparable characteristics. In the process of uncovering new drugs, enzyme inhibitors are crucial. The key cause of disorders, as discovered by molecular analysis, is the malfunction, overexpression, or excessive stimulation of the implicated enzymes. Effective enzyme inhibitors can be used to alleviate this escalated activation or overexpression of enzymes. Several enzyme inhibitors are already accessible at medical facilities [12].

14.3 Classification of Enzyme Inhibitors

14.3.1 Reversible Inhibitors

An enzyme is rendered latent by a reversible inhibitor through noncovalent interactions. A reversible inhibitor may differentiate from the enzyme as opposed to an irreversible inhibitor. Reversible inhibitors exist in various other forms. Any substance that structurally replicates an identifiable substrate and hence confronts it for binding to the center of activity of an enzyme is deemed to be a competitive inhibitor. The enzyme does not interact with the inhibitor, but it hinders the substrate from getting adjacent to the active site [13]. The relative quantities of a substrate and a competitive inhibitor dictate how much they impair an enzyme's ability to function. The inhibitor will initially block the majority of the active sites if it is present in relatively high concentrations. However, some substrate molecules will eventually adhere to the active site owing to the reversible aspect of the association and transform into a product. An elevated concentration of substrate induces the inhibitor to be shifted away from the active site. By integrating the substrate and elevating its concentration to exceed that of the inhibitor, competitive inhibition can be utterly reversed. Reversible inhibition studies have shed light on unique enzyme–substrate complexes and the interactions involving different entities at the active site. As a result, pharmaceutical companies have developed medications that suppress unique cancer cells and bacteria's metabolic activities through competition. Numerous medications are competitive inhibitors of particular enzymes [14].

14.3.2 Irreversible Inhibitors

By forging an irreversible covalent link with a specific group at the active site, an inhibitor inactivates an enzyme. Because of how strongly the inhibitor adheres to the enzyme, too much substrate will not make the inhibition disappear. By generating an enzyme–inhibitor tandem with a particular OH group of serine prevalent at the active sites of several enzymes, the nerve gases, particularly diisopropyl fluorophosphates (DIFP), irreversibly impede biological systems. DIFP suppresses trypsin and chymotrypsin, both of which have serine groups at their active sites [14].

14.3.3 Competitive Inhibitors

In the context of competitive inhibition, the inhibitor and the substrate strive to bind to the enzyme's active site. The substrate and this sort of inhibitor are structurally associated. The inability to adhere to the substrate implies diminished enzyme activity [15] (Figure 14.1).

14.3.4 Noncompetitive Inhibitors

By adhering to the enzyme at a location other than the active site, noncompetitive inhibition precludes the enzyme from executing the catalyzed reaction as envisaged. An alteration, typically conformational, which ensues from the inhibitor's adherence to the enzyme might hinder the enzyme from binding the substrate or from functioning on a substrate that has already been bound. In any situation, boosting the substrate's availability will not be able to completely counteract the inhibitor's consequences [15].

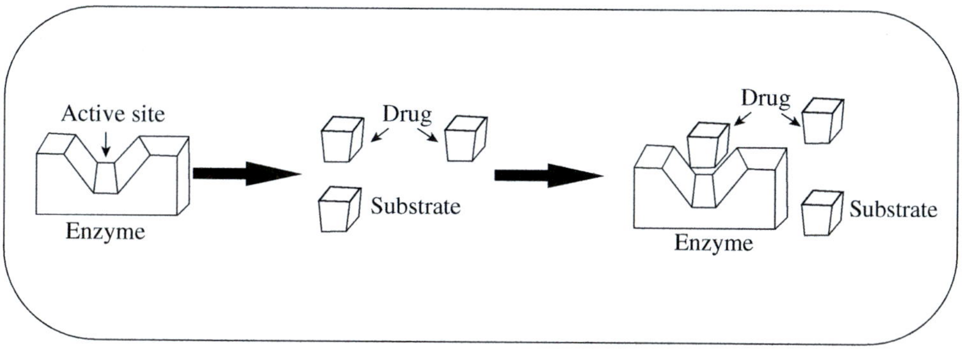

Figure 14.1 Competitive inhibition of the enzyme.

14.3.5 Allosteric Modulators

The secondary binding site implications are the core of an unusual approach to drug design. Small molecule medications are generated in this method to bind onto secondary binding sites on the targeted biomolecules rather than the major orthosteric sites. Allosterism is the term adopted to refer to the strategy and its secondary locations. A successful allosteric modulator will have the ability to attach to an allosteric site and remotely alter the conformation of the biological target's dominant orthosteric binding site [16].

The orthosteric positioning of the enzyme or receptor protein may be impacted by this conformational alteration in the following ways [17]:

- The allosteric alteration may elevate the ligand's binding affinity to the orthosteric site, enhancing the signal or boosting activity. Positive allosteric modulators are the substances that exhibit these effects.
- The orthosteric binding site may become retarded or repressed as a result of the allosteric modulation, which will attenuate the signal or diminish activity. Negative allosteric modulators are the substances that generate this result.

14.4 Strategies Employed in the Design and Development of Enzyme Inhibitors

14.4.1 Structure-Based Design

Structure-based drug design is one of the most crucial methods to generate enzyme inhibitors. This strategy utilizes the target enzyme's three-dimensional structure to pinpoint possible binding sites and generate inhibitors that perfectly complement the active site. Researchers can create inhibitor compounds that are optimized to effectively impede enzymatic activity by comprehending the connections involving the enzyme and its substrate. Pharmaceutical research and development focus a high priority on comprehending the mechanisms by which small-molecule ligands recognize and interact with macromolecules [18]. Structure-based design is the methodical implementation of structural data, either acquired experimentally or through computer homology modeling, such as macromolecular targets, referred to as receptors [19].

To obtain high receptor binding affinity, the intention is to generate ligands that possess particular electrostatic and stereochemical attributes. An in-depth investigation of the binding site design is conceivable because of the accessibility of three-dimensional macromolecular structures. Precise investigation of electrostatic attributes is also accessible. Current techniques facilitate the fabrication of ligands with the attributes needed for effective regulation of the target receptor [20]. High-affinity ligands precisely modulate a validated drug target, competing with particular cellular processes to provide the envisioned pharmacological and therapeutic outcomes [21]. To find conceivable ligands, *in silico* investigations are carried out starting with a known target structure. The most intriguing molecules are synthesized after these molecular modeling approaches. Then, utilizing an array of experimental platforms, assessments of biological parameters such as effectiveness, affinity, and efficacy are conducted [22]. The three-dimensional structure of the ligand-receptor complex can be resolved, provided that active substances are identified. Because of the available structure, it is feasible to discern an array of intermolecular attributes that assist in molecular recognition. The analysis of binding conformations, the characterization of substantial intermolecular interactions, the definition

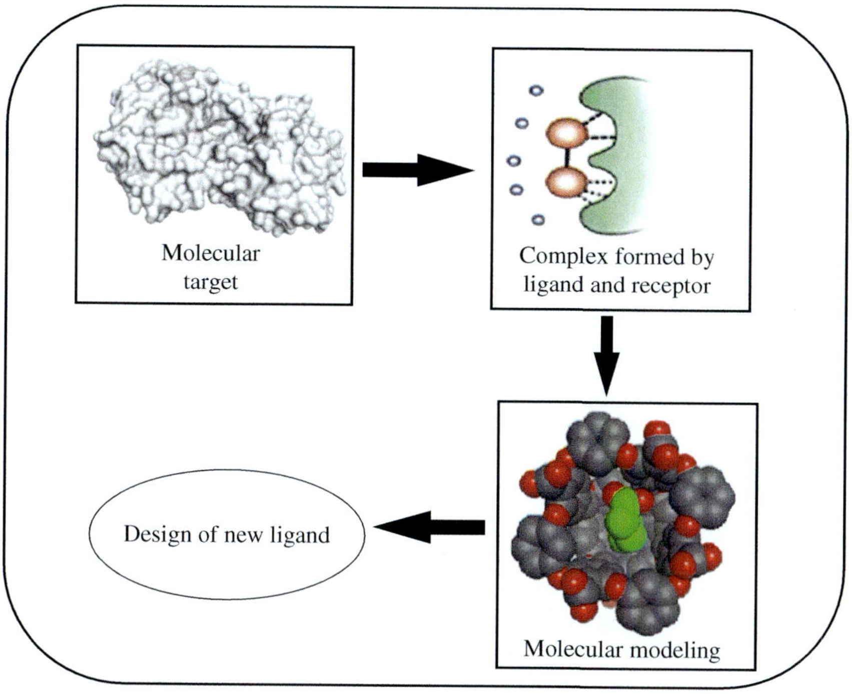

Figure 14.2 Strategies involved in structure-based design.

of speculative binding sites, mechanistic studies, and the clarification of ligand-induced conformational alterations are all made possible by structural descriptions of ligand–receptor complexes [23]. Following the recognition of a ligand–receptor complex, biological activity data corresponds with the structural details. In this fashion, the approach resumes with new phases that incorporate molecular alterations that can boost the affinity of fresh ligands for the binding site [24] (Figure 14.2).

14.4.2 Computer-Aided Design

The domain of computer-aided design integrates an array of chemical-molecular and quantum strategies with the intention of identifying, developing, and generating therapeutic chemical agents. The discovery process was facilitated by the implementation of an amalgamation of leading-edge computer strategies, biological investigation, and synthesis of chemicals, and this combinational circuitry approach broadened the scope of discovery. The acronym computer-aided design came to be employed to denote the implementation of computers in developing drugs [25]. Major accomplishments have been made leveraging these techniques, which have shown to be beneficial resources for advanced computational applications. In the specialized field of CADD, many computational techniques are utilized to model interactions between receptors and medicines to ascertain binding affinities [26]. The method has numerous additional applications besides analyzing chemical interactions and projecting binding affinities, such as generating compounds with specific physiochemical attributes and managing digital archives of chemicals (Figure 14.3).

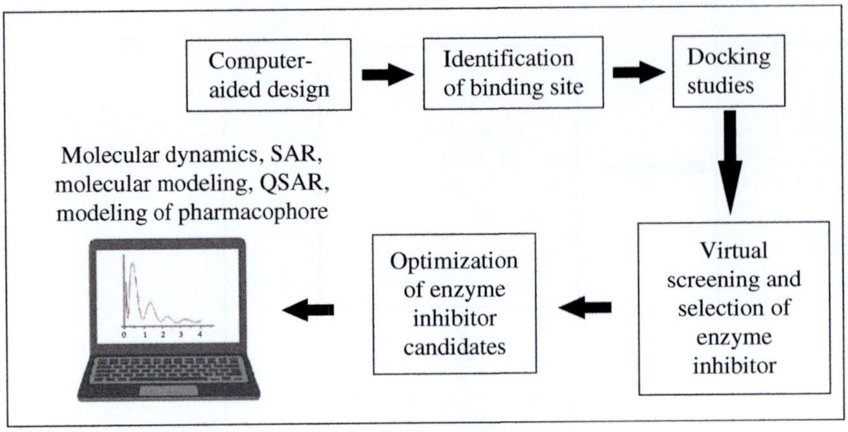

Figure 14.3 Process involved in computer-aided design.

14.4.3 Fragment-Based Design

In the development of enzyme inhibitors, fragment-based drug design is another viable method. In this technique, minuscule enzyme-binding molecule fragments are first expanded and improved to generate formidable inhibitors. Researchers can explore an expanded chemical space and uncover new inhibitors that have substantial intensity and selectivity by commencing with smaller, structurally varied fragments. Modifying the arrangement of substrates and cofactors, or emulating intermediates or phases of transition in the postulated reaction mechanism, are prominent strategies applied to the logical development of enzyme inhibitors. When this is inadequate as an initial point to start, inhibitors are frequently found by analyzing libraries of thousands of chemicals [27].

Four stages comprise a fragment-based strategy for inhibitor design: (i) designing and assembling fragment libraries; (ii) screening fragments against a compatible variant of the target enzyme; (iii) assessing the hits; and (iv) upgrading fragments into improved inhibitors. The procedure entails qualitatively selecting segments that interact with the target protein, verifying and analyzing this interaction to precisely pinpoint the fragment while outlining its binding mode, and then iteratively upgrading the fragment into a more effective, substantial inhibitor [28] (Figure 14.4).

14.4.4 Virtual Screening Method

Presently, a crucial step in the process of medication discovery and design is virtual screening (VS). VS is typically thought of as the selection of promising therapeutic candidates utilizing computational approaches from massive libraries of chemical structures [29].

14.4.4.1 Ligand Based

Using known ligands or substrates for an enzyme as a base for ligand-based drug design is an alternative approach. Inhibitors with comparable attributes can be generated by analyzing the structure and attributes of these compounds. The goal of ligand-based approaches is to locate achievable inhibitors that interact well with the enzyme through the use of computational techniques of VS. Based on the presumption that similar compounds have similar effects, ligand-based VS techniques look for molecules that possess certain relevant common properties with known binding chemicals [30]. One or more substances that are known to adhere to the biological target of

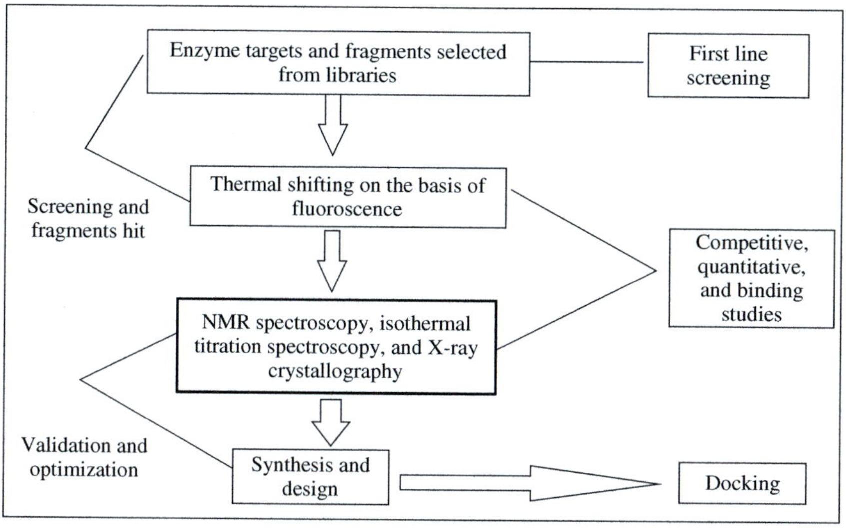

Figure 14.4 Fragment-based design.

interest must exist for a ligand-based strategy to succeed. These substances, which can be patented structures, comprehensive inhibitors, natural substrates, or even explored leads, are crucial for the screening procedure since they operate as the foundational paradigms for the search.

14.4.4.2 Receptor Based

Receptor-based approaches embrace a distinct outlook by emphasizing the connection of the ligand and the receptor rather than just the ligand itself. The main prerequisite for a receptor-based VS is the availability of a three-dimensional target structure. Such a structure may be an NMR structure, a crystallographic X-ray structure, or even a homology modeling structure [31] (Figure 14.5).

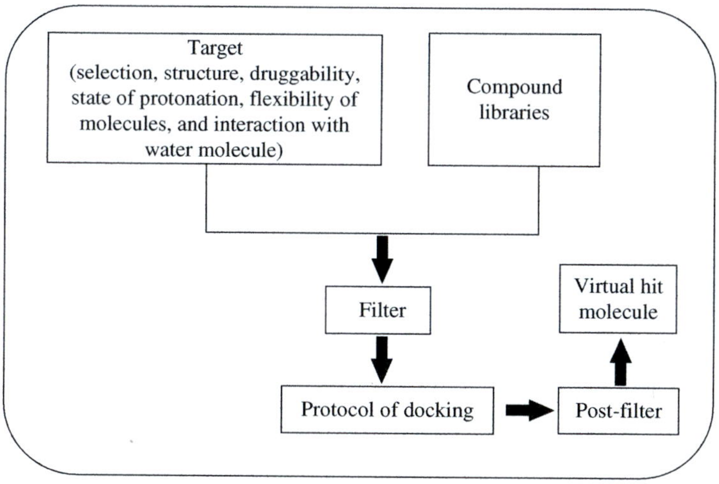

Figure 14.5 Phases of virtual screening method.

14.4.5 Natural Product-Based Discovery

Natural products are frequently touted for having an affluent framework that enables them to function as ligands for an extensive range of enzymes and receptors. The fact that natural compounds can work well as enzyme inhibitors can be attributed to a connection. Although natural products have a lot of prospects for drug discovery, there are still several issues that can impede the advancement of the pharmaceutical industry. Extracting natural products from living things is the first hurdle in the procedure [32]. The chemical classes present in the extract depend on the manner of extraction adopted. More polar solvents would boost the quantity of polar compounds in the crude extract. The biological component can be extracted using countless solvents with different polarities to elevate the variety of the retrieved natural products. The next step is incremental bioactivity-guided fractionation until the pure bioactive components are separated following the detection of a crude extract with promising pharmacological activity [33] (Figure 14.6).

14.4.6 Using Iterative Protein Crystallographic Analysis

The study of proteins' three-dimensional, atomically precise structures is recognized as protein crystallography. Over the past few decades, it has shed a great deal of light on how many different biological systems function. Over the past 10 years, the area has had a tremendous global boost in both pharmaceutical and university laboratories [34]. The fundamental impetus behind this expansion has been the hope that leads like enzyme inhibitors can be generated and existing medications' effects can be enhanced by using the three-dimensional atomic structures of proteins and other macromolecules. Understanding the complex nuances of protein structures has been essential for our comprehension of an array of biological activities, from enzymatic reactions to virus immune circumvention. Enzyme structures have supplied information about the interactions of substrates and inhibitors and the mechanism of enzyme-catalyzed reactions [35].

14.4.7 Utilization of Covalent Inhibitors

In drug discovery and treatment, covalent inhibitors are widely acknowledged as being crucial. In general, molecules that evolve a covalent link with a particular molecular target qualify as covalent inhibitors. Depending on the intended warhead, the covalent bond might be either reversible or irreversible. Cysteine, serine, threonine, tyrosine, and lysine are just a few of the individual amino acid residues that have been the target of various warheads [36].

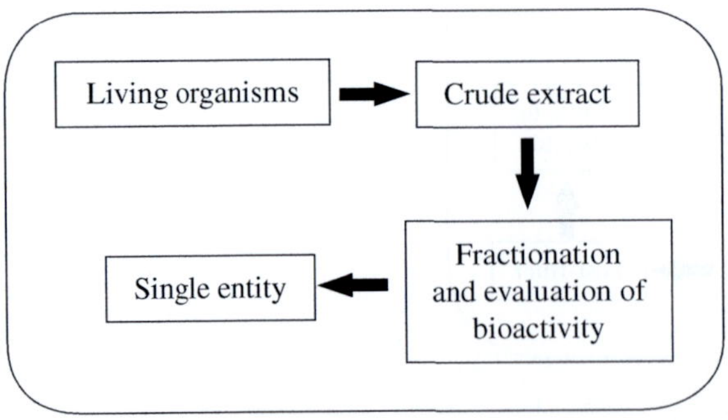

Figure 14.6 Natural product-based discovery.

By forging specific interactions between the ligand and the target protein, covalent medicines impair the normal functioning of proteins. Compared to a reversible mode of action, a covalent mechanism of action can offer multiple states pharmacological benefits, such as improved effectiveness, selectivity, and sustained duration of action. Many of the covalent medications that have been approved ended up being discovered by chance. The emergence of a new class of covalent medications known as "targeted covalent inhibitors" has been made foreseeable by computer-assisted drug design [37].

14.4.8 Encapsulation Techniques

Enzyme recycling and deterioration under harsh circumstances can be minimized by encasing the enzyme in a particulate nano- or microstructure. The size and chemoselectivity of the catalyzed reaction are altered by the small reaction space. Surface effects may boost the reaction rate, and elevating the local concentration of intermediates in a multicomponent reaction can maximize the efficiency of the reaction [38, 39]. Therefore, unlike an enzyme in solution, one that is encapsulated is frequently a better mimic of nature. The simple recovery for catalysis applications or greater resilience to harmful conditions for usage in biotechnology and medicine are just a couple of the many useful benefits of enzyme encapsulation [40].

Different matrix materials and various methods, such as physical entrapment, adsorption through noncovalent interactions, and covalent attachment to the matrix, can be used to encapsulate enzymes. Proteins that naturally form casings are used in protein-based enzyme encapsulation [41]. Capsid proteins from both viral and nonviral organisms have been used for this. The multimeric protein assemblies that make up viral capsids, also known as virus-like particles, have a high copy number and generate hollow shells with an elevated degree of rigidity [42].

14.4.9 Based on Active-Site Specificity

When substrate molecules adhere to an enzyme, a chemical reaction ensues. This area is known as the active site. The binding site, or amino acid residues, and the catalytic site, or amino acid residues, constitute the active site. The binding site initiates transient bonds with the substrate. High specificity is accomplished by each active site being developed to be best suited to bind a specific substrate and catalyze a specific reaction. The assembly of amino acids within the active site and the makeup of the substrates both impact this selectivity. To execute their job, enzymes occasionally need to bind with some cofactors [43].

Noncovalent linkages draw inhibitors and substrates to the active site. As a result, new molecules that resemble the substrates are created. It has a group that can chemically react with a protein's amino acid residue. The compound's characteristics that resemble a substrate would help it find the active site, and the reactive group would then create a covalent bond that would bind the inhibitor to the site [44]. The enzyme would subsequently become irreversibly inactive as the irreversible complex would stop any further catalytic activity. Since then, active sites of pure proteins and enzymes have been studied using active-site-directed labeling. Understanding chemical modification strategies of enzyme specificity is necessary for the rational design of site-directed enzymes. Knowing the components of the substrates that are not necessary for binding is vital when designing an inhibitor so that their positions can be used to introduce a possible connecting group. It may be possible to add linking groups at certain sites using the enzyme's tolerance to bulky substituents at specific positions of the substrate. In addition, it is crucial to create a chemical for chemotherapy that will only interact with one of the numerous comparable enzymes [45].

14.4.10 Machine Learning Inhibitor Design

The practice of generating new biocatalysts leveraging machine learning has gained prominence in recent years. This approach is data-driven, as opposed to model-driven rational design, where it seeks patterns in the current data to project the traits of a forthcoming input that has not been seen before but is similar. ML-based design can generate new, previously undetected but promising variations based on the patterns in the acquired data, as opposed to directed evolution which periodically selects the current mutants. ML is used in tandem with rational design and directed evolution [46].

The main goal of the majority of ML algorithms is to identify patterns in the provided data. These data often comprise data points with many attributes or descriptors, such as amino acid changes, enzyme secondary and tertiary structures, and sequences. Typically, there are hundreds to thousands of characteristics, making the problem high-dimensional. Unsupervised, supervised, and semi-supervised learning are the three main types of machine learning. In supervised learning, one or more target attributes, such as enzyme activity or stability, are identified as labels. The objective is to create a predictor that, utilizing a labeled training data set, would return labels for unseen data points based on their descriptors. Semi-supervised learning is the combination of these two supervised and unsupervised ML types, which happens frequently. The following are the various steps [47]:

- The data are typically transformed into tables and divided into training and test portions. The performance of the predictor will be impacted by any errors, biases, or imbalances; thus, they must be taken into consideration.
- Using the training data set, the predictor is trained.
- Using the test data set as a basis, the predictor's performance is assessed.

14.4.11 Enzyme-Templated Dynamic Combinatorial Chemistry

A new method for effectively finding novel chemical compounds that will attach to a specific enzyme is called enzyme-directed dynamic combinatorial chemistry. To create a combinatorial library, this method typically uses a library of tiny molecules that interact reversibly with one another. The combinatorial library's elements are in thermodynamic equilibrium with one another. When an enzyme is added to the equilibrium mixture, the equilibrium position will shift and any combinatorial library components that interact with the protein will be amplified. These components can then be identified by a suitable biophysical technique. This knowledge can serve as a springboard for further chemical synthesis of inventive protein ligands and enzyme inhibitors [48].

The basic building blocks of a protein-compatible combinatorial library typically consist of a variety of tiny and simple compounds that may interact with one another reversibly in aqueous solution and at physiologically relevant pH. They are frequently referred to as "building blocks" for systems. Under thermodynamic control, all of these small molecule building blocks and their reaction products are in equilibrium with one another in the mixture. Their relative distribution in the mixture indicates each component's thermodynamic stability within the equilibrium. Essentially, any enzyme-directed experiment should include four elements: a protein blueprint, an archive of components, a reversible reaction, and analytical methods [49] (Figure 14.7).

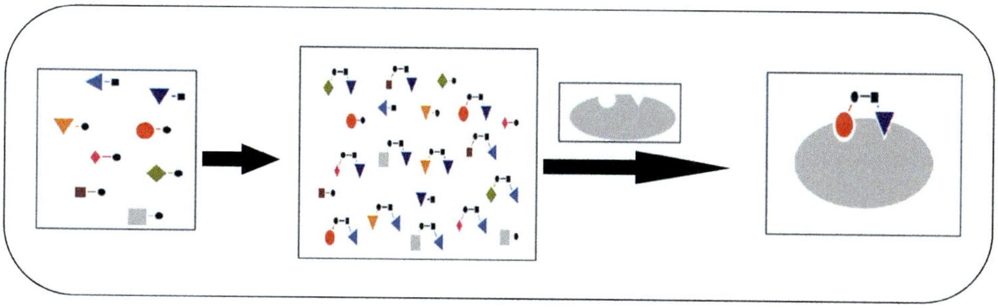

Figure 14.7 Enzyme-templated dynamic combinatorial chemistry.

14.5 Limitations and Challenges

Nevertheless, there are hurdles in generating enzyme inhibitors. Achieving selectivity is a massive challenge. For off-target effects to be kept to a minimum and treatment efficacy to be improved, inhibitors must be assured to target the desired enzyme selectively while not impacting other important proteins or enzymes. Understanding an enzyme's structure, activity, and specificity in depth is frequently crucial for accomplishing selectivity. The fabrication of enzyme inhibitors must also overcome drug resistance. The effectiveness of inhibitors can be diminished through the emergence of mechanisms by illnesses or enzymes to circumvent inhibition. The development of inhibitors that circumvent or bypass the aforementioned resistance pathways is what researchers must do to anticipate and combat these resistance mechanisms [50]. Combination treatment, which combines the use of many enzyme inhibitors or inhibitors with other therapeutic methods, can also aid in minimizing medication resistance. In the design of enzyme inhibitors, pharmacokinetic issues frequently arise. To guarantee efficient transport and action at the target location, it is essential to optimize the inhibitors' drug-like qualities, such as bioavailability, stability, and metabolism. To enhance pharmacological characteristics while preserving the desired inhibitory effect, substantial medicinal chemistry efforts are frequently required [51].

Several strategies can be used to deal with these problems. Designing inhibitors with increased selectivity and potency benefits from rational drug development, which makes use of structural data and computational modeling. Potential inhibitors that can be further optimized for desired qualities can be found by high-throughput screening of vast chemical libraries. An effective technique to investigate the chemical realm and find potential areas for inhibitor development is through fragment-based lead generation [52]. The efficacy of enzyme inhibitors may also be boosted by using combination therapy and personalized medicine strategies. To target particular enzymes and thwart disease processes, enzyme inhibitors are designed strategically for medication development. However, aspects including accomplishing selectivity, managing resistance, and strengthening pharmacokinetic traits must be tackled. Researchers may generate efficient enzyme inhibitors with great therapeutic potential for a variety of diseases by combining several techniques and continuously improving our understanding of enzyme function [53].

14.6 Future Directions

The capacity to target enzymes participating in numerous pathways as well as the creation of more strong and selective inhibitors are key to the future of enzyme inhibitor-based medication discovery. The use of enzyme inhibitors in conjunction with other therapeutic approaches, such as gene therapy, also has a lot of potential. And finally, one of the most intriguing areas of research is the creation of revolutionary computational techniques for the design and enhancement of medications based on enzyme inhibitors. Through the use of such techniques, it may be possible to find novel inhibitors that are even more potent than those that are already known, which would speed up the creation of more effective medicines for several diseases [54].

14.7 Conclusion

A crucial component of drug discovery and development is the design of enzyme inhibitors. The activity of particular enzymes implicated in disease processes is modulated by enzyme inhibitors, which provide focused therapeutic interventions. To find and improve potential inhibitors, techniques including structure-based drug design, ligand-based drug design, and fragment-based drug design are helpful. Enzyme inhibitor design does, however, provide several formidable difficulties. Selectivity is a key goal that must be attained to guarantee that inhibitors only affect the desired enzyme and leave off-target proteins unaffected. Innovative methods and combination therapies are needed to combat medication resistance, which occurs when illnesses or enzymes discover ways to circumvent inhibition. For the inhibitors to be delivered and used effectively, it is also crucial to optimize their pharmacokinetic qualities. To overcome these difficulties, researchers develop inhibitors with increased selectivity, potency, and drug-like features using rational drug design, high-throughput screening, fragment-based lead generation, and other techniques. Processes for designing and optimizing inhibitors could be sped up with the integration of computer techniques like artificial intelligence and machine learning. Exploring new target classes and newly discovered enzyme families for therapeutic intervention is where the future of enzyme inhibitor creation lies. Precision targeting and advances in personalized therapy provide considerable promise for improved efficacy and fewer adverse effects by customizing inhibitors to specific patients or disease subtypes. In general, effective medication development depends on an awareness of the approaches and difficulties in enzyme inhibitor design. Researchers can overcome obstacles and create highly potent and selective enzyme inhibitors for a variety of diseases by utilizing novel approaches and continually expanding our knowledge of enzyme biology. By doing so, they can significantly advance the field of drug design and improve patient outcomes.

References

1 Hopkins, A.L. and Groom, C.R. (2002). The druggable genome. *Nature Reviews. Drug Discovery* 1: 727–730. https://doi.org/10.1038/nrd892.

2 Consalvi, S., Biava, M., and Poce, G. (2015). COX inhibitors: a patent review (2011–2014). *Expert Opinion on Therapeutic Patents* 25: 1357–1371. https://doi.org/10.1517/13543776.2015.1090973.

3 Ng, H.L., Chen, S.Y., Chew, E.H., and Chui, W.K. (2016). Applying the designed multiple ligands approach to inhibit dihydrofolate reductase and thioredoxin reductase for antiproliferative activity. *European Journal of Medicinal Chemistry* 115: 63–74. https://doi.org/10.1016/j.ejmech.2016.03.002.

4 Nikolic, K., Mavridis, L., Djikic, T. et al. (2016). Drug design for CNS diseases: polypharmacological profiling of compounds using cheminformatic, 3D-QSAR and virtual screening methodologies. *Frontiers in Neuroscience* 10. https://doi.org/10.3389/fnins.2016.00265.

5 Hughes, R.E., Nikolic, K., and Ramsay, R.R. (2016). One for all? Hitting multiple Alzheimer's disease targets with one drug. *Frontiers in Neuroscience* 10. https://doi.org/10.3389/fnins.2016.00177.

6 Nikolic, K., Mavridis, L., Bautista-Aguilera, O.M. et al. (2015). Predicting targets of compounds against neurological diseases using cheminformatic methodology. *Journal of Computer-Aided Molecular Design* 29: 183–198. https://doi.org/10.1007/s10822-014-9816-1.

7 Saura, J., Bleuel, Z., Ulrich, J. et al. (1996). Molecular neuroanatomy of human monoamine oxidases A and B revealed by quantitative enzyme radioautography and *in situ* hybridization histochemistry. *Neuroscience* 70: 755–774. https://doi.org/10.1016/S0306-4522(96)83013-2.

8 Rojas, R.J., Edmondson, D.E., Almos, T. et al. (2015). Reversible and irreversible small molecule inhibitors of monoamine oxidase B (MAO-B) investigated by biophysical techniques. *Bioorganic & Medicinal Chemistry* 23: 770–778. https://doi.org/10.1016/j.bmc.2014.12.063.

9 Lawrence, C.M., Menon, S., Eilers, B.J. et al. (2009). Structural and functional studies of archaeal viruses. *The Journal of Biological Chemistry* 284: 12599–12603.

10 Jugder, B.E. and Watnick, P.I. (2020). Vibrio cholera sheds its coat to make itself comfortable in the gut. *Cell Host & Microbe* 27: 161–163.

11 Yang, L.W. and Bahar, I. (2005). Coupling between catalytic site and collective dynamics: a requirement for mechanochemical activity of enzymes. *Structure* 13: 893–904. https://doi.org/10.1016/j.str.2005.03.015.

12 Herschlag, D. and Natarajan, A. (2013). Fundamental challenges in mechanistic enzymology: progress toward understanding the rate enhancement of enzyme. *Biochemistry* 52: 2050–2067. https://doi.org/10.1021/bi4000113.

13 Cleland, W.W. (1963). The kinetics of enzyme-catalyzed reactions with two or more substrates or products. II. Inhibition: nomenclature and theory. *Biochimica et Biophysica Acta* 67: 173–187. https://doi.org/10.1016/0926-6569(63)90226-8.

14 Powers, J.C., Asgian, J.L., Ekici, Ö.D., and James, K.E. (2002). Irreversible inhibitors of serine, cysteine, and threonine proteases. *Chemical Reviews* 102 (12): 4639–4750. https://doi.org/10.1021/cr010182v.

15 Molla, G. (2017). Competitive inhibitors unveil structure/function relationships in human D-amino acid oxidase. *Frontiers in Molecular Biosciences* 4: 80. https://doi.org/10.3389/fmolb.2017.00080.

16 Abdel-Magid, A.F. (2015). Allosteric modulators: an emerging concept in drug discovery. *ACS Medicinal Chemistry Letters* 6 (2): 104–107. https://doi.org/10.1021/ml5005365.

17 Rosse, G. (2013). Negative allosteric modulators of metabotropic glutamate receptor subtype. *ACS Medicinal Chemistry Letters* 4 (6): 500–501. https://doi.org/10.1021/ml400138p.

18 Salum, L., Polikarpov, I., and Andricopulo, A.D. (2008). Structure-based approach for the study of estrogen receptor binding affinity and subtype selectivity. *Journal of Chemical Information and Modeling* 48: 2243–2253.

19 Kalyaanamoorthy, S. and Chen, Y.P. (2011). Structure-based drug design to augment hit discovery. *Drug Discovery Today* 16: 831–839.

20 Drwal, M.N. and Griffith, R. (2013). Combination of ligand- and structure-based methods in virtual screening. *Drug Discovery Today: Technologies* 10: e395–e401.

21 Valasani, K.R., Vangavaragu, J.R., Day, V.W., and Yan, S.S. (2014). Structure-based design, synthesis, pharmacophore modeling, virtual screening, and molecular docking studies for identification of novel cyclophilin D inhibitors. *Journal of Chemical Information and Modeling* 54: 902–912.

22 Blaney, J. (2012). A very short history of structure-based design: how did we get here and where do we need to go? *Journal of Computer-Aided Molecular Design* 26: 13–14.

23 Wilson, G.L. and Lill, M.A. (2011). Integrating structure-based and ligand-based approaches for computational drug design. *Future Medicinal Chemistry* 3: 735–750.

24 Shoichet, B.K. and Kobilka, B.K. (2012). Structure-based drug screening for G-protein-coupled receptors. *Trends in Pharmacological Sciences* 33: 268–272.

25 Song, C.M., Lim, S.J., and Tong, J.C. (2009). Recent advances in computer-aided drug design. *Briefings in Bioinformatics* 10 (5): 579–591. https://doi.org/10.1093/bib/bbp023.

26 Veselovsky, A.V., Zharkova, M.S., Poroikov, V.V., and Nicklaus, M.C. (2014). Computer-aided design and discovery of protein–protein interaction inhibitors as agents for anti-HIV therapy. *SAR and QSAR in Environmental Research* 25 (6): 457–471. https://doi.org/10.1080/1062936X.2014.898689.

27 Rees, D.C., Congreve, M.S., Murray, C.W., and Carr, R. (2004). Fragment-based lead discovery. *Nature Reviews. Drug Discovery* 3: 660–672.

28 Erlanson, D.A., McDowell, R.S., and O'Brien, T. (2004). Fragment-based drug discovery. *Journal of Medicinal Chemistry* 47: 3463–3482.

29 Sousa, S.F., Cerqueira, N.M., Fernandes, P.A., and Ramos, M.J. (2010). Virtual screening in drug design and development. *Combinatorial Chemistry & High Throughput Screening* 13 (5): 442–453. https://doi.org/10.2174/138620710791293001.

30 Stahura, F.L. and Bajorath, M. (2005). New methodologies for ligand-based virtual screening. *Current Pharmaceutical Design* 11: 1189–1202.

31 Lengauer, T., Lemmen, C., Rarey, M., and Zimmermann, M. (2004). Novel technologies for virtual screening. *Drug Discovery Today* 9: 27–34.

32 Atanasov, A.G., Waltenberger, B., Pferschy-Wenzig, E.-M. et al. (2015). Discovery and resupply of pharmacologically active plant-derived natural products: a review. *Biotechnology Advances* 33: 1582–1614.

33 Harvey, A.L., Edrada-Ebel, R., and Quinn, R.J. (2015). The re-emergence of natural products for drug discovery in the genomics era. *Nature Reviews. Drug Discovery* 14: 111–129.

34 Appelt, K., Bacquet, R.J., Bartlett, C.A. et al. (1991). Design of enzyme inhibitors using iterative protein crystallographic analysis. *Journal of Medicinal Chemistry* 34 (7): 1925–1934. https://doi.org/10.1021/jm00111a001.

35 Mehlman, T.S., Biel, J.T., Azeem, S.M. et al. (2023). Room-temperature crystallography reveals altered binding of small-molecule fragments to PTP1B. *eLife* 12: https://doi.org/10.7554/eLife.84632.

36 Potashman, M.H. and Duggan, M.E. (2009). Covalent modifiers: an orthogonal approach to drug design. *Journal of Medicinal Chemistry* 52: 1231–1246.

37 Robertson, J.G. (2005). Mechanistic basis of enzyme-targeted drugs. *Biochemistry* 44: 5561–5571.

38 Zhang, Y., Wu, H., Li, J. et al. (2008). Protamine-templated biomimetic hybrid capsules: efficient and stable carrier for enzyme encapsulation. *Chemistry of Materials* 20: 1041–1048.

39 Glasgow, J.E., Asensio, M.A., Jakobson, C.M. et al. (2015). Influence of electrostatics on small molecule flux through a protein nanoreactor. *ACS Synthetic Biology* 4: 1011–1019. https://doi.org/10.1021/acssynbio.5b00037.

40 Betancor, L. and Luckarift, H.R. (2008). Luckarift bioinspired enzyme encapsulation for biocatalysis. *Trends in Biotechnology* 26: 566–572. https://doi.org/10.1016/j.tibtech.2008.06.009.

41 Zdarta, J., Meyer, A.S., Jesionowski, T., and Pinelo, M. (2018). Pinelo A general overview of support materials for enzyme immobilization: characteristics, properties, practical utility. *Catalysts* 8: 92.

42 Zhang, Z., Zhang, R., and McClements, D.J. (2017). Lactase (β-galactosidase) encapsulation in hydrogel beads with controlled internal pH microenvironments: impact of bead characteristics on enzyme activity. *Food Hydrocolloids* 67: 85–93. https://doi.org/10.1016/j.foodhyd.2017.01.005.

43 de la Lastra, P., Manuel, J., Baca-González, V. et al. (2021). Antibodies targeting enzyme inhibition as potential tools for research and drug development. *Biomolecular Concepts* 12 (1): 215–232. https://doi.org/10.1515/bmc-2021-0021.

44 Chattopadhyay, K., Lazar-Molnar, E., Yan, Q. et al. (2009). Sequence, structure, function, immunity: structural genomics of costimulation. *Immunological Reviews* 229(1): 356–86. https://doi.org/10.1111/j.1600-065X.2009.00778.x

45 Mäntsälä, P. and Niemi, J. (2009). Enzymes: the biological catalysts of life. *Physiology and Maintenance* 2: 1–22.

46 Mazurenko, S., Prokop, Z., and Damborsky, J. (2020). Machine learning in enzyme engineering. *ACS Catalysis* 10 (2): 1210–1223. https://doi.org/10.1021/acscatal.9b04321.

47 Libbrecht, M.W. and Noble, W.S. (2015). Machine learning applications in genetics and genomics. *Nature Reviews. Genetics* 16: 321–332.

48 Huang, R. and Leung, I.K.H. (2016). Protein-directed dynamic combinatorial chemistry: a guide to protein–ligand and inhibitor discovery. *Molecules* 21 (7): 910. https://doi.org/10.3390/molecules 21070910.

49 Shi, B., Stevenson, R., Campopiano, D.J., and Greaney, M.F. (2006). Discovery of glutathione S-transferase inhibitors using dynamic combinatorial chemistry. *Journal of the American Chemical Society* 128 (26): 8459–8467. https://doi.org/10.1021/ja058049y.

50 de la Fuente, M., Lombardero, L., Gómez-González, A. et al. (2021). Enzyme therapy: current challenges and future perspectives. *International Journal of Molecular Sciences* 22 (17): 9181. https://doi.org/10.3390/ijms22179181.

51 Tjhung, K.F., Shokhirev, M.N., Horning, D.P., and Joyce, G.F. (2020). An RNA polymerase ribozyme that synthesizes its ancestor. *Proceedings of the National Academy of Sciences of the United States of America* 117: 2906–2913. https://doi.org/10.1073/pnas.1914282117.

52 Kaiser, E.T., Lawrence, D.S., and Rokita, S.E. (1985). The chemical modification of enzymatic specificity. *Annual Review of Biochemistry* 54: 565–595. https://doi.org/10.1146/annurev.bi.54.070185.003025.

53 Robertson, J.G. (2007). Enzymes as a special class of therapeutic target: clinical drugs and modes of action. *Current Opinion in Structural Biology* 17: 674–679. https://doi.org/10.1016/j.sbi.2007.08.008.

54 Kuldeep, S., Jeetendra, G.K., Devendra, P., and Shivendra, K. (2023). The use of enzyme inhibitors in drug discovery: current strategies and future prospects. *Current Enzyme Inhibition* 19 (3): https://doi.org/10.2174/1573408019666230731113105.

15

Rational Design of Peptides and Protein Molecules in Drug Discovery

Ipsa Padhy[1], *Abanish Biswas*[2], *Chandan Nayak*[3], *and Tripti Sharma*[1,4]

[1] *Department of Pharmaceutical Chemistry, School of Pharmaceutical Sciences, Siksha 'O' Anusandhan (Deemed to be University), Bhubaneswar, Odisha, India*
[2] *Department of Pharmaceutical Sciences and Technology, Birla Institute of Technology, Ranchi, Jharkhand, India*
[3] *School of Pharmaceutical Education and Research, Berhampur University, Berhampur, Odisha, India*
[4] *School of Pharmaceutical Sciences and Research, Chhatrapati Shivaji Maharaj University, Navi Mumbai, Maharashtra, India*

15.1 Introduction

The side effects associated with small molecule drugs have always created a need for an alternative therapeutic preposition. In this context, the peptides represent a unique class of therapeutic agents having features of small molecule drugs, large proteins, and other endogenous biologics [1]. All the signaling molecules regulating the metabolic pathways are enzymes, hormones, and other endogenous proteins. Consequently, the peptides emerge as an appropriate class of pharmacotherapeutic agents emulating natural entities when there is an imbalance or insufficiency in endogenous secretion levels [2].

The discovery of peptides as therapeutic entities can be dated back to 1921–1922 when insulin was first isolated from bovine and porcine pancreas, proving a boon for human civilization in the management of type 1 diabetes [3]. In the 1950s, with the development of solution phase peptide synthetic methods, several other peptides were chemically developed like synthetic oxytocin [4] and synthetic vasopressin [5]. Later, in the mid-1960s, the peptide drug discovery was fueled by the inception of solid-phase synthetic methods [6]. Implementation of more advanced techniques such as microwave irradiation in both solid and solution phase peptide syntheses, chemical ligations, and late-stage functionalizations further led to the clinical development of synthetic therapeutic peptides like recombinant insulin, gonadotropic-releasing hormones, leuprolide, and goserelin [7]. Since 2000, about 40 peptides have attained FDA approval as drugs. Peptide drugs have a vast therapeutic repertoire [8–12]. Currently, more than 200 peptides are under active clinical trials, along with several others undergoing preclinical investigations [13].

The peptide molecules can potentially and selectively bind to larger protein interfaces, thereby advocating low off-target side effects and toxicity [14]. With the advances in computational approaches, the rational design and development of therapeutic peptides have become more effective [15]. *In silico* molecular docking methods enable accurate prediction of protein structure, surface topology, and interaction affinities. Peptide docking has been divided into template-based and template-free docking depending on the requirement size of input data [16]. The physics-based

Computational Methods for Rational Drug Design, First Edition. Edited by Mithun Rudrapal.

molecular simulations are currently well integrated with emerging artificial (AI)- and alfafold (AF)-based tools for reshaping the peptide discovery landscape [17].

In this chapter, we discuss various methods utilized in the design and development of therapeutic peptides with a special focus on current technological advancements. We emphasize the therapeutic applications of peptides. A comparative analysis between the peptide and small molecule drugs respective to their pharmacodynamic and a pharmacokinetic property has also been conversed. We outline various state-of-the-art peptide docking methods. An overview of available specific peptide docking software tools is provided.

15.2 Peptides as Therapeutics

The management of diseases by small molecule drugs is impeccable. However, the side effects associated with therapeutic small molecules have also generated a need for alternative therapeutic entities. In this scenario, peptides present themselves as a distinct class of therapeutic candidates. Metabolic disorders and cancers have been the thrust disease areas driving peptide drug discoveries [18]. The emergence of antimicrobial-resistant pathogens has also irked the development of novel peptide antibiotics [19]. Gastrointestinal (GI) disorders have also provided a research area eliciting a lot of interest in the development of peptides as potential therapeutic treatment options [20].

15.2.1 Peptide Antibiotics

Antimicrobial resistance has become a global health issue basically arising due to overprescribing antibiotics for nonbacterial infections or frequent misuse of antibiotics as prophylactics [21]. With growing microbial resistance, the last-line antimicrobials have entered the pre-antibiotic era as well a very slow pace in discovery of newer broad-spectrum antimicrobials for multidrug-resistant pathogens, the therapeutic utilization of peptide antibiotics becomes the foremost important [22, 23]. Moreover, peptide antibiotics stand high as a source of new antimicrobials [24].

The peptide antibiotics or host defense peptides are front-line defense entities in both prokaryotes and eukaryotes. The antimicrobial peptides have broad-spectrum inhibitory activity against bacteria, protozoan parasites, pathogenic fungi, and viruses. The peptide antibiotics are diverse with respect to their structure, physicochemical properties, and mechanism of action. The peptide antibiotics are isolated from bacteriophages, bacteria, fungi, mollusks, amphibians, arthropods, mammals, and plants as well [25]. The first reported peptide antibiotic was gramicidin, which showed inhibitory activity against many Gram-positive bacteria. It was isolated from the soil bacterium *Bacillus brevis* [26]. The antimicrobial peptide database (APD3) has cataloged more than 3000 peptide antibiotics [27].

In general, the peptide antibiotics comprise 50 amino acid residues. The majority of the AMPs are cationic, with more than 45% hydrophobic residues [27]. The cationic nature, as well as the hydrophobic residues of these peptides, generate an amphipathic backbone facilitating the bacterial membrane infiltration [28]. Different models have been demonstrated for targeting microbial cell membranes by the antimicrobial peptides (AMPs) (Figure 15.1).

Cathelicidin LL-37 peptides induce anionic lipid clustering to permeate bacterial membranes and rupture the phospholipid bilayer by carpet/toroidal models. The proline-rich peptides target bacterial ribosomes, while lantibiotics and cyclotides can both bind to phosphatidylethanolamines (PEs) [29–32]. Rather, these peptides may work synergistically for optimal outcomes.

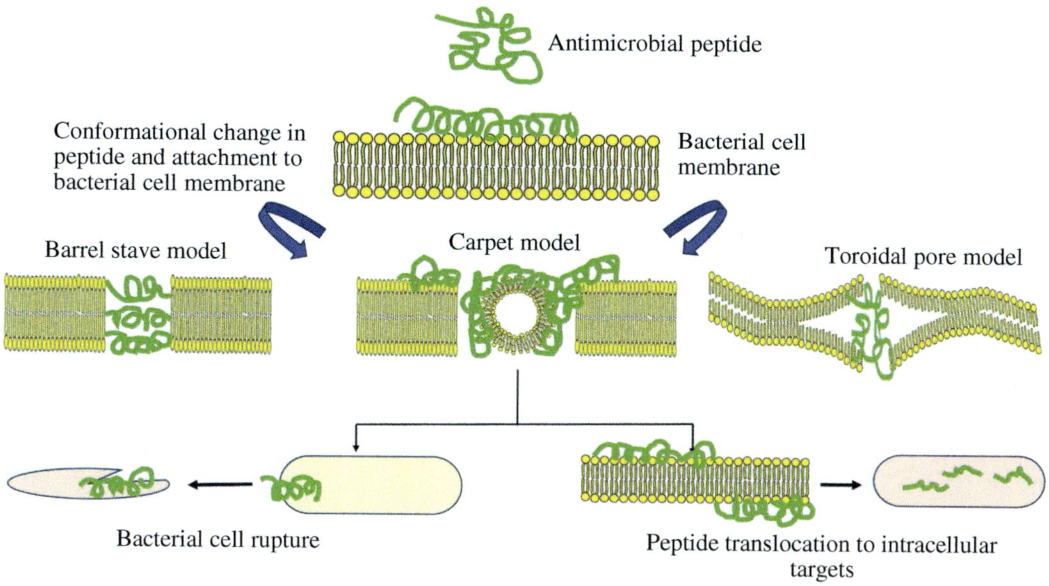

Figure 15.1 Models displaying the mechanism of membrane disruption by peptide antibiotics.

Finally, AMPs can boost the immune response to further clear invading pathogens. All of these mechanisms make it difficult for pathogens to develop resistance.

15.2.1.1 Peptides in Bone Diseases

Bone diseases are a wide group of musculoskeletal disorders that often require high drug doses for effective therapy. However, such high doses of small-molecule drugs evoke adverse effects. In such cases, bone tissue-targeted pharmacotherapy is a prerequisite peptide that has been developed in the last three decades for the treatment of bone disorders, especially degenerative bone resorption triggered by osteoporosis and bone metastasis [33–35]. Numerous peptide therapeutic candidates can be classified into four categories like bone resorption inhibitors (W9, OP3-4, RANKL inhibitor peptide), bone formation stimulators (B2A, P1, P2, P3, P24, P15, TP508, OGP, PTH), dual bone resorption inhibitors, and bone formation stimulators and bone targeting peptides [36]. The designed peptides inhibited bone resorption by antagonizing actions of TNF-α and RANKL, modulating RANK–RANKL signaling pathways, mimic Loop3 of RANKL, and blocked the RANKL-binding activity and RANKL-induced differentiation of osteoclast, blocking NEMO/IKK-β interactions and suppressing of the activity of the IKK complex inhibiting TNF-α-induced NF-κB signaling pathway for bone resorption preventing inflammatory bone resorption.

15.2.1.2 Peptides in Cancer

Immunotherapy majorly attacks the tumor cells by targeting immune checkpoints. FDA has approved monoclonal antibodies (MABs) for cancer therapy by targeting specific immune checkpoints. Although very effective, MABs have some disadvantages associated with high immunogenicity, poor solubility, and very high cost [37]. Being smaller in size, good stability, and less immunogenicity, the peptides have been seeking serious attention in the arena of cancer therapy and diagnosis. The peptides are applied for cancer theranostics in many different ways (Figure 15.2).

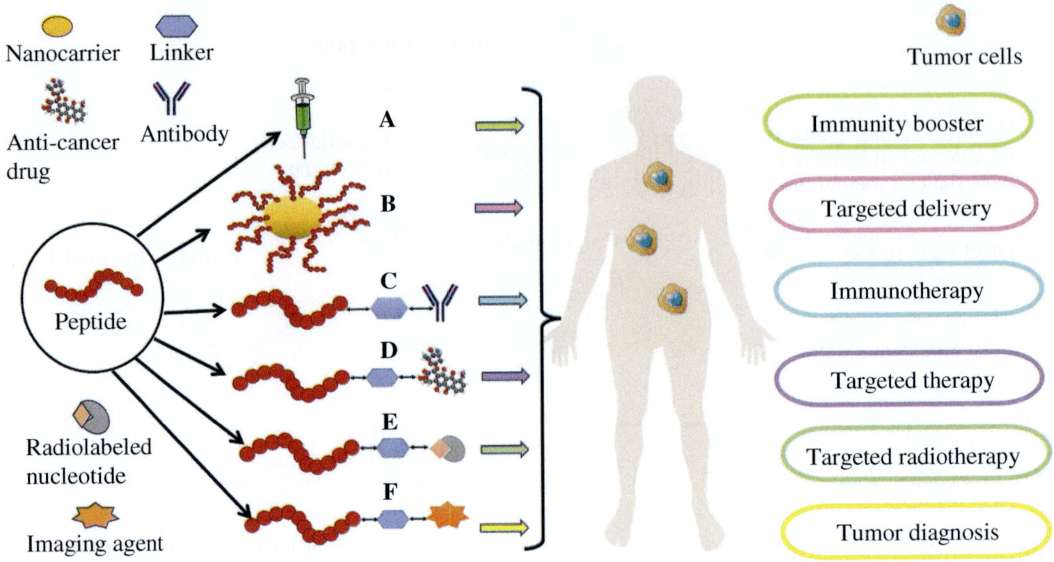

Figure 15.2 Peptide-based cancer theranostics. (A) Peptide vaccine; (B) nanocarrier-loaded peptides; (C) peptide conjugated to the antibody; (D) peptide conjugated to anticancer drug; (E) peptide conjugated to radionucleotide; and (F) peptide conjugated to the imaging agent.

The peptides can be labeled with dyes, radioisotopes, or other specific molecules for cancer diagnosis or imaging. The peptides are conjugated with nanocarriers for targeted therapy. The peptides have been developed as vaccines. Peptides are directly used as targeted pharmacotherapeutics [38, 39].

The peptide-based imaging probes bind to receptors expressed on the cell surface (α-integrins, somatostatin receptor, transferrin receptor, neurotensin receptor), within the intracellular matrix (cyclin A, cyclin kinase) or extracellular matrix (fibronectin, matrix metalloproteinases, prostate-specific antigen) [38]. The tumor distribution can be visualized by single photon emitted computed tomography imaging/scanning techniques. Octreoscan and depreotide are radiolabeled conjugates of somatostatin peptide approved by the FDA for diagnosis of neuroendocrine and lung cancers [40]. Although depreotide was later withdrawn. Tc-3PRGD2 is an iodine labeled α-integrin-based peptide probe developed for the detection of thyroid cancer [41]. On a similar note, cancer radiotherapy with isotope-labeled peptides was developed. A good example is Lutetium 177 dotatate, somatostatin labeled with radioactive lutetium 177. It is FDA-approved for the treatment of gastroenteric neuroendocrine cancers [42–44]. The major side effect of targeted radiotherapy is inevitable damage to normal tissues surrounding the metastatic tissues [45].

Peptides are also conjugated with anticancer drugs, genes, and RNAs for targeting tumor cells [45]. Adriamycin-conjugated AN-150 and AN-207 (luteinizing releasing hormone analogs) were found to be effective against endometrial and ovarian cancers in phases I and II clinical trials. The PEGylated NGR peptide conjugated to liposome polycation DNA effectively delivered the small interfering RNA to tumor cells *in vivo* and, decreased expressions of c-Myc and induced apoptosis [46]. The cell-penetrating peptides act as carriers for other peptides, nucleic acids, and drugs to target tumor sites [47]. Antigenic peptides like TERT572Y have been developed as anticancer vaccines (lung cancer), which trigger immune activities of killer T cells or helper T cells by coupling to MHC I and II complexes [48]. Peptides conjugation to specific antigens induce vaccine immunogenicity [49, 50]. Itself peptides exert anticancer effects. PD-1/PD-L1 is one of the major signaling pathways being targeted for anticancer peptide drug discovery. Notably, PD-LI peptide mimics like PL120131, DS-I & II, FITC-YT-16, and NY-12 were developed that demonstrated promising antitumor potency against preclinical cancer models [51–54].

Interestingly, natural spider venom toxins like Hanatoxin-1 from the Chilean spider [55] and Lafr26, a porogenic peptide isolated from the venom of *Lachesana* species spiders [56] have shown effective anticancer potency against colon and lung cancers *in vitro*. The natural venom-derived peptides exert anticancer activity by blocking potassium ion channels and specific transmembrane receptors Many other naturally derived and semisynthetic peptides have shown promising anticancer activities, like aurein 1.2, BMAP-27, BMAP-28, brevinine, cecropin A and B, citropin 1.1, gaegurins, HMGB 1, HNP 1, 2 and 3, hBD3, and PR-39 against leukemia, bladder cancer, breast cancer, and colon cancers *in vitro*. The potency of natural antimicrobial peptides such as LL-37, magainin 2, LfcinB, melittin, and tachyplesin 1 against cancers has also been revealed [57]. A chemically synthesized θ-defensin peptide analog exerted anti-breast cancer activity, probably by membrane rupturing activity [58].

15.2.1.3 Peptides in Metabolic Diseases

T2DM has been successfully treated with peptide drugs, including GLP-1 receptor agonists [59]. Exenatide, liraglutide, lixisenatide, dulaglutide, and semaglutide are some FDA-approved peptide drugs for clinical use in type 2 diabetes [60]. Clinical trial reports have proven that GLP-1 receptor agonists like lixisenatide are instrumental in improving diabetic nephropathy [61]. Moreover, it was reported that liraglutide and semaglutide antiatherosclerotic potency [62, 63].

The renin–angiotensin–aldosterone system (RAAS) is an exclusive target for the pharmacotherapy of cardiac disorders by peptides. Synthetic angiotensin II was approved by the FDA in 2017 for increasing blood pressure via intravenous infusion in adults with septicemia or other distributed shock [64]. Four peptides (WPRGYFL, GPDRPKFLGPF, WYGPDRPKFL, and SDWDRF) isolated from *Tetradesmus obliquus* microalgae inhibited the angiotensin-converting enzymes (ACEs) [65]. IRW, an egg white-derived tripeptide exhibited antihypertensive activity *in vivo* by upregulating angiotensin-converting enzyme (ACE2) [66]. The natriuretic peptide receptors pose as excellent targets for cardiovascular drugs [67, 68]. Nesiritide is a recombinant human BNP that was approved by the FDA in 2001 for the treatment of acutely decompensated heart failure in patients [69, 70]. NPs act mainly through NPR-A and/or NPR-B receptors, while NPR-C is mainly used for scavenging NPs [71]. Cenderitide (from green mamba snake) is a dual NPR-A/NPR-B agonist in an active clinical trial and is having a safe therapeutic index [16, 72]. Infusing vasoactive intestinal peptides increased the concentration of myocardial vasoactive intestinal peptides and reversed existing myocardial fibrosis in rats [73, 74]. Cyclopeptide RD808 attenuated β1-adrenergic induced myocardial injury *in vivo* [75]. The central adrenocorticotropin-releasing factor (CRF)-related peptide system is currently attracting increasing attention as a target for the prevention of cardiovascular disease [76, 77].

15.2.1.4 Peptides in Gastrointestinal Diseases

The gut microbiome of human secretes a plethora of antimicrobial peptides. GI diseases and excessive use of antibiotics degrade the symbiotic gut microflora that usually acts as a gut defense system. GI diseases are being now targeted by peptides. Proline-arginine-39, isolated from porcine bone marrow and lymphoid tissue, exhibited antibacterial, immunomodulatory, and intestinal epithelial repair functions and may provide a safe alternative therapy for IBD [78]. Subcutaneous injection of teduglutide a GLP-2 analogue developed by recombinant DNA technique, increased intestinal absorption in patients with short bowel syndrome (SBS) [79, 80]. EGF, erythropoietin, and hepatocyte growth factor have also shown therapeutic potential in SBS [81].

Periplanetasin-2 isolated from the American cockroach blocked the mucosal damage and inflammation induced by *Clostridium difficile* toxin A [82]. Preincubating or co-incubating (*Clostridium perfringens* endotoxin) CPE with the claudin-4 extracellular loop ECL-2 peptide significantly inhibited CPE-induced luminal fluid accumulation and histological lesions in rabbit intestinal loop [83], indicating synthetic peptide ECL-2 can be used to negate food poisoning. Cathelicidin produced from the human colon defended *Salmonella typhimurium* infection by obstructing bacteria infiltration into the colon epithelium by upregulating Toll-like receptor-4 and releasing pro-cytokines [84].

15.2.2 Advantages and Limitations of Peptide Therapeutics

The peptide drugs basically mimic endogenous hormones, growth factors, neurotransmitters, ion channel ligands, or anti-infective agents. Typically, they attach to cellular receptors and induce specific effects intracellularly. The peptide drugs, in comparison to other therapeutic biologicals, incur less production cost as well as display less immunogenetic [2]. Therapeutic peptides, in comparison to small molecule drugs, are target-specific and have minimal side effects owing to shorter half-life. The tissue accumulation by therapeutic peptides is low [85]. Therapeutic peptides have two intrinsic drawbacks, membrane impermeability and poor *in vivo* stability, which represent major stumbling blocks for peptide drug development [86, 87].

15.2.3 FDA-Approved Peptide Therapeutics

Peptide drug discovery is targeted at various diseases, such as autoimmune, endocrinological, skin, bone disorders, reproductive and cardiovascular diseases, and diseases related to metabolic dysfunction. However, the prime focus has remained on anticancer and antimicrobial therapeutics, justified by the high number of drug approvals for these indications (Table 15.1).

15.2.4 Peptide-Based Entities in Clinical Trials

Peptides currently in clinical trials (Table 15.2) are being investigated for a wide range of applications, including microbial infections, diabetes, hormonal disorders, and cancer [7]. Interestingly, some trials are being carried out simultaneously in two phases, such as phase I/II or II/III, to hasten the approval process (https://www.clinicaltrials.gov). The data highlight that FDA approvals of peptide drugs had been predominantly via parenteral rather than oral routes of administration. However, advances in chemical modifications have led to an almost equivalent presence of oral peptides in clinical trials to that of the intravenous route. This is a promising trend, given that oral administration is expected to widen the applicability of peptide drugs.

15.2.5 Peptide Synthesis and Diversification

With rapid technological innovations in the fields of synthetic biology, biotechnology, and chemical sciences, peptide drug discovery and development has achieved enormous success. In this chapter, we recapitulate the existing state of art techniques (Figure 15.3) employed for peptide synthesis and diversification.

Table 15.1 FDA-approved peptide drugs 2000–2022.

Peptide therapeutic	Approval year	Method of development	Target pharmacological action	Indication
Cetrorelix	2000	Synthetic	Gonadotropin-releasing hormone antagonist	Infertility
Bivalirudin	2000	Synthetic	Inhibitor of prothrombin	Anticoagulant
Caspofungin	2001	Semi-synthetic	Cell wall synthesis inhibitor	Aspergillosis
Nesiritide	2001	Recombinant DNA technique	Atrial natriuretic peptide-1 agonist	Chronic heart failure
Teriparatide	2002	Recombinant DNA technique	Analog of parathyroid hormone	Osteoporosis
Enfuvirtide	2003	Synthetic	gp41 protein fusion inhibitor of HIV-1 virus	HIV infections
Abarelix	2003	Synthetic	Gonadotropin-releasing hormone antagonist	Prostate cancer
Desirudin	2003	Recombinant DNA technique	Inhibitor of human thrombin	Prophylaxis of deep thrombosis
Atazanavir	2003	Synthetic	HIV-1 viral protease inhibitor	HIV infections
Ziconotide	2004	Natural peptide from the venom of *Conus magus*	Blocks calcium channels	Acute and chronic pain
Human secretin	2004	Synthetic	Agonist for secretin receptor	Pancreatic exocrine dysfunction
Pramlintide	2005	Synthetic	Agonist for calcitonin receptor	Types 1 and 2 diabetes mellitus
Micafungin	2005	Semi-synthetic	Fungal cell wall synthesis inhibitor	Candidiasis
Exenatide	2005	Synthetic	GLP-1 agonist	Type 2 diabetes
Lanreotide	2007	Synthetic	Somatostatin receptor agonist	Acromegaly
Romiplostim	2008	Phage display technique	Thrombopoietin receptor agonist	Thrombocytopenia purpura
Romidepsin	2009	Natural, from *Chromobacterium violaceum*	Inhibitor of histone deacetylase	Lymphoma of skin
Liraglutide	2010	Recombinant DNA technique	GLP-1 agonist	Type 2 diabetes and obesity
Tesamorelin	2010	Synthetic	Growth hormone-releasing hormone agonist	Lipodystrophy
Telaprevir	2011	Synthetic	NS3/4A viral protease inhibitor	Hepatitis C
Boceprevir	2011	Synthetic	NS3/4A viral protease inhibitor	Hepatitis C
Icatibant	2011	Synthetic	Bradykinin B2 receptor antagonist	Genetic angioedema

(Continued)

Table 15.1 (Continued)

Peptide therapeutic	Approval year	Method of development	Target pharmacological action	Indication
Carfilzomib	2012	Semi-synthetic	Proteasome inhibitor	Multiple myeloma
Lucinactant	2012	Synthetic	Reduce surface tension of alveolar epithelium and allow gas exchange	Respiratory distress syndrome (RDS) in premature infants
Pasireotide	2012	mRNA display	Somatostatin receptors agonist	Acromegaly
Peginesatide	2012	Synthetic	Erythropoietin receptor stimulant	For anemia in chronic kidney disease (CKD) patients on dialysis
Linaclotide	2012	Synthetic	Guanylate cyclase-C agonist	Irritable Bowel Syndrome with constipation (IBS-C) and chronic idiopathic constipation (CIC)
Teduglutide	2012	rDNA technology	GLP-2 agonist	Short bowel syndrome
Vasopressin	2014	Synthetic	Vasopressin V2 receptor agonist	Postoperative abdominal distention, in abdominal roentgenography to dispel interfering gas shadows, and in diabetes insipidus
Dalbavancin	2014	Semi-synthetic	Bacterial cell wall synthesis inhibitor	Antibacterial for acute bacterial skin and skin structure infection (ABSSSI)
Oritavancin	2014	Semi-synthetic	Inhibit transglycosylation	Acute bacterial skin and skin structure
Albiglutide	2014	rDNA technology	GLP-1 agonist	Type 2 diabetes
Dulaglutide	2014	rDNA technology	GLP-1 agonist	Type 2 diabetes
Parathyroid hormone	2015	rDNA technology	Parathyroid hormone 2 (PTH 2) receptor agonist	Hypocalcemia in patients with hypoparathyroidism
Lixisenatide	2016	rDNA technology	GLP-1 receptor agonist	Type 2 diabetes
Grazoprevir	2016	Synthetic	NS3/4A viral protease inhibitor	Hepatitis C virus in combination with elbasvir
Gallium dotatate Ga-68	2016	Synthetic	Binds to somatostatin receptors	Diagnosis of somatostatin receptor-positive neuroendocrine tumors (NETs)
Macimorelin	2017	Synthetic	Growth hormone secretagogue receptor type 1 agonist, increases growth hormone secretion	Diagnosis of growth hormone deficiency
Voxilaprevir	2017	Synthetic	NS3/4A viral protease inhibitor	Hepatitis C virus

Table 15.1 (Continued)

Peptide therapeutic	Approval year	Method of development	Target pharmacological action	Indication
Angiotensin II	2017	Synthetic	Type-1 angiotensin II receptor agonist	Increases blood pressure in adults with septic or other distributive shock
Plecanatide	2017	Synthetic	Guanylate cyclase soluble subunit alpha-2 agonist	CIC and IBS-C
Etelcalcetide	2017	Synthetic	Calcium sensing receptor agonist	Secondary hyperparathyroidism in patients with CKD on dialysis
Abaloparatide	2017	Synthetic	Parathyroid hormone receptor 1 (PTR-1) agonist	Osteoporosis
Semaglutide	2017	rDNA technology	GLP-1 receptor agonist	T2DM
Lutetium Lu 177 dotatate	2018	Synthetic	Binds to somatostatin receptors subtype-2	Gastroenteropancreatic neuroendocrine tumors
Ivosidenib	2018	Synthetic	Isocitrate dehydrogenase-1 inhibitor	Relapsed or refractory acute myeloid leukemia cholangiocarcinoma
Bremelanotide	2019	Synthetic	Melanocortin receptor (MCR) agonist	Hypoactive sexual desire disorder in premenopausal women
Afamelanotide	2019	Synthetic	Melanocortin-1 receptor (MC1R) agonist	Patients with phototoxic reactions (damage to skin) from erythropoietic protoporphyria
Gallium Dotatoc Ga-68	2019	Synthetic	Binds to somatostatin subtype 2 receptors (SSTR2)	PET imaging of neuroendocrine tumors
Gallium Ga-68 gozetotide	2020	Synthetic	Binds to prostate-specific membrane antigen (PSMA)	PET imaging of PSMA for prostate carcinoma
Copper Dotatate Cu-64	2020	Synthetic	Binds to somatostatin receptors with the highest affinity for subtype 2 receptors (SSTR2)	PET/CT imaging of neuroendocrine tumors
Setmelanotide	2020	Synthetic	Melanocortin receptor 4 agonist	Obesity
Vosoritide	2021	Synthetic	Atrial natriuretic peptide receptor 2 agonist	Growth in children with achondroplasia
Pegcetacoplan	2021	mRNA display	Complement protein C3 inhibitor	Paroxysmal nocturnal hemoglobinuria (PNH)
Melphalan flufenamide	2021	Synthetic	DNA cross-linking and alkylation	Multiple myeloma
Lonapegsomatropin-TCGD *(Skytrofa)*	2021	rDNA technology	Release somatotropin, growth hormone receptor agonist	Short stature

(Continued)

Table 15.1 (Continued)

Peptide therapeutic	Approval year	Method of development	Target pharmacological action	Indication
Difelikefalin	2021	Synthetic	Kappa-type opioid receptor agonist	Pruritus associated with CKD patients on dialysis
Odevixibat	2021	Synthetic	Ileal sodium/bile acid transporter (IBAT) inhibitor	Pruritus associated with progressive familial intrahepatic cholestasis (PFIC)
Dasiglucagon	2021	Synthetic	Glucagon receptor agonist	Hypoglycemia
Voclosporin	2021	Synthetic	Calcineurin inhibitor	Lupus nephritis
Lutetium (^{177}Lu) vipivotide tetraxetan	2022	Synthetic	DNA denaturation and apoptosis	Metastatic prostate cancer
Tirzepatide	2022	Synthetic	GIP/GLP-1 dual agonist	T2DM

Table 15.2 Peptide-based drugs in clinical trials.

Therapeutic peptide candidate	Peptide nature	Indication	Phase of trial	Trial number
Bovine lactoferin	Lactoferin-derived peptides	Infant lactose intolerance	1	NCT04738058
Dulaglutide	Analogue of GLP-1	Cessation of smoking	2	NCT03204396
Efinopegdutide	IgG4 conjugated peptide	Fatty liver	2	NCT04944992
Ghrelin	Side chain modified ghrelin	Obesity	1	NCT02953587
Somapacitan	Human growth hormone analog	Children growth hormone deficiency	3	NCT03811535
Thymalfasin	Modified prothymosin	Cancer of thymus gland	2	NCT0366764
Vosoritide	CNP cyclic derivative	Achondroplasia	2	NCT03989947
Angiotensin 1-7	Angiotensin synthetic analogue	Treatment of hypertension	1	NCT02245230
Deslorelin	Gonadotropin releasing hormone agonist	Breast cancer	2	NCT00080756
Thyrotropin	Recombinant thyrotropin	Cancers of head and neck	3	NCT00415233
K27M	Peptide vaccine	Brain tumors	2	NCT02960230
Ivosidenib	Inhibitor of IDH1	Acute myeloid leukemia, solid tumors	1	NCT04176393
LSALT peptide	Linear peptide	COVID-19	2	NCT04402957
LY3437943	Analogue of proglucagon	Obesity treatment	2	NCT0846778
IB1362	Synthetic peptide	T2DM, obesity	2	NCT04440345
AZP-3601	Analog of parathyroid hormone	Hyperthyroidism	1	NCT05239221
Metreleptin	Leptin synthetic analogue	Type 2 diabetes, hyperlipidemia	2	NCT00085982
Avexitide	GLP-1 antagonist	Obesity	2	NCT02697253

Table 15.2 (Continued)

Therapeutic peptide candidate	Peptide nature	Indication	Phase of trial	Trial number
Motixafortide	Heterodetic cyclic peptide, CXCR4 inhibitor	Multiple myeloma	1	NCT05293171
Trofinetide	Synthetic analog and terminal tripeptide of IGF1	Rett syndrome	3	NCT04279314
Cagrilintide	Novel lipidated amylin analog	T2DM	2	NCT04982575
Rusfertide	Hepcidin mimetic	Polycythemia vera	2	NCT04767802
AXT107	20-mer amino acid derived from the sequence in type IV collagen edema	Diabetic macular and neovascular age-related macular degeneration	2	NCT04279314
Zilucoplan	Synthetic macrocyclic peptide inhibitor	Amyotrophic lateral sclerosis, myasthenia gravis	3	NCT04436497
Aviptadil	Octacosapeptide vasoactive intestinal peptide analog	COVID-19	3	NCT04843761
Zetomipzomib	P1 keto-epoxide-containing peptide	Polymyositis/ dermatomyositis, Lupus nephritis, erythematosus	2	NCT04033926
Bulevirtide	Lipopeptide	Chronic hepatitis delta	3	NCT03852719
Glepaglutide	Long-acting GLP2 analog	Short bowel syndrome III 2018	3	NCT03690206
Setmelanotide	Cyclic peptide analog of pro-opiomelanocortin	Rare genetic disorders of obesity	3	NCT03013543
Adegramotide	Peptide vaccine	Renal cell carcinoma, urothelial carcinoma, glioblastoma	3	NCT0331133
Mifamuritide	Synthetic derivative of muramyl dipeptide	Osteosarcoma	2	NCT01459484
Cu-64 TP3805	Activator of pituitary cyclic adenylate	Bladder cancer	1	NCT03039413

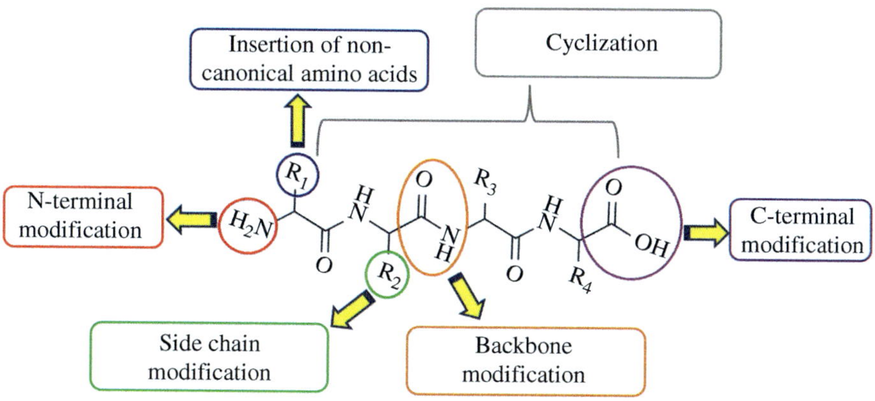

Figure 15.3 Peptide diversification strategies.

15.2.5.1 Chemical Synthesis of Peptides

Ever since the introduction of solid phase synthesis (SPSS) of peptides by Merrifield [6], the development of peptides has become easier, more accessible, and efficient enough. SPSS is one of the major methods for current day peptide synthesis in terms of methodology and synthetic materials used. SPSS was the perfect alternative approach to the time-consuming and less convenient classical solution phase synthetic methods, which necessitated purification after each cycle of amino acid coupling [88]. Precisely, in SPSS, the N-protected C-terminal amino acid is covalently attached via the carboxylic group to an insoluble amino or hydroxyl to finally yield a C-terminal amide peptide or C-terminal acid, respectively. Further, the peptide chain is elongated in the C-terminus to the N-terminus strategy through a series of coupling and deprotection steps. The α-amino groups are protected by either Boc (t-butyloxycarbonyl)/Fmoc (fluroenylmethyloxycarbonyl) groups to avoid side reactions with reagents used in the synthesis process (Figure 15.1). The protecting groups and peptidyl resins are removed by employing strong anhydrous hydrogen fluoride or trifluromethanesulfonic acid in Boc-SPSS and trifluoracetic acid in Fmoc-SPSS, respectively [89, 90]. Of note, the removal of Fmoc is achieved in milder conditions, thereby making Fmoc-SPSS a method of choice for peptide synthesis [91]. The Boc-based SPSS employs Merrifield resin for the synthesis of C-terminus acid peptides, whereas MBHA resin is used for C-terminus amide peptides. On a similar note, Fmoc-based SPSS uses Pal resin, Rink resin, and Sieber resin to synthesize C-terminus amide peptides. Trityl chloride resin, 2-chloro trityl chloride resin, HMPB resin, Wang resin, and Sasrin resin are widely used for the synthesis of C-terminus acid peptides [88]. A wide range of functional resins have been developed by coupling various linkers to resins and are being frequently used for the synthesis of longer peptides and peptide cyclization [92].

The synthesis of long peptides by SPSS still remains challenging due to interchain aggregation and formation of aspartimide segments specially in Fmoc-SPSS, which reduces the purity of crude peptides. Several methods, like the application of microwaves, using low substitution resins, using mixed solvents, and introducing pseudoproline into reaction media, have been undertaken to reduce reaction time, to separate the peptide chain, and more importantly to avoid aggregation during the synthesis phase [93–96]. The formation of aspartimide during Fmoc SPSS is negated by using microwaves for reaction, using N-α-alkyl Asp-Gly dipeptide, or by addition of 1-hydroxy benzotriazole (HOBt), Oxyma Pure to reaction medium [97–100].

Very long peptides (>50 amino acids) are difficult to synthesize via SPSS. However, several chemoselective ligation techniques were conjugated with SPSS to facilitate the smooth synthesis of long peptides. In the chemical ligation method, a C-terminus thiocarboxylic acid group is used for coupling with N-terminal amine through a peptide bond analogous linker. Alternatively, an activated thioester group at the C-terminus of one peptide can act as a strong nucleophile to attack the N-terminus amine of another peptide segment [101, 102]. Several other developed chemoselective ligation techniques for the synthesis of peptides are briefly presented in Table 15.3.

Other techniques developed is the expressed protein ligation (EPL) [109] and sequential protein ligation (SPL) [110]. Both methods are ideal for the synthesis of semisynthetic proteins. The semisynthetic proteins are synthesized chemoselectively ligating recombinant peptides and SPSS-constructed peptides. Nonnatural amino acids can be incorporated into the peptide segments via EPL and SPL. Currently, automated peptide synthesizers are used for lab scale synthesis of peptides. CEM liberty PRIME and CSBio-II are automatic peptide synthesizers that utilize microwave or infrared heat to sequentially ligate peptides [111]. Microwave and infrared heating have limited application for large-scale peptide synthesis owing to risks of overheating and lack of larger equipment [112].

Table 15.3 Chemical ligation techniques for peptide synthesis.

Ligation technique	Linker functional group	Methodology	Reference
Thioester ligation	Peptide segment 1-CO-S-peptide segment 2	Nucleophilic attack by thioacid sulfur of N-terminal peptide at alkyl halide of C-terminal peptide	[103]
Oxime ligation	Peptide segment 1-CH=NO-peptide segment 2	Condensation of aldehyde functional groups at the C-terminus of the N-terminal peptide with aminooxy groups at the N-terminus of the C-terminal peptide	[104]
Hydrazone ligation	Peptide segment 1-CH=NNH-peptide segment 2	Condensation of aldehyde functional groups at the C-terminus of the N-terminal peptide with hydrazine groups at the N-terminus of the C-terminal peptide	[105]
Thiazolidine/Oxazolidone ligation	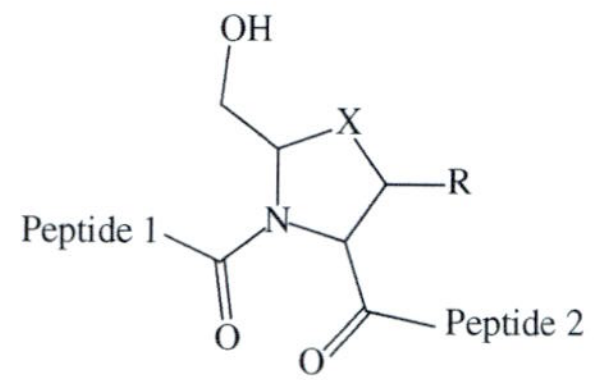 X=S/O; R=H, CH_3	Formation of thiazolidine/oxazolidone capture device between C-terminal peptide ester aldehyde and N-terminal thiol/hydroxyl amine group	[106]
Native chemical ligation	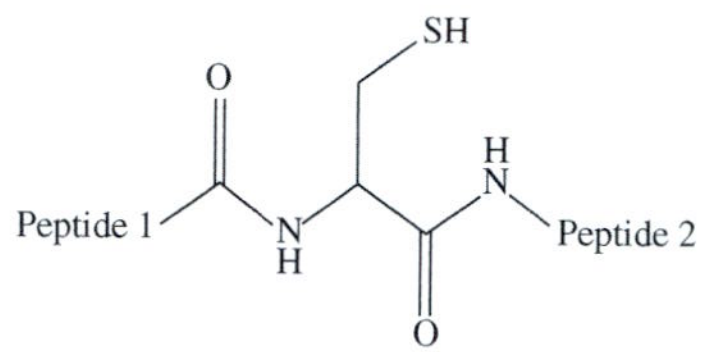	Ligation between thioester at the C-terminus of one peptide and cysteine residue at the N-terminus of another peptide following spontaneous rearrangement of thioester bond to amide bond	[107]
Thioacid capture ligation	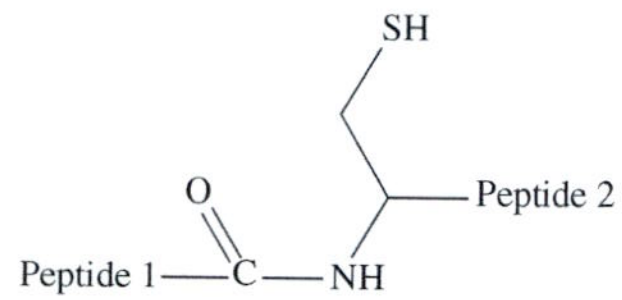	Thioacid from C-terminal peptide capture by NPYS (5-nitropyridyl-2-sulfenyl) activated cysteine thiol of N-terminal peptide leading to form ligated peptide via acyl disulfide intermediate	[108]

15.2.5.2 Chemical Modification of Peptide and Peptidomimetics

Chemical diversification of peptides generally involves late-stage modifications and peptidomimetic strategies. The modification of peptide structure is important for mimicking, stabilizing, or constructing a superlative secondary structure to enhance their biological [113]. Prior to chemical modification, an alanine scan is carried out to determine key residues that impart biological properties to lead peptide [114, 115]. Peptidomimetics structurally possess a pharmacophore that mimic peptide structures and display pharmacological action. Peptidomimetics are instrumental in abrogating pharmacokinetic disadvantages of therapeutic peptides [7]. Based on their structural features, the peptidomimetics have been divided into four groups [116]. Class A comprises peptidomimetics that have minute modifications in their backbone or amino acid sequences like voclosporin and trofinetide. Class B peptidomimetics have incorporated noncanonical amino acids in their basic sequence [117]. The incorporation of unnatural amino acids in peptides improves the inherent pharmacokinetic and pharmacodynamic properties. Odevixibat (2021) and setmelanotide (2020) are exquisite examples of FDA-approved therapeutic peptides having unnatural amino acids in their peptide sequence. Class C peptidomimetics are modified peptides wherein the amide bonds are replaced by oxetane, aza, or triazole linkages. Saquinavir is a good example of backbone-modified peptide. Class D peptidomimetics are small drug molecules that only mimic the pharmacological activity of the nascent peptide without any structural resemblance or side-chain modifications, such as navitoclax and venetoclax (BH3 mimetics).

15.2.5.3 Backbone Modification of Peptides

Backbone diversifications uplift the physicochemical properties of therapeutic peptides [118]. The backbone modifications of the peptide could be removable so as to improve solubility by inserting soluble polyarginine tags and facilitating synthesis or nonremovable where unnatural amino acids are inserted in the nascent peptide for improving the pharmacology [119, 120].

The basic strategies for nonremovable modifications involve Cα-methylation, backbone cyclizations, and Nα-methylations. The insertion of noncanonical amino acids in the peptide reduces flexibility and thereby imparting tertiary structure to the modified peptide [121–123].

Replacing an α-amino acid with a side chain bearing amino acid on β3 carbon prolongs half-lives of peptides [124, 125]. Selepressin has a prolonged plasma half-life developed by the backbone modification of vasopressin [126]. Backbone modification of cyclic peptides by N-methylations has improved their membrane permeability and oral bioavailability [127].

15.2.5.4 Side-Chain Modification of Peptides

By replacing the canonical amino acids from the parent peptides with their respective analogs, side-chain modifications can be achieved. Such side-chain modifications make the peptide target specific with improved binding affinity [128]. B-phenylalanine, homoarginine, and benzyloxytyrosine are some analogs of natural amino acids also available commercially and have been successfully used for side-chain modifications [129, 130]. Liraglutide and semaglutide are the GLP-1 agonists with side-chain modifications [131].

15.2.5.5 Peptide Cyclization

In the roadway of uplifting the biophysical characteristics of the peptides as well as broadening their applicability as biocatalysts and therapeutics, cyclization is an escalating field of research. Cyclization enhances biological activity, target binding specificity, cell permeability, and metabolic stability of therapeutic peptides [132]. The peptides can be cyclized in four different

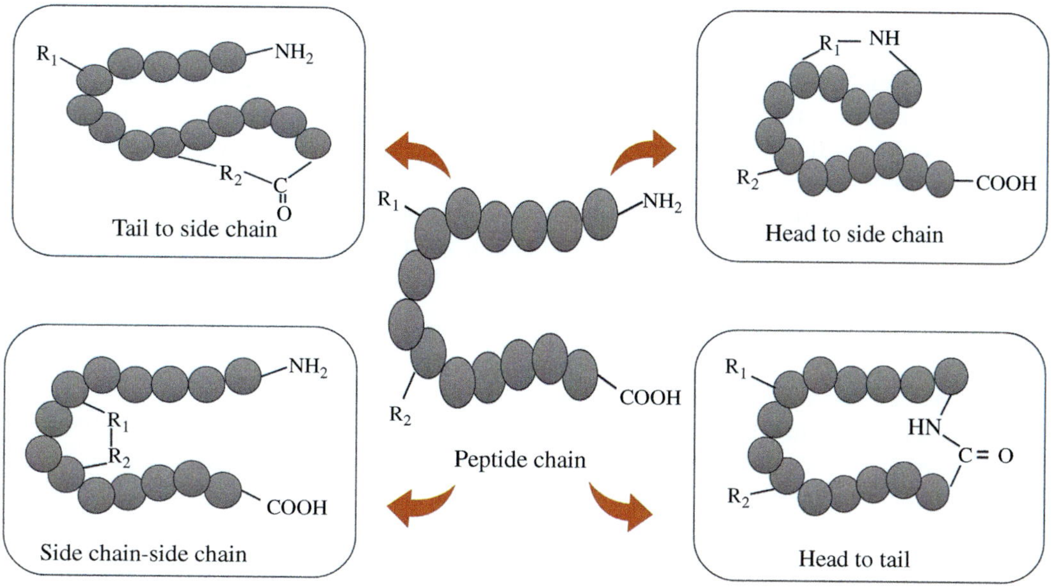

Figure 15.4 Strategies for peptide cyclization.

orientations: side chain-to-side chain, head-to-tail, head-to-side chain, and tail-to-side chain (Figure 15.4). Peptide cyclization can be achieved by chemical, enzymatic, and protein tagging techniques [133].

The chemical methods encompass direct amide bond formations, native chemical ligations, C-terminal aldehyde ligations, disulfide bond formations, and biorthogonal reactions. Direct amide bond formation is achieved by converting the C-terminal –OH into a better-leaving group using a single or combination of coupling agents along with tertiary amine base additive like DIEA [134]. Backbone peptide cyclization of peptides can be achieved by intramolecular native chemical ligations even at millimolar concentrations [135]. Several cyclic peptides like daptomycin [136], cyclomontanin B [137], and many more are developed from chemical ligations where a C-terminal glycolaldehyde reacts with N-terminal serine or threonine residue to form a benzylidene acetal amide intermediate that rearranges to form peptide bond [138]. More recently, the CyClick method has been applied for the synthesis of thermodynamically stable bicyclic peptides [139]. Biorthogonal reactions are also extensively used for the synthesis of cyclic peptides. The commonly used biorthogonal reactions are Staudinger ligations [140], KAHA ligations [141], TAMM ligations [142], CuAAC cycloadditions, and RuAAC cycloadditions [143]. Cyclization through disulfide bond formation can be achieved by the insertion of orthogonal cysteine residues in peptide chains [144]. Using disulfide stapling agents, strong thioether linkages are established between the cysteine residues which make the cyclic peptide more stable biophysically [145].

Cyclic peptides are also synthesized by employing enzymes. Subglitases [146], sortases [147], and asparaginyl transpeptidases are used for the backbone cyclization of many peptides. Amino acid side chains can be ligated using transglutaminases as well [148]. Noncatalytic protein domains also called as tags are conjugated to peptide of interest for initiating cyclization. Inteins [149], Spy tag/Spy catcher [150], and Snoop tag/Snoop catcher [151] exclusively used tags for the synthesis of cyclic peptides.

15.2.5.6 Peptide Mimicking of α-Helices and Stabilization

Hydrogen bond surrogates or stabilized α-helices are produced by replacing hydrogen bonds with covalent bonds in the parent peptide chain [152]. The α-helices can also be stabilized by inserting cross-links in the amino acid side chains. Building crosslinks at i and $i+4$ as well as i and $i+7$ positions in the α-helix side chains is very effective in producing stable peptide structures [153]. Stable α-helices are also produced by inserting lactam-based cross-links, disulfide bonds, and biselctrophillic linkers [154–156]. Stapling of peptides is currently used in innovative cross-linking approaches for stabilizing α-helices. The technique forms nucleophilic cross-links via insertion of electrophilic unnatural amino acid residues at i and $i+4$ as well as i and $i+7$ positions [157].

15.2.5.7 Peptide Mimicking of β-Strands and β-Sheets

The design of peptidomimetics and peptides for targeting secondary structures in various PPIs is a burgeoning research area. β-Strands and sheets are secondary structures of proteins. D-proline and L-proline templates are being majorly used for stabilizing β-hairpins in PPI inhibitors [158]. Some notable peptide mimetic strategies employed for targeting PPIs involving destabilized β-strands and β-sheets are showcased in Table 15.4.

15.2.5.8 Peptide Production by Recombinant Technology

Cheaper, sustainable, and scalable bioactive peptide production is enabled by recombinant DNA technology. The expression systems used are bacteria, yeast, mammalian cells, insect cell cultures, and plants [166]. The basic strategy encompasses genome modification of the host/expression system by inserting gene encoding peptide or protein of interest followed by its expression in the host. Later, the recombinant peptide/protein of interest is isolated and purified [167]. The first human peptide to be produced recombinantly was somatostatin followed by insulin in 1982 [168, 169]. Table 15.5 displays various recombinant therapeutic peptides.

Table 15.4 Peptidomimetics for stabilization of β-sheet/β strands in PPIs.

Peptidomimetic strategy	PPI aggregations	Target disorder	Reference
Modified β6 and β7 strands of SOD1 (β-strand mimics)	SOD1 (superoxide dismutase-1)	ALS (amyotrophic lateral sclerosis)	[159]
Isolated sequence recognition elements (β-strand mimics)	Amyloid β-peptide (segments 1–42)	Alzheimer's disease	[160]
Linear peptide with β-sheet breakers	Insulin aggregation	Impaired insulin absorption	[161]
Cyclic β-hairpins	Dimerization of epithelial growth factor receptor	Carcinomas	[162]
Acyclic β-hairpins	Transthyretin aggregation	Familial amyloid polyneuropathy	[163]
Tweezer molecules	Dimerization of HIV-1 protease	AIDS	[164]
Peptide antibodies	Human α-synuclein aggregation	Parkinson's disease	[165]

Table 15.5 rDNA-derived therapeutic peptides.

Therapeutic peptide	Amino acid residues	Brand name	Manufacturer	Indication
Human insulin	51	Humulin, Novolin	Eli lily, Novo Nordisk	Types 1 and 2 diabetes
Insulin detemir	50	Levemir	Novo Nordisk	Types 1 and 2 diabetes
Glucagon	28	GlucaGen	Eli lily, Novo Nordisk, Zymo Genetics	Hypoglycemia
Teduglutide	36	Revestive	Nycomed	Short bowel syndrome
Liraglutide	37	Victoza, Saxenda	Novo Nordisk	Type 2 diabetes, obesity
Albiglutide	2×30	Tanzeum	GlaxoSmithkine	Type 2 diabetes
Dulaglutide	2×30	Trulicity	Eli lily	Type 2 diabetes
Semaglutide	31	Ozempic	Novo Nordisk	Type 2 diabetes, obesity
Carperitide	28	HANP injection	Daiichi Pharmaceuticals	Acute heart failure
Neseritide	32	Natrecor	Scios	Congestive heart failure
Calcitonin	32	Acticalcin	TRB Pharma	Postmenopausal osteoporosis
Teriparatide	34	Forteo	Eli lily	Osteoporosis
Mecasemin	70	Increlex	Ipsen	Insulin-like growth factor-1 deficiency

15.2.5.9 Peptides Modification by Genetic Code Expansion

Proteins are naturally synthesized from 20 natural or canonical amino acids. The amino acid sequences in biosynthesized protein molecules are strictly conserved, thereby constraining their structural and functional diversity. Of note, evolution in genetic code expansion has played a vital role in overcoming such restriction [170]. During the translation of proteins, noncanonical amino acids with specific physiological properties can be selectively incorporated into a growing polypeptide by implementing genetic code expansion [171]. In the process, an orthogonal aminoacyl-tRNA synthetase charges a noncanonical amino acid to the orthogonal tRNA that meanwhile suppresses the unique stop codon and expresses the amber or quadruplet codon (UAG). Consequently, the noncanonical amino acids are site specifically incorporated into the growing peptide molecule [172].

A plethora of noncanonical amino acids with specific functional modules have been genetically encoded into viruses, prokaryotes, and eukaryotes that evidently provided precious tools for protein engineering [173–175]. Numerous vaccines have been developed through genetic code expansion [176]. Site-specific insertion of noncanonical immunogenic amino acids like pNO2F in m-TNFα and EGF as well posttranslational modification of Tyr residues at desired locations in IgG have broken immunological tolerances and further developed therapeutic vaccines Noncanonical amino acids with phenylalanine backbone were successfully inserted in *Mycobacterium tuberculosis*, *Mycobacterium smegmatis*, and *Bacillus Calmette Guerin* for the development of tuberculosis vaccines Genetic code expansion was also implemented for the development of live attenuated HIV-1 vaccine by incorporation of phenyl alanine analogues into the protein envelope of HIV virus by controlling its replication cycles.

15.2.5.10 PEGylation of Peptides and Proteins

PEGylation improves renal clearance of therapeutic peptides by enhancing their effective molecular weight [177]. PEGylation has been the method of choice for increasing the half-life and metabolic stability of peptide therapeutics in the last two decades [178, 179]. Consequently, PEGylation has evolved as a prime strategy for the modification of therapeutic peptides. Currently, there are >10 FDA-approved therapeutic peptides in the market and several other potential candidates under clinical investigations [180].

The majority of the approved therapeutic peptides have been synthesized by nonsite-specific PEGylation, specially at Lys or Cys residues [181]. The nonsite-specific conjugation strategy although leads to the accumulation of heterogenous products if more than one reactive Lys or Cys residue is present on the target protein and is quite difficult to separate [182]. Moreover, the activity of the protein may be lost due to PEGylation at the active site [183], entanglement in peptide chains by PEG residues [184] or conformational changes in protein [185]. Under such circumstances, the site specific PEGylation becomes an effective alternative approach where selective conjugation and positional control is achieved.

Physiochemical properties of the amino acid residues play a vital role in the selective modification of peptides. Differences in the pK_a values between amino groups of lysine, arginine, and N-terminal amino acid residue provide a good platform for the N-terminal modification of peptides [186]. N-terminal conjugation can be achieved by the use of oxidized agents, chemoselective utilization of catechol, pH-controlled acylation and alkylation, transamination, and reductive alkylation [187]. ATRP and RAFT methods (graft from) polymerization methods have been developed to grow PEG from the surface of protein [188–190]. Genetic code expansion has also emerged a promising tool for PEG conjugation. In this approach, a noncanonical amino acid is genetically encoded with bio-orthogonal chemical handle at required position into the target protein and further conjugated to PEG. IFN-a2b [191], hGH variants [192], and GLP-1 [193] have been modified by genetic code expansion coupled with site-specific PEGylation to improve pharmacokinetic and pharmacodynamic profiles. Currently, enzymes are also employed for site-specific ligation of amino acids with PEG [194].

15.3 New Technologies for Peptide-Based Drug Discovery

Several cutting-edge methods for finding therapeutic peptides have emerged in recent years. The development of novel peptide medications is being facilitated by these cutting-edge methods at a breakneck pace. Peptide-based drug discovery is benefiting from the development of new display technologies that have several advantages over more conventional approaches. Peptide libraries can now be made that are larger, manufactured more quickly, and at a lower cost than ever before because of the availability of these technologies. In this chapter, we look at some of the cutting-edge tools for peptide-based drug development

15.3.1 Phage Display

Phage display systems have improved, allowing this technology to be employed in immunology, biomedicine, material development, and more. The 2018 Nobel Prize "for the phage display of peptides and antibodies" recognized the importance of phage display systems [195]. The process of designing the library is considered to be one of the most significant aspects of the phage display experiment. This stage involves the incorporation of millions, if not more, of DNA

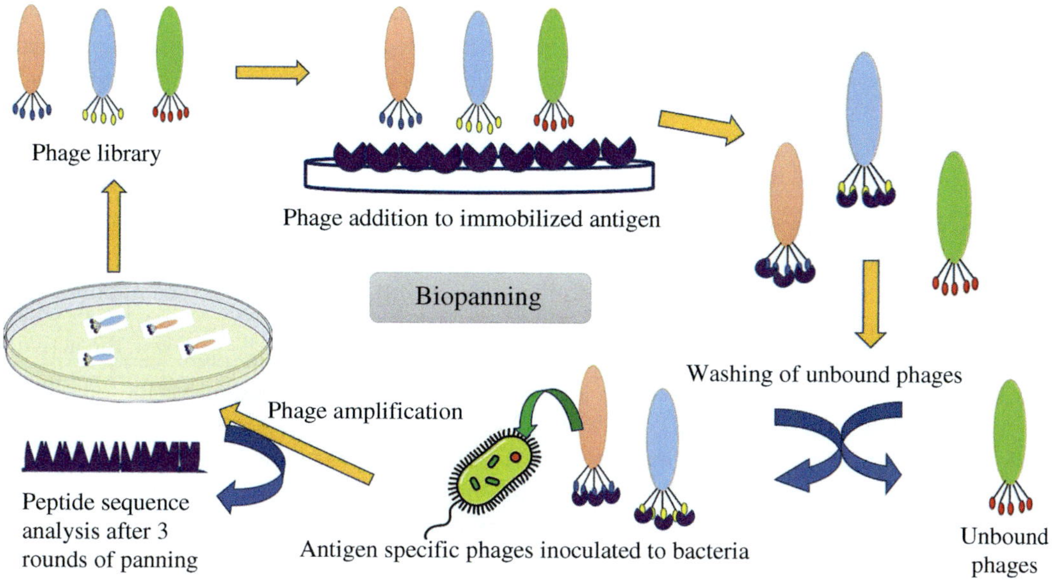

Figure 15.5 Biopanning and phage display method for peptide expression.

clones that contain target sequences that encode peptide or antibody fragments. Following that, these fragments are capable of being duplicated, copied, translated, and shown on the surface of the phage [196]. The subsequent step is to incorporate the library into the genome of the phage by means of either phagemid replication or standard vector replication. Utilizing biopanning (Figure 15.5) which is the process of selecting a specific phage particle [197], it is feasible to accomplish identification in a very basic manner. Following the verification of the functional expression of a peptide or protein, mutagenesis makes it possible to generate a large number of variations.

Several FDA-approved peptides have been designed by phage display technology, like peginesatide, romiplostim, and ecallantide. Many peptides have been identified by phage display as leads for antibiotic synthesis, antifreeze agents, and protease inhibitors of the SARS coronavirus [7].

15.3.2 mRNA Display

mRNA display represents a cutting-edge molecular technology that holds great promise in the realms of protein engineering, drug discovery, and functional genomics. This innovative technique involves the covalent linkage of a coding mRNA molecule to its encoded protein, creating a stable mRNA–protein complex. This linkage allows for the direct phenotypic screening of vast combinatorial libraries, enabling the rapid identification and optimization of peptides and proteins with desired functions or binding properties. Unlike traditional display methods, mRNA display uniquely captures the genotype–phenotype relationship in a single entity, facilitating the exploration of diverse sequence spaces. With applications ranging from the selection of therapeutic antibodies to the engineering of enzymes and the study of protein–protein interactions, mRNA display emerges as a powerful tool at the intersection of molecular biology and protein engineering, poised to redefine the landscape of functional genomics and accelerate the pace of drug discovery.

Phase II/III peptides like zilucoplan and phase III peptides like pegcetacoplan were found using mRNA display technology; a number of lead peptides have been identified as strong inhibitors of the SARS-CoV-2 spike protein [198], E6AP ubiquitin ligase [199], and influenza virus [200].

15.3.3 DNA-Encoded Libraries

In the dynamic landscape of drug discovery, the constant pursuit of innovative methodologies has significantly shaped both the pharmaceutical industry and academic research. A notable advancement in this arena is the transition from traditional approaches of accumulating compounds for robotic high-throughput screening (HTS) to the revolutionary realm of DEL technology [201]. DELs represent combinatorial collections of drug-like molecules, each uniquely labeled with a DNA barcode, offering a powerful platform for screening a vast array of compounds for their ability to interact with specific proteins.

The development of reactions compatible with DELs has been a formidable task, with stringent requirements such as compatibility with water, achieving quantitative yield, working with dilute reactants, and ensuring DNA orthogonality. These challenges underscore the complexities faced by modern synthetic organic chemistry in the pursuit of effective DEL-compatible reactions. Click reactions, with their ability to achieve diverse chemical functions with minimal reactions, align perfectly with these requirements, facilitating the creation of high-quality DELs with optimal structural diversity [202].

Recently, DELs have been successfully used in the design of respiratory syncytial virus protein inhibitors [203] and PPI inhibitors for skp1/skp2 aggregation [204].

15.3.4 Cell-Penetrating Peptides

Cell-penetrating peptides (CPPs) are sequences of diverse amino acids that facilitate cellular uptake through various mechanisms. The discovery of the transactivator of transcription (TAT) protein from HIV-1 in 1988 marked the inception of CPPs, followed by the identification of "penetratin" from the Drosophila Antennapedia-homeotic transcription factor in 1994. Subsequently, numerous natural and synthetic CPPs have been explored and refined for their cell-penetrating capabilities [205].

CPPs can be manipulated systematically to deliver cargo to specific cells or tissues, showcasing their adaptability. In addition, CPPs can be combined with other delivery systems to enhance cell-penetrating abilities [206].

15.3.5 Macrocyclic Peptides

Macrocyclic peptides, known for their ring structure with 12 or more members, emerge as a promising class for tackling complex therapeutic targets, especially protein–protein interactions. Their capacity for substantial binding interfaces, resembling natural protein–protein interactions, coupled with restricted conformational flexibility due to cyclization, imparts noteworthy selectivity and affinity compared to their noncyclic counterparts. With over 40 macrocyclic peptides in clinical use, the field sees the introduction of about one new macrocyclic peptide drug annually over the past decade [207].

Two main discovery approaches for macrocyclic peptides exist: design-led and screening/selection-led approaches. Design-led strategies leverage knowledge of the bound conformation to engineer ring-forming modifications that stabilize the active conformation. In contrast, screening/selection approaches rely on diverse candidate ligand pools or libraries without prior target structure or binding conformation knowledge. Recent technologies like DELs, split intein circular ligation of peptides and proteins (SICLOPPS), phage display, and mRNA display encode library members with unique genetic tags, facilitating the selection of molecules with desired traits using molecular biology and DNA sequencing. This differs from traditional screening, which individually assesses each library member's interaction with the target [208].

15.4 Computational Approaches in Peptide Drug Discovery

Computational methods enable an elaborate study of biomolecules in conjunction with their receptors at molecular levels. The knowledge of molecular-level protein and peptide interactions paves the way for rational peptide drug design and development by overcoming inherent shortcomings such as poor oral bioavailability and membrane permeability [209].

One of the strategies to improve oral bioavailability of peptide therapeutics is by enhancing their water solubility. The aqueous solubility of the peptides can be optimized by identifying and replacing undesired hydrophobic amino acid residues; however, the process is empirically large [210]. To expedite such optimizations, several support vector machine (SVM) learning tools like ccSOL omics [211] and PROSO II [212] were developed. The online tools facilitated in predicting peptide solubility by identifying hydrophobic amino acid residues along with their positions in the protein or peptide structure.

The CPPs can easily cross the plasma membranes owing to their hydrophobic nature. The cell penetrating ability of the CPPs are well exploited for delivering biological cargos, including antibodies, small molecule drugs, or other peptides which are otherwise impermeable across biological lipid bilayers [213]. Various robust bioinformatic tools are currently applied for predicting and optimizing peptide designing using the CPPs. Table 15.6 displays some online webservers available for predicting peptide solubility and permeability of CPPs.

Table 15.6 Online tools for CPP designing.

Bioinformatic tool	Server	Prediction	Machine learning technique	Input format
KELM-CPPpred	http://sairam.people.iitgn.ac.in/KELMCPPpred.html/	Permeability	Kernel extreme learning model	FASTA
CellPPD	http://crdd.osdd.net/raghava/cellppd/multi_pep.php	Permeability	Support vector machine	FASTA
CPPpred	http://bioware.ucd.ie/cpppred	Permeability	Artificial neural networks	FASTA
CPPpred-RF	http://server.malab.cn/CPPred-RF	Permeability	Random forest	FASTA
PROSO II	http://mbiljj45.bio.med.uni-muenchen.de:8888/prosoII/prosoII.seam	Solubility	Support vector machine	FASTA
ccSOL omics	http://s.tartaglialab.com/static_files/shared/tutorial_ccsol_omics.html	Solubility	Support vector machine	FASTA

Computational methods are extremely beneficial in predicting protein structures, surface charges, and interaction capacities. The *in silico* docking methods are cutting-edge tools for studying the protein–protein interactions that usually define disease pathologies. The peptide–protein docking methods can be classified as local, global, and template-based docking. The local docking methods screen the best binding pose of a peptide at user defined target binding site. Global docking identifies the best peptide pose and peptide binding site on the target protein. Global docking is the method of choice for analyzing protein–peptide interactions. Template docking generates the best peptide–protein interaction models by abstracting amino acid sequences from query peptide and binding site residue sequences of the target receptor [16]. The software tools used in specific peptide–protein docking are enlisted in Table 15.7.

Table 15.7 Molecular modeling tools for peptide–protein docking.

Software	Docking type	Accessibility	Features	Reference
PepCrawler	Local docking	http://bioinfo3d.cs.tau.ac.il/PepCrawler	Uses rapidly exploring random tree algorithm Sampling based on motion planning Peptide structure is flexible	[214]
Rosetta FlexPepDock	Local docking	http://flexpepdock.furmanlab.cs.huji.ac.il or http://www.rosettacommons.org/software	Monte-Carlo-based algorithm Modeling of hot spot residue Receptor is flexible Clustering and scoring are based on Rosetta energy function	[215]
HADDOCK	Local docking	http://haddock.science.uu.nl/services/HADDOCK2.2/	Homology modeling using ensemble canonical structure Docking at user defined residues in binding pocket of target protein Uses a binding energy-based scoring function Flexible docking	[216]
DynaDock	Local docking	Commercial	Performs soft core potential optimized molecular dynamic simulation Faster sampling time Flexible docking	[217]
Pepsite 2.0	Local docking	http://pepsite2.russelllab.org	Fastest identification of peptide binding site Generate low resolution peptide model Generate coarse grained peptide orientation by using spatial specific scoring matrix	[218]
ClusPro Peptidock	Global docking	https://peptidock.cluspro.org/	Fourier transform based docking Motif based structure prediction	[219]

Table 15.7 (Continued)

Software	Docking type	Accessibility	Features	Reference
pepATTRACT	Global docking	http://bioserv.rpbs.univparisdiderot.fr/services/pepATTRACT/	Predicts peptide structure by employing threading sequence	[220, 221]
HPEPDOCK	Global docking	http://huanglab.phys.hust.edu.cn/hpepdock/	Ensemble protein structure conformation Uses hierarchical algorithm	[202]
GalaxyPepDock	Template docking	http://galaxy.seoklab.org/pepdock	Motif similarity search Model building using PeptiDB database Rigid docking	[222]
SPRINT-str	Template docking	http://sparks-lab.org/server/SPRINT-Str	Predicts interacting residues on peptide–protein binding interface Uses SVM algorithm Ability to distinguish binding site on macromolecules	[223]
PBRpredict-suite	Template docking	http://cs.uno.edu/~{}tamjid/Software/PBRpredict/pbrpredict-suite.zip	Using template sequences from NCBI database the probable interacting residues are predicted Use integrated machine learning algorithm	[224]
PepComposer	Template docking	https://cassandra.med.uniroma1.it/pepcomposer/webserver/pepcomposer.php	Uses monomeric protein databases for similarity search of binding site motifs Monte-Carlo-based geometry optimization	[225]

Molecular simulations are applied to investigate therapeutic peptide–excipient interactions, assessing drug–conjugate interaction, study interactions of synthesized compounds with peptides and proteins, screening of peptide-based enzyme inhibitors as well as a rational formulation design tool. Molecular docking, in combination with genetic algorithms, expedites the virtual screening of large peptide libraries [226]. The interaction of water-soluble metal complexes with human serum albumin was well demonstrated through molecular docking analysis using HEX 6.1 software by Tabassum and group [227]. *In silico* tools were instrumental in the design of N-octyl-N-arginine chitosan as a permeation-enhancing and drug-loading polymer for the oral delivery of insulin [228]. Molecular dynamic studies favored the glycosylation of peptides, which improved their poor membrane permeability via sodium-dependent glucose transporters [229]. The stability of peptide conjugates with polymers or other molecules can also be elucidated by molecular simulations [230, 231]. Molecular docking guided formulation of bioinspired oral insulin delivery system using yeast microcapsules is yet another groundbreaking application of computational avenues in the development of therapeutic peptides [232].

Computational tools are valuable assets for providing insight into the activity of peptide antimicrobials and thereby exploiting their potential to design new analogs. The APD3 have played a

Table 15.8 Online tools for AMP designing.

Webtool	URL	Basic algorithm	Application
APD3	aps.unmc.edu/AP/	Parameter space	AMP prediction
CAMP	www.bicnirrh.res.in/antimicrobial	DA, SVM, RF	AMP prediction
CS-AMPPred	sourceforge.net/projects/csamppred/	SVM	Prediction of cytosine stabilized AMPs
PeptideLocator	bioware.ucd.ie	BRNN	Predicting bioactive peptides in proteins
AVPpred	crdd.osdd.net/servers/avppred	SVM, BLASTP	Prediction of antiviral peptides
AntiBP2	http://crdd.osdd.net/raghava/antibp/	SVM	Prediction of AMP domains in protein sequences
CPPpred	bioware.ucd.ie/cpppred	NN	Prediction of cell penetrating peptides
C-P Amp	bioserver2.bioacademy.gr/Bioserver/CPAmP/	SVM	Prediction of plant AMPs
Defensinpred	www.defensinpred.cdac.in	SVM	Prediction of defensin peptides
BAGEL	bioinformatics.biol.rug.nl/websoftware/bage	ORF prediction tools	Predicting bacteriocin ORFs in DNA sequences
iAMP2L	www.jci-bioinfo.cn/iAMP-2L	FKNN	Predicting AMP query sequences

major role in the screening of large peptide libraries and the development of computational tools for the prediction and design of novel peptide antibiotics [173]. Several existing web tools allow the predictive design of AMPs with the aid of machine learning techniques (Table 15.8).

Computational approaches have been employed for the discovery of novel antimicrobial peptides. Several novel peptides have been designed by performing database screening and *ab initio* methods in APD for MRSA infections [233]. Capsid proteins from various enveloped and nonenveloped viruses were screened for antimicrobial potency using AMPA online tools [234]. A plethora of computational tools are also available for designing therapeutic peptides.

15.5 Conclusion

In recent years, peptides have become an increasingly important therapeutic class. Their use was generally hampered by poor membrane impermeability, oral bioavailability, and poor stability *in vivo*. However, peptides have been studied extensively to overcome such limitations. The amalgamation of existing traditional discovery approaches with new advances, such as peptide diversification through unnatural amino acids, C–H functionalization, recombinant synthesis, and DELs and display technologies, has resulted in the rapid discovery and development of peptide therapeutics. The advent of new chemical and biological synthetic processes allows large-scale production of the peptides. Thus, more than 100 peptide drugs have entered the clinic, and a significant number are in clinical development. Peptide-based discovery approaches are emerging as valuable tools for addressing several unresolved disease areas. Therefore, peptides have enormous therapeutic potential and economic market value, and recent advances are expected to enable the rapid identification of drugs to combat various diseases, including cancer.

References

1 Apostolopoulos, V., Bojarska, J., Chai, T.-T. et al. (2021). A global review on short peptides: frontiers and perspectives. *Molecules* 26 (2): 430.
2 Fosgerau, K. and Hoffmann, T. (2015). Peptide therapeutics: current status and future directions. *Drug Discovery Today* 20 (1): 122–128.
3 Henninot, A., Collins, J.C., and Nuss, J.M. (2017). The current state of peptide drug discovery: back to the future? *Journal of Medicinal Chemistry* 61 (4): 1382–1414.
4 du Vigneaud, V., Ressler, C., Swan, C.J. et al. (1953). The synthesis of an octapeptide amide with the hormonal activity of oxytocin. *Journal of the American Chemical Society* 75 (19): 4879–4880.
5 du Vigneaud, V., Gish, D.T., Katsoyannis, P.G., and Hess, G.P. (1958). Synthesis of the pressor-antidiuretic hormone, arginine-vasopressin. *Journal of the American Chemical Society* 80 (13): 3355–3358.
6 Merrifield, R.B. (1963). Solid phase peptide synthesis. 1. The synthesis of a tetrapeptide. *Journal of the American Chemical Society* 85 (14): 2149–2154.
7 Sharma, K., Sharma, K.K., Sharma, A., and Jain, R. (2023). Peptide-based drug discovery: current status and recent advances. *Drug Discovery Today* 28 (2): 103464.
8 Fisher, E., Pavlenko, K., Vlasov, A., and Ramenskaya, G. (2019). Peptide-based therapeutics for oncology. *Pharmaceutical Medicine* 33 (1): 9–20.
9 Iyengar, S., Ossipov, M.H., and Johnson, K.W. (2017). The role of calcitonin gene-related peptide in peripheral and central pain mechanisms including migraine. *Pain* 158 (4): 543–559.
10 Sloan, L.A. (2019). Review of glucagon-like peptide-1 receptor agonists for the treatment of type 2 diabetes mellitus in patients with chronic kidney disease and their renal effects. *Journal of Diabetes* 11 (12): 938–948.
11 Peterson, S.C. and Barry, A.R. (2018). Effect of glucagon-like peptide-1 receptor agonists on all-cause mortality and cardiovascular outcomes: a meta-analysis. *Current Diabetes Reviews* 14 (3): 273–279.
12 Torres, M.D.T., Sothiselvam, S., Lu, T.K., and de la Fuente-Nunez, C. (2019). Peptide design principles for antimicrobial applications. *Journal of Molecular Biology* 431 (18): 3547–3567.
13 Lau, J.L. and Dunn, M.K. (2018). Therapeutic peptides: historical perspectives, current development trends, and future directions. *Bioorganic & Medicinal Chemistry* 26 (10): 2700–2707.
14 Tsomaia, N. (2015). Peptide therapeutics: targeting the undruggable space. *European Journal of Medicinal Chemistry* 94: 459–470.
15 Maurya, N.S., Kushwaha, S., and Mani, A. (2019). Recent advances and computational approaches in peptide drug discovery. *Current Pharmaceutical Design* 25 (31): 3358–3366.
16 Lee, A.C., Harris, J.L., Khanna, K.K., and Hong, J.H. (2019). A comprehensive review on current advances in peptide drug development and design. *International Journal of Molecular Sciences* 20 (10): 2383.
17 Chang, L., Mondal, A., Singh, B. et al. (2023). Revolutionizing peptide based drug discovery: advances in the post AlphaFold era. *Wiley Interdisciplinary Reviews: Computational Molecular Science* e1693.
18 Davenport, A.P., Scully, C.C., de Graaf, C. et al. (2020). Advances in therapeutic peptides targeting G protein-coupled receptors. *Nature Reviews Drug Discovery* 19 (6): 389–413.
19 Yu, L., Li, K., Zhang, J. et al. (2022). Antimicrobial peptides and macromolecules for combating microbial infections: from agents to interfaces. *ACS Applied BioMaterials* 5 (2): 366–393.
20 Elliott, J.A., Reynolds, J.V., le Roux, C.W., and Docherty, N.G. (2016). Physiology, pathophysiology and therapeutic implications of enteroendocrine control of food intake. *Expert Review of Endocrinology & Metabolism* 11 (6): 475–499.

21 Ventola, C.L. (2015). The antibiotic resistance crisis: part 1: causes and threats. *Pharmacy and Therapeutics* 40 (4): 277.

22 Zaman, S.B., Hussain, M.A., Nye, R. et al. (2017). A review on antibiotic resistance: alarm bells are ringing. *Cureus* 9 (6).

23 Luong, H.X., Thanh, T.T., and Tran, T.H. (2020). Antimicrobial peptides – advances in development of therapeutic applications. *Life Sciences* 260: 118407.

24 Lewies, A., Du Plessis, L.H., and Wentzel, J.F. (2019). Antimicrobial peptides: the Achilles' heel of antibiotic resistance? *Probiotics and Antimicrobial Proteins* 11: 370–381.

25 Huan, Y., Kong, Q., Mou, H., and Yi, H. (2020). Antimicrobial peptides: classification, design, application and research progress in multiple fields. *Frontiers in Microbiology* 11: 2559.

26 Dubos, R.J. and Hotchkiss, R.D. (1941). The production of bactericidal substances by aerobic sporulating bacilli. *The Journal of Experimental Medicine* 73 (5): 629–640.

27 Wang, G., Li, X., and Wang, Z. (2016). APD3: the antimicrobial peptide database as a tool for research and education. *Nucleic Acids Research* 44 (D1): D1087–D1093.

28 Wang, G., Epand, R.F., Mishra, B. et al. (2012). Decoding the functional roles of cationic side chains of the major antimicrobial region of human cathelicidin LL-37. *Antimicrobial Agents and Chemotherapy* 56 (2): 845–856.

29 Krizsan, A., Volke, D., Weinert, S. et al. (2014). Insect derived proline-rich antimicrobial peptides kill bacteria by inhibiting bacterial protein translation at the 70 S ribosome. *Angewandte Chemie, International Edition* 53 (45): 12236–12239.

30 Mardirossian, M., Grzela, R., Giglione, C. et al. (2014). The host antimicrobial peptide Bac71-35 binds to bacterial ribosomal proteins and inhibits protein synthesis. *Chemistry & Biology* 21 (12): 1639–1647.

31 Troeira Henriques, S. and Craik, D.J. (2017). Cyclotide structure and function: the role of membrane binding and permeation. *Biochemistry* 56 (5): 669–682.

32 Hullin-Matsuda, F., Makino, A., Murate, M., and Kobayashi, T. (2016). Probing phosphoethanolamine-containing lipids in membranes with duramycin/cinnamycin and aegerolysin proteins. *Biochimie* 130: 81–90.

33 Grigore, M.E. (2018). Drug delivery systems in hard tissue engineering. *SF Journal of Biotechnology and Biomedical Engineering* 1 (1): 1001.

34 Suva, L.J., Brander, B.E., and Makhoul, I. (2011). Update on bone-modifying agents in metastatic breast cancer. *Nature Reviews. Endocrinology* 7: 380–381.

35 Deal, C. (2009). Potential new drug targets for osteoporosis. *Nature Clinical Practice. Rheumatology* 5: 20–27.

36 Aoki, K., Alles, N., Soysa, N., and Ohya, K. (2012). Peptide-based delivery to bone. *Advanced Drug Delivery Reviews* 64 (12): 1220–1238.

37 Yang, Y. (2015). Cancer immunotherapy: harnessing the immune system to battle cancer. *The Journal of Clinical Investigation* 125: 3335–3337.

38 Zhao, N., Qin, Y., Liu, H., and Cheng, Z. (2018). Tumor-targeting peptides: ligands for molecular imaging and therapy. *Anti-Cancer Agents in Medicinal Chemistry* 18: 74–86.

39 Xiao, Y.F., Jie, M.M., Li, B.S. et al. (2015). Peptide-based treatment: a promising cancer therapy. *Journal of Immunology Research* 2015: 761820.

40 Xu, C. and Zhang, H. (2015). Somatostatin receptor based imaging and radionuclide therapy. *BioMed Research International* 2015: 917968.

41 Gao, R., Zhang, G.J., Wang, Y.B. et al. (2018). Clinical value of (99m)Tc-3PRGD2 SPECT/CT in differentiated thyroid carcinoma with negative (131)I whole-body scan and elevated thyroglobulin level. *Scientific Reports* 8: 473.

42 Strosberg, J., El-Haddad, G., Wolin, E. et al. (2017). Phase 3 trial of (177) Lu-dotatate for midgut neuroendocrine tumors. *The New England Journal of Medicine* 376: 125–135.

43 Tagawa, S.T., Vallabhajosula, S., Christos, P.J. et al. (2019). Phase 1/2 study of fractionated dose lutetium-177-labeled anti-prostate-specific membrane antigen monoclonal antibody J591 ((177) LuJ591) for metastatic castration-resistant prostate cancer. *Cancer* 125: 2561–2569.

44 Escala Cornejo, R.A., García-Talavera, P., Navarro Martin, M. et al. (2018). Large cell neuroendocrine carcinoma of the lung with atypical evolution and a remarkable response to lutetium Lu 177 dotatate. *Annals of Nuclear Medicine* 32: 568–572.

45 Hagimori, M., Fuchigami, Y., and Kawakami, S. (2017). Peptide-based cancer-targeted DDS and molecular imaging. *Chemical & Pharmaceutical Bulletin (Tokyo)* 65: 618–624.

46 Chen, Y., Wu, J.J., and Huang, L. (2010). Nanoparticles targeted with NGR motif deliver c-myc siRNA and doxorubicin for anticancer therapy. *Molecular Therapy* 18: 828–834.

47 Ruoslahti, E. (2017). Tumor penetrating peptides for improved drug delivery. *Advanced Drug Delivery Reviews* 110–111: 3–12.

48 Kotsakis, A., Papadimitraki, E., Vetsika, E.K. et al. (2014). A phase II trial evaluating the clinical and immunologic response of HLA-A2(+) non-small cell lung cancer patients vaccinated with an hTERT cryptic peptide. *Lung Cancer* 86: 59–66.

49 Kumai, T., Kobayashi, H., Harabuchi, Y., and Celis, E. (2017). Peptide vaccines in cancer-old concept revisited. *Current Opinion in Immunology* 45: 1–7.

50 Tsoras, A.N. and Champion, J.A. (2018). Cross-linked peptide nanoclusters for delivery of oncofetal antigen as a cancer vaccine. *Bioconjugate Chemistry* 29: 776–785.

51 Boohaker, R.J., Sambandam, V., Segura, I. et al. (2018). Rational design and development of a peptide inhibitor for the PD-1/PD-L1 interaction. *Cancer Letters* 434: 11–21.

52 Zhou, K., Lu, J., Yin, X. et al. (2019). Structure-based derivation and intramolecular cyclization of peptide inhibitors from PD-1/PD-L1 complex interface as immune checkpoint blockade for breast cancer immunotherapy. *Biophysical Chemistry* 253: 106213.

53 Abbas, A.B., Lin, B., Liu, C. et al. (2019). Design and synthesis of A PD-1 binding peptide and evaluation of its anti-tumor activity. *International Journal of Molecular Sciences* 20: 572.

54 Sasikumar, P.G., Ramachandra, R.K., Adurthi, S. et al. (2019). A rationally designed peptide antagonist of the PD-1 signaling pathway as an immunomodulatory agent for cancer therapy. *Molecular Cancer Therapeutics* 18: 1081–1091.

55 Mahadevappa, R., Ma, R., and Kwok, H.F. (2017). Venom peptides: improving specificity in cancer therapy. *Trends Cancer* 3: 611–614.

56 Okada, M., Ortiz, E., Corzo, G., and Possani, L.D. (2019). Pore-forming spider venom peptides show cytotoxicity to hyperpolarized cancer cells expressing K+ channels: a lentiviral vector approach. *PLoS One* 14: e0215391.

57 Ghaly, G., Tallima, H., Dabbish, E. et al. (2023). Anti-cancer peptides: status and future prospects. *Molecules* 28 (3): 1148.

58 Strzelecka, P., Czaplinska, D., Sadej, R. et al. (2017). Simplified, serine-rich theta-defensin analogues as antitumour peptides. *Chemical Biology & Drug Design* 90: 52–63.

59 Rodbard, H.W. (2018). The clinical impact of GLP-1 receptor agonists in type 2 diabetes: focus on the long-acting analogs. *Diabetes Technology & Therapeutics* 20: S233–S241.

60 George, C., Byun, A., and Howard-Thompson, A. (2018). New injectable agents for the treatment of type 2 diabetes part 2-glucagon-like peptide-1 (GLP-1) agonists. *The American Journal of Medicine* 131: 1304–1306.

61 Tonneijck, L., Muskiet, M.H., Blijdorp, C.J. et al. (2019). Renal tubular effects of prolonged therapy with the GLP-1 receptor agonist lixisenatide in patients with type 2 diabetes mellitus. *American Journal of Physiology – Renal Physiology* 316: F231–F240.

62 von Scholten, B.J., Hansen, T.W., Goetze, J.P. et al. (2015). Glucagon-like peptide 1 receptor agonist (GLP-1 RA): long-term effect on kidney function in patients with type 2 diabetes. *Journal of Diabetes and Its Complications* 29: 670–674.

63 Greco, E.V., Russo, G., Giandalia, A. et al. (2019). GLP-1 receptor agonists and kidney protection. *Medicina (Kaunas, Lithuania)* 55: 233.

64 Correa, T.D., Takala, J., and Jakob, S.M. (2015). Angiotensin II in septic shock. *Critical Care* 19 (98): 302.

65 Montone, C.M. et al. (2018). Peptidomic strategy for purification and identification of potential ACE-inhibitory and antioxidant peptides in *Tetradesmus obliquus* microalgae. *Analytical and Bioanalytical Chemistry* 410: 3573–3586.

66 Liao, W., Fan, H., Davidge, S.T., and Wu, J. (2019). Egg white-derived antihypertensive peptide IRW (Ile-Arg-Trp) reduces blood pressure in spontaneously hypertensive rats via the ACE2/Ang (1-7)/ Mas receptor axis. *Molecular Nutrition & Food Research* 63: e1900063.

67 Kuwahara, K. (2021). The natriuretic peptide system in heart failure: diagnostic and therapeutic implications. *Pharmacology & Therapeutics* 227: 107863.

68 Burnett, J.C. Jr. (2019). Atrial natriuretic peptide, heart failure and the heart as an endocrine organ. *Clinical Chemistry* 65: 1602–1603.

69 Shah, S.J. and Teerlink, J.R. (2007). Nesiritide: a reappraisal of efficacy and safety. *Expert Opinion on Pharmacotherapy* 8: 361–369.

70 Sackner-Bernstein, J.D., Kowalski, M., Fox, M., and Aaronson, K. (2005). Short-term risk of death after treatment with nesiritide for decompensated heart failure: a pooled analysis of randomized controlled trials. *JAMA* 293: 1900–1905.

71 Forte, M., Madonna, M., Schiavon, S. et al. (2019). Cardiovascular pleiotropic effects of natriuretic peptides. *International Journal of Molecular Sciences* 20: 3874.

72 Ichiki, T., Dzhoyashvili, N., and Burnett, J.C. Jr. (2019). Natriuretic peptide based therapeutics for heart failure: Cenderitide: a novel first-in-class designer natriuretic peptide. *International Journal of Cardiology* 281: 166–171.

73 Lee, C.Y., Chen, H.H., Lisy, O. et al. (2009). Pharmacodynamics of a novel designer natriuretic peptide, CDNP, in a first-in-human clinical trial in healthy subjects. *Journal of Clinical Pharmacology* 49: 668–673.

74 Duggan, K.A., Hodge, G., Chen, J., and Hunter, T. (2019). Vasoactive intestinal peptide infusion reverses existing myocardial fibrosis in the rat. *European Journal of Pharmacology* 862: 172629.

75 Dong, Y., Bai, Y., Zhang, S. et al. (2019). Cyclic peptide RD808 reduces myocardial injury induced by beta1-adrenoreceptor autoantibodies. *Heart and Vessels* 34: 1040–1051.

76 Gao, H.R. and Gao, H.Y. (2019). Cardiovascular functions of central corticotropin-releasing factor related peptides system. *Neuropeptides* 75: 18–24.

77 Barchetta, I., Ciccarelli, G., Barone, E. et al. (2019). Greater circulating DPP4 activity is associated with impaired flow-mediated dilatation in adults with type 2 diabetes mellitus. *Nutrition, Metabolism, and Cardiovascular Diseases* 29: 1087–1094.

78 Holani, R., Shah, C., Haji, Q. et al. (2016). Proline-arginine rich (PR-39) cathelicidin: structure, expression and functional implication in intestinal health. *Comparative Immunology, Microbiology and Infectious Diseases* 49: 95–101.

79 Jeppesen, P.B. (2015). Gut hormones in the treatment of short-bowel syndrome and intestinal failure. *Current Opinion in Endocrinology, Diabetes, and Obesity* 22 (1): 14–20.

80 Billiauws, L., Bataille, J., Boehm, V. et al. (2017). Teduglutide for treatment of adult patients with short bowel syndrome. *Expert Opinion on Biological Therapy* 17 (5): 623–632.

81 Lim, D.W., Levesque, C.L., Vine, D.F. et al. (2017). Synergy of glucagon-like peptide-2 and epidermal growth factor coadministration on intestinal adaptation in neonatal piglets with short bowel syndrome. *American Journal of Physiology. Gastrointestinal and Liver Physiology* 312: G390–G404.

82 Hong, J., Zhang, P., Yoon, I.N. et al. (2017). The American cockroach peptide periplanetasin-2 blocks *Clostridium difficile* toxin a-induced cell damage and inflammation in the gut. *Journal of Microbiology and Biotechnology* 27: 694–700.

83 Shrestha, A., Robertson, S.L., Garcia, J. et al. (2014). A synthetic peptide corresponding to the extracellular loop 2 region of claudin-4 protects against *Clostridium perfringens* enterotoxin *in vitro* and *in vivo*. *Infection and Immunity* 82: 4778–4788.

84 Marin, M., Holani, R., Blyth, G.A. et al. (2019). Human cathelicidin improves colonic epithelial defenses against *Salmonella typhimurium* by modulating bacterial invasion, TLR4 and proinflammatory cytokines. *Cell and Tissue Research* 376: 433–442.

85 Smith, A.J. (2015). New horizons in therapeutic antibody discovery: opportunities and challenges versus small-molecule therapeutics. *Journal of Biomolecular Screening* 20: 437–453.

86 Räder, A.F., Weinmüller, M., Reichart, F. et al. (2018). Orally active peptides: is there a magic bullet? *Angewandte Chemie International Edition* 57 (44): 14414–14438.

87 Merrifield, R.B. (1963). Solid phase peptide synthesis. I. The synthesis of a tetrapeptide. *Journal of the American Chemical Society* 85 (14): 2149–2154.

88 Hou, W., Zhang, X., and Liu, C.F. (2017). Progress in chemical synthesis of peptides and proteins. *Transactions of Tianjin University* 23: 401–419.

89 Behrendt, R., White, P., and Offer, J. (2016). Advances in Fmoc solid-phase peptide synthesis. *Journal of Peptide Science* 22 (1): 4–27.

90 Weidmann, J., Dimitrijević, E., Hoheisel, J.D., and Dawson, P.E. (2016). Boc-SPPS: compatible linker for the synthesis of peptide o-aminoanilides. *Organic Letters* 18 (2): 164–167.

91 Wołczański, G. and Lisowski, M. (2018). A general method for preparation of N-Boc-protected or N-Fmoc-protected α, β-didehydropeptide building blocks and their use in the solid-phase peptide synthesis. *Journal of Peptide Science* 24 (8–9): e3091.

92 Puentes, A.R., Morejón, M.C., Rivera, D.G., and Wessjohann, L.A. (2017). Peptide macrocyclization assisted by traceless turn inducers derived from Ugi peptide ligation with cleavable and resin-linked amines. *Organic Letters* 19 (15): 4022–4025.

93 Coin, I., Beyermann, M., and Bienert, M. (2007). Solid-phase peptide synthesis: from standard procedures to the synthesis of difficult sequences. *Nature Protocols* 2 (12): 3247–3256.

94 García-Martín, F., Quintanar-Audelo, M., García-Ramos, Y. et al. (2006). ChemMatrix, a poly (ethylene glycol)-based support for the solid-phase synthesis of complex peptides. *Journal of Combinatorial Chemistry* 8 (2): 213–220.

95 Pedersen, S.L., Tofteng, A.P., Malik, L., and Jensen, K.J. (2012). Microwave heating in solid-phase peptide synthesis. *Chemical Society Reviews* 41 (5): 1826–1844.

96 Paradís-Bas, M., Tulla-Puche, J., and Albericio, F. (2016). The road to the synthesis of "difficult peptides". *Chemical Society Reviews* 45 (3): 631–654.

97 Palasek, S.A., Cox, Z.J., and Collins, J.M. (2007). Limiting racemization and aspartimide formation in microwave-enhanced Fmoc solid phase peptide synthesis. *Journal of Peptide Science* 13: 143–148.

98 Cardona, V., Eberle, I., Barthélémy, S. et al. (2008). Application of Dmb-dipeptides in the Fmoc SPPS of difficult and aspartimide-prone sequences. *International Journal of Peptide Research and Therapeutics* 14: 285–292.

99 Michels, T., Dölling, R., Haberkorn, U., and Mier, W. (2012). Acid-mediated prevention of aspartimide formation in solid phase peptide synthesis. *Organic Letters* 14 (20): 5218–5221.

100 Subirós Funosas, R., El Faham, A., and Albericio, F. (2012). Use of Oxyma as pH modulatory agent to be used in the prevention of base driven side reactions and its effect on 2-chlorotrityl chloride resin. *Peptide Science* 98 (2): 89–97.

101 Hojo, H. and Aimoto, S. (1991). Polypeptide synthesis using the S-alkyl thioester of a partially protected peptide segment. Synthesis of the DNA-binding domain of c-Myb protein (142–193)-NH2. *Bulletin of the Chemical Society of Japan* 64 (1): 111–117.

102 Aimoto, S. (2001). Contemporary methods for peptide and protein synthesis. *Current Organic Chemistry* 5 (1): 45–87.

103 Schnölzer, M. and Kent, S.B. (1992). Constructing proteins by dove-tailing unprotected synthetic peptides: backbone-engineered HIV protease. *Science* 256 (5054): 221–225.

104 Rose, K. (1994). Facile synthesis of homogeneous artificial proteins. *Journal of the American Chemical Society* 116 (1): 30–33.

105 Fisch, I., Kunzi, G., Rose, K., and Offord, R.E. (1992). Site-specific modification of a fragment of a chimeric monoclonal antibody using reverse proteolysis. *Bioconjugate Chemistry* 3 (2): 147–153.

106 Liu, C.F. and Tam, J.P. (1994). Chemical ligation approach to form a peptide bond between unprotected peptide segments. Concept and model study. *Journal of the American Chemical Society* 116 (10): 4149–4153.

107 Dawson, P.E., Muir, T.W., Clark-Lewis, I., and Kent, S.B. (1994). Synthesis of proteins by native chemical ligation. *Science* 266 (5186): 776–779.

108 Liu, C.F., Rao, C., and Tam, J.P. (1996). Acyl disulfide-mediated intramolecular acylation for orthogonal coupling between unprotected peptide segments. Mechanism and application. *Tetrahedron Letters* 37 (7): 933–936.

109 Muir, T.W., Sondhi, D., and Cole, P.A. (1998). Expressed protein ligation: a general method for protein engineering. *National Academy of Sciences of the United States of America* 95 (12): 6705–6710.

110 Shigenaga, A., Sumikawa, Y., Tsuda, S. et al. (2010). Sequential native chemical ligation utilizing peptide thioacids derived from newly developed Fmoc-based synthetic method. *Tetrahedron* 66 (18): 3290–3296.

111 Varela, Y.F., Vanegas Murcia, M., and Patarroyo, M.E. (2018). Synthetic evaluation of standard and microwave-assisted solid phase peptide synthesis of a long chimeric peptide derived from four plasmodium falciparum proteins. *Molecules* 23 (11): 2877.

112 Hansen, A.M., Bonke, G., Hogendorf, W.F. et al. (2019). Microwave-assisted solid-phase synthesis of antisense acpP peptide nucleic acid-peptide conjugates active against colistin- and tigecycline-resistant *E. coli* and *K. pneumoniae*. *European Journal of Medicinal Chemistry* 168: 134–145.

113 Goodwin, D., Simerska, P., and Toth, I. (2012). Peptides as therapeutics with enhanced bioactivity. *Current Medicinal Chemistry* 19: 4451–4461.

114 Jamieson, A.G., Boutard, N., Sabatino, D., and Lubell, W.D. (2013). Peptide scanning for studying structure-activity relationships in drug discovery. *Chemical Biology & Drug Design* 81: 148–165.

115 Eustache, S., Leprince, J., and Tuffery, P. (2016). Progress with peptide scanning to study structure-activity relationships: the implications for drug discovery. *Expert Opinion on Drug Discovery* 11: 771–784.

116 Pelay-Gimeno, M., Glas, A., Koch, O., and Grossmann, T.N. (2015). Structure-based design of inhibitors of protein–protein interactions: mimicking peptide binding epitopes. *Angewandte Chemie, International Edition* 54 (31): 8896–8927.

117 Noisier, A.F. and Brimble, M.A. (2014). C–H functionalization in the synthesis of amino acids and peptides. *Chemical Reviews* 114 (18): 8775–8806.

118 Boto, A., González, C.C., Hernández, D. et al. (2021). Site-selective modification of peptide backbones. *Organic Chemistry Frontiers* 8: 6720–6759.

119 Zuo, C., Tang, S., Si, Y.Y. et al. (2016). Efficient synthesis of longer Aβ peptides via removable backbone modification. *Organic & Biomolecular Chemistry* 14 (22): 5012–5018.

120 Zheng, J.S., Yu, M., Qi, Y.K. et al. (2014). Expedient total synthesis of small to medium-sized membrane proteins via Fmoc chemistry. *Journal of the American Chemical Society* 136 (9): 3695–3704.

121 Weinstock, M.T., Francis, J.N., Redman, J.S., and Kay, M.S. (2012). Protease-resistant peptide design—empowering nature's fragile warriors against HIV. *Peptide Science* 98 (5): 431–442.

122 Glas, A., Bier, D., Hahne, G. et al. (2014). Constrained peptides with target-adapted cross-links as inhibitors of a pathogenic protein–protein interaction. *Angewandte Chemie, International Edition* 53 (9): 2489–2493.

123 Cheloha, R.W., Maeda, A., Dean, T. et al. (2014). Backbone modification of a polypeptide drug alters duration of action *in vivo*. *Nature Biotechnology* 32 (7): 653–655.

124 Werner, H.M., Cabalteja, C.C., and Horne, W.S. (2016). Peptide backbone composition and protease susceptibility: impact of modification type, position, and tandem substitution. *ChemBioChem* 17 (8): 712–718.

125 Urban, J., Vaisar, T., Shen, R., and Lee, M.S. (1996). Lability of N-alkylated peptides towards TFA cleavage. *International Journal of Peptide and Protein Research* 47:182–189.

126 Maybauer, M.O., Maybauer, D.M., Enkhbaatar, P. et al. (2014). The selective vasopressin type 1a receptor agonist selepressin (FE 202158) blocks vascular leak in ovine severe sepsis. *Critical Care Medicine* 42: e525–e533.

127 Räder, A.F.B., Reichart, F., Weinmüller, M., and Kessler, H. (2018). Improving oral bioavailability of cyclic peptides by N-methylation. *Bioorganic & Medicinal Chemistry* 26 (10): 2766–2773.

128 Henninot, A., Collins, J.C., and Nuss, J.M. (2018). The current state of peptide drug discovery: back to the future? *Journal of Medicinal Chemistry* 61 (4): 1382–1414.

129 Hone, A.J., Fisher, F., Christensen, S. et al. (2019). PeIA-5466: a novel peptide antagonist containing non-natural amino acids that selectively targets alpha 3 beta 2 nicotinic acetylcholine receptors. *Journal of Medicinal Chemistry* 62: 6262–6275.

130 Masri, E., Ahsanullah, Accorsi, M., and Rademann, J. (2020). Side-chain modification of peptides using a phosphoranylidene amino acid. *Organic Letters* 22: 2976–2980.

131 Del Olmo-Garcia, M.I. and Merino-Torres, J.F. (2018). GLP-1 receptor agonists and cardiovascular disease in patients with type 2 diabetes. *Journal Diabetes Research* 2018: 4020492.

132 Wang, L., Wang, N., Zhang, W. et al. (2022). Therapeutic peptides: current applications and future directions. *Signal Transduction and Targeted Therapy* 7 (1): 48.

133 Hayes, H.C., Luk, L.Y.P., and Tsai, Y.H. (2021). Approaches for peptide and protein cyclisation. *Organic & Biomolecular Chemistry* 19 (18): 3983–4001. https://doi.org/10.1039/d1ob00411e. PMID: 33978044; PMCID: PMC8114279.

134 Jin, K., Sam, I.H., Po, K.H. et al. (2016). Total synthesis of teixobactin. *Nature Communications* 7 (1): 12394.

135 Ollivier, N., Toupy, T., Hartkoorn, R.C. et al. (2018). Accelerated microfluidic native chemical ligation at difficult amino acids toward cyclic peptides. *Nature Communications* 9 (1): 2847.

136 Lam, H.Y., Zhang, Y., Liu, H. et al. (2013). Total synthesis of daptomycin by cyclization via a chemoselective serine ligation. *Journal of the American Chemical Society* 135 (16): 6272–6279.

137 Wong, C.T., Lam, H.Y., and Li, X. (2013). Effective synthesis of kynurenine-containing peptides via on-resin ozonolysis of tryptophan residues: synthesis of cyclomontanin B. *Organic & Biomolecular Chemistry* 11 (43): 7616–7620.

138 Liu, H. and Li, X. (2018). Serine/threonine ligation: origin, mechanistic aspects, and applications. *Accounts of Chemical Research* 51 (7): 1643–1655.

139 Adebomi, V., Cohen, R.D., Wills, R. et al. (2019). CyClick chemistry for the synthesis of cyclic peptides. *Angewandte Chemie* 131 (52): 19249–19256.

140 Soellner, M.B., Tam, A., and Raines, R.T. (2006). Staudinger ligation of peptides at non-glycyl residues. *The Journal of Organic Chemistry* 71 (26): 9824–9830.

141 Bode, J.W., Fox, R.M., and Baucom, K.D. (2006). Chemoselective amide ligations by decarboxylative condensations of N-alkylhydroxylamines and α-ketoacids. *Angewandte Chemie International Edition* 45 (8): 1248–1252.

142 Zheng, X., Li, Z., Gao, W. et al. (2020). Condensation of 2-((alkylthio)(aryl) methylene) malononitrile with 1,2-aminothiol as a novel bioorthogonal reaction for site-specific protein modification and peptide cyclization. *Journal of the American Chemical Society* 142 (11): 5097–5103.

143 Bottecchia, C. and Noël, T. (2019). Photocatalytic modification of amino acids, peptides, and proteins. *Chemistry – A European Journal* 25 (1): 26–42.

144 Patel, S.G., Sayers, E.J., He, L. et al. (2019). Cell-penetrating peptide sequence and modification dependent uptake and subcellular distribution of green florescent protein in different cell lines. *Scientific Reports* 9 (1): 6298.

145 Li, X., Chen, S., Zhang, W.D., and Hu, H.G. (2020). Stapled helical peptides bearing different anchoring residues. *Chemical Reviews* 120 (18): 10079–10144.

146 Weeks, A.M. and Wells, J.A. (2019). Subtiligase-catalyzed peptide ligation. *Chemical Reviews* 120 (6): 3127–3160.

147 Dai, X., Böker, A., and Glebe, U. (2019). Broadening the scope of sortagging. *RSC Advances* 9 (9): 4700–4721.

148 Touati, J., Angelini, A., Hinner, M.J., and Heinis, C. (2011). Enzymatic cyclisation of peptides with a transglutaminase. *ChemBioChem* 12 (1): 38–42.

149 Shah, N.H., Dann, G.P., Vila-Perelló, M. et al. (2012). Ultrafast protein splicing is common among cyanobacterial split inteins: implications for protein engineering. *Journal of the American Chemical Society* 134 (28): 11338–11341.

150 Zakeri, B., Fierer, J.O., Celik, E. et al. (2012). Peptide tag forming a rapid covalent bond to a protein, through engineering a bacterial adhesin. *National Academy of Sciences of the United States of America* 109 (12): E690–E697.

151 Veggiani, G., Nakamura, T., Brenner, M.D. et al. (2016). Programmable polyproteams built using twin peptide superglues. *National Academy of Sciences of the United States of America* 113 (5): 1202–1207.

152 Jochim, A.L. and Arora, P.S. (2010). Systematic analysis of helical protein interfaces reveals targets for synthetic inhibitors. *ACS Chemical Biology* 5: 919–923.

153 Sawyer, N., Watkins, A.M., and Arora, P.S. (2017). Protein domain mimics as modulators of protein–protein interactions. *Accounts of Chemical Research* 50: 1313–1322.

154 Khoo, K.K. et al. (2011). Lactam-stabilized helical analogues of the analgesic muconotoxin KIIIA. *Journal of Medicinal Chemistry* 54: 7558–7566.

155 Galande, A.K. et al. (2004). Thioether side chain cyclization for helical peptide formation: inhibitors of estrogen receptor–coactivator interactions. *The Journal of Peptide Research* 63: 297–302.

156 Muppidi, A., Zhang, H., Curreli, F. et al. (2014). Design of antiviral stapled peptides containing a biphenyl cross-linker. *Bioorganic & Medicinal Chemistry Letters* 24: 1748–1751.

157 Reguera, L. and Rivera, D.G. (2019). Multicomponent reaction toolbox for peptide macrocyclization and stapling. *Chemical Reviews* 119 (17): 9836–9860.

158 Cheng, P.N., Liu, C., Zhao, M. et al. (2012). Amyloid beta-sheet mimics that antagonize protein aggregation and reduce amyloid toxicity. *Nature Chemistry* 4: 927–933.

159 Banerjee, V., Shani, T., Katzman, B. et al. (2016). Superoxide dismutase 1 (SOD1)-derived peptide inhibits amyloid aggregation of familial amyotrophic lateral sclerosis SOD1 mutants. *ACS Chemical Neuroscience* 7 (11): 1595–1606.

160 Ryan, P., Patel, B., Makwana, V. et al. (2018). Peptides, peptidomimetics, and carbohydrate–peptide conjugates as amyloidogenic aggregation inhibitors for Alzheimer's disease. *ACS Chemical Neuroscience* 9 (7): 1530–1551.

161 Mishra, N.K., Joshi, K.B., and Verma, S. (2013). Inhibition of human and bovine insulin fibril formation by designed peptide conjugates. *Molecular Pharmaceutics* 10 (10): 3903–3912.

162 Hanold, L.E., Oruganty, K., Ton, N.T. et al. (2015). Inhibiting EGFR dimerization using triazolyl-bridged dimerization arm mimics. *PLoS One* 10 (3): e0118796.

163 Hopping, G., Kellock, J., Caughey, B., and Daggett, V. (2013). Designed Trpzip-3 β-hairpin inhibits amyloid formation in two different amyloid systems. *ACS Medicinal Chemistry Letters* 4 (9): 824–828.

164 Dufau, L., Marques Ressurreição, A.S., Fanelli, R. et al. (2012). Carbonylhydrazide-based molecular tongs inhibit wild-type and mutated HIV-1 protease dimerization. *Journal of Medicinal Chemistry* 55 (15): 6762–6775.

165 Mirecka, E.A., Shaykhalishahi, H., Gauhar, A. et al. (2014). Sequestration of a β-hairpin for control of α-synuclein aggregation. *Angewandte Chemie International Edition* 53 (16): 4227–4230.

166 Wibowo, D. and Zhao, C. (2019). Recent achievements and perspectives for large-scale recombinant production of antimicrobial peptides. *Applied Microbiology and Biotechnology* 103: 659–671.

167 Rosano, G.L., Morales, E.S., and Ceccarelli, E.A. (2019). New tools for recombinant protein production in *Escherichia coli*: a 5-year update. *Protein Science* 28 (8): 1412–1422.

168 Itakura, K. et al. (1977). Expression in *Escherichia coli* of a chemically synthesized gene for the hormone somatostatin. *Science* 198: 1056–1063.

169 Johnson, I.S. (1983). Human insulin from recombinant DNA technology. *Science* 219: 632–637.

170 Wang, L. (2017). Engineering the genetic code in cells and animals: biological considerations and impacts. *Accounts of Chemical Research* 50: 2767–2775.

171 Oller-Salvia, B. and Chin, J.W. (2019). Efficient phage display with multiple distinct noncanonical amino acids using orthogonal ribosome-mediated genetic code expansion. *Angewandte Chemie (International Ed. in English)* 58: 10844–10848.

172 Valentini, T.D., Lucas, S.K., Binder, K.A. et al. (2020). Bioorthogonal non-canonical amino acid tagging reveals translationally active subpopulations of the cystic fibrosis lung microbiota. *Nature Communications* 11: 2287.

173 Liu, J., Hemphill, J., Samanta, S. et al. (2017). Genetic code expansion in zebrafish embryos and its application to optical control of cell signaling. *Journal of the American Chemical Society* 139: 9100–9103.

174 Brown, W., Liu, J., and Deiters, A. (2018). Genetic code expansion in animals. *ACS Chemical Biology* 13: 2375–2386.

175 Liu, S., Fan, L., Sun, J. et al. (2017). Computational resources and tools for antimicrobial peptides. *Journal of Peptide Science* 23 (1): 4–12.

176 Fok, J.A. and Mayer, C. (2020). Genetic-code-expansion strategies for vaccine development. *ChemBioChem* 21: 3291–3300.

177 Veronese, F.M. and Mero, A. (2008). The impact of PEGylation on biological therapies. *BioDrugs* 22: 315–329.

178 Swierczewska, M., Han, H.S., Kim, K. et al. (2016). Polysaccharide-based nanoparticles for theranostic nanomedicine. *Advanced Drug Delivery Reviews* 99: 70–84.

179 Santos, J.H., Carretero, G., Ventura, S.P. et al. (2019). PEGylation as an efficient tool to enhance cytochrome c thermostability: a kinetic and thermodynamic study. *Journal of Materials Chemistry B* 7 (28): 4432–4439.

180 Gupta, V., Bhavanasi, S., Quadir, M. et al. (2019). Protein PEGylation for cancer therapy: bench to bedside. *Journal of Cell Communication and Signaling* 13: 319–330.

181 Zhang, C., Yang, X.L., Yuan, Y.H. et al. (2012). Site-specific PEGylation of therapeutic proteins via optimization of both accessible reactive amino acid residues and PEG derivatives. *BioDrugs* 26: 209–215.

182 Alconcel, S.N., Baas, A.S., and Maynard, H.D. (2011). FDA-approved poly(ethylene glycol)–protein conjugate drugs. *Polymer Chemistry* 2 (7): 1442–1448.

183 Schiavon, O., Caliceti, P., Ferruti, P., and Veronese, F.M. (2000). Therapeutic proteins: a comparison of chemical and biological properties of uricase conjugated to linear or branched poly(ethylene glycol) and poly(N-acryloylmorpholine). *Il Farmaco* 55 (4): 264–269.

184 Xiaojiao, S., Corbett, B., Macdonald, B. et al. (2016). Modeling and optimization of protein pegylation. *Industrial and Engineering Chemistry Research* 55 (45): 11785–11794.

185 Chiu, K., Agoubi, L.L., Lee, I. et al. (2010). Effects of polymer molecular weight on the size, activity, and stability of PEG-functionalized trypsin. *Biomacromolecules* 11 (12): 3688–3692.

186 Roberts, M.J., Bentley, M.D., and Harris, J.M. (2012). Chemistry for peptide and protein PEGylation. *Advanced Drug Delivery Reviews* 54 (4): 459–476. https://doi.org/10.1016/j.addr.2012.09.025.

187 Belén, L.H., Rangel-Yagui, C.D., Beltrán Lissabet, J.F. et al. (2019). From synthesis to characterization of site-selective PEGylated proteins. *Frontiers in Pharmacology* 10: 1450.

188 Ameringer, T., Ercole, F., Tsang, K.M. et al. (2013). Surface grafting of electrospun fibers using ATRP and RAFT for the control of biointerfacial interactions. *Biointerphases* 8 (1).

189 Wallat, J.D., Rose, K.A., and Pokorski, J.K. (2014). Proteins as substrates for controlled radical polymerization. *Polymer Chemistry* 5 (5): 1545–1558.

190 Tucker, B.S., Coughlin, M.L., Figg, C.A., and Sumerlin, B.S. (2017). Grafting-from proteins using metal-free PET–RAFT polymerizations under mild visible-light irradiation. *ACS Macro Letters* 6 (4): 452–457.

191 Zhang, B., Xu, H., Chen, J. et al. (2015). Development of next generation of therapeutic IFN-α2b via genetic code expansion. *Acta Biomaterialia* 19: 100–111.

192 Wu, L., Chen, J., Wu, Y. et al. (2017). Precise and combinatorial PEGylation generates a low-immunogenic and stable form of human growth hormone. *Journal of Controlled Release* 249: 84–93.

193 Fu, C., Chen, Q., Zheng, F. et al. (2019). Genetically encoding a lipidated amino acid for extension of protein half-life *in vivo*. *Angewandte Chemie* 131 (5): 1406–1410.

194 Wang, H.H., Altun, B., Nwe, K., and Tsourkas, A. (2017). Proximity-based sortasemediated ligation. *Angewandte Chemie, International Edition* 56 (19): 5349–5352. https://doi.org/10.1002/anie.201701419.

195 Jaroszewicz, W., Morcinek-Orłowska, J., Pierzynowska, K. et al. (2022). Phage display and other peptide display technologies. *FEMS Microbiology Reviews* 46 (2): fuab052.

196 Marintcheva, B. (2017). *Harnessing the Power of Viruses*. London: Academic Press.

197 Bábíčková, J., Tóthová, Ľ., Boor, P., and Celec, P. (2013). *In vivo* phage display: a discovery tool in molecular biomedicine. *Biotechnology Advances* 31 (8): 1247–1259.

198 Norman, A., Franck, C., Christie, M. et al. (2021). Discovery of cyclic peptide ligands to the SARS-CoV-2 spike protein using mRNA display. *ACS Central Science* 7: 1001–1008.

199 Yamagishi, Y., Shoji, I., Miyagawa, S. et al. (2011). Natural product-like macrocyclic N-methyl-peptide inhibitors against a ubiquitin ligase uncovered from a ribosome-expressed *de novo* library. *Chemistry & Biology* 18: 1562–1570.

200 Saito, M., Itoh, Y., Yasui, F. et al. (2021). Macrocyclic peptides exhibit antiviral effects against influenza virus HA and prevent pneumonia in animal models. *Nature Communications* 12: 2654.

201 Zimmermann, G. and Neri, D. (2016). DNA-encoded chemical libraries: foundations and applications in lead discovery. *Drug Discovery Today* 21 (11): 1828–1834.

202 Zhou, Y., Li, C., Peng, J. et al. (2018). DNA-encoded dynamic chemical library and its applications in ligand discovery. *Journal of the American Chemical Society* 140: 15859–15867.

203 Zhu, Z., Shaginian, A., Grady, L.C. et al. (2018). Design and application of a DNA-encoded macrocyclic peptide library. *ACS Chemical Biology* 13 (1): 53–59.

204 Shin, M.H., Lee, K.J., and Lim, H.S. (2019). DNA-encoded combinatorial library of macrocyclic peptoids. *Bioconjugate Chemistry* 30 (11): 2931–2938.

205 Habault, J. and Poyet, J.L. (2019). Recent advances in cell penetrating peptide-based anticancer therapies. *Molecules* 24: 927. https://doi.org/10.3390/MOLECULES24050927.

206 Hyun, S., Lee, Y., Jin, S.M. et al. (2018). Oligomer formation propensities of dimeric bundle peptides correlate with cell penetration abilities. *ACS Central Science* 4: 885–893.

207 Peacock, H. and Suga, H. (2021). Discovery of *de novo* macrocyclic peptides by messenger RNA display. *Trends in Pharmacological Sciences* 42 (5): 385–397.

208 Vinogradov, A.A. et al. (2019). Macrocyclic peptides as drug candidates: recent progress and remaining challenges. *Journal of the American Chemical Society* 141: 4167–4181.

209 Pandya, A.K. and Patravale, V.B. (2021). Computational avenues in oral protein and peptide therapeutics. *Drug Discovery Today* 26 (6): 1510–1520.

210 Wu, S.J., Luo, J., O'Neil, K.T. et al. (2010). Structure-based engineering of a monoclonal antibody for improved solubility. *Protein Engineering, Design & Selection* 23: 643–651.

211 Agostini, F., Cirillo, D., Livi, C.M. et al. (2014). cc SOL omics: a webserver for solubility prediction of endogenous and heterologous expression in *Escherichia coli*. *Bioinformatics* 30: 2975–2977.

212 Smialowski, P., Doose, G., Torkler, P. et al. (2012). PROSO II – a new method for protein solubility prediction. *The FEBS Journal* 279: 2192–2200.

213 Qian, Z., LaRochelle, J.R., Jiang, B. et al. (2014). Early endosomal escape of a cyclic cell-penetrating peptide allows effective cytosolic cargo delivery. *Biochemistry* 53: 4034–4046.

214 Donsky, E. and Wolfson, H.J. (2011). PepCrawler: a fast RRT-based algorithm for high-resolution refinement and binding affinity estimation of peptide inhibitors. *Bioinformatics* 27: 2836–2842.

215 Raveh, B., London, N., and Schueler-Furman, O. (2010). Sub-angstrom modeling of complexes between flexible peptides and globular proteins. *Proteins* 78: 2029–2040.

216 Trellet, M., Melquiond, A.S., and Bonvin, A.M. (2013). A unified conformational selection and induced fit approach to protein-peptide docking. *PLoS One* 8: e58769.

217 Antes, I. (2010). DynaDock: a new molecular dynamics-based algorithm for protein-peptide docking including receptor flexibility. *Proteins* 78: 1084–1104.

218 Trabuco, L.G., Lise, S., Petsalaki, E., and Russell, R.B. (2012). PepSite: prediction of peptide-binding sites from protein surfaces. *Nucleic Acids Research* 40: 423–427.

219 Porter, K.A., Xia, B., Beglov, D. et al. (2017). ClusPro PeptiDock: efficient global docking of peptide recognition motifs using FFT. *Bioinformatics* 33: 3299–3301.

220 De Vries, S.J., Rey, J., Schindler, C.E.M. et al. (2017). The pepATTRACT web server for blind, large-scale peptide–protein docking. *Nucleic Acids Research* 45: 361–364.

221 Zhou, P., Jin, B., Li, H., and Huang, S.Y. (2018). HPEPDOCK: a web server for blind peptide-protein docking based on a hierarchical algorithm. *Nucleic Acids Research* 46: 443–450.

222 Lee, H., Heo, L., Lee, M.S., and Seok, C. (2015). GalaxyPepDock: a protein–peptide docking tool based on interaction similarity and energy optimization. *Nucleic Acids Research* 43: 431–435.

223 Taherzadeh, G., Zhou, Y., Liew, A.W., and Yang, Y. (2018). Structure-based prediction of protein-peptide binding regions using random forest. *Bioinformatics* 34: 477–484.

224 Iqbal, S. and Hoque, M.T. (2018). PBRpredict-Suite: a suite of models to predict peptide-recognition domain residues from protein sequence. *Bioinformatics* 34: 3289–3299.

225 Obarska-Kosinska, A., Iacoangeli, A., Lepore, R., and Tramontano, A. (2016). PepComposer: computational design of peptides binding to a given protein surface. *Nucleic Acids Research* 44: 522–528.

226 Yagi, Y., Terada, K., Noma, T. et al. (2007). *In silico* panning for a non-competitive peptide inhibitor. *BMC Bioinformatics* 8: 1–1.

227 Tabassum, S., Al-Asbahy, W.M., Afzal, M., and Arjmand, F. (2012). Synthesis, characterization and interaction studies of copper based drug with human serum albumin (HSA): spectroscopic and molecular docking investigations. *Journal of Photochemistry and Photobiology B: Biology 114*: 132–139.

228 Zhang, Z.H., Abbad, S., Pan, R.R. et al. (2013). N-octyl-N-arginine chitosan micelles as an oral delivery system of insulin. *Journal of Biomedical Nanotechnology* 9 (4): 601–609.

229 Moradi, S.V., Hussein, W.M., Varamini, P. et al. (2016). Glycosylation, an effective synthetic strategy to improve the bioavailability of therapeutic peptides. *Chemical Science* 7 (4): 2492–2500.

230 Yang, C., Lu, D., and Liu, Z. (2011). How PEGylation enhances the stability and potency of insulin: a molecular dynamics simulation. *Biochemistry* 50 (13): 2585–2593.

231 Clardy-James, S., Allis, D.G., Fairchild, T.J., and Doyle, R.P. (2012). Examining the effects of vitamin B 12 conjugation on the biological activity of insulin: a molecular dynamic and *in vivo* oral uptake investigation. *MedChemComm* 3 (9): 1054–1058.

232 Sabu, C., Raghav, D., Jijith, U.S. et al. (2019). Bioinspired oral insulin delivery system using yeast microcapsules. *Materials Science and Engineering: C* 103: 109753.

233 Mishra, B. and Wang, G. (2012). *Ab initio* design of potent anti-MRSA peptides based on database filtering technology. *Journal of the American Chemical Society* 134: 12426–12429. https://doi.org/10.1021/ja305644e.

234 Freire, J.M., Almeida, D.S., Flores, L. et al. (2015). Mining viral proteins for antimicrobial and cell-penetrating drug delivery peptides. *Bioinformatics* 31: 2252–2256. https://doi.org/10.1093/bioinformatics/btv131.

16

Rational Design of Drugs for Neurodegenerative Disorders

Priyanka Kamaria

Department of Pharmaceutical Chemistry, KLE College of Pharmacy, Bangalore, India

16.1 Introduction

Neurodegenerative disorders are a group of chronic and progressive conditions characterized by the gradual degeneration and loss of structure or function of nerve cells (neurons) in the central nervous system. These disorders primarily affect the brain and, in some cases, the spinal cord. Neurodegenerative diseases often result in cognitive decline, movement disorders, and various neurological symptoms [1]. The exact causes of these disorders can vary, including genetic factors, environmental influences, and a combination of both.

Some common neurodegenerative disorders include

1) **Alzheimer's disease (AD):** AD disease accounts for a significant portion of age-related cognitive decline, making it the most common cause of dementia. In this disease, beta-amyloid plaques and tau tangles are among the abnormal protein aggregates found in the brain. AD primarily affects memory, cognition, and behavior [2].
2) **Parkinson's disease (PD):** In PD, dopamine-producing neurons in the brain gradually degenerate. In turn, this leads to motor symptoms such as tremors, bradykinesia, and rigidity. PD is characterized by abnormal protein deposits called Lewy bodies [3].
3) **Huntington's disorder (HD):** A mutation in the huntingtin gene results in the inherited condition known as Huntington's disease. It causes the cerebral cortex and basal ganglia's neurons to degenerate. Involuntary motions, cognitive deterioration, and emotional difficulties are among the symptoms [4].
4) **Amyotrophic lateral sclerosis (ALS):** The loss of both upper and lower motor neurons in the brain and spinal cord is a hallmark of ALS, a motor neuron disease. This eventually results in respiratory failure, paralysis, and increasing muscle weakness [5].
5) **Multiple sclerosis (MS):** The main feature of MS is the immune system targeting the protective layer of myelin that covers nerve fibers for protection. As a result, there is a disruption in the brain's ability to communicate with the body, which can cause various neurological symptoms like exhaustion, weak muscles, and poor coordination [6].
6) **Frontotemporal dementia (FTD):** A collection of illnesses known as frontal and/or temporal lobe degeneration (FTD) are distinguished by their progressive nature. It may result in modifications to behavior, personality, and linguistic skills [7].

Computational Methods for Rational Drug Design, First Edition. Edited by Mithun Rudrapal.

16.2 Common Mechanism of Neurodegeneration

Neurodegeneration in different neurodegenerative illnesses is attributed to some common pathways. Mechanistic convergence can be divided into five main areas: (1) factors related to the environment, including age, diet, and exercise; (2) metabolic stress, including mitochondrial dysfunction and elevated reactive oxygen species (ROS); (3) genetic factors, including sex-linked genetic contributions and risk alleles found through genome-wide association studies (GWAS); (4) neurovascular coupling, which includes impaired neurovascular coupling and breakdown of the blood–brain barrier; and (5) neurological inflammation, which is characterized by glial reactivity and the infiltration of peripheral immune cells. Every mechanistic facet of degeneration is influenced by environmental variables.

1) **Environmental factors**
   Neurodegenerative disease progression is influenced by environmental variables in both cellular and epigenetic aspects. Neurotoxic drugs, age, food, and exercise can exacerbate or initiate underlying events. These elements play a crucial role in common degenerative processes. Age, a primary risk factor across diseases, particularly affects postmitotic cells like neurons, making them sensitive to aging effects like genomic instability and mitochondrial dysfunction. Exercise and nutrition play a critical role in maintaining the health of the central nervous system, influencing functions like immunological responses and mitochondrial ATP synthesis. Understanding the influence of extrinsic variables on signaling pathways and their role in degeneration provides prospective opportunities for novel therapeutic techniques in neurodegenerative illnesses [8].
2) **Neuroinflammation**
   Neuroinflammation stands as a pervasive commonality in various neurodegenerative diseases. Characterized by increased glial reactivity and infiltration of peripheral immune cells, it fuels pathological processes. The inflammatory response, triggered by diverse factors like genetic predisposition and environmental influences, contributes significantly to neuronal damage and dysfunction. In conditions like Parkinson's, Alzheimer's, and MS, neuroinflammation is a shared hallmark, amplifying disease progression. Understanding and targeting this inflammatory cascade presents a promising avenue for therapeutic interventions, offering potential strategies to mitigate the impact of neuroinflammation on neural tissues and improve outcomes in the realm of neurodegenerative disorders [9, 10].
3) **Neurovascular coupling**
   It appears as a common mechanism among several neurodegenerative conditions. It entails the complex interaction between cerebral blood flow control and neuronal activity. Dysfunction in this coupling, characterized by blood–brain barrier breakdown and impaired neurovascular interactions, contributes to pathological processes. In conditions like Alzheimer's, stroke, and vascular dementia, compromised neurovascular coupling accelerates cognitive decline and exacerbates neuronal damage. Disruptions in nutrient and oxygen supply to the brain further amplify the degenerative cascade. Recognizing the pivotal role of neurovascular coupling in various disorders offers valuable insights for targeted therapeutic strategies, aiming to restore vascular integrity and alleviate the impact of impaired neurovascular interactions on neurological health [11].
4) **Metabolic stress**
   It emerges as a prevalent mechanism across various neurodegenerative diseases. Manifesting as mitochondrial dysfunction and increased levels of ROS, it inflicts cellular damage and exacerbates neurodegeneration. The compromised energy production and oxidative stress lead to a

cascade of detrimental effects on neurons. PD, AD, and HD all exhibit links to metabolic stress, further underscoring its significance. Understanding and addressing this shared mechanism provides a promising avenue for therapeutic interventions. Mitigating metabolic stress-induced damage holds the potential for alleviating the progression of neurodegenerative disorders, offering a targeted approach to enhance cellular resilience and maintain neurological well-being [12].

5) **Genetic contributors**
   It constitutes a pivotal common mechanism in the landscape of neurodegenerative diseases. GWAS unveil risk alleles associated with conditions like Parkinson's, Alzheimer's, and ALS, elucidating genetic underpinnings. Sex-linked genetic influences also play a role in disease susceptibility. The identified genetic markers contribute to altered protein functions, impaired cellular processes, and neuronal vulnerability, collectively shaping the course of neurodegeneration. Recognizing these genetic nuances across diverse disorders opens avenues for targeted therapeutic interventions, offering potential strategies to modify the underlying genetic factors and mitigate the impact of inherited predispositions on the onset and progression of neurodegenerative diseases [13].

16.3 Brief Overview of Computational Methods in Drug Design

Computational methods play a pivotal role in drug design, revolutionizing the way researchers identify and optimize potential drug candidates. These methods leverage computational power to analyze biological systems, predict molecular interactions, and streamline the drug discovery process. Here is a brief overview of computational methods in drug design:

1) **Molecular docking**
   Computational method called "molecular docking" is used in drug discovery to determine a ligand's preferred orientation when it binds to a target protein receptor. It involves exploring the potential interactions and binding conformations between the ligand and receptor to estimate the binding affinity and potential efficacy of a drug candidate. By simulating the molecular interactions at the atomic level, docking algorithms help to identify promising drug candidates with high binding affinity, aiding in the rational design of new pharmaceuticals. This approach accelerates the drug discovery process by facilitating the selection of compounds most likely to exhibit desired biological activity [14].
2) **Quantitative structure–activity relationship (QSAR)**
   QSAR is a computational modeling technique in chemistry and pharmacology used to predict the biological activity of molecules based on their chemical structure. QSAR establishes mathematical relationships between chemical descriptors representing molecular structure and experimental biological activity. By analyzing these relationships, QSAR models can forecast the activity of new compounds, aiding in drug design and environmental toxicity assessment. QSAR models are crucial in rational drug design, enabling the identification of lead compounds with desired pharmacological properties and reducing the time and resources needed for experimental testing. They play a significant role in optimizing molecular structures to enhance desired biological effects [15].
3) **Pharmacophore modeling**
   Pharmacophore modeling is a computational method employed in drug discovery to identify and characterize essential structural features of molecules that are crucial for binding to a biological target. It involves analyzing the spatial arrangement of functional groups and atoms

within active ligands to create a 3D representation of the pharmacophore, which serves as a blueprint for designing new drug candidates. By elucidating key interactions between ligands and target receptors, pharmacophore modeling guides the rational design of compounds with enhanced binding affinity and specificity. This approach accelerates the drug discovery process by facilitating the identification of lead compounds and optimization of their pharmacological properties [16, 17].

4) **Virtual screening**
Virtual screening is a computational technique employed in drug discovery to efficiently sift through large libraries of chemical compounds and predict their potential to interact with a target protein or receptor. Using molecular docking, pharmacophore modeling, or similarity searching, virtual screening evaluates the likelihood of compounds binding to the target based on their structural and chemical properties. By prioritizing the most promising candidates for experimental testing, virtual screening accelerates the drug discovery process, saving time and resources. It plays a vital role in identifying lead compounds with desired biological activity, ultimately aiding in the development of new drugs and therapeutic interventions [18].

5) **Molecular dynamics simulation**
Molecular dynamics (MD) is a computational simulation method used in physics and chemistry to study the motion and behavior of atoms and molecules over time. It employs classical mechanics principles to model the trajectories of individual atoms as they interact within a defined system, accounting for forces such as electrostatics and van der Waals interactions. By numerically solving Newton's equations of motion, MD simulations provide insight into the dynamic behavior of complex molecular systems, revealing information on structural changes, thermodynamic properties, and molecular interactions. Widely applied in fields like materials science, biochemistry, and drug design, MD simulations contribute to understanding fundamental molecular processes and guiding experimental research [19].

6) **Machine learning in drug design**
Machine learning (ML) in drug design leverages algorithms to analyze large datasets of chemical compounds and biological targets, aiding in the identification of potential drug candidates. ML models predict molecular properties, bioactivity, and toxicity, guiding the selection and optimization of lead compounds. Techniques such as QSAR, molecular docking, and virtual screening utilize ML to expedite the drug discovery process by prioritizing compounds for experimental validation. By integrating diverse sources of data, including chemical structures, genomic information, and experimental results, ML enhances the efficiency and accuracy of drug design, facilitating the development of novel therapeutics with improved efficacy and safety profiles [20].

7) **Fragment-based drug design (FBDD)**
FBDD is a strategy in medicinal chemistry that utilizes small molecular fragments as starting points for developing drug candidates. These fragments, typically low-molecular-weight compounds, bind to specific regions on target proteins with high affinity. Through techniques like X-ray crystallography and NMR spectroscopy, fragments' binding interactions are analyzed, guiding the assembly of larger compounds that retain the essential binding motifs. FBDD allows for efficient exploration of chemical space, facilitating the design of compounds with improved potency, selectivity, and pharmacokinetic properties. It has emerged as a powerful approach in drug discovery, particularly for challenging targets, offering promise in developing novel therapeutics for various diseases [21].

8) ***De novo* drug design**
De novo drug design is a computational approach in medicinal chemistry aimed at designing novel drug molecules from scratch, rather than modifying existing compounds. It involves the

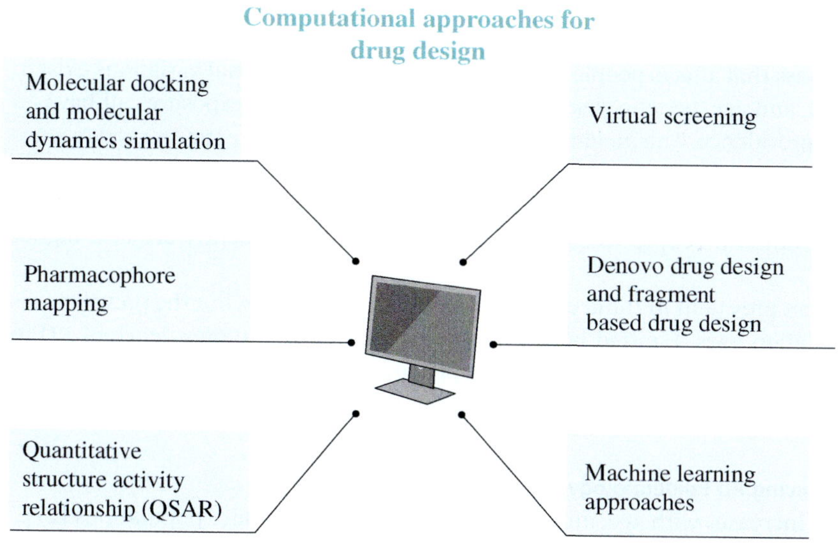

Figure 16.1 Different computational approaches commonly used in drug design process.

generation of molecular structures that possess desired pharmacological properties and are optimized for target specificity, efficacy, and safety. Using computational methods such as molecular docking, MD simulations, and quantum chemistry calculations, *de novo* drug design explores chemical space to identify lead compounds with the potential to interact favorably with target proteins. This method accelerates the drug discovery process by providing innovative solutions for addressing unmet medical needs and developing therapeutics with tailored properties [22, 23].

These computational methods significantly accelerate the drug discovery and development process, helping researchers to identify potential drug candidates more efficiently and cost-effectively. They also contribute to a deeper understanding of the molecular mechanisms underlying diseases, paving the way for more targeted and effective therapeutic interventions. Different approaches used for computer-aided drug design (CADD) are depicted in Figure 16.1.

16.4 Parkinson's Disease as Prevalent Neurodegenerative Disorder

Among the most common neurodegenerative diseases is Parkinson's disease. It is characterized by the gradual degeneration of brain neurons that produce dopamine, resulting in motor symptoms such as bradykinesia (slowness of movement), stiffness, and postural instability. Additional symptoms include sleep difficulties, gastrointestinal tract, bladder, and mental expressions. In addition to dopaminergic (DA) neurons in the substantia nigra (SNc), PD impacts additional neurological pathways. The genesis of PD can be caused by various variables, such as brain damage, environmental factors, genetic background, age, or dietary inadequacies. By 2030, it is expected that the number of adults over 65 with PD would have doubled. In the United States alone, PD is the 14th most common cause of death, with approximately USD 25 billion spent annually on treatment. While AD is generally recognized as the most common neurodegenerative disorder, PD is among the top neurodegenerative conditions in terms of prevalence [24].

16.4.1 Epidemiology of Parkinson's Disease

PD is a neurological illness that affects people all over the world today. Regional variations exist in the epidemiology of PD, and age, genetic susceptibility, and environmental exposures all have an impact on the disease's prevalence. The incidence of PD rises dramatically after age 60 and reaches over 3% in those over the age of 80. Most often, men experience PD at a higher incidence than women. Toxins in the environment can exacerbate PD symptoms, and eating habits might change the likelihood of the condition. People who smoke and drink coffee frequently are at a higher risk [25].

PD is reported to be less prevalent in India compared to Western countries, but the prevalence is increasing as the population ages. Limited population-based studies on the prevalence of PD in India are available, and the estimates vary. The reported prevalence in some studies ranges from 41 to 328 cases per 100,000 people, with variations across different regions [26].

16.4.1.1 Factors Influencing PD Epidemiology

1) **Age:** The risk of PD increases with age, and its prevalence is higher in older populations [27].
2) **Genetics:** Certain genetic factors can contribute to an increased risk of PD. Familial forms of PD, although less common, are influenced by genetic mutations [28].
3) **Environmental factors:** Exposure to certain environmental factors, such as pesticides, well water contaminants, and rural living, has been suggested to play a role in the risk of developing PD [29].

16.4.2 Pathogenesis of PD

Several factors that play roles in the pathogenesis of PD are depicted in Figure 16.2. These factors are as follows:

1) **Accumulation of Lewy bodies in substantia nigra**
PD is characterized by a number of pathophysiological alterations, one of which is the development of Lewy bodies in the SNc. Abnormal protein deposits called Lewy bodies form in the brain's nerve cells. Alpha-synuclein is the main protein that makes them up. Lewy bodies are mostly detected in the SNc, a part of the brain related to motor control, in PD. The buildup of Lewy bodies in this region aids in the degeneration of dopamine-producing neurons, which causes tremors, stiffness, and problems with balance and coordination that are characteristic of PD. Lewy body buildup is thought to cause oxidative stress, inflammation, and disruption of cellular processes, all of which can contribute to DA neuron degradation and ultimately death occurs [30].
2) **Mitochondrial dysfunction**
Mitochondrial dysfunction is linked to PD, affecting energy production and leading to oxidative stress. Mitochondria, responsible for ATP production, play a crucial role, and their impairment compromises neuronal function. Dysfunctional mitochondria generate ROS, contributing to PD's neurodegenerative process. Alpha-synuclein, associated with PD pathology, disrupts mitochondrial function. Mutations in mitochondrial DNA observed in PD brains may impair efficiency. Impaired mitophagy, responsible for removing damaged mitochondria, adds to cellular stress in PD. Environmental toxins, like pesticides, impact mitochondrial function, potentially contributing to dysfunction. Therapeutic strategies targeting mitochondrial function are explored for PD treatment in ongoing research [31].

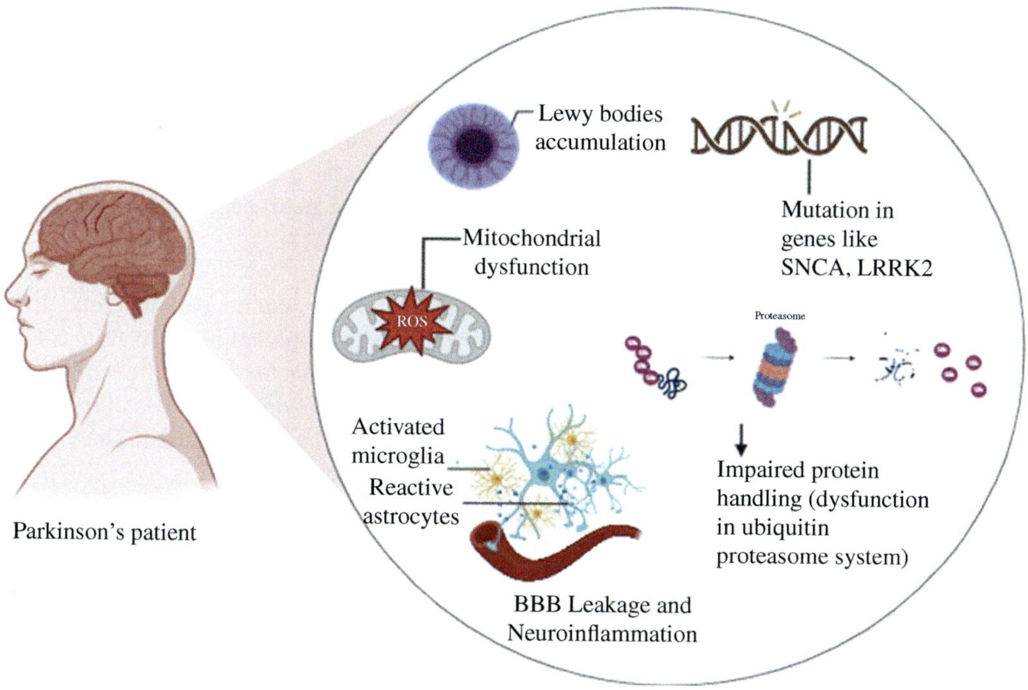

Figure 16.2 Pathogenesis of Parkinson's disease.

3) **Genetic factors**
Genetic factors significantly contribute to PD, encompassing both familial and sporadic cases. Approximately 5–10% of PD cases exhibit a familial pattern, involving specific genetic mutations. Among the genes associated with PD, SNCA (alpha-synuclein), LRRK2 (leucine-rich repeat kinase 2), PARK2 (parkin), PINK1 (PTEN-induced kinase 1), and GBA (glucocerebrosidase) are associated with an increased PD risk. Alpha-synuclein mutations influence Lewy body formation, LRRK2 mutations are prevalent in certain populations, PARK2 mutations impact protein degradation, and GBA mutations affect lipid metabolism. PD is a complex interplay of genetic and environmental factors, emphasizing the need for a comprehensive understanding in diagnosis and potential therapeutic development [32].
4) **Neuroinflammation**
Neuroinflammation plays a significant role in the pathogenesis of PD. It involves the activation of the brain's immune cells, such as microglia and astrocytes, in response to various triggers like alpha-synuclein aggregates and cellular damage. While inflammation is a normal protective response, chronic neuroinflammation in PD can become detrimental. Activated immune cells release proinflammatory molecules, leading to oxidative stress and further damage to DA neurons in the SNc. This inflammatory response may contribute to the progressive degeneration seen in PD. Understanding and targeting neuroinflammation are areas of active research for potential therapeutic interventions in PD [33].
5) **Impaired protein handling**
Impaired protein handling is integral to PD pathogenesis. Dysfunctional processes like the ubiquitin-proteasome system and autophagy, responsible for clearing damaged proteins, contribute to the accumulation of toxic protein aggregates, notably alpha-synuclein. Inefficient

clearance leads to the formation of Lewy bodies, affecting DA neurons in the SNc. The presence of these aggregates disrupts cellular function and contributes to neurodegeneration. Understanding and addressing these disruptions in protein handling mechanisms are crucial for developing therapeutic strategies to mitigate or prevent the progression of PD [34].

6) **Oxidative stress**
Increased production of ROS and oxidative stress play a role in the degeneration of DA neurons. Oxidative damage to cellular components contributes to the neurodegenerative process in PD [35].
7) **Environmental toxins**
Exposure to certain environmental toxins, including pesticides (rotenone, paraquat) and drugs like maneb and epoxomycin, have been associated with the development of PD in animal studies. This suggests a potential link between exposure to these specific toxins and the initiation or progression of PD-like symptoms in experimental research involving animals. Although the translation of these results to human implications requires further investigation. These toxins may contribute to mitochondrial dysfunction and oxidative stress [36].

16.4.3 Signaling Pathway of Parkinson's Disease

Several signaling pathways are implicated in the PD pathogenesis. Some key pathways include:

1) **DA signaling**
Neurodegeneration can be brought on by abnormalities in the signaling system, which is dependent on DA, the neurotransmitter found in DA neurons. DA neuron loss in PD results in decreased dopamine in the striatum, which affects the nigrostriatal system and causes motor abnormalities. Central to the development of PD are DA neuron loss, impaired dopamine release, and altered neurotransmission. Alpha-synuclein aggregation, oxidative stress, ferroptosis, mitochondrial failure, and neuroinflammation are among the mechanisms linked to PD. PD is caused by protein aggregation and mitochondrial failure in DA neurons. Environmental and neurotoxin exposure can also induce mitochondrial impairment, leading to DA cell loss and PD development. Thus, understanding the DA signaling pathway and associated mechanisms is crucial for comprehending and addressing PD progression [37].
2) **MAPK/ERK pathway**
Cell growth, differentiation, and survival are all regulated by the essential cell signaling mechanism known as the mitogen-activated protein kinase/extracellular signal-regulated kinase (MAPK/ERK) pathway. Within the framework of PD, the pathophysiology and advancement of the condition have been linked to the MAPK/ERK pathway. One important aspect of PD is the destruction of DA neurons in the SNc, which may be facilitated by dysregulation of the MAPK/ERK pathway. Dopamine is a neurotransmitter that is essential to motor function and is produced by DA neurons. Increased oxidative stress, inflammation, and apoptosis (programmed cell death) in DA neurons may arise from abnormal activation of the MAPK/ERK pathway. It is thought that these processes play a role in the neurodegenerative alterations seen in PD [38].
3) **PI3K/Akt/mTOR pathway**
Activation of the phosphoinositide 3-kinase (PI3K) enzyme is triggered by extracellular signals, including growth factors. Phosphatidylinositol 3,4,5-trisphosphate (PIP3) is produced when activated PI3K phosphorylates phosphatidylinositol 4,5-bisphosphate (PIP2). Akt (protein kinase B): Upon binding to PIP3, the serine/threonine kinase Akt is drawn to the cell membrane. Phosphorylation at the membrane causes Akt to become active. By phosphorylating several

downstream targets, activated Akt promotes cell growth and survival. Mammalian target of rapamycin, or mTOR, is a protein kinase that controls autophagy, cell division, and protein synthesis by combining signals from the Akt and other pathways. There are two mTOR complexes, mTORC1 and mTORC2, each of which has a different task [39].

The PI3K/Akt pathway has been linked to neuroprotection because it inhibits apoptosis, or programmed cell death, and encourages cell survival. Autophagy is a cellular process that is regulated by the mTOR pathway and is responsible for eliminating damaged or malfunctioning cellular components. Alpha-synuclein is one of the misfolded proteins that accumulate and is characteristic of PD; its buildup has been linked to dysregulation of autophagy. The control of mitochondrial function depends upon the PI3K/Akt/mTOR pathway. A common characteristic of neurodegenerative illnesses, such as PD, is mitochondrial malfunction. The PI3K/Akt/mTOR pathway is being intensively studied by researchers as a potential PD treatment approach [40].

4) **Wnt/β-catenin pathway**

The signaling molecules called Wnt proteins are released and start the Wnt pathway. Wnt ligands come in various forms that can open the pathway. Cell surface receptors such as frizzled (FZD) receptors and coreceptors such as low-density lipoprotein receptor-related protein (LRP) are bound by Wnt ligands. Signaling cascades within cells are activated by this binding. β-Catenin is typically the target of a protein complex that contains glycogen synthase kinase-3 beta (GSK-3β) for destruction in the absence of Wnt signaling. β-Catenin is stabilized and accumulates in the cytoplasm when Wnt signaling is activated. When stabilized β-catenin reaches the nucleus, it combines with transcription factors belonging to the T-cell factor/lymphoid enhancer factor (TCF/LEF) family to form a complex. Target genes implicated in various biological functions have their transcription activated by this complex.

It has been suggested that the Wnt/β-catenin pathway aids in neuroprotection and neuronal survival. It might contribute to preserving the integrity of the SNc's DA neurons, which are lost in a selective manner in PD. The formation of new neurons, or neurogenesis, has also been linked to the activation of this pathway. In the treatment of neurodegenerative illnesses, this mechanism may aid in the repair of neurons. The pathway is also connected to the control of mitochondrial activity, which is essential for the production of energy and maintenance of cellular homeostasis [41].

5) **NF-κB (nuclear factor-κB) pathway**

One important signaling mechanism that controls immunological responses, inflammation, and cell survival is the NF-κB pathway. Regarding PD, disruption of the NF-κB pathway has been connected to the neuroinflammatory mechanisms that accompany the disease's advancement. Generally, inhibitor proteins (IκBs) sequester NF-κB in the cytoplasm, rendering it inactive. Proinflammatory signals are one of the many stimuli that might cause the pathway to become active. Activation causes IκB proteins to get phosphorylated and then degrade, which permits NF-κB to translocate into the nucleus. Upon entering the nucleus, NF-κB functions as a transcription factor, controlling the expression of genes linked to immune response, inflammation, and cell viability. The brain's NF-κB pathway can be activated to enhance the production of chemokines and cytokines that promote inflammation [42].

6) **Autophagy-lysosomal pathway**

This pathway is pivotal in PD, ensuring cellular homeostasis. It begins with autophagosome formation, engulfing damaged components, including alpha-synuclein. Fusion with lysosomes forms autolysosomes for degradation, preventing toxic protein aggregation. Dysregulation leads to alpha-synuclein accumulation, contributing to Lewy bodies characteristic of Parkinson's pathology. Proper autophagy-lysosomal function is vital for neuronal health, and therapeutic strategies targeting this pathway may alleviate proteinopathy, offering a promising avenue for PD interventions [43].

7) **JNK (c-Jun N-terminal kinase) pathway**
This pathway is implicated in neurodegeneration. Activated JNK contributes to DA neuronal death, fostering inflammation, oxidative stress, and promoting alpha-synuclein aggregation. These processes collectively drive the progression of PD pathology. Modulating the JNK pathway emerges as a potential therapeutic strategy to impede neuroinflammation and apoptotic cascades, offering avenues for disease intervention and neuroprotection in PD [44].
8) **AMPK (AMP-activated protein kinase) pathway**
This pathway is implicated in PD and is essential for maintaining cellular energy balance. In PD, AMPK activation is associated with neuroprotection. Activated AMPK helps maintain cellular energy balance, regulates mitochondrial function, and suppresses oxidative stress, contributing to neuronal survival. Modulating the AMPK pathway presents a potential therapeutic avenue for PD, aiming to enhance cellular resilience, improve energy metabolism, and mitigate neurodegenerative processes. Understanding the intricate involvement of AMPK in PD pathophysiology offers insights into novel strategies for disease-modifying interventions [45].
9) **Nrf2 (nuclear factor erythroid 2-related factor 2) pathway**
The Nrf2 pathway is crucial in cellular defense against oxidative stress and inflammation, and its dysregulation is implicated in PD. In PD, diminished Nrf2 activity contributes to increased oxidative damage and neuroinflammation. Nrf2 regulates the expression of antioxidant and detoxification genes, promoting cellular resilience. Therapeutic strategies aiming to enhance Nrf2 activation show promise in mitigating PD-related neurodegeneration. Understanding the dynamics of the Nrf2 pathway offers insights into potential interventions to bolster cellular defenses, reduce oxidative stress, and counteract the progression of PD [46].

Understanding these signaling pathways provides insights into the molecular mechanisms underlying PD. Targeting specific pathways may offer therapeutic strategies to modify disease progression.

16.4.4 Enzymatic Targets in Parkinson's Disease

The targets for PD involve key enzymes or proteins associated with processes such as neurotransmitter regulation, oxidative stress, and protein degradation. Some notable enzymatic targets and their mechanism are depicted in Figure 16.3.

1) **MAO-B (monoamine oxidase B)**
MAO-B inhibition is a therapeutic strategy in PD management. In PD, DA neurons degenerate, leading to dopamine deficiency. MAO-B, an enzyme, metabolizes dopamine, exacerbating its depletion. Inhibiting MAO-B preserves dopamine levels, alleviating PD symptoms. Selegiline and rasagiline are FDA-approved MAO-B inhibitors used as adjuncts to levodopa therapy, enhancing its efficacy and extending its duration of action. By reducing oxidative stress and neuroinflammation, MAO-B inhibition may also exert neuroprotective effects. Targeting MAO-B offers symptomatic relief and potential disease-modifying benefits in PD, making it a valuable enzymatic target in the management of this neurodegenerative disorder [47].
2) **COMT (catechol-O-methyltransferase)**
COMT inhibition is a therapeutic avenue in PD treatment. In PD, DA neurons degenerate, leading to dopamine depletion. COMT, an enzyme, metabolizes levodopa, reducing its efficacy and

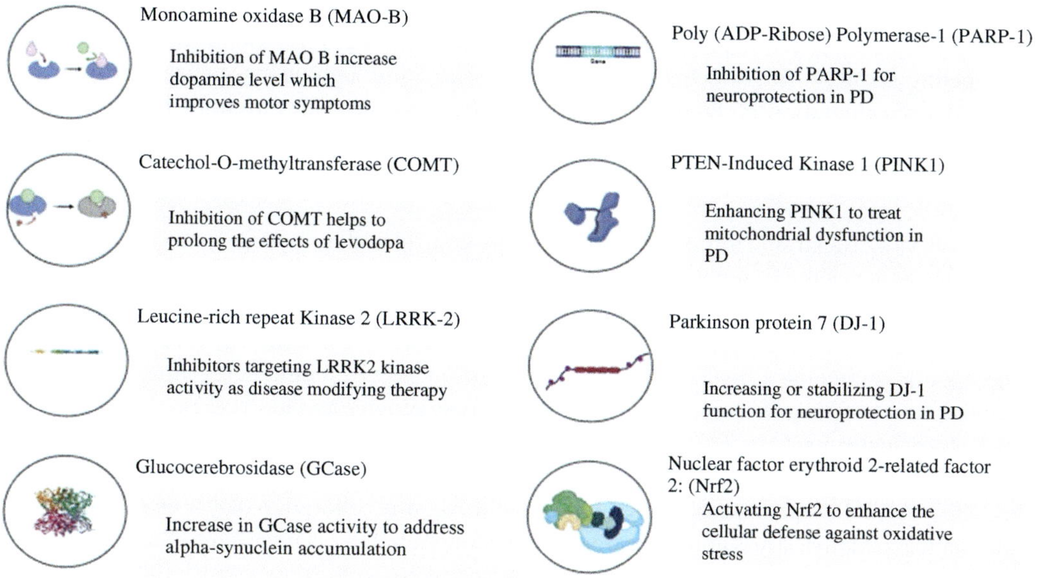

Figure 16.3 Enzymatic targets for Parkinson's disease and their mechanism.

causing motor fluctuations. Inhibiting COMT enhances levodopa's bioavailability and prolongs its therapeutic effect, reducing motor fluctuations and improving symptom control. Tolcapone and entacapone are FDA-approved COMT inhibitors used in combination with levodopa therapy. By optimizing levodopa pharmacokinetics, COMT inhibition offers symptomatic relief and enhances patient quality of life in PD. Targeting COMT represents a valuable strategy for managing motor complications and optimizing levodopa therapy in PD patients [48].

3) **LRRK2**

LRRK2 has emerged as a significant enzymatic target in PD research. Mutations in the LRRK2 gene are associated with both familial and sporadic forms of PD. LRRK2 is a kinase enzyme implicated in various cellular processes, including vesicle trafficking, cytoskeletal dynamics, and autophagy. Dysregulation of LRRK2 activity leads to aberrant phosphorylation events and cellular dysfunction, contributing to PD pathology. Targeting LRRK2 with small-molecule inhibitors is a promising therapeutic approach, aiming to modulate its kinase activity and mitigate neurodegeneration in PD. Developing effective LRRK2 inhibitors holds potential for disease-modifying treatments in PD management [49].

4) **GCase (glucocerebrosidase)**

GCase plays a role in lysosomal function and alpha-synuclein metabolism. Mutations in the GBA1 gene, encoding, are a significant genetic risk factor for PD. Reduced GCase activity leads to impaired degradation of alpha-synuclein, promoting its aggregation and contributing to PD pathogenesis. Therapeutic strategies aimed at enhancing GCase activity or promoting lysosomal function hold promise for slowing disease progression and alleviating PD symptoms. Investigating GCase as an enzymatic target offers insights into novel approaches for tackling alpha-synuclein pathology and advancing PD treatment options [50].

5) **PARP-1 [poly(ADP-ribose) polymerase-1]**

PARP-1 is involved in DNA repair, inflammation, and cell death pathways implicated in PD pathology. Dysregulated PARP-1 activity exacerbates oxidative stress and neuroinflammation, contributing to DA neuron degeneration. Inhibiting PARP-1 shows promise in preclinical studies for neuroprotection and alleviating PD symptoms. Targeting PARP-1 offers a potential therapeutic strategy for slowing disease progression and enhancing neuroprotection in PD. Further exploration of PARP-1 as an enzymatic target provides avenues for developing novel treatments that address underlying molecular mechanisms and improve outcomes for PD patients [51].

6) **PINK1**
   PINK1 plays a crucial role in mitochondrial quality control and mitophagy, processes essential for maintaining cellular homeostasis and preventing neurodegeneration. Mutations in the PINK1 gene are linked to familial forms of PD. Dysfunctional PINK1 leads to mitochondrial dysfunction and accumulation of damaged mitochondria, contributing to PD pathology. Therapeutic strategies aimed at modulating PINK1 activity or enhancing mitophagy hold promise for protecting DA neurons and slowing disease progression in PD [52].
7) **DJ-1 (Parkinson protein 7)**
   DJ-1, also known as PARK7, mutations in the DJ-1 gene are linked to familial forms of PD. Although DJ-1's precise role in PD remains elusive, it is believed to play a neuroprotective role against oxidative stress and mitochondrial dysfunction. DJ-1 is involved in various cellular processes, including antioxidative defense mechanisms and mitochondrial homeostasis. Therapeutic strategies targeting DJ-1 aim to enhance its neuroprotective functions or restore its activity in PD [53].
8) **Nrf2**
   Nrf2 is a transcription factor that regulates the expression of antioxidant and detoxifying enzymes, playing a crucial role in cellular defense against oxidative stress. Dysregulation of Nrf2 signaling is implicated in PD pathology, with decreased Nrf2 activity observed in affected brain regions. Therapeutic approaches aimed at enhancing Nrf2 activation offer promise for mitigating oxidative damage, reducing neuroinflammation, and promoting neuronal survival in PD. Targeting Nrf2 represents a potential strategy for developing disease-modifying treatments that address underlying mechanisms of neurodegeneration in PD [46].

   These enzymatic targets are part of ongoing research efforts aimed at developing disease-modifying therapies for PD. It is important to note that while progress has been made, many potential treatments are still in preclinical or early clinical stages, and further research is needed to determine their safety and efficacy in human patients.

16.4.5 Current Therapeutic Approaches to Treat PD

Several therapeutic approaches are used to manage the symptoms of PD. It is important to note that ongoing research and developments may bring new treatments, and individualized management plans are tailored by healthcare professionals based on the specific needs of each patient. Drugs for treating motor and nonmotor symptoms and disease-modifying therapy of PD are given in Table 16.1. The common therapeutic approaches include the following:

1) **Drugs to treat motor symptoms of PD**
   - **Dopamine precursor-levodopa (L-DOPA):** It is a precursor to dopamine, a neurotransmitter that is lacking in PD patients' brains. In the brain, levodopa gets converted into dopamine, which reduces motor symptoms. On the other hand, prolonged usage could result in dyskinesias and motor irregularities [54].

Table 16.1 Current treatment strategies for Parkinson's disease.

	Drugs to treat motor symptoms of PD		Drugs to treat nonmotor symptoms of PD		
S. no.	Category	Examples	Category	Examples	Disease-modifying therapies
1	Dopamine precursors	Levodopa	Cholinesterase inhibitors	Rivastigmine	Deep brain stimulation (DBS)
2	Dopamine agonist	Pramipexole, ropinirole	Antidepressants	Fluoxetine, Sertraline	Physical and occupational therapy
3	MAO-B inhibitor	Selegiline, rasagiline	Antipsychotics	Quetiapine	Speech and swallowing therapy
4	COMT inhibitors	Entacapone	Antiemetics	Ondansetron, metoclopramide	Exercise programs
5	Anticholinergic drugs	Trihexyphenidyl, benztropine	Antimuscarinic agents	Tolterodine, oxybutynin	Diet and nutrition

- **Dopamine agonists:** These drugs function in the brain similarly to dopamine. They can be used alone or in conjunction with levodopa as monotherapy. Ropinirole and pramipexole are two examples [55].
- **MAO-B inhibitors:** By preventing dopamine from being broken down, MAO-B inhibitors like rasagiline and selegiline raise dopamine levels in the brain. They are typically applied to PD in its early stages [56].
- **COMT inhibitors:** Because they stop levodopa from breaking down in the peripheral tissues, entacapone and other COMT inhibitors prolong the effects of the drug. They frequently work in combination with levodopa [57].
- **Anticholinergic medications:** These drugs, such as trihexyphenidyl and benztropine, help control tremors and rigidity by modulating the activity of acetylcholine, another neurotransmitter [58].

2) **Drugs to treat non-motor symptoms of PD**
   - **Cholinesterase inhibitors:** Medications such as rivastigmine may be prescribed to address cognitive impairment in PD, especially if there are symptoms of dementia [59].
   - **Antidepressants:** Depression and anxiety are prevalent nonmotor symptoms in PD and can be treated with selective serotonin reuptake inhibitors (SSRIs) like sertraline or ventlafaxine or duloxetine or serotonin–norepinephrine reuptake inhibitors (SNRIs) like sertraline [60].
   - **Antipsychotics:** Traditional antipsychotics are usually not prescribed due to their potential to worsen motor symptoms while atypical antipsychotics like quetiapine may be used cautiously to treat psychosis, hallucinations, or severe behavioral disturbances [61].
   - **Antiemetics:** Ondansetron and metoclopramide are specifically designed to prevent or relieve drug-induced nausea and vomiting [62].
   - **Antimuscarinic agents:** Antimuscarinic medications like tolterodine and oxybutynin can help to reduce bladder spasms and may be used to treat overactive bladder symptoms, which may help with symptoms of urinary urgency and frequency [63].
3) **Disease-modifying therapies to treat PD**
   - **Deep brain stimulation (DBS):** DBS entails implanting electrodes into particular movement-regulating brain areas. These electrodes are attached to a stimulator that modifies

aberrant neural activity by delivering electrical impulses. In cases of advanced PD, DBS is frequently taken into consideration [64].

- **Physical and occupational therapy:** Physical therapy helps maintain mobility and improve balance, while occupational therapy focuses on adapting daily activities to the individual's capabilities [65].
- **Speech and swallowing therapy:** Speech therapy helps individuals with PD overcome speech and swallowing difficulties that may arise as the disease progresses [66].
- **Exercise programs:** Frequent physical activity, such as strength training and aerobics, has been demonstrated to improve mobility and general well-being in people with PD [67].
- **Diet and nutrition:** A balanced diet, along with adequate fluid intake, is important for managing PD symptoms and potential medication side effects [68].

Research is ongoing to develop disease-modifying therapies that can slow or halt the progression of PD. Additionally, emerging therapies, including gene therapies and novel drug candidates, are being explored in clinical trials.

16.4.6 Current Therapeutic Challenges to Treat Parkinson's disease

There were several limitations associated with existing treatments for PD. Keep in mind that there may have been advancements or changes since then. Here are some of the common limitations:

1) **Symptomatic relief only**
The majority of PD medications now in use concentrate on symptom relief rather than treating the disease's fundamental cause. Although drugs like levodopa can effectively relieve motor symptoms, the disease's course is neither slowed down nor stopped by them.
2) **Motor fluctuations and dyskinesias**
Levodopa is a vital drug for treating PD symptoms; however, prolonged use of it might cause dyskinesias and motor irregularities. Patients may have periods of decreased pharmaceutical efficacy, which could result in "off" periods when symptoms reappear or "on" periods when they have dyskinesias, or uncontrollable movements.
3) **Limited efficacy in nonmotor symptoms**
Numerous nonmotor symptoms, including mood disorders, cognitive decline, and autonomic dysfunction, are linked to PD. These elements of the condition may not be adequately addressed by current treatments, and patients may need to take extra drugs or undergo other therapies to control their symptoms.
4) **Disease progression**
Although the current medications can temporarily alleviate symptoms, PD cannot be stopped from progressing. This emphasizes the need for medications that alter the course of disease and slow down or stop the brain's DA neurons from degenerating.
5) **Side effects**
PD treatments come with a number of adverse effects, and different people may react differently to different medications. It might be difficult to weigh the advantages against any possible drawbacks, particularly as the illness worsens. Some common side effects of drugs used to treat motor symptoms of PD are given in Table 16.2.
6) **Limited treatment options for advanced PD**
When DBS becomes less successful in treating symptoms, solutions such as medication may be considered in advanced stages of PD. But not every patient is a good fit for these treatments, and there could be dangers involved.

Table 16.2 Side effects of drugs used to treat motor symptoms.

S. no.	Drugs	Examples	Side effects
1	Dopamine precursors	Carbidopa/levodopa	Shortness of breath, numbness
2	Dopamine agonist	Bromocriptine, Pramipexole	Blurred vision
		Ropinirole	Chest pain or discomfort, insomnia
3	MAO-B inhibitor	Rasagiline, Selegiline, Safinamide	Sudden increase in blood pressure, depression
4	COMT inhibitors	Tolcapone, Entacapone	Dyskinesia, confusion, hallucinations
5	Anticholinergic drugs	Trihexyphenidyl and Benztropine	Drowsiness, irregular heartbeat

7) **Individual variability**
As a diverse disorder, Parkinson's affects people differently in how it shows up. There is no one-size-fits-all method of therapy, and determining the right combination of drugs and therapies can be difficult and time-consuming [69].

Researchers and clinicians continue to explore new treatment avenues, including gene therapies, neuroprotective agents, and interventions targeting specific pathways involved in PD.

16.4.7 Unmet Needs in Parkinson's Disease Therapeutics

Unmet needs in PD therapeutics include a lack of disease-modifying treatments to slow or halt progression. Nonmotor symptoms such as cognitive issues and mood disorders require targeted interventions. Improved formulations for levodopa, addressing motor fluctuations and dyskinesias, are essential. Biomarkers for early diagnosis and personalized treatments based on individual variability are crucial. Neuroprotective therapies, a deeper understanding of PD subtypes, and symptomatic relief without side effects remain pressing challenges. Integrating patient-reported outcomes into clinical assessments and ensuring global access to therapies are vital for comprehensive PD care. Ongoing research aims to address these gaps and enhance overall management [70].

16.4.8 Significance of Computational Approaches in Parkinson's Disease

- **CADD approaches to overcome therapeutic challenges in PD**
  CADD is a powerful approach that can contribute to overcoming therapeutic challenges in PD in several ways.
- **Target identification and validation**
  CADD can assist in identifying and validating potential drug targets by analyzing the complex molecular interactions involved in PD pathogenesis. This aids in the selection of targets with the highest therapeutic potential.
- **Virtual screening**
  Through virtual screening, CADD helps in identifying potential drug candidates by analyzing large chemical databases. This accelerates the drug discovery process, allowing for the efficient identification of molecules that may modulate specific targets associated with PD.
- **Predicting drug–target interactions**
  CADD employs computational models to predict the interactions between drugs and their target proteins. This helps in understanding how potential therapeutics may affect specific molecular pathways involved in PD.

- **Optimizing drug candidates**
  CADD assists in optimizing lead compounds by predicting their pharmacokinetic properties, bioavailability, and potential side effects. This reduces the likelihood of failure in later stages of drug development and ensures the selection of compounds with better chances of success.
- **Understanding molecular mechanisms**
  CADD provides insights into the detailed molecular mechanisms underlying PD. This understanding is crucial for designing drugs that can specifically target the pathological processes involved in the disease [71].
- **Repurposing existing drugs**
  CADD allows for the systematic exploration of existing drug databases to identify potential candidates that can be repurposed for PD treatment. This strategy may accelerate the development of new therapies by leveraging the safety profiles of already-approved drugs.
- **Personalized medicine**
  CADD can contribute to the development of personalized medicine by analyzing individual genetic and molecular profiles. Tailoring treatments to the specific characteristics of each patient may enhance therapeutic efficacy and minimize side effects.
- **Simulating drug effects**
  Computational simulations can predict the effects of potential drugs on cellular and molecular systems. This aids in understanding how drugs may influence complex biological processes, allowing for more informed decision-making in drug development.
- **Drug delivery optimization**
  CADD can assist in optimizing drug delivery methods, including overcoming the blood–brain barrier, to ensure that therapeutic agents reach the target areas in the brain effectively.

By integrating these CADD approaches into the drug discovery and development process, researchers can streamline the identification and optimization of potential therapeutics for PD. This computational approach enhances the efficiency of drug discovery, potentially leading to the development of more effective and targeted treatments for PD [72].

16.4.9 Use of Computational Tools in Identifying Biomarkers

Computational tools play a crucial role in identifying biomarkers for PD, contributing to early diagnosis and disease monitoring. Various approaches leverage advanced technologies:

- **Bioinformatics and data analysis**
  Computational tools analyze large-scale biological data, including genomics, transcriptomics, and proteomics, to identify potential biomarkers associated with PD. This aids in understanding the molecular mechanisms and pathways involved in the disease.
- **ML algorithms**
  ML models can analyze complex datasets to identify patterns and correlations that may not be apparent through traditional methods. This enables the discovery of potential biomarkers and the development of predictive models for early PD diagnosis.
- **Imaging analysis**
  Computational tools are used to process and analyze medical imaging data, such as functional MRI (fMRI) and positron emission tomography (PET) scans. This assists in identifying structural and functional changes in the brain that may serve as biomarkers for PD.

- **Network analysis**
  Computational tools evaluate biological networks to identify key nodes or pathways associated with PD. This systems biology approach helps uncover molecular interactions and potential biomarkers linked to disease progression.
- **Artificial intelligence (AI) in wearable technology**
  AI-driven analysis of data from wearable devices, such as smartwatches or activity trackers, can provide continuous monitoring of motor symptoms and aid in the identification of biomarkers related to movement patterns and fluctuations.
- **Clinical data integration**
  Computational tools integrate clinical data from various sources, including electronic health records and patient databases. This comprehensive analysis helps identify correlations between patient characteristics, treatment responses, and disease progression, facilitating the discovery of relevant biomarkers.
- **Predictive modeling**
  Computational models predict the risk of developing PD based on various risk factors, genetics, and environmental exposures. This aids in identifying individuals at higher risk, enabling early intervention strategies.
- **Biochemical pathway analysis**
  Computational tools assess biochemical pathways associated with PD, which help to identify potential biomarkers related to specific metabolic or signaling processes that may be dysregulated in the disease.

Integration of these computational approaches enhances the identification and validation of biomarkers, ultimately contributing to a more comprehensive understanding of PD pathophysiology. These tools expedite the translation of molecular insights into clinical applications, fostering advancements in early diagnosis, personalized treatment strategies, and the development of disease-modifying therapies for PD [73–75].

16.4.10 Neuroprotective Strategies Through Computational Insights

16.4.10.1 Computational Models for Neuroprotection

Computational models have been developed to study neuroprotection in PD, aiming to understand the mechanisms behind neurodegeneration and identify potential therapeutic targets. Some of these models include:

- **PD signaling cascade planned and implemented using a systems biology approach:** this work focused on the relationship between alpha-synuclein and protein kinase A (PKA) and used nanoparticle-mediated suppression of the disease [76].
- **A computational model of PD-related DA cell loss because of excitotoxicity induced by glutamate:** This model examined the relationship between the subthalamic nucleus (STN) and SNc, demonstrating how the loss of DA cells is aggravated when SNc cells are initially lost. This process is known as a positive-feedback loop [77].
- **A multiscale computational model of levodopa-induced toxicity in PD:** It is a multiscale computational model. The SNc-striatum system was the subject of this model, which modeled striatal neurons at the spiking level and SNc neurons at the biophysical level. The toxicity of L-DOPA in SNc, which is brought on by an energy deficit, was captured by the model. This model may provide insight into possible treatment targets and assist in understanding the process underlying neurodegeneration in PD [77].

These computational models contribute to the understanding of PD mechanisms and provide insights into potential therapeutic targets. However, more research is needed to develop effective treatments for PD.

16.4.11 Examples of Computational Successes in Parkinson's Disease Drug Development

Computational approaches have played a significant role in various aspects of PD drug development, offering insights into disease mechanisms, identifying potential drug candidates, and optimizing treatment strategies. Here are a few examples of computational successes in PD drug development:

1) **Target identification and validation**
   LRRK2 inhibitors: Computational modeling and bioinformatics analyses have been crucial in identifying LRRK2 as a potential target for PD. Inhibitors targeting LRRK2 have been explored for their neuroprotective effects, and computational tools helped in validating their therapeutic potential [78].
2) **Drug repurposing**
   Isradipine: Computational analyses and virtual screening have been used to repurpose existing drugs for PD. Isradipine, originally an antihypertensive medication, showed neuroprotective effects in preclinical models of PD, and computational methods played a role in identifying its potential [79].
3) **Alpha-synuclein aggregation inhibitors**
   Virtual screening: Computational methods, such as molecular docking and dynamics simulations, have been employed in virtual screening to identify small molecules capable of inhibiting the aggregation of alpha-synuclein, a key pathological feature in PD. These approaches aid in the discovery of potential disease-modifying agents [80].
4) **Deep learning in biomarker discovery**
   Biomarker identification: Deep learning algorithms applied to multiomics data have shown promise in identifying biomarkers associated with PD. These computational approaches enhance our understanding of disease progression and may aid in developing diagnostic and prognostic markers [81].
5) **Personalized medicine**
   Genomic medicine: Computational tools analyze genetic data to identify patient-specific variations that may influence drug responses. This personalized medicine approach aims to tailor treatments based on individual genetic profiles, optimizing therapeutic outcomes [71].
6) **Drug-induced neuroprotection**
   Network pharmacology: Computational network pharmacology approaches help elucidate the complex interactions between drugs and biological pathways. This has been applied to identify compounds that may induce neuroprotection by modulating multiple targets involved in PD pathogenesis [82].
7) **Optimizing clinical trials**
   Clinical trial simulation: Computational modeling and simulation are used to optimize clinical trial design, including patient selection criteria, dosage regimens, and expected outcomes. This helps streamline the drug development process and enhance the chances of success in clinical trials [83].

16.4.12 Future Directions and Innovations in Computational Methods for Parkinson's Disease

Some potential future directions and innovations in computational methods for PD are as follows:

1) **ML and AI-based diagnostics**
ML algorithms, particularly deep learning models, can analyze large datasets such as neuroimaging, genetic, and clinical data to identify patterns associated with PD. These models can assist in early diagnosis, prognosis prediction, and subtype classification of PD [84].
2) **Wearable technology integration**
Wearable sensors, such as smartwatches and accelerometers, provide continuous monitoring of movement patterns and symptoms in PD patients. Integrating data from wearable devices with computational algorithms can offer insights into disease progression, medication response, and motor fluctuations, enabling personalized treatment strategies [85].
3) **Multimodal data fusion**
Integrating heterogeneous data sources, including genetic, imaging, clinical, and environmental data, can provide a more comprehensive understanding of PD etiology and progression. Computational methods for multimodal data fusion, such as fusion algorithms and network-based approaches, can uncover complex relationships between different data modalities and also identify biomarkers for early detection and disease monitoring [86].
4) **Predictive modeling of disease progression**
Computational models, including mathematical modeling and simulation techniques, can simulate the progression of PD based on individual patient characteristics and disease-related factors. These predictive models can aid clinicians in treatment planning, clinical trial design, and assessing the long-term impact of interventions [87].
5) **Network analysis of brain connectivity**
Graph theory-based approaches can analyze structural and functional brain networks derived from neuroimaging data to characterize alterations in brain connectivity associated with PD. Network-based biomarkers can help to identify disease-related changes in brain organization and facilitate the development of targeted therapies aimed at modulating dysfunctional neural circuits [88].
6) **Personalized treatment optimization**
Computational methods, such as reinforcement learning and optimization algorithms, can optimize treatment strategies for individual PD patients based on their clinical profiles, medication responses, and disease progression trajectories. Personalized therapeutic interventions, including medication dosing schedules and DBS parameters, can improve symptom management and enhance the quality of life for PD patients [89].
7) **Data sharing and collaboration platforms**
Collaborative platforms and data repositories enable researchers to share and analyze large-scale datasets, fostering interdisciplinary collaboration and accelerating scientific discoveries in PD research. Open-access data initiatives and standardized data formats facilitate the development and validation of computational methods across different research groups and institutions [90].

Overall, the integration of computational methods with advances in technology and data analytics holds great promise for advancing our understanding of PD pathophysiology, improving diagnostic accuracy, and optimizing personalized treatment approaches. Continued innovation and collaboration in this field are essential for translating computational research findings into clinical practice and ultimately improving outcomes for PD patients.

16.5 Conclusion

In conclusion, the rational design of drugs for neurodegenerative disorders represents a promising avenue for addressing the urgent need for effective therapies. By leveraging computational methods such as molecular docking, QSAR modeling, and network pharmacology analysis, researchers can identify and validate novel drug targets, design small molecules or biologics with improved potency and selectivity, and repurpose existing drugs for therapeutic intervention. The use of CADD approaches enables researchers to understand the intricate molecular mechanisms underlying neurodegenerative disorders, allowing for the targeted design of therapeutic agents that can modulate specific biological targets implicated in disease pathology. By leveraging the vast amount of available structural and chemical data, CADD facilitates the rational selection and optimization of lead compounds, thereby increasing the likelihood of successful clinical translation. Additionally, ML algorithms facilitate the analysis of large-scale omics data and clinical datasets to identify disease biomarkers, predict drug responses, and optimize treatment strategies. Overall, the rational design of drugs for neurodegenerative disorders holds great promise for accelerating the drug discovery process, advancing personalized medicine approaches, and ultimately improving patient outcomes in the treatment of these debilitating conditions.

References

1 Lamptey, R.N.L., Chaulagain, B., Trivedi, R. et al. (2022). A review of the common neurodegenerative disorders: current therapeutic approaches and the potential role of nanotherapeutics. *International Journal of Molecular Sciences* 23: 1851. https://doi.org/10.3390/ijms230318512.

2 Breijyeh, Z. and Karaman, R. (2020). Comprehensive review on Alzheimer's disease: causes and treatment. *Molecules* 25: 5789. https://doi.org/10.3390/molecules25245789.

3 Goyal, V. and Radhakrishnan, D. (2018). Parkinson's disease: a review. *Neurology India* 66: 26. https://doi.org/10.4103/0028-3886.226451.

4 McColgan, P. and Tabrizi, S.J. (2017). Huntington's disease: a clinical review. *European Journal of Neurology* 25: 24–34. https://doi.org/10.1111/ene.13413.

5 Masrori, P. and Van Damme, P. (2020). Amyotrophic lateral sclerosis: a clinical review. *European Journal of Neurology* 27: 1918–1929. https://doi.org/10.1111/ene.14393.

6 Dobson, R. and Giovannoni, G. (2018). Multiple sclerosis – a review. *European Journal of Neurology* 26: 27–40. https://doi.org/10.1111/ene.13819.

7 Sivasathiaseelan, H., Marshall, C., Agustus, J. et al. (2019). Frontotemporal dementia: a clinical review. *Seminars in Neurology* 39: 251–263. https://doi.org/10.1055/s-0039-1683379.

8 Nabi, M. and Tabassum, N. (2022). Role of environmental toxicants on neurodegenerative disorders. *Frontiers in Toxicology* 4: https://doi.org/10.3389/ftox.2022.837579.

9 Niranjan, R. (2018). Recent advances in the mechanisms of neuroinflammation and their roles in neurodegeneration. *Neurochemistry International* 120: 13–20. https://doi.org/10.1016/j.neuint.2018.07.003.

10 Kwon, H.S. and Koh, S.H. (2020). Neuroinflammation in neurodegenerative disorders: the roles of microglia and astrocytes. *Translational Neurodegeneration* 9: https://doi.org/10.1186/s40035-020-00221-2.

11 Zhu, W.M., Neuhaus, A., Beard, D.J. et al. (2022). Neurovascular coupling mechanisms in health and neurovascular uncoupling in Alzheimer's disease. *Brain* 145: 2276–2292. https://doi.org/10.1093/brain/awac174.

12 De Felice, F.G. and Lourenco, M.V. (2015). Brain metabolic stress and neuroinflammation at the basis of cognitive impairment in Alzheimer's disease. *Frontiers in Aging Neuroscience* 7: https://doi.org/10.3389/fnagi.2015.00094.

13 Pihlstrøm, L., Wiethoff, S., and Houlden, H. (2018). Genetics of neurodegenerative diseases: an overview. *Handbook of Clinical Neurology* 309–323. https://doi.org/10.1016/B978-0-12-802395-2.00022-5.

14 Meng, X.Y., Zhang, H.X., Mezei, M., and Cui, M. (2011). Molecular docking: a powerful approach for structure-based drug discovery. *Current Computer Aided-Drug Design* 7: 146–157. https://doi.org/10.2174/157340911795677602.

15 Muhammad, U., Uzairu, A., and Ebuka Arthur, D. (2018). Review on: quantitative structure activity relationship (QSAR) modeling. *Journal of Analytical & Pharmaceutical Research* 7: https://doi.org/10.15406/japlr.2018.07.00232.

16 Voet, A., Qing, X., Lee, X.Y. et al. (2014). Pharmacophore modeling: advances, limitations, and current utility in drug discovery. *Journal of Receptor, Ligand and Channel Research* 81: https://doi.org/10.2147/JRLCR.S46843.

17 Yang, S.Y. (2010). Pharmacophore modeling and applications in drug discovery: challenges and recent advances. *Drug Discovery Today* 15: 444–450. https://doi.org/10.1016/j.drudis.2010.03.013.

18 Maia, E.H.B., Assis, L.C., de Oliveira, T.A. et al. (2020). Structure-based virtual screening: from classical to artificial intelligence. *Frontiers in Chemistry* 8: https://doi.org/10.3389/fchem.2020.00343.

19 De Vivo, M., Masetti, M., Bottegoni, G., and Cavalli, A. (2016). Role of molecular dynamics and related methods in drug discovery. *Journal of Medicinal Chemistry* 59: 4035–4061. https://doi.org/10.1021/acs.jmedchem.5b01684.

20 Dara, S., Dhamercherla, S., Jadav, S.S. et al. (2021). Machine learning in drug discovery: a review. *Artificial Intelligence Review* 55: 1947–1999. https://doi.org/10.1007/s10462-021-10058-4.

21 Kirsch, P., Hartman, A.M., Hirsch, A.K.H., and Empting, M. (2019). Concepts and core principles of fragment-based drug design. *Molecules* 24: 4309. https://doi.org/10.3390/molecules24234309.

22 Hartenfeller, M. and Schneider, G. (2010). *De novo* drug design. *Methods in Molecular Biology* 299–323. https://doi.org/10.1007/978-1-60761-839-3_12.

23 Aarsland, D., Batzu, L., Halliday, G.M. et al. (2021). Parkinson disease-associated cognitive impairment. *Nature Reviews Disease Primers* 7: https://doi.org/10.1038/s41572-021-00280-3.

24 Khan, A.U., Akram, M., Daniyal, M., and Zainab, R. (2018). Awareness and current knowledge of Parkinson's disease: a neurodegenerative disorder. *International Journal of Neuroscience* 129: 55–93. https://doi.org/10.1080/00207454.2018.1486837.

25 Tysnes, O.B. and Storstein, A. (2017). Epidemiology of Parkinson's disease. *Journal of Neural Transmission* 124: 901–905. https://doi.org/10.1007/s00702-017-1686-y.

26 Ascherio, A. and Schwarzschild, M.A. (2016). The epidemiology of Parkinson's disease: risk factors and prevention. *The Lancet Neurology* 15: 1257–1272. https://doi.org/10.1016/S1474-4422(16)30230-7.

27 Verma, A.K., Raj, J., Sharma, V. et al. (2017). Epidemiology and associated risk factors of Parkinson's disease among the north Indian population. *Clinical Epidemiology and Global Health* 5: 8–13. https://doi.org/10.1016/j.cegh.2016.07.003.

28 Klein, C. and Westenberger, A. (2012). Genetics of Parkinson's disease. *Cold Spring Harbor Perspectives in Medicine* 2: a008888. https://doi.org/10.1101/cshperspect.a008888.

29 Ball, N., Teo, W.P., Chandra, S., and Chapman, J. (2019). Parkinson's disease and the environment. *Frontiers in Neurology* 10: https://doi.org/10.3389/fneur.2019.00218.

30 Mahul-Mellier, A.L., Burtscher, J., Maharjan, N. et al. (2020). The process of Lewy body formation, rather than simply α-synuclein fibrillization, is one of the major drivers of neurodegeneration.

Proceedings of the National Academy of Sciences 117: 4971–4982. https://doi.org/10.1073/pnas.1913904117.

31 Moon, H.E. and Paek, S.H. (2015). Mitochondrial dysfunction in Parkinson's disease [Internet]. *Experimental Neurobiology* 24: 103–116. https://doi.org/10.5607/en.2015.24.2.103.

32 Ye, H., Robak, L.A., Yu, M. et al. (2023). Genetics and pathogenesis of Parkinson's syndrome [Internet]. *Annual Review of Pathology: Mechanisms of Disease* 18: 95–121. https://doi.org/10.1146/annurev-pathmechdis-031521-034145.

33 Troncoso-Escudero, P., Parra, A., Nassif, M., and Vidal, R.L. (2018). Outside in: unraveling the role of neuroinflammation in the progression of Parkinson's disease [Internet]. *Frontiers in Neurology* 9: https://doi.org/10.3389/fneur.2018.00860.

34 Cook, C., Stetler, C., and Petrucelli, L. (2012). Disruption of protein quality control in Parkinson's disease. *Cold Spring Harbor Perspectives in Medicine* 2: a009423. https://doi.org/10.1101/cshperspect.a009423.

35 Dias, V., Junn, E., and Mouradian, M.M. (2013). The role of oxidative stress in Parkinson's disease. *Journal of Parkinson's Disease* 3: 461–491. https://doi.org/10.3233/JPD-130230.

36 Pan-Montojo, F., Loewenbrück, K.F., and Reichmann, H. (2020). The role of environmental toxins and inflammation in Parkinson's disease pathophysiology: a historical perspective and research-based evidence. *Diagnosis and Management in Parkinson's Disease* 39–55. https://doi.org/10.1016/B978-0-12-815946-0.00003-X.

37 Dong-Chen, X., Yong, C., Yang, X. et al. (2023). Signaling pathways in Parkinson's disease: molecular mechanisms and therapeutic interventions. *Signal Transduction and Targeted Therapy* 8: https://doi.org/10.1038/s41392-023-01353-3.

38 Bohush, A., Niewiadomska, G., and Filipek, A. (2018). Role of mitogen activated protein kinase signaling in Parkinson's disease. *International Journal of Molecular Sciences* 19: 2973. https://doi.org/10.3390/ijms19102973.

39 Xu, F., Na, L., Li, Y., and Chen, L. (2020). Retracted article: roles of the PI3K/AKT/mTOR signalling pathways in neurodegenerative diseases and tumours. *Cell & Bioscience* 10: https://doi.org/10.1186/s13578-020-00416-0.

40 Goyal, A., Agrawal, A., Verma, A., and Dubey, N. (2023). The PI3K-AKT pathway: a plausible therapeutic target in Parkinson's disease. *Experimental and Molecular Pathology* 129: 104846. https://doi.org/10.1016/j.yexmp.2022.104846.

41 Serafino, A. and Cozzolino, M. (2023). The Wnt/β-catenin signaling: a multifunctional target for neuroprotective and regenerative strategies in Parkinson's disease. *Neural Regeneration Research* 18: 306. https://doi.org/10.4103/1673-5374.343908.

42 Singh, S.S., Rai, S.N., Birla, H. et al. (2019). NF-κB-mediated neuroinflammation in Parkinson's disease and potential therapeutic effect of polyphenols. *Neurotoxicity Research* 37: 491–507. https://doi.org/10.1007/s12640-019-00147-2.

43 Bonam, S.R., Tranchant, C., and Muller, S. (2021). Autophagy-lysosomal pathway as potential therapeutic target in Parkinson's disease. *Cells* 10: 3547. https://doi.org/10.3390/cells10123547.

44 Peng, J. and Andersen, J. (2003). The role of c-Jun N-terminal kinase (JNK) in Parkinson's disease. *IUBMB Life* 55: 267–271. https://doi.org/10.1080/1521654031000121666.

45 Curry, D.W., Stutz, B., Andrews, Z.B., and Elsworth, J.D. (2018). Targeting AMPK signaling as a neuroprotective strategy in Parkinson's disease. *Journal of Parkinson's Disease* 8: 161–181. https://doi.org/10.3233/JPD-171296.

46 Yang, X.X., Yang, R., and Zhang, F. (2022). Role of Nrf2 in Parkinson's disease: toward new perspectives. *Frontiers in Pharmacology* 13: https://doi.org/10.3389/fphar.2022.919233.

47 Dezsi, L. and Vecsei, L. (2017). Monoamine oxidase B inhibitors in Parkinson's disease. *CNS & Neurological Disorders – Drug Targets* 16: https://doi.org/10.2174/1871527316666170124165222.

48 Rivest, J., Barclay, C.L., and Suchowersky, O. (1999). COMT inhibitors in Parkinson's disease. *Canadian Journal of Neurological Sciences/Journal Canadien des Sciences Neurologiques* 26: S34–S38. https://doi.org/10.1017/s031716710000007x.

49 Wojewska, D.N. and Kortholt, A. (2021). LRRK2 targeting strategies as potential treatment of Parkinson's disease. *Biomolecules* 11: 1101. https://doi.org/10.3390/biom11081101.

50 Martínez-Bailén, M., Clemente, F., Matassini, C., and Cardona, F. (2022). GCase enhancers: a potential therapeutic option for Gaucher disease and other neurological disorders. *Pharmaceuticals* 15: 823. https://doi.org/10.3390/ph15070823.

51 Salemi, M., Mazzetti, S., De Leonardis, M. et al. (2021). Poly (ADP-ribose) polymerase 1 and Parkinson's disease: a study in post-mortem human brain. *Neurochemistry International* 144: 104978. https://doi.org/10.1016/j.neuint.2021.104978.

52 Vizziello, M., Borellini, L., Franco, G., and Ardolino, G. (2021). Disruption of mitochondrial homeostasis: the role of PINK1 in Parkinson's disease. *Cells* 10: 3022. https://doi.org/10.3390/cells10113022.

53 Heremans, I.P., Caligiore, F., Gerin, I. et al. (2022). Parkinson's disease protein PARK7 prevents metabolite and protein damage caused by a glycolytic metabolite. *Proceedings of the National Academy of Sciences* 119: https://doi.org/10.1073/pnas.2111338119.

54 Haddad, F., Sawalha, M., Khawaja, Y. et al. (2017). Dopamine and levodopa prodrugs for the treatment of Parkinson's disease. *Molecules* 23: 40. https://doi.org/10.3390/molecules23010040.

55 Luo, D., Reith, M., and Dutta, A.K. (2020). Dopamine agonists in treatment of Parkinson's disease: an overview. *Diagnosis and Management in Parkinson's Disease* 445–460. https://doi.org/10.1016/B978-0-12-815946-0.00026-0.

56 Tan, Y.Y., Jenner, P., and Chen, S.D. (2022). Monoamine oxidase-B inhibitors for the treatment of Parkinson's disease: past, present, and future. *Journal of Parkinson's Disease* 12: 477–493. https://doi.org/10.3233/JPD-212976.

57 Fabbri, M., Ferreira, J.J., and Rascol, O. (2022). COMT inhibitors in the management of Parkinson's disease. *CNS Drugs* 36: 261–282. https://doi.org/10.1007/s40263-021-00888-9.

58 Katzenschlager, R., Sampaio, C., Costa, J., and Lees, A. (2002). Anticholinergics for symptomatic management of Parkinson's disease. *Cochrane Database of Systematic Reviews* 2010: https://doi.org/10.1002/14651858.CD003735.

59 Pagano, G., Rengo, G., Pasqualetti, G. et al. (2014). Cholinesterase inhibitors for Parkinson's disease: a systematic review and meta-analysis. *Journal of Neurology, Neurosurgery & Psychiatry* 86: 767–773. https://doi.org/10.1136/jnnp-2014-308764.

60 Zhuo, C., Xue, R., Luo, L. et al. (2017). Efficacy of antidepressive medication for depression in Parkinson disease. *Medicine* 96: e6698. https://doi.org/10.1097/MD.0000000000006698.

61 Shotbolt, P., Samuel, M., and David, A. (2010). Quetiapine in the treatment of psychosis in Parkinson's disease. *Therapeutic Advances in Neurological Disorders* 3: 339–350. https://doi.org/10.1177/1756285610389656.

62 Tarsy, D., Parkes, J.D., and Marsden, C.D. (1975). Metoclopramide and pimozide in Parkinson's disease and levodopa-induced dyskinesias. *Journal of Neurology, Neurosurgery & Psychiatry* 38: 331–335. https://doi.org/10.1136/jnnp.38.4.331.

63 Barrett, M.J., Sargent, L., Nawaz, H. et al. (2021). Antimuscarinic anticholinergic medications in Parkinson disease: to prescribe or deprescribe? *Movement Disorders Clinical Practice* 8: 1181–1188. https://doi.org/10.1002/mdc3.13347.

64 Groiss, S.J., Wojtecki, L., Südmeyer, M., and Schnitzler, A. (2009). Review: deep brain stimulation in Parkinson's disease. *Therapeutic Advances in Neurological Disorders* 2: 379–391. https://doi.org/10.1177/1756285609339382.

65 Radder, D.L.M., Sturkenboom, I.H., van Nimwegen, M. et al. (2017). Physical therapy and occupational therapy in Parkinson's disease. *International Journal of Neuroscience* 127: 930–943. https://doi.org/10.1080/00207454.2016.1275617.

66 Tjaden, K. (2008). Speech and swallowing in Parkinson's disease. *Topics in Geriatric Rehabilitation* 24: 115–126. https://doi.org/10.1097/01.TGR.0000318899.87690.44.

67 Oliveira de Carvalho, A., Filho, A.S.S., Murillo-Rodriguez, E. et al. (2018). Physical exercise for Parkinson's disease: clinical and experimental evidence. *Clinical Practice & Epidemiology in Mental Health* 14: 89–98. https://doi.org/10.2174/1745017901814010089.

68 Knight, E., Geetha, T., Burnett, D., and Babu, J.R. (2022). The role of diet and dietary patterns in Parkinson's disease. *Nutrients* 14: 4472. https://doi.org/10.3390/nu14214472.

69 Pires, A.O., Teixeira, F.G., Mendes-Pinheiro, B. et al. (2017). Old and new challenges in Parkinson's disease therapeutics. *Progress in Neurobiology* 156: 69–89. https://doi.org/10.1016/j.pneurobio.2017.04.006.

70 LeWitt, P.A. and Chaudhuri, K.R. (2020). Unmet needs in Parkinson disease: motor and non-motor. *Parkinsonism & Related Disorders* 80: S7–S12. https://doi.org/10.1016/j.parkreldis.2020.09.024.

71 Shalaby, K.E. and El-Agnaf, O.M.A. (2022). Gene-based therapeutics for Parkinson's disease. *Biomedicines* 10: 1790. https://doi.org/10.3390/biomedicines10081790.

72 Glaab, E. (2017). Computational systems biology approaches for Parkinson's disease. *Cell and Tissue Research* 373: 91–109. https://doi.org/10.1007/s00441-017-2734-5.

73 Yousef, M., Najami, N., Abedallah, L., and Khalifa, W. (2014). Computational approaches for biomarker discovery. *Journal of Intelligent Learning Systems and Applications* 6: 153–161. https://doi.org/10.4236/jilsa.2014.64012.

74 Kumar, R., Malik, M.Z., Thanaraj, T.A. et al. (2023). A computational biology approach to identify potential protein biomarkers and drug targets for sporadic amyotrophic lateral sclerosis. *Cellular Signalling* 112: 110915. https://doi.org/10.1016/j.cellsig.2023.110915.

75 Zhang, F., Wu, X., and Chen, J.Y. (2013). Computational biomarker discovery. *Approaches in Integrative Bioinformatics* 355–386. https://doi.org/10.1007/978-3-642-41281-3_13.

76 Kaushik, A.C., Bharadwaj, S., Kumar, S., and Wei, D.Q. (2018). Nano-particle mediated inhibition of Parkinson's disease using computational biology approach. *Scientific Reports* 8: https://doi.org/10.1038/s41598-018-27580-1.

77 Muddapu, V.R., Mandali, A., Chakravarthy, V.S., and Ramaswamy, S. (2019). A computational model of loss of dopaminergic cells in Parkinson's disease due to glutamate-induced excitotoxicity. *Frontiers in Neural Circuits* 13: https://doi.org/10.3389/fncir.2019.00011.

78 Tan, S., Lu, R., Yao, D. et al. (2023). Identification of LRRK2 inhibitors through computational drug repurposing. *ACS Chemical Neuroscience* 14: 481–493. https://doi.org/10.1021/acschemneuro.2c00672.

79 Vaz, R.L., Sousa, S., Chapela, D. et al. (2020). Identification of antiparkinsonian drugs in the 6-hydroxydopamine zebrafish model. *Pharmacology Biochemistry and Behavior* 189: 172828. https://doi.org/10.1016/j.pbb.2019.172828.

80 Kurnik, M., Sahin, C., Andersen, C.B. et al. (2018). Potent α-synuclein aggregation inhibitors, identified by high-throughput screening, mainly target the monomeric state. *Cell Chemical Biology* 25: 1389–1402.e9. https://doi.org/10.1016/j.chembiol.2018.08.005.

81 Zhang, H., Deng, K., Li, H. et al. (2020). Deep learning identifies digital biomarkers for self-reported Parkinson's disease. *Patterns* 1: 100042. https://doi.org/10.1016/j.patter.2020.100042.

82 Liu, Y.Y., Yu, L.H., and Zhang, J. (2021). Network pharmacology-based and molecular docking-based analysis of Suanzaoren decoction for the treatment of Parkinson's disease with sleep disorder. (Alatas, B., ed.). *BioMed Research International* 2021: 1–12. https://doi.org/10.1155/2021/1752570.

83 Prasad, E.M. and Hung, S.Y. (2021). Current therapies in clinical trials of Parkinson's disease: a 2021 update. *Pharmaceuticals* 14: 717. https://doi.org/10.3390/ph14080717.

84 Mei, J., Desrosiers, C., and Frasnelli, J. (2021). Machine learning for the diagnosis of Parkinson's disease: a review of literature. *Frontiers in Aging Neuroscience* 13: https://doi.org/10.3389/fnagi.2021.633752.

85 Yadav, R., Bagrodia, V., Holla, V. et al. (2022). Parkinson's disease and wearable technology: an Indian perspective. *Annals of Indian Academy of Neurology* 25: 817. https://doi.org/10.4103/aian.aian_653_22.

86 Morgan, C., Tonkin, E.L., Masullo, A. et al. (2023). A multimodal dataset of real world mobility activities in Parkinson's disease. *Scientific Data* 10: https://doi.org/10.1038/s41597-023-02663-5.

87 Ahmed, S., Komeili, M., and Park, J. (2022). Predictive modelling of Parkinson's disease progression based on RNA-Sequence with densely connected deep recurrent neural networks. *Scientific Reports* 12: https://doi.org/10.1038/s41598-022-25454-1.

88 Bergamino, M., Keeling, E.G., Ray, N.J. et al. (2023). Structural connectivity and brain network analyses in Parkinson's disease: a cross-sectional and longitudinal study. *Frontiers in Neurology* 14: https://doi.org/10.3389/fneur.2023.1137780.

89 Mishima, T., Fujioka, S., Morishita, T. et al. (2021). Personalized medicine in Parkinson's disease: new options for advanced treatments. *Journal of Personalized Medicine* 11: 650. https://doi.org/10.3390/jpm11070650.

90 Cubo, E. and Delgado-López, P.D. (2022). Telemedicine in the management of Parkinson's disease: achievements, challenges, and future perspectives. *Brain Sciences* 12: 1735. https://doi.org/10.3390/brainsci12121735.

17

Rational Design of Anti-inflammatory Therapeutics

Kratika Singh[1], Anmol Gupta[2], Irum Siddiqui[3], Ashapurna Sinha[2], Mukesh Kumar Patwa[1], and Urmila Singh[1]

[1] *Department of Microbiology, King George Medical University, Lucknow, Uttar Pradesh, India*
[2] *Department of Biosciences, Integral University, Lucknow, Uttar Pradesh, India*
[3] *IIRC-1, Department of Bioengineering, Integral University, Lucknow, Uttar Pradesh, India*

17.1 Introduction

Inflammatory disease (ID) is recognized as a leading cause of morbidity within the population and its term comprises a comprehensive spectrum of diseases involved in chronic inflammation for the progression of disease progression [1]. These conditions are linked to an engaged immune system, encompassing activated immune cells and various biomolecules [2]. Inflammation serves as a defensive response of an organism to counteract the intrusion of foreign entities such as viruses, bacteria, and parasites [3]. IDs exert a profound impact on the physiology of organs, whether deeply situated or superficially located. There are many conditions are present that affect various organs, such as the intestine (inflammatory bowel disease; IBD), lungs (acute respiratory distress syndrome; ARDS), liver (hepatitis), arteries (atherosclerosis), pancreas (pancreatitis), kidney (nephritis), and heart (myocardial infarction). Similarly, psoriasis (influencing the skin), arthritis (affecting joints), periodontitis (impacting dental health), and uveitis (affecting the eyes) are examples of IDs involving superficially located organs. It includes both "autoimmune diseases" and "auto-IDs" [4]. The attack of the immune system on self-tissues like psoriasis, ulcerative, and rheumatoid arthritis is called an autoimmune disease in many diseases, whereas diseases like hepatitis, uveitis, atherosclerosis, pulmonary diseases, myocardial infarction, pancreatitis, nephritis involves uncontrolled inflammation called as auto-IDs [5]. In aggregate, IDs significantly contribute to the global disease burden, commonly assessed through disability-adjusted life year (DALY) statistics. For example, in 2017, the average DALY rate for both genders with chronic obstructive pulmonary disease (COPD) was 1028.8 per 100,000 age-standardized years [6]. Over the decades, some small molecules, like nonsteroidal anti-inflammatory drugs (NSAID), glucocorticoids, and antioxidants have been widely used for the treatment of various ID, and other than their effect they have many side effects like aseptic joint necrosis, gastrointestinal bleeding cardiovascular risk, liver/kidney injury [7]. Certain monoclonal antibodies that target different pro-inflammatory cytokines, chemokines, or other bio-macromolecules have been used in clinical settings or studied in preclinical studies for the treatment of numerous conditions like rheumatoid arthritis (RA) [8], IBD [9],

Computational Methods for Rational Drug Design, First Edition. Edited by Mithun Rudrapal.

atherosclerotic disease [10], and asthma [11]. These antibodies regulate molecular and cellular procedures closely linked to inflammatory responses (IRs). Still, the clinical application of these agents has been accompanied by various limitations and side effects like cost effectiveness, a primary or secondary lack of reaction, as well as an increased risk of serious infections [12]. Undesirable effects in large part result from the systemic delivery in nontarget cells and tissue, irrespective of whether it's small molecule drugs or biological therapies. In addition, for certain newly developed therapies like specialized pro-resolving mediators, which exhibit potent inflammation-resolving activity and are highly effective in treating the inflammatory disorder, it also becomes challenging to provide a controlled release and site-specific distribution in inflammatory cells/tissues [7]. Over the past decades, many cutting-edge drug delivery strategies have been explored to overcome the drawbacks accompanying traditional formulations of anti-inflammatory therapies.

17.2 Navigating Inflammation and its Microenvironment

Inflammation is a completely natural biological response to internal and/or external stimuli of a biophysical or chemical nature. This reaction is distinguished by the coordinated involvement of various inflammatory cells and molecular mediators. Based on the duration of inflammation, it is possible to classify it into two types: chronic and acute inflammatory responses. Severe inflammation, which is characterized by the prominent clinical symptoms of redness, swelling, pain, and heat, typically remains used for some hours to a few days. In response to invading tissue injuries, acute inflammation is generally accompanied by a quick and remarkable accumulation of inflammatory cells (especially neutrophils), cytokines, fluid, and chemokines in the affected tissues/organs [13, 14]. Acute IR typically resolves on its own, restoring tissue balance (Figure 17.1). Nevertheless, should a pathological insult persist or resolution be unattainable, tissue injury and an ongoing IR may occur, resulting in chronic inflammation that can endure for months or even years. Monocytes/macrophages are likely to play a significant role in this process by releasing numerous inflammatory cytokines, which are closely linked to the development of chronic diseases. While the specific triggers and body responses determine the initiation and inflammation progression, their shared physiological and pathological characteristics are important targets for diagnosing and treating various inflammatory conditions. In the subsequent part, we will briefly discuss the typical pathophysiological traits of the inflammatory microenvironment, which provide valuable insights for designing drug delivery systems and bio-responsive materials to manage inflammation-related diseases.

17.2.1 Inflammatory Cell Infiltration and Vascular Permeability

The altered structure and function of the microvasculature are the most important characteristics of initial inflammation. This results in the widening of blood vessels, reduced permeability of the blood vessels, and an increase in white blood cells in the tissue [15]. Substances in the plasma and cells that cause blood vessels to react, such as complement components, histamine, serotonin, fibrin, kinins, prostaglandins, and platelet-activating factors, are unlikely to have the main influence on the immediate responses. These substances can bind to specific receptors on the cells that line the blood vessels, causing them to contract and form gaps. As a result, the permeability of the blood vessels increases, and fluids from the blood leak out within 30 minutes of an injury.

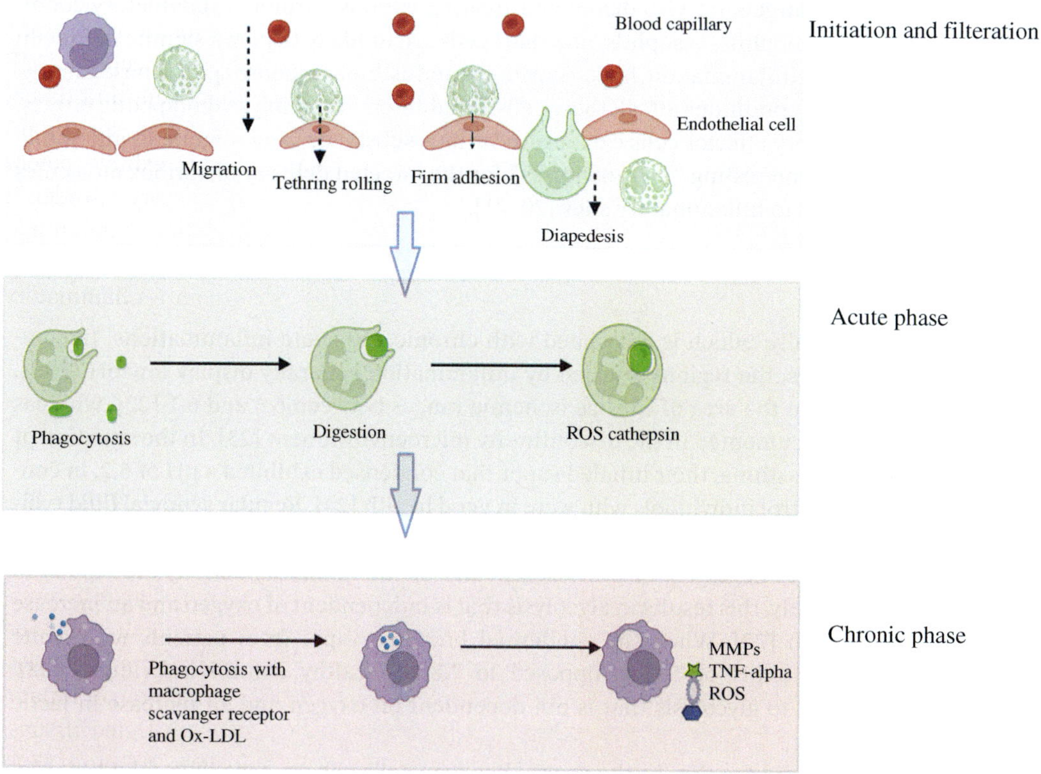

Figure 17.1 Inflammation and environment of inflammation with acute and chronic phases in connective tissues.

Temporarily, the activation of cells that line the interior of blood vessels is initiated by these substances that induce inflammation to increase the synthesis of diverse molecules that promote cellular adhesion. These molecules can be categorized into four primary groups, characterized by their function and structure: immunoglobulins, integrins, selectins, and cadherins. The cells lining the inner walls of blood vessels display the expression of these molecules that promote adhesion. Subsequently, these molecules aid the movement of white blood cells from the bloodstream into the damaged tissues via a meticulously controlled process that encompasses various stages of attachment, rolling, strong adhesion, and migration. The body's IR involves diverse subgroups of white blood cells, such as monocytes/macrophages, basophils, eosinophils, lymphocytes, and primarily neutrophils, as well as mast cells. Neutrophils, which constitute a significant proportion of white blood cells in the bloodstream (50–70%), play a significant role in the immediate response to inflammation. They perform a number of tasks, including creating extracellular neutrophil traps, generating reactive oxygen species (ROSs), and triggering and controlling the immune systems of the body [16, 17]. Monocytes can migrate toward inflamed areas and transform into macrophages, which depends on the severity and duration of the injury and different substance activations. These macrophages, along with invading macrophages and their cytokines, are present in the tissues and play a crucial role in maintaining tissue and host equilibrium during the acute phase response of inflammation.

Furthermore, peripheral acute inflammatory and immune reactions rarely lead to acute inflammation in affected tissues, as short and controlled macrophage responses are linked to various noninfectious and infectious diseases [18]. Therefore, macrophages have been extensively researched as potential targets for visualizing and treating various chronic inflammatory conditions [19]. However, eosinophils, basophils, and mast cells are unlikely to play a significant role in allergies, parasite-related inflammation, hypersensitivity, and asthma responses [20]. Consequently, associated inflammatory conditions attack these cells. In addition to serving as therapeutic targets, these unique inflammatory effector cells can absorb various substances and migrate to inflammatory sites, making them interesting "Trojan Horses" for the targeted delivery of various molecules and particulate therapies to inflammatory sites [20, 21].

17.2.2 Acidosis

Acidification occurs locally, which is associated with chronic and acute inflammations. In comparison to healthy tissues, the regions affected by inflammation generally display low-pH levels. As an example, the pH in the area of cardiac ischemia ranges between 6.6 and 6.1 [22], whereas levels ≥4.2 have been documented in the discontinuity microenvironment [23]. In the situation of patients who have acute asthma, their inhaled vapor that condensed exhibited a pH of 5.2, in contrast to a pH of 7.8 in control individuals who were in good health [24]. Regular synovial fluid typically has a pH ranging from 7.4 to 7.8, whereas synovial fluid in arthritic joints decreases to a range of 6.7–7.3 [25, 26]. The degree of the sickness and its inflammatory activity are related to this pH drop. Consequently, this results in glycolysis that is independent of oxygen and an increase in lactic acid production [27]. When the condensed breathed vapor from patients with acute asthma cleared, it had a pH of 5.2 as opposed to 7.8 in healthy control individuals [28]. Consequently, this leads to glycolysis that is not dependent on oxygen and an increase in lactic acid production [27].

During an infection, acidification in the spaces between cells can be contributed by bacterial metabolites like lactate [29, 30]. Inflammatory lung diseases, including asthma and COPD, can result in respiratory acidosis due to accidental inhalation of acids present in the air and pollution from fog. Extended acid exposure can exacerbate acidosis due to varying airway pH controls, which is a major risk factor for the development of bronchitis and asthma. By acting as warning signs and regulating the production and release of extracellular acidosis, inflammatory mediators alter the intrinsic properties of different immune and inflammatory cells [31, 32]. Furthermore, delivery systems that display acid-triggered release behaviors for various medicines may be designed using acidosis as an internal signal. It is important to note that, apart from these pathological situations, various cells, organs, and tissues in the body have different levels of acidic environments under regular physiological circumstances. For example, the gastrointestinal tract's pH gradient, which ranges from 5.6 to 7.6 in the colon to 1.1–3.6 in the stomach, is well-regulated [33]. To treat inflammatory disorders, a variety of pH-responsive biomaterials and delivery devices have been created.

17.2.3 Increased Oxidative Stress in Tissues

ROS are chemical compounds that contain oxygen and display reactivity. Some examples of typical ROS consist of hydrogen peroxide (H_2O_2), hypochlorous acid (HOCl), singlet oxygen ($^1O_2$), hydroxyl radical ($^{\bullet}OH$), and superoxide anion radical ($\bullet O_2^-$). Physiological levels of ROS play a significant role in the maintenance and regulation of cellular functions, like proliferation, survival, migration, and cell differentiation. Furthermore, ROS are important for the defense against pathogens. However, extreme production of ROS can lead to oxidative damage of lipids, proteins, and DNA. As a result of this, apoptosis or necrosis may eventually cause cell death [34].

Oxidative stress, which is caused by ROS, has been directly associated with a number of diseases, including diabetes, asthma, acute ischemic stroke, acute lung or liver injury, acute gastroenteritis, acute bronchitis, neurodegenerative diseases, IBD, cardiovascular diseases, and RA. When compared to healthy entities that have *Helicobacter pylori* or IBD, the ROS levels are more than 10–100 times greater [35]. Extracellular H_2O_2 concentration rose to $100\,\mu M$ during reperfusion and ischemia in several neurological disorders. This eventually resulted in the development of Alzheimer's and Parkinson's diseases due to changes in structural elements and neuronal proteins as well as loss of cognitive function [36]. Respiratory lining cells with inflammatory lung diseases had 20 times greater H_2O_2 levels than in healthy tissues [37]. Research carried out on animal models of colitis, acute lung damage, and alcoholic liver injury demonstrated that the amount of ROS often follows a parabolic trajectory as different IDs proceed. The kind of illness and whether the inflammatory stimuli are endogenous or external dictate when the maximal level occurs [38, 39]. Therefore, endogenous biochemical stimuli are treated by oxidative stress at the site of inflammation.

17.3 The Demand for Advanced Anti-inflammatory Medications

A major healthcare problem worldwide is IDs, which include different types of rheumatic diseases in which inflammation disorder badly affects human health. There are various agents available to treat the disorder but they have adverse side effects on human health, in traditional practice there are two types of drugs either steroidal anti-inflammatory drugs or NSAIDs are available to manage the ID. Through the cyclooxygenase (COX) inhibition in the prostaglandins biosynthesis, NSAIDs can inhibit in the early steps, and these are the main drugs for the reduction of inflammation consequences [40]. The use of NSAIDs has many side effects which are related to renal, gastrointestinal, and toxicities [41]. The use of corticosteroids is central to hypertension, hyperglycemia, osteoporosis, and growth arrest [42]. The significant challenge associated with presently utilized synthetic drugs is their potential for toxicity and the recurrence of symptoms upon discontinuation [40]. There is a growing interest in addressing this issue by exploring the advances of safer anti-inflammatory agents [43]. A rational and productive strategy toward the treatment of inflammatory conditions involves the progress of anti-inflammatory drugs consequent from normal sources [44]. For the treatment of IDs natural products are the biocompatible, safe, and cost-effective alternatives, efficacious.

Since ancient times, there have been many countries, such as India, Sri Lanka, China, and Brazil called hubs of natural products used as traditional medicine. Recently, research has been delving into the pharmacologic and molecular mechanisms that underlie the health benefits and therapeutic potential of many compounds. The integration of traditional knowledge and indigenous resources is deemed essential for advancing the development of new anti-inflammatory leads [41]. Numerous reports focus on traditional plants employed in folk medicine for their anti-inflammatory properties, providing valuable insights that contribute to the potential use of natural products as anti-inflammatory drugs [45]. The impact of phytoconstituents is manifested through their influence on pivotal regulatory molecules such as cytokines, inducible nitric oxide synthase (iNOS), and COX [46, 47].

17.4 Natural Products Used for Anti-inflammatory Drug Development: Systematic Approach in Use of Different Animal Models for Evaluations

Pharmacology plays an essential role in modern medicine in the search for new drugs with the use of appropriate models and identifying the mechanism of the selected molecules. For the prediction of clinically and physiologically suitable models the trial models are used, based on pharmacological ethics, the model is selected based on their results which should be matched with the clinical results, for the preclinical assessment of drugs Vogel et al. [48] designated various methods, i.e., *in vivo* and *in vitro*. For the screening of anti-inflammatory agents, many models are used they have used different mechanisms for the screening of activity, some examples are enlisted in Table 17.1.

Before beginning any experimental technique, a number of preparatory criteria such as medication dosage, sample size, toxicity, and others should be taken into account [60]. There are several research on the anti-inflammatory properties of phytoconstituents that demonstrate their safety, toxicity, and mode of action. During the drug discovery process, improper work planning leads to the rejection of phytoconstituents; thus, program selection must be done carefully. The main cause of the failure of phytoconstituents-based drug development programs appears to be program selection.

17.5 Rational Design of Anti-inflammatory Agents

17.5.1 Creating Anti-inflammatory Polymers Through Phosphoramidite Chemistry Inspired by Apoptotic Processes

The polymers are designed with immunosuppressive effects for anti-inflammatory therapies, which results in the modification of the morphology of immune cells. The researcher used the phosphoramidite method in which phosphoramidite chemistry was utilized to synthesize the apoptotic cell-membrane-inspired polymers. The macrophages employed in the *in vitro* cell viability assays of PS-modified polymers demonstrated their nontoxicity at concentrations of 1×10^5 μg/mL. The expression of NF–κB has been decreased, which showed that the macrophages are protected from lipopolysaccharide-induced inflammation [72].

17.5.2 New Anti-inflammatory Agent with Indoyl-imidazole Hybrids

For designing new agents researchers used *in vivo* analgesic and anti-inflammatory activity as well as ulcerogenicity of the prepared hybrids were evaluated in experimental animals. COX-2 enzyme was employed in the computational studies, which produced two new hybrid compounds with the lowest binding energy [73].

17.5.3 Rational Design of Novel Aminopiperidinyl Amide

In this study, the C-X-C chemokine receptor type 4 has been used as a targeted enzyme for ID. In this study, the ZINC chemical database has been screened, and after molecular dynamic simulation, the Z7R was identified as a novel compound with nanomolar affinity and 78.5% chemotaxis inhibition which shows potent anti-inflammatory activity [74].

Table 17.1 Methods for evaluating anti-inflammatory activity in animal models.

S. no.	Inflammation	Model	Mechanism	Advantage	Disadvantage	References
1.	Acute-inflammation	Carrageenan-induced paw edema	Initiate the inflammation phase from the beginning; cyclooxygenase pathway activation dilates post-capillary venules	Inflammation is reproducible, acute and nonimmune Susceptible to inhibitors of cyclooxygenase Adequate for NSAIDs evaluation	The researcher was instructed in recording the consistent and dependable paw volume	[49, 50]
		Histamine/5-HT-induced paw edema	It causes an increase in lymph flow, which causes edema to occur Histamine injected subplantarly stimulates the release of fluid and plasma proteins from the cell	Used as a secondary model 5-HT inhibitor drugs assessed	Prostaglandin inhibitor drugs cannot be checked	[51, 52]
		Paw edema induced by bradykinin	Bradykinin-stimulated human endothelial cell cultures produce metabolites of arachidonic acid	Drugs suppressing prostaglandins are successful in this model	It only causes moderate, temporary edema	[53]
		Paw edema induced by dextran	Increased permeability of blood vessels, activation of kinin, and release of mediators, including histamine and serotonin, which produce osmotic edema with insufficient neutrophils and proteins	This can be used to bolster the results of the paw model caused by carrageenan	Unsuitable for drugs acting through mechanisms other than antiserotonin or antihistamine	[53–55]
		LPS-induced paw edema	TNF-α-mediated inflammation drugs assisted	The simultaneous assessment of anti-inflammatory and analgesic can be done	NA	[54, 56]
		Arachidonic acid-induced ear edema	Fast transformation into lipoxygenase and cyclooxygenase products	Anti-inflammatory for eicosanoid inhibition recognized	Animals are sacrificed	[57]
		Croton oil/TPA-induced ear edema	Topical administration of croton oil causes stimulates neutrophil influx, vascular permeability, synthesis of eicosanoids, vasodilatation, and liberation of serotonin and histamine	Most suitable for steroidal and NSAID	Animals are sacrificed	[58, 59, 60, 61]

(*Continued*)

Table 17.1 (Continued)

S. no.	Inflammation	Model	Mechanism	Advantage	Disadvantage	References
		Oxazolone-induced ear edema	Oxazolone dramatically raises the levels of IFN-γ while having a negligible effect on IL-4 levels Interferon-γ can cause the activation of various inflammatory cell	Recognize the model of delayed-type hypersensitivity	Animals are sacrificed	[62, 63]
		Acetic acid/ Compound 48/80-induced vascular permeability	Compound 48/80 is an activator of histamine release and mast cell degranulation	Appropriate model for evaluating the acute-inflammatory effect	Severe irritation	[64]
		Pleurisy model	Exudes, the overall protein composition and the inhibition of leukocyte migration demonstrate the acute anti-inflammatory effect	This test enables the evaluation of the inflammatory phenomenon	Severe pain, animal sacrifices	[65, 66]
2.	Sub-acute inflammation	Granuloma pouch model	Carcinogenic and mutagenic tissues grown Maintain direct contact of a target with the compounds	Test substances into air pouches	Animal sacrifices	[67, 68]
3.	Chronic inflammation	Cotton pellet-induced granuloma	The transudatory and proliferative elements evaluated	Give information about proliferative changes	It needs surgical skills May cause localized sepsis	[64, 69]
		Formalin-induced paw edema	Bradykinin and substance-p mediate the early neurogenic phase, while histamine, 5-HT, prostaglandins, and Bradykinin are implicated in the later inflammatory phase	It resembles human arthritis	Irritating and painful for animals	[64]
		Arthritis induced by CFA	Leukocyte infiltration, cytokine, and chemokine levels elevated with TNF-α and IL-1β destruction of bone and cartilage accompanied by edema and deformity and the release of ROS	The best model for arthritis alteration	It may affect the severity of arthritis, and it needs attention	[70, 71]

TNF, tumor necrosis factor; IL, interleukin; IFN, interferon; NSAID, nonsteroidal anti-inflammatory drugs; LPS, lipopolysaccharide; 5-HT, hydroxytryptamine; CFA, complete Freund's adjuvant.

17.5.4 Lipid Nanoparticles (LNPs) as Anti-inflammatory Agents

LNPs are the most used as a delivery agent for RNA therapeutics and have the ability to trigger an innate immune response and cause inflammation. It has been found that Dexamethasone (Dex) is used as an anti-inflammatory medication which is directly related to cholesterol, which is the main component of LNPs. In a study, it was found that the expression of TNF-α has been reduced in *in vivo* and *in vitro* experiments after the treatment of Dex-incorporated LNPs [75].

17.6 Conclusion and Future Perspectives

In conclusion, IDs place a significant burden on global health, affecting various organs and systems within the body. The complex interaction between immune cells and molecular mediators defines the inflammatory environment, resulting in either acute or chronic responses. The altered small blood vessels, acidosis, and increased oxidative stress contribute to the complexity of these conditions. There are drawbacks and possible adverse effects to the available treatments, which include glucocorticoids and NSAIDs. Although monoclonal antibodies that target proinflammatory cytokines are promising, there are drawbacks as well, including expense, the possibility of serious infections and cancers, and primary or secondary unresponsiveness. The demand for advanced anti-inflammatory medications has led to the exploration of natural products as potential alternatives. Traditional knowledge and resources from indigenous populations play a crucial role in identifying new leads for the treatment of inflammatory conditions. However, the challenge lies in overcoming the toxicity and recurrence of symptoms associated with current synthetic drugs. To achieve effective anti-inflammatory results, it is essential to follow a systematic approach that incorporates the use of various animal models for evaluation. The rational design of new compounds, including polymers inspired by apoptotic cell death, indoyl-imidazole hybrids, aminopiperidinyl amide, and LNPs, offers promising avenues for therapy development. In general, the development of anti-inflammatory treatments mandates a multidisciplinary approach that incorporates insights from pharmacology, natural products, and innovative methods of drug delivery, also the future of anti-inflammatory drug development lies in a multidisciplinary approach that leverages technological advancements, novel therapeutic strategies, and a deep understanding of the complex biology of inflammation. By addressing current limitations and exploring new frontiers, we can develop safer, more effective treatments that significantly improve the quality of life for individuals affected by IDs. The objective is to overcome the limitations of current treatments and provide individuals struggling with IDs with safer and more effective options.

Authors' Contribution

Conceptualization: K.S. and A.G.; writing: original draft preparation, K.S., A.G., I.S., A.S., and U.S.; manuscript review and editing: K.S., A.G., I.S., A.S., and U.S.; Supervision: K.S. and A.G.; All the authors have read and agreed to the published version of the manuscript.

References

1 Dewanjee, S., Dua, T.K., and Sahu, R. (2013). Potential anti-inflammatory effect of *Leea macrophylla* Roxb. leaves: a wild edible plant. *Food and Chemical Toxicology* 59: 514–520.

2 Kulkarni, R., Achaiah, G., and Narahari, S.G. (2006). Novel targets for antiinflammatory and antiarthritic agents. *Current Pharmaceutical Design* 12 (19): 2437–2454.

3 Calixto, J.B., Otuki, M.F., and Santos, A.R.S. (2003). Anti-inflammatory compounds of plant origin. Part I. Action on arachidonic acid pathway, nitric oxide and nuclear factor kB (NF-kB). *Planta Medica* 69 (11): 973–983.

4 Ahamad, N., Kar, A., Mehta, S. et al. (2021). Immunomodulatory nanosystems for treating IDs. *Biomaterials* 274: 120875.

5 Park, H., Bourla, A.B., Kastner, D.L. et al. (2012). Lighting the fires within: the cell biology of autoIDs. *Nature Reviews Immunology* 12 (8): 570–580. Available from: https://www.nature.com/articles/nri3261.

6 Kassebaum, N.J., Arora, M., Barber, R.M. et al. (2016). Global, regional, and national disability-adjusted life-years (DALYs) for 315 diseases and injuries and healthy life expectancy (HALE), 1990–2015: a systematic analysis for the global burden of disease study 2015. *The Lancet* 388 (10053): 1603–1658.

7 Dou, Y., Li, C., Li, L. et al. (2020). Bioresponsive drug delivery systems for the treatment of IDs. *Journal of Controlled Release* 327: 641–666.

8 Taylor, P.C. and Feldmann, M. (2009). Anti-TNF biologic agents: still the therapy of choice for rheumatoid arthritis. *Nature Reviews Rheumatology* 5 (10): 578–582. Available from: https://www.nature.com/articles/nrrheum.2009.181.

9 Tephan, T., Argan, S.R., Tephen, H. et al. (1997). A short-term study of chimeric monoclonal antibody cA2 to tumor necrosis factor α for Crohn's disease. *The New England Journal of Medicine [Internet]* 337 (15): 1029–1036. https://doi.org/10.1056/nejm199710093371502.

10 Kojima, Y., Volkmer, J.P., McKenna, K. et al. (2016). CD47-blocking antibodies restore phagocytosis and prevent atherosclerosis. *Nature* 536 (7614): 86–90. Available from: https://www.nature.com/articles/nature18935.

11 Leckie, M.J., Ten Brinke, A., Khan, J. et al. (2000). Effects of an interleukin-5 blocking monoclonal antibody on eosinophils, airway hyper-responsiveness, and the late asthmatic response. *The Lancet* 356 (9248): 2144–2148.

12 Khilfeh, I., Guyette, E., Watkins, J. et al. (2019). Adherence, persistence, and expenditures for high-cost anti-inflammatory drugs in rheumatoid arthritis: an exploratory study. *Journal of Managed Care & Specialty Pharmacy* 25 (4): 461–467. https://doi.org/10.18553/jmcp.2019.25.4.461.

13 Rock, K.L. and Kono, H. (2008). The inflammatory response to cell death. *Annual Review of Pathology: Mechanisms of Disease* 3: 99–126. https://doi.org/10.1146/annurev.pathmechdis.3.121806.151456.

14 Zhang, J.M. and An, J. (2007). Cytokines, inflammation, and pain. *International Anesthesiology Clinics* 45 (2): 27–37. Available from: https://scholar.google.com/scholar_lookup?title=Cytokines%2C%20inflammation%2C%20and%20pain&author=J.-M.%20Zhang&publication_year=2007&pages=27-37.

15 da Silva, S.B., Ferreira, D., Pintado, M., and Sarmento, B. (2016). Chitosan-based nanoparticles for rosmarinic acid ocular delivery – *in vitro* tests. *International Journal of Biological Macromolecules* 84: 112–120.

16 Shields, C.W., Wang, L.L., Evans, M.A., and Mitragotri, S. (2020). Materials for immunotherapy. *Advanced Materials* 32: 13, 1901633.

17 Ahamad, N., Prabhakar, A., Mehta, S. et al. (2020). Trigger-responsive engineered-nanocarriers and image-guided theranostics for rheumatoid arthritis. *Nanoscale* 12 (24): 12673–12697.

18 Yi, B.G., Park, O.K., Jeong, M.S. et al. (2017). *In vitro* photodynamic effects of scavenger receptor targeted-photoactivatable nanoagents on activated macrophages. *International Journal of Biological Macromolecules* 97: 181–189.

19 Yang, H., Sun, Y., Wei, J. et al. (2019). The effects of ultrasound-targeted microbubble destruction (UTMD) carrying IL-8 monoclonal antibody on the inflammatory responses and stability of atherosclerotic plaques. *Biomedicine & Pharmacotherapy* 118: 109161.

20 Schett, G., Elewaut, D., McInnes, I.B. et al. (2013). How cytokine networks fuel inflammation: toward a cytokine-based disease taxonomy. *Nature Medicine* 19 (7): 822–824.

21 Schwartz, D.M., Kanno, Y., Villarino, A. et al. (2017). JAK inhibition as a therapeutic strategy for immune and IDs. *Nature Reviews Drug Discovery* 16 (12): 843–862. Available from: https://www.nature.com/articles/nrd.2017.201.

22 Corbo, C., Cromer, W.E., Molinaro, R. et al. (2017). Engineered biomimetic nanovesicles show intrinsic anti-inflammatory properties for the treatment of inflammatory bowel diseases. *Nanoscale* 9 (38): 14581–14591.

23 Ziraldo, C., Vodovotz, Y., Namas, R.A. et al. (2013). Central role for MCP-1/CCL2 in injury-induced inflammation revealed by *in vitro*, *in silico*, and clinical studies. *PLoS One* 8 (12): e79804.

24 Leuschner, F., Dutta, P., Gorbatov, R. et al. (2011). Therapeutic siRNA silencing in inflammatory monocytes in mice. *Nature Biotechnology* 29 (11): 1005–1010. Available from: https://www.nature.com/articles/nbt.1989.

25 Ponzoni, M., Pastorino, F., Di Paolo, D. et al. (2018). Targeting macrophages as a potential therapeutic intervention: impact on IDs and cancer. *International Journal of Molecular Sciences* 19 (7): 1953.

26 Fromen, C.A., Kelley, W.J., Fish, M.B. et al. (2017). Neutrophil-particle interactions in blood circulation drive particle clearance and alter neutrophil responses in acute inflammation. *ACS Nano [Internet]* 11 (11): 10797–10807. https://doi.org/10.1021/acsnano.7b03190.

27 Liu, Y.C., Zou, X.B., Chai, Y.F., and Yao, Y.M. (2014). Macrophage polarization in IDs. *International Journal of Biological Sciences* 10 (5): 520–529.

28 Leuschner, F., Dutta, P., Gorbatov, R. et al. (2011). Therapeutic siRNA silencing in inflammatory monocytes in mice. *Nature Biotechnology* 29 (11): 1005–1010. Available from: https://www.nature.com/articles/nbt.1989.

29 Murray, P.J. and Wynn, T.A. (2011). Protective and pathogenic functions of macrophage subsets. *Nature Reviews Immunology* 11 (11): 723–737. Available from: https://www.nature.com/articles/nri3073.

30 Saito, E., Kuo, R., Pearson, R.M. et al. (2019). Designing drug-free biodegradable nanoparticles to modulate inflammatory monocytes and neutrophils for ameliorating inflammation. *Journal of Controlled Release* 300: 185–196.

31 Kwon, D., Cha, B.G., Cho, Y. et al. (2017). Extra-large pore mesoporous silica nanoparticles for directing *in vivo* M2 macrophage polarization by delivering IL-4. *Nano Letters [Internet]* 17 (5): 2747–2756. https://doi.org/10.1021/acs.nanolett.6b04130.

32 Brown, B.N., Ratner, B.D., Goodman, S.B. et al. (2012). Macrophage polarization: an opportunity for improved outcomes in biomaterials and regenerative medicine. *Biomaterials* 33 (15): 3792–3802.

33 Taratummarat, S., Sangphech, N., Vu, C.T.B. et al. (2018). Gold nanoparticles attenuates bacterial sepsis in cecal ligation and puncture mouse model through the induction of M2 macrophage polarization. *BMC Microbiology [Internet]* 18 (1): 1–10. https://doi.org/10.1186/s12866-018-1227-3.

34 Thomas, T.P., Goonewardena, S.N., Majoros, I.J. et al. (2011). Folate-targeted nanoparticles show efficacy in the treatment of inflammatory arthritis. *Arthritis & Rheumatology [Internet]* 63 (9): 2671–2680. https://doi.org/10.1002/art.30459.

35 van der Veeken, J., Gonzalez, A.J., Cho, H. et al. (2016). Memory of inflammation in regulatory T cells. *Cell* 166 (4): 977–990.

36 Chen, Z., Vong, C.T., Gao, C. et al. (2020). Bilirubin nanomedicines for the treatment of reactive oxygen species (ROS)-mediated diseases. *Molecular Pharmaceutics [Internet]* 17 (7): 2260–2274. https://doi.org/10.1021/acs.molpharmaceut.0c00337.

37 Zeng, F., Wu, Y., Li, X. et al. (2018). Custom-made ceria nanoparticles show a neuroprotective effect by modulating phenotypic polarization of the microglia. *Angewandte Chemie* 130 (20): 5910–5914.

38 Gou, S., Huang, Y., Wan, Y. et al. (2019). Multi-bioresponsive silk fibroin-based nanoparticles with on-demand cytoplasmic drug release capacity for CD44-targeted alleviation of ulcerative colitis. *Biomaterials* 212: 39–54.

39 Jung, E., Lee, J., Jeong, L. et al. (2019). Stimulus-activatable echogenic maltodextrin nanoparticles as nanotheranostic agents for peripheral arterial disease. *Biomaterials* 192: 282–291.

40 Jo, W.S., Yang, K.M., Choi, Y.J. et al. (2010). *In vitro* and *in vivo* anti-inflammatory effects of pegmatite. *Molecular & Cellular Toxicology [Internet]* 6 (2): 195–202. https://doi.org/10.1007/s13273-010-0027-0.

41 Dewanjee, S., Dua, T.K., and Sahu, R. (2013). Potential anti-inflammatory effect of Leea macrophylla Roxb. leaves: a wild edible plant. *Food and Chemical Toxicology* 59: 514–520.

42 Gautam, R. and Jachak, S.M. (2009). Recent developments in anti-inflammatory natural products. *Medicinal Research Reviews* 29 (5): 767–820.

43 Gorzalczany, S., López, P., Acevedo, C., and Ferraro, G. (2011). Anti-inflammatory effect of Lithreamolleoides extracts and isolated active compounds. *Journal of Ethnopharmacology* 133 (3): 994–998.

44 Sofidiya, M.O., Imeh, E., Ezeani, C. et al. (2014). Antinociceptive and anti-inflammatory activities of ethanolic extract of *Alafia barteri*. *Revista Brasileira de Farmacognosia* 24 (3): 348–354.

45 Patil, K.R., Mahajan, U.B., Unger, B.S. et al. (2019). Animal models of inflammation for screening of anti-inflammatory drugs: implications for the discovery and development of phytopharmaceuticals. *International Journal of Molecular Sciences* 20 (18): 4367.

46 de Oliveira RG, Mahon CPAN, Ascêncio PGM, Ascêncio SD, Balogun SO, Martins DT de O. Evaluation of anti-inflammatory activity of hydroethanolic extract of *Dilo dendron bipinnatum* Radlk. *Journal of Ethnopharmacology* 2014;155(1): 387–95.

47 Bellik, Y., Boukraâ, L., Alzahrani, H.A. et al. (2013). Molecular mechanism underlying anti-inflammatory and anti-allergic activities of phytochemicals: an update. *Molecules* 18 (1): 322–353. Available from: https://www.mdpi.com/1420-3049/18/1/322/htm.

48 Han, J., Kim, Y.S., Lim, M.Y. et al. (2018). Dual roles of graphene oxide to attenuate inflammation and elicit timely polarization of macrophage phenotypes for cardiac repair. *ACS Nano* 12 (2): 1959–1977.

49 Boominathan, R., Parimaladevi, B., Mandal, S.C., and Ghoshal, S.K. (2004). Anti-inflammatory evaluation of *Ionidium suffruticosam* Ging. in rats. *Journal of Ethnopharmacology* 91 (2–3): 367–370.

50 Sarkhel, S. (2016). Evaluation of the anti-inflammatory activities of *Quillaja saponaria* Mol. saponin extract in mice. *Toxicology Reports* 3: 1–3.

51 Singh, B., Bani, S., Gupta, D.K. et al. (2003). Anti-inflammatory activity of 'TAF' an active fraction from the plant *Barleria prionitis* Linn. *Journal of Ethnopharmacology* 85 (2–3): 187–193.

52 Vasudevan, M., Gunnam, K.K., and Parle, M. (2007). Antinociceptive and anti-inflammatory effects of *Thespesia populnea* bark extract. *Journal of Ethnopharmacology* 109 (2): 264–270.

53 De Araújo Vilar, M.S., De Souza, G.L., De Araújo, V.D. et al. (2016). Assessment of phenolic compounds and anti-inflammatory activity of ethyl acetate phase of Anacardium occidentale L. Bark. *Molecules* 21 (8): 1087. Available from: https://www.mdpi.com/1420-3049/21/8/1087/htm.

54 Coura, C.O., Souza, R.B., Rodrigues, J.A.G. et al. (2015). Mechanisms involved in the anti-inflammatory action of a polysulfated fraction from *Gracilaria cornea* in rats. *PLoS One [Internet]* 10 (3): e0119319. https://doi.org/10.1371/journal.pone.0119319.

55 Babu, N.P., Pandikumar, P., and Ignacimuthu, S. (2009). Anti-inflammatory activity of *Albizia lebbeck* Benth., an ethnomedicinal plant, in acute and chronic animal models of inflammation. *Journal of Ethnopharmacology* 125 (2): 356–360.

56 Calil, I.L., Zarpelon, A.C., Guerrero, A.T.G. et al. (2014). Lipopolysaccharide induces inflammatory hyperalgesia triggering a TLR4/MyD88-dependent cytokine cascade in the mice paw. *PLoS One [Internet]* 9 (3): e90013. https://doi.org/10.1371/journal.pone.0090013.

57 Boller, S., Soldi, C., Marques, M.C.A. et al. (2010). Anti-inflammatory effect of crude extract and isolated compounds from *Baccharis illinita* DC in acute skin inflammation. *Journal of Ethnopharmacology* 130 (2): 262–266.

58 Sadeghi, H., Parishani, M., AkbartabarTouri, M. et al. (2017). Pramipexole reduces inflammation in the experimental animal models of inflammation. *Immunopharmacology and Immunotoxicology [Internet]* 39 (2): 80–86. https://doi.org/10.1080/08923973.2017.1284230.

59 Vignali, D.A.A., Collison, L.W., and Workman, C.J. (2008). How regulatory T cells work. *Nature Reviews Immunology* 8 (7): 523–532. Available from: https://www.nature.com/articles/nri2343.

60 Taratummarat, S., Sangphech, N., Vu, C.T.B. et al. (2018). Gold nanoparticles attenuates bacterial sepsis in cecal ligation and puncture mouse model through the induction of M2 macrophage polarization. *BMC Microbiology* 18 (1): 85.

61 Nonato, F.R., Nogueira, T.M.O., De Almeida Barros, T.A. et al. (2011). Antinociceptive and antiinflammatory activities of *Adiantum latifolium* lam.: evidence for a role of IL-1β inhibition. *Journal of Ethnopharmacology* 136 (3): 518–524.

62 Bralley, E.E., Greenspan, P., Hargrove, J.L. et al. (2008). Topical anti-inflammatory activity of Polygonum cuspidatum extract in the TPA model of mouse ear inflammation. *Journal of Inflammation [Internet]* 5 (1): 1–7. https://doi.org/10.1186/1476-9255-5-1.

63 Bas, E., Recio, M.C., Máñez, S. et al. (2007). New insight into the inhibition of the inflammatory response to experimental delayed-type hypersensitivity reactions in mice by scropolioside A. *European Journal of Pharmacology* 555 (2–3): 199–210.

64 Segawa, S., Takata, Y., Kaneda, H., and Watari, J. (2007). Effects of a hop water extract on the compound 48/80-stimulated vascular permeability in ICR mice and histamine release from OVA-sensitized BALB/c mice. *Bioscience, Biotechnology, and Biochemistry [Internet]* 71 (6): 1577–1581. https://doi.org/10.1271/bbb.70047.

65 Wu, Y., Zhou, C., Song, L. et al. (2006). Effect of total phenolics from *Laggera alata* on acute and chronic inflammation models. *Journal of Ethnopharmacology* 108 (2): 243–250.

66 Rachmawati, H., Safitri, D., Pradana, A., and Adnyana, I. (2016). TPGS-stabilized curcumin nanoparticles exhibit superior effect on carrageenan-induced inflammation in Wistar rat. *Pharmaceutics* 8 (3): 24.

67 Gupta, M., Mazumder, U.K., Sambath Kumar, R. et al. (2005). Anti-inflammatory, analgesic and antipyretic effects of methanol extract from *Bauhinia racemosa* stem bark in animal models. *Journal of Ethnopharmacology* 98 (3): 267–273.

68 Neto, A.G., Costa, J.M.L.C., Belati, C.C. et al. (2005). Analgesic and anti-inflammatory activity of a crude root extract of *Pfaffia glomerata* (Spreng) Pedersen. *Journal of Ethnopharmacology* 96 (1–2): 87–91.

69 Lalrinzuali, K., Vabeiryureilai, M., and Jagetia, G.C. (2016). Investigation of the anti-inflammatory and analgesic activities of ethanol extract of stem bark of Sonapatha *Oroxylum indicum in vivo*. *International Journal of Inflammation* 2016 (1): 8247014.

70 Cascão, R., Vidal, B., Raquel, H. et al. (2014). Potent anti-inflammatory and antiproliferative effects of gambogic acid in a rat model of antigen-induced arthritis. *Mediators of Inflammation* 2014 (1): 195327.

71 Mbiantcha, M., Almas, J., Shabana, S.U. et al. (2017). Anti-arthritic property of crude extracts of *Piptadeniastrum africanum* (Mimosaceae) in complete Freund's adjuvant-induced arthritis in rats. *BMC Complementary and Alternative Medicine [Internet]* 17 (1): 1–16. https://doi.org/10.1186/s12906-017-1623-5.

72 Nakagawa, Y., Saitou, A., Aoyagi, T., and Ebara, M. (2018). Rational design of anti-inflammatory polymers inspired by apoptotic cell death using phosphoramidite chemistry. *Polymer (Guildf)* 134: 85–93.

73 Husain, A., AlAsmari, A.F., Azmi, S.N.H. et al. (2022). Rational drug design, synthesis, and *in vivo* biological activity of new indolyl-imidazolone hybrids as potential and safer non-steroidal anti-inflammatory agents. *Journal of King Saud University, Science* 34 (4): 102023.

74 Oum, Y.H., Kell, S.A., Yoon, Y. et al. (2020). Discovery of novel aminopiperidinyl amide CXCR4 modulators through virtual screening and rational drug design. *European Journal of Medicinal Chemistry* 201: 112479.

75 Zhang, H., Han, X., Alameh, M.G. et al. (2022). Rational design of anti-inflammatory lipid nanoparticles for mRNA delivery. *Journal of Biomedical Materials Research. Part A* 110 (5): 1101–1108.

18

Rational Design of Antibacterial Agents for Multidrug-Resistant Infections

Sathish Kumar Konidala[1], Podila Naresh[1], Risy Namratha Jamullamudi[2], Kamma Harsha Sri[3], Richie Rashmin Bhandare[4,5], and Afzal Basha Shaik[1]

[1] *Department of Pharmaceutical Sciences, School of Biotechnology and Pharmaceutical Sciences, Vignan's Foundation for Science, Technology and Research, Guntur, AP, India*
[2] *Department of Pharmacy, Koneru Lakshmaiah Education Foundation, Vaddeswaram, AP, India*
[3] *Department of Pharmaceutical Sciences, Vignan Pharmacy College, Guntur, AP, India*
[4] *Department of Pharmaceutical Sciences, College of Pharmacy and Health Sciences, Ajman University, Ajman, United Arab Emirates*
[5] *Centre of Medical and Bio-allied Health Sciences Research, Ajman University, Ajman, United Arab Emirates*

18.1 Introduction

Single-celled microbes, known as bacteria, are capable of surviving in various settings, including the human body. A wide range of diseases known as "bacterial infections" are brought on by different bacterial species. Some bacteria can cause illnesses when they enter the body and multiply, although many of them are benign or even helpful [1, 2]. These infections can range in severity from moderate to severe and can affect various body areas. Bacterial infections are conditions, diseases, or ill health caused by bacteria and their toxins, which stand in second place as endogenous causes of death throughout the globe. According to current statistics, bacterial infections may be the cause of one out of every eight fatalities globally [3]. Pathogenic bacteria, which can result in disease, are the main source of bacterial infections. Through various routes, including ingestion, inhalation, cuts, and wounds, these bacteria can enter the body. Person-to-person contact, contaminated food or drink, insect bites, and exposure to polluted surfaces are common ways for the disease to spread. Some infections can be more difficult to treat because bacteria can develop drug resistance [4].

Bacterial infections are of different types based on which part of the host is infected, the transmitted route or medium, etc., for example, pneumonia caused by bacteria, food poisoning by bacteria (gastroenteritis), topical infections, and sexually transmitted infections. These infections can be prevalent through carriers like mosquitoes or ticks, contact with an infected person, either direct or indirect, airborne particles or droplets, contaminated water or food, etc. [5]. Depending on the type of bacteria involved and the area of the body affected, the symptoms of a bacterial infection can change. The most typical signs of bacterial infections include fever, pain or discomfort, fatigue, cough, diarrhea, skin changes, urinary symptoms, and neurological symptoms. Depending on factors like sanitation, access to healthcare, and immunization rates, the incidence of particular bacterial illnesses might differ by region [6]. Some bacterial infections, like the common cold or

Computational Methods for Rational Drug Design, First Edition. Edited by Mithun Rudrapal.

urinary tract infections, are rather typical and frequently not serious. Others can be more difficult to manage and treat, such as tuberculosis or bacterial strains resistant to antibiotics. There are many reasons why new bacterial illnesses can appear, including bacterial mutations, environmental changes, increasing human interaction, and travel. Significant public health issues could be caused by these newly developing infections. The rise of antibiotic-resistant bacteria, also known as "superbugs," has become a significant concern in recent years. Because these bacteria are resistant to many widely used medicines, illnesses are more challenging to treat [7].

New bacterial strains or antibiotic-resistant strains may eventually appear, and bacterial infections have long been a public health problem. Carbapenem-resistant *Enterobacteriaceae* (CRE) is a well-known example of a recently discovered bacterial illness in recent years. Several antibiotics, including carbapenems, which are thought to be the last choice for treating such infections, are ineffective against CRE bacteria. *Clostridium difficile*, or *C. difficile*, is another noteworthy example. It is frequently linked to healthcare environments and produces extremely severe diarrhea. Reports of a rise in the frequency and severity of *C. difficile* infections have surfaced over time [8].

Center for Disease Control and Prevention reported the outbreak of some bacterial infections like *Salmonella* infections (outbreak in 2023) most often in children younger than five years due to the handling of small turtles and backyard poultry birds. Various environmental samples obtained from two of the ice cream house locations were discovered to contain an outbreak strain of *Listeria* (outbreak in 2023) [9]. Consumption of raw milk or its products contaminated with *Brucella* may result in brucellosis (outbreak in 2019). A strain of *Brucella* known as *Brucella melitensis* or *Brucella abortus* is responsible for the majority of brucellosis infections linked to raw milk in people who have gone to nations where these strains are prevalent and have consumed tainted cow, sheep, or goat milk [10].

18.2 Treatment

Antibiotics, which are drugs intended to either kill or stop the growth of bacteria, are commonly used in the treatment of bacterial illnesses. The precise type of bacteria causing the infection and its susceptibility to medications determine which antibiotic is appropriate and how long to treat the infections [11]. The classification of antibacterial agents, commonly referred to as antibiotics, takes into account several factors, such as their chemical composition, mode of action, range of activity, and mode of delivery [12]. The antibacterial agents are categorized here in general as follows, based on chemical structure: β-lactam antibiotics including penicillins, cephalosporins, carbapenems, and monobactams contain a β-lactam ring in their chemical structure. Macrolides include erythromycin, clarithromycin, and azithromycin, which have a macrocyclic lactone ring. Quinolones include ciprofloxacin and levofloxacin, which have a quinolone core structure. Tetracyclines like doxycycline and tetracycline have four linearly fused rings. Aminoglycosides like gentamicin and amikacin are composed of amino-modified sugars [13, 14].

Though a wide range of antibiotics are available and newer antibiotic agents were discovered timely for treating bacterial infections, the outbreak and prevalence of bacterial infections that are already faced or controlled by the health system might be due to the development of antibacterial resistance (ABR) [15], incomplete treatment courses, hospital-acquired infections (HAIs), global travel and trade, poor sanitation and hygiene, immunocompromised populations, emerging infectious diseases, climate change, lack of access to healthcare, inadequate surveillance and reporting, etc. [16, 17].

18.3 Antibacterial Resistance

ABR is one of the leading problems in controlling bacterial infections. The development of the ability of bacteria to withstand exposure to antibiotics intended to either kill them or slow their growth is known as antibacterial resistance or antibiotic resistance, and it is a global public health concern [18–20]. Due to this condition, normal treatments become ineffective, which increases the danger of spreading the virus to others and results in persistent infections. The development of ABR is due to overuse and misuse of antibiotics, incomplete treatment courses, self-medication, use in agriculture, lack of new antibiotics, globalization and travel, poor infection control, environmental factors, and a lack of surveillance and monitoring [21]. In healthcare settings like hospitals, where antimicrobial agent usage is increasing, multidrug resistance (MDR) is frequently been found in recent days. Resistance may arise as a result of the selection pressure that using several medications exerts. The scarcity of appropriate medications makes treating infections brought on by multidrug-resistant organisms more difficult. As a result, combined treatments, multitargeted drugs, or the use of antibiotics as a last resort may be necessary. Resistant infections can be encountered by promoting antibiotic stewardship programs, global cooperation, increasing public awareness about the usage and regulation of antibiotics, and conducting research and development of new antibiotic agents [22, 23].

Therefore, developing a multitargeted medication design is essential for treating infections with MDR. Understanding the process of resistance development is necessary to design and discover the right drug for treating infections that are resistant to treatment [24]. There are a few possible mechanisms shown in Figure 18.1 for the development of resistance in bacteria or microbes against antibiotics, which are classified as follows [25, 26].

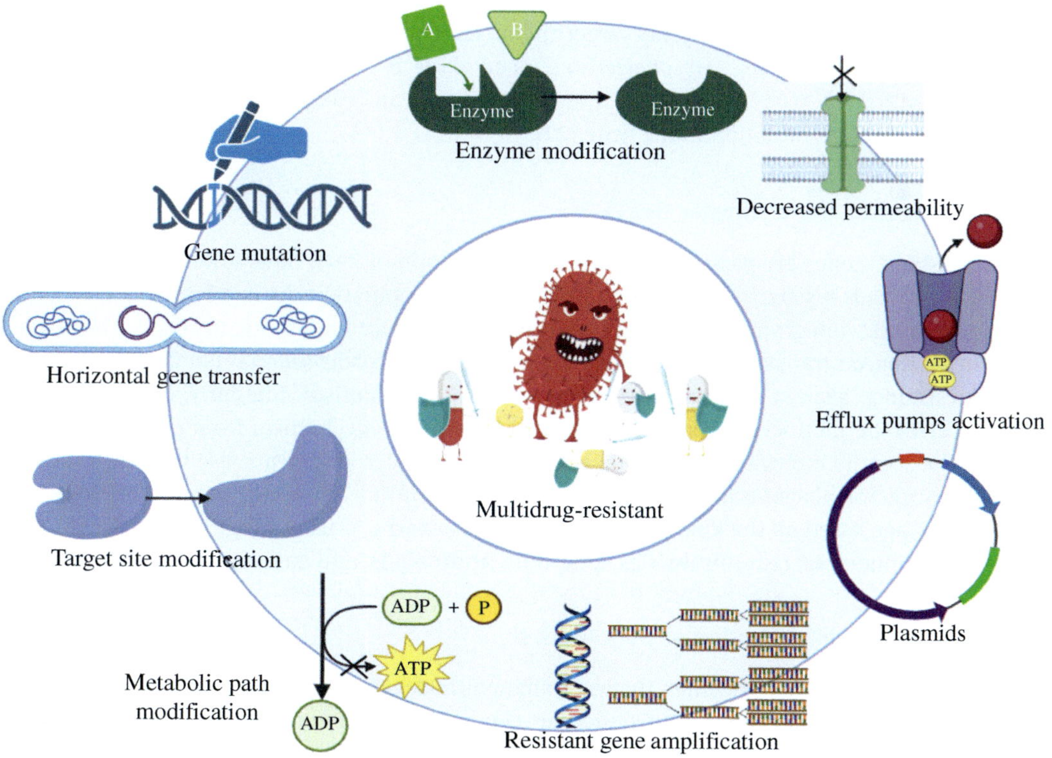

Figure 18.1 Multidrug resistance mechanisms.

I) The genetic basis of antimicrobial resistance
   a) Mutational resistance
   b) Horizontal gene transfer
II) Mechanistic basis of antimicrobial resistance
   a) Modification of the antibiotic molecule
      i) Chemical modification of antibiotics
      ii) Destruction of the antibiotic
   b) Decreased antibiotic penetration and efflux
      i) Decreased permeability
      ii) Efflux pumps
   c) Changes in target sites
      i) Target protection
      ii) Modification of the target site
         - Mutation of the target site
         - Enzymatic alterations of the target site
         - Complete replacement or bypass of the target site
   d) Resistance due to global cell adaptations

The microbial genome serves as the primary source of the genetic basis of antimicrobial resistance, which can develop through various pathways.

18.3.1 Mutation

Unintentional changes in a microbial genome can result in modifications to vital genes, changing the sites at which antimicrobial agents target them. This may lessen the medications' ability to combat the mutant microbes. *Mycobacterium tuberculosis* is one the best examples showing resistance to different types of antibiotics such as rifampicin, streptomycin, ethambutol, and pyrazinamide, through mutation in the gene [27].

18.3.2 Horizontal Gene Transfer (HGT)

Several resistance genes are passed from one species or strain of bacteria to another horizontally. This process spreads resistance genes throughout bacterial populations by involving the exchange of genetic material (integrons, transposons, and plasmids) between microorganisms. For instance, *Enterobacter cloacae* transfer OXA-48 plasmid for other *Enterobacteriaceae* family members to develop resistance against carbapenem through a HGT mechanism. Similarly, *Staphylococcus aureus* receives the methicillin resistance gene mecA by HGT mechanism from other species of bacteria. The specific molecular and cellular strategies that microorganisms use to withstand the effects of antimicrobial agents are the mechanistic basis of antimicrobial resistance. These mechanisms can change based on the kind of microorganism (bacteria, viruses, fungi, and parasites) and the type of antimicrobial (e.g., antibiotics, antivirals, antifungals, and antiparasitics) [28].

18.3.3 Enzymatic Modification or Degradation

Certain microbes generate enzymes that can change or degrade antimicrobial agents, making them less effective. β-Lactamases, for instance, are produced by bacteria and are responsible for the breakdown of β-lactam antibiotics like cephalosporins and penicillins. The interaction between the serine hydroxyl group in the active site of penicillin-binding protein (PBP) and the β-lactams carbonyl group makes the PBP inactive for antibacterial activity. Carbapenemase enzymes produced by *Klebsiella pneumoniae* species can degrade the carbapenems and other β-lactam drugs [29].

18.3.4 Target Site Modification

Microbes can alter the locations where antimicrobial agents bind, making the medications less effective. This modification can be brought about by other mechanisms that change the structure of the drug-binding site or by mutations in the genes that encode the target proteins. *Escherichia coli* with the mcr-1 gene can change its cell wall structure by adding compounds so that the colistin antibiotic cannot bind to the cell wall of *Escherichia coli* having other types of genes. Similarly, triazoles cannot bind to the cyp1A gene of the *Aspergillus fumigatus* due to target site change in this type of gene than the other types [30].

18.3.5 Decreased Permeability

Certain microbes may evolve defense mechanisms to lessen the absorption of antibiotics, preventing the medications from reaching their intended targets. This may entail modifications to the permeability of the cell wall or membrane of the microorganism. Most common species such as *Klebsiella*, *Enterobacter*, *Serratia*, and *Salmonella* can show resistance to antibiotics through reduction of permeability across the cell membrane [31].

18.3.6 Efflux Pumps

Antimicrobial agents can be actively pumped out of bacteria's cells by these pumps, which lowers the drug's intracellular concentration. Increased efflux and resistance can result from upregulating the genes that encode these pumps. *S. aureus* develops intrinsic resistance against most antibiotics through norA, mepA, etc., types of chromosomally encoded efflux pumps [32].

18.3.7 Plasmids

Small, circular DNA molecules called plasmids are distinct from bacterial chromosomes. They are easily transmitted between bacteria and frequently carry genes encoding resistance mechanisms. The spread of resistance is greatly aided by this horizontal transfer of resistance plasmids.

18.3.8 Transposons

DNA sequences known as transposons can travel throughout the genome and between plasmids and chromosomes. Resistance genes may be carried by transposons, which would facilitate the transfer of these genes between bacteria and aid in the spread of resistance.

18.3.9 Gene Amplification

Certain microbes can multiply particular genes, such as those that confer resistance. Higher expression levels of resistance mechanisms may arise from this amplification, increasing the microorganism's resistance to the antimicrobial agent. Resistance against the sulfonamides and trimethoprim by bacterial species might be due to gene amplification [33].

18.3.10 Formation of Biofilms

Communities of microorganisms encased in a matrix known as biofilms can increase resistance to antimicrobials. The genes responsible for the formation of biofilms can tangentially enhance resistance by shielding microbes from the effects of medications.

18.3.11 Modified Metabolic Pathways

The metabolic pathways of certain resistant microbes may differ. By allowing the microbes to redirect resources away from processes that antimicrobials target, these changes may aid in developing resistance.

18.3.12 Adaptive Evolution

When exposed to antibiotics, microorganisms may go through an adaptive evolutionary process. This entails the accumulation of genetic alterations that give rise to a survival benefit when the antimicrobial is present. The ongoing evolution of resistance mechanisms facilitates the persistence of resistant strains.

18.4 Medicinal Chemistry Strategies for the Design of Antibacterials Combating Multidrug-Resistant Bacterial Infections

The design and development of antibacterials to fight bacterial infections resistant to multiple drugs is a difficult task that calls for creative solutions to deal with the changing face of antibiotic resistance.

18.4.1 Structure-Based Drug Design

Making use of a thorough understanding of the bacterial target's structure to create medication that binds more successfully and might get around current resistance mechanisms. To understand the mechanism of resistance of mutated targets, the study of both normal and mutated target structures is essential, which might also be essential and helpful in predicting the better target feature to overcome the resistance through a modeling approach. The topoisomerases are the target for fluoroquinolones to exert potential antibacterial action, but over time, resistance to the fluoroquinolones was developed due to the accumulation of mutants in the gyrA subunit of TopoIV and gyrase. To overcome this resistance problem, scientists developed the compound GSK299423, which will act differently than fluoroquinolones by binding to the active site in between two gyrA subunits [34].

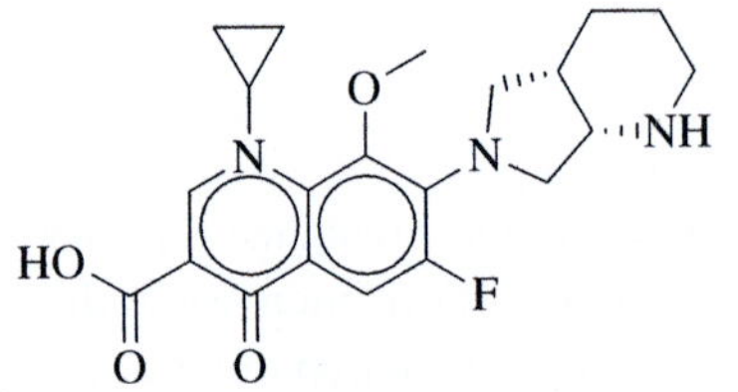

Moxifloxacin

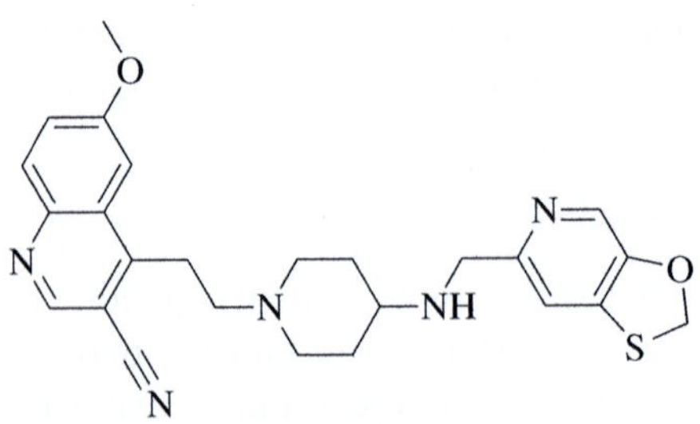

GSK299423

Other research groups have reported the different heterocyclic lead molecules [35–38] to combat the resistant bacterial strains based on the structure-based drug design.

18.4.2 Modification of Existing Antibiotics

Modifying the structure of reported antibiotics improves their activity against resistant strains or prevents the development of resistance mechanisms. This mode of drug design was used in the early stages of antibacterial development like the development of semisynthetic penicillins from penicillin; still, this mode of antibacterial drug development is in use. For instance, the development of doripenem forms a structural modification of imipenem to overcome the resistance to β-lactamase and is prone to elimination by dehydropeptidase-1 [39].

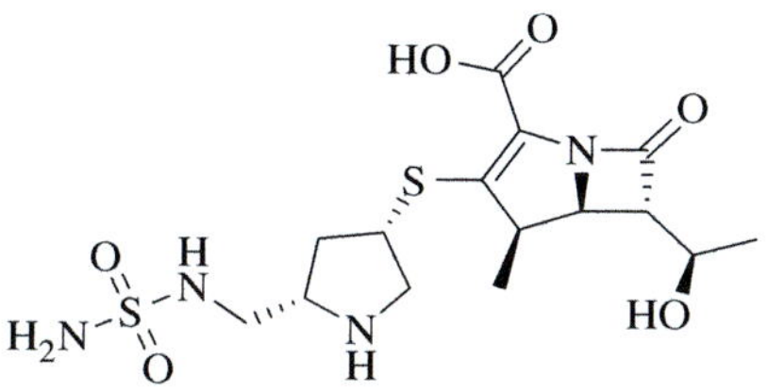
Doripenem

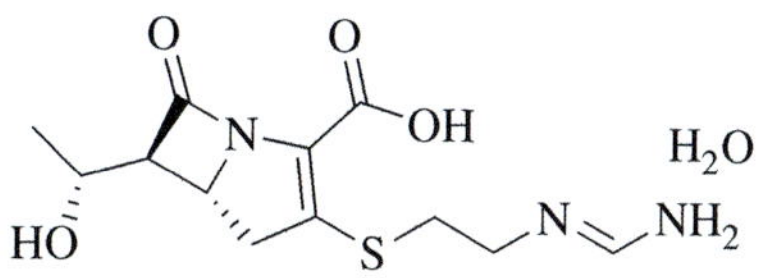
Imipenem

Ridinilazole, a novel medication with a special structure that targets the bacterial cell directly, was created specially to treat infections caused by *Clostridium difficile*. Delafloxacin is a modified chemical structure of fluoroquinolone that increases its potency at varying pH values, making it adaptable to diverse bacterial strains.

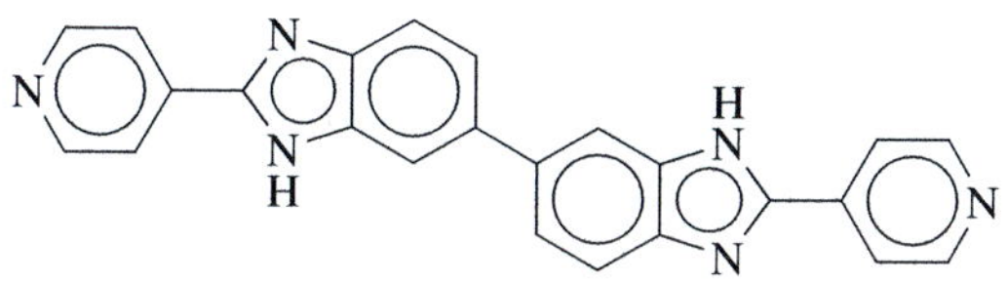
Ridinilazole

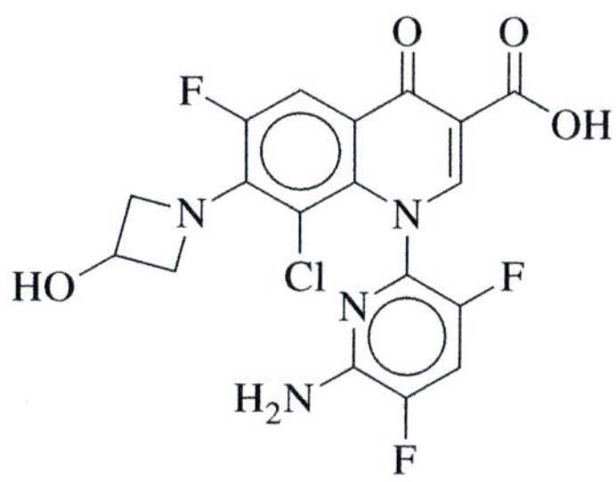
Delafloxacin

18.4.3 Bioisosterism

Enhancing a drug's potency, stability, or bioavailability by substituting structurally identical but chemically distinct groups for certain portions of the molecule is known as bioisosterism. The replacement of the hydroxyl group in phenol with the amine group (resulting in aniline) is a classical isosteric modification that can alter the physicochemical properties, pharmacokinetic profile, and nature and nullify the activity, whereas nonclassic isosteric modification like the replacement of the hydroxyl group in phenol in (±)-Phenylephrine with sulfonamide leads to formation of methanesulfonic acid; though they are different in electronic and steric properties they can augment the biological activity [40].

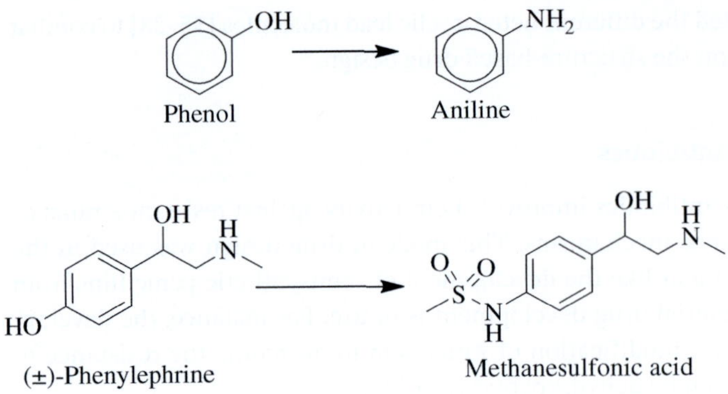

18.4.4 Prodrug Strategies

Prodrugs are the dormant precursors of active medications that become active inside the biological system through biotransformation shown in Figure 18.2, leading to improving selectivity and minimizing negative effects. When designing a prodrug, specific parent drug components can be added or removed to change the drug's permeability, bioavailability, and absorption without changing the drug's pharmacological activity. Three categories exist for prodrugs: (i) prodrugs connected to carriers, in which the active ingredient is connected to a promoiety that is eliminated through a chemical or enzymatic process, thereby releasing the active ingredient; (ii) bioprecursor prodrugs, in which the active drug is molecularly modified and can be released through oxidation or reduction reactions; and (iii) double prodrugs, which are joined by two linkers and can be broken down by distinct mechanisms, such as a codrug, which is a combination of two physiologically active drugs in a single molecule [41].

The poor absorption problem of β-lactams was solved by the prodrug approach, and the bioavailability of ampicillin was increased by ester prodrugs like pivampicillin (the first prodrug),

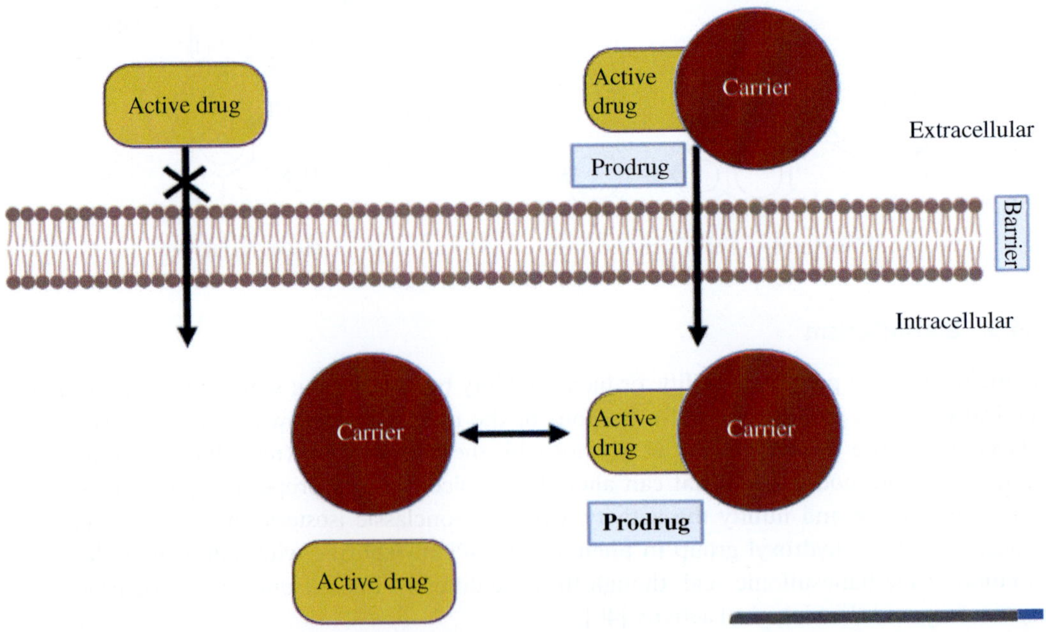

Figure 18.2 Biotransformation of prodrug at the site of action.

talampicillin, hetacillin, and bacampicillin; these prodrugs can be biotransformed to ampicillin by the esterases through enzymatic hydrolysis. A prodrug cephalosporin-fluoroquinolone of ciprofloxacin was developed to overcome the effect of the β-lactamase enzyme on ciprofloxacin and its selective delivery into the ciprofloxacin-resistant bacteria.

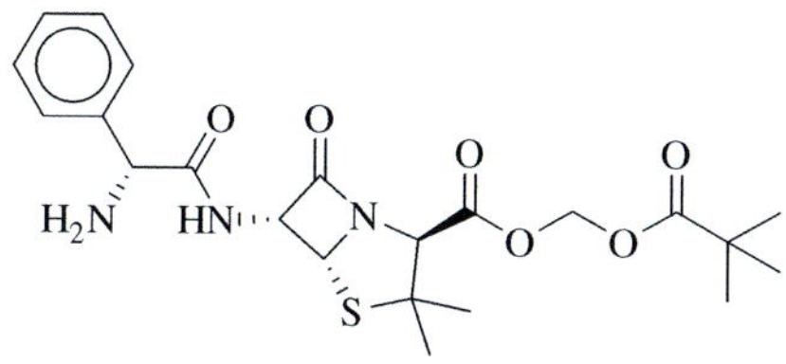

Pivampicillin

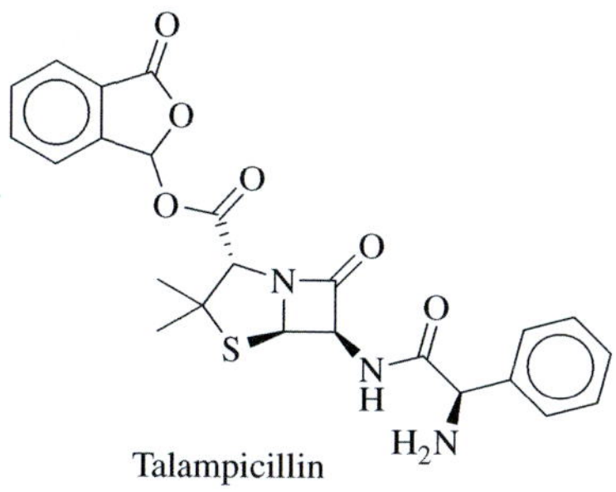

Talampicillin

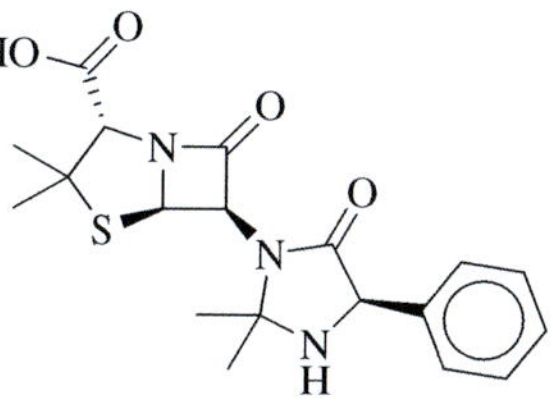

Hetacillin

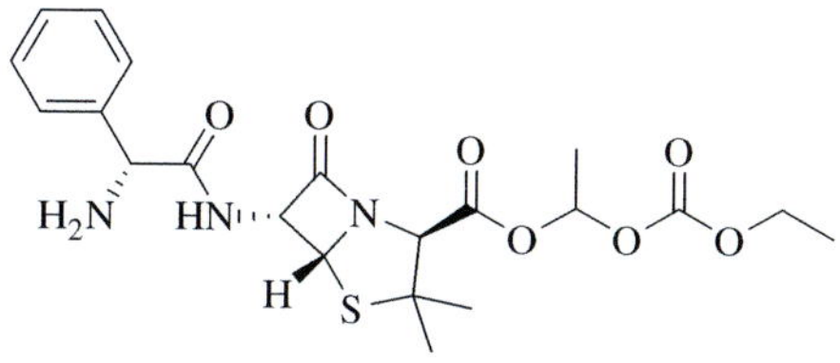

Bacampicillin

18.4.5 Similar Bacterial Components Target

Determining and targeting the components of bacteria that are different from human cells but necessary for their survival is another strategy in novel antibacterial drug design. Targets could include things like bacterial cell walls, proteins, or enzymes that are absent or different in human cells. The drugs used to fight against bacteria by targeting the regular regulatory targets are not responsive since most of the antibacterials are developed with resistance so novel alternatives or similar targets need to be selected to develop the antibacterial drug development. Targets like C55-PP and lipid II, enzymes in the peptidoglycan (PG) biosynthesis, Mur enzymes (GlmS, GlmM, and GlmU), etc., for instance, cyclic lipopeptide Friulimicin B forms a Ca^{2+}-dependent complex with bactoprenol phosphate carrier C55-P, suppress the formation of cell wall precursors [42].

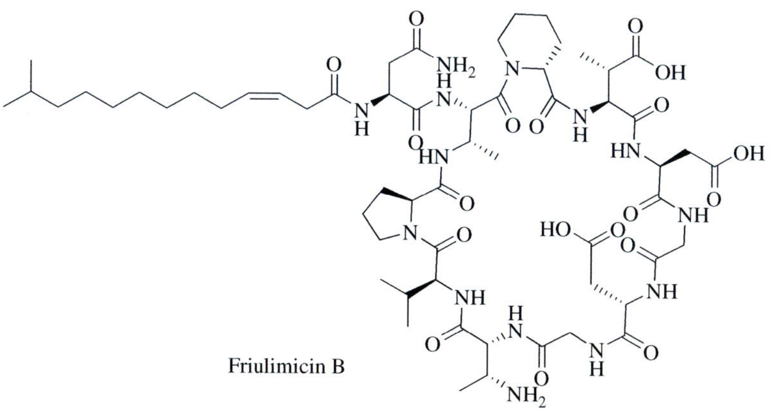

Friulimicin B

18.4.6 Combine or Combination Therapy

Employ combination treatments simultaneously targeting several bacterial pathways. This strategy can improve treatment efficacy and lessen the chance that resistance will develop. Combination therapy of Ceftazidime + Avibactam and Meropenem + Vaborbactam with β-lactamase inhibitors overcomes resistance mechanisms. Imipenem + Cilastatin + Relebactam combined therapy along with β-lactamase inhibitors is designed to increase the range of β-lactam antibiotics [43].

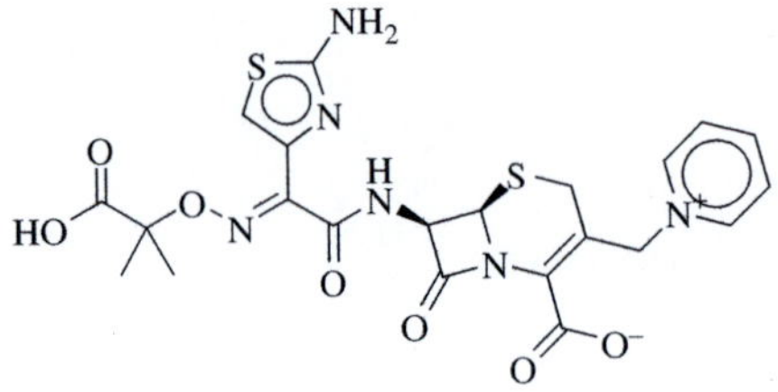

Ceftazidime

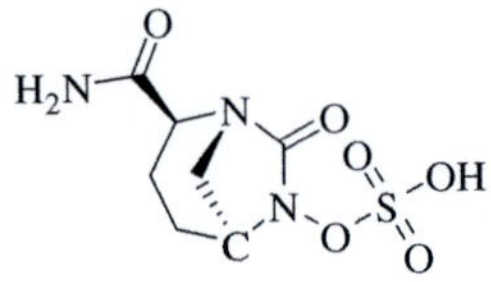

Avibactam

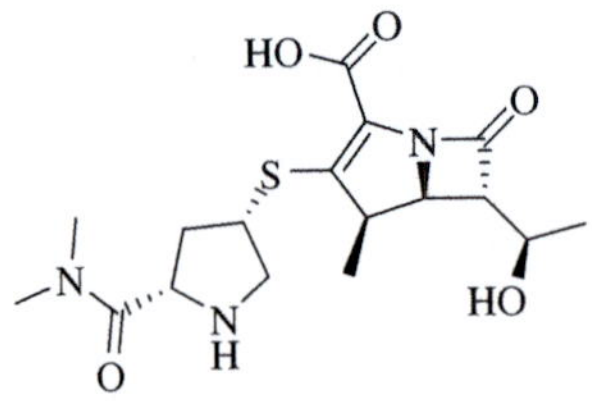

Meropenem

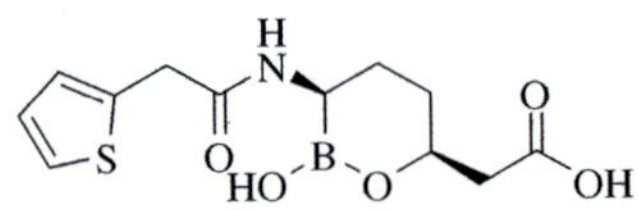

Vaborbactam

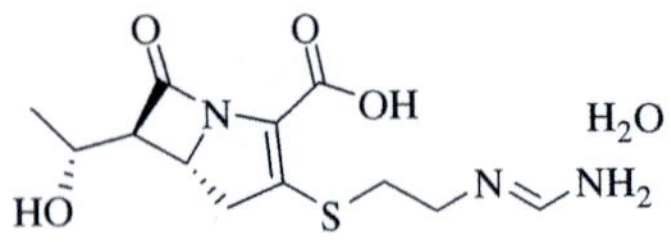

Imipenem

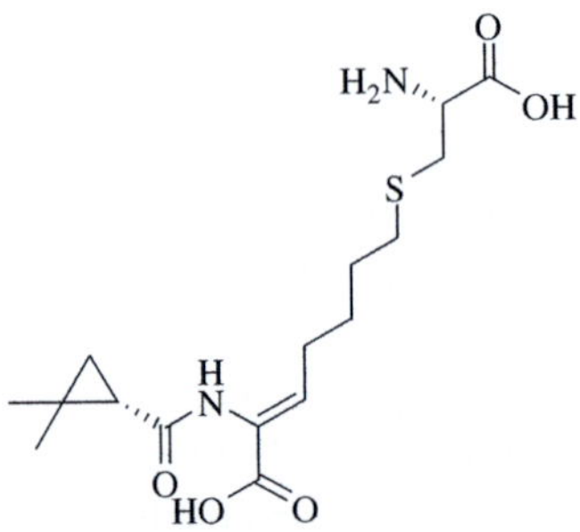

Cilastatin

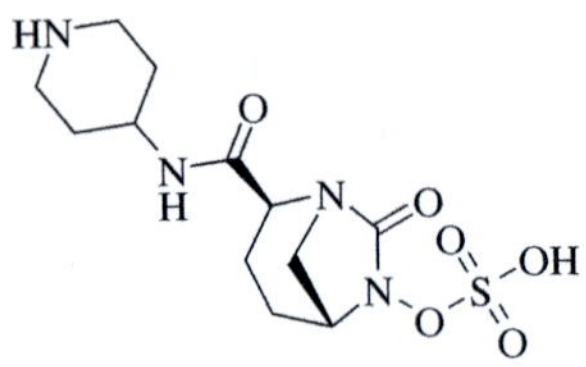

Relebactam

In the treatment against Gram-negative bacterial infections, the combination of outer membrane permeabilises and antibacterial drugs can synergize the treatment efficiency. For instance, pentamidine along with novobiocin, colistin along with rifampicin, and thanatin along with meropenem improve the antibacterial activity by increasing the permeability of antibacterial agents through the Gram-negative bacteria cell membrane by outer membrane permeabilizes.

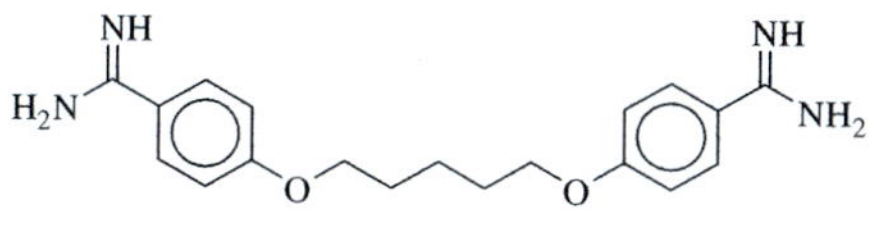

Pentamidine

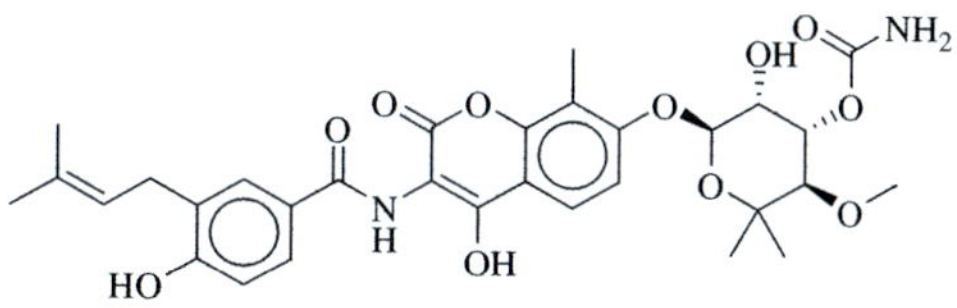

Novobiocin

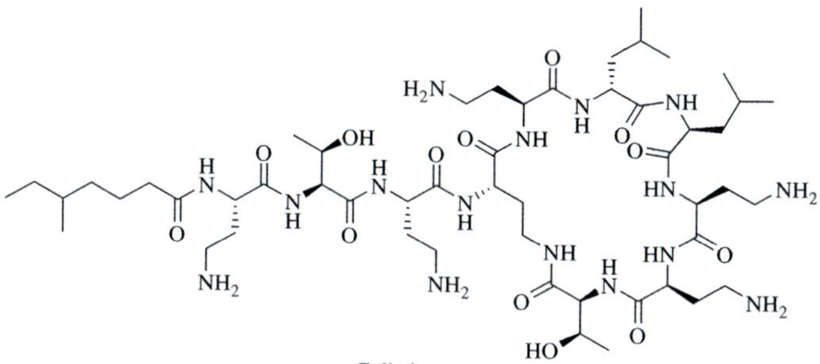

Colistin

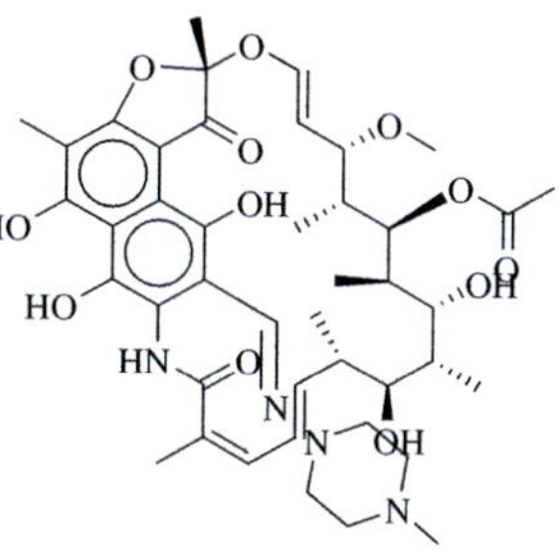

Rifampicin

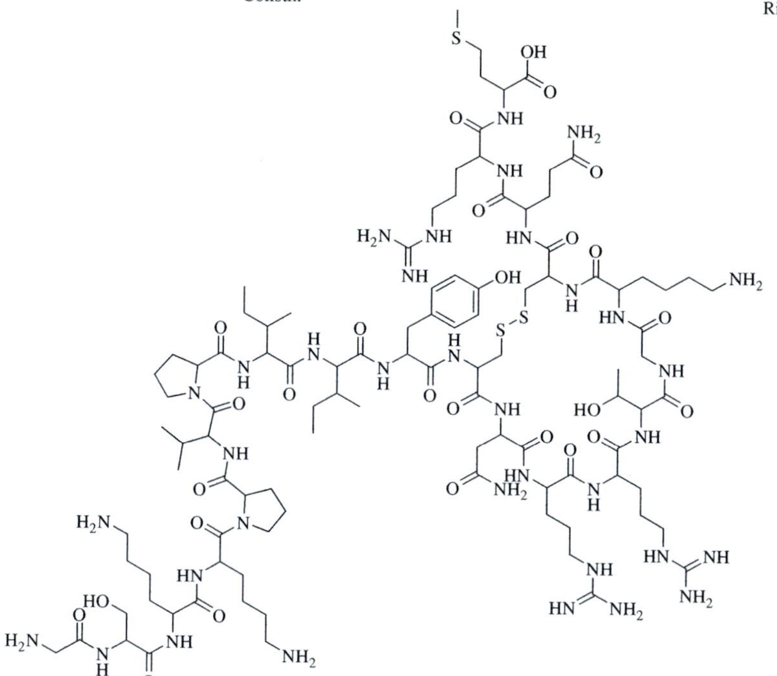

Thanatin

Similarly, the combination therapy of antibacterial drugs along with efflux pump inhibitors can make the antimicrobial agent available for a longer period at the target site to exert its potential action. Loperamide along with minocycline, econazole along with colistin, tobramycin-lysine hybrid along with novobiocin, etc., work efficiently in this way. In recent years, the combination of biofilm damages along with antimicrobial agents is also used in combined therapy.

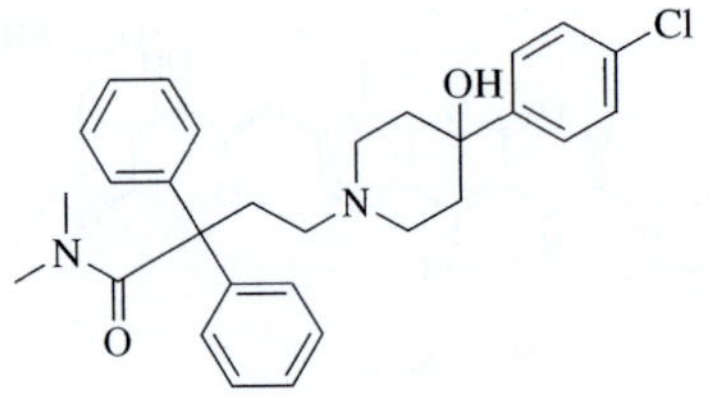

Loperamide

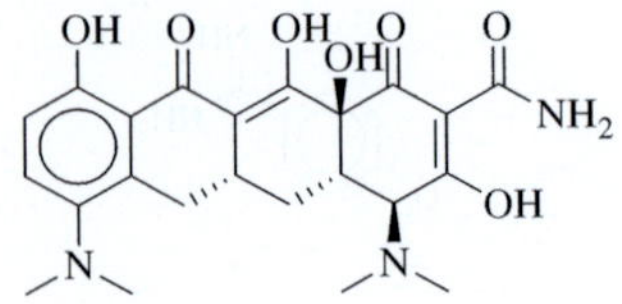

Minocycline

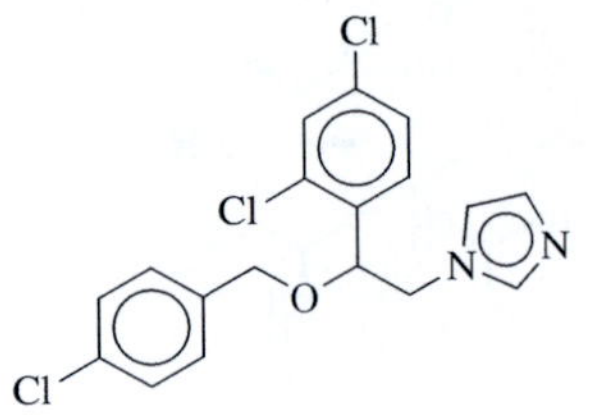

Econazole

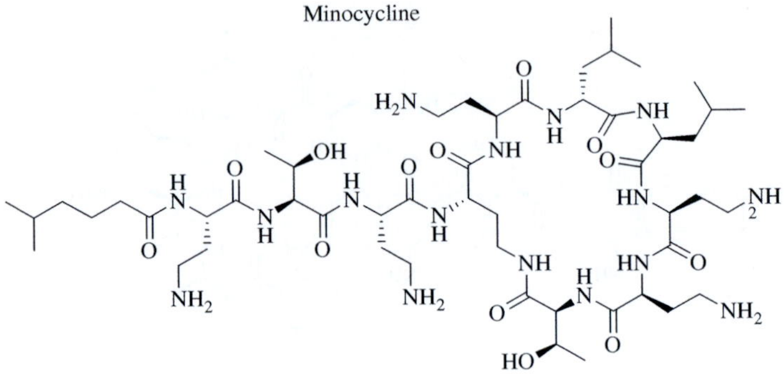

Colistin

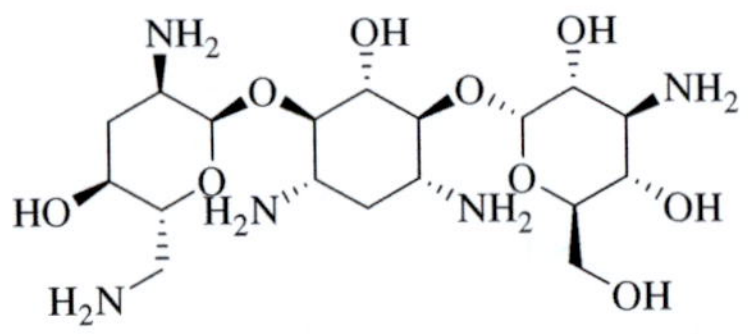

Tobramycin

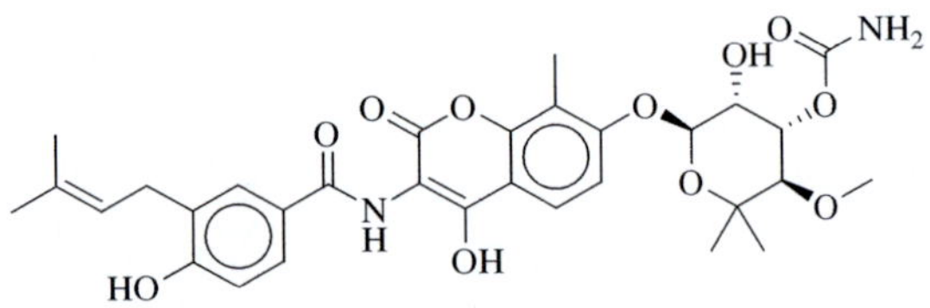

Novobiocin

18.4.7 Drug Repurposing

It includes examining current medications for any novel antibacterial qualities. Finding new applications for medications that were first created for different purposes is known as drug repurposing. This strategy can hasten the process of developing new drugs. The phosphoinositide 3-kinase inhibitors are repurposed to use as adjuvants in the treatment of bacterial infections; repurposing of novel β-lactams and polymyxins along with adjuvants (nonantibiotics) like antibiotic inactivating enzyme inhibitors, membrane permeabilizes, efflux pump inhibitors, and biofilm damages are some of the reported examples of drug-repurposing approaches in MDR antibacterial drug development [44].

18.4.8 Resistant Mechanism Blocking

Create medications that block particular resistance pathways, like efflux pumps, decreased permeability, or the enzymes that render a medication inactive. Combining these inhibitors with existing antibiotics can enhance their effectiveness. Cefiderocol is a siderophore cephalosporin that improves entry into bacterial cells by taking advantage of the iron transport system in bacteria, hence circumventing some resistance mechanisms [45].

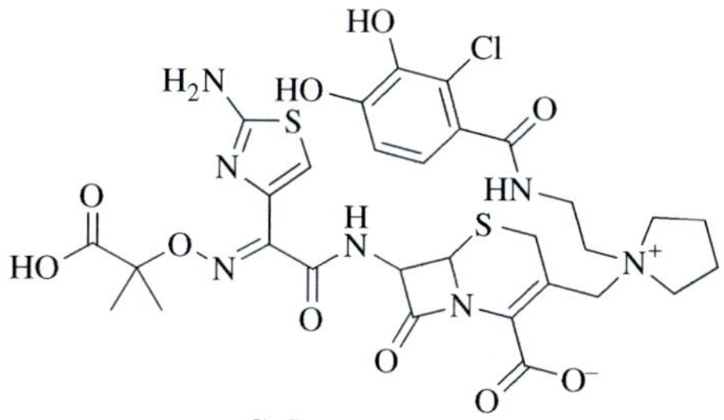

Cefiderocol

18.4.9 Improving Drug Delivery by Nanotechnology

Combining antibiotics with nanomaterial compounds also known as nano-antibiotics is another innovative way to boost the efficacy of antibiotics. For instance, vancomycin nanoformulation has increased antibacterial action against multidrug-resistant pathogens such as vancomycin-resistant *Enterococcus* and methicillin-resistant *Staphylococcus aureus* when coupled with silver nanoparticles. Nanotechnology was employed to reduce adverse effects on host cells. Nanoparticles and nanocarriers can maximize drug stability, bioavailability, and targeted delivery to bacterial cells. Recently, ampicillin silver (AgNPs) and gold nanoparticles (AuNPs) were proven as broad-spectrum antibacterial agents to overcome the resistance against multidrug-resistant strains of bacterial species functionalized with vancomycin and were developed to combat vancomycin-resistant *enterococci*. The antibacterial activity of alginate nanoparticles comprising colistin showed more potent activity against colistin-resistant bacterial strains. Similarly, linezolid bound to silver/gold bimetallic nanoparticles (Au@Ag@Lz) showed significant efficacy even against methicillin-resistant *Staphylococcus aureus* [46].

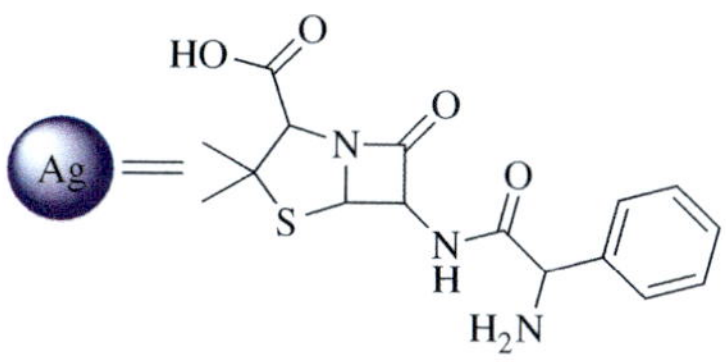

Ampicillin silver nanoparticles

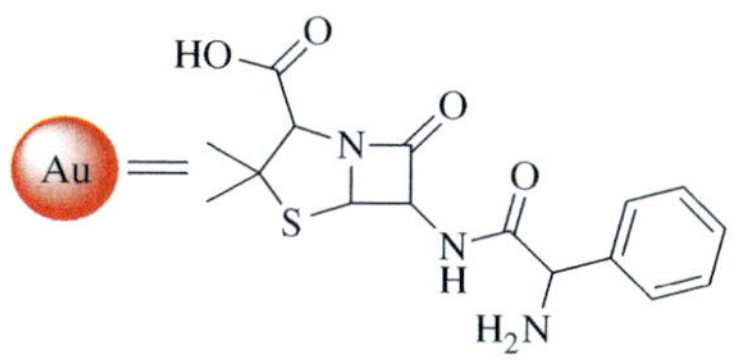

Ampicillin gold nanoparticles

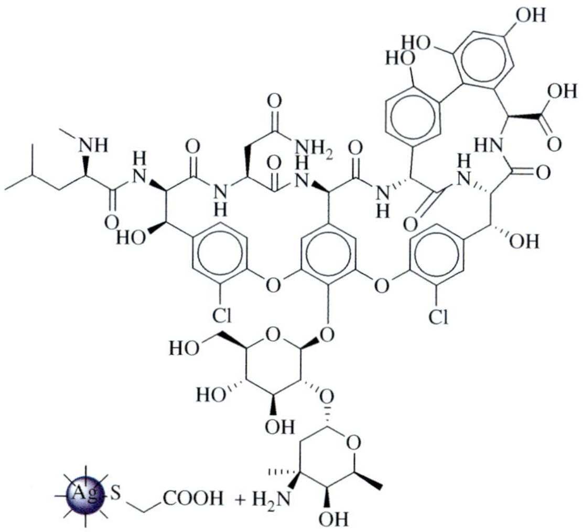

Vancomycin-Ag Nanoparticle

18.4.10 Phage Intervention

Examine the potential applications of bacteriophages and viruses that infect bacteria as a medical strategy. Phage therapy offers a precise and focused method of eliminating bacteria, and phages can change over time to combat resistance in bacteria. The host range of lytic phages is determined by the fact that the majority of them are only infectious to bacteria that also carry their corresponding receptor. Phages differ in their host specificity; some are strain specific, while others can infect various bacterial strains and even species. Bacteria have developed a multitude of defense mechanisms against lytic phage infection, while phages have amassed an equally remarkable array of defense mechanisms against this resistance. This can involve the deletion or mutation of receptors in bacteria as well as the integration of phage DNA into the CRISPR/Cas system, which is a clustered regularly interspaced palindromic repeats system. For phages, this can involve the detection of anti-CRISPR genes and new or altered receptors. For the first time, bacterial phases were used to treat pediatric dysentery in 1919. Phase therapy was continued in the 20th century for treating cholera, dysentery, and bubonic plague through the production of bacteriophage preparations on a commercial scale. Recent works revealed that therapy with phase preparations significantly improved *P. aeruginosa* gut-derived sepsis. In animal models, phage cocktails have also been utilized to treat resistant *P. aeruginosa* infections that affect the skin, lungs, and digestive system [47].

18.4.11 Host Targeting

Bacteria such as *Mycobacterium tuberculosis* and *Coxiella burnetii* can delay or avoid the fusion of their vacuoles with lysosomes and develop a replicate niche that helps to escape from phagosomes thereby entering into the cytosol. Designing potential drugs that specifically target host cell elements like the resolution of signaling pathways, enhancing the autophagy activity, stabilizing and improving the innate immune pathways, adapting the neutralizing mechanism for reactive species, and reducing the host proinflammatory response are the ways through which the bacterial life cycle can be disturbed, and the likelihood of bacterial resistance development is reduced by altering host factors. For instance, when monophosphoryl lipid A, a chemically altered derivative of the lipid A moiety of lipopolysaccharide, was inhaled, the amount of *Moraxella catarrhalis* and *Haemophilus influenzae* bacteria that were collected from the nasopharynx was significantly lower. The *Escherichia coli* infection in the bladder was greatly reduced (10-fold) by administration of an HIF-1α-stabilizing agent (AKB-4924) [48].

18.4.12 CRISPR-Cas Technique

Make use of CRISPR-Cas technology to target and remove specific DNA from bacteria. This method can target particular bacterial strains that are prone to ABR precisely because of their high selectivity. For instance, the viability of *E. coli* was drastically reduced by targeting the virulence factor gene eae, which is essential for the pathology and colonization of these bacteria by CRISPR-Cas technique. Similarly, this technique was employed to potentiate the bactericidal activity of carbapenem against *E. coli* by targeting resistance genes such as blaOXA-48 and bla IMP-1 and colistin against *E. coli* by targeting resistance genes like mcr-1&2 [49].

18.4.13 Peptides as Antibacterials

The application of antimicrobial peptides (AMPs) can damage bacterial membranes or obstruct vital bacterial processes. These peptides might be less likely to develop resistance and exhibit broad-spectrum activity. The first AMP discovered and isolated in 1939 from soil *Bacillus brevis* is gramicidin; it exhibits its antimicrobial action by affecting the strength or disruption of bacterial cell walls through different types of interactions. For instance, the biofilms formed in *K. pneumonia*-resistant drugs like imipenem, meropenem, and cefepime were efficiently encountered by the protease-resistant peptide. Similarly, the peptide and its derivatives isolated from the frog skin secretions showed better therapeutic activity against biofilms of *Enterococcus faecalis* and methicillin-resistant *S. aureus* [50].

18.4.14 Immunizations and Immunotherapy

Make investments in the creation of vaccines to combat bacterial strains resistant to antibiotics. As an addition to traditional antimicrobial treatments, immunotherapies that strengthen the host immune response against bacterial infections may be investigated. For instance, Raxibacumab toxin a monoclonal antibody was proven to be effective in the treatment of anthrax caused by *B. anthraci*, the monoclonal antibody Bexlotoxumab was developed for treating recurrent *Clostridium difficile* infections [51].

18.4.15 Natural Product Derivatives

Since many of the antibiotics currently in use are derived from natural products, exploring and modifying compounds derived from natural sources might be hopeful in this drug discovery. So many natural compounds were reported for antibacterial activity against different bacteria, Piperine is potent against *Lactobacillus* and *Micrococcus*, and anthemic acid is potent against *M. tuberculosis*, *Staphylococcus aureus*, etc. [52].

18.4.16 Fragment-Based Drug Discovery (FBDD)

This method entails determining the interactions that low-molecular-weight molecules (100–300 Da) have with their potential targets, frequently at low affinity (KD ~0.1–1 mM). Beginning with tiny chemical fragments and working your way up to more complex compounds can bind effectively to bacterial targets. The fragments with low molecular weight to high molecular weight biologically active compounds were screened for potential antibacterial activity against bacterial targets like Biotin Carboxylase, DNA Gyrase, Cell Division Protein FtsZ, and β-lactamase [53, 54].

18.4.17 Metabolomics and Genetics

Integrating information from bacterial genetics with metabolomics uncovers new targets and pathways that can be exploited for antibiotic development.

18.4.18 Cheminformatics

The study of chemical data analysis and the prediction of novel compounds' antibacterial activity through the use of algorithms and data analysis tools such as SwissADMET, Drugmint, SwissBioisostere, pkCSM, DataWarrior, Galaxy, BioTransformer, and Knime for determining drug-likeness, and ADMET properties. [55, 56].

18.5 Summary and Conclusion

In summary, bacterial infections, with their various origins, symptoms, and developing treatment approaches, continue to pose a serious threat to global health. Current outbreaks highlight the enduring danger presented by bacterial infections and the pressing need to create efficient defenses against them. Treatment attempts are made more difficult by the emergence of antimicrobial resistance, which calls for a deeper comprehension of the fundamental mechanisms causing resistance.

Antimicrobial resistance has been shown to occur through multiple methods, such as genetic changes, HGT, and biofilm formation. These mechanisms pose significant challenges to conventional antibiotic therapy. Additionally, the emergence of antibiotic resistance makes clinical management of bacterial infections even more difficult, restricting available treatments and raising death rates.

A multidisciplinary strategy that includes creative antibacterial drug creation techniques and coordinated attempts to tackle resistance mechanisms is needed to address these issues. Novel therapeutic targets are necessary, and improvements in combination therapy and drug delivery methods present intriguing paths to overcome resistance.

However, creating antibacterial medications that work against resistant types of bacteria is still a difficult endeavor with many unknowns in science. The intricate relationship that exists between host–pathogen interactions, environmental variables, and bacterial physiology highlights the importance of interdisciplinary collaboration and integrated research efforts.

In light of these factors, a number of approaches, such as target validation, structure-based drug design, and repurposing of already-existing molecules, have enormous potential in the search for next-generation antibacterial medicines. Adopting a comprehensive strategy that combines proteomic, genomic, and computational techniques can spur the creation of novel treatments that can evade resistance mechanisms and protect public health.

In conclusion, the fight against bacterial infections and antibiotic resistance necessitates persistent attention to detail, creative thinking on the part of scientists and cooperative efforts from a wide range of stakeholders. Through the utilization of state-of-the-art technologies and the promotion of an innovative culture, it is possible to reduce the effects of pathogen resistance and maintain the effectiveness of antibacterial treatments for future generations. This chapter explained the bacterial infections: causes, symptoms, symptoms, and recent outbreaks of bacterial infections, elaborated treatment strategies, development of resistance against the antimicrobial agents, i.e., antimicrobial resistance, and mechanisms involved in developing antimicrobial resistance or MDR. This chapter also briefed about the essentiality and challenges in antibacterial drug design against resistant bacteria shown through different mechanisms. Further, it described the various strategies that can be effective in antibacterial drug design against multidrug-resistant strains. The information provided in this chapter might be made knowledgeable and enlighten them with different thought processes. Hence, this information might be helpful in novel antibacterial drug design and development against vulnerable and endangered MDR bacterial infections in the future.

References

1 Drexler, M. (2010). How infection works. In: *What You Need to Know About Infectious Disease*. Washington, DC: National Academies Press (US).

2 Doron, S. and Gorbach, S.L. (2008). Bacterial infections: overview. *International Encyclopedia of Public Health* 2008: 273–282.

3 Mancuso, G., Midiri, A., Gerace, E., and Biondo, C. (2021). Bacterial antibiotic resistance: the most critical pathogens. *Pathogens* 10 (10): 1310.

4 Murray, P.R., Rosenthal, K.S., Pfaller, M.A. et al. (2020). *Medical Microbiology*, 9e. Elsevier.

5 Jawetz, E., Adelberg, E.A., and Brooks, G. F. (Eds.). (2019). *Jawetz, Melnick & Adelberg's Medical Microbiology*, 28e. McGraw-Hill Education.

6 Brooks, G.F., Carroll, K.C., Butel, J.S., and Morse, S.A. (ed.) (2018). *Jawetz, Melnick, & Adelberg's Medical Microbiology*, 27e. McGraw-Hill Education.

7 World Health Organization (WHO) (2020). *Antimicrobial Resistance: Global Report on Surveillance*. Geneva: World Health Organization.

8 Centers for Disease Control and Prevention (CDC) (2021). *Antibiotic Resistance Threats in the United States, 2019*. U.S. Department of Health and Human Services.

9 Centers for Disease Control and Prevention (n.d.). *Outbreaks of Salmonella Infections Linked to Backyard Poultry*. Atlanta, GA: Centers for Disease Control and Prevention https://www.cdc.gov/salmonella/backyardpoultry-06-22/index.html (accessed 11/02/2024).

10 Islam, M.S., Islam, M.A., Rahman, M.M. et al. (2023). Presence of Brucella spp. in milk and dairy products: a comprehensive review and its perspectives. *Journal of Food Quality* 2023.

11 Hutchings, M.I., Truman, A.W., and Wilkinson, B. (2019). Antibiotics: past, present and future. *Current Opinion in Microbiology* 51: 72–80.

12 Leekha, S., Terrell, C.L., and Edson, R.S. (2011). General principles of antimicrobial therapy. *Mayo Clinic Proceedings* 86 (2): 156–167.

13 Pancu, D.F., Scurtu, A., Macasoi, I.G. et al. (2021). Antibiotics: conventional therapy and natural compounds with antibacterial activity – a pharmaco-toxicological screening. *Antibiotics* 10 (4): 401.

14 Bush, K. and Bradford, P.A. (2016). β-Lactams and β-lactamase inhibitors: an overview. *Cold Spring Harbor Perspectives in Medicine* 6 (8).

15 Aslam, B., Wang, W., Arshad, M.I. et al. (2018). Antibiotic resistance: a rundown of a global crisis. *Infection and Drug Resistance* 1645–1658.

16 Monegro AF, Muppidi V, Regunath H. *Hospital Acquired Infections*. https://www.ncbi.nlm.nih.gov/books/NBK441857/ (accessed 11/02/2024).

17 Alamer, A., Alharbi, F., Aldhilan, A. et al. (2022). Healthcare-associated infections (HAIs): challenges and measures taken by the radiology department to control infection transmission. *Vaccine* 10 (12): 2060.

18 Prestinaci, F., Pezzotti, P., and Pantosti, A. (2015). Antimicrobial resistance: a global multifaceted phenomenon. *Pathogens and Global Health* 109 (7): 309–318.

19 Antimicrobial Resistance Collaborators (2022). Global burden of bacterial antimicrobial resistance in 2019: a systematic analysis. *The Lancet* 399 (10325): P629–P655. https://doi.org/10.1016/S0140-6736(21)02724-0.

20 *Drug-resistant Infections: A Threat to our Economic Future*. https://www.worldbank.org/en/topic/health/publication/drug-resistant-infections-a-threat-to-our-economic-future (accessed 11/02/2024)

21 Ventola, C.L. (2015). The antibiotic resistance crisis: part 1: causes and threats. *Pharmacy and Therapeutics* 40 (4): 277.

22 Aslam, B., Wang, W., Arshad, M.I. et al. (2018). Antibiotic resistance: a rundown of a global crisis. *Infection and Drug Resistance* 1645–1658.

23 Uddin, T.M., Chakraborty, A.J., Khusro, A. et al. (2021). Antibiotic resistance in microbes: history, mechanisms, therapeutic strategies and future prospects. *Journal of Infection and Public Health* 14 (12): 1750–1766.

24 Suckling, C.J., Hunter, I.S., and Scott, F.J. (2022). Multitargeted anti-infective drugs: resilience to resistance in the antimicrobial resistance era. *Future Drug Discovery* 4 (1): FDD73.

25 Nikaido, H. (2009). Multidrug resistance in bacteria. *Annual Review of Biochemistry* 78: 119–146.

26 Reygaert, W.C. (2018). An overview of the antimicrobial resistance mechanisms of bacteria. *AIMS Microbiology* 4 (3): 482.

27 Woodford, N. and Ellington, M.J. (2007). The emergence of antibiotic resistance by mutation. *Clinical Microbiology and Infection* 13 (1): 5–18.

28 Tao, S., Chen, H., Li, N. et al. (2022). The spread of antibiotic resistance genes in vivo model. *Canadian Journal of Infectious Diseases and Medical Microbiology* 2022.

29 Egorov, A.M., Ulyashova, M.M., and Rubtsova, M.Y. (2018). Bacterial enzymes and antibiotic resistance. *Acta Naturae (англоязычная версия)* 10 (4(39)): 33–48.

30 https://www.cdc.gov/drugresistance/about/how-resistance-happens.html#:~:text=Germs%20change%20the%20antibiotic's%20target,colistin%20cannot%20latch%20onto%20it (accessed 11/02/2024).

31 Van, J.C.N. and Gutmann, L. (1994). Resistance to antibiotics caused by decrease of the permeability in gram-negative bacteria. *Presse Medicale (Paris, France: 1983)* 23 (11): 522–527.

32 Sharma, A., Gupta, V.K., and Pathania, R. (2019). Efflux pump inhibitors for bacterial pathogens: from bench to bedside. *The Indian Journal of Medical Research* 149 (2): 129.

33 Davies, J. and Davies, D. (2010). Origins and evolution of antibiotic resistance. *Microbiology and Molecular Biology Reviews* 74 (3): 417–433.

34 Reeve, S.M., Lombardo, M.N., and Anderson, A.C. (2015). Understanding the structural mechanisms of antibiotic resistance sets the platform for new discovery. *Future Microbiology* 10 (11): 1727–1733.

35 Verma, S.K., Verma, R., Kumar, K.S. et al. (2021). A key review on oxadiazole analogs as potential methicillin-resistant *Staphylococcus aureus* (MRSA) activity: structure–activity relationship studies. *European Journal of Medicinal Chemistry* 219: 113442.

36 Lagu, S.B., Yejella, R.P., Bhandare, R.R., and Shaik, A.B. (2020). Design, synthesis, and antibacterial and antifungal activities of novel trifluoromethyl and trifluoromethoxy substituted chalcone derivatives. *Pharmaceuticals* 13 (11): 375.

37 Konidala, S.K., Kotra, V., Danduga, R.C.S.R. et al. (2021). Design, multistep synthesis and in-vitro antimicrobial and antioxidant screening of coumarin clubbed chalcone hybrids through molecular hybridization approach. *Arabian Journal of Chemistry* 14 (6): 103154.

38 Bhandare, R.R., Munikrishnappa, C.S., Kumar, G.S. et al. (2022). Multistep synthesis and screening of heterocyclic tetrads containing furan, pyrazoline, thiazole and triazole (or oxadiazole) as antimicrobial and anticancer agents. *Journal of Saudi Chemical Society* 26 (3): 101447.

39 Hutchings, M.I., Truman, A.W., and Wilkinson, B. (2019). Antibiotics: past, present and future. *Current Opinion in Microbiology* 51: 72–80.

40 Oladipupo, A.R. (2020). Toxin to medicine and bioisosterism in drug development: a study of the discovery and development of ACE inhibitors from snake venom. *Macedonian Pharmaceutical Bulletin* 66 (2): 15–33.

41 Jubeh, B., Breijyeh, Z., and Karaman, R. (2020). Antibacterial prodrugs to overcome bacterial resistance. *Molecules* 25 (7): 1543.

42 Belete, T.M. (2019). Novel targets to develop new antibacterial agents and novel alternatives to antibacterial agents. *Human Microbiome Journal* 11: 100052.

43 Si, Z., Pethe, K., and Chan-Park, M.B. (2023). Chemical basis of combination therapy to combat antibiotic resistance. *JACS Au* 3 (2): 276–292.

44 Gurajala, S. (2023). Antimicrobial drug repurposing. *Journal of Population Therapeutics and Clinical Pharmacology* 30 (17): 14–39.

45 Laws, M., Shaaban, A., and Rahman, K.M. (2019). Antibiotic resistance breakers: current approaches and future directions. *FEMS Microbiology Reviews* 43 (5): 490–516.

46 Hetta, H.F., Ramadan, Y.N., Al-Harbi, A.I. et al. (2023). Nanotechnology as a promising approach to combat multidrug-resistant bacteria: a comprehensive review and future perspectives. *Biomedicine* 11 (2): 413.

47 Lin, D.M., Koskella, B., and Lin, H.C. (2017). Phage therapy: an alternative to antibiotics in the age of multi-drug resistance. *World Journal of Gastrointestinal Pharmacology and Therapeutics* 8 (3): 162.

48 Chiang, C.Y., Uzoma, I., Moore, R.T. et al. (2018). Mitigating the impact of antibacterial drug resistance through host-directed therapies: current progress, outlook, and challenges. *mBio* 9 (1): 10–128.

49 Kundar, R. and Gokarn, K. (2022). CRISPR-Cas system: a tool to eliminate drug-resistant gram-negative bacteria. *Pharmaceuticals* 15 (12): 1498.

50 Mba, I.E. and Nweze, E.I. (2022). Focus: antimicrobial resistance: antimicrobial peptides therapy: an emerging alternative for treating drug-resistant bacteria. *The Yale Journal of Biology and Medicine* 95 (4): 445.

51 Seixas, A.M., Sousa, S.A., and Leitão, J.H. (2022). Antibody-based immunotherapies as a tool for tackling multidrug-resistant bacterial infections. *Vaccine* 10 (11): 1789.

52 Qadri, H., Shah, A.H., Ahmad, S.M. et al. (2022). Natural products and their semi-synthetic derivatives against antimicrobial-resistant human pathogenic bacteria and fungi. *Saudi Journal of Biological Sciences* 29 (9): 103376.

53 Konaklieva, M.I. and Plotkin, B.J. (2023). Fragment-based lead discovery strategies in antimicrobial drug discovery. *Antibiotics* 12 (2): 315.

54 Lamoree, B. and Hubbard, R.E. (2018). Using fragment-based approaches to discover new antibiotics. *SLAS Discovery: Advancing Life Sciences R&D* 23 (6): 495–510.

55 Oselusi, S.O., Christoffels, A., and Egieyeh, S.A. (2021). Cheminformatic characterization of natural antimicrobial products for the development of new lead compounds. *Molecules* 26 (13): 3970.

56 Bakchi, B., Krishna, A.D., Sreecharan, E. et al. (2022). An overview on applications of SwissADME web tool in the design and development of anticancer, antitubercular and antimicrobial agents: a medicinal chemist's perspective. *Journal of Molecular Structure* 1259: 132712.

19

Rational Design of Antiviral Therapeutics

Sneha Dokhale[1], Samiksha Garse[2], Shine Devarajan[2], Vaishnavi Thakur[2], and Shaunak Kolhapure[2]

[1] *Department of Biotechnology, B. K. Birla College of Arts, Science & Commerce, Kalyan, Maharashtra, India*
[2] *School of Biotechnology and Bioinformatics, D Y Patil Deemed to be University, Navi Mumbai, Maharashtra, India*

19.1 Introduction to Antiviral Therapeutics

19.1.1 Overview

Viral infections are a significant category of diseases caused by viruses, which are tiny, infectious agents composed of genetic material (either DNA or RNA) surrounded by a protein coat [1]. Viruses do not possess cellular structures and do not have metabolic processes, which make them intracellular parasites, unlike bacteria. The world of antiviral therapeutics is closely connected to the building blocks of life we know as biomolecules. These majorly include carbohydrates, lipids, nucleic acids, and proteins, which play a crucial role in our defense against viruses [2]. An efficient way to fight viruses is by designing novel target-specific bioactive compounds that interact with these biomolecules to alter their function in favoring human health. The major focus for designing these compounds is on DNA and RNA [1, 2], as these nucleic acids are like instruction manuals for the cells, including the ones that viruses use. By changing how these instructions are read, it is possible to disrupt the viral processes, inhibiting their activity and probably eradicating them from the host cell [3]. Similarly, proteins that replicate viruses can also be targeted by designing compounds that inhibit the viral spread. Moreover, Pillay et al. [4] have summarized key points for drug resistance, approximately 25 years later drug resistance remains a major challenge for researchers and clinicians across the globe [4, 5].

However, the rational design of bioactive agents for antiviral activity represents a critical aspect in the development of effective antiviral drugs and treatments. The approach leverages the prior knowledge of the viral target and its mechanism to computationally screen and develop multigenerational drug-combating viruses [5, 6]. Advances in various scientific disciplines, including organic physical chemistry, theoretical chemistry, computational chemistry, and pharmacochemistry, have significantly contributed to the implementation of sophisticated algorithms aimed at predicting the interactions between organic ligands and their biological receptors [5]. Computer-aided drug design (CADD) and virtual screening (VS) have gained prominence in designing antivirals. These computational approaches, particularly molecular docking techniques employed by software like Autodock and Rosetta Ligand (details are mentioned

Computational Methods for Rational Drug Design, First Edition. Edited by Mithun Rudrapal.

in Section 19.3.1 [6], play a fundamental role in identifying optimal binding modes of candidate molecules within the binding cavities of viral targets [6]. These methods enable researchers to assess the potential efficacy of various compounds and prioritize them with their favorable binding interactions. The bioinformatics tools empower researchers in the rational design of novel antiviral agents based on the three-dimensional structure of drug targets. In addition, it allows for the creation of molecular structures with enhanced affinity and specificity toward viral targets, optimizing the potential antiviral activity of the designed molecules. Structural biology information provides an invaluable understanding of the precise interactions between bioactive molecules and viral targets by resolving X-ray crystallography structures of target–ligand complexes [7]. These structural insights prove essential for the development of potent antiviral drugs. Moreover, the integration of CADD with artificial intelligence, machine learning, and systems biology (SB) approaches has proven fruitful in the development of antiviral therapeutics with a personalized touch.

19.1.2 Blueprints for Antiviral Drug Interventions

Structural informatics is the foundation of antiviral research, providing a glimpse into the intricate architecture and dynamics of viral biomolecules. Through advanced techniques like X-ray crystallography, cryo-electron microscopy (cryo-EM), and nuclear magnetic resonance (NMR) spectroscopy, structural biology reveals the atomic complexity of these biomolecules [8]. This data is instrumental in identifying precise viral targets for antiviral drug development. It unveils the spatial arrangement of viral proteins, nucleic acids, and lipids, shedding light on their active sites and binding pockets [9]. Researchers can then rationally design bioactive molecules to precisely interact with these viral targets. This enables the creation of antiviral compounds that disrupt essential viral functions with exceptional specificity and continuously optimize bioactive molecules to enhance their potency and reduce side effects [10]. The data also aids in predicting potential drug resistance mutations by highlighting regions of viral biomolecules that are less amenable to binding [11]. This guides the development of drugs resistant to viral mutation. Ultimately, the utilization of structural data in antiviral research is paramount, as it empowers scientists in the pursuit of effective antiviral therapeutics, bringing us one step closer to combating viral infections on a global scale [4, 5]. Figure 19.1 illustrates the structural blueprint with a generic description of each component's structural function with its inhibition actions on viral assembly.

Apart from these, there are some additional factors that are taken into consideration for drug intervention as follows.

19.1.2.1 Protein Folding and Binding Sites

Viral proteins have unique three-dimensional structures that are crucial for their functions. These structures are determined by the protein sequence and are often stabilized by hydrogen bonds, disulfide bridges, and hydrophobic interactions [12]. Understanding protein folding is important because it provides researchers with insights into potential drug-binding sites and regions vulnerable to disruption [7, 13]. Moreover, many antiviral drugs work by binding to specific sites on viral proteins. These sites can be active sites where chemical reactions occur or allosteric sites that modulate protein function [13, 14]. In some cases, drugs mimic the natural substrates of viral enzymes and compete for binding, inhibiting the enzymatic activity [13].

19.1.2.2 Conformational Changes

Enveloped viruses often rely on conformational changes within the fusion proteins that are located on their envelopes. These changes fuse the viral envelope with the host's membrane, wherein the genetic material reaches out into the host's cytoplasm [15]. Many viruses have protective protein

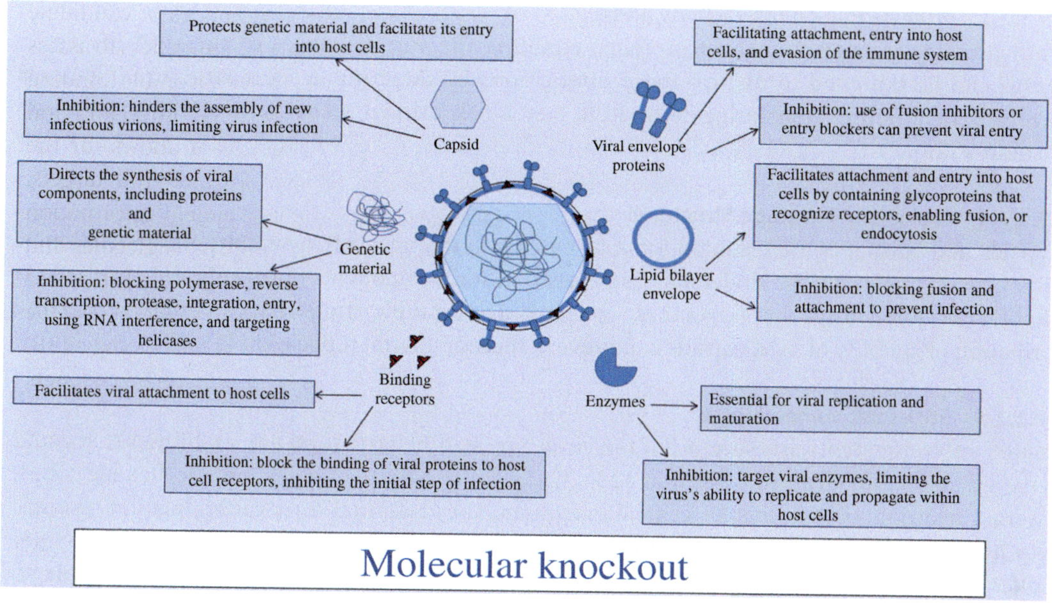

Figure 19.1 Molecular knockouts of viruses.

coats or capsids surrounding their genetic material. During the assembly of new viral particles, conformational changes in viral structural proteins are required to facilitate the packaging of the viral genetic material into newly formed capsids [16]. This step is crucial for the production of infectious virions. The alterations of surface proteins pose as a viral strategy to escape recognition by antibodies or immune cells, contributing to their ability to persist in the host [17]. Similarly, these changes play a major role in viral enzymes (e.g., polymerases) [18]; they enable the replication of viral genetic material or transcribe it into mRNA [19].

Some antiviral drugs can bring about conformational change in the structure of viral proteins. The changes induced by antiviral drugs, in the virus family Picornaviridae, more specifically poliovirus [20] underline mechanisms that involve the interactions of the ligand-induced conformational drugs and the viral capsid proteins. Crystal structures of drug complexes, particularly with the compounds R80633 and R77975 [21], have been studied to assess the structural alterations and understand their impact on viral infection [22]. The unique structure of the poliovirus capsid, with its characteristic pocket and surface features, makes it susceptible to interactions with specific antiviral drugs. The drugs either slightly exceed or fall short of the optimal size for the binding pocket. This results in imperfect packing at both ends of the pocket, influencing the interaction between capsid proteins [21, 23]. The recognized amino acids (Met1132, Leu1134, and Leu1261) in the drug-binding pocket are critical for determining the efficiency of drug binding [23]. The mechanical responses and structural shifts observed near the fivefold axis provide insights into how the poliovirus family responds to antiviral drug-induced stress. These antiviral compounds, which prevent viral uncoating by impeding conformational changes required for entering the host, have been scrutinized in complex with the Mahoney strain of type 1 poliovirus [21, 24]. Notably, the Mahoney strain displays relative insensitivity toward the antivirals, demanding substantial spatial changes to allow drug binding. These alterations not only affect the nearby binding site of the drug but also farther off the loops that lie within the viral particle's fivefold axis. The shifts correlated with drug effectiveness provide insights into how antiviral drugs interfere with the normal conformational repertoire of the virus, ultimately inhibiting productive cell entry [24, 25].

19.1.2.3 Protein–Protein Interactions (PPIs)

Viral proteins interact with host proteins, enabling the various stages of the viral life cycle. Drugs can be designed to disrupt these interactions by targeting specific structural elements involved in binding. Interfering with PPIs can block critical steps in viral replication and assembly [26].

19.1.2.4 Capsid and Envelope Structures

Capsids and envelopes are essential for viral entry into the host cell, assembly, and protection of viral genetic material [22]. Understanding the structural components of capsids and envelopes is vital for designing drugs that target these elements. For example, drugs that can interfere with the formation or stability of viral capsids can prevent successful viral replication [15, 27].

19.1.2.5 Structural Vulnerabilities

Some viral components, irrespective of the virus type, exhibit structural vulnerabilities that could serve as targets for drug intervention [28]. These vulnerabilities might include flexible loops, exposed hydrophobic regions, or critical enzymatic sites essential for viral replication or entry into host cells. For example, in addition to the fusion peptide (FP) domain [29] in the SARS-CoV-2 spike (S) protein, other viruses possess analogous regions or domains susceptible to inhibition.

In general, these structural vulnerabilities represent potential sites for small-molecule intervention. By targeting these regions with specific inhibitors, such as chlorcyclizine (CCZ) [28, 30] or other compounds, it becomes feasible to disrupt viral replication, entry, or other essential processes crucial for viral survival. Importantly, these vulnerabilities often differ between viruses or even among strains of the same virus, offering a unique opportunity for the development of selective drug therapies tailored to combat specific viral infections [28]. HIV-1 Env (envelope protein), similar to other viral surface glycoproteins, exhibits a higher degree of varied sequence and glycosylation that facilitates evade strategies against host immune responses. However, recent advances in structural characterization have revealed vulnerable sites on the HIV-1 envelope, responsible for receptor binding and fusion machinery [16]. These vulnerabilities have been targeted broadly by neutralizing antibodies (nAbs) [31], leading to the development of potential strategies for vaccine design or the creation of antiviral therapeutics.

19.1.2.6 Enzymatic Activities

Viral enzymes are essential components of the intricate machinery employed by viruses to infect host cells, replicate their genetic material, and evade host defenses. Among these enzymes, viral proteases, polymerases, and deubiquitinases (DUBs) play pivotal roles in viral pathogenesis [32]. Viral proteases, exemplified by the M48USP, a protease found in cytomegalovirus (CMV) [33], exhibit unique catalytic triads enabling them to cleave specific host and viral proteins, further facilitating viral replication and evasion of host immune responses [32, 33]. Similarly, DUBs like BPLF1 in Epstein–Barr virus (EBV) and UL36USP in human cytomegalovirus (HCMV) exert a multifunctional role in viral replication and immune evasion by modulating the ubiquitination status of proteins and interfering with host innate immune pathways [32]. Structural studies of these viral enzymes have provided valuable insights into their catalytic mechanisms and substrate specificities, facilitating the development of antiviral drugs targeting their activities [12]. Small-molecule inhibitors and engineered viral mutants deficient in enzyme activity represent promising strategies for combating viral infections by disrupting essential enzymatic functions crucial for viral replication and immune evasion [23, 34].

19.1.2.7 Viral Attachment

Many viruses attach to host cell receptors through specific interactions between viral proteins and receptor molecules. Enveloped viruses often employ fusion proteins to facilitate the merger of viral and host cell membranes. Membrane fusion is a crucial step in the life cycle of enveloped viruses involving key structural elements and mechanisms [15]. Enveloped viruses possess a lipid bilayer membrane derived from the host cell, with embedded viral envelope proteins, that include fusion proteins undergoing conformational changes typically containing hydrophobic fusion peptides or domains [35, 36]. Once the fusion process begins, the attachment of viral envelope proteins to host cell receptors triggers conformational changes exposing FPs. These peptides insert into the host cell membrane, forming fusion intermediates and transient fusion pores [36]. The viral and host cell membranes merge, as the fusion pores expand, allowing viral content release into the host cell [35]. Understanding these structural elements and mechanisms is crucial for designing antiviral strategies that target specific steps in the fusion process, potentially preventing viral entry and infection.

19.1.2.8 Viral Assembly and Replication Machinery

Viral assembly and release of the genetic components represent key stages in the viral lifecycle, where new virions are formed and then released from host cells. These processes are highly orchestrated and offer valuable targets for antiviral therapeutics. The replication machinery is a critical cellular process, which is responsible for the accurate duplication of DNA or RNA during cell division [1]. In situations like viral infections, inhibiting this machinery becomes essential to prevent uncontrolled viral replication. Biomolecules offer promising avenues for developing antiviral therapeutics that target the replication machinery [18, 27].

19.1.2.9 The Host's Immune Response

The host immune response is a complex and highly regulated system that plays a critical role in defending the body against infections and diseases. It involves various components, including white blood cells, antibodies, and cytokines, working together to recognize and eliminate pathogens like bacteria, viruses, and fungi [17]. While the immune system is essential for maintaining health, some pathogens have evolved mechanisms to evade or interfere with the host's immune response. This includes mechanisms like antigenic variation, where pathogens change their surface proteins to avoid recognition, and where they can mimic host molecules to avoid immune attack [31]. Suppression of immune activation: Some pathogens produce bioactive molecules that actively suppress the host's immune response. For example, certain viruses can inhibit the production of interferons, which are essential signaling molecules for antiviral defense. Researchers are exploring strategies to block or counteract these immune-suppressing bioactive molecules, allowing the immune system to mount a more effective response [37, 38].

HCMV has evolved several proteins that directly target key components of the host's immune system and often mimic or inhibit their immune responses. For example, the HCMV protein pUL83, also known as pp65, can interfere with antigen presentation by inhibiting the transporter associated with antigen processing (TAP), thereby reducing the presentation of viral antigens to cytotoxic T cells [39].

19.2 Targets for Antiviral Therapeutics and Inhibition Strategies

Bioactive molecules employ sophisticated strategies to inhibit viral entry by involving receptor blockade compounds that mimic host cell receptors or bind directly to viral attachment proteins, preventing initial virus–cell interactions. Fusion inhibition targets conformational changes in viral fusion proteins postreceptor binding, halting membrane fusion, and viral entry [27, 36]. Coreceptor interference involves blocking coreceptor binding sites on viral proteins and impeding their effective interaction with primary receptors and coreceptors. By targeting viral receptors, such as CD4, CXCR4, and CCR5 for HIV [16, 34, 40], or salicylic acid analogs for influenza [27], promising avenues for antiviral intervention are revealed. The discovery of novel receptors, exemplified by chANXA2 for ALV-J, underscores the potential of host-targeted approaches in combating viral infections, with identified host receptors serving as valuable markers for diverse antiviral strategies. pH modulation comes into play with viruses entering via endocytosis, by raising endosomal pH, this strategy disrupts fusion events essential for cytoplasmic entry [41]. Allosteric inhibition affects distant structural sites on viral proteins, inducing conformational changes that impair viral entry [7, 13, 14]. Preventing membrane fusion targets viral fusion proteins or interferes with fusion pore formation, effectively blocking virus-cell membrane merging. Intracellular signaling interference disrupts host cell signaling pathways used by some viruses for entry. RNA interference (RNAi) can specifically target and degrade viral RNA sequences, inhibiting viral replication and entry [42].

19.2.1 Enzyme Inhibitors

For catalyzing transcription and replication, viruses often use their own RNA and DNA polymerase that reflects diversity in the replication machinery [18]. Studying the crystal structures of viral polymerases provides insights into their active sites and binding pockets. Inhibition of viral polymerases can be achieved through competitive binding of small molecules or biomolecules, disrupting RNA and DNA synthesis. Helicases [43] and nucleoproteins [44] play pivotal roles in unwinding viral RNA or DNA, coating and protecting viral genomes. In measles and Ebola viruses, helicase is absent where nucleoprotein functions as helicase [45]. In Herpes Simplex virus type 1 (HSV-1), a helicase-primase complex poses as a potential alternative for the development of antiviral drugs [46]; for example, Amenamevir (AMNV) disrupts the complex and halts viral replication [47]. Viral proteases encode proteases for processing viral polyproteins and are essential for the maturation of viral particles [48]. Structural insights provide the basis for rationally developing compounds that fit snugly into the active site, blocking protease activity, and disrupting the maturation of viral proteins, thereby inhibiting viral assembly and release [12]. There are enzyme inhibitors targeting reverse transcriptase [49], which are either nucleoside or nucleotide specific [50]. Table 19.1 discusses some inhibitors targeting viral pathogens.

19.2.2 Antiviral Peptides

Antiviral peptides are designed based on intricate structural insights into viral proteins involved in assembly and release [35]. By precisely targeting the interaction sites, antiviral peptides can obstruct the assembly of viral components or interfere with PPIs, disrupting viral assembly. For example, the FP of the HIV-1 envelope glycoprotein, structural studies have unveiled its interaction with host cell membranes during viral entry [35, 36]. Antiviral peptides designed to mimic this FP can competitively bind to host cell membranes [27], thwarting the viral envelope's fusion process with the host cell membrane [36]. This disruption profoundly impedes the final stages of viral entry, assembly, and release [19]. Enfuvirtide prevents fusion and blocks the formation of new virions of HIV further hindering the viral infection. Boceprevir and telaprevir are synthesized peptides that inhibit protease to restrict hepatitis C replication [35, 36].

Table 19.1 Target inhibitors of viral structure.

Inhibitors	Type	Drug name	Effects on target disease	Reference
Attachment inhibitors		Fostemsavir	Reduces HIV-1 growth by blocking a virus's ability to bind with its CD4 receptor	[34]
	Neuraminidase inhibitors	Oseltamivir (Tamiflu)	Attaches to and inhibits the neuraminidase enzymes' active site, which is necessary for all influenza viruses to release their progeny virions from infected host cells	[51]
		Zanamivir	By its antagonist action, it inhibits neuraminidase protein, which also prevents the virus from spreading and infecting neighboring cells	[52]
Entry inhibitors	Fusion inhibitors	Enfuvirtide	Inhibits various types of encapsulated viruses, especially HIV strains	[34]
		Maraviroc	Blocks the CCR5 coreceptor to prevent HIV-1 from entering host cells	[34]
Protease inhibitors		Nelfinavir	Inhibits cytochrome P450 3A and HIV protease	[53]
		Darunavir	It maintains to have antiretroviral efficacy against certain HIV strains that are resistant to multiple drugs	[54]
Polymerase inhibitors	DNA polymerase inhibitors	Valacyclovir	It is used to treat cytomegaloviruses, varicella-zoster, and herpes	[55]
	RNA polymerase inhibitors	Ribavirin	By attaching to the substrate-binding site of the IMPDH enzyme and preventing it from accessing its endogenous substrate, Ribavirin inhibits the enzyme, lowering GTP levels and production	[56]
	Nucleoside polymerase inhibitors	Sofosbuvir	It effectively inhibits the SARS Virus-CoV-2 RdRp	[53]
	Nonnucleoside polymerase inhibitors	Rilpivirine	HCV-NS5B polymerase inhibitor of hepatitis C virus (HCV)	[57]
Nucleoside and nucleotide reverse transcriptase inhibitors		Abacavir	Abacavir's intracellular anabolite carbovir-triphosphate inhibits HIV viral RNA-dependent DNA polymerase (reverse transcriptase), which inhibits viral replication and has an antiviral effect	[49]
		Tenofovir	Prevent HIV replication by posing a rival to deoxyadenosine 5′-triphosphate, which is naturally occurring and the substrate used by HIV transcription, for DNA insertion	[50]

(Continued)

Table 19.1 (Continued)

Inhibitors	Type	Drug name	Effects on target disease	Reference
Nonnucleoside reverse-transcriptase inhibitors		Efavirenz	It inhibits the HIV reverse transcription enzyme's activity by attaching to a noncatalytic location on the enzyme. As a result, this activity prevents HIV replication and other DNA polymerase-related activities	[16]
		Etravirine	It causes direct inhibition of the human immunodeficiency virus type 1 (HIV-1) reverse transcriptase enzyme. It limits the action of polymerase that is dependent on both DNA and RNA by directly binding to reverse transcriptase	[58]
Integrase inhibitors		Raltegravir	Prevents HIV-1 DNA from being inserted into the genome of the host cell by inhibiting the action of HIV-1 integrase	[40]
		Dolutegravir	Inhibits integrase binding and blocks the HIV DNA transfer into the host	[59]

19.2.3 Antiviral Antibodies

Monoclonal antibodies (mAbs) represent a potent class of biomolecules designed to recognize and bind to specific viral proteins, particularly those involved in assembly and release [31]. Structural biology provides critical insights into epitopes on viral proteins that are susceptible to antibody binding. Detailed structural studies show the conformational changes and PPIs central to viral assembly and release, enabling the selection or design of mAbs that can precisely target these epitopes, thereby interfering with the viral lifecycle [60, 61]. Studies on SARS-CoV-2 have unveiled the structure of its spike protein, which plays a pivotal role in viral entry, assembly, and release [29]. mAbs are engineered to bind specific regions of the spike protein that can disrupt viral attachment, assembly, and release by blocking interactions with host cell receptors, thus offering an avenue for antiviral intervention [31]. mAbs like Bamlanivimab, Etesevimab, and Casirivimab pose as a passive immunotherapy for mild to moderate infection that acts as ACE2 blockers inhibiting viral entry [31].

19.2.4 Lipid-Mimicking Compounds

The design of compounds that mimic the lipid composition of host cell membranes depends on an understanding of the structural properties of lipid bilayers and the interactions that occur between the viral envelope proteins and lipids [62]. This understanding guides the development of molecules that can effectively compete with host cell lipids, thereby disrupting viral envelope formation and impeding viral budding [36]. Some lipid-mimicking compounds imitate the lipid rafts present in host cell membranes. Because of this, there is a disruption in the assembly of viral envelope proteins into lipid rafts, preventing the proper formation of viral envelopes [42]. This interference halts the release of mature virions and significantly reduces viral infectivity [22].

19.2.5 Vaccines

Another aspect of antiviral therapeutics is vaccines that mimic the natural immunity against the virus. Antigens, adjuvants, or genetic materials like RNA or DNA (vectors) are commonly used in these types of vaccinations to activate and identify the viral proteins and trigger the defense processes [42]. Antigens are like key players that imitate the viral protein and produce immunological responses through the synergistic effect of creating antibodies (by B cells) and enhancing immunity (by T cells) [39, 44]. Adjuvants can stimulate the production of cytokines and other immune mediators, leading to a stronger and longer-lasting immune response. For example, the oral polio vaccine (OPV) and the measles, mumps, and rubella (MMR) vaccine [63], to reduce its virulence and prevent it from infecting healthy individuals with disease, the live pathogen is modified. Bioactive molecule-based immunizations use specific molecules (mRNA, viral vectors, or purified proteins) that encode or imitate a pathogen component [64]. When provided, they stimulate cells to produce pathogenic components, which sets off an immune reaction without causing sickness. Vector development process can be viral (harmless viruses) and nonviral (nanoparticles, liposomes, etc.) that carry and deliver genes to the host cells that decode for antigens to enhance immune recognition [42]. The human papillomavirus (HPV) vaccine (Gardasil and Cervarix) neutralizes antibody responses thereby averting HPV infection and reducing the chances of cervical cancer [65]. Some of the nanoparticle-based vectors are discussed in Section 19.3.5.

19.2.6 Immunomodulation

Bioactive molecules can also be used to modulate the immune response in cases of autoimmune diseases or chronic inflammation. Immunomodulatory therapies aim to balance the immune system, either by suppressing an overactive response or by enhancing a weakened one [39, 44]. These strategies often involve the use of bioactive compounds like mAbs or cytokine inhibitors [31].

Looking at the example of HCMV employs a range of clever tactics to manipulate the human immune system, particularly through its viral cytokine-like molecules [39]. These molecules, such as cmvIL-10 and LAcmvIL-10, are key players in altering the body's immune responses. CmvIL-10 acts as an immune suppressor by decreasing the activity of MHC-II, which is an essential component for recognizing and responding to infections [66]. Simultaneously, it enhances the immune system's ability to identify and eliminate infected cells by boosting the expression of receptors that aid in recognizing and engulfing these cells. In contrast, LAcmvIL-10's role involves hiding HCMV from immune cells, particularly CD4+ cells, which are vital for detecting and combating infections. By making it more challenging for CD4+ cells to identify HCMV-infected cells, LAcmvIL-10 aids the virus in evading immune surveillance and establishing long-term persistence within the host. These modulation strategies highlight HCMV's adaptability in manipulating the immune system to its advantage [39, 66].

19.3 Rational Strategies for Antiviral Therapeutics

With the revolutionalized technologies and integration of Omics, rational drug designing strategies have overpowered the traditional and serendipitous discoveries, allowing researchers to overcome drug resistance challenges and understand the complexity of host–pathogen interaction. Figure 19.2 illustrates the multidisciplinary integrated pipeline for the development of antiviral therapeutics including CADD, artificial Intelligence (AI) and machine learning (ML) algorithms, and SB. Antiviral therapeutics based on these approaches are mentioned in Table 19.2.

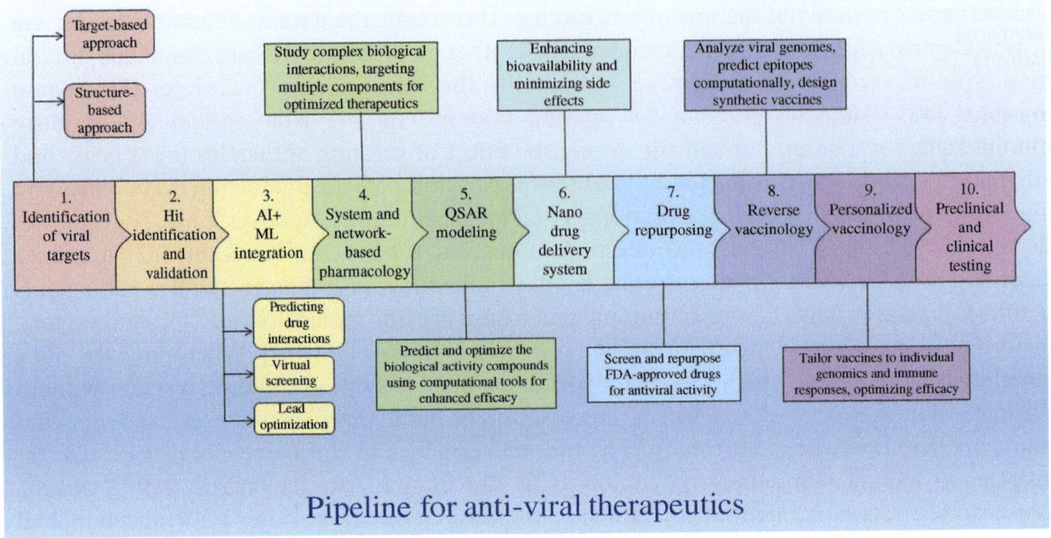

Figure 19.2 Rational pipeline for the development of antiviral therapeutics.

19.3.1 CADD and QSAR (Quantitative Structure–Activity Relationship)

CADD is based on either target- or drug-oriented pipelines, which are also known as structure- and ligand-based approaches, respectively. In the structure-based (target-oriented) approach, the structural data of the target is known; these could be viral enzymes or structural proteins of interest. In the case of an unknown structure, protein modeling allows structure prediction based on the templates that are experimentally validated [77]. AlphaFold, an AI-generated tool to predict the 3D structure against the primary structure of protein [14], can be further used to determine the function of the target. Once this information is availed, docking and molecular dynamics (MD) are performed to investigate the molecular features like binding affinity and conformational modifications, leading to probable therapeutics of various drugs [78, 79]. In the ligand-based (drug-oriented) approach, the 3D structure of the target is unknown, but with the prior structural and functional features of the drugs, the targets can be identified by preparing ligands through VS. Lipinski Rule of 5 approach can also be used to filter compounds based on basic properties like molecular weight, log$P$, and the number of H-bond acceptors and donors [79]. Nowadays with the advances in technology, the CADD approach integrates AI and ML models for the VS [30, 73]. The pharmacophore-based ligand searching plays a crucial role in identifying structural key features that mimic and comprehend biological activity. Pharmacophore docking can also be done if the 3D structure of the target is available [11, 37]. These features can be ranked according to electrostatic and steric descriptors. Hence, ligand- and structure-based mapping and searching are very important in selecting appropriate candidate compounds. Dai et al. identified two promising drugs with the same structural approach, to inhibit SARS-CoV protease, an enzyme that mediates the replication of virus [29]. Similarly, Ahmed et al. (2024) used homology modeling for building the nonstructural protein (nsP4) of RdRp of the Chikungunya virus and ProdDrug server for predicting active sites that serve as druggable pockets. Pharmacophore modeling and VS filtered thousands of compounds with the highest likelihood for the predicted pockets. Docking and MD result in three ligands with optimal interactions with three subdomains of the nsP4, i.e., catalytic, thumb, and palm triggering specific inhibition responses against the virus [11].

Table 19.2 Antiviral therapeutics based on various drug discovery pipelines.

Based drug discovery	Drug name	Mechanism	Reference
Structure-based drug discovery	Remdesivir	By entering human cells through the ACE2 receptor, SARS-CoV-2 transforms remdesivir into triphosphate and forms a complex with ATP, viral RNA, and RNA-dependent RNA polymerase that prevents the synthesis of new viral RNA	[67]
	Lopinavir	The HIV-1 protease, which is necessary for the development of a mature, infectious virus, is inhibited by lopinavir. It functions by inhibiting HIV-1's development, which prevents the virus from spreading	[53]
Ligand-based drug discovery	Maraviroc	Maraviroc (MVC) is an entrance inhibitor that inhibits gp120 from binding to the chemokine receptor CCR5 (CCR5), preventing HIV from infecting human cells	[68]
	Imiquimod	The immune response modifier imiquimod activates Toll-like receptor 7, which in turn causes the release of cytokines such as tumor necrosis factor-alpha and interferons. The release of interferon-alpha and a Th1 response are two examples of the innate and adaptive immune responses that are triggered by this. In addition to its antiviral properties, imiquimod also promotes apoptosis in skin cancer cells and demonstrates antitumoral effect	[38]
Pharmacophore drug discovery	Sofosbuvir	Sofosbuvir is a nucleotide prodrug targeted at the liver, which functions as a chain terminator in HCV genotypes 1–4 by blocking the RNA-dependent NS5B RNA polymerase	[69]
	Oseltamivir (Tamiflu)	Oseltamivir blocks the neuraminidase enzyme of influenza virus, which is essential for viral growth. It limits the spread of infection by stopping the release of fresh virus particles from infected cells. The prodrug oseltamivir phosphate is converted into the active form oseltamivir carboxylate in the body	[37]
QSAR-based drug discovery	Ritonovir	Indinavir is an antiretroviral protease inhibitor that works by binding to the HIV protease's catalytic site and blocking the cleavage of viral polyprotein precursors into mature, functional proteins that are required for viral replication	[70]
	Lamivudine (3TC)	Lamivudine functions as a cytosine analog to prevent the synthesis of viral DNA. After entering the cell, it undergoes phosphorylation to its active forms lamivudine monophosphate (L-MP) and lamivudine triphosphate (L-TP), which block the replication of viral DNA. Since lamivudine is not effectively recognized by human polymerases, unlike natural cytidine triphosphate, it can compete and inhibit the synthesis of reverse transcriptase DNA in HIV-1 and HBV infections. It can, however, weakly inhibit mitochondrial DNA polymerase and mammalian DNA polymerases	[71]

(Continued)

Table 19.2 (Continued)

Based drug discovery	Drug name	Mechanism	Reference
AL and ML-based drug discovery	Acyclovir	By operating as an analog of deoxyguanosine triphosphate (dGTP), acyclovir triphosphate competitively inhibits viral DNA polymerase. As acyclovir triphosphate does not have a 3′ hydroxyl group, it does not allow other nucleosides to bond to DNA, which causes chain termination	[72]
	Zidovudine (AZT)	Zidovudine is categorized as a nucleoside reverse transcriptase inhibitor (NRTI) and is a synthetic analog of the nucleoside thymidine. By replacing thymidine in freshly synthesized viral DNA and functioning as a viral DNA chain terminator, zidovudine serves as an antiviral drug. This disrupts the HIV-1 life cycle by preventing HIV-1 reverse transcriptase from creating viral DNA from the RNA template	[73]
Systems biology or network-based drug discovery	Baloxavir marboxil (Xofluza)	This prodrug is hydrolyzed to produce Baloxavir, the active form. To prevent the influenza virus from replicating, it blocks the endonuclease activity of the polymerase acidic (PA) protein	[74]
	Favipiravir (Avigan)	Favipiravir inhibits viral RNA polymerase, which blocks viral protein synthesis, once it has been metabolized to its active form, favipiravir-RTP. To prevent future viral replication, it can also be integrated into the RNA of the virus. Furthermore, it is uncertain if it is active against SARS-CoV-2 because it causes deadly mutagenesis in some viruses	[75]
Reverse vaccinology-based drug discovery	REGN-EB3	The vaccine neutralizes the virus by hindering its entry and recruiting additional immune cells to seek out and eliminate contaminated cells from the body	[76]
	Gardasil 9 (9vHPV)	The exact mechanism of the 9-valent human papillomavirus (9vHPV) vaccine is still unclear; it is thought to function by triggering the humoral immune response. According to studies, the vaccination stimulates the development of neutralizing antibodies against different HPV strains, resulting in a potent immune response that protects against diseases linked to HPV and dysplastic lesions. The vaccine's humoral response mechanisms have demonstrated efficacy, as seen by the much higher antibody levels created by the vaccine compared to those produced by natural infection	[65]

One of the important ligand-based modeling approaches is QSAR, which models and highlights 2D and 3D descriptors of the ligand that correlate with biological activity like IC50, GL50, and LD50 [37, 70]. The diverse set of descriptors like topological, electronic, constitutional, and geometrical can be calculated and applied for constructing the models. Programming languages like R (Rcpi) [80] and Python (PyL3dMD) [81] possess simulation packages that underpin molecular informatics by merging bioinformatics and chemoinformatics knowledge of a molecule or a protein. The prediction of the biological activity ($\log 1/c$) is used to correlate against the pharmacological properties of descriptors [37]. The best-correlated QSAR model reflects R^2, nearer to 0.8 or 0.9 and the quality of the QSAR model can be further checked by using the cross-validation (Q^2) based on the LOO (leave one compound out) approach [81]. However, the error rate of the QSAR model can be observed as root mean square error (RMSE) [11]. Multidimensional QSAR can be used in 4D, 5D, and 6D and are a few popular computational methods for antiviral drug design apart from the traditional QSAR modeling. These are some emerging tools for incorporating different ligands such as nanoparticles and small molecules [34, 82].

19.3.2 AI and ML

Compared to the conventional methods, the integration of AI and Omics has crossed the barriers associated with time, cost, precision, and accuracy. Systems can be trained with the preexisting data of viral targets (e.g., HIV and COVID-19) to screen the primary hits (antiviral targets) to be further docked using algorithms like protein–ligand scoring functions. A cloud-based platform VS integrated with multiple databases to generate datasets of potential antiviral compounds that can be tested against viruses using *in vitro* and *in vivo* methods for validation [83]. Algorithms of ANN and deep learning (DL) train the dataset from different repositories, where the neural network can identify and classify antiviral peptides for their antiactivity. The particle swarm optimization (PSO) algorithm optimally classifies the viral peptides that are critical for the innate immune system with an ability to inhibit viral growth [84]. Deep-AVPpred [85], AVPIden [86], D3AI-CoV [87], Meta-iAVP [88], and iAVPs-ResBi [89] are a few of the state-of-the-art classifiers that adopt DL and ML algorithms to model based on the multiple descriptors and feature selections specific to the antiviral mechanism. These interfaces classify and predict novel antiviral compounds with drug-like properties that can be chemically synthesized, optimized, and scientifically verified. Emami et al. (2022) have discussed some of the optimization algorithms that use heuristic approaches integrating evolutionary theory to model the best features that can mimic the viral behavior leading to novel drug discovery [84]. Accuracy of various scoring functions is implemented in VS and prediction tools to attain the properties or descriptors that show similar biological activity [34, 81]. Geometric DL and the knowledge of graph theory use SMILES notations of drug and target to fed in a form node features to train the neural network for predicting PPI networks to identify highly potential antiviral drug targets [10].

19.3.3 Systems Biology and Network Pharmacology

The SB approach also known as "Network Pharmacology" integrates networks of proteins, genes, and host–pathogen interactions. Large datasets with different file formats retrieved from databases such as GEO, Uniprot, GeneBank, and STRING are used to construct networks to identify viral interactions with host proteins [90]. Tools like CytoScape have many plugins to cluster the nodes on topological parameters, which are calculated based on qualitative data, quantitative data, and numerical modeling of biological systems. Kyoto Encyclopedia of Genes and Genomics (KEGG), gene ontology (GO), and enrichment analysis determine potential pathways and genes that can alter biomolecular processes [26]. In addition, this approach can identify several host variables that are common between any two infectious pathways [91]. For example, Zhuang et al. [92] investigated bioactive compounds and target genes of Shuanghuanglian, a traditional Chinese medicine (TCM). These findings indicated that four key hub genes have been enriched in pathways of HCMV and papillomavirus with potential pharmacological responses of the TCM against these viruses [92]. Farooq et al. curated protein datasets of the influenza A virus (IAV) and identified potent therapeutic targets that are linked not only with the IAV but also with hepatitis B and C viruses, measles, and cancer-triggering viruses [26]. The SB models can hypothesize based on the comparative analysis of molecular mechanisms to unravel the insights of pathogenesis and replication of the viruses [93]. Recently, Liu et al. [94] study screened target genes of Mulberry leaves for antiviral therapeutics against EV71 virus (Enterovirus 71). The network identified Quercetin to possess drug-likeness to inhibit NF-κB signaling pathway and other associated pathways eventually targeting the effectiveness of EV71 [94].

19.3.4 CRISPR Systems

CRISPR-Cas systems can be engineered to target and cleave essential viral genes involved in the replication of the viral cycle. The CRISPR-based antiviral therapy has the potential for selective inhibition of viral replication. Antiviral therapeutics targeting the replication machinery face challenges such as viral mutations causing drug resistance, delivery methods, and potential off-target effects [95]. The combination of more than one inhibitory strategy, such as using a combination of small molecules and CRISPR-Cas systems, may be a promising approach to overcome these challenges [96]. CRISPR technology has evolved beyond SpCas9, introducing innovations like SaCas9, FnCas9, C2c1/2/3, and Cas13 in the antiviral realm [95]. SaCas9, a compact alternative to SpCas9, facilitates versatile genome editing using adeno-associated virus (AAV) delivery [97], overcoming limitations associated with the size and immunogenicity of other vectors. SaCas9 variants, such as SaCas9 (KKH), broaden target recognition sites, offering flexibility in biomedical applications across various organisms [95, 97]. Cpf1, a type V CRISPR system, recognizes 5′-TTN-3′ or 5′-CTA-3′ motifs and processed crRNA design, which has high efficiency in inhibiting SAR-COV-2 and influenza viruses [98, 99]. Notably, Cpf1 also exhibits RNase activity, extending its potential to target these viral RNA genomes through genome editing provoking antimechanism [100]. FnCas9, derived from *F. novicida*, uniquely modulates bacterial virulence by degrading endogenous antibacterial lipoprotein mRNA [95]. It demonstrates its versatility by targeting RNA viruses, for example, in the hepatitis C virus (HCV), the system weakens the phosphorylation of p38, an essential factor for cellular functions (like cell entry, replication, and virion assembly) [15]. Eventually showcasing its potential as an antiviral defense tool by suppressing the replication and translation process of the virus.

Class 2 candidate (1/2/3) and Cas13 [95] cleave the RNA restricting the replication and disrupting the virus [100]. The targeted host factors and restriction mechanisms employed by CRISPR-Cas systems present a dynamic approach to antiviral research. The alteration of the genetic makeup of viruses such as HIV, hepatitis (B and C) SARS, MERS, and influenza, through these systems, has shed some light on innovative strategies for antiviral targets by minimizing on off-target activity and drug resistance.

19.3.5 Nanotechnology-Based Design and Delivery Systems

Nonviral vectors, such as lipid-based or polymer-based nanocarriers, show promising alternatives to viral vectors, addressing integration-related issues and immune responses [101]. Nanoparticle-based inhibition employs functionalized nanoparticles to block viral entry by binding to viral particles, disrupting receptor interactions, or interfering with fusion processes [102]. The use of nanocarriers for targeted drug delivery represents a cutting-edge approach to antiviral drug development. The nanotechnology-driven systems circumvent biological barriers by potentially minimizing side effects; enhancing drug delivery specificity, drug stability, and circulating time [101, 103]. Nanoparticles (NPs) such as gold, silver, and certain metallic oxides of zinc, copper, titanium, silicon, and ceric fused with antiviral peptides modify the surface proteins of a broad spectrum of viruses such as dengue, papilloma, influenza, hepatitis, and HIV [103]. Virus-like particles (VLPs) and virosomes directly target the infectious sites by attaching to the envelope proteins, which boosts the bioavailability by modulating the burst release of the drug [104]. siRNAs conjugated with gold NPs reduced viral duplication of dengue [105] and SARS-CoV-2 virus [82]. In order to prevent and control viral infections, research on improving immune response and the use of nanomaterials as an adjuvant to antiviral vaccinations have shown promising outcomes.

19.3.6 Reverse Vaccinology

In the relentless pursuit of effective antiviral therapeutics amidst the perpetual threat of viral infections, the conventional landscape of drug discovery has encountered formidable challenges in keeping stride with the rapid mutational dynamics of viruses and the emergence of resilient drug-resistant strains. However, amid this, an approach emerges in the form of reverse vaccinology, heralding a transformative paradigm in the quest for novel antiviral interventions. During the prescreening phase, primary information on membrane proteins and their localized and secreted virulence factors is retrieved from various databases that are associated with the pathogen of interest [60]. Vaccine candidates are selected based on the PPI networks, molecular parameters of proteins like molecular weight, associated antigens, and various B-cell and T-cell epitopes conjugated with their adjuvants [61]. The rigid interaction of the vaccine with its receptor is simulated and predicted through docking and MD results further predict the immunological response. Overall, the computational methods discussed in the above sections are used for antiviral vaccines depending on the structural properties of the pathogens [64].

The development of vaccines is hampered by the diversity of viruses because each one interacts differently with a standardized medication because of its structure and replication cycle. For example, Arboviruses represent a diverse group of viruses belonging to different families, including Flaviviridae, Togaviridae, Bunyaviridae, and Reoviridae [106]. Developing vaccines against individual arboviruses may not provide sufficient protection, especially in regions where multiple arboviruses cocirculate. A multivalent vaccine targeting several diverse viruses simultaneously offers the potential for cross-protection and broad-spectrum immunity, enhancing its efficacy in diverse epidemiological settings [61, 106]. Hence, instead of designing a specific drug, designing a multivalent vaccine against Herpes [107], Arboviruses [106], Monkeypox [108], and Ebola virus [76] through the innovative approach of reverse vaccinology is more viable. This is done by employing bioinformatics tools and methodologies to potential vaccine targets and epitopes across various viruses to identify. Advances in high-throughput sequencing technologies have enabled researchers to rapidly sequence the genomes of a wide range of viruses, providing valuable insights into their genetic composition, structure, and evolution.

19.4 Conclusion

Rational drug designing are multifaceted strategies that offer diverse approaches for inhibiting viral entry, tailored to the unique mechanisms employed by different viruses, and hold significant promise for antiviral development. Blocking attachment sites involves compounds that directly bind to these sites and competitively hinder the virus's attachment to host cells. Bioactive molecules that interfere with these evasion mechanisms can be valuable in enhancing the immune response against such pathogens. These approaches involve assigning confidence scores to candidate compounds, with higher scores indicating a greater likelihood of efficacy. Additionally, nanoparticle-based inhibition employs functionalized nanoparticles to block viral entry by binding to viral particles, disrupting receptor interactions, or interfering with fusion processes. By leveraging the number of drug targets associated with each candidate, researchers can effectively gauge the potential therapeutic utility of these compounds across multiple biological pathways or molecular targets. The "drug–target–drug" computational approach showcases its robustness. However, as computational techniques evolve, integrating diverse approaches such as molecular simulations, RNAseq, and drug network analyses could provide a more comprehensive understanding of drug-repurposing strategies.

Conversely, the removal of pharmaceuticals from categories that include prohibited, discontinued, or nutraceuticals helps mitigate potential safety or regulatory concerns, ensuring that repurposed candidates meet stringent regulatory standards and can proceed smoothly through the drug development pipeline. The retention of approved drugs in the candidate pool is pivotal, as it ensures that repurposed candidates have already undergone rigorous preclinical and clinical evaluation, possess a proven safety profile, and exhibit known pharmacokinetics. This accelerates the drug development process by bypassing the need for extensive safety and toxicity studies, thereby expediting the translation of promising candidates into clinical trials. This approach not only enhances the efficiency of drug-repurposing efforts but also contributes to the rapid identification and development of novel treatments for a wide range of medical conditions. As drug repurposing continues to gain traction as a viable strategy for therapeutic innovation, the utilization of such innovative approaches will be instrumental in addressing unmet medical needs and advancing patient care on a global scale.

In the future, precision medicine and personalized immunotherapy approaches may become more prominent, tailoring bioactive interventions to an individual's unique immune profile. Additionally, ongoing research in immunology and bioactive molecule design will likely uncover new targets and strategies for optimizing the host immune response against various diseases, ultimately improving healthcare outcomes for patients. In summary, the process of prioritizing repurposed drug candidates through the calculation of confidence scores which is based on the number of drug targets represents a crucial aspect of drug-repurposing endeavors, particularly in the context of addressing emerging infectious diseases and pandemics. Advances in bioactive molecule design are contributing to the emergence of personalized antiviral therapies. By taking into account the individual's genetic makeup and viral strain characteristics, researchers can tailor antiviral treatments to maximize efficacy and minimize side effects.

References

1 Payne, S. (2022). *Viruses: From Understanding to Investigation*. Elsevier.

2 Singh, S., Kola, P., Kaur, D. et al. (2021). Therapeutic potential of nutraceuticals and dietary supplements in the prevention of viral diseases: a review. *Frontiers in Nutrition* 8: 679312.

3 Liang, L., Ahamed, A., Ge, L. et al. (2020). Advances in antiviral material development. *ChemPlusChem* 85 (9): 2105–2128.

4 Pillay, D. and Zambon, M. (1998). Antiviral drug resistance. *BMJ* 317 (7159): 660–662.

5 Matthew, A.N., Leidner, F., Lockbaum, G.J. et al. (2021). Drug design strategies to avoid resistance in direct-acting antivirals and beyond. *Chemical Reviews* 121 (6): 3238–3270.

6 Rudrapal, M. and Khan, J. (2023). *CADD and Informatics in Drug Discovery*. Springer Nature.

7 Rupp, B. (2009). *Biomolecular Crystallography: Principles, Practice, and Application to Structural Biology*. Garland Science.

8 Dobson, C.M. (2019). Biophysical techniques in structural biology. *Annual Review of Biochemistry* 88: 25–33.

9 Duelfer, J., Kadek, A., Kopicki, J.D. et al. (2019). Structural mass spectrometry goes viral. *Advances in Virus Research* 105: 189–238.

10 Das, B., Kutsal, M., and Das, R. (2022). A geometric deep learning model for display and prediction of potential drug–virus interactions against SARS-CoV-2. *Chemometrics and Intelligent Laboratory Systems* 229: 104640.

11 Ahmed, M.H., Singh, G., Castrosanto, M. et al. (2024). A structure-based drug design approach for the identification of antiviral compounds targeting the chikungunya virus RdRp protein. *Chemical Physics Impact* 8: 100450.

12 Li, Q. and Kang, C. (2021). Structure and dynamics of Zika virus protease and its insights into inhibitor design. *Biomedicine* 9 (8): 1044.

13 Lim, L., Gupta, G., Roy, A. et al. (2019). Structurally- and dynamically-driven allostery of the chymotrypsin-like proteases of SARS, Dengue and Zika viruses. *Progress in Biophysics and Molecular Biology* 143: 52–66.

14 Nussinov, R., Zhang, M., Liu, Y., and Jang, H. (2022). AlphaFold, artificial intelligence (AI), and allostery. *The Journal of Physical Chemistry B* 126 (34): 6372–6383.

15 Mohamed, A., Lin, Q.F., Eaton, H.E., and Shmulevitz, M. (2023). p38 mitogen-activated protein kinase signaling enhances reovirus replication by facilitating efficient virus entry, capsid uncoating, and postuncoating steps. *Journal of Virology* 97 (2): e00009–e00023.

16 Vanangamudi, M., Nair, P.C., Engels, S.M. et al. (2021). Structural insights to human immunodeficiency virus (HIV-1) targets and their inhibition. *Antiviral Drug Discovery and Development* 63–95.

17 Thakur, A., Mikkelsen, H., and Jungersen, G. (2019). Intracellular pathogens: host immunity and microbial persistence strategies. *Journal of Immunology Research* 2019.

18 Menéndez-Arias, L. and Andino, R. (2017). Viral polymerases. *Virus Research* 234: 1.

19 Wang, Y., Zirbel, C.L., Leontis, N.B., and Ding, B. (2018). RNA 3-dimensional structural motifs as a critical constraint of viroid RNA evolution. *PLoS Pathogens* 14 (2): e1006801.

20 Adeyemi, O.O., Sherry, L., Ward, J.C. et al. (2019). Involvement of a nonstructural protein in poliovirus capsid assembly. *Journal of Virology* 93 (5): https://doi.org/10.1128/jvi.01447-18.

21 Tsang, S.K.-Y. (2001). *The Poliovirus N to A Transition: A Kinetic Analysis and Strategies for Drug Discovery*. Harvard University.

22 Fry, E.E. and Stuart, D.I. (2010). Virion structure. *The Picornaviruses* 57–71.

23 Joseph-McCarthy, D., Tsang, S.K., Filman, D.J. et al. (2001). Use of MCSS to design small targeted libraries: application to picornavirus ligands. *Journal of the American Chemical Society* 123 (51): 12758–12769.

24 Soleimani, S., Kamalzadeh, M., and Lotfi, M. (2019). Preparation and characterization of a neutralizing monoclonal antibody against poliovirus type 1 (Mahoney strain). *Medical Laboratory Journal* 13 (1).

25 Lancaster, K.Z. and Pfeiffer, J.K. (2011). Mechanisms controlling virulence thresholds of mixed viral populations. *Journal of Virology* 85 (19): 9778–9788. https://doi.org/10.1128/JVI.00355-11.

26 Farooq, Q.u.A., Shaukat, Z., Aiman, S. et al. (2020). A systems biology-driven approach to construct a comprehensive protein interaction network of influenza A virus with its host. *BMC Infectious Diseases* 20: 1–10.

27 Liu, C., Hu, L., Dong, G. et al. (2023). Emerging drug design strategies in anti-influenza drug discovery. *Acta Pharmaceutica Sinica B* 13 (12): 4715–4732.

28 Hu, Z., Rolt, A., Hu, X. et al. (2020). Chlorcyclizine inhibits viral fusion of hepatitis C virus entry by directly targeting HCV envelope glycoprotein 1. *Cell Chemical Biology* 27 (7): 780–792. e5.

29 Dai, W., Zhang, B., Jiang, X.M. et al. (2020). Structure-based design of antiviral drug candidates targeting the SARS-CoV-2 main protease. *Science* 368 (6497): 1331–1335.
30 Santos, F.R., Nunes, D.A., Lima, W.G. et al. (2019). Identification of Zika virus NS2B-NS3 protease inhibitors by structure-based virtual screening and drug repurposing approaches. *Journal of Chemical Information and Modeling* 60 (2): 731–737.
31 Kumar, S., Chandele, A., and Sharma, A. (2021). Current status of therapeutic monoclonal antibodies against SARS-CoV-2. *PLoS Pathogens* 17 (9): e1009885.
32 Proulx, J., Borgmann, K., and Park, I.-W. (2021). Role of virally-encoded deubiquitinating enzymes in regulation of the virus life cycle. *International Journal of Molecular Sciences* 22 (9): 4438.
33 Gredmark-Russ, S., Isaacson, M.K., Kattenhorn, L. et al. (2009). A gammaherpesvirus ubiquitin-specific protease is involved in the establishment of murine gammaherpesvirus 68 infection. *Journal of Virology* 83 (20): 10644–10652.
34 Lai, Y.-T. (2021). Small molecule HIV-1 attachment inhibitors: discovery, mode of action and structural basis of inhibition. *Viruses* 13 (5): 843.
35 Agarwal, G. and Gabrani, R. (2021). Antiviral peptides: identification and validation. *International Journal of Peptide Research and Therapeutics* 27 (1): 149–168.
36 Lozada, C., Barlow, T.M., Gonzalez, S. et al. (2021). Identification and characteristics of fusion peptides derived from enveloped viruses. *Frontiers in Chemistry* 9: 689006.
37 Inamdar, P., Bhandari, S., Sonawane, B. et al. (2014). Structure optimization of neuraminidase inhibitors as potential anti-influenza (H1N1Inhibitors) agents using QSAR and molecular docking studies. *Iranian Journal of Pharmaceutical Research: IJPR* 13 (1): 49.
38 Nanda, J. and Bermudez, R. (2023). Imiquimod. In: *StatPearls [Internet]* (ed. J. Nanda and R. Bermudez). StatPearls Publishing.
39 Long, X., Qiu, Y., Zhang, Z., and Wu, M. (2021). Insight for immunotherapy of HCMV infection. *International Journal of Biological Sciences* 17 (11): 2899.
40 Liang, G., He, Y., Zhao, L. et al. (2022). CTNNBL1 restricts HIV-1 replication by suppressing viral DNA integration into the cell genome. *Cell Reports* 38 (11).
41 Mei, M., Ye, J., Qin, A. et al. (2015). Identification of novel viral receptors with cell line expressing viral receptor-binding protein. *Scientific Reports* 5 (1): 7935.
42 Lundstrom, K. (2020). Viral vectors applied for RNAi-based antiviral therapy. *Viruses* 12 (9): 924.
43 Du Pont, K.E., McCullagh, M., and Geiss, B.J. (2022). Conserved motifs in the flavivirus NS3 RNA helicase enzyme. *Wiley Interdisciplinary Reviews: RNA* 13 (2): e1688.
44 Šantak, M. and Matić, Z. (2022). The role of nucleoprotein in immunity to human negative-stranded RNA viruses – not just another brick in the viral nucleocapsid. *Viruses* 14 (3): 521.
45 Guseva, S., Milles, S., Blackledge, M., and Ruigrok, R.W. (2019). The nucleoprotein and phosphoprotein of measles virus. *Frontiers in Microbiology* 10: 460035.
46 Bermek, O. and Williams, R.S. (2021). The three-component helicase/primase complex of herpes simplex virus-1. *Open Biology* 11 (6): 210011.
47 Sato, Y., Suenaga, T., Kobayashi, M. et al. (2021). Characteristics of helicase-primase inhibitor Amenamevir-resistant herpes simplex virus. *Antimicrobial Agents and Chemotherapy* 65 (10): https://doi.org/10.1128/aac.00494-21.
48 Amin, S.A., Banerjee, S., Ghosh, K. et al. (2021). Protease targeted COVID-19 drug discovery and its challenges: insight into viral main protease (Mpro) and papain-like protease (PLpro) inhibitors. *Bioorganic & Medicinal Chemistry* 29: 115860.
49 Amblard, F., Patel, D., Michailidis, E. et al. (2022). HIV nucleoside reverse transcriptase inhibitors. *European Journal of Medicinal Chemistry* 240: 114554.

50 Wonganan, P., Limpanasithikul, W., Jianmongkol, S. et al. (2020). Pharmacokinetics of nucleoside/nucleotide reverse transcriptase inhibitors for the treatment and prevention of HIV infection. *Expert Opinion on Drug Metabolism & Toxicology* 16 (7): 551–564.
51 Davies, B.E. (2010). Pharmacokinetics of oseltamivir: an oral antiviral for the treatment and prophylaxis of influenza in diverse populations. *Journal of Antimicrobial Chemotherapy* 65 (suppl_2): ii5–ii10.
52 Sehnert, B., Mietz, J., Rzepka, R. et al. (2021). Neuraminidase inhibitor zanamivir ameliorates collagen-induced arthritis. *International Journal of Molecular Sciences* 22 (3): 1428.
53 Hossain, M.I., Asha, A.T., Hossain, M.A. et al. (2024). Investigating the role of hypothetical protein (AAB33144. 1) in HIV-1 virus pathogenicity: a comparative study with FDA-approved inhibitor compounds through *in silico* analysis and molecular docking. *Heliyon* 10 (1).
54 Kogawa, A.C. and Salgado, H.R.N. (2014). Characteristics, complexation and analytical methods of darunavir. *British Journal of Pharmaceutical Research* 1276–1286.
55 Vardanyan, R. and Hruby, V. (2016). *Synthesis of Best-Seller Drugs*. Academic Press.
56 Thomas, E., Ghany, M.G., and Liang, T.J. (2012). The application and mechanism of action of ribavirin in therapy of hepatitis C. *Antiviral Chemistry and Chemotherapy* 23 (1): 1–12.
57 Beg, M. and Athar, F. (2020). Anti-HIV and Anti-HCV drugs are the putative inhibitors of RNA-dependent-RNA polymerase activity of NSP12 of the SARS CoV-2 (COVID-19). *Pharmacy & Pharmacology International Journal* 8 (3): 163–172.
58 Sluis-Cremer, N. (2021). Retroviral reverse transcriptase: structure, function and inhibition. *Enzymes* 50: 179–194. https://doi.org/10.1016/bs.enz.2021.06.006.
59 Underwood, M., Horton, J., Nangle, K. et al. (2022). Integrase inhibitor resistance mechanisms and structural characteristics in antiretroviral therapy-experienced, integrase inhibitor-naive adults with HIV-1 infection treated with dolutegravir plus two nucleoside reverse transcriptase inhibitors in the DAWNING study. *Antimicrobial Agents and Chemotherapy* 66 (1): e01643–e01621.
60 Rashid, M.I., Rehman, S., Ali, A., and Andleeb, S. (2019). Fishing for vaccines against *Vibrio cholerae* using *in silico* pan-proteomic reverse vaccinology approach. *PeerJ* 7 (22): e6223.
61 Procopio, A.C. et al. (2024). Integrated one health strategies in dengue. *One Health* 100684.
62 Fischer, M., Müller, P., Scheidt, H.A., and Luck, M. (2022). Drug–membrane interactions: effects of virus-specific RNA-dependent RNA polymerase inhibitors remdesivir and favipiravir on the structure of lipid bilayers. *Biochemistry* 61 (13): 1392–1403.
63 Di Pietrantonj, C., Rivetti, A., Marchione, P. et al. (2020). Vaccines for measles, mumps, rubella, and varicella in children. *Cochrane Database of Systematic Reviews* 4.
64 Mohammadi, Y., Nezafat, N., Negahdaripour, M. et al. (2023). In silico design and evaluation of a novel mRNA vaccine against BK virus: a reverse vaccinology approach. *Immunologic Research* 71 (3): 422–441.
65 Soliman, M., Oredein, O., and Dass, C.R. (2021). Update on safety and efficacy of HPV vaccines: focus on gardasil. *International Journal of Molecular and Cellular Medicine* 10 (2): 101.
66 Potts, M. (2023). *Characterisation of Novel Secreted Antiviral Factors to Human Cytomegalovirus*. University of Cambridge.
67 Khanal, P. (2020). Remdesivir for COVID-19 treatment: mechanism OF action, synthesis, and clinical trials. *Journal of Pharmacy & Pharmaceutical Sciences* 9: 1062–1068.
68 Schaechter, M. (2009). *Encyclopedia of Microbiology*. Academic Press.
69 Gritsenko, D. and Hughes, G. (2015). Ledipasvir/Sofosbuvir (harvoni): improving options for hepatitis C virus infection. *Pharmacy and Therapeutics* 40 (4): 256.
70 Olotu, F.A., Agoni, C., Soremekun, O., and Soliman, M.E. (2020). The recent application of 3D-QSAR and docking studies to novel HIV-protease inhibitor drug discovery. *Expert Opinion on Drug Discovery* 15 (9): 1095–1109.

71 Patel, P.H. and Zulfiqar, H. (2023). Reverse transcriptase inhibitors. In: *StatPearls [Internet]* (ed. P.H. Patel and H. Zulfiqar). StatPearls Publishing.

72 Khan, T. and Raza, S. (2023). Exploration of computational aids for effective drug designing and management of viral diseases: a comprehensive review. *Current Topics in Medicinal Chemistry* 23 (17): 1640–1663.

73 Kumar, N. and Acharya, V. (2023). Machine intelligence-guided selection of optimized inhibitor for human immunodeficiency virus (HIV) from natural products. *Computers in Biology and Medicine* 153: 106525.

74 Todd, B., Tchesnokov, E.P., and Götte, M. (2021). The active form of the influenza cap-snatching endonuclease inhibitor baloxavir marboxil is a tight binding inhibitor. *Journal of Biological Chemistry* 296.

75 Gider, V. and Budak, C. (2022). Instruction of molecular structure similarity and scaffolds of drugs under investigation in ebola virus treatment by atom-pair and graph network: a combination of favipiravir and molnupiravir. *Computational Biology and Chemistry* 101: 107778.

76 Ullah, M.A., Sarkar, B., and Islam, S.S. (2020). Exploiting the reverse vaccinology approach to design novel subunit vaccines against Ebola virus. *Immunobiology* 225 (3): 151949.

77 Yang, W., Wang, Y., Han, D. et al. (2024). Recent advances in application of computer-aided drug design in anti-COVID-19 virials drug discovery. *Biomedicine & Pharmacotherapy* 173: 116423.

78 Diakou, I., Papakonstantinou, E., Papageorgiou, L. et al. (2022). Novel computational pipelines in antiviral structure-based drug design. *Biomedical Reports* 17 (6): 1–5.

79 Ferreira, L.G., Dos Santos, R.N., Oliva, G., and Andricopulo, A.D. (2015). Molecular docking and structure-based drug design strategies. *Molecules* 20 (7): 13384–13421.

80 Xiao, N., D. Cao, and Q. Xu, *Rcpi: R/Bioconductor Package as an Integrated Informatics Platform in Drug Discovery*. 2015.

81 Panwar, P., Yang, Q., and Martini, A. (2023). PyL3dMD: Python LAMMPS 3D molecular descriptors package. *Journal of Cheminformatics* 15 (1): 69.

82 Ullah, A., Qazi, J., Rahman, L. et al. (2020). Nanoparticles-assisted delivery of antiviral-siRNA as inhalable treatment for human respiratory viruses: a candidate approach against SARS-COV-2. *Nano Select* 1 (6): 612–621.

83 Namba-Nzanguim, C.T., Turon, G., Simoben, C.V. et al. (2022). Artificial intelligence for antiviral drug discovery in low resourced settings: a perspective. *Frontiers in Drug Discovery* 2: 1013285.

84 Emami, H. (2022). Anti-coronavirus optimization algorithm. *Soft Computing* 26 (11): 4991–5023.

85 Sharma, R., Shrivastava, S., Singh, S.K. et al. (2021). Deep-AVPpred: artificial intelligence driven discovery of peptide drugs for viral infections. *IEEE Journal of Biomedical and Health Informatics* 26 (10): 5067–5074.

86 Pang, Y., Yao, L., Jhong, J.H. et al. (2021). AVPIden: a new scheme for identification and functional prediction of antiviral peptides based on machine learning approaches. *Briefings in Bioinformatics* 22 (6): bbab263.

87 Yang, Y., Zhou, D., Zhang, X. et al. (2022). D3AI-CoV: a deep learning platform for predicting drug targets and for virtual screening against COVID-19. *Briefings in Bioinformatics* 23 (3): bbac147.

88 Schaduangrat, N., Nantasenamat, C., Prachayasittikul, V., and Shoombuatong, W. (2019). Meta-iAVP: a sequence-based meta-predictor for improving the prediction of antiviral peptides using effective feature representation. *International Journal of Molecular Sciences* 20 (22): 5743.

89 Ma, X., Liang, Y., and Zhang, S. (2023). iAVPs-ResBi: identifying antiviral peptides by using deep residual network and bidirectional gated recurrent unit. *Mathematical Biosciences and Engineering: MBE* 20 (12): 21563–21587.

90 Sherman, A.C., Mehta, A., Dickert, N.W. et al. (2019). The future of flu: a review of the human challenge model and systems biology for advancement of influenza vaccinology. *Frontiers in Cellular and Infection Microbiology* 9: 107.

91 Nasrin, T., Hoque, M., and Ali, S. (2023). Systems biology of the genomes' microsatellite signature of Orthopoxvirus including the Monkeypox virus. *Comparative Immunology, Microbiology and Infectious Diseases* 98: 102002.

92 Zhuang, Z., Wen, J., Zhang, L. et al. (2020). Can network pharmacology identify the anti-virus and anti-inflammatory activities of Shuanghuanglian oral liquid used in Chinese medicine for respiratory tract infection? *European Journal of Integrative Medicine* 37: 101139.

93 Shah, P.S., Beesabathuni, N.S., Fishburn, A.T. et al. (2022). Systems biology of virus-host protein interactions: from hypothesis generation to mechanisms of replication and pathogenesis. *Annual Review of Virology* 9: 397–415.

94 Liu, T., Li, Y., Wang, L. et al. (2023). Network pharmacology-based exploration identified the antiviral efficacy of Quercetin isolated from mulberry leaves against enterovirus 71 via the NF-κB signaling pathway. *Frontiers in Pharmacology* 14: 1260288.

95 Chen, S., Yu, X., and Guo, D. (2018). CRISPR-Cas targeting of host genes as an antiviral strategy. *Viruses* 10 (1): 40.

96 Hryhorowicz, M., Lipiński, D., and Zeyland, J. (2023). Evolution of CRISPR/cas systems for precise genome editing. *International Journal of Molecular Sciences* 24 (18): 14233.

97 Bijlani, S., Pang, K.M., Sivanandam, V. et al. (2022). The role of recombinant AAV in precise genome editing. *Frontiers in Genome Editing* 3: 799722.

98 Abbott, T.R., Dhamdhere, G., Liu, Y. et al. (2020). Development of CRISPR as an antiviral strategy to combat SARS-CoV-2 and influenza. *Cell* 181 (4): 865–876.e12.

99 Park, H.M., Liu, H., Wu, J. et al. (2018). Extension of the crRNA enhances Cpf1 gene editing *in vitro* and *in vivo*. *Nature Communications* 9 (1): 3313.

100 Rajan, A., Shrivastava, S., Janhawi et al. (2022). CRISPR-Cas system: from diagnostic tool to potential antiviral treatment. *Applied Microbiology and Biotechnology* 106 (18): 5863–5877.

101 Delshadi, R., Bahrami, A., McClements, D.J. et al. (2021). Development of nanoparticle-delivery systems for antiviral agents: a review. *Journal of Controlled Release* 331: 30–44.

102 Yang, G., Chen, S., and Zhang, J. (2019). Bioinspired and biomimetic nanotherapies for the treatment of infectious diseases. *Frontiers in Pharmacology* 10: 751.

103 Chakravarty, M. and Vora, A. (2021). Nanotechnology-based antiviral therapeutics. *Drug Delivery and Translational Research* 11: 748–787.

104 Cordeiro, A.S., Patil-Sen, Y., Shivkumar, M. et al. (2021). Nanovaccine delivery approaches and advanced delivery systems for the prevention of viral infections: from development to clinical application. *Pharmaceutics* 13 (12): 2091.

105 Maus, A., Strait, L., and Zhu, D. (2021). Nanoparticles as delivery vehicles for antiviral therapeutic drugs. *Engineered Regeneration* 2: 31–46.

106 Dhanushkumar, T., Kumar Selvam, P., Santhosh, M.E. et al. (2024). Rational design of a multivalent vaccine targeting arthropod-borne viruses using reverse vaccinology strategies. *International Journal of Biological Macromolecules* 258: 128753.

107 Sarkar, B. and Ullah, M.A. (2020). Designing novel subunit vaccines against herpes simplex virus-1 using reverse vaccinology approach. *BioRxiv* 2020.01.10.901678.

108 Kharisma, V.D., Ansori, A.N.M., Murtadlo, A.A.A. et al. (2024). Construction of peptide vaccine candidate based on β-cell epitopes of Indonesian Monkeypox virus (MPXV) virulence protein: a reverse vaccinology. *Research Journal of Pharmacy and Technology* 17 (1): 291–296.

20

Rational Design of Anticancer Therapeutics

Debarupa Dutta Chakraborty and Prithviraj Chakraborty

Royal School of Pharmacy, The Assam Royal Global University, Guwahati, Assam, India

20.1 Introduction

Cancer ranks as the second most common cause of mortality after cardiovascular disorders worldwide. Its therapy is therefore quite challenging [1, 2]. Globally, the rising rate of cancer incidence and death is among the most significant public health issues with a significant socioeconomic impact. Decades of worldwide research efforts have yielded no cure for cancer since effective and targeted medications are still lacking. This leads to high rates of recurrence and relapse, which in turn perpetuates cancer's leading role in global mortality. One of the primary obstacles to overcoming the high incidence and prevalence of the disease in this environment is the design of fresh, effective anticancer medicines that can lessen the unwanted effects of current treatments [3, 4].

A lot of ongoing research are being done on cancer therapy and prevention. Radiation, chemotherapy, and surgery are the clinical methods of cancer treatment that are most frequently employed. As a systemic treatment, chemotherapy can be beneficial for certain malignancies that have metastasized to more advanced forms of the tumor or have a tendency to spread throughout the body [5, 6]. Creating a new medication is difficult, costly, time consuming, and hazardous. The traditional drug discovery process is thought to require up to 15 years and more than $1 billion USD before a new medication is ready for the market. Fortunately, new techniques have recently emerged, altering this situation. To enhance the effectiveness of the discovery stage, numerous innovative tools and approaches have been created, and computational techniques are now an essential part of many drug discovery initiatives. Many discovery efforts employ strategies like ligand- or structure-based virtual screening (VS), which range from hit identification to lead optimization. In the case of developing possible anticancer medications and drug candidates, these computational methods have shown to be quite influential throughout time and have yielded valuable insightful findings in the field of cancer research [7].

Utilizing disease-related screens for a wide range of chemical types, followed by systematic chemical alteration of active compounds, is the traditional method for finding new anticancer medications. Conversely, identified specific macromolecular targets are becoming a more common starting point for novel drug research. As the findings from research on the molecular basis of oncogenesis are used to treat and prevent cancer, the significance of this trend will increase. Specifically, the involvement of specific oncogenic proteins as targets for chemotherapeutic intervention is highlighted by their participation in several malignancies, either as mutations or as overexpressed levels [8].

Computational Methods for Rational Drug Design, First Edition. Edited by Mithun Rudrapal.

Rational drug design entails using molecular modeling tools like pharmacophore modeling, molecular dynamics, VS, and molecular docking in order to describe molecular factors for drug target contact, explain biomolecule function, and create more effective drug candidates [9]. The Nobel Prizes given to George H. Hitchings and Gertrude Elion for their contributions to the study of nucleic acid metabolism in both normal and malignant human cells can be used to trace the origins of rational drug design. By introducing concepts from biochemistry and physiology to the science of drug discovery, they transformed the field and made it easier to comprehend the molecular basis of many disorders [10, 11].

Three components make up an excellent candidate for rational drug design: an imaging reporter, a nanoparticle carrier, and a therapeutic DNA/RNA component. DNA/RNA-targeted methods may play a crucial role in rational therapeutics. This crucial component utilizes the genome and transcriptome "coded" nature. Therefore, DNA/RNA-targeted techniques provide the perfect framework for a completely rational design of therapeutic and diagnostic agents based on the complementarity phenomenon. An additional element of the rational treatment that is envisaged is nanotechnology. More specifically, nanoparticles are carriers capable of incorporating all three of the proposed therapeutic components. Without affecting the oligos' ability to function, they are readily functionalized with oligonucleotides. Standard synthetic chemistry, such as liposomes and iron oxide nanoparticles, can be used to fine-tune their design. Finally, an imaging reporter, such as a radionuclide, can be used to label them. The third component of the envisioned drug design is image guidance. One can monitor the therapeutic agent's transport to the target tissue by labeling the drug with an imaging reporter. Delivery to target tissues can be evaluated and controlled with the use of image-guided delivery. Imaging can assist in determining the best drug design, delivery schedule, route, and therapeutic dose on an individual basis and suggest alternatives in the event that a patient's medication is unsuccessful [12].

With the advent of precision medicine, researchers have discovered that changes in key intracellular biomolecules – which are often found at the subcellular level – play a crucial role in carcinogenesis and the progression of cancer. The use of molecular-level pathogenesis to design therapeutic candidates has emerged as a new paradigm in drug discovery. In addition to delivering therapeutic compounds to specific tissues and cells, efficient nanoparticle-based drug delivery systems (NDDSs) should precisely target organelles as discrete subcellular locations. They are thought to be among the most promising methods for treating cancer. Through targeted modifications and proper design, tumor cells are enriched in subcellular-targeting nanoformulations. They are internalized by endocytosis across subcellular barriers and target-specific subcellular structures. Subsequently, the therapeutic compounds are released under controlled conditions at the desired locations, thereby improving their anticancer efficacy, decreasing their deleterious effects, and surmounting the most significant obstacle to intracellular drug delivery-multidrug resistance (MDR) [13–18].

This study describes the rational design of nanomedicine, the drug delivery mechanism required for cancer treatment, and focuses on the essential components based on the most current research advancements over the previous few years. Furthermore, various methods of rational drug design in anticancer theranostics will be summarized.

20.2 Rational Design of Nanomedicine for Cancer Treatment

Cancer nanomedicines have been designed to overcome the pharmacokinetic restrictions connected with conventional drugs. The biggest benefit is their capacity to accumulate preferentially in tumor tissues by taking advantage of the enhanced permeability and retention (EPR) effect that tumors exhibit due to their inadequate lymphatic drainage and pathologically leaky vasculature.

Clinical data has accumulated evidence demonstrating the nanoparticle localization in solid human tumors following systemic delivery to cancer patients [19–28].

These nanomedicines have demonstrated numerous benefits over traditional chemotherapeutics, including increased drug solubility, extended circulation in the blood compartments, improved bioavailability, and significantly decreased adverse effects from the parent drugs, all of which have significantly improved patients' quality of life. However, current cancer nanomedicines have failed to improve the therapeutic benefits – these nanomedicines cannot or only modestly improve patients' overall survival [19, 29–33].

Although these nanomedicines were able to accumulate more in tumor tissues than free pharmaceuticals with less adverse effects, detailed clinical assessments revealed that larger drug concentrations did not improve therapeutic efficacy. Therefore, the secret to creating next-generation nanomedicines with great therapeutic efficacy lies in understanding how to construct them so that they can both effectively target tumor tissues and exert the drugs' therapy [19, 34].

20.3 The CAPIR Cascade: A Nanomedicine Strategy for Administering Cancer Medications

Delivering medications into cancer cells as free molecules so they can have their therapeutic effects is the ultimate aim of cancer drug delivery. The process of administering a nanomedicine intravenously usually involves five steps: the drug circulates through the blood compartments, accumulates in the tumor due to its poorly constructed and leaky vasculature, penetrates deep into the tumor tissue to reach the tumor cells, internalizes within those cells, and finally releases the drug intracellularly: CAPIR cascade, to put it briefly (Figure 20.1). Thus, the overall delivery efficiency, the percentage of the drug as free molecules inside targeted cells over the injected, of a system is a product of efficiencies of the five steps [19, 35–37].

20.4 Rational Regulation of Nanoparticle's Physicochemical Characteristics

Particle size, shape, surface decoration, and intelligent change under particular conditions are examples of physicochemical features that can be the main determinants of nanoparticle tumor accumulation and deep penetration capabilities [38].

20.4.1 Particle Size

One of the most crucial aspects of nanoparticles is their size, as it has a direct bearing on the *in vivo* circulation, distribution, tumor accumulation, and elimination. Compelling evidence from multiple groups has led to an agreement that improved nanoparticle penetration into the tumor mass requires smaller particle sizes.

20.4.2 Shape

Another crucial aspect of nanoscale vehicles is particle shape, which plays a key role in tumor penetration. Elongated particles have demonstrated deeper penetration in multiple tumor models compared to spherical particles.

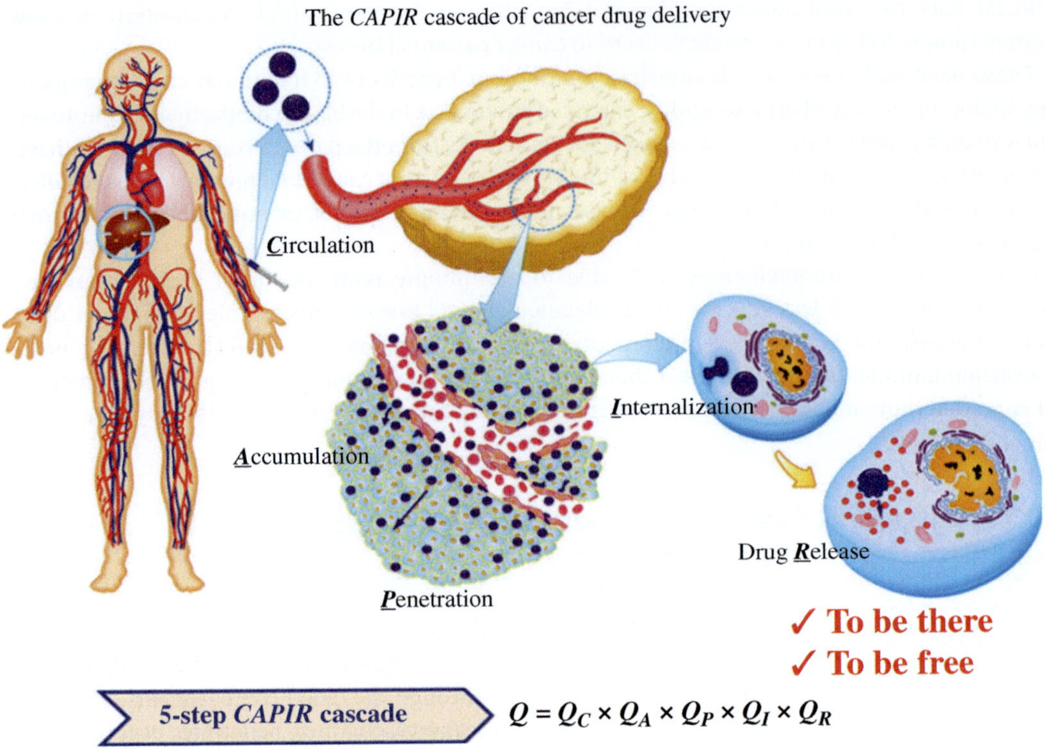

Figure 20.1 A 5-step CAPIR cascade of nanomedicines. *Source:* Sun et al. (19)/with permission of John Wiley & Sons.

20.4.3 Surface Modification

Another essential component of nanoparticles that influences their *in vivo* destiny and penetration effectiveness in primary tumors and metastases is surface modification. Most significantly, active targeting moieties added to the surface of nanoparticles are widely used to enhance selective accumulation and therapeutic effects.

A wide range of biological, organic, and inorganic materials can be used to create nanoscale particles, which can then be designed with a variety of qualities to aid with tumour penetration and medication administration. Smaller nanoparticles have superior penetration effects in primary tumours or lymph node metastases, according to all of the above data from a range of materials and structures, especially when the diameter is less than 50 nm. To treat metastatic tumours in the lung and brain, for example, nonspherical nanoparticles can gather preferentially in those organs. However, the shape of these particles needs to be carefully regulated. Surface modification is essential for improving detection and penetration in primary or metastatic tumours, in addition to optimised particle size or shape.

20.5 Some Approaches of Rational Drug Design in Anticancer Theranostics

In contemporary molecular pharmacology, the discovery of novel, pharmacologically active small molecules is a significant and quickly developing field. Since there are just a few proteins that can be drugged, it is critical to find as many chemical effectors as possible in order to determine the most effective anticancer treatment plan for each unique situation. An interesting target for small-molecule inhibitors is Mdm2, an E3 ubiquitin ligase that regulates ubiquitin-dependent degradation of the critical tumor suppressor p53. Numerous novel inhibitors of the p53-Mdm2 interaction were found by employing a hybrid technique that combines high-content screening utilizing cancer cell lines with the rational creation of small compounds picked from the virtual library. At rates greater than the well-known Mdm2 inhibitor Nutlin-3, these substances were able to activate and stabilize the p53 protein, resulting in extensive apoptosis, preferably in p53-positive cells [39].

A product of glutamic acid, glutamine is a necessary growth factor for tumor cells that proliferate quickly. Studies have shown that tumors significantly alter the host's glutamine metabolism, causing the nitrogen metabolism to adjust to the increased glutamine requirements of the tumor. Overall, glutamine promotes the rapid growth and proliferation of cancer cells, pointing to a potential positive correlation between glutamine and glutamic acid in cancer. Moreover, the reintroduction of thalidomide, a synthetic derivative of glutamic acid, in clinical studies for the management of different malignant tumors provides strong evidence linking glutamine and glutamic acid to cancer. Since thalidomide is a synthetic derivative of glutamic acid with a chiral stereocenter, it experiences *in vivo* racemization resulting in the production of *R* and *S* enantiomers. It is well known that thalidomide has antiangiogenic, anticytokine, anti-integrin, and immunomodulatory properties. Additionally, it inhibits the synthesis of proinflammatory cytokine TNF-α and initiates the synthesis of IL-2 and IFN-γ. Furthermore, it is commonly recognized that thalidomide has proapoptotic and antiproliferative effects on tumor cells. It was thought that thalidomide would be extremely important in the treatment of multiple myeloma and other cancers since it has positive effects on the immune system and inhibits the angiogenesis process. Owing to thalidomide's low solubility in water and its propensity for spontaneous hydrolytic cleavage, a number of chemically altered thalidomide analogs were created, produced, and studied for use in treating different cancer types. In comparison to thalidomide, Lenalidomide (CC-5013), a structural thalidomide analog, exhibited immunomodulatory effects and significantly greater potency in the proliferation and tube formation assays of human umbilical vein endothelial cells. It has been noted that Pomalidomide (CC-4047) triggers protective, long-lasting, and tumor-specific *in vivo* Th1-type responses. It is significantly more effective than thalidomide *in vitro* and *in vivo* at inhibiting the angiogenesis and proliferation of myeloma malignancies. Thalidomide's small-molecule analog is ENMD-0995. It is interesting to note that there was no evidence of hazardous side effects and that this thalidomide analog exhibited improved angiogenesis-inhibiting activity. ENMD-0995 was designated as an orphan drug by the FDA in 2002 in order to treat multiple myeloma patients. Once outlawed and removed from the market, thalidomide is now being used as a model for the design of new anticancer medications and as the preferred anticancer medication for treating a number of solid tumors [1, 40–52].

Another novel class of potential cancer treatment is represented by membranolytic anticancer peptides (ACPs). Their receptor-independent mechanism of action might prevent the development of cellular resistance. In the 1990s, "Simulated Molecular Evolution (SME)" – a stochastic optimization method – was created for computational peptide design. SME is a member of the evolutionary algorithm class, which also includes genetic algorithms, and can be used to optimize peptide properties encoded in a theoretical fitness function or in conjunction with an experimental fitness evaluation in situations where structure–activity relationships cannot be beforehand. The computational approach established by Gabernet et al. [53] is a step toward the rational design of effective and targeted ACPs. By merging a machine learning model with the SME algorithm, ACPs with low-micromolar efficacy against various cancer cells and selectivity against nontransformed cells (HDMEC, human dermal microvascular endothelial cells) and human erythrocytes were generated in this study. The machine-learning classifier by itself was able to recognize active peptides, but it was not able to recognize peptides that were specific to cancer cells. Four new ACPs were found using VS of computationally built peptide libraries using an integrated machine-learning classifier. These ACPs will provide as the foundation for selectivity optimization by SME. SME is applicable to all kinds of experimental readouts and avoids using more traditional peptide optimization techniques like alanine scanning. Simultaneously, the findings imply that further enhanced cancer cell selectivity of membranolytic ACPs might come at the price of reduced peptide potency. This working hypothesis serves as a foundation for future research [53–59].

20.6 Artificial Intelligence's Progress in Anticancer Drug Development

The simulation of human intelligence in machines that are designed to act and think like humans is known as artificial intelligence, or AI. These days, it is applied to a broad variety of cancer research, including the prediction of drug–protein interactions, the prediction of target protein structures, and the image classification of aberrant cancer cells. These findings show how the application of AI methods can significantly alter the development of anticancer medications. Figure 20.2 shows some applications where AI is used in the design of anticancer drugs [60–65].

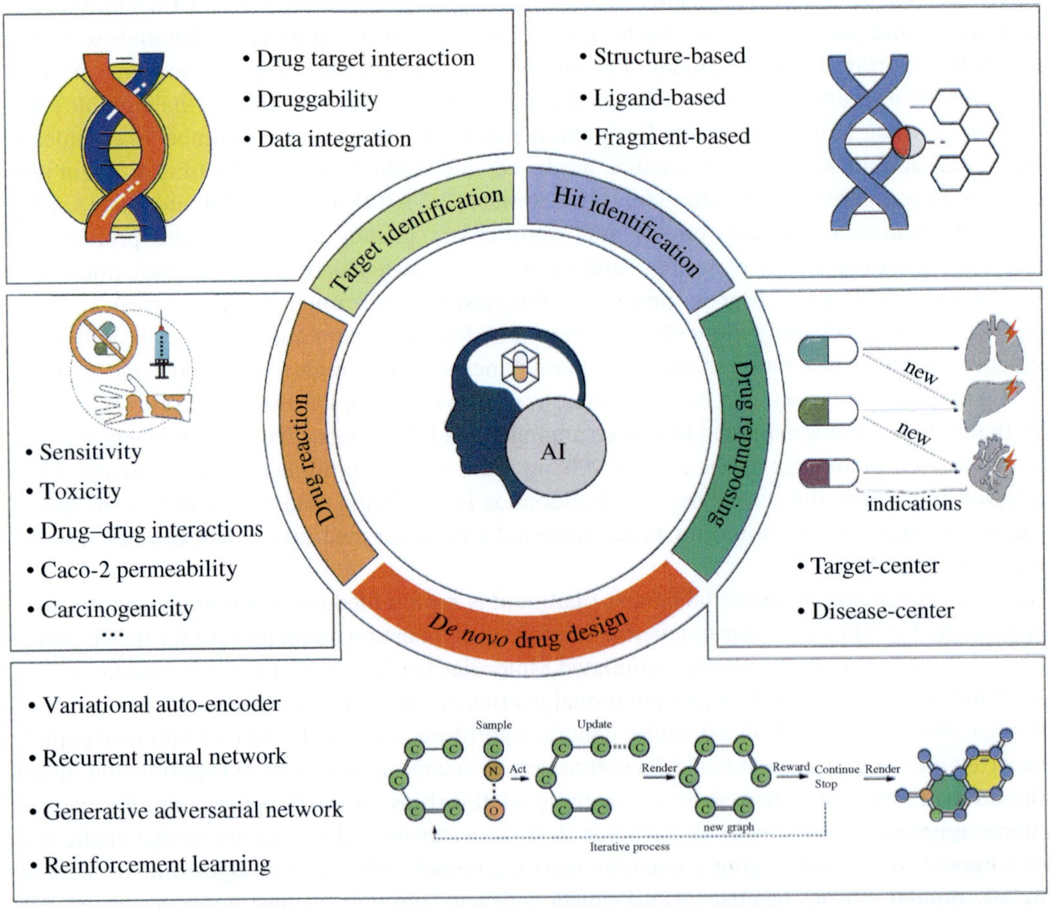

Figure 20.2 Application of AI in anticancer drug design. *Source:* Xuan et al. [60]/MDPI/CC BY 4.0.

20.6.1 Identification of Anticancer Drug Targets Using Artificial Intelligence

The first stage in designing an anticancer drug is to identify drug–target interactions (DTIs). Binding affinity constants, which include indicators like a dissociation constant (K_d), an inhibition constant (K_i), and a half-maximal inhibitory concentration (IC50), are frequently used to characterize the strength of drug–target binding. The computational prediction of DTIs is quite interesting because experimentally determining them is expensive and time-consuming. Precise and effective DTI forecasts can greatly aid in the development of new drugs and hasten up the search for lead or hit compounds [64, 66].

20.6.2 Artificial Intelligence for Hit Compound Screening of Anticancer Drugs

The next step after determining the therapeutic targets for anticancer drugs is to search for anticancer drug hit compounds, which are molecules with initial activity against a particular target or mechanism of action. The primary method for finding computer-assisted hit compounds is high-throughput screening (HTS). HTS can be accomplished using two different techniques: ligand-based screening and structure-based screening. Structure-based VS uses docking and scoring to find compounds with high binding affinities for target proteins. This technique is an important tool for the discovery of anticancer drugs, despite the fact that many of the existing docking approaches are time-consuming and pose challenges for large-scale VS. Ligand-based screening works by taking small compounds with known activity and using a compound library to find structures as candidates that have comparable chemical or physical properties [64, 67–69].

20.6.3 Artificial Intelligence-Based *De Novo* Anticancer Drug Design

The estimated number of drug-like compounds in the chemical space is 10^{23}~10^{60}. Consequently, mining the full chemical space with computational approaches is almost impossible. In this setting, identifying particular lead molecules within the large chemical space is a significant issue. HTS and VS technologies are able to efficiently evaluate compounds in huge compound libraries using a broad variety of filters due to the fast advancements of computational power and experimental procedures. *De novo* drug design is a molecular generation technique that uses AI to create and optimize a molecule [64, 70–73].

20.6.4 Artificial Intelligence for Repurposing Anticancer Drugs

A key component of drug development is the efficient identification of novel indications from approved or well-established clinical medications. One term for this procedure is "drug repositioning." Theoretically, repurposing is less expensive, quicker, safer, and easier than the established obstacles to the creation of novel molecules. Possibilities for repurposing drugs are frequently determined by incidental findings or time-consuming, nonhypothesis-driven preclinical drug screening [64].

20.6.5 Reactions to Anticancer Drugs Accurately Predicted with Artificial Intelligence Support

Drug sensitivity, drug toxicity, and drug–drug interactions are all influenced by the ADMET characteristics of the drugs. Predicting drug reactions accurately may enhance patient outcomes and boost clinical trial success rates. A rising number of relevant research employing AI approaches during the drug design stage are being proposed as a result of the AI technologies' quick development [64, 74, 75].

The incorporation of AI techniques in various stages of anticancer drug development offers promising solutions to obstacles like identifying DTIs, screening for hit compounds, designing novel drugs, repurposing old ones, and predicting drug reactions accurately. AI-powered approaches hold the potential to significantly expedite the drug discovery process, reduce costs, and improve patient outcomes in cancer treatment. As AI technologies continue to evolve, their application in drug development is anticipated to further advance, leading to more effective and targeted therapies for cancer patients.

20.7 Conclusion

In conclusion, the fight against cancer continues to be one of the biggest obstacles in modern medicine, given its devastating impact on global health and the lack of a definitive cure. Despite decades of research and significant advancements in therapeutic modalities including chemotherapy, radiation, and surgery, the high incidence and prevalence of cancer persist, necessitating the creation of new and more potent anticancer treatments. The amalgamation of rational drug design principles, nanomedicine approaches, and AI technologies offers promising avenues for revolutionizing cancer therapy. Rational drug design strategies leverage molecular insights into cancer pathogenesis to develop targeted therapies with enhanced efficacy and reduced side effects. Nanomedicine platforms enhance drug delivery to tumor tissues while minimizing systemic toxicity, potentially improving patient outcomes. Moreover, AI-driven approaches enable the efficient identification of drug targets, screening of potential hit compounds, designing new drugs, repurposing of existing medications, and forecasting drug reactions, thereby expediting the drug discovery process. By combining these innovative approaches, researchers are poised to usher in a new era of precision medicine, offering personalized and more effective treatment options for cancer patients worldwide.

References

1 Ali, I., Wani, W.A., Haque, A., and Saleem, K. (2013). Glutamic acid and its derivatives: candidates for rational design of anticancer drugs. *Future Medicinal Chemistry* 5 (8): 961–978.

2 Ali, I., Wani, W.A., and Saleem, K. (2011). Cancer scenario in India with future perspective. *Cancer Therapy* 8: 56–70.

3 Morais, T.S., Fernandes, A.R., Baptista, P.V., and Gambino, D. (2022). Editorial: rational drug design of metal complexes for cancer therapy. *Frontiers in Chemistry* 10.

4 Debela, D.T., Muzazu, S.G., Heraro, K.D. et al. (2021). New approaches and procedures for cancer treatment: current perspectives. *SAGE Open Medicine* 9: 205031212110343.

5 Ellis, C.N., Ellis, M.B., and Blakemore, W.S. (1987). Effect of adriamycin on heart mitochondrial DNA. *Biochemical Journal* 245 (1): 309–312.

6 Meng, Q.Y., Cong, H.L., Hu, H., and Xu, F.J. (2020). Rational design and latest advances of codelivery systems for cancer therapy. *Materials Today Bio* 7: 100056.

7 Prada-Gracia, D., Huerta-Yépez, S., and Moreno-Vargas, L.M. (2016). Application of computational methods for anticancer drug discovery, design, and optimization. *Boletín Médico del Hospital Infantil de México* 73 (6): 411–423.

8 Neidle, S. (1994). Discovery of new anticancer drugs by computer-aided drug design. *Annals of Oncology* 5: S51–S54.

9 Gagic, Z., Ruzic, D., Djokovic, N. et al. (2020). *In silico* methods for design of kinase inhibitors as anticancer drugs. *Frontiers in Chemistry* 7.

10 Magalhaes, L.G., Ferreira, L.L.G., and Andricopulo, A.D. (2018). Recent advances and perspectives in cancer drug design. *Anais da Academia Brasileira de Ciências* 90 (1 suppl 2): 1233–1250.

11 Baguley, B.C. and Kerr, D.J. (ed.) (2001). *Anticancer Drug Development*. Elsevier.

12 Yoo, B., Billig, A.M., and Medarova, Z. (2017). Guidelines for rational cancer therapeutics. *Frontiers in Oncology* 7.

13 Fu, X., Shi, Y., Qi, T. et al. (2020). Precise design strategies of nanomedicine for improving cancer therapeutic efficacy using subcellular targeting. *Signal Transduction and Targeted Therapy* 5 (1): 262.

14 Stratton, M.R. (2011). Exploring the genomes of cancer cells: progress and promise. *Science (1979)* 331 (6024): 1553–1558.

15 Rozenblatt-Rosen, O., Regev, A., Oberdoerffer, P. et al. (2020). The human tumor atlas network: charting tumor transitions across space and time at single-cell resolution. *Cell* 181 (2): 236–249.

16 Murciano-Goroff, Y.R., Taylor, B.S., Hyman, D.M., and Schram, A.M. (2020). Toward a more precise future for oncology. *Cancer Cell* 37 (4): 431–442.

17 Liu, J., Li, M., Luo, Z. et al. (2017). Design of nanocarriers based on complex biological barriers *in vivo* for tumor therapy. *Nano Today* 15: 56–90.

18 Rajendran, L., Knölker, H.J., and Simons, K. (2010). Subcellular targeting strategies for drug design and delivery. *Nature Reviews. Drug Discovery* 9 (1): 29–42.

19 Sun, Q., Zhou, Z., Qiu, N., and Shen, Y. (2017). Rational design of cancer nanomedicine: nanoproperty integration and synchronization. *Advanced Materials* 29 (14): 1606628.

20 Blanco, E., Shen, H., and Ferrari, M. (2015). Principles of nanoparticle design for overcoming biological barriers to drug delivery. *Nature Biotechnology* 33 (9): 941–951.

21 Chrastina, A., Massey, K.A., and Schnitzer, J.E. (2011). Overcoming *in vivo* barriers to targeted nanodelivery. *WIREs Nanomedicine and Nanobiotechnology* 3 (4): 421–437.

22 Farokhzad, O.C. and Langer, R. (2009). Impact of nanotechnology on drug delivery. *ACS Nano* 3 (1): 16–20.

23 Matsumura, Y. and Maeda, H. (1986). A new concept for macromolecular therapeutics in cancer chemotherapy: mechanism of tumoritropic accumulation of proteins and the antitumor agent smancs. *Cancer Research* 46 (12 Pt 1): 6387–6392.

24 Gerlowski, L.E. and Jain, R.K. (1986). Microvascular permeability of normal and neoplastic tissues. *Microvascular Research* 31 (3): 288–305.

25 Maeda, H. (2015). Toward a full understanding of the EPR effect in primary and metastatic tumors as well as issues related to its heterogeneity. *Advanced Drug Delivery Reviews* 91: 3–6.

26 Clark, A.J., Wiley, D.T., Zuckerman, J.E. et al. (2016). CRLX101 nanoparticles localize in human tumors and not in adjacent, nonneoplastic tissue after intravenous dosing. *Proceedings of the National Academy of Sciences* 113 (14): 3850–3854.

27 Davis, M.E., Zuckerman, J.E., Choi, C.H.J. et al. (2010). Evidence of RNAi in humans from systemically administered siRNA via targeted nanoparticles. *Nature* 464 (7291): 1067–1070.

28 Eliasof, S., Lazarus, D., Peters, C.G. et al. (2013). Correlating preclinical animal studies and human clinical trials of a multifunctional, polymeric nanoparticle. *Proceedings of the National Academy of Sciences* 110 (37): 15127–15132.

29 Stirland, D.L., Nichols, J.W., Miura, S., and Bae, Y.H. (2013). Mind the gap: a survey of how cancer drug carriers are susceptible to the gap between research and practice. *Journal of Controlled Release* 172 (3): 1045–1064.

30 Cabral, H. and Kataoka, K. (2014). Progress of drug-loaded polymeric micelles into clinical studies. *Journal of Controlled Release* 190: 465–476.

31 Rifkin, R.M., Gregory, S.A., Mohrbacher, A., and Hussein, M.A. (2006). Pegylated liposomal doxorubicin, vincristine, and dexamethasone provide significant reduction in toxicity compared with doxorubicin, vincristine, and dexamethasone in patients with newly diagnosed multiple myeloma. *Cancer* 106 (4): 848–858.

32 O'Brien, M.E.R., Wigler, N., Inbar, M. et al. (2004). Reduced cardiotoxicity and comparable efficacy in a phase III trial of pegylated liposomal doxorubicin HCl(CAELYX^{TM}/Doxil®) versus conventional doxorubicin for first-line treatment of metastatic breast cancer. *Annals of Oncology* 15 (3): 440–449.

33 Iwamoto, T. (2013). Clinical application of drug delivery systems in cancer chemotherapy: review of the efficacy and side effects of approved drugs. *Biological & Pharmaceutical Bulletin* 36 (5): 715–718.

34 Lammers, T., Kiessling, F., Ashford, M. et al. (2016). Cancer nanomedicine: is targeting our target? *Nature Reviews Materials* 1 (9): 16069.

35 Sun, Q., Sun, X., Ma, X. et al. (2014). Integration of nanoassembly functions for an effective delivery cascade for cancer drugs. *Advanced Materials* 26 (45): 7615–7621.

36 Brannon-Peppas, L. and Blanchette, J.O. (2004). Nanoparticle and targeted systems for cancer therapy. *Advanced Drug Delivery Reviews* 56 (11): 1649–1659.

37 Brannon-Peppas, L. and Blanchette, J.O. (2012). Nanoparticle and targeted systems for cancer therapy. *Advanced Drug Delivery Reviews* 64: 206–212.

38 Zhang, Z., Wang, H., Tan, T. et al. (2018). Rational design of nanoparticles with deep tumor penetration for effective treatment of tumor metastasis. *Advanced Functional Materials* 28 (40).

39 Barlev, N., Fedorova, O., Lezina, L., and Piletsky, S. (2015). T86. *European Journal of Cancer Supplements* 13 (1): 3–4.

40 Singhal, S., Mehta, J., Desikan, R. et al. (1999). Antitumor activity of thalidomide in refractory multiple myeloma. *New England Journal of Medicine* 341 (21): 1565–1571.

41 Hideshima, T., Chauhan, D., Shima, Y. et al. (2000). Thalidomide and its analogs overcome drug resistance of human multiple myeloma cells to conventional therapy. *Blood* 96 (9): 2943–2950.

42 Corral, L.G. and Kaplan, G. (1999). Immunomodulation by thalidomide and thalidomide analogues. *Annals of the Rheumatic Diseases* 58 (Suppl 1): i107–i113.

43 Sampaio, E.P., Sarno, E.N., Galilly, R. et al. (1991). Thalidomide selectively inhibits tumor necrosis factor alpha production by stimulated human monocytes. *The Journal of Experimental Medicine* 173 (3): 699–703.

44 B de Sanctis, J., Mijares, M., Suarez, A. et al. (2010). Pharmacological properties of thalidomide and its analogues. *Recent Patents on Inflammation & Allergy Drug Discovery* 4 (2): 144–148.

45 Jacobson, J.M. (2000). Thalidomide: a remarkable comeback. *Expert Opinion on Pharmacotherapy* 1 (4): 849–863.

46 Medina, M.Á. (2001). Glutamine and cancer. *The Journal of Nutrition* 131 (9): 2539S–2542S.

47 Melchert, M. and List, A. (2007). The thalidomide saga. *The International Journal of Biochemistry & Cell Biology* 39 (7–8): 1489–1499.

48 Streetly, M.J., Gyertson, K., Daniel, Y. et al. (2008). Alternate day pomalidomide retains anti-myeloma effect with reduced adverse events and evidence of *in vivo* immunomodulation. *British Journal of Haematology* 141 (1): 41–51.

49 Lentzsch, S., Rogers, M.S., LeBlanc, R. et al. (2002). S-3-amino-phthalimido-glutarimide inhibits angiogenesis and growth of B-cell neoplasias in mice. *Cancer Research* 62 (8): 2300–2305.

50 Dredge, K., Marriott, J.B., Todryk, S.M. et al. (2002). Protective antitumor immunity induced by a costimulatory thalidomide analog in conjunction with whole tumor cell vaccination is mediated by increased Th1-type immunity. *The Journal of Immunology* 168 (10): 4914–4919.

51 Tohnya, T.M., Hwang, K., Lepper, E.R. et al. (2004). Determination of CC-5013, an analogue of thalidomide, in human plasma by liquid chromatography–mass spectrometry. *Journal of Chromatography B* 811 (2): 135–141.

52 Juliusson, G., Celsing, F., Turesson, I. et al. (2000). Frequent good partial remissions from thalidomide including best response ever in patients with advanced refractory and relapsed myeloma. *British Journal of Haematology* 109 (1): 89–96.

53 Gabernet, G., Gautschi, D., Müller, A.T. et al. (2019). *In silico* design and optimization of selective membranolytic anticancer peptides. *Scientific Reports* 9 (1): 11282.

54 Papo, N. and Shai, Y. (2005). Host defense peptides as new weapons in cancer treatment. *CMLS Cellular and Molecular Life Sciences* 62 (7–8): 784–790.

55 Mader, J.S. and Hoskin, D.W. (2006). Cationic antimicrobial peptides as novel cytotoxic agents for cancer treatment. *Expert Opinion on Investigational Drugs* 15 (8): 933–946.

56 Schweizer, F. (2009). Cationic amphiphilic peptides with cancer-selective toxicity. *European Journal of Pharmacology* 625 (1–3): 190–194.

57 Schneider, G., Schrödl, W., Wallukat, G. et al. (1998). Peptide design by artificial neural networks and computer-based evolutionary search. *Proceedings of the National Academy of Sciences* 95 (21): 12179–12184.

58 Schneider, G., Schuchhardt, J., and Wrede, P. (1995). Peptide design in machina: development of artificial mitochondrial protein precursor cleavage sites by simulated molecular evolution. *Biophysical Journal* 68 (2): 434–447.

59 Schneider, G. and Wrede, P. (1994). The rational design of amino acid sequences by artificial neural networks and simulated molecular evolution: *de novo* design of an idealized leader peptidase cleavage site. *Biophysical Journal* 66 (2): 335–344.

60 Xuan, P., Zhang, Y., Cui, H. et al. (2021). Integrating multi-scale neighbouring topologies and cross-modal similarities for drug–protein interaction prediction. *Briefings in Bioinformatics* 22 (5).

61 Pandiyan, S. and Wang, L. (2022). A comprehensive review on recent approaches for cancer drug discovery associated with artificial intelligence. *Computers in Biology and Medicine* 150: 106140.

62 Cao, L., Yang, J., Rong, Z. et al. (2021). A novel attention-guided convolutional network for the detection of abnormal cervical cells in cervical cancer screening. *Medical Image Analysis* 73: 102197.

63 Abriata, L.A. and Dal Peraro, M. (2021). State-of-the-art web services for *de novo* protein structure prediction. *Briefings in Bioinformatics* 22 (3).

64 Wang, L., Song, Y., Wang, H. et al. (2023). Advances of artificial intelligence in anti-cancer drug design: a review of the past decade. *Pharmaceuticals* 16 (2): 253.

65 You, J., Liu, B., Ying, Z. et al. (2018). Graph convolutional policy network for goal-directed molecular graph generation. *Advances in Neural Information Processing Systems* 31.

66 Ghimire, A., Tayara, H., Xuan, Z., and Chong, K.T. (2022). CSatDTA: prediction of drug–target binding affinity using convolution model with self-attention. *International Journal of Molecular Sciences* 23 (15): 8453.

67 Chen, H. and Engkvist, O. (2019). Has drug design augmented by artificial intelligence become a reality? *Trends in Pharmacological Sciences* 40 (11): 806–809.

68 Lu, S.H., Wu, J.W., Liu, H.L. et al. (2011). The discovery of potential acetylcholinesterase inhibitors: a combination of pharmacophore modeling, virtual screening, and molecular docking studies. *Journal of Biomedical Science* 18 (1): 8.

69 Sheng, C., Dong, G., Miao, Z. et al. (2015). State-of-the-art strategies for targeting protein–protein interactions by small-molecule inhibitors. *Chemical Society Reviews* 44 (22): 8238–8259.

70 Yang, T., Li, Z., Chen, Y. et al. (2021). DrugSpaceX: a large screenable and synthetically tractable database extending drug space. *Nucleic Acids Research* 49 (D1): D1170–D1178.

71 Vamathevan, J., Clark, D., Czodrowski, P. et al. (2019). Applications of machine learning in drug discovery and development. *Nature Reviews. Drug Discovery* 18 (6): 463–477.

72 Paul, D., Sanap, G., Shenoy, S. et al. (2021). Artificial intelligence in drug discovery and development. *Drug Discovery Today* 26 (1): 80–93.

73 Mouchlis, V.D., Afantitis, A., Serra, A. et al. (2021). Advances in *de novo* drug design: from conventional to machine learning methods. *International Journal of Molecular Sciences* 22 (4): 1676.

74 Hodgson, J. (2001). ADMET – turning chemicals into drugs. *Nature Biotechnology* 19 (8): 722–726.

75 Niu, J., Straubinger, R.M., and Mager, D.E. (2019). Pharmacodynamic drug–drug interactions. *Clinical Pharmacology and Therapeutics* 105 (6): 1395–1406.

21

PROTAC and ProTide Strategies in Drug Design

Maitreyee Mukherjee

Department of Pharmaceutical Technology, NSHM Knowledge Campus – Group of Institutions, Kolkata 700053, West Bengal, India

21.1 Introduction

The origin of the word "drug" was from the French word "drogue," which means dry herb [1]. Drug is an external molecule that has specific effects on biological systems and is applied to diagnose, prevent, or cure a disease. The origin of a drug can be natural or can be produced synthetically. An "ideal" drug must follow some specific features like particular action, should be nontoxic, without or may have minimal side effects, stable, and soluble in water as well as in lipids. Drugs interact with specific targets in the human body to exert its effect and these interactions create two different kinds of effects, i.e., the impact of the drug on the human body and the impact of the human body on the drug, which are determined as pharmacodynamics and pharmacokinetics, respectively. The physicochemical properties of drug molecules influence the kinetics and dynamics of the body [2]. The discovery and development of drugs are time consuming and costly processes. To accelerate drug discovery, design, development, and optimization, there is an increasing endeavor to apply computational resources to the integrated chemical and biological domain. The ligand-based drug design, structure-based drug design (drug–target docking), and quantitative structure–activity and quantitative structure–property relationships are some examples of frequently used computational approaches. Although there are a few drawbacks related to computer-aided drug design, these are time-taking process of knowing how to run software, high production and purchase costs for software, and the work can be lost because of the sudden breakdown of computers [3].

Due to obtained drug resistance, inaccessible targets for drugs, and low selectivity within target family members, the discovery of potent small-molecule drugs for specific therapies comes with numerous significant obstacles. Novel small-molecule alternatives have come to light and acquired clinical importance [4].

Proteolysis targeting chimeras, or PROTACs, is a novel technique that has appeared as a standout treatment paradigm in recent years. Recently, focusing on the enzymatic activities of 20–25% approximately of every protein target have been studied. PROTACs inhibit the entire biological function of specific target proteins by binding to them and generating following proteasomal degradation, unlike small molecules. Besides all the benefits, the notable limitations of PROTACs are transcription factors, nuclear proteins, and many other scaffolding proteins, which are difficult to maneuver along with

Computational Methods for Rational Drug Design, First Edition. Edited by Mithun Rudrapal.

conventional small-molecule inhibitors. Recently, diverse proteins, such as BTK, BRD4, AR, STAT3, IRAK4, and tau, are degraded by PROTAC [5]. ARV-110 and ARV-471 proteins showed outstanding efficacy in the clinical II study. It is unidentified, which targets are preferable for PROTAC technology to achieve benefits compared to small-molecule inhibitors and for researchers, it is more challenging to rationally develop the design of effective PROTACs and optimize it for oral effectiveness [6].

ProTide is an approach in which an active chemical entity is masked with the help of a cleavable group (promoiety) that, in the human body, breaks off under certain conditions, exposing the active species [7]. ProTide is a phosphate or phosphonate prodrug technology designed to deliver nucleoside analog monophosphates and monophosphonates efficiently into the cells. In this method, the hydroxyls of the monophosphate or monophosphonate groups are masked by an aromatic group and an amino acid ester moiety. These are then enzymatically cleaved off inside the cells to liberate free nucleoside monophosphate and monophosphonate entities. From a perspective of structure, this is the culmination of nearly 30 years of medicinal chemistry research. It began with simple alkyl groups masking nucleoside monophosphate and monophosphonate groups, and it developed into the complex ProTide system that is used today. Two FDA-approved ProTides (antiviral) have already been discovered, thanks to the widespread use of this method in research. The invention of nucleoside-based ProTides for Parkinson's disease was made possible by this method [8]. Furthermore, as a prodrug strategy for the intracellular delivery of monophosphorylated non-nucleoside compounds like glucosamine, sphingosine 1-phosphate (S1P) [9], 4-phospho-D-erythronohydroxamic acid, and 5-phospho erythronohydroxamic acid, the ProTide approach is gaining more interest [9]. In this chapter we discussed drug design strategies, the ubiquitin–proteasome system, chemical formulations of PROTAC, ProTide technology for designing drugs, implementation of ProTides as nucleoside analogs, and applications of PROTACs and ProTide [10].

21.2 Drug Design: Past to Present

Drug design has a rich history in the pharmaceutical field. At the end of the 19th century, scientist Emil Fischer proposed the significance of drug–receptor interaction that is similar to the interaction between key and lock, an incredible progress has been made in the area of drug design. Drug design, now the most sophisticated method for drug discovery, has grown into a rational and well-planned discipline based on strong theoretical aspects and practical implications [2, 11]. In drug discovery, the generation of novel drugs with potential interactions with specific targets plays an important role. Discovery of a drug and development (identification of lead molecule to an FDA-approved marketed drug) is an immensely challenging, cost-effective, and prolonged process [12]. A new as well as advanced approach was introduced to overcome these limitations, which is referred to as computer-aided drug design (CADD). The rational drug design is cost-effective and time-saving compared to the conventional drug discovery process. The rational drug design technique is also recognized as reverse pharmacology as the first step is to detect promising target proteins and these identified target proteins are further involved in the screening of small-molecule libraries [13]. These computational methods are also relevant in cutting down the excessive use of animal models in pharmacological research [14].

CADD, a part of modern drug discovery, plays a crucial role in the selection of drugs with acceptable bioactivity and side-effect profiles. This is applicable in both designing novel drugs and replacing existing ones. CADD research has led to promising results in the treatment of inflammation and pain related to musculoskeletal conditions [15].

Significant breakthroughs have been achieved in structural and molecular biology, as well as biomolecular spectroscopic structure identification techniques and the 3D structure of more than 100,000 proteins provided by these methods. In addition to storing and organizing such types of information, a lot of discussion are there regarding the development of advanced and robust computational techniques. The rate of drug development is enhanced due to the advancement of bioinformatics. The availability of a large number of target proteins that are responsible for the completion of the Human Genome is the foundation for structure-based drug design (SBDD). This is an important aspect of industry drug development programs and academic research [16]. SBDD is a more precise, effective, and rapid method for lead identification and optimization since it incorporates the 3D structure of a target protein as well as molecular information about the illness [17]. The most prevalent computational approaches used in SBDD encompass structure-based virtual screening (SBVS), molecular docking, and molecular dynamics (MD) simulations. These approaches have several uses, including the analysis of binding energetics, interaction between ligand–protein, and evaluation of conformational modifications that occur during the docking process [18]. In recent years, the software industry has experienced a significant increase in applications for effective drug discovery procedures.

Computational tools are an effective technology for speeding up the process of drug discovery, which comprises numerous screening processes, combinatorial chemistry, and calculations of properties like absorption, distribution, metabolism, excretion, and toxicity (ADMET) [13]. SBDD is an iterative process that goes through multiple steps to get an improved drug to clinical trials. The identification of a possible therapeutic target as well as active ligands is the first phase of activity. The initial step is to clone the target gene, which is then extracted, and purified, and the protein's 3D structure determined. Many computational methods can be utilized to dock large databases of small molecules or fractions of compounds into the binding site of the target protein [13].

21.3 PROTAC Strategy in Drug Design

Numerous diseases are the target of proteins. As a result, different types of strategies for blocking target proteins associated with disease have been developed [19]. The protein homeostasis is regulated by the ubiquitin–proteasome system with the help of proteasome as the main element of eukaryotic protein degradation. To mark proteins for proteasomal breakdown, ubiquitin covalently binds to surface lysines [18]. An abnormal protein functioning is the main cause of inherited as well as acquired diseases, which is now targeted by applying an occupancy-based pharmaceutical approach, where the disease-implicated proteins are the target for binding of the inhibitors and inhibitors can block the longer protein functions and achieve a higher clinical benefit. To achieve optimal treatment efficacy, high local inhibitory concentrations (IC90–95) must be maintained regularly. But frequently, this causes off-target binding to follow the side effects. Event-driven techniques offer an alternative, in which drug binding enhances an action that decreases the cellular levels of the protein implicated in the disease. There are several nucleotide-based approaches with low *in vivo* stability and low bioavailability, such as genome editing strategies like CRISPR-Cas9, antisense oligonucleotides, and small interfering RNA (siRNA). However, several proteins for specific targets can be selectively degraded by small molecules. This technique is named as PROTACs. Ligases such as cereblon (CRBN), Von Hippel–Lindau (VHL), IAP, and MDM2 have been the focus of extensive studies on PROTACs across a wide range of disease domains. PROTACs have been found effective for attacking cancer targets such as the androgen receptor, estrogen receptor, BTK,

CDK8, BCL2, and c-MET [6e11]. Multiple ligases are used to construct several types of BET family like BRD2, BRD3, and BRD4-PROTACs. The membrane permeability of dBET6, and dBET57 PROTAC are enhanced by MDM2-based BRD4 PROTAC [20], CRBN-based dBET1 [21], and BETd-24-6 for triple-negative breast cancer [22]. PROTACs have been tested for viral, immunological, and neurodegenerative diseases, respectively, for the protease of the hepatitis C virus (HCV), IRAK4, and Tau [17e19] [23].

A protein of interest (POI) ligand, E3 ligase ligand, and linker are the components of a PROTAC, a kind of heterobifunctional molecule that can be developed and synthesized to meet a wide range of requirements. The availability in cells for POI with a high degree of selectivity is decreased by the new mechanism of PROTACs which may lower the side effects in comparison to conventional small-molecular inhibitors [24]. The first PROTAC, i.e., PROTAC-1 was discovered in 2001, since then the target protein degradation (TPD), a novel approach and innovative suggestion for therapeutics is growing rapidly. TPD offers the modern explanation of the concepts of classical drug discovery and is driven by target activity that is event-based as compared to activity that is occupancy-driven. The ovalicin and a 10-amino acid phosphopeptide from nuclear factor-κB inhibitor-α (NF-κBIα; also known as IκBα), which was targeted by Protac-1, has been identified by the E3 ligase β-transducing repeat-containing E3 ubiquitin–protein ligase (β-TRCP) [24]. The methionyl aminopeptidase 2 (METAP2) was targeted by a newly developed Protac-1. In between METAP2 and β-TRCP, the ligase facilitated by Protac-1 to ubiquitylate METAP2 in extracts from unfertilized eggs of *Xenopus laevis*. Additionally, with the emergence of these early PROTAC compounds, the immunomodulatory imide drugs (IMiDs) thalidomide and its analogs lenalidomide and pomalidomide [6] were discovered to target the E3 ligase cereblon (CRBN) in the field of cancer therapy [25]. For targeting the IKAROS family zinc finger 1 (IKZF1) and IKZF3 for degradation [3, 4, 23], these drugs cooperate CRBN and are presently regarded as breakthrough examples of molecular glue [24]. The small-molecule mimetics of the HIF1α peptide were introduced by the rational design of PROTACs. Systematic detection of degrader compounds, which are termed as molecular glues, can be utilized in conjunction with the modular proteolysis-targeting chimera (PROTAC) technique to promote rational degrader discovery. Molecular glues mediate protein–protein interfaces to promote the formation of a complex between an E3 ligase and a target. There are many examples of intramolecular and intermolecular glues which work outside the ubiquitin–proteasome system (UPS), like the allosteric protein tyrosine phosphatase nonreceptor type 11 (PTPN11; another name SHP2) inhibitor SHP099 [26], which maintains a closed conformation of SHP2 (intramolecular); and cyclosporin, which causes for the proximity of calcineurin and cyclophilin [24].

Besides a biological chemistry tool, PROTAC is a widely used technology for the drug design of new therapies. The E3 ligase binding peptide ALAPYIP to recruit von Hippel–Lindau (VHL) is carried by PROTAC, which can lead to the degradation of FK506 binding protein 12 (FKBP12) and AR [6].

As the early PROTACs are not complete small-molecule structures but also include peptide ligands for the E3 ligase, these are regarded as "bioPROTACs." The rational design of PROTACs was introduced by the small-molecule mimetics of HIF1α peptide, e.g., the first PROTACs, which utilized VHL to degrade bromodomain-containing protein 4 (BRD4), were based on the bromodomain protein inhibitor JQ1. One popular biochemical modification involves attaching one or more ubiquitins through covalent bonds to one or more acceptor lysines located on the substrate protein. This results in the degradation of the ubiquitinated protein by the proteasome. The UPS breaks down about 85% of the protein in the body [27].

21.3.1 Ubiquitin Proteasome System and PROTACs

The ubiquitin-activating enzymes (E1), ubiquitin-conjugating enzymes (E2), and ubiquitin–protein ligases (E3) have the sequential action to enhance the binding of ubiquitin to substrate proteins and lead to ubiquitination. The modification of protein is introduced to the proteasome for degradation which is caused by K48 polyubiquitination. The E3 ubiquitin–protein ligases bind directly to the substrate and proceed to ubiquitination. Craig M. Crews and Raymond J. Deshaies came up with the idea of seizing the ubiquitin-dependent proteolysis pathway, which might be an effective way to modulate the protein abundance of normal diseased cells as the degradation of ubiquitinated proteins occurs rapidly in cells [17]. They established a method for targeting proteins to the ubiquitin/proteasome pathway (Figure 21.1). The development of small molecules that modulate the interactions has specific medicinal potential as protein–protein interactions (PPIs) ruling substrate identification by E3 ubiquitin ligases are crucial for cellular function. Occupancy-driven pharmacology is primarily utilized in therapy, while PROTAC technology relies on the principle of event-driven pharmacology. The idea of bringing together two special proteins which generally do not generate a complex, has a role in the development of the compounds created by the PROTAC [28, 29]. The protein of interest (POI) ligand first binds to the pathogenic protein; following that, the E3 ligand attached to the other terminus of the linker recruits an appropriate E3 ligase to the nearby location of POI, which causes the target protein to become polyubiquitinated. Ultimately, the 26S proteasome breaks down the recognized polyubiquitinated protein. A study report that first peptide-based PROTACs have the potency to degrade methionine aminopeptidase 2 (MetAP-2). The effective approach of PROTACs is enabled in various areas of drug

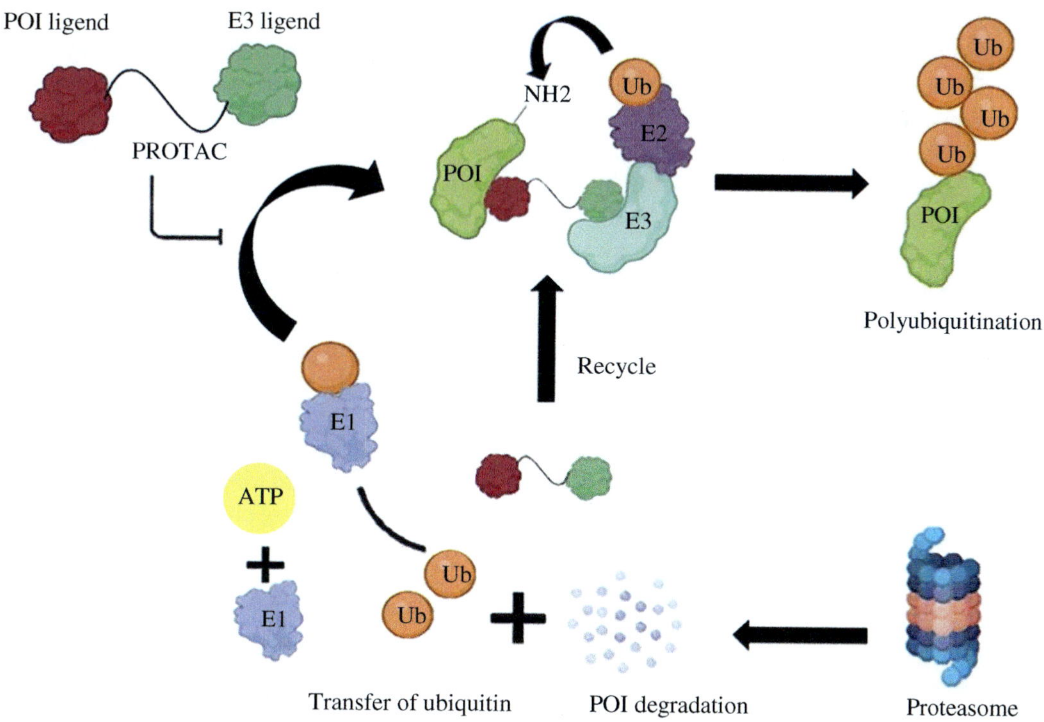

Figure 21.1 A systematic diagram stating the mechanism of action of PROTACs.

development by the mechanism of action of complete target degradation and corresponding PROTAC compounds are targeted more than 130 proteins, which are associated with leukemia, cancer, cardiovascular disease, virus infection (e.g., hepatitis C virus (HCV) NS3/4A PROTAC DGY-08-097) and neurodegenerative disease for degradation. Several PROTACs for the treatment of malignant tumors have been created as candidates for clinical trials in the development of antitumor drugs; these consist of ARV-110 [2], an androgen receptor-targeted PROTAC for prostate cancer, and ARV-471 [3], an estrogen receptor-targeted PROTAC for breast cancer, both developed by Arvinas; these are currently in phase II clinical trials; furthermore, DT2216 [4], a BCL-XL degrader for hematomas and solid tumors, produced by Dialectic Therapeutics, is now in phase I clinical trial.

The rational design of PROTACs was introduced by the small-molecule mimetics of HIF1α peptide. In the year 2001, the Deshaies laboratory developed and produced Protac-1, a functional molecule that recruits the UPS to maintain intracellular protein homeostasis. This process helps in the degradation of methionine aminopeptidase-2 (MetAP-2). Three covalently linked parts make up Protac-1 as a domain with ovalicin (MetAP-2 inhibitor), which has been identified by the Skp1-Cullin-F-box complex (SCF, an E3 ligase that starts protein ubiquitination and degradation by UPS), and a linker that connects these two domains [30].

21.3.2 Chemical Formulations of PROTACs

By mediating PROTACs passive diffusion, active transport processes, or incorporation into nanodelivery vehicles to overcome the obstacles for reaching the target site to gain the wanted activity, researchers have adopted various approaches that improve the cellular uptake and solubility of PROTACs. The conformational flexibility in the PROTACs structure by approving the creation of intramolecular H-bonds (IMHB) for the formation of a more polar and soluble form yielding their increased permeability was introduced by Alex et al. Further the concept of click chemistry for the production of bifunctional PROTACs with better permeability that described simultaneous degradation of two key oncogenic targets was applied by Lebraud et al. The specific peptides (p-PROTAC) are the basic component for the development of PROTACs to achieve specific as well as effective deterioration of undruggable target proteins [31]. Several different pathways in cancer progression and neurodegenerative disorders and cancer progression have been targeted by applying this technology including PI3K, ERa, CREPT, AKT, X-protein, and Tau-protein for the treatment of different types of cancers like breast cancer, ovarian cancer, pancreatic cancer, and neuroblastoma. Although this process still has drawbacks as it possesses low stability and poor membrane permeability barrier tough to reach the specific target tissue. A trastuzumab–PROTAC conjugate (Ab-PROTAC 3) was introduced by Maneiro and colleagues which has target degradation potency only in HER2-positive breast cancer cell lines, without showing any effect on HER2-negative cells [31]. A number of PROTAC–antibody conjugates comprising two specific Era targeting degrader entities along with three separate ADC linker modalities for targeted delivery of chimeric ERa degrader molecules developed by Dragovich et al. They showed that integrated degrader payloads may be delivered to MCF7-neo/HER2 in a target–antigen-dependent manner. On the other hand, creating antibody–PROTAC combines for targeted PROTAC delivery may be an expensive and time-consuming process. Nanotechnology is possibly used to avoid these hurdles along with achieving particular accumulation of PROTACs in tumor target sites and nonspecific deterioration of target proteins [32].

There are several new concepts have emerged regarding PROTAC, such as HaloPROTACs, HomoPROTACs, autophagy-targeted chimera (AUTAC), lysosome-targeting chimera (LYTAC), and autophagosome-tethering compound (ATTEC) [33]. HaloPROTACs is a novel PROTAC ligand produced by exchanging POI ligands with chloroalkanes. The chloroalkane-labeled moieties can covalently bind to HaloTag protein. If HaloTag is binded to the specific protein, indirectly the PROTAC molecules bound to them enhance their degradation process. Recently, HaloPROTACs have been used by a few researchers for the degradation purpose of different types of HaloTag fusion proteins. The POI ligand in PROTAC along with E3 ligase ligand is replaced by a few groups that are similar to dimerize E3 ligase ligand. The ternary compounds are commonly referred to as homo-PROTACs, and here, the E3 ligase enzymes transform into their substrates and are susceptible to UPS degradation [34]. Apart from the ubiquitin–proteasome pathway, there exists an additional protein degradation mechanism named the lysosomal degradation pathway, which comprises the process of autophagy and the endosome/lysosome pathway. Another approach that researchers have found to counteract the off-target impacts of nanocarriers and encapsulated chemotherapeutics is the active targeting of PROTAC molecules via simulation of nanoformulations targeted to overexpressed receptors on the surface of cancer cells. The incorporated MZ1 PROTAC into PLA-based polymeric nanoparticles (ACNPs), which have received FDA approval and were conjugated with trastuzumab to target HER2 receptor-overexpressing breast cancer cells and improve both the antitumor efficaciousness and pharmacokinetic profile. At low dosages, these ACNPs had a potential cytotoxic effect that guided the NPs toward cancer cells that expressed HER2, without causing any additional adverse effects. To target liver tumors specifically, researchers developed asialoglycoprotein receptor (ASGPR)-directed nanoliposomes (GALARV) which comprised a BRD4-PROTAC: ARV-825. The aqueous solubility increased by 140-fold when ARV-825 encapsulated in GALARV and prevention of its microsomal metabolism was observed to improve its *in vitro* half-life by 100-fold compared with only ARV-825 alone. For the purpose of treating nonsmall-cell lung cancer, Vartak et al. developed nanostructured lipid carriers (AP-NLCs) that integrate ARV-825 and greatly increased its solubility [19]. Polymeric PROTAC (POLY-PROTAC) nanomedicine was recently developed by Gao et al. to accomplish tumor-specific breakdown of BRD4 proteins. In order to enhance intratumoral accumulation and retention of azide-modified POLY-PROTAC nanoparticles (NPs) through an *in situ* bioorthogonal click reaction, these POLY-PROTACs are loaded with dibenzocyclooctyne (DBCO). In response to an acidic pH, a reductive tumor microenvironment and extracellular matrix metalloproteinase-2, POLY-PROTACs assemble themselves to create polymeric micellar NPs, which then emit the PROTAC payload. Additionally, acid-activatable bioorthogonal POLY-PROTAC NPs worked in conjunction with photodynamic treatment to promote tumor-specific BRD4 decomposition, which significantly regressed the growth of tumors and increased survival by 40% in a mouse model of MDA-MB-231 breast cancer [35]. A significant obstacle in the delivery of a large anticancer molecule such as PROTACs to tumor cells is getting through the dense and fibrotic tumor stroma to the target tumor cells. In the past, scientists have boosted the delivery of medications, macromolecules, and nanomedicine to tumor cells by using stromal-disrupting substances. For vemurafenib-resistant melanoma, Fu et al. recommended the use of nintedanib as an antifibrotic drug in a liposomal nanocarrier (ARNIPL) alongside ARV-825. In order to explore the antiangiogenic and antivasculogenic mimicry impact of palmitoyl-DL-carnitine chloride, a protein kinase C inhibitor, in combination with ARV-825, co-loaded in PEGylated nanoliposomes (LARPC), by a group of researchers mixed the two substances [27, 35]. For the delivery of PROTAC through two main routes (i.e., parenteral and oral) formulation and route of administration are the main focus. Without a doubt, injectables offer the fastest route for drug molecules across physiological barriers, particularly the GIT environment, improving the concentration of drugs in blood circulation. The parenteral route: intravenous or intraperitoneal – is still used for most of the significant *in vivo* studies on PROTACs that have been published in peer-reviewed literature [36]. Parenteral administration of PROTACs is often achieved by solubilizing these molecules in an appropriate solvent or surfactant, yet this method may have disadvantages of its own [37]. As a result, oral PROTAC administration would boost the attraction of the novel approaches. Also, in the majority of chronic illnesses, both doctors and patients always favor the oral route [38]. The proportion of active escaping intestinal and/or hepatic metabolism, as well as the fraction absorbed where PROTACs have difficulties at each of these parameters, are closely associated with oral bioavailability [39]. Lipid-based drug delivery systems provide attractive platforms for delivering complicated compounds where parenteral delivery is the goal. Chen and his colleagues also published a study showing the creation of LNPs for parenteral delivery of pre-fused PROTACs, which exhibited better therapeutic efficacy [19].

21.3.3 Advent of PROTACs as Antiviral

Hepatitis B viral defense using peptide-based PROTACs, hepatitis B virus (HBV) X-protein (154 amino acids, 17 kDa) is necessary for virus infection [36]. It has a role in the development of cirrhosis caused by HBV or even hepatocellular carcinoma (HCC) in patients who suffer from persistent HBV infection. In 2014, Montrose et al. gave the first report on the use of PROTAC technology to reduce the amount of protein associated with a virus [40]. They also demonstrated how the HBV X-protein was neutralized and eliminated by a new cell-permeable PROTAC [41]. The N-terminal oligomerization motif of the X-protein was attached to the C-terminal instability domains to form the PROTAC, which was then made cell-permeable by N-terminal fusing to a polyarginine cell-penetrating peptide. In a study, the X-protein and the oligomerization motif linked together, and the unstable portion of the X-protein turned it into a target that was misfolded or damaged and could be broken down by the UPS. Also, it has been confirmed that the PROTAC's oligomerization domain opposes the X-protein's proapoptotic activity. The applicability of this peptide-based PROTAC in the field of anti-HBV therapy is limited since it has not been further studied for the reduction of HBV replication and the progression of chronic hepatitis [36, 42].

21.3.4 NS3/4A-Targeting PROTACs Against HCV

As a result of its protracted latent phase, the positive single-strand RNA virus known as HCV can cause several progressive infection symptoms, including cirrhosis, liver fibrosis, and even cancer. As a result, the virus poses a risk to public health. There are 10 structural and nonstructural proteins produced by the proteolytic breakdown of an HCV polyprotein [43]. The NS3/4A complex protein is a multifunctional enzyme that has a C-terminal helicase region and an N-terminal serine protease domain among the viral proteins implicated in HCV replication. Numerous essential mature nonstructural proteins (NSPs), including RNA-dependent RNA polymerases (RdRp), helicases/nucleotide triphosphatases, and serine proteases, are produced by this enzyme through the cleavage of HCV polyprotein precursors. Additionally, the NS3/4A protease cleaves host proteins implicated in the beginning of the cellular immune response, such as TRIF and MAVS [42]. The only protease that has been effectively employed as a DTA target at this point is NS3/4A. Yang et al. demonstrated in 2019 that small-molecule PROTACs against HCV promoted NS3/4A degradation *in vitro*. This was the first instance of combining PROTAC technology as a ligand of a POI with an antiviral small-molecule inhibitor. The pyrazine group of the HCV NS3/4A-Telaprevir co-crystal structural complex (PDB: 3SV6) proved to be a solvent-exposed fragment in this research, which allowed it to function as a tethering site [44]. A range of heterobifunctional compounds were created and produced by conjugating telaprevir to various CRBN ligands through the usage of alkyl or PEG polymers. This was done to verify if the PROTACs technology may be utilized as a viable and innovative virus-fighting tactic. These conjugators effectively engaged the CRL4CRBN complex in addition to retaining the ability to inhibit the NS3/4A protease with IC50 values ranging from 247 to 385 nM. Through a concentration-dependent way, these bifunctional degraders markedly stimulated rapid and sustained proteasome-mediated NS3 protein degradation in a cultured cell line. With a DC50 value of 50 nM, compound DGY-08-097 demonstrated the highest efficacy in inducing NS3 degradation via the UPS pathway. However, a decrease in the neo-substrate abundance of IMiDs (IKZF1 and IKZF3) has not been observed, likely due to the application of a novel tricyclic imide molecule as the highly particular CRBN ligand. The compound DGY-08-097 showed inhibition activity against the NS3-V55A and NS3-A156S protease mutants, with IC50 values of 508 and 1561 nM, respectively, which was three times higher than that of wild-type NS3. This is consistent with several PROTACs that can overcome mutational variation in cancers [44].

21.3.4.1 Neuraminidase-Targeting PROTACs

A hydrophobic, single-stranded mosaic of 29 amino acids encases influenza neuraminidase (NA), a tetramer with mushroom-like structure, is a membrane glycoprotein that is present inside the envelope of the influenza virus. By interfering with the interaction between the glycoprotein HA on the surface of virion and the sialic acid receptor on the host cell membrane, NA releases progeny virions from the surface of infected cells during influenza virus replication. This enables the virus to infect additional host cells and multiply faster [45]. However, numerous investigations have been conducted to find the protein degraders for HCV or SARS-CoV-2, and this has been motivated by the severe acquired drug resistance of clinical IAV NA inhibitors. Oseltamivir's cocrystal structure complex (PDB: 2HU0) with NA showed that the amino and carboxylic acid groups were suitable locations for additional modification [38]. This is compatible with the findings of earlier medicinal chemistry investigations intended to increase the binding affinity among NA and oseltamivir through SAR [42].

21.4 Emergence of ProTide Technology in Drug Design

Since its formal introduction in 1958, the applicability of a prodrug approach to drug discovery has been introduced. Analogs of nucleosides have several disadvantages, including inadequate absorption by cells because membrane transporters are not expressed enough, immediate breakdown, and a delayed conversion to triphosphate because of the initial phosphorylation step that limits the rate of conversion [7]. Prodrugs of NA with a phosphate group or an isosteric and isoelectronic phosphonate moiety have been explored to get around these limitations. A few of these methods have proven effective, such as prodrugs that are now being used as antiviral treatments in clinics. ProTide (Pro Nucleotide) is a phosphoramidate technique developed over 20 years ago by Cardiff University's Professor Chris McGuigan and his colleagues [7]. When nucleoside analogs were initially developed several decades ago, they proved to be an efficient treatment for a wide range of illnesses, including cancer and viral infections such as herpes simplex virus (HSV), human cytomegalovirus (HCMV), HIV, hepatitis B virus (HBV), more recently, hepatitis C virus (HCV). At present, the clinic offers over 20 distinct nucleoside analogs for the treatment of cancer and viral infections. These chemical compounds either passively diffuse into the cell or are assisted by transporters like peptide and concentrative nucleoside transporters [46]. Several nucleoside and nucleotide kinases activate the nucleoside analogs when they are within the cell. They then phosphorylate the analogs step-by-step, forming the mono-, di-, and triphosphorylated nucleoside analog metabolites. The therapeutic effects of activated (phosphorylated) antiviral nucleoside analogs are achieved by inhibition of intracellular enzymes, which are typically virus-encoded DNA or RNA polymerases, and/or incorporation in the viral nucleic acid chains, which ceases the elongation process. Nevertheless, there are plenty of limitations accompanied by the administration of nucleoside analog drugs. The structural dissimilarity of nucleoside analogs to genuine nucleosides renders their phosphorylation by cellular or viral kinases usually inefficient. This eventually restricts the synthesis of the active triphosphate metabolites, typically nucleoside analog active metabolite. The low intestinal permeability of these compounds, which prevents them from being carried over the cell border via the paracellular route, also frequently results in low oral bioavailability. Furthermore, resistance restricts the therapeutic application of many nucleoside analogs with antiviral and anticancer properties. Resistance arises from a variety of mechanisms, including (i) downregulating the nucleoside kinases that activate nucleoside analogs; (ii) depleting transporters that impact the effective trafficking of these analogs into target cells; (iii) activating phosphatases, or 5′-nucleotidases, which dephosphorylate the active metabolites of nucleoside analogs; and (iv) increased catabolic bioconversion of the nucleoside analogs, including (deoxy)cytidine deamination, which may result in inactive versions of some of the analogs [47]. It was first proposed that some of these deficiencies could be addressed by supplying phosphorylated metabolites of the nucleoside analog to get around these restrictions. Numerous investigations that revealed the slow phosphorylation of nucleoside analogs into their nucleoside monophosphates as the rate-limiting step in their *in situ* activation further supported this idea [10].

The ProTides were first developed in the late 1980s. The goal was to hide the phosphate group's oxygen atoms in nucleoside monophosphate analogs so that, at physiological pH, they are neutral and more readily absorbed by cells. Initially, the phosphate groups were simply hidden by alkyl groups. However, over time, the technique developed into cleverly hidden phosphate groups using aryl groups and amino acid esters, which can be broken down inside intact cells. Alkyl and haloalkyl phosphate esters, alkyloxy and haloalkyloxy phosphoramidates, phosphorodiamidates, lactyl-derived systems, diaryl phosphates, and aryloxy phosphoramidates, referred to as ProTides, are the six phases of this development process, which lasted at least two decades [48].

21.5 Approaches of ProTides in Drug Development

Numerous nucleoside monophosphate or monophosphonate ProTides have been synthesized using this method since in the early 1990s, the McGuigan laboratory developed the ProTide technology. While pyrimidine nucleosides like AZT (3′-azidothymidine) and d4T (2′,3′-didehydro-2′, 3′-dideoxythymidine) were the initial subjects of structure–activity relationship (SAR) investigations, a wide range of additional ProTide derivatives have also been produced and tested for antiviral effectiveness. ProTides with phosphoramidate technology have been observed to have significantly higher antiviral potencies when applied to purine nucleoside analogs, such as 2′,3′-dideoxyadenosine (ddA) and 2′,3′-didehydro-2′,3′-dideoxyadenosine (D4A), carbocyclic adenosine derivatives, [49], and carbovir/abacavir [50]. Recently, it was also demonstrated that the antiherpetic acyclovir ProTides produced antiherpetic and anti-HIV activity allowing the antiviral range of the medication to be increased. It should be noted in this regard that ProTide technology not only makes it possible to significantly optimize the currently accessible antiviral medications, but it also makes it possible to transform molecules that initially appear inactive into active antivirals through enhanced absorption and direct delivery of their activated (monophosphate) species.

Some ProTides have made it to clinical trials, and two of them have received approval for usage in clinic. The FDA-approved ProTides and important ProTides that are reportedly undergoing clinical trials will be discussed in order to demonstrate the value of this prodrug technology in drug discovery [51].

Due to the emergence of resistance to this anticancer treatment drug, gemcitabine became the preferred choice for the ProTide technology. The inability of gemcitabine to be converted into its pharmacologically active diphosphate and triphosphate forms because of deoxycytidine kinase downregulation, the deamination of gemcitabine into its very less active uridine derivative, and mutations in the membrane transporter that mediates its active uptake into cells which is responsible for a few of the mechanisms of gemcitabine resistance. The cytostatic properties of certain gemcitabine ProTides were investigated when they were developed. These investigations produced the lead structure, which showed remarkable antiproliferative efficacy, major *in vivo* stability, and an outstanding PK profile [52]. One significant finding from the ProTide was its ability to successfully overcome resistance to gemcitabine. Three key explanations were offered for this: first, gemcitabine is known to be catabolized by the enzyme cytidine deaminase, which is present in both serum and cells. Unprotected hydroxyl groups on the 3′- and 5′-positions of the nucleoside analog are necessary for deamination to take place. Because the phosphoramidate moiety was present on the ribose moiety's 5′-*O*-position, ProTide 1 showed resistance to deamination. Being resistant to deaminase breakdown also improves safety since less harmful difluorodeoxyuridine is generated. Secondly, it was discovered that the ProTide was lipophilic enough to get inside the cells by passive diffusion and that the cellular uptake was not dependent on nucleoside transporters. An experiment that involves the downregulation of nucleoside transport in cancer cells was mimicked using dipyridamole, a known inhibitor of the human equilibrative nucleoside transporter 1 (hENT1) was deduced by ProTide. When pancreatic PANC1 cancer cells were treated with gemcitabine in the presence of dipyridamole, the cytotoxic activity of the drug was seen to be dramatically reduced, and the cytostatic effect decreased from 611 nM to >2000 nM (CC50) [53]. In spite of hENT1 blockade, the activity stayed rather constant (CC50 = 162 μM), indicating that cellular absorption is not dependent on nucleoside transporters. The ProTide's structure enables it to bypass the initial phosphorylation, which restricts the rate and is mediated by deoxycytidine kinase, which is reduced in certain carcinomas. As the competing natural substrate for gemcitabine activation to its 5′-monophosphate by 2′-deoxycytidine kinase, 2′-deoxycytidine kinase (EC50 = 0.7 μM in the presence of 2′-deoxycytidine compared to 0.2 μM in the absence of dCyd) when incubated with RT112 cancer cells, indicated a minor change in its activity level 1. On the other hand, when 2′-deoxcytidine was administered concurrently with gemcitabine, its efficacy was reduced as the two drugs were competing for the same active site on the deoxycytidine kinase [54]. ProTide 1 was studied *in vivo* in animals, with a special emphasis on pancreatic tumors that were partially resistant to gemcitabine. The preliminary findings were encouraging. In mice models using human tumor xenografts, shrank the tumor more rapidly than gemcitabine, and on day seven after ProTide delivery, the tumors shrank significantly. In order to determine the animals' tolerance to the substance, their body weight was also noted. ProTide 1 was related to a decrease in body weight of less than 4%, and it was determined that, in comparison to mice treated with gemcitabine, the mean body weight loss was less noticeable in the former group for the duration of the treatment. This suggested that the parent nucleoside analog was not as well tolerated as the gemcitabine ProTide [52]. Compared to the pharmacokinetics of ProTide were better than gemcitabines' as its half-life was 7.9 hours as opposed to 1.5 hours for gemcitabine. ProTide additionally caused intracellular concentrations of the active metabolite of gemcitabine triphosphate to rise to 13 times greater levels than those of gemcitabine itself [52]. A total number of 68 patients participated in phase I/II clinical studies, which demonstrated that 1 was beneficial for several types of cancer, most notably ovarian, pancreatic, and biliary malignancies. Phase III trials with a specific emphasis on certain cancers will be carried out. The stability observed from the *in vivo* study of ProTide was with a plasma half-life of 8.3 hours. The active metabolite obtained its maximum intracellular concentrations fast, peaked at 475 μM, and remained there for 24 hours. And 33 patients had stabilized their condition, while many individuals had seen a tumor reduction of more than 30% [10].

The FDA has approved two ProTide medications for application in clinical settings. The FDA authorized sofosbuvir (Sovaldi) in December 2013 for the treatment of HCV. The product known as sofosbuvir is a phosphoramidate derivative of the parent nucleoside 2′-deoxy-2′-fluoro-2′-C-methyluridine. After surpassing 50 other ProTides in expansion, 35 sofosbuvir was acquired by Gilead Sciences, where it took on the name sofosbuvir and underwent furthermore, thorough clinical testing. It has been found that sofosbuvir acts by preventing the HCV NS5B polymerase, an enzyme necessary for the virus's genome replication. When this enzyme is inhibited, RNA chain termination occurs, which eventually stops HCV from replicating [55]. Sofosbuvir is absorbed by hepatocytes and then hydrolyzed sequentially by CES1 and cathepsin A to produce active nucleoside monophosphate, again hydrolysis by HINT1, the pharmacologically active triphosphate metabolite emerged, which acts on NS5B polymerase through phosphorylations. The results of sofosbuvir's initial animal trials were positive. In dogs with portal vein cannulation, ProTide was successfully absorbed after oral administration, and its bioavailability was 9.9%, or 36% of the total percentage absorbed. Over 80% of the dose administered to human volunteers was absorbed; however, the exact bioavailability was not confirmed [53]. The sofosbuvir maximum concentration (C_{max}) was attained 0.5–2 hours after oral administration, while the C_{max} of the inactive parent nucleoside metabolite was reached 2–4 hours afterward administration, with a half-life of 0.48–0.75 hours. The absorption was fast. *Ex vivo* analysis of the anti-HCV agent's distribution demonstrated that 82% of healthy subjects and 85% of subjects with end-stage renal disease bound plasma proteins. According to the typical *in vivo* degradation of the ProTides, tests involving the diastereomeric mixture of sofosbuvir in people demonstrated that it was quickly absorbed and degraded in the liver. The parent nucleoside monophosphate has been created by cleaving off the P–N bond of the phosphoramidate group after the isopropyl ester motif was hydrolyzed into the exposed carboxylate group [56]. After the 100 and 800 mg doses of the diastereomeric combination of sofosbuvir, the parent nucleoside that is an inactive metabolite, found in the plasma, and it is possible to detect for at least 24 hours. The main component found in the urine (77.7%) was the inactive parent nucleoside, indicating renal excretion, while sofosbuvir showed little evidence of renal excretion (3.5% recovered in urine; renal clearance = 0.238 L/min). Further, this was supported by research done on the pharmacokinetics of sofosbuvir in patients who had decreased renal function, which proved that the metabolites of sofosbuvir were actively excreted by the kidneys. The renal clearance of the parent nucleoside was approximately two times higher than the glomerular filtration rate. A minimal effect with minimization of any adverse effects was provided by the combination of pegylated interferon and ribavirin with sofosbuvir in the phase II clinical trial. Initial study results showed that the ideal sofosbuvir dosages were 200 or 400 mg in combination with pegylated interferon and ribavirin [57]. That was because the 100 mg dose was found to be less efficient than the 200 and 400 mg doses in terms of viral load reduction. Fatigue and nausea were the most common side effects at these dosages, while there was a low prevalence of other side effects. In addition to sofosbuvir's brief half-life, once-daily dosing was advised because there was minimal drug buildup. Phase II clinical trials later verified sofosbuvir's safety. It was hypothesized that there might be fewer drug–drug interactions since sofosbuvir is not metabolized by the cytochrome P450–3A4 enzymes. Additionally, precautions were recommended when co-administering carbazepine and rifabutin, though, as sofosbuvir concentrations could decline significantly because of the apparent induction of the *P*-glycoprotein efflux pump [58]. The high efficacy of sofosbuvir was demonstrated in a single-group, open-label phase III trial of treatment-naive individuals with HCV genotypes 1, 4, 5, and 6 who received sofosbuvir plus pegylated interferon and ribavirin for 12 weeks, and 12 weeks after the end of HCV treatment, 80% of cirrhotic patients and 92% of noncirrhotic patients had viral counts below the lower limit of quantifications (SVR12). These earlier findings showed the effectiveness of sofosbuvir in treating different HCV genotypes. A sustained viral response was obtained in 78% of treated patients, with a 93% response rate in genotype 2 patients and a 61% response rate in genotype 3 patients, in a phase III placebo-controlled study comparing sofosbuvir and ribavirin for 12 weeks with matched interferon ineligible regulates in patients with genotypes 2 and 3 [59]. The response rates displayed a 73% SVR12 in the 16-week arm compared to 50% in the 12-week arm in a study that involved patients who had previously received interferon treatment and were treated with either sofosbuvir and ribavirin for 12 weeks followed by placebo for 4 weeks or sofosbuvir and ribavirin for 16 weeks. Remarkably, after receiving sofosbuvir and ribavirin for 12 weeks, 94% of genotype 2 patients reached SVR12, and within 16 weeks 87% achieved it [60]. Among patients with genotype 3, 30% attained SVR12 in 12 weeks as opposed to 62% in 16 weeks. Remarkably, compared to 27% of patients in the 16-week therapy arm, 46% of patients getting sofosbuvir and ribavirin for 12 weeks experienced a viral relapse. The effectiveness of giving genotype 3 patients an extended 24-week course of sofosbuvir along with ribavirin was assessed in another study [61]. The findings suggested that whereas 85% of the 250 patients who had genotype 3, received sofosbuvir and ribavirin treatment for 24 weeks reached SVR12, 93% of the 73 patients who had genotype 2, took part in the study achieved SVR12 within 12 weeks of treatment. This study was the first extensive trial establishing the effectiveness of long-term interferon-free therapy for patients having genotype 3. Depending on the HCV genotype to be treated, sofosbuvir has been approved for clinical usage in combination with pegylated interferon, ribavirin, and the NS5A inhibitors ledipasvir and velpatasvir because of its exceptional safety profile and pan-genotypic efficiency. Each of these factors has combined to make sofosbuvir a mainstay of HCV treatment (Table 21.1) [7].

FDA gives approval for the new ProTide-based drug, i.e., tenofovir alafenamide (TAF) for the treatment of HIV [63]. TAF is the latest ProTide of the acyclic nucleoside phosphonate medication tenofovir, developed by Gilead Sciences Inc. Another prodrug of tenofovir is Tenofovir disoproxil fumarate (TDF), had long been an FDA-approved treatment for HIV; nevertheless, ProTide

Table 21.1 Example of ProTides that are already approved or are in a clinical study.

Chemical compound	Nucleoside	Application	Clinical trial/Approved
Sofosbuvir	PSI-6206	Antiviral (HCV)	Approved
Remdisivir	GS441524	COVID-19	Approved
Stampidine	Stavudine	Antiviral (HIV)	Phase I
TAF	Tenofovir	Antiviral (HIV and HBV)	Approved
Acelarin	Gemcitabine	Anticancer	Phase III
Thymectacin	Brivudine	Anticancer	Phase I
NUC3373	2′-Deoxy-5-fluorouridine	Anticancer	Phase II
NUC7738	3′-Deoxy-adenosine	Anticancer	Phase I

Source: Adapted from [62].

TAF exhibited better anti-HIV efficacy with greater *in vivo* stability, and fewer side effects compare to both tenofovir and TDF [63]. TAF is the latest ProTide of the acyclic nucleoside phosphonate medication tenofovir, developed by Gilead Sciences Inc. Tenofovir disoproxil fumarate (TDF), another prodrug of tenofovir, had long been an FDA-approved treatment for HIV; nevertheless, ProTide TAF displayed better anti-HIV efficacy with greater *in vivo* stability, and fewer side effects than both tenofovir and TDF (Table 21.1). After being absorbed by the cell, the medication becomes phosphorylated and pharmacologically active in the diphosphate form, inhibiting reverse transcriptase. Whenever this enzyme becomes blocked, the DNA chain breaks, and viral replication is stopped [10, 63, 64].

21.6 Implementation of ProTides as Nucleoside Analogs

21.6.1 Antiviral Applications of ProTides

The development of the phosphoramidate prodrugs started in the 1990s when McGuigan and his colleagues assessed several azidothymidine (AZT; zidovudine) phosphate triester derivatives for their ability to inhibit the human immunodeficiency virus (HIV-1). Substituted dialkyl phosphates were effective against HIV, while simple AZT and 2′,3′-dideoxythymidine (d4T, stavudine) dialkyl phosphates were ineffective. These results confirmed that the biological impacts of strategically changed phosphates were mediated by the intracellular release of their nucleotide form [60]. This resulted in McGuigan and Professor Jan Balzarini working together to modify the prodrug's structure with the aim of maximizing activation and potency [65]. Following numerous adjustments, aryloxyphosphoramidates containing AZT were created. In CEM cells, these compounds exhibited strong, selective anti-HIV activity; the biological effect's strength varied significantly depending on the type of phosphate-masking groups used. As per the literature, the first known examples of ProTides are AZT aryloxy phosphoramidates, which provide the McGuigan group recognition for creating this group of prodrugs. Throughout several years, the collaboration between the laboratories of McGuigan and Balzarini gifted the progress of ProTide technology. The ProTide method was successfully utilized for several more antiviral NAs after the AZT studies. The examples of ProTides for treatment of various types of viral infections are 2′,3′-dideoxyadenosine and 2′,3′-didehydro-2′,3′-dideoxyadenosine,2′,3′-didehydro-2′,3′-dideoxythymidine (d4T, Stavudine), 2′,3′-dideoxyadenosine-3′-fluoroadenosine [42], 2′,3′-dideoxydrouridine, and 2′,3′-didehydro-2′3′-dideoxyuridine, carbocyclic nucleoside Abacavir (ABC, Ziagen), carbocyclic adenosine derivatives, and 2′,3′-dideoxy-2′3′-didehydro-7-deazaadenosine [66, 67].

Although the ProTide technique was only partially successful, attempts to target influenza viruses were attempted. When McGuigan's team and Inhibitex worked to develop INX189, Pharmasset worked on developing PSI-7977, another nucleoside phosphoramidate. After that Gilead advanced this drug, and the FDA approved it as Sofosbuvir, in December 2013 for the therapy of HCV infections. The antiviral drug 2C-deoxy-2C-α-fluoro-β-C-methyluridine-5′-monophosphate, whose triphosphate class is a potent inhibitor of the HCV ribonucleic acid- (RNA) dependent RNA polymerase (NS5B), is produced by the liver metabolism of sofosbuvir. Phase I trials of sofosbuvir were conducted using the diastereoisomeric mixture; phase II and III trials were conducted using the pure Sp diastereoisomer. Its Rp isomer was found to be eight times less powerful than it [47]. Eventually, sofosbuvir emerged as the medicine with the quickest global sales history. Patients with genotype 1, 2, 3, or 4 infections, including those with hepatocellular carcinoma, showed its effectiveness [68].

Gilead contributed to the development of GS-6620, a further anti-HCV proTide, alongside sofosbuvir. This drug exhibited significant anti-HCV efficiency in phase I clinical study, but its substantial intra- and interpatient pharmacokinetic and pharmacodynamic variability impeded future development ProTide GS5734 (Remdesevir), which is being tested in clinical trials for the treatment of Ebola virus infections, was found as a result of ongoing studies into it as a possible treatment for other viral disorders [10].

21.7 Conclusion

Targeted protein degradation might grow to become a crucial therapeutic approach given the advancements made over the previous 20 years as well as the recent degree of interest and funding from both academia and industry. It has been rewarding to see how it has evolved from an intriguing hypothesis to a clinically validated idea, first with IMiDs and molecular glues and, recently, with heterobifunctional PROTACs in phase I and phase II clinical trials. The effective use of PROTACs in conjunction with antivirals has the potential to alleviate the current drug discovery bottleneck and offer a more readily available means of combating the pandemic. PROTACs do, however, come with a lot of drawbacks, and applying them in the antiviral field presents a lot of difficulties. According to current reports, 15 different PROTAC individuals are reportedly undergoing clinical trials at this time; most of them are aimed at treating tumors; however, one is also looking into treating autoimmune illnesses. PROTACs attach to the target proteins using a reversible noncovalent arrangement, which leads to the target protein being rapidly and completely eliminated within the cell and ceases all activities related to it. This is because the primary mechanism of action of PROTACs is the recruitment of the cell's own ubiquitin–proteasome protein degradation pathway. Because the protein needs to be resynthesized to restore its function and distribution after breakdown, the method of protein degradation is more robust and effective. Through this, the disadvantages of small-molecule drugs that merely inhibit the active site are reduced [19]. On the other hand, ProTide is an approach in which an active chemical entity is masked with a cleavable group (promoiety) that, in the human body, breaks off under certain conditions, exposing the active species [7]. The development and implementation of multiple nucleoside monophosphate and monophosphate analogs in clinical applications have been made easier by ProTide technology. The successful ProTide method that we utilize today was developed over many years of trial and error, especially by the McGuigan group, to maximize the masking groups of the phosphate. This technology was subsequently utilized for the early discovery of nucleoside-based ProTides for Parkinson's disease, even though it is used in the discovery of nucleoside-based ProTides as antiviral and anticancer agents. As a prodrug strategy for intracellular administration of monophosphorylated non-nucleoside compounds such as glucosamine, Sphingosine 1-phosphate (S1P), 4-phospho-D-erythronohydroxamic acid, and 5-phospho erythronohydroxamic acid, there is growing interest regarding the ProTide approach. There is hope these methods could provide patients with novel therapeutic alternatives for a variety of conditions in the near future.

References

1 Singh, V. and Kumar, A. (ed.) (2021). *Advances in Bioinformatics*. Springer Singapore: Singapore.

2 Doytchinova, I. (2022). Drug design—past, present, future. *Molecules* 27.

3 Sinha, S. and Vohora, D. (2018). Drug discovery and development. *Pharmaceutical Medicine and Translational Clinical Research* 19–32.

4 Han, X. and Sun, Y. (2023). PROTACs: a novel strategy for cancer drug discovery and development. *MedComm* 4.

5 Yao, T., Xiao, H., Wang, H., and Xu, X. (2022). Recent advances in PROTACs for drug targeted protein research. *International Journal of Molecular Sciences* 23 (18): 10328.

6 Liu, Z., Hu, M., Yang, Y. et al. (2022). An overview of PROTACs: a promising drug discovery paradigm. *Molecular Biomedicine* 3.

7 Serpi, M. and Pertusati, F. (2021). An overview of ProTide technology and its implications to drug discovery. *Expert Opinion on Drug Discovery* 16: 1149–1161.

8 Osgerby, L., Lai, Y.C., Thornton, P.J. et al. (2017). Kinetin riboside and its ProTides activate the Parkinson's disease associated PTEN-induced putative kinase 1 (PINK1) independent of mitochondrial depolarization. *Journal of Medicinal Chemistry* 60 (8): 3518–3524.

9 McGuigan, C., Serpi, M., Bibbo, R. et al. (2008). Phosphate prodrugs derived from *N*-acetylglucosamine have enhanced chondroprotective activity in explant cultures and represent a new lead in antiosteoarthritis drug discovery. *Journal of Medicinal Chemistry* 51 (18): 5807–5812.

10 Mehellou, Y., Rattan, H.S., and Balzarini, J. (2018). The ProTide prodrug technology: from the concept to the clinic. *Journal of Medicinal Chemistry* 61: 2211–2226.

11 Carracedo-Reboredo, P., Liñares-Blanco, J., Rodríguez-Fernández, N. et al. (2021). A review on machine learning approaches and trends in drug discovery. *Computational and Structural Biotechnology Journal* 19: 4538–4558.

12 Song, C.M., Lim, S.J., and Tong, J.C. (2009). Recent advances in computer-aided drug design. *Briefings in Bioinformatics* 10 (5): 579–591.

13 Batool, M., Ahmad, B., and Choi, S. (2019). A structure-based drug discovery paradigm. *International Journal of Molecular Sciences* 20.

14 Brogi, S., Ramalho, T.C., Kuca, K. et al. (2020). Editorial: *in silico* methods for drug design and discovery. *Frontiers in Chemistry* 8.

15 Tsagkaris, C., Corriero, A.C., Rayan, R.A. et al. (2023). Success stories in computer-aided drug design. In: *Computational Approaches in Drug Discovery, Development and Systems Pharmacology*, 237–253. Elsevier.

16 Kliza, K. and Husnjak, K. (2020). Resolving the complexity of ubiquitin networks. *Frontiers in Molecular Biosciences* 7.

17 Konstantinidou, M., Li, J., Zhang, B. et al. (2019). PROTACs – a game-changing technology. *Expert Opinion on Drug Discovery* 14 (12): 1255–1268.

18 Cromm, P.M. and Crews, C.M. (2017). Targeted protein degradation: from chemical biology to drug discovery. *Cell Chemical Biology* 24 (9): 1181–1190.

19 Saraswat, A.L., Vartak, R., Hegazy, R. et al. (2023). Drug delivery challenges and formulation aspects of proteolysis targeting chimera (PROTACs). *Drug Discovery Today* 28.

20 Belcher, B.P., Ward, C.C., and Nomura, D.K. (2023). Ligandability of E3 ligases for targeted protein degradation applications. *Biochemistry* 62 (3): 588–600.

21 Finley, D. (2009). Recognition and processing of ubiquitin-protein conjugates by the proteasome. *Annual Review of Biochemistry* 78 (1): 477–513.

22 Weng, G., Cai, X., Cao, D. et al. (2023). PROTAC-DB 2.0: an updated database of PROTACs. *Nucleic Acids Research* 51 (D1): D1367–D1372.

23 Gross, P.H., Sheets, K.J., Warren, N.A. et al. (2022). Accelerating PROTAC drug discovery: establishing a relationship between ubiquitination and target protein degradation. *Biochemical and Biophysical Research Communications* 628: 68–75.

24 Békés, M., Langley, D.R., and Crews, C.M. (2022). PROTAC targeted protein degraders: the past is prologue. *Nature Reviews Drug Discovery* 21: 181–200.

25 Kannt, A. and Đikić, I. (2021). Expanding the arsenal of E3 ubiquitin ligases for proximity-induced protein degradation. *Cell Chemical Biology* 28 (7): 1014–1031.

26 Chen, Y.N.P., LaMarche, M.J., Chan, H.M. et al. (2016). Allosteric inhibition of SHP2 phosphatase inhibits cancers driven by receptor tyrosine kinases. *Nature* 535 (7610): 148–152.

27 Li, J. and Liu, J. (2020). PROTAC: a novel technology for drug development. *ChemistrySelect* 5: 13232–13247.

28 Guedeney, N., Cornu, M., Schwalen, F. et al. (2023). PROTAC technology: a new drug design for chemical biology with many challenges in drug discovery. *Drug Discovery Today* 28.

29 Lee, J., Lee, Y., Jung, Y.M. et al. (2022). Discovery of E3 ligase ligands for target protein degradation. *Molecules* 27 (19): 6515.

30 Li, X., Pu, W., Zheng, Q. et al. (2022). Proteolysis-targeting chimeras (PROTACs) in cancer therapy. *Molecular Cancer* 21.

31 Musharrafieh, R., Ma, C., Zhang, J. et al. (2019). Validating enterovirus D68-2A^{pro} as an antiviral drug target and the discovery of Telaprevir as a potent D68-2A^{pro} inhibitor. *Journal of Virology* 93 (7).

32 Bond, M.J., Chu, L., Nalawansha, D.A. et al. (2020). Targeted degradation of oncogenic KRAS ^{G12C} by VHL-recruiting PROTACs. *ACS Central Science* 6 (8): 1367–1375.

33 Zhang, X., Crowley, V.M., Wucherpfennig, T.G. et al. (2019). Electrophilic PROTACs that degrade nuclear proteins by engaging DCAF16. *Nature Chemical Biology* 15 (7): 737–746.

34 Lee, H., Puppala, D., Choi, E. et al. (2007). Targeted degradation of the aryl hydrocarbon receptor by the PROTAC approach: a useful chemical genetic tool. *ChemBioChem* 8 (17): 2058–2062.

35 Tran, H.T.L., Morikawa, K., Anggakusuma, Z.R. et al. (2020). OCIAD1 is a host mitochondrial substrate of the hepatitis C virus NS3-4A protease. *PLoS One* 15 (7): e0236447.

36 Xie, M., Yang, Z., Liu, Y., and Zheng, M. (2018). The role of HBV-induced autophagy in HBV replication and HBV related-HCC. *Life Sciences* 205: 107–112.

37 Sincere, N.I., Anand, K., Ashique, S. et al. (2023). PROTACs: emerging targeted protein degradation approaches for advanced druggable strategies. *Molecules* 28 (10): 4014.

38 Gamblin, S.J. and Skehel, J.J. (2010). Influenza hemagglutinin and neuraminidase membrane glycoproteins. *Journal of Biological Chemistry* 285 (37): 28403–28409.

39 Chen, Y., Tandon, I., Heelan, W. et al. (2022). Proteolysis-targeting chimera (PROTAC) delivery system: advancing protein degraders towards clinical translation. *Chemical Society Reviews* 51 (13): 5330–5350.

40 Chigbu, D., Loonawat, R., Sehgal, M. et al. (2019). Hepatitis C virus infection: host–virus interaction and mechanisms of viral persistence. *Cells* 8 (4): 376.

41 Papatheodoridi, M., Tampaki, M., Lok, A.S., and Papatheodoridis, G.V. (2022). Risk of HBV reactivation during therapies for HCC: a systematic review. *Hepatology* 75 (5): 1257–1274.

42 Liang, J., Wu, Y., Lan, K. et al. (2023). Antiviral PROTACs: opportunity borne with challenge. *Cell Insight* 2 (3): 100092.

43 Sonntag, E., Hahn, F., Bertzbach, L.D. et al. (2019). *In vivo* proof-of-concept for two experimental antiviral drugs, both directed to cellular targets, using a murine cytomegalovirus model. *Antiviral Research* 161: 63–69.

44 Lahm, A., Yagnik, A., Tramontano, A., and Koch, U. (2002). Hepatitis C virus proteins as targets for drug development: the role of bioinformatics and modelling. *Current Drug Targets* 3 (4): 281–296.

45 Ferraris, O. and Lina, B. (2008). Mutations of neuraminidase implicated in neuraminidase inhibitors resistance. *Journal of Clinical Virology* 41 (1): 13–19.

46 Löffler, M., Morote-Garcia, J.C., Eltzschig, S.A. et al. (2007). Physiological roles of vascular nucleoside transporters. *Arteriosclerosis, Thrombosis, and Vascular Biology* 27 (5): 1004–1013.

47 De, C.E. and Holý, A. (2005). Acyclic nucleoside phosphonates: a key class of antiviral drugs. *Nature Reviews. Drug Discovery* 4 (11): 928–940.

48 Cobb, D.A., Smith, N., Deodhar, S. et al. (2021). Transformation of tenofovir into stable ProTide nanocrystals with long-acting pharmacokinetic profiles. *Nature Communications* 12 (1): 5458.

49 McGuigan, C., Hassan-Abdallah, A., Srinivasan, S. et al. (2006). Application of phosphoramidate ProTide technology significantly improves antiviral potency of carbocyclic adenosine derivatives. *Journal of Medicinal Chemistry* 49 (24): 7215–7226.

50 McGuigan, C., Harris, S.A., Daluge, S.M. et al. (2005). Application of phosphoramidate pronucleotide technology to abacavir leads to a significant enhancement of antiviral potency. *Journal of Medicinal Chemistry* 48 (10): 3504–3515.

51 Siccardi, D., Kandalaft, L.E., Gumbleton, M., and McGuigan, C. (2003). Stereoselective and concentration-dependent polarized epithelial permeability of a series of phosphoramidate triester prodrugs of d4T: an *in vitro* study in Caco-2 and Madin-Darby canine kidney cell monolayers. *Journal of Pharmacology and Experimental Therapeutics* 307 (3): 1112–1119.

52 Slusarczyk, M., Lopez, M.H., Balzarini, J. et al. (2014). Application of ProTide technology to Gemcitabine: a successful approach to overcome the key cancer resistance mechanisms leads to a new agent (NUC-1031) in clinical development. *Journal of Medicinal Chemistry* 57 (4): 1531–1542.

53 Procházková, E., Navrátil, R., Janeba, Z. et al. (2019). Reactive cyclic intermediates in the ProTide prodrugs activation: trapping the elusive pentavalent phosphorane. *Organic & Biomolecular Chemistry* 17 (2): 315–320.

54 McGuigan, C., Daverio, F., Nájera, I. et al. (2009). The application of the phosphoramidate ProTide approach confers micromolar potency against Hepatitis C virus on inactive agent 4′-azidoinosine: Kinase bypass on a dual base/sugar modified nucleoside. *Bioorganic & Medicinal Chemistry Letters* 19 (11): 3122–3124.

55 Alanazi, A.S., James, E., and Mehellou, Y. (2019). The ProTide prodrug technology: where next? *ACS Medicinal Chemistry Letters* 10: 2–5.

56 Chapman, H., Kernan, M., Rohloff, J. et al. (2001). Purification of PMPA amidate prodrugs by SMB chromatography and X-ray crystallography of the diastereomerically pure GS-7340. *Nucleosides, Nucleotides & Nucleic Acids* 20 (4–7): 1085–1090.

57 Pradere, U., Garnier-Amblard, E.C., Coats, S.J. et al. (2014). Synthesis of nucleoside phosphate and phosphonate prodrugs. *Chemical Reviews* 114 (18): 9154–9218.

58 Van Rompay, A.R., Johansson, M., and Karlsson, A. (2000). Phosphorylation of nucleosides and nucleoside analogs by mammalian nucleoside monophosphate kinases. *Pharmacology & Therapeutics* 87 (2–3): 189–198.

59 Jacobson, I.M., Gordon, S.C., Kowdley, K.V. et al. (2013). Sofosbuvir for hepatitis C genotype 2 or 3 in patients without treatment options. *New England Journal of Medicine* 368 (20): 1867–1877.

60 Bordoni, C., Cima, C.M., Azzali, E. et al. (2019). Microwave-assisted organic synthesis of nucleoside ProTide analogues. *RSC Advances* 9 (35): 20113–20117.

61 Foster, G.R., Afdhal, N., Roberts, S.K. et al. (2015). Sofosbuvir and Velpatasvir for HCV genotype 2 and 3 infection. *New England Journal of Medicine* 373 (27): 2608–2617.

62 McGuigan, C., Shackleton, J.M., Tollerfield, S.M., and Riley, P.A. (1989). Synthesis and evaluation of some novel phosphate and phosphonate derivatives of araA. Studies on the mechanism of action of phosphate triesters. *Nucleic Acids Research* 17 (24): 10171–10177.

63 De Clercq, E. (2016). Tenofovir alafenamide (TAF) as the successor of tenofovir disoproxil fumarate (TDF). *Biochemical Pharmacology* 119: 1–7.

64 Menéndez-Arias, L., Álvarez, M., and Pacheco, B. (2014). Nucleoside/nucleotide analog inhibitors of hepatitis B virus polymerase: mechanism of action and resistance. *Current Opinion in Virology* 8: 1–9.

65 McGuigan, C., Davies, M., Pathirana, R. et al. (1994). Synthesis and anti-HIV activity of some novel diaryl phosphate derivatives of AZT. *Antiviral Research* 24 (1): 69–77.

66 McGuigan, C., Turner, S., Nicholls, S.R. et al. (1994). Haloalkyl phosphate derivatives of AZT as inhibitors of HIV: studies in the phosphate region. *Antiviral Chemistry & Chemotherapy* 5 (3): 162–168.

67 McGuigan, C., Pathirana, R.N., Balzarini, J., and De Clercq, E. (1993). Intracellular delivery of bioactive AZT nucleotides by aryl phosphate derivatives of AZT. *Journal of Medicinal Chemistry* 36 (8): 1048–1052.

68 Simmons, B., Liu, Z., Klapars, A. et al. (2017). Mechanism-based solution to the ProTide synthesis problem: selective access to Sofosbuvir, Acelarin, and INX-08189. *Organic Letters* 19 (9): 2218–2221.

22

Advancing Lung Cancer Treatment Through ALK Receptor-targeted Drug Metabolism and Pharmacokinetics

Vivek Yadav[1], *Shikha Goswami*[2], *Rajiv Kumar Tonk*[1], *and Mithun Rudrapal*[3]

[1] *Department of Pharmaceutical Chemistry, Delhi Pharmaceutical Sciences and Research University, New Delhi, India*
[2] *Department of Pharmacology, Delhi Pharmaceutical Sciences and Research University, New Delhi, India*
[3] *Department of Pharmaceutical Sciences, School of Biotechnology and Pharmaceutical Sciences, Vignan's Foundation for Science, Technology & Research, Guntur, India*

22.1 Introduction

Around 80% of lung malignancies are NSCLC, with 3–7% of these being ALK+NSCLC [1]. This receptor belongs to the insulin receptor superfamily and has a few crucial roles in mammal's normal tissues [2]. ALK, however, was revealed to be carcinogenic in different types of malignancies due to abnormal expression and hyperactivation. The people with ALK-positive nonsmall cell lung cancer (NSCLC) are usually younger, nonsmokers, and often present with the histological subtype adenocarcinoma [3]. From the published meta-analysis report it was revealed that the ALK-positive NSCLC patients have a higher prevalence of thromboembolism compared to non-ALK-positive patients [4]. The proto-oncogene ALK encodes for anaplastic lymphoma kinase, which is predominantly expressed in the nervous system [5, 6].

ALK signaling is primarily stimulated in cancer cells via gene fusions, gene amplification, and activating point mutations [7]. ALK gene fusion with the 5′ sequence of the echinoderm microtubule-associated protein-like 4 (EML4) gene resulted in the first ALK rearrangement discovered in NSCLC in 2007 [8]. The result of this rearrangement is the production of the EML4-ALK fusion protein. Since then, several cancer types have been shown to harbor different ALK fusion gene variants. When two ALK proteins fuse, the resulting protein gains control over both the degree of gene expression and the multimerization of the ALK kinase domain, possibly affecting a variety of biological processes, including cell differentiations [9]. ALK mutations were originally discovered as a result of the discovery of the nucleophosmin-ALK fusion gene in 1994, which was a turning point in the study of cancer [10].

Crizotinib, the first ALK inhibitor launched in 2011, was first used to treat NSCLC patients with ALK rearrangements [11]. In comparison to other therapies, it showed noticeably greater median progression-free survival and objective response rates. However, the majority of patients exhibited primary or secondary resistance to crizotinib after a year of treatment. Second-generation ALK inhibitors including ceritinib, brigatinib, and alectinib were created by researchers to address the problem of crizotinib-resistant mutations [12]. To address resistance and improve treatment results, the third-generation ALK inhibitor lorlatinib with a macrocyclic structure was also released.

Computational Methods for Rational Drug Design, First Edition. Edited by Mithun Rudrapal.

22.2 ALK Receptor and Its Role

22.2.1 ALK Structure

An extracellular ligand-binding domain, a transmembrane domain, and an intracellular tyrosine kinase domain make up the three primary structural elements of the ALK target [13]. Despite having a great deal in common with the insulin receptor (IR), the human ALK's tyrosine kinase domain, the ALK extracellular domain (ECD), is unique within the receptor tyrosine kinase (RTK) family. It has a meprin A-5 protein, MAM domain which is a receptor protein tyrosine phosphatase mu, LDLa domain (low-density lipoprotein receptor class A) and a glycine-rich region (Figure 22.1).

22.2.1.1 ALK Extracellular Domain

The similarity between the ECDs of ALK and the RTK places them in a distinctive subgroup within the insulin receptor superfamily. The human ALK RTK subfamily gets activated by specific small secreted peptide/protein ligands known as FAM150A (AUGβ) and FAM150B (AUGα), which effectively trigger ALK signaling.

Two MAM domains, one low-density lipoprotein domain (LDLa) and a glycine-rich region, make up the ALK ECD. This ECD may be split into a number of domains, each of which may have roles in dimerization, ligand binding, and communication with co-receptors and secreted regulatory proteins. The activation of the intracellular kinase domain is hypothesized to be triggered by conformational changes that are relayed by these domains. The proteolytic cleavage of the full-length ALK receptor ECD results in the release of a 140 kDa highly tyrosine-phosphorylated shortened receptor and an ECD fragment of around 80 kDa. Uncertainty persists regarding the precise physiological importance and molecular processes driving this cleavage event. According to a previous study, cleaved ALK is more stable and active than intact ALK, and the cleavage process may also contribute to ligand-mediated activation process.

22.2.1.2 ALK Kinase Domain

A "hinge" region connects the well-conserved amino-terminal and carboxy-terminal lobes of the ALK kinase domain, forming a cleft that acts as the ATP binding site. This makes it possible for the catalytic kinase process to occur. Furthermore, the kinase domain's internal structures support its adaptability and dynamic infrastructure, providing crucial allosteric modulation within and between the lobes. The transition between the conformations of an active and inactive kinase is facilitated by this regulatory mechanism [2].

The important tyrosine kinase receptor known as the ALK receptor controls cellular signaling networks involved in cell growth and survival. Oncogenic factors have been identified as ALK gene rearrangements in a subset of nonsmall cell lung tumors, which result in persistent ALK activation.

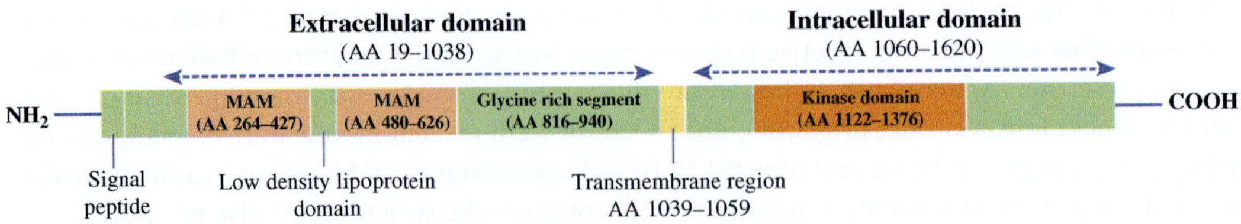

Figure 22.1 Structure of the anaplastic lymphoma kinase (ALK) gene.

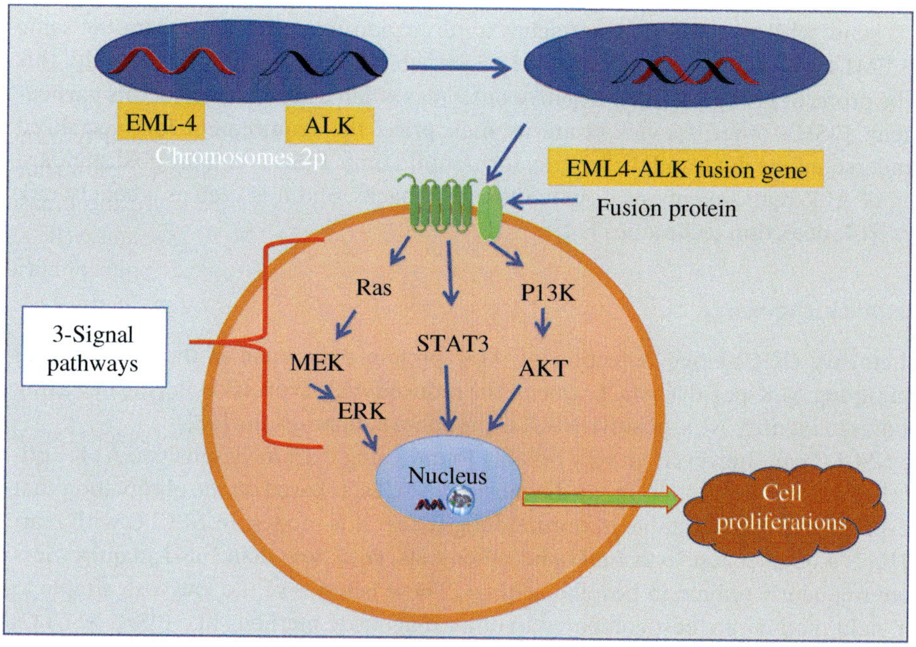

Figure 22.2 Mechanism of EML4-ALK rearrangement.

22.2.2 Mechanism (EML4-ALK)

The echinoderm microtubule-associated protein-like 4 (EML4) gene's 5′ sequence fused with the 3′ region of the ALK gene. Through the multimerization of the ALK kinase domain, this rearrangement caused the production of the EML4-ALK fusion protein, which supports a number of biological processes including cell differentiation, proliferation, and antiapoptosis. ALK has the ability to activate several signaling cascades; some of them are the mitogen-activated protein kinase (MAPK), phosphatidylinositol 3-kinase (PI3K), Janus kinase/signal transducer, and activator of transcription (JAK/STAT) pathways which stimulates the nucleus resulting in the proliferation of the cancerous cells (Figure 22.2).

22.3 Diagnostic Methods for ALK Rearranged NSCLC

Detection of ALK rearrangements in tumor tissue can be accomplished using various methods such as:

- Fluorescence *in situ* hybridization (FISH)
- Immunohistochemistry (IHC)
- Reverse transcription-polymerase chain reaction (RT-PCR)
- Next-generation sequencing (NGS)

22.3.1 Fluorescence *In Situ* Hybridization (FISH)

The Food and Drug Administration (FDA) approved the Vysis Dual Color break-apart fluorescence *in situ* hybridization (FISH) by Abbot Molecular Inc. as the pioneering detection method for NSCLC with a positive ALK test result. In this method, a red probe binds to the region located just

3′ from the ALK gene, while a green probe attaches to the region immediately 5′ from the same gene. Although EML4 and ALK are situated only 12.5 megabases apart on chromosome 2p, this technique may be prone to producing false negative outcomes when utilized to detect this particular rearrangement. FISH's other drawbacks are its high price, the requirement for specialized knowledge to understand the data, and the lengthy turnaround time. FISH is still the gold standard for the detection of ALK rearrangements despite these drawbacks, and it is used as a benchmark to validate other ALK detection techniques [14].

22.3.2 Immunohistochemistry

Immunohistochemistry (IHC)-based detection of ALK protein expression is the current gold standard for identifying ALK-positive ALCL. Due of the reduced ALK expression in NSCLC, using the same antibodies to identify ALK-positive NSCLC produces subpar results [15].

ALK-positive NSCLC can, however, be very reliably found using extremely sensitive ALK antibodies. The idea behind employing IHC in the diagnosis of NSCLC is based on the observation that whereas ALK expression is undetectable in normal lung tissue, it is modest in NSCLC with rearranged ALK. IHC is a quicker test than FISH and other tests, costs less than FISH, requires less skill, and is more frequently offered in hospital settings. These benefits of the low cost, simple to use, and higher yield method are best compared to other diagnostic methods like FISH. NSCLCs that were negative for ALK by IHC in certain cases, however, were reported to be positive by FISH, and like FISH, IHC does not allow for the identification of the fusion partner.

ALK IHC has been regarded as the main diagnostic test for NSCLC due to this test's validation in two well-known clinical trials with ALK inhibitors and the benefits of IHC over FISH that have been stated above. However, most laboratories with substantial experience in NSCLC and ALK testing advise IHC first, followed by confirmation by FISH, due to the risk of a false negative with IHC.

22.3.3 Reverse Transcription-Polymerase Chain Reaction (RT-PCR)

Different ALK fusion partners might have various dimerization and signaling potentials, which would also influence the tumor biology. Therefore, while deciding on the best course of action, fusion partner identification can be crucial. Identification of ALK fusion partners can be achieved through the utilization of specific primers designed for known ALK fusion partners. By employing RT-PCR, researchers can ascertain the fusion partner with precision. The drawback of this method is that numerous primers had to be used before the ALK fusion partner variant could be successfully identified, and unidentified fusion variants could not be found. The development of new assays, such as the ALK RGQ RT-PCR kit manufactured by Qiagen in Manchester, UK, offers a promising solution to address this problem. This advanced one-step quantitative RT-PCR (qRT-PCR) assay is designed to detect all ALK rearrangements, irrespective of the fusion partner or variant. It accomplishes this by measuring the expression of mRNA that encodes the ALK tyrosine kinase domain after qualification through an endogenous control reaction.

A study was conducted to compare the ALK RGQ RT-PCR assay with FISH and IHC using FFPE specimens from a group of 95 patients. The results showed that the qRT-PCR successfully identified all cases (21 patients) with ALK rearrangements, as determined by FISH. Additionally, the qRT-PCR detected some discordant cases that were initially classified as ALK-negative by FISH and IHC, but were later confirmed to have ALK rearrangements through next-generation sequencing. This indicates the potential of implementing qRT-PCR as a routine diagnostic tool for ALK in NSCLC. Furthermore, the qRT-PCR offers additional advantages, such as rapid turnaround time, ease of analysis, and the ability to use biopsy or cytology specimens with smaller tumor content, which is not feasible with precise FISH and IHC methods.

22.3.4 Next-Gen Sequencing

By resolving conflicting or borderline instances, the development of molecular methods for the identification of ALK fusions, such as qRT-PCR, can improve the diagnostic precision. However, one of the key drawbacks for clinical use is its low precision between 3′/5′ gene imbalance value, which can lead to the wrong diagnosis results. An alternate method to solve this issue is amplicon-based next-generation sequencing (NGS). Examples of these approaches are the Ion AmpliSeq RNA Lung Cancer Research Fusion Panel and the Archer® FusionPlex® ALK, RET, and ROS1 v2 kit. Moreover, the Archer FusionPlex® kit not only provides correct identified the fusions but also able to detect novel ALK fusions (DCTN1 – dynactin subunit 1, CLIP1 – CAP-GLY domain-containing linker protein-1, and GCC2 – GRIP and coiled-coil domain-containing protein-2) according to a recent study comparing these kits to IHC and FISH in a subset of 37 patients with NSCLC. It is interesting to note that two out of three patients with these uncommon and new rearrangements responded well to crizotinib treatment. This shows that the Archer® FusionPlex® could offer a precise, useful substitute for FISH testing for the identification of known and novel ALK fusions to direct NSCLC diagnosis and treatment [16].

22.4 ALK Inhibitors Drug Development

USFDA-approved ALK inhibitors for advanced ALK+NSCLC have been categorized into first, second, and third generations, as depicted in Figure 22.3. Among them, Crizotinib was a groundbreaking development in 2011, being the first-in-class, first-generation ALK TKI (tyrosine kinase inhibitor). This innovation developed the treatment approach for ALK-positive NSCLC.

22.4.1 First-generation ALK Inhibitor

22.4.1.1 Crizotinib

Crizotinib is a type of small molecule known as a first-generation ALK inhibitor. Through phase I and II studies, crizotinib has shown promising and long-lasting responses in advanced ALK-positive NSCLC patients. The Crizotinib drug got accelerated FDA approval based on the results obtained from (PROFILE 1007), phase III clinical trials, which demonstrated enhanced progression-free survival from the previous chemotherapy techniques, both in patients receiving it as a first-line treatment and those who had been treated previously [17].

However, one limitation of crizotinib is its poor penetration of the blood–brain barrier, and as a result, it may not effectively treat cancer that has spread to the central nervous system (CNS). Additionally, patients undergoing crizotinib treatment may eventually develop acquired resistance. This resistance is often associated with specific mutations, such as L1196M, G1269A, and C1156Y, which cause structural changes in the ATP-binding pocket of ALK, preventing crizotinib from effectively binding to the target [18].

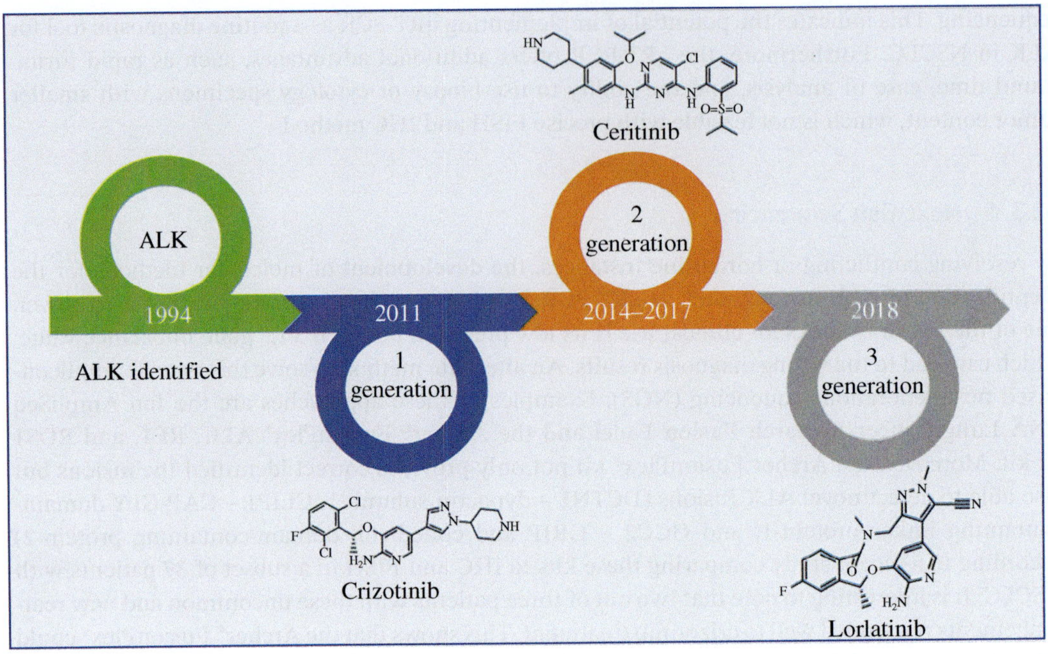

Figure 22.3 ALK timeline development process.

22.4.2 Second-Generation ALK Inhibitor

Second-generation ALK-TKIs such as alectinib, ceritinib, ensartinib, and brigatinib were specifically designed to address the issue of crizotinib resistance in ALK + NSCLC patients. These drugs have demonstrated strong efficacy in treating individuals who developed resistance to crizotinib.

22.4.2.1 Alectinib

The next-generation inhibitor alectinib targets the ALK (anaplastic lymphoma kinase) protein with excellent specificity. Alectinib has higher blood-brain barrier penetration than crizotinib and is not a *P*-glycoprotein substrate. Based on two single-arm studies (NP28761 and NP28673) that comprised 225 patients receiving an oral dosage of 600 mg alectinib twice daily, the FDA authorized alectinib for second-line therapy in 2015 [19]. Alectinib, a second-generation ALK inhibitor, shows improvements in progression-free survival (PFS) and superior tolerability when compared to crizotinib, particularly at doses of 300 mg twice daily, according to the J-ALEX study, which was a significant milestone. Based on the encouraging findings from the phase III ALEX study, which used a 600 mg twice daily dose, the FDA subsequently authorized alectinib for first-line therapy of ALK + NSCLC in 2017 [20]. It is crucial to remember that the J-ALEX study's final analysis failed to show that alectinib had a superior overall survival (OS) rate over crizotinib. This result raises the possibility that switching to second-line therapies after first-line therapy may have had a major impact on overall survival (OS), especially in patients with ALK + NSCLC who would have benefited greatly from all ALK tyrosine kinase inhibitors (TKIs). The occurrence of G1202R and I1171N/S/T mutations can also be linked to on-target resistance to alectinib [21].

22.4.2.2 Brigatinib

Brigatinib (AP26113) has showed potential in overcoming resistance to ALK tyrosine kinase inhibitors (TKIs) in preclinical studies. Clinical trials such as the ALTA trial in crizotinib-treated patients and the ALTA-1L trial in crizotinib-naïve patients with ALK + NSCLC demonstrated encouraging antitumor activity of brigatinib, particularly in CNS metastases cases [22]. The final analysis of ALTA-1L revealed that brigatinib's efficacy was superior to crizotinib, irrespective of the specific ALK fusion variant or TP53 mutation status, particularly in patients with pre-existing brain metastases. A network meta-analysis highlighted brigatinib's high efficacy against CNS metastases when compared to alectinib, although alectinib ranked higher for overall population efficacy. Overall, brigatinib is well-tolerated, although concerns have been raised about its early-onset pulmonary toxicity [23].

22.4.2.3 Ceritinib

The FDA approved crizotinib in 2014 for the treatment of ALK-positive patients who had advanced or were intolerant to it. Additionally, it received the first-line therapeutic approval in 2017. Based on the results of the ASCEND-1 and ASCEND-2 investigations, the permission was granted. Patients with ALK + NSCLC who had previously had crizotinib treatment were given a standard dosage of 750 mg of ceritinib daily as part of the phase II ASCEND-2 research. 38.6% was the study's stated objective response rate (ORR) [24]. Ceritinib 450 mg, when given with meals, had equivalent effectiveness to the 750-mg fasting dosage in the phase I ASCEND-8 research, which assessed three distinct treatment arms [25]. Furthermore, the reduced gastrointestinal toxicity was linked to the lower dosage. Ceritinib's efficacy against conventional chemotherapy in the first-line and second-line scenarios was examined in two randomized Phase III studies, ASCEND-4 and ASCEND-5, respectively [26]. When compared to chemotherapy, ceritinib's toxicity profile in these trials showed a greater frequency of dosage interruptions and adjustments brought on by adverse events [27].

22.4.3 Third-Generation ALK Inhibitor

A third-generation ALK TKI, lorlatinib is a compact macrocyclic inhibitor with increased metabolic stability and less *P*-glycoprotein-mediated efflux *in vitro*. Following successive ALK-TKI therapies, different ALK mutations have been found in cells or patient samples that are resistant to lorlatinib. Lorlatinib successfully blocks the G1202R mutation, although it performs less well against compound mutations. As a second- or third-line therapy for ALK + NSCLC, lorlatinib received FDA approval in 2018. Lorlatinib received first-line FDA clearance in March 2021 as a result of the phase III CROWN trial, which compared lorlatinib to crizotinib and showed greater differential PFS advantages [28, 29]. Crossover was not allowed in the trial, which may have altered the first-line standard of care. However, the absence of direct comparisons between lorlatinib and second-generation ALK TKIs has sparked debates about the most suitable first-line treatment for ALK-positive NSCLC patients. Due to the increased lipophilicity and CNS penetration ability, it showed higher incidences of adverse effects as compared to the previous ALK generation drugs [30].

22.4.4 Fourth-generation ALK TKIs Under Investigation

Sequentially using ALK TKIs that target ALK "single mutants" can result in the development of double ALK resistance mutations. To address this issue, fourth-generation ALK TKIs were developed like TPX-0131 and NVL-655, specifically targeting "double mutants." TPX-0131, a compact macrocyclic inhibitor, suitable for the perfect ATP-binding pocket cavity, which potentially reduces susceptibility to various ALK TKI-resistant mutations, including those in the solvent front, hinge region, gatekeeper, and compound mutations [31]. TPX-0131 not only remains effective against most single resistant mutations but also shows promise in combating compound mutations like G1202R + L1198F, G1202R + L1196M, L1196M + L1198F, and G1202R + C1156F [31].

NVL-655, another fourth-generation ALK tyrosine kinase inhibitor, is a brain-penetrant small-molecule inhibitor capable of targeting solvent front drug-resistance mutations, such as G1202R, G1202R + L1196M, and G1202R + G1269A. Additionally, NVL-655 demonstrates the ability to penetrate the brain, offering good possible treatment options for brain metastases with reduced CNS adverse effects.

22.5 Drug Metabolism of Reported ALK Inhibitor

Drug metabolism in ALK inhibitors has been extensively investigated and documented. These inhibitors undergo various metabolic processes in the body, which include oxidation, reduction, hydrolysis, and conjugation. Enzymes like cytochrome P450 (CYP) enzymes and UDP-glucuronosyltransferases (UGTs) play a significant role in mediating these metabolic pathways. ALK inhibitors are essential in treating NSCLC associated with ALK gene rearrangement. Patients with ALK gene rearrangement have shown sensitivity and clinical benefits when treated with ALK tyrosine kinase inhibitors (ALK-TKIs). Different types of ALK-TKIs drugs, such as crizotinib, ceritinib, alectinib, brigatinib, lorlatinib, and entrectinib, have been approved by the USFDA and the European Medicines Agency for the treatment of ALK gene-rearranged forms of NSCLC [32]. While the oral administration of these inhibitors is convenient and enhances patient conformity, it can be influenced by various factors, such as food, intragastric pH, cytochrome P450 enzymes, transporters, and *P*-glycoprotein. All ALK-TKIs are administered orally, providing flexibility and convenience and improving the quality of life for patients. However, it is important to acknowledge that the oral administration of the drugs may be responsible for the potential drug–drug interactions (DDIs) and lower therapeutic effectiveness or higher toxicities. These pharmacokinetic (PK) interactions, which affect the drug's absorption, distribution, metabolism, and excretion, can alter its bioavailability and potentially result in unfavorable outcomes.

The metabolism of different ALK-TKIs and how it may lead to potential DDIs and it is very crucial to be aware that when taking crizotinib, ceritinib, brigatinib, lorlatinib, and entrectinib, concurrent usage of strong CYP3A inhibitors or inducers should be avoided since these medications can be influenced by such interactions. If it is not possible to avoid using these substances together, it is advisable to adjust the doses of these five ALK-TKIs accordingly. Besides hepatic CYP enzymes, intestinal CYP enzymes also play a significant role in drug metabolism. The inhibition or induction of intestinal CYP enzymes can affect drug absorption. For instance, grapefruit is known to inhibit intestinal CYP enzymes, potentially increasing the bioavailability of ALK-TKIs. PK absorption of the drug is mainly reduced due to the interactions among food and orally administered drugs. Concerning ceritinib and alectinib, having food while administering these drugs significantly enhances their absorption, and patients generally tolerate the associated side effects well. As a result, it is recommended to take these medications with food. However, the presence of food does not appear to have clinically relevant implications for the absorption of crizotinib, brigatinib, lorlatinib, and entrectinib.

22.5.1 Pharmacokinetic Parameters of ALK Inhibitor

Crizotinib belongs to the first generation of ALK-TKIs, while ceritinib, alectinib, and brigatinib belong to the second generation, and lorlatinib belongs to the third generation. Contrastingly, Entrectinib demonstrates robust efficacy as an orally active inhibitor targeting tropomyosin receptor kinases (TRK) A/B/C, c-Ros oncogene 1 (ROS1), and ALK, the tyrosine kinases. Upon oral intake, crizotinib, ceritinib, alectinib, and entrectinib reach their median maximum plasma concentration (C_{max}) between 4 and 6 hours, while brigatinib and lorlatinib achieve their C_{max} in 1–4 hours and 1.2–2 hours, respectively. Among the available ALK-TKIs, lorlatinib exhibits the highest bioavailability at 81%. Approximately 66% of lorlatinib attaches to proteins in the bloodstream, while more than 90% of the other five ALK-TKIs bind to plasma proteins. All existing ALK-TKIs are widely distributed in the body's tissues. Crizotinib and ceritinib have limited access to the central nervous system (CNS), whereas alectinib, brigatinib, lorlatinib, and entrectinib can cross the blood–brain barrier (BBB) and reach CNS tissues.

Crizotinib, ceritinib, alectinib, brigatinib, and entrectinib undergo primarily metabolic processes via the cytochrome P450 (CYP450) pathway, while lorlatinib undergoes metabolism through both CYP3A4 and UDP glucuronosyltransferase (UGT1A4). The elimination half-lives ($t_{1/2}$) of all six drugs exceed 20 hours. With the exception of lorlatinib, these drugs are primarily eliminated through fecal excretion. Refer to Table 22.1 for a summary of the pharmacokinetic (PK) parameters of these six agents (Table 22.1).

To understand the metabolism of ALK inhibitors, it is essential to consider drug transporters present throughout the body, particularly in the intestinal gut, liver, kidney, and the blood–brain barrier (BBB). These transporters show crucial roles in facilitating the membrane transport of

Table 22.1 The list of the pharmacokinetic parameters for ALK inhibitors.

	Absorption		Distribution		Metabolism			Excretion	
PK	T_{max} (h)	F (%)	V_d/V_{ss} (L)	PB (%)	Enzymes	$T_{1/2}$ (h)	CL/F (L/h)	Feces (%)	Urine (%)
Crizotinib	4 6^a	43	1772^b	91	CYP3A	42^a	100^a 60^c	63	22
Ceritinib	4 6^a	NK	4230	97	CYP3A	41^a	88.5^a 33.2^c	92	1.3
Alectinib	4^c	37	475^b	>99	CYP3A4	33	81.9	98	<0.5
Brigatinib	1 4^a	NK	153^b	91	CYP3A4 CYP2C8	25^c	12.7^c	65	25
Lorlatinib	1.2^a 2^c	81	305^b	66	CYP3A4 UGT1A4	24^a	11^a 18^c	41	48
Entrectinib	4 6^a	>50	551	>99	CYP3A4	20	19.6	83	3

T_{max}, time to achieve C_{max}; V_d, apparent volume of distribution; PB, plasma protein binding; $T_{1/2}$, elimination half-life; CL/F, apparent clearance; F, absolute bioavailability; NK, not known.
[a] Single dose administration.
[b] Steady state volume of distribution (V_s).
[c] Multiple dose administrations at steady state.
Source: Zhao et al. [32]/with permission of Taylor & Francis.

various drugs. Broadly classified into two families, drug transporters include the ATP-binding cassette transporters (ABCs) and the solute carrier transporters (SLCs). The ABC transporters, such as P-gp, multidrug resistance-associated protein 2 (MRP2), and breast cancer resistance protein (BCRP), function as efflux transporters for drugs. Inhibition or induction of these drug transporters can significantly affect the pharmacokinetics of drugs, influencing their absorption and metabolism. For instance, the use of elacridar, a P-gp and ABCG2 inhibitor, can enhance the oral bioavailability of crizotinib [33]. Consequently, it is crucial to carefully consider DDIs that occur during the metabolism process mediated by drug transporters.

22.5.2 PK DDIs: Metabolism

The majority of DDIs involving PK (pharmacokinetic) processes take place as a result of the liver's processing of medicines, particularly those metabolized by CYP enzymes [34]. The liver is the main location for drug metabolism, while the small intestine may also play a role in rare cases. DDIs can happen when strong CYP inducers or inhibitors change how CYP substrate medicines are exposed. DDIs may also develop when two substrates bind to the same CYP enzyme-binding site in a competitive manner. As CYP enzyme substrates, the five ALK-TKIs (anaplastic lymphoma kinase tyrosine kinase inhibitors) are vulnerable to DDIs whether supplied with inducers, substrates, or inhibitors of CYP enzymes [35]. Drug transporters, which are found in the liver and kidney and are linked to hepatic CYP enzymes, are essential for the removal of drugs. The liver is the main clearance mechanism for the six ALK-TKIs, with renal elimination making only small contributions. As a result, the excretion of these medicines might be impacted by the inhibition or activation of hepatic transporters and CYP enzymes, resulting in clinically significant DDIs [36]. These theories are supported by a small amount of clinical evidence, but further investigation is required to determine how alterations in hepatic transporters affect the pharmacokinetics of crizotinib, ceritinib, brigatinib, and entrectinib. Being nonsubstrates for transporters, alectinib and lorlatinib should not be impacted by coadministration with inducers or inhibitors of transporters. ALK-TKIs can potentially affect the pharmacokinetics of other medications by altering the expression, enzymes action and transporters. Researching how ALK-TKIs affect the clearance of other substrate medicines is crucial, for this reason. Hepatic (HI) or renal (RI) impairment may be the cause of DDIs that have to do with elimination [37]. HI can dramatically affect the pharmacokinetics and result in DDIs in the case of medications that are mostly removed by the liver. Similar to this, RI can drastically alter the pharmacokinetics and result in DDIs for medications that are largely removed by the kidney. RI may also affect how medications are removed through the liver but not by the kidneys. Therefore, understanding the impact of RI and HI on drug exposure is crucial, including their effects on ALK-TKIs.

22.5.2.1 Crizotinib

The primary metabolic pathway of crizotinib involves the enzyme CYP3A, which plays a significant role in the formation of crizotinib lactam and O-desalkyl metabolites [32].

22.5.2.2 Ceritinib

CYP3A plays a crucial role in the breakdown of ceritinib, being the primary enzyme responsible for its metabolism. When a radiolabeled dose of ceritinib is administered, the predominant component found in the bloodstream is unchanged ceritinib, accounting for approximately 82% of the total. Additionally, 11 metabolites of ceritinib were detected, but in relatively small amounts, with each metabolite representing no more than 2.3% of the original quantity. Ceritinib has demonstrated its ability to inhibit CYP3A, implying that the nonlinear pharmacokinetics of ceritinib could be due to its self-inhibition of CYP3A [38].

22.5.2.3 Alectinib

Alectinib undergoes extensive metabolism primarily mediated by the enzyme CYP3A4, leading to the formation of M4 product. Further metabolism of M4 product also occurs via CYP3A4, resulting in the production of M6 product. When a radiolabeled dose is administered, the unchanged form of alectinib is the main component present in the bloodstream, accounting for 61% of the total radioactivity. In contrast, M4 product represents 15% of the total radioactivity detected in plasma [39].

22.5.2.4 Brigatinib

Brigatinib undergoes metabolic processes primarily mediated by CYP2C8 and CYP3A4 enzymes. When administered as a single radiolabeled dose, the predominant circulating radioactive element is unchanged brigatinib, making up 92% of the total radioactivity. On the other hand, its primary metabolite, known as AP26123, contributes to 3.5% of the observed radioactivity [40].

22.5.2.5 Lorlatinib

Lorlatinib undergoes primary metabolism through CYP3A4 and UGT1A4 enzymes, while CYP2C8, CYP2C19, CYP3A5, and UGT1A3 enzymes play a minor role in its metabolism. When a radiolabeled dose is administered, approximately 21% of the radioactivity present in plasma is attributed to a metabolite called benzoic acid (M8), which is pharmacologically inactive [41].

22.5.2.6 Entrectinib

The primary metabolic pathway of Entrectinib involves CYP3A4. In terms of active metabolites, M5 stands out as the predominant circulating metabolite. During steady-state conditions, the exposure of circulating M5 in patients accounted for 40% of the corresponding entrectinib exposure.

22.6 Resistance and Mutations

ALK inhibitors are particularly made to prevent cancer cells from acting abnormally when the ALK receptor is activated. These inhibitors can bind to the ALK kinase domain, inhibiting both the enzyme's enzymatic activity and the signaling pathways that it activates downstream. The therapeutic management of ALK-positive lung cancer is complicated significantly by the establishment of resistance mechanisms. Changes in drug metabolism and pharmacokinetics are other possible causes of resistance, as are modifications to the ALK receptor itself. ALK resistance mechanisms fall into two basic kinds. TKIs, namely off-target modifications like bypassing signaling pathways and on-target modifications like ALK mutation/gene amplification. The ALK signaling pathway may be activated by substitution with ALK-destabilizing mutations, conferring inhibitor medication resistance. For first- and second-generation ALK-TKIs, only a small percentage of patients with acquired resistance to ALK-TKIs had inherent ALK resistance mutations. The major mechanism of resistance is caused by ALK mutations such somatic kinase domain mutations. L1196M (7%) and G1269A (4%) were two significant ALK mutations following the first-generation ALK TKI crizotinib, which affects 3D conformation and impairs tyrosine kinase inhibitor binding. Specific mutations including G1202R, I1171N, S1206Y, and E1201K, for which not all ALK inhibitors are equally effective, are linked to resistance to second-generation ALK TKIs. Following the

Table 22.2 Resistance and mutations reported of the ALK TKIs.

Drugs	Mutations
Crizotinib	1151Tins, L1152R, C1156Y, F1174L/V, L1196M, G1202R, S1206Y, F1245C, G1269A
Ceritinib	G1123S, F1174C/V, G1202R
Alectinib	I1171N/T/S, G1202R
Lorlatinib	L1198F

Source: Adapted from Gristina et al. [42].

administration of cetitinib, secondary mutations were found in 56% of patient samples, with 17% exhibiting double mutations, including G1202R (21%), F1174 C/L (17%), and C1156Y (8%). In approximately 53% of patients, acquired alectinib mutations were detected, comprising G1202R (29%), I1171T/S (12%), V11180L (6%), and L1196M (6%) (Table 22.2). G1269A (6.6%) was found additional and more frequently than G1202R (2.8%) among 14.2% of patients associated with secondary ALK mutations after second-line ensartinib therapy, although G1202R (2.8%) was not the most common ALK mutation in ensartinib-resistant patients. Compound ALK mutations are a crucial factor in the development of on-target resistance to lorlatinib, a third-generation ALK inhibitor. ALK kinase domain secondary mutations are linked to around 50% of cases of resistance to second-generation ALK TKIs.

When the patients develop secondary ALK mutations and become resistant to the initial ALK TKI treatment, other ALK TKIs can be employed. The concept of resensitization has led to the sequential or alternate use of several ALK TKIs. The emergence of diverse and numerous resistance mechanisms against novel generation ALK TKIs reflects the selective genetic pressure imposed by these medications.

22.7 Conclusion

In this chapter, we have provided an overview of the ALK targets and PK (pharmacokinetics) parameters of ALK inhibitor drugs from the first generation to the fourth generation. However, ALK-targeted therapies have transformed the treatment of ALK-driven cancers, but there are still some persistent challenges, which include resistance and mutations, the need for precise diagnostic biomarkers, and managing potential side effects. The information provided in this chapter aims to assist healthcare professionals in optimizing treatment effectiveness while reducing the likelihood of adverse reactions when prescribing ALK-TKIs (anaplastic lymphoma kinase tyrosine kinase inhibitors). Future perspectives for ALK-targeted treatments involve overcoming resistance through innovative therapies, precision medicine, and combinational therapy approaches to enhance drug's efficacy, providing optimism for improved outcomes and personalized care in the treatments associated with ALK-related cancers.

Conflict of Interest

The authors declare no conflict of interest.

References

1 Peng, L., Zhu, L., Sun, Y. et al. (2022). Targeting ALK rearrangements in NSCLC: current state of the art. *Frontiers in Oncology* 12 (April): 1–13.

2 Hallberg, B. and Palmer, R.H. (2016). The role of the ALK receptor in cancer biology. *Annals of Oncology [Internet]* 27: iii4–iii15. https://doi.org/10.1093/annonc/mdw301.

3 Riudavets, M., Garcia de Herreros, M., Besse, B., and Mezquita, L. (2022). Radon and lung cancer: current trends and future perspectives. *Cancers (Basel)* 14 (13): 1–17.

4 Zhu, V.W., Zhao, J.J., Gao, Y. et al. (2021). Thromboembolism in ALK+ and ROS1+ NSCLC patients: a systematic review and meta-analysis. *Lung Cancer* 157: 147–155.

5 Yadav, V., Tonk, R.K., and Khatri, R. (2020). Molecular docking, 3D-QSAR, fingerprint-based 2D-QSAR, analysis of pyrimidine, and analogs of ALK (anaplastic lymphoma kinase) inhibitors as an anticancer agent. *Letters in Drug Design & Discovery* 18 (5): 509–521.

6 Johnson, T.W., Richardson, P.F., Bailey, S. et al. (2014). Discovery of (10R)-7-amino-12-fluoro-2,10,16-trimethyl-15-oxo-10,15,16,17-tetrahydro-2H-8,4-(metheno)pyrazolo[4,3-h][2,5,11]-benzoxadiazacyclotetradecine-3-carbonitrile (PF-06463922), a macrocyclic inhibitor of anaplastic lymphoma kinase (ALK) and c-ros O. *Journal of Medicinal Chemistry [Internet]* 57 (11): 4720–4744. https://doi.org/10.1021/jm500261q.

7 Wen, S., Dai, L., Wang, L. et al. (2019). Genomic signature of driver genes identified by target next-generation sequencing in Chinese non-small cell lung cancer. *The Oncologist* 24 (11): e1070–e1081.

8 Xia, B., Nagasaka, M., Zhu, V.W. et al. (2020). How to select the best upfront therapy for metastatic disease? Focus on ALK-rearranged non-small cell lung cancer (NSCLC). *Translational Lung Cancer Research* 9 (6): 2521–2534.

9 Morris, S.W., Naeve, C., Mathew, P. et al. (1997). ALK, the chromosome 2 gene locus altered by the t(2;5) in non-Hodgkin's lymphoma, encodes a novel neural receptor tyrosine kinase that is highly related to leukocyte tyrosine kinase (LTK). *Oncogene [Internet]* 14 (18): 2175–2188. https://doi.org/10.1038/sj.onc.1201062.

10 Kasprzycka, M., Marzec, M., Liu, X. et al. (2006). Nucleophosmin/anaplastic lymphoma kinase (NPM/ALK) oncoprotein induces the T regulatory cell phenotype by activating STAT3. *Proceedings of the National Academy of Sciences of the United States of America* 103 (26): 9964–9969. https://doi.org/10.1073/pnas.0603507103.

11 Abdelgalil, A.A. and Alkahtani, H.M. (2023). Crizotinib: a comprehensive profile. *Profiles of Drug Substances, Excipients and Related Methodology* 48: 39–69.

12 Wu, J., Savooji, J., and Liu, D. (2016). Second- and third-generation ALK inhibitors for non-small cell lung cancer. *Journal of Hematology & Oncology* 9: 19.

13 Iwahara, T., Fujimoto, J., Wen, D. et al. (1997). Molecular characterization of ALK, a receptor tyrosine kinase expressed specifically in the nervous system. *Oncogene [Internet]* 14 (4): 439–449. https://doi.org/10.1038/sj.onc.1200849.

14 Gao, X., Sholl, L.M., Nishino, M. et al. (2015). Clinical implications of variant ALK FISH rearrangement patterns. *The International Association for the Study of Lung Cancer Lung Cancer* 10 (11): 1648–1652.

15 Kim, H., Yoo, S.-B., Choe, J.-Y. et al. (2011). Detection of ALK gene rearrangement in non-small cell lung cancer: a comparison of fluorescence *in situ* hybridization and chromogenic *in situ* hybridization with correlation of ALK protein expression. *The International Association for the Study of Lung Cancer Lung Cancer* 6 (8): 1359–1366.

16 Beadling, C., Wald, A.I., Warrick, A. et al. (2016). A multiplexed amplicon approach for detecting gene fusions by next-generation sequencing. *Molecular Diagnostics Journal [Internet]* 18 (2): 165–175. Available from: https://www.sciencedirect.com/science/article/pii/S1525157815002573.

17 Costa, D.B., Shaw, A.T., Ou, S.H.I. et al. (2015). Clinical experience with crizotinib in patients with advanced ALK-rearranged non-small-cell lung cancer and brain metastases. *Journal of Clinical Oncology* 33 (17): 1881–1888.

18 Barrows, S.M., Wright, K., Copley-Merriman, C. et al. (2019). Systematic review of sequencing of ALK inhibitors in ALK-positive non-small-cell lung cancer. *Lung Cancer: Targets and Therapy* 10: 11–20.

19 Noé, J., Lovejoy, A., Ou, S.H.I. et al. (2019). ALK mutation status before and after alectinib treatment in locally advanced or metastatic ALK-positive NSCLC: pooled analysis of two prospective trials. *Journal of Thoracic Oncology [Internet]* https://doi.org/10.1016/j.jtho.2019.10.015.

20 Nakagawa, K., Hida, T., Nokihara, H. et al. (2020). Final progression-free survival results from the J-ALEX study of alectinib versus crizotinib in ALK-positive non-small-cell lung cancer. *Lung Cancer [Internet]* 139 (November): 195–199. https://doi.org/10.1016/j.lungcan.2019.11.025.

21 Peters, S., Camidge, D.R., Shaw, A.T. et al. (2017). Alectinib versus crizotinib in untreated ALK-positive non-small-cell lung cancer. *The New England Journal of Medicine* 377 (9): 829–838.

22 Camidge, D.R., Kim, H.R., Ahn, M.J. et al. (2020). Brigatinib versus crizotinib in advanced ALK inhibitor–naive ALK-positive non-small cell lung cancer: second interim analysis of the phase III ALTA-1L trial. *Journal of Clinical Oncology* 38 (31): 3592–3603.

23 Spencer, S.A., Riley, A.C., Matthew, A., and Di Pasqua, A.J. (2019). Brigatinib: novel ALK inhibitor for non-small-cell lung cancer. *The Annals of Pharmacotherapy* 53 (6): 621–626.

24 Crinò, L., Ahn, M.-J., De Marinis, F. et al. (2016). Multicenter phase II study of whole-body and intracranial activity with ceritinib in patients with ALK-rearranged non-small-cell lung cancer previously treated with chemotherapy and crizotinib: results from ASCEND-2. *Journal of Clinical Oncology: Official Journal of the American Society of Clinical Oncology* 34 (24): 2866–2873.

25 Cho, B.C., Kim, D., Bearz, A. et al. (2017). ASCEND-8: a randomized phase 1 study of ceritinib, 450 mg or 600 mg, taken with a low-fat meal versus 750 mg in fasted state in patients with anaplastic lymphoma kinase (ALK) –rearranged metastatic non-small cell lung cancer (NSCLC). *Journal of Thoracic Oncology [Internet]* 12 (9): 1357–1367. https://doi.org/10.1016/j.jtho.2017.07.005.

26 Claxton, L., O'Connor, J., Woolacott, N. et al. (2019). Ceritinib for untreated anaplastic lymphoma kinase-positive advanced non-small-cell lung cancer: an evidence review group evaluation of a NICE single technology appraisal. *Pharmacoeconomics [Internet]* 37 (5): 645–654. https://doi.org/10.1007/s40273-018-0720-8.

27 Santarpia, M., Daffinà, M.G., D'Aveni, A. et al. (2017). Spotlight on ceritinib in the treatment of ALK+ NSCLC: design, development and place in therapy. *Drug Design, Development and Therapy* 11: 2047–2063.

28 Shaw, A.T., Bauer, T.M., de Marinis, F. et al. (2020). First-line lorlatinib or crizotinib in advanced ALK-positive lung cancer. *The New England Journal of Medicine* 383 (21): 2018–2029.

29 Solomon, B.J., Bauer, T.M., Ignatius Ou, S.-H. et al. (2022). Post hoc analysis of lorlatinib intracranial efficacy and safety in patients with ALK-positive advanced non-small-cell lung cancer from the phase III CROWN study. *Journal of Clinical Oncology: Official Journal of the American Society of Clinical Oncology* 40 (31): 3593–3602.

30 Akamine, T., Toyokawa, G., Tagawa, T. et al. (2019). Lorlatinib for the treatment of patients with non-small cell lung cancer. *Drugs of Today* 55 (2): 107–116.

31 Ou, S.I., Nagasaka, M., Brazel, D. et al. (2021). Translational oncology will the clinical development of 4th-generation "double mutant active" ALK TKIs (TPX-0131 and NVL-655) change the future treatment paradigm of ALK + NSCLC? *Translational Oncology [Internet]* 14 (11): 101191. https://doi.org/10.1016/j.tranon.2021.101191.

32 Zhao, D., Chen, J., Chu, M. et al. (2020). Pharmacokinetic-based drug–drug interactions with anaplastic lymphoma kinase inhibitors: a review. *Drug Design, Development and Therapy* 14: 1663–1681.

33 König, J., Müller, F., and Fromm, M.F. (2013). Transporters and drug–drug interactions: important determinants of drug disposition and effects. *Pharmacological Reviews* 65 (3): 944–966.

34 Haddad, A., Davis, M., and Lagman, R. (2007). The pharmacological importance of cytochrome CYP3A4 in the palliation of symptoms: review and recommendations for avoiding adverse drug interactions. *Supportive Care in Cancer [Internet]* 15 (3): 251–257. https://doi.org/10.1007/s00520-006-0127-5.

35 Nakanishi, T. and Tamai, I. (2015). Interaction of drug or food with drug transporters in intestine and liver. *Current Drug Metabolism* 16 (9): 753–764.

36 Keiser, M.J., Setola, V., Irwin, J.J. et al. (2009). Predicting new molecular targets for known drugs. *Nature* 462 (7270): 175–181.

37 Krens, S.D., Lassche, G., Jansman, F.G.A. et al. (2019). Dose recommendations for anticancer drugs in patients with renal or hepatic impairment. *The Lancet Oncology* 20 (4): e200–e207.

38 Otoukesh, S., Sanchez, T., Mirshahidi, S. et al. (2019). ASCEND-8 pharmacokinetic, safety, and efficacy data for ceritinib 450 mg with food in patients with anaplastic lymphoma kinase-positive non-small cell lung cancer: a clinical perspective. *Cancer Treatment and Research Communications* 20: 100149.

39 Morcos, P.N., Yu, L., Bogman, K. et al. (2017). Absorption, distribution, metabolism and excretion (ADME) of the ALK inhibitor alectinib: results from an absolute bioavailability and mass balance study in healthy subjects. *Xenobiotica* 47 (3): 217–229.

40 Backman, J.T., Filppula, A.M., Niemi, M., and Neuvonen, P.J. (2016). Role of cytochrome P450 2C8 in drug metabolism and interactions. *Pharmacological Reviews* 68 (1): 168–241.

41 Stypinski, D., Fostvedt, L., Lam, J.L. et al. (2020). Metabolism, excretion, and pharmacokinetics of lorlatinib (PF-06463922) and evaluation of the impact of radiolabel position and other factors on comparability of data across 2 ADME studies. *Journal of Clinical Pharmacology* 60 (9): 1254–1267.

42 Gristina, V., La Mantia, M., Iacono, F. et al. (2020). The emerging therapeutic landscape of ALK inhibitors in non-small cell lung cancer. *Pharmaceuticals [Internet]* 13 (12): Available from: https://www.mdpi.com/1424-8247/13/12/474.

23

Targeting Intrinsically Disordered Proteins (IDPs) in Drug Discovery

Opportunities and Challenges

Sridhar Vemulapalli

Department of Pharmaceutical Sciences, University of Nebraska Medical Center, Omaha, NE, United States

23.1 Introduction

Intrinsically disordered proteins (IDPs) represent a fascinating category of biomolecules characterized by their lack of a fixed three-dimensional structure in physiological conditions. Instead, they exist as dynamic ensembles capable of assuming various conformational states [1]. Their pervasive presence in biological systems underscores their pivotal roles in cellular processes and regulatory networks [2]. IDPs encompass crucial signaling molecules and transcription factors, emphasizing their significance in orchestrating intricate cellular functions. Understanding their role is paramount, as unlike traditional proteins, IDPs contribute to cellular signaling, transcriptional regulation, and molecular recognition due to their dynamic nature [3]. This inherent flexibility enables IDPs to engage in complex interactions, making them indispensable in various physiological and pathological processes.

IDPs significantly influence crucial cellular processes, affecting pathways involved in growth, differentiation, and response to external stimuli [4]. Moreover, their implication in diseases like neurodegeneration and cancer underscores their potential as therapeutic targets [1]. However, the dynamic nature of IDPs poses challenges to conventional drug discovery approaches, which typically target well-defined binding sites [5]. This limitation prompts the development of specialized drug design strategies tailored for IDPs [6]. These innovative approaches aim to harness the unique characteristics of IDPs, thereby opening new avenues for therapeutic intervention (Figure 23.1).

23.2 Properties and Significance of IDPs

23.2.1 Structural Characteristics of IDPs

IDPs constitute a captivating class of biomolecules distinguished by their structural dynamics, devoid of a fixed three-dimensional conformation [7]. Unlike traditional proteins, IDPs exist as dynamic ensembles, showcasing remarkable flexibility and adaptability in their structural states. This intrinsic dynamism underpins their pivotal roles in various cellular processes, enabling IDPs to swiftly alternate between conformations to interact with diverse binding partners [8]. An exemplary illustration of IDPs' structural plasticity is evident in the tumor suppressor protein p53, a crucial regulator of cell cycle progression. P53 harbors disordered regions, facilitating interactions

Computational Methods for Rational Drug Design, First Edition. Edited by Mithun Rudrapal.

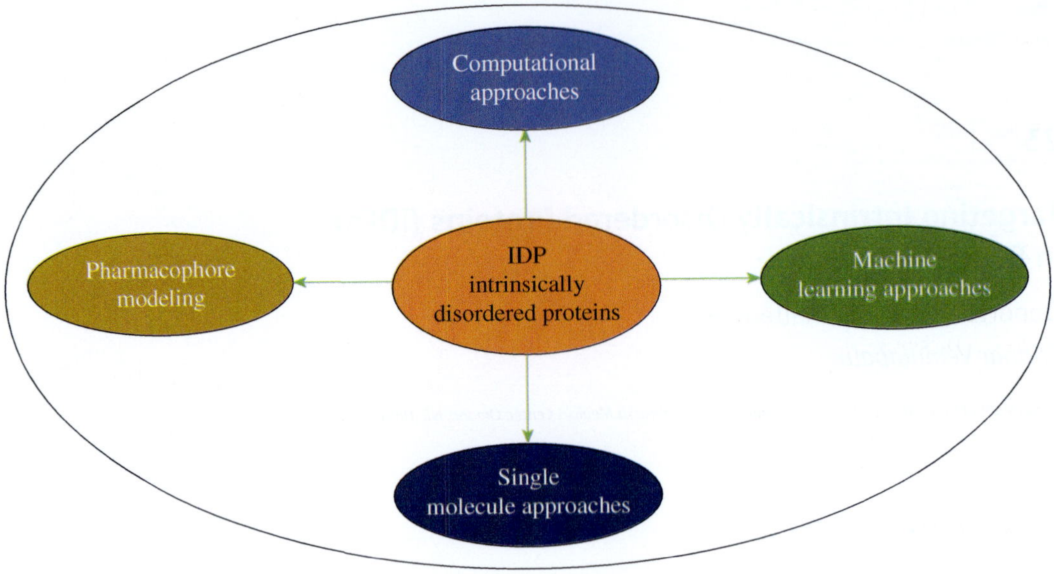

Figure 23.1 Intrinsically disordered proteins drug discovery approaches.

with multiple partners, underscoring the versatility of IDPs in molecular recognition [9]. Moreover, IDPs serve as molecular hubs within signaling pathways, mediating intricate protein–protein interactions and fostering the assembly of protein complexes essential for cellular function [10]. For example, the kinase-inducible domain of the transcription factor CREB (cyclic AMP response element-binding protein) undergoes a disorder-to-order transition upon phosphorylation, enabling its binding to the transcriptional coactivator CBP. This dynamic interplay highlights the indispensable role of IDPs in orchestrating molecular events crucial for cellular homeostasis and function [5].

23.2.2 Relationship Between IDPs and Diseases

The intricate interplay between IDPs and diseases, including cancer and neurodegenerative disorders, highlights the pivotal role of IDPs in pathophysiology [10]. Dysregulated IDPs can instigate uncontrolled cell proliferation and circumvent apoptosis in cancer, contributing to tumor development and progression. For instance, MDM2, an oncoprotein involved in the negative regulation of p53, harbors disordered regions crucial for its interaction with p53, leading to its subsequent degradation [11]. Moreover, in neurodegenerative conditions like Alzheimer's and Parkinson's diseases, aberrant folding and aggregation of IDPs, such as tau and α-synuclein, are implicated in disease pathogenesis [12]. Tau protein's misfolding contributes to the formation of neurofibrillary tangles in Alzheimer's disease (AD), while α-synuclein aggregates are characteristic of Parkinson's disease pathology [13]. Understanding these intricate associations provides valuable insights into potential therapeutic avenues targeting IDPs, offering promise for the development of innovative strategies to mitigate disease progression and improve patient outcomes [1] (Table 23.1).

Table 23.1 List of intrinsically disordered proteins and their role in cellular/regulatory networks and diseases.

S. no.	IDP	Role in cellular/Regulatory networks	Diseases involved	Reference
1	p53	Transcription factor; regulates cell cycle and apoptosis	Various cancers	[14]
2	Retinoblastoma protein (pRb)	Regulates cell cycle by inhibiting E2F transcription factors	Retinoblastoma, osteosarcoma	[15]
3	Disheveled	Component of Wnt signaling pathway; regulates embryonic development	Cancer, developmental disorders, neurodegenerative diseases	[16]
4	α-Synuclein	Regulates synaptic function and neurotransmitter release	Parkinson's disease, Alzheimer's disease	[17]
5	Tau	Microtubule-associated protein; regulates microtubule stability	Alzheimer's disease, frontotemporal dementia	[18]
6	Prion protein (PrP)	Involved in neuronal function and synaptic transmission	Prion diseases	[19]
7	Androgen receptor	Transcription factor; regulates gene expression in response to androgens	Prostate cancer	[20]
8	Huntingtin	Associated with vesicular trafficking and intracellular transport	Huntington's disease	[21]
9	β-Catenin	Transcriptional coactivator in Wnt signaling pathway	Various cancers	[22]
10	NF-κB	Transcription factor; regulates immune response and inflammation	Inflammatory diseases, cancer	[23]
11	c-Myc	Transcription factor; regulates cell proliferation and apoptosis	Various cancers	[24]
12	Estrogen receptor	Transcription factor; mediates estrogen signaling	Breast cancer	[3]
13	PTEN	Tumor suppressor; regulates cell growth and survival	Various cancers	[25]
14	STAT3	Transcription factor; regulates cell growth and survival	Cancer, inflammatory diseases	[20]
15	HIF-1α	Transcription factor; regulates cellular response to hypoxia	Cancer, cardiovascular diseases	[26]
16	SIRT1	Histone deacetylase; regulates gene expression and metabolism	Aging-related diseases, metabolic disorders	[27]
17	Bcl-2	Antiapoptotic protein; regulates mitochondrial membrane permeability	Cancer, neurodegenerative diseases	[28]
18	Nrf2	Transcription factor; regulates antioxidant response	Cancer, neurodegenerative diseases, metabolic disorders	[29]

(Continued)

Table 23.1 (Continued)

S. no.	IDP	Role in cellular/Regulatory networks	Diseases involved	Reference
19	E-cadherin	Cell adhesion molecule; maintains tissue integrity	Cancer, developmental disorders	[30]
20	FOXO	Transcription factor; regulates cell cycle and apoptosis	Cancer, aging-related diseases	[31]
21	Hsp90	Molecular chaperone; facilitates protein folding and stability	Cancer, neurodegenerative diseases	[32]
22	BRCA1	Tumor suppressor; involved in DNA repair	Breast cancer, ovarian cancer	[33]
23	PPARγ	Transcription factor; regulates adipocyte differentiation	Obesity, diabetes	[34]
24	Sonic Hedgehog	Morphogen; regulates embryonic development	Developmental disorders, cancer	[35]
25	SOD1	Antioxidant enzyme; protects against oxidative stress	Amyotrophic lateral sclerosis (ALS)	[36]
26	Annexin A1	Regulates inflammation and immune response	Inflammatory diseases, cancer	[37]
27	Keap1	RegKeaulator of Nrf2; controls antioxidant response	Cancer, neurodegenerative diseases	[29]
28	NFAT	Transcription factor; regulates immune response	Autoimmune diseases, cancer	[38]
29	DJ-1	Oxidoreductase; regulates oxidative stress response	Parkinson's disease, cancer	[39]
30	IRF3	Transcription factor; mediates innate immune response	Viral infections, autoimmune diseases	[40]
31	YAP/TAZ	Transcriptional coactivators; regulate cell proliferation and tissue growth	Cancer, tissue regeneration	[41]
32	Cyclin D	Regulatory subunit of cyclin-dependent kinases	Cancer	[42]

23.3 Challenges in Targeting IDPs

Targeting IDPs in drug discovery presents inherent challenges due to their dynamic structure [1]. Unlike well-structured proteins with defined binding sites, IDPs lack a stable conformation, posing obstacles to identifying specific points for drug interaction [43]. These challenges are exemplified by the oncogenic protein c-Myc, shedding light on the complexities of designing drugs for IDPs.

23.3.1 Structural Heterogeneity

c-Myc exhibits structural heterogeneity, lacking a fixed three-dimensional structure due to its disordered regions [1]. This inherent flexibility complicates drug design efforts, as it becomes challenging to pinpoint specific binding sites for drug molecules [44].

23.3.1.1 Dynamic Conformations

IDPs like c-Myc can adopt multiple conformations, facilitating interactions with diverse binding partners [45]. This dynamic behavior adds another layer of complexity to drug design, as targeting a single conformation may not effectively inhibit its activity [44].

23.3.1.2 Lack of Well-Defined Binding Pockets

Traditional drug design relies on targeting well-defined binding pockets on protein structures. However, IDPs often lack such pockets, making it difficult to design drugs that selectively bind to and inhibit their activity [9]. This challenge necessitates innovative approaches to drug design tailored to the unique characteristics of IDPs.

23.3.1.3 Functional Implications

The disordered regions of c-Myc play critical roles in its function, including protein–protein interactions and transcriptional regulation [46]. Targeting these regions without disrupting essential functions poses a significant challenge in drug development, requiring precise molecular understanding and careful consideration of functional implications [6].

Addressing the challenges associated with the dynamic structure of IDPs, as exemplified by c-Myc, requires innovative approaches and a deep understanding of molecular interactions. Overcoming these obstacles holds the key to unlocking the therapeutic potential of targeting IDPs in drug discovery.

Traditional drug discovery approaches encounter significant limitations when targeting IDPs. Unlike structured proteins, IDPs lack a rigid, well-defined structure, challenging the classical "lock-and-key" model of drug–target interaction [47]. This poses several obstacles to drug discovery, particularly evident in the case of the tumor suppressor protein p53.

23.3.1.4 Dynamic Conformations

The disordered regions of p53 exhibit dynamic conformations, allowing the protein to adopt multiple shapes and interact with diverse binding partners [48, 49]. This inherent flexibility complicates drug design efforts, as targeting a specific conformation may not effectively modulate p53's activity.

23.3.1.5 Lack of Well-Defined Binding Sites

Unlike structured proteins with defined binding pockets, p53's disordered regions lack specific sites for small-molecule binding [50]. This absence of well-defined binding sites hampers the identification of suitable targets for drug development, limiting the efficacy of traditional drug discovery approaches.

23.3.1.6 Functional Importance

The disordered regions of p53 are crucial for mediating protein–protein interactions essential for its tumor suppressor functions [51]. Consequently, targeting these regions without disrupting vital interactions presents a formidable challenge in drug development, as any interference could compromise p53's tumor-suppressive role.

23.3.1.7 Flexibility and Plasticity

The inherent flexibility and plasticity of p53's disordered regions allow it to adapt to various cellular contexts and environmental stimuli [52]. This adaptability further complicates drug design efforts, as the protein's behavior may vary depending on its surroundings, making it challenging to develop drugs that effectively modulate its activity.

Navigating these limitations necessitates innovative approaches that acknowledge the unique characteristics of IDPs. Integrating advanced computational modeling, structural biology techniques, and high-throughput screening methods can enhance our understanding of IDP dynamics and facilitate the design of targeted therapeutics [53]. By overcoming these obstacles, researchers can unlock the full therapeutic potential of IDPs, paving the way for novel treatments for a wide range of diseases.

Targeting IDPs presents multifaceted challenges that extend beyond structural complexities, encompassing intricate interactions and functional implications. For instance, the disordered region of α-synuclein, implicated in Parkinson's disease, interacts promiscuously with various cellular membranes, posing significant hurdles for targeted drug interventions [54].

23.3.1.8 Promiscuous Interactions

IDPs like α-synuclein engage in promiscuous interactions, binding to multiple partners through diverse binding regions. This promiscuity complicates drug development, as targeting a specific interaction may inadvertently disrupt other cellular processes vital for normal function [10, 55]. In the case of α-synuclein, its amphipathic nature enables interactions with various cellular membranes, including the plasma membrane and synaptic vesicles [56].

23.3.1.9 Challenges in Selective Modulation

Modulating α-synuclein's pathological interactions without affecting essential cellular functions is challenging due to nonspecific binding and functional importance. α-Synuclein's interactions with cellular membranes are not only implicated in disease progression but also play crucial roles in synaptic transmission and vesicle trafficking [55–57]. Developing therapies that selectively target pathological interactions while preserving normal cellular functions requires careful consideration of these dual roles.

23.3.1.10 Dynamic Interactions

The dynamic nature of α-synuclein's interactions further complicates drug design efforts. Its conformational flexibility allows for multiple binding modes with cellular membranes, adding another layer of complexity to targeted interventions [55, 58–60]. Designing specific inhibitors that selectively interfere with pathological interactions while sparing normal cellular functions necessitates a nuanced understanding of α-synuclein's membrane interactions and their implications for cellular physiology.

23.3.1.11 Comprehensive Impact

α-Synuclein's interactions with various cellular membranes, including the plasma membrane, mitochondrial membrane, and endoplasmic reticulum (ER) membrane, disrupt fundamental cellular processes such as neurotransmission, energy metabolism, and protein folding [5, 55, 61, 62]. Targeting these interactions selectively requires a holistic approach that considers the intricate balance between pathological and physiological functions.

Addressing the complexities of targeting IDPs like α-synuclein demands innovative strategies that transcend traditional drug discovery paradigms. By integrating advanced computational modeling, structural biology techniques, and systems-level analyses, researchers can unravel the intricacies of IDP interactions and pave the way for targeted therapeutic interventions with enhanced efficacy and specificity [59].

23.4 Computational Tools for IDP Analysis

23.4.1 Exploring Computational Methods for IDP Analysis

Advancements in computational methods have significantly contributed to unraveling the intricacies of IDPs [10]. Various bioinformatics tools aid in predicting IDP regions, understanding their dynamic behavior, and deciphering their interactions with binding partners [63]. Notably, the disorder prediction algorithm utilizes sequence information to predict regions likely to lack a stable structure. Tools like IUPred and DISOPRED are widely used to identify disordered regions within protein sequences, providing valuable insights into the disorder propensity of IDPs [64].

23.4.2 Role of Computational Tools in Rational Drug Design

Computational tools play an indispensable role in the rational design of drugs targeting IDPs. Molecular dynamics (MD) simulations enable researchers to simulate the dynamic behavior of IDPs and their interactions with potential drug candidates [6]. This approach allows for the exploration of binding pockets and the identification of allosteric sites crucial for designing drugs that can modulate IDP activity [65]. For instance, in the case of the Tau protein associated with AD, MD simulations have been employed to study the conformational changes of the disordered regions and guide drug design efforts [66].

23.4.3 Molecular Dynamics Simulations for Tau Protein in Alzheimer's Disease

MD simulations offer a powerful tool to elucidate the conformational dynamics of IDPs like the Tau protein implicated in AD.

Mechanism: MD simulations model the behavior of atoms and molecules over time, providing insights into the dynamic behavior of IDPs [5, 6]. In the case of Tau protein, MD simulations capture the transient conformations of its disordered regions, revealing how they interact with other molecules and undergo structural changes implicated in AD pathology [54].

23.4.3.1 Relevance to Alzheimer's Disease

Aggregation mechanism: MD simulations reveal how Tau protein undergoes conformational changes, leading to the formation of aggregates, a hallmark of AD.

Oligomerization studies: Recent MD simulations focus on Tau oligomers and fibrils, providing insights into their structural properties and stability, which is crucial for understanding AD progression [67].

Destabilization mechanisms: MD simulations elucidate how factors like posttranslational modifications or ligand interactions destabilize Tau protein structures, contributing to AD pathology [68, 69].

Drug design insights: *In silico* drug design studies, coupled with MD simulations, identify potential small molecules targeting Tau aggregation, facilitating the development of novel therapeutics for AD [70, 71].

Structure-based discoveries: MD simulations aid in discovering small molecules that interfere with Tau fibril formation, offering promise for AD treatment [72, 73].

Inhibition mechanisms: Computational studies explore how inhibitors prevent Tau fibril formation, providing insights into their inhibitory mechanisms at the molecular level.

MD simulations, in conjunction with experimental data, offer invaluable insights into the Tau protein's role in AD and facilitate the development of targeted therapeutic strategies [74, 75].

23.4.4 Examples of Computational Tools in Action

The application of computational tools is showcased in the identification of binding sites on IDPs [1]. ANCHOR, a bioinformatics tool, specializes in predicting regions of IDPs that transition from disorder to order upon binding to a partner. This predictive capability is invaluable for designing drugs that can stabilize these interactions, thereby inhibiting pathological processes [76]. Moreover, machine learning approaches like DeepIdP have emerged to predict IDP–protein interactions, offering insights into the specificity and strength of these interactions. These computational tools have revolutionized the study of IDPs by enabling the prediction of disordered regions and facilitating rational drug design [77]. The integration of bioinformatics and molecular simulations has significantly advanced our comprehension of IDP structure and interactions, laying the groundwork for innovative therapeutic interventions [54] (Table 23.2).

23.5 Rational Design Approaches for IDP Inhibition

In the pursuit of effective drug development against IDPs, rational design strategies have emerged as indispensable. This section elucidates the complexities of allosteric and orthosteric binding site identification methods, underscoring their significance in crafting targeted therapies.

23.5.1 Examination of Allosteric and Orthosteric Binding Site Identification Methods

23.5.1.1 Allosteric Binding Sites

Allosteric sites, situated distally from the active site, present a unique avenue for targeted drug design. Computational tools such as FTMap facilitate the pinpointing of allosteric pockets on IDPs by predicting regions with energetically favorable binding [94]. Allostery, marked by conformational changes upon ligand binding, affords precise modulation of IDP function.

23.5.1.2 Orthosteric Binding Sites

Orthosteric sites, the classical active sites of proteins, pose challenges owing to the dynamic nature of IDPs [95]. Advanced techniques like MD simulations and NMR spectroscopy, and single-molecule approaches play pivotal roles in characterizing these sites [96–101]. Understanding the structural dynamics of orthosteric sites is paramount for designing drugs capable of effectively competing with endogenous binding partners.

23.5.2 Case Studies Illustrating Successful Rational Design Campaigns

23.5.2.1 Pin1–Par14 Interaction

A prominent case study revolves around the Pin1–Par14 interaction, which has implications for cancer pathogenesis. Through meticulous rational design endeavors, researchers pinpointed an allosteric binding pocket on Pin1. This breakthrough led to the development of small molecules capable of disrupting the Pin1–Par14 interaction, thereby inhibiting cancer cell proliferation [102]. Such targeted interventions highlight the efficacy of rational drug design in modulating critical protein–protein interactions associated with disease progression.

Table 23.2 Computational approaches and tools for IDP analysis.

S. no.	Computational approach/Tools	Computational tools	Advantages	Disadvantages	Reference
1	Scoring function-based methods	DISOclust, PrDOS	Rapid prediction	Limited accuracy	[78]
			Applicable to large datasets	Sensitive to sequence variations	
			Moderate computational requirements	May struggle with highly flexible regions	
2	Machine learning-based methods	IUPred2A, SPOT-disorder	Improved accuracy compared to scoring function-based methods	Training data biases may affect predictions	[79]
			Robust performance across diverse protein sequences	Limited interpretability of predictions	
			Effective for identifying short disordered regions	Computational intensity for model training	
3	Sequence composition methods	FoldIndex, DisEMBL	Simple and computationally efficient	May overlook context-dependent disorder	[80]
			Suitable for identifying long disordered regions	Sensitive to variations in protein length and composition	
			Tolerant to missing structural information	May struggle with highly flexible regions	
4	Comparative sequence analysis	DICHOT, ANCHOR	Utilizes evolutionary conservation for improved predictions	Limited applicability to highly divergent sequences	[76]
			Effective for identifying functional motifs within disordered regions	Sensitivity to the quality of multiple sequence alignments	
			Provides insights into potential binding regions within IDPs	Requires homologous sequences for accurate predictions	
5	Scoring function-based methods	DISOclust, PrDOS	Rapid prediction	Limited accuracy	[81]
			Applicable to large datasets	Sensitive to sequence variations	
			Moderate computational requirements	May struggle with highly flexible regions	
6	Structural ensemble methods	Ensemble disorder prediction (EDP), VSL2	Accounts for conformational dynamics in predicting disorder	Limited by the availability of experimentally determined structures	[82]
			Improved accuracy for proteins with multiple conformations	May struggle with large-scale structural ensembles	
			Provides insights into disorder-to-order transitions	Limited application to highly flexible regions	

(Continued)

Table 23.2 (Continued)

S. no.	Computational approach/Tools	Computational tools	Advantages	Disadvantages	Reference
7	Sequence-based methods		Simple and computationally efficient	Limited by sequence information	[1]
			Applicable to a wide range of proteins	May struggle with highly flexible regions	
			Suitable for large-scale analysis	Sensitive to sequence variations	
8	DisEMBL		Predicts disorder regions based on sequence information	Limited accuracy for highly flexible regions	[1]
			Provides disorder prediction for specific residue positions	Sensitivity to variations in input sequences	
			Web-based tool for user-friendly access and analysis	May not capture conformational changes	
9	IUPred2A		Employs machine learning for disorder prediction	Limited accuracy for long disordered regions	[1]
			Allows prediction of disorder in specific regions	Dependent on the quality and diversity of training data	
			Web server and API for user-friendly and programmatic access	May not capture disorder-to-order transitions	
10	PONDR®		Utilizes neural networks for disorder prediction	Limited accuracy for proteins with complex folding patterns	[1]
			Provides disorder scores and graphical representation	May require additional experimental validation	
			Utilizes neural networks for disorder prediction	Limited accuracy for proteins with complex folding patterns	
11	SPOT-Disorder2		Employs deep learning techniques for disorder prediction	Requires computational resources for model training	[1]
			Achieves high accuracy for long disordered regions	May not capture disorder-to-order transitions	
			Web server for user-friendly access and analysis	Dependent on the quality and diversity of training data	

12	D^2P^2	Integrates multiple disorder prediction methods	Ensemble predictions may not cover all possible conformations	[1]
		Provides disorder prediction scores and sequence annotations	Limited to available prediction methods	
		Facilitates the comparison of predictions from various tools	May not capture context-specific disorder	
13	Molecular dynamics simulations	Captures dynamic behavior of IDPs	Computationally expensive	[1]
		Provides atomistic details of interactions	Limited timescale for certain systems	
		Allows investigation of conformational ensembles	Requires expertise in simulation methodologies	
14	NMR spectroscopy	Determines dynamic structural information	Limited for large and highly disordered proteins	[1]
		Provides site-specific information on IDP motions	Sensitivity to sample conditions	
		Suitable for studying transient interactions	May require isotopic labeling for certain studies	
15	Single-molecule techniques	Investigates individual IDP molecules	Limited spatial and temporal resolution	[1]
		Provides insights into folding and binding processes	Challenging for highly dynamic systems	
		Enables observation of heterogeneous behavior	Technical challenges in data analysis	
16	X-ray crystallography	Provides high-resolution structural information	Limited applicability for IDPs	[83]
		Captures well-structured regions of IDPs	Requires protein crystallization	
		Ideal for understanding ordered regions within IDPs	Limited information on dynamic behavior	
17	Bioinformatics and computational prediction	Provides rapid and large-scale disorder predictions	Limited accuracy for certain protein regions	[83–93]
		Facilitates screening of potential IDPs	May not capture context-specific disorder	
		Integrates various prediction algorithms	Dependent on training datasets and algorithms	

23.5.2.2 MDM2-p53 Interaction

In the context of cancer therapeutics, the MDM2-p53 interaction represents a pivotal target. Notably, Nutlin-3 was strategically engineered to bind orthosterically, effectively blocking the MDM2-p53 interaction. By disrupting this interaction, Nutlin-3 promotes apoptosis specifically in cancer cells [103]. This exemplifies the success of rational drug design strategies in precisely modulating IDP interactions, thereby offering promising avenues for therapeutic intervention in cancer. These case studies underscore the significance of rational design campaigns in targeting protein–protein interactions implicated in disease pathogenesis, thereby paving the way for the development of innovative therapeutics.

Targeting the disordered Tau protein in AD showcases innovative approaches in neurodegenerative disorders. Small molecules like TRx0237 have emerged as promising candidates capable of modulating Tau aggregation, thus offering potential therapeutic avenues [70, 104, 105]. The rational design approach applied in developing such compounds underscores the importance of precision medicine in combating AD.

23.5.2.3 IDP-Targeting PROTACs

Proteolysis-targeting chimeras (PROTACs) represent a paradigm shift in drug development. Notably, ARV-110 targeting the androgen receptor exemplifies how PROTACs induce selective degradation of IDPs, presenting a novel therapeutic strategy [106, 107]. This innovative approach holds promise in addressing diseases characterized by aberrant protein function, including cancer and neurodegeneration.

23.5.2.4 Examples of IDP Targeting

23.5.2.4.1 Small Molecules Inducing Dynamic Effects It was previously demonstrated that small molecules can exert substantial effects by modulating dynamic interactions with IDPs. This observation highlights the potential for designing drugs that target dynamic behaviors crucial for cellular functions, opening new avenues for therapeutic intervention [108].

23.5.2.4.2 Low-Molecular-Weight Inhibitors Clinically relevant IDPs and their low-molecular-weight inhibitors. These inhibitors are designed to regulate transient interactions involving IDPs, offering innovative therapeutic approaches with the potential to address various diseases [109].

23.5.2.4.3 High-Throughput Screening Emphasized the discovery of IDP-binding ligands through high-throughput experimental screening of chemical compound libraries. This approach has led to the identification of compounds with potential clinical applications, accelerating drug discovery efforts [110].

23.5.3 Evaluation of Clinical Impact

23.5.3.1 Phosphorylation-mediated Effects

Posttranslational modifications (PTMs), such as phosphorylation, profoundly impact IDP behavior [111]. MD simulations have been instrumental in studying these effects, providing insights into the clinical implications of PTMs on IDP function and disease pathology [112].

23.5.3.2 Consideration of Interactions in Drug Design

Staid et al. [5] highlighted the importance of considering interactions between RNA-binding proteins (RBPs), microRNAs (miRs), and IDPs in drug design [19]. This comprehensive approach accounts for potential clinical effects, emphasizing the need for holistic strategies in drug development.

23.5.4 Cancer Therapeutics

Santofimia-Castaño et al. [109] summarized the successful targeting of IDPs involved in various cancer types. These advancements hold significant clinical promise, offering innovative approaches to cancer therapeutics by targeting dysregulated protein–protein interactions. The multifaceted approaches to targeting IDPs, including small molecules, low-molecular-weight inhibitors, and high-throughput screening, have led to significant advancements. The comprehensive evaluation of clinical impact spans from molecular interactions to therapeutic outcomes, driving progress in drug discovery and precision medicine.

In conclusion, rational design approaches, whether targeting allosteric or orthosteric sites, have yielded transformative results in IDP drug development. Case studies underscore the efficacy of these strategies, paving the way for innovative and precise therapeutic interventions.

23.6 Case Studies

An in-depth exploration of case studies highlights the pivotal role of computational methods in designing compounds targeting IDPs, showcasing their profound impact on clinical efficacy.

23.6.1 Case Study 1: p53-MDM2 Interaction Inhibition

Computational elucidation of the p53-MDM2 binding interface, as demonstrated, has driven the design of small molecules disrupting this interaction. These compounds show promising clinical implications, particularly in cancer therapeutics, by reactivating the p53 tumor suppressor pathway [113]. This exemplifies how MD simulations offer insights into protein–protein interactions, facilitating the development of targeted therapies.

23.6.2 Case Study 2: N-terminal Tau Binding Pocket

Utilizing computational tools like molecular docking, identified a novel N-terminal Tau binding pocket. Small molecules targeting this site show potential in mitigating Tau aggregation, offering a promising avenue for intervention in neurodegenerative disorders [70]. This underscores the importance of computational methods in uncovering druggable sites on IDPs, thus expanding the repertoire of therapeutic targets.

23.6.3 Case Study 3: Disordered Protein–Protein Interactions

Metadynamics simulations to analyze disordered protein–protein interactions, paving the way for designing compounds that selectively modulate these interactions [65]. This approach presents innovative therapeutic strategies across various diseases by precisely targeting dynamic protein-protein interfaces. Computational simulations enable a detailed understanding of the complex interplay between IDPs, facilitating the development of tailored interventions.

23.6.4 Case Study 4: AlphaFold in Drug Repurposing

AlphaFold, an advanced AI algorithm, extends its utility beyond structural prediction to identify potential IDP targets for existing drugs. This highlights how computational methods contribute not only to novel compound design but also to repurposing existing therapeutics for IDP-related diseases [114]. By leveraging AI-driven approaches, drug repurposing becomes a cost-effective and efficient strategy in IDP-targeted drug discovery.

These case studies underscore the transformative impact of computational methods in IDP-targeting compound design, from elucidating binding interfaces to repurposing existing drugs, ultimately advancing precision medicine.

23.6.5 Evaluation of Clinical Impact

23.6.5.1 Cardiovascular Diseases: Troponin I–Actin Interaction

Computational analysis delved into the interaction between Troponin I and Actin, pivotal in cardiovascular function. Utilizing computational insights, researchers designed small molecules that effectively modulate this interaction, hinting at their potential therapeutic utility in cardiovascular diseases [115, 116]. This highlights the profound impact of computational methodologies in uncovering intricate protein interactions and devising tailored interventions for complex cardiovascular conditions, underscoring their significance in advancing treatment modalities.

23.6.5.2 Infectious Diseases: IDPs in Pathogen–Host Interactions

Computational approaches play a crucial role in deciphering the involvement of IDPs in pathogen–host interactions, as noted in previous research [1]. Leveraging computational insights enables the targeted design of compounds aimed at disrupting these interactions, offering a promising avenue for developing therapeutics against infectious diseases. By elucidating the molecular mechanisms underlying pathogen–host interplay, computational methods contribute to the development of innovative strategies to combat infectious pathogens, emphasizing their potential in addressing global health challenges [117–119]. These case studies underscore the versatility and efficacy of computational methods in IDP drug discovery, demonstrating their broad applicability across diverse disease contexts. The clinical impact of compounds designed through computational insights underscores the transformative potential of these approaches in revolutionizing medical treatment paradigms.

23.6.6 In-Depth Analysis of Computational Methods in IDP-targeting Compounds Design

Understanding the intricate world of IDPs and their potential in drug design necessitates a comprehensive exploration of computational methodologies. Here is a detailed analysis of diverse case studies showcasing the transformative impact of computational methods in IDP-targeting compound design, elucidating the computational methods employed in each case for a deeper understanding.

23.6.6.1 Case Study 1: Disruption of c-Myc–Max Interaction

Computational Method: Molecular Modeling

Rationale: Molecular modeling techniques, enable the visualization and analysis of protein–protein interaction interfaces. By employing computational simulations, researchers unravel the intricate details of the c-Myc–Max interaction interface, which is crucial in cancer progression. Understanding these structural nuances facilitates the rational design of small molecules capable of disrupting this interaction, offering potential therapeutic avenues in cancer treatment [120].

23.6.6.2 Case Study 2: Tau Protein Aggregation in Alzheimer's Disease

Computational Method: Molecular Dynamics Simulations

Rationale: MD simulations, delve into the dynamic behavior of Tau proteins implicated in AD pathology. By simulating the movement of atoms over time, these simulations provide insights into the structural changes driving Tau aggregation [121, 122]. Armed with this knowledge, researchers can design compounds targeting specific regions involved in Tau aggregation, offering promise in mitigating AD progression.

23.6.6.3 Case Study 3: Disordered Proteins in Neurodegenerative Disorders

Computational Method: Metadynamics Simulations

Rationale: Metadynamics simulations, offer a dynamic view of protein–protein interactions in neurodegenerative disorders. These simulations capture the complex interplay between disordered proteins, shedding light on potential therapeutic targets. By manipulating these interactions through computational modeling, researchers can design compounds that modulate disease-associated pathways, presenting innovative treatment strategies for neurological conditions like Parkinson's and Huntington's disease [5, 78, 79].

23.6.7 Computational Methods in IDP-targeting Compound Design: Clinical Impact Evaluation

The integration of computational methodologies in designing compounds targeting IDPs has revolutionized drug discovery, offering innovative solutions to complex disease challenges. Here, we delve into the clinical impact of computational approaches through a series of case studies, highlighting their efficacy in diverse therapeutic contexts.

23.6.7.1 Breast Cancer Treatment: HER2–HER3 Disruption

Computational insights: MD simulations elucidate the dynamic interface of the HER2–HER3 heterodimer, guiding the rational design of small molecules disrupting this interaction. Computational techniques such as docking simulations and molecular mechanics calculations facilitate compound optimization. Clinical trials demonstrate the efficacy of these compounds, providing a personalized therapeutic option for breast cancer patients. The targeted disruption of HER2–HER3 heterodimer highlights the translational potential of computational methodologies in oncology [123–125].

23.6.7.2 Alzheimer's Disease: Aβ Oligomerization Inhibitors

Computational Insights: Molecular docking and dynamics simulations elucidate the structural determinants of Aβ oligomerization, informing the design of inhibitors targeting Aβ aggregation. Free energy calculations optimize drug binding affinities, enhancing therapeutic efficacy. Clinical assessments reveal promising results, underscoring the disease-modifying potential of computationally designed inhibitors in AD management. These findings signify a paradigm shift in neurodegenerative disorder therapeutics [126–128].

23.6.7.3 Metabolic Syndrome Intervention

Computational insights: Protein–ligand docking, and virtual screening identify IDPs implicated in metabolic disorders, guiding compound design. MD simulations optimize binding affinity and pharmacokinetic properties, enhancing therapeutic effectiveness. Clinical trials validate the clinical benefits of computationally designed therapeutics in mitigating metabolic dysregulation [129–133]. The successful translation of computational insights underscores their utility in addressing complex metabolic conditions.

23.6.7.4 Antiviral Strategies: Disruption of Viral–Host Interactions

Computational insights: Protein–protein docking, and network analysis identify IDPs involved in viral–host interactions, guiding compound design. *In silico* screening optimizes drug-likeness and ADMET properties, facilitating clinical translation. Preclinical and clinical evaluations demonstrate the efficacy of computationally designed compounds in antiviral therapy. These findings highlight the versatility of computational methods in combating infectious diseases [108, 117, 124, 134].

In conclusion, computational methodologies play a pivotal role in designing IDP-targeting compounds with significant clinical impact across diverse therapeutic areas. Through a synthesis of computational insights and clinical trials, these case studies underscore the transformative potential of computational approaches in precision medicine, offering tailored therapeutic interventions for complex diseases.

23.7 Future Directions

In recent years, the field of IDPs has experienced a paradigm shift in drug design strategies, marked by significant advancements and promising avenues for future exploration.

Leveraging computational approaches: One notable trend involves harnessing computational methods such as MD simulations and bioinformatics tools to delve into the dynamic nature of IDPs. These techniques enable researchers to unravel the complex conformational ensembles of IDPs, providing insights into their structure–function relationships. By identifying potential binding sites and employing rational drug design strategies, computational approaches offer a powerful toolset for developing targeted therapeutics tailored to IDPs [64, 94].

Exploration of unconventional binding modes: Traditional drug design paradigms often struggle with the inherent flexibility and lack of fixed structure exhibited by IDPs. However, recent advancements have paved the way for exploring unconventional binding modes for IDP-targeted drugs. Innovative strategies focus on designing molecules capable of exploiting the transient conformations of IDPs or inducing folding upon binding. These approaches showcase the adaptability of drug design methodologies in addressing the unique challenges posed by IDPs, promising breakthroughs in therapeutic interventions [65].

23.7.1 Addressing Ongoing Challenges and Avenues for Improvement

Limited understanding of structure–function relationships: Despite progress, gaps persist in comprehensively understanding the intricate structure–function relationships of IDPs. Precisely predicting their behavior remains a challenge, hampering drug discovery efforts. Integration of multimodal techniques, including NMR spectroscopy, mass spectrometry, and computational predictions, holds promise in enhancing our understanding of IDP dynamics and refining drug design strategies [1].

Standardization of experimental and computational methodologies: The absence of standardized experimental and computational methodologies presents a significant obstacle in obtaining reliable data for drug discovery. Efforts to establish robust protocols and benchmarks for experimental and computational studies are crucial for ensuring reproducibility and validity in IDP-targeted drug design endeavors [135]. *Exploration of novel screening approaches*: Researchers are actively exploring novel screening approaches, such as fragment-based drug design and library screening, to identify small molecules capable of interacting with specific IDPs. These methods, coupled with advancements in high-throughput techniques, streamline the screening process and accelerate the discovery of potential drug candidates tailored to IDPs [1, 109].

In conclusion, the future of IDP-targeted drug design holds immense promise, driven by advancements in computational methodologies, innovative binding modes exploration, and ongoing efforts to address persistent challenges and enhance screening approaches.

23.7.2 Discussion of the Possibility of Combination Therapies Involving IDP-targeting Drugs

The complexity of diseases linked to IDPs necessitates innovative therapeutic strategies, with combination therapies emerging as a promising avenue.

Simultaneous targeting of multiple IDPs: By concurrently addressing multiple IDPs implicated in a signaling pathway or disease network, combination therapies offer the potential to enhance treatment efficacy while mitigating the development of resistance. This synergistic approach leverages the interplay between different targets, leading to more robust therapeutic outcomes [54, 109, 136].

Integration with conventional and immunotherapies: Combining IDP-targeting drugs with conventional therapies or immunotherapies presents an opportunity for synergistic effects. By complementing each other's mechanisms of action, these combinations can offer a comprehensive treatment strategy for diseases such as cancer and neurodegenerative disorders. This integration taps into the strengths of different modalities to address the multifaceted nature of IDP-associated diseases [137].

Combining small molecules with biologics: Integrating small molecules with biologics, such as monoclonal antibodies, represents a pioneering approach in IDP-targeted combination therapies. Monoclonal antibodies, with their high specificity, can selectively bind to IDPs, modulating their activity. Meanwhile, small molecules can target other nodes in the pathway, amplifying the therapeutic impact. This combined approach holds promise for achieving enhanced efficacy with minimized side effects, highlighting the potential of synergistic interactions between different therapeutic modalities [131].

In conclusion, the exploration of combination therapies in IDP-targeted drug design reflects the dynamic and evolving nature of this field. These innovative approaches offer new avenues for developing effective treatments for diseases associated with IDPs.

23.8 Conclusions

In conclusion, the significance of IDPs as viable drug targets in contemporary drug discovery cannot be overstated. IDPs play pivotal roles in diverse cellular processes, contributing both to normal physiological functions and the onset of diseases. Their unique ability to adopt multiple conformations and engage in dynamic interactions renders them attractive targets for therapeutic interventions [1].

The evolution of drug discovery has undergone a paradigmatic shift with the increasing recognition of IDPs as druggable entities. Implicated in a spectrum of diseases, including cancer, neurodegenerative disorders, and cardiovascular diseases, IDPs present novel avenues for therapeutic exploration, expanding the repertoire of potential drug targets beyond conventionally structured proteins [65].

Computational methodologies have emerged as indispensable assets in elucidating the intricacies of IDP-targeted drug design. Techniques such as MD simulations, bioinformatics tools, and machine learning algorithms stand at the vanguard of computational approaches. These tools offer insights into the dynamic behavior of IDPs, facilitating the identification of binding sites and rational drug design. Notably, hierarchical computational strategies for IDP drug virtual screening have yielded compounds with promising therapeutic potential [6].

Despite notable advancements, challenges persist in targeting IDPs, including the absence of well-defined structures and the requisite for innovative drug design strategies. Computational techniques, adept at navigating the inherent flexibility of IDPs, rise to meet these challenges. Progress in computational drug discovery streamlines the identification of lead compounds and optimizes their interactions with IDPs. For instance, rational drug design targeting IDPs has culminated in the development of small-molecule and peptide drugs advancing into clinical trials [64]. In conclusion, the convergence of understanding the indispensable role of IDPs, harnessing computational techniques, and surmounting challenges in drug discovery underscores the promise of IDPs as compelling drug targets. This holistic approach fosters avenues for therapeutic innovation, offering substantial contributions to the progression of modern medicine.

References

1 Wang, H., Xiong, R., and Lai, L. (2023). Rational drug design targeting intrinsically disordered proteins. *WIREs Computational Molecular Science* 13 (6): e1685.

2 Chen, J. and Kriwacki, R.W. (2018). Intrinsically disordered proteins: structure, function and therapeutics. *Journal of Molecular Biology* 430 (16): 2275–2277.

3 Guillien, M., Le Maire, A., Mouhand, A. et al. (2020). IDPs and their complexes in GPCR and nuclear receptor signaling. *Progress in Molecular Biology and Translational Science [Internet]* 105–155.

4 Bondos, S.E., Dunker, A.K., and Uversky, V.N. (2022). Intrinsically disordered proteins play diverse roles in cell signaling. *Cell Communication and Signaling: CCS* 20 (1): 20.

5 Fatafta, H., Samantray, S., Sayyed-Ahmad, A. et al. (2021). Molecular simulations of IDPs: from ensemble generation to IDP interactions leading to disorder-to-order transitions. *Progress in Molecular Biology and Translational Science [Internet]* 135–185.

6 Salo-Ahen, O.M.H., Alanko, I., Bhadane, R. et al. (2020). Molecular dynamics simulations in drug discovery and pharmaceutical development. *Processes* 9 (1): 71.

7 Chowdhury, A., Nettels, D., and Schuler, B. (2023). Interaction dynamics of intrinsically disordered proteins from single-molecule spectroscopy. *Annual Review of Biophysics* 52 (1): 433–462.

8 Mukhopadhyay, S. (2020). The dynamism of intrinsically disordered proteins: binding-induced folding, amyloid formation, and phase separation. *The Journal of Physical Chemistry B* 124 (51): 11541–11560.

9 Chakrabarti, P. and Chakravarty, D. (2022). Intrinsically disordered proteins/regions and insight into their biomolecular interactions. *Biophysical Chemistry* 283: 106769.
10 Trivedi, R. and Nagarajaram, H.A. (2022). Intrinsically disordered proteins: an overview. *International Journal of Molecular Sciences* 23 (22): 14050.
11 Pan, M. and Blattner, C. (2021). Regulation of p53 by E3s. *Cancers* 13 (4): 745.
12 Tsoi, P.S., Quan, M.D., Ferreon, J.C., and Ferreon, A.C.M. (2023). Aggregation of disordered proteins associated with neurodegeneration. *International Journal of Molecular Sciences* 24 (4): 3380.
13 Shim, K.H., Kang, M.J., Youn, Y.C. et al. (2022). Alpha-synuclein: a pathological factor with Aβ and tau and biomarker in Alzheimer's disease. *Alzheimer's Research & Therapy* 14 (1): 201.
14 Zou, R., Zhou, Y., Wang, Y. et al. (2020). Free energy profile and kinetics of coupled folding and binding of the intrinsically disordered protein p53 with MDM2. *Journal of Chemical Information and Modeling* 60 (3): 1551–1558.
15 Dyson, H.J. (2023). Vital for viruses: intrinsically disordered proteins. *Journal of Molecular Biology* 435 (11): 167860.
16 Vazquez, D.S., Toledo, P.L., Gianotti, A.R., and Ermácora, M.R. (2022). Protein conformation and biomolecular condensates. *Current Research in Structural Biology* 4: 285–307.
17 Johnson, D.H. and Zeno, W.F. (2024). Synergy between the structured and disordered regions of alpha-synuclein in membrane remodeling. *Biophysical Journal* 123 (3): 74a.
18 Ukmar-Godec, T., Fang, P., Ibáñez De Opakua, A. et al. (2020). Proteasomal degradation of the intrinsically disordered protein tau at single-residue resolution. *Science Advances* 6 (30): eaba3916.
19 Do Amaral, M.J. and Cordeiro, Y. (2021). Intrinsic disorder and phase transitions: pieces in the puzzling role of the prion protein in health and disease. *Progress in Molecular Biology and Translational Science [Internet]* 1–43.
20 Sadar, M.D. (2020). Discovery of drugs that directly target the intrinsically disordered region of the androgen receptor. *Expert Opinion on Drug Discovery* 15 (5): 551–560.
21 Mohanty, P., Phan, T.M., and Mittal, J. (2024). Dynamic interdomain interactions in oligomeric and higher-order assemblies can modulate the intrinsic structural features of huntingtin exon 1. *Biophysical Journal* 123 (3): 26a–27a.
22 Rowling, P.J.E., Murton, B.L., Du, Z., and Itzhaki, L.S. (2022). Multivalent interaction of beta-catenin with its intrinsically disordered binding partner adenomatous polyposis coli. *Frontiers in Molecular Biosciences* 9: 896493.
23 Dyson, H.J. and Wright, P.E. (2023). From immunogenic peptides to intrinsically disordered proteins. *Israel Journal of Chemistry* 63 (10–11): e202300051.
24 Gunasinghe, K.K.J., Rahman, T., and Chee, W.X. (2024). Unraveling the behavior of intrinsically disordered protein c-Myc: a study utilizing Gaussian-accelerated molecular dynamics. *ACS Omega* 9 (2): 2250–2262.
25 Djulbegovic, M.B., Uversky, V.N., Harbour, J.W. et al. (2021). Structural protein analysis of driver gene mutations in conjunctival melanoma. *Genes* 12 (10): 1625.
26 Meyers, L.M., Krawic, C., Luczak, M.W., and Zhitkovich, A. (2022). Vulnerability of HIF1α and HIF2α to damage by proteotoxic stressors. *Toxicology and Applied Pharmacology* 445: 116041.
27 Bui, H.L., Prado, J., Wang, N., and Grazioli, G. (2024). Investigating intracellular SIRT1 regulation mechanisms using molecular dynamics simulations and machine learning. *Biophysical Journal* 123 (3): 551a.
28 Lan, Y.J., Yeh, P.S., Kao, T.Y. et al. (2020). Anti-apoptotic BCL-2 regulation by changes in dynamics of its long unstructured loop. *Communications Biology* 3 (1): 668.
29 Karunatilleke, N.C., Fast, C.S., Ngo, V. et al. (2021). Nrf2, the major regulator of the cellular oxidative stress response, is partially disordered. *International Journal of Molecular Sciences* 22 (14): 7434.

30 Morris, O.M., Torpey, J.H., and Isaacson, R.L. (2021). Intrinsically disordered proteins: modes of binding with emphasis on disordered domains. *Open Biology* 11 (10): 210222.

31 Gui, T. (2023). *The Structure–Function Paradigm Reshaped by Intrinsically Disordered Regions* [Internet] [dr.]. Utrecht University; p. 0003028520231030. [cited 2024 Feb 15]; Available from: https://dspace.library.uu.nl/handle/1874/432850.

32 Kolhe, J.A., Babu, N.L., and Freeman, B.C. (2023). The Hsp90 molecular chaperone governs client proteins by targeting intrinsically disordered regions. *Molecular Cell* 83 (12): 2035–2044.e7.

33 Witus, S.R., Tuttle, L.M., Li, W. et al. (2023). BRCA1/BARD1 intrinsically disordered regions facilitate chromatin recruitment and ubiquitylation. *The EMBO Journal* 42 (15): e113565.

34 Li, Z., Luo, L., Yu, W. et al. (2022). PPARγ phase separates with RXRα at PPREs to regulate target gene expression. *Cell Discovery* 8 (1): 37.

35 Csermely, P., Kunsic, N., Mendik, P. et al. (2020). Learning of signaling networks: molecular mechanisms. *Trends in Biochemical Sciences* 45 (4): 284–294.

36 Eleutherio, E.C.A., Silva Magalhães, R.S., De Araújo, B.A. et al. (2021). SOD1, more than just an antioxidant. *Archives of Biochemistry and Biophysics* 697: 108701.

37 De Souza Ferreira, L.P., Da Silva, R.A., Gil, C.D., and Geisow, M.J. (2023). Annexin A1, A2, A5, and A6 involvement in human pathologies. *Proteins: Structure, Function, and Bioinformatics* 91 (9): 1191–1204.

38 Kumar, R., DuMond, J.F., Khan, S.H. et al. (2020). NFAT5, which protects against hypertonicity, is activated by that stress via structuring of its intrinsically disordered domain. *Proceedings of the National Academy of Sciences* 117 (33): 20292–20297.

39 Jimenez-Harrison, D., Huseby, C.J., Hoffman, C.N. et al. (2023). DJ-1 molecular chaperone activity depresses tau aggregation propensity through interaction with monomers. *Biochemistry* 62 (5): 976–988.

40 Tsimbalyuk, S., Cross, E., Hoad, M. et al. (2020). The intrinsically disordered W protein is multifunctional during henipavirus infection, disrupting host signalling pathways and nuclear import. *Cells* 9 (8): 1913.

41 Wu, D., Luo, L., Yang, Z. et al. (2021). Targeting human hippo TEAD binding interface with YAP/TAZ-derived, flexibility-reduced peptides in gastric cancer. *International Journal of Peptide Research and Therapeutics* 27 (1): 119–128.

42 Jiao, X., Di Sante, G., Casimiro, M.C. et al. (2024). A cyclin D1 intrinsically disordered domain accesses modified histone motifs to govern gene transcription. *Oncogenesis* 13 (1): 4.

43 Mitra, G. (2021). Emerging role of mass spectrometry-based structural proteomics in elucidating intrinsic disorder in proteins. *Proteomics* 21 (3–4): 2000011.

44 Sivakumar, K.C., Haixiao, J., Naman, C.B., and Sajeevan, T.P. (2020). Prospects of multitarget drug designing strategies by linking molecular docking and molecular dynamics to explore the protein–ligand recognition process. *Drug Development Research* 81 (6): 685–699.

45 Arai, M., Suetaka, S., and Ooka, K. (2024). Dynamics and interactions of intrinsically disordered proteins. *Current Opinion in Structural Biology* 84: 102734.

46 Beaulieu, M.E., Castillo, F., and Soucek, L. (2020). Structural and biophysical insights into the function of the intrinsically disordered Myc oncoprotein. *Cells* 9 (4): 1038.

47 Avery, C., Patterson, J., Grear, T. et al. (2022). Protein function analysis through machine learning. *Biomolecules* 12 (9): 1246.

48 Bakker, M.J., Sørensen, H.V., and Skepö, M. (2024). Exploring the role of globular domain locations on an intrinsically disordered region of p53: a molecular dynamics investigation. *Journal of Chemical Theory and Computation* 20 (3): 1423–1433.

49 Gomes, A.S., Ramos, H., Inga, A. et al. (2021). Structural and drug targeting insights on mutant p53. *Cancers* 13 (13): 3344.

50 Ruan, H., Yu, C., Niu, X. et al. (2021). Computational strategy for intrinsically disordered protein ligand design leads to the discovery of p53 transactivation domain I binding compounds that activate the p53 pathway. *Chemical Science* 12 (8): 3004–3016.

51 Kuusk, A., Boyd, H., Chen, H., and Ottmann, C. (2020). Small-molecule modulation of p53 protein–protein interactions. *Biological Chemistry* 401 (8): 921–931.

52 Moses, D., Ginell, G.M., Holehouse, A.S., and Sukenik, S. (2023). Intrinsically disordered regions are poised to act as sensors of cellular chemistry. *Trends in Biochemical Sciences* 48 (12): 1019–1034.

53 Luo, S., Wohl, S., Zheng, W., and Yang, S. (2023). Biophysical and integrative characterization of protein intrinsic disorder as a prime target for drug discovery. *Biomolecules* 13 (3): 530.

54 Chen, J., Liu, X., and Chen, J. (2020). Targeting intrinsically disordered proteins through dynamic interactions. *Biomolecules* 10 (5): 743.

55 Khare, S.D., Chinchilla, P., and Baum, J. (2023). Multifaceted interactions mediated by intrinsically disordered regions play key roles in alpha synuclein aggregation. *Current Opinion in Structural Biology* 80: 102579.

56 Runwal, G. and Edwards, R.H. (2021). The membrane interactions of synuclein: physiology and pathology. *Annual Review of Pathology: Mechanisms of Disease* 16 (1): 465–485.

57 Man, W.K., Tahirbegi, B., Vrettas, M.D. et al. (2021). The docking of synaptic vesicles on the presynaptic membrane induced by α-synuclein is modulated by lipid composition. *Nature Communications* 12 (1): 927.

58 Metskas, L.A. and Rhoades, E. (2020). Single-molecule FRET of intrinsically disordered proteins. *Annual Review of Physical Chemistry* 71 (1): 391–414.

59 Kaur, U. and Lee, J.C. (2020). Unroofing site-specific α-synuclein–lipid interactions at the plasma membrane. *Proceedings of the National Academy of Sciences* 117 (32): 18977–18983.

60 Liu, C., Zhao, Y., Xi, H. et al. (2021). The membrane interaction of alpha-synuclein. *Frontiers in Cellular Neuroscience* 15: 633727.

61 Oliveira, L.M.A., Gasser, T., Edwards, R. et al. (2021). Alpha-synuclein research: defining strategic moves in the battle against Parkinson's disease. *npj Parkinson's Disease* 7 (1): 65.

62 Vasquez, V., Mitra, J., Wang, H. et al. (2020). A multi-faceted genotoxic network of alpha-synuclein in the nucleus and mitochondria of dopaminergic neurons in Parkinson's disease: emerging concepts and challenges. *Progress in Neurobiology* 185: 101729.

63 Lindorff-Larsen, K. and Kragelund, B.B. (2021). On the potential of machine learning to examine the relationship between sequence, structure, dynamics and function of intrinsically disordered proteins. *Journal of Molecular Biology* 433 (20): 167196.

64 Kumar, P., Bhardwaj, A., Uversky, V.N. et al. (2022). Computational methods to study intrinsically disordered proteins. *Advances in Protein Molecular and Structural Biology Methods [Internet]* 489–504.

65 Saurabh, S., Nadendla, K., Purohit, S.S. et al. (2023). Fuzzy drug targets: disordered proteins in the drug-discovery realm. *ACS Omega* 8 (11): 9729–9747.

66 El Khatabi, K., El-Mernissi, R., Aanouz, I. et al. (2021). Identification of novel acetylcholinesterase inhibitors through 3D-QSAR, molecular docking, and molecular dynamics simulation targeting Alzheimer's disease. *Journal of Molecular Modeling* 27 (10): 302.

67 Leonard, C., Phillips, C., and McCarty, J. (2021). Insight into seeded tau fibril growth from molecular dynamics simulation of the Alzheimer's disease protofibril core. *Frontiers in Molecular Biosciences* 8: 624302.

68 Penke, B., Szűcs, M., and Bogár, F. (2020). Oligomerization and conformational change turn monomeric β-amyloid and tau proteins toxic: their role in Alzheimer's pathogenesis. *Molecules* 25 (7): 1659.

69 Kanchi, P.K. and Dasmahapatra, A.K. (2021). Destabilization of the Alzheimer's amyloid-β protofibrils by THC: a molecular dynamics simulation study. *Journal of Molecular Graphics & Modelling* 105: 107889.

70 Mouchlis, V.D., Melagraki, G., Zacharia, L.C., and Afantitis, A. (2020). Computer-aided drug design of β-secretase, γ-secretase and anti-tau inhibitors for the discovery of novel Alzheimer's therapeutics. *International Journal of Molecular Sciences* 21 (3): 703.

71 Job, N., Thimmakondu, V.S., and Thirumoorthy, K. (2023). *In silico* drug design and analysis of dual amyloid-beta and tau protein-aggregation inhibitors for Alzheimer's disease treatment. *Molecules* 28 (3): 1388.

72 Gholami, A. (2023). Alzheimer's disease: the role of proteins in formation, mechanisms, and new therapeutic approaches. *Neuroscience Letters* 817: 137532.

73 Seidler, P.M., Murray, K.A., Boyer, D.R. et al. (2022). Structure-based discovery of small molecules that disaggregate Alzheimer's disease tissue derived tau fibrils in vitro. *Nature Communications* 13 (1): 5451.

74 Zhu, L. and Qian, Z. (2022). Recent studies of atomic-resolution structures of tau protein and structure-based inhibitors. *Quantitative Biology* 10 (1): 17–34.

75 Tarika, J.D.D., Dexlin, X.D.D., Arun Kumar, A. et al. (2022). Computational insights on charge transfer and non-covalent interactions of antibacterial compound 4-dimethylaminopyridinium pyridine-2-carboxylate pentahydrate. *Journal of Molecular Structure* 1256: 132525.

76 Zhang, J., Ghadermarzi, S., and Kurgan, L. (2020). Prediction of protein-binding residues: dichotomy of sequence-based methods developed using structured complexes versus disordered proteins. Elofsson A, editor. *Bioinformatics* 36 (18): 4729–4738.

77 Ferruz, N., Schmidt, S., and Höcker, B. (2022). ProtGPT2 is a deep unsupervised language model for protein design. *Nature Communications* 13 (1): 4348.

78 Kurgan, L. (2022). Resources for computational prediction of intrinsic disorder in proteins. *Methods* 204: 132–141.

79 Barik, A., Katuwawala, A., Hanson, J. et al. (2020). DEPICTER: intrinsic disorder and disorder function prediction server. *Journal of Molecular Biology* 432 (11): 3379–3387.

80 Sinha, S., Sharma, S., Johari, S. et al. (2023). Design and development of hydrophobicity and net charge based artificial neural network model for IDP/IDPR prediction. *Procedia Computer Science* 218: 438–448.

81 Zhao, B. and Kurgan, L. (2023). Machine learning for intrinsic disorder prediction. *Machine Learning in Bioinformatics of Protein Sequences [Internet]* 205–236. https://doi.org/10.1142/9789811258589_0008.

82 Schor, M., Mey, A.S.J.S., and MacPhee, C.E. (2016). Analytical methods for structural ensembles and dynamics of intrinsically disordered proteins. *Biophysical Reviews* 8 (4): 429–439.

83 Lermyte, F. (2020). Roles, characteristics, and analysis of intrinsically disordered proteins: a mini review. *Life* 10 (12): 320.

84 Han, B., Ren, C., Wang, W. et al. (2023). Computational prediction of protein intrinsically disordered region related interactions and functions. *Genes* 14 (2): 432.

85 Su, C.T., Chen, C.Y., and Ou, Y.Y. (2006). Protein disorder prediction by condensed PSSM considering propensity for order or disorder. *BMC Bioinformatics* 7 (1): 319.

86 Mészáros, B., Erdős, G., and Dosztányi, Z. (2018). IUPred2A: context-dependent prediction of protein disorder as a function of redox state and protein binding. *Nucleic Acids Research* 46 (W1): W329–W337.

87 Dayhoff, G.W. and Uversky, V.N. (2022). Rapid prediction and analysis of protein intrinsic disorder. *Protein Science* 31 (12): e4496.

88 Hanson, J., Paliwal, K.K., Litfin, T., and Zhou, Y. (2019). SPOT-Disorder2: improved protein intrinsic disorder prediction by ensembled deep learning. *Genomics, Proteomics & Bioinformatics* 17 (6): 645–656.

89 Oates, M.E., Romero, P., Ishida, T. et al. (2012). D2P2: database of disordered protein predictions. *Nucleic Acids Research* 41 (D1): D508–D516.

90 Henriques, J., Cragnell, C., and Skepö, M. (2015). Molecular dynamics simulations of intrinsically disordered proteins: force field evaluation and comparison with experiment. *Journal of Chemical Theory and Computation* 11 (7): 3420–3431.

91 Stangler, T., Hartmann, R., Willbold, D., and Koenig, B.W. (2006). Modern high resolution NMR for the study of structure, dynamics and interactions of biological macromolecules. *Zeitschrift für Physikalische Chemie* 220 (5): 567–613.

92 Alston, J.J., Soranno, A., and Holehouse, A.S. (2021). Integrating single-molecule spectroscopy and simulations for the study of intrinsically disordered proteins. *Methods* 193: 116–135.

93 Doerge, R.W. (2006). Bioinformatics and computational biology solutions using R and bioconductor. Edited by Gentleman, R., Carey, V., Huber, W., Irizarry, R., and Dudoit, S. *Biometrics* 62 (4): 1270–1271.

94 Sheik Amamuddy, O., Veldman, W., Manyumwa, C. et al. (2020). Integrated computational approaches and tools for allosteric drug discovery. *International Journal of Molecular Sciences* 21 (3): 847.

95 Kulkarni, P., Leite, V.B.P., Roy, S. et al. (2022). Intrinsically disordered proteins: ensembles at the limits of Anfinsen's dogma. *Biophysical Reviews* 3 (1): 011306.

96 Camacho-Zarco, A.R., Schnapka, V., Guseva, S. et al. (2022). NMR provides unique insight into the functional dynamics and interactions of intrinsically disordered proteins. *Chemical Reviews* 122 (10): 9331–9356.

97 Vemulapalli, S., Hashemi, M., and Lyubchenko, Y.L. (2021). Site-search process for synaptic protein-DNA complexes. *International Journal of Molecular Sciences* 23 (1): 212.

98 Vemulapalli, S., Hashemi, M., Kolomeisky, A.B., and Lyubchenko, Y.L. (2021). DNA looping mediated by site-specific SfiI–DNA interactions. *The Journal of Physical Chemistry B* 125 (18): 4645–4653.

99 Vemulapalli, S., Hashemi, M., Kolomeisky, A.B., and Lyubchenko, Y.L. (2023). Assembly of synaptic protein–DNA complexes: critical role of non-specific interactions. *International Journal of Molecular Sciences* 24 (12): 9800.

100 Vemulapalli, S., Hashemi, M., Chen, Y. et al. (2024). Nanoscale interaction of endonuclease APE-1 with DNA characterized by atomic force microscopy [Internet]. *Biophysics* https://doi.org/10.1101/2024.02.11.579811.

101 Stormberg, T., Vemulapalli, S., Filliaux, S., and Lyubchenko, Y.L. (2021). Effect of histone H4 tail on nucleosome stability and internucleosomal interactions. *Scientific Reports* 11 (1): 24086.

102 Chang, W., Hsieh, C., Kuo, I. et al. (2023). Nutlin-3 acts as a DNA methyltransferase inhibitor to sensitize esophageal cancer to chemoradiation. *Molecular Carcinogenesis* 62 (2): 277–287.

103 Zerihun, M. and Qvit, N. (2023). Selective inhibitors targeting Fis1/Mid51 protein–protein interactions protect against hypoxia-induced damage in cardiomyocytes. *Frontiers in Pharmacology* 14: 1275370.

104 Dar, K.B., Bhat, A.H., Amin, S. et al. (2020). Elucidating critical proteinopathic mechanisms and potential drug targets in neurodegeneration. *Cellular and Molecular Neurobiology* 40 (3): 313–345.

105 Burns, S., Selman, A., Sehar, U. et al. (2022). Therapeutics of Alzheimer's disease: recent developments. *Antioxidants* 11 (12): 2402.

106 Zhang, G., Zhang, J., Gao, Y. et al. (2022). Strategies for targeting undruggable targets. *Expert Opinion on Drug Discovery* 17 (1): 55–69.

107 Zhang, W., Roy Burman, S.S., Chen, J. et al. (2022). Machine learning modeling of protein-intrinsic features predicts tractability of targeted protein degradation. *Genomics, Proteomics & Bioinformatics* 20 (5): 882–898.

108 Bhattarai, A. and Emerson, I.A. (2020). Dynamic conformational flexibility and molecular interactions of intrinsically disordered proteins. *Journal of Biosciences* 45 (1): 29.

109 Santofimia-Castaño, P., Rizzuti, B., Xia, Y. et al. (2020). Targeting intrinsically disordered proteins involved in cancer. *Cellular and Molecular Life Sciences* 77 (9): 1695–1707.

110 Hu, X., Xu, Y., Wang, C. et al. (2023). Combined prediction and design reveals the target recognition mechanism of an intrinsically disordered protein interaction domain. *Proceedings of the National Academy of Sciences* 120 (39): e2305603120.

111 Spolaore, B., Secco, L., Rocca, G. et al. (2023). Proteomic tools to study phosphorylation of intrinsically disordered proteins. *Expert Review of Proteomics* 20 (4–6): 93–107.

112 Newcombe, E.A., Delaforge, E., Hartmann-Petersen, R. et al. (2022). How phosphorylation impacts intrinsically disordered proteins and their function. Mukhopadhyay S, editor. *Essays in Biochemistry* 66 (7): 901–913.

113 Fang, Y., Liao, G., and Yu, B. (2020). Small-molecule MDM2/X inhibitors and PROTAC degraders for cancer therapy: advances and perspectives. *Acta Pharmaceutica Sinica B* 10 (7): 1253–1278.

114 Nussinov, R., Zhang, M., Liu, Y., and Jang, H. (2022). AlphaFold, artificial intelligence (AI), and allostery. *The Journal of Physical Chemistry B* 126 (34): 6372–6383.

115 Zhan, C., Tang, T., Wu, E. et al. (2023). From multi-omics approaches to personalized medicine in myocardial infarction. *Frontiers in Cardiovascular Medicine* 10: 1250340.

116 Asiri, F., Haque Siddiqui, M.I., Ali, M.A. et al. (2023). Mathematical modeling of active contraction of the human cardiac myocyte: a review. *Heliyon* 9 (9): e20065.

117 Blundell, T.L., Gupta, M.N., and Hasnain, S.E. (2020). Intrinsic disorder in proteins: relevance to protein assemblies, drug design and host–pathogen interactions. *Progress in Biophysics and Molecular Biology* 156: 34–42.

118 Sudhakar, P., Machiels, K., Verstockt, B. et al. (2021). Computational biology and machine learning approaches to understand mechanistic microbiome–host interactions. *Frontiers in Microbiology* 12: 618856.

119 Gao, C., Ma, C., Wang, H. et al. (2021). Intrinsic disorder in protein domains contributes to both organism complexity and clade-specific functions. *Scientific Reports* 11 (1): 2985.

120 Poluri, K.M., Gulati, K., Tripathi, D.K., and Nagar, N. (2023). Drug design methods to regulate protein–protein interactions. *Protein–Protein Interactions [Internet]* 265–341. https://doi.org/10.1007/978-981-99-2423-3_6.

121 Candelise, N., Scaricamazza, S., Salvatori, I. et al. (2021). Protein aggregation landscape in neurodegenerative diseases: clinical relevance and future applications. *International Journal of Molecular Sciences* 22 (11): 6016.

122 Rahman, A., Saikia, B., Gogoi, C.R., and Baruah, A. (2022). Advances in the understanding of protein misfolding and aggregation through molecular dynamics simulation. *Progress in Biophysics and Molecular Biology* 175: 31–48.

123 Zuo, K., Kranjc, A., Capelli, R. et al. (2023). Metadynamics simulations of ligands binding to protein surfaces: a novel tool for rational drug design. *Physical Chemistry Chemical Physics* 25 (20): 13819–13824.

124 Wang, X., Ni, D., Liu, Y., and Lu, S. (2021). Rational design of peptide-based inhibitors disrupting protein–protein interactions. *Frontiers in Chemistry* 9: 682675.

125 Murali, R., Zhang, H., Cai, Z. et al. (2021). Rational design of constrained peptides as protein interface inhibitors. *Antibodies* 10 (3): 32.

126 Arar, S., Haque, M.A., and Kayed, R. (2023). Protein aggregation and neurodegenerative disease: structural outlook for the novel therapeutics. *Proteins: Structure, Function, and Bioinformatics* prot.26561.

127 Mishra, P., Basak, S., Mukherjee, A., and Basu, A. (2022). Design and study of *in silico* binding dynamics of certain isoxazole bearing leads against Aβ-42 and BACE-1 loop in protein fibrillation. *Letters in Drug Design & Discovery* 19 (3): 192–213.

128 Ugbaja, S.C., Lawal, I.A., Kumalo, H.M., and Lawal, M.M. (2022). Alzheimer's disease and β-secretase inhibition: an update with a focus on computer-aided inhibitor design. *Current Drug Targets* 23 (3): 266–285.

129 Das, B., Mathew, A.T., Baidya, A.T.K. et al. (2023). Artificial intelligence assisted identification of potential tau aggregation inhibitors: ligand- and structure-based virtual screening, in silico ADME, and molecular dynamics study. *Molecular Diversity [Internet]* https://doi.org/10.1007/s11030-023-10645-3.

130 Varma, D.A., Singh, M., Wakode, S. et al. (2023). Structure-based pharmacophore mapping and virtual screening of natural products to identify polypharmacological inhibitor against c-MET/EGFR/VEGFR-2. *Journal of Biomolecular Structure & Dynamics* 41 (7): 2956–2970.

131 Pei, H., Guo, W., Peng, Y. et al. (2022). Targeting key proteins involved in transcriptional regulation for cancer therapy: current strategies and future prospective. *Medicinal Research Reviews* 42 (4): 1607–1660.

132 Rehman, A.U., Khurshid, B., Ali, Y. et al. (2023). Computational approaches for the design of modulators targeting protein–protein interactions. *Expert Opinion on Drug Discovery* 18 (3): 315–333.

133 Delaunay, M. and Ha-Duong, T. (2022). Computational tools and strategies to develop peptide-based inhibitors of protein–protein interactions. In: *Computational Peptide Science [Internet]*, vol. 2405 (ed. T. Simonson), 205–230. New York, NY: Springer US https://doi.org/10.1007/978-1-0716-1855-4_11.

134 Selvaraj, C., Pravin, M.A., Alhoqail, W.A. et al. (2022). Intrinsically disordered proteins in viral pathogenesis and infections. In: *Advances in Protein Chemistry and Structural Biology [Internet]*, 221–242. Elsevier.

135 Schaduangrat, N., Lampa, S., Simeon, S. et al. (2020). Towards reproducible computational drug discovery. *Journal of Cheminformatics* 12 (1): 9.

136 Choudhary, S., Lopus, M., and Hosur, R.V. (2022). Targeting disorders in unstructured and structured proteins in various diseases. *Biophysical Chemistry* 281: 106742.

137 Radaeva, M., Ton, A.T., Hsing, M. et al. (2021). Drugging the 'undruggable'. Therapeutic targeting of protein–DNA interactions with the use of computer-aided drug discovery methods. *Drug Discovery Today* 26 (11): 2660–2679.

Index

Note: *Italicized* and **bold** page numbers refer to figures and tables, respectively.

Computational Methods for Rational Drug Design, First Edition. Edited by Mithun Rudrapal.

b

d

h

i

n

o

p

q

r

Printed and bound by CPI Group (UK) Ltd, Croydon, CR0 4YY
17/09/2026
14964334-0003